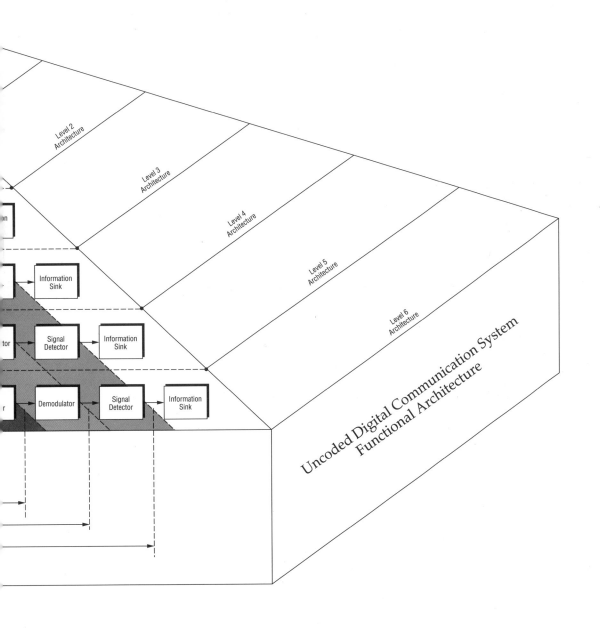

Level 2
Architecture

Level 3
Architecture

Level 4
Architecture

Level 5
Architecture

Level 6
Architecture

Information
Sink

Signal
Detector

Information
Sink

Demodulator

Signal
Detector

Information
Sink

Uncoded Digital Communication System
Functional Architecture

Digital Communication Techniques

Signal Design and Detection

Marvin K. Simon
Jet Propulsion Laboratory

Sami M. Hinedi
Jet Propulsion Laboratory

William C. Lindsey
LinCom Corporation
University of Southern California

PTR Prentice Hall
Upper Saddle River, New Jersey 07458
http://www.phptr.com

Library of Congress Cataloging-in-Publication Data

Simon, Marvin Kenneth,
    Digital communication techniques / Marvin K. Simon, Sami M.
Hinedi, William C. Lindsey.
      p. cm.
    Includes bibliographical references and index.
    Contents: v. 1. Signal design and detection
    ISBN 0-13-200610-3
    1. Digital communciations. I. Hinedi, Sami M. II. Lindsey,
William C. III. Title.
TK5103.7.S52   1994
621.382—dc20   93-9053
                                          CIP

Editorial/production supervision and interior design: *Harriet Tellem*
Cover design: *Aren Graphics*
Manufacturing buyer: *Alexis Heydt*
Acquisitions editor: *Paul W. Becker*
Editorial assistant: *Maureen Diana*

© 1995 Prentice Hall PTR
Prentice-Hall, Inc.
A Simon and Schuster Company
Upper Saddle River, New Jersey 07458

The publisher offers discounts on this book when ordered in bulk quantities:

    Corporate Sales Department
    PTR Prentice Hall
    1 Lake Street
    Upper Saddle River, New Jersey 07458

    Phone: 201-592-2863
    FAX: 201-592-2249

Printed in the United States of America
10 9 8 7 6 5 4

ISBN: 0-13-200610-3

Prentice-Hall International (UK) Limited, *London*
Prentice-Hall of Australia Pty. Limited, *Sydney*
Prentice-Hall Canada Inc., *Toronto*
Prentice-Hall Hispanoamericana, S.A., *Mexico*
Prentice-Hall of India Private Limited, *New Delhi*
Prentice-Hall of Japan, Inc., *Tokyo*
Simon & Schuster Asia Pte. Ltd., *Singapore*
Editora Prentice-Hall do Brasil, Ltda., *Rio de Janeiro*

Dedication

Marvin K. Simon would like to dedicate this book to his wife Anita in appreciation for her consultation at every step of the way through the preparation of this manuscript. Indeed this author could have never completed his portion of this book nor for that matter any of his previous books without the support, encouragement, guidance, and sharing in the many anxieties that accompany a project of this magnitude. This book is also dedicated to his children Brette and Jeff who have made him very proud through their development and accomplishments both as human beings and in their respective fields of interest.

Sami M. Hinedi would like to thank his father Mounir and his mother Hikmat for their constant support, encouragement and guidance and for the various sacrifices they made in order to provide their children with the best education one can hope for. He also would like to thank his sister Fadwa, his brother Taha and last but not least, his dear uncle Abdul Kader for their help and for being there when he needed them most. This book is dedicated to all of them.

William C. Lindsey would like to dedicate this book to Dorothy, my wife, and to John, my son, for the support system they have provided me throughout by professional career. Also, to Corinne Leslie, my secretary for 27 years, for her loyal, continuous, and dedicated support; and to my sisters, Loretta and Jean for their lifelong inspiration and encouragement.

Contents

Contents

Preface

Future applications of **Digital Communication Techniques** to architecting and implementing global information transportation and computing systems have never been brighter. This outlook is driven by social, economical, political, and technological reasons. From a technical perspective, it is recognized by most communication engineers that communications is required to accomplish computing while computers are required to accomplish communications. From a technology viewpoint, this technical perspective is rapidly being accomplished using emerging digital microelectronic technologies (DSP and VLSI) to implement digital communication systems.

    Digital Communication Techniques are exciting and are of vital importance to all societies. Countries have failed to be competitive simply because they did not succeed in establishing good communication infrastructures. Consequently, one major purpose of this textbook is to present, in a unique and innovative way, a functional architecture and a theory for use in the design of uncoded and coded digital communication systems. The system architecture is pyramidal and the theoretical development is unique in that it is presented, for the first time, from a **Systems Engineering** perspective for both bandlimited and power limited communication channels. This perspective adopts the point of view that **coding** and **modulation** are both components of the **signal design** problem and that **demodulation** and **decoding** are both components of the **signal detection** problem. Beginning with Chapter 1, the subject matter progresses top down and systematically in a hierarchial way. The geometric concepts, first introduced by Shannon and Kotelnikov, and later documented by Wozencraft and Jacobs in their book *Principles of Communication Engineering*, are used in Chapter 3 to set the foundations for signal design and detection. Starting with Chapter 4 and ending in Chapter 8, **coherent, noncoherent, partially coherent, and differentially coherent** detection techniques are treated for numerous uncoded modulation techniques, such as BPSK, QPSK, MPSK, and MFSK. In Chapter 10, these same detection techniques are applied to more advanced forms of uncoded modulation such as QAM, CPFSK, MSK, QFPM, and CPQFPM. As opposed to the M-ary error probability criterion used in designing uncoded systems, the R_0-criterion is introduced for use in optimizing the design of coded systems. Since R_0 is a function of the codec and modem choice, this criterion leads to a combined codec-modem design that employs the most effective coding and modulation technique. In fact, soft-decision demodulators can be systematically designed using R_0. Chapters 11, 12, and 13 consider **block**, **convolutional**, and **concatenated coding** techniques from the systems perspective. In addition, the counterparts of maximum-likelihood (ML) decoding and ML decoding using the Viterbi algorithm are given. A variety of **interleaving-deinterleaving** techniques (block, convolutional, helical, hybrid) are presented. To understand the connection among the various coding techniques presented, a Venn diagram for error correcting codes is constructed; emphasis is placed on presenting the communciations efficiency achieved using **Hamming, Golay, Bose-Chaudhuri-Hocquenghem (BCH), Reed Solomon (RS),** and **convolutional** codes.

There are a vast number of textbooks on the market today that deal with the subject of digital communications. In fact, in a quick survey of the textbook literature, we were able to come up with at least ten books that *include* the words "digital communications" in their titles. At least three bear these words as their *entire* title implying, somewhat, that they are an all-inclusive treatment of the subject. Many of these books are quite broad in scope, but also quite shallow in detail. Striking a proper balance between these two attributes, yet maintaining a high level of readability, is no simple task. We believe that our book accomplishes this important goal and sets it apart from all other digital communication texts currently on the market. Several key features that distinguish our book from the others are as follows:

- A top-down perspective of digital communication system design, using a pyramid structure to describe the system functional architecture.
- A top-down presentation of the theory needed to perform uncoded and coded system design.
- Includes R_0 criterion for use in the design of coded systems.
- Includes more recent developments in the field that have occurred over the last 20 or 30 years.
- A universal appeal to graduate students as well as system architects and engineers.
- Written by authors whose combined industrial and university experience exceeds 60 years.

There are many specific features that make this book unique and beneficial to its readers. With the advent of today's advances in the solid state microelectronic technologies, a variety of novel digital communication systems are appearing on the market and are serving as motivation for the introduction of new telecommunication and information services. Chapter 1 of this book provides the reader with examples of such services and top level system architectures thereby indicating the highly complex nature of these systems. We believe that Chapter 2, which discusses the computation of power spectral density of digital modulations, is the best treatment of spectrum efficiency evaluation found anywhere. This computation is essential to assessing the bandwidth (spectral) occupancy requirement of a digital modulation, yet it is ignored in many books. Another key feature of this book is the organization and order of presentation of the material in Chapters 3 through 7. By first describing coherent detection and then successively following with noncoherent, partially coherent, and differentially coherent detection, the reader is provided with a logical flow starting with the conceptually simplest technique and proceeding top down to the more complex techniques. The discussion of double differentially coherent detection in Chapter 8 is unique to our book.[1] Here the reader will learn how to design differentially coherent communication systems that are robust to frequency offsets due, for example, to Doppler and oscillator instabilities. Chapter 9 treats the voluminous subject of bandlimited communications in a condensed and unified way. Included here are the important subjects of Nyquist

---

1. There are a few Russian-authored textbooks that discuss double differentially coherent detection, but as yet, they have not been translated into English.

and partial response signaling, maximum-likelihood detection in the presence of intersymbol interference, and equalization. To the authors' knowledge, Chapter 10 of this book is the most complete and up-to-date treatment of advanced modulation techniques. It guides the reader through the most recent modulation techniques described in the literature and how they compare in terms of such properties as modulation type, pulse shaping, continuity of phase, variation of envelope, I and Q channel data rates, and parameter offsets. Still another key feature is the proper identification of the important link (mapping function) between the modulation and coding functions in coded digital communication systems. Most books that discuss both modulation and coding treat these two functions as separate and independent entities. In some books where modulation and coding are treated as combined, the treatment is strictly limited to trellis coded modulation. This book is unique in that it presents a general formulation for coded communication systems by properly defining the key parameters and interfaces between the various modulation and coding functions. This discussion is presented in Chapter 11 which includes many examples to clearly elucidate this often overlooked but all important aspect of system design. Chapter 12 discusses the use of the R_0 criterion in the design of coded systems. Finally, Chapter 13 presents a compact yet authoritative discussion of the design of convolutionally-coded communication systems, a subject that, by itself, can occupy an entire textbook.

We recognize that a complete study of reliable and efficient communication of information requires a full and detailed treatment of the two important disciplines: information theory and communication theory. Since the main focus in this book is on the latter, we do not treat the problem of efficient packaging of information (data compression of text, voice, video, etc.) nor do we treat the important problem of designing the ultimate error control coding-decoding technique which achieves the ultimate transmission speed (channel capacity). The solution to these problems are best treated in separate books on information theory and error control coding, and, indeed, there are such books available.

This book has been written for use as a textbook at universities involved in teaching Communication Sciences. It has also been designed to accommodate certain needs of the systems architect, systems engineer, the professor, and communication sciences researcher. The lecture material has been organized and written in a form whereby theory and practice are continuously emphasized. Most of the problems suggested at the conclusion of each chapter have evolved from teaching graduate level courses to students at the University of Southern California and the California Institute of Technology. Through homework assignments, most of the problems have been field tested, corrected, and enhanced over the years.

The architecture for this book is predicated upon two graduate level Communication Theory courses (EE564 and EE664) taught at the University of Southern California's Communication Sciences Institute over the past 25 years. In this sense, two semesters are required to cover its contents. The organization and presentation of the material is largely based upon the academic and course design work of Professors William C. Lindsey and Robert A. Scholtz. The authors' approach to presenting the solution to the problem of vector communications in the presence of colored noise and the representation of bandpass random processes are largely due to Professor Scholtz. The Digital Communication System architecture presented in the pyramids of Chapter 1 was created by Professor Lindsey. In addition, certain exercises given in and at the conclusion of Chapters 1, 3, 4, 5, and 6 were taken with permission from problems created and used by Professors Robert M. Gagliardi, Vijay Kumar, William C. Lindsey, Andreas Polydoros, Charles L. Weber, and Robert A. Scholtz all of the University of Southern California's Communication Sciences Institute. The authors wish to thank these individuals for permission to use these challenging problems as homework exercises.

The authors are grateful for the help of many colleagues and students during the preparation of this manuscript, and are particularly indebted to: Dr. Michael Mandel, formerly a graduate student at the California Institute of Technology; Dr. Debbie Van Alphen, Professor of Electrical Engineering at California State University, Northridge; and Dr. Jorge M. N. Pereira of GTE; for their comments and criticisms. Furthermore, we thank our colleagues at the Jet Propulsion Laboratory, Pasadena, CA, namely: Dr. Tien Nguyen, Dr. Haiping Tsou, Dr. Daniel Raphaeli, Mr. Biren Shah, Mr. Samson Million, Mr. Alexander Mileant, Mr. Mazen Shihabi, Dr. Ramin Sadr, Mr. Roland Bevan, and Dr. Victor Vilnrotter for reading and reviewing various parts of the manuscript.

Marvin K. Simon
Sami M. Hinedi
William C. Lindsey

Pasadena, CA

Chapter 1

Introduction to Telecommunications

Today's advances in solid state electronic technologies are impacting the design and implementation of various digital communication systems, and specifically, terrestrial and space-related telecommunication networks and services. Many systems/networks that once were considered impractical due to their complexities are now well within the reach of current technology and are presently being designed, implemented, and operated. Many microelectronic technologies are maturing so fast that what is considered impractical today may well be implementable tomorrow. As a result, novel telecommunication and information services are continuously being introduced into the market.

At the system component level, microprocessors are operating faster and faster, front-end amplifiers are achieving gains in frequency regions that were once considered unattainable with noise figures that approach the theoretical limit, antenna surfaces are becoming "smoother," thus enabling higher frequency transmission and reception. Various other advanced technologies, especially wireless, are affecting and sometimes dictating the system design architecture and strategy. At the systems level, satellites are moving into an era of multiple beams and high-gain spot beams allowing for frequency reuse, supplementing global communications coverage and even allowing the use of "very" small antenna apertures. Cellular terrestrial and proposed space radio systems around the world are testing digital systems to replace their analog FM equipment, while cordless telephones are moving outside the boundaries of the home and may merge with cellular systems, enabling users to carry phones with them. Mobile satellite experiments [1] are currently being carried out to test the next generation system in which a mobile user, located in a car, on a plane or a ship, can directly communicate with a satellite, bypassing the need for a relay station. Existing telephone networks are being upgraded using fiber-optic technology [2] to allow for higher data rates and better transmission quality enabling novel information services. These advances and others are resulting in improved system performances, higher reliability, easier operability and maintainability. Consequently, systems life cycles are increased dramatically and life cycle costs decreased. Global mobile and personal communication systems are the wave of the future, and their applications are endless.

Telecommunication engineering and more specifically, coding, modulation, demodulation, decoding, detection, and synchronization techniques are the technical topics to be covered in the chapters to follow. Fundamentals of signal detection, being *coherent*, *non-coherent*, *partially* and *differentially coherent*, are the key methodologies to the design of optimum receivers. The choice of the modulation scheme from the "simple" *binary phase shift keying* (BPSK) to the more "elaborate" *quadrature amplitude modulations* (QAM) is essential in trading channel bandwidth and achievable bit error rate for a given and fixed transmitted signal power. The *synchronization function* is fundamental in operating any communication link. At the receiving site, various timing and phase references are needed and are derived from the incoming noisy signal; these include carrier phase and frequency estimation, possibly subcarrier phase and frequency, and definitely symbol (bit) timing to recover the transmitted information bits. Synchronization precludes communication and includes these two functions: acquisition and tracking. Each of the carrier, subcarrier, and symbol acquisition processes consists of first frequency, then phase acquisition, and the total lock-up (or acquisition time) depends on the specific structures or algorithms employed. The transition from acquisition to tracking function is nonuniquely defined and is typically said to have occurred when the instantaneous phase error decreases and remains below a predetermined threshold. Automatic gain control (AGC) circuitry is essential in maintaining reasonable received power levels in the receiver and in providing some protection against "large" intentional or accidental interferences. Finally, signal reference generators at both the transmitter and receiver require some degree of time and frequency stability to maintain a fixed reference throughout the system.

This chapter introduces the reader to the architecture of a digital communication system and introduces the various terms and key parameters used throughout the book. Section 1.2 presents practical "telecommunication networks" and discusses future networks. The various elements and key functions of an end-to-end communication link are discussed in Section 1.3. The key performance parameters such as signal-to-noise ratios, bandwidth, and so on, are defined along with the various losses that need to be accounted for in a link budget analysis. The "information" capacity of a communication channel as defined by Shannon is discussed in Section 1.4 and in Chapter 11. Communication with subcarriers and data formatting of various signals is the topic of Section 1.5.

1.1 DIGITAL COMMUNICATION SYSTEM FUNCTIONAL ARCHITECTURE

For the most part, the functional architecture of a digital communication system [3] has evolved over the past 35 years; with few exceptions, the functional architecture matches the physical structure. This evolution has been driven by two major forces, viz., the development of communication and information theory (communication sciences) and the development of communications, computer, and information technology. In presenting the functional architecture of a digital communication system, the top-down approach will be used. This approach takes advantage of the hierarchical nature of any system, in particular, an *information transportation system* which transports information using a *digital communication system*. In addition, the top-down approach fits well with the presentation and development of the digital communication techniques and theory presented in the chapters that follow.

Our top-down approach begins with the simplest architectural level, *Level 0*: The *Conceptual Level,* and proceeds downward until the bottom level, viz., the *Physical Level*, is reached. Figure 1.1 demonstrates that the basic building blocks of an *Information Transportation System (ITS)* in level one of the hierarchy are: (1) an information source to be transported, (2) a communication channel or information pipe, and (3) an information user or a sink. In order to connect the information source to the channel, a *transducer* is needed. In level 2 of the hierarchy in Fig. 1.1, this transducer is identified as the *communication transmitter*. Simultaneously, a transducer is also needed at the channel output of level one, and this is accomplished in level 2 by adding a *communication receiver* to interface with the information *user*. Thus, we conclude that the basic building blocks of a digital communication system are a transmitter, a physical channel (communications pipe), and a receiver. These blocks will be further partitioned and discussed.

As we shall see, the system engineering and design of a digital communication system best proceeds top-down, while implementation and system integration proceeds best bottom-up, that is, top-down design driven by *system requirements* followed by bottom-up integration of the system building blocks yields a system that meets the *system user requirements*. From the system requirements, system engineers generate *design specifications* and from these specifications, implementations (software and hardware) are perceived. Executing this top-down design process, which is driven by the system requirements and bottom-up integration, constitutes what is frequently referred to in practice as *Systems Engineering and Integration* (SE&I).

The interfaces between the functional blocks are control signals and are specified by *interface control documents* (ICDs). In what follows, a further breakdown of the functional architecture of the information transportation system will be presented in the form of a hierarchy. All systems are hierarchical in nature—one engineer's part is another's system. The transmitter, channel, and receiver can easily be identified at each level in the hierarchy. Figure 1.2 serves to demonstrate the building blocks for the *communications transmitter* and the *communication receiver* in level 2 of a digital communication system architecture in Fig. 1.1. This figure shows that each system building block is a distinct architecture within a larger architecture in the hierarchy. These building blocks will be modeled and elaborated upon below, as well as in Chapters 3 through 13.

1.1.1 Communications Channel

We observe from Fig. 1.1 that communication system architecture centers or pivots around the *communications channel*. The architecture of this channel can best be demonstrated by introducing the *channel hierarchy*. The architecture of this channel hierarchy can be conveniently represented using six levels; these levels are identified in Fig. 1.3. In this chapter and Chapter 2, we are primarily concerned with modeling and analysis of levels 4, 5, and 6. In the remaining chapters, we will be incorporating channel levels 4, 5, and 6 as a part of the IF channel in level 3. In other words, the mathematical characterization of the IF channel must serve to mimic the features of the RF channel, the Antenna-to-Antenna Channel, and the Physical Channel. In a sense, the IF channel model will serve as a "*virtual channel*" model that mimics the physics of the hardware and the physical channel at levels 4, 5, and 6. The concept of a *virtual channel* is key to understanding the modeling of the *communications process*. A *virtual channel* is a set of resources necessary to complete the *communication link* between information source and user. In this book, we elect to model their effects, for the most part, below level 3 in the *Information Transportation System (ITS)* hierarchy. This model will consider the effects of power (energy) reduction as a function of the building blocks and source-user separation primarily below level 3 of the system hierarchy. In the following section, we further identify the building blocks and features at each channel level in the hierarchy demonstrated in Fig. 1.3. In essence, the building blocks of the communication channel at any level are always defined by those building blocks in the architectural levels below.

Figure 1.4 demonstrates the communication systems engineering education pyramid.

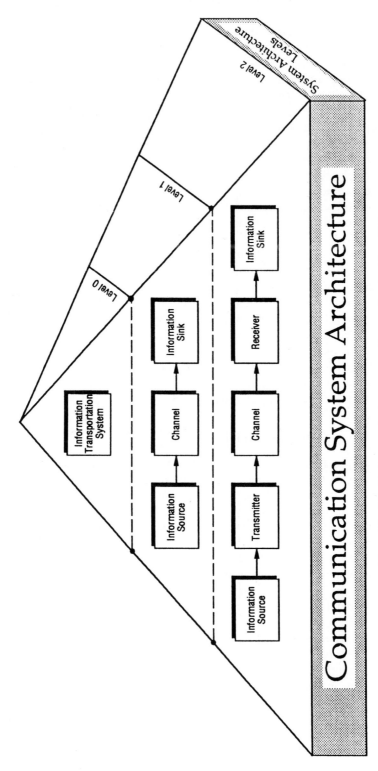

Figure 1.1 Simplified communication system architecture (Reprinted from [38])

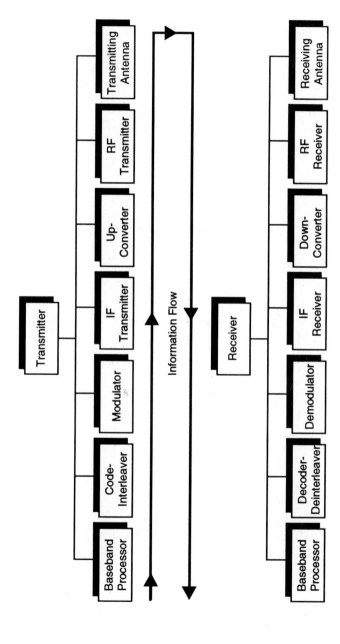

Figure 1.2   Transmitter and receiver building blocks (Reprinted from [38])

5

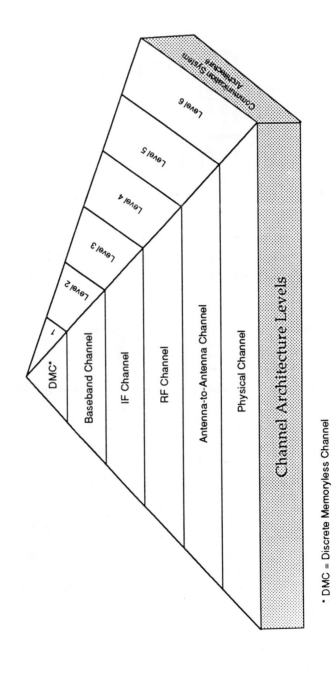

Figure 1.3 Channel architecture versus communication system architecture (Reprinted from [38])

* DMC = Discrete Memoryless Channel

This pyramid summarizes the relationships between the theory needed to perform analysis and/or system simulation at various channel levels in the overall architecture. This figure demonstrates that the lower the level to be worked, the larger the knowledge base required from a system engineering viewpoint.

1.1.2 Uncoded Digital Communication System Architecture

In considering the more detailed breakdown of the architecture at the lower levels in the hierarchy, it is convenient to break this into two components, viz., (1) the functional architecture of an *uncoded digital communication system* and (2) the functional architecture of a *coded digital communication system.* Chapters 3 through 10 will be concerned with the theory and techniques associated with the design and optimization of an *Uncoded Digital Communication System*; the performance criterion will be "*bit error probability.*" Chapters 10 through 13 will provide the necessary interface that will lead to presenting error control coding techniques associated with the design and optimization of a *Coded Digital Communication System*; the performance criteria will be the "R_0 criterion" and bit error probability.

Figure 1.5 demonstrates the *Information Transportation System (ITS) Pyramid* for the architecture of an uncoded digital communications system. At each level in the architecture, there exists a channel that serves to characterize and model the physical features associated with the signal flow path below it. This channel hierarchy serves to define (one-to-one) the elements associated with the channel layers illustrated in Fig. 1.3. The frequency up converter present in practice between the IF and RF transmitter and the down converter between the RF receiver and IF receiver (see Fig. 1.2) have been included respectively in the RF transmitter and receiver building blocks. In this chapter, we are primarily concerned with modeling and characterizing the power transfer and channel effects in channel layers 5 and 6.

1.1.3 Coded Digital Communication System Architecture

When information source and error control coding techniques are used to mitigate the deleterious effects of channel interference and memory on communication transmission, the functional architecture changes. Figure 1.6 demonstrates the *ITS Pyramid* when source and error control coding techniques are assumed. Comparing the pyramids in Figures 1.5 and 1.6, we see that a *source encoder* and a *channel encoder* are required on the transmit side and a *channel decoder* and a *source decoder* are required on the receive side. In practice, these elements are frequently referred to as the *source codec* and the *channel codec*. The informed reader will note that the *interleaver* and *deinterleaver* functions demonstrated in Fig. 1.2 have been omitted in Fig. 1.6; these functions, designed to alleviate channel memory effects, are assumed to be incorporated in the channel coder and decoder of Fig. 1.6, respectively. The details of these functions are covered in Chapters 11 through 13 from a system design and engineering perspective. As before, the channel at any level of the coded system architecture serves to incorporate the features of the channel below it with regard to any deleterious effects on signal (information) flow.

The *ITS Pyramids* serve to demonstrate the feature of the signal (information) flow path (the communication channel) that constitute hardware *versus* those that constitute the communication medium (that is, cable, waveguide, atmospheric, and so on.) In essence, the *ITS Pyramids* serve to answer the question: "*What functional elements constitute the*

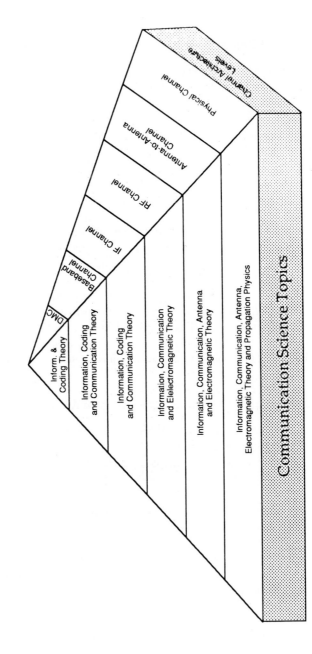

Figure 1.4 Communication systems engineering education pyramid (Reprinted from [38])

The following labels appear within the figure:

- DMC
- Baseband Channel
- IF Channel
- RF Channel
- Antenna-to-Antenna Channel
- Physical Channel
- Channel Architecture Levels

- Inform. & Coding Theory
- Information, Coding and Communication Theory
- Information, Coding and Communication Theory
- Information, Communication and Elelectromagnetic Theory
- Information, Communication, Antenna and Electromagnetic Theory
- Information, Communication, Antenna, Electromagnetic Theory and Propagation Physics

Communication Science Topics

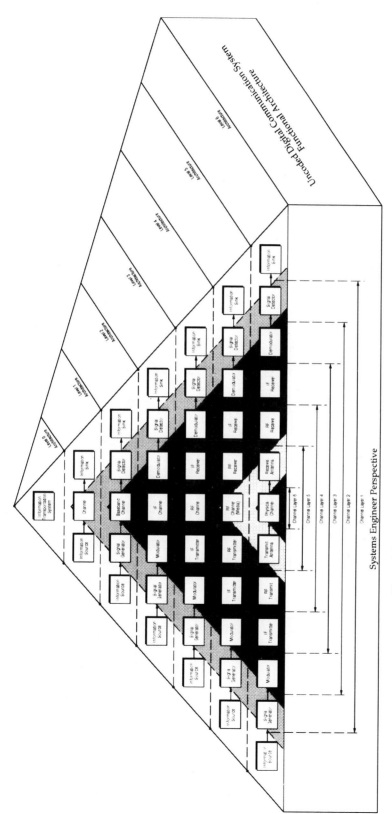

Figure 1.5 Information Transportation System (ITS) pyramid; uncoded digital communication system functional architecture (Reprinted from [38]) (In color on front end paper.)

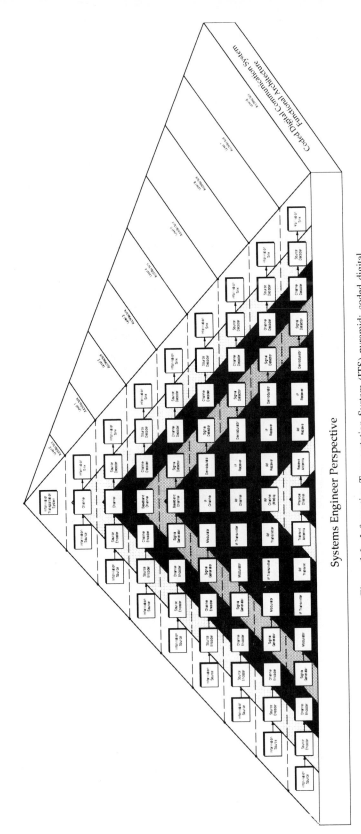

Figure 1.6 Information Transportation System (ITS) pyramid; coded digital communication system functional architecture (Reprinted from [38]) (In color on back end paper.)

10

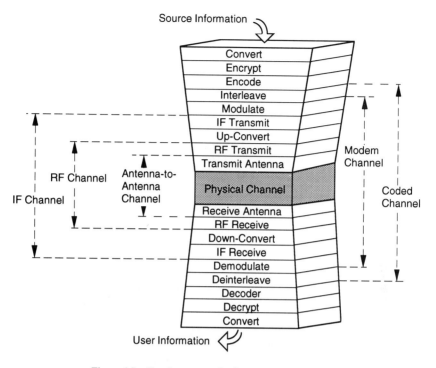

Figure 1.7   Simplex communication channel architecture

communication channel from the theoretical perspective?" The digital communication system building blocks elements which define the channel, the signal processing functions, the building blocks and hardware/software interfaces are integrated in the *communications pipe* illustrated in Fig. 1.7. It is suggested that the reader carefully examines Fig. 1.7 to understand this important communications pipe.

1.1.4 ISO-OSI Network Architecture

In digital communication networks, data and voice can be sent from one point to another in one of two fashions. A circuit connection can be set up in the network between the source and the user. Then the message is transmitted, and when it finishes, the circuit connection is deleted. Another method for communication is to break up the message into smaller packets and transmit the packets through the network (possibly through different paths) to the destination, and then reconstruct the message at the user site. The latter technique results in packet-switched data networks while the former technique results in circuit-switched networks.

Figure 1.8 demonstrates the seven layers that make up the ISO-OSI Network Architecture for packet-switched networks and shows how the architecture demonstrated by the *ITS Pyramids* fits into the Physical Layer (Layer 1) of the ISO-OSI Architecture: Open System Interconnection (OSI) network architecture is an approach to interprocessor communications developed by the International Organization for Standardization (ISO) and is the focus of that research effort. For details of this important and detailed standard, the reader is referred to the basic material contained in [4–10]. The functions of each of the layers are summarized in the following list.

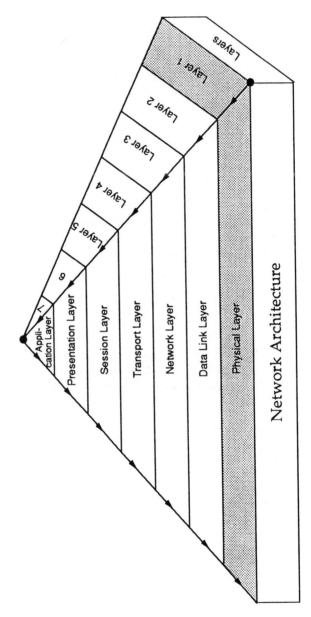

Figure 1.8   ISO-OSI network architecture model

1. *The Physical Layer:* The architecture of this layer is presented in Figs. 1.5 and 1.6. It is further detailed in Fig. 1.7. This layer is the physical communication medium that supports the transmission of bits and is concerned with the modulation and demodulation process and functions such as synchronization and framing.

2. *The Data-Link Layer:* This layer provides the basic link frame, including information packet, start and end headers, and an error check field. It ensures that packets are reliably transmitted through the network and it shields higher layers from any concerns about the physical transmission medium. The service provided by the *data link layer* is the reliable transfer of a packet over a single link in the network. This is accomplished through positive and negative acknowledgments for error checking in the link.

3. *The Network Layer:* This layer provides the networkwide (or internetwork) services required to transmit, route, and relay data between computers (or nodes) without regard for the communications medium. This layer generates (at the source) and reads (at the destination) the address information and control fields for purposes of flow control. It is responsible for establishing, maintaining, and terminating the network connection between two users and for transferring data along that connection.

4. *The Transport Layer:* This layer is responsible for providing data transfer between two users at an agreed on *quality of service (QOS)*. When a connection is established between two users, the layer is responsible for selecting a particular class of service, to monitor transmissions to ensure the appropriate service quality is maintained, and to notify the users if it is not.

5. *The Session Layer:* This layer provides dialogue services for those functions that require the establishment of connection and synchronization between any two machines prior to the exchange of information. This layer provides the "Are you there"—"Yes I am" exchange prior to the exchange of application data. A primary concern of the session layer is controlling when users can send and receive, based on whether they can send concurrently or alternately.

6. *The Presentation Layer:* This layer provides a common representation of application data that is communicated between application entities. Common representation refers to the encoding of data and the order of bits and bytes exchanged between processors. This may include character code translation, data conversion, or data compression and expansion. For example, the exchange of data between a processor using American Standard Code for Information Interchange (ASCII) encoded characters and a processor using Extended Binary Coded Decimal Interchange Code (EBCDIC) encoded characters requires a data translation before the information can be utilized.

7. *The Application Layer:* This layer provides the user program with an interface to an OSI system. In this case, the user is any computer program or application program that requires interprocessor communications. The layer includes services used to establish and terminate the connections between users and to monitor and manage the systems being interconnected and the various resources they employ. A number of common communication functions have been identified and grouped into so-called application entities with standard services and protocols. The application entities are the key to open systems.

1.2 TELECOMMUNICATION NETWORKS AND SERVICES

Telecommunication networks are used extensively in the modern world to interconnect geographically separated nodes or "terminals" for the purpose of providing a specific service. The elements of a network can be spread over a relatively small area, such as a building or a university campus, or over cities or even continents. The elements or nodes themselves can be stationary or dynamic depending on the application. The number of nodes typically ranges from a few elements to a few thousand or more. Each node can consist of a simple telephone or a computer terminal to a more elaborate "receiver" or signal processing center. The network performs a function or provides a service from which its name is typically derived. Basically, any interconnected set of elements that communicate with each other in some manner qualify as a telecommunication network. A very simple network can consist of a set of two personal computers that communicate with each other to share a common data base. It is worthwhile pointing out at this time the technology forecast made in 1945 by Arthur C. Clarke [11] about using rocket stations to provide worldwide radio coverage. It ranks among the most accurate and history-changing of all technology forecasts in modern times. The majority of current networks rely on satellites or "rocket stations" to communicate the information necessary to provide their respective services. Examples of such networks are the National Aeronautics Space Administration's (NASA's) Deep Space Network, the Telecommunication and Data Relay Satellite System, and the Global Positioning System.

1.2.1 Deep Space Network (DSN)

The Deep Space Network (DSN) [12, 13] is the largest, most sensitive scientific telecommunications and radio navigation network in the world. Its principal responsibilities are to support automated interplanetary spacecraft missions and radio and radar astronomy observations in the exploration of the solar system and the universe. The DSN also supports high-Earth orbiters, lunar and Space Shuttle missions.

The DSN is managed, technically directed, and operated for the National Aeronautics and Space Administration (NASA) by the Jet Propulsion Laboratory (JPL), California Institute of Technology, Pasadena, California. The DSN is a collection of three communication complexes located in California (USA), Madrid (Spain), and Canberra (Australia) that support deep space missions. Each complex consists of basically three or more "large" antennas that receive and transmit commands or data to the space probes. The location of the complexes was chosen to provide global support and maximize coverage availability. Figure 1.9 depicts the DSN's largest antenna, which is 70 meters in diameter. Each node of this network consists of either a complete communication complex or a deep space spacecraft. The network is nonstationary in the sense that the relative distance between some of the nodes is changing as a function of time. The use of large antennas in addition to supercooled front-end receivers has enabled the DSN to receive some of the weakest signals (-185 dBm) used in communication systems.

1.2.2 Tracking and Data Relay Satellite System (TDRSS)

The Tracking and Data Relay Satellite System (TDRSS) [14] consists of a space network (SN) and a ground network (GN) and provides two-way relay of communications between a user spacecraft and the ground terminal located at White Sands, New Mexico. The system has evolved in two phases, TDRSS-I and TDRSS-II. In TDRSS-I, the two operational space

Figure 1.9 The Deep Space Network (DSN) 70-meter antenna (Courtesy of Jet Propulsion Laboratory)

segment elements, TDRS-East and TDRS-West, are shown in Fig. 1.10a and are located respectively at about 41° and 171° west longitude in geosynchronous orbits. A spare satellite (TDRS-Spare), located at 61° west longitude, can be repositioned at either location in case of a satellite malfunction. The SN provides tracking and data acquisition with more than 85% coverage for orbit altitudes of 200 to 1,200 km, and 100% coverage for orbit altitudes of 1,200 to 12,100 km. Return data rates of up to 300 Mb/sec from user spacecraft are supported, as are a number of different services. In TDRSS-II depicted in Fig. 1.10b, the Zone of Exclusion (ZOE) is removed using the Out-Of-View (OOV) TDRS-II. Return data rates of up to 600 Mbits/sec can be supported. TDRSS supports the primary communication links to the space shuttle, as illustrated in Fig. 1.11.

1.2.3 Global Positioning System (GPS)

The third network to be discussed is NAVSTAR's Global Positioning System (GPS) [15, 16]. Its purpose is to provide three-dimensional position and velocity information as well as time transfer to users anywhere in the world at any time. The position determination is based on measuring the delay of radio frequency (RF) signals from four satellites of a total constellation of 24. A minimum of four measurements are necessary to solve for the four unknowns—three-dimensional position in addition to the user's clock offset. The 24 satellites are shown in Fig. 1.12 and operate in 12-hour orbits at an altitude of 20,183 km. The constellation provides visibility of six to eleven satellites at five degrees or more above the horizon to users located anywhere in the world at any time. The four satellites offering the best geometry can be selected automatically by the receivers using ephemeris

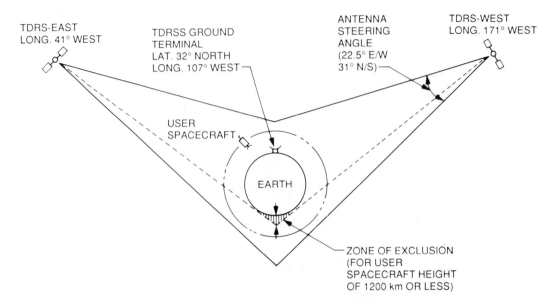

Figure 1.10a   Tracking and Data Relay Satellite System-I (TDRSS-I)

information transmitted by the satellites. Accuracies on the order of 10 meters in position are anticipated using commercial receivers and 1 meter or less using military receivers. Comparable position location accuracies can also be achieved using the GLONASS system, which has been designed and implemented by the former Soviet Union.

GPS has revolutionized position location and navigation in this century. For example, all commercial jets are being equipped with GPS receivers to allow for more efficient navigation and routing between two cities. GPS is also allowing for smarter highways and transportation systems where cars get information on traffic and adaptive routing is employed to minimize congestion [17].

1.2.4 Integrated Services Digital Network (ISDN)

The Integrated Services Digital Network (ISDN) and its broadband version have been under development for several years and are currently undergoing field tests [18,19]. The three fundamental principles embodied in ISDN and broadband ISDN are digitization with high bandwidth, worldwide standards, and integrated functionality and services. ISDN will finally integrate voice and data services enabling novel applications limited only by the human imagination [20]; for example, personalized newspapers in which an individual, using his home "terminal," can retrieve the sports section from the *Los Angeles Times*, the food section from the *New York Times*, the world news from the *Christian Science Monitor*, and so on, and print all sections in the form of one "personal" newspaper. Another example is the cost-effective home office in which an employee can communicate with his employers through voice, facsimile, and video simultaneously, thus enabling him or her to hold meetings with fellow employees also located at home or in different company locations.

1.2.5 Military Satellite Communications

In the military domain, the next generation satellite communications (MILSATCOM) system, designed to serve both the nation's strategic and tactical forces, is currently in the

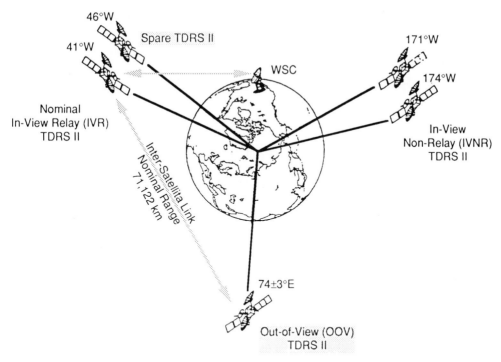

Figure 1.10b   Tracking and Data Relay Satellite System-II (TDRSS-II)

deployment phase. The new system, called MILSTAR [21] and depicted in Fig. 1.13, will provide a worldwide, highly jam-resistant, survivable and enduring MILSATCOM capability under conditions of both electronic warfare and physical attack. Through the use of extremely high frequency (EHF) and other advanced techniques, the system will recover very quickly from the propagation degradation caused by a high-altitude nuclear detonation. MILSTAR satellite crosslinks will operate in the W-band (60 GHz) to prevent eavesdropping and interference from enemy ground stations. They will also enable message relay around the world above the atmosphere without reliance on intermediate downlinks to vulnerable ground stations. This capability would ensure, for instance, that a presidential jet flying over the Pacific would be able to communicate with a tactical nuclear force in Europe even if all ground stations had been destroyed.

1.2.6 Mobile and Personal Communications

Satellite systems for mobile and personal communications and their interconnection with the Public Switched Telephone Network (PSTN) will no doubt be the next layer of telecommunication infrastructure developed for purposes of global communications and navigation. Mobile satellite communication systems (MSCS) are capable of delivering a range of services to a wide variety of terminal types. The architecture of a future MSCS is illustrated in Fig. 1.14a with objectives depicted in Fig. 1.14b; namely, communication from one hand-held terminal to another hand-held terminal anywhere in the world. Examples of mobile satellite terminal platforms include land vehicles, aircraft, marine vessels, remote data collection and control sites. Additionally, services can be provided to portable terminals, which are currently the size of a briefcase and no doubt will be reduced to hand-held sizes for future systems. Many future services will include position reporting to determine the position

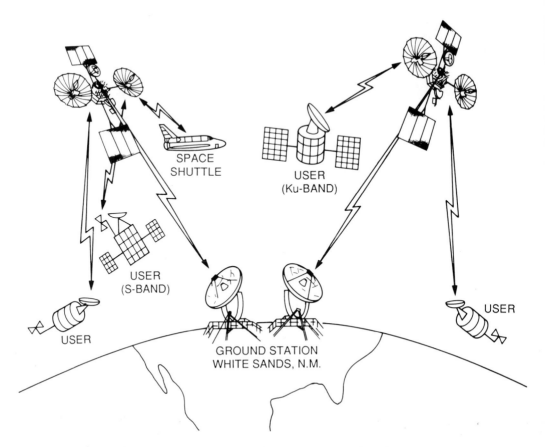

Figure 1.11 TDRSS/User spacecraft configuration

of the mobile user, with transmission of information via the mobile satellite communications system. In the future, *Mobile Satellite Services* and *Radio Determination Satellite Services* (RDSS) will provide the ultimate in *Global Personal Communications* when interoperated with the *Public Telephone Switched Network*. Ultimately, mobile satellite systems should be capable of providing *personal communications services* (PCS) such as voice and data (low-rate) to very small terminals including hand-held units. Applications requiring much higher data rates should be possible with somewhat larger mobile and portable terminals. Mobility and high capacity are the major new requirements that these future systems must satisfy.

At least three approaches have been proposed to the Federal Communication Commission (FCC). These include different network architectures consisting of satellite constellations in low Earth orbits (LEO), medium Earth orbits (MEO), and geosynchronous orbits (GEO). In addition, Europe is considering satellite constellations in highly eccentric (elliptic) orbits (HEO). These various orbits are depicted in Fig. 1.15. Presently, the International Maritime Satellite (Inmarsat) [22] Organization and the American Mobile Satellite Corporation (AMSC) systems utilize GEO.

Currently, there are several proposed systems: (1) Motorola's proposed*Iridium* [23] and Loral/Qualcomm's proposed *GlobalStar* [24] consisting of a satellite constellation in LEO, (2) TRW's proposed *Odyssey* [25] consisting of a satellite constellation in MEO, and

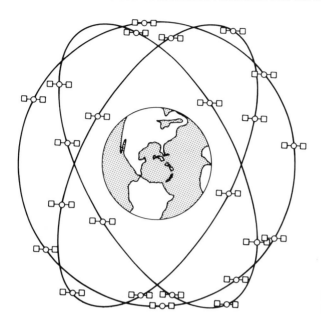

Figure 1.12   The Global Positioning
System (GPS) orbit configuration

(3) Europe's proposed *Archimedes* [26] consisting of satellites in HEO. It is clear that the multinational development of *global wireless communications* will be a ubiquitous reality in the twenty-first century. Table 1.1 summarizes the characteristics of the proposed systems in terms of number of satellites, number of orbit planes, orbit altitude, and so on.

The final system/network architecture associated with this wireless world is difficult to speculate on at this time; however, it seems safe to predict that before the year 2000, small shirt-pocket size, low-weight, full-featured personal communication hand-held terminals will be available with long talk time and high quality voice, data and position location capabilities that can be used anywhere in the world—either at home, in the office, on the street, or while on travel with either terrestrial and/or satellite facilities. This personal communications network (PCN) will be implemented such that any of the network facilities will be totally transparent to the end user. In the next few sections, we will give a brief overview of several systems proposed as the infrastructure for future generation wireless, mobile, personal communications.

Inmarsat.   Inmarsat [22] is currently the world's largest mobile satellite communications organization, backed by more than 60 national telecommunications enterprises, as well as a worldwide network of equipment manufacturers, distributors, and service providers. Established in 1979, Inmarsat has its headquarters in London and has 64 member countries as of May 1992.

The worldwide services that are currently available through Inmarsat include: direct-dial telephone, facsimile, telex, electronic mail, data transmission, flight deck voice and data services, position reporting, and fleet management. In the near future, global paging, briefcase telephones, and voice and data for the business traveler will be available. Inmarsat currently uses a total of 10 satellites in its global system. While some of these satellites are carrying commercial communications traffic, others are in orbit as backup spares to ensure continuity of service. The satellites, as shown in Fig. 1.16, are arranged in geostationary orbit to provide communications coverage of all areas of the globe with the exception

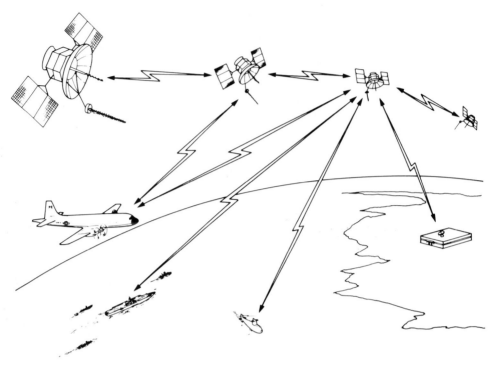

Figure 1.13 The MILSTAR system

of the extreme polar regions. To provide enough capacity to meet the huge expansion in demand, Inmarsat has acquired Inmarsat-2 satellites for the 1990s. A new generation of Inmarsat-3 satellites with dedicated spot beams is in production for launch in 1994–95, and its characteristics are shown in Fig. 1.17.

Iridium.   Motorola's proposed *Iridium* [23] is a LEO satellite-based system designed to provide global telecommunication services on a continuous basis with coverage of all units from the North Pole to the South Pole. A system overview is depicted in Fig. 1.18. Five types of basic service are presently planned, each with a separate terminal or family of terminals:

1. Radio determination service and two-way messaging: All subscriber units, except some pagers, will contain this capability, which includes automatic location reporting. Units with GPS or GLONASS services will also be available, which will provide greatly increased geolocation speed and accuracy.
2. Digital voice communications service: The system will provide duplex high-quality 4.8 kb/s voice communications.
3. Paging: An alphanumeric pager for instantaneous direct-satellite paging in any region of the world is being designed in a package only slightly larger than today's pagers.
4. Facsimile: Two types of mobile facsimile units are planned, a stand-alone facsimile unit and a unit to be used with an *Iridium* telephone.
5. Data: A 2.4 kb/s modem is being developed for use within the *Iridium* network.

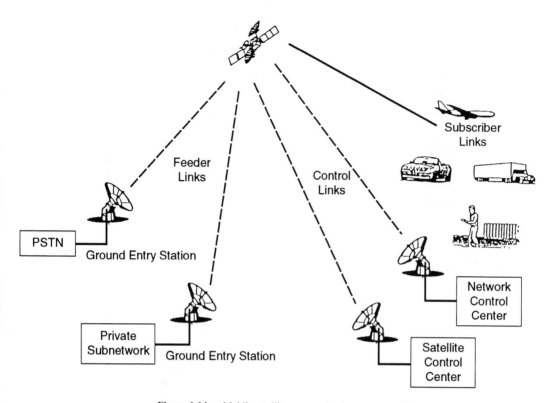

Figure 1.14a   Mobile satellite communication system architecture

The backbone of the *Iridium* system is the constellation of 77 small satellites in polar Low-Earth-Orbit (LEO) at an altitude of 420 nautical miles (778 km). This relatively low altitude assures that a user will never have to communicate further than about 1,250 nautical miles (2,315 km) to a satellite. This allows the use of lightweight low-power radiotelephones with small low-profile antennas compared to those required to utilize geostationary satellites at ranges of more than 19,000 nautical miles (35,188 km). The constellation comprises seven planes of 11 satellites, each plane equally spaced in longitude. The orbital period of each satellite is approximately 100 minutes.

The satellites are interconnected via microwave crosslinks, thus forming a global network in space, with linkage to the ground via gateways. Each gateway interfaces with the local Public Switched Telephone Network (PSTN), providing local area user/customer record keeping and interconnection capability between *Iridium* and *non-Iridium* users. The decision to locate a gateway in any given area can be made by local governments and/or servers solely out of a desire to offer local service, rather than to satisfy a functional need by the system for the gateway to be a node in the network to assure global coverage. Satellite telemetry, tracking and control, and overall system control will be accompanied by one or more centrally located System Control Facilities utilizing earth terminals and microwave links.

The radio links between users and satellites will operate somewhere in the 1,600 MHz to 1,700 MHz region of L-Band, while the gateway feeder links and satellite crosslinks will operate in small portions of the 18 GHz to 30 GHz region of K-Band. L-Band is best

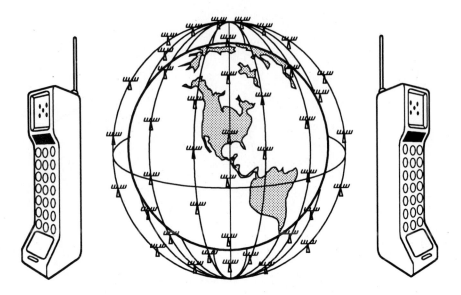

Communicate from one hand-held terminal to another hand-held terminal anywhere in the world

Figure 1.14b   Mobile Satellite Communication System (MSCS) objective

suited for direct satellite-to-portable-user links because it offers a good mix of relatively short wavelengths and low propagation losses, and it includes the upper frequency limits of economical commercial hardware using available technologies. Each satellite employs an L-Band phased array antenna complex that forms 37 interlaced spot beams (cells). Each beam operates independently. Each carrier of each beam has its downlink transmit power electronically controlled to maintain link quality for each user, allowing for variations in range and blockage/foliage. Further design details can be found in [23].

Odyssey.   TRW's proposed *Odyssey* [25] is a MEO satellite system consisting of (as few as) 12 satellites in MEO for personal communications as depicted in Fig. 1.19. *Odyssey* will provide high-quality, wireless communications from satellites worldwide. The telecommunication services include voice, data, paging, radio, determination (position and navigation), and messaging. It will use *Code Division Multiple Access* (CDMA) and *Frequency Division Multiple Access* (FDMA) techniques to achieve compatibility with other systems.

The medium altitude at which *Odyssey* satellites will operate is 10,240 km, and they will be capable of communicating with hand-held transceivers. The system's space segment consists of 12 satellites. Each satellite has two 19-beam, steerable antennas that are used to communicate with the system users. The satellite/user links are implemented in the RDSS Band. The satellites also communicate with the Earth stations, in Ka-band via a high gain steerable antenna. In addition to providing the user with RDSS functions, the system provides access to the public switched network by use of digital voice and spread spectrum techniques. The smallest hand-held user transceiver is proposed to use 0.5 Wattts of RF transmit power and is capable of carrying a digital voice signal at 4.8 kbps. The satellites operate as simple frequency translating transponders.

<u>LEO</u>: LOW EARTH ORBIT

(Circular: 900 –1,400 km Altitude
Elliptic: 400 –3,000 km Altitude Range)

<u>MEO</u>: MEDIUM EARTH ORBIT

(Circular: 10,000 km Altitude)

<u>GEO</u>: GEOSTATIONARY ORBIT

(Circular: 35,800 km Altitude Over
Equator)

<u>HEO</u>: HIGHLY ECCENTRIC ORBIT

(E.G.: 1,000 –26,555 km)

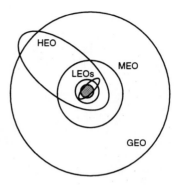

The Lower the Orbit the Faster its Period of Rotation Around the Earth

Figure 1.15   Classification of satellite orbits

Table 1.1   Characteristics of Proposed Systems

	Odyssey	GlobalStar	Iridium
Number of satellites	12	48	66
Number of orbit planes	3	8	6
Orbit inclination (deg)	55	52	87
Orbit altitude (km)	10354	1389	765
Intersatellite links	no	no	yes
Repeater type	bent-pipe	bent-pipe	processing
Frequency Bands (GHz)			
• User to satellite	1.6	1.6	1.6
• Satellite to user	2.5	2.5	1.6
Multiple-access mode	CDMA	CDMA	TDMA

High-elevation angle operation, that is, greater than 30°, reduces the possibility of signal blockage for a mobile user. The use of steerable antennas permits the coverage of selected areas of the Earth for extended periods of time, as the satellite passes overhead. This significantly reduces the amount of beam-to-beam switching that must be done and simplifies the overall system design. For RDSS purposes, the system employs a combination of ranging measurements and Doppler measurements to determine the location of a user.

For the user/satellite links, the 16.5 MHz frequency bands are divided into three sub-bands of 5.5 MHz each. Each beam, of the 19 beam user/satellite antennas, transmits or receives in only one of the three sub-bands. The antennas are designed to have similar spot beam sizes and beam positions in both the 1.6 GHz and 2.5 GHz bands. For satellite/Earth station link, each of the 19 received or transmitted sub-bands are frequency shifted to a different frequency requiring 16.5 MHz total bandwidth. This value includes data channels,

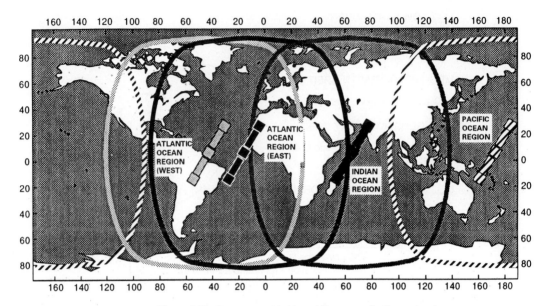

Figure 1.16   Inmarsat worldwide mobile communications network

order-wire channel, and guard bands. Because of the relatively high values of rain attenuation in the 20 and 30 GHz frequency bands, diversity Earth station techniques will be
employed. Further detailed discussion of the design and performance can be found in [25].

GlobalStar.   Loral/Qualcomm's proposed *GlobalStar* system, with characteristics
shown in Fig. 1.20, is a LEO satellite-based mobile communications system providing radio
determination, mobile voice, and data communications services. The *GlobalStar* system is
interoperable with the current and future Public Switched Telephone Network (PSTN) and
Public Land Mobile Networks (PLMN), such as cellular telephone systems, Personal Communications Networks (PCNs), or Personal Communications Services (PCS). The *Global-
Star* system also is designed to interface with private networks and provide mobile services
to private network users.

The *GlobalStar* space segment consists of a constellation of 48 LEO satellites in
circular orbits with 750 nautical miles (1389 km) altitude to provide 100% global coverage.
In this constellation, there are eight orbital planes and each plane has six satellites, which
are equally phased within the orbital plane. Each orbital plane has an inclination angle
of 52 degrees. Each satellite has a 7.5 degree phase shift to the satellite in the adjacent
orbital plane. Initially, a constellation of 24 operating satellites will be launched to provide
maximum U.S. coverage. As worldwide traffic demand develops, 24 (or more) additional
satellites will be launched to provide continuous global coverage. Further design details can
be found in [24].

Archimedes.   *Archimedes* is a proposed mobile satellite system conceived by the
European Space Agency (ESA) to effectively serve the European market for Mobile Radio
Services (MRS). *Archimedes* [26] was conceived to effectively serve the growing demands
for Mobile Radio Services (MRS). The communications systems architecture is very similar
to that for any GEO satellite communications system, the major differences being due to

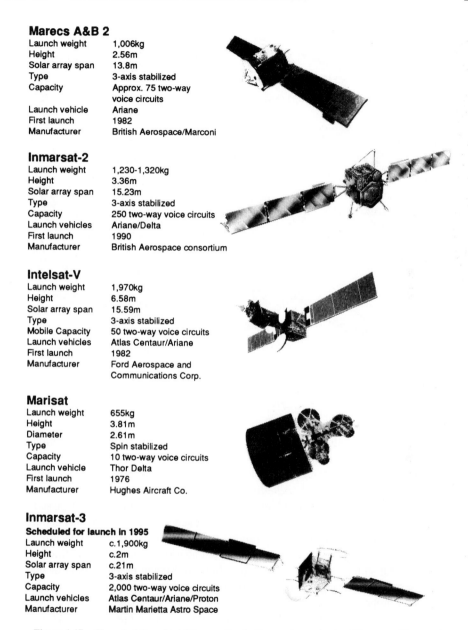

Marecs A&B 2

Launch weight	1,006kg
Height	2.56m
Solar array span	13.8m
Type	3-axis stabilized
Capacity	Approx. 75 two-way voice circuits
Launch vehicle	Ariane
First launch	1982
Manufacturer	British Aerospace/Marconi

Inmarsat-2

Launch weight	1,230-1,320kg
Height	3.36m
Solar array span	15.23m
Type	3-axis stabilized
Capacity	250 two-way voice circuits
Launch vehicles	Ariane/Delta
First launch	1990
Manufacturer	British Aerospace consortium

Intelsat-V

Launch weight	1,970kg
Height	6.58m
Solar array span	15.59m
Type	3-axis stabilized
Mobile Capacity	50 two-way voice circuits
Launch vehicles	Atlas Centaur/Ariane
First launch	1982
Manufacturer	Ford Aerospace and Communications Corp.

Marisat

Launch weight	655kg
Height	3.81m
Diameter	2.61m
Type	Spin stabilized
Capacity	10 two-way voice circuits
Launch vehicle	Thor Delta
First launch	1976
Manufacturer	Hughes Aircraft Co.

Inmarsat-3

Scheduled for launch in 1995

Launch weight	c.1,900kg
Height	c.2m
Solar array span	c.21m
Type	3-axis stabilized
Capacity	2,000 two-way voice circuits
Launch vehicles	Atlas Centaur/Ariane/Proton
Manufacturer	Martin Marietta Astro Space

Figure 1.17   Characteristics of satellites used by the Inmarsat organization (Courtesy of Inmarsat)

the fact that the *Archimedes* satellites move with respect to a point on the Earth's surface. These differences are: the requirement to regularly hand over from one satellite to the next, differential path length, and differential Doppler shift. These factors add to the complexity of both fixed and mobile terminals as circuitry has to be built in to mitigate these effects.

Services are provided in much the same way as for a GEO satellite system. Broadcasters uplink their channels via high frequency fixed feeder links to the satellite as shown in

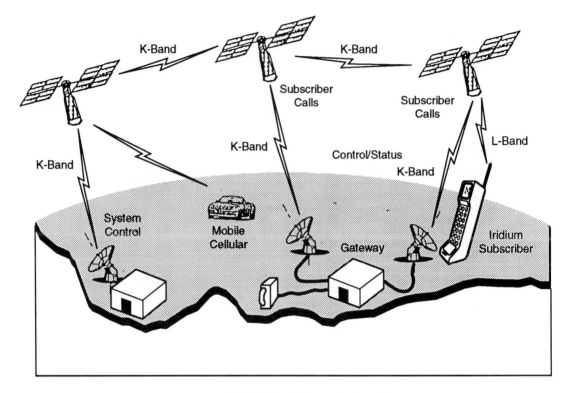

Figure 1.18a   *Iridium* system overview

Fig. 1.21. The satellite then transmits the broadcast signal at the lower mobile communications frequencies to enable reception by mobiles throughout the coverage area. Figure 1.21 illustrates the communications system architecture showing various potential user types. The overall coverage of the satellite system is Europe, but to reduce RF power per channel, it is advantageous to use high-gain spot beams. The formation of these spot beams is intrinsically linked to the choice of frequency due to the laws of antenna geometry.

A critical parameter in determining the cost per channel of the *Archimedes* system is the capacity of the system, in terms of the number of channels the satellite communication system can simultaneously support. Satellites are limited in the amount of power that they can supply to their communications payloads, and therefore, the capacity of the system is directly proportional to the power requirement of each channel onboard the satellite. The first generation *Archimedes* system is intended to serve Europe. Europe has a relatively high latitude and therefore, the elevation angle to a GEO satellite is relatively low. The propagation environment is very important for mobile communications and can be significantly improved if higher elevation angles are given. This drives the consideration of alternative orbits, particularly highly eccentric (elliptic) orbits (HEO). The increased elevation angles given for HEO can be converted into decreased RF power requirements by considering propagation margins. This is the most severe propagation environment *Archimedes* is intended to be served by terrestrial retransmission. These measurements show the minimum advantage of HEO over GEO to be 7 dB.

The mobile terminals are intended to be very simple and hence have nonsteerable antennas. This means the satellite must be within the antenna beam at all times. For HEO,

- Supports Hand-Held & Moblie Users (4800 BPS Digital Voice and Other Services)

- Constallation of 77 Satellites in 7 Polar Planes Provides Global Coverage

- Low Altitude Enables Hand-Held Users—at 900 km Min. Elevation Angle to Satellite is 10 deg.

- Each Satellite Has a 37-Cell Terrestrial Antenna Pattern Fixed Relative to Satellite but Moves Quickly Accross Earth

- Satellites Networked by Using Space Crosslinks

- System is Mirror Image of Cellular Telephone System

- TDMA/FDMA Combination

- L-Band for Both Up- & Down-Link from User

- Complex Networking

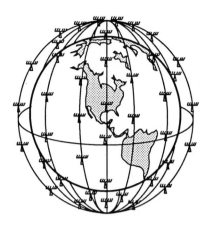

77 Processing Satellites
488 KG; 1429 W

Figure 1.18b *Iridium* characteristics

the satellite is always within a cone directly overhead and therefore a highly directional antenna can be used, whereas for GEO, the satellite can be anywhere within a fairly large toroid and therefore, a low-gain toroidal coverage antenna is needed. For further details of the proposed system, see [26].

There are too many telecommunication networks proposed or in existence to enumerate or discuss, but the examples described above are representative of the majority. In conclusion, a network is an interconnected set of "elements" or "terminals" that communicate with each other to provide a particular service.

1.3 PERFORMANCE CRITERION AND LINK BUDGETS

Characterizing the performance of a digital communication system is a key step for comparison of different systems and techniques. In this section, we first define power, bandwidth, and their relation to power spectral density. Then, the various elements and design parameters of a link are examined and unified in the link budget equation, followed by a discussion of signal power-to-noise ratio as a performance criterion.

1.3.1 Relation between Power and Bandwidth

Both power and bandwidth play a key role in system design and analysis. Often, they must be traded off against each other. The average power of a wide-sense stationary (WSS) random process $x(t)$ is given by

$$P_x = E\left\{x^2(t)\right\} = R_{xx}(0) = \int_{-\infty}^{+\infty} S_{xx}(f)df \qquad (1.1)$$

where $R_{xx}(\tau)$ is the autocorrelation function[1] of the random process and $S_{xx}(f)$ its respective *Power Spectral Density (PSD)* [27] in units of Watts/Hz. By replacing statistical

---

1. The autocorrelation function $R_{xx}(\tau)$ and its corresponding power spectral density $S_{xx}(f)$ are often denoted using one subscript, such as $R_x(\tau)$ and $S_x(f)$.

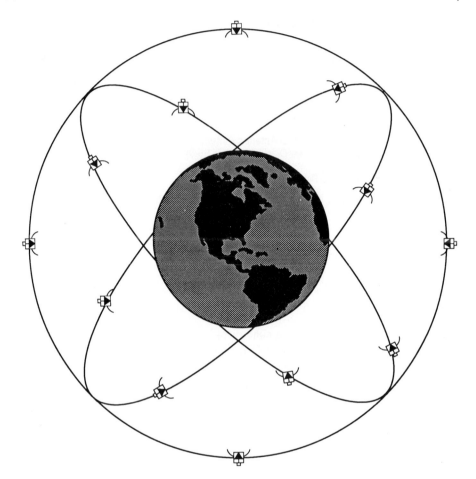

Figure 1.19 *Odyssey* satellite constellation

averaging with time averaging, the above definition can be extended to include ergodic processes. Although the power of a waveform can be defined precisely, the notion of waveform bandwidth is difficult to define and is sometimes vague and misleading. The basic objective for specifying bandwidth is to indicate the frequency range over which most of the waveform power is distributed. For random processes,[2] bandwidth can be derived from the PSD $S_{xx}(f)$ which is the Fourier transform of the autocorrelation function $R_{xx}(\tau)$. The most commonly used bandwidths (in Hertz) are the "3-dB bandwidth," the "$p\%$ fractional power bandwidth," the "*Gabor* or *root-mean-square* bandwidth," and the "*equivalent noise* bandwidth," defined as follows (one-sided definitions):[3]

1. *3-dB Bandwidth ($B_{3-\,\text{dB}}$):* Most useful for unipodal PSDs, it is the width along the positive frequency axis between the frequencies where $S_{xx}(f)$ has dropped to half its maximum value, that is,

2. When discussing bandwidth of a filter $H(f)$, replace the PSD $S_{xx}(f)$ with the magnitude squared of the filter transfer function, that is, $|H(f)|^2$, and all subsequent equations are valid.

3. We assume that the PSD is symmetric, that is, $S_{xx}(f) = S_{xx}(-f)$.

- Relies technically on Qualcomm's Cellular CDMA

- 48 Bentpipe Satellite Architecture that maximizes use of terrestrial infrastructures (mobile and fixed)

- 6 satellites in each of 8 52-deg. inclined orbits at 1,389 km; Bol Mass = 263 kg; Power = 875 w

- 6 isoflux spot beams at L-& S Bands with frequency reuse of 2

- Services include voice, data, & RDSS to hand-held & mobile

- CDMA with 1.25 MCPS spread spectrum

- Variable voice coding rate: 1.2 - 9.6 KBPS

- Minimum elevation angle: 5 deg.

- Rake reception for coherent combining of simultaneous signals through 2 or 3 satellites

Figure 1.20   *GlobalStar* overview

$$S_{xx}(B_{3-\text{dB}}) = \frac{1}{2}S_{xx}(f)\text{max} \tag{1.2}$$

This definition has the advantage that it can be computed directly from the spectral density function. It has the disadvantage that it may be misleading for spectral densities with slowly decreasing tails and it may become ambiguous and nonunique for spectral densities with multiple peaks, that is, there may be several frequencies at which the spectral density is half its maximum value.

2. *p% fractional Bandwidth (B_p):* It is the frequency width along the positive frequency axis within which $p\%$ of the one-sided spectrum lies, that is,

$$\frac{\int_{f_1}^{f_2} S_{xx}(f)df}{\int_0^\infty S_{xx}(f)df} = \frac{p}{100} \tag{1.3}$$

where $f_2 - f_1 = B_p$. This definition has the advantage of being unique. However, it can be misleading for spectral densities with somehow slowly decaying tails. Also, it has the disadvantage that it must be evaluated by integration of the power spectral density.

3. *Gabor Bandwidth (B_{rms}):* It is defined as the *root-mean-square* (rms) bandwidth of the PSD, that is,

$$B_{rms} = \left\{ \int_0^\infty f^2 S_{xx}(f)df \right\}^{1/2} \tag{1.4}$$

This bandwidth is mostly used in radar systems and it also has the disadvantage of being evaluated by integration.

4. *Equivalent Noise Bandwidth (B_n):* It is the most common definition used in studies of systems with stochastic inputs. It is defined as the width along the positive frequency axis within which a flat spectral density gives a value of power equal to that of the given power spectral density where the value of the flat spectral density corresponds to the maximum of the given spectral density (see Fig. 1.22). In mathematical terms

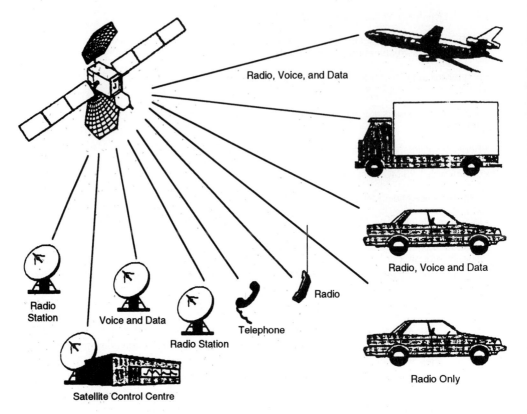

Figure 1.21 *Archimedes* communications architecture

$$B_n = \frac{\int_0^\infty S_{xx}(f)\,df}{S_{xx}(f)_{\max}} \tag{1.5}$$

Often, a two-sided noise bandwidth is also defined. In that case

$$W_n = 2B_n = \frac{\int_{-\infty}^{+\infty} S_{xx}(f)\,df}{S_{xx}(f)_{\max}} \tag{1.6}$$

An alternate interpretation of noise bandwidth is obtained by passing white noise $n(t)$ of two-sided spectral density level $N_0/2$ through a filter $H(f)$. Then, the PSD of the output process $x(t)$ is

$$S_{xx}(f) = \frac{N_0}{2}\,|\,H(f)\,|^2 \tag{1.7}$$

with average power

$$P_x = \int_{-\infty}^\infty S_{xx}(f)\,df = \int_{-\infty}^\infty \frac{N_0}{2}\,|\,H(f)\,|^2\,df \tag{1.8a}$$

If we were to pass the white noise through another brick-wall filter with gain

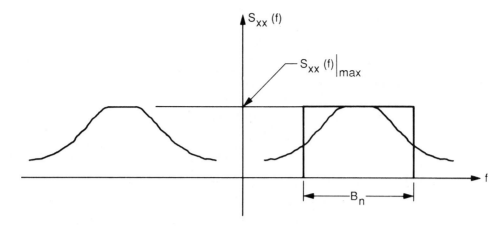

Figure 1.22   The noise bandwidth

$|H(f)|_{\max}$ and two-sided bandwidth W_n, then the output process $y(t)$ has power

$$P_y = W_n \frac{N_0}{2} |H(f)|_{\max}^2 \tag{1.8b}$$

For the two processes $x(t)$ and $y(t)$ to have identical powers, we require $P_x = P_y$ resulting in

$$W_n = \frac{\int_{-\infty}^{\infty} |H(f)|^2 \, df}{|H(f)|_{\max}^2} = 2B_n \tag{1.9}$$

In this context, W_n is the total (positive and negative) spectral width of an ideal (brick-wall) filter whose amplitude characteristic is equal to the maximum of $|H(f)|$ and which yields the same power at its output as does $H(f)$ when white noise is present at the input. For any of these bandwidths, the fractional out-of-band power, P_{ob}, is given by

$$P_{ob} = 1 - \frac{\int_0^B S_{xx}(f)df}{\int_0^{\infty} S_{xx}(f)df} \tag{1.10}$$

Example 1.1

The transfer function of a single-pole Butterworth filter is given by

$$H(f) = \frac{a}{a + j2\pi f} = \frac{1}{1 + j\frac{2\pi f}{a}}$$

It is clear that

$$H(f)_{\max} = H(0) = 1$$

The 3-dB bandwidth is thus obtained by

$$|H(B_{3-\text{dB}})|^2 = \frac{1}{1 + \left(\frac{2\pi B_{3-\text{dB}}}{a}\right)^2} = \frac{1}{2}; \quad B_{3-\text{dB}} = \frac{a}{2\pi}$$

On the other hand, the 90% bandwidth is the solution to

$$\frac{\int_0^{B_{90}} |H(f)|^2\, df}{\int_0^\infty |H(f)|^2\, df} = 0.9$$

which gives

$$\frac{a \tan^{-1}\left(\frac{2\pi f}{a}\right)\Big|_0^{B_{90}}}{a \tan^{-1}\left(\frac{2\pi f}{a}\right)\Big|_0^\infty} = 0.9$$

and

$$\tan^{-1}\left(\frac{2\pi B_{90}}{a}\right) = 0.9\frac{\pi}{2}; \quad B_{90} = \frac{6.31a}{2\pi} = 6.31 B_{3-\text{dB}}$$

Finally, the noise bandwidth is evaluated using

$$B_n = \int_0^\infty |H(f)|^2\, df = a \tan^{-1}\left(\frac{2\pi f}{a}\right)\Big|_0^\infty = \frac{a}{4}$$

Therefore,

$$B_n = \frac{a}{4} = \frac{\pi}{2} B_{3-\text{dB}}$$

Example 1.2

For an N-pole Butterworth filter, we have

$$|H(f)|^2 = \frac{1}{1 + \left(\frac{2\pi f}{a}\right)^{2N}}$$

and the 3-dB bandwidth is still given by $a/2\pi$. However, the noise bandwidth becomes

$$B_n = \int_0^\infty \frac{1}{1 + \left(\frac{2\pi f}{a}\right)^{2N}}\, df$$

and simplifies to

$$B_n = \frac{(\pi/2N)}{\sin(\pi/2N)} B_{3-\text{dB}}$$

As N gets larger, the noise bandwidth approaches the 3-dB bandwidth, as expected.

Fortunately, most of the waveform spectra encountered in typical communication systems have a single peak with relatively fast spectral roll-off. As such, all the preceding definitions yield approximately the same value and none of the above-mentioned ambiguities occur. For a more elaborate treatment, Scholtz [28] discusses cases in which discrepancies among the various definitions do occur and thus a precise bandwidth definition is essential.

1.3.2 Link Budgets

A key characteristic of a communication channel is the physical transmission media employed; this constitutes the physical channel (the media) in Figs. 1.5, 1.6 and 1.7. Depending on the nature of the information and the propagation distance, a variety of transmission media can be employed ranging from a pair of wires to waveguides or even optical fibers. The

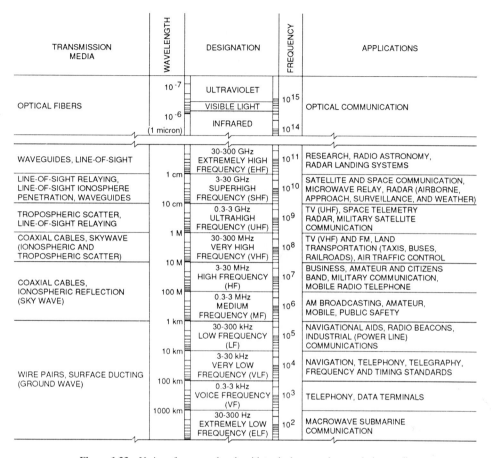

Figure 1.23   Various frequency bands with typical uses and transmission media

transmitting antenna transforms the electric signal into an electromagnetic wave for propagation while the receiving antenna performs the reciprocal process. Depending on its frequency, the wave can take on various forms ranging from a star light, to a television broadcast or even a laser beam. The electromagnetic spectrum has been divided into different bands which are assigned to serve specific types of communication as shown in Fig. 1.23. As an example, television broadcast occupies either the 30–300 MHz range (VHF band) or the 0.3–3 GHz range (UHF band).

In certain applications, such as terrestrial mobile communications, the radiated energy from one transmitting antenna may reach the receiving antenna by direct propagation or by reflections by the ground or the sky as depicted in Fig. 1.24. This phenomenon is referred to as *multipath,* which indicates that the total received electromagnetic energy is the weighted sum of "many" delayed versions of the transmitted signal where each component is randomly attenuated and delayed according to the propagation path it took to the receiver. Depending on the carrier frequency, sky reflections can result from ionospheric or tropospheric scattering (reflection and refraction). The ionosphere is a collection of gases, atoms, water droplets, pollutants, and so on, that absorb and reflect large quantities of radiated energy. It is located in the upper part of the atmosphere at an altitude of about 100–300 km

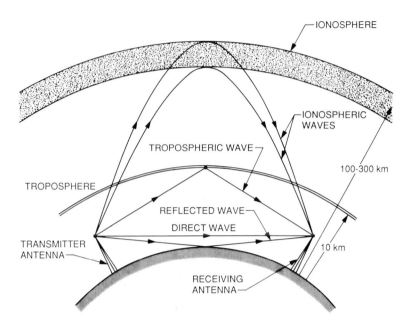

Figure 1.24 Multipath propagation in the troposhere and ionosphere

and is known to reflect signals in the VLF band. On the other hand, the troposphere is about 10 km from the earth's surface and is known to reflect carriers in the SHF band.

The design of any telecommunication link makes use of the so-called *communication equation* which contains all of the basic elements and design parameters inherent in the communication system and makes possible the specification of the received signal power-to-noise spectral density ratio (units of Hz) with a high degree of accuracy, viz.,

$$\frac{P_r}{N_0} = P_t G_t \left(\frac{\lambda}{4\pi d}\right)^2 L_\theta L_p L_a L_r \frac{G_r}{kT^\circ} \tag{1.11}$$

Here, P_r is the total received signal power (Watts), N_0 the one-sided power spectral noise density (Watts/Hz), P_t the transmitted total power (Watts), G_t the peak transmitting antenna gain (unitless), $\lambda = c/f$ the carrier wavelength (meters) and c is the speed of light (meters/sec), d the distance between transmitter and receiver (meters), L_θ the pointing loss (unitless), L_p the polarization loss (unitless), L_a the atmospheric loss (unitless), L_r the antenna surface tolerance loss (unitless), G_r the peak receiving antenna gain (unitless), k Bolzmann's constant ($1.38\ 10^{-23}$ Watts/Hz-$^\circ K$) and finally, T° the receiver temperature (Kelvin, $^\circ K$). Each element of the "power budget equation" given in (1.11) is defined and discussed here as to its effect on the system design.

A typical communication link is shown in Fig. 1.25 where various "virtual" channels are identified. At the transmitter, each information bit or block of bits is transformed into a unique baseband waveform by the baseband modulator. This signal is then upconverted twice to IF and then again to RF. Following power amplification, the signal is fed to the antenna for propagation. The *physical channel* is the *propagation media* between the transmitting and the receiving antenna. It can be a waveguide or propagation through rain as discussed earlier. Modeling the effects of the physical channel on the propagating electromagnetic field in different frequency bands and under various scenarios is a continuous and ongoing task [29]. At the receiving antenna output, the RF signal undergoes filtering and amplification using a low noise amplifier (LNA). The physics of these operations and their equivalent set at the transmitter are referred to as the *RF channel or subsystem*. Some of these operations are highly nonlinear and might introduce signal amplitude and phase distortion. Examples of such include Travelling Wave Tube (TWT) amplifiers [30] used prior to transmission which can introduce severe AM/PM effects, depending on their operating point. At the receiver, the output of the LNA is then downconverted to an appropriate IF for further filtering and processing. Communication system engineers deal mainly with modeling the IF signal in both the transmitter and receiver, or commonly designated the *IF channel or subsystem*. At the receiver, the signal model in the IF subsystem incorporates the various effects of both the physical and the RF channel such as fading due to shadowing, cross-polarization effects, AM/AM and AM/PM conversion in TWT amplifiers, amplitude and phase scintillation due to plasmas, and so on. Typically, the IF signal is again downconverted to baseband using inphase and quadrature references for further processing including bit detection. In modern receivers, the latter stage is typically implemented digitally as long as the data rates are well within the technology capabilities. The baseband signals or equivalently the *baseband channel or subsystem* are equivalent to their IF counterparts and are dealt with by communication system engineers interchangeably as the IF conversion to baseband constitutes a mere frequency shift with no additional effects.

In this section, we concentrate on the link budget for the additive white Gaussian noise (AWGN) channel. All effects in the physical channel are frequently modeled as power degradation in the received signal. This assumption is valid for the majority of the communication links encountered in practice. Let P_a denote the maximum power at the amplifier output. Then, the actual transmitted power P_t is equal to $P_a L_c$ where L_c denotes the circuit coupling loss and is given by

$$L_c = \frac{4Z_a Z_t}{(Z_a + Z_t)^2} \tag{1.12}$$

Here, Z_a and Z_t denote the output impedance of the amplifier and the input impedance of the antenna respectively. Only when the impedances are matched can maximum power transfer occur (that is, $P_t = P_a$). The transmitted electromagnetic signal is characterized by the *Effective Isotropic Radiated Power (EIRP)*, defined by

$$EIRP = P_t G_t L_\theta L_r \tag{1.13}$$

where G_t is the transmitting antenna gain in the receiver direction and L_θ, L_r the pointing and antenna surface tolerance losses, respectively. A rough surface can cause energy scattering and loss of gain in the desired direction. The antenna surface tolerance loss is approximately given by

$$L_r \simeq e^{-(4\pi\sigma/\lambda)^2} \tag{1.14}$$

where σ is the rms roughness of the antenna surface in wavelength dimensions [31]. For $\sigma/\lambda = 0.02$, the degradation is about 0.27 dB.

Antenna gain is defined as the ratio of the capture area of the antenna to that of an isotropic antenna which radiates the same power in all directions. The peak antenna gain G_t is given by

$$G_t = \max g_t(\phi_z, \phi_l) \tag{1.15}$$

where $g_t(\phi_z, \phi_l)$ is the antenna gain as a function of both azimuth angle ϕ_z and elevation

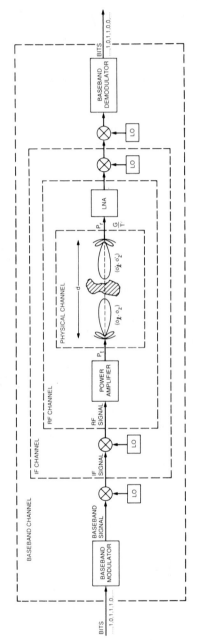

Figure 1.25   A simplified digital communication system

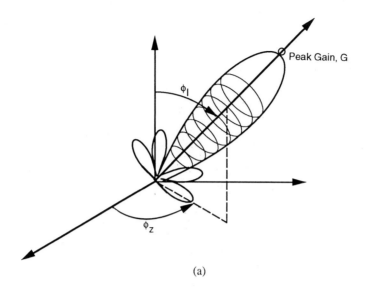

(a)

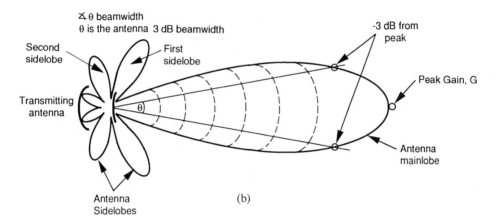

(b)

Figure 1.26   Antenna gain pattern

angle ϕ_l and G_t is the maximum gain taken over all angles (see Fig. 1.26a). Assuming perfect pointing, the gain of the transmitting antenna in the direction of the receiver would be exactly G_t. However, in more realistic scenarios, the antenna is not pointed perfectly and as a result, the gain in the receiver direction is slightly less than the peak gain G_t. That degradation is accounted for in the pointing loss L_θ. Another important parameter of an antenna gain pattern is its "beamwidth," which reflects the width of the gain in space as illustrated in Fig. 1.26b. For a circular reflector of diameter d_a, the maximum (peak) gain G_t and the half power beamwidth ϕ_b (radians) are given by

$$G_t \simeq \rho_a \left(\frac{\pi d_a}{\lambda}\right)^2 \tag{1.16}$$

$$\phi_b \simeq \frac{\lambda}{d_a \sqrt{\rho_a}} \tag{1.17}$$

where ρ_a is the antenna efficiency factor which is a function of the aperture field intensity distribution over the dish. This ρ_a is typically on the order of 55%. Note from (1.16) and (1.17) that the larger the antenna or the higher the frequency, the higher is the gain and the narrower the beamwidth. For a parabolic antenna with pointing error ϕ_e, the antenna pointing loss L_θ is approximately given by

$$L_\theta \simeq e^{-2.76(\phi_e/\phi_b)^2} \tag{1.18}$$

which indicates an exponential decrease in gain as a function of pointing error ϕ_e. Typical pointing errors are about $0.1° - 0.3°$ for geostationary satellites depending on the tracking capabilities of the antenna. Spatial pointing is maintained in the course of a track by using several techniques, one of which is conical scan in which the antenna forms small circles around the target while measuring the received power to determine the "best" spatial coordinates.

After propagating a distance d, the field flux density (FD—in units of Watts/meter^2) at the receiver becomes

$$FD(d) = \frac{(EIRP)L_a}{4\pi d^2} \tag{1.19}$$

where L_a is the atmospheric loss due to rainfall, ionized atoms in the ionosphere, and so on. The atmospheric losses are caused by absorption and scattering of the electromagnetic field by the various molecules in the atmosphere. The problem becomes more severe as the wavelength of the carrier approaches the size of these molecules. For example, severe absorption due to water vapor molecules occur at 22 GHz while oxygen absorption occurs at about 60 GHz. Other effects include degradation due to rainfall which becomes severe when the wavelength approaches the water drop size, which is dependent on the type of rainfall. For example, at a rainfall rate of 10 mm/hr, the attenuation is about 0.04 dB/km at 6 GHz but increases to 0.25 dB/km at 11 GHz. That degradation is also a function of the antenna elevation angle as the latter affects the path length of the propagation through the rain [29].

Assuming a receiving antenna with area A_r perpendicular to the direction of the transmitter, the collected signal power becomes [32]

$$P_r = FD(d)A_r L_p = \frac{(EIRP)L_a A_r L_p}{4\pi d^2} \tag{1.20}$$

where L_p accounts for any polarization loss due to misalignment in the received field polarization. Polarization is the orientation in space of the electromagnetic field and it is typically linear or circular. In the first, the electromagnetic field is aligned in one planar direction, whereas in the latter, it is radiated with two orthogonal components that when combined, appear to rotate on a circle or on an ellipse. In both cases, the receiving antenna must have a matching orientation. Circular polarization has the advantage of not being affected by an erroneous phase rotation in the electromagnetic field during propagation, unlike linear polarization, which suffers a degradation.

Using the reciprocity principle [32], the effective receiving antenna area can be expressed in terms of the antenna gain as

$$A_r = \frac{\lambda^2}{4\pi} g_r(\phi'_z, \phi'_l) \tag{1.21}$$

where $g_r(\phi'_z, \phi'_l)$ denotes the receiving antenna gain in the direction of the transmitter. Here again, the receiving antenna is typically not pointed perfectly and the gain in the transmitter direction might not be the peak gain G_r. Using (1.21) in (1.20), we have

$$P_r = P_t G_t L_s L_\theta L_p L_a L_r G_r \tag{1.22}$$

where L_s denotes the space loss and is given by $(\lambda/4\pi d)^2$. Both coupling as well as pointing losses can also occur in the receiver and need to be accounted for in (1.22).

The parameter N_0 in (1.11) represents the normalized noise power spectral density in Watts per Hertz (W/Hz). This factor is usually referenced to the input stage of the receiver since the signal-to-noise ratio is established at that point. It is given by

$$N_0 = kT^\circ \tag{1.23}$$

where k is Boltzmann's constant and T° the system noise temperature in Kelvin ($^\circ K$). This equivalent system noise temperature can be expressed as

$$T^\circ = T_b^\circ + T_e^\circ \tag{1.24}$$

where T_b° is the background noise temperature and T_e° the temperature of the receiver electronics. The latter is equivalently given in terms of the receiver *noise figure F* [33]

$$T_e^\circ = (F - 1)290^\circ \tag{1.25}$$

The standard technique employed to measure the noise temperature of a system is to compare, in a fixed bandwidth B, a measurement of system noise power P_n with the noise figure from a resistor at a known temperature. Since P_n and B are related by $P_n = kT^\circ B$, the unknown system noise temperature can be determined quite simply. The quantity B is the single-sided noise bandwidth, and noise power is always measured with respect to it.

The link budget of (1.11) is often rewritten as

$$\frac{P_r}{N_0} = \left[\frac{EIRP}{k}\right][L_p L_a]\left[\frac{G_r}{T^\circ}\right] \tag{1.26}$$

where the transmitter parameters are lumped in the first bracket, propagation parameters in the second, and the receiver figure of merit in the third.

Example 1.3

Table 1.2 depicts a link budget for an RF communication link between Mars and Earth as shown in Fig. 1.27. Two different link options are considered: a direct link and an indirect link. The first option consists of communicating from Mars to a Mars relay satellite (MRS) and then to Earth directly at a frequency of 8.4 GHz (X-band) with a data rate of 20 Mbits/sec [Table 1.2(a)]. The second option is similar except that the link to Earth will be relayed through a satellite (Earth relay satellite—ERS) and the carrier frequency is at 32.4 GHz (Ka-band) and the data rate is reduced to 10 Mbits/sec [Table 1.2(b)]. Since Ka-band suffers large degradations in bad weather, the ERS is essential in increasing link availability.

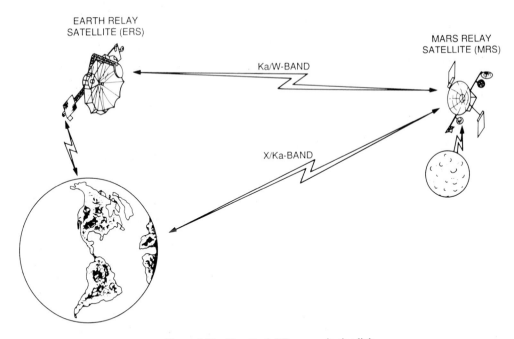

EARTH RELAY
SATELLITE (ERS)

MARS RELAY
SATELLITE (MRS)

Ka/W-BAND

X/Ka-BAND

Figure 1.27   Mars-Earth RF communication link

1.3.3 Signal-to-Noise Ratio as a Performance Criterion

Any performance criterion should assess the capability of a communication system to produce at its output a desired signal, that is, it should relate the desired operation to the actual
operation in the presence of the channel disturbances.

One of the most widely used measures of performance, particularly in systems with
stochastic additive disturbances, is the signal-to-noise ratio (SNR). The scenario consists of
a desired waveform (baseband or modulated carrier) contaminated by an additive interfering
waveform. When this combined waveform appears at any point in the system, the SNR at
that point (typically at the output of the receiver) is defined as

$$\text{SNR} = \frac{\text{Power in desired signal waveform}}{\text{Power in interfering waveform}} \tag{1.27}$$

SNR is an indication of how much stronger the desired signal is than the interfering signal
at the point in question. It is common to express SNR in dB, that is,

$$(\text{SNR})_{dB} = 10\log_{10}(\text{SNR}) = 10\log_{10}(P) - 10\log_{10}(N) \tag{1.28}$$

where P and N are the powers[4] of the desired signal $s(t)$ and interference (noise) $n(t)$,
respectively. Thus, SNR is useful as a performance criterion when the faithfulness of the
reconstructed waveshape is not the primary concern as in digital communication and radar
systems. In comparing two systems on the basis of SNR performance, one must compute
the SNR at the same point in both systems (typically, the receiver output where the final
reconstructed information waveform is to be extracted) when $s(t) + n(t)$ is input to both.
The system with the higher SNR would be considered the more favorable.

---

4. From here on in the book, P denotes the received signal power and the subscript "r" will be omitted.

Table 1.2(a)   Mars Return Link Direct to Earth (20 Mbps at $f = 8.4$ GHz)

Frequency	GHz	8.4	
Data Rate	Mbps	20.0	
(1) Mars relay EIRP	dBw		73.4
TX output power	watts	200.0	
Circuit losses	dB	0.8	
Antenna gain	dBi	51.3	
Diameter	m	5.0	
Efficiency	%	70.0	
Pointing loss	dB	0.1	
Pointing error	degrees	0.04	
(2) Space loss	dB		282.4
Range	AU	2.5	
(3) Clear sky loss	dB		0.1
(4) Rain attenuation	dB		0.0
(5) Polarization loss	dB		0.2
(6) Ground G/T	dB/K		60.7
Antenna gain	dBi	73.6	
Antenna diameter	m	70.0	
Antenna efficiency	%	60.0	
Pointing loss	dB	0.1	
Pointing error	degrees	0.002	
System noise temp	K	19.2	
Space background	K	3.0	
Atmospheric noise	K	6.3	
Maser noise temp	K	10.0	
Maser noise figure	dB	0.15	
(7) Boltzmann's constant	dBw/Hz-K		−228.4
(8) P_r/N_0	dB-Hz		80.0
(9) Mars-to-MRS degradation	dB		0.2
(10) Bandwidth	dB-Hz		73.0
(11) E_b/N_0 into demodulator	dB		6.8
(12) Ground equipment degradation	dB		1.5
(13) Required E_b/N_0 (with coding)	dB		2.4
Required E_b/N_0 (without coding)	dB	9.6	
Coding gain	dB	7.2	
(14) Link margin	dB		2.9

Often, we require that a system be designed to guarantee that the SNR performance at a given point is above a specified level, often called *threshold SNR*. Then the ratio of the actual SNR at that point to the threshold SNR is called the system or design margin at that point, that is,

$$\text{Design margin} = \frac{\text{SNR}_{\text{actual}}}{\text{SNR}_{\text{threshold}}} \qquad (1.29)$$

which is commonly expressed in decibels. Design margins are extremely important since

Table 1.2(b) Mars Return Link through the Earth Relay Satellite
(10 Mbps at $f = 32.4$ GHz)

Frequency	GHz	32.4	
Data Rate	Mbps	10.0	
(1) Mars-to-MRS degradation	dB		0.2
(2) MRS EIRP	dBw		81.0
TX output	watts	90	
Circuit losses	dB	0.8	
Antenna gain	dBi	62.7	
Diameter	m	5.0	
Efficiency	%	65.0	
Pointing loss		0.4	
Pointing error	degrees	0.04	
(3) Space loss (2.5 AU)	dB		294.2
(4) Polarization loss	dB		0.2
(5) ERS G/T	dB/K		62.2
Antenna gain	dBi	76.0	
Antenna diameter	m	25.0	
Antenna efficiency	%	55.0	
Pointing loss	dB	0.2	
Pointing error	degrees	0.0025	
System noise temp	K	23.0	
Space background temp	K	3.0	
RX noise temp	K	20.0	
Circuit loss	dB	0.1	
LNA noise figure	dB	0.25	
(6) Boltzmann's constant	dBw/Hz-K		−228.6
(7) P_r/N_0 at ERS	dB-Hz		77.3
(8) ERS-to-Earth degradation	dB		0.2
(9) P_r/N_0 at ground	dB-Hz		77.1
(10) Data rate	dB-bps		70.0
(11) E_b/N_0 into demodulator	dB		7.1
(12) Ground equipment degradation	dB		1.5
(13) Required E_b/N_0 (with coding)	dB		2.4
Required E_b/N_0 (without coding)	dB	9.6	
Coding gain	dB	7.2	
(14) Link margin	dB		3.2

they indicate the amount by which a system can be tolerant to unexpected events or additional degradation. The links of Example 1.3 have design margins of 2.9 and 3.2 dBs respectively.

1.4 SHANNON'S CAPACITY THEOREM

It is important to note here the premise theorized by Shannon [34–36] (also see Chapter 12) with regard to the efficiency of a communication system. Shannon shows that for the additive white Gaussian noise channel, coding schemes exist for which it is possible to communicate data *error free* over the channel provided the system data rate $\mathcal{R}_b$ (in bits per second) is less than or equal to what he called the *channel capacity* C. If $\mathcal{R}_b \geq C$, then error-free transmission is not possible. The channel capacity C, in bits per second, is defined by

$$C = B \log_2(1 + \frac{P}{N_0 B}) \tag{1.30}$$

where P represents the *received signal data power* in watts, B the one-sided channel bandwidth in Hz, and N_0 the one-sided PSD level of the noise. Now $P/N_0 B$ can be written as

$$\frac{P}{N_0 B} = \frac{P T_b \mathcal{R}_b}{N_0 B} = \frac{P T_b / N_0}{B / \mathcal{R}_b} = \delta_b \frac{E_b}{N_0} \tag{1.31}$$

where T_b is the time duration of a data bit in seconds, that is, the inverse of the data rate, $\delta_b = \mathcal{R}_b / B$ is the ratio of the data rate to the one-sided channel noise bandwidth in bits/Hz and $E_b / N_0 = P T_b / N_0$ is the energy per bit-to-noise spectral density ratio. Frequently, the parameter E_b / N_0 is referred to as the *relative efficiency* or *communication efficiency* of a particular communication scheme.

Substituting (1.31) and $B = \mathcal{R}_b / \delta_b$ into (1.30) and assuming communication at channel capacity ($\mathcal{R}_b = C$), we get

$$\delta_b = \log_2(1 + \delta_b \frac{E_b}{N_0}) \tag{1.32}$$

Equation (1.32) is plotted in Fig. 1.28 as a function of $1/\delta_b$ and E_b/N_0. Notice that as the value of E_b/N_0 decreases, the transmission bandwidth necessary to achieve 100% efficiency, operation at C, increases. It is therefore instructive to allow B to approach infinity in (1.30) or equivalently let δ_b go to zero in (1.32). Taking the limit gives (see also Chapter 12)

$$\frac{E_b}{N_0} = \ln(2) = -1.6 \text{ dB} \tag{1.33}$$

Consequently, coding schemes exist for which error-free transmission is possible in the *unconstrained bandwidth*, additive, white Gaussian noise channel provided that $E_b/N_0 \geq -1.6$ dB. Unfortunately, Shannon's theorem does not specify the data encoding method to accomplish that.

The relative efficiency of any practical communication system can be compared with the most efficient on the basis of how closely it approaches the *Shannon limit*, (1.33), for a given error probability. In the following chapters, we consider various modulation, demodulation, coding, decoding, and data-detection methods that attempt to approach the limit expressed by Shannon's theorem. Typical conclusions taken from later studies of various methods are compared with the Shannon limit in Fig. 1.28. The Shannon capacity is made more rigorous in Chapter 12.

1.5 DIGITAL COMMUNICATION WITH SUBCARRIERS AND DATA FORMATTING

Let P denote the total power of an RF signal, whether at the receiver or at the transmitter, then it is typically divided among the carrier and the data according to

$$P_d = (1 - L_m) P \tag{1.34}$$

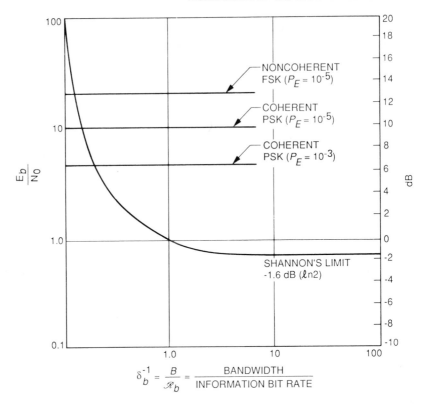

Figure 1.28   Relative efficiency versus the ratio of bandwidth to information bit rate

and

$$P_c = L_m P \tag{1.35}$$

where the parameter L_m represents the transmitter *modulation loss*. Essentially, it specifies the power division between the carrier and information sidebands resulting from a particular modulation process. If the basic information carrier is a microwave, continuous-wave (CW) sinusoidal voltage, the modulation can be impressed on the transmitted carrier by angle modulation [phase (PM) or frequency (FM)], amplitude modulation (AM), or a simultaneous combination of the two. Analytically speaking, the instantaneous value of a modulated RF carrier can be expressed by

$$s(t) = \sqrt{2P} a(t) \cos(\omega_c t + \theta(t)) \tag{1.36}$$

where $a(t)$ characterizes the amplitude modulation, $\theta(t)$ characterizes the angle modulation, ω_c is the carrier radian frequency, and $\sqrt{2P}$ denotes the peak carrier amplitude when $a(t)$ is unity. The modulation signals, the information to be transmitted, can be classified as digital, analog or a combination of the two.

One typical telemetry modulation is to *phase-shift-key* the data onto a square-wave subcarrier and then phase-modulate onto a sinusoidal carrier (PCM/PSK/PM) as depicted in Fig. 1.29. However, an alternate scheme phase-modulates the data directly on the carrier. In this latter case, the data being phase-shift keyed is called PCM/PM. In either case, the modulated carrier can be represented mathematically by

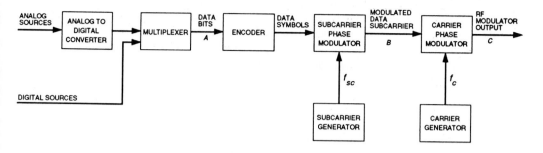

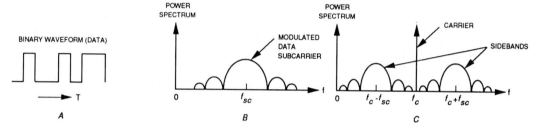

Figure 1.29   Typical modulation spectrum involving subcarriers

$$s(t) = \sqrt{2P} \sin\left(\omega_c t + \sum_{i=1}^{K} \Delta_i D_i(t)\right) \tag{1.37}$$

where Δ_i denotes modulation index (in degrees) for the ith "data" signal, $i = 1, 2, \ldots, K$, K the total number of data channels and $D_i(t)$ the normalized data sequence (in the case of PCM/PM) or a normalized modulated square-wave subcarrier (in the case of PCM/PSK/PM), that is,

$$D_i(t) = \begin{cases} d_i(t), & \text{PCM/PM} \\ d_i(t)Sin(\omega_{sc_i}t), & \text{PCM/PSK/PM} \end{cases} \tag{1.38}$$

where $d_i(t) = \pm 1$ is the normalized data sequence, $Sin(\omega_{sc_i}t) = \pm 1$ the square-wave subcarrier, and ω_{sc_i} the ith subcarrier frequency. Using trigometric identities with the fact that $D_i(t) = \pm 1$, $i = 1, 2, \ldots, K$, (1.37) may be expanded for $K = 1$ as

$$s(t) = \sqrt{2P} \cos \Delta_1 \sin(\omega_c t) + \sqrt{2P} D_1(t) \sin \Delta_1 \cos(\omega_c t) \tag{1.39}$$

If $0° < \Delta_1 < 90°$, this phase-modulated carrier comprises a pilot tone (residual carrier) and a Double-Sideband (DSB) modulated carrier. A system with $\Delta_1 < 90°$ is called a *residual* carrier system. A system with $\Delta_1 = 90°$ is called a *suppressed* carrier system. A residual carrier employs a phase-locked loop to track the pilot tone and provide a coherent carrier phase reference for demodulating the DSB-modulated carrier. A Costas loop may be used as a suppressed carrier receiver. For $K = 2$, (1.37) may be expanded as

$$\begin{aligned} s(t) = \sqrt{2P}\{&\cos \Delta_1 \cos \Delta_2 \sin(\omega_c t) + D_1(t) \sin \Delta_1 \cos \Delta_2 \cos(\omega_c t) \\ &+ D_2(t) \cos \Delta_1 \sin \Delta_2 \cos(\omega_c t) - D_1(t) D_2(t) \sin \Delta_1 \sin \Delta_2 \sin(\omega_c t)\} \end{aligned} \tag{1.40}$$

When we have a single data channel, as in (1.39), the first term is the carrier, while the second term is the data channel. Hence the modulation index Δ_1 has allocated the total power P in the transmitted signal $s(t)$ to the carrier and to the data channel, where the carrier power and the data power are, respectively

$$P_c = P \cos^2 \Delta_1 \tag{1.41}$$

$$P_d = P \sin^2 \Delta_1 \tag{1.42}$$

Comparing (1.41) to (1.35), it is clear that the modulation loss L_m is equal to $\cos^2 \Delta_1$. When we have two data channels as in (1.40), the first term is the carrier component, the second term is the modulated subcarrier $D_1(t)$, the third term is the modulated subcarrier $D_2(t)$ component, and the fourth term is the cross-modulation loss component. Hence, the corresponding powers in the four components are allocated by both modulation indices Δ_1 and Δ_2 with

$$P_c = P \cos^2 \Delta_1 \cos^2 \Delta_2 \tag{1.43}$$

$$P_{d1} = P \sin^2 \Delta_1 \cos^2 \Delta_2 \tag{1.44}$$

$$P_{d2} = P \cos^2 \Delta_1 \sin^2 \Delta_2 \tag{1.45}$$

$$P_m = P \sin^2 \Delta_1 \sin^2 \Delta_2 \tag{1.46}$$

Similar expressions may be carried out for the cases $K \geq 3$.

In some applications, the subcarrier is a sine-wave, in which case

$$D_i(t) = d_i(t) \sin(\omega_{sc_i} t) \tag{1.47}$$

Using the identity

$$\exp[j\beta \sin(\omega t + \phi)] = \sum_{n=-\infty}^{\infty} J_n(\beta) \exp[jn(\omega t + \phi)] \tag{1.48}$$

we have from (1.37) for $K = 1$ (letting $\omega_{sc} = \omega_{sc_1}$)

$$s(t) = \sqrt{2P} \sin(\omega_c t + \Delta_1 d_1(t) \sin \omega_{sc} t)$$

$$= \sqrt{2P} \{\cos(\Delta_1 \sin \omega_{sc} t) \sin \omega_c t + d_1(t) \sin(\Delta_1 \sin \omega_{sc} t) \cos \omega_c t\}$$

$$= \sqrt{2P} J_0(\Delta_1) \sin \omega_c t + \sqrt{2P} \sin \omega_c t \left\{ \sum_{n=1}^{\infty} 2 J_{2n}(\Delta_1) \cos(2n\omega_{sc} t) \right\} \tag{1.49}$$

$$+ \sqrt{2P} d_1(t) \cos \omega_c t \left\{ \sum_{n=0}^{\infty} 2 J_{2n+1}(\Delta_1) \sin((2n + 1)\omega_{sc} t) \right\}$$

where $J_n(x) = (-1)^n J_n(-x)$ and $J_n(x)$ is the nth order Bessel function [37]. The first term of (1.49) is the discrete carrier, the second term is other discrete components at $\omega_c \pm 2n\omega_{sc}$, and the last term consists of suppressed carrier modulations centered at $\omega_c \pm (2n + 1)\omega_{sc}$.

The baseband data $d_i(t)$ can have different formats as illustrated in Fig 1.30. With a Non-Return to Zero-Level (NRZ-L) data format, a logical "one" is represented by one level and a logical "zero" by the other. With NRZ-M (Mark), a logical "one" is represented by a change in level and a logical "zero" by no change. Two other formats, Biphase and Miller, are also defined in Fig. 1.30. It is worthwhile noting that Bi-ϕ-L is identical to NRZ-L when the latter is modulated on a square-wave subcarrier with a cycle period equal to

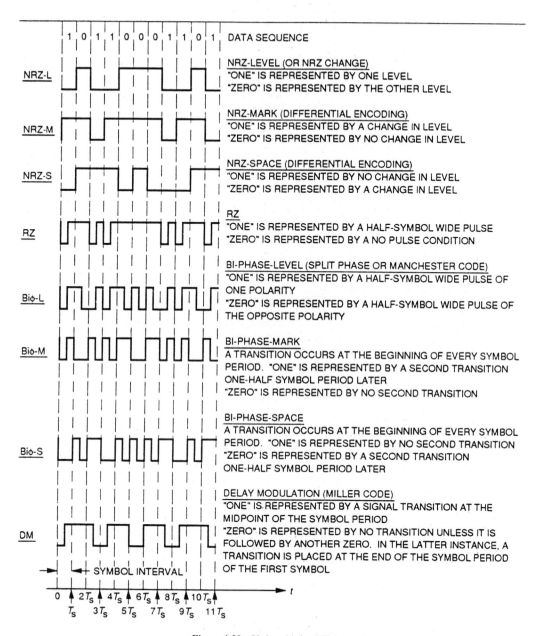

Figure 1.30 Various binary PCM waveforms

the bit period, so that a transition occurs at midpoint. Thus, PCM/PM with Bi-ϕ-L data formatting is a special case of PCM/PSK/PM with NRZ-L data formatting.

PROBLEMS

1.1 Let Z_1 and Z_2 be two independent Gaussian random variables with moments

$$E\{Z_1\} = A_1, \quad \text{Var}\{Z_1\} = \sigma_1^2$$

$$E\{Z_2\} = A_2, \quad \text{Var}\{Z_2\} = \sigma_2^2$$

It is desired to find the probability density function of the *product* random variable $Z = Z_1 Z_2$. One way of doing this is to represent Z in the following form:

$$Z \triangleq Z_1 Z_2 = \left(\frac{Z_1 + Z_2}{2}\right)^2 - \left(\frac{Z_1 - Z_2}{2}\right)^2 \triangleq Z_+^2 - Z_-^2$$

(a) Find the probability density functions (pdf's) of Z_+ and Z_-.
(b) Are Z_+ and Z_- correlated? Are they dependent?
(c) Find the pdf's of Z_+^2 and Z_-^2.
(d) Based on your results in (a), (b), and (c), find the characteristic function of Z and indicate how you would find (do not attempt to carry out the evaluation) its pdf.

1.2 Let a time-invariant system be given by the following differential equation:

$$\frac{dy(t)}{dt} + 2y(t) = x(t)$$

Let the input to the system $x(t)$ be a wide sense stationary (WSS) random process with mean $m_x = 1$ and covariance function $K_x(\tau) = 3 \exp(-|\tau|)$. Compute the following quantities:
(a) The mean of $y(t)$,
(b) The autocorrelation function of $y(t)$ and its PSD,
(c) The cross-correlation and cross-power spectral density of the input $x(t)$ and output $y(t)$.

1.3 Consider the discrete-time system represented by the linear constant coefficient difference equation [$x(n)$ is the input and $y(n)$ is the output]

$$y(n) - \frac{3}{8}y(n-1) + \frac{1}{4}y(n-2) = x(n) + 2x(n-1)$$

(a) Draw a block diagram representation of the system in terms of adders, delays, and coefficient multipliers.
(b) Let the input $x(n)$ be a white WSS random sequence with PSD equal to σ^2. Compute the PSD of the output $y(n)$.
(c) Design a digital causal filter (specify its impulse response) which operates on $y(n)$ (that is, cascaded with the first filter) such that its output $z(n)$ is white.
(d) Suppose that a third filter $H(z)$ is cascaded at the output of the whitening filter. Is the output of $H(z)$ still white if

$$H(z) = \frac{r - z^{-1}e^{j\theta}}{1 - rz^{-1}e^{j\theta}}, \quad |r| < 1$$

What did the insertion of $H(z)$ do?

1.4 Consider the system illustrated in Fig. P1.4. Assume that the input is a WSS random process with autocorrelation $R_x(\tau)$ and corresponding PSD $S_x(f)$.
(a) Find the autocorrelation function of $y(t)$ in terms of $R_x(\tau)$.

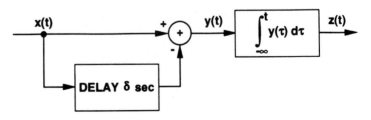

Figure P1.4

(b) Compute the PSD of $y(t)$ in terms of $S_x(f)$.

(c) Compute the PSD of $y(t)$ in terms of $S_x(f)$. If $S_x(f)$ is flat for all frequencies, that is, $S_x(f) = 1$, sketch $S_z(f)$ and its autocorrelation $R_z(\tau)$. What is the power of the signal $z(t)$?

(d) Find the cross-correlation function $R_{xy}(\tau)$ and the corresponding cross-power spectral density $S_{xy}(f)$.

1.5 A WSS random process $x(t)$ has PSD $S_x(f)$ given by

$$S_x(f) = \begin{cases} 1, & 0 \le |f| \le f_a(1-\alpha) \\ \cos^2\left\{\frac{\pi}{4\alpha}[\frac{|f|}{f_a} - 1 + \alpha]\right\}, & f_a(1-\alpha) \le |f| \le f_a(1+\alpha) \\ 0, & \text{otherwise} \end{cases}$$

This is referred to as a *raised cosine* roll-off characteristic with excess bandwidth factor α where $0 \le \alpha \le 1$.

(a) Suppose that $x(t)$ is sampled at a rate $f_s = 1/T_s = 2f_a(1+\alpha)$ producing the sampled function

$$x^*(t) = \sum_{n=-\infty}^{\infty} x(nT_s)\delta(t - nT_s)$$

Are the samples uncorrelated? (Explain.)

(b) Suppose now that $x(t)$ is sampled at a rate $f_s = 1/T_s = 2f_a$ (note that this corresponds to the case of undersampling). Determine and sketch the PSD of the sampled process $x^*(t)$ (show all scaling of both axes). Are the samples uncorrelated?

(c) Assume now that $x^*(t)$ of (b) is passed through the ideal filter

$$H(f) = \begin{cases} \frac{1}{2f_a}, & |f| \le f_a \\ 0, & \text{otherwise} \end{cases}$$

producing $\hat{x}(t)$. Write a sampling expansion appropriate to the filter output $\hat{x}(t)$. Are the samples of this expansion uncorrelated?

1.6 An autocorrelation function which frequently arises in practical problems is

$$R_x(\tau) = e^{-\alpha|\tau|} \cos \beta\tau$$

(a) Calculate the corresponding PSD $S_x(f)$.

(b) For what value of f does $S_x(f)$ assume its maximum value?

(c) Show that $S_x(f)$ has a maximum at a frequency unequal to zero if $\alpha < \sqrt{3}\beta$.

(d) Compute the maximum value of $S_x(f)$.

(e) Show that the half-power point occurs at $f = r/(2\pi)$ (where $r^2 = \alpha^2 + \beta^2$) and is always to the right of the peak value.

1.7 Consider a random process $x(t)$ consisting of an ensemble of rectangular waves that alternate between the values E and $-E$. Assume that the distribution of zero crossings is Poisson, that is, the probability of n zero crossings (transitions) occurring in a duration of time T is given by

$$P_T(n) = \frac{(\lambda T)^n}{n!} e^{-\lambda T}$$

where λ is the average number of zero crossings per unit time. Also, assume that $x(t)$ takes on the values E and $-E$ with equal a priori probability.

(a) Sketch a sample waveform of the random process $x(t)$.

(b) Compute the autocorrelation function of $x(t)$.

(c) Compute the PSD corresponding to the result in (b).

1.8 Consider a random data sequence with PSD given by

$$S_x(f) = E\left(\frac{\sin(\pi Tf)}{(\pi Tf)}\right)^2 \quad \text{Watts/Hz}$$

    (a) Find the rms bandwidth.
    (b) Find the equivalent noise bandwidth.
    (c) Find the 3-dB bandwidth.
    (d) Find the $p\%$ fractional bandwidth.

1.9 Repeat Problem 1.8 if the PSD is given by

$$S_x(f) = T\frac{\sin^4(\pi fT/2)}{(\pi fT/2)^2}$$

1.10 Repeat Problem 1.8 if the PSD is given by

$$S_x(f) = T\frac{\pi^2 T(\cos \pi fT)^2}{4\left[(\pi fT)^2 - (\frac{\pi}{2})^2\right]^2}$$

1.11 (a) Find the 3-dB and equivalent noise bandwidth of the Butterworth spectra of order n if

$$S_n(\omega) = |H_n(\omega)H_n(-\omega)| = |H_n(\omega)|^2$$

    where

$$H_n(\omega) \triangleq \frac{1}{\prod_{i=1}^{n}\left[j\omega - e^{j\pi[(2i+n-1)/2n]}\right]}$$

    (b) What is the limiting equivalent noise bandwidth as n approaches infinity?

1.12 The impulse response of a filter corresponding to a raised-cosine spectrum is

$$p(t) = (\frac{\sin \pi t/T}{\pi t/T})\frac{\cos \alpha\pi t/T}{1 - (2\alpha t/T)^2}$$

    (a) Sketch the pulse shapes for $\alpha = 0.25, \ 0.5$, and 1.
    (b) Determine and sketch the raised cosine spectra for $\alpha = 0.25, \ 0.5$, and 1.

1.13 Define the frequency range involved in: (a) HF, (b) UHF, (c) VHF, (d) EHF, (e) ELF, (f) L-band, (g) S-band, (h) X-band, and (i) K-band.

1.14 (a) Determine the space loss at a distance of 22,000 miles (35,420 km), typical of a satellite in synchronous orbit, when the carrier frequency is 150 MHz, 1.2 GHz, 2.3 GHz, 8.6 GHz, and 32 GHz; that is, VHF, L-band, S-band, X-band, and Ka-band.
    (b) Repeat (a) for a distance of 870 miles (1,400 km) which is typical of a satellite in a low Earth orbit (LEO).
    (c) Repeat (a) for a distance of 6,211 miles (10,000 km) which is typical of a satellite in a medium Earth orbit (MEO).

1.15 In order to maintain the antenna surface tolerance loss below 0.3 dB, what is the required rms roughness of the antenna at 1.2 GHz, 2 GHz, 8 GHz, and 32 GHz?

1.16 In the deep space network (DSN), parabolic antennas with 26, 34, and 70 m in diameter are used.
    (a) Compute the peak gain for each antenna at L-band (1.2 GHz), X-band (8.4 GHz), and Ka-band (32 GHz), assuming an antenna efficiency of 45%.
    (b) Compute the half-power beamwidths for each antenna in the various frequency bands.

1.17 A transmitting rectangular antenna of width w and length l produces the gain function

$$g_t(\phi_z, \phi_l) = 100\left[\frac{\sin[(\pi w \sin \phi_l)/\lambda]\sin[(\pi l \sin \phi_z)/\lambda]}{[(\pi w \sin \phi_l)/\lambda][\pi l \sin \phi_z)/\lambda]}\right]^2$$

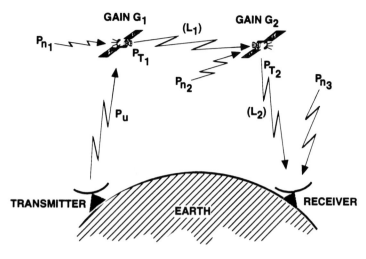

GAIN G_1 (L_1) GAIN G_2

P_{n_1} P_{T_1} P_{n_2} P_{T_2} P_{n_3}

P_u (L_2)

TRANSMITTER   EARTH   RECEIVER

Figure P1.19

when operating with wavelength λ. Determine the half-power beamwidth in the $\phi_l = 0$ and in the $\phi_z = 0$ plane.

1.18 A communication satellite is to transmit to Earth at L-band (1.2 GHz).

    (a) Assuming a LEO (1,400 km altitude) satellite, compute the required diameter of a parabolic transmitting antenna so that the beam covers a 1,000 km^2 area. Assume an antenna efficiency of 40%. What is the resulting antenna gain?

    (b) Repeat (a) for a MEO (10,000 km altitude) satellite and a desired ground footprint of 2,000,000 km^2 [roughly the size of continental U.S. (CONUS)].

1.19 In the Iridium network, a phone call might require routing through various satellites to get it to its destination. Consider the 2-satellite system depicted in Fig. P1.19. Assume ideal frequency-translating satellites, white noise, and equal bandwidths.

    (a) Compute the received signal power at the receiver as a function of uplink power P_u, satellite gains G_1 and G_2, and path losses L_1 and L_2.

    (b) Compute the received noise power at the receiver as a function of P_{n1}, P_{n2}, P_{n3}, G_1, G_2, L_1, and L_2.

    (c) Compute the signal-to-noise ratio at the receiver.

1.20 A satellite at 1,000 km altitude is to use multiple downlink spot beams to achieve frequency re-use, while covering a distance on Earth of 1,200 km (use one-dimensional antenna patterns here). The beams are to have center-to-center separation of 0.1 radians with the gain pattern depicted in Fig. P1.20. A receiving station in any beam must have 50 dB isolation from any other beam (the received carrier power at any ground station from any other beam after front end filtering is to be less than 50 dB from the carrier power at peak gain in its own beam). The downlink signal is transmitted with bandwidth B, but can be located at any carrier frequency f_c, $f_c + B$, $f_c + 2B$, etc. Neglecting dual polarization, determine the amount a frequency can be reused with this system.

1.21 In Table 1.2(a), assume the frequency is moved to Ku-band (12 GHz) and that the antenna diameter is reduced from 5 to 3 m.

    (a) Compute P_r/N_0 in dB-Hz.

    (b) What is the link margin?

    (c) Assume that a 34-m antenna is used on the ground instead of a 70-m antenna, compute P_r/N_0 and the link margin.

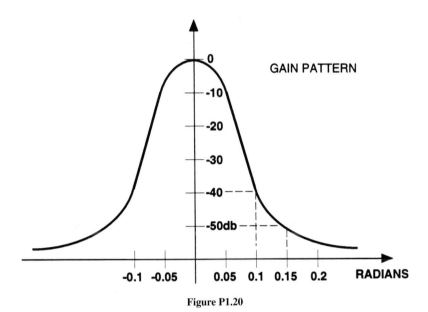

Figure P1.20

(d) With a 6 dB link margin and a 34-m antenna on the ground, what is the maximum data rate that can be supported?

1.22 In Table 1.2(b), assume that the Mars Relay Satellite (MRS) transmitted power is reduced from 90 to 50 watts but the Earth Relay Satellite (ERS) antenna is increased to 30 m in diameter.

(a) Compute P_r/N_0 in dB-Hz at ERS.

(b) Compute P_r/N_0 in dB-Hz on the ground.

(c) What is the link margin with 10 Mbps?

(d) Given a desirable link margin of 6 dB, what is the maximum data rate (in bps) that can be supported?

1.23 Consider the case in which 100 kbps is to be transmitted over a 3 kHz voice grade circuit. Is it possible to achieve error-free transmission with a SNR of 10 dB? If not, what modifications would you suggest?

1.24 Consider a voice grade telephone circuit with a bandwidth of 3 kHz. Assume the circuit can be modeled as an additive white Gaussian noise (AWGN) channel.

(a) What is the capacity of such a circuit if the SNR is 30 dB?

(b) What is the minimum SNR required for a data rate of 4800 bps on such a voice grade circuit?

1.25 (a) Find the average capacity in bps that would be required to transmit a high-resolution black-and-white TV signal at the rate of 32 pictures per sec if each picture is made up of 2×10^6 picture elements (pixels) and 16 different brightness levels. All picture elements are assumed to be independent and all levels have equal likelihood of occurrence.

(b) For color TV, this system additionally provides for 64 different shades of color. How much more system capacity is required for a color system compared to the black and white system?

(c) Find the required capacity if 100 of the possible brightness-color combinations occur with probability of 0.003 each, 300 of the combinations occur with probability 0.001 each, and 624 of the combinations occur with probability 0.00064 each.

1.26 Assume $E_b/N_0 = 3.4$ dB at $T = 28°$K and the data rate is $R_b = 1/T_b = 115{,}000$ bps.

(a) Find the power-to-noise energy ratio, P/N_0 in dB-Hz, at $T = 28°$K.

(b) For $T = 0°$K, $28°$K, and $300°$K, what is the noise power?

(c) What are the communication efficiencies at the temperatures given in (b)?

(d) Find the equivalent bit mass, say m_μ, that satisfies Einstein's result relating energy to mass, that is, $E = mc^2$, if $T = 28°$K.

(e) Find the equivalent energy associated with an electron and the ratio of the bit mass given in (d) to that of an electron.

(f) If each spacecraft TV picture transmitted from Jupiter consists of 2,000,000 bits, find how many electron masses make up one TV picture. (Assume $T = 28°$K.)

(g) What is the channel capacity if the single-sided bandwidth is B $= 200,000$ Hz? (Assume $T = 28°$K.)

(h) If a picture is worth a thousand words, how many bits/word are there in a picture transmitted from Jupiter?

(i) If each word contains 5 bits of information, how many bits are there in a "1,000-word" picture?

1.27 Consider *phase-shift keying* with a carrier component which can be written as

$$x_c(t) = A \sin\left(2\pi f_c t + \cos^{-1}(ad(t))\right)$$

Assume that $d(t)$ is a ± 1-valued binary message signal. Show that the ratio of power in the carrier and modulation components of $x_c(t)$ is (for $a \neq 0$)

$$\frac{P_c}{P_m} = \frac{1 - a^2}{a^2}$$

REFERENCES

1. NASA's Mobile Satellite Experiment, Pasadena, CA: Jet Propulsion Laboratory, *MSAT-X Quarterly*, no. 1, October 1984.

2. Special Issue on Optical Communications, *Proceedings of the IEEE*, vol. 58, no. 10, October 1970.

3. Lindsey, W. C. and M. K. Simon, *Telecommunication Systems Engineering*, Englewood Cliffs, N.J.: Prentice Hall, 1973. Reprinted by Dover Press, New York, NY, 1991.

4. International Organization for Standardization, *Information Processing Systems—Open Systems Interconnection—Basic Reference Model, ISO 7498*, New York: American National Standards Institute, October 1984.

5. Henshall, J. and S. Shaw, *OSI Explained, End-to-End Computer Communication Standards*, West Sussex, England: Ellis-Horwood Limited, pp. 10–20, 1988.

6. Knightson, K. G., J. Larmouth and T. Knowles, *Standards for Open Systems Interconnection*, New York: McGraw-Hill, pp. 12–19, 20–54, 259–301, 1988.

7. International Organization for Standardization, *Information Processing Systems—Open Systems Interconnection—Manufacturing Message Specification, ISO 9506*, New York: American National Standards Institute, 1989.

8. International Organization for Standardization, *Information Processing Systems—Open Systems Interconnection—Basic Reference Model, Part 4: Management Framework, ISO 7498-4*, New York: American National Standards Institute, 1988.

9. International Organization for Standardization, *Information Processing Systems—Open Systems Interconnection—Systems Management Overview, ISO 10164*, New York: American National Standards Institute, 1990.

10. International Organization for Standardization, *Information Processing Systems–Open Systems*

Interconnection—Structure of Management Information, ISO 10165, New York: American National Standards Institute, 1990.

11. Clarke, A. C., "Extra-Terrestrial Relays," *MSN & CT*, August 1985, pp. 59–67, also in *Wireless World*, 1945.

12. *Deep Space Network, Flight Projects Interface Design Handbook, Vol I: Existing DSN Capabilities*, Pasadena, CA: Jet Propulsion Laboratory, Document 810-5, Rev. D, March 1990.

13. *Deep Space Network, Flight Projects Interface Design Handbook, Vol II: Proposed DSN Capabilities*, Pasadena, CA: Jet Propulsion Laboratory, Document 810-5, Rev. D, March 1990.

14. *Space Network (SN) User's Guide*, National Aeronautics and Space Administration, Goddard Space Flight Center, Revision 6, September 1988.

15. *Global Positioning System, Vol. I*, Washington, D.C.: The Institute of Navigation, 1980.

16. *Global Positioning System, Vol. II*, Washington, D.C.: The Institute of Navigation, 1984.

17. Kirson, A. M.,"RF Data Communication Considerations in Advanced Driver Information Systems," *IEEE Transactions on Vehicular Technology*, vol. 40, no. 1, February 1991, pp. 51–55.

18. Décima, M., W. S. Gifford, R. Potter and A. A. Robrock, eds., "Special Issue on Integrated Services Digital Network: Recommendations and Field Trials I," *IEEE Journal On Selected Areas in Communications*, vol. SAC-4, no. 3, May 1986.

19. Décima, M., W. S. Gifford, R. Potter and A. A. Robrock, eds., "Special Issue on Integrated Services Digital Network: Technology and Implementations II," *IEEE Journal On Selected Areas In Communications*, vol. SAC-4, no. 8, November 1986.

20. "ISDN: A Means Towards A Global Information Society," *IEEE Communications Magazine*, vol. 25, no. 12, December 1987.

21. Ricci, F. and D. Schutzer, *U.S. Military Communications*, Maryland: Computer Science Press, 1986.

22. Wood, P., "Mobile Satellite Services for Travellers," *IEEE Communications Magazine*, vol. 29, no. 11, November 1991.

23. Grubb, J. L., "The Traveller's Dream Come True," *IEEE Communications Magazine*, vol. 29, no. 11, November 1991.

24. Kwan, R. K. and R. A. Wiedeman, "GlobalStar: Linking the World via Mobile Connections," *International Symposium on Personal, Indoor and Mobile Communications*, London, U.K., September 1991, pp. 318–23.

25. Rusch, R. J., P. Cress, M. Horstein, R. Huang and E. Wismell, "Odyssey, A Constellation For Personal Communications," *Fourteenth AIAA International Communication Satellite System Conference*, Washington, D.C., March 1992.

26. Taylor, S. C. and W. C. Shurvinton, "The Archimedes Satellite System," *American Institute of Aeronautics and Astronautics*, March 1992.

27. Papoulis, A., *Probability, Random Variables and Stochastic Processes*, New York: McGraw-Hill, 1965.

28. Scholtz, R. A., "How Do You Define Bandwidth," *Proceedings of the International Communication Conference*, Los Angeles, October 1972, pp. 281– 88.

29. Ippolito, L., et al., *Propagation Handbook for Satellite Systems Design*, NASA Reference Publication 1082, December 1981; 1083, June 1983.

30. Strauss, R., J. Bretting and R. Metivier, "TWT for Communication Satellites," *Proceedings of the IEEE*, vol. 65, March 1977, pp. 387– 400.

31. Ruze, R., Effect of Surface Error Tolerance on Efficiency and Gain of Parabolic Antennas, *Proceedings of the IEEE*, vol. 54, April 1966.

32. Rusch, W. and P. Potter, *Analysis of Reflector Antennas*, New York: Academic Press, 1970.

33. Bennett, W., *Electrical Noise*, New York: McGraw-Hill, 1960.

34. Shannon, C. E. and W. Weaver, *The Mathematical Theory Of Communication*, Urbana, IL: University of Illinois Press, 1959.

35. Gallager, R. G., *Information Theory and Reliable Communication*, New York: John Wiley and Sons, 1968.

36. Wozencraft, J. M. and I. M. Jacobs, *Principles of Communication Engineering*, New York: John Wiley and Sons, 1965.

37. Abromowitz, M. and I. Stegun, *Handbook of Mathematical Functions*, Washington, D.C.: National Bureau of Standards, 1965.

38. Lindsey, W. C., "System Architecting and Engineering Digital Communication Systems," *LinCom Corporation*, Technical report 0850-100-09.01, LinCom Corporation, 5110 Goldleaf Circle, Los Angeles, CA, 90056, September 1992.

Chapter 2

⊓⊔

Power Spectral Density
of Digital Modulations

⊓⊔

In Chapter 1, several definitions of *bandwidth* were given as measures of the spectral occupancy of a chosen signaling or modulation technique. All of these were in some way related to the *power spectral density (PSD)* of the signaling scheme. Thus, the purpose of this chapter is to characterize the PSD of various types of synchronous data pulse streams which include as special cases the data formats introduced in Chapter 1.

2.1 POWER SPECTRAL DENSITY OF A SYNCHRONOUS DATA PULSE STREAM

In this section, we derive a general result for the power spectral density of a synchronous data pulse stream where the underlying data sequence $\{a_n\}$ that generates the pulse stream has known statistical properties and the transmitted waveform is a single known pulse shape, $p(t)$. We shall assume a data model that covers a wide class of applications, namely, the generating sequence is wide sense stationary (WSS) with known autocorrelation function $R(m)$. Such a model includes the commonly encountered cases of Markov and purely random data sources.

In the former case, each data bit, for example, a_n, in the sequence takes on M possible values (states) $\{\alpha_i; i = 1, 2, \ldots, M\}$ with probabilities $\{p_i; i = 1, 2, \ldots, M\}$ (often called the *stationary* probabilities). Furthermore, the transitions between states occur with *transition* probabilities $\{p_{ik}; i, k = 1, 2, \ldots, M\}$ where p_{ik} denotes the probability that α_k is transmitted in a given transmission interval following the occurrence of α_i in the previous transmission interval. These transition probabilities are conveniently arranged in an $M \times M$ *transition matrix* $\mathbf{P} = \{p_{ik}\}$. A special case of the above is a purely random source, that is, one that emits a data bit in a given signaling interval independent of those emitted in previous signaling intervals. Such a source can be modeled as a degenerate case of the above M-ary Markov source (that is, $\mathbf{P}^{-1}$ does not exist) whose transition matrix $\mathbf{P}$ has identical rows. When $M = 2$, and the two states α_1 and α_2 are equal and opposite, we get the familiar results for antipodal binary signaling.

Occasionally, one finds situations where the generating data sequence is not WSS but, however, is *cyclostationary*, that is, its first two moments are periodic. One such case occurs at the output of a convolutional encoder where the period (in code symbols) is equal to the reciprocal of the code rate (more about this later on in the chapter). Thus, in order to characterize the spectral properties of the output of such an encoder, one must develop an expression for the power spectral density of a synchronous data pulse stream whose generating sequence is cyclostationary. We begin with the derivation of the power spectral density of a synchronous data pulse stream generated by a WSS sequence.

2.1.1 Power Spectral Density of a Synchronous Data Pulse Stream Generated by a Binary, Zero Mean, WSS Sequence

Consider the binary (± 1), zero mean, WSS sequence $\{a_n\}$ for which it is known that

$$\overline{a_n} = 0; \quad \overline{a_n a_m} = R_a(m-n) \tag{2.1}$$

and the overbar denotes statistical expectation. From this sequence, we form the data pulse stream

$$m(t) = \sum_{n=-\infty}^{\infty} a_n p(t - nT) \tag{2.2}$$

where $p(t)$ is the elementary signal (pulse shape) and is not necessarily restricted to be time-limited to only a single signaling interval, that is, the pulse train $m(t)$ can contain overlapping pulses. Irrespective of the properties of the generating sequence $\{a_n\}$, the pulse stream $m(t)$ is itself cyclostationary since the expected value of the product $m(t)m(t+\tau)$ is, in addition to being a function of τ, a periodic function of t. Thus, to compute the PSD of $m(t)$, namely, $S_m(f)$, we must first average the statistical correlation function

$$R_m(t; \tau) \triangleq \overline{m(t)m(t+\tau)} \tag{2.3}$$

over t [the average is performed over the period of $R_m(t; \tau)$] and then take the Fourier transform of the result. Thus,

$$S_m(f) \triangleq \mathcal{F}\{\langle R_m(t; \tau)\rangle\} \tag{2.4}$$

where $<>$ denotes time average and $\mathcal{F}$ denotes Fourier transform. Substituting (2.2) into (2.3) and making use of (2.1) and the definition of Fourier transform, we get[1]

$$S_m(f) = \int_{\tau} \left\langle \sum_m \sum_n R_a(m-n)p(t-nT)p(t+\tau-mT) \right\rangle e^{-j2\pi f\tau} d\tau \tag{2.5}$$

Since

$$p(t) = \int_{-\infty}^{\infty} P(f)e^{j2\pi ft} df \tag{2.6}$$

where $P(f)$ is the Fourier transform of $p(t)$, then substituting (2.6) in (2.5) gives

$$S_m(f) = \int_y \int_z \sum_m \sum_n R_a(m-n)P(y)P^*(z)e^{-j2\pi ynT}$$

$$\times \ e^{j2\pi zmT} \left\langle e^{j2\pi(y-z)t} \right\rangle \int_{\tau} e^{-j2\pi(f+z)\tau} d\tau\, dy\, dz \tag{2.7}$$

---

1. Unless otherwise noted, all summations and integrals range from $-\infty$ to ∞.

Recalling that

$$\int_{\tau} e^{-j2\pi x\tau}\,d\tau = \delta(x) \tag{2.8}$$

then (2.7) simplifies to

$$S_m(f) = \int_y \sum_m \sum_n R_a(m-n)P(y)P^*(-f)e^{-j2\pi f(m-n)T}\,e^{-j2\pi(y+f)nT}$$

$$\times \left\langle e^{j2\pi(y+f)t}\right\rangle dy \tag{2.9}$$

Recalling that the time average is performed over the period of $R_m(t;\tau)$, that is, T, then

$$\left\langle e^{j2\pi(y+f)t}\right\rangle \triangleq \frac{1}{T}\int_{-T/2}^{T/2} e^{j2\pi(y+f)t}\,dt = \frac{\sin\left[\pi(y+f)T\right]}{\pi(y+f)T} \tag{2.10}$$

Substituting (2.10) in (2.9) and letting $l = m - n$ gives

$$S_m(f) = \sum_l R_a(l)e^{-j2\pi flT}\int_y P(y)P^*(-f)\frac{\sin\left[\pi(y+f)T\right]}{\pi(y+f)T}$$

$$\times \sum_n e^{-j2\pi(y+f)nT}\,dy \tag{2.11}$$

Finally, from Poisson's formula, we have that

$$\sum_n e^{-j2\pi xnT} = \frac{1}{T}\sum_k \delta\left(x - \frac{k}{T}\right) \tag{2.12}$$

which when substituted in (2.11) yields

$$S_m(f) = \left[\frac{1}{T}\sum_l R_a(l)\,e^{-j2\pi flT}\right]\frac{1}{T}\sum_k P\left(-f + \frac{k}{T}\right)P^*(-f)\frac{\sin\pi k}{\pi k} \tag{2.13}$$

Since

$$\frac{\sin\pi k}{\pi k} = \begin{cases} 1; & k = 0 \\ 0; & k \neq 0 \end{cases} \tag{2.14}$$

and $|P(f)|^2$ is an even function of f, we get the desired result[2]

$$S_m(f) = S_p(f)S_a(f) \tag{2.15}$$

where

$$S_p(f) \triangleq \frac{1}{T}|P(f)|^2 \tag{2.16}$$

---

2. The result in (2.15) can also be obtained (see Problem 2.1) directly from the basic definition of the PSD of an arbitrary random process $m(t)$, namely, $S_m(f) = \lim_{T_0\to\infty} E\left\{\frac{1}{T_0}|M_0(f)|^2\right\}$ where $M_0(f)$ is the Fourier transform of a T_0-sec observation of an ensemble member of $m(t)$.

is the power spectral density of the pulse shape $p(t)$ and

$$S_a(f) \triangleq \sum_{l=-\infty}^{\infty} R_a(l) e^{-j2\pi f l T} \tag{2.17}$$

is the power spectral density of the sequence, that is, the discrete Fourier transform of its autocorrelation function. Note that if the data sequence is purely random, that is,

$$\overline{a_n a_m} = R_a(m-n) = \begin{cases} 1; & m=n \\ 0; & m \neq n \end{cases} \tag{2.18}$$

then, from (2.17)

$$S_a(f) = 1 \tag{2.19}$$

and hence

$$S_m(f) = S_p(f) \tag{2.20}$$

which is a commonly used result.

2.1.2 Power Spectral Density of a Synchronous Data Pulse Stream Generated by a Binary, Zero Mean, Cyclostationary Sequence

Suppose now that we have a binary sequence $\{a_n\}$ that has the properties

$$\overline{a_n} = 0; \qquad \overline{a_n a_m} = R_a(n; m-n) \tag{2.21}$$

Furthermore

$$R_a(n; m-n) = R_a(n+kN; m-n); \quad k = 0, \pm1, \pm2, \ldots \tag{2.22}$$

where N denotes the period of the autocorrelation function $R_a(n; m-n)$. Then, following the steps leading up to (2.9), we see that the analogous result is now

$$S_m(f) = \int_y \sum_n \sum_m R_a(n; m-n) P(y) P^*(-f) e^{-j2\pi f(m-n)T}$$

$$\times e^{-j2\pi(y+f)nT} \left\langle e^{j2\pi(y+f)t} \right\rangle dy \tag{2.23}$$

Thus far, we have not made use of the cyclostationary property given in (2.22). If now we use this property to evaluate the sum on n in (2.23), and as before let $l = m-n$, then it can be easily shown that

$$\sum_n R_a(n; l) e^{-j2\pi(y+f)nT} = \left[\sum_{n=1}^{N} R_a(n; l) e^{-j2\pi(y+f)nT} \right]$$

$$\times \left[\sum_k e^{-j2\pi(y+f)kNT} \right] \tag{2.24}$$

or making use of (2.12)

$$\sum_n R_a(n; l) e^{-j2\pi(y+f)nT} = \left[\sum_{n=1}^{N} R_a(n; l) e^{-j2\pi(y+f)nT} \right]$$

$$\times \left[\frac{1}{NT} \sum_k \delta\left(y + f - \frac{k}{NT} \right) \right] \tag{2.25}$$

Substituting (2.25) into (2.23) and performing the integration on y gives

$$S_m(f) = \frac{1}{NT} \sum_k \sum_l \left[\sum_{n=1}^{N} R_a(n; l) e^{-j2\pi nk/N} \right]$$

$$\times e^{-j2\pi flT} P\left(-f + \frac{k}{NT} \right) P^*(-f) \left\langle e^{j2\pi kt/NT} \right\rangle \tag{2.26}$$

Since $R_m(t; \tau)$ is now periodic in t with period NT, then

$$\left\langle e^{j2\pi kt/NT} \right\rangle \triangleq \frac{1}{NT} \int_{-NT/2}^{NT/2} e^{j2\pi kt/NT} dt = \frac{\sin \pi k}{\pi k} = \begin{cases} 1; & k = 0 \\ 0; & k \neq 0 \end{cases} \tag{2.27}$$

Finally, substituting (2.27) into (2.26) gives the desired result

$$S_m(f) = S_p(f) S_{\bar{a}}(f) \tag{2.28}$$

where $S_p(f)$ is still given by (2.16) and

$$S_{\bar{a}}(f) = \sum_{l=-\infty}^{\infty} \left[\frac{1}{N} \sum_{n=1}^{N} R_a(n; l) \right] e^{-j2\pi flT} \tag{2.29}$$

is the power spectral density of the equivalent stationary data sequence, that is, the periodicity of the correlation function caused by the cyclostationary behavior of the sequence $\{a_n\}$ must be "averaged out" before taking the discrete Fourier transform.

2.1.3 Power Spectral Density of a Synchronous Data Pulse Stream Generated by a Binary, Nonzero Mean, Cyclostationary Sequence

When the generating sequence $\{a_n\}$ has nonzero mean, then the spectrum of the corresponding synchronous data pulse stream will have a discrete component in addition to the customary continuous component. An example of a situation where this might occur is at the output of a convolutional encoder whose input is random but not equiprobable binary data. The procedure for handling this case is as follows:

Define the zero mean cyclostationary sequence $\{A_n\}$ by

$$A_n = a_n - \overline{a_n} \tag{2.30}$$

which has the properties

$$\overline{A_n} = 0; \quad \overline{A_n A_m} = R_A(n; m - n) \tag{2.31}$$

Then, using the results of the previous section, the *continuous* component of the power spectrum $(S_m(f))_c$ for the synchronous data pulse stream generated by $\{a_n\}$ is given by

$$(S_m(f))_c = S_p(f) S_{\bar{A}}(f) \tag{2.32}$$

where

$$S_{\bar{A}}(f) = \sum_{l=-\infty}^{\infty} \left[\frac{1}{N} \sum_{n=1}^{N} R_A(n; l) \right] e^{-j2\pi flT} \tag{2.33}$$

The discrete spectral component $(S_m(f))_d$ is found from

$$(S_m(f))_d = \mathcal{F} \left\{ \left\langle \sum_m \sum_n \overline{a_n \, a_m} \, p(t - nT) p(t + \tau - mT) \right\rangle \right\} \tag{2.34}$$

Making use of the Fourier transform relation of (2.6) gives the alternate form

$$(S_m(f))_d = \int_y \sum_n \sum_m \overline{a_n} e^{-j2\pi ynT} \overline{a_m} e^{-j2\pi fmT} P(y) P^*(-f) \left\langle e^{j2\pi(y+f)t} \right\rangle dy \tag{2.35}$$

Since, for a cyclostationary sequence, $\overline{a_n}$ and $\overline{a_m}$ are both periodic with period N, then analogous to (2.25) we have

$$\sum_n \overline{a_n} e^{-j2\pi ynT} = \left[\sum_{n=1}^{N} \overline{a_n} e^{-j2\pi ynT} \right] \frac{1}{NT} \sum_l \delta \left(y - \frac{l}{NT} \right)$$

$$\sum_m \overline{a_m} e^{-j2\pi fmT} = \left[\sum_{m=1}^{N} \overline{a_m} e^{-j2\pi fmT} \right] \frac{1}{NT} \sum_k \delta \left(f - \frac{k}{NT} \right) \tag{2.36}$$

Substituting (2.36) into (2.35) and integrating on y yields after simplification

$$(S_m(f))_d = \frac{1}{NT} \sum_k \delta \left(f - \frac{k}{NT} \right) \sum_{m=1}^{N} \overline{a_m} e^{-j2\pi mk/N}$$

$$\times \frac{1}{NT} \sum_l P \left(\frac{l}{NT} \right) P^* \left(-\frac{k}{NT} \right) \sum_{n=1}^{N} \overline{a_n} e^{-j2\pi nl/N} \left\langle e^{j2\pi(k+l)t/NT} \right\rangle \tag{2.37}$$

Performing the time average over the period NT gives as before

$$\left\langle e^{j2\pi(k+l)t/NT} \right\rangle = \frac{\sin \pi(k+l)}{\pi(k+l)} = \begin{cases} 1; & k = -l \\ 0; & k \neq -l \end{cases} \tag{2.38}$$

Finally, substituting (2.38) into (2.37) and recalling that $|P(f)|^2$ is an even function of f gives the desired result

$$(S_m(f))_d = \frac{1}{(NT)^2} \sum_k \left| P \left(\frac{k}{NT} \right) \right|^2 \left(\sum_{m=1}^{N} \overline{a_m} e^{-j2\pi mk/N} \right)$$

$$\times \left(\sum_{n=1}^{N} \overline{a_n} e^{j2\pi nk/N} \right) \delta \left(f - \frac{k}{NT} \right) \tag{2.39}$$

Note that when $\{a_n\}$ is a WSS sequence (that is, $N = 1$), then (2.39) reduces to

$$(S_m(f))_d = \frac{(\overline{a_1})^2}{T^2} \sum_k \left| P \left(\frac{k}{T} \right) \right|^2 \delta \left(f - \frac{k}{T} \right) \tag{2.40}$$

Example 2.1

Consider a binary, zero mean Markov source characterized by

$$\Pr\{a_{n+1} \neq a_n\} = p_t$$
$$\Pr\{a_{n+1} = a_n\} = 1 - p_t \tag{2.41}$$

The autocorrelation function for such a source is given by

$$R_a(l) = (1 - 2p_t)^{|l|} \tag{2.42}$$

Substituting (2.42) into (2.17) gives

$$S_a(f) = 1 + 2 \sum_{l=1}^{\infty} (1 - 2p_t)^l \cos 2\pi f l T \tag{2.43}$$

From [1; p. 84, Eq. (454)], we have that

$$\sum_{k=1}^{n} a^k \cos k\theta = \frac{a \cos \theta - a^2 + a^{n+2} \cos n\theta - a^{n+1} \cos [(n+1)\theta]}{1 - 2a \cos n\theta + a^2} \tag{2.44}$$

which for $a < 1$ and $n \to \infty$ becomes

$$\sum_{k=1}^{\infty} a^k \cos k\theta = \frac{a \cos \theta - a^2}{1 - 2a \cos \theta + a^2} \tag{2.45}$$

Furthermore

$$1 + 2 \sum_{k=1}^{\infty} a^k \cos k\theta = \frac{1 - a^2}{1 - 2a \cos \theta + a^2} \tag{2.46}$$

Applying (2.46) to (2.43) with $a = 1 - 2p_t$ and $q = 2\pi f T$ immediately gives the desired result, namely

$$S_a(f) = \frac{1 - (1 - 2p_t)^2}{1 + (1 - 2p_t)^2 - 2(1 - 2p_t) \cos 2\pi f T}$$
$$= \frac{4p_t (1 - p_t)}{2(1 - 2p_t)(1 - \cos 2\pi f T) + 4p_t^2} \tag{2.47}$$

Furthermore, if the data stream generated by this sequence uses rectangular pulses, that is,

$$p(t) = \begin{cases} A; & 0 \leq t \leq T \\ 0; & \text{otherwise} \end{cases} \tag{2.48}$$

then, using (2.16) and (2.48) we get

$$S_m(f) = A^2 T \frac{\sin^2(\pi f T)}{(\pi f T)^2} \left[\frac{4p_t(1 - p_t)}{2(1 - 2p_t)(1 - \cos 2\pi f T) + 4p_t^2} \right] \tag{2.49}$$

Example 2.2

Consider the sequence formed by interleaving N independent first order Markov sources with respective transition probabilities p_{t_n}; $n = 1, 2, \ldots, N$. Then, the resulting sequence is cyclostationary with correlation function

$$R_a(n; l) = \begin{cases} (1 - 2p_{t_n})^{|l|/N}; & l = 0, \pm N, \pm 2N, \ldots \\ 0; & \text{all other integer } l \end{cases} \tag{2.50}$$

The power spectral density $S_{\bar{a}}(f)$ of (2.29) is computed as (letting $l = kN, k = 0, \pm 1, \pm 2, \ldots$)

$$S_{\bar{a}}(f) = \frac{1}{N} \sum_{n=1}^{N} \sum_{k} \left(1 - 2p_{t_n}\right)^{|k|} e^{-j2\pi f kNT}$$

$$= \frac{1}{N} \sum_{n=1}^{N} \left\{ 1 + 2 \sum_{k=1}^{\infty} \left(1 - 2p_{t_n}\right)^{k} \cos 2\pi f kNT \right\}$$

(2.51)

Noting the similarity between (2.43) and (2.51) (for fixed n), we can immediately write down the result, namely

$$S_{\bar{a}}(f) = \frac{1}{N} \sum_{n=1}^{N} \left[\frac{4p_{t_n}(1 - p_{t_n})}{2(1 - 2p_{t_n})(1 - \cos 2\pi f NT) + 4p_{t_n}^2} \right]$$

(2.52)

2.2 POWER SPECTRAL DENSITY OF A GENERALIZED *M*-ARY MARKOV SOURCE

Consider a random *M*-ary source which every T seconds emits an elementary signal from the set $\{s_i(t); i = 1, 2, \ldots, M\}$ with probability p_i (the *stationary* probabilities). Note that, in contrast to the data model in Section 2.1, here it is the *waveshape* that changes with the data rather than merely the *amplitude* of a fixed waveshape (pulse). For a Markov source, the sequence of waveforms is further characterized by the *transition* probability matrix $\mathbf{P} = \{p_{ik}\}$ where now p_{ik} denotes the probability that signal $s_k(t)$ is transmitted in a given transmission interval after the occurrence of the signal $s_i(t)$ in the previous transmission interval.

From the above statistical description of the source, it can be shown [2], by an approach analogous to that taken in Section 2.1, that the power spectral density of a modulation $m(t)$ generated by this source is given by[3]

$$S_m(f) = \frac{1}{T} \sum_{i=1}^{M} p_i \left| S_i'(f) \right|^2$$

$$+ \frac{1}{T^2} \sum_{n=-\infty}^{\infty} \left| \sum_{i=1}^{M} p_i S_i \left(\frac{n}{T}\right) \right|^2 \delta \left(f - \frac{n}{T}\right)$$

(2.53)

$$+ \frac{2}{T} \mathrm{Re} \left[\sum_{i=1}^{M} \sum_{k=1}^{M} p_i S_i'^*(f) S_k'(f) p_{ik}(e^{-j2\pi fT}) \right]$$

where $S_i(f)$ is the Fourier transform of the ith elementary signal $s_i(t)$ and

$$p_{ik}(z) \triangleq \sum_{n=1}^{\infty} p_{ik}^{(n)} z^n; \quad s_i'(t) = s_i(t) - \sum_{k=1}^{M} p_k s_k(t)$$

(2.54)

Also, $S_i'(f)$ is the Fourier transform of $s_i'(t)$. The quantity $p_{ik}^{(n)}$ is defined as the probability that the elementary signal $s_k(t)$ is transmitted n signaling intervals after the occurrence of $s_i(t)$. Hence, from the properties of Markov sequences, $p_{ik}^{(n)}$ is the ik^{th} element of the matrix $\mathbf{P}^n$. Also, by definition, $p_{ik}^{(1)} = p_{ik}$.

---

3. Note that this result is consistent with those in Section 2.1 when $s_i(t) = a_i p(t)$.

Notice that the second term (the *discrete* or *line* spectrum) of (2.53) vanishes when

$$\sum_{i=1}^{M} p_i S_i \left(\frac{n}{T}\right) = 0 \tag{2.55}$$

which implies that a necessary and sufficient condition for the absence of a line spectrum is that

$$\sum_{i=1}^{M} p_i s_i(t) = 0 \tag{2.56}$$

Many special classes of signals exist for which the general PSD of (2.53) can be simplified. The first class is generated by the source defined by the conditions:

(i) for each signal waveform $s_i(t)$ of the set of elementary signals, the negative $-s_i(t)$ is also in the set;

(ii) the stationary probabilities on $s_i(t)$ and $-s_i(t)$ are equal, and

(iii) the transition probability $p_{ik} = p_{rs}$ whenever $s_i(t) = \pm s_r(t)$ and $s_k(t) = \pm s_s(t)$.

Such a signaling source is said to be *negative equally probable* (NEP). Its overall spectrum is characterized by the absence of a line spectrum and furthermore is independent of the transition probabilities themselves. For this special case, the PSD reduces to the first term of (2.53), namely

$$S_m(f) = \frac{1}{T} \sum_{i=1}^{M} p_i \, |S_i(f)|^2 \tag{2.57}$$

which is the weighted sum of the energy spectrum of the elementary signaling set.

Another class of signals of interest is those generated by a purely random source whose properties were discussed in Section 2.1. Such a source, that is, one whose transition matrix $\mathbf{P}$ has identical rows, has the property that $\mathbf{P}^n = \mathbf{P}$ for all $n \geq 1$. In this case, the PSD of (2.53) simplifies to

$$S_m(f) = \frac{1}{T} \sum_{i=1}^{M} p_i(1 - p_i) \, |S_i(f)|^2$$

$$+ \frac{1}{T^2} \sum_{n=-\infty}^{\infty} \left| \sum_{i=1}^{M} p_i S_i \left(\frac{n}{T}\right) \right|^2 \delta \left(f - \frac{n}{T}\right) \tag{2.58}$$

$$- \frac{2}{T} \sum_{\substack{i=1 \\ i \neq k,\ i < k}}^{M} \sum_{\substack{k=1 \\ }}^{M} p_i p_k \operatorname{Re} \left[S_i(f) S_k^*(f) \right]$$

When $M = 2$ and $p_1 = p$, $p_2 = 1 - p$, (2.58) reduces to

$$S_m(f) = \frac{1}{T} p(1 - p) \, |S_1(f) - S_2(f)|^2$$

$$+ \frac{1}{T^2} \sum_{n=-\infty}^{\infty} \left| p S_1 \left(\frac{n}{T}\right) + (1 - p) S_2 \left(\frac{n}{T}\right) \right|^2 \delta \left(f - \frac{n}{T}\right) \tag{2.59}$$

Obviously, if $s_1(t) = -s_2(t) = p(t)$ and $p = 1/2$, then the signaling format becomes NEP and (2.59) reduces to (2.57), which is then identical to (2.16).

At this point we turn to an evaluation of the PSDs of the various PCM signaling formats discussed in Chapter 1.

2.2.1 NRZ Baseband Signaling

Referring to Fig. 1.30 and assuming that the binary waveform levels are $\pm A$, the NRZ signaling format falls into the class of signals whose spectrum is given by (2.59). Since the elementary signal is a rectangular pulse of width T, its Fourier transform is

$$S_1(f) = -S_2(f)AT \exp(-j\pi f T)\frac{\sin(\pi f T)}{\pi f T} \tag{2.60}$$

Substituting (2.60) into (2.59) and letting $E = A^2 T$, we get

$$\frac{S_m(f)}{E} = \frac{1}{T}(1 - 2p)^2 \delta(f) + 4p(1 - p)\left[\frac{\sin^2(\pi f T)}{(\pi f T)^2}\right] \tag{2.61}$$

This result could also be obtained from (2.32) and (2.40) by recognizing that

$$\overline{a_n} = (1 - 2p); \quad \overline{A_n A_m} = \begin{cases} 1 - (1 - 2p)^2 = 4p(1 - p); & m = n \\ 0; & m \neq n \end{cases} \tag{2.62}$$

$$P\left(\frac{k}{T}\right) = \begin{cases} AT; & k = 0 \\ 0; & k \neq 0 \end{cases}$$

When $p = 1/2$, the dc spike at the origin disappears and the NRZ signaling format falls into the NEP class with

$$\frac{S_m(f)}{E} = \frac{\sin^2(\pi f T)}{(\pi f T)^2} \tag{2.63}$$

2.2.2 RZ Baseband Signaling

Here we have a situation where $S_1(f) = 0$ and $S_2(f)$ corresponds to the Fourier transform of a half-symbol wide pulse, that is,

$$S_2(f) = \frac{AT}{2} \exp\left(-\frac{j\pi f T}{2}\right)\left[\frac{\sin(\pi f T/2)}{\pi f T/2}\right] \tag{2.64}$$

Since the source is again purely random, then substituting (2.64) into (2.59) gives

$$\frac{S_m(f)}{E} = \frac{1}{4T}(1 - p)^2 \delta(f) + \frac{1}{4T}(1 - p)^2 \sum_{\substack{n=-\infty \\ n \text{ odd}}}^{\infty} \left(\frac{2}{n\pi}\right)^2 \delta\left(f - \frac{n}{T}\right)$$

$$+ \frac{1}{4}p(1 - p)\left[\frac{\sin^2(\pi f T/2)}{(\pi f T/2)^2}\right] \tag{2.65}$$

2.2.3 Biphase (Manchester) Baseband Signaling

Here the two elementary signals are defined by

$$
s_1(t) = \begin{cases} A; & 0 \le t \le T/2 \\ -A; & T/2 \le t \le T \end{cases}
$$

$$
s_2(t) = -s_1(t)
$$

(2.66)

Substituting the Fourier transform of (2.66) into (2.59) gives

$$
\frac{S_m(f)}{E} = \frac{1}{T}(1 - 2p)^2 \sum_{\substack{n=-\infty \\ n \text{ odd}}}^{\infty} \left(\frac{2}{n\pi}\right)^2 \delta\left(f - \frac{n}{T}\right)
$$

$$
+ 4p(1 - p)\left[\frac{\sin^4(\pi fT/2)}{(\pi fT/2)^2}\right]
$$

(2.67)

For $p = 1/2$, the line spectrum disappears and

$$
\frac{S_m(f)}{E} = \frac{\sin^4(\pi fT/2)}{(\pi fT/2)^2}
$$

(2.68)

2.2.4 Delay Modulation or Miller Coding

As indicated by Hecht and Guida [3], the Miller coding scheme can be modeled as a Markov source with four states whose stationary probabilities are all equal to 1/4 and whose transition matrix is given by

$$
\mathbf{P} = \begin{pmatrix} 0 & 1/2 & 0 & 1/2 \\ 0 & 0 & 1/2 & 1/2 \\ 1/2 & 1/2 & 0 & 0 \\ 1/2 & 0 & 1/2 & 0 \end{pmatrix}
$$

(2.69)

Another property of the Miller code is that it satsifies the recursion relation

$$
\mathbf{P}^{4+i}\mathbf{\Gamma} = -\frac{1}{4}\mathbf{P}^i\mathbf{\Gamma} \quad i \ge 0
$$

(2.70)

where $\mathbf{\Gamma}$ is the signal correlation matrix whose ik^{th} element is defined by

$$
\gamma_{ik} \triangleq \frac{1}{\sqrt{E_i E_j}} \int_0^T s_i(t)s_k(t)dt \quad i, k = 1, 2, 3, 4
$$

(2.71)

For the Miller code, the four elementary signals are defined by

$$
s_1(t) = -s_4(t) = A; \quad 0 \le t \le T
$$

$$
s_2(t) = -s_3(t) = \begin{cases} A; & 0 \le t \le T/2 \\ -A; & T/2 \le t \le T \end{cases}
$$

(2.72)

and $E_i = A^2T; \quad i = 1, 2, 3, 4$. Substituting (2.72) into (2.71) and arranging the results in

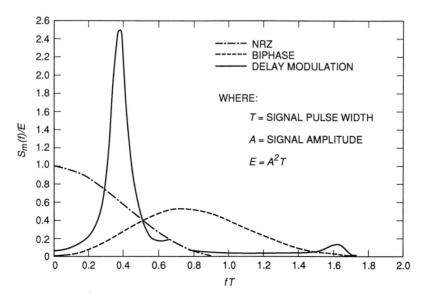

Figure 2.1   Power spectral densities: NRZ, biphase, and delay modulation (Reprinted from [15])

the form of a matrix, we get

$$
\Gamma = \begin{pmatrix} 1 & 0 & 0 & -1 \\ 0 & 1 & -1 & 0 \\ 0 & -1 & 1 & 0 \\ -1 & 0 & 0 & 1 \end{pmatrix}
\tag{2.73}
$$

Finally, using (2.69), (2.70), and (2.73) in the general PSD result of (2.53) yields the result for the Miller code:

$$
\frac{S_m(f)}{E} = \frac{1}{2\theta^2 (17 + 8 \cos 8\theta)} (23 - 2 \cos \theta - 22 \cos 2\theta - 12 \cos 3\theta
$$
$$
+ 5 \cos 4\theta + 12 \cos 5\theta + 2 \cos 6\theta - 8 \cos 7\theta + 2 \cos 8\theta)
\tag{2.74}
$$

where $\theta \triangleq \pi f T$.

The power spectral densities of the NRZ, biphase, and delay modulation signaling schemes as given by (2.63), (2.68), and (2.74), respectively, are plotted in Fig. 2.1.

Spectral properties of the Miller code that make it attractive for magnetic tape recording as well as phase-shift-keyed signaling include:

1. The majority of the signaling energy lies in frequencies less than one-half of the data rate, $R = 1/T$.

2. The spectrum is small in the vicinity of $f = 0$. This spectral minimum facilitates carrier tracking, which because of property 1 can also be more efficiently achieved than Manchester coding. A reduced spectral density in the vicinity of $f = 0$ is an important consideration in tape recording, primarily because of the poor dc response of the channel.

3. As a result of property 2, a lower magnetic tape recording speed (higher packing density) can be used.

4. The Miller code is insensitive to the $180°$ phase ambiguity common to NRZ-L and Manchester coding.

5. Bandwidth requirements are approximately one-half those needed by Manchester coding.

Other well-known techniques for reducing spectral bandwidth rely on ternary rather than binary alphabets [4]; however, this is achieved at the expense of an increase in error probability performance and a complication of implementation.

2.3 POWER SPECTRAL DENSITY FOR DATA PULSE STREAMS WITH DATA ASYMMETRY

In many applications [5–7], the data pulse stream experiences *data asymmetry* in that the rising and falling transitions do not occur at the nominal data transition time instants. Such asymmetry in the data waveform causes the presence of a line spectrum in the PSD as well as distortion of the continuous component of the PSD. Both of these ultimately cause degradation of the error probability of the receiver. In order to assess this degradation, it is desirable to evaluate the PSD of NRZ and Manchester streams with data asymmetry.

2.3.1 NRZ Data

In this section, we use the data asymmetry model proposed in [5] which assumes that $+1$ NRZ symbols are elongated by $\Delta T/2$ (relative to their nominal value of T sec) when a negative-going data transition occurs and -1 symbols are shortened by the same amount when a positive-going data transition occurs. When no data transition occurs, the symbols maintain their nominal T-sec value. As such, we can use the generalized M-ary source model of Section 2.2 where now $M = 4$ with

$$s_1(t) = \begin{cases} A; & -T/2 \le t \le T(1+\Delta)/2 \\ 0; & \text{otherwise} \end{cases}$$

$$s_2(t) = \begin{cases} -A; & -T/2 \le t \le T(1-\Delta)/2 \\ 0; & \text{otherwise} \end{cases}$$

$$s_3(t) = \begin{cases} A; & -T/2 \le t \le T/2 \\ 0; & \text{otherwise} \end{cases} \tag{2.75}$$

$$s_4(t) = \begin{cases} -A; & -T/2 \le t \le T/2 \\ 0; & \text{otherwise} \end{cases}$$

The stationary probabilities associated with these four elementary waveforms are, respectively

$$p_1 = p p_t; \quad p_2 = (1-p) p_t;$$

$$p_3 = p(1-p_t); \quad p_4 = (1-p)(1-p_t) \tag{2.76}$$

where p_t is the transition probability which for a purely random source is related to the a priori probability of a $+1$ NRZ symbol, p, by

$$p_t = 2p(1-p) \tag{2.77}$$

Taking the Fourier transform of the four elementary signals in (2.75) and substituting the results in (2.58) gives after much simplification [6]

$$S_m(f) = A^2 T \frac{\sin^2(\pi f T)}{(\pi f T)^2} [A_1(p_t) + A_2(p, p_t, \eta)] + A^2 T \frac{\sin^2(\pi f T \eta)}{(\pi f T)^2} A_3(p_t, \eta)$$

$$+ A^2 T \frac{\sin(2\pi f T)}{(\pi f T)^2} [A_4(p, p_t, \eta) - A_5(p, p_t)] \tag{2.78}$$

$$+ A^2 [2p - (1 - \eta p_t)]^2 \delta(f) + \frac{2A^2}{\pi^2} p_t^2 \sum_{n=1}^{\infty} \frac{1}{n^2} C(n, p, \eta) \delta\left(f - \frac{n}{T}\right)$$

where

$$A_1(p_t) = p_t(1 - p_t)[1 + 2(1 - p_t)] - p_t^3$$

$$A_2(p, p_t, \eta) = \left(3p_t^3 + p_t(1 - p_t)[1 + 2(1 - 2p)]\right) \cos^2(\pi f T \eta)$$

$$A_3(p_t, \eta) = p_t \left(1 + p_t^2 - p_t\right) \cos^2 \pi f T + p_t^3 \cos(2\pi f T \eta)$$

$$A_4(p, p_t, \eta) = p_t(1 - p_t)(1 - 2p) \left[\frac{1}{2} \cos(2\pi f T \eta) - p \sin(2\pi f T \eta)\right] \tag{2.79}$$

$$A_5(p, p_t) = \frac{1}{2} p_t(1 - p_t)(1 - 2p)$$

$$C(n, p, \eta) = \sin^2(n\pi\eta) \left[\cos^2(n\pi\eta) + (1 - 2p)^2 \sin^2(n\pi\eta)\right]$$

$$\eta = \frac{\Delta}{2}$$

Figure 2.2 is a plot of the continuous component of (2.78), $(S_m(f))_c$, normalized by the energy $E = A^2 T$, for various values of data asymmetry η in percent.

2.3.2 Manchester Data

In this section, we use the data asymmetry model proposed in [5] which assumes that for a $+1$ data bit, the first half of the Manchester symbol is elongated by $\Delta T/4$ (relative to its nominal value of $T/2$ sec) and for a -1 symbol, the first half of the Manchester symbol is shortened by the same amount. When a data transition follows, the second half of the Manchester symbol retains its nominal value (i.e., $T/2$). When no data transition occurs, the second half of the Manchester symbol adjusts itself so that the total Manchester symbol maintains its nominal T-sec value. In view of the above asymmetry model, we can use the generalized M-ary source model of Section 2.3 where now $M = 4$ with

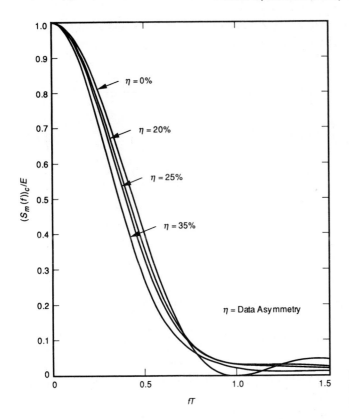

Figure 2.2 Normalized power spectra versus data asymmetry—NRZ data (Reprinted from [6] © IEEE)

$$s_1(t) = \begin{cases} A; & -T/2 \le t \le \Delta T/4 \\ -A; & \Delta T/4 \le t \le (T/2)(1 + \Delta/2) \\ 0; & \text{otherwise} \end{cases}$$

$$s_2(t) = \begin{cases} -A; & -T/2 \le t \le -\Delta T/4 \\ A; & -\Delta T/4 \le t \le (T/2)(1 - \Delta/2) \\ 0; & \text{otherwise} \end{cases}$$

$$s_3(t) = \begin{cases} A; & -T/2 \le t \le \Delta T/4 \\ -A; & \Delta T/4 \le t \le T/2 \\ 0; & \text{otherwise} \end{cases}$$ (2.80)

$$s_4(t) = \begin{cases} -A; & -T/2 \le t \le -\Delta T/4 \\ A; & -\Delta T/4 \le t \le T/2 \\ 0; & \text{otherwise} \end{cases}$$

The stationary probabilities associated with these four elementary waveforms are again given by (2.76). Taking the Fourier transform of the four elementary signals in (2.80) and substituting the results in (2.58) gives after much simplification

$$S_m(f) = (S_m(f))_c + (S_m(f))_d$$ (2.81)

where for $p = p_t = 1/2$, the discrete component $(S_m(f))_d$ is given by [7]

$$(S_m(f))_d = \frac{9}{4}A^2\eta^2\delta(f)$$

$$+ \frac{2A^2}{\pi^2} \sum_{m=1}^{\infty} \frac{1}{m^2} [H_1(m, \eta) + H_2(m, \eta) + H_3(m, \eta)]\,\delta\left(f - \frac{m}{T}\right) \tag{2.82}$$

with

$$\eta = \Delta/4$$

$$H_1(m, \eta) = \frac{\sin^2(m\pi\eta)\,[1 + 2h_1(m, \eta)]^2}{4}$$

$$H_2(m, \eta) = \sin^2(2m\pi\eta) \tag{2.83}$$

$$H_3(m, \eta) = 2\sin^2(m\pi\eta)\,(\cos m\pi\eta)\,[1 + 2h_1(m, \eta)]$$

The parameter $h_1(m, \eta)$ in (2.83) is defined as

$$h_1(m, \eta) = \begin{cases} \cos^2\left(\frac{m\pi\eta}{2}\right); & m \text{ odd} \\ \sin^2\left(\frac{m\pi\eta}{2}\right); & m \text{ even} \end{cases} \tag{2.84}$$

Similarly, for $p = p_t = 1/2$ the continuous component of (2.81) is given by

$$(S_m(f))_c = \frac{A^2T}{4} \frac{\sin^4\left(\frac{\pi fT}{2}\right)}{\left(\frac{\pi fT}{2}\right)^2} - A^2T\,[C_1(\eta) + C_2(\eta) + C_3]\frac{\sin^2(\pi fT\eta)}{\left(\frac{\pi fT}{2}\right)^2}$$

$$- A^2TC_4(\eta)\frac{\sin^2\left(\frac{\pi fT\eta}{2}\right)}{\left(\frac{\pi fT}{2}\right)^2} + A^2TC_5(\eta)\frac{\sin^2\left(\frac{\pi fT(1+\eta)}{2}\right)}{\left(\frac{\pi fT}{2}\right)^2} \tag{2.85}$$

$$+ A^2TC_6(\eta)\frac{\sin^2\left(\frac{\pi fT(1-\eta)}{2}\right)}{\left(\frac{\pi fT}{2}\right)^2}$$

where

$$C_1(\eta) = \frac{1}{4} \sin^2\left[\frac{\pi f T(1+\eta)}{2}\right] \left\{ \sin^2\left[\frac{\pi f T(1-\eta)}{2}\right] + \cos \pi f T \eta \right\}$$

$$C_2(\eta) = \frac{1}{8} \cos \pi f T \eta \left\{ 2 \sin^2\left[\frac{\pi f T(1-\eta)}{2}\right] - \sin^2\left[\frac{\pi f T \eta}{2}\right] \right\}$$

$$C_3 = \frac{1}{8}\left[1 - 4\cos\left(\frac{\pi f T}{4}\right)\right]$$

$$C_4(\eta) = \frac{\sin \pi f T \eta}{8} \left\{ \sin\left(\frac{\pi f T \eta}{2}\right) \right.$$

$$\left. + \sin\left(\frac{5\pi f T \eta}{2}\right)[1 - \cos \pi f T \cos \pi f T \eta] \right\}$$

$$- \frac{3}{8} \sin\left(\frac{\pi f T}{4}\right) \sin\left(\frac{\pi f T \eta}{4}\right) + \frac{1}{8}[2\cos 3\pi f T \eta + \cos 2\pi f T \eta]$$

$$C_5(\eta) = \frac{1}{8}\left\{ \sin^2\left[\frac{\pi f T(1-\eta)}{2}\right] - \frac{3}{2}\sin^2\left[\frac{\pi f T(1+\eta)}{2}\right]\right\}$$

$$C_6(\eta) = \frac{3}{16}\sin^2\left[\frac{\pi f T(1-\eta)}{2}\right] + \frac{1}{4}\sin^2\left[\frac{\pi f T}{2}\right]\cos \pi f T \eta$$

(2.86)

For $\eta = 0$, $(S_m(f))_d = 0$ and $(S_m(f))_c$ of (2.85) reduces to (2.68). Figure 2.3 is a plot of $(S_m(f))_c$ of (2.85) normalized by the energy $E = A^2 T$ for various values of data asymmetry η in percent. Note that as η increases, the null at $f = 0$ disappears.

2.4 POWER SPECTRAL DENSITY OF CONVOLUTIONALLY CODED MODULATIONS

In this section, we apply the results of Section 2.1 to the case where the data sequence $\{a_n\}$ corresponds to the output of a convolutional encoder (see Chapter 13 for a more detailed discussion of convolutional codes). For most of the optimum performance convolutional codes, a purely random input gives a purely random output with a bandwidth expansion equal to the reciprocal of the code rate. As such, the PSD of the encoder output would merely be a frequency scaled version of the input spectrum. There is, however, a class of codes (to be defined shortly) for which the above is not a true statement, and indeed some of the optimum performance codes fall into this category. In these instances, the bandwidth expansion produced by the encoder can be considerably less than the reciprocal of the code rate. A discussion of the implications of this statement will be given later on. Another consideration is the case where the input to the code is not a purely random source. Along these lines, we shall again consider two data models, namely, an unbalanced NRZ source where the two signaling states of the binary waveform are not assumed to occur with equal probability, that is, p and $(1 - p)$, and a first-order Markov source which allows transition probabilities, p_t, between 0 and 1.

We begin by noting that the symbol sequence at the output of a convolutional encoder is a cyclostationary process (not necessarily zero mean) with period equal to the number of modulo-2 summers, n. As such the results of Section 2.1.3, in particular, are the ones that apply.

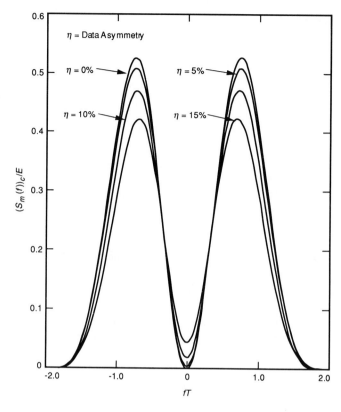

η = Data Asymmetry

η = 0%

η = 5%

η = 10%

η = 15%

Figure 2.3 Normalized power spectra versus data asymmetry—Manchester data (Reprinted from [7] © IEEE)

2.4.1 Convolutional Encoder Model

Consider a convolutional encoder with constraint length K and rate b/n, as shown in Fig. 2.4. In the mth information interval, b information symbols a_{mb+j}; $j = 0, 1, 2, \ldots, b-1$ enter the encoder and n channel symbols x_{mn+v}; $v = 1, 2, \ldots, n$ exit the encoder. The structure of such a convolutional encoder can also be defined by a generator (connection) matrix, namely

$$\mathbf{G} = \begin{bmatrix} g_{1,1} & g_{1,2} & \cdot & \cdot & g_{1,Kb} \\ g_{2,1} & g_{2,2} & \cdot & \cdot & g_{2,Kb} \\ \cdot & & & & \cdot \\ \cdot & & & & \cdot \\ g_{n,1} & g_{n,2} & \cdot & \cdot & g_{n,Kb} \end{bmatrix} \tag{2.87}$$

where $g_{i,j}$ is either one or zero depending, respectively, on whether or not the ith modulo summer is connected to the jth shift register stage. For mathematical convenience, we shall assume that both the input symbols $\{a_{mb+j}\}$ and the output symbols $\{x_{mn+v}\}$ take on values plus and minus one. This allows modulo-2 summation operations to be replaced by algebraic products. Thus, the encoder of Fig. 2.4 has the input output relation

$$x_{mn+v} = \prod_{i=1}^{Kb} \left[a_{(m+1)b-i} \right]^{g_{v,i}} \tag{2.88}$$

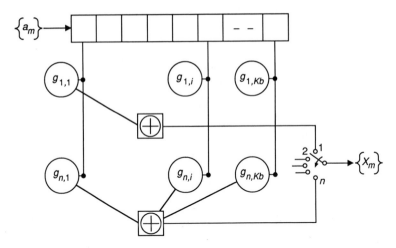

Figure 2.4   A general constraint length K, rate b/n convolutional encoder (Reprinted from [8], [16] © IEEE)

Clearly, the output sequence $\{x_{mn+v}\}$ is cyclostationary with period n as previously mentioned.

If, analogous to (2.2), a modulation $m(t)$ with pulse shape $p(t)$ is formed from the sequence $\{x_{mn+v}\}$, then the spectrum of this modulation is given by (2.32) and (2.34). In terms of the present notation, the continuous component of the spectrum $(S_m(f))_c$ would be

$$(S_m(f))_c = S_p(f) S_{\bar{X}}(f) \tag{2.89}$$

where $S_p(f)$ is given by (2.16) and

$$S_{\bar{X}}(f) = \frac{1}{n} \sum_{v=1}^{n} \left[\sum_{\kappa=-\infty}^{\infty} R_X(v; \kappa) e^{-j2\pi \kappa f T} \right] \tag{2.90}$$

where

$$R_X(v; \kappa) = \overline{(x_v - \bar{x}_v)(x_{v+\kappa} - \bar{x}_{v+\kappa})} \triangleq \overline{X_v X_{v+\kappa}} \tag{2.91}$$

Letting

$$\kappa = nl + q - v; \quad v = 1, 2, \ldots, n \tag{2.92}$$

we can rewrite (2.90) as

$$S_{\bar{X}}(f) = \frac{1}{n} \sum_{v=1}^{n} \sum_{q=1}^{n} \sum_{l=-\infty}^{\infty} R_X(v; nl + q - v) e^{-j2\pi(nl+q-v)fT} \tag{2.93}$$

Equation (2.93) will be more useful later on since it shows that x_v is the symbol generated from the vth interleaving tap output and x_{nl+q} is the symbol generated from the qth interleaving tap output.

From the definition of a cyclostationary sequence, we have that

$$\overline{X_v X_{nl+q}} = \overline{X_{v-nl} X_q} \tag{2.94}$$

Thus, using (2.94), we can rewrite (2.93) in the more compact form

$$S_{\bar{X}}(f) = \frac{1}{n} \sum_{v=1}^{n} \sum_{q=1}^{n} \sum_{l=0}^{\infty} \varepsilon_l R_X(v; nl + q - v) \cos[2\pi(nl + q - v)fT] \quad (2.95)$$

where ε_l is the Neumann factor defined by

$$\varepsilon_l = \begin{cases} 1; & l = 0 \\ 2; & l \neq 0 \end{cases} \quad (2.96)$$

We shall define the "memory" of the cyclostationary process as the smallest integer l^* such that $\overline{X_v X_{nl+q}} = 0$ for $l > l^*$.

If $\{x_i\}$ is not zero mean, then in addition to the continuous PSD component of (2.89), we will also have a discrete spectral component $(S_m(f))_d$ given by

$$(S_m(f))_d = \frac{1}{(nT)^2} \sum_{k=-\infty}^{\infty} \left| P\left(\frac{k}{nT}\right) \right|^2 \left\{ \left[\sum_{v=1}^{n} \bar{x}_v \cos\left(\frac{2\pi vk}{n}\right) \right]^2 \right.$$

$$\left. + \left[\sum_{v=1}^{n} \bar{x}_v \sin\left(\frac{2\pi vk}{n}\right) \right]^2 \right\} \delta\left(f - \frac{k}{nT}\right) \quad (2.97)$$

2.4.2 Encoder Output Spectrum for Independent Binary Symbol Input

Consider the input to the encoder of Fig. 2.4 to be a sequence of independent binary symbols $\{a_j\}$ that take on values ± 1 with probabilities p and $(1 - p)$. Let $m = 0$ in (2.88) with no loss in generality; then since the symbols are independent, the mean of x_v is given by

$$\bar{x}_v = (\bar{a})^{\alpha_v} = (1 - 2p)^{\alpha_v}; \quad v = 1, 2, \ldots, n \quad (2.98)$$

where

$$\alpha_v \triangleq \sum_{i=1}^{Kb} g_{v,i} \quad (2.99)$$

that is, the algebraic sum of the number of $+1$'s in the vth row of **G**. Furthermore, it can be shown that

$$R_X(v; nl + q - v) = \begin{cases} (\bar{a})^{\alpha_v + \alpha_q + 2\beta_{vq}(l)} - (\bar{a})^{\alpha_v + \alpha_q}; & 0 \leq l \leq K - 1 \\ (\bar{a})^{\alpha_v + \alpha_q + 2\beta_{qv}(-l)} - (\bar{a})^{\alpha_v + \alpha_q}; & -(K - 1) \leq l \leq 0 \quad (2.100) \\ 0; & \text{otherwise} \end{cases}$$

where

$$\beta_{vq}(l) \triangleq \sum_{i=1}^{(K-l)b} g_{v,i} g_{q,i+lb} \quad (2.101)$$

that is, the cross-correlation of the vth row[4] of **G** with the qth row shifted by lb elements. Finally, substituting (2.98) and (2.100) into (2.95) and (2.97) gives the desired results for

---

4. The cross-correlation is performed only on the parts of the vth and shifted q^{th} rows whose elements overlap.

the continuous and discrete components of the encoder output spectrum, namely

$$
S_{\bar{X}}(f) = \frac{1}{n} \sum_{v=1}^{n} \sum_{q=1}^{n} \sum_{l=0}^{K-1} \varepsilon_l \left[(\bar{a})^{\alpha_v + \alpha_q + 2\beta_{vq}(l)} - (\bar{a})^{\alpha_v + \alpha_q} \right] \cos \left[2\pi (nl + q - v) fT \right]
$$

$$(2.102)$$

$$
(S_m(f))_c = T \left(\frac{\sin \pi f T}{\pi f T} \right)^2 S_{\bar{X}}(f)
$$

where a unit rectangular pulse has been assumed for $p(t)$ and

$$
(S_m(f))_d = \frac{1}{n^2} \sum_{k=-\infty}^{\infty} \left(\frac{\sin \frac{\pi k}{n}}{\frac{\pi k}{n}} \right)^2 \left\{ \left[\sum_{v=1}^{n} (\bar{a})^{\alpha_v} \cos \left(\frac{2\pi v k}{n} \right) \right]^2 \right.
$$

$$(2.103)$$

$$
\left. + \left[\sum_{v=1}^{n} (\bar{a})^{\alpha_v} \sin \left(\frac{2\pi v k}{n} \right) \right]^2 \right\} \delta \left(f - \frac{k}{nT} \right)
$$

Several interesting points can be deduced from (2.102). First, note from (2.99) and (2.103) that for $v = q$ and $l = 0$, the exponent $\alpha_v + \alpha_q - 2\beta_{vq}(l) = 0$ *independent of the code.* Thus, for these terms

$$
(\bar{a})^{\alpha_v + \alpha_q - 2\beta_{vq}(l)} = 1
$$

$$(2.104)$$

for all $\bar{a}$ (including zero). It is also possible that the above exponent can be zero for other combinations of v, q, and l. The conditions on the elements of **G** under which this is possible are derived in [8] and are given by

$$
g_{v,i} = 0; \quad i = (K - l)b + 1, (K - l)b + 2, \ldots, Kb
$$

$$
g_{q,i} = 0; \quad i = 1, 2, \ldots, lb
$$

$$(2.105)$$

$$
g_{v,i} = g_{q,i+lb}; \quad i = 1, 2, \ldots, (K - l)b
$$

which are graphically illustrated in Fig. 2.5. This defines a class of convolutional codes whose significance will become apparent shortly. In the meantime, it is sufficient to remark that if the code is such that (2.105) can be satisfied for any values of v, q, and l (other than $v = q$ and $l = 0$), then (2.104) is satisfied for all $\bar{a}$ (again including zero).

Returning to (2.102), we further note that independent of the code, the exponent $\alpha_v + \alpha_q$ is never equal to zero. Thus, the term $(\bar{a})^{\alpha_v + \alpha_q}$ will be equal to one only when $\bar{a}$ is equal to one and will vanish when $\bar{a}$ is equal to zero (that is, $p = 1/2$). This last statement is crucial to the results that now follow.

The case of a purely random data input (p = 1/2).   Consider the special case of a purely random data input with rectangular pulses and $p = 1/2$ ($\bar{a} = 0$). If the code is such that (2.105) is not satisfied for any combination of v, q, and l (other than $v = q$ and $l = 0$), then the only nonzero terms in (2.102) will be the n terms correponding to $v = q$ and $l = 0$ for which (2.104) applies. Thus, $S_{\bar{X}}(f) = 1$ and the continuous component of the spectrum is just given by (2.63). Furthermore, the discrete component vanishes. Thus, the conclusion to be reached is that *for all codes that do not satisfy (2.105), a random NRZ data input results in a random NRZ data output scaled (expanded) in frequency by the reciprocal*

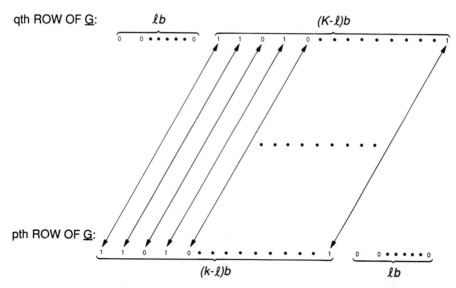

Figure 2.5   An illustration of the constraints of Equation (2.105) (Reprinted from [8], [16] © IEEE)

of the code rate, that is, n/b. We define this class of codes as *uncorrelated convolutional codes.*[5]

If, on the other hand, the code is such that (2.105) can be satisfied for at least one combination of v, q, and l (other than $v = q$ and $l = 0$), then $S_{\tilde{X}}(f)$ of (2.102) will be frequency dependent and the encoder output spectrum will not be merely a frequency scaled version of the input. In fact, let $\{v, q, l\}$ denote the set of combinations of v, q, and l (other than $v = q$ and $l = 0$) for which (2.105) can be satisfied; then (2.102) simplifies to

$$S_{\tilde{X}}(f) = 1 + \frac{1}{n} \sum_{\{v,q,l\}} \sum \sum \varepsilon_l \cos\left[2\pi(nl + q - v)fT\right] \tag{2.106}$$

Note that the discrete component still vanishes. We define this class of codes as *correlated convolutional codes* and observe that for such codes the encoder output spectrum will differ (in form) from the input spectrum.

An example of a correlated convolutional code is one whose generator matrix is such that two or more rows are identical since it satisfies (2.105) with $l = 0$. Larsen [9] tabulates short constraint length codes (up to and including $K = 14$) noncatastrophic codes with maximal free distance for rates 1/2, 1/3, and 1/4. Included in this tabulation is a rate 1/3, $K = 3$ code with generator matrix

$$\mathbf{G} = \begin{bmatrix} 1 & 0 & 1 \\ 1 & 1 & 1 \\ 1 & 1 & 1 \end{bmatrix} \tag{2.107}$$

This code was first found by Odenwalder [10]. The two combinations of v, q, and l that satisfy (2.105) are $v = 2, q = 3, l = 0$, and $v = 3, q = 2, l = 0$ in which case (2.106) becomes

$$S_{\tilde{X}}(f) = 1 + \frac{2}{3} \cos 2\pi fT \tag{2.108}$$

---

5. Note that all rate 1/2, noncatastrophic convolutional codes are uncorrelated codes.

CODE GENERATOR MATRIX

1 0 1
1 1 1
1 1 1

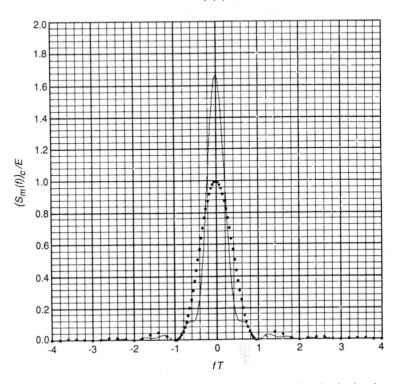

Figure 2.6   Spectrum for best rate $1/3$, constraint length 3 convolutional code; dotted curve is spectrum of uncoded NRZ (Reprinted from [8], [16] © IEEE)

Multiplying (2.108) by (2.63) gives the encoder output spectrum illustrated in Fig. 2.6.

Another interesting result may be gleaned from the example just presented. Since codes whose generator matrix have at least two identical rows are a subclass of correlated convolutional codes for which $l = 0$, then for these codes (2.99) simplifies still further to

$$S_{\bar{X}}(f) = 1 + \frac{1}{n} \sum_{\{v,q\}} \sum \cos[2\pi(q-v)fT] \tag{2.109}$$

where the set $\{v, q\}$ now represents all pairs of integers corresponding to pairs of identical rows in **G**. Since (2.109) only depends on the *difference* $v - q$, and small values of $v - q$ correspond to less oscillatory behavior of $S_{\bar{X}}(f)$, then it is clearly desirable from a spectral concentration standpoint to have the identical rows packed close together in **G**. For example, if the bottom row of **G** in (2.107) is moved to the top, then $v - q = 2$ and

$$S_{\bar{X}}(f) = 1 + \frac{2}{3} \cos 4\pi fT \tag{2.110}$$

which is illustrated in Fig. 2.7.

CODE GENERATOR MATRIX

1 1 1
1 0 1
1 1 1

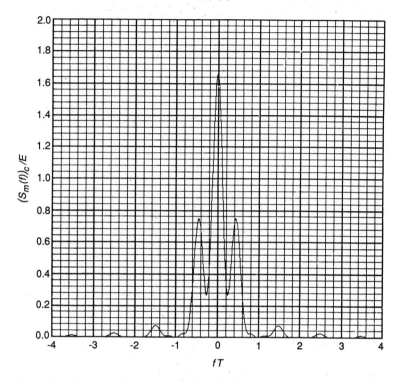

Figure 2.7   Spectrum for best rate 1/3, constraint length 3 convolutional code with permuted code generator matrix (Reprinted from [8], [16] © IEEE)

The important point to realize from the above is that *permuting the rows of the generator matrix of a convolutional code can be used to change the encoder output spectrum without any effect on the bit error probability performance of the code.*

The case of an unbalanced NRZ input (p ≠ 1/2).   Consider the optimum rate 1/2, constraint length 3 code whose generator matrix is given by the first two rows of (2.107). For this code $\alpha_1 = 2$, $\alpha_2 = 3$ and the sets of exponents $\alpha_v + \alpha_q$ and $\alpha_v + \alpha_q - 2\beta_{v,q}(l)$; $l = 0, 1, 2$; $v, q = 1, 2$ are tabulated below:

l	v	q	$2l + q - v$	$\alpha_v + \alpha_q$	$\alpha_v + \alpha_q - 2\beta_{v,q}(l)$
0	1	1	0	4	0
0	1	2	1	5	1
0	2	1	−1	5	1
0	2	2	0	6	0
1	1	1	2	4	4
1	1	2	3	5	3

l	ν	q	$2l+q-\nu$	$\alpha_\nu + \alpha_q$	$\alpha_\nu + \alpha_q - 2\beta_{\nu,q}(l)$
1	2	1	1	5	3
1	2	2	2	6	2
2	1	1	4	4	2
2	1	2	5	5	3
2	2	1	3	5	3
2	2	2	4	6	4

Using these results in (2.102) gives the closed form result

$$S_{\bar{X}}(f) = 1 - \frac{1}{2}\left(\bar{a}^4 - \bar{a}^6\right) + \left(\bar{a} + \bar{a}^3 - 2\bar{a}^5\right)\cos 2\pi fT$$

$$+ \left(\bar{a}^2 - \bar{a}^6\right)[\cos 4\pi fT + \cos 8\pi fT] \tag{2.111}$$

$$+ \left(\bar{a}^3 - \bar{a}^5\right)[2\cos 6\pi fT + \cos 10\pi fT]$$

Since this code is an uncorrelated code, for $\bar{a} = 0$ (2.111) reduces to $S_{\bar{X}}(f) = 1$ as it should.

Figure 2.8 is an illustration of the corresponding continuous spectrum $(S_m(f))_c$ [obtained by multiplying (2.111) by (2.63)] for several values of $p = (1 - \bar{a})/2$. We note that the output encoder spectrum narrows as p is decreased despite the fact that the continuous component of the input spectrum remains unaltered in shape as p is varied [see (2.61)].

The discrete component of the output spectrum is easily found from (2.103) to be

$$(S_m(f))_d = \left(\bar{a}^3 - \bar{a}^2\right)^2 \sum_{k=1,3,5,\dots}^{\infty} \left(\frac{1}{\pi k}\right)^2 \delta\left(f - \frac{k}{2T}\right) \tag{2.112}$$

2.4.3 Encoder Output Spectrum for First-Order Markov Input

For the first-order Markov source, we first transform the information sequence $\{a_n\}$ into a transition sequence $\{t_j\}$ where $t_j = a_j a_{j-1};\ j = 1, 2, \dots$ The sequence $\{t_j\}$ is a stationary sequence of independent ± 1 random variables with transition probability p_t. For any l and $n \geq 1$, we have

$$a_{l-n} = a_l \prod_{i=1}^{n} t_{l-i+1}; \quad a_{l+n} = a_l \prod_{i=1}^{n} t_{l+i} \tag{2.113}$$

Defining

$$h_{\nu,i} = g_{\nu,i} \oplus g_{\nu,i+1} \oplus \dots \oplus g_{\nu,Kb} \tag{2.114}$$

where $\oplus$ denotes modulo-2 summation, and

$$\rho_\nu = \sum_{i=2}^{Kb} h_{\nu,i} \tag{2.115}$$

then after considerable algebraic manipulation, it can be shown [8] that

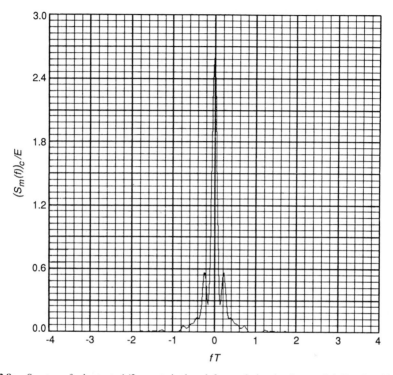

CODE GENERATOR MATRIX

1 0 1
1 1 1

Figure 2.8a   Spectrum for best rate $1/2$, constraint length 3 convolutional code; $p = 0.1$ (Reprinted from [16])

$$S_{\bar{X}}(f) = \frac{1}{n} \sum_{v=1}^{n} \sum_{q=1}^{n} \left(1 - h_{v,1} - h_{q,1} + 2\dot{h}_{v,1}h_{q,1}\right)$$

$$\times \sum_{l=0}^{K-1} \varepsilon_l \left[(\bar{t})^{\rho_v + \rho_q - 2\xi_{v,q}(l)} - \left(1 - h_{q,1}\right)(\bar{t})^{\rho_v + \rho_q} \right]$$

$$\times \cos\left[2\pi(nl + q - v)fT\right] \tag{2.116}$$

$$+ \frac{2}{n} \sum_{v=1}^{n} \sum_{q=1}^{n} h_{v,1}h_{q,1} (\bar{t})^{\rho_v + \rho_q + Kb}$$

$$\times \frac{\cos\left[2\pi(nK + q - v)fT\right] - \bar{t}^b \cos\left[2\pi(n(K-1) + q - v)fT\right]}{1 - 2\bar{t}^b \cos 2\pi nfT + \bar{t}^{2b}}$$

where

$$\xi_{v,q}(l) = \sum_{i=2}^{(K-l)b} h_{v,i}h_{q,i+lb} - \frac{1}{2}h_{q,1}\left(2 + lb - 2\sum_{i=1}^{lb+1} h_{q,i}\right) \tag{2.117}$$

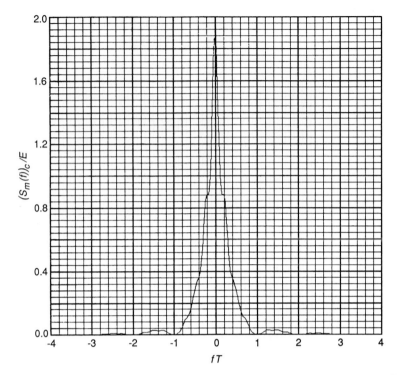

Figure 2.8b   Spectrum for best rate 1/2, constraint length 3 convolutional code; $p = 0.3$ (Reprinted from [16])

Equation (2.117) is used in (2.89) to get the continuous component of the encoder output spectrum.

The discrete spectrum is obtained as

$$(S_m(f))_d = \frac{1}{n^2} \sum_{k=-\infty}^{\infty} \left(\frac{\sin \frac{\pi k}{n}}{\frac{\pi k}{n}} \right)^2 \left\{ \left[\sum_{v=1}^{n} (1 - h_{v,1}) \, \bar{i}^{\rho_v} \cos \left(\frac{2\pi vk}{n} \right) \right]^2 \right.$$

$$\left. + \left[\sum_{v=1}^{n} (1 - h_{v,1}) \, \bar{i}^{\rho_v} \sin \left(\frac{2\pi vk}{n} \right) \right]^2 \right\} \delta \left(f - \frac{k}{nT} \right)$$

(2.118)

Note that a discrete spectrum can potentially exist at the encoder output despite the fact that the encoder has only a continuous spectrum at its input such as that in (2.49) with $p_t = (1 - \bar{i})/2$. Note that if the code is *transparent*, that is, each row of **G** has an odd number of ones, then from (2.114)

$$h_{v,i} = 1; \quad v = 1, 2, \ldots, n \tag{2.119}$$

which when substituted in (2.118) results in $(S_m(f))_d = 0$.

As an example of the application of (2.116), again consider the optimum rate 1/2 constraint length 3 convolutional code for which the modified generator matrix **H** whose elements are defined in (2.114) is given by

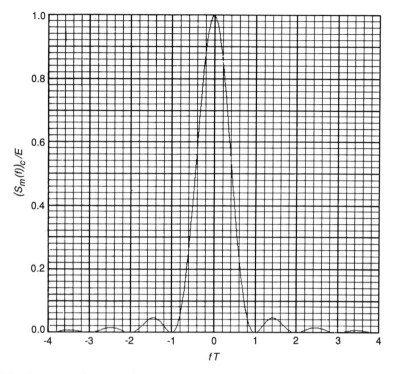

Figure 2.8c   Spectrum for best rate $1/2$, constraint length 3 convolutional code; $p = 0.5$ (Reprinted from [16])

$$\mathbf{H} = \begin{bmatrix} 0 & 1 & 1 \\ 1 & 0 & 1 \end{bmatrix} \tag{2.120}$$

The corresponding continuous component of the encoder output spectrum may be obtained from the following:

$$S_{\bar{X}}(f) = 1 - \frac{1}{2}\bar{i}^4 \tag{2.121}$$

$$+ \frac{-\bar{i}^3(1 + \bar{i} - \bar{i}^2) + \bar{i}^2(1 + \bar{i}^3 - \bar{i}^4)\cos 4\pi fT + \bar{i}^2(1 - \bar{i} - \bar{i}^2 + \bar{i}^3)\cos 8\pi fT}{1 - 2\bar{i}\cos 4\pi fT + \bar{i}^2}$$

Figure 2.9 is an illustration of $(S_m(f))_c$ [using (2.121)] for several values of transition density $p_t = (1 - \bar{i})/2$.

2.5 COMPARISON OF THE SPECTRAL OCCUPANCY OF VARIOUS I-Q MODULATION TECHNIQUES

Thus far, we have considered evaluation of the PSD only for *baseband* modulations. In a large part of the remainder of this book, we shall address ourselves to a form of *bandpass* modulation in which two lowpass modulations are impressed on inphase and quadrature carriers. Such an inphase-quadrature (I-Q) signal has the generic form

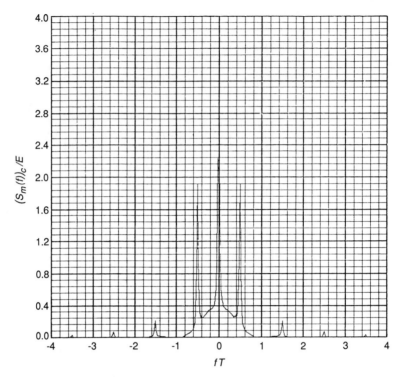

Figure 2.9a Spectrum for best rate 1/2, constraint length 3 convolutional code; first-order Markov source; $p_t = 0.1$ (Reprinted from [8], [16] © IEEE)

$$s(t) = A_I m_I(t) \cos \omega_c t - A_Q m_Q(t) \sin \omega_c t$$

$$= \text{Re} \left\{ \tilde{s}(t) e^{j\omega_c t} \right\}; \quad \tilde{s}(t) = A_I m_I(t) + j A_Q m_Q(t) \tag{2.122}$$

where $\omega_c = 2\pi f_c$ is the radian carrier frequency, $m_I(t)$ and $m_Q(t)$ are baseband modulations of the form in (2.2), that is,

$$m_I(t) = \sum_{n=-\infty}^{\infty} a_n p_I(t - nT_I); \quad m_Q(t) = \sum_{n=-\infty}^{\infty} b_n p_Q(t - nT_Q) \tag{2.123}$$

and $\tilde{s}(t)$ is the equivalent complex lowpass signal. Assuming that $\{a_n\}$ and $\{b_n\}$ are ± 1, zero mean WSS sequences as in (2.1) but are independent of each other, then the PSD of $s(t)$ is related to the PSD's of $m_I(t)$ and $m_Q(t)$ by

$$S_s(f) = \frac{A_I^2}{4} \left[S_{m_I}(f - f_c) + S_{m_I}(f + f_c) \right]$$

$$+ \frac{A_Q^2}{4} \left[S_{m_Q}(f - f_c) + S_{m_Q}(f + f_c) \right] \tag{2.124}$$

One popular carrier modulation to be discussed in Chapter 4 that falls in the class defined by (2.122) is *quadrature phase-shift-keying (QPSK)* in which $\{a_n\}$ and $\{b_n\}$ are

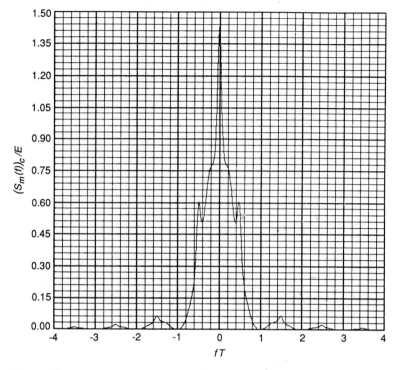

Figure 2.9b Spectrum for best rate 1/2, constraint length 3 convolutional code; first-order Markov source; $p_t = 0.3$ (Reprinted from [16])

random data sequences, $A_I = A_Q = A$, $T_I = T_Q = 2T_b$ (T_b is the bit time), and $p_I(t) = p_Q(t) = p(t)$ is a unit rectangular pulse of duration $2T_b$. Using the results of Section 2.1.1, we find that the PSD of QPSK is given by

$$S_s(f) = PT_b \left[\frac{\sin^2 2\pi (f - f_c)T_b}{(2\pi (f - f_c)T_b)^2} + \frac{\sin^2 2\pi (f + f_c)T_b}{(2\pi (f + f_c)T_b)^2} \right] \tag{2.125}$$

where $P = A^2$ is the power in $s(t)$.

Another popular carrier modulation technique to be discussed in Chapter 10 is *minimum-shift-keying (MSK)* [11] which has an I-Q representation in which $\{a_n\}$ and $\{b_n\}$ are random data sequences, $A_I = A_Q = A$, $T_I = T_Q = 2T_b$, and

$$p_I(t) = \begin{cases} \cos \frac{\pi t}{2T_b}; & -T_b \leq t \leq T_b \\ 0; & \text{otherwise} \end{cases} , \quad p_Q(t) = \begin{cases} \sin \frac{\pi t}{2T_b}; & 0 \leq t \leq 2T_b \\ 0; & \text{otherwise} \end{cases} \tag{2.126}$$

Again using the results of Section 2.1.1, the PSD of MSK is found to be

$$S_s(f) = \frac{8PT_b}{\pi^2} \left[\frac{\cos^2 2\pi (f - f_c)T_b}{(1 - 16(f - f_c)^2 T_b^2)^2} + \frac{\cos^2 2\pi (f + f_c)T_b}{(1 - 16(f + f_c)^2 T_b^2)^2} \right] \tag{2.127}$$

Here $P = A^2/2$. Note that, asymptotically, the PSD of MSK rolls off at a rate f^{-4} as compared with QPSK which rolls off as f^{-2}.

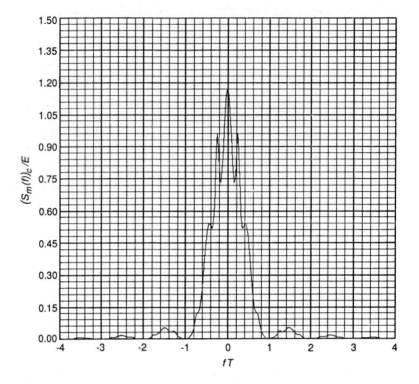

Figure 2.9c Spectrum for best rate $1/2$, constraint length 3 convolutional code; first-order Markov source; $p_t = 0.7$ (Reprinted from [16])

Even faster spectral roll-offs can be achieved by further smoothing of the pulse shape. For example, so called *sinusoidal frequency-shift-keying (SFSK)* [12] (see Chapter 10) has an I-Q representation in the form of (2.122) where $\{a_n\}$ and $\{b_n\}$ are random data sequences, $A_I = A_Q = A$, $T_I = T_Q = 2T_b$, and

$$p_I(t) = \begin{cases} \cos\left[\frac{\pi t}{2T_b}\left(1 - \frac{\sin 2\pi t/T_b}{2\pi t/T_b}\right)\right]; & -T_b \leq t \leq T_b \\ 0; & \text{otherwise} \end{cases}$$

$$p_Q(t) = \begin{cases} \sin\left[\frac{\pi t}{2T_b}\left(1 - \frac{\sin 2\pi t/T_b}{2\pi t/T_b}\right)\right]; & 0 \leq t \leq T_b \\ \sin\left[\frac{\pi(2T_b-t)}{2T_b}\left(1 - \frac{\sin 2\pi(2T_b-t)/T_b}{2\pi(2T_b-t)/T_b}\right)\right]; & T_b \leq t \leq 2T_b \\ 0; & \text{otherwise} \end{cases} \tag{2.128}$$

For SFSK, the PSD can be obtained in the form of an infinite series, in particular [12]

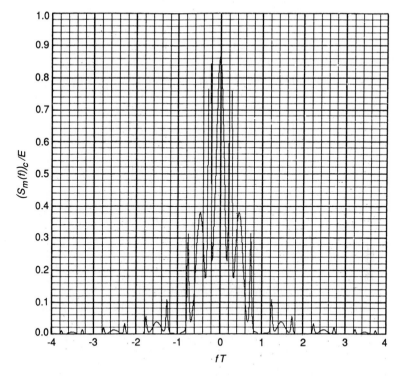

Figure 2.9d Spectrum for best rate 1/2, constraint length 3 convolutional code; first-order Markov source; $p_t = 0.9$ (Reprinted from [8], [16] © IEEE)

$$S_s(f) = \frac{PT_b}{2} \left\{ \left[J_0\left(\frac{1}{4}\right) A_0(f - f_c) + 2\sum_{n=1}^{\infty} J_{2n}\left(\frac{1}{4}\right) B_{2n}(f - f_c) \right. \right.$$

$$\left. + 2\sum_{n=1}^{\infty} J_{2n-1}\left(\frac{1}{4}\right) B_{2n-1}(f - f_c) \right]^2$$

$$+ \left[J_0\left(\frac{1}{4}\right) A_0(f + f_c) + 2\sum_{n=1}^{\infty} J_{2n}\left(\frac{1}{4}\right) B_{2n}(f + f_c) \right.$$

$$\left. \left. + 2\sum_{n=1}^{\infty} J_{2n-1}\left(\frac{1}{4}\right) B_{2n-1}(f + f_c) \right]^2 \right\}$$

(2.129)

where $J_n(x)$ is the nth order Bessel function of argument x and

$$A(f) = 2T_b \frac{\sin 2\pi f T_b}{2\pi f T_b}$$

$$A_0(f) = \frac{1}{2}A\left(f + \frac{1}{4T_b}\right) + \frac{1}{2}A\left(f - \frac{1}{4T_b}\right) = \frac{4T_b}{\pi} \frac{\cos 2\pi f T}{1 - 16f^2 T_b^2}$$

$$A_{2n}(f) = \frac{1}{2}A\left(f + \frac{2n}{T_b}\right) + \frac{1}{2}A\left(f - \frac{2n}{T_b}\right);$$

$$A_{2n-1}(f) = \frac{1}{2}A\left(f + \frac{2n-1}{T_b}\right) - \frac{1}{2}A\left(f - \frac{2n-1}{T_b}\right)$$

$$B_{2n}(f) = \frac{1}{2}A_{2n}\left(f + \frac{1}{4T_b}\right) + \frac{1}{2}A_{2n}\left(f - \frac{1}{4T_b}\right);$$

$$B_{2n-1}(f) = -\frac{1}{2}A_{2n-1}\left(f + \frac{1}{4T_b}\right) + \frac{1}{2}A_{2n-1}\left(f - \frac{1}{4T_b}\right)$$

(2.130)

It can be shown that (2.130) rolls off asymptotically as f^{-6}.

For all three of the above modulations, the bandpass PSD can be expressed in the form

$$S_s(f) = \frac{1}{4}[G(f - f_c) + G(f + f_c)] \tag{2.131}$$

where $G(f)$ is the equivalent two-sided lowpass PSD, that is, the PSD of the complex lowpass signal $\tilde{s}(t)$ defined in (2.122). Figure 2.10 is a plot of $G(f)/2PT_b$ (in dB) versus fT_b for the three modulations. This figure clearly illustrates the trade-off between the width of the main lobe and rate of roll-off of the spectrum. Figure 2.11 is a plot of fractional out-of-band power, η (in dB), for the same three modulations versus BT_b where the equivalent lowpass bandwidth, B, is defined, as in Chapter 1, by

$$\eta = 1 - \frac{\int_{-B}^{B} G(f)df}{\int_{-\infty}^{\infty} G(f)df} \tag{2.132}$$

2.6 POWER SPECTRAL DENSITY OF CONTINUOUS PHASE MODULATION (CPM)

Not all digital modulations can be expressed in I-Q form. One very popular example of this is continuous phase modulation (CPM) in which a carrier is frequency modulated with a digital pulse train of the form in (2.2). Mathematically speaking, a CPM signal has the form[6]

$$s(t) = A \operatorname{Re}\left\{ e^{j\left(\omega_c t + 2\pi h \sum_{n=-\infty}^{\infty} a_n q(t - nT)\right)} \right\}$$

(2.133)

$$= A \cos\left(\omega_c t + 2\pi h \sum_{n=-\infty}^{\infty} a_n q(t - nT)\right)$$

6. See Chapter 10 for a more detailed discussion of CPM.

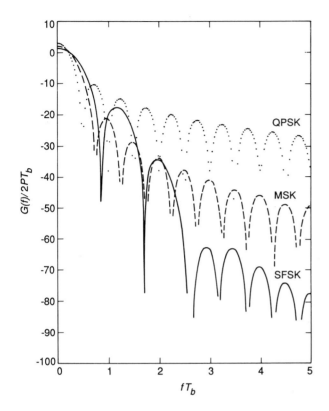

Figure 2.10   Equivalent lowpass power spectral densities of various carrier modulations (Reprinted from [17] © IEEE)

where h is the *modulation index*, and the phase shaping pulse, $q(t)$, is given by

$$q(t) = \int_{-\infty}^{t} r(\tau)d\tau \tag{2.134}$$

with $r(t)$ the frequency pulse, and T the signaling interval. Typically, h is chosen as a rational number, that is, $h = J/M$ in which case the data symbols $\{a_n\}$ take on values ± 1, $\pm 3, \ldots, \pm(M-1)$ and the number of states in CPM is finite. Furthermore, $r(t)$ is assumed to be of finite duration, that is, $r(t)$ is nonzero only in the interval $0 \le t \le LT$ where L is the pulse length in symbols. The case $L = 1$ is referred to as *full response CPM* whereas the more general case of $L > 1$ wherein the data symbols overlap is called *partial response CPM*. Also $r(t)$ is assumed normalized such that

$$\int_{-\infty}^{\infty} r(\tau)d\tau = \frac{1}{2} \tag{2.135}$$

The special case where $L = 1$ and the frequency pulse $r(t)$ is rectangular, namely

$$r(t) = \begin{cases} \frac{1}{2T}; & 0 \le t \le T \\ 0; & \text{otherwise} \end{cases} \tag{2.136}$$

is referred to as *continuous phase frequency-shift-keying (CPFSK)*. The special case of CPFSK for which $h = 0.5$ corresponds to MSK which, as we have already seen, has an I-Q representation.

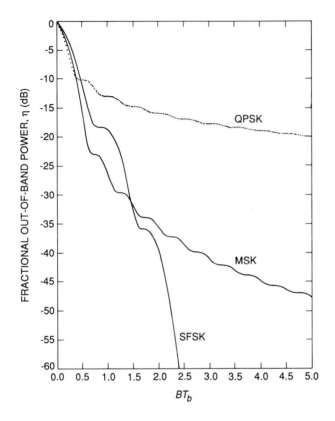

Figure 2.11   Fractional out-of-band power of various carrier modulations (Reprinted from [17] © IEEE)

To compute the power spectrum of a full response CPM signal with rational modulation index, we make use of the signal representation due to Rimoldi [13] which can be written in the form

$$s(t) = A \operatorname{Re} \left\{ e^{j(\omega_c t + 2\pi f_H t + \psi(t; u_n, \sigma_n))} \right\} = A \cos(\omega_c t + 2\pi f_H t + \psi(t; u_n, \sigma_n)) \quad (2.137)$$

where

$$f_H = -h \frac{M-1}{2T} \quad (2.138)$$

and $\psi(t; u_n, \sigma_n)$ is referred to as the *physical tilted phase* which, for M even and $0 \le t \le T$, is given by

$$\psi(t; u_n, \sigma_n) = 2\pi h \sigma_n + 4\pi h u_n q(t) + w(t) \quad (2.139)$$

Here u_n is a translated and scaled version of a_n, that is, $u_n = (a_n + M - 1)/2$ so that u_n takes on only nonnegative values, namely, $0, 1, 2, \ldots, M - 1$ and

$$w(t) = \pi h(M-1) \frac{t}{T} - 2\pi h(M-1)q(t) \quad (2.140)$$

hence, the notion of *tilted* phase. Finally, σ_n denotes the *state* of the modulator in the nth transmission interval and the transition between states is described by

$$\sigma_{n+1} = R_S [u_n + \sigma_n] \quad (2.141)$$

where $R_M[X]$ denotes the remainder of X after division by M, and

$$S = \frac{\text{l.c.m.}(J, M)}{J} \tag{2.142}$$

is the number of modulator states with l.c.m. denoting "least common multiple."

The state sequence $\{\sigma_n\}$ behaves as a Markov sequence with S states and thus is characterized by an $S \times S$ probability transition matrix $\mathbf{P} = \{p_{ik}\}$. It can be shown [14] that all of the elements of $\mathbf{P}$ have value $1/S$ and thus the stationary probabilities, $\{p_i\}$, also all have value $1/S$. Following a procedure analogous to that in Section 2.2, Biglieri and Visintin [14] have computed the equivalent lowpass PSD, $G(f)$. Omitting the details, we present the following result:

$$G(f) = \frac{1}{T} \left[\frac{1}{M} \sum_{k=0}^{M-1} |\alpha_k(f)|^2 \right]$$

$$+ \frac{2}{T} \text{Re} \left\{ e^{-j2\pi(f-f_H)T} \frac{1}{M} \sum_{k=0}^{M-1} e^{j2\pi hk} \alpha_k^*(f) \frac{1}{M} \sum_{l=0}^{M-1} \alpha_l(f) \right\} \tag{2.143}$$

where

$$\alpha_k(f) = \mathcal{F} \left\{ A e^{j[2\pi f_H t + 4\pi hk q(t) + w(t)]} \right\} \tag{2.144}$$

Note the absence of a discrete spectrum.

For the special case of CPFSK, $q(t) = t/2T$ and consequently, $w(t) = 0$. Thus, (2.144) becomes

$$\alpha_k(f) = A_0 (fT - hk) e^{-j\pi((f-f_H)T - hk)} \tag{2.145}$$

where

$$A_0(fT) = AT \frac{\sin \pi(f - f_H)T}{\pi(f - f_H)T} \tag{2.146}$$

Substituting (2.145) together with (2.146) into (2.143) gives

$$G(f) = \frac{1}{T} \left[\frac{1}{M} \sum_{k=0}^{M-1} A_0^2(fT - hk) \right]$$

$$+ \frac{2}{T} \text{Re} \left\{ e^{-j2\pi(f-f_H)T} \frac{1}{M} \sum_{k=0}^{M-1} e^{j2\pi hk} A_0(fT - hk) \right.$$

$$\left. \times \frac{1}{M} \sum_{l=0}^{M-1} e^{j\pi(l-k)h} A_0(fT - hl) \right\} \tag{2.147}$$

Finally, letting $h = 1/2$ (i.e., $J = 1$ and $M = 2$), $T = T_b$, and $P = A^2/2$, (2.147) reduces, after much manipulation, to

$$G(f) = \frac{32 P T_b}{\pi^2} \frac{\cos^2 2\pi f T_b}{\left(1 - 16 f^2 T_b^2\right)^2} \tag{2.148}$$

which together with (2.131) yields (2.127).

PROBLEMS

2.1 Starting with the basic definition of the PSD for an arbitrary random process $m(t)$, namely

$$S_m(f) = \lim_{T_0 \to \infty} E\left\{\frac{1}{T_0}|M_0(f)|^2\right\}$$

where $M_0(f)$ is the Fourier transform of a T_0-sec observation of an ensemble member of $m(t)$, derive Eq. (2.15) corresponding to $m(t)$ in the form of a binary pulse train as in (2.2). For convenience, let T_0 correspond to an integer number of bits over the interval $-NT \le t \le (N+1)T$, that is, $T_0 = (2N+1)T$.

2.2 Consider a binary random pulse train, $m(t)$, such as that characterized by (2.2) together with (2.18). Suppose now that $m(t)$ is periodically pulsed on (for T_d sec) and off (for $T_p - T_d$ sec) as in Figure P2.2, thus producing the random process, $z(t)$. Assume that T_d and T_p are integer multiples of T. Using the approach of Problem 2.1, find the PSD of $z(t)$. How does it compare to the PSD of the nonpulsed modulation as given by (2.15)?

2.3 Consider a deterministic ± 1 periodic waveform $w(t)$ as shown in Figure P2.3a. Suppose now that $w(t)$ is periodically pulsed on (for T_d sec) and off (for $T_p - T_d$ sec) as shown in Figure P2.3b, thus producing the random process, $z(t)$. Assume that T_d and T_p are integer multiples of T. Using the approach of Problem 2.1, find the PSD of $z(t)$.

2.4 For the purpose of frame synchronization, a deterministic ± 1 waveform (sync pattern) $w(t)$ is periodically inserted into a binary data pulse stream $m(t)$, thus producing the waveform $z(t)$ as shown in Figure P2.4. Evaluate the PSD of the synchronized data stream $z(t)$. Note that $z(t)$ can be expressed as the sum of two random processes, one in the form of $z(t)$ in Problem 2.2 and one in the form of $z(t)$ in Problem 2.3.

2.5 An FSK signal is characterized in the nth transmission interval $nT \le t \le (n+1)T$ by

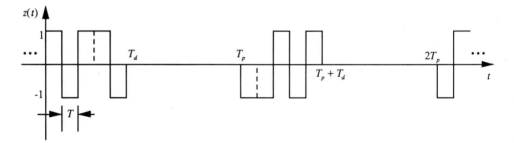

Figure P2.2

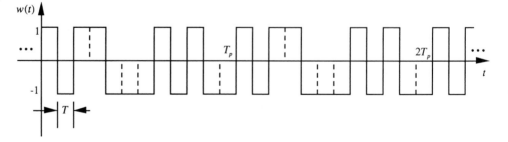

Figure P2.3a

$$m(t) = A \cos \left[\left(\omega_c + a_n \frac{\Delta \omega}{2} \right) t + \theta_n \right]$$

where a_n takes on values of ± 1 with equal probability and the θ_n's are uniformly distributed in $(-\pi, \pi)$ and are independent of each other.

(a) Find the time-averaged autocorrelation $R_m(\tau) \triangleq \overline{\langle m(t)m(t+\tau) \rangle}$ where the statistical average is taken over the θ_n's and the data sequence $\{a_n\}$.

(b) Compute the PSD corresponding to your answer in part (a).

2.6 Repeat Problem 2.5 for the case of an M-FSK signal where now a_n takes on values of ± 1, $\pm 3, \ldots, \pm (M-1)$ with equal probability.

2.7 A binary-valued (± 1) periodic sequence $\{d_n\}$ is defined by the property $d_{p+k} = d_k$, $k = 1, 2, \ldots$ where p is the number of sequence positions in a single period.

(a) Show that the autocorrelation function of $\{d_n\}$, namely, $R_d(k) \triangleq \frac{1}{p} \sum_{n=1}^{p} d_n d_{n+k}$, is given by

$$R_d(k) = \frac{A(k) - D(k)}{A(k) + D(k)}$$

where $A(k)$ is the number of sequence positions in which the two sequences agree in polarity and $D(k)$ is the number of sequence positions in which they disagree.

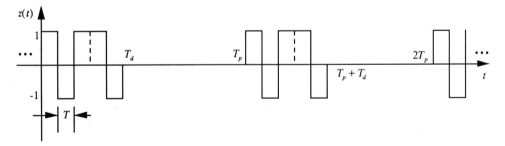

Figure P2.3b

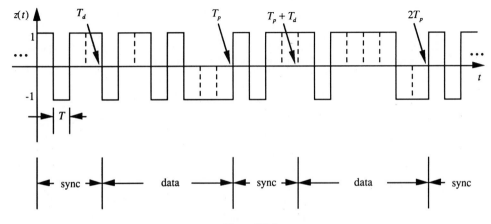

Figure P2.4

(b) A pseudonoise (PN) sequence is a particular type of binary-valued periodic sequence that has the following additional properties:
1. The period p is equal to $2^N - 1$ (N integer).
2. In a single period of the sequence, the number of -1's exceeds the number of 1's by one.
3. The product of a PN sequence and a cyclic shift of itself produces another cyclic shift of the original PN sequence. Mathematically speaking, $\{d_k\} \times \{d_{k+m}\} = \{d_{k+l}\}$ where m is in general unequal to l.

Show that the autocorrelation function of a PN sequence is given by

$$R_d(k) = \begin{cases} 1; & k = 0 \\ -\frac{1}{p}; & k = \pm 1, \pm 2, \ldots \pm (p-1) \end{cases}$$

together with $R_d(k) = R_d(k + np)$, $n = \pm 1, \pm 2, \ldots$ That is, the autocorrelation of a PN sequence is a periodic (with period p) two-level discrete function of the discrete shift variable, k, possessing one value for $k = 0$ and another value for all other $k \neq 0$ within the period.

(c) Consider a binary pulse stream $d(t)$ generated by the PN sequence defined in part (b) where the pulse shape $p(t)$ is a unit rectangular pulse of width T. Find the autocorrelation function $R_d(\tau)$ of the waveform $d(t)$.

(d) Find the PSD of the PN waveform of part (c). Is it a continuous spectrum, a discrete spectrum, or a combination of both?

2.8 Consider a binary random pulse train, $m(t)$, such as that characterized by Eq. (2.2) together with (2.18). After each N data bits, a *parity check bit* bit, b_i, is now inserted into the sequence. Thus, the sequence that generates $m(t)$ would appear as $\ldots a_{-N}, a_{-N+1}, \ldots, a_{-1}, b_{-1}, a_0, a_1,$ $\ldots, a_N, b_0, a_{N+1}, a_{N+2}, \ldots, a_{2N}, b_1, \ldots$ The ith parity check bit, b_i, is related to the previous N data bits by

$$b_i = -\prod_{k=1}^{N} a_{iN+k}$$

Compute the PSD of the resulting $m(t)$. Does it have any discrete components?

2.9 Consider a binary random pulse train, $m(t)$, such as that characterized by Eq. (2.2) together with (2.18). This modulation is transmitted over a multipath channel that produces a single delayed and attenuated reflection which adds to $m(t)$. Thus, the received modulation $r(t)$ can be expressed as $r(t) = m(t) + \alpha m(t - T_d)$ where α denotes the attenuation and T_d denotes the delay. Compute the PSD of $r(t)$.

2.10 A digital pulse stream generated by a binary, zero mean, WSS sequence is impulse sampled at a rate $1/T^*$. The resulting modulation, $m^*(t)$, can be written as

$$m^*(t) = m(t)i(t) = m(t) \sum_{i=-\infty}^{\infty} \delta(t - iT^* - t_0)$$

where $m(t)$ takes the form of (2.2) together with (2.1) and the epoch t_0 is a uniformly distributed random variable in the interval $(-T^*/2, T^*/2)$ which is independent of the data sequence.

(a) Find the time-averaged statistical autocorrelation function of $m^*(t)$, namely $R_{m^*}(\tau) \triangleq \overline{\langle m^*(t)m^*(t+\tau) \rangle}$ where the statistical average is taken both over the data sequence and the random epoch t_0. Hint: Use the Poisson sum relation of (2.12) to express the doubly infinite train of impulse functions as a doubly infinite train of exponentials.

(b) Show that the PSD corresponding to $R_{m^*}(\tau)$ of part (a) can be expressed as

$$S_{m^*}(f) = \frac{1}{(T^*)^2} \sum_{i=-\infty}^{\infty} S_m\left(f - i\frac{1}{T^*}\right)$$

where $S_m(f)$ is the PSD of the unsampled modulation as in (2.15). That is, the PSD of a uniform impulse sampled digital pulse stream is a doubly infinite sum of translations of the PSD of the unsampled modulation where the frequency interval between translations is equal to the sampling frequency.

2.11 Suppose now that in Problem 2.10 the sampling pulse train is nonideal in that $i(t)$ is now given by

$$i(t) = \sum_{i=-\infty}^{\infty} f(t - iT^* - t_0)$$

where $f(t)$ is a narrow pulse with unit area. Show that analogous to the result in part (b) of Problem 2.7, the PSD is now given by

$$S_{m^*}(f) = \frac{1}{(T^*)^2} \sum_{i=-\infty}^{\infty} \left| F\left(i\frac{1}{T^*}\right) \right|^2 S_m\left(f - i\frac{1}{T^*}\right)$$

where $F(f)$ is the Fourier transform of $f(t)$.

2.12 A sinusoidal carrier is phase modulated with a binary data-modulated subcarrier thus forming the signal

$$s(t) = \sqrt{2P} \sin\left(\omega_c t + \beta_m m(t) \sin(\omega_{sc} t + \theta_{sc}) + \theta_c\right)$$

where $0 \leq \beta_m \leq \pi/2$ is the modulation angle, ω_c and ω_{sc} are the carrier and subcarrier radian frequencies, respectively, and θ_c, θ_{sc} are the carrier and subcarrier phases which are uniformly distributed in $(-\pi, \pi)$ and independent of each other.

(a) Using simple trigonometry, express $s(t)$ in terms of its unmodulated and modulated components at the carrier frequency, and even and odd harmonics of the subcarrier frequency above and below the carrier frequency.

(b) Using the form of $s(t)$ found in part (a), find its PSD. Identify the various components of the PSD as being discrete or continuous and the average power of each.

2.13 Suppose that in Problem 2.12 a unit square-wave subcarrier were used instead of the sinusoidal one. As such the signal is now of the form

$$s(t) = \sqrt{2P} \sin\left(\omega_c t + \beta_m m(t) Sq(\omega_{sc} t + \theta_{sc}) + \theta_c\right)$$

where $Sq(\omega_{sc} t + \theta_{sc})$ is as illustrated in Figure P2.13. How do your answers to parts (a) and (b) of Problem 2.10 change?

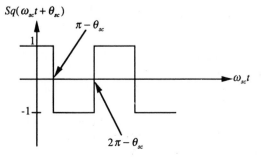

Figure P2.13

2.14 Consider a sinusoidal carrier that is phase modulated by *two* binary data modulations. One is directly modulated onto the carrier with a modulation angle of $\pi/2$ radians. The other is first multiplied by the first one and then applied to a sinusoidal subcarrier which is then phase modulated onto the carrier with a modulation angle β_m. The resulting signal is of the form

$$s(t) = \sqrt{2P} \sin\left(\omega_c t + \frac{\pi}{2} m_1(t) + \beta_m m_1(t) m_2(t) \sin(\omega_{sc} t + \theta_{sc}) + \theta_c\right)$$

(a) Using simple trigonometry, express $s(t)$ in terms of its various components at the carrier frequency, and even and odd harmonics of the subcarrier frequency above and below the carrier frequency. Are there any discrete (unmodulated) components in this signal? If so, at what frequencies?

(b) Using the form of $s(t)$ found in part (a), find its PSD. Identify the various components of the PSD as being discrete or continuous and the average power of each.

2.15 Consider a staggered quadriphase pseudonoise (PN) modulation where identical (except for the 1/2-chip delay between them) PN waveforms (see Problem 2.7) modulate each of the quadrature carriers. The mathematical form of such a modulation is

$$m(t) = \sqrt{P}\ [d(t) \cos \omega_c t - d(t - T/2) \sin \omega_c t]$$

where ω_c is the radian carrier frequency, T is the chip time interval, and $d(t)$ is a PN waveform that has a statistical autocorrelation function $R_d(\tau)$ as determined in Problem 2.7. For simplicity, assume that the code sequence period, p, is very much greater than unity.

(a) Find the statistical autocorrelation function $R_m(\tau)$ of the cyclostationary process $m(t)$ defined above.

(b) Find the corresponding PSD $S_m(f)$.

(c) Discuss how the PSD found in part (b) is different from that corresponding to a staggered quadriphase PN modulation with different and independent PN sequences modulating the quadrature carriers, that is,

$$m(t) = \sqrt{P}\ [d_1(t) \cos \omega_c t - d_2(t - T/2) \sin \omega_c t]$$

where $d_1(t)$ and $d_2(t)$ are still assumed to have the same chip period T.

2.16 Starting with (2.53) and the properties of the Miller Code as described by (2.69), (2.70) and (2.73), derive its PSD, namely, (2.74).

2.17 Starting with the general form of the equivalent lowpass PSD for a CPFSK signal as given by (2.147), derive the equivalent lowpass PSD for the special case of MSK as given by (2.148).

REFERENCES

1. Jolley, L. B. W., *Summation of Series*, New York: Dover Publications, 1961.

2. Titsworth, R. C. and L. R. Welch, "Power Spectra of Signals Modulated by Random and Pseudorandom Sequences," Pasadena, CA: Jet Propulsion Laboratory, Technical Report No. 32-140, October 1961.

3. Hecht, M. and A. Guida, "Delay Modulation," *Proceedings of the IEEE*, vol. 57, no. 7, July 1969, pp. 1314–16.

4. Deffebach, H. L. and W. O. Frost, "A Survey of Digital Baseband Signaling Techniques," NASA No. N71-37703, June 1971.

5. Simon, M. K., K. Tu and B. H. Batson, "Effects of Data Asymmetry on Shuttle Ku-Band Communications Link Performance," *IEEE Transactions on Communications*, vol. COM-26, no. 11, November 1978, pp. 1639–51.

6. Nguyen, T. M., "The Impact of NRZ Data Asymmetry on the Performance of a Space Telemetry System," *IEEE Transactions on Electromagnetic Compatibility*, vol. 33, no. 4, November 1991, pp. 343–50.

7. Nguyen, T. M., "Space Telemetry Degradation due to Manchester Data Symmetry Induced Carrier Tracking Phase Error," *IEEE Transactions on Electromagnetic Compatibility*, vol. 33, no. 3, August 1991, pp. 262–68.

8. Simon, M. K. and D. Divsalar, "Spectral Characteristics of Convolutionally Coded Digital Signals," *IEEE Transactions on Communications*, vol. COM-28, no. 2, February 1980, pp. 173–86.

9. Larsen, K. J., "Short Convolutional Codes with Maximal Free Distance for Rates 1/2, 1/3, and 1/4," *IEEE Transactions on Information Theory*, vol. IT-19, no. 3, May 1973, pp. 371–72.

10. Odenwalder, J. P., "Optimal Decoding of Convolutional Codes," Ph.D. dissertation, University of California, Los Angeles, 1970.

11. Doelz, M. L. and E. H. Heald, "Minimum-Shift Data Communication System," U. S. Patent no. 2,977,417, March 28, 1961 (assigned to Collins Radio Company).

12. Amoroso, F., "Pulse and Spectrum Manipulation in the MSK Format," *IEEE Transactions on Communications*, vol. COM-24, no. 3, March 1976, pp. 381–84.

13. Rimoldi, B. E., "A Decomposition Approach to CPM," *IEEE Transactions on Information Theory*, vol. 34, no. 3, March 1988, pp. 260–70.

14. Biglieri, E. and M. Visintin, "A Simple Derivation of the Power Spectrum of Full-Response CPM and Some of Its Properties," *IEEE Transactions on Communications*, vol. 38, no. 3, March 1990, pp. 267–69.

15. Lindsey, W. C. and M. K. Simon, *Telecommunication Systems Engineering*, Englewood Cliffs, NJ: Prentice Hall, 1973. Reprinted by Dover Press, New York, NY, 1991.

16. Simon, M. K. and D. Divsalar, "Spectral Characteristics of Convolutionally Coded Digital Signals," Pasadena, CA: Jet Propulsion Laboratory, JPL Publication 79-93, August 1, 1979.

17. Simon, M. K., "A Generalization of Minimum-Shift-Keying (MSK)–Type Signaling Based Upon Input Data Symbol Pulse Shaping," *IEEE Transactions on Communications*, vol. COM-24, no. 8, August 1976, pp. 845–56.

Chapter 3

Scalar and Vector Communications Over the Memoryless Channel

In the previous chapters, a system level description of a communication link was presented and the power spectral densities of various modulation techniques were computed. Modulation and demodulation were discussed without much detail into their respective operations. The various tasks performed in an IF demodulator can be grouped into two main functions, namely, synchronization and detection. For the purpose of this chapter, we assume perfect synchronization (or timing) and concentrate on the basics of detection and hypothesis testing. The degradation due to imperfect synchronization on the detection performance will be the topic of a future chapter.

Information about the world is acquired by observation and physical measurement. All measurements are subject to error. All efforts to completely eliminate this error finally reach a bound set by nature's underlying chaos, which infects all observations of physical phenomenon with an innate uncertainty (error). Acquiring information about any physical system involves making *decisions*. Generally, we must decide which of a set of *hypotheses* (statements) best describe the system insofar as observations permit one to judge. Such hypotheses exist and, in the context of some theory, signify the state of the system. Such a theory requires modeling the sources of uncertainty (error) that corrupt the observations. The irreducible minimum component of error in making decisions is characterized by analyzing the *decision rules* that minimize some convenient measure of the average amount of error.

The transportation of information requires a communication system. Such a system corrupts the information by the introduction of random background noise. The need to ferret weak signals out of such noise has called for the development of a *Theory of Signal Detection*. This need arose during World War II in the context of determining the presence or absence of an enemy emitter. At that time, it was realized that decisions made in the face of uncertainty could be treated by the *Theory of Signal Decisions*, commonly known as *Hypothesis Testing*. Such an elegant theory had been fully developed in the 1930s by Neyman and Pearson [1,2], Wald [3] and others. The motivation of this theory came from "An Essay Toward Solving a Problem in the Doctrine of Chances" which appeared in the Philosophical Transactions of The Royal Society of London in 1763 [4]. Therein, the insightful Reverend Thomas Bayes treated the problem of making decisions under conditions of uncertainty, and he recommended selecting that hypothesis for which the *a posteriori probability* is greatest. This is called *Bayes strategy,* and the general formulation of hypothesis testing is often called "Bayesian."

In the context of communication of information, the abstract set of hypotheses $\{H_m; \ m = 0, \ldots, M - 1\}$ takes on the physical meaning of messages. For convenience, we denote this message set as $\{m_m, \ m = 0, 1, \ldots, M - 1\}$ such that hypothesis H_m has a one-to-one correspondence with message m_m. We will apply the theory of hypothesis testing to the problem of communication of information with minimum probability of error. Our top-down approach to demonstrating the theory of signal design (message representation) and *signal detection* (message decision) is to apply *Statistical Decision Theory* to solve this communication problem. As we shall see, it is rich in results of great practical interest.

In the communication of information, the message-bearing signal is observed as values of the voltage $r(t)$ appearing at the terminals of an antenna during a known interval $[0, T]$ of observation. This voltage waveform is a random process described by probability density functions that embody what is known about the message-bearing signals to be detected and the statistical properties of the noise that corrupt them. Statistical decision theory, as applied to the communication problem, shows how best to process that input voltage $r(t)$ in order to decide among several hypotheses (messages) about what $r(t)$ contains. As we shall see, it prescribes certain kinds of processing, manipulations, and appropriate logical operations that end in *decisions*. The probabilities of error that these decisions incur reveal the ultimate limits on the *detectability* (*Probability of Message Error*) of the message embedded in $r(t)$.

Using the top-down approach to decision making depicted in Fig. 3.1, we first consider the scalar communication problem in which the observation is a discrete random variable taking on values $r = r_0, r_1, \ldots, r_{J-1}$. At first, this may seem to be a gross oversimplification of the problem; ironically enough, we shall see, in Chapters 4, 11 through 14, that it is indeed the model that one must ultimately characterize in order to evaluate the performance of an uncoded (quantized) and of a coded digital communication system. Next, we model our observation as a continuous random variable taking on values $r = \rho \epsilon R$ (R is the set of real numbers). A generalization of this model allows for our observation to be a random vector taking on values $\mathbf{r} = \boldsymbol{\rho} \epsilon R^N$ (R^N is the set of N-tuples of real numbers). This model allows for evaluation of the performance of most uncoded communication systems, as will be discussed in future chapters. Moreover, the vector model offers great insight into the detection problem wherein the transmitted messages are represented by waveforms. It will also be shown that the waveform communication problem can be reduced to a corresponding vector problem with the appropriate transformation, and hence the vector problem serves as a platform from which the waveform problem can be approached. Figure 3.2 illustrates the detection problem for all cases of interest. The information source outputs one of M messages, which is transmitted using a scalar, a vector, or a waveform. The channel is represented by the probabilistic transition mechanism which outputs an element of the observation space. Finally, the decision rule maps the observed element into one of M hypotheses (messages), according to a certain prescribed criterion. A communication link from a systems engineer perspective is shown in Fig. 3.3. This perspective has been discussed in great detail in Chapter 1. For our purpose, it suffices to note that depending on the application, an appropriate architectural level can be chosen to model the channel as a discrete memoryless channel (DMC) or a continuous memoryless channel (CMC).

Top Down

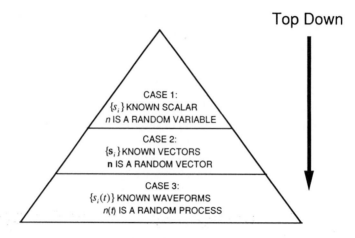

Figure 3.1   Scalar, vector, and waveform communications (top-down approach)

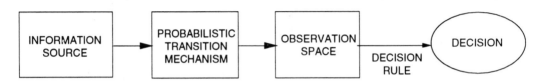

Figure 3.2   Elements of the detection system

3.1 SCALAR COMMUNICATIONS OVER THE MEMORYLESS CHANNEL

In scalar communications, the transmitter transforms each message m into a distinct voltage s for transmission. The channel is assumed to corrupt the transmitted voltage s by adding a random variable n representing the noise. The receiver observes $r = s + n$ and estimates the transmitted message m. For an *additive Gaussian noise channel*, the random variable n has a normal (or Gaussian) distribution. The channel can be modeled differently depending on the application. A more general channel model is given by $r = f(s, n)$ in which the received voltage is a nonlinear two-dimensional function of both the signal and noise. An example of a nonlinear channel is the case where the signal is amplified by a Travelling Wave Tube (TWT) amplifier operating in the saturation mode. Another example is the fading channel where the receiver observes $r = as + n$ and a is a random variable that models signal attenuation due to the fade. The attenuation itself can be time-varying and frequency dependent.

As previously mentioned, the detection problem is best formulated as a *hypothesis testing* problem in which the function of the receiver is to decide which of the M hypotheses (messages) is most likely. Letting H_i denote the ith hypothesis, then

$$
\begin{aligned}
m_0 &\longrightarrow H_0 : r = s_0 + n \\
m_1 &\longrightarrow H_1 : r = s_1 + n \\
&\ \ \vdots \\
m_{M-1} &\longrightarrow H_{M-1} : r = s_{M-1} + n
\end{aligned}
$$

(3.1)

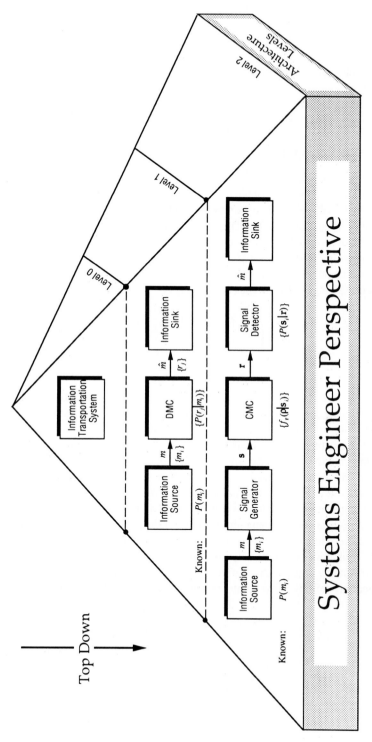

Figure 3.3 Systems engineer perspective

for the additive noise channel where s_i is a known deterministic voltage representing the ith message and n is in general a zero-mean noise with a known probability density function. The function of the receiver is to operate a decision rule that maps the observed random variable r into one of the M hypotheses. Typically, the optimization criterion will be the minimization of the message error probability, but another criteria such as minimizing the average cost, as in Bayesian detection, can be chosen and is discussed in more detail in Appendix 3A in the case of vector communications.

3.1.1 Scalar Communications Over the Discrete Memoryless Channel

Let's consider first the discrete memoryless channel (DMC) which corrupts the signal by adding a discrete random variable n. In this case, the output r is constrained to assume one of a set of J discrete values. We assume that the M a priori (before reception) message probabilities $P(m_i)$ are known. These are the probabilities with which each message gets transmitted over the channel. In this case, the channel is described by MJ conditional (transition) probabilities $p_{i,j} = P(r_j/m_i)$ where M is the number of messages and J the number of channel observable outputs, as depicted in Fig. 3.4. Often, these probabilities are grouped into a single M by J matrix $\mathcal{P}$ given by

$$\mathcal{P} = \begin{pmatrix} p_{0,0} & p_{0,1} & \cdots & p_{0,J-1} \\ p_{1,0} & p_{1,1} & \cdots & p_{1,J-1} \\ \vdots & \vdots & \ddots & \vdots \\ p_{M-1,0} & p_{M-1,1} & \cdots & p_{M-1,J-1} \end{pmatrix} \tag{3.2}$$

Examples of such channels are depicted in Fig. 3.5. In the binary channel (BC), $M = J = 2$ and the two messages m_0 and m_1 experience in general different transition probabilities. On the other hand, they experience identical transition probabilities in the binary symmetric channel (BSC). In other channels, the message either is received correctly or is erased. Such is the case in the binary erasure channel (BEC). Other examples include the *noiseless channel* in which $p_{i,j} = 1$ for $i = j$ and 0 for $i \neq j$, the *useless channel* in which $p_{i,j} = p_j$ for all i and j, and finally, the *M-ary symmetric channel* defined by $p_{i,j} = p_{j,i}$.

For any discrete memoryless channel (DMC), we would like to derive the optimum receiver that minimizes the probability of error. That is, upon observing the received symbol r_j, the receiver should decide which message m_i was transmitted in such a way as to maximize the probability of correct decision, $P(C)$. We therefore expect the optimum receiver to outperform any other receiver after a "long" series of transmissions. Using the method of conditional probabilities, the joint probability of transmitting m_i and receiving r_j is given by

$$P(m_i, r_j) = P(m_i)P(r_j/m_i) \tag{3.3}$$

where $P(m_i)$ depends solely on the message source. The total probability of receiving r_j is the sum of all joint probabilities $P(m_i, r_j)$ over all possible transmitted messages, that is

$$P(r_j) = \sum_{i=0}^{M-1} P(m_i, r_j) \tag{3.4}$$

Using Bayes rule [5,6], the *a posteriori (after reception) probability* that m_i was transmitted

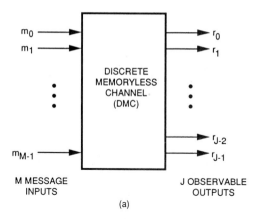

M MESSAGE
INPUTS

J OBSERVABLE
OUTPUTS

(a)

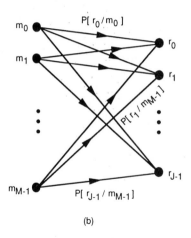

(b)

Figure 3.4   The discrete memoryless channel (DMC)

given that r_j is received (observed) is

$$P(m_i/r_j) = \frac{P(r_j/m_i)P(m_i)}{P(r_j)} \tag{3.5}$$

Before transmission, each message m_i has an a priori probability $P(m_i)$ of being transmitted. Therefore, if the receiver were to decide on which message was transmitted before making any observation, the message with the highest probability of transmission [that is, highest a priori probability $P(m_i)$] would be chosen. After observing r_j, the probability that m_i was transmitted is altered from the a priori probability $P(m_i)$ to the a posteriori probability $P(m_i/r_j)$. We expect that a "smart" receiver will exploit this change to make a better decision on the transmitted message.

On the basis of this set of known a posteriori probabilities, the goal is to find out how does the optimum receiver make a decision (that minimizes the probability of error) regarding the transmitted message? That is, how does the receiver map the observed symbol r_j onto the message input set $\{m_i\}$ such that each possible received symbol is assigned one and only one element (but not necessarily a distinct element) from the message set? Let

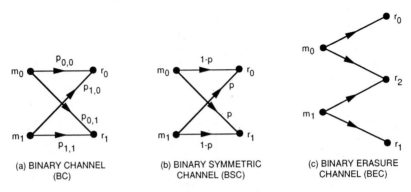

Figure 3.5   Examples of discrete memoryless channels: (a) Binary channel (BC), (b) Binary symmetric channel (BSC), (c) Binary erasure channel (BEC)

$\widehat{m}(r_j)$ denote the transmitted symbol attributed to observing r_j; for example, the receiver decides that m_1 was transmitted if r_3 is received, that is, $\widehat{m}(r_3) = m_1$. Then, the conditional probability of being correct given r_j is observed is just the probability that $\widehat{m}(r_j)$ was actually transmitted, that is

$$P(C/r_j) = P(\widehat{m}(r_j)/r_j) \tag{3.6}$$

The conditional probability $P(C/r_j)$ is then maximized by choosing $\widehat{m}(r_j)$ to be that member of the set $\{m_i\}$ with the largest a posteriori probability. This *maximum a posteriori* (MAP) probability decision rule determines the optimum receiver. The unconditional probability of a correct decision, $P(C)$, is the weighted average of the above conditional probabilities, that is

$$P(C) = \sum_{j=0}^{J-1} P(C/r_j)P(r_j) \tag{3.7}$$

and the probability of error is therefore

$$P(E) = 1 - P(C) \tag{3.8}$$

Since the probabilities $P(r_j)$ are positive and independent of the decision rule, the sum in (3.7) is maximized if each of the $P(C/r_j)$ terms is maximum. Since $P(r_j)$ is independent of the transmitted message, then using (3.5), the decision rule simplifies to: set $\widehat{m}(r_j) = m_k$ if and only if (iff)

$$P(m_k)P(r_j/m_k) \geq P(m_i)P(r_j/m_i) \quad \text{for all } i \tag{3.9}$$

and is referred to as the maximum a posteriori (MAP) decision rule. In the particular case where the a priori probabilities are equal, the decision rule simplifies further to: set $\widehat{m}(r_j) = m_k$ if and only if

$$P(r_j/m_k) \geq P(r_j/m_i) \quad \text{for all } i \tag{3.10}$$

This special case of the MAP decision rule is referred to as the *maximum likelihood* (ML) decision rule. Thus, we conclude that the optimum receiver is a probability computer, in particular, an a posteriori probability computer.

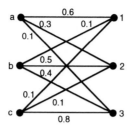

Figure 3.6   Channel of Example 3.1

Example 3.1

Consider the mathematical model of a discrete communication channel having 3 possible input messages $\{a, b, c\}$ and 3 possible output symbols $\{1, 2, 3\}$ as shown in Fig. 3.6. The channel model is completely described by the set of nine conditional probabilities

$$P(1/a) = 0.6 \quad P(2/a) = 0.3 \quad P(3/a) = 0.1$$
$$P(1/b) = 0.1 \quad P(2/b) = 0.5 \quad P(3/b) = 0.4$$
$$P(1/c) = 0.1 \quad P(2/c) = 0.1 \quad P(3/c) = 0.8$$

which specify the probability of receiving each output symbol conditioned on each input message. Assume that we know the set of the three a priori probabilities $P(a)$, $P(b)$, and $P(c)$ with which the input messages are transmitted

$$P(a) = 0.3 \quad P(b) = 0.5 \quad P(c) = 0.2$$

1. Given that the received symbol is 1, what is the decision of the optimum receiver regarding which message is transmitted?
Using Bayes rule, we have

$$P(a/1) = \frac{P(1/a)P(a)}{P(1)}$$

where

$$P(1) = P(1/a)P(a) + P(1/b)P(b) + P(1/c)P(c)$$
$$= (0.6)(0.3) + (0.1)(0.5) + (0.1)(0.2) = 0.25$$

Therefore

$$P(a/1) = \frac{(0.6)(0.3)}{(0.25)} = 0.72$$

Similarly

$$P(b/1) = \frac{P(1/b)P(b)}{P(1)} = \frac{(0.1)(0.5)}{(0.25)} = 0.2$$

$$P(c/1) = \frac{P(1/c)P(c)}{P(1)} = \frac{(0.1)(0.2)}{(0.25)} = 0.08$$

Since $P(a/1)$ is greater than both $P(b/1)$ and $P(c/1)$, the optimum receiver will decide that message a was transmitted given that the received symbol is 1.

2. Repeat (1) for the received symbol 2.
Similarly

$$P(a/2) = \frac{P(2/a)P(a)}{P(2)}$$

where

$$P(2) = P(2/a)P(a) + P(2/b)P(b) + P(2/c)P(c)$$
$$= (0.3)(0.3) + (0.5)(0.5) + (0.1)(0.2) = 0.36$$

Therefore

$$P(a/2) = \frac{(0.3)(0.3)}{(0.36)} = 0.25$$

and

$$P(b/2) = \frac{P(2/b)P(b)}{P(2)} = \frac{(0.5)(0.5)}{(0.36)} = 0.6944\ldots$$

$$P(c/2) = \frac{P(2/c)P(c)}{P(2)} = \frac{(0.1)(0.2)}{(0.36)} = 0.0555\ldots$$

If the received symbol is 2, the receiver decides that message b was transmitted.

3. Repeat (1) for the received symbol 3.

$$P(a/3) = \frac{P(3/a)P(a)}{P(3)}$$

where

$$P(3) = P(3/a)P(a) + P(3/b)P(b) + P(3/c)P(c)$$
$$= (0.1)(0.3) + (0.4)(0.5) + (0.8)(0.2) = 0.39$$

Therefore

$$P(a/3) = \frac{(0.1)(0.3)}{(0.39)} = 0.0769\ldots$$

$$P(b/3) = \frac{P(3/b)P(b)}{P(3)} = \frac{(0.4)(0.5)}{(0.39)} = 0.5128\ldots$$

$$P(c/3) = \frac{P(3/c)P(c)}{P(3)} = \frac{(0.8)(0.2)}{(0.39)} = 0.4102\ldots$$

The receiver will decide that message b is transmitted if the received symbol is 3. Note from (2) that message b is also decided when symbol 2 is received. Therefore, message "c" is never chosen by the MAP receiver. This is acceptable since each received symbol is assigned a unique transmitted message, but not necessarily a distinct message. This problem illustrates that in some DMCs, the MAP decision rule, which minimizes the average probability of error, does not always decide on all messages in the transmitted set.

4. Given that the received symbol is 2 and the receiver decides that message b was sent, what is the probability that the decision is correct, that is, that message b was actually transmitted?

$$P(\text{correct decision}/2) = P(b/2) = \frac{0.25}{0.36} = 0.6944\ldots$$

5. What is the probability of error $P(E)$ of the optimum receiver?

$$P(\text{correct decision}/1) = P(C/1) = P(a/1) = \frac{0.18}{0.25} = 0.72$$

$$P(\text{correct decision}/2) = P(b/2) = 0.6944\ldots$$

$$P(\text{correct decision}/3) = P(b/3) = 0.5128\ldots$$

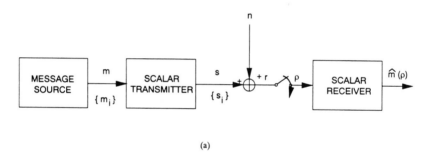

(a)

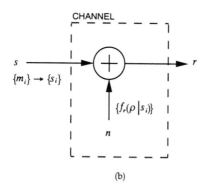

(b)

Figure 3.7   Continuous memoryless channel (CMC) model for scalar communications

Therefore, the probability of correct decision is

$$P(C) = P(C/1)P(1) + P(C/2)P(2) + P(C/3)P(3)$$
$$= (0.72)(0.25) + (0.69)(0.36) + (0.51)(0.39) = 0.6273\ldots$$

and the probability of error $P(E)$ is thus given by

$$P(E) = 1 - P(C) = 1 - 0.6273 = 0.3727\ldots$$

3.1.2 Scalar Communications Over the Continuous Memoryless Channel

In most channels, the noise n is a continuous random variable with probability density function (pdf) $f_n(\alpha)$.[1] As a result, the received random variable r is also of the continuous type and the continuous memoryless channel (CMC) in the scalar case is modeled as shown in Fig. 3.7. For a particular observed value of r, say $r = \rho$, assume that the receiver sets $\widehat{m}(\rho) = m_i$. Then, the conditional probability of correct detection is just the probability that m_i was actually transmitted given that $r = \rho$ is observed, that is

$$P(C/r = \rho) = P(\widehat{m}(\rho) = m_i / r = \rho) = P(m_i / r = \rho) \tag{3.11}$$

So the probability of being correct, $P(C)$, is thus

$$P(C) = \int_{-\infty}^{\infty} P(C/r = \rho) f_r(\rho) d\rho \tag{3.12}$$

1. Strictly speaking, the subscript n in $f_n(\alpha)$ denotes the random variable while the argument is an independent variable (in this case, α). Sometimes, we will denote the independent variable by the same symbol as the random variable, i.e., $f_n(n)$.

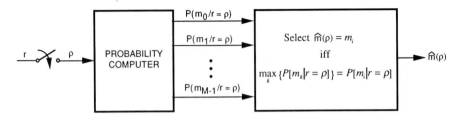

Figure 3.8   Optimum receiver for scalar communications

where $f_r(\rho)$ is the probability density function of the received voltage r. Here again, $P(C)$ is maximized when $P(C/r = \rho)$ is maximized since $f_r(\rho) \geq 0$. But, $P(C/r = \rho) = P(m_i/r = \rho)$ is maximized when m_i is actually transmitted. As in the discrete memoryless channel, the optimum receiver maps ρ into m_i for which the a posteriori probability $P(m_i/r = \rho)$ is maximum. Therefore, the decision rule is: set $\widehat{m}(\rho) = m_i$ if and only if

$$P(m_i/r = \rho) = \max_k \{P(m_k/r = \rho)\} \quad \text{for all } k \tag{3.13}$$

Thus, in cases of both discrete and continuous noise, the optimum receiver is a "probability computer" as depicted in Fig. 3.8. The optimum receiver computes the a posteriori probability, given that $r = \rho$, for all messages in the set and decides on the message with the largest computed a posteriori probability.

3.1.3 Optimum Scalar Receivers

So far, we have shown that the optimum receiver is a probability computer that computes all the a posteriori probabilities. Indeed, this is the case for any channel with arbitrary signal and noise statistics. We would like to know if the receiver can be simplified for some specific channels and how it can be mechanized in a relatively "simple" manner. We will first consider the case of binary communications to understand the basic fundamentals, and then generalize the results to M-ary communications.

Binary scalar receivers.   In the binary case ($M = 2$), the optimum detector sets $\widehat{m}(\rho) = m_0$ if and only if

$$P(m_0/r = \rho) > P(m_1/r = \rho) \tag{3.14}$$

Using Bayes rule in mixed form, we obtain

$$P(m_i/r = \rho) = \frac{f_r(\rho/m_i)P(m_i)}{f_r(\rho)} \tag{3.15}$$

where $f_r(\rho/m_i)$ is the conditional probability density of r given that m_i was transmitted. Inserting (3.15) into (3.14), we have the following decision rule: decide that m_0 was transmitted if and only if

$$\frac{f_r(\rho/m_0)P(m_0)}{f_r(\rho)} > \frac{f_r(\rho/m_1)P(m_1)}{f_r(\rho)} \tag{3.16}$$

Cancelling out $f_r(\rho)$ from both sides, we have

$$f_r(\rho/m_0)P(m_0) \underset{m_1}{\overset{m_0}{\gtrless}} f_r(\rho/m_1)P(m_1) \tag{3.17}$$

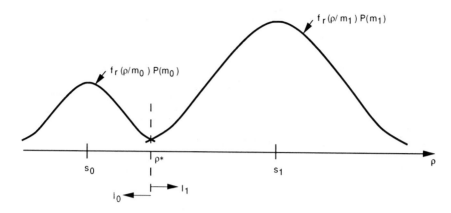

Figure 3.9   Division of observation space in binary scalar communications

The above notation indicates that the receiver decides that message m_0 was transmitted if the left-hand side of (3.17) is greater than its right-hand side. However, message m_1 is chosen if the reverse is true. In the case of equality, either message can be chosen randomly. Figure 3.9 depicts a typical division of the observation space into two disjoint decision regions, say I_0 and I_1. If $\rho \epsilon I_1$, then $\widehat{m}(\rho) = m_1$, that is, m_1 is decided. On the other hand, if $\rho \epsilon I_0$, then $\widehat{m}(\rho) = m_0$. For $\rho = \rho^*$ (that is, $\rho \epsilon I_0$ and $\rho \epsilon I_1$ simultaneously), both messages are equally likely to have been transmitted and the receiver can choose either of the two randomly. Sometimes, (3.17) is expressed as a ratio, that is

$$\Lambda_r(\rho) \triangleq \frac{f_r(\rho/m_1)}{f_r(\rho/m_0)} \underset{m_0}{\overset{m_1}{\gtrless}} \frac{P(m_0)}{P(m_1)} = \lambda^* \tag{3.18}$$

This is referred to as the *likelihood ratio* and it measures which of the two messages is more likely (in terms of larger a posteriori probability) to have been transmitted. Note that the threshold λ^* depends only on the a priori probabilities of the messages.

To further evaluate the rule, we need to make additional assumptions on the channel and the statistics of the noise and the signal. Assuming that the signal is disturbed by additive noise and that the latter is independent of the signal source, then if s_i is transmitted, $r = \rho$ is received if and only if the noise takes on the value $n = \rho - s_i$, that is

$$f_r(\rho/m_i) = f_n(\rho - s_i/m_i) = f_n(\rho - s_i) \tag{3.19}$$

where the latter equality is due to the independence of the noise from the signal. Therefore, the decision rule of (3.17) simplifies to: set $\widehat{m}(\rho) = m_0$ if and only if

$$f_n(\rho - s_0)P(m_0) > f_n(\rho - s_1)P(m_1) \tag{3.20}$$

Assuming further that the additive noise is Gaussian with probability density function (pdf)

$$f_n(x) = \frac{\exp(-x^2/2\sigma^2)}{\sqrt{2\pi\sigma^2}} \tag{3.21}$$

where σ^2 is the noise variance (and is assumed to be known at the receiver), (3.20) reduces

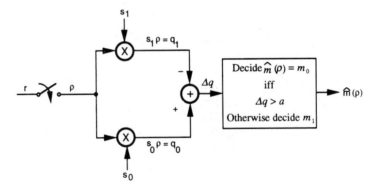

Figure 3.10   Implementation of binary scalar receivers

to

$$P(m_0)\frac{\exp\left[-\frac{(\rho-s_0)^2}{2\sigma^2}\right]}{\sqrt{2\pi\sigma^2}} > P(m_1)\frac{\exp\left[-\frac{(\rho-s_1)^2}{2\sigma^2}\right]}{\sqrt{2\pi\sigma^2}} \tag{3.22}$$

Note that when the noise has a nonzero mean, the latter can be added to the signals s_0 and s_1 and (3.22) still holds for the modified signals. Taking the natural logarithm of both sides, (3.22) can be simplified to yield

$$-(\rho - s_0)^2 > -(\rho - s_1)^2 + 2\sigma^2\ln\frac{P(m_1)}{P(m_0)} \tag{3.23}$$

or equivalently

$$\Delta q \triangleq \rho(s_0 - s_1) > a \tag{3.24}$$

where the threshold a is given by

$$a \triangleq \frac{s_0^2 - s_1^2}{2} + \sigma^2\ln\frac{P(m_1)}{P(m_0)} \tag{3.25}$$

This rule suggests the mechanization depicted in Fig. 3.10 in which the received signal, ρ, is multiplied (*correlated*) with the two expected signals s_0 and s_1 or the difference signal $s_0 - s_1$ to form Δq, and the result is compared to a threshold which is a function of the signal powers s_0^2, s_1^2, the noise power σ^2, and the a priori probabilities $P(m_0)$ and $P(m_1)$. In the special case where the signals have equal power (that is, $s_0 = -s_1$) and equal a priori probabilities, the threshold a becomes zero and the receiver reduces to multiplying $r = \rho$ with s_0 and s_1 and choosing the larger of the two.

Example 3.2

Consider a binary communication link where message m_0 is transmitted by sending $s_0 = 0$ and message m_1 is transmitted by sending a random variable s with a known probability density function. To be more specific, consider the following binary hypothesis testing problem:

$$H_0: \quad m_0 \longrightarrow r = n$$
$$H_1: \quad m_1 \longrightarrow r = s + n$$

where s and n are independent random variables with respective pdf's (assume $a > 0$, $b > 0$ and $a \neq b$)

$$f_s(S) = \begin{cases} ae^{-aS}, & S \geq 0 \\ 0, & S < 0 \end{cases}$$

and

$$f_n(N) = \begin{cases} be^{-bN}, & N \geq 0 \\ 0, & N < 0 \end{cases}$$

Find the decision rule that minimizes the error probability if m_0 and m_1 are equally likely. Using the likelihood ratio test of (3.18), the decision rule becomes

$$\Lambda(\rho) = \frac{f_r(\rho/m_1)}{f_r(\rho/m_0)} \underset{m_0}{\overset{m_1}{\gtrless}} 1$$

as the a priori probabilities are equal. Since r is the sum of two independent random variables when message m_1 is transmitted, the conditional pdf $f_r(\rho/m_1)$ is the convolution of the individual pdf's, that is

$$f_r(\rho/m_1) = f_s(\rho) * f_n(\rho)$$

$$= \int_0^\rho ae^{-a(\rho-N)}be^{-bN}dN$$

$$= \frac{ab}{a-b}\{e^{-b\rho} - e^{-a\rho}\} \quad a \neq b$$

The pdf of r, conditioned on m_0, is simply

$$f_r(\rho/m_0) = f_n(\rho) = be^{-b\rho}$$

Hence, the decision rule becomes

$$\Lambda(\rho) = \frac{a}{a-b}\{1 - e^{(b-a)\rho}\} \underset{m_0}{\overset{m_1}{\gtrless}} 1, \quad a \neq b$$

Rearranging and then taking the natural logarithm of both sides yields

$$\rho \underset{m_1}{\overset{m_0}{\gtrless}} \frac{1}{b-a}\ln[\frac{b}{a}], \quad a \neq b$$

M-ary scalar receivers. We have shown that the optimum M-ary receiver sets $\widehat{m}(\rho) = m_i$ if and only if the a posteriori probability $P(m_i/r = \rho)$ is the largest. Using Bayes rule in mixed form as given in (3.15) and cancelling $f_r(\rho)$, the decision rule becomes: set $\widehat{m}(\rho) = m_i$ if and only if

$$f_r(\rho/m_i)P(m_i) \geq f_r(\rho/m_k)P(m_k) \text{ for all } k \qquad (3.26)$$

Again assuming Gaussian noise with pdf given by (3.21), we get (ignoring the $1/\sqrt{2\pi\sigma^2}$ factor)

$$P(m_i)\exp\left[-\frac{(\rho - s_i)^2}{2\sigma^2}\right] \geq P(m_k)\exp\left[-\frac{(\rho - s_k)^2}{2\sigma^2}\right] \qquad (3.27)$$

Simplifying, we have

$$2\sigma^2\ln P(m_i) - \rho^2 + 2\rho s_i - s_i^2 \geq 2\sigma^2\ln P(m_k) - \rho^2 + 2\rho s_k - s_k^2 \qquad (3.28)$$

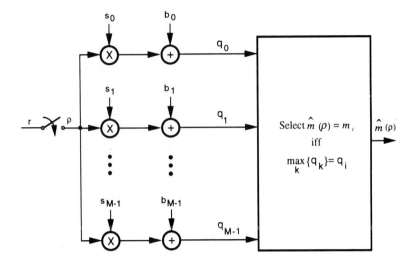

Figure 3.11   Implementation of M-ary scalar cross-correlation receivers

So, the optimum receiver sets $\hat{m}(\rho) = m_i$ if and only if

$$\rho s_i + b_i \geq \rho s_k + b_k \text{ for all } k \tag{3.29}$$

where the "bias" b_k is given by

$$b_k \triangleq \frac{1}{2}\left[2\sigma^2\ln P(m_k) - s_k^2\right] \tag{3.30}$$

This rule suggests the functional diagram (mechanization) shown in Fig. 3.11, where $q_i = \rho s_i + b_i$. Note that the optimum decision rule is to multiply the received voltage with each of the signals, add a bias to account for different signal powers and a priori probabilities, and finally choose the largest. This receiver is linear since the decision rule is a linear function of ρ and this is the result of the Gaussian assumption on the noise. For other noise pdf's, the receiver may indeed be a nonlinear function of ρ. In the special case where the a priori probabilities are equal, the $\sigma^2 \ln(1/M)$ term can be ignored and the bias reduces to

$$b_k = -\frac{1}{2}s_k^2 \tag{3.31}$$

Adding $(-\rho^2/2)$ to both sides of (3.29) and multiplying by (-2) yields

$$(\rho - s_i)^2 \leq (\rho - s_k)^2 \tag{3.32}$$

or equivalently, the receiver decides $\hat{m}(\rho) = m_i$ if and only if

$$|\rho - s_i| \leq |\rho - s_k| \text{ for all } k \tag{3.33}$$

In other words, the receiver decides on the message m_i whose voltage s_i is " closest" to the received voltage ρ. This rule is referred to as *minimum distance decoding* and it will be used again when considering vector communications.

Example 3.3

A system is used to transmit M equally likely messages using the voltages $s_i = i\Delta$, $i = 0, 1, \ldots, M - 1$. The received voltage, when the ith message is transmitted, is $r = s_i + n$ where

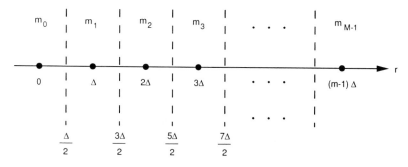

Figure 3.12   Decision regions of Example 3.3

n is a zero-mean Gaussian noise with variance σ^2. Draw the decision regions of the optimum receiver.

Since the noise is Gaussian, the optimum rule is minimum distance decoding as shown in Fig. 3.12. Hence, if $r \leq \Delta/2$, decide m_0. If $\Delta/2 < r \leq 3\Delta/2$, decide m_1, and so on.

3.1.4 Error Probability Performance

In this section, the error probability is used to assess the performance of the optimum receivers for both binary and M-ary scalar signaling. In other fields such as radar, signal acquisition, antenna pointing, and so on, other performance criteria are sometimes used in detection and these are discussed in Appendix 3A.

Binary scalar receiver performance.   We now evaluate the probability of message error, $P(E)$, for additive Gaussian noise as a function of the transmitter voltage levels (s_0, s_1) and the a priori probabilities $[P(m_0), P(m_1)]$. Without loss of generality, assume that $s_0 > s_1$ and that the threshold a given by (3.25) is positive. When m_i is transmitted for either $i = 0$ or 1, the receiver is correct if and only if r lies in region I_i, depicted in Fig. 3.9. The probability of this occurrence is

$$P(C/m_i) = P(r \epsilon I_i/m_i) = \int_{I_i} f_r(\rho/m_i)d\rho \qquad (3.34)$$

and therefore

$$P(C) = \sum_{i=0}^{1} P(m_i)P(C/m_i) = \sum_{i=0}^{1} P(m_i) \int_{I_i} f_r(\rho/m_i)d\rho \qquad (3.35)$$

But $f_r(\rho/m_i) = f_n(\rho - s_i)$; hence

$$P(C) = P(m_0) \int_{a'}^{\infty} \frac{\exp[-\frac{(\rho-s_0)^2}{2\sigma^2}]}{\sqrt{2\pi\sigma^2}} d\rho + P(m_1) \int_{-\infty}^{a'} \frac{\exp[-\frac{(\rho-s_1)^2}{2\sigma^2}]}{\sqrt{2\pi\sigma^2}} d\rho \qquad (3.36)$$

where $a' \triangleq a/(s_0 - s_1)$. The probability of error, $P(E)$, is thus given by

$$P(E) = 1 - P(C) = P(m_0) + P(m_1) - P(C) \qquad (3.37)$$

Using $P(C)$ from (3.36), we have

$$P(E) = P(m_0)(1 - \int_{a'}^{\infty} \frac{\exp[-\frac{(\rho-s_0)^2}{2\sigma^2}]}{\sqrt{2\pi\sigma^2}} d\rho) + P(m_1)(1 - \int_{-\infty}^{a'} \frac{\exp[-\frac{(\rho-s_1)^2}{2\sigma^2}]}{\sqrt{2\pi\sigma^2}} d\rho) \qquad (3.38)$$

Substituting $\alpha = \frac{\rho - s_0}{\sigma}$ and $\beta = \frac{\rho - s_1}{\sigma}$, (3.38) becomes

$$
P(E) = P(m_0)\left(1 - \int_{\frac{a'-s_0}{\sigma}}^{\infty} \frac{\exp(-\alpha^2/2)}{\sqrt{2\pi}}d\alpha\right)
$$

$$
+ P(m_1)\left(1 - \int_{-\infty}^{\frac{a'-s_1}{\sigma}} \frac{\exp(-\beta^2/2)}{\sqrt{2\pi}}d\beta\right)
\tag{3.39}
$$

If we define the Gaussian probability integral as

$$
Q(x) \triangleq \int_x^{\infty} \frac{\exp(-y^2/2)}{\sqrt{2\pi}}dy
\tag{3.40}
$$

then using the property $Q(-x) = 1 - Q(x)$, (3.39) reduces to

$$
P(E) = P(m_0)Q(\frac{s_0 - a'}{\sigma}) + P(m_1)Q(\frac{a' - s_1}{\sigma})
\tag{3.41}
$$

A table of the Gaussian integral is given in Appendix 3B along with an nth order series approximation. If the signals were equally probable, that is, $P(m_0) = P(m_1) = 1/2$, then $a' = \frac{s_0 + s_1}{2}$ and

$$
P(E) = Q(\frac{|s_1 - s_0|}{2\sigma})
\tag{3.42}
$$

where $|x|$ denotes the absolute value of x. The absolute value in (3.42) is necessary to handle both cases $s_0 > s_1$ and $s_1 > s_0$. Hence, the probability of error depends only on the ratio of the "distance" or "potential difference" between the signals to the noise standard deviation σ. If we were to choose the transmitted voltages s_0 and s_1 with the additional constraint that $s_0^2 + s_1^2$ is fixed (that is, total power is constraint), then the optimum signal design [7] that will minimize the probability of error is to set the voltages the furthest apart using

$$
s_1 = -s_0 = s
\tag{3.43}
$$

which results in

$$
P_{min}(E) = Q\left(\frac{|s|}{\sigma}\right) = Q\left(\sqrt{\frac{s^2}{\sigma^2}}\right)
\tag{3.44}
$$

Thus, the error performance depends only on the signal-to-noise power ratio (SNR) given by s^2/σ^2. The larger this ratio, the smaller is the probability of message error.

_M_-ary scalar receiver performance.   The optimum receiver in the M-ary case divides the observation space (a line in the scalar communication problem) into disjoint regions denoted $I_0, I_1, \ldots, I_{M-1}$ respectively, as shown in Fig. 3.13. If the received signal $\rho \in I_i$, then $\widehat{m}(\rho) = m_i$ and the probability of being correct is given by

$$
P(C) = \sum_{i=0}^{M-1} P(m_i)P(C/m_i) = \sum_{i=0}^{M-1} P(m_i)\int_{I_i} f_r(\rho/m_i)d\rho
\tag{3.45}
$$

Example 3.4
    Compute the probability of error of the receiver in Example 3.3.

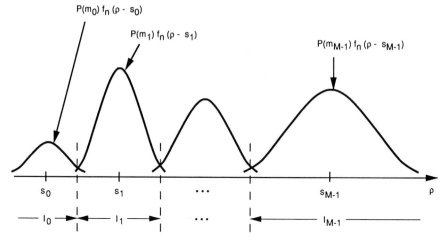

Figure 3.13   Partitioning of observation space in M-ary scalar communications

The probability of error is the weighted sum of the conditional probabilities of error, that is

$$P(E) = \frac{1}{M} \sum_{i=0}^{M-1} P(E/m_i)$$

But due to symmetry, we have

$$P(E/m_0) = P(E/m_{M-1}) = \int_{\Delta/2}^{\infty} \frac{\exp(-r^2/2\sigma^2)}{\sqrt{2\pi\sigma^2}} dr = Q\left(\frac{\Delta}{2\sigma}\right)$$

and

$$P(E/m_1) = P(E/m_2) = \cdots = P(E/m_{M-2})$$

where

$$P(E/m_1) = \int_{-\infty}^{\Delta/2} \frac{\exp\left(-(r-\Delta)^2/2\sigma^2\right)}{\sqrt{2\pi\sigma^2}} dr + \int_{3\Delta/2}^{\infty} \frac{\exp\left(-(r-\Delta)^2/2\sigma^2\right)}{\sqrt{2\pi\sigma^2}} dr = 2Q\left(\frac{\Delta}{2\sigma}\right)$$

Hence

$$P(E) = \left(\frac{2}{M}\right) Q\left(\frac{\Delta}{2\sigma}\right) + 2\left(\frac{M-2}{M}\right) Q\left(\frac{\Delta}{2\sigma}\right) = 2\left(\frac{M-1}{M}\right) Q\left(\frac{\Delta}{2\sigma}\right)$$

3.2 VECTOR COMMUNICATIONS OVER THE CONTINUOUS MEMORYLESS CHANNEL

In communication systems, each message m_i is transmitted using a unique waveform $s_i(t)$. We will later map the waveform communication problem into an equivalent vector problem and show that the mapping does indeed preserve the inherent properties. Note that the signal vectors are defined with respect to an orthonormal basis. Depending on the choice of the latter, the dimensionality N of the problem can be reduced as discussed in Appendix 3C.

In this section, we will proceed with the vector problem and consider the vector communication system model shown in Fig. 3.14. Here again, the source outputs one of M messages which is mapped into its respective vector $\mathbf{s}_i$ by the vector transmitter. We assume

Figure 3.14   Vector communication system

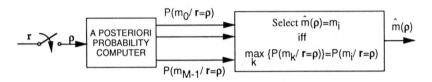

Figure 3.15   Optimum receiver for vector communications

that the a priori probabilities $\{P(m_i)\}$, the signal set $\{s_i\}$, and the statistical characterization of the channel $\{f_{\mathbf{r}}(\boldsymbol{\rho}/s = s_i)\}$ are all known. The receiver observes the vector $\boldsymbol{\rho}$ and decides which message m_i was transmitted.

3.2.1 A Posteriori Probability Computer

Let $P(C/\mathbf{r} = \boldsymbol{\rho})$ denote the conditional probability of correct decision given that $\mathbf{r} = \boldsymbol{\rho}$ is observed and assume that the receiver sets $\widehat{m}(\boldsymbol{\rho}) = m_k$. Then the probability of correct decision is the probability that m_k was indeed transmitted, that is

$$P(C/\mathbf{r} = \boldsymbol{\rho}) = P(\widehat{m}(\boldsymbol{\rho}) = m_k/\mathbf{r} = \boldsymbol{\rho}) = P(m_k/\mathbf{r} = \boldsymbol{\rho}) \tag{3.46}$$

The unconditional probability of correct detection is then

$$P(C) = \underbrace{\int \cdots \int}_{N-fold} P(C/\mathbf{r} = \boldsymbol{\rho}) f_{\mathbf{r}}(\boldsymbol{\rho}) d\boldsymbol{\rho} \tag{3.47}$$

where N is the dimensionality of the observation space. Since $f_{\mathbf{r}}(\boldsymbol{\rho}) \geq 0$, then $P(C)$ is maximized when the conditional probability $P(C/\mathbf{r} = \boldsymbol{\rho})$ is indeed maximum. Hence, the optimum receiver sets $\widehat{m}(\boldsymbol{\rho}) = m_k$ if and only if

$$P(m_k/\mathbf{r} = \boldsymbol{\rho}) \geq P(m_i/\mathbf{r} = \boldsymbol{\rho}) \text{ for all } i \tag{3.48}$$

Here again, the optimum receiver is an a posteriori probability computer as depicted in Fig. 3.15. Using Bayes rule in mixed form, we have

$$P(m_i/\mathbf{r} = \boldsymbol{\rho}) = \frac{P(m_i) f_{\mathbf{r}}(\boldsymbol{\rho}/s = s_i)}{f_{\mathbf{r}}(\boldsymbol{\rho})} \tag{3.49}$$

Thus, the optimum receiver sets $\widehat{m}(\boldsymbol{\rho}) = m_k$ when the *decision function* (rule)

$$P(m_i) f_{\mathbf{r}}(\boldsymbol{\rho}/s = s_i) \tag{3.50}$$

is maximum for $i = k$. This defines the **maximum a posteriori probability vector receiver**. Another related receiver is the **maximum likelihood vector receiver** which sets $\widehat{m}(\boldsymbol{\rho}) = m_k$ if and only if

$$f_{\mathbf{r}}(\boldsymbol{\rho}/s = s_k) = \max_i \{f_{\mathbf{r}}(\boldsymbol{\rho}/s = s_i)\} \tag{3.51}$$

Note that the maximum likelihood receiver does not depend on the knowledge of the a priori probabilities $\{P(m_i)\}$ and that if all messages are equally probable, then the maximum likelihood vector receiver is equivalent to the maximum a posteriori probability vector receiver.

3.2.2 Optimum Vector Correlation and Minimum Distance Receivers

When the channel transition probability densities $f_\mathbf{r}(\boldsymbol{\rho}/\mathbf{s} = \mathbf{s}_i)$ are known, the maximum a posteriori probability vector receiver can be simplified further. We will first consider the optimum receivers in the presence of additive white Gaussian noise (AWGN) and then investigate the receiver structures when the noise is colored.

When the signal $\mathbf{s}$ is disturbed by an additive N-dimensional noise vector $\mathbf{n}$ with zero mean, equal variance and uncorrelated components, the noise vector is defined as "white." In the special case when it is Gaussian in addition to being white, the noise components are independent and we define the channel to be additive white Gaussian noise (AWGN). Letting $\mathbf{n} = [n_1, n_2, \ldots, n_N]^T$, we have

$$f_{n_j}(\alpha_j) = \frac{1}{\sqrt{2\pi\sigma^2}} \exp\left[-\alpha_j^2/2\sigma^2\right] \tag{3.52}$$

resulting in

$$f_\mathbf{n}(\boldsymbol{\alpha}) = \prod_{j=1}^{N} f_{n_j}(\alpha_j) = \left(\frac{1}{2\pi\sigma^2}\right)^{N/2} \exp\left[-\frac{1}{2\sigma^2}\boldsymbol{\alpha}^T\boldsymbol{\alpha}\right] \tag{3.53}$$

where $\boldsymbol{\alpha} = [\alpha_1, \alpha_2, \ldots, \alpha_N]^T$. For the AWGN channel, $\sigma = \sqrt{N_0/2}$ where $N_0 = kT^\circ$ Joules.

Binary vector receivers in the AWGN channel.   In the case of binary signaling, the optimum receiver sets $\widehat{m}(\boldsymbol{\rho}) = m_0$ if and only if

$$P(m_0) f_\mathbf{r}(\boldsymbol{\rho}/\mathbf{s} = \mathbf{s}_0) > P(m_1) f_\mathbf{r}(\boldsymbol{\rho}/\mathbf{s} = \mathbf{s}_1) \tag{3.54}$$

But since $\mathbf{r} = \mathbf{s}_i + \mathbf{n}$, $f_\mathbf{r}(\boldsymbol{\rho}/\mathbf{s} = \mathbf{s}_i) = f_\mathbf{n}(\boldsymbol{\rho} - \mathbf{s}_i)$ as the signals and the noise are independent. Using (3.53), the decision rule becomes (since $\boldsymbol{\alpha}^T\boldsymbol{\alpha} = |\boldsymbol{\alpha}|^2$ and ignoring the $1/\sqrt{2\pi\sigma^2}$ factor)

$$P(m_0) \exp\left[-\frac{|\boldsymbol{\rho} - \mathbf{s}_0|^2}{2\sigma^2}\right] \underset{m_1}{\overset{m_0}{\gtrless}} P(m_1) \exp\left[-\frac{|\boldsymbol{\rho} - \mathbf{s}_1|^2}{2\sigma^2}\right] \tag{3.55}$$

Taking the natural logarithm gives

$$-\frac{|\boldsymbol{\rho} - \mathbf{s}_0|^2}{2\sigma^2} + \ln P(m_0) \underset{m_1}{\overset{m_0}{\gtrless}} -\frac{|\boldsymbol{\rho} - \mathbf{s}_1|^2}{2\sigma^2} + \ln P(m_1) \tag{3.56}$$

or equivalently, the receiver sets $\widehat{m}(\boldsymbol{\rho}) = m_0$ if and only if

$$|\boldsymbol{\rho} - \mathbf{s}_0|^2 < |\boldsymbol{\rho} - \mathbf{s}_1|^2 - 2\sigma^2 \ln\left[\frac{P(m_1)}{P(m_0)}\right] \tag{3.57}$$

For equally probable messages, the rule becomes: set $\widehat{m}(\boldsymbol{\rho}) = m_0$ if and only if

$$|\boldsymbol{\rho} - \mathbf{s}_0|^2 < |\boldsymbol{\rho} - \mathbf{s}_1|^2 \tag{3.58}$$

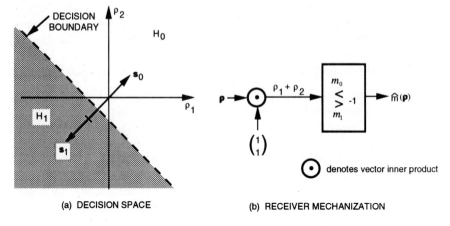

(a) DECISION SPACE   (b) RECEIVER MECHANIZATION

Figure 3.16   Decision space and receiver mechanization of Example 3.5

This rule decides on the message whose corresponding signal $\mathbf{s}$ is "closest" to $\boldsymbol{\rho}$. This rule leads to the notion of **Minimum Distance Decoding** encountered earlier. Coding theorists work from this perspective.

Example 3.5

One of two equally likely messages is to be communicated over the AWGN channel. Assuming that the transmitter uses the signal vectors

$$\mathbf{s}_0 = \begin{pmatrix} 1 \\ 1 \end{pmatrix} \quad \text{and} \quad \mathbf{s}_1 = \begin{pmatrix} -2 \\ -2 \end{pmatrix}$$

draw the decision space and show a mechanization for the optimum receiver.

Since the messages are equally likely and the channel is AWGN, the optimum receiver employs minimum distance decision and the observation space can be divided into two distinct regions as shown in Fig. 3.16(a). Let $\boldsymbol{\rho}^T = (\rho_1 \; \rho_2)$, then the decision rule becomes

$$2\boldsymbol{\rho}^T\mathbf{s}_0 - |\mathbf{s}_0|^2 \underset{m_1}{\overset{m_0}{\gtrless}} 2\boldsymbol{\rho}^T\mathbf{s}_1 - |\mathbf{s}_1|^2$$

or

$$2(\rho_1 + \rho_2) - 2 \underset{m_1}{\overset{m_0}{\gtrless}} 2(-2\rho_1 - 2\rho_2) - 8$$

or

$$\rho_1 + \rho_2 \underset{m_1}{\overset{m_0}{\gtrless}} -1$$

Hence, the receiver can be mechanized as shown in Fig. 3.16(b). The vector operation $\boldsymbol{\rho}^T\mathbf{s}_0$ is referred to as *cross-correlation* in practice.

M-ary vector receivers in the AWGN channel.   Simplifying (3.50) in AWGN channels, the receiver in the M-ary case sets $\widehat{m}(\boldsymbol{\rho}) = m_k$ if and only if

$$\boldsymbol{\rho}^T\mathbf{s}_k + B_k \geq \boldsymbol{\rho}^T\mathbf{s}_i + B_i \quad \text{for all } i \tag{3.59}$$

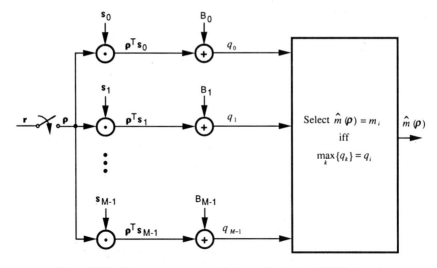

Figure 3.17   *M*-ary vector cross-correlation receiver for the AWGN channel

where

$$B_i \triangleq \frac{1}{2}\left(2\sigma^2 \ln P(m_i) - |\mathbf{s}_i|^2\right) \tag{3.60}$$

The vector cross-correlation receiver can be mechanized as shown in Fig. 3.17, where $q_i = \boldsymbol{\rho}^T \mathbf{s}_i + B_i$. The received signal $\boldsymbol{\rho}$ is correlated with each possible signal $\mathbf{s}_i$ (that is, form the inner product $\boldsymbol{\rho}^T \mathbf{s}_i$) and a bias B_i is added to compensate for different signal powers and a priori probabilities. When all the signals have equal lengths and a priori probabilities, the bias can be ignored.

M-ary vector receivers in the colored Gaussian noise channel.   So far, we have considered the case when the channel noise is Gaussian and white. In this section, we consider the case when the noise is still Gaussian, but not necessarily white. In general, the N-dimensional Gaussian pdf is given by

$$f_{\mathbf{n}}(\boldsymbol{\alpha}) \triangleq G(\boldsymbol{\alpha}; \mathbf{m_n}; \mathcal{K_n})$$
$$= \frac{1}{(2\pi)^{N/2}|\det \mathcal{K_n}|^{1/2}} \exp\left\{-\frac{1}{2}(\boldsymbol{\alpha} - \mathbf{m_n})^T \mathcal{K_n}^{-1}(\boldsymbol{\alpha} - \mathbf{m_n})\right\} \tag{3.61}$$

where N is the dimension of the vector $\mathbf{n}$, $\mathbf{m_n}$ is the mean vector of $\mathbf{n}$ (that is, $\mathbf{m_n} = E\{\mathbf{n}\}$ where $E(.)$ denotes statistical expectation), $\mathcal{K_n}$ is the $N \times N$ symmetric *covariance matrix* of $\mathbf{n}$ given by

$$\mathcal{K_n} = E\left\{(\mathbf{n} - \mathbf{m_n})(\mathbf{n} - \mathbf{m_n})^T\right\} \tag{3.62}$$

and $\det \mathcal{K_n}$ denotes the determinant of the square matrix $\mathcal{K_n}$. Note that if the Gaussian vector $\mathbf{n}$ is white and each component has variance σ^2, then $\mathcal{K_n} = \sigma^2 \mathcal{I}$ ($\mathcal{I}$ denotes the $N \times N$ identity matrix), $\det \mathcal{K_n} = (\sigma^2)^N$, and (3.61) reduces to (3.53) as expected. From (3.50), the MAP receiver sets $\widehat{m}(\rho) = m_i$ if $P(m_i)f_{\mathbf{r}}(\boldsymbol{\rho}/\mathbf{s} = \mathbf{s}_i)$ is maximum. Since $f_{\mathbf{r}}(\boldsymbol{\rho}/\mathbf{s} = \mathbf{s}_i) =$

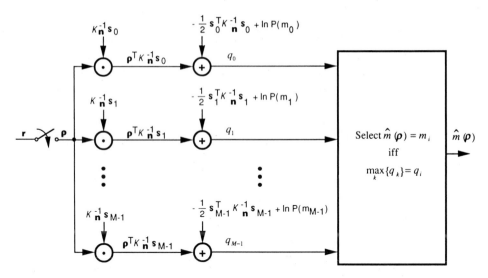

Figure 3.18 Optimum cross-correlation receiver mechanization in the presence of colored Gaussian noise

$f_{\mathbf{n}}(\boldsymbol{\rho} - \mathbf{s}_i)$ and taking the natural logarithm, the decision rule in the binary case becomes

$$\ln P(m_0) - \frac{1}{2}(\boldsymbol{\rho} - \mathbf{s}_0)^T K_{\mathbf{n}}^{-1}(\boldsymbol{\rho} - \mathbf{s}_0) \underset{m_1}{\overset{m_0}{\gtrless}} \ln P(m_1) - \frac{1}{2}(\boldsymbol{\rho} - \mathbf{s}_1)^T K_{\mathbf{n}}^{-1}(\boldsymbol{\rho} - \mathbf{s}_1) \quad (3.63)$$

To arrive at the above equation, we assumed that the noise has zero-mean. The nonzero mean case can be handled by incorporating the means into new modified vectors, given by $\mathbf{s}'_i = \mathbf{s}_i + \mathbf{m_n}$. Expanding the products and simplifying, the rule reduces to

$$\boldsymbol{\rho}^T K_{\mathbf{n}}^{-1}(\mathbf{s}_0 - \mathbf{s}_1) \underset{m_1}{\overset{m_0}{\gtrless}} b \quad (3.64)$$

where

$$b \triangleq \frac{1}{2}\{\mathbf{s}_0^T K_{\mathbf{n}}^{-1}\mathbf{s}_0 - \mathbf{s}_1^T K_{\mathbf{n}}^{-1}\mathbf{s}_1\} + \ln \frac{P(m_1)}{P(m_0)} \quad (3.65)$$

Note that in arriving at (3.64), we used the fact that $\boldsymbol{\rho}^T K_{\mathbf{n}}^{-1}\mathbf{s}_0 = \mathbf{s}_0^T K_{\mathbf{n}}^{-1}\boldsymbol{\rho}$ since both quantities are scalars and are the transpose of each other (recall that $K_{\mathbf{n}}^{-1}$ is symmetric since $K_{\mathbf{n}}$ is symmetric and hence $K_{\mathbf{n}}^{-1} = (K_{\mathbf{n}}^{-1})^T$). In the M-ary case, the decision rule can be generalized to give: select $\widehat{m}(\boldsymbol{\rho}) = m_i$ if and only if

$$\boldsymbol{\rho}^T K_{\mathbf{n}}^{-1}\mathbf{s}_i - \frac{1}{2}\mathbf{s}_i^T K_{\mathbf{n}}^{-1}\mathbf{s}_i + \ln P(m_i) \quad (3.66)$$

is maximum and the resulting cross-correlation receiver mechanization is depicted in Fig. 3.18. In the case of AWGN channel, $K_{\mathbf{n}}^{-1} = (1/\sigma^2)\mathcal{I}$ and (3.66) reduces to $(1/\sigma^2)(\boldsymbol{\rho}^T\mathbf{s}_i + B_i)$ where B_i is given by (3.60).

Let's consider an alternative derivation of the optimum receiver in the presence of colored Gaussian noise. Knowing how to perform signal detection in the presence of white Gaussian noise, a natural extension would be to perform an operation on the received vector $\boldsymbol{\rho}$, which contains colored noise, such that the resulting observation is corrupted by white noise only. This is referred to as a *whitening* operation and assuming that it is linear, the output noise vector, in addition to being white, would remain Gaussian, since it is a by-product of a linear operation on a Gaussian vector. Following the whitening process, the optimum detection rule (assuming equally likely messages) is minimum distance decoding as we have seen before. However, the detection has to be performed on the modified set of transmitted vectors at the output of the whitening operation. This two-stage approach to the detection problem is very intuitive, but it is not clear if it is equivalent to the optimum MAP receiver.

Let's start by drawing on a key result from the field of linear algebra on the factorization of matrices, namely that any nonnegative definite symmetric matrix of dimension $N \times N$ has N associated orthogonal eigenvectors $\mathbf{e}_i$, $i = 1, \ldots, N$ [8,9]. As a result of this, the covariance matrix of any random vector, being nonnegative definite symmetric, can be decomposed as

$$\mathcal{K}_{\mathbf{n}} = \Omega \Lambda \Omega^T \tag{3.67}$$

where Ω is the matrix of orthonormal eigenvectors given by

$$\Omega = [\mathbf{e}_1 \ \mathbf{e}_2 \ \ldots \ \mathbf{e}_N] \tag{3.68}$$

and Λ is a diagonal matrix of nonnegative eigenvalues given by

$$\Lambda = \begin{pmatrix} \lambda_1 & 0 & \ldots & 0 \\ 0 & \lambda_2 & 0 & \vdots \\ \vdots & 0 & \ddots & 0 \\ 0 & \ldots & 0 & \lambda_N \end{pmatrix} \tag{3.69}$$

The eigenvectors and eigenvalues are solutions to the following sets of linear equations:

$$\mathcal{K}_{\mathbf{n}} \mathbf{e}_i = \lambda_i \mathbf{e}_i, \text{ for } \ i = 1, 2, \ldots, N \tag{3.70}$$

Factoring the matrix Λ as

$$\Lambda = \Lambda^{1/2} \Lambda^{1/2} \tag{3.71}$$

where

$$\Lambda^{1/2} = \begin{pmatrix} \sqrt{\lambda_1} & 0 & \ldots & 0 \\ 0 & \sqrt{\lambda_2} & 0 & \vdots \\ \vdots & 0 & \ddots & 0 \\ 0 & \ldots & 0 & \sqrt{\lambda_N} \end{pmatrix} \tag{3.72}$$

then, (3.67) can be rewritten as

$$\mathcal{K}_{\mathbf{n}} = \Omega \Lambda^{1/2} (\Lambda^{1/2})^T \Omega^T = (\Omega \Lambda^{1/2})(\Omega \Lambda^{1/2})^T = \mathcal{Q} \mathcal{Q}^T \tag{3.73}$$

where the matrix $\mathcal{Q}$ is defined as

$$\mathcal{Q} = \Omega \Lambda^{1/2} \tag{3.74}$$

Now in order to whiten the zero-mean vector $\mathbf{n}$ with a covariance matrix $\mathcal{K}_{\mathbf{n}}$, suppose we form the linear transformation

$$\mathbf{w} = \mathcal{P} \mathbf{n} \tag{3.75}$$

Figure 3.19   Vector detection in colored noise

where P is an $N \times N$ matrix chosen such that the resulting vector $\mathbf{w}$ is white, that is, $\mathcal{K}_{\mathbf{w}} = \mathcal{I}$. From (3.75), we have

$$\mathcal{K}_{\mathbf{w}} = E\left\{\mathbf{w}\mathbf{w}^T\right\} = E\left\{P\mathbf{n}\mathbf{n}^T P^T\right\} = P\mathcal{K}_{\mathbf{n}}P^T \tag{3.76}$$

But, we have shown that $\mathcal{K}_{\mathbf{n}}$ can be factored as $\mathcal{Q}\mathcal{Q}^T$, hence

$$\mathcal{K}_{\mathbf{n}} = P\mathcal{Q}\mathcal{Q}^T P^T = (P\mathcal{Q})(P\mathcal{Q})^T \tag{3.77}$$

The matrix P could be chosen as $P = \mathcal{Q}^{-1}$ because then $\mathcal{K}_{\mathbf{w}} = \mathcal{I}$ and the noise has been whitened. Therefore, in conclusion, the noise vector $\mathbf{n}$ can be whitened using the linear transformation $\mathcal{Q}^{-1}$ where $\mathcal{K}_{\mathbf{n}} = \mathcal{Q}\mathcal{Q}^T$ and $\mathcal{Q} = \Omega\Lambda^{1/2}$.

Returning to the detection problem in the presence of colored noise, our approach would be to whiten the observation first and then perform signal detection in the presence of white noise, as shown in Fig. 3.19. Hence, the problem (in the binary case) reduces to deciding between the two modified hypotheses

$$\begin{aligned} m_0 &\longrightarrow H_0 : \mathbf{y} = \mathcal{Q}^{-1}\mathbf{s}_0 + \mathbf{w} \\ m_1 &\longrightarrow H_1 : \mathbf{y} = \mathcal{Q}^{-1}\mathbf{s}_1 + \mathbf{w} \end{aligned} \tag{3.78}$$

where the new observation $\mathbf{y} = \mathcal{Q}^{-1}\boldsymbol{\rho}$. From (3.58), the optimal decision rule for white noise is

$$|\mathbf{y} - \mathcal{Q}^{-1}\mathbf{s}_0|^2 \underset{m_0}{\overset{m_1}{\gtrless}} |\mathbf{y} - \mathcal{Q}^{-1}\mathbf{s}_1|^2 \tag{3.79}$$

or equivalently

$$\mathbf{y}^T\left(\mathcal{Q}^{-1}\mathbf{s}_1 - \mathcal{Q}^{-1}\mathbf{s}_0\right) \underset{m_0}{\overset{m_1}{\gtrless}} \frac{1}{2}\left\{|\mathcal{Q}^{-1}\mathbf{s}_1|^2 - |\mathcal{Q}^{-1}\mathbf{s}_0|^2\right\} \tag{3.80}$$

The left-hand side (LHS) of the above equation reduces to

$$\begin{aligned} \text{LHS} &= \left(\mathcal{Q}^{-1}\boldsymbol{\rho}\right)^T \mathcal{Q}^{-1}(\mathbf{s}_1 - \mathbf{s}_0) = \boldsymbol{\rho}^T\left(\mathcal{Q}^T\right)^{-1}\mathcal{Q}^{-1}(\mathbf{s}_1 - \mathbf{s}_0) \\ &= \boldsymbol{\rho}^T\left(\mathcal{Q}\mathcal{Q}^T\right)^{-1}(\mathbf{s}_1 - \mathbf{s}_0) = \boldsymbol{\rho}^T\mathcal{K}_{\mathbf{n}}^{-1}(\mathbf{s}_1 - \mathbf{s}_0) \end{aligned} \tag{3.81}$$

Similarly, the right-hand side (RHS) becomes

$$\text{RHS} = \frac{1}{2}\left\{\left(\mathcal{Q}^{-1}\mathbf{s}_1\right)^T\left(\mathcal{Q}^{-1}\mathbf{s}_1\right) - \left(\mathcal{Q}^{-1}\mathbf{s}_0\right)^T\left(\mathcal{Q}^{-1}\mathbf{s}_0\right)\right\}$$

$$= \frac{1}{2}\left\{\mathbf{s}_1^T\left(\mathcal{Q}^{-1}\right)^T\mathcal{Q}^{-1}\mathbf{s}_1 - \mathbf{s}_0^T\left(\mathcal{Q}^{-1}\right)^T\mathcal{Q}^{-1}\mathbf{s}_0\right\} \qquad (3.82)$$

$$= \frac{1}{2}\left\{\mathbf{s}_1^T\mathcal{K}_\mathbf{n}^{-1}\mathbf{s}_1 - \mathbf{s}_0^T\mathcal{K}_\mathbf{n}^{-1}\mathbf{s}_0\right\}$$

and hence, the decision rule reduces to

$$\boldsymbol{\rho}^T\mathcal{K}_\mathbf{n}^{-1}(\mathbf{s}_1 - \mathbf{s}_0) \underset{m_0}{\overset{m_1}{\gtrless}} \frac{1}{2}\left(\mathbf{s}_1^T\mathcal{K}_\mathbf{n}^{-1}\mathbf{s}_1 - \mathbf{s}_0^T\mathcal{K}_\mathbf{n}^{-1}\mathbf{s}_0\right) \qquad (3.83)$$

which is identical to (3.64) for equally likely signals. Hence, we have shown that the optimum MAP receiver, in the presence of colored noise, does indeed whiten the observation and then perform signal detection in the presence of white noise. Again, the cross-correlation process is involved in signal detection.

Example 3.6

Consider transmitting the equally likely messages m_0, m_1 using the signals

$$\mathbf{s}_0 = \begin{pmatrix} 0 \\ 10 \end{pmatrix} \quad \text{and} \quad \mathbf{s}_1 = \begin{pmatrix} 0 \\ -10 \end{pmatrix}$$

over the additive vector Gaussian noise channel with covariance matrix

$$\mathcal{K}_\mathbf{n} = \begin{pmatrix} 4 & 2 \\ 2 & 7 \end{pmatrix}$$

Derive the optimum receiver structure and draw the decision space.
Using (3.64), we have

$$\mathcal{K}_\mathbf{n}^{-1}(\mathbf{s}_0 - \mathbf{s}_1) = \frac{1}{24}\begin{pmatrix} 7 & -2 \\ -2 & 4 \end{pmatrix}\begin{pmatrix} 0 \\ 20 \end{pmatrix} = \frac{5}{3}\begin{pmatrix} -1 \\ 2 \end{pmatrix}$$

and the threshold $b = 0$. Hence, the decision rule becomes

$$\frac{5}{3}(-\rho_1 + 2\rho_2) \underset{m_1}{\overset{m_0}{\gtrless}} 0$$

In this case, the factor $5/3$ can be dropped without changing the rule. Also, note that the first component of the observation, ρ_1, is being used in the detection process, even though the transmitted signals have identical first components. This is because the noise is not white and the observation of ρ_1 offers information on the noise component of ρ_2. If the noise vector were white, the decision rule would have been

$$\rho_2 \underset{m_0}{\overset{m_1}{\gtrless}} 0$$

and ρ_1 would not be used in the detection process. The decision spaces of both cases of white and colored noises are shown in Fig. 3.20.

In all derivations so far, we have assumed that the covariance matrix $\mathcal{K}_\mathbf{n}$ is nonsingular (or equivalently, that the inverse does exist). What if $\mathcal{K}_\mathbf{n}^{-1}$ does not exist? In this case, the noise vector lies in a space of dimension less than N, even though it has N components (similar to three vectors lying in a plane even though each vector can have three components that are nonzero). Expanding the noise vector in terms of the eigenvectors of $\mathcal{K}_\mathbf{n}$, it can be

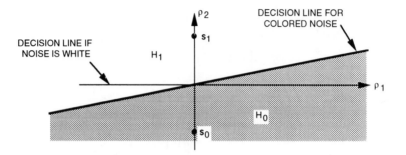

Figure 3.20   Decision space in Example 3.6

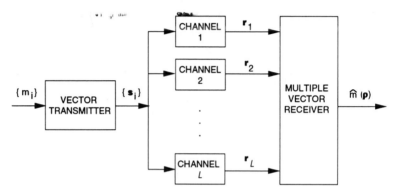

Figure 3.21   Diversity communications

shown that **n** has at least one noise-free dimension or direction. Perfect detection is then possible in this case, provided that the projections of the transmitted vectors in these noise-free dimensions are distinct. This is referred to as *singular* detection. If the projections are not distinct, then the detection should be performed in the reduced dimension space, spanned by the noise vector [8].

M-ary multiple (diversity) vector receivers.   In some applications, the signal is received through various communications channels as depicted in Fig. 1.9 and the receiver has to jointly process all the received vectors, in order to minimize the probability of error. This is referred to as *diversity reception* and is depicted in Fig. 3.21. Such is the case in multipath channels which create signal fading as a consequence of scattering and delay in the physical media. In practice, messages are transmitted using waveforms and diversity can be accomplished in time, frequency, spatially, or through electromagnetic field polarity. In all cases, the total receiver input can be modeled by an augmented received vector **r** given by

$$\mathbf{r}^T = \left(\mathbf{r}_1^T, \mathbf{r}_2^T, \ldots, \mathbf{r}_L^T\right) \tag{3.84}$$

where L denotes the number of received vectors (or diversity channels) and

$$\mathbf{r}_i^T = (r_{i1}, r_{i2}, \ldots, r_{iN}), \quad i = 1, 2, \ldots, L \tag{3.85}$$

Given that $\mathbf{r}^T = \boldsymbol{\rho}^T = (\boldsymbol{\rho}_1{}^T, \boldsymbol{\rho}_2{}^T, \ldots, \boldsymbol{\rho}_L{}^T)$ is observed, the MAP receiver sets $\widehat{m}(\boldsymbol{\rho}) = m_i$

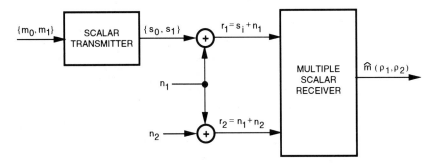

Figure 3.22   Diversity channel of Example 3.7

if and only if

$$P(m_j/\mathbf{r}_1 = \boldsymbol{\rho}_1, \mathbf{r}_2 = \boldsymbol{\rho}_2, \ldots, \mathbf{r}_L = \boldsymbol{\rho}_L) \qquad (3.86)$$

is maximum for $j = i$, or equivalently if

$$P(m_i) f_{\mathbf{r}}(\boldsymbol{\rho}/\mathbf{s} = \mathbf{s}_i) \qquad (3.87)$$

is maximum. Using the fact that

$$f_{\mathbf{r}}(\boldsymbol{\rho}_1, \boldsymbol{\rho}_2, \ldots, \boldsymbol{\rho}_L) = f_{\mathbf{r}_1/\mathbf{r}_2, \ldots, \mathbf{r}_L}(\boldsymbol{\rho}_1/\boldsymbol{\rho}_2, \ldots, \boldsymbol{\rho}_L) f_{\mathbf{r}_2, \ldots, \mathbf{r}_L}(\boldsymbol{\rho}_2, \ldots, \boldsymbol{\rho}_L) \qquad (3.88)$$

(3.87) simplifies to

$$P(m_i) f_{\mathbf{r}_1/\mathbf{r}_2, \ldots, \mathbf{r}_L, \mathbf{s}}(\boldsymbol{\rho}_1/\boldsymbol{\rho}_2, \ldots, \boldsymbol{\rho}_L, \mathbf{s} = \mathbf{s}_i) f_{\mathbf{r}_2, \ldots, \mathbf{r}_L/\mathbf{s}}(\boldsymbol{\rho}_2, \ldots, \boldsymbol{\rho}_L/\mathbf{s} = \mathbf{s}_i) \qquad (3.89)$$

If $\mathbf{r}_1$ conditioned on $\mathbf{r}_2, \ldots, \mathbf{r}_L$ is statistically independent of $\mathbf{s}$, the decision rule simplifies to: set $\widehat{m}(\boldsymbol{\rho}) = m_i$ if and only if

$$P(m_i) f_{\mathbf{r}_2, \ldots, \mathbf{r}_L/\mathbf{s}}(\boldsymbol{\rho}_2, \ldots, \boldsymbol{\rho}_L/\mathbf{s} = \mathbf{s}_i) \qquad (3.90)$$

is maximum. In this case, the knowledge that $\mathbf{r}_1 = \boldsymbol{\rho}_1$ is ignored by the MAP receiver as it does not enter into the maximization of (3.87) with respect to i. Hence, $\mathbf{r}_1 = \boldsymbol{\rho}_1$ is irrelevant to the decision process and the performance of the receiver is not degraded [10]. This reasoning can be applied iteratively to (3.90) to ignore all irrelevant observations and simplify the MAP multivector receiver to the extent possible.

Example 3.7

In the multiscalar communication system depicted in Fig. 3.22, the transmitted signals ($s_1 = -s_0 = s$) and the noise n_1, n_2 are all statistically independent random voltages. Assuming that the messages are equally likely and that the noises are Gaussian with zero-mean and variance σ^2, derive the structure of the optimum multiscalar receiver.

The decision rule is given by (since $L = 2$)

$$f_{r_1, r_2}(\rho_1, \rho_2/m_0) \overset{m_0}{\underset{m_1}{\gtrless}} f_{r_1, r_2}(\rho_1, \rho_2/m_1)$$

Applying Bayes rule, we have

$$f_{r_2/r_1}(\rho_2/\rho_1, m_0) f_{r_1}(\rho_1/m_0) \overset{m_0}{\underset{m_1}{\gtrless}} f_{r_2/r_1}(\rho_2/\rho_1, m_1) f_{r_1}(\rho_1/m_1)$$

But

$$f_{r_2/r_1}(\rho_2/\rho_1, m_0) = f_{n_2}(r_2 - r_1 + s_0)$$

Similarly

$$f_{r_2/r_1}(\rho_2/\rho_1, m_1) = f_{n_2}(r_2 - r_1 + s_1)$$

$$f_{r_1}(\rho_1/m_0) = f_{n_1}(r_1 - s_0) \quad \text{and} \quad f_{r_1}(\rho_1/m_1) = f_{n_1}(r_1 - s_1)$$

Hence, the decision rule becomes

$$\exp\left[-\frac{1}{2\sigma^2}\left((\rho_1 + s)^2 + (\rho_2 - \rho_1 - s)^2\right)\right] \underset{m_1}{\overset{m_0}{\gtrless}} \exp\left[-\frac{1}{2\sigma^2}\left((\rho_1 - s)^2 + (\rho_2 - \rho_1 + s)^2\right)\right]$$

Taking the logarithm and rearranging, we get (assuming $s > 0$)

$$\rho_1 - \frac{1}{2}\rho_2 \underset{m_0}{\overset{m_1}{\gtrless}} 0$$

3.2.3 Decision Rules, Functions, Regions, and Boundaries

The optimum decision rule partitions the N-dimensional observation space into M disjoint decision regions $R_0, R_1, R_2, \ldots, R_{M-1}$ corresponding to the M messages. The decision regions are defined by the *decision rule* which sets $\widehat{m} = m_k$ if the *decision function* $|\mathbf{r} - \mathbf{s}_i|^2 - 2\sigma^2 \ln P(m_i)$ is minimum for $i = k$.

To gain geometric insight into this decision rule, consider first the case of equiprobable messages. In this case, the second term in the above decision function can be dropped and the receiver decides $\widehat{m} = m_k$ if the reduced decision function $|\mathbf{r} - \mathbf{s}_i|^2$ is minimum for $i = k$. Noting that $|\mathbf{r} - \mathbf{s}_i|$ is the Euclidean distance between $\mathbf{r}$ and $\mathbf{s}_i$, then the decision rule has a very simple geometric interpretation, namely, the decision is made in favor of that signal (message) that is closer to $\mathbf{r}$. This is called *minimum distance decoding* which is a familiar concept in coding theory.

When the messages are not equiprobable, then a particular message m_i will be most probable. In this case, a decision is made more often in favor of m_i and the decision regions will be biased in favor of m_i. This point is illustrated by the appearance of the term $\ln P(m_i)$ in the decision function.

Binary communication with equal a priori probabilities. To be more specific, let us assume that $M = N = 2$. Consider a two-dimensional space and the two equiprobable vectors $\mathbf{s}_0$ and $\mathbf{s}_1$ shown in Fig. 3.23a. From this figure, it is clear that when the decision is made in favor of that message corresponding to the signal vector that is closest to $\mathbf{r}$, the decision boundary that separates the two decision regions is the perpendicular bisector of the line that joins the vectors $\mathbf{s}_0$ and $\mathbf{s}_1$. This is true since any point on the boundary is equidistant from $\mathbf{s}_0$ and $\mathbf{s}_1$. If $\mathbf{r}$ falls on the boundary itself, one flips an unbiased coin to decide whether to select m_0 or m_1.

Binary communication with unequal a priori probabilities. Consider now the binary case with unequal a priori probabilities for the two messages. Let $d_0 = |\mathbf{r} - \mathbf{s}_0|$ and $d_1 = |\mathbf{r} - \mathbf{s}_1|$ denote the Euclidean distance of $\mathbf{r}$ from $\mathbf{s}_0$ and $\mathbf{s}_1$, respectively. From (3.57), we see that a decision is made in favor of m_0 if $d_0^2 - d_1^2 < c^2 \triangleq 2\sigma^2 \ln[P(m_0)/P(m_1)]$ and in favor of m_1 if $d_0^2 - d_1^2 > c^2$. Along the decision boundary, $d_0^2 - d_1^2 = c^2$. We shall now show that this boundary is a straight line perpendicular to (but

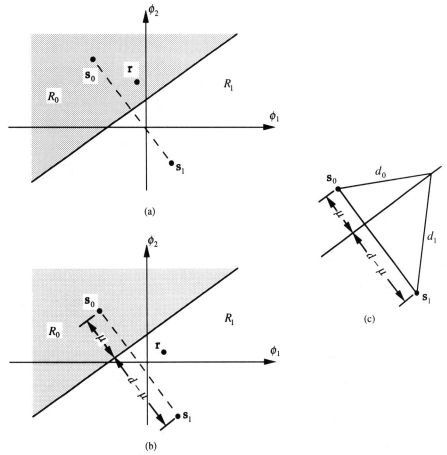

Figure 3.23   Decision regions and decision boundaries for binary communication

not the bisector of) the line that joins the vectors s_0 and s_1 at a distance μ from s_0 (see Fig. 3.23b). Here, $\mu = (c^2 + d^2)/2d = (\sigma^2/d) \ln[P(m_0)/P(m_1)] + d/2$ where d is the distance between s_0 and s_1.

To prove the above, we make use of the geometry in Fig. 3.23c. It is clear from this figure that $d_0^2 = \alpha^2 + \mu^2$ and $d_1^2 = \alpha^2 + (d - \mu)^2$, so that $d_0^2 - d_1^2 = 2d\mu - d^2 = c^2$. This is the desired result since along the decision boundary, $d_0^2 - d_1^2$ is constant and equal to c^2. Note that for equal a priori probabilities, $c = 0$, and along the boundary, $d_0 = d_1$.

_M_-ary communication with $N = 2$.   The decision boundaries for the decision regions are determined along similar lines to those discussed above for the binary case. In particular, consider the case of $M = 3$ equiprobable messages with $N = 2$. The decision regions are as illustrated in Fig. 3.24. The decision boundaries are constructed by taking the vectors s_0, s_1, and s_2 in pairs and constructing the three perpendicular bisectors of the lines joining the vector pairs. When the signals are not equiprobable, then the decision boundaries will still be lines drawn perpendicular to the lines joining the vector pairs. However, they will be shifted away from their midpoint positions in accordance with the ratios of a priori probabilities as above.

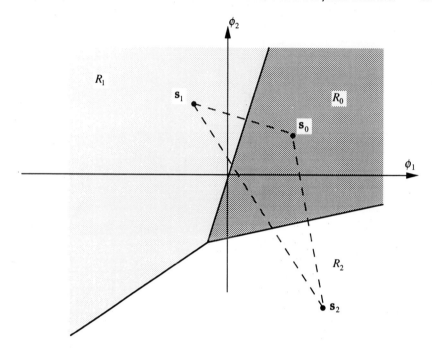

Figure 3.24   Decision regions and decision boundaries for ternary communication

M-ary communication in N-dimensions.   Finally, for M signal vectors in N-dimensional space, the decision regions will be N-dimensional polyhedra and the decision boundaries will be $(N-1)$-dimensional hyperplanes. Proof of this fact is left as an analytic geometry exercise for the reader.

3.2.4 Error Probability Performance

In this section, we derive the message error probability of the optimum MAP vector receiver in AWGN channels. First, we consider the case of binary signaling and then generalize to M-ary communications. The performance in the presence of colored noise can also be derived following similar steps.

Binary vector receiver performance.   In the binary case, the probability of message error can be found using the theorem on total probability, that is

$$P(E) = P(m_0)P(E/m_0) + P(m_1)P(E/m_1) \qquad (3.91)$$

where $P(E/m_i)$ is the conditional probability of error given m_i. The MAP decision rule is given by (3.57) and assuming that m_0 is actually transmitted, the probability of correct decision is then the probability that

$$|\boldsymbol{\rho} - \mathbf{s}_0|^2 < |\boldsymbol{\rho} - \mathbf{s}_1|^2 - 2\sigma^2 \ln\left(\frac{P(m_1)}{P(m_0)}\right) \qquad (3.92)$$

Expanding the squares and simplifying, one obtains

$$\boldsymbol{\rho}^T(\mathbf{s}_1 - \mathbf{s}_0) < \frac{1}{2}\left(|\mathbf{s}_1|^2 - |\mathbf{s}_0|^2\right) - \sigma^2 \ln\left(\frac{P(m_1)}{P(m_0)}\right) \qquad (3.93)$$

or equivalently (since $\boldsymbol{\rho} = \mathbf{s}_0 + \mathbf{n}$ when m_0 is transmitted)

$$(\mathbf{s}_0 + \mathbf{n})^T (\mathbf{s}_1 - \mathbf{s}_0) < \frac{1}{2}\left(|\mathbf{s}_1|^2 - |\mathbf{s}_0|^2\right) - \sigma^2 \ln\left(\frac{P(m_1)}{P(m_0)}\right) \tag{3.94}$$

Performing the inner products and simplifying, we have

$$\mathbf{n}^T (\mathbf{s}_1 - \mathbf{s}_0) < \frac{1}{2}|\mathbf{s}_1 - \mathbf{s}_0|^2 - \sigma^2 \ln\left(\frac{P(m_1)}{P(m_0)}\right) \tag{3.95}$$

The inner product $\mathbf{n}^T (\mathbf{s}_1 - \mathbf{s}_0)$ is a zero-mean Gaussian random variable $z(u)$, with variance $\sigma_z^2 = E\{z^2(u)\} = (\mathbf{s}_1 - \mathbf{s}_0)^T \mathcal{K}_\mathbf{n}(\mathbf{s}_1 - \mathbf{s}_0)$. Since $\mathcal{K}_\mathbf{n} = \sigma^2 \mathcal{I}$, the variance reduces to $\sigma_z^2 = \sigma^2 |\mathbf{s}_1 - \mathbf{s}_0|^2$. Using the transformation $z(u) = \sigma |\mathbf{s}_1 - \mathbf{s}_0| x(u)$, where $x(u)$ is a normalized (zero-mean, unit-variance) Gaussian random variable, (3.95) can be rewritten as

$$x(u) < \frac{d_{10}}{2\sigma} - \frac{\sigma}{d_{10}} \ln\left(\frac{P(m_1)}{P(m_0)}\right) \tag{3.96}$$

where $d_{10} = |\mathbf{s}_1 - \mathbf{s}_0|$ denotes the distance between vectors $\mathbf{s}_0$ and $\mathbf{s}_1$. Using the Gaussian tail integral, the probability of error, conditioned on m_0, is then given by

$$P(E/m_0) = Q\left(\frac{d_{10}}{2\sigma} - \frac{\sigma}{d_{10}} \ln\left(\frac{P(m_1)}{P(m_0)}\right)\right) \tag{3.97}$$

Similarly, it can be shown that

$$P(E/m_1) = Q\left(\frac{d_{10}}{2\sigma} + \frac{\sigma}{d_{10}} \ln\left(\frac{P(m_1)}{P(m_0)}\right)\right) \tag{3.98}$$

Hence, the total probability of error becomes

$$P(E) = P(m_0)Q\left(\frac{d_{10}}{2\sigma} - \frac{\sigma}{d_{10}} \ln\left(\frac{P(m_1)}{P(m_0)}\right)\right)$$
$$+ P(m_1)Q\left(\frac{d_{10}}{2\sigma} + \frac{\sigma}{d_{10}} \ln\left(\frac{P(m_1)}{P(m_0)}\right)\right) \tag{3.99}$$

When the signals are equally likely, the expression reduces to

$$P(E) = Q\left(\frac{d_{10}}{2\sigma}\right) \tag{3.100}$$

which is a function of the ratio of the distance between the signals to the noise standard deviation. This is a fundamental result which suggests selecting the binary modulation technique that maximizes the "distance" between the vectors that represent the messages in the channel.

M-ary vector receiver performance.   In the M-ary case, the receiver partitions the observation space into M disjoint decision regions R_i as depicted in Fig. 3.13 [since the figure deals with scalar communications, R_i (a multidimensional region) is replaced with I_i (a one-dimensional region)] and the conditional probability of correct decision given that m_i was transmitted is given by

$$P(C/m_i) = P(\boldsymbol{\rho} \epsilon R_i/m_i) = \underbrace{\int \cdots \int_{R_i}}_{N-fold} f_\mathbf{r}(\boldsymbol{\rho}/m_i)d\boldsymbol{\rho} \tag{3.101}$$

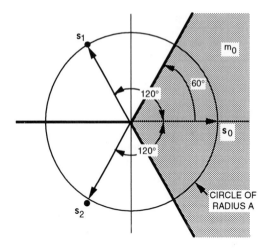

Figure 3.25   Signal set of Example 3.8

and

$$P(C) = \sum_{i=0}^{M-1} P(m_i)P(C/m_i) \tag{3.102}$$

For some specific signal constellations, the multidimensional integral can be reduced to the product of N one-dimensional integrals when the noise is white and Gaussian. However, in the general case, the integral cannot be simplified any further.

Example 3.8

A ternary communication system employs the vector signal set illustrated in Fig. 3.25 to transmit equally probable messages m_0, m_1, and m_2 over the additive white Gaussian noise channel with zero-mean and covariance matrix $\mathcal{K}_n = \sigma^2 \mathcal{I}$. Write a mathematical expression for the exact probability of error $P(E)$.

Noting that $\cos 60° = 1/2$ and $\sin 60° = \sqrt{3}/2$, the signals vectors are given by

$$\mathbf{s}_0 = A \begin{pmatrix} 1 \\ 0 \end{pmatrix} \quad \mathbf{s}_1 = A \begin{pmatrix} -1/2 \\ \sqrt{3}/2 \end{pmatrix} \quad \text{and} \quad \mathbf{s}_2 = A \begin{pmatrix} -1/2 \\ -\sqrt{3}/2 \end{pmatrix}$$

Since the noise is white and Gaussian, minimum distance decoding is the optimal rule, as shown in Fig. 3.25. Since the signals are symmetric and equally probable

$$P(E) = 1 - P(C/m_0)$$

But

$$P(C/m_0) = P(\boldsymbol{\rho} \epsilon R_0 / m_0)$$

Assuming that m_0 is transmitted, $\boldsymbol{\rho} = \mathbf{s}_0 + \mathbf{n}$ resulting in

$$f_{\mathbf{r}}(\boldsymbol{\rho}/m_0) = f_{\mathbf{n}}(\boldsymbol{\rho} - \mathbf{s}_0)$$

Hence

$$P(C/m_0) = \int_{R_0} f_{\mathbf{n}}(\boldsymbol{\rho} - \mathbf{s}_0) d\boldsymbol{\rho}$$

$$= \int_{R_0} \frac{1}{2\pi\sigma^2} \exp\left\{ -\frac{1}{2\sigma^2}((\rho_1 - A)^2 + \rho_2^2) \right\} d\rho_2 d\rho_1$$

or

$$P(C/m_0) = \int_0^\infty \int_{-\sqrt{3}\rho_1}^{\sqrt{3}\rho_1} \frac{1}{2\pi\sigma^2} \exp\left\{ -\frac{1}{2\sigma^2}((\rho_1 - A)^2 + \rho_2^2) \right\} d\rho_2 d\rho_1$$

and

$$P(E) = 1 - P(C/m_0)$$

The above integral cannot be simplified any further and needs to be evaluated numerically for performance prediction purposes.

3.2.5 Effects of Translation and Rotation of the Signal Vectors on Message Error Probability

To facilitate analysis and/or to minimize the average required transmitted energy, it is often advantageous to translate and rotate the signal constellation $\{s_0, s_1, \ldots, s_{M-1}\}$.

Consider the conditional probability of correct decision given that m_i is transmitted, that is, $P(C/m_i)$. This conditional probability is the probability that the received vector $\mathbf{r} = \mathbf{s}_i + \mathbf{n}$ lies in region R_i, as depicted in Fig. 3.26a. This is also the probability that the noise vector drawn from the point $\mathbf{s}_i$ lies within R_i. Since this probability does not depend on the origin of the coordinate system, we can translate the coordinate system to facilitate the analysis. Such a translation is equivalent to translating the signal point and the corresponding decision region. Consequently, $P(C/m_i) = P(C/\mathbf{s}_i)$ for the translated system shown in Fig. 3.26b is identical to that of the system in Fig. 3.26a. We conclude that translations of the entire signal constellation and the corresponding decision regions do not alter message error probability. This in turn suggests that "different" signal sets yield equivalent performances.

In the case of spherically symmetric additive white Gaussian noise, we make another important observation, namely, the rotation of the coordinate system does not affect the message error probability because the pdf $f_{\mathbf{n}}(\boldsymbol{\alpha})$ possesses spherical symmetry. To prove this, consider Fig. 3.26c which is generated from Fig. 3.26a by translation and rotation. Notice that rotation of the coordinate system is equivalent to rotating the signal constellation in the opposite sense. The probability that the noise vector $\mathbf{n}$ (drawn from $\mathbf{s}_i$) lies in R_i is the same in Fig. 3.26a as the probability that the noise vector (drawn from $\mathbf{s}_i$) belongs to R_i in Fig. 3.26c. This is because the probability $P(C/m_i)$ is characterized by the integral over the region R_i of the noise pdf $f_{\mathbf{n}}(\boldsymbol{\alpha} - \mathbf{s}_i)$ in Fig. 3.26a. If one translates $\mathbf{s}_i$ to the origin, the probability in region R_i is not affected by any rotation of the region R_i because symmetric Gaussian noise (same variance in all directions) possesses spherical (circular) symmetry.

Example 3.9

One of sixteen equally likely messages is to be communicated over a vector channel which adds a statistically independent zero-mean Gaussian random variable with variance $N_0/2$ to each transmitted vector component. Assuming that the transmitter uses the signal vectors shown in Fig. 3.27(a), express the $P(E)$ produced by an optimum receiver in terms of the Gaussian tail integral.

Since the channel is additive white Gaussian, the decision rule is the minimum distance decoding rule as shown in Fig. 3.27(b). Dividing the observation space into three major types as depicted in Fig. 3.27(c), there are 4 regions of type ①, 4 regions of type ②, and 8 regions of type ③. Since translation and rotation does not affect $P(E)$ and hence $P(C)$, we have

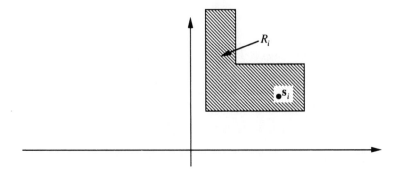

(a) Signal point $\mathbf{s}_i$ and decision region R_i

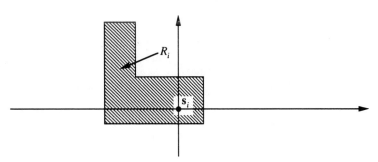

(b) Translation of $\mathbf{s}_i$ and R_i

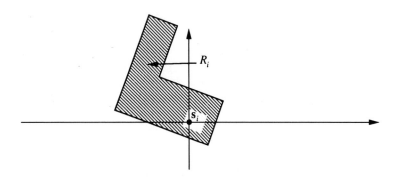

(c) Translation and rotation of $\mathbf{s}_i$ and R_i

Figure 3.26   Translation and rotation of signal vectors and decision regions

$$P(C) = \frac{1}{4} \int_{-d/2}^{\infty} \int_{-d/2}^{\infty} \frac{1}{\pi N_0} \exp\left[-\frac{1}{N_0}\left(x^2 + y^2\right)\right] dxdy$$

$$+ \frac{1}{2} \int_{-d/2}^{d/2} \int_{-d/2}^{\infty} \frac{1}{\pi N_0} \exp\left[-\frac{1}{N_0}\left(x^2 + y^2\right)\right] dxdy$$

$$+ \frac{1}{4} \int_{-d/2}^{d/2} \int_{-d/2}^{d/2} \frac{1}{\pi N_0} \exp\left[-\frac{1}{N_0}\left(x^2 + y^2\right)\right] dxdy$$

or

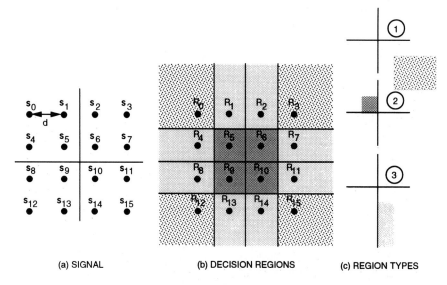

(a) SIGNAL   (b) DECISION REGIONS   (c) REGION TYPES

Figure 3.27   Signals and decision regions of Example 3.9

$$P(C) = \frac{1}{4}(1-p)^2 + \frac{1}{2}(1-p)(1-2p) + \frac{1}{4}(1-2p)^2$$

$$= \left(1 - \frac{3}{2}p\right)^2$$

where $p = Q(d/\sqrt{2N_0})$. Hence

$$P(E) = 1 - P(C) = 1 - \left(1 - \frac{3}{2}p\right)^2$$

3.2.6 Translation to Minimize the Required Signal Energy

Based upon our arguments presented in the previous section, there exists infinitely many signal constellations which yield the same message error probability. From a system design prespective, it makes sense to choose the set of signal vectors that minimizes the mean energy required to yield a specified message error probability. This suggests choosing the signals which are closest to the origin. By translating and rotating the signal set, one can always find an arrangement of signals which require minimum energy. For example, translating a set of orthogonal vectors to minimize the required energy yields the transorthogonal (simplex) set.

Consider the mean energy

$$\overline{E} \triangleq E\left\{|\mathbf{s}_i|^2\right\} = \sum_{i=0}^{M-1} P(m_i)|\mathbf{s}_i|^2 \tag{3.103}$$

of the signal set. Translation of the signal set to minimize the required energy does not affect the message error probability and is equivalent to substracting some vector **a** from each vector $\mathbf{s}_i$ to create a new vector $\mathbf{s}'_i = \mathbf{s}_i - \mathbf{a}$ for all i. In other words, we wish to find the vector $\mathbf{a}$ which minimizes the new mean energy

$$\overline{E'} = \sum_{i=0}^{M-1} P(m_i)|\mathbf{s}'_i|^2 \tag{3.104}$$

One can easily show (see Problem 3.42) that the vector $\mathbf{a}$ specifies the center of gravity of the M vectors $\mathbf{s}_0, \mathbf{s}_1, \ldots, \mathbf{s}_{M-1}$, that is

$$\mathbf{a} = \sum_{i=0}^{M-1} P(m_i)\mathbf{s}_i = E\{\mathbf{s}_i\} \tag{3.105}$$

This result readily follows from the fact that the energy E' is equivalent to the moment of inertia about a point $\mathbf{a}$ of the "M particles of probability masses" $P(m_0), P(m_1), \ldots, P(m_{M-1})$, respectively.

Example 3.10

Consider the equilikely binary orthogonal signal vectors $\mathbf{s}_0^T = (\sqrt{E}, 0)$ and $\mathbf{s}_1^T = (0, \sqrt{E})$. Creating the mean signal vector $\mathbf{a}$ and subtracting it from $\mathbf{s}_0$ and $\mathbf{s}_1$ yields the new signal vectors $\mathbf{s}_0'^T = (-\sqrt{E/2}, 0)$ and $\mathbf{s}_1'^T = (\sqrt{E/2}, 0) = -\mathbf{s}_0'$. The message error probability for both signal sets is $Q(\sqrt{E/N_0})$; however, the latter requires a smaller mean energy. This is one of the reasons why BPSK (antipodal vectors) is selected in practice over BFSK (orthogonal vectors).

Example 3.11

Transorthogonal (simplex) signal vectors are the minimum energy equivalent set of equilikely orthogonal signal vectors. This is illustrated in Fig. 3.28 for the case of two- and three-dimensional signal space. Note that energy minimization effectively reduces the number of dimensions (bandwidth) of the signal space by one. For two-dimensional orthogonal signal vectors (BFSK), the corresponding transorthogonal signal vectors lie on the same straight line, that is, the one-dimensional antipodal signals (BPSK) as shown in Fig. 3.28a. For three-dimensional orthogonal vectors, the plane formed by the three signals is constructed where the center of gravity of the signal vectors on this plane becomes the new origin. It is obvious that the equivalent transorthogonal signal vectors are reduced to two dimensions as depicted in Fig. 3.28b. Similarly for the four-dimensional case, the minimum energy set lies at the vertices of a tetrahedron (as shown in Fig. 4.21 in Chapter 4) reducing it to a three-dimensional space. It can be shown (in Chapter 12) that for a given mean energy, the equilikely transorthogonal signal vectors are optimum in the sense of minimum error probability in white Gaussian noise. For the orthogonal signal vectors $\mathbf{s}_i = \sqrt{E}\boldsymbol{\phi}_i$, the mean signal vector $\mathbf{a}$ is given by

$$\mathbf{a} = \left(\frac{\sqrt{E}}{M}\right) \sum_{i=0}^{M-1} \boldsymbol{\phi}_i$$

and

$$\mathbf{s}'_i = \sqrt{E}\boldsymbol{\phi}_i - \mathbf{a}$$

so that

$$E' = |\mathbf{s}'_i|^2 = E\left(1 - \frac{1}{M}\right)$$

Hence for the same message error probability, the mean energy of the transorthogonal set is $(1 - 1/M)$ times that of the orthogonal set. For large M, orthogonal and transorthogonal signal sets have the same message error probability.

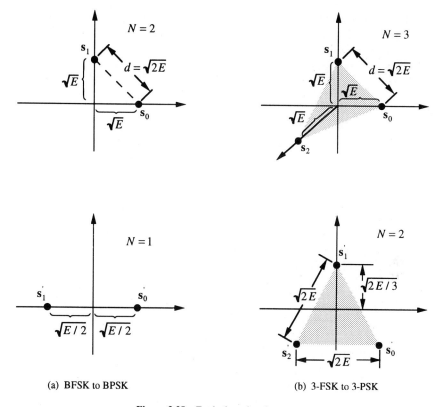

Figure 3.28 Equivalent signal vector sets

3.2.7 Union Bound on *M*-ary Vector Receiver Performance

For most signaling schemes, the computation of the exact probability of error is tedious and an upper bound on the performance might be sufficient. In this section, we will derive such a bound for vector communications over the additive white Gaussian noise channel. The bound is general and can be applied to any set of vectors.

We know from the previous discussion that the optimum a posteriori probability receiver sets $\widehat{m}(\boldsymbol{\rho}) = m_k$ if and only if

$$\boldsymbol{\rho}^T \mathbf{s}_k + B_k \geq \boldsymbol{\rho}^T \mathbf{s}_i + B_i \quad \text{for all } i \tag{3.106}$$

where B_i is a constant given by (3.60). When the signals are of equal length and are equally probable, the constants can be ignored and the rule simplifies to evaluating the cross-correlations

$$\boldsymbol{\rho}^T \mathbf{s}_k \geq \boldsymbol{\rho}^T \mathbf{s}_i \tag{3.107}$$

Hence, the conditional probability of error, given that m_k was transmitted, is

$$P(E/m_k) = P\left(\boldsymbol{\rho}^T \mathbf{s}_k < \boldsymbol{\rho}^T \mathbf{s}_j \quad \text{for some } j \neq k/m_k\right) \tag{3.108}$$

Let E_j $(j \neq k)$ denote the event that $\boldsymbol{\rho}^T \mathbf{s}_j > \boldsymbol{\rho}^T \mathbf{s}_k$. Then, an error can result if only one event E_j occurs for $j \neq k$, or if any two events E_j, E_m $(j \neq k, m \neq k)$ occur simultaneously, or if any three events E_j, E_m, E_n ($j \neq k$, $m \neq k$ and $n \neq k$) occur jointly, . . . , or if all $M-1$ events $E_0, \ldots, E_{k-1}, E_{k+1}, \ldots, E_{M-1}$ occur together. Mathematically, this can be denoted using a union of events as

$$P(E/m_k) = P(E_0/m_k) + \cdots + P(E_{k-1}/m_k) + P(E_{k+1}/m_k) + \cdots + P(E_{M-1}/m_k)$$
$$\pm P(E_0, E_1/m_k) + \cdots + P(E_{M-2}, E_{M-1}/m_k)$$
$$\pm \cdots$$
$$\pm P(E_0, \ldots, E_{k-1}, E_{k+1}, \ldots, E_{M-1}/m_k) \tag{3.109}$$
$$= P\left(\bigcup_{\substack{j=0 \\ j \neq k}}^{M-1} E_j/m_k \right)$$

The signs in (3.109) depend on M since $P(A+B) = P(A) + P(B) - P(AB)$ but $P(A+B+C) = P(A) + P(B) + P(C) - P(AB) - P(AC) - P(BC) + P(ABC)$. Since $P(A+B) \leq P(A) + P(B)$, applying this inductively to the union in (3.109), we obtain a bound on $P(E/m_k)$, namely

$$P(E/m_k) \leq \sum_{\substack{j=0 \\ j \neq k}}^{M-1} P(E_j/m_k) \tag{3.110}$$

or, equivalently

$$P(E/m_k) \leq \sum_{\substack{j=0 \\ j \neq k}}^{M-1} P\left(\boldsymbol{\rho}^T \mathbf{s}_k < \boldsymbol{\rho}^T \mathbf{s}_j / m_k \right) \tag{3.111}$$

This bound is referred to as the **Union bound** and is very useful in communication problems. Consider the event that $\{\boldsymbol{\rho}^T \mathbf{s}_k < \boldsymbol{\rho}^T \mathbf{s}_j / m_k\}$. Assuming that the noise is white Gaussian, the probability of this event is given in (3.100) as

$$P(E_j/m_k) = Q\left(\frac{d_{kj}}{2\sigma} \right) \tag{3.112}$$

where d_{kj} is the distance between the vectors $\mathbf{s}_k$ and $\mathbf{s}_j$. Hence, the Union bound simplifies to

$$P(E/m_k) \leq \sum_{\substack{j=0 \\ j \neq k}}^{M-1} Q\left(\frac{d_{kj}}{2\sigma} \right) \tag{3.113}$$

The probability of error $P(E)$ can then be bounded using (3.113) with the appropriate a priori probabilities. This is a key result that will be used in the chapters on coded communications.

Example 3.12

Assuming that a communication system uses M orthogonal signals to communicate over an AWGN channel with $\mathcal{K_n} = \sigma^2 \mathcal{I}$, derive a bound on $P(E)$ when $|\mathbf{s}_i|^2 = E$, $i = 0, 1, \ldots, M-1$. The signal can be modeled by

$$\mathbf{s}_i = \sqrt{E} \boldsymbol{\phi}_i \quad i = 0, 1, \ldots, M-1$$

where

$$\phi_i^T \phi_j = \begin{cases} 1 & i = j \\ 0 & i \neq j \end{cases}$$

for all i and j in the set $\{0, 1, \ldots, M-1\}$. The distance between any two vectors is given by

$$d = \sqrt{E + E} = \sqrt{2E}$$

and is independent of which pair we choose due to the orthogonality. Using (3.113), we get

$$P(E) \leq (M-1)Q\left(\sqrt{\frac{E}{N_0}}\right)$$

where we have assumed that $\sigma^2 = N_0/2 = kT^\circ/2$ Joules. At room temperature, $T^\circ = 300K = 28^\circ C$ and $kT^\circ = k300K = 1/40$ eV (electron-volt).

3.2.8 Error Probability Performance of Two-Dimensional, *M*-ary Vector Receivers—A New Approach

In Section 3.2.4 we described the traditional approach for evaluating the probability of correct decision for N-dimensional, M-ary vector receivers wherein the observation space is partitioned into M disjoint N-dimensional regions $R_i; i = 0, 1, 2, \ldots, M-1$ surrounding the signal vectors. The origin of coordinates for defining these decision regions is assumed to be the same for each of them and is located in the N-dimensional space defined by the *received* vector **r**. In particular, for the ith signal transmitted, the conditional probability of correct decision is described by the N-dimensional integral in (3.101) which, in the general case, cannot be simplified any further. Even for the case of a two-dimensional signal vector set and an AWGN channel (such as in Example 3.8), we saw that the expression for the conditional probability of correct decision still took the form of a double integral. Thus, in general, the average probability of correct decision, or equivalently, the average probability of error, would be expressed as the sum of M such double integrals.

Recently, Craig [16] cleverly showed that the evaluation of average probability of error for the two-dimensional, AWGN channel case could be considerably simplified by choosing the origin of coordinates for each decision region in the two-dimensional space defined by the corresponding *signal* vector. This shift of the coordinate system origin from the received vector space to the signal vector space allows the integrand of the two-dimensional integral describing the conditional probability of error to be independent of the transmitted signal. In particular, this integrand is simply the pdf of the two-dimensional noise vector **n** which is characterized by a two-dimensional zero mean Gaussian pdf with independent components. When expressed in polar coordinates, that is, (r, θ), this pdf can be integrated on r to yield a closed form result, that is, a single exponential, *regardless of the complexity of the integration limits or their dependence on* θ, the latter being dependent on the shape of the decision boundaries. Thus, at worst, the average probability of error would now be expressible as the sum of M single integrals. In short, this new approach of computing average error probability has shifted the complexity of the computation from the integrand to the integration limits.

To illustrate the above approach, we consider a generalization of Example 3.8 corresponding to an M-ary communication system with a vector signal set having M points

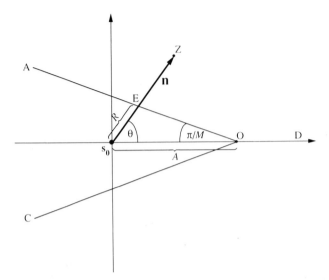

Figure 3.29 The geometry for the decision region corresponding to the signal point s_0

uniformly distributed around a circle of radius A. The geometry for the decision region associated with the signal point

$$s_0 = A \begin{pmatrix} -1 \\ 0 \end{pmatrix}$$

is illustrated in Fig. 3.29. An error is made if the noise vector $\mathbf{n}$ lies outside the wedge-shaped decision region bounded by the semi-infinite lines OA and OC, as for example, the point Z. In particular, the probability of an error given that signal s_0 is transmitted is twice the probability that Z lies above the boundary AOD, namely

$$P\,(E\,|m_0) = 2 \int_0^{(M-1)\pi/M} d\theta \int_R^\infty f_{\mathbf{n}}\,(r,\theta)\,dr \tag{3.114}$$

where R is the distance from the signal point to the boundary point E (in general, a function of θ), and $f_{\mathbf{n}}\,(r,\theta)$ is the above-mentioned bivariate Gaussian pdf of the noise vector expressed in polar coordinates, that is,

$$f_{\mathbf{n}}\,(r,\theta) = \frac{r}{2\pi\sigma^2} \exp\left(-\frac{r^2}{2\sigma^2}\right) \tag{3.115}$$

which is clearly independent of θ. Substituting (3.115) into (3.114) and carrying out the integration on r gives

$$P\,(E\,|m_0) = \frac{1}{\pi} \int_0^{(M-1)\pi/M} \exp\left[-\frac{A^2 \sin^2 \frac{\pi}{M}}{2\sigma^2 \sin^2 \left(\theta + \frac{\pi}{M}\right)}\right] d\theta \tag{3.116}$$

where, from the law of sines

$$R = \frac{A \sin \frac{\pi}{M}}{\sin \left(\theta + \frac{\pi}{M}\right)} \tag{3.117}$$

Making the change of variables $\Phi \triangleq [(M-1)\pi/M] - \theta$, the final form for $P(E \,|m_0)$ is

$$P(E\,|m_0) = \frac{1}{\pi} \int_0^{(M-1)\pi/M} \exp\left[-\frac{A^2 \sin^2 \frac{\pi}{M}}{2\sigma^2 \sin^2 \Phi}\right] d\Phi \qquad (3.118)$$

Finally, since from the symmetry of the vector signal set, we see that $P(E\,|m_i)$ is independent of i, then the average error probability is also given by (3.118), that is,

$$P(E) = \frac{1}{\pi} \int_0^{(M-1)\pi/M} \exp\left[-\frac{A^2 \sin^2 \frac{\pi}{M}}{2\sigma^2 \sin^2 \Phi}\right] d\Phi \qquad (3.119)$$

For the ternary case ($M = 3$) as in Example 3.8, (3.119) becomes

$$P(E) = \frac{1}{\pi} \int_0^{2\pi/3} \exp\left[-\frac{A^2 \sin^2 \frac{\pi}{3}}{2\sigma^2 \sin^2 \Phi}\right] d\Phi \qquad (3.120)$$

which is a considerable simplification of the result previously obtained for this case in the form of one minus a double integral. Furthermore, for small values of $P(E)$, the direct numerical evaluation of (3.120) will yield improved accuracy relative to the previous indirect result where a number close to unity (the probability of correct decision) is first evaluated and then subtracted from unity.

For the binary case ($M = 2$), (3.119) evaluates to

$$P(E) = \frac{1}{\pi} \int_0^{\pi/2} \exp\left[-\frac{A^2}{2\sigma^2 \sin^2 \Phi}\right] d\Phi \qquad (3.121)$$

However, for this case, the result as obtained from the traditional evaluation approach in Section 3.2.4 is expressible as a Gaussian tail integral, namely

$$P(E) = Q\left(\frac{A}{\sigma}\right) \qquad (3.122)$$

Thus, we see that the Gaussian tail integral, which is normally defined as the integral over a semi-infinite interval of a Gaussian pdf, can be expressed as the finite integral

$$Q(x) = \frac{1}{\pi} \int_0^{\pi/2} \exp\left[-\frac{x^2}{2\sin^2 \Phi}\right] d\Phi \qquad (3.123)$$

which is useful for numerical evaluation purposes.

It is straightforward to generalize the above results to signal sets whose decision region boundaries are polygons surrounding each of the signal points. A typical example of such is described by the geometry in Fig. 3.30 where the nth signal point s_n is illustrated inside its correct decision region described by the four-sided polygon ABCDA. If signal s_n (corresponding to message m_n) is transmitted, then an error occurs if the tip of the noise vector lies outside of this polygon. The region outside of this polygon is made up of the four disjoint regions ①, ②, ③, and ④. The probability that the tip of the noise vector falls in region ① is described by an integral analogous to (3.116), namely

$$P_1(E\,|m_n) = \frac{1}{2\pi} \int_0^{\theta_1} \exp\left[-\frac{A_1^2 \sin^2 \psi_1}{2\sigma^2 \sin^2(\theta + \psi_1)}\right] d\theta \qquad (3.124)$$

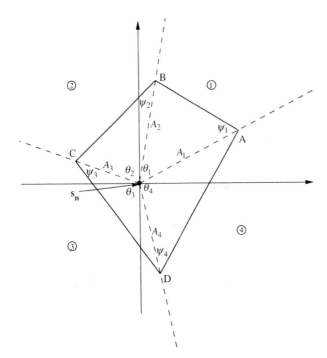

Figure 3.30 The geometry for the decision region corresponding to the signal point s_n

The probability that the tip of the noise vector falls in any of the other three regions is given by an expression similar to (3.124). Since the four regions are disjoint and comprise the entire error space, then the conditional probability of error given that message m_n was transmitted is

$$P(E \,|m_n) = \sum_{i=1}^{4} P_i(E \,|m_n) = \sum_{i=1}^{4} \frac{1}{2\pi} \int_0^{\theta_i} \exp\left[-\frac{A_i^2 \sin^2 \psi_i}{2\sigma^2 \sin^2(\theta + \psi_i)} \right] d\theta \quad (3.125)$$

The conditional probability of error for any other point in the signal constellation is given by an expression analogous to (3.125) except that the number of terms in the sum might be different depending on the number of sides in the polygon that defines its decision region. Finally, the average symbol error probability would be obtained by averaging all of these conditional error probabilities over the a priori probabilities of the message set. As such, the average symbol error probability is expressible as a weighted sum of single integrals with finite limits and an exponential integrand.

Appendix 3A

Other Detection Criteria

So far, we have considered the probability of error detection as the only performance measure for the receivers. Depending on the application (for example, in radar system design), other criteria [11] might be more suitable and these make use of differing amounts of information and problem specification. The two criteria for the decision rule that we have encountered are the maximum likelihood (ML) criterion and the maximum a posteriori (MAP) criterion. In this appendix, we will consider three more criteria: the Bayes decision rule, the minimum-maximum rule, and finally, the Neyman-Pearson criterion. Only the basics of each rule are discussed assuming a binary hypothesis and the reader is referred to appropriate references for a more detailed treatment.

3A.1 THE BAYES CRITERION

In both detection criteria considered so far, we have assumed that all errors and all correct decisions are equally important and thus, are treated equally by maximizing the a posteriori probabilities. In some applications, some errors are more costly than others and as a result, some kind of weighting is necessary to reflect that. In Bayesian detection, each course of action is assigned a cost which is a function of both the chosen and true hypothesis. These costs are denoted by C_{ij}, where in the binary case, we have

$$C_{00} = \text{Cost of deciding hypothesis } H_0 \text{ when } H_0 \text{ is true}$$
$$C_{10} = \text{Cost of deciding hypothesis } H_1 \text{ when } H_0 \text{ is true}$$
$$C_{01} = \text{Cost of deciding hypothesis } H_0 \text{ when } H_1 \text{ is true}$$
$$C_{11} = \text{Cost of deciding hypothesis } H_1 \text{ when } H_1 \text{ is true}$$

The expected value of the cost, referred to as the risk R, is given by

$$
\begin{aligned}
R = C_{00}P(H_0)P(H_0/H_0) + C_{10}P(H_0)P(H_1/H_0) \\
+ C_{01}P(H_1)P(H_0/H_1) + C_{11}P(H_1)P(H_1/H_1)
\end{aligned}
\tag{3A.1}
$$

where $P(H_i)$ is the a priori probability of hypothesis H_i and $P(H_i/H_j)$ is the probability of deciding on hypothesis H_i given that hypothesis H_j is true. In the binary case, the observation region is divided into two distinct regions, R_0 and R_1, where we decide H_0 if $\rho \in R_0$ and decide H_1 if $\rho \in R_1$. Hence

$$
P(H_i/H_j) = \int_{R_i} f_{\mathbf{r}}(\boldsymbol{\rho}/H_j)d\boldsymbol{\rho}
\tag{3A.2}
$$

However, note that since the received vector belongs to either R_0 or to R_1 (that is, $R_0 \cup R_1$ = complete observation space), then

$$
\int_{R_1} f_{\mathbf{r}}(\boldsymbol{\rho}/H_0)d\boldsymbol{\rho} = 1 - \int_{R_0} f_{\mathbf{r}}(\boldsymbol{\rho}/H_0)d\boldsymbol{\rho}
\tag{3A.3}
$$

and

$$
\int_{R_1} f_{\mathbf{r}}(\boldsymbol{\rho}/H_1)d\boldsymbol{\rho} = 1 - \int_{R_0} f_{\mathbf{r}}(\boldsymbol{\rho}/H_1)d\boldsymbol{\rho}
\tag{3A.4}
$$

Hence, the risk can be expressed as

$$R = C_{10}P(H_0) + C_{11}P(H_1)$$
$$+ \int_{R_0} \left\{ P(H_1)(C_{01} - C_{11})f_{\mathbf{r}}(\boldsymbol{\rho}/H_1) - P(H_0)(C_{10} - C_{00})f_{\mathbf{r}}(\boldsymbol{\rho}/H_0) \right\} d\boldsymbol{\rho} \tag{3A.5}$$

Since correct decisions are less costly than erroneous ones, we can safely assume that

$$C_{01} > C_{11} \quad \text{and} \quad C_{10} > C_{00} \tag{3A.6}$$

so that $C_{01} - C_{11} > 0$ and $C_{10} - C_{00} > 0$. Bayes decision criterion selects the decision region R_0 in order to minimize the risk. Since the first two terms of (3A.5) are independent of R_0, they are not involved in the minimization. The integral, on the other hand, is minimized by selecting R_0 to contain all the observations $\boldsymbol{\rho}$ that satisfy

$$P(H_0)(C_{10} - C_{00})f_{\mathbf{r}}(\boldsymbol{\rho}/H_0) > P(H_1)(C_{01} - C_{11})f_{\mathbf{r}}(\boldsymbol{\rho}/H_1) \tag{3A.7}$$

This results in the most negative possible value of the integral and as a result, the risk is at its minimum. Since R_0 and R_1 are disjoint sets, all observed vectors $\boldsymbol{\rho}$ for which

$$P(H_0)(C_{10} - C_{00})f_{\mathbf{r}}(\boldsymbol{\rho}/H_0) < P(H_1)(C_{01} - C_{11})f_{\mathbf{r}}(\boldsymbol{\rho}/H_1) \tag{3A.8}$$

are included in R_1. Hence, the Bayes decision rule is given by

$$P(H_0)(C_{10} - C_{00})f_{\mathbf{r}}(\boldsymbol{\rho}/H_0) \underset{H_1}{\overset{H_0}{\gtrless}} P(H_1)(C_{01} - C_{11})f_{\mathbf{r}}(\boldsymbol{\rho}/H_1) \tag{3A.9}$$

or equivalently

$$\Lambda_{\mathbf{r}}(\boldsymbol{\rho}) \triangleq \frac{f_{\mathbf{r}}(\boldsymbol{\rho}/H_1)}{f_{\mathbf{r}}(\boldsymbol{\rho}/H_0)} \underset{H_0}{\overset{H_1}{\gtrless}} \frac{P(H_0)(C_{10} - C_{00})}{P(H_1)(C_{01} - C_{11})} = \lambda \tag{3A.10}$$

which is a likelihood ratio test. Note that when $C_{01} = C_{10} = 1$ and $C_{00} = C_{11} = 0$, the Bayes rule reduces to the MAP decision rule, as given in (3.18) for the scalar case. All results derived herein can be generalized to M-ary decisions as given in [12–14].

Example 3A.1

A decisioning system must decide between hypothesis H_0 and H_1, where H_0 is the hypothesis that an observed random variable $x(u)$ has a pdf

$$f_x(x/H_0) = \frac{1}{\sqrt{2\pi}} \exp\left(-\frac{x^2}{2}\right)$$

and H_1 is the hypothesis that $x(u)$ has a pdf

$$f_x(x/H_1) = \begin{cases} \frac{1}{3}, & \text{if } 0 \leq x \leq 3 \\ 0, & \text{otherwise} \end{cases}$$

The a priori probabilities of H_0 and H_1 are 3/4 and 1/4, respectively. The costs are given by

$$C_{00} = 0 \quad C_{10} = 1$$

$$C_{01} = 2 \quad C_{11} = 0$$

Determine and describe the optimal Bayes decision procedure.

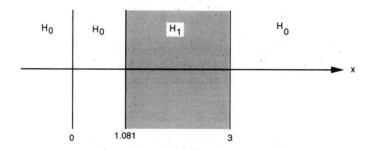

Figure 3A.1 Decision regions in Example 3A.1

Since $f_x(x/H_1) = 0$ for $x \le 0$ and $x > 3$, then decide H_0 in this region. For $0 \le x \le 3$, use the ratio in (3A.10) to get

$$\frac{\sqrt{2\pi}(1/3)}{\exp(-x^2/2)} \underset{H_0}{\overset{H_1}{\gtrless}} \frac{(3/4)(1)}{(1/4)(2)}, \quad \text{for } 0 \le x \le 3$$

or

$$\exp(x^2/2) \underset{H_0}{\overset{H_1}{\gtrless}} \frac{9}{2\sqrt{2\pi}}, \quad \text{for } 0 \le x \le 3$$

or

$$x \underset{H_0}{\overset{H_1}{\gtrless}} \sqrt{2\ln(\frac{9}{2\sqrt{2\pi}})} = 1.081, \quad \text{for } 0 \le x \le 3$$

Hence, the decision regions are as shown in Fig. 3.A1.

3A.2 THE MIN-MAX CRITERION

In some applications, the a priori probabilities $P(H_0)$ and $P(H_1)$ are not known, in which case, the minimum-maximum (min-max) criterion can be employed. The average cost in (3A.5) can be rewritten as

$$R = C_{10}(1 - P(H_1)) + C_{11}P(H_1) \tag{3A.11}$$
$$+ \int_{R_0} \left\{ P(H_1)(C_{01} - C_{11})f_{\mathbf{r}}(\boldsymbol{\rho}/H_1) - (1 - P(H_1))(C_{10} - C_{00})f_{\mathbf{r}}(\boldsymbol{\rho}/H_0) \right\} d\boldsymbol{\rho}$$

Since $P(H_1)$ is unknown, the min-max rule maximizes the risk with respect to $P(H_1)$ and then minimizes that maximum with the choice of R_0. Hence, the rule is given by

$$\min_{R_0} \max_{P(H_1)} R \tag{3A.12}$$

In most applications, the minimization and maximization operations can be interchanged resulting in

$$\min_{R_0} \max_{P(H_1)} R = \max_{P(H_1)} \min_{R_0} R \tag{3A.13}$$

The minimization of R, with respect to R_0, is simply the Bayes rule associated with $P(H_1)$. Hence, R_0 contains the observed vectors $\boldsymbol{\rho}$ for which

$$P(H_1)(C_{01} - C_{11})f_{\mathbf{r}}(\boldsymbol{\rho}/H_1) < (1 - P(H_1))(C_{10} - C_{00})f_{\mathbf{r}}(\boldsymbol{\rho}/H_0) \tag{3A.14}$$

giving the decision rule

$$\frac{f_{\mathbf{r}}(\boldsymbol{\rho}/H_0)}{f_{\mathbf{r}}(\boldsymbol{\rho}/H_1)} \underset{H_1}{\overset{H_0}{\gtrless}} \frac{P(H_1)(C_{01} - C_{11})}{(1 - P(H_1))(C_{10} - C_{00})} \tag{3A.15a}$$

associated with $P(H_1)$, or equivalently

$$\frac{f_{\mathbf{r}}(\boldsymbol{\rho}/H_1)}{f_{\mathbf{r}}(\boldsymbol{\rho}/H_0)} \underset{H_0}{\overset{H_1}{\gtrless}} \frac{(1 - P(H_1))(C_{10} - C_{00})}{P(H_1)(C_{01} - C_{11})} \tag{3A.15b}$$

This is followed by a maximization to find the least favorable $P(H_1)$.

Example 3A.2

Assuming that the a priori probabilities are unknown, repeat Example 3A.1 for the min-max rule using the pdf's

$$f_x(x/H_0) = \exp(-x), \quad x > 0$$

and

$$f_x(x/H_1) = 2 \exp(-2x), \quad x > 0$$

Using (3A.15a), we get the likelihood ratio test

$$\frac{\exp(-x)}{2 \exp(-2x)} \underset{H_1}{\overset{H_0}{\gtrless}} 2 \left(\frac{P(H_1)}{1 - P(H_1)} \right)$$

or

$$x \underset{H_1}{\overset{H_0}{\gtrless}} \ln \left(\frac{4P(H_1)}{1 - P(H_1)} \right)$$

or

$$x \underset{H_1}{\overset{H_0}{\gtrless}} \ln \left(4\frac{1 - P(H_0)}{P(H_0)} \right) = T^*$$

The decision rule is valid as long as T^* is positive (that is, $P(H_0) \leq 4/5$). In this case, the risk is given by [using (3A.1)]

$$R = P(H_0)P(H_1/H_0) + 2(1 - P(H_0))P(H_0/H_1)$$

But

$$P(H_1/H_0) = \int_0^{T^*} \exp(-z)dz$$

and

$$P(H_0/H_1) = \int_{T^*}^{\infty} 2 \exp(-2z)dz$$

Substituting for T^* and solving the integrals, we get

$$R = \begin{cases} \frac{9P^2(H_0) - 8P(H_0)}{8(P(H_0) - 1)} & \text{for } P(H_0) \leq 4/5 \\ 2(1 - P(H_0)) & \text{for } P(H_0) \geq 4/5 \end{cases}$$

Now, we need to solve for $P(H_0)$ which maximizes the risk R. For $P(H_0) \leq 4/5$, the maximum is obtained by setting $dR/dP(H_0) = 0$ and solving for $P(H_0)$ to get $P(H_0) = 2/3$ (with corresponding $R = 1/2$). For $P(H_0) \geq 4/5$, the maximum R is $2/5$ achieved at $P(H_0) = 4/5$. Hence

$P(H_0) = 2/3$ maximizes the risk and results in $T^* = \ln 2$. Therefore, the min-max decision rule reduces to

$$x \underset{H_1}{\overset{H_0}{\gtrless}} \ln 2$$

3A.3 THE NEYMAN-PEARSON CRITERION

In radar, signal acquisition, and antenna pointing, a decision between one of two hypotheses must be made. Here, hypothesis H_0 denotes target (signal) absent while H_1 denotes target (signal) present. The performance of the radar detector (signal detector) is characterized by the probability of each of these four possible events:

1. decide target present when target is present (H_1/H_1),
2. decide target not present when target is present (H_0/H_1),
3. decide target present when target is not present (H_1/H_0),
4. decide target not present when target is not present (H_0/H_0)

The first event is called "target detection" and the probability associated with this event is called the probability of detection, P_D. The second event always occurs if event (1) does not occur; therefore, the target is missed and the *miss* event occurs with probability $P_M = 1 - P_D$. Event (3) is called a *false alarm* since a target presence is declared even though it is not actually present; its probability is denoted by P_{FA}. Using the same reasoning as in events (1) and (2), the probability of event (4) must be the complement of event (3), or $P_C = 1 - P_{FA}$, and is sometimes called the probability of correct dismissal. Since the decision process just described is binary, only two probabilities are needed to fully specify the radar (signal presence indicator) performance. Notice that the error probability is given by $P(E) = P(H_0)P_{FA} + P(H_1)(1 - P_D)$, when the prior probabilities are known. In terms of the channel transition densities, we have

$$P_{FA} \triangleq \text{Probability of false alarm} = P(H_1/H_0) = \int_{R_1} f_{\mathbf{r}}(\boldsymbol{\rho}/H_0)d\boldsymbol{\rho} \qquad (3A.16)$$

$$P_C \triangleq \text{Probability of correct dismissal} = P(H_0/H_0) = \int_{R_0} f_{\mathbf{r}}(\boldsymbol{\rho}/H_0)d\boldsymbol{\rho} = 1 - P_{FA} \qquad (3A.17)$$

$$P_D \triangleq \text{Probability of detection} = P(H_1/H_1) = \int_{R_1} f_{\mathbf{r}}(\boldsymbol{\rho}/H_1)d\boldsymbol{\rho} \qquad (3A.18)$$

$$P_M \triangleq \text{Probability of miss} = P(H_0/H_1) = \int_{R_0} f_{\mathbf{r}}(\boldsymbol{\rho}/H_1)d\boldsymbol{\rho} = 1 - P_D \qquad (3A.19)$$

The Neyman-Pearson criterion maximizes the probability of detection, P_D, under the constraint that the probability of false alarm, P_{FA}, is equal to β. The solution is obtained by using Lagrange multipliers to reformulate the problem as an unconstrained one. Construct

$$\Gamma = P_D - \lambda(P_{FA} - \beta)$$

$$= \int_{R_1} \left\{ f_{\mathbf{r}}(\boldsymbol{\rho}/H_1) - \lambda f_{\mathbf{r}}(\boldsymbol{\rho}/H_0) \right\} d\boldsymbol{\rho} + \lambda\beta \qquad (3A.20)$$

Clearly, if $P_{FA} = \beta$, then maximizing Γ maximizes P_D. As before, the integral is maximized if R_1 is set to contain all observed vectors $\boldsymbol{\rho}$, for which

$$f_{\mathbf{r}}(\boldsymbol{\rho}/H_1) > \lambda f_{\mathbf{r}}(\boldsymbol{\rho}/H_0) \tag{3A.21}$$

resulting in the likelihood ratio test

$$\Lambda_{\mathbf{r}}(\boldsymbol{\rho}) = \frac{f_{\mathbf{r}}(\boldsymbol{\rho}/H_1)}{f_{\mathbf{r}}(\boldsymbol{\rho}/H_0)} \underset{H_0}{\overset{H_1}{\gtrless}} \lambda \tag{3A.22}$$

The parameter λ is then selected such that the constraint

$$P_{FA} = \beta = \int_{R_1} f_{\mathbf{r}}(\boldsymbol{\rho}/H_0) d\boldsymbol{\rho} \tag{3A.23}$$

is satisfied. Note the similarity of (3A.22) to the maximum likelihood criterion, in which the threshold λ is unity.

Example 3A.3

Consider the binary problem with the following conditional pdf's

$$f_x(x/H_0) = \frac{1}{\sqrt{2\pi}} \exp(-x^2/2)$$

$$f_x(x/H_1) = \frac{1}{\sqrt{2\pi}} \exp(-(x-m)^2/2)$$

We require that $P_{FA} = \beta$. Derive the Neyman-Pearson test.
From (3A.22), we have

$$\frac{\exp(-(x-m)^2/2)}{\exp(-x^2/2)} \underset{H_0}{\overset{H_1}{\gtrless}} \lambda$$

or equivalently

$$x \underset{H_0}{\overset{H_1}{\gtrless}} \frac{\ln \lambda}{m} + \frac{m}{2}$$

The false-alarm probability is therefore

$$P_{FA} = P(H_1/H_0) = \int_{\frac{\ln \lambda}{m} + \frac{m}{2}}^{\infty} \frac{1}{\sqrt{2\pi}} \exp\left(-x^2/2\right) dx = Q\left(\frac{\ln \lambda}{m} + \frac{m}{2}\right) = \beta$$

When $\beta = 0.1$ and $m = 1$, we have

$$Q\left(\ln \lambda + \frac{1}{2}\right) = 0.1$$

From the table in Appendix 3B, we find that

$$Q(1.282) = 0.1$$

and hence, $\ln \lambda + 1/2 = 1.282$ resulting in $\lambda = 2.18$. The decision rule reduces to

$$x \underset{H_0}{\overset{H_1}{\gtrless}} 1.282$$

Appendix 3B

The Gaussian Integral $Q(x)$

In this appendix, we present a numerical table and a series approximation for $Q(x)$, which are useful in decision-theory problems. The Gaussian tail integral is defined by

$$Q(x) \triangleq \int_x^\infty \frac{1}{\sqrt{2\pi}} \exp\left(-\frac{y^2}{2}\right) dy = 1 - Q(-x) \tag{3B.1}$$

We also define the error function $\text{erf}(x)$ as

$$\text{erf}(x) \triangleq \frac{2}{\sqrt{\pi}} \int_0^x \exp\left(-u^2\right) du \tag{3B.2}$$

and the complementary error function $\text{erfc}(x)$ as

$$\text{erfc}(x) \triangleq 1 - \text{erf}(x) \tag{3B.3}$$

The Gaussian tail integral $Q(x)$ is related to $\text{erfc}(x)$ through

$$\text{erfc}(x) = 2Q\left(\sqrt{2}x\right) \tag{3B.4}$$

or, inversely

$$Q(y) = \frac{1}{2}\text{erfc}\left(\frac{y}{\sqrt{2}}\right) \tag{3B.5}$$

3B.1 TABLE OF $Q(x)$

A numerical table of $Q(x)$ is given in Table 3B.1.

3B.2 SERIES APPROXIMATION FOR $Q(x)$

In this section, we present an nth order series approximation for $Q(x)$, along with a bound on the percentage error. For $x > 0$, $Q(x)$ is approximately given by

$$Q(x) \simeq Q_n(x) = \frac{1}{\sqrt{2\pi}x} \exp\left(-\frac{x^2}{2}\right) \left\{1 + \sum_{k=1}^{n}(-1)^k \frac{(1)(3)(5)\ldots(2k-1)}{x^{2k}}\right\} \tag{3B.6}$$

where

$$\% \text{ error} < \left|\frac{(1)(3)(5)\ldots(2n+1)}{x^{2n+2}(1-\frac{1}{x^2})}\right|, \quad x > 0 \tag{3B.7}$$

In order to derive that bound, rewriting $Q(x)$ as

$$Q(x) = \frac{1}{\sqrt{2\pi}} \int_x^\infty \frac{1}{z}z \exp\left(-\frac{z^2}{2}\right) dz \tag{3B.8}$$

Table 3B.1 $Q(x)$. *Source:* Abramowitz, M. and I. A. Stegun, *Handbook of Mathematical Functions*, New York: Dover Publications, 1970.

x	$1 - Q(x)$	x	$1 - Q(x)$
0.00	0.50000 00000 00000	1.00	0.84134 47460 68543
0.02	0.50797 83137 16902	1.02	0.84613 57696 27265
0.04	0.51595 34368 52831	1.04	0.85083 00496 69019
0.06	0.52392 21826 54107	1.06	0.85542 77003 36091
0.08	0.53188 13720 13988	1.08	0.85992 89099 11231
0.10	0.53982 78372 77029	1.10	0.86433 39390 53618
0.12	0.54775 84260 20584	1.12	0.86864 31189 57270
0.14	0.55567 00048 05907	1.14	0.87285 68494 37202
0.16	0.56355 94628 91433	1.16	0.87697 55969 48657
0.18	0.57142 37159 00901	1.18	0.88099 98925 44800
0.20	0.57925 97094 39103	1.20	0.88493 03297 78292
0.22	0.58706 44226 48215	1.22	0.88876 75625 52166
0.24	0.59483 48716 97796	1.24	0.89251 23029 25413
0.26	0.60256 81132 01761	1.26	0.89616 53188 78700
0.28	0.61026 12475 55797	1.28	0.89972 74320 45558
0.30	0.61791 14221 88953	1.30	0.90319 95154 14390
0.32	0.62551 58347 23320	1.32	0.90658 24910 06528
0.34	0.63307 17360 36028	1.34	0.90987 73275 35548
0.36	0.64057 64332 17991	1.36	0.91308 50380 52915
0.38	0.64802 72924 24163	1.38	0.91620 66775 84986
0.40	0.65542 17416 10324	1.40	0.91924 33407 66229
0.42	0.66275 72731 51751	1.42	0.92219 61594 73454
0.44	0.67003 14463 39407	1.44	0.92506 63004 65673
0.46	0.67724 18897 49653	1.46	0.92785 49630 34106
0.48	0.68438 63034 83778	1.48	0.93056 33766 66669
0.50	0.69146 24612 74013	1.50	0.93319 27987 31142
0.52	0.69846 82124 53034	1.52	0.93574 45121 81064
0.54	0.70540 14837 84302	1.54	0.93821 98232 88188
0.56	0.71226 02811 50973	1.56	0.94062 00594 05207
0.58	0.71904 26911 01436	1.58	0.94294 65667 62246
0.60	0.72574 68822 49927	1.60	0.94520 07083 00442
0.62	0.73237 11065 31017	1.62	0.94738 38615 45748
0.64	0.73891 37003 07139	1.64	0.94949 74165 25897
0.66	0.74537 30853 28664	1.66	0.95154 27737 33277
0.68	0.75174 77695 46430	1.68	0.95352 13421 36280
0.70	0.75803 63477 76927	1.70	0.95543 45372 41457
0.72	0.76423 75022 20749	1.72	0.95728 37792 08671
0.74	0.77035 00028 35210	1.74	0.95907 04910 21193
0.76	0.77637 27075 62401	1.76	0.96079 60967 12518
0.78	0.78230 45624 14267	1.78	0.96246 20196 51483
0.80	0.78814 46014 16604	1.80	0.96406 96808 87074
0.82	0.79389 19464 14187	1.82	0.96562 04975 54110
0.84	0.79954 58067 39551	1.84	0.96711 58813 40836
0.86	0.80510 54787 48192	1.86	0.96855 72370 19248
0.88	0.81057 03452 23288	1.88	0.96994 59610 38800
0.90	0.81593 98746 53241	1.90	0.97128 34401 83998
0.92	0.82121 36203 85629	1.92	0.97257 10502 96163
0.94	0.82639 12196 61376	1.94	0.97381 01550 59548
0.96	0.83147 23925 33162	1.96	0.97500 21048 51780
0.98	0.83645 69406 72308	1.98	0.97614 82356 58492
1.00	0.84134 47460 68543	2.00	0.97724 98680 51821

Table 3B.1 $Q(x)$. *Source:* Abramowitz, M. and I. A. Stegun, *Handbook of Mathematical Functions*, New York: Dover Publications, 1970.

x	$1 - Q(x)$	x	$1 - Q(x)$
2.00	0.97724 98680 51821	3.00	0.99865 01020
2.02	0.97830 83062 32353	3.05	0.99885 57932
2.04	0.97932 48371 33930	3.10	0.99903 23968
2.06	0.98030 07295 90623	3.15	0.99918 36477
2.08	0.98123 72335 65062	3.20	0.99931 28621
2.10	0.98213 55794 37184	3.25	0.99942 29750
2.12	0.98299 69773 52367	3.30	0.99951 65759
2.14	0.98382 26166 27834	3.35	0.99959 59422
2.16	0.98461 36652 16075	3.40	0.99966 30707
2.18	0.98537 12692 24011	3.45	0.99971 97067
2.20	0.98609 65524 86502	3.50	0.99976 73709
2.22	0.98679 06161 92744	3.55	0.99980 73844
2.24	0.98745 45385 64054	3.60	0.99984 08914
2.26	0.98808 93745 81453	3.65	0.99986 88798
2.28	0.98869 61557 61447	3.70	0.99989 22003
2.30	0.98927 58899 78324	3.75	0.99991 15827
2.32	0.98982 95613 31281	3.80	0.99992 76520
2.34	0.99035 81300 54642	3.85	0.99994 09411
2.36	0.99086 25324 69428	3.90	0.99995 19037
2.38	0.99134 36809 74484	3.95	0.99996 09244
2.40	0.99180 24640 75404	4.00	0.99996 83288
2.42	0.99223 97464 49447	4.05	0.99997 43912
2.44	0.99265 63690 44652	4.10	0.99997 93425
2.46	0.99305 31492 11376	4.15	0.99998 33762
2.48	0.99343 08808 64453	4.20	0.99998 66543
2.50	0.99379 03346 74224	4.25	0.99998 93115
2.52	0.99413 22582 84668	4.30	0.99999 14601
2.54	0.99445 73765 56918	4.35	0.99999 31931
2.56	0.99476 63918 36444	4.40	0.99999 45875
2.58	0.99505 99842 42230	4.45	0.99999 57065
2.60	0.99533 88119 76281	4.50	0.99999 66023
2.62	0.99560 35116 51879	4.55	0.99999 73177
2.64	0.99585 46986 38964	4.60	0.99999 78875
2.66	0.99609 29674 25147	4.65	0.99999 83403
2.68	0.99631 88919 90825	4.70	0.99999 86992
2.70	0.99653 30261 96960	4.75	0.99999 89829
2.72	0.99673 59041 84109	4.80	0.99999 92067
2.74	0.99692 80407 81350	4.85	0.99999 93827
2.76	0.99710 99319 23774	4.90	0.99999 95208
2.78	0.99728 20550 77299	4.95	0.99999 96289
2.80	0.99744 48696 69572	5.00	0.99999 97133
2.82	0.99759 88175 25811		
2.84	0.99774 43233 08458		
2.86	0.99788 17949 59596		
2.88	0.99801 16241 45106		
2.90	0.99813 41866 99616		
2.92	0.99824 98430 71324		
2.94	0.99835 89387 65843		
2.96	0.99846 18047 88262		
2.98	0.99855 87580 82660		
3.00	0.99865 01019 68370		

let's integrate by parts using $\int u\,dv = uv - \int v\,du$ where

$$u = \frac{1}{z}, \quad du = -\frac{1}{z^2}dz$$

$$dv = z\exp\left(-\frac{z^2}{2}\right)dz, \quad v = -\exp\left(-\frac{z^2}{2}\right)$$

and get

$$Q(x) = \frac{1}{\sqrt{2\pi}}\left\{\frac{1}{x}\exp\left(-\frac{x^2}{2}\right) - \int_x^\infty \frac{1}{z^2}\exp\left(-\frac{z^2}{2}\right)dz\right\} \tag{3B.9}$$

Since $\int_x^\infty \frac{1}{z^2}\exp\left(-\frac{z^2}{2}\right)dz > 0$, then for $x > 0$, we have

$$Q(x) < \frac{1}{\sqrt{2\pi}x}\exp\left(-\frac{x^2}{2}\right) \tag{3B.10}$$

Also integrating by parts the second term in (3B.9) with

$$u = \frac{1}{z^3}, \quad du = -\frac{3}{z^4}dz$$

$$dv = z\exp\left(-\frac{z^2}{2}\right)dz, \quad v = -\exp\left(-\frac{z^2}{2}\right)$$

we obtain

$$\int_x^\infty \frac{1}{z^2}\exp\left(-\frac{z^2}{2}\right)dz = \frac{1}{x^3}\exp\left(-\frac{x^2}{2}\right) - \int_x^\infty \frac{3}{z^4}\exp\left(-\frac{z^2}{2}\right)dz \tag{3B.11}$$

Thus

$$Q(x) = \frac{1}{\sqrt{2\pi}}\left\{\frac{1}{x}\exp\left(-\frac{x^2}{2}\right) - \frac{1}{x^3}\exp\left(-\frac{x^2}{2}\right) + \int_x^\infty \frac{3}{z^4}\exp\left(-\frac{z^2}{2}\right)dz\right\} \tag{3B.12}$$

Ignoring the third term which is always positive, we have for $x > 0$

$$Q(x) > \frac{1}{\sqrt{2\pi}x}\left(1 - \frac{1}{x^2}\right)\exp\left(-\frac{x^2}{2}\right) \tag{3B.13}$$

Combining (3B.10) and (3B.13), we have

$$\frac{1}{\sqrt{2\pi}x}\left(1 - \frac{1}{x^2}\right)\exp\left(-\frac{x^2}{2}\right) < Q(x) < \frac{1}{\sqrt{2\pi}x}\exp\left(-\frac{x^2}{2}\right) \tag{3B.14}$$

This is an upper and lower bound on $Q(x)$ for $x > 0$. This bound can be improved by rewriting (3B.12) as

$$Q(x) = \frac{1}{\sqrt{2\pi}x}\left\{\left(1 - \frac{1}{x^2}\right)\exp\left(-\frac{x^2}{2}\right) + x\int_x^\infty \frac{3}{z^4}\exp\left(-\frac{z^2}{2}\right)dz\right\} \tag{3B.15}$$

Integrating by parts the second term with

$$u = \frac{3}{z^5}, \quad du = -\frac{(3)(5)}{z^6} dz$$

$$dv = z \exp\left(-\frac{z^2}{2}\right) dz, \quad v = -\exp\left(-\frac{z^2}{2}\right)$$

we have

$$Q(x) = \frac{1}{\sqrt{2\pi x}} \left\{ \left(1 - \frac{1}{x^2} + \frac{3}{x^4}\right) \exp\left(-\frac{x^2}{2}\right) - x \int_x^\infty \frac{(3)(5)}{z^6} \exp\left(-\frac{z^2}{2}\right) dz \right\} \quad \text{(3B.16)}$$

Again, integrating the second term by parts with

$$u = \frac{(3)(5)}{z^7}, \quad du = -\frac{(3)(5)(7)}{z^8} dz$$

$$dv = z \exp\left(-\frac{z^2}{2}\right) dz, \quad v = -\exp\left(-\frac{z^2}{2}\right)$$

we have

$$Q(x) = \frac{\exp(-\frac{x^2}{2})}{\sqrt{2\pi x}} \left\{ 1 - \frac{1}{x^2} + \frac{3}{x^4} - \frac{(3)(5)}{x^6} \right.$$

$$\left. + x \exp\left(\frac{x^2}{2}\right) \int_x^\infty \frac{(3)(5)(7)}{z^8} \exp\left(-\frac{z^2}{2}\right) dz \right\} \quad \text{(3B.17)}$$

From above, we obtain after integrating by parts n times

$$Q(x) = \frac{1}{\sqrt{2\pi x}} \exp\left(-\frac{x^2}{2}\right) \left\{ 1 + \sum_{k=1}^n (-1)^k \frac{(1)(3)(5)\ldots(2k-1)}{x^{2k}} + R_n \right\} \quad \text{(3B.18)}$$

where R_n denotes the residue after n integrations and is given by

$$R_n = (-1)^{n+1} x \exp\left(\frac{x^2}{2}\right) \int_x^\infty \frac{(1)(3)(5)\ldots(2n+1)}{z^{2n+2}} \exp\left(-\frac{z^2}{2}\right) dz \quad \text{(3B.19)}$$

To simplify R_n, we write it as

$$R_n = (-1)^{n+1}(1)(3)\ldots(2n+1) x \int_x^\infty \frac{\exp\left\{(x^2 - z^2)/2\right\}}{z^{2n+2}} dz \quad \text{(3B.20)}$$

In the integral, let $t = (1/2)(z^2 - x^2)$. Then $dz = (2t + x^2)^{-1/2} dt$ and

$$R_n = (-1)^{n+1}(1)(3)(5)\ldots(2n+1) x \int_0^\infty \frac{1}{(2t + x^2)^{n+1}} \exp(-t)(2t + x^2)^{-1/2} dt$$

$$= (-1)^{n+1} \frac{(1)(3)(5)\ldots(2n+1)}{x^{2n+2}} \int_0^\infty \left(1 + \frac{2t}{x^2}\right)^{-n-\frac{3}{2}} \exp(-t) dt \quad \text{(3B.21)}$$

or

$$R_{n-1} = \left\{ (-1)^n \frac{(1)(3)(5)\ldots(2n-1)}{x^{2n}} \right\} \Phi(x) \quad \text{(3B.22)}$$

where

$$\Phi(x) = \int_0^\infty \left(1 + \frac{2t}{x^2}\right)^{-n-\frac{1}{2}} \exp(-t)dt \tag{3B.23}$$

Note that $\Phi(x) \leq 1$ since

$$\int_0^\infty \left(1 + \frac{2t}{x^2}\right)^{-n-\frac{1}{2}} \exp(-t)dt < \int_0^\infty \exp(-t)dt = 1$$

Let

$$Q_n(x) = \frac{1}{\sqrt{2\pi}x} \exp\left(-\frac{x^2}{2}\right)\left\{1 + \sum_{k=1}^n (-1)^k \frac{(1)(3)(5)\ldots(2k-1)}{x^{2k}}\right\} \tag{3B.24}$$

be the nth order approximation to $Q(x)$, then the percentage error is given by

$$\% \text{ error} = \left|\frac{Q(x) - Q_n(x)}{Q(x)}\right| = \left|\frac{\exp\left(-\frac{x^2}{2}\right)R_n}{\sqrt{2\pi}x Q(x)}\right| \tag{3B.25}$$

Since $\Phi(x) \leq 1$, we have

$$\% \text{ error} < \left|\frac{\exp\left(-\frac{x^2}{2}\right)}{\sqrt{2\pi}x Q(x)} \frac{(1)(3)(5)\ldots(2n+1)}{x^{2(n+1)}}\right| \tag{3B.26}$$

Using the lower bound in (3B.13), we obtain

$$\% \text{ error} < \left|\frac{(1)(3)(5)\ldots(2n+1)}{x^{2n+2}\left(1 - \frac{1}{x^2}\right)}\right|, \quad x > 0 \tag{3B.27}$$

Example 3B.1

Assume that $x = 3$. Calculate a simple bound on the percentage error when $Q(x)$ is approximated by the first n terms in the asymptotic series, for $n = 2, 3$. Repeat for $x = 5$ and $n = 2, 3$, and 4.

(a) For $x = 3$

$$n = 2, \% \text{ error} < \frac{(1)(3)(5)}{3^6\left(1 - \frac{1}{3^2}\right)} = 0.0231 = 2.31\%$$

$$n = 3, \% \text{ error} < \frac{(1)(3)(5)(7)}{3^8\left(1 - \frac{1}{3^2}\right)} = 0.018 = 1.8\%$$

(b) For $x = 5$

$$n = 2, \text{ \% error} < \frac{(1)(3)(5)}{5^6 \left(1 - \frac{1}{5^2}\right)} = 0.001 = 0.1\%$$

$$n = 3, \text{ \% error} < \frac{(1)(3)(5)(7)}{5^8 \left(1 - \frac{1}{5^2}\right)} = 0.0003 = 0.03\%$$

$$n = 4, \text{ \% error} < \frac{(1)(3)(5)(7)(9)}{5^{10} \left(1 - \frac{1}{5^2}\right)} = 0.0001 = 0.01\%$$

Appendix 3C

Gram-Schmidt Orthogonalization Procedure

The Gram-Schmidt orthogonalization procedure [15] is a useful tool in constructing a suitable coordinate system to define a decision space with the least possible coordinates or equivalently, with the smallest dimensionality. Given a set $S = \{s_0(t), s_1(t), \ldots, s_{M-1}(t)\}$ of M signals, we wish to find a corresponding set of orthogonal signals, such that all signals in the original set S can be represented as a linear combination of the orthogonal signals, such that the number of the orthogonal signals is minimized. The procedure is very simple, and it consists of choosing the first orthogonal signal $\beta_1(t)$ as the first signal, that is

$$\beta_1(t) = s_0(t) \tag{3C.1}$$

The second orthogonal signal $\beta_2(t)$ is that component of the second signal $s_1(t)$ which is linearly independent of $\beta_1(t)$, that is

$$\beta_2(t) = s_1(t) - \left(\frac{\int_0^{T_s} \beta_1(t)s_1(t)dt}{\int_0^{T_s} \beta_1^2(t)dt} \right) \beta_1(t) \tag{3C.2}$$

The third orthogonal signal $\beta_3(t)$ is again that component of $s_2(t)$ which is linearly independent of $\beta_1(t)$ and $\beta_2(t)$, that is

$$\begin{aligned} \beta_3(t) = s_2(t) &- \left(\frac{\int_0^{T_s} \beta_1(t)s_2(t)dt}{\int_0^{T_s} \beta_1^2(t)dt} \right) \beta_1(t) \\ &- \left(\frac{\int_0^{T_s} \beta_2(t)s_2(t)dt}{\int_0^{T_s} \beta_2^2(t)dt} \right) \beta_2(t) \end{aligned} \tag{3C.3}$$

In general, we have

$$\beta_k(t) = s_{k-1}(t) - \sum_{j=1}^{k-1} \left(\frac{\int_0^{T_s} \beta_j(t)s_{k-1}(t)dt}{\int_0^{T_s} \beta_j^2(t)dt} \right) \beta_j(t) \tag{3C.4}$$

for all $k = 2, \ldots, M$. To get these signals normalized, we simply define [for $\beta_i(t) \neq 0$]

$$\phi_i(t) = \frac{\beta_i(t)}{\int_0^{T_s} \beta_i^2(t)dt}, \quad i = 1, \ldots, M \tag{3C.5}$$

to obtain an orthonormal (orthogonal and normalized) set of signals. The same procedure is applicable to vectors by noting the $\int_0^{T_s} s_i(t)s_j(t)dt = \mathbf{s}_i^T \mathbf{s}_j$.

Given the orthonormal set of signals (or vectors), each signal $s_i(t)$ (or vector $\mathbf{s}_i$) can then be written as a linear combination of the orthonormal functions (or vectors) as

$$s_i(t) = \sum_{j=1}^N s_{ij}\phi_j(t) \tag{3C.6}$$

for $0 \leq t \leq T_s$ and $s_i(t) = 0$ elsewhere. Here

$$s_{ij} = \int_0^{T_s} s_i(t)\phi_j(t)dt \tag{3C.7}$$

Note that the number of orthonormal elements N is always less than or equal to the number of original signals M. If the set S consists of linearly independent signals, then $M = N$ and the number of orthonormal elements is identical to the number of signals. Once the set of orthonormal functions $\{\phi_i(t)\}$ are identified, each transmitted waveform given in (3C.6) is completely determined by the vector transpose $\mathbf{s}_i^T = (s_{i1}, s_{i2}, \ldots, s_{iN})$ for all $i = 0, 1, \ldots, M - 1$. In vector notation, each N-tuple denotes the vector

$$\mathbf{s}_i = \sum_{j=1}^{N} s_{ij} \boldsymbol{\phi}_j \qquad (3C.8)$$

where the inner product $\boldsymbol{\phi}_i^T \boldsymbol{\phi}_j = \delta_{ij}$ and where $\boldsymbol{\phi}_j$ denotes the unit vector along the jth-axis, $j = 1, 2, \ldots, N$. In this sense, we can visualize the M vectors $\{\mathbf{s}_i\}$ as defining M points in the N-dimensional Euclidean space, the *signal space*. The coefficients s_{ij} in (3C.7) are equivalently found from the vector $\mathbf{s}_i$ by evaluating the projection $s_{ij} = \mathbf{s}_i^T \boldsymbol{\phi}_j$ for all i and j.

Example 3C.1

Consider the following three vectors:

$$\mathbf{s}_0^T = (1, 1, 1), \mathbf{s}_1^T = (1, 0, 1), \mathbf{s}_2^T = (0, 1, 0)$$

Then

$$\boldsymbol{\beta}_1^T = \mathbf{s}_0^T = (1, 1, 1)$$

$$\boldsymbol{\beta}_2^T = \mathbf{s}_1^T - \left(\frac{\boldsymbol{\beta}_1^T \mathbf{s}_1}{|\boldsymbol{\beta}_1|^2}\right) \boldsymbol{\beta}_1^T = (1, 0, 1) - \frac{2}{3}(1, 1, 1) = \frac{1}{3}(1, -2, 1)$$

$$\boldsymbol{\beta}_3^T = \mathbf{s}_2^T - \left(\frac{\boldsymbol{\beta}_1^T \mathbf{s}_2}{|\boldsymbol{\beta}_1|^2}\right) \boldsymbol{\beta}_1^T - \left(\frac{\boldsymbol{\beta}_2^T \mathbf{s}_2}{|\boldsymbol{\beta}_2|^2}\right) \boldsymbol{\beta}_2^T$$

$$= (0, 1, 0) - \frac{1}{3}(1, 1, 1) + \frac{1}{3}(1, -2, 1) = (0, 0, 0)$$

Note that $\boldsymbol{\beta}_3$ can not be in the new set since by definition, the elements of a basis are a set of nonzero vectors. Check orthogonality property of the new vectors:

$$\boldsymbol{\beta}_1^T \boldsymbol{\beta}_2 = \frac{1 - 2 + 1}{3} = 0$$

Normalizing $\boldsymbol{\beta}_1$ and $\boldsymbol{\beta}_2$, we have

$$\boldsymbol{\phi}_1^T = \frac{\boldsymbol{\beta}_1^T}{|\boldsymbol{\beta}_1|} = \frac{1}{\sqrt{3}}(1, 1, 1)$$

$$\boldsymbol{\phi}_2^T = \frac{\boldsymbol{\beta}_2^T}{|\boldsymbol{\beta}_2|} = \frac{1}{\sqrt{6}}(1, -2, 1)$$

Expressing the original signals in terms of the new orthonormal vectors, we have

$$s_0 = (s_0^T \phi_1)\phi_1 + (s_0^T \phi_2)\phi_2 = \sqrt{3}\phi_1$$

$$s_1 = (s_1^T \phi_1)\phi_1 + (s_1^T \phi_2)\phi_2 = \frac{2}{\sqrt{3}}\phi_1 + \frac{2}{\sqrt{6}}\phi_2$$

$$s_2 = s_0 - s_1 = \frac{1}{\sqrt{3}}\phi_1 - \frac{2}{\sqrt{6}}\phi_2$$

In terms of the new orthonormal base (ϕ_1, ϕ_2), we can write

$$s_0^T = (\sqrt{3}, 0), \quad s_1^T = (\frac{2}{\sqrt{3}}, \frac{2}{\sqrt{6}}), \quad s_2^T = (\frac{1}{\sqrt{3}}, -\frac{2}{\sqrt{6}})$$

We have therefore defined a new observation space ϕ_1 and ϕ_2 such that the signals can be represented in two dimensions. Note that in terms of the standard base $\phi_1^T = (1, 0, 0)$, $\phi_2^T = (0, 1, 0)$, and $\phi_3^T = (0, 0, 1)$, the signals were represented using three coordinates even though they lied in two dimensions (that is, a plane). As can be seen from this example, choosing the appropriate base can reduce the dimensionality of the observation space and could greatly simplify the detection problem and reduce the receiver's operations.

Appendix 3D

Connecting the Linear Vector Space to the Physical Waveform Space

The relationship between signals and vectors will be needed; this suggests the possibility of a geometrical representation of signals and an application of analytic geometry. Consider an N-dimensional vector space with unit vectors $\boldsymbol{\phi}_1, \boldsymbol{\phi}_2, \ldots, \boldsymbol{\phi}_N$ along the coordinate axis. These vectors are called the *basis vectors* and form a complete set. The set of N basis vectors is not unique. One can choose any N independent vectors; they may or may not be mutually orthogonal. In our work, we shall assume the set of basis vectors to be mutually orthogonal and each normalized to a unit length. Such a set is called an *orthonormal set*.

Once the N basis vectors are chosen, any vector $\mathbf{r}$ in this space can be specified by N numbers $\mathbf{r}^T = (r_1, r_2, \ldots, r_N)$, which represent the magnitude of components of $\mathbf{r}$ along the N basis vectors, respectively. The vector $\mathbf{r}$ can be expressed as

$$\mathbf{r} = \sum_{n=1}^{N} r_n \boldsymbol{\phi}_n \tag{3D.1}$$

where, for normalized basis vectors, the projection of $\mathbf{r}$ onto $\boldsymbol{\phi}_n$ is the scalar product

$$r_n = \mathbf{r}^T \boldsymbol{\phi}_n \tag{3D.2}$$

for all n. The vector $\mathbf{r}$ can be represented geometrically by a point $(r_1, r_2, \ldots, r_N)$ in the coordinate system formed by the basis vectors.

We have an analogous situation for signals which evolve with the passage of time. Analogous to the base vectors, suppose we have N waveforms $\phi_1(t), \phi_2(t), \ldots, \phi_N(t)$ which form an orthonormal set of *basis signals*, N may be finite or infinite. Then , an arbitrary signal $r(t)$ in this N-dimensional Euclidean space can be expressed as

$$r(t) = \sum_{n=1}^{N} r_n \phi_n(t), \quad 0 \le t \le T \tag{3D.3}$$

where, for orthonormal basis signals, the *projections* of $r(t)$ onto the nth basis signal is

$$r_n = \int_0^T r(t) \phi_n(t) dt \tag{3D.4}$$

for all n. The set of basis signals $\{\phi_n(t)\}$ must form a complete set of signals for the space. In practice, the basis signals are usually generated using a local oscillator to generate the inphase and quadrature components $\sqrt{2/T} \cos \omega_c t$ and $\sqrt{2/T} \sin \omega_c t$.

Once the basic signals $\{\phi_n(t)\}$ are specified, one can also represent the signal $r(t)$ by an N-tuple $(r_1, r_2, \ldots, r_N)$. Alternatively, we may represent this signal geometrically by a point $(r_1, r_2, \ldots, r_N)$ in an N-dimensional space. We can now associate a vector $\mathbf{r}$ with signal $r(t)$. Note that the basis signal $\phi_1(t)$ is represented by the corresponding basis vector $\boldsymbol{\phi}_1$ and $\phi_2(t)$ is represented by $\boldsymbol{\phi}_2$, and so on. The vector $\mathbf{r}$ is called the *projection* of $r(t)$ onto the vector space. In the applications that follow, the linear vector space needed by the optimum vector receiver is created from the physically observable **waveform** $r(t) = s(t) + n(t)$ by projecting the waveform onto that space. Such projections are equivalent to filtering and sampling the received waveform, namely, the operation of *cross-correlation*.

The scalar product.   In a certain signal space, let $s(t)$ and $r(t)$ be two signals represented by their projection vectors $\mathbf{s}^T = (s_1, s_2, \ldots, s_N)$ and $\mathbf{r}^T = (r_1, r_2, \ldots, r_N)$. If $\{\phi_n(t)\}$ are the orthonormal basis signals, then

$$s(t) = \sum_{n=1}^{N} s_n \phi_n(t) \tag{3D.5a}$$

$$r(t) = \sum_{n=1}^{N} r_n \phi_n(t) \tag{3D.5b}$$

Hence,

$$\int_0^T r(t)s(t)dt = \sum_{n=1}^{N} r_n s_n \tag{3D.6}$$

since

$$\int_0^T \phi_n(t)\phi_m(t)dt = \begin{cases} 1, & n = m \\ 0, & n \neq m \end{cases} \tag{3D.7}$$

However, the right-hand side of the above equation is by definition the *scalar product* or *inner product* of vectors $\mathbf{r}$ and $\mathbf{s}$

$$\mathbf{r}^T \mathbf{s} = \int_0^T r(t)s(t)dt \tag{3D.8}$$

We conclude that *the integral of the product of two signals is equal to the scalar product of the corresponding vectors.*

Energy of a signal.   For an arbitrary signal $s(t)$, the energy E is given by

$$E = \int_0^T s^2(t)dt \text{ Joules} \tag{3D.9}$$

It follows from (3D.5a) that

$$E = \mathbf{s}^T \mathbf{s} = |\mathbf{s}|^2 \tag{3D.10}$$

where $|\mathbf{s}|^2$ is the square of the length of the vector $\mathbf{s}$ and $|\mathbf{s}|$ is called *the norm of* $\mathbf{s}$. Hence, *the signal energy is given by the square of the length of the corresponding vector*.

Example 3D.1

Consider the two-dimensional signal vectors depicted in Fig. 3D.1. For *Binary On-Off Keying* (OOK), we use the inphase component generated by a local oscillator to write

$$\mathbf{s}_0 = \mathbf{0} \rightarrow s_0(t) = 0$$

$$\mathbf{s}_1 = \sqrt{E}\boldsymbol{\phi}_1 \rightarrow s_1(t) = \sqrt{2E/T} \cos \omega_c t$$

for $0 \leq t \leq T$. For *Binary Frequency Shift Keying* (BFSK), we use the inphase and quadrature components of a local oscillator to write

$$\mathbf{s}_0 = \sqrt{E}\boldsymbol{\phi}_1 \rightarrow s_0(t) = \sqrt{2E/T} \cos \omega_c t$$

$$\mathbf{s}_1 = \sqrt{E}\boldsymbol{\phi}_2 \rightarrow s_1(t) = \sqrt{2E/T} \sin \omega_c t$$

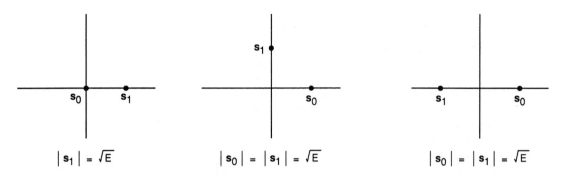

(a) On-off keying (OOK)   (b) Binary frequency shift keying (BFSK)   (c) Binary phase shift keying (BPSK)

Figure 3D.1   Signals in Example 3D.1

for $0 \le t \le T$. For *Binary Phase Shift Keying* (BPSK), we have

$$\mathbf{s}_0 = -\mathbf{s}_1 = \sqrt{E}\boldsymbol{\phi}_1 \rightarrow s_0(t) = -s_1(t) = \sqrt{2E/T}\cos\omega_c t$$

for $0 \le t \le T$.

Appendix 3E

The Pairwise Probability of Error (PPE): A Tool for System Performance Evaluation and Signal Design

One of the most powerful analytical tools used in the development of communication theory is that of linear vector spaces, needed in the representation of signals (codes). In *vector communications*, the *Pairwise Probability of Error (PPE)* depends on the Euclidean distance (a true mathematical distance measure induced by the inner product or cross-correlation) and the signal space with this distance forms a Hilbert space. Analysis and synthesis of vector communication system techniques can be reduced, in many practical cases, to consideration of the geometrical features of this Hilbert space.

A basic building block in the development of M-ary vector communications is the so-called PPE. It also serves as the basic building block in the development of the Union bound, which is used to characterize the symbol error probability (system performance) and is composed of $M - 1$ terms, each of which is a PPE. This bound is used to develop the ensemble symbol error probability arising in the Shannon theory, which demonstrates that a code exists that exhibits a zero asymptotic error probability in the presence of noise, under some conditions.

The *pairwise error event* is defined as

$$E_{jk} \triangleq \left[\mathbf{r} : |\mathbf{r} - \mathbf{s}_j|^2 \leq |\mathbf{r} - \mathbf{s}_k|^2 / m_k \right] \tag{3E.1}$$

where we have assumed that message m_k (signal $\mathbf{s}_k$) was transmitted and $\mathbf{s}_j$ is any other signal (code) vector in the transmitted set. The geometry of this event is illustrated in Fig. 3.E1. With $\mathbf{r} = \mathbf{s}_k + \mathbf{n}$, then the event can be equivalently written as

$$E_{jk} = \left[\mathbf{r} : 2\mathbf{n}^T (\mathbf{s}_k - \mathbf{s}_j) + |\mathbf{s}_k - \mathbf{s}_j|^2 \leq 0 / m_k \right]$$

$$= \left[\mathbf{n}^T (\mathbf{s}_j - \mathbf{s}_k) \geq \frac{d_{jk}^2}{2} \right] \tag{3E.2}$$

where d_{jk} denotes the distance between the vectors. Defining the unit vector $\mathbf{e}_{jk} \triangleq (\mathbf{s}_j - \mathbf{s}_k)/|\mathbf{s}_j - \mathbf{s}_k|$ and introducing the random variable $z(u) = \mathbf{n}^T \mathbf{e}_{jk}$, one can write the pairwise error event as

$$E_{jk} = \left[z(u) \geq \frac{d_{jk}}{2} \right] \tag{3E.3}$$

The probability of this event is just one minus the value of the probability distribution function of the random variable $z(u)$ evaluated at $d_{jk}/2$, that is

$$P(E_{jk}) \triangleq P_2(\mathbf{s}_j, \mathbf{s}_k)$$

$$= P \left(z(u) \geq \frac{d_{jk}}{2} / m_k \right) \tag{3E.4}$$

which says that the pairwise error event occurs if the projection $z(u)$ of the noise vector onto the *unit difference signal vector* $\mathbf{e}_{jk}$ is greater than one-half the distance between signal (code) vectors $\mathbf{s}_j$ and $\mathbf{s}_k$ as shown in Fig. 3E.1. It is easy to show that $z(u)$ is a zero-mean

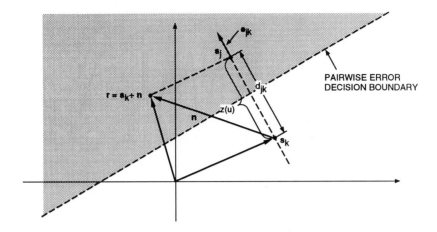

PAIRWISE ERROR EVENT: $[\, z(u) \geq d_{jk}/2 \,]$

$$\Pr\,[z(u) \geq d_{jk}/2] = \Pr\,[\, \mathbf{r} : |\mathbf{r} - \mathbf{s}_j| \leq |\mathbf{r} - \mathbf{s}_k|/m_k \,]$$

$$= \int_{\frac{d_{jk}}{2}}^{\infty} \frac{1}{\sqrt{\pi N_0}}\, \exp\left[-z^2/N_0\right] dz \;=\; Q\left[\frac{d_{jk}}{\sqrt{2N_0}}\right] \;<\; e^{-d_{jk}^2/4N_0}$$

- Characterizes exact performance of a binary communication system employing two equally likely messages

- Holds for two arbitrary vectors in any normed vector space

- Basic building block in developing:

  - Shannon's random coding theorem

  - Union bound (which characterizes performance for an arbitrary set of signal vectors)

  - The ensemble symbol error probability (R_0-criterion)

  - Performance of systems employing sequence estimation

Figure 3E.1   Pairwise probability of error (PPE)

Gaussian random variable with variance $\sigma^2 = N_0/2$. Using this fact in (3E.4), we obtain

$$P_2(\mathbf{s}_j, \mathbf{s}_k) = \frac{1}{\sqrt{\pi N_0}} \int_{d_{jk}/2}^{\infty} \exp(-z^2/N_0)\, dz \tag{3E.5}$$

which is just the Gaussian integral

$$P_2(\mathbf{s}_j, \mathbf{s}_k) = Q\!\left(\frac{d_{jk}}{\sqrt{2N_0}}\right) \tag{3E.6}$$

for all $j \neq k$. This PPE has numerous applications. For example, it characterizes (1) the exact error probability performance of all binary communication systems transmitting equal energy, equal probable signals, (2) the symbol error probability performance of an M-ary system is Union bounded by a linear combination of PPE's, that is

$$P(E/m_k) \le \sum_{\substack{k=0 \\ j \ne k}}^{M-1} P_2(\mathbf{s}_j, \mathbf{s}_k) \tag{3E.7}$$

This suggests that the performance of an M-ary vector communication system is bounded by a linear combination of the performances of $M-1$ binary systems employing signal vectors $\mathbf{s}_j$ and $\mathbf{s}_k$, that is

$$P(E) \le \sum_{k=0}^{M-1} \sum_{\substack{j=0 \\ j \ne k}}^{M-1} P(m_k) P_2(\mathbf{s}_j, \mathbf{s}_k) \tag{3E.8}$$

Finally, there is no loss in generality in allowing $\mathbf{s}_j$ and $\mathbf{s}_k$ to be any two arbitrary encoded data sequences (vectors) so that (3E.6) can be used in evaluating the performance of coding-decoding algorithms of the sequential type, for example, decoding of convolutional codes using the Viterbi algorithm. This theory building block will be applied in Chapters 4, 12, and 13. If one identifies the vectors $\mathbf{s}_m^T = (s_{m1}, \ldots, s_{mN})$ with the coded message m of N symbols, then the PPE that message m' (consisting of another coded message of N symbols) represented by $\mathbf{s}_{m'}^T = (s_{m'1}, \ldots, s_{m'N})$ is just

$$P(m \to m') = Q\left(\frac{d_{m'm}^2}{\sqrt{2N_0}}\right) \tag{3E.9}$$

where $d_{mm'}^2 \triangleq |\mathbf{s}_m - \mathbf{s}_{m'}|^2$ denotes the squared Euclidean distance between the two vectors $\mathbf{s}_m$ and $\mathbf{s}_{m'}$. This result will be used numerous times in later chapters.

PROBLEMS

3.1 An observation x is defined as follows:

$$H_1: \quad x = mb + n$$
$$H_0: \quad x = n$$

where b and n are two independent, zero-mean Gaussian random variables with variances σ_b^2 and σ_n^2, respectively and m is a constant. Assuming equiprobable hypotheses,
(a) Find and sketch the optimum (MAP) receiver.
(b) Calculate the minimum probability of error for the receiver found in (a).
(c) Repeat (a) and (b) when the random variable b has mean value m_b.
(d) Using (c), find the error probability when $\sigma_b \to 0$ and discuss your result.

3.2 An observation x is characterized under two hypotheses according to the following pdf's:

$$H_0: \quad f_x(x/H_0) = \begin{cases} \sqrt{\frac{2}{\pi}} \exp\{-\frac{x^2}{2}\}, & x \ge 0 \\ 0, & x < 0 \end{cases}$$

$$H_1: \quad f_x(x/H_1) = \begin{cases} \exp\{-x\}, & x \ge 0 \\ 0, & x < 0 \end{cases}$$

The probabilities of the two hypotheses H_0 and H_1 are 5/8 and 3/8, respectively.

(a) Find the minimum error probability test to choose between H_0 and H_1.

(b) Find the minimum error probability that results from the test in (a).

3.3 Consider the following scalar communication problem:

$$H_i: \quad f_x(x/H_i) = \begin{cases} \frac{x^{m_i}}{m_i!} \exp\{-x\}, & x \geq 0 \\ 0, & x < 0 \end{cases} \quad i = 0, 1$$

Assuming equiprobable hypotheses,

(a) Find a likelihood ratio test that minimizes the probability of error.

(b) For $m_0 = 0$ and $m_1 = 1$, it is convention to refer to the probability of choosing H_1 when indeed H_0 is true as the *false alarm probability*, P_{FA}. Similarly, the probability of choosing H_1, when indeed H_1 is true is referred to as the *probability of detection*, P_D. Evaluate P_{FA} and P_D numerically.

3.4 Consider the following scalar communication problem:

$$H_i: \quad x = s_i + n \quad i = 0, 1$$

where $s_0 = 1$, $s_1 = -1$ and

$$f_n(n) = \frac{1}{2}\left[\frac{1}{\sqrt{2\pi\frac{a^2}{16}}}\exp\left\{-\frac{(n-a)^2}{2\frac{a^2}{16}}\right\} + \frac{1}{\sqrt{2\pi\frac{a^2}{16}}}\exp\left\{-\frac{(n+a)^2}{2\frac{a^2}{16}}\right\}\right]$$

and the hypotheses are equiprobable.

(a) Assuming that $a < 1$, describe the optimum receiver that will yield the minimum probability of error. Repeat for $a > 1$.

(b) Find an appropriate expression for the probability of error in each case.

3.5 Consider the scalar communication problem where one of two voltage levels $s_0 = 0$ and $s_1 = 1$ volt is transmitted over an additive noise channel where the noise pdf is characterized by

$$f_n(n) = \frac{1}{\pi(1+n^2)}; \quad -\infty \leq n \leq \infty$$

(a) Determine the likelihood ratio test that specifies the maximum-likelihood receiver.

(b) Evaluate the false alarm probability $P_{FA} = \Pr\{H_1 \text{ is chosen } / H_0 \text{ is true}\}$ and the probability of detection $P_D = \Pr\{H_1 \text{ is chosen } / H_1 \text{ is true}\}$ if the threshold voltage is set to 0.5 volts. Also calculate the probability of error.

3.6 (a) Using just a single observation, what is the optimum receiver to choose between the hypotheses

$$H_0: \quad \text{the sample is zero-mean Gaussian, with variance } \sigma_0^2$$

$$H_1: \quad \text{the sample is zero-mean Gaussian, with variance } \sigma_1^2 \ (\sigma_1^2 > \sigma_0^2)$$

if the a priori probability of H_1 is twice that of H_0.

(b) In terms of the observation, what are the decision regions I_0 and I_1?

(c) What is the probability of choosing H_1 when H_0 is true?

(d) Determine the probability of error if $\sigma_1^2/\sigma_0^2 = e^5$ and $\sigma_1 = 1$.

3.7 Based on N independent samples, design a test to choose between

$$H_0: \quad r = n$$
$$H_1: \quad r = 1 + n$$
$$H_2: \quad r = -1 + n$$

where n is a zero-mean Gaussian random variable with variance σ^2. Assume equal a priori probabilities for each hypothesis. Show that the test statistic may be chosen to be

$$\bar{r} = \frac{1}{N} \sum_{i=1}^{N} r_i$$

Find the decision region for $\bar{r}$.

3.8 Given the Poisson distributed random variable K with parameter x, that is

$$\text{Prob}\,\{K = k\} = \frac{x^k e^{-x}}{k!}; \quad k = 0, 1, \cdots$$

Assume that x takes on one of two known values, say

$$H_0: \quad x = n$$

$$H_1: \quad x = a + n$$

with equal probability.

(a) Develop the decision rule which minimizes the probability of error in announcing the value of x given the observable k.

(b) Assume now that $x = 4$ if H_1 is true and $x = 2$ if H_0 is true, simplify the decision rule found in (a).

3.9 A binary hypothesis test is made on the basis of the observed statistics:

$$H_0 \to f_r(\rho/H_0) = \frac{\rho \exp\left(-\rho^2/2\sigma^2\right)}{\sigma^2}, \quad \rho \geq 0$$

$$H_1 \to f_r(\rho/H_1) = \frac{\rho}{\sigma^2} \exp\left(-\frac{\rho^2 + \alpha^2}{2\sigma^2}\right) I_0\left(\frac{\alpha\rho}{\sigma^2}\right), \quad \rho \geq 0$$

where α and σ are known and $I_0(x)$ is the zero-order modified Bessel function of the first kind.

(a) Specify the likelihood ratio test if

$$\text{Prob}\{H_0\} = \text{Prob}\{H_1\} = \frac{1}{2}$$

(b) Write an expression for the probability of false alarm, P_{FA}.

(c) Write an expression for the probability of detection, P_D.

(d) Write an expression for the probability of missed detection, P_M.

(e) Write an expression for the probability of error, P_E.

(f) Table P3.9 lists approximate values for the Marcum Q-function for integer arguments. That is, it lists $Q(\alpha, \beta)$ where α is as shown in the left margin and β is given in the top row. More complete tables can be found in papers by Marcum. What is the threshold setting which gives $P_{FA} = 0.01$ (assume $\sigma = 1$ volt)?

(g) For this threshold setting, determine the sensitivity of the false alarm probability to increases and decreases of σ by $\pm 25\%$ about one volt. (Plot results.)

(h) Repeat (g) for P_D as a function of α/σ.

Note: Marcum Q-function[1]

$$Q(\alpha, \beta) \triangleq \int_{\beta}^{\infty} x \exp\left(-\frac{x^2 + \alpha^2}{2}\right) I_0(\alpha x)\, dx$$

3.10 Consider the following binary decision problem representing the detection of a known signal s

---

1. See M. S. Roden, "Digital and Data Communication Systems," Englewood Cliffs, N.J.: Prentice Hall, 1982.

Table P3.9

α	1	2	3	4	5	β 6	7	8	9	10
1	0.875	0.268	0.044	0.003	0	0	0	0	0	0
2	0.915	0.603	0.213	0.034	0.002	0	0	0	0	0
3	0.991	0.885	0.595	0.195	0.031	0.002	0	0	0	0
4	0.999	0.986	0.875	0.575	0.186	0.029	0.002	0	0	0
5	0.999	0.999	0.984	0.865	0.564	0.180	0.027	0.002	0	0
6	0.999	0.999	0.999	0.984	0.860	0.543	0.176	0.027	0.002	0
7	0.999	0.999	0.999	0.999	0.981	0.855	0.523	0.173	0.026	0.002
8	0.999	0.999	0.999	0.999	0.999	0.980	0.850	0.511	0.170	0.026
9	0.999	0.999	0.999	0.999	0.999	0.999	0.977	0.845	0.497	0.167
10	0.999	0.999	0.999	0.999	0.999	0.999	0.999	0.975	0.840	0.482

in additive white noise n:

$$\text{Message } m_0: \quad r = n$$

$$\text{Message } m_1: \quad r = s + n$$

The pdf which characterizes the additive random noise voltage is the generalized pdf

$$f_n(\alpha) = [c\eta(\sigma, c)/2\Gamma(1/c)] \exp\left\{-\eta(\sigma, c)|\alpha|^c\right\}$$

where

$$\eta(\sigma, c) \triangleq \sigma^{-1}[\Gamma(3/c)/\Gamma(1/c)]^{1/2}$$

for $|\alpha| \leq \infty$. Here c is a known positive parameter controlling the rate of decay, $\Gamma(.)$ is the Gamma function, and σ^2 is the noise variance. Note for $c = 2$, this density reduces to the Gaussian density whereas for $c = 1$, it becomes the Laplace density.

(a) Find the optimum decision rule which minimizes the probability of error. Assume that $\text{Prob}\{m_0\} = \text{Prob}\{m_1\} = 1/2$.

(b) Sketch the optimum detector suggested by the rule found in (a).

(c) Find an expression for the error probability.

(d) Repeat (b) for $c = 1$ and $c = 2$.

3.11 Consider the following binary hypothesis testing problem:

$$H_0: \quad r = n$$

$$H_1: \quad r = s + n$$

where s and n are independent random variables with

$$f_s(S) = a \exp\{-aS\}, \quad S \geq 0$$

$$f_n(N) = b \exp\{-bN\}, \quad N \geq 0$$

Find the decision region which minimizes the error probability if H_0 and H_1 are equally probable.

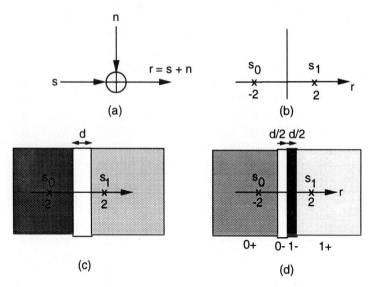

Figure P3.15

3.12 The two equally likely hypotheses are

$$H_1: \quad f_r(R) = \frac{1}{2}\exp(-|R|)$$

$$H_0: \quad f_r(R) = \frac{1}{\sqrt{2\pi}}\exp(-\frac{1}{2}R^2)$$

(a) Find the optimum decision rule.
(b) Compute the decision regions.

3.13 Let the conditional densities for a binary detection problem be given by the Cauchy densities

$$f_r(r/m_i) = \frac{1}{\pi b}\frac{1}{1 + (\frac{r-a_i}{b})^2} \quad i = 0, 1$$

If the two messages m_0 and m_1 are equally likely and if a_0 and a_1 are known constants with $a_1 > a_0$,
(a) Find the a posteriori probabilities for the messages, $P(m_0/r)$ and $P(m_1/r)$;
(b) State the decision rule that will result in minimum probability of error;
(c) Compute the resulting probability of error.

3.14 The likelihood ratio is defined as

$$\Lambda_r(\rho) = \frac{f_r(\rho/H_1)}{f_r(\rho/H_0)}$$

If we replace the observation ρ in $\Lambda_r(\rho)$ by the random variable r, then $\Lambda_r(r)$ is itself a random variable with special properties. Prove:
(a) $E\{\Lambda_r^n(r) / H_1\} = E\{\Lambda_r^{n+1}(r)/H_0\}$
(b) $E\{\Lambda_r(r)/H_0\} = 1$
(c) $E\{\Lambda_r(r)/H_1\} - E\{\Lambda_r(r)/H_0\} = \text{Var}\{\Lambda_r(r)/H_0\}$

3.15 The random variable n in Fig. P3.15a is Gaussian, with zero-mean. If one of two equally likely messages is transmitted, using the signals of Fig. P3.15b, an optimum receiver yields $P(E) = 0.001$.
(a) Compute the variance σ^2 of n.

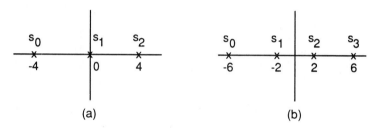

Figure P3.16

(b) Erasures are sometimes employed in receivers to allow them to avoid a decision when it seems that a decision might contribute an error. Consider the decision rule of Fig. P3.15c:

if $r > \frac{d}{2}$, decide s_1
if $r < -\frac{d}{2}$, decide s_0
if $|r| < \frac{d}{2}$, decide an erasure

Compute d such that the erasure probability is 10^{-4}. Compute all necessary probabilities $P(r_i/m_j)$ to reduce the continuous channel to a discrete memoryless channel (DMC) with 2 inputs ($M = 2$) and 3 outputs ($J = 3$).

(c) In some applications, it is better to quantify the quality of the received symbol. This is typically accomplished by measuring the confidence level of the observed voltage. Consider the decision rule depicted in Fig. P3.15d. In this case, the receiver distinguishes between a soft "1" (denoted by 1–) and a strong "1" (denoted by 1+). The new decision rule is as follows:

if $r < -\frac{d}{2}$, decide a strong "0" (0+)
if $-\frac{d}{2} < r < 0$, decide a soft "0" (0–)
if $0 < r < \frac{d}{2}$, decide a soft "1" (1–)
if $r > \frac{d}{2}$, decide a strong "1" (1+)

Compute d such that the probability of any soft symbol is 10^{-4}. Compute all necessary probabilities $P(r_i/m_j)$ to reduce the continuous channel to a discrete memoryless channel (DMC) with 2 inputs ($M = 2$) and 4 outputs ($J = 4$).

3.16 The random variable n in Fig. P3.15a is Gaussian, with zero-mean. If one of two equally likely messages is transmitted, using the signals of Fig. P3.15b, an optimum receiver yields $P(E) = 0.001$.

(a) What is the minimum attainable probability of error when the channel of Fig. P3.15a is used with the three equilikely signals of Fig. P3.16a?

(b) How about with the four equilikely signals of Fig. P3.16b?

(c) How do the answers of (b) change if it is known that the noise has mean m.

3.17 A communication system employs two channels to transmit a voltage s_i, i=0,1, as shown in Fig. P3.17. The voltages s_0 and s_1 correspond, respectively, to the messages m_0 and m_1. Two received voltages r_1 and r_2 are available to the decision device, that is, the receiver makes its decisions based on a joint observation of r_1 and r_2. Assume that the additive noises n_1 and n_2 are zero-mean Gaussian random variables with variances σ_1^2 and σ_2^2, respectively. Also, assume that s_i, n_1, and n_2 are statistically independent. Letting $s_0 = \sqrt{E}$ with probability $P(m_0)$ and $s_1 = -\sqrt{E}$ with probability $P(m_1)$,

(a) Determine the structure of the optimum receiver that results in the minimum probability of error.

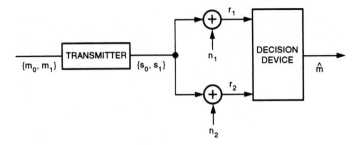

Figure P3.17

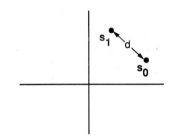

Figure P3.19

(b) Calculate the average probability of error for the case $P(m_0) = P(m_1)$ and $\sigma_1^2 = \sigma_2^2$.

3.18 (a) Sketch the optimum vector receiver for each of the signal sets illustrated in Fig. 3D.1.

(b) Sketch the decision regions and boundaries.

(c) Compute the minimum probability of error for each case.

3.19 (a) Given two equilikely messages, m_0 and m_1, to be transmitted (using the signals illustrated in Fig. P3.19) over a channel that adds white Gaussian noise, find an expression for the minimum error probability.

(b) What can one conclude regarding the energy of the two signals from the result found in (a)?

3.20 Consider the following N-dimensional decision problem: the received vector $\mathbf{r} = \mathbf{s}_i + \mathbf{n}$ where

$$\mathbf{s}_i = (i, i, \cdots, i)^T \quad i = 1, 2, 3, 4$$

$$\mathbf{r} = (r_1, r_2, \cdots, r_N)^T$$

Assume that the noise vector is Gaussian with mean $\mathbf{0}$ and covariance matrix $\sigma^2 \mathcal{I}$, where $\mathcal{I}$ is the $N \times N$ identity matrix.

(a) Design a decision rule to choose between the four equally likely hypotheses.

(b) For the case $N = 2$, assume the received vector is $\mathbf{r} = (4, 0)^T$. Which hypothesis would the receiver choose?

3.21 The two equally likely hypotheses are

$$H_0: \quad f(x_1, x_2/H_0) = \frac{1}{2\pi\sigma_0^2} \exp\left(-\frac{x_1^2}{2\sigma_0^2} - \frac{x_2^2}{2\sigma_0^2}\right), \quad -\infty < x_1, x_2 < \infty$$

$$H_1: \quad f(x_1, x_2/H_1) = \frac{1}{4\pi\sigma_0\sigma_1}\left[\exp\left(-\frac{x_1^2}{2\sigma_1^2} - \frac{x_2^2}{2\sigma_0^2}\right) + \exp\left(-\frac{x_1^2}{2\sigma_0^2} - \frac{x_2^2}{2\sigma_1^2}\right)\right],$$

$$-\infty < x_1, x_2 < \infty$$

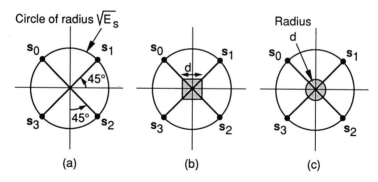

Figure P3.22

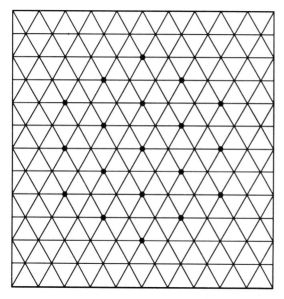

Figure P3.23

Find the optimum decision rule.

3.22 One of four equilikely messages is to be communicated over a vector channel which adds to each transmitted vector component a statistically independent zero-mean Gaussian random variable with variance $N_0/2$.

   (a) Assuming that the transmitter uses the signal vectors shown in Fig. P3.22a, express the probability of error in terms of the function $Q(x)$ defined in (3.40).

   (b) Assume that the optimum receiver is modified such that any received vector in the square centered at the origin and of diameter d is declared an erasure as shown in Fig. P3.22b. Compute the probability of erasures as a function of d and E_s/N_0. Compute all necessary probabilities $P(r_i/m_j)$ to reduce the continuous channel to a discrete memoryless channel (DMC) with 4 inputs ($M = 4$) and 5 outputs ($J = 5$).

   (c) Repeat (b) if the square is replaced by a circle of radius d as shown in Fig. P3.22c.

3.23 For the equilikely signals illustrated in Fig. P3.23, sketch the optimum decision boundaries in an AWGN channel.

3.24 When the two signal vectors s_0 and s_1 shown in Fig. P3.24a are transmitted over a channel

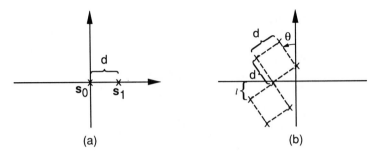

Figure P3.24

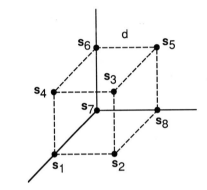

Figure P3.25

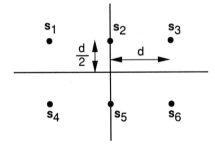

Figure P3.26

disturbed by additive white Gaussian noise, an optimum receiver will achieve $P(E) = q$ for equilikely messages. Assuming that the seven vectors indicated by ×'s in Fig. P3.24b are used as signals over the same channel, compute the probability of error achieved by an optimum receiver in terms of q, θ, and l, assuming equilikely messages.

3.25 Consider the PCM (pulse code modulation) signaling system where each PCM word has a length of 3 bits, that is, there are $2^3 = 8$ possible symbols. The signal vectors are placed on the vertices of a "3-dimensional hypercube" as shown in Fig. P3.25. Assume all signals are equilikely and that the noise is an AWGN with PSD of $N_0/2$ watts/Hz. Find an expression for the probability of symbol error.

3.26 Consider the 6-QAM constellation set depicted in Fig. P3.26. Assume all signal vectors are equilikely to be transmitted and that the channel is disturbed by AWGN with PSD $N_0/2$.

(a) Find the boundaries for the decision regions.

(b) Find the probability of symbol error.

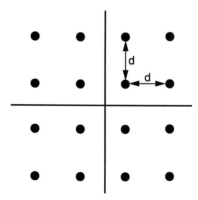

Figure P3.27

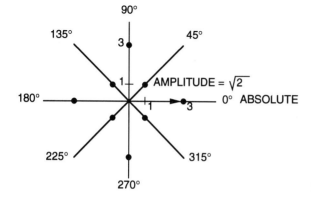

Figure P3.28

3.27 Consider the 4×4 QAM signal constellation depicted in Fig. P3.27. Assume an AWGN channel and equilikely signals.
  (a) Draw the optimum decision boundaries and illustrate the decision regions.
  (b) Using an appropriate basis, write an expression for each of the 16 vector signals.
  (c) Compute the average energy for this signal constellation.
  (d) Draw the optimum detector.
  (e) Write an expression for the probability of symbol error.

3.28 Consider the V.29 signal constellation depicted in Fig. P3.28, which transmits 3 bits/symbol. Assume an AWGN channel and equilikely signals.
  (a) Draw the optimum decision boundaries and illustrate the decision regions.
  (b) Using an appropriate basis, write an expression for each of the eight vector signals.
  (c) Compute the average energy for this signal constellation.
  (d) Draw the optimum detector.
  (e) Write an expression for the probability of symbol error.

3.29 Consider the signal constellation shown in Fig. P3.29. This signaling scheme is referred to as 8-QAM (quadrature amplitude modulation). Note that the energy of each signal is not equal.
  (a) Find the average energy of the signaling set. Assume all signals are equilikely.
  (b) Find an upper bound for the probability of symbol error.

3.30 Consider a 3-dimensional "biorthogonal" signaling scheme as shown in Fig. P3.30. Assume all these signals are equilikely and that they are over an AWGN channel with PSD $N_0/2$. Using the union bound, find an upper bound for the probability of error as a function of E_s/N_0 where E_s denotes signal energy.

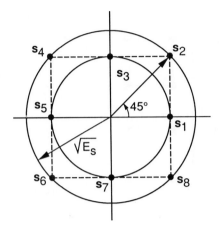

Figure P3.29

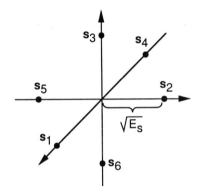

Figure P3.30

3.31 Consider the V.29 signal constellation depicted in Fig. P3.31, which transmits 4 bits/symbol. Assume an AWGN channel and equilikely signals.
   (a) Draw the optimum decision boundaries and illustrate the decision regions.
   (b) Using an appropriate basis, write an expression for each of the 16 vector signals.
   (c) Compute the average energy for this signal constellation.
   (d) Draw the optimum detector.
   (e) Write an expression for the probability of symbol error.

3.32 Consider a 32-QAM signal set as shown in Fig. P3.32. Assume all signals are equilikely and are transmitted over an AWGN channel with PSD $N_0/2$.
   (a) Calculate the "average" energy of this set.
   (b) Find an upper bound for the probability of symbol error.

3.33 Consider the problem in which you must decide which of two hypotheses is true, based on the observation of x:

$$H_0: \quad x = s_0 + n_0$$

$$H_1: \quad x = s_1 + n_1$$

where s_0 and s_1 are known constants and n_0, n_1 are zero-mean independent Gaussian random variables with respective variances σ_0^2 and σ_1^2.

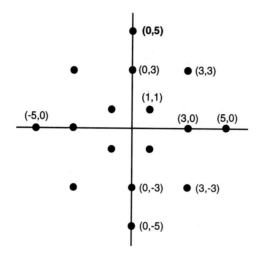

Figure P3.31

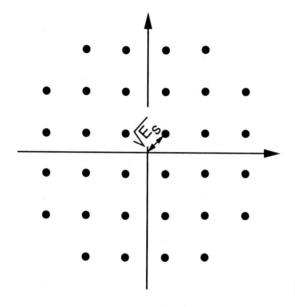

Figure P3.32

(a) Derive the decision rule that minimizes the probability of error (assuming equilikely probabilities) and write expressions to characterize $P(\text{error}/H_0)$ and $P(\text{error}/H_1)$.

(b) Suppose that the problem is reformulated as follows: decide which of two hypotheses is true based on the observation of the vector $\mathbf{x}$ where

$$H_0: \quad \mathbf{x} = \mathbf{s}_0 + \mathbf{n}$$

$$H_1: \quad \mathbf{x} = \mathbf{s}_1 + \mathbf{n}$$

where $\mathbf{s}_0 = (s_0, 0)^T$, $\mathbf{s}_1 = (0, s_1)^T$ and $\mathbf{n}$ is a zero-mean Gaussian random vector with covariance matrix

$$K_n = \begin{pmatrix} \sigma_0^2 & 0 \\ 0 & \sigma_1^2 \end{pmatrix}$$

What is the receiver that minimizes the probability of error?

(c) Write expressions to characterize $P(\text{error}/H_0)$ and $P(\text{error}/H_1)$ for the receiver in (b).

(d) Are the two receivers equivalent in performance? Explain your answer.

3.34 Given a random vector $\mathbf{x}$ with zero-mean and covariance matrix

$$K_{\mathbf{x}} = \begin{pmatrix} 1 & -1/2 & -1/2 \\ -1/2 & 1 & -1/2 \\ -1/2 & -1/2 & 1 \end{pmatrix}$$

(a) Let the random variable y be formed by the inner product $y = \mathbf{h}^T \mathbf{x}$, where $\mathbf{h}$ is constrained to have unity magnitude. Find the $\mathbf{h}$ vector which maximizes the mean-squared value of y.

(b) Determine the mean and variance of the random variable y with $\mathbf{h}$ chosen as in (a).

(c) Let $\mathbf{w} = \mathcal{G}\mathbf{x}$. Find the matrix $\mathcal{G}$ such that the resulting vector $\mathbf{w}$ has a covariance matrix equal to the identity matrix (that is, perform a whitening transformation on $\mathbf{x}$).

(d) Express $\mathbf{x}$ as

$$\mathbf{x} = \sum_{i=1}^{n} \sqrt{\lambda_i} \omega_i \mathbf{e}_i$$

where the ω_i's are zero-mean, uncorrelated random variables with unit variance (that is, find the λ_i's and the $\mathbf{e}_i$'s).

3.35 Let $\mathcal{R}_{\mathbf{x}}$ be a correlation matrix with zero determinant. Prove that the matrix $\mathcal{E}_m \mathcal{E}_m^T$ has eigenvalues zero or one where $\mathcal{E}_m$ is the matrix whose columns are the eigenvectors corresponding to nonzero eigenvalues of $\mathcal{R}_{\mathbf{x}}$.

3.36 Based on the observation of the random vector $\mathbf{x}$, determine the detector that will minimize the probability of error between the two equilikely hypotheses:

$$H_0: \quad \mathbf{x} = (1, 1, 0, 0)^T + \mathbf{n}$$

$$H_1: \quad \mathbf{x} = (0, 0, 1, -1)^T + \mathbf{n}$$

where the covariance matrix of the Gaussian zero-mean noise vector is given by

$$K_{\mathbf{n}} = \begin{pmatrix} 1 & 1 & 0 & 0 \\ 1 & 1 & 0 & 0 \\ 0 & 0 & 1 & -1 \\ 0 & 0 & -1 & 1 \end{pmatrix}$$

Hint: Is the covariance matrix singular? Is perfect decisions possible? How about preprocessing the observation $\mathbf{x}$ with a matrix $\mathcal{H}$ such that $\mathcal{H}\mathbf{x}$ has a nonsingular covariance matrix?

3.37 You are asked to design an optimum decision rule to decide between the following two hypotheses:

$$H_0: \quad x(n) = s_0(n) + \mu(n)$$

$$H_1: \quad x(n) = s_1(n) + \mu(n)$$

where $s_0(n) = \cos(n\pi)$, $s_1(n) = \cos((n+1)\pi))$ and $\mu(n)$ is a Gaussian zero-mean wide sense stationary (WSS) sequence.

(a) Based on observing $x(10)$ and $x(11)$, design the optimum decision rule assuming $\mu(n)$ is white with PSD $S_\mu(f) = \sigma^2$.

(b) Bound the probability of error of the receiver in (a).

(c) Derive an exact expression for the probability of error for the receiver in (a).

(**d**) Assuming that the noise $\mu(n)$ is no longer white, but has PSD

$$S_\mu(f) = \frac{\sigma^2(1 - \rho^2)}{1 - 2\rho\cos(2\pi f) + \rho^2} \quad \text{with } 0 \le |\rho| < 1$$

derive the optimum decision rule. Check to see that when $\rho = 0$, the decision rule reduces to that in (a).

(**e**) Bound the probability of error of the receiver in (d).

(**f**) Derive an exact expression of the receiver in (d). Check to see that the expression reduces to that in (c) when $\rho = 0$.

3.38 A decisioning system must decide between the hypothesis H_0 and H_1 where

$$H_0: \quad f_r(x) = \frac{1}{\sqrt{2\pi}} \exp\left(-\frac{x^2}{2}\right)$$

$$H_1: \quad f_r(x) = \begin{cases} \frac{1}{3}, & 0 \le x \le 3 \\ 0, & \text{otherwise} \end{cases}$$

The a priori probability of H_0 is 3/4 and H_1 is 1/4. The cost of decisioning is

$$C_{00} = 0 \quad C_{10} = 1$$

$$C_{01} = 1 \quad C_{11} = 0$$

Determine and describe the optimal Bayes decision procedure for this problem.

3.39 Consider the hypotheses

$$H_0: \quad r = n$$

$$H_1: \quad r = s + n$$

where s and n are independent random variables with pdf's

$$f_s(S) = \exp(-S), \quad S \ge 0$$

$$f_n(N) = 10\exp(-10N), \quad N \ge 0$$

(**a**) Compute the conditional pdf's $f_r(\rho/H_0)$ and $f_r(\rho/H_1)$.

(**b**) Find the likelihood ratio test, $\Lambda_r(\rho)$.

(**c**) If H_0 has a priori probability 1/4 and H_1 a priori probability 3/4, $C_{01} = C_{10} = 3$ and $C_{00} = C_{11} = 0$, find the threshold for a Bayes test.

(**d**) Show that the likelihood ratio test of (c) can be expressed as

$$\rho \underset{H_1}{\overset{H_0}{\gtrless}} \gamma$$

Find the numerical value for γ for the Bayes test of (c).

(**e**) Compute the risk for the Bayes test in (c).

(**f**) Compute the threshold for a Neyman-Pearson test with P_{FA} less than or equal to 10^{-4}. Find P_D for this threshold.

(**g**) Reducing the Neyman-Pearson test of (f) to

$$\rho \underset{H_1}{\overset{H_0}{\gtrless}} \gamma$$

find P_{FA} and P_D for an arbitrary γ.

3.40 A decisioning system must decide between two hypotheses with pdf's

$$f_r(\rho/H_0) = 2\exp(-2\rho), \quad \rho \geq 0$$

$$f_r(\rho/H_1) = 1/2, \quad -1 \leq \rho \leq 1$$

Assuming that the a priori probabilities are unknown, formulate a decision rule using the min-max criterion.

3.41 A decisioning system must decide between two hypotheses with pdf's

$$f_r(\rho/H_0) = \begin{cases} 1 - |\rho|, & |\rho| < 1 \\ 0, & |\rho| > 1 \end{cases}$$

$$f_r(\rho/H_1) = \begin{cases} (2 - |\rho|)/4, & |\rho| < 2 \\ 0, & |\rho| > 2 \end{cases}$$

Choosing H_0 when H_1 is true costs three times as much as choosing H_1 when H_0 is true. Correct choices cost nothing. Formulate a decision rule using the min-max criterion.

3.42 Show that the vector

$$\mathbf{a} = \sum_{i=0}^{M-1} P(m_i)\mathbf{s}_i$$

minimizes the required mean energy

$$\overline{E'} = \sum_{i=0}^{M-1} P(m_i)|\mathbf{s}_i - \mathbf{a}|^2$$

without altering the message error probability.

REFERENCES

1. Neyman, J. and E. Pearson, "The Testing of Statistical Hypotheses in Relation to Probability A Priori," *Proc. Comb. Phil Soc.*, vol. 29, (4), pp. 492–510 (1933).

2. Neyman, J. and E. Pearson, "On The Problem of the Most Efficient Tests of Statistical Hypothesis," *Phil. Trans. Roy Soc. (London)*, Ser. A, 231(9), pp. 289–337 (1933).

3. Wald, A., *Statistical Decision Functions*, New York: John Wiley and Sons, 1950.

4. Bayes, T., "An Essay Toward Solving a Problem in the Doctrine of Chances," *Phil. Trans. Roy. Soc. (London)* , pp. 370–418 (1763). Reprinted in *Biometrika* 45, pp. 293–315 (1958).

5. Papoulis, A., *Probability, Random Variables and Stochastic Processes*, New York: McGraw-Hill, 1965.

6. Larson, H. J. and B. O. Shubert, *Probabilistic Models in Engineering Sciences, Vol. I: Random Variables and Stochastic Processes*, New York: John Wiley and Sons, 1979.

7. Weber, C. L., *Elements of Detection and Signal Design*, New York: Springer-Verlag, 1987.

8. Strang, G., *Linear Algebra and Its Applications*, 3rd ed., San Diego: Harcourt Brace Jovanovich, 1988.

9. Friedberg, S. H., A. J. Insel and L. E. Spence, *Linear Algebra*, Englewood Cliffs, N.J.: Prentice-Hall, 1979.

10. Wozencraft, J. M. and I. M. Jacobs, *Principles of Communication Engineering*, New York: John Wiley and Sons, 1965.

11. Blackwell, D. and M. A. Girshick, *Theory of Games and Statistical Decisions*, New York: Dover Publications, 1954.

12. Van Trees, H. L., *Detection, Estimation and Modulation Theory, Part I*, New York: John Wiley and Sons, 1968.

13. Helstrom, C. W., *Statistical Theory of Signal Detection*, 2nd ed., Oxford: Pergamon Press, 1968.

14. Melsa, J. L. and D. L. Cohn, *Decision and Estimation Theory*, New York: McGraw-Hill, 1978.

15. Friedberg, S. H., A. J. Insel and L. E. Spence, *Linear Algebra*, Englewood Cliffs, N.J.: Prentice-Hall, 1979.

16. Craig, J. W., "A New, Simple and Exact Result for Calculating the Probability of Error for Two-Dimensional Signal Constellations, " *IEEE MILCOM'91 Conference Record*, Boston, MA, pp. 25.5.1–25.5.5.

Chapter 4

Coherent Communications with Waveforms

In the previous chapter, we investigated the problem of scalar and vector communications and derived decision rules that minimize the probability of message error. Other performance criteria were also considered to illustrate alternative parameter optimization. In the presence of white Gaussian noise, the decision rules resulted in optimum receiver structures that correlate the received vector with all possible transmitted vectors (with the addition of the appropriate bias), and then decide on the message with the maximum correlation. In this chapter, we consider the problem of waveform communications in which each message is transmitted using a distinct voltage function. We start with a baseband signal model which assumes perfect carrier phase coherency. We will first derive the optimum receiver structures in the presence of white Gaussian noise and then evaluate the resulting performance for various sets of transmitted waveforms. The approach is then generalized to handle nonwhite Gaussian noise, using various techniques, namely, the sampling approach, the Karhunen-Loève (K-L) expansion, a direct derivation using a sufficient statistic, and a "whitening" operation on the observed waveform. Finally, inphase (I) and quadrature (Q) modulation and demodulation are introduced to establish the link between the derived receiver structures and communication signals. We will then relax our assumptions regarding the carrier phase coherency and the effects of imperfect carrier and symbol synchronization are investigated for the simple model of a slow time-varying phase error process. Finally, the deleterious effects of data asymmetry and suboptimum detection on system performance are treated.

Our approach to waveform receiver design [1] is to first approximate the waveforms using finite samples to provide a quick but nevertheless correct derivation of the receiver structures. The more rigorous approach, which does not require the finite sample approximation, uses the K-L expansion. It will be shown that the results obtained are the same; however, it is felt that time sampling is more appealing from a practical viewpoint. In the current case, the transmitter maps each message m into a distinct voltage waveform $s(t)$ for transmission as shown in Fig. 4.1. It is assumed that each transmitted message lasts for T seconds and that the system is perfectly synchronized in the sense that the receiver has full knowledge of the start and end epoch of each signal. The receiver observes a sample function $\rho(t)$, $0 \leq t \leq T$, of the random process $r(t)$, and maps the observation into one of the messages in the transmitted set. Formulating the detection problem as a hypothesis testing problem, then in mathematical terms

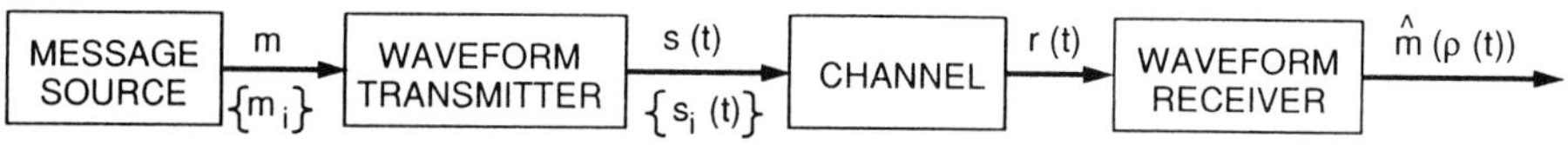

Figure 4.1   Waveform communications model

$$m_0 \to H_0 \; : \; r(t) = s_0(t) + n(t)$$
$$m_1 \to H_1 \; : \; r(t) = s_1(t) + n(t)$$
$$\vdots$$
$$m_{M-1} \to H_{M-1} \; : \; r(t) = s_{M-1}(t) + n(t)$$

$$(4.1)$$

for the additive noise channel where $s_i(t)$ is a known deterministic time function, of duration T seconds, used to transmit message m_i, $i = 0, 1, \ldots, M - 1$. It is also assumed that a full statistical description of the additive noise process $n(t)$ is available, that is, the joint pdfs $f_{n(t_1), n(t_2), \ldots, n(t_k)}(n_1, n_2, \ldots, n_k)$ are known for any finite k and for any set $\{t_1, t_2, \ldots, t_k\}$.

4.1 OPTIMUM WAVEFORM RECEIVERS IN WHITE GAUSSIAN NOISE

To gain insight into the waveform detection problem, we will first consider the case when the channel noise is white and Gaussian. The received signal, assuming that message m_i is transmitted, is given by

$$r(t) = s_i(t) + n(t), \quad 0 \le t \le T \tag{4.2}$$

where the Gaussian random process $n(t)$ has a symmetric power spectral density $N_0/2$ Watts/Hz double-sided in the frequency range $-B$ to B Hz and $BT \gg 1$. Now according to the sampling theorem [2], $n(t)$ is completely characterized by its samples taken every $1/2B$ seconds, that is,

$$n(t) = \sum_{k=-\infty}^{\infty} n(t_k) \, \text{sinc}(2Bt - k) \tag{4.3}$$

where $\text{sinc}(x) \triangleq (\sin \pi x)/(\pi x)$ and $t_k \triangleq k/2B$. Moreover, we shall have occasion not only to represent time functions as samples but also reconvert from sums of samples to time integrals. Thus, we note that

$$\int_{-\infty}^{\infty} n^2(t) dt = \frac{1}{2B} \sum_{k=-\infty}^{\infty} n^2(t_k) \tag{4.4}$$

represents the energy in the sample function $n(t)$. Now the joint probability density function of the vector $\mathbf{n}^{\mathbf{T}} = (n(t_1), n(t_2), \cdots, n(t_k))$ created from sampling a T-second strip of $n(t)$ is given by

$$f_{n(t_1), \ldots, n(t_k)}(\mathbf{n}) = C \exp \left\{ -\frac{\sum_{k=1}^{2BT} n^2(t_k)}{2N_0 B} \right\} \tag{4.5}$$

as the $2BT$ samples are uncorrelated and Gaussian. Here, C represents a normalization constant. Using (4.4) to convert from sums of samples to a time integral, we readily approximate

(4.5) via

$$f_{n(t_1),\cdots,n(kT)}(\mathbf{n}) \simeq C \exp\left\{-\frac{\int_{T_k} n^2(t)dt}{N_0}\right\} \tag{4.6}$$

for the kth time interval $T_k = [(k-1)T, kT]$. Without loss in generality, we set $k=1$ and consider the time interval $[0, T]$. Hence

$$f_{n(0),\cdots,n(T)}(\mathbf{n}) = C \exp\left\{-\frac{\int_0^T n^2(t)dt}{N_0}\right\} \tag{4.7}$$

Assuming that message m_i is transmitted in that interval, then $\mathbf{r}^T = (r(t_1), r(t_2), \cdots, r(T))$ and $\mathbf{s}_i^T = (s_i(t_1), s_i(t_2), \cdots, s_i(T))$ and hence

$$f_\mathbf{r}(\boldsymbol{\rho}/m_i) = f_\mathbf{n}(\boldsymbol{\rho} - \mathbf{s}_i/m_i) = f_\mathbf{n}(\boldsymbol{\rho} - \mathbf{s}_i)$$

$$= C \exp\left\{-\frac{\int_0^T (\rho(t) - s_i(t))^2 dt}{N_0}\right\} \tag{4.8}$$

This will be the basis from which the optimum receivers in white Gaussian noise will be derived.

4.1.1 Binary Cross-Correlation Receivers

In the binary case, the receiver needs to decide between messages m_0 and m_1. Using MAP decision theory, the receiver sets $\widehat{m}(\boldsymbol{\rho}) = m_0$ iff

$$P(m_0/\mathbf{r} = \boldsymbol{\rho}) > P(m_1/\mathbf{r} = \boldsymbol{\rho}) \tag{4.9}$$

or equivalently iff

$$\frac{P(m_0) f_\mathbf{r}(\boldsymbol{\rho}/m_0)}{f_\mathbf{r}(\boldsymbol{\rho})} > \frac{P(m_1) f_\mathbf{r}(\boldsymbol{\rho}/m_1)}{f_\mathbf{r}(\boldsymbol{\rho})} \tag{4.10}$$

Using (4.8) in (4.10) and assuming equilikely messages, the decision rule becomes: set $\widehat{m}(\rho(t)) = m_0$ iff

$$\exp\left\{-\frac{\int_0^T (\rho(t) - s_0(t))^2 dt}{N_0}\right\} > \exp\left\{-\frac{\int_0^T (\rho(t) - s_1(t))^2 dt}{N_0}\right\} \tag{4.11}$$

Taking natural logarithm and simplifying, the rule becomes: set $\widehat{m}(\rho(t)) = m_0$ iff

$$\int_0^T (\rho(t) - s_0(t))^2 dt < \int_0^T (\rho(t) - s_1(t))^2 dt \tag{4.12}$$

or equivalently, choose the message corresponding to the waveform $s_i(t)$ that is "closest" to $\rho(t)$ where distance is defined through

$$d(s_i(t), \rho(t)) \triangleq \sqrt{\int_0^T (\rho(t) - s_i(t))^2 dt} \tag{4.13}$$

This is referred to as *minimum distance decoding* and it was encountered earlier in vector communications. Expanding (4.12), we have

$$\int_0^T s_0^2(t)dt - 2\int_0^T \rho(t)s_0(t)dt < \int_0^T s_1^2(t)dt - 2\int_0^T \rho(t)s_1(t)dt \tag{4.14}$$

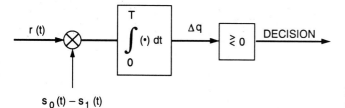

$s_0(t) - s_1(t)$

Figure 4.2   Binary waveform receiver (equal energy, equiprobable signals)

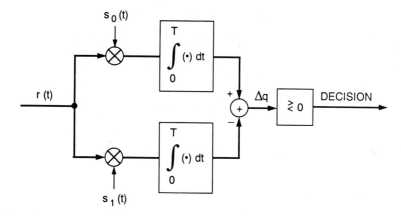

Figure 4.3   Alternative implementation of the binary waveform receiver (equal energy, equiprobable signals)

Letting E_i denote the energy of the ith signal, that is

$$E_i \triangleq \int_0^T s_i^2(t)dt \qquad (4.15)$$

then the rule reduces to: set $\widehat{m}(\rho(t)) = m_0$ iff

$$\Delta q \triangleq \int_0^T \rho(t)\,(s_0(t) - s_1(t))\,dt > \frac{1}{2}(E_0 - E_1) \qquad (4.16)$$

For equal energy signals, $E_0 = E_1$ and the test reduces to

$$\Delta q = \int_0^T \rho(t)\,(s_0(t) - s_1(t)) \underset{m_1}{\overset{m_0}{\gtrless}} 0 \qquad (4.17)$$

Hence, the receiver correlates $\rho(t)$ with the difference signal $s_0(t) - s_1(t)$ over $(0, T)$ and compares the outcome to zero as shown in Fig. 4.2. If the correlation is positive, the receiver selects message m_1. Otherwise, it selects m_0. Alternatively, the receiver can be implemented using two correlators as depicted in Fig. 4.3 and selects the signal corresponding to maximum correlation. Note that the decision rule operates linearly on the received signal and that no other rule (linear or nonlinear) can produce a smaller error probability. The zero threshold is due to the assumption made on the binary signal. If the signals are not equal energy or equally probable, the threshold will be biased away from zero.

4.1.2 Matched Filter Receivers

To gain further insight into the structure of the optimum waveform receiver, let's consider the problem of detecting the presence of a known signal $s(t)$ using the block diagram shown

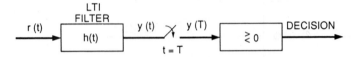

Figure 4.4   Detection of the presence of a signal

in Fig. 4.4, where the received signal is given by

$$r(t) = s(t) + n(t) \tag{4.18}$$

and $s(t)$ is nonzero for $t \epsilon [0, T]$ and $n(t)$ is AWGN with two-sided PSD $N_0/2$ Watts/Hz. The output $y(t)$ of the linear time-invariant (LTI) filter $h(t)$ is sampled at the end of the signal duration, that is, at $t = T$, and the resulting voltage compared to zero. The goal is to design the filter $h(t)$ such that the signal-to-noise ratio (SNR) of the sample $y(T)$ is maximized, where SNR is defined to be the ratio of the mean squared to the variance of the sample, that is

$$\text{SNR} = \frac{E^2\{y(T)\}}{\sigma_{y(T)}^2} \tag{4.19}$$

Using convolution, $y(t)$ can be written as

$$y(t) = \int_{-\infty}^{\infty} r(\lambda)h(t - \lambda)d\lambda \tag{4.20}$$

Since the noise is zero-mean, we have

$$E\{y(T)\} = \int_{-\infty}^{\infty} s(\lambda)h(T - \lambda)d\lambda \tag{4.21}$$

Similarly, the variance becomes

$$\sigma_{y(T)}^2 = \int_{-\infty}^{\infty} \int_{-\infty}^{\infty} R_n(\lambda_1 - \lambda_2)h(T - \lambda_1)h(T - \lambda_2)d\lambda_1\lambda_2 \tag{4.22}$$

Since the noise is white, $R_n(\tau) = (N_0/2)\delta(\tau)$ and (4.22) reduces to

$$\sigma_{y(T)}^2 = \frac{N_0}{2} \int_{-\infty}^{\infty} h^2(\lambda)d\lambda \tag{4.23}$$

Therefore, the SNR as given by (4.19) becomes

$$\text{SNR} = \frac{\left[\int_{-\infty}^{\infty} s(\lambda)h(T - \lambda)d\lambda\right]^2}{(N_0/2)\int_{-\infty}^{\infty} h^2(\lambda)d\lambda} \tag{4.24}$$

or alternatively

$$\text{SNR} = \frac{\left(\int_{-\infty}^{\infty} s(T - \beta)h(\beta)d\beta\right)^2}{(N_0/2)\int_{-\infty}^{\infty} h^2(\alpha)d\alpha} \tag{4.25}$$

Using Schwartz's inequality [3], which states that

$$\left(\int_{-\infty}^{\infty} h_1(t)h_2(t)dt\right)^2 \leq \int_{-\infty}^{\infty} h_1^2(t)dt \int_{-\infty}^{\infty} h_2^2(t)dt \tag{4.26}$$

with equality if and only if $h_2(t) = ch_1(t)$ (c being any constant), (4.25) can be upper bounded by

$$\text{SNR} \leq \left(\frac{2}{N_0}\right) \frac{\int_{-\infty}^{\infty} s^2(T - \lambda)d\lambda \int_{-\infty}^{\infty} h^2(\beta)d\beta}{\int_{-\infty}^{\infty} h^2(\alpha)d\alpha} \tag{4.27}$$

or since $E = PT = \int_0^T s^2(t)dt$, then

$$\text{SNR} \leq \frac{2}{N_0} \int_{-\infty}^{\infty} s^2(T - \lambda)d\lambda = \frac{2E}{N_0} = 2(\text{PG})(\text{CNR}) \tag{4.28}$$

where PG denotes the processing gain of the matched filter and is given by PG=WT and CNR $= P/(N_0W)$ where P is the signal power and W is the front-end one-sided bandwidth of the receiver. Here, CNR denotes the signal-to-noise ratio in the front-end bandwidth W where as $E/N_0 = PT/N_0$ denotes the signal-to-noise ratio in the data rate bandwidth $1/T$. The ratio of these two bandwidths is the processing gain PG of the matched filter. Exact equality in (4.28) occurs only when (c any constant)

$$h(t) = cs(T - t) \tag{4.29}$$

Therefore, SNR is maximized when the filter $h(t)$ is matched to the incoming signal $s(t)$, thus the terminology *matched filter*. Of course, such a filter with a finite impulse response is not physically realizable. However, as a practical matter, the filter becomes both physically realizable and essentially optimum whenever almost all of the energy in the signal $s(t)$ is located within $0 \leq t \leq T$ and $-B \leq f \leq B$ where B is the one-sided baseband bandwidth of the receiver and is related to the IF bandwidth of the receiver through $W = 2B$. When implementing a matched filter, the output $y(t)$ can be expressed as

$$y(t) = \int_{-\infty}^{\infty} r(\lambda)h(t - \lambda)d\lambda$$
$$= \int_{-\infty}^{\infty} r(\lambda)s(T - t + \lambda)d\lambda \tag{4.30}$$

When sampled at $t = T$, then

$$y(T) = \int_{-\infty}^{\infty} r(\lambda)s(\lambda)d\lambda \tag{4.31}$$

which is exactly the correlation between the incoming waveform $r(t)$ and the signal $s(t)$. From Fig. 4.3, the optimum binary waveform receiver can alternatively be implemented using matched filters as shown in Fig. 4.5. In this case, two filters matched to the possible incoming waveforms are used to filter the observed signal and their outputs sampled appropriately at $t = T$ to provide maximum SNR. The resulting voltages are fed into a comparator for selection of maximum value. Note that the signal component of the matched filter output is given by

$$E[y(t)] = \int_{-\infty}^{\infty} s(\lambda)s(T - t + \lambda)d\lambda \tag{4.32}$$

Recalling that the time-autocorrelation function [4] of any real function $g(t)$ is given by

$$R_g(\tau) \triangleq \int_{-\infty}^{\infty} g(t)g(t + \tau)dt \tag{4.33}$$

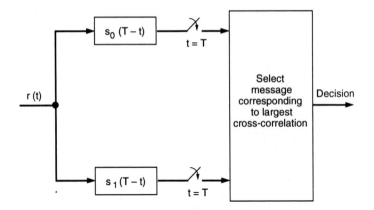

Figure 4.5   Matched filter implementation of the binary waveform receiver of Figure 4.3

Hence

$$E[y(t)] = R_s(T - t) \tag{4.34}$$

and at the sampling instant $t = T$, the mean becomes

$$E[y(T)] = R_s(0) = E \tag{4.35}$$

which is the maximum value of $R_s(\tau)$. Thus, the time-correlation operation can be equivalently performed using a matched filter. Note that when the output of the matched filter is sampled at $t = T + \hat{\delta}$ due to a timing error $\hat{\delta}$, then $E\{y(T + \hat{\delta})\} = R_s(\hat{\delta})$ and the signal-to-noise ratio is degraded accordingly.

4.1.3 *M*-ary Waveform Receivers

In the case of M-ary signals, the optimum receiver sets $\hat{m}(\rho(t)) = m_i$ iff

$$P(m_i) \exp \left\{ -\frac{\int_0^T (\rho(t) - s_i(t))^2 dt}{N_0} \right\}$$

$$= \max_k \left\{ P(m_k) \exp \left\{ -\frac{\int_0^T (\rho(t) - s_k(t))^2 dt}{N_0} \right\} \right\} \tag{4.36}$$

Simplifying and taking the logarithm, the receiver reduces to maximizing

$$\ln P(m_i) - \frac{1}{N_0} \int_0^T s_i^2(t) dt + \frac{2}{N_0} \int_0^T \rho(t) s_i(t) dt \tag{4.37}$$

or equivalently, the optimum receiver sets $\hat{m}(\rho(t)) = m_i$ iff

$$q_i \triangleq \int_0^T \rho(t) s_i(t) dt + B_i = \max_k \left\{ \int_0^T \rho(t) s_k(t) dt + B_k \right\} \tag{4.38}$$

where the bias term B_k is given by

$$B_k \triangleq \frac{N_0}{2} \ln P(m_k) - \frac{E_k}{2} \tag{4.39}$$

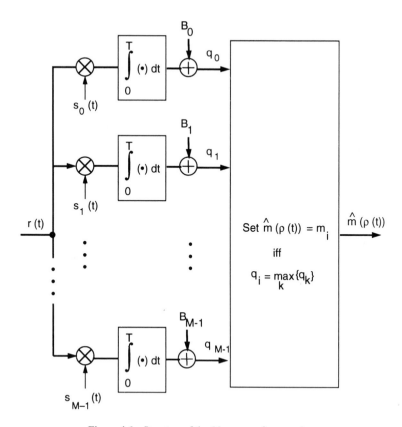

Figure 4.6   Structure of the M-ary waveform receiver

where E_k is the signal energy given in (4.15). In the case of equally likely messages, the bias reduces to $-E_k/2$. In addition, if the signals had equal energy, the bias term can be ignored. Note that the bias term is equivalent to that given in (3.60) when $\sigma^2 = N_0/2$. Hence, the receiver should correlate the observed signal with all possible transmitted waveforms (add the appropriate bias for unequal energy or unequal a priori probabilities) and select that message corresponding to maximum correlation as shown in Fig. 4.6. This implementation requires M correlators and hence, the receiver complexity increases linearly with the number of transmitted signals. If the signals were designed such that (see Appendix 3C)

$$s_i(t) = \sum_{j=1}^{N} s_{ij}\phi_j(t) \tag{4.40}$$

then

$$\int_0^T \rho(t)s_i(t)dt = \int_0^T \rho(t) \sum_{j=1}^{N} s_{ij}\phi_j(t)dt$$

$$= \sum_{j=1}^{N} s_{ij} \int_0^T \rho(t)\phi_j(t)dt = \sum_{j=1}^{N} s_{ij}r_j \tag{4.41}$$

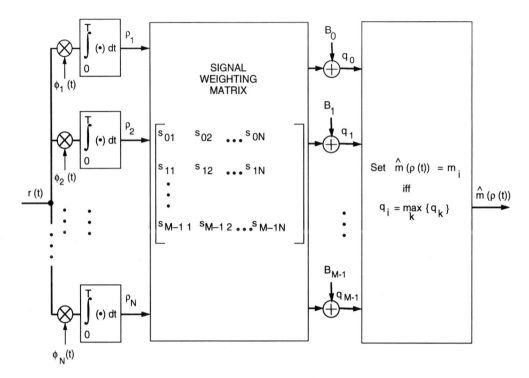

Figure 4.7   Alternative implementation of the M-ary waveform receiver using N correlators

Hence, the correlation can be achieved by correlating $\rho(t)$ with the N orthogonal functions and then weighting the output with the signal coefficients to form the required signal correlation. This processing is depicted in Fig. 4.7 and requires only N correlators for any signal set. When the number of signals M is greater than the dimensionality N, the second configuration requires fewer correlators and is more suitable for implementation. Since the correlation operation can also be implemented using a filter and a sampler, the optimum receiver can alternatively be implemented using either M or N matched filters (with the appropriate weighting) as illustrated in Fig. 4.8. Note that when the signal waveforms satisfy (4.40), we can map the waveform problem into an equivalent vector problem where $\mathbf{s}_i^T = (s_{i1}, s_{i2}, \cdots, s_{iN}), i = 0, 1, \cdots, M - 1$ and use all the tools developed in Chapter 3.

4.1.4 Error Probability Performance

In this section, the performances of the various coherent receivers are derived for both binary and M-ary signaling. First, binary signals are considered to gain insight into the problem [5], and then the results are generalized to the M-ary case. Equiprobable signals are considered as they reflect most communication systems. But the results can also incorporate unequiprobable signals with the appropriate weights.

Binary waveform receiver performance.   For equiprobable, equal energy signals, the optimum receiver computes $\Delta q = \int_0^T \rho(t)(s_0(t) - s_1(t))dt$ and compares the resulting random variable to zero. Then, the conditional mean and variance of Δq become

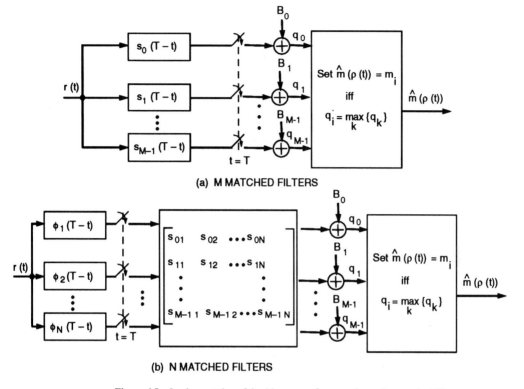

Figure 4.8   Implementation of the M-ary waveform receiver using matched filters

$$E[\Delta q / m_i] = \int_0^T s_i(t)\,(s_0(t) - s_1(t))\,dt \qquad (4.42)$$

and

$$\sigma_{\Delta q}^2 = E\left\{ \left(\int_0^T n(t)(s_0(t) - s_1(t))dt \right) \left(\int_0^T n(u)(s_0(u) - s_1(u))du \right) \right\}$$

$$= \left(\frac{N_0}{2}\right) \left(\int_0^T (s_0(t) - s_1(t))^2 dt \right) \qquad (4.43)$$

Because of the equal energy assumption, $\int_0^T s_1^2(t)dt = \int_0^T s_0^2(t)dt = E$ and (4.43) reduces to

$$\sigma_{\Delta q}^2 = N_0 \left\{ E - \int_0^T s_1(t)s_0(t)dt \right\} \qquad (4.44)$$

If we denote

$$\gamma_{ij} \triangleq \int_0^T \frac{s_i(t)}{\sqrt{E_i}} \frac{s_j(t)}{\sqrt{E_j}} dt \qquad (4.45)$$

as the correlation coefficient between signal $s_i(t)$ and $s_j(t)$, then

$$\overline{\Delta q_0} \triangleq E[\Delta q / m_0] = E(1 - \gamma_{10}) \qquad (4.46)$$

$$\overline{\Delta q_1} \triangleq E[\Delta q/m_1] = E(\gamma_{10} - 1) \tag{4.47}$$

and

$$\sigma_{\Delta q}^2 = N_0 E(1 - \gamma_{10}) \tag{4.48}$$

Note that

$$|\gamma_{ij}| = \left| \int_0^T \frac{s_i(t)}{\sqrt{E_i}} \frac{s_j(t)}{\sqrt{E_j}} dt \right| \leq \left| \int_0^T \left(\frac{s_i(t)}{\sqrt{E_i}} \right)^2 dt \right|^{1/2} \left| \int_0^T \left(\frac{s_j(t)}{\sqrt{E_j}} \right)^2 dt \right|^{1/2} = 1 \tag{4.49}$$

where the upper bound is obtained using Schwartz's inequality of (4.26). Due to the equal probable, equal energy assumption, the probability of error, conditioned on m_0, $P(E/m_0)$, is identical to $P(E/m_1)$. As a result, the unconditional probability of error, $P(E)$, for equilikely signals is given by

$$P(E) = \int_{-\infty}^{0} \frac{1}{\sqrt{2\pi\sigma_{\Delta q}^2}} \exp\left\{ -\frac{(\Delta q - \overline{\Delta q_0})^2}{2\sigma_{\Delta q}^2} \right\} d(\Delta q) \tag{4.50}$$

which, using the substitution $x = (\Delta q - \overline{\Delta q_0})/\sigma_{\Delta q}$, can be simplified to

$$P(E) = Q\left(\frac{\overline{\Delta q_0}}{\sigma_{\Delta q}} \right) \tag{4.51}$$

Substituting (4.47) and (4.48) into (4.51), we have

$$P(E) = Q\left(\sqrt{\frac{E}{N_0}(1 - \gamma_{10})} \right) \tag{4.52}$$

where E/N_0 denotes the signal energy-to-noise ratio. Alternatively, (4.52) can be expressed as

$$P(E) = Q\left(\frac{d_{10}}{\sqrt{2N_0}} \right) \tag{4.53}$$

where d_{ij} denotes the distance between signals $s_i(t)$ and $s_j(t)$ and is given by

$$d_{ij}^2 \triangleq \int_0^T (s_i(t) - s_j(t))^2 dt$$
$$= E_i + E_j - 2\sqrt{E_i E_j}\gamma_{ij} \tag{4.54}$$

For the two binary signals with equal energy, $d_{10}^2 = 2E(1 - \gamma_{10})$ and since the most negative value γ_{10} can take is minus one, the corresponding $P(E)$ is then

$$P(E) = Q\left(\sqrt{\frac{2E}{N_0}} \right) \quad \text{Antipodal (BPSK) signals} \tag{4.55}$$

which occurs when $s_0(t) = -s_1(t)$ and the pair is referred to as **antipodal** signals. Note that any pair of antipodal signals will generate the same $P(E)$ for a given energy E. Examples of three pairs of antipodal signals are given in Fig. 4.9, namely: (a) pulse code modulation

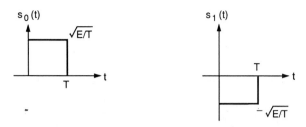

(a) PULSE CODE MODULATED (PCM) SIGNALS

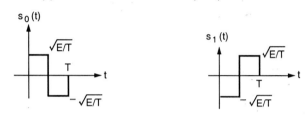

(b) MANCHESTER OR BI-φ SIGNALS

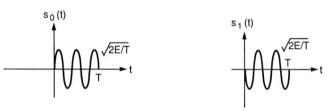

(c) BINARY PHASE SHIFT KEYED (BPSK) SIGNALS

Figure 4.9   Examples of antipodal signals: (a) Pulse code modulated (PCM) signals, (b) Manchester or Bi-φ signals, (c) Binary phase-shift-keyed (BPSK) signals

(PCM), (b) coded PCM typically referred to as Manchester or Bi-ϕ, and (c) binary phase-shift-keyed (BPSK) signals. The various signals are given by

PCM:

$$s_0(t) = -s_1(t) = \begin{cases} \sqrt{E/T} & 0 \le t \le T \\ 0 & \text{otherwise} \end{cases} \tag{4.56}$$

Bi-φ:

$$s_0(t) = -s_1(t) = \begin{cases} \sqrt{E/T} & 0 \le t \le T/2 \\ -\sqrt{E/T} & T/2 \le t \le T \\ 0 & \text{otherwise} \end{cases} \tag{4.57}$$

BPSK:

$$s_0(t) = -s_1(t) = \sqrt{2E/T} \sin(\omega_c t + \theta) \quad 0 \le t \le T, \quad f_c T = \text{integer} \tag{4.58}$$

The latter is referred to as a phase shift keyed signal since $s_0(t) = -s_1(t) = \sqrt{2E/T} \sin(\omega_c t + \theta \pm \pi)$ and hence, $s_0(t)$, $s_1(t)$ are transmitted by assigning the phase of the signal to 0 or $\pm\pi$, respectively. Since the phase is changing between two defined states, this scheme is referred to as binary phase-shift-keying (BPSK). Later, we will consider M-PSK signals where the phase is varied among M states, equally spaced around the unit circle.

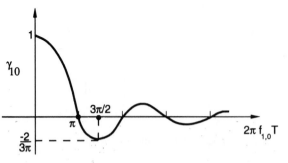

Figure 4.10 Correlation coefficient of coherent frequency shift signals as a function of $2\pi f_{1,0}T$

For orthogonal signals, $\int_0^T s_1(t)s_0(t)dt = 0$ (that is, $\gamma_{10} = 0$) and $P(E)$ of (4.52) reduces to

$$P(E) = Q\left(\sqrt{\frac{E}{N_0}}\right) \quad \text{Orthogonal signals} \tag{4.59}$$

which is a factor of two (or 3 dB) worse in E/N_0 than the performance of antipodal signals. Another signaling scheme is coherent frequency-shift-keying (FSK) where the messages are transmitted using two distinct frequencies, that is

$$s_1(t) = \sqrt{\frac{2E}{T}} \sin(\omega_1 t + \theta), \quad 0 \leq t \leq T \tag{4.60}$$

$$s_0(t) = \sqrt{\frac{2E}{T}} \sin(\omega_0 t + \theta), \quad 0 \leq t \leq T \tag{4.61}$$

and $\omega_1 T, \omega_0 T \gg 1$. In this case, the correlation coefficient is given by

$$\gamma_{10} = \frac{1}{E}\int_0^T s_1(t)s_0(t)dt = \frac{\sin(\omega_1 - \omega_0)T}{(\omega_1 - \omega_0)T} = \frac{\sin 2\pi f_{1,0}T}{2\pi f_{1,0}T} \tag{4.62}$$

where the correlation is plotted in Fig. 4.10 as a function of the frequency separation $f_{1,0} = f_1 - f_0$. The minimum possible correlation is given by $\gamma_{10} = -\frac{2}{3\pi}$, resulting in

$$P(E) = Q\left(\sqrt{\frac{E}{N_0}\left(1 + \frac{2}{3\pi}\right)}\right) \quad \text{Coherent BFSK} \tag{4.63}$$

which is $2/(1 + \frac{2}{3\pi}) = 2.1$ dB worse in SNR than antipodal signals. The corresponding frequency separation is $(f_1 - f_0)T = f_{1,0}T = 3/4$. On the other hand, when $f_{1,0}T = 0.5$, that is , the frequency separation is equal to half the message rate, orthogonal binary signaling is obtained and coherent FSK with this minimum frequency separation is sometimes referred to as minimum shift keying (MSK).

In all cases so far, we have considered equiprobable, equal energy signals. For unequal energy and unequiprobable signals, it can be shown (see Problem 4.10) that

$$P(E) = P(m_0)Q\left(\frac{d_{10}}{\sqrt{2N_0}} - \frac{1}{d_{10}}\sqrt{\frac{N_0}{2}}\ln\frac{P(m_1)}{P(m_0)}\right)$$

$$+ P(m_1)Q\left(\frac{d_{10}}{\sqrt{2N_0}} + \frac{1}{d_{10}}\sqrt{\frac{N_0}{2}}\ln\frac{P(m_1)}{P(m_0)}\right)$$

(4.64)

For the special case of equiprobable, unequal energy signals, (4.64) simplifies to

$$P(E) = Q\left(\frac{d_{10}}{\sqrt{2N_0}}\right)$$

(4.65)

as expected. An example of a scheme with unequal powers is "On-Off" keying where $s_0(t)$ is sent by turning the power off and $s_1(t)$ is sent by turning the power on, that is

$$s_1(t) = \begin{cases} \sqrt{E/T} & 0 \le t \le T \\ 0 & \text{otherwise} \end{cases}$$

(4.66)

and $s_0(t) = 0$ for $0 \le t \le T$. In this case, $\gamma_{10} = 0$ since $s_0(t) = 0$ and $d_{10}^2 = E$. Hence

$$P(E) = Q\left(\sqrt{\frac{E}{2N_0}}\right) \quad \text{On-Off keying}$$

(4.67)

Note that the maximum energy delivered is E, but the average energy is $E/2$ since both signals are equally likely. Hence, (4.67) can be expressed in terms of the average energy E_{avg} as

$$P(E) = Q\left(\sqrt{\frac{E_{\text{avg}}}{N_0}}\right)$$

(4.68)

which is identical to (4.59).

M-ary waveform receiver performance.   The performance of an M-ary MAP waveform receiver is first derived for any signal set of equally probable, equal energy signals. Then, the results are applied to signals of great interest, namely orthogonal ($\perp$), biorthogonal (Bi-$\perp$), transorthogonal, L-orthogonal, multiple-phase-shift-keyed (M-PSK) and finally, multiple amplitude modulated signals (M-AM).

Referring to Fig. 4.6, which depicts the structure of the optimum M-ary waveform receiver in an AWGN channel, the output of each integrate-and-dump filter is denoted by q_i and the receiver selects the message or signal corresponding to the largest q_i. At this point, we need to differentiate between symbols and bits as both will be used in this section. In binary communications ($M = 2$), each message or signal is transmitting one binary bit of information, that is, a binary "0" or a binary "1." Hence, the signal duration T and energy E are the bit duration and bit energy, denoted more appropriately by T_b and E_b, respectively. In M-ary communications (with $M = 2^n$), each signal is transmitting $n = \log_2 M$ binary bits. We will thus refer to each message or signal as a symbol and to the duration and energy of the signal as symbol duration T_s and symbol energy E_s. Since in every T_s sec, $\log_2 M$ bits are transmitted, $(T_s/T_b) = (E_s/E_b) = \log_2 M$. We will also differentiate between symbol and bit error probabilities, denoted respectively by $P_s(E)$ and $P_b(E)$. The bit error probability $P_b(E)$ will be used throughout this section to compare the performances of binary and M-ary communication systems. The probability of message (symbol) error, $P_s(E)$, can in general be expressed as

$$P_s(E) = \sum_{i=0}^{M-1} P_s(E/m_i) P(m_i) \tag{4.69}$$

where $P_s(E/m_i)$ is the probability of message or symbol error assuming that m_i is transmitted. In order to evaluate $P_s(E/m_i)$, the outputs of the integrate-and-dump filters have to be jointly characterized through the M-dimensional pdf. Recalling that

$$q_k = \int_0^{T_s} \rho(t) s_k(t) dt \tag{4.70}$$

where $\rho(t) = s_i(t) + n(t)$ when m_i is transmitted, the output of the kth integrator becomes

$$q_{k/i} = E_s \gamma_{ki} + \int_0^{T_s} n(t) s_k(t) dt \tag{4.71}$$

with conditional mean

$$\overline{q_{k/i}} \triangleq E[q_{k/i}] = E_s \gamma_{ki} \tag{4.72}$$

Similarly, any pair of outputs q_k and q_j will have conditional covariance (that is, subtract the means)

$$q_{kj} = E\left\{ \int_0^{T_s} n(t) s_k(t) dt \int_0^{T_s} n(u) s_j(u) du \right\} = \frac{N_0}{2} \int_0^{T_s} s_k(t) s_j(t) dt = \frac{N_0}{2} E_s \gamma_{kj} \tag{4.73}$$

Hence, the outputs $q_0, q_1, \ldots, q_{M-1}$, conditioned on m_i, are Gaussian random variables with nonzero means and nonzero correlations. Their joint M-dimensional pdf is given by (3.61) as

$$f_{\mathbf{q}}(\mathbf{q}/m_i) = G\left(\mathbf{q}; \bar{\mathbf{q}}_i; \mathcal{K}_{\mathbf{q}}\right)$$

$$= \frac{1}{(2\pi)^{M/2} |\det \mathcal{K}_{\mathbf{q}}|^{\frac{1}{2}}} \exp\left\{ -\frac{1}{2} (\mathbf{q} - \bar{\mathbf{q}}_i)^T \mathcal{K}_{\mathbf{q}}^{-1} (\mathbf{q} - \bar{\mathbf{q}}_i) \right\} \tag{4.74}$$

where $\mathbf{q} = (q_0, q_1, \ldots, q_{M-1})^T$, $\bar{\mathbf{q}}_i$ denotes the conditional mean vector given by

$$\bar{\mathbf{q}}_i = (\overline{q_{0/i}}, \overline{q_{1/i}}, \ldots, \overline{q_{M-1/i}})^T = E_s(\gamma_{0i}, \gamma_{1i}, \ldots, \gamma_{M-1i})^T \tag{4.75}$$

and $\mathcal{K}_{\mathbf{q}}$ denotes the covariance matrix given by

$$\mathcal{K}_{\mathbf{q}} = \begin{pmatrix} q_{00} & q_{01} & \cdots & q_{0M-1} \\ q_{10} & q_{11} & \cdots & q_{1M-1} \\ \vdots & \vdots & \ddots & \vdots \\ q_{M-10} & q_{M-11} & \cdots & q_{M-1M-1} \end{pmatrix}$$

$$= \frac{N_0 E_s}{2} \begin{pmatrix} \gamma_{00} & \gamma_{01} & \cdots & \gamma_{0M-1} \\ \gamma_{10} & \gamma_{11} & \cdots & \gamma_{1M-1} \\ \vdots & \vdots & \ddots & \vdots \\ \gamma_{M-10} & \gamma_{M-11} & \cdots & \gamma_{M-1M-1} \end{pmatrix} \tag{4.76}$$

Equation (4.74) assumes that $\mathcal{K}_{\mathbf{q}}$ is invertible. If $\mathcal{K}_{\mathbf{q}}$ is singular, the dimensionality of the observed vector can be reduced without any loss in generality and can result in singular detection. The conditional probability of symbol error $P_s(E/m_i) = 1 - P_s(C/m_i)$, where $P_s(C/m_i)$ is the probability that all q_k's, except $k = i$, are less than q_i, that is

$$P_s(C/m_i) = \text{Prob}\, \{q_i > (q_0, q_1, \cdots, q_{i-1}, q_{i+1}, \cdots, q_{M-1})/m_i\} \qquad (4.77)$$

$$= \int_{-\infty}^{\infty} \underbrace{\int_{-\infty}^{q_i} \cdots \int_{-\infty}^{q_i}}_{M-1 \text{ integrals}} f_{\mathbf{q}}(\mathbf{q}/m_i) dq_0 dq_1 \cdots dq_{i-1} dq_{i+1} \cdots dq_{M-1} dq_i$$

Note the order of integration where the last integral is over q_i. Since $P(C/m_i)$ depends on γ_{ji}, $j = 0, 1, \ldots, M-1$, the resulting probability of symbol error, $P_s(E)$, depends solely on the symbol energy E_s, the noise spectral level N_0, and the signal correlation matrix Γ[1] given by

$$\Gamma = \begin{pmatrix} \gamma_{00} & \gamma_{01} & \cdots & \gamma_{0M-1} \\ \gamma_{10} & \gamma_{11} & \cdots & \gamma_{1M-1} \\ \vdots & \vdots & \ddots & \vdots \\ \gamma_{M-10} & \gamma_{M-11} & \cdots & \gamma_{M-1M-1} \end{pmatrix} \qquad (4.78)$$

If we normalize the q_i's by introducing the change of variables

$$z_i = (q_i - \overline{q_i})/\sqrt{q_{ii}} \qquad (4.79)$$

then (4.77) may be rewritten as

$$P_s(C/m_i) = \int_{-\infty}^{\infty} \int_{-\infty}^{z_i + \sqrt{2E_s/N_0}(1-\gamma_{i1})} \cdots \int_{-\infty}^{z_i + \sqrt{2E_s/N_0}(1-\gamma_{iM-1})} f_{\mathbf{z}}(\mathbf{z}/m_i) d\mathbf{z} \qquad (4.80)$$

where $\mathbf{z}^T = (z_0, z_1, z_2, \cdots, z_{M-1})$ and $f_{\mathbf{z}}(\mathbf{z}/m_i) = G(\mathbf{z}; \mathbf{0}; \Gamma)$, independent of m_i. Using a similar procedure as in Chapter 3, a Union bound can be derived for waveform communications resulting in

$$P(E/m_k) \le \sum_{\substack{j=0 \\ j \ne k}}^{M-1} Q\left(\frac{d_{kj}}{\sqrt{2N_0}}\right) \qquad (4.81)$$

where d_{kj} is the distance given by (4.54).

The bit error probability is what a design engineer needs in a link budget. In order to compare binary and M-ary communications, we will therefore use the bit error probability for both signaling schemes as a measure. In binary communications, each signal transmits one binary bit and hence, the probability of signal error is identical to the probability of bit error. In M-ary communications, let $M = 2^n$ so that each signal or symbol is transmitting $\log_2 M = n$ binary bits. Let's consider the transmission of a sequence of K bits where K is very large. When employing M-ary signaling, we can write $K = hn$ where h is the number of symbols and n is the number of bits per symbol. The probability of bit error $P_b(E)$ can be written as

1. The covariance matrix $\mathcal{K}_{\mathbf{q}}$ is given by $\mathcal{K}_{\mathbf{q}} = \frac{N_0 E_s}{2}\Gamma$.

$$P_b(E) = \frac{\text{Average number (\#) of erroneous bits}}{\text{Total \# of bits}}$$

$$= \frac{(\text{Avg. \# of erroneous bits per symbol})(\text{\# of symbols})}{\text{Total \# of bits}} \qquad (4.82)$$

$$= \frac{(\text{Avg. \# of erroneous bits per symbol})h}{h \log_2 M}$$

which is independent of h and hence of K. But

$$(\text{Avg. \# of erroneous bits per symbol}) = \sum_{i=0}^{M-1} (\text{Avg. \# of erroneous bits}/m_i) \, P(m_i)$$

$$= \frac{1}{M} \sum_{i=0}^{M-1} (\text{Avg. \# of erroneous bits}/m_i) \qquad (4.83)$$

The last equality is due to the fact that we are transmitting a "large" number of symbols and as a result, all messages occur equally likely. Hence

$$P_b(E) = \frac{1}{M \log_2 M} \sum_{i=0}^{M-1} (\text{Avg. \# of erroneous bits}/m_i) \qquad (4.84)$$

Now let $\eta_{i,j}$ denote the event that message i is transmitted but message j is decided (that is, vector $\mathbf{s}_j$ is closest to observed vector $\boldsymbol{\rho}$) and (i, j) the number of bits that symbol i differs from symbol j; then

$$(\text{Avg. \# of erroneous bits given symbol } i) = \sum_{j=0}^{M-1} P(\eta_{i,j})(i, j) \qquad (4.85)$$

Hence

$$P_b(E) = \frac{1}{M \log_2 M} \sum_{i=0}^{M-1} \sum_{j=0}^{M-1} P(\eta_{i,j})(i, j) \qquad (4.86)$$

The probability of bit error, $P_b(E)$, depends on the symbol-to-bit mapping, otherwise known as the code. The computation of $P(\eta_{i,j})$ is very difficult if not impossible for a general M and signal constellation. It is therefore desirable to obtain an upper bound on $P_b(E)$. Let $\xi_{i,j}$ denote the event that message i is transmitted but signal j is closer to the received vector, ignoring all other signals. The probability of the event $\xi_{i,j}$ is referred to as *the pairwise probability of error (PPE)* and it will be used extensively in future chapters to bound the probability of bit error. Figure 4.11 illustrates the regions where $\eta_{0,1}$ and $\xi_{0,1}$ occur for an example of four signals. In all cases, the region where the event $\xi_{i,j}$ is true is greater than the corresponding region where $\eta_{i,j}$ is true. As a result

$$P(\xi_{i,j}) \geq P(\eta_{i,j}) \quad \text{for all } i \text{ and } j \qquad (4.87)$$

where $P(\xi_{i,j}) = P_2(\mathbf{s}_i, \mathbf{s}_j) = Q(d_{ij}/\sqrt{2N_0})$ from Appendix 3E and hence, (4.86) can be upper bounded by

$$P_b(E) \leq \frac{1}{M \log_2 M} \sum_{i=0}^{M-1} \sum_{j=0}^{M-1} (i, j) P(\xi_{i,j}) \qquad (4.88)$$

Equations (4.86) and (4.88) are exact and upper bound expressions for the bit error probability for any signaling format and any code (that is, symbol-to-bit mapping). Both expressions

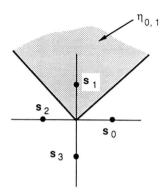

IN $\eta_{0,1}$, ρ IS CLOSER TO
s_1, THAN ANY OTHER
SIGNAL
$\therefore$ s_1 IS THE CLOSEST TO ρ

IN $\varepsilon_{0,1}$, ρ IS CLOSER TO s_1,
THAN s_0 BUT s_2 MIGHT BE
THE CLOSEST TO ρ (CONSIDER Ⓧ)

Figure 4.11   Example of $\eta_{0,1}$ and $\xi_{0,1}$

can be further simplified by noting that $(i, j) = (j, i)$ for all i and j. However, the events $\eta_{i,j}$ and $\eta_{j,i}$ are not always identical and depend on the signal set of interest.

Example 4.1

Consider the four signals illustrated in Fig. 4.12(a) with the corresponding bit mapping. This set corresponds to quadrature phase-shift-keyed (QPSK) signals, which will be discussed in more detail when considering M-PSK signals. Using (4.88), we have ($M = 4$)

$$P_b(E) \le \frac{1}{(2)(4)} \sum_{i=0}^{3} \sum_{j=0}^{3} (i, j) P(\xi_{i,j})$$

$$= \frac{1}{8} \{ (0, 1) P(\xi_{0,1}) + (0, 2) P(\xi_{0,2}) + (0, 3) P(\xi_{0,3}) + (1, 0) P(\xi_{1,0})$$

$$+ (1, 2) P(\xi_{1,2}) + (1, 3) P(\xi_{1,3}) + (2, 0) P(\xi_{2,0}) + (2, 1) P(\xi_{2,1})$$

$$+ (2, 3) P(\xi_{2,3}) + (3, 1) P(\xi_{3,1}) + (3, 2) P(\xi_{3,2}) + (3, 0) P(\xi_{3,0}) \}$$

But

$$(0, 1) = (0, 3) = (1, 0) = (1, 2) = (2, 1) = (2, 3) = (3, 0) = (3, 2) = 1$$

and

$$(0, 2) = (2, 0) = (1, 3) = (3, 1) = 2$$

Also from Fig. 4.12(b) and (c), it is clear that

$$P(\xi_{0,1}) = P(\xi_{0,3}) = P(\xi_{1,0}) = P(\xi_{1,2}) = P(\xi_{2,1}) = P(\xi_{2,3}) = P(\xi_{3,2}) = P(\xi_{3,0}) = Q\left(\sqrt{\frac{E_s}{N_0}}\right)$$

and

$$P(\xi_{0,2}) = P(\xi_{1,3}) = P(\xi_{2,0}) = P(\xi_{3,1}) = Q\left(\sqrt{\frac{2E_s}{N_0}}\right)$$

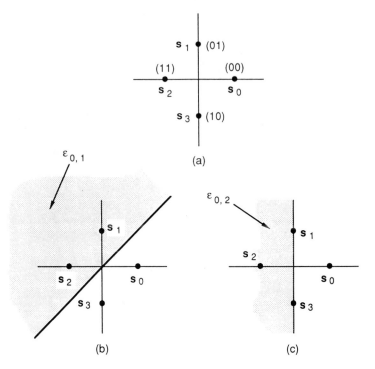

Figure 4.12 Signals and bit mapping of Example 4.1

Hence

$$P_b(E) \le Q\left(\sqrt{\frac{E_s}{N_0}}\right) + Q\left(\sqrt{\frac{2E_s}{N_0}}\right)$$

and $E_s/N_0 = 2E_b/N_0$. Note that $P_b(E)$ can be made arbitrarily small by making E_s/N_0 larger and larger. This is not the case in binary or bit-by-bit signaling where the probability of a bit error occurring approaches unity for any E_b/N_0 when transmitting many bits. To illustrate that, consider the probability of a bit error in a single transmission [that is, $Q(\sqrt{2E_b/N_0})$] assuming antipodal signals. Then, the probability of a bit error in K transmissions is $1 - [1 - Q(\sqrt{2E_b/N_0})]^K$ and approaches unity as $K \to \infty$ (for a fixed E_b/N_0). Note that this comparison is between block transmission as the block length increases and bit-by-bit transmission as the number of bits grows. The key advantage is obtained by varying the block length in one case versus a fixed block (or bit) in the other.

Orthogonal Signals.   Orthogonal signals are characterized by the fact that any two distinct signals have zero correlation, that is

$$\gamma_{ij} = \begin{cases} 0 & i \ne j \\ 1 & i = j \end{cases} \tag{4.89}$$

Hence, their signal correlation matrix Γ_O is given by

$$\Gamma_O = \begin{pmatrix} 1 & 0 & \cdots & 0 \\ 0 & 1 & 0 & \cdots \\ \vdots & 0 & \ddots & \vdots \\ 0 & \cdots & 0 & 1 \end{pmatrix} = \mathcal{I}_M \ (M \times M \text{ identity matrix}) \tag{4.90}$$

In this case, the integrate-and-dump filter outputs $q_0, q_1, \cdots, q_{M-1}$ are Gaussian, i.i.d., with

$$\overline{q_{k/i}} = \begin{cases} 0 & k \neq i \\ E_s & k = i \end{cases} \tag{4.91}$$

and variance $\sigma^2 = N_0 E_s/2$. From (4.77), we have

$$P_s(E/m_i) = 1 - \int_{-\infty}^{\infty} \left[\int_{-\infty}^{q_i} \frac{1}{\sqrt{2\pi\sigma^2}} \exp\left(-\frac{q_k^2}{2\sigma^2}\right) dq_k \right]^{M-1}$$

$$\times \frac{1}{\sqrt{2\pi\sigma^2}} \exp\left(-\frac{(q_i - E_s)^2}{2\sigma^2}\right) dq_i \tag{4.92}$$

Since $P_s(E/m_i)$ is independent of hypothesis i, the unconditional probability of symbol error $P_s(E)$ becomes (using Appendix 3B)

$$P_s(E) = 1 - \int_{-\infty}^{\infty} \left[\frac{1}{2} \text{erfc}\left(-q - \sqrt{\frac{E_s}{N_0}}\right) \right]^{M-1} \frac{e^{-q^2}}{\sqrt{\pi}} dq \tag{4.93}$$

and is depicted in Fig. 4.13 as a function of E_s/N_0 for $M = 2, 4, 8$, and 16. The performance degrades for increasing M because even though each M-ary system is transmitting at the same symbol rate $\mathcal{R}_s$ (symbols/sec), the corresponding bit rate $\mathcal{R}_b$ (bits/sec) is changing and is given by

$$\mathcal{R}_b = \frac{1}{T_s} \log_2 M \quad \text{(bits/sec)} \tag{4.94}$$

Hence, as M increases, the bit rate also increases. In order to compare all M-ary systems at the same bit rate, we need to assess $P_s(E)$ as a function of bit SNR, E_b/N_0, given by

$$\frac{E_b}{N_0} = \frac{E_s}{N_0} \frac{1}{\log_2 M} \tag{4.95}$$

Figure 4.14 depicts $P_s(E)$ versus E_b/N_0 when $M = 2^n$. It is clear that the bit rate adjustment leads to an improvement in performance as M is increased.

To convert from symbol to bit errors, note that with orthogonal signals, the incorrectly decided symbol is equally likely to be any of the remaining $M - 1$ symbols since each symbol (equivalent signal vector) is equidistant from all others. Consider Fig. 4.15, which depicts the bits for eight symbols (that is, $M = 2^3 = 8$). In any column of bits, there are exactly 2^{n-1} ones and 2^{n-1} zeros (four of each in Fig. 4.15). Thus, if a symbol in incorrectly decided, then for any given bit position (or column), there are 2^{n-1} out of a possible $2^n - 1$ ways in which that bit can be in error. Thus, the probability of bit error, $P_b(E)$, is given by

$$P_b(E) = \frac{2^{n-1}}{2^n - 1} P_s(E) \tag{4.96}$$

and is plotted in Fig. 4.16 as a function of E_b/N_0 for several values of n. Using the union bound of (4.81), $P_s(E)$ can be upper bounded by

$$P_s(E) \leq (M - 1) Q\left(\sqrt{\frac{E_s}{N_0}}\right) \tag{4.97}$$

which is easily computable.

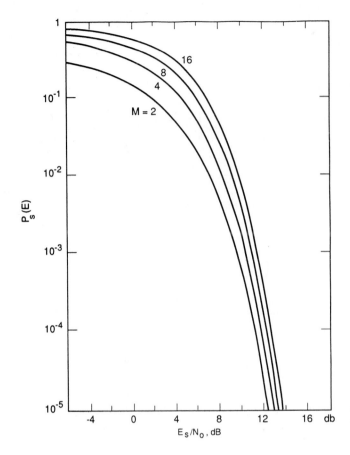

Figure 4.13 Symbol error probability for M orthogonal signals as a function of symbol SNR (Reprinted from [1])

Example 4.2

A typical set of signals that may be generated in practice for M-ary transmissions is the set of M nonoverlapping sinusoidal pulses

$$s_i(t) = \begin{cases} \sqrt{2PM} \sin \omega_c t, & \frac{iT}{M} \leq t \leq \frac{(i+1)T}{M} \\ 0, & \text{otherwise} \end{cases}$$

This set is orthogonal and is useful in Pulse Position Modulation (PPM) systems. Another orthogonal set of signals consists of sinusoidal pulses of different frequencies

$$s_i(t) = \sqrt{2P} \sin(\omega_c t + \frac{i\pi}{T}t), \quad 0 \leq t \leq T$$

where the frequency separation is $1/2T$ Hz to ensure orthogonality. Such a set is referred to as multiple frequency-shift-keyed (M-FSK) signals.

Biorthogonal Signals. Biorthogonal (Bi-$\perp$) signals can be constructed from orthogonal signals by augmenting the set with the negative of the signals. Let's start with $\frac{M}{2}$ orthogonal signals, denoted by $s_0(t), s_1(t), \cdots, s_{\frac{M}{2}-1}(t)$. Adding their negatives to the set, we have $s_{\frac{M}{2}}(t) = -s_0(t), s_{\frac{M}{2}+1}(t) = -s_1(t), \cdots, s_{M-1}(t) = -s_{\frac{M}{2}-1}(t)$. Hence, the signal correlation coefficients are given by

1
10-1
10-2
10-3
10-4
10-5
Ps(E)
n = 1
n = 2
n = 3
n = 4
n = 5
n = 6
n = 7
n = 8
n = 9
n = 10
n = 11
n = 12
n = ∞
-1.6 dB
-10
-5
0
5
10
15
20
Eb/No, dB

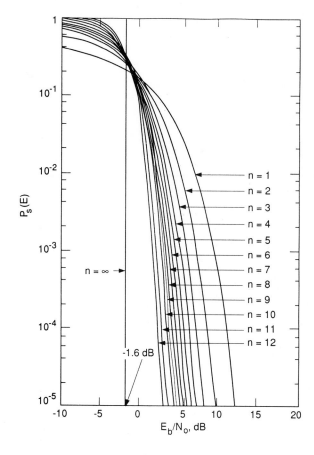

Figure 4.14 Symbol error probability for M orthogonal signals ($M = 2^n$) as a function of bit SNR (Reprinted from [12])

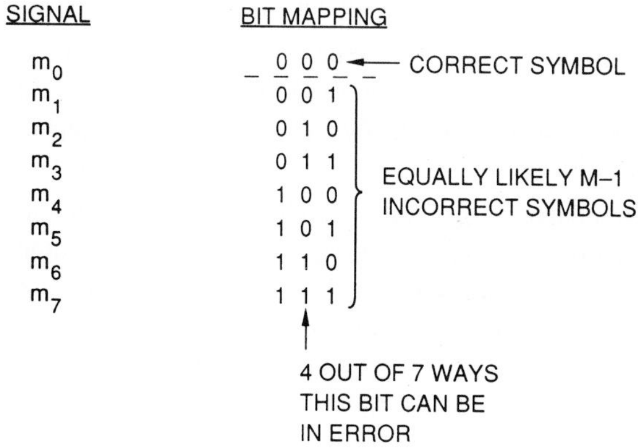

SIGNAL
BIT MAPPING
m0 0 0 0 ← CORRECT SYMBOL
m1 0 0 1
m2 0 1 0
m3 0 1 1
m4 1 0 0
m5 1 0 1
m6 1 1 0
m7 1 1 1
EQUALLY LIKELY M–1 INCORRECT SYMBOLS
4 OUT OF 7 WAYS THIS BIT CAN BE IN ERROR

Figure 4.15 Symbol-to-bit mapping for orthogonal signals ($M = 8$)

$$\gamma_{ij} = \begin{cases} 0 & i \neq j, |i-j| \neq \frac{M}{2} \\ 1 & i = j \\ -1 & |i-j| = \frac{M}{2} \end{cases} \tag{4.98}$$

and as a result

$$\Gamma_B = \begin{pmatrix} \mathcal{I}_{\frac{M}{2}} & -\mathcal{I}_{\frac{M}{2}} \\ -\mathcal{I}_{\frac{M}{2}} & \mathcal{I}_{\frac{M}{2}} \end{pmatrix} \tag{4.99}$$

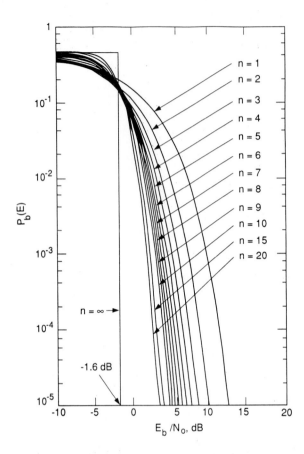

Figure 4.16   Bit error probability for M orthogonal signals ($M = 2^n$) as a function of bit SNR (Reprinted from [12])

where $\mathcal{I}_{\frac{M}{2}}$ is the $\frac{M}{2} \times \frac{M}{2}$ identity matrix. Since the signals and their negatives are in the signal set, the receiver can be simplified and implemented using $\frac{M}{2}$ correlators as shown in Fig. 4.17. In the decision rule, the signal is first chosen to correspond to the maximum correlation in absolute value. Then, the sign of the signal is chosen to correspond to the sign of the correlation. The joint pdf of $q_0, q_1, \cdots, q_{\frac{M}{2}-1}$ is the multivariate Gaussian pdf of (4.74) with an identity covariance matrix corresponding to the set of $\frac{M}{2}$ orthogonal signals. As a result, the probability of correct symbol detection, when m_i is transmitted, is the probability that $|q_i| > |q_k|$ for all $k \neq i$ and $i = 0, 1, \cdots, \frac{M}{2} - 1$ and has the correct sign (assumed positive without any loss in generality), that is

$$P_s(C/m_i) = \text{Prob}\left\{q_i > 0, |q_i| > (|q_0|, \cdots, |q_{i-1}|, |q_{i+1}|, \cdots, |q_{\frac{M}{2}-1}|)/m_i\right\}$$

$$= \int_0^{\infty} \underbrace{\int_{-q_i}^{q_i} \cdots \int_{-q_i}^{q_i}}_{\frac{M}{2}-1 \text{ integrals}} \tag{4.100}$$

$$\times f_{\mathbf{q}}(\mathbf{q}/m_i) dq_0 dq_1 \cdots dq_{i-1} dq_{i+1} \cdots dq_{\frac{M}{2}-1} dq_i$$

Since the q_i's, $i = 0, 1, \cdots, \frac{M}{2} - 1$ are i.i.d., then (with change of variables)

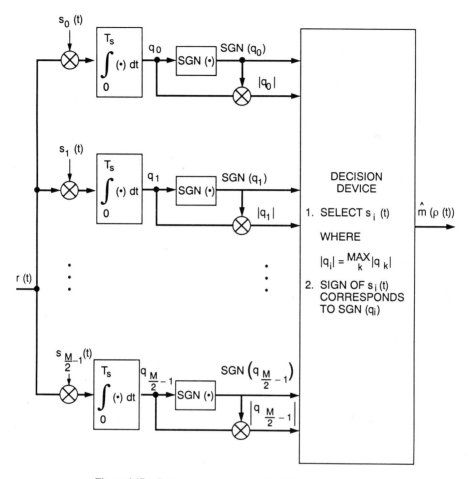

Figure 4.17   Cross-correlation receiver for M biorthogonal signals

$$P_s(C/m_i) = \int_{-\sqrt{\frac{E_s}{N_0}}}^{\infty} \frac{e^{-q^2}}{\sqrt{\pi}} \left[\int_{-q-\sqrt{\frac{E_s}{N_0}}}^{q+\sqrt{\frac{E_s}{N_0}}} \frac{e^{-z^2}}{\sqrt{\pi}} dz \right]^{\frac{M}{2}-1} dq$$

(4.101)

$$= \int_{-\sqrt{\frac{E_s}{N_0}}}^{\infty} \frac{e^{-q^2}}{\sqrt{\pi}} \left[1 - \operatorname{erfc}\left(q + \sqrt{\frac{E_s}{N_0}} \right) \right]^{\frac{M}{2}-1} dq$$

Again, since $P_s(C/m_i)$ is independent of m_i, $P_s(E)$ becomes

$$P_s(E) = 1 - \int_{-\sqrt{\frac{E_s}{N_0}}}^{\infty} \frac{e^{-q^2}}{\sqrt{\pi}} \left[1 - \operatorname{erfc}\left(q + \sqrt{\frac{E_s}{N_0}} \right) \right]^{\frac{M}{2}-1} dq$$

(4.102)

The symbol error probability, $P_s(E)$, is plotted in Fig. 4.18 as a function of bit SNR, E_b/N_0. Since for any choice of signals, one of the remaining $M - 1$ is antipodal (distance $= 2\sqrt{E_s}$)

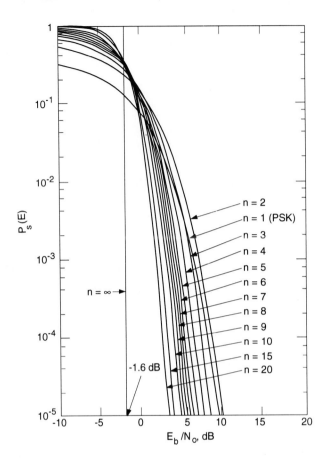

Figure 4.18 Symbol error probability for M biorthogonal signals ($M = 2^n$) as a function of bit SNR (Reprinted from [12])

while the rest are orthogonal (distance $= \sqrt{2E_s}$), $P_s(E)$ can be bounded using (4.81) to produce

$$P_s(E) \le (M - 2)Q\left(\sqrt{\frac{E_s}{N_0}}\right) + Q\left(\sqrt{\frac{2E_s}{N_0}}\right) \tag{4.103}$$

In order to compute the bit error probability, there are two types of symbol errors that can occur—when the negative $-s_i(t)$ of the transmitted signal $s_i(t)$ is selected or when any one of the $M - 2$ remaining orthogonal signals is decided. We will refer to the former event as errors of the first type and to the latter event as errors of the second type. The probability of symbol error of the first type is given by

$$P_{s1}(E) = \text{Prob}\left\{q_i < 0, |q_i| > (|q_0|, \cdots, |q_{i-1}|, |q_{i+1}|, \cdots, |q_{\frac{M}{2}-1}|)/m_i\right\}$$

$$= \int_{-\infty}^{0} \underbrace{\int_{-q_i}^{q_i} \cdots \int_{-q_i}^{q_i}}_{\frac{M}{2}-1 \text{ integrals}} \tag{4.104}$$

$$\times f_{\mathbf{q}}(\mathbf{q}/m_i)dq_0 dq_1 \cdots dq_{i-1} dq_{i+1} \cdots dq_{\frac{M}{2}-1} dq_i$$

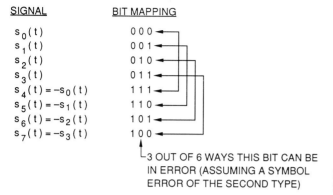

SIGNAL   BIT MAPPING

$s_0(t)$ 0 0 0
$s_1(t)$ 0 0 1
$s_2(t)$ 0 1 0
$s_3(t)$ 0 1 1
$s_4(t) = -s_0(t)$ 1 1 1
$s_5(t) = -s_1(t)$ 1 1 0
$s_6(t) = -s_2(t)$ 1 0 1
$s_7(t) = -s_3(t)$ 1 0 0

3 OUT OF 6 WAYS THIS BIT CAN BE
IN ERROR (ASSUMING A SYMBOL
ERROR OF THE SECOND TYPE)

Figure 4.19   Complementary bit encoding for $M = 8$ biorthogonal signals

Substituting for the joint pdf and simplifying, $P_{s1}(E)$ becomes

$$P_{s1}(E) = \int_{-\infty}^{-\sqrt{\frac{E_s}{N_0}}} \frac{e^{-q^2}}{\sqrt{\pi}} \left[-1 + \text{erfc}(q + \sqrt{\frac{E_s}{N_0}}) \right]^{\frac{M}{2}-1} dq \qquad (4.105)$$

The probability of error of the second type, $P_{s2}(E)$, is greater than $P_{s1}(E)$ since the orthogonal signals are "closer" to the transmitted signal than its negative version. Since both error events are disjoint and form all possibilities of symbol errors, then $P_{s1}(E) + P_{s2}(E) = P_s(E)$ and as a result

$$P_{s2}(E) = P_s(E) - P_{s1}(E) \qquad (4.106)$$

In order to proceed ahead and compute the probability of bit error, we need to assume a symbol to bit mapping. Since the distance between a signal and its negative is the maximum it can be, we will encode a symbol and its negative with complementary bit patterns as shown in Fig. 4.19. This will ensure that the event when all bits are decided in error occurs with the least possible probability, that is, a symbol error of the first type. Hence, if a symbol error of the first type is made, all bits are in error and the conditional bit error probability is exactly one. For a symbol error probability of the second type, in any bit position (or column) of the remaining $2^n - 2$ possible incorrect symbols, there are an equal number, $2^{n-1} - 1$, of ones and zeros. Thus, the probability of a bit being in error is one-half given that a symbol error of the second type occurred. Thus, the bit error probability for biorthogonal signals is given by

$$P_b(E) = P_{s1}(E) + \frac{1}{2} P_{s2}(E) \qquad (4.107)$$

and is plotted in Fig. 4.20 versus bit SNR, E_b/N_0.

Transorthogonal Signals.   Transorthogonal signals are equally correlated signals with the most negative correlation that can be achieved with M signals. Recall that in binary waveform communications, antipodal signals, with minus one signal correlation, offer the best performance. The motivation for transorthogonal signals is that their performance would be superior compared with orthogonal signals, because they are constructed with the

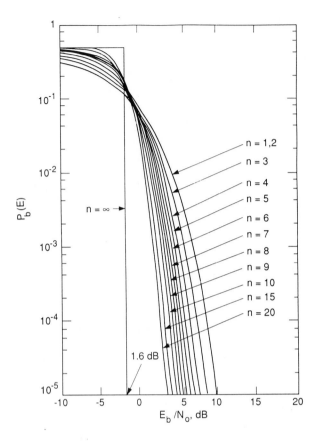

Figure 4.20 Bit error probability for M biorthogonal signals ($M = 2^n$) as a function of bit SNR (Reprinted from [12])

largest achievable negative correlation [6]. The value of the signal correlation is limited as follows:

$$\int_0^{T_s} (\sum_{i=0}^{M-1} s_i(t))^2 dt = \int_0^{T_s} \sum_{i=0}^{M-1} s_i^2(t) dt + \int_0^{T_s} \sum_{i=0}^{M-1} \sum_{\substack{j=0 \\ i \neq j}}^{M-1} s_i(t) s_j(t) dt \qquad (4.108)$$

$$= M E_s + M(M - 1) E_s \gamma$$

where γ is the correlation between any pair of distinct signals. Since $\int_0^{T_s} (\sum_{i=0}^{M-1} s_i(t))^2 dt \geq 0$ must be true for any signal set, then

$$M E_s + M(M - 1) E_s \gamma \geq 0 \qquad (4.109)$$

or

$$\gamma \geq -\frac{1}{M - 1} \qquad (4.110)$$

This is the maximum negative correlation achievable with M equally correlated signals. If we choose a set with $\gamma = -1/(M - 1)$, the signal set is called a **simplex** signal set with corresponding signal correlation matrix

$$\Gamma_{TO} = \begin{pmatrix} 1 & -\frac{1}{M-1} & \cdots & -\frac{1}{M-1} \\ -\frac{1}{M-1} & 1 & \ddots & \vdots \\ \vdots & \ddots & \ddots & -\frac{1}{M-1} \\ -\frac{1}{M-1} & \cdots & -\frac{1}{M-1} & 1 \end{pmatrix} \tag{4.111}$$

One method to construct simplex signals is to start with M orthogonal signals, $s_i(t)$, $i = 0, 1, \cdots, M-1$, each with energy E_s, and translate each signal in the set by the average of the signals, that is, let

$$a(t) \triangleq \frac{1}{M} \sum_{i=0}^{M-1} s_i(t) \tag{4.112}$$

Then, form the new signals

$$s_i'(t) = s_i(t) - a(t) \quad i = 0, 1, \cdots, M-1 \tag{4.113}$$

The signal energy of the new set is given by

$$\begin{aligned} E_s' &= \int_0^{T_s} (s_i'(t))^2 dt = \int_0^{T_s} s_i^2(t)dt + \int_0^{T_s} a^2(t)dt - 2\int_0^{T_s} s_i(t)a(t)dt \\ &= E_s + \frac{1}{M^2}(ME_s) - \frac{2}{M}E_s = E_s\left(1 - \frac{1}{M}\right) \end{aligned} \tag{4.114}$$

and the new signal normalized correlation coefficients become

$$\begin{aligned} \gamma_{ij} &= \frac{1}{E_s'}\int_0^{T_s} s_i'(t)s_j'(t)dt = \frac{1}{E_s'}\int_0^{T_s} (s_i(t) - \frac{1}{M}\sum_{k=0}^{M-1} s_k(t))(s_j(t) - \frac{1}{M}\sum_{l=0}^{M-1} s_l(t))dt \\ &= \frac{1}{E_s'}\left(0 - \frac{E_s}{M} - \frac{E_s}{M} + \frac{ME_s}{M^2}\right) = -\frac{1}{M-1} \quad \text{independent of } i \text{ and } j \end{aligned} \tag{4.115}$$

Hence, the new signal set achieves the minimum correlation possible for M signals and the correlation is independent of the signal pair. The simplex signals for $M = 2, 3$ and 4 are depicted in Fig. 4.21. When $M = 2$, the simplex set reduces to two antipodal signals as expected. Since the simplex set is obtained from an orthogonal set by translating the signals, the symbol error rate remains the same as it is insensitive to signal translation or rotation in AWGN channels. Using (4.93) and realizing that the signal energy is scaled by the factor $(1 - \frac{1}{M})$, then

$$P_s(E) = 1 - \int_{-\infty}^{\infty} \left[\frac{1}{2}\text{erfc}\left(-q - \sqrt{\frac{E_s}{N_0}\left(\frac{M}{M-1}\right)}\right)\right]^{M-1} \frac{e^{-q^2}}{\sqrt{\pi}}dq \tag{4.116}$$

where we let E_s denote the energy of the simplex signal (that is, replaced E_s' by E_s). Since all the signals are equidistant from each other, a symbol error is equally likely to be in favor of any incorrect symbol and as a result, the bit error probability is related to the symbol error probability in the same way it was for orthogonal signals. The Union bound for simplex

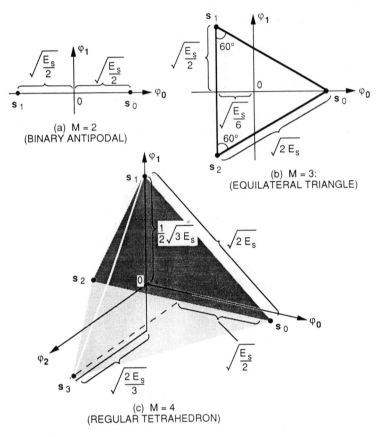

Figure 4.21   Simplex signals $M = 2, 3$, and 4 (Reprinted from [1])

signals is identical to that of orthogonal signals of (4.97) except for the adjustment in energy, that is

$$P_s(E) \le (M - 1)Q\left(\sqrt{\frac{E_s}{N_0}\left(\frac{M}{M - 1}\right)}\right) \tag{4.117}$$

Multiple Phase-Shift-Keyed (M-PSK) Signals.   This signal set corresponds to M equally spaced vectors around the unit circle. The elements of the covariance matrix that define this set of signals are given by

$$\gamma_{ij} = \begin{cases} \cos\left[\frac{2\pi}{M}(i - j)\right] & i \ne j \\ 1 & i = j \end{cases} \tag{4.118}$$

so that

$$\Gamma_{M-PSK} = \begin{pmatrix} 1 & \cos(\frac{2\pi}{M}) & \cdots & \cos\left(\frac{2\pi(M-1)}{M}\right) \\ \cos(\frac{2\pi}{M}) & 1 & \ddots & \vdots \\ \vdots & & \ddots & \ddots & \cos(\frac{2\pi}{M}) \\ \cos\left(\frac{2\pi(M-1)}{M}\right) & \cdots & \cdots & 1 \end{pmatrix} \qquad (4.119)$$

For $M = 2$, the signal set reduces to antipodal signals, while for $M = 4$, we obtain a biorthogonal signal set. Notice that this matrix is singular since the sum of elements in every column is zero; hence, the matrix has zero determinant.

For multiple phase-shift-keying (M-PSK), a convenient set of signals that may be generated is the polyphase signal set

$$s_i(t) = \sqrt{2P} \sin(\omega_c t + \theta_i), \quad 0 \le t \le T_s \qquad (4.120)$$

where ω_c is the carrier frequency and $\theta_i = 2\pi i/M$ for $i = 0, 1, \cdots, M - 1$. Notice that for $M = 2$, we get antipodal signals or binary-phase-shift-keying (BPSK), the case $M = 4$ produces quadriphase-shift-keying (QPSK), and $M = 8$ yields octaphase-shift-keying (8PSK). The three signal sets are depicted in Fig. 4.22, along with the decision regions of an optimum receiver (which employs minimum distance decoding).

Since the matrix Γ_{M-PSK} is singular, the dimensionality of the integration in (4.77) must be reduced in order to evaluate $P_s(E)$. A simpler method is to transform the problem into polar coordinates by representing the ith integrator output in terms of its amplitude and phase. The optimum receiver computes for all i the quantities

$$\begin{aligned} q_i &= \int_0^{T_s} r(t)s_i(t)dt = \sqrt{2P} \int_0^{T_s} r(t)\sin(\omega_c t + \theta_i)dt \\ &= \sqrt{2P}\left[\cos\theta_i \int_0^{T_s} r(t)\sin\omega_c t\, dt + \sin\theta_i \int_0^{T_s} r(t)\cos\omega_c t\, dt\right] \\ &= A\cos[\theta_i - \eta] \end{aligned} \qquad (4.121)$$

where defining

$$A_s \triangleq \int_0^{T_s} r(t)\left[\sqrt{2P}\sin\omega_c t\right]dt \qquad (4.122)$$

$$A_c \triangleq \int_0^{T_s} r(t)\left[\sqrt{2P}\cos\omega_c t\right]dt \qquad (4.123)$$

we have

$$A \triangleq \sqrt{A_s^2 + A_c^2} \qquad (4.124)$$

and

$$\eta \triangleq \tan^{-1}\frac{A_c}{A_s} \qquad (4.125)$$

On the basis of this representation of q_i, it is important to note that the receiver of Fig. 4.6 requires M correlators, whereas an equivalent receiver (see Fig. 4.23) requires only two

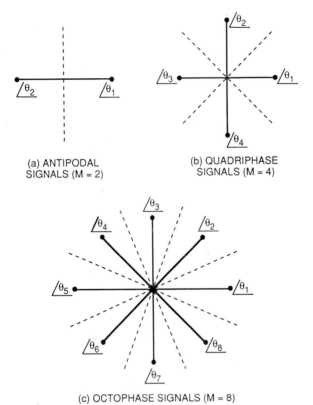

(a) ANTIPODAL
SIGNALS (M = 2)

(b) QUADRIPHASE
SIGNALS (M = 4)

(c) OCTOPHASE SIGNALS (M = 8)

THE DOTTED LINES REPRESENT DECISION
THRESHOLDS IN THE RECEIVER

Figure 4.22   The signal vectors for various polyphase signal sets: (a) Antipodal signals $(M = 2)$, (b) Quadriphase signals $(M = 4)$, (c) Octaphase signals $(M = 8)$

correlators. Since for any i, q_i represents a sample of a Gaussian process, the joint pdf of A and η, given that $s_i(t)$ was transmitted, is

$$f(A, \eta) = \begin{cases} \frac{A}{\pi E_s N_0} \exp\left[-\frac{A^2 - 2E_s A \cos(\eta - \theta_i) + E_s^2}{E_s N_0}\right], & 0 \le A \le \infty, 0 \le \eta \le 2\pi \\ 0, & \text{elsewhere} \end{cases} \quad (4.126)$$

Letting $r \triangleq A/\sqrt{E_s N_0}$ and averaging over the random variable r gives

$$f(\eta) = \begin{cases} \int_0^\infty \frac{r}{\pi} \exp\left\{-\left[r^2 - 2r\sqrt{\frac{E_s}{N_0}} \cos(\eta - \theta_i) + \frac{E_s}{N_0}\right]\right\} dr, & 0 \le \eta \le 2\pi \\ 0, & \text{elsewhere} \end{cases} \quad (4.127)$$

Since q_i is a maximum when the distance $|\theta_i - \eta|$ is a minimum, then for η in the region $(\theta_i - \pi/M, \theta_i + \pi/M)$, the inequality $|\theta_j - \eta| > |\theta_i - \eta|$ is valid for all $j \ne i$ and hence $q_i > q_j$ for all $j \ne i$. Based on this result, the probability of correctly detecting the ith signal is the probability that η is in the region $[(2i - 1)\pi/M, (2i + 1)\pi/M]$, that is

$$P_s(C/m_i) = \int_{(2i-1)\pi/M}^{(2i+1)\pi/M} f(\eta) d\eta \quad (4.128)$$

Letting $\psi \triangleq \eta - 2\pi i/M$, we see that $P_s(C/m_i)$ is independent of i; hence, the probability

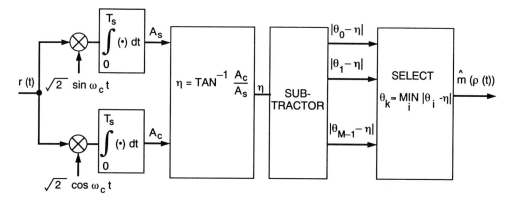

Figure 4.23   *M*-PSK receiver requiring two correlators

of correct detection becomes (see Problem 4.22)

$$P_s(C) = \frac{2}{\pi} \exp\left(-\frac{E_s}{N_0}\right) \int_0^\infty r \exp\left(-r^2\right) \left[\int_0^{\pi/M} \exp\left(2r\sqrt{\frac{E_s}{N_0}} \cos\psi\right) d\psi \right] dr \quad (4.129)$$

Introducing the change of variables $u = r\cos\psi$, $v = r\sin\psi$ and noting that $v = u\tan\psi$, (4.129) reduces to

$$P_s(C) = \frac{2}{\pi} \int_0^\infty \exp\left\{-(u - \sqrt{\frac{E_s}{N_0}})^2\right\} \left[\int_0^{u\tan\pi/M} \exp(-v^2) dv \right] du \quad (4.130)$$

The corresponding symbol error probability is derived in Appendix 4A and can be written as

$$P_s(E) = \frac{M-1}{M} - \frac{1}{2}\mathrm{erf}\left[\sqrt{\frac{E_s}{N_0}} \sin\frac{\pi}{M}\right]$$

$$- \frac{1}{\sqrt{\pi}} \int_0^{\sqrt{E_s/N_0}\sin\pi/M} \exp(-y^2)\mathrm{erf}\left(y\cot\frac{\pi}{M}\right) dy \quad (4.131)$$

For the case of BPSK, $M = 2$ and the performance given by (4.131) reduces to (4.55). It is also possible to evaluate the integral in (4.131) in closed form for $M = 4$, that is, QPSK. The result is given by

$$P_s(E)|_{M=4} = \mathrm{erfc}\left(\sqrt{\frac{E_s}{2N_0}}\right) - \frac{1}{4}\mathrm{erfc}^2\left(\sqrt{\frac{E_s}{2N_0}}\right) \quad (4.132)$$

In the region where $E_s/N_0 \gg 1$, (4.132) is well approximated by

$$P_s(E) \simeq \mathrm{erfc}\left(\sqrt{\frac{E_s}{2N_0}}\right) \quad (4.133)$$

This implies that the symbol error probability performance using four signals is about 3 dB worse than that of orthogonal signals discussed earlier. For values of $P_s(E)$ less than 10^{-3},

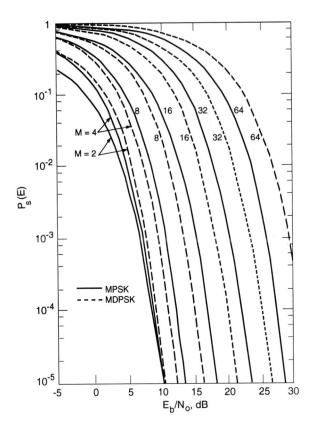

Figure 4.24 Symbol error probability for M-PSK signals as a function of bit SNR (Reprinted from [12])

the region of great interest in practice, the error probability for $M > 2$ is well approximated by

$$P_s(E) \simeq \mathrm{erfc}\left(\sqrt{\frac{E_s}{N_0}}\sin\frac{\pi}{M}\right) \qquad (4.134)$$

since when an error occurs, one of the two neighbors of the transmitted signal is chosen with high probability. This approximation becomes extremely tight for fixed M as E_s/N_0 increases and serves as an upper bound on the probability of symbol error for any E_s/N_0 (see Problem 4.23). The performance for various values of M is depicted in Fig. 4.24 for $M = 2, 4, 8, 16, 32$, and 64, as a function of bit SNR.

The computation of the bit error rate for M-PSK signals when $M = 2^n$ is considered for Gray code mapping, as given in Table 4.1. The Gray code has the property that in going from one symbol to an adjacent symbol, only one bit out of the $\log_2 M$ bits changes. As a result, an error in an adjacent symbol is accompanied by one and only one bit error. At large symbol SNRs, we can assume that the only significant symbol errors are those whose corresponding signal phases are adjacent to that of the transmitted signal. Hence, the bit error probability can be approximated by

$$P_b(E) \simeq \frac{P_s(E)}{\log_2 M} \qquad (4.135)$$

Table 4.1   Gray Code Bit Mapping and Hamming Weight (Reprinted from [7])

MESSAGE	BITS	W_m	MESSAGE	BITS	W_m
n = 2 (M = 4)			n = 5 (M = 32)		
0	0 0	0	0	0 0 0 0 0	0
1	0 1	1	1	0 0 0 0 1	1
2	1 1	2	2	0 0 0 1 1	2
3	1 0	1	3	0 0 0 1 0	1
n = 3 (M = 8)			4	0 0 1 1 0	2
			5	0 0 1 1 1	3
0	0 0 0	0	6	0 0 1 0 1	2
1	0 0 1	1	7	0 0 1 0 0	1
2	0 1 1	2	8	0 1 1 0 0	2
3	0 1 0	1	9	0 1 1 0 1	3
4	1 1 0	2	10	0 1 1 1 1	4
5	1 1 1	3	11	0 1 1 1 0	3
6	1 0 1	2	12	0 1 0 1 0	2
7	1 0 0	1	13	0 1 0 1 1	3
n = 4 (M = 16)			14	0 1 0 0 1	2
			15	0 1 0 0 0	1
0	0 0 0 0	0	16	1 1 0 0 0	2
1	0 0 0 1	1	17	1 1 0 0 1	3
2	0 0 1 1	2	18	1 1 0 1 1	4
3	0 0 1 0	1	19	1 1 0 1 0	3
4	0 1 1 0	2	20	1 1 1 1 0	4
5	0 1 1 1	3	21	1 1 1 1 1	5
6	0 1 0 1	2	22	1 1 1 0 1	4
7	0 1 0 0	1	23	1 1 1 0 0	3
8	1 1 0 0	2	24	1 0 1 0 0	2
9	1 1 0 1	3	25	1 0 1 0 1	3
10	1 1 1 1	4	26	1 0 1 1 1	4
11	1 1 1 0	3	27	1 0 1 1 0	3
12	1 0 1 0	2	28	1 0 0 1 0	2
13	1 0 1 1	3	29	1 0 0 1 1	3
14	1 0 0 1	2	30	1 0 0 0 1	2
15	1 0 0 0	1	31	1 0 0 0 0	1

At lower symbol SNRs where the approximation of (4.135) is not valid, it was not until recently that the bit error rates were able to be evaluated exactly. Only the key results are presented here and the reader is referred to [7, 8] and its references for additional detail. Let R_m denote the decision region when message m is transmitted. Let S_m be the probability that the received signal vector $\mathbf{r}$ falls into R_m when message "0," corresponding to zero phase, is transmitted (no loss in generality is obtained by considering message "0"). Then, it can be shown that

$$
S_m = \begin{cases}
\int_0^\infty \frac{e^{-(z-d_n)^2/2}}{\sqrt{2\pi}} \{Q\,[z\tan((2m-1)a_n)] \\
\quad - Q\,[z\tan((2m+1)a_n)]\}\,dz & m = 0, 1, \cdots, \frac{M}{4} - 1 \\
\int_0^\infty \frac{e^{-(z+d_n)^2/2}}{\sqrt{2\pi}} \{Q\,[z\tan((M-2m-1)a_n)] \\
\quad - Q\,[z\tan((M-2m+1)a_n)]\}\,dz & m = \frac{M}{4} + 1, \cdots, \frac{M}{2}
\end{cases} \qquad (4.136)
$$

where $d_n = \sqrt{2n\frac{E_b}{N_0}}$, $a_n = \pi/2^n$ with $M = 2^n$ and

$$S_m = S_{M-m} \text{ for } m = \frac{M}{2} + 1, \cdots, M - 1 \tag{4.137}$$

Let W_m be the Hamming weight of the bits assigned to symbol m (see Table 4.1). Since the probability of a bit error is $W_m / \log_2 M$ when message "0" (all zero bit pattern) is transmitted and symbol m is decided (that is, $r \epsilon R_m$), then

$$P_b(E) = \frac{1}{\log_2 M} \sum_{m=1}^{M-1} W_m S_m \tag{4.138}$$

For $M = 2, 4$, it can be shown using integration by parts that

$$P_b(E) = Q(d_1 \sin a_1) \quad M = 2 \tag{4.139}$$

$$P_b(E) = \frac{1}{2}\{S_1 + 2S_2 + S_3\} = Q(d_2 \sin a_2) \quad M = 4 \tag{4.140}$$

Since $d_1 \sin a_1 = d_2 \sin a_2$, the bit error rates for BPSK and QPSK are identical. For $M = 8$, we have

$$
\begin{aligned}
P_b(E) &= \frac{1}{3}\{S_1 + 2S_2 + S_3 + 2S_4 + 3S_5 + 2S_6 + S_7\} \\
&= \frac{2}{3}\{Q(d_3 \sin a_3) + Q(d_3 \sin 3a_3)[1 - Q(d_3 \sin a_3)]\}
\end{aligned} \tag{4.141}
$$

For $M = 16$,

$$P_b(E) = \frac{2}{4}\left\{\sum_{m=1}^{8} S_m + \sum_{m=2}^{5} S_m + S_5 + 2S_6 + S_7\right\} \tag{4.142}$$

The exact BERs for M-PSK are shown in Table 4.2 for $M = 2, 4, 8, 16, 32$, and 64 and plotted in Fig. 4.25.

     A more recent study [8] presented a technique for evaluating the bit error probability with any bit mapping for M-PSK signals. A simple technique for evaluating S_m is suggested which uses closed form expressions for the half-plane and quadrant plane probabilities in terms of the Q-function. Bit error probabilities for M-PSK with Gray, natural binary, and folded binary bit mapping are derived.

     L-Orthogonal Signals. L-orthogonal signal sets [10] are generalizations of biorthogonal signal sets and consist of $M = NL$ equal energy, equal time-duration signals. A given L-orthogonal signal set has the following two properties: (1) the signal set may be divided into N disjoint subsets, each subset containing L signals; and (2) signals from different subsets are orthogonal. One instance of an L-orthogonal signal set is realized when each signal subset is chosen as a set of polyphase signals. A construction of such an L-orthogonal signal set is given in Reed and Scholtz [10]. The signal correlation coefficients are given by[2]

$$\gamma_{ij} = \begin{cases} 0, & \left[\frac{i-1}{L}\right] \neq \left[\frac{j-1}{L}\right] \\ 1, & i = j \\ \cos\left(\frac{2\pi}{L}(i-j)\right), & \left[\frac{i-1}{L}\right] = \left[\frac{j-1}{L}\right] \end{cases} \tag{4.143}$$

2. $[y] = $ integer part of y.

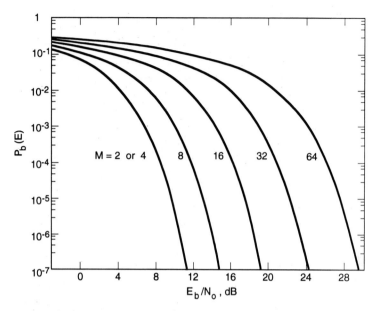

Figure 4.25   Bit error rate of M-PSK over AWGN channel (Reprinted from [7])

The corresponding correlation matrix $\Gamma_{L-\text{Orth}}$ can be partitioned in the following form:

$$\Gamma_{L-\text{Orth}} = \begin{bmatrix} \Gamma_{L-\text{PSK}} & \mathcal{O} & \cdots & \mathcal{O} \\ \mathcal{O} & \Gamma_{L-\text{PSK}} & & \vdots \\ \vdots & & \ddots & \vdots \\ \mathcal{O} & \cdots & \cdots & \Gamma_{L-\text{PSK}} \end{bmatrix} \tag{4.144}$$

where $\Gamma_{L-\text{PSK}}$ is the matrix defined in (4.119) with M replaced by L, and $\mathcal{O}$ is the $L \times L$ null matrix whose elements are all zero. Again, the special case $L = 2$ yields biorthogonal signal sets. Other special cases appear of little practical importance in the implementation of phase-coherent systems.

A set of M signals that have the correlation properties of (4.143) can be realized by combining the orthogonal M-FSK signal discussed in Example 4.2 with the polyphase signal representation of (4.120). Specifically, the set of $M = NL$ signals

$$s_i(t) = \begin{cases} \sqrt{2P} \sin\left[\omega_c t + \frac{n\pi t}{T} + \theta_l\right], & L = 1, 2 \\ \sqrt{2P} \sin\left[\omega_c t + \frac{2n\pi t}{T} + \theta_l\right], & L > 2 \end{cases} \tag{4.145}$$

is L-orthogonal (see Problem 4.24), where $\theta_l = 2l\pi/L$ and $i = nL + l$ with $n = 0, 1, \cdots,$ $N - 1$ and $l = 0, 1, \cdots, L - 1$. Several other examples of L-orthogonal signals sets are given in Problem 4.25. Note that the minimum frequency separation between the signals for $L > 2$ is $1/T$ as opposed to $1/2T$ required for $L = 1, 2$. We will see in a later chapter that $1/T$ is the minimum signal separation required for noncoherent reception to guarantee orthogonality. When the phase takes on more than two values, it can be argued that it becomes "unknown" or random and hence, the frequency separation for noncoherent signals is more appropriate.

Since for the above set of signals, the correlation matrix of (4.144) is singular, the

Table 4.2   Exact BER Values of M-ary PSK with Gray Code Mapping (Reprinted from [7])

E_b/N_0, dB	P_b (E) ($M = 2, 4$)	P_b (E) ($M = 8$)	P_b (E) ($M = 16$)	P_b (E) ($M = 32$)	P_b (E) ($M = 64$)
−5.0	2.132E-01	2.468E-01	2.867E-01	3.185E-01	3.376E-01
−4.0	1.861E-01	2.217E-01	2.646E-01	3.003E-01	3.242E-01
−3.0	1.584E-01	1.961E-01	2.420E-01	2.817E-01	3.099E-01
−2.0	1.306E-01	1.708E-01	2.191E-01	2.629E-01	2.951E-01
−1.0	1.038E-01	1.461E-01	1.965E-01	2.442E-01	2.800E-01
0.0	7.865E-02	1.227E-01	1.745E-01	2.256E-01	2.649E-01
1.0	5.628E-02	1.008E-01	1.535E-01	2.073E-01	2.497E-01
2.0	3.751E-02	8.061E-02	1.338E-01	1.892E-01	2.345E-01
3.0	2.288E-02	6.225E-02	1.155E-01	1.714E-01	2.194E-01
4.0	1.250E-02	4.589E-02	9.865E-02	1.538E-01	2.043E-01
5.0	5.954E-03	3.186E-02	8.292E-02	1.368E-01	1.895E-01
6.0	2.388E-03	2.048E-02	6.816E-02	1.207E-01	1.747E-01
7.0	7.727E-04	1.195E-02	5.429E-02	1.055E-01	1.599E-01
8.0	1.909E-04	6.181E-03	4.145E-02	9.147E-02	1.452E-01
9.0	3.363E-05	2.748E-03	2.998E-02	7.840E-02	1.307E-01
10.0	3.872E-06	1.011E-03	2.025E-02	6.614E-02	1.165E-01
11.0	2.613E-07	2.937E-04	1.256E-02	5.451E-02	1.030E-01
12.0	9.006E-09	6.338E-05	7.010E-03	4.349E-02	9.027E-02
13.0		9.417E-06	3.427E-03	3.325E-02	7.848E-02
14.0		8.756E-07	1.421E-03	2.406E-02	6.752E-02
15.0		4.516E-08	4.789E-04	1.627E-02	5.724E-02
16.0			1.246E-04	1.010E-02	4.747E-02
17.0			2.342E-05	5.642E-03	3.819E-02
18.0			2.925E-06	2.763E-03	2.950E-02
19.0			2.187E-07	1.147E-03	2.163E-02
20.0			8.573E-09	3.876E-04	1.486E-02
21.0				1.011E-04	9.417E-03
22.0				1.907E-05	5.394E-03
23.0				2.393E-06	2.725E-03
24.0				1.799E-07	1.177E-03
25.0				7.099E-09	4.176E-04
26.0					1.159E-04
27.0					2.361E-05
28.0					3.264E-06

dimensionality of the integration in (4.77) must be reduced. A procedure for doing so is described in Reed and Scholtz. The details of the derivation are quite involved, and hence we merely present the results. For $L > 2$, the probability of correct detection is given by[3]

$$P_s(C) = \int\int_B \frac{\exp\{-(x_1^2 + x_2^2)\}}{\pi}$$

$$\times \left[\int\int_{A(x_1)} \frac{\exp\{-(y_1^2 + y_2^2)\}}{\pi} dy_1 dy_2 \right]^{N-1} dx_1 dx_2 \tag{4.146}$$

where $A(x_1)$ and B are regions of integration as shown in Fig. 4.26. In particular, B is the set of points (x_1, x_2) in the wedge bounded by the straight lines

---

3. For $L = 1$ and $L = 2$, we have the well-known results of orthogonal and biorthogonal signaling respectively.

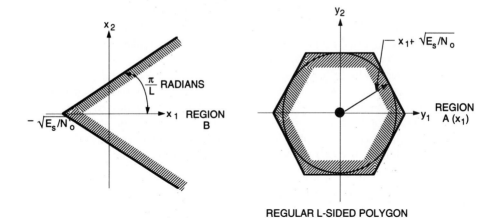

Figure 4.26   Areas of integration (Examples drawn for $L = 6$)

$$x_2 = \left(x_1 + \sqrt{\frac{E_s}{N_0}}\right) \tan \frac{\pi}{L} \tag{4.147}$$

$$x_2 = -\left(x_1 + \sqrt{\frac{E_s}{N_0}}\right) \tan \frac{\pi}{L} \tag{4.148}$$

and $A(x_1)$ is the set of points within an L-sided regular polygon circumscribed on a circle of radius $x_1 + \sqrt{E_s/N_0}$. Substituting (4.147) and (4.148) in (4.146) and making the change of variables $u_1 = x_1 + \sqrt{E_s/N_0}$ gives (for $L > 2$)

$$P_s(C) = \frac{2}{\pi} \int_0^\infty \exp\left[-\left(u_1 - \sqrt{\frac{E_s}{N_0}}\right)^2\right]\left[\int_0^{u_1 \tan(\pi/L)} \exp\left(-x_2^2\right) dx_2\right]$$

$$\times \left[\iint_{A(u_1 - \sqrt{E_s/N_0})} \frac{\exp\left\{-\left(y_1^2 + y_2^2\right)\right\}}{\pi} dy_1 dy_2\right]^{N-1} du_1 \tag{4.149}$$

Since any L-sided regular polygon can be subdivided into L equal area triangles, then from the symmetry of the double integral, we get

$$P_s(C) = \frac{2}{\pi} \int_0^\infty \exp\left[-\left(u_1 - \sqrt{\frac{E_s}{N_0}}\right)^2\right]\left[\int_0^{u_1 \tan(\pi/L)} \exp\left(-x_2^2\right) dx_2\right]$$

$$\times \left[L \int_0^{u_1} \int_{-y_1 \tan(\pi/L)}^{y_1 \tan(\pi/L)} \frac{\exp\left\{-\left(y_1^2 + y_2^2\right)\right\}}{\pi} dy_1 dy_2\right]^{N-1} du_1 \tag{4.150}$$

$$= \frac{1}{\sqrt{\pi}} \int_0^\infty \exp\left[-\left(u_1 - \sqrt{\frac{E_s}{N_0}}\right)^2\right] \operatorname{erf}\left(u_1 \tan \frac{\pi}{L}\right)$$

$$\times \left[\frac{L}{\sqrt{\pi}} \int_0^{u_1} \exp\left(-y_1^2\right) \operatorname{erf}\left(y_1 \tan \frac{\pi}{L}\right) dy_1\right]^{N-1} du_1$$

The special case of $L = 2$ corresponds to the biorthogonal signal set and hence (4.150) reduces to (4.102) (see Problem 4.26). When $N = 1$, we merely have a single set of polyphase signals, and by inspection (4.150) reduces to (4.130). For other combinations of N and L, the integrals in (4.150) must be evaluated numerically.

Upper and lower bounds on the probability of symbol error $P_s(E) = 1 - P_s(C)$ may be readily calculated by circular approximation to the area of integration $A(u_1)$. In particular, an upper bound on $P_s(E)$ [lower bound on $P_s(C)$] is obtained by approximating $A(u_1)$ by its inscribed circle of radius u_1. Similarly, a lower bound on $P_s(E)$ may be calculated by approximating $A(u_1)$ by its circumscribed circle of radius $u_1 \sec(\pi/L)$. Considering first the upper bound on $P_s(E)$ and letting $y_1 = r \cos \psi$ and $y_2 = r \sin \psi$ gives

$$
P_s(C) \geq \frac{2}{\pi} \int_0^\infty \exp\left[-\left(u_1 - \sqrt{\frac{E_s}{N_0}}\right)^2\right] \left[\int_0^{u_1 \tan(\pi/L)} \exp\left(-x_2^2\right) dx_2\right]
$$

$$
\times \left[\int_0^{2\pi} \int_0^{u_1} \frac{r \exp\left(-r^2\right)}{\pi} dr d\psi\right]^{N-1} du_1
$$

(4.151)

$$
= \frac{2}{\pi} \int_0^\infty \exp\left[-\left(u_1 - \sqrt{\frac{E_s}{N_0}}\right)^2\right] \left[\int_0^{u_1 \tan(\pi/L)} \exp\left(-x_2^2\right) dx_2\right]
$$

$$
\times \left[1 - \exp\left(-u_1^2\right)\right]^{N-1} du_1
$$

Recall the binomial expansion

$$
\left\{1 - \exp\left[-u_1^2\right]\right\}^{N-1} = \sum_{k=0}^{N-1} {}_{N-1}C_k (-1)^k \exp\left(-k u_1^2\right)
$$

(4.152)

where ${}_n C_k$ is the binomial coefficient given by

$$
{}_n C_k \triangleq \frac{n!}{k!(n-k)!}
$$

(4.153)

Substituting (4.152) into (4.151) and making the changes of variables $u = u_1\sqrt{1+k}$, $v = x_2\sqrt{1+k}$ gives

$$
P_s(C) \geq \sum_{k=0}^{N-1} \frac{(-1)^k}{1+k} {}_{N-1}C_k \exp\left\{-\frac{kE_s/N_0}{1+k}\right\}
$$

(4.154)

$$
\times \left\{\frac{2}{\pi} \int_0^\infty \exp\left[-\left(u - \sqrt{\frac{E_s/N_0}{1+k}}\right)^2\right] \left[\int_0^{u \tan \pi/L} \exp\left\{-\frac{v^2}{1+k}\right\} dv\right] du\right\}
$$

Recognizing that the $k = 0$ term in (4.154) corresponds to $P_s(C)|_{M-PSK}$ as given by (4.130), we write the upper bound on $P_s(E)$ as[4]

---

4. $P_s(E)|_{L-PSK}$ is the symbol error probability of $M - PSK$ signals with M replaced by L, hence the notation $L - PSK$.

$$P_s(E) \leq P_s(E)|_{L-PSK} - \sum_{k=1}^{N-1} \frac{(-1)^k}{1+k} {}_{N-1}C_k \exp\left\{-\frac{kEs/N_0}{1+k}\right\} \tag{4.155}$$

$$\times \left\{ \frac{2}{\pi} \int_0^\infty \exp\left[-\left(u - \sqrt{\frac{E_s/N_0}{1+k}}\right)^2\right] \left[\int_0^{u\tan\pi/L} \exp\left\{-\frac{v^2}{1+k}\right\} dv\right] du \right\}$$

where $P_s(E)|_{L-\text{PSK}} \triangleq 1 - P_s(C)|_{L-\text{PSK}}$. Similarly, one can derive a lower bound on $P_s(E)$ with the result

$$P_s(E) \geq P_s(E)|_{L-PSK} - \sum_{k=1}^{N-1} \frac{(-1)^k}{1+k'} {}_{N-1}C_k \exp\left\{-\frac{k'Es/N_0}{1+k'}\right\} \tag{4.156}$$

$$\times \left\{ \frac{2}{\pi} \int_0^\infty \exp\left[-\left(u - \sqrt{\frac{E_s/N_0}{1+k'}}\right)^2\right] \left[\int_0^{u\tan\pi/L} \exp\left\{-\frac{v^2}{1+k'}\right\} dv\right] du \right\}$$

where $k' \triangleq k \sec^2(\pi/L)$. The bounds of (4.155) and (4.156) can be simplified by reducing the double integration involving an infinite limit to a single integration having a finite limit. The procedure for doing this reduction parallels that given in Appendix 4A. The results are

$$P_s(E) \leq P_{s_u}(E) \triangleq P_s(E)|_{L-PSK} - \sum_{k=1}^{N-1} \frac{(-1)^k}{1+k} {}_{N-1}C_k \exp\left\{-\frac{kEs/N_0}{1+k}\right\}$$

$$\times \left[\frac{\theta}{\pi} + \frac{1}{2}\text{erf}\left(\sqrt{\frac{E_s/N_0}{1+k}}\sin\theta\right) \right. \tag{4.157}$$

$$\left. + \frac{1}{\sqrt{\pi}} \int_0^{\sqrt{(E_s/N_0)/(k+1)}\sin\theta} \exp\left(-y^2\right)\text{erf}(y\cot\theta)dy \right]$$

with $\theta \triangleq \tan^{-1}\left[\frac{\tan(\pi/L)}{\sqrt{1+k}}\right]$ and

$$P_s(E) \geq P_{s_l}(E) \triangleq P_s(E)|_{L-PSK} - \sum_{k=1}^{N-1} \frac{(-1)^k}{1+k'} {}_{N-1}C_k \exp\left\{-\frac{k'Es/N_0}{1+k'}\right\}$$

$$\times \left[\frac{\theta'}{\pi} + \frac{1}{2}\text{erf}\left(\sqrt{\frac{E_s/N_0}{1+k}}\sin\theta'\right) \right. \tag{4.158}$$

$$\left. + \frac{1}{\sqrt{\pi}} \int_0^{\sqrt{(E_s/N_0)/(k+1)}\sin\theta'} \exp\left(-y^2\right)\text{erf}\left(y\cot\theta'\right) dy \right]$$

with $\theta' \triangleq \tan^{-1}\left[\frac{\tan(\pi/L)}{\sqrt{1+k'}}\right]$. We note from (4.157) and (4.158) that for $N = 1$, the two bounds become equal and give the exact performance; that is $P_s(E)|_{L-\text{PSK}}$. Also for large L (small N), the bounds are very tight since both the inscribed and circumscribed circles are good approximations to the L-sided polygon. Figure 4.27 illustrates the upper and lower bounds of (4.157) and (4.158) respectively as functions of E_b/N_0 in dB for $M = 64$ and various combinations of N and L.

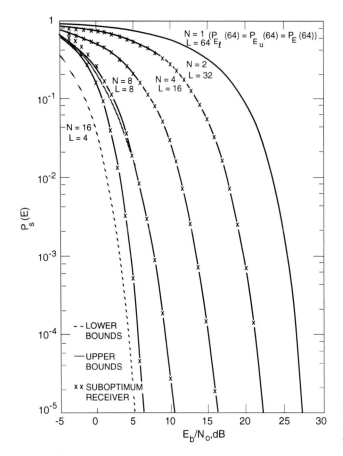

Figure 4.27 Symbol error probability of L-orthogonal ($M = 64$) signals as a function of bit SNR (Reprinted form [12])

Optimum detection of the set of signals in (4.145) requires the correlation receiver of Fig. 4.6 and yields the performance as given by (4.150). When M becomes large, the number of correlators required to implement the receiver of Fig. 4.6 becomes prohibitive. Hence, we investigate a suboptimum receiver that requires only $2N$ correlators and whose error probability performance is quite close to that of the optimum receiver.

The suboptimum receiver illustrated in Fig. 4.28 uses a noncoherent detector to determine the most probable index n characterizing the received signal [see (4.145)] and simultaneously makes a maximum-likelihood decision on the phase θ_l, or equivalently, the index l. We note from (4.145) that the index i is correct if and only if n and l are correctly chosen. The probability that n is correct is the probability of the event $\{V_n^2 = \max_k V_k^2\}$ or equivalently, the probability of choosing the correct frequency from a total of N choices. This probability $P_s(C)|_{NON}$ is evaluated in Chapter 5 and can be obtained by taking one minus the error probability performance as given by (5.77) with M replaced by N. Having selected the correct frequency, the conditional probability $P_s(C)$ of correctly choosing l (or equivalently, the phase θ_l from a total of L choices) is given by (4.130) with M replaced by L. Thus, the probability of correct reception is

$$P_s(C)|_{M \text{ signals}} = P_s(C)|_{NON,N \text{ signals}} \ P_s(C)|_{L-\text{PSK}} \qquad (4.159)$$

or equivalently, the symbol error probability for the suboptimum receiver of Fig. 4.28 is given by

$$
\begin{aligned}
P_s(E)|_{M \text{ signals}} = {}& P_s(E)|_{NON,N \text{ signals}} + P_s(E)|_{L-\text{PSK}} \\
& - P_s(E)|_{NON,N \text{ signals}} \, P_s(E)_{L-\text{PSK}}
\end{aligned}
\tag{4.160}
$$

The error probability performance of the suboptimum receiver as given by (4.160) is superimposed in Fig. 4.27 on the performance bounds for the optimum receiver in Fig. 4.6. For large E_b/N_0, the difference in performance between the optimum and suboptimum receivers is virtually negligible. Finally, note that the receiver implementation of Fig. 4.28 requires N noncoherent frequency detectors and L phase estimators. Since the phase estimate is obtained after the correct frequency has been chosen, an alternate implementation could employ only a single phase estimator whose input terminals are switched to those of the noncoherent detector corresponding to the chosen frequency.

M-ary Amplitude Modulation (M-AM).   Unlike previous signal sets, M-ary amplitude modulated signals are not equal energy signals. In this case, the messages are transmitted using M distinct amplitudes[5], that is

$$
s_i(t) = \sqrt{E_i}\phi(t), \quad 0 \le t \le T_s
\tag{4.161}
$$

for $i = 0, 1, \cdots, M-1$ where $\phi(t)$ is any unit energy function of duration T_s seconds, that is, $\int_0^{T_s} \phi^2(t)dt = 1$. Since the energies are unequal, the normalized correlation coefficients become [see (4.45)]

$$
\gamma_{ij} = \int_0^{T_s} \left(\frac{s_i(t)}{\sqrt{E_i}} \right) \left(\frac{s_j(t)}{\sqrt{E_j}} \right) dt = 1 \text{ for all } i \text{ and } j
\tag{4.162}
$$

Hence, the signal correlation matrix Γ has unity entries and is noninvertible. The optimum receiver is the minimum distance decision rule and since all the signals are expressed in terms of one function $\phi(t)$, the waveform problem can be reduced to an M-ary scalar equivalent problem. When $\sqrt{E_i} = i\Delta$, the optimum receiver is given in Example 3.3 and its performance is evaluated in Example 3.4. Noting that $\sigma^2 = N_0/2$, we obtain

$$
P_s(E) = 2 \left(\frac{M-1}{M} \right) Q \left(\sqrt{\frac{\Delta^2}{2N_0}} \right)
\tag{4.163}
$$

If we further assume that the transmitter is subject to an average power limitation of E_{avg}/T_s (Watts), it follows that

$$
\sum_{i=0}^{M-1} \left(\frac{E_i}{T_s} \right) \frac{1}{M} = \frac{E_{\text{avg}}}{T_s}
\tag{4.164}
$$

Making use of the inequality

$$
\sum_{j=0}^{M-1} j^2 = \frac{M(M-1)(2M-1)}{6}
\tag{4.165}
$$

---

5. Quite often, M-AM is defined by a set of amplitudes that are uniformly spaced and symmetrically located around zero.

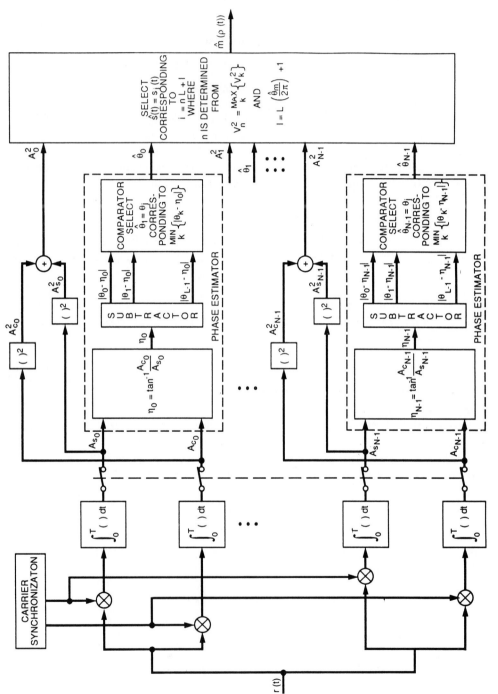

Figure 4.28 A suboptimal receiver for coherent detection of L-orthogonal signals

it is readily seen that the quantity Δ is constrained to equal

$$\Delta = \sqrt{\frac{6E_{avg}}{(M-1)(2M-1)}} \tag{4.166}$$

Consequently, the average probability of error may be written as

$$P_s(E) = 2\left(\frac{M-1}{M}\right) Q\left(\sqrt{\frac{3E_{avg}}{N_0(M-1)(2M-1)}}\right) \tag{4.167}$$

When $M = 2$, (4.167) reduces to the "on-off" keying performance of (4.68) as expected. Given a symbol-to-bit mapping, the bit error probability can be computed exactly in closed form using (4.86) and the probabilities of Example 3.4.

4.1.5 Limiting Error Probability for Block Codes as $M \to \infty$

We are now in a position to relate the limiting error probability performance of block codes as the number of codewords M approaches infinity to the notions of channel capacity, communication efficiency, and the Shannon limit introduced in Chapter 1. We shall deal with orthogonal signal sets, although the conclusions drawn apply to a more general class of block codes, including biorthogonal and transorthogonal sets.

Consider the probability of correct symbol detection for orthogonal signals as given in (4.93), that is

$$P_s(C) = \int_{-\infty}^{\infty} \left[\frac{1}{2}\text{erfc}\left(-q - \sqrt{\frac{E_s}{N_0}}\right)\right]^{M-1} \frac{\exp(-q^2)}{\sqrt{\pi}} \, dq \tag{4.168}$$

With a simple change of variables, (4.168) can be expressed as

$$P_s(C) = \int_{-\infty}^{\infty} \frac{\exp(-v^2/2)}{\sqrt{2\pi}} \left[Q\left(-v - \sqrt{\frac{2E_b}{N_0}\log_2 M}\right)\right]^{M-1} \, dv \tag{4.169}$$

Consider the limiting behavior of $P_s(C)$ as M approaches infinity, that is

$$\lim_{M\to\infty} P_s(C) = \lim_{M\to\infty} \int_{-\infty}^{\infty} \frac{\exp(-v^2/2)}{\sqrt{2\pi}} \left[Q\left(-v - \sqrt{\frac{2E_b}{N_0}\log_2 M}\right)\right]^{M-1} \, dv$$

$$= \int_{-\infty}^{\infty} \frac{\exp(-v^2/2)}{\sqrt{2\pi}} \lim_{M\to\infty} \left[Q\left(-v - \sqrt{\frac{2E_b}{N_0}\log_2 M}\right)\right]^{M-1} \, dv \tag{4.170}$$

If we define $\beta \triangleq E_b/N_0$ and consider the limit of the logarithm of the expression in brackets, we obtain

$$\lim_{M\to\infty} \ln\left[Q\left(-v - \sqrt{2\beta \log_2 M}\right)\right]^{M-1} = \lim_{M\to\infty} \frac{\ln Q\left(-v - \sqrt{2\beta \log_2 M}\right)}{1/(M-1)} \tag{4.171}$$

As $M \to \infty$, the limit approaches $\ln Q(-\infty)/0 = 0/0$. Treating M as a continuous variable and using l'Hospital's rule, we have[6]

$$\lim_{M\to\infty} \frac{\left(\frac{1}{Q(-v-\sqrt{2\beta\log_2 M})}\right)\left(\frac{1}{\sqrt{2\pi}}\right)\left(\exp\left\{-\frac{1}{2}\left(v+\sqrt{2\beta\log_2 M}\right)\right\}^2\right)\frac{d}{dM}\left(v+\sqrt{2\beta\log_2 M}\right)}{-1/(M-1)^2}$$

(4.172)

Defining the constants

$$c_1 \triangleq \sqrt{\frac{\beta}{2\ln 2}}\,\frac{1}{\sqrt{2\pi}}\exp\left(-\frac{v^2}{2}\right)$$

(4.173)

$$c_2 \triangleq v\sqrt{\frac{2\beta}{\ln 2}}$$

(4.174)

and noting that

$$\frac{d}{dM}\left(v+\sqrt{2\beta\log_2 M}\right) = \frac{1}{M\sqrt{\ln M}}\sqrt{\frac{\beta}{2\ln 2}}$$

(4.175)

$$\exp\left(-\beta\frac{\ln M}{\ln 2}\right) = M^{-(\beta/\ln 2)}$$

(4.176)

(4.172) can be expressed as

$$\lim_{M\to\infty}(-c_1)\frac{(M-1)^2\exp\left(-c_2\sqrt{\ln M}\right)}{M^{(1+\frac{\beta}{\ln 2})}}\frac{1}{\sqrt{\ln M}}\frac{1}{Q\left(-v-\sqrt{2\beta\log_2 M}\right)}$$

(4.177)

For large M, $Q(-v-\sqrt{2\beta\log_2 M}) \to 1$ and $M-1 \simeq M$. Hence, (4.177) gives

$$\lim_{M\to\infty}(-c_1)M^{(1-\frac{\beta}{\ln 2})}\frac{\exp\left(-c_2\sqrt{\ln M}\right)}{\sqrt{\ln M}}$$

(4.178)

Letting $w \triangleq \sqrt{\ln M}$, then $M = \exp(w^2)$ and (4.178) reduces to

$$\lim_{w\to\infty}(-c_1)\frac{\exp\left(-c_2 w\right)\exp\left(\left(1-\frac{\beta}{\ln 2}\right)w^2\right)}{w} = \begin{cases} 0, & \beta/\ln 2 > 1 \\ -\infty, & \beta/\ln 2 < 1 \end{cases}$$

(4.179)

The limit is independent of the sign of c_2 as the exponential in w^2 always dominates the exponential in w as $w \to \infty$. Since this is the limit of the logarithm in brackets, the limit of the expression itself is

$$\lim_{M\to\infty}\left[Q\left(-v-\sqrt{\frac{2E_b}{N_0}\log_2 M}\right)\right]^{M-1} = \begin{cases} 1, & E_b/N_0 > \ln 2 \\ 0, & E_b/N_0 < \ln 2 \end{cases}$$

(4.180)

6. To compute the derivative of $Q(x)$, use Leibnitz's rule

$$\frac{d}{da}\int_p^q f(x,a)dx = \int_p^q \frac{\partial}{\partial a}f(x,a)dx + f(q,a)\frac{dq}{da} - f(p,a)\frac{dp}{da}$$

so that the limit of (4.168) becomes

$$\lim_{M \to \infty} P_s(C) = \begin{cases} 1, & E_b/N_0 > \ln 2 \\ 0, & E_b/N_0 < \ln 2 \end{cases} \tag{4.181}$$

or, in terms of error probability

$$\lim_{M \to \infty} P_s(E) = \begin{cases} 0, & E_b/N_0 > \ln 2 \\ 1, & E_b/N_0 < \ln 2 \end{cases} \tag{4.182}$$

Allowing M to approach infinity is equivalent to considering the case of unconstrained bandwidth. Hence, an orthogonal coding scheme produces error-free transmission in the unconstrained bandwidth, additive white Gaussian noise channel provided that $E_b/N_0 > \ln 2$. The limiting behavior is particularly significant from an information-theoretic point of view. Shannon's formula for the capacity of a channel of bandwidth B perturbed by additive white Gaussian noise with one-sided PSD N_0 is given by (1.30) and repeated here for convenience:

$$C = B \log_2 \left(1 + \frac{P}{N_0 B} \right) \text{ bits/sec} \tag{4.183}$$

This represents the maximum bit rate for error-free transmission. Letting the bandwidth B approach infinity and using the expansion $\ln(1 + x) = x - x^2/2 + x^3/3 - \cdots, |x| < 1$, then

$$\lim_{B \to \infty} C = \lim_{B \to \infty} \frac{B}{\ln 2} \ln \left(1 + \frac{P}{N_0 B} \right) = \left(\frac{B}{\ln 2} \right) \left(\frac{P}{N_0 B} \right) = \frac{P}{N_0 \ln 2} \tag{4.184}$$

Thus, from (4.182) and (4.184), it follows that orthogonal signals achieve error-free transmission in the limit as $M \to \infty$, provided that the bit rate $\mathcal{R}_b$ is less than

$$\mathcal{R}_b < \frac{P}{N_0 \ln 2} = \lim_{B \to \infty} C \quad \text{bits/sec} \tag{4.185}$$

4.1.6 Matrix Generation of Binary Signal Sets

In this section, we shall briefly discuss methods of generating orthogonal, biorthogonal, and transorthogonal signal sets for the binary case, that is

$$s_i(t) = \sum_{j=1}^{N} s_{ij} \phi_j(t) \quad i = 0, 1, \cdots, M - 1 \tag{4.186}$$

where $s_{ij} = \pm 1$ for all i and j and $\{\phi_j(t), j = 1, 2, \cdots\}$ is any set of orthonormal functions over the interval $(0, T_s)$. The signals as defined by (4.186) are normalized with energy N. For any energy E_s, the signals should be scaled by $\sqrt{E_s/N}$. The signals can alternatively be represented by their respective vectors $\mathbf{s}_i = (s_{i1}, s_{i2}, \cdots, s_{iN})^T$. Let $\mathcal{H}_N$ be an $N \times N$ matrix defined inductively by

$$\mathcal{H}_N = \mathcal{H}_{\frac{N}{2}} \otimes \mathcal{H}_{\frac{N}{2}} \triangleq \begin{pmatrix} \mathcal{H}_{\frac{N}{2}} & \mathcal{H}_{\frac{N}{2}} \\ \mathcal{H}_{\frac{N}{2}} & -\mathcal{H}_{\frac{N}{2}} \end{pmatrix} \tag{4.187}$$

where

$$\mathcal{H}_2 = \begin{pmatrix} 1 & 1 \\ 1 & -1 \end{pmatrix} \tag{4.188}$$

Such a matrix is called "Sylvester-type Hadamard" matrix and satisfies the property that

$$\mathcal{H}_N \mathcal{H}_N^T = N \mathcal{I}_N \tag{4.189}$$

By definition, the construction of such matrices is known when $N = 2^k$. It is also conjectured that when $N = 0 \mod 4$, such matrices exist [11]. As an example, $\mathcal{H}_{12}$ is given by

$$\mathcal{H}_{12} = \begin{pmatrix} 1 & 1 & 1 & 1 & 1 & 1 & 1 & 1 & 1 & 1 & 1 & 1 \\ 1 & -1 & 1 & -1 & 1 & 1 & 1 & -1 & -1 & -1 & 1 & -1 \\ 1 & -1 & -1 & 1 & -1 & 1 & 1 & 1 & -1 & -1 & -1 & 1 \\ 1 & 1 & -1 & -1 & 1 & -1 & 1 & 1 & 1 & -1 & -1 & -1 \\ 1 & -1 & 1 & -1 & -1 & 1 & -1 & 1 & 1 & 1 & -1 & -1 \\ 1 & -1 & -1 & 1 & -1 & -1 & 1 & -1 & 1 & 1 & 1 & -1 \\ 1 & -1 & -1 & -1 & 1 & -1 & -1 & 1 & -1 & 1 & 1 & 1 \\ 1 & 1 & -1 & -1 & -1 & 1 & -1 & -1 & 1 & -1 & 1 & 1 \\ 1 & 1 & 1 & -1 & -1 & -1 & 1 & -1 & -1 & 1 & -1 & 1 \\ 1 & 1 & 1 & 1 & -1 & -1 & -1 & 1 & -1 & -1 & 1 & -1 \\ 1 & -1 & 1 & 1 & 1 & -1 & -1 & -1 & 1 & -1 & -1 & 1 \\ 1 & 1 & -1 & 1 & 1 & 1 & -1 & -1 & -1 & 1 & -1 & -1 \end{pmatrix} \tag{4.190}$$

Since $\mathcal{H}_N \mathcal{H}_N^T = N \mathcal{I}_N$, the columns of these matrices can be used to construct M orthogonal signals each with N coordinates with $M \leq N$. To construct M biorthogonal signals, use any $M/2$ columns and augment the set with the negative of the columns to obtain the complete set of M signals. To obtain M simplex signals, delete the top row (or first column) of $\mathcal{H}_N$ and use the remaining columns as vectors. The number of coordinates will be one less than the number of constructed vectors as expected.

Example 4.3

Design binary vectors representing eight orthogonal, biorthogonal, and simplex signal sets with $E_s = 8$. Use Sylvester-type Hadamard matrices to obtain the vectors.

Since $M = 8$, we need to compute $\mathcal{H}_8$ as

$$\mathcal{H}_8 = \begin{pmatrix} \mathcal{H}_4 & \mathcal{H}_4 \\ \mathcal{H}_4 & -\mathcal{H}_4 \end{pmatrix} = \begin{pmatrix} 1 & 1 & 1 & 1 & 1 & 1 & 1 & 1 \\ 1 & -1 & 1 & -1 & 1 & -1 & 1 & -1 \\ 1 & 1 & -1 & -1 & 1 & 1 & -1 & -1 \\ 1 & -1 & -1 & 1 & 1 & -1 & -1 & 1 \\ 1 & 1 & 1 & 1 & -1 & -1 & -1 & -1 \\ 1 & -1 & 1 & -1 & -1 & 1 & -1 & 1 \\ 1 & 1 & -1 & -1 & -1 & -1 & 1 & 1 \\ 1 & -1 & -1 & 1 & -1 & 1 & 1 & -1 \end{pmatrix}$$

$$= (s_0 \ s_1 \ s_2 \ \cdots s_7)$$

To obtain the orthogonal vectors ($E_s = 8$), use

$$s_0 \ s_1 \ s_2 \ \cdots s_7$$

For biorthogonal signals, use

$$s_0 \ s_1 \ s_2 \ s_3 \ -s_0 \ -s_1 \ -s_2 \ -s_3$$

For simplex vectors, delete first row from $\mathcal{H}_8$ and use remaining columns to obtain

$$\mathbf{s}_0' = \begin{pmatrix} 1 \\ 1 \\ 1 \\ 1 \\ 1 \\ 1 \\ 1 \end{pmatrix}, \mathbf{s}_1' = \begin{pmatrix} -1 \\ 1 \\ -1 \\ 1 \\ -1 \\ 1 \\ -1 \end{pmatrix}, \mathbf{s}_2' = \begin{pmatrix} 1 \\ -1 \\ -1 \\ 1 \\ 1 \\ -1 \\ -1 \end{pmatrix}, \cdots, \mathbf{s}_7' = \begin{pmatrix} -1 \\ -1 \\ 1 \\ -1 \\ 1 \\ 1 \\ -1 \end{pmatrix}$$

We can check to see that

$$\mathbf{s}_1'^T \mathbf{s}_1' = 7 = E_s\left(1 - \frac{1}{M}\right) = 8\left(1 - \frac{1}{8}\right)$$

and

$$\mathbf{s}_1'^T \mathbf{s}_2' = -1 = -\frac{E_s}{M} = -\frac{8}{8}$$

4.1.7 Coherent Detection of Differentially Encoded M-PSK Signals

Thus far in our discussion of optimum coherent detection of M-PSK signals, we have tacitly assumed that the receiver was perfectly synchronized to (had exact knowledge of) the frequency and phase of the received signal's carrier. In practice, the receiver derives its frequency and phase demodulation references from a carrier synchronization loop which, depending on its type,[7] may exhibit an M-fold phase ambiguity. That is, the receiver is capable of establishing a phase reference only to within an angle ϕ_a of the exact phase. Since the phase ambiguity indeed takes on values equal to those of the M-PSK data phases themselves, then any value of ϕ_a other than zero will cause a correctly detected phase to be erroneously mistaken for one of the other possible transmitted phases, *even in the absence of noise!* Clearly, something must be done to resolve this phase ambiguity else the receiver will suffer a performance degradation. For example, for a phase ambiguity corresponding to $m = M/2$, the receiver will be $180°$ out of phase with the received signal and all data decisions will be inverted resulting in very poor error probability performance.

One simple solution to the above is to employ *differential encoding* at the transmitter yet still maintain coherent detection at the receiver. The theory of differential encoding and how it resolves the above-mentioned phase ambiguity is described as follows. Let $s_i(t) = \sqrt{2P}\cos(\omega_c t + \theta_i)$ represent[8] the M-PSK signal to be transmitted during the ith transmission interval $iT_s \leq t \leq (i+1)T_s$ where θ_i ranges over the set of allowable values $\beta_m = 2\pi m/M; m = 0, 1, \ldots, M-1$. Ordinarily, θ_i would be exactly equal to the information phase, say $\Delta\theta_i$, to be communicated in that same transmission interval. At the conclusion of the ith transmission interval, that is, at time $t = (i+1)T_s$, the receiver makes its M-ary decision $\hat{\theta}_i$ on θ_i (or equivalently, $\Delta\theta_i$). Because of the ambiguity associated with the demodulation reference phase, however, this decision consists of an unambiguous part $\hat{\theta}_{u,i}$ plus the ambiguity ϕ_a itself, that is, $\hat{\theta}_i = \hat{\theta}_{u,i} + \phi_a$. Suppose now that instead of assigning θ_i identically equal to $\Delta\theta_i$, we instead encode $\Delta\theta_i$ as the difference between two adjacent

---

7. A complete discussion of the various types of carrier synchronization loops for M-PSK modulation is presented in Volume II of this book.

8. In this section, it is necessary to introduce a notation to characterize discrete time. For simplicity, we shall use a subscript, e.g., i, on a variable to denote the value of that variable in the ith transmission interval. To avoid confusion with the previous notation of θ_i to denote the ith message phase $2\pi i/M$ (i^{th} hypothesis), we introduce a new Greek symbol, i.e., β_i, to denote the ith message phase. Whenever it is necessary to represent a variable *both* in discrete time and/or by an associated hypothesis, we shall use a parenthetical superscript for the latter.

transmitted phases, that is, $\Delta\theta_i = \theta_i - \theta_{i-1}$. If the receiver still performs decisions on the transmitted phases θ_{i-1} and θ_i but arrives at its decision on the *information phase* based on the difference decision variable $\hat{\theta}_i - \hat{\theta}_{i-1} = \left(\hat{\theta}_{u,i} + \phi_a\right) - \left(\hat{\theta}_{u,i-1} + \phi_a\right) = \hat{\theta}_{u,i} - \hat{\theta}_{u,i-1}$, then clearly the phase ambiguity has been eliminated. Inherent in the above ambiguity resolution is the assumption that the phase ambiguity ϕ_a introduced by the carrier synchronization loop remains constant over the two transmission intervals of interest, that is, for a duration of $2T_s$ seconds. If we continue to encode the information phases in the above manner and base our decisions on the difference of adjacent decisions made on the actual transmitted phases, then clearly a sequence of $N_s + 1$ transmitted phases is needed to convey N_s symbols of information. That is, the first symbol in the transmitted sequence is used to resolve the phase ambiguity.

As a first example, consider a BPSK modulation ($M = 2$) and suppose that we desire to transmit the information sequence $\Delta\theta_1 = \pi, \Delta\theta_2 = \pi, \Delta\theta_3 = 0, \Delta\theta_4 = \pi$, and $\Delta\theta_5 = 0$. Assuming $\theta_0 = 0$ as the reference phase, then the remainder of the transmitted (that is, the differentially encoded) phases reduced modulo 2π would be $\theta_1 = \theta_0 + \Delta\theta_1 = \pi, \theta_2 = \theta_1 + \Delta\theta_2 = 0, \theta_3 = \theta_2 + \Delta\theta_3 = 0, \theta_4 = \theta_3 + \Delta\theta_4 = \pi, \theta_5 = \theta_4 + \Delta\theta_5 = \pi$. In the absence of noise and assuming no phase ambiguity introduced by the carrier synchronization loop, the receiver would make perfect decisions on the received sequence, that is, $\hat{\theta}_i = \theta_i$; $i = 0, 1, 2, \ldots, 5$ and would then *differentially decode* the modulo 2π difference of these decisions as $\hat{\theta}_1 - \hat{\theta}_0 = \pi, \hat{\theta}_2 - \hat{\theta}_1 = \pi, \hat{\theta}_3 - \hat{\theta}_2 = 0, \hat{\theta}_4 - \hat{\theta}_3 = \pi$, and $\hat{\theta}_5 - \hat{\theta}_4 = 0$ which is identical to the information sequence. If now the carrier synchronization loop introduced a phase ambiguity of π radians thus converting the received phase symbols into the sequence $\theta'_0 = \pi, \theta'_1 = 0, \theta'_2 = \pi, \theta'_3 = \pi, \theta'_4 = 0$, and $\theta'_5 = 0$, then in the absence of noise we would make perfect decisions, that is, $\hat{\theta}'_i = \theta'_i$; $i = 0, 1, 2, \ldots, 5$ and after differential decoding (modulo 2π) obtain $\hat{\theta}'_1 - \hat{\theta}'_0 = \pi, \hat{\theta}'_2 - \hat{\theta}'_1 = \pi, \hat{\theta}'_3 - \hat{\theta}'_2 = 0, \hat{\theta}'_4 - \hat{\theta}'_3 = \pi, \hat{\theta}'_5 - \hat{\theta}'_4 = 0$, which is once again identical to the information sequence.

As a second example, consider a QPSK modulation ($M = 4$) and suppose that we desire to transmit the information sequence $\Delta\theta_1 = \pi/2, \Delta\theta_2 = \pi, \Delta\theta_3 = 0, \Delta\theta_4 = 3\pi/2, \Delta\theta_5 = 3\pi/2$. Again assuming $\theta_0 = 0$ as the reference phase, then the remainder of the transmitted (that is, the differentially encoded) phases reduced modulo 2π would be $\theta_1 = \theta_0 + \Delta\theta_1 = \pi/2, \theta_2 = \theta_1 + \Delta\theta_2 = 3\pi/2, \theta_3 = \theta_2 + \Delta\theta_3 = 3\pi/2, \theta_4 = \theta_3 + \Delta\theta_4 = \pi, \theta_5 = \theta_4 + \Delta\theta_5 = \pi/2$. In the absence of noise and phase ambiguity, the receiver would differentially decode (modulo 2π) its perfect decisions, that is, $\hat{\theta}_i = \theta_i$ into $\hat{\theta}_1 - \hat{\theta}_0 = \pi/2, \hat{\theta}_2 - \hat{\theta}_1 = \pi, \hat{\theta}_3 - \hat{\theta}_2 = 0, \hat{\theta}_4 - \hat{\theta}_3 = 3\pi/2$, and $\hat{\theta}_5 - \hat{\theta}_4 = 3\pi/2$ which is identical to the information sequence. In the presence of a phase ambiguity, say $\phi_a = \pi/2$, the received phase symbols would be converted to the sequence $\theta'_0 = \pi/2, \theta'_1 = \pi, \theta'_2 = 0, \theta'_3 = 0, \theta'_4 = 3\pi/2, \theta'_5 = \pi$ and in the absence of noise the receiver would make perfect decisions, that is, $\hat{\theta}'_i = \theta'_i$; $i = 0, 1, 2, \ldots, 5$ and after differential decoding (modulo 2π) obtain $\hat{\theta}'_1 - \hat{\theta}'_0 = \pi/2, \hat{\theta}'_2 - \hat{\theta}'_1 = \pi, \hat{\theta}'_3 - \hat{\theta}'_2 = 0, \hat{\theta}'_4 - \hat{\theta}'_3 = 3\pi/2, \hat{\theta}'_5 - \hat{\theta}'_4 = 3\pi/2$ which is once again identical to the information sequence.

Figures 4.29a and 4.29b are illustrations of a differential encoder and a differential decoder, respectively, where the inputs and outputs of these devices are the phase symbols themselves. Alternately, one can represent differential encoding and decoding in terms of the complex baseband representation of the signals which results in the configurations of

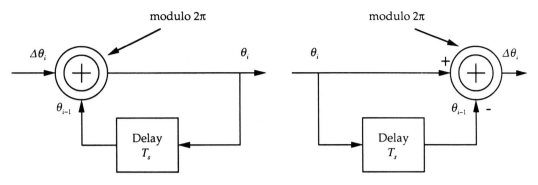

Figure 4.29   Differential encoder and decoder (real form)

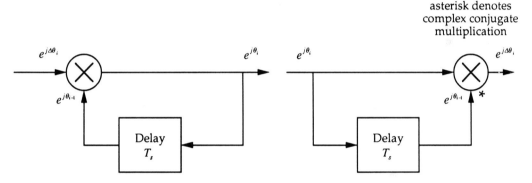

Figure 4.30   Differential encoder and decoder (complex form)

Figures 4.30a and 4.30b. Which differential encoder/decoder representation is used in a particular situation is a function of whether the receiver is represented in its real or complex form.

A digital communication system that employs differential encoding of the information symbols and coherent detection of successive differentially encoded M-PSK signals is called a *differentially encoded coherent M-PSK system*. We now derive the optimum (MAP) receiver for such a system followed by an evaluation of its error probability performance. We shall show that the structure of this receiver takes the form of the optimum receiver for M-PSK signals without differential encoding (see Fig. 4.23) followed by a differential decoder.

Optimum (MAP) Receiver for Differentially Encoded M-PSK.   Based on the above discussion of differential encoding, we observe that the desire to convey the information phase $\Delta\theta_i$ in the transmission interval $iT_s \leq t \leq (i+1)T_s$ results in the actual transmission over the channel of θ_{i-1} in the interval $(i-1)T_s \leq t \leq iT_s$ and θ_i in the interval $iT_s \leq t \leq (i+1)T_s$. Hence, a receiver that makes an optimum coherent decision on $\Delta\theta_i$ should observe the received signal plus noise, $r(t)$, in the $2T_s$-sec interval $(i-1)T_s \leq t \leq (i+1)T_s$ and make a joint optimum coherent decision on $\theta'_{i-1} = \theta_{i-1} + \phi_a$ and $\theta'_i = \theta_i + \phi_a$, namely, $\left(\hat{\theta}'_{i-1}, \hat{\theta}'_i\right)$. The ultimate decision on $\Delta\theta_i$ would then be obtained from $\Delta\hat{\theta}_i = \hat{\theta}'_i - \hat{\theta}'_{i-1}$ modulo 2π.

The design of an optimum MAP coherent receiver for jointly detecting θ'_{i-1} and θ'_i follows the general likelihood function approach described in Section 4.1.3 and illustrated in Fig. 4.6 and later applied to conventional (without differential encoding) M-PSK in Fig. 4.23. Here however, for fixed ϕ_a, there are M^2 hypotheses (pairs of possible values for θ'_{i-1} and θ'_i) and a joint decision is made based on a maximum correlation over the $2T_s$-sec interval $(i-1)T_s \le t \le (i+1)T_s$. In particular, analogous to (4.38), the receiver computes

$$x^{(k,l)} \triangleq \int_{(i-1)T_s}^{(i+1)T_s} \rho'(t)s^{(k,l)}(t)dt \tag{4.191}$$

where the prime on $\rho(t)$ denotes the fact that the sample function of $r(t)$ includes the phase ambiguity ϕ_a and

$$s^{(k,l)}(t) = \begin{cases} s^{(k)}(t + T_s); & (i-1)T_s \le t \le iT_s \\ s^{(l)}(t); & iT_s \le t \le (i+1)T_s \\ 0; & \text{otherwise} \end{cases} \tag{4.192}$$

with $s^{(l)}(t) = \sqrt{2P}\cos(\omega_c t + \beta_l)$; $iT_s \le t \le (i+1)T_s$. The decision rule on the transmitted phases θ_{i-1} and θ_i in the presence of the phase ambiguity is then made as follows. Choose $\hat{\theta}'_{i-1} = \beta_{\hat{k}}$ and $\hat{\theta}'_i = \beta_{\hat{l}}$ where $x^{\left(\hat{k},\hat{l}\right)}$ is the maximum value of $x^{(k,l)}$. Finally, the decision on $\Delta\theta_i$ would be $\Delta\hat{\theta}_i = \hat{\theta}'_i - \hat{\theta}'_{i-1} = \beta_{\hat{l}} - \beta_{\hat{k}}$. Since (4.191) can be obviously written as

$$x^{(k,l)} \triangleq \int_{(i-1)T_s}^{iT_s} \rho'(t)s^{(k)}(t+T_s)dt + \int_{iT_s}^{(i+1)T_s} \rho'(t)s^{(l)}(t)dt \tag{4.193}$$

a receiver that implements the above decision rule is illustrated in Fig. 4.31. Furthermore, because of the white noise model and the assumption of uncoded data [that is, $s_{i-1}(t)$ is independent of $s_i(t)$], then

$$\max_{k,l} x^{(k,l)} = \max_k \left\{ \int_{(i-1)T_s}^{iT_s} \rho'(t)s^{(k)}(t+T_s)dt \right\} + \max_l \left\{ \int_{iT_s}^{(i+1)T_s} \rho'(t)s^{(l)}(t)dt \right\} \tag{4.194}$$

$$\triangleq \max_k x_{i-1}^{(k)} + \max_l x_i^{(l)}$$

and the *joint* decision on θ'_{i-1} and θ'_i can be visualized as successive decisions on θ'_{i-1} and θ'_i independently. In view of this, the receiver of Fig. 4.31 can be redrawn in simpler form (M correlators as opposed to M^2 correlators) as in Fig. 4.32a. Furthermore, transforming the problem to polar coordinates as was done in arriving at Fig. 4.23 from Fig. 4.6, we obtain the familiar form (see [12], Fig. 5-20) illustrated in Fig. 4.32b which requires only two correlators. We emphasize that the equivalence between Figures 4.31 and 4.32a or 4.32b would not be valid for coded (correlated) data despite the white noise assumption.

Error Probability Performance.   To evaluate the symbol error probability performance of the receiver of Fig. 4.32, we begin by considering the possible ways of making a correct decision on $\Delta\theta_i$. Clearly, if both $\hat{\theta}'_{i-1}$ and $\hat{\theta}'_i$ correspond to correct decisions,

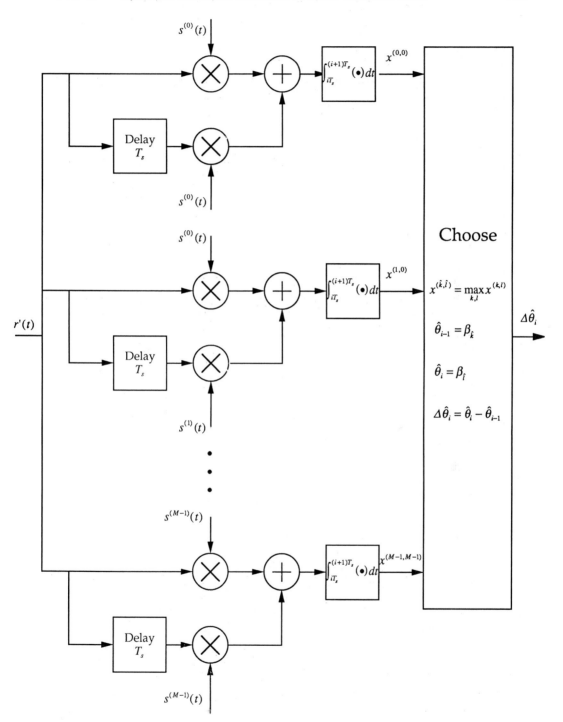

Figure 4.31   Optimum receiver for differentially encoded M-PSK

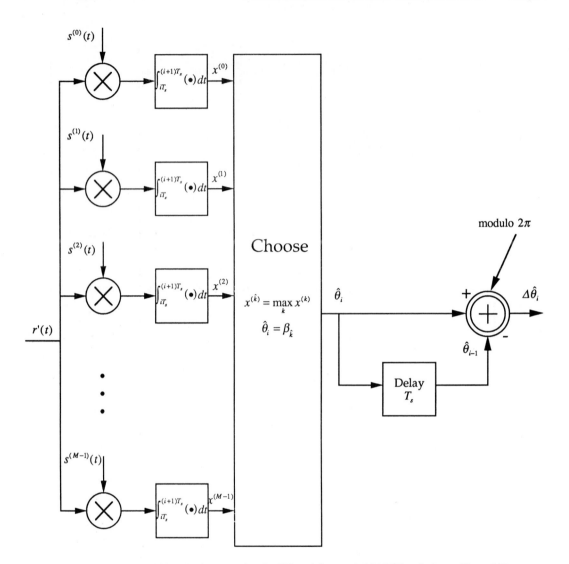

Figure 4.32a   Optimum receiver for differentially encoded M-PSK equivalent to Figure 4.31

that is, $\hat{\theta}'_{i-1} = \theta_{i-1} + \phi_a$ and $\hat{\theta}'_i = \theta_i + \phi_a$, then $\Delta\theta_i$ will be detected correctly. The probability of this event occurring is $P^{(0)}_{s,i}(C) \triangleq \left[P_s(C)|_{M-\text{PSK}} \right]^2$ where $P_s(C)|_{M-\text{PSK}}$ is given by (4.130). Furthermore, if both $\hat{\theta}'_{i-1}$ and $\hat{\theta}'_i$ are in error but by the same amount, that is, $\hat{\theta}'_{i-1} = \theta_{i-1} + \phi_a + 2k\pi/M$ and $\hat{\theta}'_i = \theta_i + \phi_a + 2k\pi/M$, then $\Delta\theta_i$ will still be detected correctly. Denoting the probability of this event by $P^{(k)}_{s,i}(C)$, then referring to Fig. 4.32, we have

$$P^{(k)}_{s,i}(C) = \Pr\left\{(2k-1)\pi/M \le \eta_{i-1} - \theta_{i-1} - \phi_a \le (2k+1)\pi/M,\right.$$

$$\left.(2k-1)\pi/M \le \eta_i - \theta_i - \phi_a \le (2k+1)\pi/M\right\}$$

(4.195)

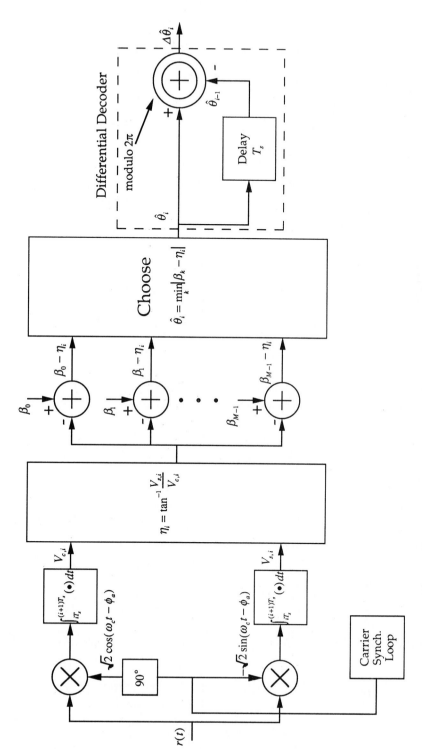

Figure 4.32b Optimum receiver for differentially encoded M-PSK equivalent to Figure 4.32a

231

where η_{i-1} and η_i are the receiver's noisy estimates of the phases in the $i-1^{\text{th}}$ and i^{th} transmission intervals, respectively. Analogous to (4.127), the pdf's of η_{i-1} and η_i are

$$
f(\eta_{i-1}) =
\begin{cases}
\int_0^\infty \frac{r}{\pi} \exp\left\{-\left[r^2 - 2r\sqrt{\frac{E_s}{N_0}}\right.\right. \\
\qquad \left.\left. \times \cos(\eta_{i-1} - \theta_{i-1} - \phi_a) + \frac{E_s}{N_0}\right]\right\} dr; & 0 \le \eta_{i-1} \le 2\pi \\
0; & \text{otherwise}
\end{cases}
\tag{4.196}
$$

$$
f(\eta_i) =
\begin{cases}
\int_0^\infty \frac{r}{\pi} \exp\left\{-\left[r^2 - 2r\sqrt{\frac{E_s}{N_0}}\right.\right. \\
\qquad \left.\left. \times \cos(\eta_i - \theta_i - \phi_a) + \frac{E_s}{N_0}\right]\right\} dr; & 0 \le \eta_i \le 2\pi \\
0; & \text{otherwise}
\end{cases}
$$

Since η_{i-1} and η_i are statistically independent, then

$$
P_{s,i}^{(k)}(C) = \int_{\theta_{i-1}+\phi_a+(2k-1)\pi/M}^{\theta_{i-1}+\phi_a+(2k+1)\pi/M} f(\eta_{i-1})\, d\eta_{i-1} \int_{\theta_i+\phi_a+(2k-1)\pi/M}^{\theta_i+\phi_a+(2k+1)\pi/M} f(\eta_i)\, d\eta_i
$$

$$
= \left[\int_{(2k-1)\pi/M}^{(2k+1)\pi/M} \int_0^\infty \frac{r}{\pi} \exp\left\{-\left[r^2 - 2r\sqrt{\frac{E_s}{N_0}}\cos\Theta + \frac{E_s}{N_0}\right]\right\} dr\, d\Theta\right]^2
\tag{4.197}
$$

$$
\triangleq \left(P_i^{(k)}\right)^2
$$

Using the same change of variables as in going from (4.129) to (4.130) and following the approach taken in Appendix 4A, we arrive at the following result:

$$
P_i^{(k)} = \frac{1}{\pi} \int_0^\infty \exp\left[-\left(u - \sqrt{\frac{E_s}{N_0}}\right)^2\right] \int_{u\tan[(2k-1)\pi/M]}^{u\tan[(2k+1)\pi/M]} \exp\left(-v^2\right) dv\, du
$$

$$
= \frac{1}{M} + \frac{1}{4}\operatorname{erf}\left[\sqrt{\frac{E_s}{N_0}}\sin\frac{(2k+1)\pi}{M}\right] - \frac{1}{4}\operatorname{erf}\left[\sqrt{\frac{E_s}{N_0}}\sin\frac{(2k-1)\pi}{M}\right]
\tag{4.198a}
$$

$$
+ \frac{1}{2\sqrt{\pi}} \int_0^{\sqrt{E_s/N_0}\sin[(2k+1)\pi/M]} \exp\left(-y^2\right)\operatorname{erf}\left[y\cot\frac{(2k+1)\pi}{M}\right] dy
$$

$$
- \frac{1}{2\sqrt{\pi}} \int_0^{\sqrt{E_s/N_0}\sin[(2k-1)\pi/M]} \exp\left(-y^2\right)\operatorname{erf}\left[y\cot\frac{(2k-1)\pi}{M}\right] dy
$$

Alternately, using the approach of Section 3.2.8, we obtain the simpler result (see Problem 4.27)

$$
P_i^{(k)} = \frac{1}{2\pi} \int_0^{\pi(1-(2k-1)/M)} \exp\left\{-\frac{E_s}{N_0}\left(\frac{\sin^2\frac{(2k-1)\pi}{M}}{\sin^2\phi}\right)\right\} d\phi
$$

$$
- \frac{1}{2\pi} \int_0^{\pi(1-(2k+1)/M)} \exp\left\{-\frac{E_s}{N_0}\left(\frac{\sin^2\frac{(2k+1)\pi}{M}}{\sin^2\phi}\right)\right\} d\phi
\tag{4.198b}
$$

Since the preceding events are mutually exclusive and exhaustive of the ways to make a correct decision, the probability of a correct decision on the ith symbol is

$$P_{s,i}(C) = \sum_{k=0}^{M-1} P_{s,i}^{(k)}(C) \tag{4.199}$$

Furthermore, since (4.199) is independent of i and since we have assumed equiprobable symbols in arriving at Fig. 4.32, the average symbol error probability is

$$P_s(E) = 1 - \frac{1}{M} \sum_{i=0}^{M-1} P_{s,i}(C)$$

$$= 2\, P_s(E)|_{M-\text{PSK}} \left[1 - \frac{1}{2}\, P_s(E)|_{M-\text{PSK}} - \frac{1}{2} \frac{\sum_{k=1}^{M-1} P_{s,i}^{(k)}(C)}{P_s(E)|_{M-\text{PSK}}} \right] \tag{4.200}$$

where $P_s(E)|_{M-\text{PSK}} = 1 - P_s(C)|_{M-\text{PSK}}$ is the average symbol error probability for M-PSK without differential encoding and is given by (4.131).

Some special cases of (4.200) are of interest. For $M = 2$, we have coherent detection of differentially encoded BPSK for which (4.200) reduces to

$$P_s(E) = 2\, P_s(E)|_{M-\text{PSK}} \left[1 - P_s(E)|_{M-\text{PSK}} \right]$$

$$= \left(\text{erfc}\sqrt{\frac{E_s}{N_0}} \right) \left[1 - \frac{1}{2}\text{erfc}\sqrt{\frac{E_s}{N_0}} \right] \tag{4.201}$$

For coherent detection of differentially encoded QPSK ($M = 4$), (4.200) simplifies to

$$P_s(E) = 2\,\text{erfc}\sqrt{\frac{E_s}{2N_0}} - 2\,\text{erfc}^2\sqrt{\frac{E_s}{2N_0}} + \text{erfc}^3\sqrt{\frac{E_s}{2N_0}} - \frac{1}{4}\text{erfc}^4\sqrt{\frac{E_s}{2N_0}} \tag{4.202}$$

Analogous to the formation of a QPSK signal from two independent BPSK signals, a differentially encoded QPSK signal can also be formed from two differentially encoded BPSK signals. The optimum receiver would then differentially decode each BPSK signal independently and then combine these two binary decisions to form the resulting decision on the QPSK information symbol. The receiver for this system is illustrated in Fig. 4.33 where the symbol phase decision is made in accordance with the bit to symbol mapping of Fig. 4.12a. To demonstrate that the symbol error probability performance of Fig. 4.33 is identical to that of Fig. 4.32, we proceed as follows. For Fig. 4.33, the probability of a correct symbol decision is equal to the probability that *both* binary decisions are correct. Since each binary decision is the result of differential decoding a BPSK signal, then the probability of a correct binary decision is one minus the probability of error given in (4.201) and hence the probability of a correct QPSK symbol decision is equal to $\left[1 - P_s(E)|_{\text{BPSK}} \right]^2$. Finally, the probability of error for the QPSK symbol is

$$P_s(E)|_{\text{D.E. QPSK}} = 1 - \left[1 - P_s(E)|_{\text{D.E. BPSK}} \right]^2$$

$$= 2\, P_s(E)|_{\text{D.E. BPSK}} - \left(P_s(E)|_{\text{D.E. BPSK}} \right)^2 \tag{4.203}$$

Substituting (4.201) into (4.203) and simplifying gives the identical result to (4.202).

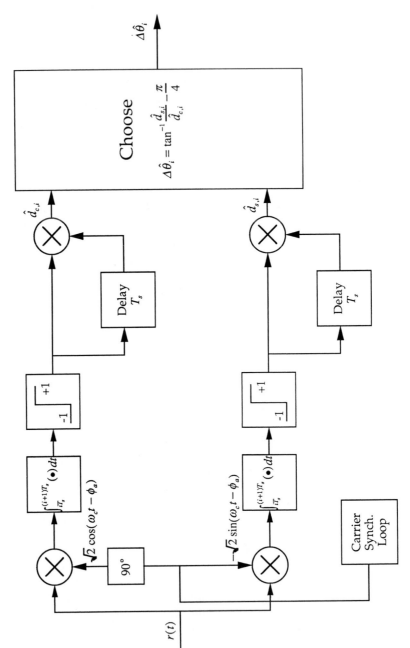

Figure 4.33 Optimum receiver for differentially encoded QPSK

4.2 OPTIMUM WAVEFORM RECEIVERS IN COLORED GAUSSIAN NOISE

The problem of detecting signals in colored Gaussian noise can be approached in several ways, namely, using the sampling approach, the K-L expansion, by performing an intuitive whitening operation on the received signal or by a direct derivation with a sufficient statistic. All four methods will be explored in this section to derive the optimum receiver in colored Gaussian noise. Unlike detection in white Gaussian noise, the detection problem in colored Gaussian noise and the performance of the optimum receiver depend on the length of the observation interval. In white noise, the observation of the received signal over the interval $(0, T)$, that is, over the signal duration, is sufficient to perform optimum detection. In colored noise, the observation of the received signal outside the signal duration (and hence, the observation of the corrupting noise) is of key importance as it provides information on the noise present inside the signal interval. Limiting the observation interval to the signal duration will in general result in a receiver performance inferior to that of observing a larger interval. There are two key issues that will be addressed in this section: (1) the structure and performance of the optimum receiver assuming a finite observation interval, and (2) the structure and performance of the optimum receiver assuming an infinite observation interval. The problem is then to decide between M messages where $r(t) = s_i(t) + n(t)$ under m_i, $i = 0, 1, \cdots, M - 1$, over some observation period (finite or infinite) and $n(t)$ is Gaussian, zero-mean, nonwhite with autocorrelation function $R_n(t_1, t_2) \triangleq E\{n(t_1)n(t_2)\}$.

4.2.1 Time-Sampling Approach

Let's first start by investigating the time-sampling approach. In this case, the received signal $r(t)$, observed over the interval $[0, T]$, is sampled to produce the N-dimensional vector

$$\mathbf{r}_N^T = (r(\Delta t), r(2\Delta t), \cdots, r(N\Delta t)) \tag{4.204}$$

consisting of samples taken Δt seconds apart. This provides a finite dimensional approximation to the waveform problem and we will then let N become large, improving the accuracy of the approximation. Define the transmitted signal and noise vectors as

$$\mathbf{s}_i^T = (s_i(\Delta t), s_i(2\Delta t), \cdots, s_i(N\Delta t)) \tag{4.205}$$

and

$$\mathbf{n}^T = (n(\Delta t), n(2\Delta t), \cdots, n(N\Delta t)) \tag{4.206}$$

The noise correlation matrix of the noise vector $\mathbf{n}$ of dimension N is thus given by

$$\mathcal{R}_N = \begin{pmatrix} R_n(\Delta t, \Delta t) & \cdots & R_n(\Delta t, N\Delta t) \\ \vdots & \ddots & \vdots \\ R_n(N\Delta t, \Delta t) & \cdots & R_n(N\Delta t, N\Delta t) \end{pmatrix} \tag{4.207}$$

The density of the received vector $\mathbf{r}_N$, assuming message m_i is transmitted, is thus Gaussian with mean $\mathbf{s}_i$ and correlation matrix $\mathcal{R}_N$ as given by (3.61), that is

$$f_{\mathbf{r}_N}(\boldsymbol{\rho}_N / m_i) = G\left(\boldsymbol{\rho}_N; \mathbf{s}_i; \mathcal{R}_N\right) \tag{4.208}$$

and the unconditional density $f_{\mathbf{r}_N}(\boldsymbol{\rho}_N)$ therefore becomes

$$f_{\mathbf{r}_N}(\boldsymbol{\rho}_N) = \sum_{i=0}^{M-1} P(m_i) G\left(\boldsymbol{\rho}_N; \mathbf{s}_i; \mathcal{R}_N\right) \tag{4.209}$$

The MAP receiver decides on that message with maximum a posteriori probability, that is

$$P(m_i/\mathbf{r}_N = \boldsymbol{\rho}_N) = \frac{P(m_i) f_{\mathbf{r}_N}(\boldsymbol{\rho}_N/m_i)}{f_{\mathbf{r}_N}(\boldsymbol{\rho}_N)} \tag{4.210}$$

$$= \frac{P(m_i) \frac{1}{(2\pi)^{N/2}} \frac{1}{|\mathcal{R}_N|^{1/2}} \exp\left\{-\frac{1}{2}(\boldsymbol{\rho}_N - \mathbf{s}_i)^T \mathcal{R}_N^{-1}(\boldsymbol{\rho}_N - \mathbf{s}_i)\right\}}{\sum_{j=0}^{M-1} P(m_j) \frac{1}{(2\pi)^{N/2}} \frac{1}{|\mathcal{R}_N|^{1/2}} \exp\left\{-\frac{1}{2}(\boldsymbol{\rho}_N - \mathbf{s}_j)^T \mathcal{R}_N^{-1}(\boldsymbol{\rho}_N - \mathbf{s}_j)\right\}}$$

Taking the limit to improve the accuracy of the approximation and simplifying, (4.210) becomes

$$P(m_i/r(t) = \rho(t)) = \lim_{N \to \infty} \frac{P(m_i) \exp\left\{\boldsymbol{\rho}_N^T \mathcal{R}_N^{-1} \mathbf{s}_i - \frac{1}{2}\mathbf{s}_i^T \mathcal{R}_N^{-1} \mathbf{s}_i\right\}}{\sum_{j=0}^{M-1} P(m_j) \exp\left\{\boldsymbol{\rho}_N^T \mathcal{R}_N^{-1} \mathbf{s}_j - \frac{1}{2}\mathbf{s}_j^T \mathcal{R}_N^{-1} \mathbf{s}_j\right\}} \tag{4.211}$$

In arriving at (4.211), we used the fact that $\boldsymbol{\rho}_N^T \mathcal{R}_N^{-1} \mathbf{s}_i$ is a scalar and equal to its transpose $\mathbf{s}_i^T \mathcal{R}_N^{-1} \boldsymbol{\rho}_N$ as $(\mathcal{R}_N^{-1})^T = \mathcal{R}_N^{-1}$ due to the symmetry property of the correlation matrix $\mathcal{R}_N$. Note that the denominator is a constant $C(N)$ independent of i. If the limit exists and is finite, then

$$\lim_{N \to \infty} \boldsymbol{\rho}_N^T \mathcal{R}_N^{-1} \mathbf{s}_i = \lim_{N \to \infty} \left(\Delta t \boldsymbol{\rho}_N^T\right) \left(\frac{1}{\Delta t} \mathcal{R}_N^{-1} \mathbf{s}_i\right) \tag{4.212}$$

Define

$$\mathbf{g}_i \triangleq \frac{1}{\Delta t} \mathcal{R}_N^{-1} \mathbf{s}_i \tag{4.213}$$

or

$$\mathbf{s}_i = \mathcal{R}_N \mathbf{g}_i \Delta t \tag{4.214}$$

Hence, the jth element of $\mathbf{s}_i$ is given by

$$s_i(j\Delta t) = \sum_{k=1}^{N} R_n(j\Delta t, k\Delta t) g_i(k\Delta t) \Delta t, \quad \text{for all } j = 1, 2, \cdots, N \tag{4.215}$$

By continuity of $s_i(t)$ and $R_n(t, s)$, we write in the limit as $N \to \infty$, $s_i(j\Delta t) \to s_i(t)$ and

$$s_i(t) = \int_0^T R_n(t, s) g_i(s) ds, \quad \forall t \epsilon [0, T] \tag{4.216}$$

and $g_i(t)$ is determined by solving the integral equation. From (4.213), we find

$$\lim_{N \to \infty} \boldsymbol{\rho}_N^T \mathcal{R}_N^{-1} \mathbf{s}_i = \lim_{N \to \infty} (\Delta t) \boldsymbol{\rho}_N^T \mathbf{g}_i$$

$$= \lim_{N \to \infty} \sum_{k=1}^{N} \rho(k\Delta t) g_i(k\Delta t) \Delta t \tag{4.217}$$

$$= \int_0^T r(s) g_i(s) ds$$

Similarly

$$\lim_{N\to\infty} s_i^T R_N^{-1} s_i = \lim_{N\to\infty} (\Delta t) s_i^T g_i$$

$$= \int_0^T s_i(s) g_i(s) ds \tag{4.218}$$

Thus, the MAP test reduces to the following: select $\widehat{m}(\rho(t)) = m_i$ iff

$$P(m_i/r(t) = \rho(t)) = \max_k \{P(m_k/r(t) = \rho(t))\} \tag{4.219}$$

$$= \max_k \left\{ C^{-1} P(m_k) \exp\left\{ \int_0^T r(t) g_k(t) dt - \frac{1}{2} \int_0^T s_k(t) g_k(t) dt \right\} \right\}$$

where $C = \lim_{N\to\infty} C(N)$. Assuming that C exists in the limit, we have derived the optimum receiver. The convergence of C will be treated later when dealing with the K-L expansion. Simplifying (4.219) by taking logarithm, the decision rule becomes: set $\widehat{m}(\rho(t)) = m_i$ iff

$$\int_0^T r(t) g_i(t) dt + B_i' = \max_k \left\{ \int_0^T r(t) g_k(t) dt + B_k' \right\} \tag{4.220}$$

where

$$B_k' = \ln P(m_k) - \frac{1}{2} \int_0^T s_k(t) g_k(t) dt \tag{4.221}$$

and $g_k(t)$ is the solution to the integral equation given by (4.216). This test reduces to the mechanization given in Fig. 4.34. In the special case of white noise, $R_n(t, s) = (N_0/2)\delta(t - s)$ and $g_i(t) = (2/N_0)s_i(t)$. The bias B_k' in the presence of AWGN is related to B_k of (4.39) through $B_k' = (2/N_0) B_k$ and the receiver reduces to that in Fig. 4.6 as expected.

4.2.2 Karhunen-Loève (K-L) Expansion Approach

Let $\{\phi_i(t), \ i = 1, 2, \cdots, \infty\}$ be a sequence of orthonormal functions defined on the interval $[0, T]$. Assume that the set $\{s_i(t), \ i = 0, 1, \cdots, M - 1\}$ of known signals can be expanded in terms of the $\phi_i(t)$'s, that is

$$s_i(t) = \sum_{j=1}^{\infty} s_{ij}\phi_j(t), \quad i = 0, 1, \cdots, M - 1 \tag{4.222}$$

where

$$s_{ij} \triangleq \int_0^T s_i(t)\phi_j(t) dt \tag{4.223}$$

Here the definition of orthonormality implies

$$\int_0^T \phi_i(t)\phi_j(t) dt = \begin{cases} 0, & \text{if } i \neq j \\ 1, & \text{if } i = j \end{cases} \tag{4.224}$$

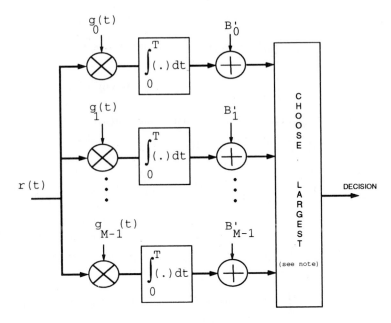

Note: From here on, **"choose largest"** implies choose
message corresponding to largest input voltage or value.

Figure 4.34   Waveform receiver in colored Gaussian noise (derived using sampling approach)

The Gaussian noise random process $n(t)$ can also be expanded in terms of the basis functions to yield

$$n(t) = \sum_{j=1}^{\infty} n_j \phi_j(t) \tag{4.225}$$

where

$$n_j \triangleq \int_0^T n(t)\phi_j(t)dt \tag{4.226}$$

are random variables. For these expansions or sequences to converge, we require that

$$\lim_{N \to \infty} \left| s_i(t) - \sum_{j=1}^{N} s_{i,j}\phi_j(t) \right|^2 = 0 \tag{4.227}$$

for the signal and

$$\lim_{N \to \infty} E\left\{ \left| n(t) - \sum_{j=1}^{N} n_j\phi_j(t) \right|^2 \right\} = 0 \tag{4.228}$$

for the noise. The latter is referred to as mean-squared convergence. Let $\mathbf{r}_N^T = (r_1, r_2, \cdots,$

r_N) where

$$r_j = \int_0^T r(t)\phi_j(t)dt \tag{4.229}$$

and

$$r_j = s_{i,j} + n_j, \quad j = 1, 2, \cdots, N \tag{4.230}$$

when m_i is transmitted. The random variables n_i, $i = 1, 2, \cdots, N$ are Gaussian (since they are the result of a linear operation on a Gaussian random process) with

$$E\{n_j\} = E\left\{\int_0^T n(t)\phi_j(t)dt\right\} = 0 \tag{4.231}$$

since $n(t)$ is zero-mean and

$$E\{n_i n_j\} \triangleq \lambda_{ij} = E\left\{\int_0^T \int_0^T n(t)\phi_i(t)n(s)\phi_j(s)dtds\right\}$$

$$= \int_0^T \int_0^T \phi_i(t)R_n(t,s)\phi_i(s)dtds \tag{4.232}$$

The covariances can be thought of as the coefficients in a double orthonormal function expansion of

$$R_n(t,s) = \sum_{i=1}^{\infty} \sum_{j=1}^{\infty} \lambda_{i,j}\phi_i(t)\phi_j(s) \tag{4.233}$$

If we set

$$\mathcal{R}_N = \begin{pmatrix} \lambda_{11} & \cdots & \lambda_{1N} \\ \vdots & \ddots & \vdots \\ \lambda_{N1} & \cdots & \lambda_{NN} \end{pmatrix} \tag{4.234}$$

then, when m_i is transmitted

$$E\{\mathbf{r}_N/m_i\} = \mathbf{s}_i \tag{4.235}$$

where $\mathbf{s}_i^T = (s_{i1}, s_{i2}, \cdots, s_{iN})$. Hence, from (4.210), the a posteriori probability $P(m_i/\mathbf{r}_N)$ is

$$P(m_i/\mathbf{r}_N = \boldsymbol{\rho}_N) = \frac{P(m_i)\exp\left\{-\frac{1}{2}(\boldsymbol{\rho}_N - \mathbf{s}_i)^T \mathcal{R}_N^{-1}(\boldsymbol{\rho}_N - \mathbf{s}_i)\right\}}{\sum_{j=0}^{M-1} P(m_j)\exp\left\{-\frac{1}{2}(\boldsymbol{\rho}_N - \mathbf{s}_j)^T \mathcal{R}_N^{-1}(\boldsymbol{\rho}_N - \mathbf{s}_j)\right\}} \tag{4.236}$$

To eliminate the problem associated with $\mathcal{R}_N^{-1}$ as $N \to \infty$, we use a special case of $\phi_i(t)$'s, which render all $\mathcal{R}_N$ to be invertible, that is, let λ_i and $\phi_i(t)$ (eigenvalue and eigenfunction) be solutions to

$$\lambda_i \phi_i(t) = \int_0^T R_n(t,s)\phi_i(s)ds \tag{4.237}$$

where $\lambda_{ij} = \lambda_i \delta_{ij}$. Thus, the sample function expansion of the noise $n(t) = \sum_{i=1}^{\infty} n_i \phi_i(t)$ is the Karhunen-Loève (K-L) expansion. This implies that the n_j's are uncorrelated and

$$\lambda_{i,j} = \lambda_i \delta_{i,j} = \int_0^T \int_0^T \phi_i(t) R_n(t,s) \phi_j(s) ds dt \tag{4.238}$$

In this case, (4.233) reduces to

$$R_n(t,s) = \sum_{i=1}^{\infty} \lambda_i \phi_i(t) \phi_i(s) \tag{4.239}$$

resulting in Mercer's theorem [13]. Thus, for a fixed N, we have

$$\mathcal{R}_N = \begin{pmatrix} \lambda_1 & 0 & \cdots & 0 \\ 0 & \lambda_2 & \cdots & 0 \\ \vdots & \ddots & \cdots & \vdots \\ 0 & \cdots & 0 & \lambda_N \end{pmatrix} \text{ or } \mathcal{R}_N^{-1} = \begin{pmatrix} \lambda_1^{-1} & 0 & \cdots & 0 \\ 0 & \lambda_2^{-1} & \cdots & 0 \\ \vdots & \ddots & \cdots & \vdots \\ 0 & \cdots & 0 & \lambda_N^{-1} \end{pmatrix} \tag{4.240}$$

and therefore (4.236) becomes

$$\lim_{N \to \infty} P(m_i / \mathbf{r}_N) = \frac{P(m_i) \exp\left\{ \sum_{j=1}^{\infty} \lambda_j^{-1} \left(r_j s_{ij} - \frac{1}{2} s_{ij}^2 \right) \right\}}{\sum_{k=0}^{M-1} P(m_k) \exp\left\{ \sum_{j=1}^{\infty} \lambda_j^{-1} \left(r_j s_{kj} - \frac{1}{2} s_{kj}^2 \right) \right\}} \tag{4.241}$$

The denominator is independent of i and is a constant, say C^{-1}. Hence

$$\lim_{N \to \infty} P(m_i / \mathbf{r}_N) = C P(m_i) \exp\left\{ \sum_{j=1}^{\infty} \lambda_j^{-1} \left(r_j s_{ij} - \frac{1}{2} s_{ij}^2 \right) \right\} \tag{4.242}$$

or

$$\ln P(m_i / r(t)) = \ln C + \ln P(m_i) + \sum_{j=1}^{\infty} \lambda_j^{-1} \left(r_j s_{ij} - \frac{1}{2} s_{ij}^2 \right) \tag{4.243}$$

for all $i = 0, 1, \cdots, M - 1$. Therefore, the optimum receiver sets $\widehat{m}(r(t)) = m_i$ iff

$$\sum_{j=1}^{\infty} \lambda_j^{-1} r_j s_{ij} + B_i'' = \max_k \left\{ \sum_{j=1}^{\infty} \lambda_j^{-1} r_j s_{kj} + B_k'' \right\} \tag{4.244}$$

where

$$B_k'' = \ln P(m_i) - \frac{1}{2} \sum_{j=1}^{\infty} s_{ij}^2 \lambda_j^{-1} \tag{4.245}$$

represents a bias term. In the presence of AWGN, $\lambda_j = \frac{N_0}{2} \, \forall j$ and B_k'' reduces to B_k' of (4.221). This decision rule admits the interpretation depicted in Fig. 4.35 in terms of signal processing. Implementation of the optimum receiver using the K-L expansion requires knowledge of the eigenfunctions and eigenvalues, that is, $\phi_i(t)$ and λ_i for all i. These are obtained by solving the integral equation given by (4.237). The procedure for doing so involves differentiating the integral equation to transform it into a differential equation which

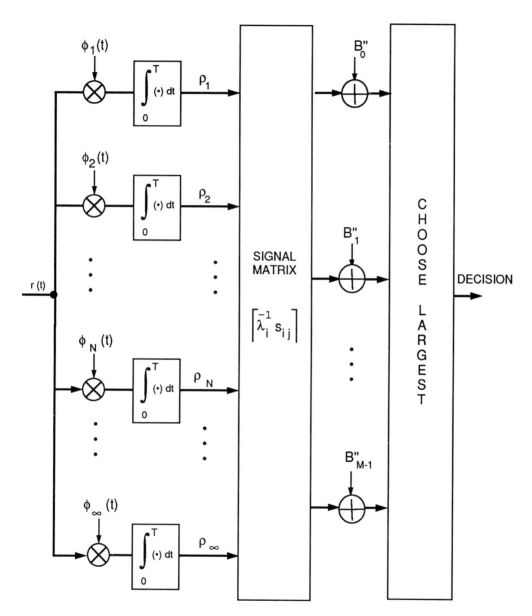

Figure 4.35   Waveform receiver in colored Gaussian noise (derived using K-L expansion)

is easier to solve. The eigenvalues are solved for simultaneously by imposing certain constraints on the form of the solution of the differential equation (see Problem 4.31).

Interconnection of K-L and sampling approaches.   Let's interconnect the K-L approach to the sampling approach derived in the previous section. Note that

$$\sum_{j=1}^{\infty} \lambda_j^{-1} s_{ij} r_j = \sum_{j=1}^{\infty} \lambda_j^{-1} s_{ij} \int_0^T r(t)\phi_j(t)dt$$

(4.246)

$$= \int_0^T r(t)\tilde{g}_i(t)dt$$

where

$$\tilde{g}_i(t) \triangleq \sum_{j=1}^{\infty} \lambda_j^{-1} s_{ij} \phi_j(t)$$

(4.247)

Multiplying both sides of (4.247) by $R_n(t, s)$ and integrating, we have

$$\int_0^T R_n(t, s)\tilde{g}_i(t)dt = \sum_{j=1}^{\infty} \lambda_j^{-1} s_{ij} \left(\int_0^T R_n(t, s)\phi_j(t)dt \right)$$

$$= \sum_{j=1}^{\infty} \lambda_j^{-1} s_{ij} \left(\lambda_j \phi_j(s) \right)$$

(4.248)

$$= \sum_{j=1}^{\infty} s_{ij} \phi_j(s)$$

$$= s_i(s)$$

Recall $g_i(t)$ in the sampling approach defined through (4.216). Thus, the K-L and the sampling approaches are equivalent if

$$g_i(t) = \tilde{g}_i(t)$$

(4.249)

Convergence of receiver.   Let's now focus on the convergence of the receiver and more specifically, on the convergence of the constant C. First note that the convergence of $\lim_{N\to\infty} \sum_{i=1}^{N} n_i \phi_i(t)$ to $n(t)$ is guaranteed by the K-L theorem. The convergence of the signal expansion $\lim_{N\to\infty} \sum_{j=1}^{N} s_{ij} \phi_j(t)$ to $s_i(t)$ is guaranteed whenever the signals are of finite energy, that is

$$\int_0^T s_i^2(t)dt < \infty$$

(4.250)

or equivalently

$$\lim_{N\to\infty} \sum_{i=1}^{N} s_{ij}^2 < \infty$$

(4.251)

The convergence of the random sums computed in the receiver is handled in the following way; let S_N denote

$$S_N \triangleq \sum_{j=1}^{N} \lambda_j^{-1} r_j s_{ij}$$

(4.252)

This sequence of random variables converges in the mean-square sense as $N \to \infty$ whenever

$$\lim_{N \to \infty} E\left\{|S_N - S_{N+M}|^2\right\} = 0 \text{ for any } M > 0 \tag{4.253}$$

This is the Cauchy convergence criterion [14] for random sequences. Suppose that m_k is transmitted. Then, the r_j's are independent Gaussian random variables with

$$r_j = \int_0^T r(t)\phi_j(t)dt = s_{kj} + n_j \tag{4.254}$$

and

$$E\{r_j/m_k\} = s_{kj} \text{ and } \text{Var}\{r_j\} = \lambda_j \tag{4.255}$$

Thus

$$E\left\{|S_N - S_{N+M}|^2 / m_k\right\} = E\left\{\left|\sum_{j=N+1}^{N+M} \lambda_j^{-1} r_j s_{ij}\right|^2 / m_k\right\}$$

$$= E\left\{\left|\sum_{j=N+1}^{N+M} \lambda_j^{-1}(s_{kj} + n_j)s_{ij}\right|^2 / m_k\right\} \tag{4.256}$$

$$= \left|\sum_{j=N+1}^{N+M} \frac{s_{kj}s_{ij}}{\lambda_j}\right|^2 + \sum_{j=N+1}^{N+M} \frac{s_{ij}^2}{\lambda_j}$$

Suppose we number the eigenfunctions such that $\lambda_1 \geq \lambda_2 \geq \cdots$. Noting that

$$\text{Average noise power} = E\left\{\int_0^T n^2(t)dt\right\} = \int_0^T R_n(t,t)dt$$

$$= \sum_{i=1}^{\infty} \lambda_i \tag{4.257}$$

then for finite noise power in the observation interval, we require that

$$\lim_{n \to \infty} \lambda_n = 0 \tag{4.258}$$

For convergence of original sum in (4.252), we require from (4.256) that

$$\lim_{N \to \infty} \sum_{j=N+1}^{N+M} \frac{s_{ij}^2}{\lambda_j} = 0 \text{ for any } M \tag{4.259}$$

Since the denominator elements are going to zero, we not only require that

$$\lim_{j \to \infty} s_{ij} = 0 \tag{4.260}$$

but additionally, we require

$$\lim_{j \to \infty} \frac{s_{ij}^2}{\lambda_j} = 0 \tag{4.261}$$

If the above is satisfied for all i, then it is guaranteed that the remaining term in (4.256) also converges because of Schwartz's inequality, that is

$$\left| \sum_{j=N+1}^{N+M} \frac{s_{kj}s_{ij}}{\lambda_j} \right| \leq \left(\sum_{j=N+1}^{N+M} \frac{s_{kj}^2}{\lambda_j} \right)^{1/2} \left(\sum_{j=N+1}^{N+M} \frac{s_{ij}^2}{\lambda_j} \right)^{1/2} \tag{4.262}$$

In summary for the sums to be convergent, we require three conditions, namely

(a) signal waveform energy to be finite, that is

$$\int_0^T s_i^2(t)dt < \infty$$

(b) noise waveform energy to be finite, that is

$$E\left\{ \int_0^T n^2(t)dt \right\} < \infty$$

(c) let s_{ij}^2/λ_j be the signal energy-to-noise ratio for signal i in the direction of $\phi_j(t)$, then

$$\lim_{N\to\infty} \sum_{j=1}^N \frac{s_{i,j}^2}{\lambda_j} < \infty \text{ for all } i$$

When condition (c) is not satisfied, we arrive at the case of *singular detection* in which there exists at least one direction or dimension that is noise free (that is, zero eigenvalue). In this case, perfect detection is possible provided that the projections of the various transmitted signals onto that coordinate are distinct, that is, unequal. If the projections are not distinct, the dimensionality of the problem can then be reduced by the number of zero eigenvalues and the problem reworked in a less dimensional vector space.

Receiver truncation and theorem of irrelevance.   Recall in additive Gaussian noise, the optimum receiver sets $\widehat{m}(\rho(t)) = m_i$ iff

$$\sum_{j=1}^\infty \lambda_j^{-1} r_j s_{ij} + B_i'' = \max_k \left\{ \sum_{j=1}^\infty \lambda_j^{-1} r_j s_{kj} + B_k'' \right\} \tag{4.263}$$

where B_k'' is given by (4.245) and the rule results in the functional block discussed previously in Fig. 4.35. Assume now that the transmitted signals are designed such that they occupy the first N dimensions, that is

$$s_i(t) = \sum_{j=1}^N s_{ij}\phi_j(t) \tag{4.264}$$

then $r_j = \int_0^T r(t)\phi_j(t)dt = s_{ij} + n_j$, where the last equality assumes that m_i is transmitted. Decompose the vector $\mathbf{r}$ into

$$\mathbf{r}^T = \left(\mathbf{r}_1^T, \mathbf{r}_2^T \right) \tag{4.265}$$

where

$$\mathbf{r}_1^T = (r_1, r_2, \cdots, r_N) \tag{4.266}$$

and

$$\mathbf{r}_2^T = (r_{N+1}, r_{N+2}, \cdots, r_\infty) \tag{4.267}$$

where $r_j = n_j$ for $j > N$ because $s_{ij} = 0$ for $j > N$. Thus

$$r(t) = \sum_{j=1}^{\infty} r_j \phi_j(t)$$

$$= \underbrace{\sum_{j=1}^{N} r_j \phi_j(t)}_{\text{depends on } i} + \underbrace{\sum_{j=N+1}^{\infty} n_j \phi_j(t)}_{\substack{\text{does not depend on } i \\ \text{and } \mathbf{r}_2 \text{ is irrelevant}}} \tag{4.268}$$

Since $\mathbf{r}_2$ is independent of i and is uncorrelated with $\mathbf{r}_1$, it is irrelevant to the detection process. Mathematically, we have

$$P(m_i/\mathbf{r} = \boldsymbol{\rho}) = \frac{P(m_i) f_{\mathbf{r}}(\boldsymbol{\rho}/m_i)}{f_{\mathbf{r}}(\boldsymbol{\rho})}$$

$$= \frac{P(m_i) f_{\mathbf{r}_1, \mathbf{r}_2}(\boldsymbol{\rho}_1, \boldsymbol{\rho}_2/m_i)}{f_{\mathbf{r}}(\boldsymbol{\rho})} \tag{4.269}$$

Since $\mathbf{r}_1, \mathbf{r}_2$ are uncorrelated, then

$$P(m_i/\mathbf{r} = \boldsymbol{\rho}) = \frac{P(m_i) f_{\mathbf{r}_1}(\boldsymbol{\rho}_1/m_i) f_{\mathbf{r}_2}(\boldsymbol{\rho}_2/m_i)}{f_{\mathbf{r}}(\boldsymbol{\rho})} \tag{4.270}$$

But $f_{\mathbf{r}_2}(\boldsymbol{\rho}_2/m_i) = f_{\mathbf{r}_2}(\boldsymbol{\rho}_2)$ independent of i; hence

$$P(m_i/\mathbf{r} = \boldsymbol{\rho}) = \frac{P(m_i) f_{\mathbf{r}_1}(\boldsymbol{\rho}_1/m_i) f_{\mathbf{r}_2}(\boldsymbol{\rho}_2)}{f_{\mathbf{r}}(\boldsymbol{\rho})} \tag{4.271}$$

Therefore, optimum receiver sets $\widehat{m}(\rho(t)) = m_i$ whenever

$$P(m_i) f_{\mathbf{r}_1}(\boldsymbol{\rho}_1/m_i) = \max_k \left\{ P(m_k) f_{\mathbf{r}_1}(\boldsymbol{\rho}_1/m_k) \right\} \tag{4.272}$$

or alternatively iff

$$\sum_{j=1}^{N} \lambda_j^{-1} r_j s_{ij} + B_i'' = \max_k \left\{ \sum_{j=1}^{N} \lambda_j^{-1} r_j s_{kj} + B_k'' \right\} \tag{4.273}$$

Thus, the minimum number of correlators required to implement optimum receivers is N as shown in Fig. 4.36.

In the special case of AWGN with $R_n(t, s) = (N_0/2)\delta(t - s)$, any orthonormal set of N waveforms can be used in the receiver resulting in $\lambda_j = N_0/2$ for all j. Hence, the optimum receiver sets $\widehat{m}(\rho(t)) = m_i$ iff

$$\frac{2}{N_0} \sum_{j=1}^{N} r_j s_{ij} + B_i'' = \max_k \left\{ \frac{2}{N_0} \sum_{j=1}^{N} r_j s_{kj} + B_k'' \right\} \tag{4.274}$$

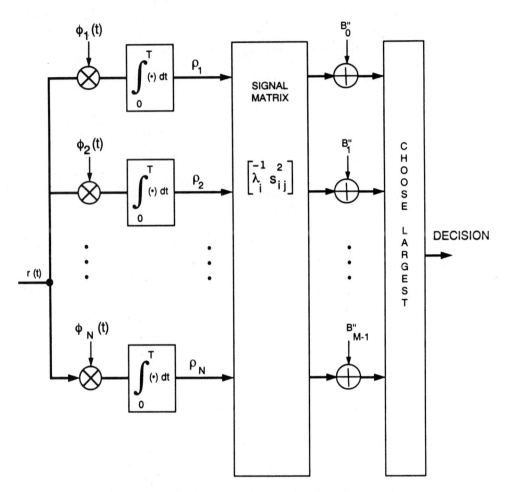

Figure 4.36   Waveform receiver truncation in colored Gaussian noise

In this case

$$\mathbf{r}_1^T \mathbf{s}_i = \mathbf{r}^T \mathbf{s}_i = \sum_{j=1}^{N} r_j s_{ij} = \int_0^T r(t) s_i(t) dt \text{ for all } i \tag{4.275}$$

and $\int_0^T s_i^2(t) dt = \sum_{j=1}^{N} s_{ij}^2 = E_i$. Hence, the receiver reduces to selecting $\widehat{m}(\rho(t)) = m_i$ iff

$$\int_0^T r(t) s_i(t) dt + B_i = \max_k \left\{ \int_0^T r(t) s_k(t) dt + B_k \right\} \tag{4.276}$$

where $B_k = (N_0/2) \ln P(m_k) - E_k/2$ as given in (4.39).

4.2.3 Whitening Approach

Since the optimum MAP structure of Fig. 4.36 whitens the received signal by projecting it on the appropriate eigenfunctions, let's rederive the receiver by imposing a whitening filter up front.

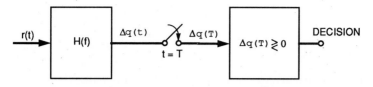

Figure 4.37   Antipodal receiver in colored Gaussian noise

Infinite observation interval.   Let's first consider the detection of antipodal binary signals in colored wide sense stationary (WSS) Gaussian noise assuming infinite $(-\infty, \infty)$ observation interval. Consider the structure of Fig. 4.37 where the received signal is first filtered and then sampled at the end of the signal duration. The idea is to whiten the noise and then perform matched filtering on the new signal. The filter $H(f)$ combines both the whitening and the matched filtering operations. Recall that for binary signaling in white Gaussian noise, the bit error probability is given by (4.51) as

$$P(E) = Q\left(\frac{\overline{\Delta q_0}}{\sigma_{\Delta q}}\right) \tag{4.277}$$

where $\overline{\Delta q_0}$ is the mean conditioned on m_0 and $\sigma_{\Delta q}^2$ the noise variance at the output of the matched filter. Since the noise at the output of $H(f)$ is Gaussian in Fig. 4.37, (4.277) is still applicable with the appropriate mean and variance. Since the output $\Delta q(t) = \int_{-\infty}^{\infty} r(\tau)h(t-\tau)d\tau$, then

$$\overline{\Delta q_0} = \int_{-\infty}^{\infty} s(\tau)h(T-\tau)d\tau = \int_{-\infty}^{\infty} S(f)H(f)e^{j2\pi fT}df \tag{4.278}$$

and

$$\sigma_{\Delta q}^2 = \int_{-\infty}^{\infty} S_n(f)|H(f)|^2 df \tag{4.279}$$

where $S_n(f)$ is the PSD of $n(t)$ and $S(f)$ the Fourier transform of the signaling pulse $s(t)$. Hence

$$\Lambda \triangleq \frac{\overline{\Delta q_0}}{\sigma_{\Delta q}} = \frac{\int_{-\infty}^{\infty} S(f)H(f)e^{j2\pi fT}df}{\left(\int_{-\infty}^{\infty} S_n(f)|H(f)|^2 df\right)^{1/2}} \tag{4.280}$$

Defining

$$G_1(f) \triangleq H(f)\sqrt{S_n(f)} \tag{4.281}$$

$$G_2(f) \triangleq \frac{S(f)e^{j2\pi fT}}{\sqrt{S_n(f)}} \tag{4.282}$$

then

$$\Lambda = \frac{\int_{-\infty}^{\infty} G_1(f)G_2(f)df}{\left(\int_{-\infty}^{\infty} |G_1(f)|^2 df\right)^{1/2}} \tag{4.283}$$

Using Schwartz's inequality, that is

$$\left|\int_{-\infty}^{\infty} G_1(f)G_2(f)df\right|^2 \le \int_{-\infty}^{\infty} |G_1(f)|^2 df \int_{-\infty}^{\infty} |G_2(f)|^2 df \tag{4.284}$$

we have

$$\Lambda \leq \left(\int_{-\infty}^{\infty} \frac{|S(f)|^2}{S_n(f)} df \right)^{1/2} \tag{4.285}$$

with equality if and only if $G_1(f) = cG_2^*(f)$ (c any constant) or

$$H(f)\sqrt{S_n(f)} = \frac{S^*(f)e^{-j2\pi fT}}{\sqrt{S_n(f)}} \tag{4.286}$$

or equivalently

$$H(f) = \frac{S^*(f)e^{-j2\pi fT}}{S_n(f)} \tag{4.287}$$

This is the optimal filter for antipodal signals with colored Gaussian noise with PSD $S_n(f)$. The filter can be decomposed as a series of two filters, that is

$$H(f) = H_1(f)H_2(f) \tag{4.288}$$

where

$$H_1(f) = \frac{1}{\sqrt{S_n(f)}} \tag{4.289}$$

and

$$H_2(f) = \frac{S^*(f)e^{-j2\pi fT}}{\sqrt{S_n(f)}} \tag{4.290}$$

The filter $H_1(f)$ whitens the noise[9] and the filter $H_2(f)$ is matched to the "whitened" signals as shown in Fig. 4.38.

Assuming optimal filtering, the bit error probability is given by $P(E) = Q(\Lambda_{max})$ where

$$\Lambda_{max} = \int_{-\infty}^{\infty} \frac{|S(f)|^2}{S_n(f)} df \tag{4.291}$$

Hence, $P(E)$ is now dependent on the choice of $S(f)$, that is, it depends on the form of the actual signaling pulse $s(t)$, not simply its energy E as in AWGN channels. Then, what

---

9. When $S_n(f)$ is a rational function, then it can be expressed as

$$S_n(f) = c \frac{\prod_{i=1}^{2n}(f - z_i)}{\prod_{j=1}^{2m}(f - p_j)}$$

where z_i and p_j are the zeros and the poles, respectively. Since $S_n(f)$ is real and nonnegative, then the complex roots of $S_n(f)$ have to occur in complex conjugate pairs and the real roots occur with even order. In this case, the whitening filter can be constructed using

$$H_1(f) = c^{1/2} \frac{\prod_{i=1}^{n}(f - z_i)}{\prod_{j=1}^{m}(f - p_j)}$$

where $\text{Im}(z_i) \geq 0$, $1 \leq i \leq n$ and $\text{Im}(p_j) > 0$, $1 \leq j \leq m$, that is, $H_1(f)$ is formed with the right-hand side zeros and poles of $S_n(f)$. In this case, $H_1(f)$ is guaranteed to be a causal filter.

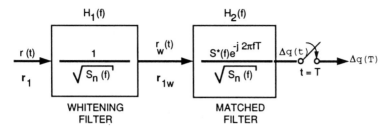

Figure 4.38   Decomposition of $H(f)$ into a whitening and a matched filter

would be the optimal signal shape to transmit for a given channel $S_n(f)$ or equivalently, what is the choice of $s(t)$ that maximizes $\Lambda_{\max}$? That is

$$\max_{s(t)} \int_{-\infty}^{\infty} \frac{|S(f)|^2}{S_n(f)} df \text{ subject to } \int_{-\infty}^{\infty} |S(f)|^2 df = E \tag{4.292}$$

Consider expanding the received signal $r(t)$ using a K-L expansion as in the previous section (remember that only the first N components are sufficient for detection), that is

$$r_1(t) = \sum_{i=1}^{N} r_i \phi_i(t) \text{ and } r_i = s_{ki} + n_i \tag{4.293}$$

assuming m_k is transmitted with

$$E\{n_i n_j\} = \begin{cases} 0, & i \neq j \\ \lambda_i, & i = j \end{cases} \tag{4.294}$$

Then, $r(t)$ can be mapped into the vector $\mathbf{r}_1$ and the whitening operation can be represented by the equivalent vector transformation[10]

$$\mathbf{r}_{1w} = \begin{pmatrix} \lambda_1^{-1/2} & 0 & \cdots & 0 \\ 0 & \lambda_2^{-1/2} & \cdots & \vdots \\ \vdots & \cdots & \ddots & 0 \\ 0 & \cdots & 0 & \lambda_N^{-1/2} \end{pmatrix} \mathbf{r}_1 \tag{4.295}$$

since

$$E\{\mathbf{n}_w \mathbf{n}_w^T\} = \begin{pmatrix} \lambda_1^{-1/2} & 0 & \cdots & 0 \\ 0 & \lambda_2^{-1/2} & \ddots & \vdots \\ \vdots & \ddots & \ddots & 0 \\ 0 & \cdots & 0 & \lambda_N^{-1/2} \end{pmatrix} \begin{pmatrix} \lambda_1 & 0 & \cdots & 0 \\ 0 & \lambda_2 & \ddots & \vdots \\ \vdots & \ddots & \ddots & 0 \\ 0 & \cdots & 0 & \lambda_N \end{pmatrix}$$

$$\times \begin{pmatrix} \lambda_1^{-1/2} & 0 & \cdots & 0 \\ 0 & \lambda_2^{-1/2} & \ddots & \vdots \\ \vdots & \ddots & \ddots & 0 \\ 0 & \cdots & 0 & \lambda_N^{-1/2} \end{pmatrix} = \mathcal{I}_N \tag{4.296}$$

10. $\mathbf{r}_{1w}$ stands for the whitened received vector and is given by $\mathbf{r}_{1w} = \mathbf{s}_w + \mathbf{n}_w$.

and the noise components of $\mathbf{r}_{1w}$ are actually white. The signal component of $\mathbf{r}_{1w}$, $\mathbf{s}_w$, becomes

$$\mathbf{s}_w = \begin{pmatrix} \lambda_1^{-1/2} & 0 & \cdots & 0 \\ 0 & \lambda_2^{-1/2} & \ddots & \vdots \\ \vdots & \cdots & \ddots & \vdots \\ 0 & \cdots & 0 & \lambda_N^{-1/2} \end{pmatrix} \mathbf{s} \tag{4.297}$$

and

$$\sum_{i=1}^{N} \frac{s_i^2}{\lambda_i} = \int_{-\infty}^{\infty} s_w^2(t) dt = \int_{-\infty}^{\infty} \frac{|S(f)|^2}{S_n(f)} df \tag{4.298}$$

where the first equality is due to the fact that a signal and its expansion have the same energy and the second to Parseval's theorem [15]. The optimization problem can therefore be restated in terms of the vector components as

$$\max_{\mathbf{s}} \sum_{i=1}^{N} \frac{s_i^2}{\lambda_i} \text{ subject to } \sum_{i=1}^{N} s_i^2 = E \tag{4.299}$$

Let $\lambda_{\min}$ denote the smallest eigenvalue, then

$$\sum_{i=1}^{N} \frac{s_i^2}{\lambda_i} \le \sum_{i=1}^{N} \frac{s_i^2}{\lambda_{\min}} = \frac{E}{\lambda_{\min}} \tag{4.300}$$

with exact equality if we place all the signal energy E in the dimension with the smallest noise variance $\lambda_{\min}$. Therefore, the optimum performance is obtained by transmitting all the signal power in that coordinate where the noise is weakest. Thus, the optimal antipodal signal pair in nonwhite Gaussian noise is

$$\pm s(t) = \pm \sqrt{E_b} \phi_{\min}(t) \tag{4.301}$$

where $\phi_{\min}(t)$ is the eigenfunction of the K-L expansion of the noise having the smallest eigenvalue. Here again, perfect detection (that is, $P(E) = 0$) can be achieved if $\lambda_{\min} = 0$ (case of singular detection).

Finite observation interval. When the noise is still WSS but the observation interval is finite, the derivation of the optimum receiver is not as straightforward and requires solving integral equations. Let $(0, T)$ denote the observation interval, then the output of the whitening filter in Fig. 4.39, when message m_i is transmitted, becomes (From here on, we will let $H(f)$ denote the whitening filter. Note that in Figures 4.37 and 4.38, $H(f)$ denoted the combination of whitening and matched filters)

$$r_w(t) = \int_0^T h(t - \tau) r(\tau) d\tau = s_{w,i}(t) + n_w(t) \tag{4.302}$$

where

$$s_{w,i}(t) = s_i(t) * h(t) = \int_0^T s_i(t) h(t - \tau) d\tau \tag{4.303}$$

$$n_w(t) = n(t) * h(t) = \int_0^T n(t) h(t - \tau) d\tau \tag{4.304}$$

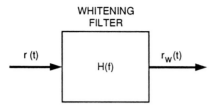

Figure 4.39   Whitening filter

Since the noise at the output of the filter, $n_w(t)$, is white, the optimum receiver for equally likely messages computes

$$\ln\Lambda(r_w(t)/m_i) = \int_0^T r_w(t)s_{w,i}(t)dt - \frac{1}{2}\int_0^T s_{w,i}^2(t)dt \tag{4.305}$$

for every transmitted signal and decides on the message with the largest $\ln(r_w(t)/m_i)$ as shown in Fig. 4.40(a).[11] Substituting for $s_{w,i}(t)$ and $n_w(t)$, we have

$$\ln\Lambda(r_w(t)/m_i) = \int_0^T \left\{ \int_0^T h(t-\tau)r(\tau)d\tau \int_0^T h(t-\tau')s_i(\tau')d\tau' \right\} dt$$
$$- \frac{1}{2}\int_0^T \left\{ \int_0^T h(t-\tau)s_i(\tau)d\tau \int_0^T h(t-\tau')s_i(\tau')d\tau' \right\} dt \tag{4.306}$$

Defining[12]

$$Q(\tau,\tau') \triangleq \int_0^T h(t-\tau)h(t-\tau')dt \tag{4.307}$$

then (4.306) simplifies to

$$\ln\Lambda(r_w(t)/m_i) = \int_0^T \int_0^T r(\tau)Q(\tau,\tau')s_i(\tau')d\tau d\tau'$$
$$- \frac{1}{2}\int_0^T \int_0^T s_i(\tau)Q(\tau,\tau')s_i(\tau')d\tau d\tau' \tag{4.308}$$

and can be implemented as shown in Fig. 4.40(b) with

$$C_i \triangleq -\frac{1}{2}\int_0^T \int_0^T s_i(\tau)Q(\tau,\tau')s_i(\tau)d\tau d\tau' \tag{4.309}$$

Defining

$$g_i(\tau) \triangleq \int_0^T Q(\tau,\tau')s_i(\tau')d\tau' \tag{4.310}$$

---

11. $E_{w,i}$ denotes the energy of the ith whitened signal, that is, $E_{w,i} \triangleq \int_0^T s_{w,i}^2(t)dt$.

12. Note that with a simple change of variables $u = t - \tau$ and when the observation interval becomes $(-\infty, \infty)$, then

$$Q(\tau,\tau') = \int_{-\infty}^{\infty} h(u+\tau'-\tau)h(u)du \triangleq Q(\tau-\tau')$$

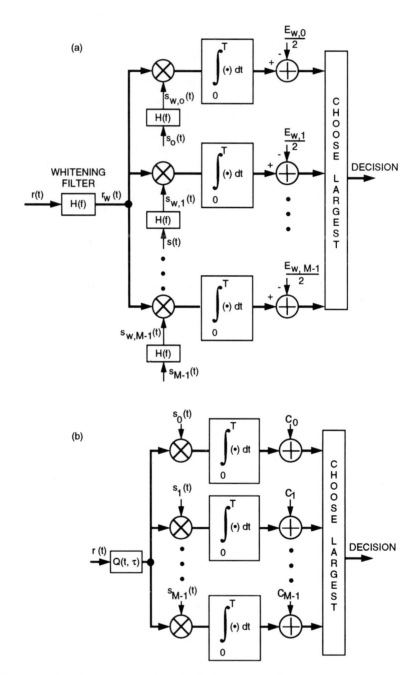

Figure 4.40a, b   Alternate structures for colored noise receiver (derived using the whitening approach)

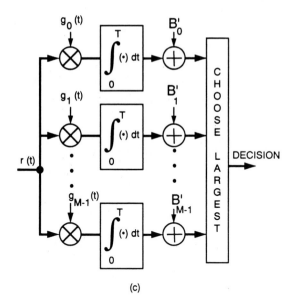

Figure 4.40c   Alternate structures for colored noise receiver (derived using the whitening approach)

then (4.308) becomes

$$\ln \Lambda(r_w(t)/m_i) = \int_0^T r(\tau)g_i(\tau)d\tau + B_i' \tag{4.311}$$

with

$$B_i' = -\frac{1}{2}\int_0^T s_i(\tau)g_i(\tau)d\tau \tag{4.312}$$

which is identical to (4.221) for equilikely signals. This suggests the alternative implementation shown in Fig. 4.40(c). The computation of the functions $g_i(t)$, $i = 0, 1, \cdots, M - 1$ require solving (4.310) which depend on $Q(\tau, \tau')$ which is a function of the whitening filter $h(t)$. The $g_i(t)$ functions can also be expressed directly in terms of the transmitted signals and noise autocorrelation function $R_n(\tau)$ as follows: since $n_w(t)$ is white, then

$$E\{n_w(t)n_w(s)\} = \int_0^T \int_0^T R_n(\tau' - \tau'')h(t - \tau')h(s - \tau'')d\tau'd\tau'' \tag{4.313}$$
$$= \delta(t - s)$$

Multiplying the right-hand side by $h(t - \tau)$ and integrating over t, then

$$\int_0^T \delta(t - s)h(t - \tau)dt = \int_0^T \int_0^T R_n(\tau' - \tau'')$$
$$\times \int_0^T h(t - \tau')h(t - \tau)dth(s - \tau'')d\tau'd\tau'' \tag{4.314}$$

or

$$h(s - \tau) = \int_0^T \int_0^T R_n(\tau' - \tau'')Q(\tau, \tau')h(s - \tau'')d\tau'd\tau'' \tag{4.315}$$

The above equation is satisfied only if

$$\int_0^T R_n(\tau' - \tau'')Q(\tau, \tau')d\tau' = \delta(\tau - \tau'') \tag{4.316}$$

This equation relates $Q(t, \tau)$ directly to the noise autocorrelation function $R_n(\tau)$. Recall from (4.310) that $g_i(\tau) = \int_0^T Q(\tau, \tau')s_i(\tau')d\tau'$. Multiply both sides by $R_n(t - \tau)$ and integrate over τ to get

$$\int_0^T g_i(\tau)R_n(t - \tau)d\tau = \int_0^T \int_0^T Q(\tau, \tau')R_n(t - \tau)s_i(\tau')d\tau'd\tau \tag{4.317}$$

Using (4.316), then

$$\int_0^T g_i(\tau)R_n(t - \tau)d\tau = \int_0^T \delta(t - \tau')s_i(\tau')d\tau' \tag{4.318}$$

or

$$s_i(t) = \int_0^T g_i(\tau)R_n(t - \tau)d\tau \tag{4.319}$$

This is an integral equation in $g_i(\tau)$ as a function of the observation interval $(0, T)$, the signal $s_i(t)$, and the noise autocorrelation function $R_n(\tau)$. Note that when the observation interval is $(-\infty, \infty)$, then $s_i(t) = g_i(t) * R_n(t)$ and $G_i(f) = S_i(f)/S_n(f)$, which is the filter obtained earlier[13] when discussing antipodal signals.

Example 4.4

Consider the binary communication problem over the colored Gaussian noise channel with $S_n(f) = (N_0/2)(f^2 + 1)/(f^2 + 4)$. Derive the optimum receiver assuming infinite observation interval and antipodal signals

$$s(t) = s_0(t) = -s_1(t) = \begin{cases} \sqrt{E}, & |t| \leq 1/2 \\ 0, & \text{otherwise} \end{cases}$$

From Fig. 4.40(c), we know that the optimum decision rule is

$$\int_{-\infty}^{\infty} r(t)g(t) \underset{m_1}{\overset{m_0}{\gtrless}} 0$$

where

$$G(f) = \frac{S(f)}{S_n(f)}$$

In this case, $S_n(f) = \frac{N_0}{2}\frac{f^2+1}{f^2+4}$ and hence

$$G(f) = \frac{2}{N_0}\left[1 + \frac{3}{f^2 + 1}\right]S(f)$$

Since the Fourier transform of $e^{-\alpha|t|}$ is $\frac{2\alpha}{\alpha^2+(2\pi f)^2}$, then

$$g(t) = \frac{2}{N_0}\left[s(t) + 6\pi e^{-2\pi|t|} * s(t)\right]$$

---

13. The filter obtained earlier is denoted by $H(f)$ and given by (4.287) as $H(f) = S^*(f)e^{-j2\pi fT}/S_n(f)$. The filter $h(t)$ and the pulse $g(t)$ are related through $h(t) = g^*(T - t)$ since $h(t)$ is actually the filter matched to $g(t)$.

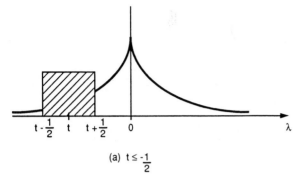

(a) $t \leq -\dfrac{1}{2}$

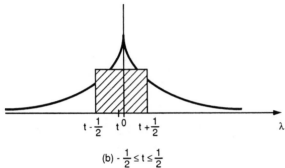

(b) $-\dfrac{1}{2} \leq t \leq \dfrac{1}{2}$

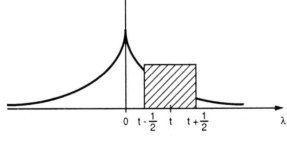

(c) $t \geq \dfrac{1}{2}$

Figure 4.41   Computation of $f(t)$ in Example 4.4

or

$$g(t) = \frac{2}{N_0} \left[s(t) + 6\pi \sqrt{E_s} f(t) \right]$$

where $f(t) = f_1(t) * f_2(t)$ and

$$f_1(t) = e^{-2\pi |t|}$$

$$f_2(t) = \begin{cases} 1 & |t| \leq 1/2 \\ 0 & \text{otherwise} \end{cases}$$

In order to compute $f(t)$, we need to consider three cases as shown in Fig. 4.41, namely

For $t \leq -1/2$

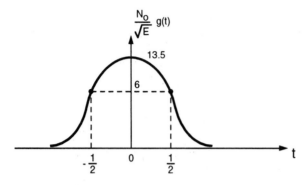

Figure 4.42   The $g(t)$ function in Example 4.4

$$f(t) = \int_{1-1/2}^{t+1/2} e^{2\pi\lambda} d\lambda = e^{2\pi t}\frac{\sinh\pi}{\pi}$$

For $-1/2 \le t \le 1/2$

$$f(t) = \int_{t-1/2}^{0} e^{2\pi\lambda} d\lambda + \int_{0}^{t+1/2} e^{-2\pi\lambda} d\lambda = \frac{1}{\pi} - \frac{e^{-\pi}}{\pi}\cosh(2\pi t)$$

For $t \ge 1/2$

$$f(t) = \int_{t-1/2}^{t+1/2} e^{-2\pi\lambda} d\lambda = e^{-2\pi t}\frac{\sinh\pi}{\pi}$$

Therefore

$$g(t) = \begin{cases} \frac{12}{N_0}\sqrt{E}\sinh(\pi)e^{2\pi t} & t \le -1/2 \\ \frac{12}{N_0}\sqrt{E}[7 - 6e^{-\pi}\cosh 2\pi t] & -1/2 \le t \le 1/2 \\ \frac{12}{N_0}\sqrt{E}\sinh(\pi)e^{-2\pi t} & t \ge 1/2 \end{cases}$$

Note that

$$g(\frac{1}{2}) = g(-\frac{1}{2}) = 12(\frac{e^{\pi} - e^{-\pi}}{2})e^{-\pi}\frac{\sqrt{E}}{N_0} \simeq 6\frac{\sqrt{E}}{N_0}$$

$$g(0) = (14 - 12e^{-\pi})\frac{\sqrt{E}}{N_0} \simeq 13.5\frac{\sqrt{E}}{N_0}$$

The function $g(t)$ is plotted in Fig. 4.42 as a function of t. Since it rolls off exponentially, the function can be truncated when it is implemented in a receiver. A "good" engineering judgment would be to neglect $(N_0/\sqrt{E})g(t)$ when it is less than 0.1 (which is less than 1% of its peak value of 13.5), that is

$$12\sinh(\pi)e^{-2\pi|t|} \le 0.1 \longrightarrow |t| \ge -\frac{1}{2\pi}\ln\frac{0.1}{12\sinh(\pi)} \simeq 1.15 \text{ sec}$$

Therefore, the receiver for all practical purposes observes $r(t)$ over the interval

$$-1.15 \le t \le 1.15$$

Recall that the optimum receiver requires solving for $g_i(t)$ which is given by the integral equation in (4.319). No attempt is made to derive the solutions to these equations, but the key results are given and the reader is referred to Van Trees [13] or Helstrom [16] for additional details. When the observation interval is finite, the noise autocorrelation $R_n(t_1, t_2) = E\{n(t_1)n(t_2)\}$ can be classified as either rational or separable. In the rational case, $R_n(t_1, t_2) = R_n(t_1 - t_2) = R_n(\tau)$ and assuming that the noise $n(t)$ can be written as the sum of two independent noises, that is, $n(t) = n_w(t) + n_c(t)$ to separate out the white from the colored noise contribution, the PSD becomes

$$S_n(f) = \frac{N_0}{2} + S_c(f) \tag{4.320}$$

where $S_c(f)$ is the PSD of $n_c(t)$ and is given by the ratio of two polynomials

$$S_c(f) \triangleq \frac{N(f^2)}{D(f^2)} \tag{4.321}$$

The numerator is a polynomial of order $2n$ and the denominator a polynomial of order $2m$. We assume that $m - n \geq 1$ so that $n_c(t)$ (colored noise contribution) has finite power. When $N_0 = 0$, that is, no white component, (4.319) reduces to

$$s_i(t) = \int_0^T R_c(t - \tau) g_i(\tau) d\tau \tag{4.322}$$

which is a **Fredholm equation of the first kind**. The general solution is given by

$$g_i(t) = g_\infty(t) + \sum_{l=1}^{2n} a_l g_{h,l}(t) + \sum_{k=0}^{m-n-1} \left[b_k \frac{d^k}{dt^k} \delta(t) + c_k \frac{d^k}{dt^k} \delta(t - T) \right] \tag{4.323}$$

where $g_\infty(t)$ is the solution to the integral equation assuming an infinite observation interval (obtained using Fourier transform techniques), $g_{h,l}(t)$ is the lth homogeneous solution to[14] (let $p \triangleq d/dt$)

$$0 = N(-p^2) g_{h,l}(t), \quad l = 1, 2, \cdots, 2n \tag{4.324}$$

The various constants are solved by substituting $g_i(t)$ back into the integral equations and equating terms.

When the white noise component is nonzero (that is, $N_0 \neq 0$), (4.319) reduces to

$$s_i(t) = \frac{N_0}{2} g_i(t) + \int_0^T R_c(t - \tau) g_i(\tau) d\tau \tag{4.325}$$

which is a **Fredholm equation of the second kind**. Its general solution is given by

$$g_i(t) = g_\infty(t) + \sum_{l=1}^{2n} a_l g_{h,l}(t) \tag{4.326}$$

where the various constants are again obtained by equating terms.

As a final category when the noise is not stationary, assume that the kernel of the colored noise is separable, that is

$$R_c(t_1, t_2) = \sum_{i=1}^{K} \lambda_i \phi_i(t_1) \phi_i(t_2), \quad 0 \leq t_1, t_2 \leq T \tag{4.327}$$

---

14. The equation is written in operator's notation.

Table 4.3   Colored Noise Receiver Solutions

OBSERVATION INTERVAL	KERNEL	WHITE NOISE COMPONENT	TECHNIQUE	SOLUTION GIVEN BY EQUATION
INFINITE	-	-	FOURIER TRANSFORM	(4.287)
FINITE	SEPARABLE	-	EIGENFUNCTIONS	(4.328)
	RATIONAL	PRESENT	FREDHOLM EQUATION OF THE SECOND KIND	(4.326)
		ABSENT	FREDHOLM EQUATION OF THE FIRST KIND	(4.223)

where λ_i and $\phi_i(t)$ are the eigenvalues and eigenfunctions of $R_c(t_1, t_2)$, then it can be shown that

$$g_i(t) = \begin{cases} \frac{2}{N_0}\left[s_i(t) - \sum_{i=1}^{K} \frac{s_i \lambda_i}{\lambda_i + N_0/2} \phi_i(t) \right], & 0 \le t \le T \\ 0 & \text{elsewhere} \end{cases} \tag{4.328}$$

where $s_i = \int_0^T s_i(t)\phi_i(t)dt$. Note that if we had allowed $K = \infty$, all kernels would be considered separable since we can always write

$$R_c(t_1, t_2) = \sum_{i=1}^{\infty} \lambda_i \phi_i(t_1)\phi_i(t_2), \quad 0 \le t_1, t_2 \le T \tag{4.329}$$

All the techniques are summarized in Table 4.3. The reader is referred to references on integral equations for additional information.

Example 4.5

Consider detecting the presence or absence of a signal $s(t)$, $t\epsilon(0, T)$ when the observed waveform is

$$m_1 \to r(t) = s(t) + n(t) \quad t\epsilon(0, T)$$

$$m_0 \to r(t) = n(t) \quad t\epsilon(0, T)$$

and $n(t)$ is a zero-mean Gaussian noise with $R_n(t_1, t_2) = \exp(-\alpha|t_1 - t_2|)$ for all $t_1, t_2\epsilon(0, T)$. Derive the decision statistic Λ.

Since $R_n(\tau) = \exp(-\alpha|\tau|)$, the PSD of the noise becomes

$$S_n(f) = \frac{2\alpha}{\alpha^2 + (2\pi f)^2}$$

The function $G_\infty(f)$ is given by the equation

$$G_\infty(f) = \frac{S(f)}{S_n(f)} = \left[\frac{\alpha^2 + (2\pi f)^2}{2\alpha} \right] S(f)$$

or

$$g_\infty(t) = \frac{1}{2\alpha}\left[\alpha^2 s(t) - s''(t)\right]$$

Since $n = 0$ and $m = 1$, (4.323) becomes

$$g(t) = g_\infty(t) + b_0\delta(t) + c_0\delta(t - T)$$

Substituting back in (4.322), one obtains

$$s(t) = \int_0^T \exp(-\alpha|t - \tau|)q(\tau)d\tau = b_0 e^{-\alpha t} + c_0 e^{\alpha(t-T)}$$

$$+ \frac{1}{2\alpha}e^{-\alpha t}\int_0^t e^{\alpha\tau}\left[\alpha^2 s(\tau) - s''(\tau)\right]d\tau$$

$$+ \frac{1}{2\alpha}e^{\alpha t}\int_t^T e^{-\alpha\tau}\left[\alpha^2 s(\tau) - s''(\tau)\right]d\tau$$

Integrating the terms containing $s''(\tau)$ twice by parts, we get

$$s(t) = b_0 e^{-\alpha t} + c_0 e^{\alpha(t-T)} - \frac{1}{2\alpha}\{ e^{-\alpha t}[\alpha s(0) - s'(0)]$$

$$+ e^{\alpha(t-T)}\left[\alpha s(T) + s'(T)\right] - 2\alpha s(t) \}$$

In order for this equation to hold, the coefficients of $e^{\alpha t}$ and $e^{-\alpha t}$ must vanish. Hence, b_0 and c_0 must satisfy

$$b_0 = [\alpha s(0) - s'(0)]/2\alpha$$

$$c_0 = [\alpha s(T) + s'(T)]/2\alpha$$

The decision statistic becomes

$$\Lambda = \frac{1}{2\alpha}\{ [\alpha s(0) - s'(0)]r(0) + [\alpha s(T) + s'(T)]r(T)$$

$$+ \int_0^T [\alpha^2 s(t) - s''(t)]r(t)dt \}$$

4.2.4 Sufficient Statistic Approach

We rederive the optimum receiver in the presence of colored Gaussian noise using a fourth separate approach.[15] A received waveform $r(t)$ is observed on a finite interval $(0, T)$ and is known to have either the form[16] $r(t) = n(t)$ (message m_0) or $r(t) = s(t) + n(t)$ (message m_1) and $s(t)$ is a known real function and $n(t)$ is a zero-mean Gaussian noise with autocorrelation $R_n(t, \tau)$. Consider those signals $s(t)$ that can be represented as

$$s(t) = \int_0^T R_n(t, \tau)g(\tau)d\tau, \quad 0 \le t \le T \tag{4.330}$$

Consider an arbitrary set of orthonormal functions $\xi_1(t), \xi_2(t), \cdots$ on the interval $(0, T)$. Add $s(t)$ to the set. The augmented set $\{s(t), \xi_1(t), \xi_2(t), \cdots\}$ is still complete and spans the same space as the original set. Orthonormalize the augmented set using the Gram-Schmidt

---

15. This approach is in several references [13, 16, 17]. We follow the paper presented by Brown [18] in 1987.

16. The more general binary detection problem ($m_0 \to r(t) = s_0(t) + n(t)$ and $m_1 \to r(t) = s_1(t) + n(t)$) can be handled by defining $r'(t) = r(t) - s_1(t)$ since $s_1(t)$ is a known signal and $s(t) = s_0(t) - s_1(t)$.

procedure outlined in Appendix 3C to obtain $\{\chi_1(t) = as(t), \chi_2(t), \cdots\}$ where $\chi_n(t)$ is orthogonal to $s(t)$ for $n \geq 2$ and a is a constant to normalize $\chi_1(t)$. Next, replace $\chi_1(t)$ by $g(t)$ and call the resulting set $\{\psi_1(t) = g(t), \psi_2(t) = \chi_2(t), \cdots\}$. Note that $\psi_n(t)$ is still orthogonal to $s(t)$ for $n \geq 2$. The altered sequence is still complete since $s(t)$ and $g(t)$ are not orthogonal since

$$\int_0^T s(t)g(t)dt = 0 \tag{4.331}$$

would imply that [using (4.330)]

$$\int_0^T \int_0^T g(t)R_n(t, \tau)g(\tau)dtd\tau = 0 \tag{4.332}$$

contradicting the positive definite property of $R_n(t, \tau)$ as an autocorrelation function. Projecting $r(t)$ onto the set $\{\psi_1(t), \psi_2(t), \cdots\}$ results in

$$r_i \triangleq \int_0^T r(t)\psi_i(t)dt, \quad i = 1, 2, \cdots \tag{4.333}$$

Note that since the set is complete, knowledge of the r_i's allows approximation of $r(t)$ in the mean-squared error sense with arbitrary accuracy. For $j \geq 2$, we have

$$E\{r_1 r_j/m_0\} = E\left\{\int_0^T n(t)\psi_j(t)dt \int_0^T n(\tau)g(\tau)d\tau\right\}$$

$$= \int_0^T \int_0^T R_n(t, \tau)g(\tau)\psi_j(t)d\tau dt = \int_0^T s(t)\psi_j(t)dt = 0 \tag{4.334}$$

since $\psi_j(t)$ is orthogonal to $s(t)$ for $j \geq 2$. Similarly

$$E\{r_1 r_j/m_1\} = \int_0^T \int_0^T s(t)s(\tau)\psi_j(t)g(\tau)dtd\tau$$

$$+ \int_0^T \int_0^T R_n(t, \tau)g(\tau)\psi_j(t)dtd\tau = 0, \quad j \geq 2 \tag{4.335}$$

Also

$$E\{r_j/m_0\} = E\left\{\int_0^T n(t)\psi_j(t)dt\right\} = 0, \text{ for all } j \geq 1 \tag{4.336}$$

and

$$E\{r_j/m_1\} = E\left\{\int_0^T (s(t) + n(t))\psi_j(t)dt\right\} = 0, \quad j \geq 2 \tag{4.337}$$

so

$$E\{r_j/m_1\} = E\{r_j/m_0\} = 0, \quad j \geq 2 \tag{4.338}$$

Next consider the cross-covariance terms of the sequence $\{r_2, r_3, \cdots\}$ when either m_0 or m_1 is transmitted:

$$E\{r_i r_j/m_0\} = \int_0^T \int_0^T R_n(t, \tau)\psi_i(t)\psi_j(\tau)dtd\tau, \quad i, j \geq 2 \tag{4.339}$$

$$E\{r_i r_j / m_1\} = E\left\{\int_0^T \int_0^T s(t)s(\tau)\psi_i(t)\psi_j(\tau)dtd\tau\right.$$

$$\left. + \int_0^T \int_0^T n(t)n(\tau)\psi_i(t)\psi_j(\tau)dtd\tau\right\} \qquad (4.340)$$

$$= \int_0^T \int_0^T R_n(t,\tau)\psi_i(t)\psi_j(\tau)dtd\tau, \quad i,j \geq 2$$

Hence

$$E\{r_i r_j / m_0\} = E\{r_i r_j / m_1\}, \quad i,j \geq 2 \qquad (4.341)$$

Therefore, we have shown that the projections $\{r_2, r_3, \cdots\}$ are statistically independent of r_1 when m_0 or m_1 is transmitted and that the means and cross-correlations among the terms of the sequence $\{r_2, r_3, \cdots\}$ are identical under m_0 and m_1.

Forming the likelihood ratio test based on the finite set $\{r_1, r_2, \cdots, r_N\}$, we have

$$\Lambda_N = \frac{f(r_1, r_2, \cdots, r_N / m_1)}{f(r_1, r_2, \cdots, r_N / m_0)} \qquad (4.342)$$

where the joint densities are Gaussian under either hypothesis. Since r_1 and $\{r_2, r_3, \cdots, r_N\}$ are independent under both m_0 and m_1, then

$$\Lambda_N = \frac{f(r_1 / m_1)f(r_2, r_3, \cdots, r_N / m_1)}{f(r_1 / m_0)f(r_2, r_3, \cdots, r_N / m_0)} \qquad (4.343)$$

However, the joint densities of $\{r_2, r_3, \cdots, r_N\}$ are multivariate normal with identical means and cross-covariances independent of which message was transmitted. Hence, $f(r_2, r_3, \cdots, r_N / m_0) = f(r_2, r_3, \cdots, r_N / m_1)$ and

$$\Lambda_N = \frac{f(r_1 / m_1)}{f(r_1 / m_0)} \qquad (4.344)$$

Since Λ_N is independent of N, the same test is obtained in the limit as $N \to \infty$. Hence, we conclude that $r_1 = \int_0^T r(t)g(t)dt$ is a sufficient statistic for deciding between m_0 and m_1. This technique can be expanded to M messages and results in the optimum receiver depicted in Figures 4.34 and 4.40(c) as expected.

4.3 INPHASE AND QUADRATURE MODULATION AND DEMODULATION

So far, we have considered detection of baseband signals represented by $s_i(t)$, $i = 0, 1, \cdots,$ $M - 1$. In real communication channels, the signals are transmitted at radio frequencies (RF) and the receiver downconverts the received RF signal to baseband, using a coherent carrier reference, before performing any message detection.

4.3.1 Real and Complex Signal Models

In most communication systems, the transmitted signal can in general be represented by

$$s(t) = \sqrt{P}\{I(t)\cos\omega_c t - Q(t)\sin\omega_c t\} \qquad (4.345)$$

where P represents the transmitted signal power, $I(t)$ and $Q(t)$ the inphase and quadrature baseband data modulations, and ω_c the carrier frequency. The $\cos\omega_c t$ term is typically referred to as the inphase (I) carrier and the $\sin \omega_c t$ as the quadrature (Q) carrier. As an example, antipodal signals are constructed by setting $Q(t) = 0$ and $I(t) = \pm p_I(t)$ for $0 \le t \le T$. Alternatively, $I(t)$ can be written as

$$I(t) = \sum_{k=-\infty}^{\infty} a_k p_I(t - kT) \tag{4.346}$$

where $a_k = \pm 1$ representing the binary random data and $p_I(t)$ is the baseband inphase pulse of duration T seconds. A specific case of antipodal signals is BPSK where

$$p_I(t) = \begin{cases} 1 & 0 \le t \le T \\ 0 & \text{otherwise} \end{cases} \tag{4.347}$$

In this case, we have[17]

$$s(t) = \sqrt{2P} \cos(\omega_c t + \theta_k), \quad kT \le t \le (k+1)T \tag{4.348}$$

where $\theta_k = 0$ or π radians, depending on the data polarity. Another example is QPSK where both $I(t) = \pm p_I(t)$ and $Q(t) = \pm p_I(t)$ for $kT \le t \le (k+1)T$, independent of each other. Alternatively, QPSK can be represented by (4.345) with $I(t)$ given by (4.346) and

$$Q(t) = \sum_{k=-\infty}^{\infty} b_k p_I(t + kT) \tag{4.349}$$

where $b_k = \pm 1$ representing an independent binary data stream. For $kT \le t \le (k+1)T$, $s(t)$ can be written as

$$s(t) = \sqrt{2P} \cos(\omega_c t + \theta_k) \tag{4.350}$$

where $\theta_k = \tan^{-1}(b_k/a_k) = \pi/4, 3\pi/4, 5\pi/4, 7\pi/4$ radians.[18] This approach can be easily generalized to accommodate M-PSK signals with $I(t)$ and $Q(t)$ as given by (4.346) and (4.349) with $a_k = \cos \theta_k$, $b_k = \sin \theta_k$, and $\theta_k = (2k+1)\pi/M, k = 0, 1, \cdots, M - 1$. A more general signal model allows for $Q(t)$ to have a pulse $p_Q(t)$ different from $p_I(t)$, as depicted in Fig. 4.43(a).

   Another example is M orthogonal signals, which can be obtained from (4.345) by one of two methods: (1) set $Q(t) = 0$ and let $I(t) = s_i(t), i = 0, 1, \cdots, M - 1$ for $0 \le t \le T$; or (2) divide the signal into two groups and let $I(t)$ be any signal from the first group and $Q(t)$ any signal from the second group. A specific set of M orthogonal signals is M-FSK where

$$s_i(t) = \sqrt{2P} \cos(\omega_c t + \frac{i \pi t}{T} + \theta), \quad 0 \le t \le T \tag{4.351}$$

and the frequency separation is $f_{i,j} = f_i - f_j$ between any two signals is $1/2T$ (independent of i and j), which is the minimum separation required for coherent detection.

   At the receiver, the observed signal is given by

$$r(t) = \sqrt{P} \{I(t) \cos(\omega_c t + \theta) - Q(t) \sin(\omega_c t + \theta)\} + n(t) \tag{4.352}$$

where θ is a random carrier phase that has to be estimated for coherent reception and $n(t)$ is a bandpass Gaussian noise process as described in Appendix 4B. Assuming that the receiver

---

17. Since $Q(t) = 0$, we replace $\sqrt{P}$ by $\sqrt{2P}$ to maintain a constant signal power of P.

18. The $\tan^{-1}(x)$ is a four quadrant inverse tangent.

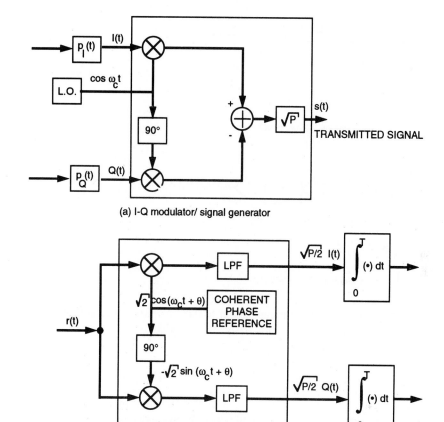

(a) I-Q modulator/ signal generator

(b) I-Q demodulator/ signal detector

Figure 4.43 Functional diagram of the I-Q modulator/demodulator using real signals: (a) I-Q modulator/signal generator, (b) I-Q demodulator/signal detector

generates the coherent references (perfect carrier phase estimate)

$$r_I(t) = \sqrt{2}\cos(\omega_c t + \theta) \tag{4.353}$$

and

$$r_Q(t) = -\sqrt{2}\sin(\omega_c t + \theta) \tag{4.354}$$

then ignoring the noise contribution, we have[19]

$$r(t)r_I(t)|_{LP} = \sqrt{P/2}\,I(t)$$
$$r(t)r_Q(t)|_{LP} = \sqrt{P/2}\,Q(t) \tag{4.355}$$

which for $kT \le t \le (k+1)T$, reduce to the ith signaling baseband pulse $s_i(t)$, assuming that m_i is transmitted. The corresponding I-Q modulator/demodulator is depicted in Fig. 4.43(b) using real signals.

---

19. The notation $r(t)r_I(t)|_{LP}$ denotes the lowpass component of $r(t)r_I(t)$.

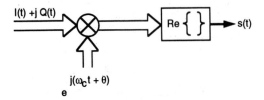

(a) Complex I-Q modulator

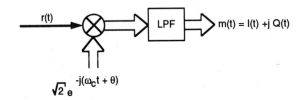

(b) Complex I-Q demodulator

Figure 4.44   Functional diagram of the I-Q modulator/demodulator using complex signals: (a) Complex I-Q modulator, (b) Complex I-Q demodulator

Sometimes, it is more convenient to use complex notation to denote the I-Q signal. Letting $m(t)$ denote the complex baseband modulation, that is, $m(t) = I(t) + jQ(t)$, then $s(t)$ can be written as

$$s(t) = \sqrt{P}\,\mathrm{Re}\left\{m(t)e^{j\omega_c t}\right\} \tag{4.356}$$

The process of coherent demodulation consists of multiplying the received signal by the complex conjugate of the signal (and hence, knowledge of the carrier phase is necessary) as shown in Fig. 4.44. The reader should interchange the real and complex models as they are equivalent. In the presence of noise, the demodulated I & Q signals become

$$r(t)r_I(t)|_{LP} = \sqrt{P/2}\,I(t) + n_c(t)$$
$$r(t)r_Q(t)|_{LP} = \sqrt{P/2}\,Q(t) + n_s(t) \tag{4.357}$$

where $n_c(t)$ and $n_s(t)$ are white Gaussian lowpass processes with one-sided bandwidth B Hz. The optimum receiver structure derived so far assumed white Gaussian noise with infinite bandwidth. It can be shown that the derived structures are still optimum provided that $B \gg 1/T$. One approach to verify the latter is to compute the performance of the integrate-and-dump receiver in the presence of the lowpass noise $n_c(t)$ and show that it reduces to the ideal performance as $BT \to \infty$.

Another more intuitive approach is to consider the detection when implemented using matched filters. In an AWGN channel, the matched filter has a bandwidth equal to roughly $1/T$ Hz, which is approximately the bandwidth of the baseband pulse. When the corrupting noise is white (that is, infinite bandwidth), the output of the matched filter is corrupted by noise with nonzero power in the filter band, that is, the bandwidth of its corresponding power spectral density (PSD) is about $1/T$ Hz. On the other hand, when the input noise is bandlimited to B Hz and $B \gg 1/T$, the output of the matched filter is still corrupted by noise with nonzero power in the filter band only, that is, a bandwidth of $1/T$ Hz. Hence, in either the case of infinite or bandlimited noise, the output of the matched filter is corrupted

Table 4.4 Coding Rule ($N = 4$)

d_0	d_1	d_2	k	a	b	ω
1	1	1	0	1	-3	$\omega_c - 3\omega_d$
1	1	-1	1	1	-1	$\omega_c - \omega_d$
1	-1	1	2	1	1	$\omega_c + \omega_d$
1	-1	-1	3	1	3	$\omega_c + 3\omega_d$
-1	1	1	0	-1	-3	$\omega_c - 3\omega_d$
-1	1	-1	1	-1	-1	$\omega_c - \omega_d$
-1	-1	1	2	-1	1	$\omega_c + \omega_d$
-1	-1	-1	3	-1	3	$\omega_c + 3\omega_d$

by similar noises and as a result, the receiver provides the same performance in both cases as long as $B \gg 1/T$.

Example 4.6

One realization of an I-Q modulation is a combined frequency/phase of a biorthogonal set of M signals is a combined frequency/phase modulation (FPM) wherein $N = M/2$ frequencies uniformly separated by $2f_d = k/2T_s$ (k an integer) are each BPSK modulated, that is, N-FSK/BPSK. Typically, $N = 2^n$ (n integer) and the frequencies are symmetrically placed around a carrier, that is, they are located at $f_c \pm f_d$, $f_c \pm 3f_d, \ldots, f_c \pm (N-1)f_d$ where $f_c = \omega_c/2\pi$ is the carrier frequency in Hertz. Such a signaling set is generated as follows. Consider as a data sequence a binary (± 1) random sequence $\{d_i\}$ of rate $1/T_b$ [equivalently a rectangular pulse data stream $d(t)$]. Blocks of $\lambda = \log_2 M = 1 + \log_2 N$ bits of duration T_b from sequence $\{d_i\}$ are assigned to FPM signals of duration $T_s = \lambda T_b$. We specifically use T_s to denote the duration of a symbol so as to be consistent with what is done later in Chapter 10 where this example is treated as a special case of a more general form of modulation. The first bit in each of these blocks is assigned to a sequence $\{a_i\}$ that is used to define the BPSK portion of the biorthogonal modulation. The remaining $\lambda - 1 = \log_2 N$ bits in each block are assigned to an N-ary $[\pm 1, \pm 3, \ldots, \pm(N-1)]$ sequence $\{b_i\}$ which is used to define the N-FSK portion of the biorthogonal modulation. The coding rule that maps $\{d_i\}$ into the binary amplitude sequence $\{a_i\}$ and the N-ary frequency sequence $\{b_i\}$, respectively, is described as follows.

For a given block of λ bits of the input sequence, say $\{d_0, d_1, \ldots, d_n\}$, the first bit ($d_0$) is assigned directly to the amplitude level a, that is, $a = d_0$. The remaining $\lambda - 1$ bits in the block $(d_1, \ldots, d_n)$ are represented as an N-ary number, say k, in accordance with the natural mapping rule which is then associated with the frequency number $b = 2k - (N-1)$ and results in the actual transmission of the frequency $f_c + bf_d$. Note that if a block of λ bits corresponds to b, then its complement (negative) also corresponds to b. An illustration of the above coding rule for $N = 4$ ($M = 8$) is given in Table 4.4.

Defining the random pulse trains $a(t)$ and $b(t)$ from the sequences $\{a_i\}$ and $\{b_i\}$, respectively, for example, $a(t) = \sum\limits_{i=-\infty}^{\infty} a_i p(t - iT_s)$ with $p(t)$ a unit rectangular pulse, then an FPM signal is characterized by

$$s(t) = \sqrt{2P}a(t) \cos\left[(\omega_c + b(t)\omega_d)\,t\right]$$

Using the methods presented in Chapter 2, it is straightforward to show that the equivalent lowpass two-sided power spectral density of FPM with minimum frequency spacing $1/2T_s$ is given by

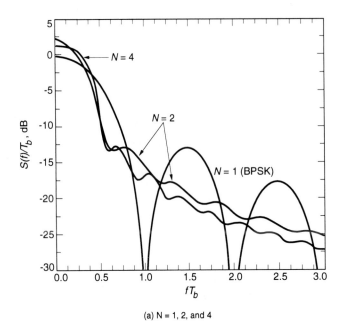

(a) N = 1, 2, and 4

Figure 4.45a   Power spectral density of N-FSK/BPSK, $N = 1, 2,$ and 4

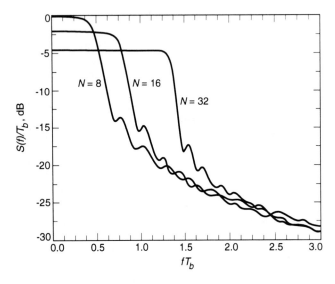

(b) N = 8, 16, and 32

Figure 4.45b   Power spectral density of N-FSK/BPSK, $N = 8, 16,$ and 32

$$S(f) = \frac{PT_s}{N} \sum_{i=0}^{N-1} \text{sinc}^2 \left(fT_s + \frac{i}{2} - \frac{N-1}{4} \right)$$

$$= \frac{(1 + \log_2 N)\, PT_b}{N} \sum_{i=0}^{N-1} \text{sinc}^2 \left((1 + \log_2 N)\, fT_b + \frac{i}{2} - \frac{N-1}{4} \right)$$

which is illustrated in Figures 4.45a,b for various values of N.

4.3.2 Effect of Imperfect Carrier Synchronization

In coherent demodulation, the carrier phase should be known precisely. Assuming that the references $r_I(t)$ and $r_Q(t)$ are not ideal, that is

$$r_I(t) = \sqrt{2}\cos(\omega_c t + \widehat{\theta})$$
$$r_Q(t) = -\sqrt{2}\sin(\omega_c t + \widehat{\theta}) \tag{4.358}$$

where $\widehat{\theta}$ is the receiver's estimate of the incoming carrier phase θ, then (4.355) becomes in the absence of noise

$$r(t)r_I(t)|_{LP} = \sqrt{P/2}[I(t)\cos\phi - Q(t)\sin\phi] \triangleq I_0(t)$$
$$r(t)r_Q(t)|_{LP} = \sqrt{P/2}[I(t)\sin\phi + Q(t)\cos\phi] \triangleq Q_0(t) \tag{4.359}$$

where $\phi = \theta - \widehat{\theta}$ denotes the carrier phase error, which is the phase difference between the received carrier and that generated locally by the carrier synchronization circuitry. Note that a nonzero carrier phase error results in cross-talk between the I and Q arms, that is, the observed I channel, $I_0(t)$, is a combination of the desired I and Q channels. Similarly for the observed Q channel, $Q_0(t)$. If $\phi = \pi/2$, then

$$I_0(t) = -\sqrt{P/2}Q(t)$$
$$Q_0(t) = \sqrt{P/2}I(t) \tag{4.360}$$

and the I and Q channels are reversed in addition to a polarity inversion in $I_0(t)$. If on the other hand $\phi = \pi$, then

$$I_0(t) = -\sqrt{P/2}I(t)$$
$$Q_0(t) = -\sqrt{P/2}Q(t) \tag{4.361}$$

and both I and Q channels are inverted, but not reversed. Using complex notation, this phenomenon can be easily expressed using

$$I_0(t) + jQ_0(t) = \sqrt{P/2}[I(t) + jQ(t)]e^{j\phi} \tag{4.362}$$

which clearly indicates an effective rotation by ϕ radians of the desired baseband I-Q modulation. The performance of the detector in the presence of carrier phase error depends on both the modulation itself and the model associated with the carrier phase error ϕ. This topic will be addressed in great detail in Volume II of this book when considering synchronization loops. Here, we will consider the simplest modulation (antipodal signals) and the easiest model of ϕ, namely, a random variable over the symbol duration. In this case, it is straightforward to show that the conditional error probability is given by

$$P(E/\phi) = Q\left(\sqrt{\frac{2E}{N_0}}\cos\phi\right) \tag{4.363}$$

The true measure of degradation is the average probability of error $P(E)$ obtained from

$$P(E) = \int_{-\pi}^{\pi} P(E/\phi)f(\phi)d\phi \tag{4.364}$$

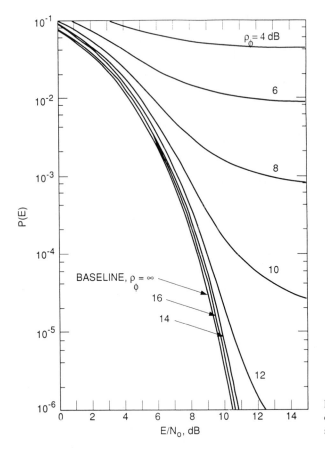

Figure 4.46 Error probability performance of antipodal system in the presence of carrier sync error

where $f(\phi)$ is the pdf of the phase error that depends on the carrier synchronization (sync) algorithm employed. Assuming that $f(\phi)$ is of the form

$$f(\phi) = \begin{cases} \frac{\exp(\rho_\phi \cos \phi)}{2\pi I_0(\rho_\phi)}, & |\phi| \leq \pi \\ 0 & \text{otherwise} \end{cases} \qquad (4.365)$$

where $\rho_\phi \triangleq 1/\sigma_\phi^2$, then using (4.365) in (4.364), $P(E)$ can be evaluated as a function of E/N_0 for various values of ρ_ϕ. The results are plotted in Fig. 4.46 illustrating the noisy reference effect on system performance. For $\rho_\phi = 12$ dB at $P(E) = 10^{-6}$, the required increase in E/N_0 is about 2 dB.

4.3.3 Effect of Imperfect Bit Synchronization

Assuming that perfect carrier synchronization is achieved with BPSK signaling, the baseband received signal in the presence of noise is given by (4.357) with P replaced by $2P$ [since $Q(t) = 0$]

$$r(t)r_I(t)|_{LP} = \sqrt{P}I(t) + n_c(t) = \sum_{k=-\infty}^{\infty} a_k p(t - (k-1)T - \delta) + n_c(t) \qquad (4.366)$$

where δ denotes the bit epoch which needs to be estimated by the bit synchronization circuitry, $a_k = \pm 1$, and $p(t)$ is the baseband pulse (with power P). [20] Consider the situation where the bit sync is in error by $\epsilon = \delta - \widehat{\delta} = \lambda T$ sec, where $\widehat{\delta}$ denotes the estimate of the bit epoch. The detector statistic on which a decision is made for the nth transmitted bit now depends on the received data in the time interval $(n-1)T + \widehat{\delta} \le t \le nT + \widehat{\delta}$. Stated mathematically

$$\widehat{a}_n = \text{sgn} \left\{ \int_{(n-1)T + \widehat{\delta}}^{nT + \widehat{\delta}} r(t) r_I(t)|_{LP} \, p[t - (n-1)T - \widehat{\delta}] dt \right\} \tag{4.367}$$

where $\widehat{a}_n$ is the nth detected bit. Making the change of variables $t' = t - (n-1)T - \widehat{\delta}$ in (4.367) and separating the signal and noise correlation components, we get

$$\widehat{a}_n = \text{sgn } Q \tag{4.368}$$

and

$$Q \triangleq \int_0^T p(t') \{ a_{n-1} p(t' + T - \lambda T) + a_n p(t' - \lambda T)$$

$$+ a_{n+1} p(t' - T - \lambda T) \} dt' + \int_0^T p(t') n_c(t' + (n-1)T + \widehat{\delta}) dt' \tag{4.369}$$

We note that in general, three input bits are involved in estimating the nth transmitted bit. Actually, if λ is known to be positive (or negative), then only two bits are involved, namely the nth and one of its adjacent bits. Denoting the signal and noise components of Q by Q_s and Q_n, respectively, then

$$Q_s = a_{n-1} R_s[-T(1-\lambda)] + a_n R_s(\lambda T) + a_{n+1} R_s[T(1+\lambda)] \tag{4.370}$$

where $R_s(\tau)$ is the signal pulse autocorrelation function given by (4.33) and repeated here for convenience

$$R_s(\tau) = \int_\tau^T p(t') p(t' - \tau) dt'$$

$$= \int_0^{T-\tau} p(t' + \tau) p(t') dt' \tag{4.371}$$

With regard to the input bits affecting the calculation of Q_s, one of four situations can occur: either the nth bit and its adjacent bit are alike (both positive or both negative) or they are unlike (negative followed by positive, or vice versa). Hence, the overall error probability conditioned on knowing λ is given by

$$P(E/\lambda) = \sum_{k=1}^{2} [\text{Prob } \{n^{\text{th}} \text{ and adjacent bits} = (-1)^k\}$$

$$\times \text{Prob } \{\text{error}/ n^{\text{th}} \text{ and adjacent bits} = (-1)^k\} \tag{4.372}$$

$$+ \text{Prob } \{n^{\text{th}} \text{ bit} = (-1)^k \text{ and adjacent bit} = (-1)^{k+1}\}$$

$$\times \text{Prob } \{\text{error}/ n^{\text{th}} \text{ bit} = (-1)^k \text{ and adjacent bit} = (-1)^{k+1}\}]$$

20. In this section, the $\sqrt{P}$ factor in (4.366) is absorbed in $p(t)$.

Since the random variable Q is conditionally Gaussian on a_{n-1}, a_n, a_{n+1} and fixed λ, with variance

$$\sigma_Q^2 = E\left\{Q_n^2\right\} = \frac{N_0 PT}{2} \tag{4.373}$$

and the input bits occur with equal probability and are independent, then

$$
\begin{aligned}
P(E/\lambda) = \frac{1}{4}\text{erfc}\left\{\sqrt{E/N_0}\left[r_s(|\lambda|T) + r_s(T - |\lambda|T)\right]\right\} \\
+ \frac{1}{4}\text{erfc}\left\{\sqrt{E/N_0}\left[r_s(|\lambda|T) - r_s(T - |\lambda|T)\right]\right\}
\end{aligned}
\tag{4.374}
$$

where

$$r_s(\tau) \triangleq \frac{R_s(\tau)}{PT} \tag{4.375}$$

denotes the normalized correlation function and is pulse shape dependent. For NRZ data, that is $p(t) = \sqrt{P};\quad 0 \leq t \leq T$

$$r_s(\tau) = \begin{cases} 1 - \frac{|\tau|}{T} & |\tau| \leq T \\ 0 & |\tau| \geq T \end{cases} \tag{4.376}$$

Substituting (4.376) into (4.374), we get (recall for BPSK, $E = E_b = E_s$)

$$P(E/\lambda) = \frac{1}{4}\text{erfc}(\sqrt{E/N_0}) + \frac{1}{4}\text{erfc}\left[\sqrt{E/N_0}(1 - 2|\lambda|)\right], \quad |\lambda| \leq 1/2 \tag{4.377}$$

If the data is Manchester coded, then the conditional error probability is given by (see Problem 4.38)

$$P(E/\lambda) = \frac{1}{4}\text{erfc}\left[\sqrt{E/N_0}(1 - 2|\lambda|)\right] + \frac{1}{4}\text{erfc}\left[\sqrt{E/N_0}(1 - 4|\lambda|)\right], \quad |\lambda| \leq 1/4 \tag{4.378}$$

When the data format is either RZ or Miller coding, then the expression for $P(E/\lambda)$ as given by (4.374) is no longer valid. The reasons for this are as follows. From Chapter 1, we recall that an RZ data sequence contains two basic pulse waveforms, $p_1(t)$ and $p_2(t)$; thus, the decision rule of (4.374) must be altered to accommodate this fact (see Problem 4.39). For Miller coding, there are four elementary pulse waveforms; furthermore, the occurrence of these waveforms in a data sequence is not independent from bit interval to bit interval (see Problem 4.40). Despite these differences, one can take an approach similar to that leading up to (4.374) and produce expressions for the conditional error probability associated with RZ and Miller-coded data. For RZ data, we obtain

$$P(E/\lambda) = \frac{1}{4}\text{erfc}\left[\sqrt{E/2N_0}\right] + \frac{1}{4}\text{erfc}\left[\sqrt{E/2N_0}(1 - 4|\lambda|)\right], \quad |\lambda| \leq 1/4 \tag{4.379}$$

and for Miller codes, we have

$$
\begin{aligned}
P(E/\lambda) = \frac{1}{2} &+ \frac{3}{16}\text{erf}^2(a) + \frac{1}{16}\text{erf}^2(b) \\
&- \frac{3}{4\sqrt{\pi}}\int_0^{\sqrt{2}c} \text{erf}(x)\exp[-(x - \sqrt{2}b)^2]dx \\
&- \frac{3}{4\sqrt{\pi}}\int_0^{\sqrt{2}c} \text{erf}(x)\exp[-(x - \sqrt{2}a)^2]dx, \quad |\lambda| \leq 1/4
\end{aligned}
\tag{4.380}
$$

where

$$a \triangleq \sqrt{E/2N_0}(1 - 4|\lambda|)$$
$$b \triangleq \sqrt{E/2N_0} \tag{4.381}$$
$$c \triangleq \sqrt{E/2N_0}(1 - 2|\lambda|)$$

The average unconditional error probability $P(E)$ is determined from

$$P(E) = \int_{-\lambda_{\max}}^{\lambda_{\max}} f(\lambda) P(E/\lambda) d\lambda \tag{4.382}$$

where, as before, $f(\lambda)$ is the probability distribution of the normalized sync error and $\lambda_{\max}$ is the maximum value for which λ is defined; for example, $\lambda_{\max} = 1/2$ for NRZ and $\lambda_{\max} = 1/4$ for Manchester, RZ, and Miller-coded data. The pdf, $f(\lambda)$, depends on the way in which bit sync is derived at the receiver. For the purpose of comparing the average error probability performance of the various data formats, we postulate a Tikhonov pdf for $f(\lambda)$ which is characterized entirely in terms of the variance σ_λ^2 of the sync error. Thus, for NRZ data [let $\rho_\lambda \triangleq 1/(2\pi\sigma_\lambda)^2$]

$$f(\lambda) = \frac{\exp(\rho_\lambda \cos(2\pi\lambda))}{I_0(\rho_\lambda)}, \quad |\lambda| \le 1/2 \tag{4.383}$$

whereas for baseband signaling that employs transitions in the middle of the symbol interval, $f(\lambda)$ is characterized by

$$f(\lambda) = \frac{2\exp\left[(\cos 4\pi\lambda)/(4\pi\sigma_\lambda)^2\right]}{I_0\left[(1/4\pi\sigma_\lambda)^2\right]}, \quad |\lambda| \le 1/4 \tag{4.384}$$

Substituting (4.383) or (4.384) into (4.382) together with the appropriate expression for $P(E/\lambda)$ and performing numerical integration provides the graphical results of Figures 4.47 through 4.50. In these curves, the average error probability is plotted versus E/N_0 with the standard deviation σ_λ of the normalized sync error λ as a parameter. The degradation in system performance relative to a perfectly synchronized system can be expressed in terms of the amount by which the signal-to-noise ratio must be increased to achieve the same error probability. As an example, at $P(E) = 10^{-5}$ for an NRZ pulse with $\sigma_\lambda = 0.06$, the degradation is about 1.4 dB.

4.3.4 Effect of Data Asymmetry

We now discuss yet another system imperfection which causes a degradation of performance relative to that of the ideal coherent receiver. This imperfection, namely *data asymmetry*, occurs when the rising and falling transitions of the baseband waveform do not occur at the nominal data transition instants [19, 20]. In Chapter 2, we considered the effects of such asymmetry on the power spectral density of the data waveform for NRZ and Manchester formats. Here we assess the degradation of this type of waveform distortion on the average error probability performance of the receiver.

NRZ Data.   The signal model (referred to as Model 1 in [20]) used to characterize NRZ data with asymmetry assumes that $+1$ NRZ bits are elongated by $\Delta T/2$ (relative to their nominal value of T sec) when a negative-going data transition occurs and -1 bits are shortened by the same amount when a positive-going data transition occurs. When

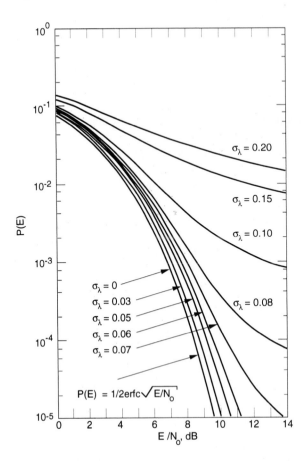

Figure 4.47 Average probability of error versus signal-to-noise ratio with standard deviation of the symbol sync error as a parameter (NRZ) (Reprinted from [12])

no data transition occurs, the bits maintain their nominal T-sec value. An illustration of this distortion for a typical sequence is given in Fig. 4.51. Assuming an integrate-and-dump (matched) filter, then for random NRZ data (50% transitions) with asymmetry, the integrate-and-dump output depends, in general, on the polarity of the bit over which it is integrating and that of the preceding and succeeding bits. As such, one must compute the integrate-and-dump output for each of the eight possible three-bit sequences and their corresponding conditional (on the particular sequence) error probabilities. Then, averaging these conditional error probabilities over the eight equally likely three-bit sequences gives the error probability performance in the presence of asymmetry.

Defining the measure of data asymmetry, η, by the difference in length between the shortest and longest pulses in the sequence divided by their sum, that is,

$$\eta \triangleq \frac{T\left(1 + \frac{\Delta}{2}\right) - T\left(1 - \frac{\Delta}{2}\right)}{T\left(1 + \frac{\Delta}{2}\right) + T\left(1 - \frac{\Delta}{2}\right)} = \frac{\Delta}{2} \tag{4.385}$$

then the average error probability performance described above is given by [20]

$$P\left(E\right) = \frac{5}{16}\,\text{erfc}\,\sqrt{\frac{E}{N_0}} + \frac{1}{8}\,\text{erfc}\left(\sqrt{\frac{E}{N_0}}\,(1 - \eta)\right) + \frac{1}{16}\,\text{erfc}\left(\sqrt{\frac{E}{N_0}}\,(1 - 2\eta)\right) \tag{4.386}$$

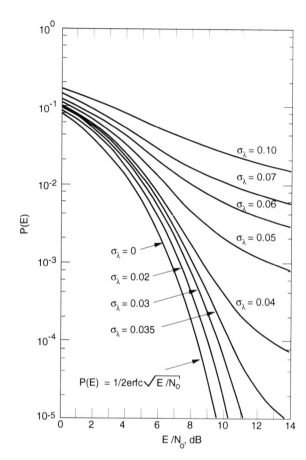

Figure 4.48   Average probability of error versus signal-to-noise ratio with standard deviation of the symbol sync error as a parameter (Manchester code) (Reprinted from [12])

For a given value of η (typically measured in percent), one can evaluate the SNR degradation by computing the additional E/N_0 required ($\Delta E/N_0$) due to asymmetry to produce the same value of bit error probability in the absence of asymmetry ($\eta = 0$). This degradation (in dB) is given by

$$\Delta E/N_0 = 10\log_{10}\left(\frac{E/N_0}{(E/N_0)_0}\right) \tag{4.387}$$

where for a given error probability $P(E)$ and data asymmetry η, E/N_0 is computed from (4.386) and $(E/N_0)_0$ is computed from (4.386) with $\eta = 0$, namely, the value corresponding to ideal coherent BPSK performance. Table 4.5 gives these degradations for error probability values between 10^{-2} and 10^{-6}, and asymmetries in the range of 5% to 25%.

Inherent in the above discussion was the assumption of perfect bit synchronization which in the presence of data asymmetry is tantamount to having a clock with a misalignment equal to half the asymmetry. If in addition to data asymmetry we include the effect of imperfect bit synchronization (additional clock misalignment) discussed in Section 4.3.3, then the conditional error probability in the presence of both imperfections is given by [20]

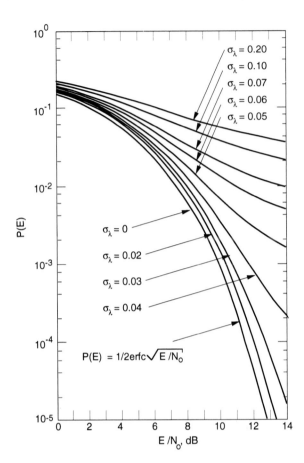

$\sigma_\lambda = 0.20$
$\sigma_\lambda = 0.10$
$\sigma_\lambda = 0.07$
$\sigma_\lambda = 0.06$
$\sigma_\lambda = 0.05$

$\sigma_\lambda = 0$
$\sigma_\lambda = 0.02$
$\sigma_\lambda = 0.03$
$\sigma_\lambda = 0.04$

$P(E) = 1/2 \, \text{erfc} \sqrt{E/N_o}$

Figure 4.49 Average probability of error versus signal-to-noise ratio with standard deviation of the symbol sync error as a parameter (RZ) (Reprinted from [12])

$$P(E|\lambda) = \begin{cases} \frac{5}{16} \, \text{erfc} \sqrt{\frac{E}{N_0}} + \frac{1}{16} \, \text{erfc} \left(\sqrt{\frac{E}{N_0}} (1 - \eta - 2|\lambda|) \right) \\ \quad + \frac{1}{16} \, \text{erfc} \left(\sqrt{\frac{E}{N_0}} (1 - \eta + 2|\lambda|) \right) \\ \quad + \frac{1}{16} \, \text{erfc} \left(\sqrt{\frac{E}{N_0}} (1 - 2\eta) \right); & 0 \leq |\lambda| \leq \frac{\eta}{2} \quad (4.388) \\ \frac{1}{4} \, \text{erfc} \sqrt{\frac{E}{N_0}} + \frac{1}{8} \, \text{erfc} \left(\sqrt{\frac{E}{N_0}} (1 - \eta - 2|\lambda|) \right) \\ \quad + \frac{1}{8} \, \text{erfc} \left(\sqrt{\frac{E}{N_0}} (1 - \eta + 2|\lambda|) \right); & \frac{\eta}{2} \leq |\lambda| \leq \frac{1}{2} \end{cases}$$

where as before λT denotes the misalignment of the bit sync clock relative to its nominal position. Note that for $\eta = 0$, (4.388) reduces to (4.377) as it should.

The average probability of error, $P(E)$, is obtained by averaging (4.388) over the pdf of the bit sync misalignment. Postulating, as in Section 4.3.3, a Tikhonov pdf for λ, which is entirely characterized by the single parameter σ_λ^2 [see (4.383)], then $P(E)$ is given by (4.382) where (4.388) is now used for $P(E|\lambda)$. Figure 4.52 illustrates $P(E)$ versus E/N_0 in dB with σ_λ as a parameter and a fixed value of asymmetry equal to 6%.

Manchester Data. The signal model (Model 1 in [20]) used to characterize Manchester data with asymmetry assumes that for a +1 data bit, the first half is elongated by $\Delta T/4$ (relative to its nominal value of $T/2$ sec) and for a −1 data bit, the first half is shortened

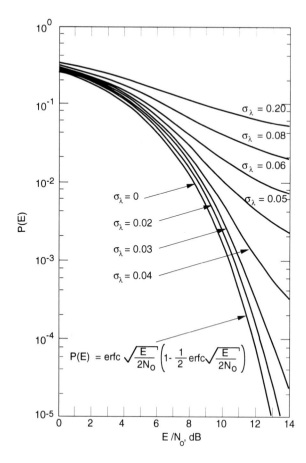

Figure 4.50 shown with curves labeled $\sigma_\lambda = 0.20$, $\sigma_\lambda = 0.08$, $\sigma_\lambda = 0.06$, $\sigma_\lambda = 0.05$, $\sigma_\lambda = 0$, $\sigma_\lambda = 0.02$, $\sigma_\lambda = 0.03$, $\sigma_\lambda = 0.04$.

$$P(E) = \text{erfc}\sqrt{\frac{E}{2N_0}}\left(1 - \frac{1}{2}\text{erfc}\sqrt{\frac{E}{2N_0}}\right)$$

Axis labels: P(E) (vertical), E/N_0, dB (horizontal)

Figure 4.50   Average probability of error versus signal-to-noise ratio with standard deviation of the symbol sync error as a parameter (Miller code) (Reprinted from [12])

by the same amount. When a data transition follows, the second half of the Manchester bit retains its nominal ($T/2$-sec) value. When no data transition occurs, the second half of the Manchester bit adjusts itself so that the total bit maintains its nominal T-sec value. An illustration of this distortion for a typical sequence is given in Fig. 4.53. Assuming an integrate-and-dump (matched) filter, then for random Manchester data (50% data transitions) with asymmetry, the integrate-and-dump output again depends, in general, on the polarity of the bit over which it is integrating and that of the preceding and succeeding bits. Thus, as for NRZ data, one must compute the integrate-and-dump output for each of the eight possible three-bit sequences and their corresponding conditional (on the particular sequence) error probabilities. Then, averaging these conditional error probabilities over the eight equally likely three-bit sequences gives the bit error probability performance in the present of asymmetry.

Defining the measure of data asymmetry, η, relative to the bit time T as was done for NRZ data[21] then analogous to (4.385) we have

---

21. By defining data asymmetry relative to the symbol time for both NRZ and Manchester data, equal percent asymmetry implies equal amounts of asymmetry (in seconds) as measured by the actual time displacements of the waveform transitions.

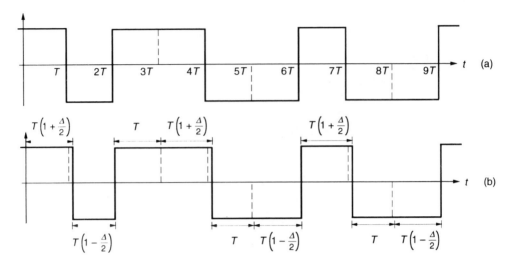

Figure 4.51 (a) Symmetric NRZ data stream; (b) Asymmetric NRZ data stream (Reprinted from [20])

Table 4.5 Degradation Due to Asymmetry for Uncoded Random NRZ Data

$P(E)$ \ η	$\Delta E/N_0$, dB				
	5%	10%	15%	20%	25%
10^{-2}	0.2	0.6	1.1	1.7	2.7
10^{-3}	0.2	0.7	1.4	2.5	2.9
10^{-4}	0.3	0.9	1.8	3.0	4.6
10^{-5}	0.3	1.0	2.1	3.4	5.0
10^{-6}	0.4	1.2	2.3	3.6	5.2

$$\eta \triangleq \frac{T\left(1+\frac{\Delta}{4}\right) - T\left(1+\frac{\Delta}{4}\right)}{T\left(1+\frac{\Delta}{4}\right) + T\left(1-\frac{\Delta}{4}\right)} = \frac{\Delta}{4} \tag{4.389}$$

and the average bit error performance becomes [20]

$$P(E) = \frac{1}{4}\,\text{erfc}\left(\sqrt{\frac{E}{N_0}}\,(1-2\eta)\right) + \frac{1}{4}\,\text{erfc}\left(\sqrt{\frac{E}{N_0}}\,(1-\eta)\right) \tag{4.390}$$

Figure 4.54 illustrates a comparison between the SNR degradation due to data asymmetry for NRZ and Manchester signals. We see that for analogous definitions of asymmetry as in (4.385) and (4.389), the Manchester signal always yields a larger SNR degradation for a given percent asymmetry as one might anticipate since, on the average, the Manchester waveform has 3/2 as many transitions as the NRZ waveform.

When both data asymmetry and imperfect bit sync timing are present, then the analogous result to (4.388) becomes

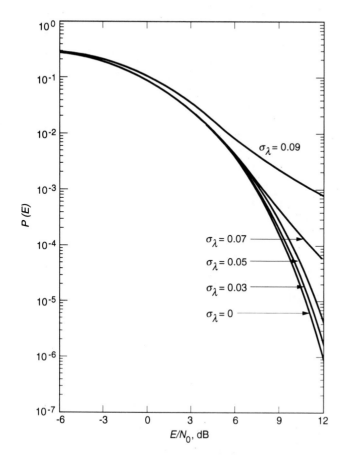

Figure 4.52   Average error probability with symbol synchronization error as a parameter (6% asymmetry), NRZ data (Reprinted from [20])

$$P(E\mid\lambda)=\begin{cases}\frac{1}{4}\,\text{erfc}\!\left(\sqrt{\frac{E}{N_0}}(1-2\eta)\right)+\frac{1}{8}\,\text{erfc}\!\left(\sqrt{\frac{E}{N_0}}(1-\eta-2\,|\lambda|)\right)\\[4pt]\quad+\frac{1}{8}\,\text{erfc}\!\left(\sqrt{\frac{E}{N_0}}(1-\eta+2\,|\lambda|)\right);\qquad\quad 0\le|\lambda|\le\frac{\eta}{2}\\[8pt]\frac{1}{4}\,\text{erfc}\!\left(\sqrt{\frac{E}{N_0}}(1-4\,|\lambda|)\right)+\frac{1}{8}\,\text{erfc}\!\left(\sqrt{\frac{E}{N_0}}(1-\eta-2\,|\lambda|)\right)\\[4pt]\quad+\frac{1}{8}\,\text{erfc}\!\left(\sqrt{\frac{E}{N_0}}(1+\eta-2\,|\lambda|)\right);\qquad\quad \frac{\eta}{2}\le|\lambda|\le\frac{1}{4}\end{cases}\tag{4.391}$$

which reduces to (4.378) when $\eta=0$. Averaging (4.391) over the pdf of the bit sync alignment[22] modeled as a Tikhonov pdf [see (4.384)], $P(E)$ is given by (4.382) where (4.391) is now used for $P(E\mid\lambda)$. Figure 4.55 illustrates $P(E)$ versus E/N_0 in dB with σ_λ as a parameter and a fixed value of asymmetry equal to 6%. Comparing Fig. 4.55 with Fig. 4.52, we observe that for fixed values of η and λ, the error probability performance of Manchester data is always worse than that for NRZ.

4.3.5 Suboptimum Detection

We conclude this chapter with a discussion of antipodal signaling with rectangular pulses but using an arbitrary filter $H(f)$ instead of a matched filter (I&D) as in Fig. 4.56. The

---

22. The nominal bit sync misalignment is now $\Delta T/8$.

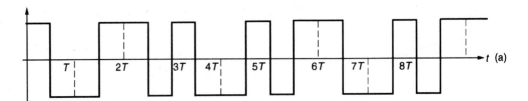

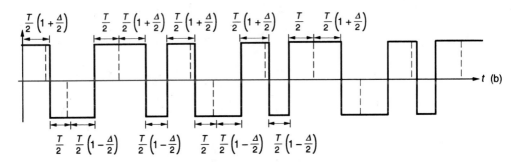

Figure 4.53 (a) Symmetric Manchester data stream; (b) Asymmetric Manchester data stream (Reprinted from [20])

input modulation takes the form

$$m(t) = \sum_{k=-\infty}^{\infty} a_k p(t - kT) \tag{4.392}$$

Here the a_k's are the independent ± 1 data bits and $p(t)$ is the rectangular pulse

$$p(t) = \begin{cases} \sqrt{E/T}, & 0 \le t \le T \\ 0, & \text{otherwise} \end{cases} \tag{4.393}$$

Zero Signal Distortion.   Assume that $|H(f)|_{\max} = 1$ and that $H(f)$ has a single-sided noise bandwidth B_N that is sufficiently wide so as to pass $m(t)$ "undistorted." Typically, a B_N of 3 to 4 times $1/T$ is sufficient. Thus, after passing through the filter $H(f)$, the signal plus noise is

$$y(t) = m(t) + \widehat{n}(t) \tag{4.394}$$

where the "hat" denotes filtering by $H(f)$ and

$$\sigma_{\widehat{n}}^2 = \sigma_y^2 = \int_{-\infty}^{\infty} S_n(f)|H(f)|^2 df = N_0 B_N \tag{4.395}$$

The mean-square value of the filter output sample is

$$E^2\{y(T)\} = \frac{E}{T} \tag{4.396}$$

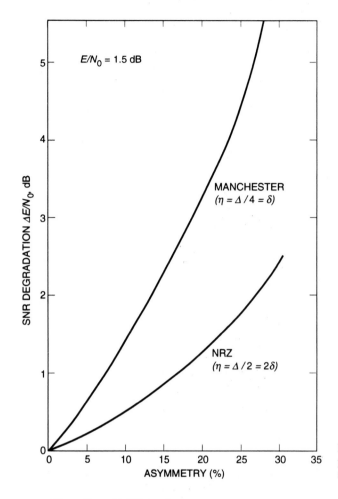

Figure 4.54 SNR degradation versus percent asymmetry for random Manchester and NRZ data (coded) (Reprinted from [20])

Thus, from (4.51), we have

$$P(E) = Q\left(\sqrt{\frac{E}{T N_0 B_N}}\right) \tag{4.397}$$

Since for $H(f)$ a matched filter, $B_N = 1/2T$ and $\text{SNR}_{\text{matched}} = 2E/N_0$, then the SNR penalty of using other than a matched filter (assuming undistorted signal) is

$$10 \log_{10}\left(\frac{\text{SNR}_{\text{matched}}}{\text{SNR}_{H(f)}}\right) = 10 \log_{10} 2B_N T \quad \text{(dB)} \tag{4.398}$$

Clearly as B_N get smaller, the distortion of $m(t)$ increases and the mean of $y(T)$ becomes less than $\sqrt{E/T}$. The value of the mean of $y(T)$ for any specific sample time $t = nT$ depends on the past sequence of a_k's, that is, the previous bits interfere with the mean of the sample corresponding to the present bit. This phenomenon is called *intersymbol interference* (ISI). In general, the effect of ISI is difficult to compute since the distortion depends on the infinite past. Typically, we make some approximation, that is, we assume

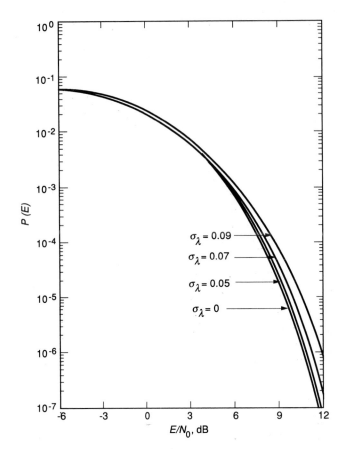

Figure 4.55 Average error probability with symbol synchronization error as a parameter (6% asymmetry), Manchester data (Reprinted from [20])

that the filter has *finite memory*, that is, the distortion of the mean of the filter output sample for a given bit depends on a finite number of previous bits. We now present an example to illustrate this point.

Worst Case Design—Upper Bound on Performance. Let $H(f)$ be a single pole Butterworth filter, that is

$$H(f) = \frac{1}{1 + j\frac{2\pi f}{2\pi f_c}}; \quad B_N = \left(\frac{\pi}{2}\right) f_c \tag{4.399}$$

Assume now that a sequence of all -1 data bits preceding and succeeding the data bit of interest which is a $+1$. This input modulation $m(t)$ (assuming rectangular pulses) is illustrated in Fig. 4.57. Assuming that $y(t)$ has value -1 just prior to the bit interval of interest, then during this interval we have

$$E\{y(t)\} = \sqrt{\frac{E}{T}}\left\{-1 + 2(1 - e^{-\omega_c t})\right\} = \sqrt{\frac{E}{T}}\left(1 - 2e^{-\omega_c t}\right) \tag{4.400}$$

The value of the mean of $y(t)$ at the end of the pulse interval of interest is

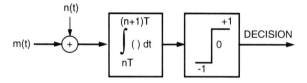

(a) Optimum detection

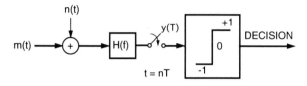

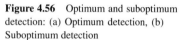

(b) Suboptimum detection

Figure 4.56   Optimum and suboptimum detection: (a) Optimum detection, (b) Suboptimum detection

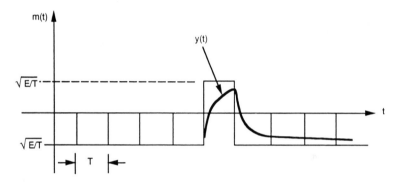

Figure 4.57   Worst case ISI sequence

$$E\{y(nT)\} = \sqrt{\frac{E}{T}}\left(1 - 2e^{-\omega_c T}\right) = \sqrt{\frac{E}{T}}\left(1 - 2e^{-4B_N T}\right) \tag{4.401}$$

and the variance is as in (4.395). Thus, the error probability is given by

$$P(E) = Q\left(\frac{E\{y(nT)\}}{\sigma_y}\right) = Q\left(\sqrt{\frac{E}{T N_0 B_N}}\left(1 - 2e^{-4B_N T}\right)\right)$$

$$= Q\left(\sqrt{\frac{E}{N_0}}\frac{\left(1 - 2e^{-4B_N T}\right)}{\sqrt{B_N T}}\right) \tag{4.402}$$

We wish to choose B_N so as to maximize the argument of the Gaussian tail integral, that is, produce the best trade-off between ISI and noise power in so far as minimizing error probability. Differentiating the argument of the Q function in (4.402) with respect to $B_N T$ gives

$$\sqrt{B_N T}\left(8e^{-4B_N T}\right) - \frac{1}{2\sqrt{B_N T}}\left(1 - 2e^{-4B_N T}\right) = 0 \tag{4.403}$$

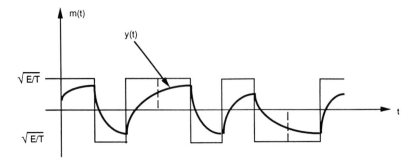

Figure 4.58   Random data sequence

which when solved results in

$$B_N T \simeq 0.62 \tag{4.404}$$

The corresponding worst case error probability is determined by substituting (4.404) into (4.402), resulting in

$$P(E) = Q\left(\sqrt{1.118\frac{E}{N_0}}\right) \tag{4.405}$$

which when compared with the matched filter result represents a $-10\log_{10}(1.118/2) = 2.5$ dB SNR loss. We now reexamine the design of the same receiver structure in light of a minimum average error probability criterion.

    Average Error Probability Performance Design—Finite Memory of Past Two Symbols.   Instead of considering the worst case ISI sequence, we now assume a random data sequence as shown in Fig. 4.58. We shall make the following approximations to the response of $H(f)$ to $m(t)$: At any potential data transition point, there is either a transition or no transition, each occurring with equal probability $1/2$.

1. Assume that after two pulse intervals, the response achieves its maximum value. Thus, if we have $a_k = a_{k-1}$, then the mean of $y(T)$ equals $\sqrt{E/T}$.

2. If there is a transition in the waveform, that is, $a_k \neq a_{k-1}$, then the preceding bit, that is, a_{k-2} will either be identical to a_{k-1} or opposite to a_{k-1}:

   (a) $a_{k-2} \neq a_{k-1}$

       This case is illustrated in Fig. 4.59(a) and we shall assume that the sequence has alternated continuously into the infinite past. Then, to determine the mean of $y(T)$, we have

$$-E\{y(T)\} + \left(\sqrt{\frac{E}{T}} + E\{y(T)\}\right)\left(1 - e^{-\omega_c T}\right) = E\{y(T)\} \tag{4.406}$$

   or

$$E\{y(T)\} = \sqrt{\frac{E}{T}}\frac{(1 - e^{-\omega_c T})}{(1 + e^{-\omega_c T})} \tag{4.407}$$

   (b) $a_{k-2} = a_{k-1}$

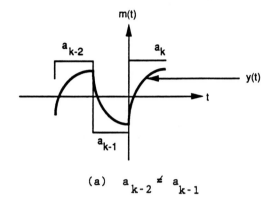

(a) $a_{k-2} \neq a_{k-1}$

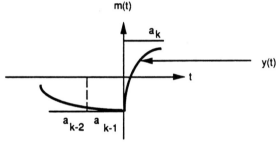

(b) $a_{k-2} = a_{k-1}$

Figure 4.59   Finite memory of past two symbols: (a) $a_{k-2} \neq a_{k-1}$, (b) $a_{k-2} = a_{k-1}$

This case is illustrated in Fig. 4.59(b). Here, we assume that the response starts from $-\sqrt{E/T}$. Thus

$$E\{y(T)\} = -\sqrt{\frac{E}{T}} + 2\sqrt{\frac{E}{T}}\left(1 - e^{-\omega_c T}\right) = \sqrt{\frac{E}{T}}\left(1 - 2e^{-\omega_c T}\right) \qquad (4.408)$$

Case 1 occurs with probability 1/2 (half the time, there is no transition). Cases 2a and 2b each occur with probability 1/4 (half the time, there is a transition. Half of that time, the two preceding bits are identical and half of that time, the two preceding bits are opposite). Thus, the average error probability is

$$P(E) = \frac{1}{2}Q\left(\sqrt{\frac{E}{N_0 T B_N}}\right) + \frac{1}{4}Q\left(\sqrt{\frac{E}{N_0 T B_N}}\left\{\frac{1 - e^{-4B_N T}}{1 + e^{-4B_N T}}\right\}\right)$$
$$+ \frac{1}{4}Q\left(\sqrt{\frac{E}{N_0 T B_N}}\left(1 - 2e^{-4B_N T}\right)\right) \qquad (4.409)$$

Again, one can choose $B_N T$ to minimize $P(E)$ by first rewriting (4.409) as

$$P(E) = \frac{1}{2} Q \left(\sqrt{\frac{E}{N_0}(\frac{1}{B_N T})} \right) + \frac{1}{4} Q \left(\sqrt{\frac{E}{N_0}(\frac{1}{B_N T})} \left\{ \frac{1 - e^{-4B_N T}}{1 + e^{-4B_N T}} \right\}^2 \right)$$

$$+ \frac{1}{4} Q \left(\sqrt{\frac{E}{N_0}(\frac{1}{B_N T})}(1 - 2e^{-4B_N T})^2 \right)$$

(4.410)

Note that the optimum $B_N T$ depends on E/N_0. Some typical values for the optimum $B_N T$ are as follows: $E/N_0 = 7$ dB, $(B_N T)_{\text{opt}} = .51$, $E/N_0 = 10$ dB, $(B_N T)_{\text{opt}} = .562$, $E/N_0 = 13$ dB, $(B_N T)_{\text{opt}} = .61$. Again for sufficiently large $B_N T$, the above reduces to (4.397) corresponding to the zero ISI result.

Appendix 4A

Derivation of the Symbol Error Probability for Polyphase Signals

The probability of error for polyphase signals is obtained by taking one minus the probability of correct detection as given by (4.130), that is

$$
P_s(E) = 1 - \frac{2}{\pi} \int_0^\infty \exp\left\{-\left(u - \sqrt{E_s/N_0}\right)^2\right\} \left[\int_0^{u \tan \pi/M} \exp\left(-v^2\right) dv\right] du \quad (4A.1)
$$

Letting $z = u - \sqrt{E_s/N_0}$

$$
P_s(E) = 1 - \frac{2}{\pi} \int_{-\sqrt{E_s/N_0}}^\infty f\left(\sqrt{E_s/N_0}, z\right) dz \quad (4A.2)
$$

where

$$
f(x, z) = \exp(-z^2) \int_0^{(z+x) \tan \pi/M} \exp(-v^2) dv \quad (4A.3)
$$

Taking the partial derivative of $P_s(E)$ with respect to $\sqrt{E_s/N_0}$, we get

$$
\frac{\partial P_s(E)}{\partial \sqrt{E_s/N_0}} = -\frac{2}{\pi} \left[\int_{-\sqrt{E_s/N_0}}^\infty \frac{\partial f(\sqrt{E_s/N_0}, z)}{\partial \sqrt{E_s/N_0}} dz + f(\sqrt{E_s/N_0}, -\sqrt{E_s/N_0})\right]
$$

$$
= -\frac{2}{\pi} \left[\int_{-\sqrt{E_s/N_0}}^\infty \frac{\partial f(\sqrt{E_s/N_0}, z)}{\partial \sqrt{E_s/N_0}} dz\right] \quad (4A.4)
$$

From (4A.3)

$$
\frac{\partial f(\sqrt{E_s/N_0}, z)}{\partial \sqrt{E_s/N_0}} = \exp\left(-z^2\right) \tan \frac{\pi}{M} \exp\left[-\left\{\left(z + \sqrt{E_s/N_0}\right) \tan \frac{\pi}{M}\right\}^2\right] \quad (4A.5)
$$

$$
= \tan \frac{\pi}{M} \exp\left[-\left\{z^2 \sec^2 \frac{\pi}{M} + 2z\sqrt{E_s/N_0} \tan^2 \frac{\pi}{M} + \frac{E_s}{N_0} \tan^2 \frac{\pi}{M}\right\}\right]
$$

Completing the square in (4A.5) gives

$$
\frac{\partial f(\sqrt{E_s/N_0}, z)}{\partial \sqrt{E_s/N_0}} = \tan \frac{\pi}{M} \exp\left\{-\frac{E_s}{N_0} \sin^2 \frac{\pi}{M}\right\}
$$

$$
\times \exp\left\{-\sec^2 \frac{\pi}{M} \left(z + \sqrt{E_s/N_0} \sin^2 \frac{\pi}{M}\right)^2\right\} \quad (4A.6)
$$

Substituting (4A.6) into (4A.4), we get

$$
\frac{\partial P_s(E)}{\partial \sqrt{E_s/N_0}} = -\left(\frac{2}{\pi}\right) \tan\left(\frac{\pi}{M}\right) \exp\left\{-\frac{E_s}{N_0} \sin^2 \frac{\pi}{M}\right\}
$$

$$
\times \int_{-\sqrt{E_s/N_0}}^\infty \exp\left\{-\sec^2 \frac{\pi}{M} \left(z + \sqrt{\frac{E_s}{N_0}} \sin^2 \frac{\pi}{M}\right)^2\right\} dz \quad (4A.7)
$$

Letting $t = (z + \sqrt{E_s/N_0} \sin^2 \pi/M) \sec(\pi/M)$

$$\frac{\partial P_s(E)}{\partial \sqrt{E_s/N_0}} = -(\frac{2}{\pi}) \sin(\frac{\pi}{M}) \exp\left\{-\frac{E_s}{N_0} \sin^2 \frac{\pi}{M}\right\} \int_{-\sqrt{E_s/N_0}\cos \pi/M}^{\infty} \exp(-t^2) dt$$

$$= -(\frac{1}{\sqrt{\pi}}) \sin \frac{\pi}{M} \exp\left\{-\frac{E_s}{N_0} \sin^2 \frac{\pi}{M}\right\} \left[1 + \text{erf}(\sqrt{E_s/N_0} \cos \frac{\pi}{M})\right]$$

(4A.8)

Integrating (4A.8) with respect to $\sqrt{E_s/N_0}$ between 0 and $\sqrt{E_s/N_0}$, we get

$$P_s(E) - P_s(E)|_{\sqrt{E_s/N_0}=0} = -\frac{1}{2}\text{erf}\left(\sqrt{\frac{E_s}{N_0}} \sin \frac{\pi}{M}\right) - \frac{\sin(\pi/M)}{\sqrt{\pi}} \int_0^{\sqrt{E_s/N_0}}$$

(4A.9)

$$\times \exp\left\{-\frac{E_s}{N_0} \sin^2 \frac{\pi}{M}\right\} \text{erf}\left(\sqrt{E_s/N_0} \cos \pi/M\right) d\sqrt{E_s/N_0}$$

Letting $y = \sqrt{E_s/N_0} \sin(\pi/M)$ and noting that $P_s(E)|_{\sqrt{E_s/N_0}=0} = (M-1)/M$, (4A.9) simplifies to the desired result

$$P_s(E) = \frac{M-1}{M} - \frac{1}{2}\text{erf}\left[\sqrt{E_s/N_0} \sin \frac{\pi}{M}\right]$$

$$- \frac{1}{\sqrt{\pi}} \int_0^{\sqrt{E_s/N_0} \sin \pi/M} \exp(-y^2) \, \text{erf}(y \cot \frac{\pi}{M}) dy$$

(4A.10)

Appendix 4B

Bandpass Random Process Representation

Let $x(t)$ be a real zero-mean bandpass random process given by

$$x(t) = \text{Re}\,\{m(t)\exp(j2\pi f_c t)\} \tag{4B.1}$$

where $m(t)$ denotes the complex baseband modulation function, consisting of both amplitude $a(t)$ and phase $\theta(t)$ modulations, that is

$$m(t) = a(t)\exp(j\theta(t)) \tag{4B.2}$$

We can also define $\tilde{x}(t)$ as the complex random process given by

$$\tilde{x}(t) = m(t)\exp(j2\pi f_c t) \tag{4B.3}$$

such that

$$x(t) \triangleq \text{Re}\,\{\tilde{x}(t)\} = a(t)\cos(2\pi f_c t + \theta(t)) \tag{4B.4}$$

Assume that the random process $x(t)$ is wide sense stationary (WSS) with PSD $S_x(f)$ as depicted in Fig. 4B.1 and expressed as

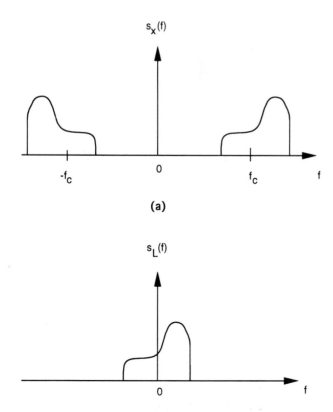

(a)

(b)

Figure 4B.1 Bandpass and baseband power spectral densities

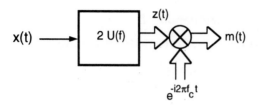

Figure 4B.2   IF to baseband demodulation

$$S_x(f) = S_L(f - f_c) + S_L(-f - f_c) \tag{4B.5}$$

with the condition that

$$S_L(f) = 0 \text{ for } f \leq -f_c \tag{4B.6}$$

Consider the processing depicted in Fig. 4B.2 where the output $z(t)$ of the linear time-invariant (LTI) filter consists solely of the positive frequency content of the input $x(t)$ by defining $U(f)$ as

$$U(f) \triangleq \begin{cases} 1, & f \geq 0 \\ 0, & \text{otherwise} \end{cases} \tag{4B.7}$$

Since the filter is LTI, the output $z(t)$ is also WSS with PSD

$$\begin{aligned} S_z(f) &= |2U(f)|^2 S_x(f) \\ &= 4 S_L(f - f_c) \end{aligned} \tag{4B.8}$$

The autocorrelation function of the modulation $m(t)$ is defined as $R_{mm^*}(t_1, t_2) = E\{m(t_1) m^*(t_2)\}$. Using $m(t) = z(t) \exp(-j2\pi f_c t)$, then

$$\begin{aligned} R_{mm^*}(t_1, t_2) &= E\{z(t_1) z^*(t_2)\} \exp\{-j2\pi f_c(t_1 - t_2)\} \\ &= R_{zz^*}(t_1 - t_2) \exp\{-j2\pi f_c(t_1 - t_2)\} \end{aligned} \tag{4B.9}$$

Therefore, $m(t)$ is also WSS and as a result, we can write $R_{mm^*}(\tau) = R_z(\tau) \exp(-j2\pi f_c \tau)$ or equivalently

$$S_m(f) = S_z(f + f_c) = 4 S_L(f) \tag{4B.10}$$

Expressing $m(t)$ in terms of its real and imaginary parts, we have

$$m(t) = m_R(t) + j m_I(t) \tag{4B.11}$$

where both $m_R(t)$ and $m_I(t)$ are real functions. In order to compute second-order statistics of $m_R(t)$ and $m_I(t)$, we need an expression for $R_{zz}(t_1, t_2) \triangleq E\{z(t_1) z(t_2)\}$, which is different from the autocorrelation function $R_{zz^*}(t_1, t_2)$ for complex processes. Consider Fig. 4B.3, which depicts a real WSS process $x(t)$ through a complex LTI filter $h(t)$. The output $z(t)$ becomes

$$z(t) = \int_{-\infty}^{\infty} h(\alpha) x(t - \alpha) d\alpha \tag{4B.12}$$

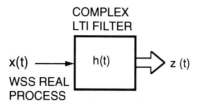

COMPLEX
LTI FILTER

$x(t) \longrightarrow$ | $h(t)$ | $\Rightarrow z(t)$

WSS REAL
PROCESS

Figure 4B.3   Complex LTI filter

resulting in

$$R_{zz}(t_1, t_2) = \int_{-\infty}^{\infty} \int_{-\infty}^{\infty} h(\alpha)h(\beta)E\{x(t_1 - \alpha)x(t_2 - \beta)\}d\alpha d\beta$$
$$= \int_{-\infty}^{\infty} \int_{-\infty}^{\infty} h(\alpha)h(\beta)R_{xx}(t_1 - t_2 - \alpha + \beta)d\alpha d\beta$$

(4B.13)

Note that $R_{zz}(t_1, t_2)$ is a function of $t_1 - t_2$ only. Therefore

$$R_{zz}(\tau) = \int_{-\infty}^{\infty} \int_{-\infty}^{\infty} h(\alpha)h(\beta) \int_{-\infty}^{\infty} S_x(f)e^{j2\pi f(\tau - \alpha + \beta)}df\, d\alpha d\beta$$
$$= \int_{-\infty}^{\infty} S_x(f)H(f)H(-f)e^{j2\pi f\tau}df$$

(4B.14)

or

$$S_{zz}(f) \triangleq \text{F.T}\{R_{zz}(\tau)\} = S_x(f)H(f)H(-f)$$

(4B.15)

where F.T.$\{f(t)\}$ denotes the Fourier transform of $f(t)$. Referring back to Fig. 4B.2, $H(f) = 2U(f)$ and

$$S_{zz}(f) = S_x(f)2U(f)2U(-f) = 0$$

(4B.16)

since $U(f)U(-f)$ is nonzero only at $f = 0$, but $S_x(0) = 0$ since it is a bandpass function. Hence, $R_{zz}(\tau) = 0$ and

$$R_{mm}(\tau) = 0 \text{ for all } \tau$$

(4B.17)

Noting that

$$R_{mm^*}(t_1, t_2) = R_{m_R}(t_1, t_2) + R_{m_I}(t_1, t_2) + jR_{m_I m_R}(t_1, t_2) - jR_{m_R m_I}(t_1, t_2)$$

(4B.18)

$$R_{mm}(t_1, t_2) = R_{m_R}(t_1, t_2) - R_{m_I}(t_1, t_2) + jR_{m_I m_R}(t_1, t_2) + jR_{m_R m_I}(t_1, t_2) = 0$$

(4B.19)

we have by adding (4B.18) and (4B.19)

$$R_{m_R}(t_1, t_2) = \frac{1}{2}\text{Re}\{R_{mm^*}(t_1, t_2)\}$$

(4B.20)

$$R_{m_I m_R}(t_1, t_2) = \frac{1}{2}\text{Im}\{R_{mm^*}(t_1, t_2)\}$$

(4B.21)

where $\text{Re}(x)$ and $\text{Im}(x)$ denote the real and imaginary parts of x, respectively. Subtracting both equations, we obtain

$$R_{m_I}(t_1, t_2) = \frac{1}{2}\text{Re}\{R_{mm^*}(t_1, t_2)\}$$

(4B.22)

$$R_{m_R m_I}(t_1, t_2) = -\frac{1}{2}\text{Im}\{R_{mm^*}(t_1, t_2)\}$$

(4B.23)

Hence

$$R_{m_R}(\tau) = R_{m_I}(\tau) = \frac{1}{2}\text{Re}\{R_{mm^*}(\tau)\} \tag{4B.24}$$

and

$$R_{m_R m_I}(\tau) = -R_{m_I m_R}(-\tau) = -\frac{1}{2}\text{Im}\{R_{mm^*}(\tau)\} \tag{4B.25}$$

From (4B.10), $S_m(f) = 4S_L(f)$ and since $R_{mm^*}(\tau) = \int_{-\infty}^{\infty} S_m(f)\exp(j2\pi f\tau)df$, (4B.22) and (4B.23) become

$$R_{m_R}(\tau) = R_{m_I}(\tau) = 2\int_{-\infty}^{\infty} S_L(f)\cos(2\pi f\tau)df \tag{4B.26}$$

and

$$R_{m_R m_I}(\tau) = -R_{m_I m_R}(-\tau) = -2\int_{-\infty}^{\infty} S_L(f)\sin(2\pi f\tau)df \tag{4B.27}$$

Using (4B.11) in (4B.1), the bandpass process $x(t)$ can be written as[1]

$$x(t) = m_R(t)\cos(2\pi f_c t) - m_I(t)\sin(2\pi f_c t) \tag{4B.28}$$

where $m_R(t)$ and $m_I(t)$ are zero-mean jointly WSS real random processes with second order statistics given by (4B.26) and (4B.27). Note that if $S_L(f)$ is symmetric about $f = 0$, then $R_{m_R m_I}(\tau) = 0$ for all τ and $m_R(t)$, $m_I(t)$ are uncorrelated. Furthermore, if we add the constraint that

$$S_L(f) = 0 \text{ for } |f| > |f_c| \tag{4B.29}$$

then $m(t)$ can be recovered by mixing $x(t)$ with the complex exponential $\exp(-j2\pi f_c t)$ and then filtering the product with an ideal lowpass filter of bandwidth f_c Hz as depicted in Fig. 4B.4.

An application of great importance is the representation of bandpass random noise processes. In any receiver, the signal is first amplified and then bandpass filtered (BPF) to reduce the noise. The noise at the output of the BPF has PSD $S_n(f)$ as shown in Fig. 4B.5(a) with a two-sided level $N_0/2$ Watts/Hz and bandwidth $2B$ Hz with $B \ll f_c$. In this case, the noise can be represented by

$$n(t) = n_c(t)\cos(\omega_c t + \psi) - n_s(t)\sin(\omega_c t + \psi) \tag{4B.30}$$

where $n_c(t)$, $n_s(t)$ are uncorrelated random processes with PSD as depicted in Fig. 4B.5(b), that is

$$S_{n_c}(f) = S_{n_s}(f) = \begin{cases} N_0 & |f| < B \\ 0 & \text{otherwise} \end{cases} \tag{4B.31}$$

and ψ is any arbitrary phase. Since the integral of the spectral density is the power of the process, we have

$$P_{n_c} = P_{n_s} = P_n = 2N_0 B \tag{4B.32}$$

---

1. By writing $m(t) = m_R(t) - jm_I(t)$, $x(t)$ can be expressed as $x(t) = m_r(t)\cos(2\pi f_c t) + m_I(t)\sin(2\pi f_c t)$ with $R_{m_R m_I}(\tau) = 2\int_{-\infty}^{\infty} S_L(f)\sin(2\pi f\tau)df$.

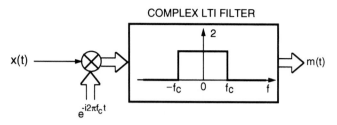

(a) Complex implementation

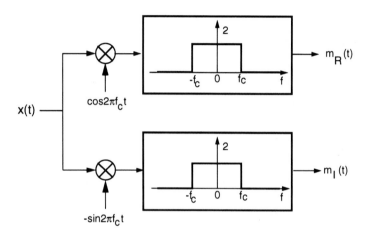

(b) Real implementation

Figure 4B.4   Baseband recovery: (a) Complex implementation, (b) Real implementation

In some instances, it is more convenient to represent the noise $n(t)$ with an additional $\sqrt{2}$ factor as

$$n(t) = \sqrt{2}\left\{n'_c(t)\cos(\omega_c t + \psi) - n'_s(t)\sin(\omega_c t + \psi)\right\} \tag{4B.33}$$

Since both representations must have equal power, then we need

$$S_{n'_c}(f) = S_{n'_s}(f) = \begin{cases} \frac{N_0}{2}, & |f| < B \\ 0, & \text{otherwise} \end{cases} \tag{4B.34}$$

to produce $P_n = 2N_0B$ as in (4B.32). It is often convenient to express the noise $n(t)$ in its complex form as

$$\begin{aligned} n(t) &= \operatorname{Re}\left\{a(t)e^{j\theta(t)}e^{j(\omega_c t + \psi)}\right\} \\ &= a(t)\cos(\omega_c t + \theta(t) + \psi) \end{aligned} \tag{4B.35}$$

where $a(t)$ and $\theta(t)$ are the envelope and phase noise processes given by

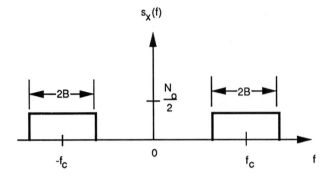

(a) Bandpass noise PSD

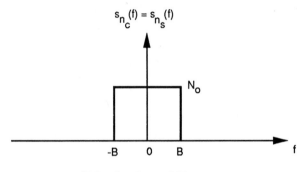

(b) Baseband noise PSD

Figure 4B.5 Receiver noise PSD: (a) Bandpass noise PSD, (b) Baseband noise PSD

$$a(t) = \sqrt{n_c^2(t) + n_s^2(t)} \tag{4B.36}$$

$$\theta(t) = \tan^{-1}\left\{ \frac{n_s(t)}{n_c(t)} \right\} \tag{4B.37}$$

PROBLEMS

4.1 Consider a binary "on-off" communication system which is used to send one of two equally probable messages in the face of AWGN (with one-sided PSD N_0) during the time interval $(0, T)$. Assume the "on-signal" has energy E.

   (a) Assume $N = 2$ (dimension of vector space). Construct the vector representation and illustrate the optimum vector receiver. Find the probability of error.

   (b) What waveform design is compatible with the vector in (a)? Illustrate the optimum waveform receiver.

4.2 Show that in an AWGN channel, when the average signal energy is constrained not to exceed E, that is

$$P(m_0) \int_0^T s_0^2(t)\,dt + P(m_1) \int_0^T s_1^2(t)\,dt \leq E$$

that the optimum signals are given by

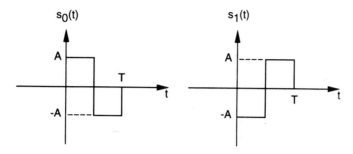

Figure P4.3

$$s_0(t) = \sqrt{\frac{P(m_1)}{P(m_0)}}s(t), \quad 0 \le t \le T$$

$$s_1(t) = -\sqrt{\frac{P(m_0)}{P(m_1)}}s(t), \quad 0 \le t \le T$$

Here $s(t)$ is any signal having energy

$$E = \int_0^T s^2(t)dt$$

4.3 Consider the following binary communication system. The information is sent using one of the two signals shown in Fig. P4.3, which are antipodal, nonreturn-to-zero (NRZ) waveforms. Assuming equilikely signals and an AWGN channel with two-sided PSD $N_0/2$, design the optimum receiver, using the minimum number of correlators, and compute the probability of error for the case $A = 2$ volts, $T = 1$ sec, and $N_0 = 2$ watts/Hz.

4.4 Consider the pulses depicted in Fig. P4.4.

(a) What is the impulse response for filters matched to pulses (a) and (b)?

(b) What is the output of the matched filters if they are excited by the signals to which they are matched?

(c) Make comments on the difference in results. Suggest a practical implementation.

4.5 Consider an amplitude-weighted binary FSK system described by the pair of signals

$$s_i(t) = A(t)\cos\left\{\left[\omega_c + (-1)^{i+1}\frac{\Delta\omega}{2}\right]t\right\}, \quad i = 0, 1 \quad 0 \le t \le T$$

(a) Find the signal cross-correlation coefficient

$$\gamma_{01} \triangleq \frac{\int_0^T s_0(t)s_1(t)dt}{\sqrt{\int_0^T s_0^2(t)dt \int_0^T s_1^2(t)dt}}$$

(b) If the weighting is exponential, that is, $A(t) = A\exp(-Wt)$, evaluate γ_{01} for the case where the frequency separation $\Delta f = \Delta\omega/2\pi$ is an integer multiple of the data rate $1/T$.

(c) Assuming an AWGN channel and an optimum matched filter receiver, what is the error probability of the system for the amplitude weighting of (b)?

(d) Is the error probability found in (c) better or worse than that for an *orthogonal* binary FSK system with $A(t) = A$?

4.6 For coherent FSK, the transmitted signal in the nth transmission interval $nT \le t \le (n + 1)T$ is

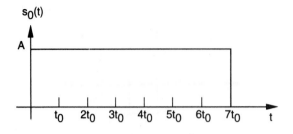

(a) Single pulse

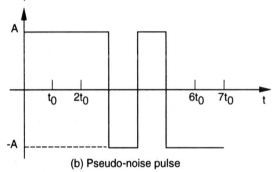

(b) Pseudo-noise pulse   **Figure P4.4**

given by

$$s(t) = A \cos\left[\omega_c t + a_n p(t - nT)\frac{\Delta\omega}{2}t\right]$$

with a_n taking on values ± 1 with equal probability. Conventionally, $p(t)$ is the unit rectangular pulse

$$p(t) = \begin{cases} 1, & 0 \leq t \leq T \\ 0, & \text{otherwise} \end{cases}$$

in which case the two possible signals (corresponding to $a_n = -1$ and $a_n = 1$, respectively), are given by

$$s_0(t) = A \cos\left[\left(\omega_c - \frac{\Delta\omega}{2}\right)t\right]$$

$$s_1(t) = A \cos\left[\left(\omega_c + \frac{\Delta\omega}{2}\right)t\right]$$

Suppose now that the pulse shape $p(t)$ is the unit square wave

$$p(t) = \begin{cases} 1, & 0 \leq t \leq T/2 \\ -1, & T/2 \leq t \leq T \end{cases}$$

which results in so-called *Manchester coded* FSK.

(a) Write the equations analogous to the above for $s_0(t)$ and $s_1(t)$.

(b) Find the signal cross-correlation coefficient γ_{01} (assume $n = 0$ for convenience). Does it differ from that found for conventional coherent FSK ($p(t)$, a unit rectangular pulse)? If so, how?

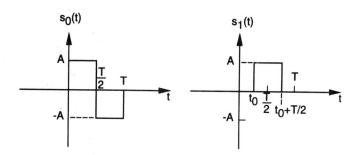

Figure P4.11

4.7 The received signal in a binary baseband communication system takes on values $\pm A$ in each T-sec interval. The probability of transmitting $+A$ is $1/4$ and the probability of transmitting $-A$ is $3/4$. An integrate-and-dump filter is used for detection with the comparison threshold set to V_0. Assume an AWGN channel with single-sided PSD N_0.
   (a) Derive an expression for the average error probability, $P(E)$, as a function of the detection threshold V_0.
   (b) What is the value of V_0 that minimizes $P(E)$?
   (c) What is the corresponding minimum value of $P(E)$?

4.8 A binary communication system (AWGN with 2-sided PSD $N_0/2$) transmits a binary one by sending either a waveform $f_1(t)$ or $f_2(t)$ (energy E, time T), selected by tossing a coin, and sending nothing for a binary zero. Assuming equally likely bits, sketch the decision mechanism that minimizes bit error probability.

4.9 **(a)** Obtain the maximum output signal-to-noise ratio of a matched filter for the following input signals: A and T_c are constants. Assume that the two-sided noise PSD at the input is $N_0/2$.
   (1) $x_1(t) = A\{u(t) - u(t - 5T_c)\}$
   (2) $x_2(t) = A\{u(t) - 2u(t - 3T_c) + 2u(t - 4T_c) - u(t - 5T_c)\}$
   where $u(t)$ denotes the unit step function.
   (b) Find the amplitude of an input signal pulse of duration T_c that will give the same maximum output signal-to-noise ratio as the signals in part (a).
   (c) Obtain the autocorrelation functions of the signal in parts (a) and (b). Comment on the utility of each for resolving delay or relative time differences.

4.10 For the binary coherent communication problem, show that the performance of the optimum receiver for the unequal energy, unequiprobable signal case is given by (4.64), that is

$$P(E) = P(m_0)Q\left(\frac{d_{10}}{\sqrt{2N_0}} - \frac{1}{d_{10}}\sqrt{\frac{N_0}{2}}\ln\frac{P(m_1)}{P(m_0)}\right)$$

$$+ P(m_1)Q\left(\frac{d_{10}}{\sqrt{2N_0}} + \frac{1}{d_{10}}\sqrt{\frac{N_0}{2}}\ln\frac{P(m_1)}{P(m_0)}\right)$$

4.11 Either of the two signals shown in Fig. P4.11 is transmitted with equal probability in a binary communication system.
   (a) Find the matched filter impulse response $h(t)$ that would be required to implement the optimum receiver in an AWGN channel.
   (b) Calculate the optimum threshold and the error probability performance of the receiver as a function of t_0.
   (c) What is the best choice for t_0 such that the error probability is minimized?
   (d) Sketch a correlator receiver for these signals.

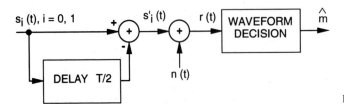

4.12 Consider the binary communication system shown in Fig. P4.12. Assume the signals are equi-likely and that $n(t)$ is AWGN with PSD $N_0/2$ watts/Hz.

(a) Illustrate the optimum correlator receiver and the optimum matched filter receiver.

(b) If $s_i(t)$, $i = 0, 1$ were antipodal signals, evaluate the probability of symbol error.

4.13 A ternary communication system transmits equally likely messages m_0, m_1, and m_2 using three orthogonal signals with equal energy. The channel adds white Gaussian noise with one-sided PSD N_0.

(a) Define vector and waveform signal models.

(b) Illustrate the optimum vector and waveform receivers.

(c) What is the message error probability in terms of E_s/N_0?

(d) What is the minimum number of correlators required to implement the optimum receiver?

(e) Illustrate the optimum waveform receiver in (d) and define all waveforms.

4.14 Consider the following hypotheses testing problem:

$$H_i : \quad r(t) = s_i(t) + n(t), \quad 0 \le t \le T$$

for $i = 1, \cdots, 8$ where

$$s_i(t) = \sqrt{2E} \sin\left(t + i\frac{\pi}{4}\right)$$

This signal set is known as 8-PSK. Assume $n(t)$ is an AWGN with PSD $N_0/2$ watts/Hz.

(a) Choose a suitable orthonormal set for the signal space. What is the dimensionality of this space?

(b) Indicate the optimum decision regions and use the union bound to find an upper bound on the probability of symbol error.

4.15 A waveform signal set is described by

$$s_i(t) = A \cos\left(\omega_c t + i\frac{2\pi}{6}\right), \quad 0 \le t \le T$$

for $i = 1, \cdots, 6$. Assume all signals are equilikely and that the signals are transmitted over an AWGN channel.

(a) Find a suitable set of orthonormal basis using minimum number of dimensions. Sketch the signal vectors and indicate the decision regions for optimum reception.

(b) Use the union bound to find an upper bound for the probability of symbol error.

4.16 Consider the problem of communicating over white Gaussian noise channel ($N_0 = 1/2$). One of four possible waveforms is transmitted every second. The source messages are equilikely and independently chosen. The transmitter has a maximum energy constraint defined by:

$$|E_s| \le 3$$

(a) What is the optimum choice for the set of four signals in this situation? Describe the set as vectors.

(b) Convert the set of vectors in (a) to waveforms.

(c) Draw a block diagram for the optimum waveform receiver. Use the smallest number of correlators (matched filters).

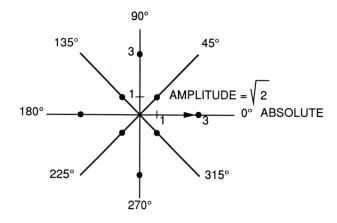

Figure P4.18

(d) Using the result from the union bound, what is an upper bound on symbol error probability?

4.17 Consider the *4*-ary hypotheses testing problem. Each observation is a *3*-dimensional vector

$$H_i: \quad \mathbf{r} = \mathbf{s}_i + \mathbf{n} \quad i = 0, 1, 2, 3$$

where

$$\mathbf{s}_0 = \begin{pmatrix} A \\ 0 \\ B \end{pmatrix} \quad \mathbf{s}_1 = \begin{pmatrix} 0 \\ A \\ B \end{pmatrix} \quad \mathbf{s}_2 = \begin{pmatrix} -A \\ 0 \\ B \end{pmatrix} \quad \mathbf{s}_3 = \begin{pmatrix} 0 \\ -A \\ B \end{pmatrix}$$

The components of the noise vector are independent identically distributed Gaussian random variables, with variance $N_0/2$.

(a) Choosing an appropriate basis, express the signal vectors as waveforms.

(b) Sketch the optimum receiver assuming equilikely signals using the least number of correlators.

(c) Compute the probability of symbol error.

4.18 Consider the V.29 signal constellation depicted in Fig. P4.18, which transmits 3 bits/symbol. Assume an AWGN channel and equilikely signals.

(a) Write an I-Q waveform model using rectangular coordinates for square and arbitrary I-Q pulses.

(b) Write an I-Q waveform model using polar coordinates for square and arbitrary I-Q pulses.

4.19 Consider an *M*-FSK system that employs the signal set (*k* an integer)

$$s_i(t) = \sqrt{2P} \sin\left[\left(\frac{k\pi}{T} + 2\pi i \Delta f\right)t\right], \quad i = 0, 1, \cdots, M - 1$$

(a) Find the minimum frequency separation spacing that produces orthogonality and the effective bandwidth occupancy of this set.

(b) Using this orthogonal set, define a biorthogonal signal set and find its effective occupancy.

(c) Repeat for a transorthogonal signal set.

4.20 Consider the V.29 signal constellation depicted in Fig. P4.20, which transmits 4 bits/symbol. Assume an AWGN channel and equilikely signals.

(a) Write an I-Q waveform model using rectangular coordinates for square and arbitrary I-Q pulses.

(b) Write an I-Q waveform model using polar coordinates for square and arbitrary I-Q pulses.

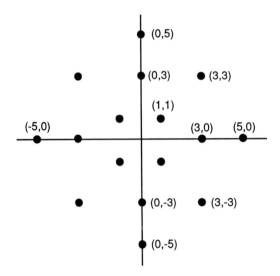

Figure P4.20

4.21 Consider a ternary communication system which transmits one of three equally likely signals

$$H_0: \quad r(t) = n(t)$$

$$H_1: \quad r(t) = A \sin \omega_c t + n(t)$$

$$H_2: \quad r(t) = -A \sin \omega_c t + n(t)$$

during $0 \le t \le T$. Assume that $n(t)$ is a white Gaussian noise process with PSD $N_0/2$ watts/Hz.
   (a) What is the optimum receiver which minimizes the error probability? Sketch this mechanization and show all operations which must be performed in making a decision.
   (b) What are the conditional probabilities of correct reception when H_0, H_1, and H_2 are true?
   (c) Using the results found in (b), write an expression for the average probability of error.
   (d) Compare this ternary system with the binary PSK system which uses $A \sin \omega_c t$ and $-A \sin \omega_c t$ to transmit data.

4.22 Establish (4.129) by substituting (4.127) into (4.128) and averaging over the equal *a priori* probabilities of the transmitted signals. Show that (4.129) reduces to (4.130) and the average symbol error probability is given by (4.131).

4.23 For large signal-to-noise conditions, Gilbert[2] has found an approximate formula for the average symbol error probability of a maximum likelihood receiver in terms of the geometrical configuration of the transmitted signals in signal space. Applying his formula to an M-PSK system, the M signals are represented as uniformly spaced points on the circumference of a circle of radius R and

$$P_s(E) \simeq \frac{N_s}{2} \operatorname{erfc} \left\{ \sqrt{\frac{E_s}{N_0}} \left(\frac{r_{min}}{2R} \right) \right\}$$

where r_{min} is the smallest of the set of $M(M-1)/2$ Euclidean distances between pairs of signal

2. E. N. Gilbert, "A Comparison of Signaling Alphabets," *Bell System Technical Journal*, vol. 31, May 1952, pp. 504–22.

points, and N_s is determined as follows: For each signal $s_i(t)$, let N_{s_i} denote the number of signals that are at the minimum distance r_{min}. Then

$$N_s = \frac{\sum_{i=1}^{M} N_{s_i}}{M}$$

(a) Show that $r_{min} = 2R \sin \pi/M$.

(b) For $M > 2$, show that $N_s = 2$.

(c) From the results of (a) and (b), verify (4.134).

4.24 A general representation for the L-orthogonal signal set is given by[3]

$$s_i(t) = \sqrt{P} \left[x_n(t) \cos \theta_l + y_n(t) \sin \theta_l \right] \quad 0 \leq t \leq T$$

for $n = 1, 2, \cdots, N, l = 1, 2, \cdots, L, i = (n-1)L + 1$ with $\theta_l = 2(l-1)\pi/L$ and

$$\frac{1}{T} \int_0^T x_j(t) x_k(t) dt = \frac{1}{T} \int_0^T y_j(t) y_k(t) dt = \begin{cases} 1, & j = k \\ 0, & j \neq k \end{cases}$$

$$\frac{1}{T} \int_0^T x_j(t) y_k(t) dt = 0 \text{ for all } j \text{ and } k$$

If

$$x_n(t) = \sqrt{2} \sin \left[\frac{\pi(j+n)t}{T} \right] \quad L = 1, 2$$

$$y_n(t) = \sqrt{2} \cos \left[\frac{\pi(j+n)t}{T} \right] \quad n = 1, 2, \cdots, N$$

and

$$x_n(t) = \sqrt{2} \sin \left[\frac{\pi(j+2n)t}{T} \right] \quad L > 2$$

$$y_n(t) = \sqrt{2} \cos \left[\frac{\pi(j+2n)t}{T} \right] \quad n = 1, 2, \cdots, N$$

Show that the signal set $\{s_i(t)\}$ is L-orthogonal. Assume that j is any integer.

4.25 (a) In terms of the L-orthogonal signal representation of Problem 4.24, let $\{x_n(t)\}$ be a set of N nonoverlapping sinusoidal pulses defined over the interval $0 \leq t \leq T/2$ and $\{y_n(t)\}$ be a set of N nonoverlapping sinusoidal pulses defined over the interval $T/2 \leq t \leq T$, that is

$$x_n(t) = 2\sqrt{N} \sin \left(\frac{2\pi j N t}{T} \right); \quad \frac{(n-1)T}{2N} \leq t \leq \frac{nT}{2N}$$

$$y_n(t) = 2\sqrt{N} \sin \left(\frac{2\pi j N t}{T} \right); \quad \frac{(N+n-1)T}{2N} \leq t \leq \frac{(N+n)T}{2N}$$

for $n = 1, 2, \cdots, N$. Show that the resulting set $\{s_i(t)\}$ is L-orthogonal.

(b) If N is a power of 2, then another pair of orthogonal signals for $\{x_n(t)\}$ and $\{y_n(t)\}$ can be

---

3. See [10], and J. J. Stiffler, A. J. Viterbi, "Performance of N-Orthogonal Codes," *IEEE Transactions on Information Theory*, vol. IT-13, no. 3, July 1967, pp. 521–22.

modeled after the Hadamard design for an orthogonal code. In particular, we define

$$x_n(t) = \sum_{i=1}^{2N} (-1)^{b_{i,n}} \, w\left(t - \frac{(i-1)T}{2N}\right) \qquad 0 \le t \le T$$

$$y_n(t) = \sum_{i=1}^{2N} (-1)^{b_{i,N+n}} \, w\left(t - \frac{(i-1)T}{2N}\right) \qquad 0 \le t \le T$$

for $n = 1, 2, \cdots, N$ where $w(t)$ is an arbitrary waveform of energy $T/2N$ and nonzero only over the interval $0 \le t \le T/2N$, and $b_{i,n}$ is the ith element in the nth row of a $2N \times 2N$ Hadamard matrix. Show that the resulting signal set $\{s_i(t)\}$ is L-orthogonal.

4.26 Show that for the special case of $L = 2$ and $L = 4$, the error probability performance of an L-orthogonal signal set of $M = NL$ signals defined by (4.145) is equivalent to that of a biorthogonal signal set of M signals.

4.27 Using the approach of Section 3.2.8, derive (4.198b) which represents the probability that the receiver's noisy estimate of the phase in the ith transmission interval lies in the kth sector of the unit circle, that is, $(\theta_i + \phi_a + (2k-1)\pi/M \le \eta_i \le \theta_i + \phi_a + (2k+1)\pi/M)$.

4.28 Consider the two sets of orthogonal functions

$$\phi_n(t) = \frac{\omega_1}{\pi} \frac{\sin(\omega_1 t - n\pi)}{\omega_1 t - n\pi}$$

$$\theta_n(t) = \frac{\omega_2}{\pi} \frac{\sin(\omega_2 t - n\pi)}{\omega_2 t - n\pi}$$

(a) Show that

$$\int_{-\infty}^{\infty} \frac{\omega_1}{\pi} \frac{\sin(\omega_1 \tau - n\pi)}{\omega_1 \tau - n\pi} \frac{\omega_2}{\pi} \frac{\sin(\omega_2(t - \tau) + m\pi)}{\omega_2(t - \tau) + m\pi} d\tau = \frac{\omega_1}{\pi} \frac{\sin\left[\omega_1 t - \left(n - m\frac{\omega_1}{\omega_2}\right)\pi\right]}{\omega_1 t - \left(n - m\frac{\omega_1}{\omega_2}\right)\pi}$$

Hint: Consider the product of the Fourier transform of $\phi_n(t)$ and $\theta_m(-t)$.

(b) Suppose that a WSS process $x(t)$ has the correlation function

$$R_x(\tau) = \frac{\omega_2}{\pi} \frac{\sin \omega_2 \tau}{\omega_2 \tau}$$

Does this process have a K-L expansion in terms of the orthogonal functions $\phi_n(t)$? If so, do the coefficients in this expansion represent samples of the random process at the sampling rate $2f_1 = 2\omega_1/2\pi$?

4.29 The covariance function of the additive Gaussian noise process present in a communication channel is characterized by

$$K_n(t, s) = \frac{1}{4} \exp\{-2|t - s|\}, \qquad |t - s| \le \infty$$

(a) Find the K-L expansion for this process.
(b) Draw the optimum receiver if

$$s_i(t) = \sum_{j=1}^{N} s_{ij}\phi_j(t), \qquad i = 0, 1, \cdots, M - 1$$

and specify the appropriate $\phi_j(t)$'s.

4.30 Let $x(t)$ be a zero-mean random process with correlation function given by

$$R_x(\tau) = \frac{\exp(-c|\tau|)}{2c}$$

where $c > 0$ is a real constant.

(a) Write the integral equation that gives the eigenvalues and the eigenfunctions in the K-L expansion of $x(t)$ over the time interval $-T/2 \leq t \leq T/2$, where T is some positive number.

(b) The usual approach to solve this integral equation is to convert it into a differential equation and solve that equation. Show that the related differential equation for this case is:

$$\lambda\ddot{\phi}(t) + (1 - c^2\lambda)\phi(t) = 0, \quad \text{for } |t| \leq T/2$$

where $\phi(t)$ denotes the eigenfunctions, λ the eigenvalues, and dots denote differentiation with respect to t.

Hint: Take derivatives on both sides of your answer in (a) two times. Use the Leibniz's theorem for differentiation of an integral

$$\frac{d}{dy}\int_{a(y)}^{b(y)} f(x, y)dx = \int_{a(y)}^{b(y)} \left[\frac{\partial}{\partial y} f(x, y)\right] dx + f(b(y), y)\frac{d}{dy}b(y) - f(a(y), y)\frac{d}{dy}a(y)$$

4.31 Consider the integral equation

$$\lambda_i\phi_i(t) = \int_0^T R_n(t, s)\phi_i(s)ds, \quad 0 \leq t \leq T$$

where[4]

$$R_n(t, s) = \begin{cases} 1 - |t - s|, & |t - s| \leq 1 \\ 0, & \text{elsewhere} \end{cases}$$

Find the eigenvalues and eigenfunctions over the interval $(0, T)$ when $T \leq 1$.

4.32 Consider the integral equation

$$\lambda_i\phi_i(t) = \int_0^T R_n(t, s)\phi_i(s)ds, \quad 0 \leq t \leq T$$

where

$$R_n(t, s) = \sum_{i=1}^{6} \sigma_i^2 \cos\left(\frac{2\pi it}{T}\right)\cos\left(\frac{2\pi is}{T}\right)$$

Find the eigenvalues and eigenfunctions over the interval $(0, T)$.

4.33 Consider the binary radar detection problem:

$$H_1: \quad r(t) = s(t) + n(t), \quad -\infty < t < \infty$$

$$H_0: \quad r(t) = n(t), \quad -\infty < t < \infty$$

where the signal $s(t)$ is nonzero in the interval $(0, T)$ and the noise $n(t)$ is Gaussian with PSD given by

$$S_n(f) = \frac{1}{2} + \frac{24}{(2\pi f)^2 + 16} \quad \text{Watts/Hz}$$

4. T. Kailath, "Some Integral Equations with "Nonrational" Kernels," *IEEE Transactions on Information Theory*, vol. IT-12, no. 4, October 1966, pp. 442–47.

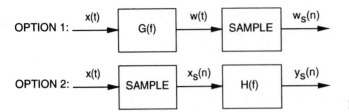

Figure P4.34

(a) Determine a suitable realizable whitening filter for the minimum probability of error receiver.

(b) The optimum receiver processes only that portion of the received signal lying within the range: (i) $[0, T]$, (ii) $[0, \infty)$, (iii) $(-\infty, 0]$, and finally (iv) $(-\infty, \infty)$. Indicate your choice and explain your answer.

4.34 A noise generator produces a WSS Gaussian random process $x(t)$, whose correlation function is

$$R_x(\tau) = \exp(-3|\tau|)$$

You must construct a random number generator from $x(t)$, which produces a new independent random number generator every tenth of a second. You have two options for doing this, as indicated in Fig. P4.34. Here the samples are instantaneous, ideal, and operate at ten samples per sec. The variables indexed with the integer n occur at times $0.1n$ sec.

(a) Compute the correlation coefficient between $X_s(0)$ and $X_s(1)$.

(b) In option 1, what filter system function $G(f)$ would you use to make $w(t)$ a white Gaussian random process with PSD $S_w(f) = 1$?

(c) In option 2, $X_s(n)$ is defined on the integers. Is $X_s(n)$ a Gaussian random process? Compute $E\{X_s(n)\}$ and $E\{X_s(n+m)X_s(n)\}$.

(d) In option 2, what causal filter system function $H(f)$ would you employ to whiten $X_s(n)$ and produce $Y_s(n)$?

(e) Are there any problems (either physical or theoretical) with these designs? Which design do you prefer?

4.35 Consider the detection of two equilikely signals in additive zero-mean colored Gaussian noise with autocorrelation $R_n(t, s)$ where

$$H_0: \quad r(t) = s_0(t) + n(t), \quad 0 \le t \le T$$

$$H_1: \quad r(t) = s_1(t) + n(t), \quad 0 \le t \le T$$

Suppose that the signals are given by

$$s_0(t) = \sum_{n=1}^{N} \sqrt{E_n}\phi_n(t)$$

$$s_1(t) = -s_0(t) = -\sum_{n=1}^{N} \sqrt{E_n}\phi_n(t)$$

where $\phi_n(t)$'s are eigenfunctions with respective eigenvalues λ_n satisfying

$$\int_0^T R_n(t, s)\phi_n(s)ds = \lambda_n\phi_n(t), \quad 0 \le t \le T \ n = 1, 2\cdots, N$$

(a) Show that the optimum decision rule is

$$\sum_{n=1}^{N} r_n\left(\frac{\sqrt{E_n}}{\lambda_n}\right) \underset{m_1}{\overset{m_0}{\gtrless}} 0$$

where $r_n = \int_0^T r(t)\phi_n(t)dt$.

(b) Find an expression for the error probability as a function of $\lambda_1, \lambda_2, \cdots, \lambda_N$ and $E_1, E_2, \cdots, E_N$. Simplify your answer for the special case where $\lambda_n = N_0/2$ for all n and $E = E_1 + E_2 + \cdots + E_N$ denotes the total signal energy.

4.36 A QPSK signal is characterized by

$$s(t) = \sqrt{P}\left[d_I p_I(t)\cos\omega_c t - d_Q p_Q(t)\sin\omega_c t\right]$$

for $0 \le t \le T$ where f_c is the center frequency of the channel and T represents the symbol duration.

(a) Determine the PSD if d_I and d_Q are independent, identically, distributed random variables taking on values $+1$ and -1 with equal probability.

(b) Repeat (a) if

$$p_I(t) = p_Q(t) = \begin{cases} 1, & 0 \le t \le T \\ 0, & \text{elsewhere} \end{cases}$$

(c) Repeat (a) if

$$p_I(t) = p_Q(t) = \begin{cases} 1, & 0 \le t \le T/2 \\ -1, & T/2 \le t \le T \\ 0, & \text{elsewhere} \end{cases}$$

(d) Repeat (a) if

$$p_I(t)\begin{cases} 1, & 0 \le t \le T \\ 0, & \text{elsewhere} \end{cases} \quad p_Q(t) = \begin{cases} 1, & 0 \le t \le T/2 \\ -1, & T/2 \le t \le T \\ 0, & \text{elsewhere} \end{cases}$$

4.37 An unbalanced QPSK (UQPSK) signal is characterized by

$$s(t) = \sqrt{P_I}d_I p_I(t)\cos\omega_c t - \sqrt{P_Q}d_Q p_Q(t)\sin\omega_c t$$

(a) Assume $P_I/P_Q = 4$ and repeat Problem 4.36.

(b) Specify the transmission bandwidth required and discuss your answer from a system engineering point of view if necessary.

4.38 Consider a data sequence which is Manchester coded.

(a) Show that the correlation function as defined by (4.375) is given by

$$r_s(\tau) = \begin{cases} 1 - 3\frac{|\tau|}{T}, & |\tau| \le \frac{T}{2} \\ \frac{|\tau|}{T} - 1, & \frac{T}{2} \le |\tau| \le T \end{cases}$$

(b) By direct substitution of the result of part (a) into (4.374), verify that the conditional error probability for Manchester coding is given by (4.378).

4.39 Consider a data sequence formed by transmitting with equal probability one of two equal energy pulse waveforms, that is, $p_1(t)$ and $p_2(t)$ independently in each symbol interval. It is easily shown that the decision rule [analogous to (4.367)] for this type of sequence is given by

$$\text{choose } p_1(t) \text{ if } Q > 0$$

$$\text{choose } p_2(t) \text{ if } Q < 0$$

where

$$Q \triangleq \int_{(n-1)T+\widehat{\delta}}^{nT+\widehat{\delta}} r(t)\left\{p_1\left[t - (n-1)T - \widehat{\delta}\right] - p_2\left[t - (n-1)T - \widehat{\delta}\right]\right\}dt$$

(a) Show that the conditional error probability for this type of sequence is given by

$$P(E/\lambda) = \frac{1}{8}\mathrm{erfc}\left\{\sqrt{(E/N_0)'}\left[r_{11}(\lambda T) + r_{11}(T - \lambda T) - r_{21}(\lambda T) - r_{12}(T - \lambda T)\right]\right\}$$

$$+ \frac{1}{8}\mathrm{erfc}\left\{\sqrt{(E/N_0)'}\left[r_{11}(\lambda T) + r_{21}(T - \lambda T) - r_{21}(\lambda T) - r_{22}(T - \lambda T)\right]\right\}$$

$$+ \frac{1}{8}\mathrm{erfc}\left\{\sqrt{(E/N_0)'}\left[r_{22}(\lambda T) + r_{22}(T - \lambda T) - r_{12}(\lambda T) - r_{21}(T - \lambda T)\right]\right\}$$

$$+ \frac{1}{8}\mathrm{erfc}\left\{\sqrt{(E/N_0)'}\left[r_{22}(\lambda T) + r_{12}(T - \lambda T) - r_{12}(\lambda T) - r_{11}(T - \lambda T)\right]\right\}$$

for $\lambda > 0$

and

$$P(E/\lambda) = \frac{1}{8}\mathrm{erfc}\left\{\sqrt{(E/N_0)'}\left[r_{11}(|\lambda|T) + r_{11}(T - |\lambda|T) - r_{12}(|\lambda|T) - r_{21}(T - |\lambda|T)\right]\right\}$$

$$+ \frac{1}{8}\mathrm{erfc}\left\{\sqrt{(E/N_0)'}\left[r_{11}(|\lambda|T) + r_{12}(T - |\lambda|T) - r_{12}(|\lambda|T) - r_{22}(T - |\lambda|T)\right]\right\}$$

$$+ \frac{1}{8}\mathrm{erfc}\left\{\sqrt{(E/N_0)'}\left[r_{22}(|\lambda|T) + r_{22}(T - |\lambda|T) - r_{21}(|\lambda|T) - r_{12}(T - |\lambda|T)\right]\right\}$$

$$+ \frac{1}{8}\mathrm{erfc}\left\{\sqrt{(E/N_0)'}\left[r_{22}(|\lambda|T) + r_{21}(T - |\lambda|T) - r_{21}(|\lambda|T) - r_{11}(T - |\lambda|T)\right]\right\}$$

for $\lambda < 0$

where

$$r_{ij}(\tau) \triangleq \frac{\int_0^{T-\tau} p_i(t + \tau)p_j(t)dt}{PT}$$

$$(E/N_0)' \triangleq \frac{(E/N_0)}{2[1 - r_{12}(0)]}$$

and we have made use of the fact that $r_{ij}(\tau) = r_{ji}(-\tau)$ for all $i, j = 1, 2$.

(b) Show that if $p_1(t) = -p_2(t)$, the expression for the conditional error probability reduces to (4.374).

(c) For RZ data, evaluate $r_{ij}(\tau)$ for $i, j = 1, 2$, and, by direct substitution of these results into the expression for $P(E/\lambda)$ given in part (a), verify (4.379).

4.40 For a Miller-coded data format, the appropriate decision rule, in the presence of symbol sync error, for the nth symbol is

$$\widehat{a}_n = \mathrm{sgn}\left[|Q_1| - |Q_2|\right]$$

where

$$Q_1 = \int_{(n-1)T+\widehat{\delta}}^{nT+\widehat{\delta}} r(t)p_1\left[t - (n-1)T - \widehat{\delta}\right]dt$$

$$Q_2 = \int_{(n-1)T+\widehat{\delta}}^{nT+\widehat{\delta}} r(t)p_2\left[t - (n-1)T - \widehat{\delta}\right]dt$$

and

$$p_1(t) = \begin{cases} \sqrt{P} & 0 \le t \le T \\ 0 & \text{elsewhere} \end{cases}$$

$$p_2(t) = \begin{cases} \sqrt{P} & 0 \le t \le T/2 \\ -\sqrt{P} & T/2 \le t \le T \\ 0 & \text{elsewhere} \end{cases}$$

Derive the expression for conditional error probability as given by (4.380).

4.41 Suppose the error probability is a function of random variable x, say $P(E/x)$. Show that the average $P(E) = E\{P(E/x)\}$ can be well approximated by

$$P(E) = P\,(E/\overline{x}) + \frac{\sigma_x^2}{2} P''\,(E/\overline{x})$$

where $\overline{x}$ is the mean of x.

4.42 Some BPSK modulators exhibit both imperfect phase switching and amplitude imbalance simultaneously. In this case, assume that the signal model is given by

$$s(t) = \sqrt{\frac{2E}{T}}[1 + \epsilon d(t)] \sin[\omega_c t + (1-\alpha)\frac{\pi}{2}d(t)]$$

where $d(t) = \pm 1$ and ϵ and α denote amplitude and phase imbalance, respectively.

(a) Assuming an AWGN channel, equally likely signals and ideal coherent demodulation/detection, find the probability of bit error.

(b) Plot (on one graph) three curves showing the loss in effective E/N_0 as a function of α for three values of ϵ, 0.1, 0.2, and 0.3, for α ranging from 0 to 1.

4.43 A QPSK system operates in AWGN with an $E_b/N_0 = 9.4$ dB (resulting in bit error probability of 10^{-5}).

(a) Find the bit error probability if there is a phase mismatch between the transmitted carrier phase and that stored in the receiver of ϕ degrees. Assume the interval $[0, T]$ is known at the receiver.

(b) Assume ϕ is 3°, 6°, and 10° and evaluate the necessary increase in E_b/N_0 such that the bit error probability remains at 10^{-5}.

4.44 (a) Repeat Problem 4.43 if there is a misalignment of the transmit and receive clocks by ϵ sec.

(b) Assume $\epsilon/T = 0.05$ and 0.1 and evaluate the required increase in E_b/N_0 needed to continue to operate at a bit rate of 10^{-5}.

4.45 As an approximation to the integrate-and-dump detector of Problem 4.7, consider replacing it with a lowpass single-pole Butterworth (RC) filter with transfer function

$$H(f) = \frac{1}{1 + j\left(\frac{2\pi f}{2\pi f_{3-\text{dB}}}\right)}$$

where $f_{3-\text{dB}}$ is the 3 − dB cutoff frequency.

(a) If $y(t)$ is the output of this filter, find the signal-to-noise ratio SNR $= [E\{y(T)\}]^2/\sigma_y^2$ of a sample of this output taken at time $t = T$. Assume that either a single positive or negative pulse of amplitude A was transmitted and that the filter capacitor is initially discharged. Also, assume an AWGN channel with single-sided PSD N_0.

(b) Find the value of the time-bandwidth product $f_{3-dB}T$ that maximizes the SNR found in (a).

4.46 A binary communication system has an input modulation characterized by an antipodal pulse stream $m(t) = \sum_{n=-\infty}^{\infty} a_n p(t - nT)$ where the a_n's take on the values $\pm A$ with equal probability and $p(t) = \mathcal{F}^{-1}\{P(f)\}$ is a known waveform, where $\mathcal{F}^{-1}\{.\}$ denotes the inverse Fourier transform. The modulation $m(t)$ is transmitted over an AWGN channel with single-sided PSD

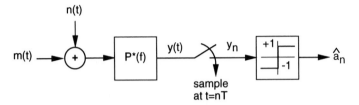

Figure P4.46

N_0 and the data is to be detected by the receiver configuration shown in Fig. P4.46. Suppose that $P(f)$ is an ideal rectangular filter with bandwidth equal to half the data rate, that is,

$$P(f) = \begin{cases} \sqrt{T}, & |f| \leq 1/(2T) \\ 0, & \text{otherwise} \end{cases}$$

(a) What is the average power of $m(t)$?

(b) What are the mean and variance (under each of the two hypotheses) of $y_n = y(nT)$? Do these quantities depend on n, that is, is there any intersymbol interference (ISI)?

(c) What is the average error probability of the receiver?

4.47 Suppose that in Problem 4.46, the transmitted pulse shape corresponds to a cosine roll-off characteristic, that is

$$P(f) = \begin{cases} \sqrt{T}, & 0 \leq |f| \leq \frac{1}{2T}(1-\alpha) \\ \sqrt{T} \cos\left\{\frac{\pi}{4\alpha}[2|f|T - 1 + \alpha]\right\}, & \frac{1}{2T}(1-\alpha) \leq |f| \leq \frac{1}{2T}(1+\alpha) \\ 0, & \text{otherwise} \end{cases}$$

Repeat (a), (b), and (c) of Problem 4.46 for this filter.

4.48 A binary communication system, with signals $s_0(t)$ and $s_1(t)$ with average power of P watts, is observed in AWGN (N_0 watts/Hz single-sided). The data rate is $1/T$ bits/sec and the binary messages are independent and equally probable.

(a) Draw the optimum coherent waveform receiver.

(b) What is the bit error probability?

(c) Assuming an additive interference characterized by $I(t) = \sqrt{2A}\cos(\omega_c t + \phi)$ (where ϕ is a uniformly distributed random variable and A is known), find the optimum receiver for $(k-1)T \leq t \leq kT$ in AWGN.

(d) What is the bit error probability of the receiver used in (a) subjected to AWGN and the interference characterized in (c) if $s_0(t) = -s_1(t)$?

(e) What is the bit error probability of the receiver found in (c) subjected to AWGN and the interference characterized in (c) if $s_0(t) = -s_1(t)$?

4.49 Consider the bandpass filter illustrated in Fig. P4.49(a), which exhibits an approximate symmetry about the center frequency f_c; the approximation improving as the Q of the resonant circuit is increased.

(a) Determine the transfer function $H(f)$.

(b) Determine the symmetric part of $H(f)$ about $f_c = \sqrt{1 - (1/4Q^2)}f_r$.

(c) Determine the antisymmetric part of $H(f)$ about f_c.

(d) Given the I-Q modulation-demodulation system illustrated in Fig. P4.49(b), develop expressions for $I_0(t)$ and $Q_0(t)$ when $\theta = \widehat{\theta}$ and $H(f) = 1$.

(e) Repeat (d) when $H(f) = 1$ but $\theta \neq \widehat{\theta}$. Comment on the system engineering meaning of your result.

(f) Repeat (d) when $\theta = \widehat{\theta}$ and the bandpass characteristic $H(f)$ is symmetrical about f_c (that is, $Q \gg 1$).

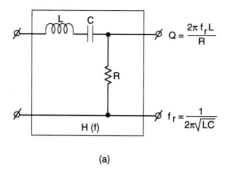

(a)

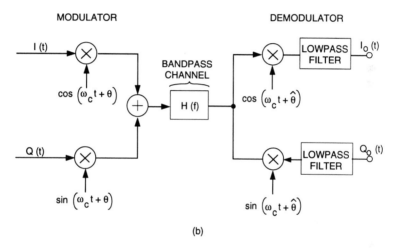

(b)

Figure P4.49

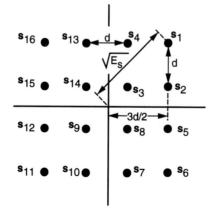

Figure P4.50

4.50 Consider the 16-QAM signal set shown in Fig. P4.50. Assume all signals are equilikely and that the noise is AWGN with PSD $N_0/2$.

(a) Write an I-Q waveform model using rectangular coordinates for square I-Q pulses.

(b) Write an I-Q waveform model using polar coordinates for square I-Q pulses.

(c) Find the probability of error in terms of d and N_0.

(d) Find the probability of error in terms of the "peak energy" E_s and N_0.

REFERENCES

1. Wozencraft, J. M. and I. M. Jacobs, *Principles of Communication Engineering*, New York: John Wiley and Sons, 1965.

2. Papoulis, A., *Signal Analysis*, New York: McGraw-Hill, 1977.

3. Strang, G., *Linear Algebra and Its Applications*, 3rd ed., San Diego: Harcourt Brace Jovanovich, 1988.

4. Bracewell, R., *The Fourier Transform and Its Applications*, 2nd ed., New York: McGraw-Hill, 1978.

5. Gagliardi, R., *Introduction to Communications Engineering*, New York: John Wiley and Sons, 1978.

6. Weber, C. L., *Elements of Detection and Signal Design*, New York: Springer-Verlag, 1987.

7. Lee, P. J., "Computation of the Bit Error Rate of Coherent M-ary PSK with Gray Code Bit Mapping," *IEEE Transactions on Communications*, vol. COM-34, no. 5, May 1986, pp. 488–91.

8. Irshid, M. I. and I. S. Salous, "Bit Error Probability for Coherent M-ary PSK Systems," *IEEE Transactions on Communications*, vol. COM-39, no. 3, March 1991, pp. 349–52.

9. Korn, I and L. M. Hii, "Relation Between Bit and Symbol Error Probabilities for DPSK and FSK with Differential Phase detection," submitted to *IEEE Transactions on Communications*.

10. Reed, I. S. and R. A. Scholtz, "N-Orthogonal Phase-Modulated Codes," *IEEE Transactions on Information Theory*, vol. IT-12, no. 3, July 1966, pp. 388–95.

11. MacWilliams, F. J. and N. A. Sloane, *The Theory of Error-Correcting Codes*, New York: North-Holland, 1977.

12. Lindsey, W. C. and M. K. Simon, *Telecommunication Systems Engineering*, Englewood Cliffs, N.J.: Prentice-Hall, 1973. Reprinted by Dover Press, New York, NY, 1991.

13. Van Trees, H. L., *Detection, Estimation and Modulation Theory—Part I*, New York: John Wiley and Sons, 1968.

14. Papoulis, A. , *Probability, Random Variables and Stochastic Processes*, New York: McGraw-Hill, 1965.

15. Cooper, G. R. and C. D. McGillen, *Methods of Signals and System Analysis*, New York: Holt, Rinehart and Winston, 1967.

16. Helstrom, C. W., *Statistical Theory of Signal Detection*, 2nd ed., Oxford: Pergamon Press, 1968.

17. Wainstein, L. A. and V. D. Zubavov, *Extraction of Signals from Noise*, translated from the Russian by R. A. Silverman, Englewood Cliffs, N.J.: Prentice-Hall, 1962.

18. Brown, J. L., "Detection of a Known Signal in Colored Gaussian Noise—A New Coordinate Approach," *IEEE Transactions on Information Theory*, vol. IT-33, no. 2, March 1987, pp. 296–98.

19. Nguyen, T., "The Impact of NRZ Data Asymmetry on the Performance of a Space Telemetry System," *IEEE Transactions on Electromagnetic Compatibility*, vol. 33, no. 4, November 1991, pp. 343–50.

20. Simon, M. K., K. Tu and B. H. Batson, "Effects of Data Asymmetry on Shuttle Ku-Band Communications Link Performance," *IEEE Transactions on Communications* (Special Issue on Space Shuttle Communications and Tracking), vol. COM-26, no. 11, November 1978, pp. 1639–51.

Chapter 5

Noncoherent Communication
with Waveforms

So far, we have concentrated on communication scenarios where a message is transmitted using a distinct signal that is precisely known to the receiver, except for the corrupting noise. In Chapter 4, it was shown that coherent demodulation involved mixing the received signal with inphase (I) and quadrature (Q) sinusoidal references, whose phases were exactly equal to the incoming received phase. It was shown that lack of knowledge of the incoming phase can result in cross-talk between the I and Q arms, the amount of which depends on the value of the carrier phase error ϕ, that is, the phase error between the locally generated and the incoming phase.

In many applications, it is not possible to perform coherent detection due to the inability of the receiver to produce meaningful carrier phase estimates, due, for example, to significant oscillator phase noise instability. In some scenarios, coherent systems are undesirable due to the additional complexity imposed by the phase estimation circuitry. In this chapter, we investigate noncoherent waveform communications in AWGN channels. First, we derive the structure and analyze the performance of optimum and suboptimum receivers in random phase channels. Both orthogonal and nonorthogonal signals are considered and the effects of imperfect time and/or frequency synchronization are assessed. We later extend the concept to both random amplitude and phase channels in which the amplitude of the incoming signal is also random due to fading in the link. Communications in both Rayleigh and Rician channels are examined and their respective performances evaluated.

In noncoherent M-ary communications, the received signal can be expressed by

$$
\begin{aligned}
m_0 \to H_0: \quad & r(t) = s_0(t, \mathbf{\Psi}) + n(t), \quad 0 \leq t \leq T \\
m_1 \to H_1: \quad & r(t) = s_1(t, \mathbf{\Psi}) + n(t), \quad 0 \leq t \leq T \\
& \quad \vdots \\
m_{M-1} \to H_{M-1}: \quad & r(t) = s_{M-1}(t, \mathbf{\Psi}) + n(t), \quad 0 \leq t \leq T
\end{aligned}
\tag{5.1}
$$

where the vector $\mathbf{\Psi}$ denotes the vector of unknown parameters [1] that are not or cannot be estimated in the receiver. As before, the noise is white Gaussian with one-sided PSD N_0 Watts/Hz and T denotes the duration of the transmitted message. We assume for now that perfect symbol synchronization has been achieved. Later, we will consider and assess the error probability in the presence of symbol sync error. Note that due to the randomness of the vector $\mathbf{\Psi}$, each message is transmitted using a sample waveform of a random process rather than a deterministic and known signal. As far as the receiver is concerned, it does not distinguish between an unknown random phase introduced in the channel or a random phase inserted at the transmitter. A message m_i can be conveyed using a set or a family of known waveforms, rather than a fixed and unique waveform. Note that for a fixed vector $\mathbf{\Psi}$, the received waveforms are known precisely to the receiver. Hypothesis testing as described by (5.1) is typically referred to as "nonsimple" detection as opposed to the "simple" detection in coherent communications.

5.1 NONCOHERENT RECEIVERS IN RANDOM PHASE CHANNELS

In random phase channels, the phase θ of the received signal is not known, nor estimated by the receiver. All the other parameters such as amplitude and frequency are assumed known and hence, the vector $\mathbf{\Psi}$ of unknown parameters consists solely of the scalar θ, that is, $\mathbf{\Psi} = \theta$. The received signal, assuming m_i is transmitted, is thus given by[1]

$$r(t) = \sqrt{2}a_i(t) \cos(\omega_c t + \phi_i(t) + \theta) + n(t), \quad 0 \le t \le T \tag{5.2}$$

where ω_c is the carrier frequency, $a_i(t)$ the amplitude information, $\phi_i(t)$ the phase information[2], θ the unknown phase, and $n(t)$ the bandpass Gaussian noise process given by (4B.33) as

$$n(t) = \sqrt{2}\{n_c(t) \cos \omega_c t - n_s(t) \sin \omega_c t\} \tag{5.3}$$

The baseband noise components $n_c(t)$ and $n_s(t)$ are independent Gaussian processes with PSDs given by

$$S_{n_c}(f) = S_{n_s}(f) = \begin{cases} \frac{N_0}{2} & |f| \le B \\ 0 & \text{otherwise} \end{cases} \tag{5.4}$$

Now according to the bandpass sampling theorem [2], $n(t)$ is completely characterized by samples of $n_c(t)$ and $n_s(t)$ taken every $1/2B$ sec, that is

$$n(t) = \sqrt{2} \sum_{k=-\infty}^{\infty} \text{sinc}(2Bt - k)\{n_c(t_k) \cos \omega_c(t - t_k) - n_s(t_k) \sin \omega_c(t - t_k)\} \tag{5.5}$$

Moreover, we note that

$$\int_{-\infty}^{\infty} n^2(t)dt = \frac{1}{2B} \sum_{k=-\infty}^{\infty} \left\{ n_c^2(t_k) + n_s^2(t_k) \right\} \tag{5.6}$$

represents the energy in the sample function $n(t)$. The joint pdf of the vector $\mathbf{n} = (n_c(t_1),$ $n_c(t_2), \ldots, n_c(T), n_s(t_1), n_s(t_2), \ldots, n_s(T))^T$ created from sampling a T-second segment of $n(t)$ is given by

$$f_{n_c(t_1),\cdots,n_s(T)}(\mathbf{n}) = C \exp\left\{ -\frac{\sum_{k=1}^{2BT} (n_c^2(t_k) + n_s^2(t_k))}{2N_0 B} \right\} \tag{5.7}$$

---

1. As in Chapter 4, we denote the received random process by $r(t)$ and a sample function of that random process by $\rho(t)$. Similarly, the received random vector is denoted by $\mathbf{r}$ and a sample vector of that random vector by $\boldsymbol{\rho}$.

2. In Chapter 5, $\phi_i(t)$ denotes the phase information when message m_i is transmitted, but it also denotes the ith basis function of the K-L expansion (consistent with the notation used in Chapter 4). The reader is therefore alerted to the dual use of $\phi_i(t)$ and special care is taken to redefine $\phi_i(t)$ whenever it is used.

since the $2 \times 2BT$ samples are uncorrelated, Gaussian each with variance $N_0 B$. The constant C is the normalization factor such that the multidimentional integral of the density is unity. Using (5.6) to convert from sums of samples to a time integral, we have

$$f_{n_c(t_1),\ldots,n_s(T)}(\mathbf{n}) \simeq C \exp\left\{-\frac{1}{N_0}\int_0^T n^2(t)dt\right\} \tag{5.8}$$

which is identical to (4.7). Assuming that message m_i is transmitted and conditioning on the random phase θ, then

$$f_{\mathbf{r}}(\boldsymbol{\rho}/m_i, \theta) = f_{\mathbf{n}}(\boldsymbol{\rho} - \mathbf{s_i}/m_i, \theta) = f_{\mathbf{n}}(\boldsymbol{\rho} - \mathbf{s_i})$$

$$= C \exp\left\{-\frac{\int_0^T \left(\rho(t) - \sqrt{2}a_i(t)\cos(\omega_c t + \phi_i(t) + \theta)\right)^2 dt}{N_0}\right\} \tag{5.9}$$

and

$$f_{\mathbf{r}}(\boldsymbol{\rho}/m_i) = \int_{-\pi}^{\pi} f_{\mathbf{r}}(\boldsymbol{\rho}/m_i, \theta) f_{\theta/m_i}(\theta/m_i)d\theta \tag{5.10}$$

Assuming that θ is random and independent of which message is transmitted, $f_{\theta/m_i}(\theta/m_i) = f_\theta(\theta)$ and (5.10) simplifies to

$$f_{\mathbf{r}}(\boldsymbol{\rho}/m_i) = \int_{-\pi}^{\pi} f_{\mathbf{r}}(\boldsymbol{\rho}/m_i, \theta) f_\theta(\theta)d\theta \tag{5.11}$$

Since the receiver does not attempt to estimate the incoming phase θ, it can be assumed random with a uniform pdf, that is

$$f_\theta(\theta) = \begin{cases} \frac{1}{2\pi} & -\pi \le \theta \le \pi \\ 0 & \text{otherwise} \end{cases} \tag{5.12}$$

Expanding (5.9), $f_{\mathbf{r}}(\boldsymbol{\rho}/m_i, \theta)$ can be expressed as

$$f_{\mathbf{r}}(\boldsymbol{\rho}/m_i, \theta) = C' \exp\left\{\frac{2}{N_0}(z_{ci}\cos\theta - z_{si}\sin\theta)\right\}\exp\left(-\frac{E_i}{N_0}\right) \tag{5.13}$$

where $E_i = \int_0^T a_i^2(t)dt$ is the energy of $a_i(t)$ and the I-Q projections generated by means of a local oscillator are

$$z_{ci} = \int_0^T \rho(t)a_i(t)\sqrt{2}\cos(\omega_c t + \phi_i(t))dt \tag{5.14a}$$

$$z_{si} = \int_0^T \rho(t)a_i(t)\sqrt{2}\sin(\omega_c t + \phi_i(t))dt \tag{5.14b}$$

As a special case when the transmitted information is contained in the amplitude only, then $r(t) = \sqrt{2}s_i(t)\cos(\omega_c t + \theta) + n(t)$, that is, $\phi_i(t) = 0 \; \forall i$ and $s_i(t) = a_i(t)$ and (5.14) can be expressed as the I-Q projections

$$z_{ci} = \int_0^T \rho_c(t)s_i(t)dt \tag{5.15a}$$

$$z_{si} = \int_0^T \rho_s(t)s_i(t)dt \tag{5.15b}$$

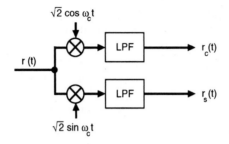

Figure 5.1 Demodulation with random phase

where

$$\rho_c(t) \triangleq \rho(t)\sqrt{2}\cos\omega_c t \tag{5.16a}$$

$$\rho_s(t) \triangleq \rho(t)\sqrt{2}\sin\omega_c t \tag{5.16b}$$

are the projections [inphase and quadrature (I and Q)] of the received waveform $\rho(t)$ onto the orthonormal basis $\{\sqrt{2}\cos\omega_c t, \sqrt{2}\sin\omega_c t\}$, as depicted in Fig. 5.1.

Transforming z_{ci} and z_{si} of (5.14) into polar coordinates using the envelope ξ_i and the phase ϕ_i as

$$z_{ci} = \xi_i \cos\phi_i \tag{5.17a}$$

$$z_{si} = \xi_i \sin\phi_i \tag{5.17b}$$

then

$$z_{ci}\cos\theta - z_{si}\sin\theta = (\xi_i \cos\phi_i)\cos\theta - (\xi_i \sin\phi_i)\sin\theta$$
$$= \xi_i \cos(\theta + \phi_i) \tag{5.18}$$

and (5.13) becomes

$$f_{\mathbf{r}/\theta}(\boldsymbol{\rho}/m_i, \theta) \sim \exp\left\{\frac{2\xi_i}{N_0}\cos(\theta + \phi_i)\right\}\exp\left(-\frac{E_i}{N_0}\right) \tag{5.19}$$

Averaging over the density of θ, we have from (5.11) and (5.13)

$$f_{\mathbf{r}}(\boldsymbol{\rho}/m_i) \sim \frac{1}{2\pi}\int_{-\pi}^{\pi}\exp\left\{\frac{2\xi_i}{N_0}\cos(\theta + \phi_i)\right\}d\theta \exp\left(-\frac{E_i}{N_0}\right) \tag{5.20}$$

Such an integral has been encountered in several engineering applications and is referred to as the "zero-order modified Bessel function of the first kind" [3, 4]. Specifically, it is defined through

$$I_0(x) \triangleq \frac{1}{2\pi}\int_{-\pi}^{\pi}e^{x\cos\beta}d\beta \tag{5.21}$$

and is plotted in Fig. 5.2 as a function of x. For positive x, it is a monotonically increasing function and because of the periodicity of the cosine, an arbitrary phase ϕ can be added to its argument without altering the value of $I_0(x)$, that is

$$I_0(x) = \frac{1}{2\pi}\int_{-\pi}^{\pi}e^{x\cos(\beta+\phi)}d\beta \tag{5.22}$$

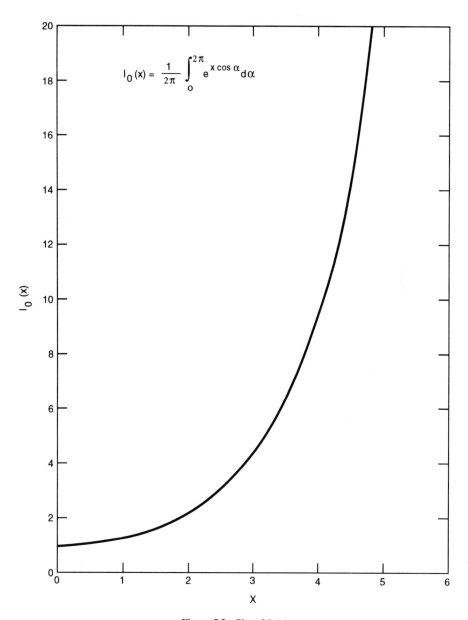

$$I_0(x) = \frac{1}{2\pi} \int_0^{2\pi} e^{x \cos \alpha} d\alpha$$

Figure 5.2   Plot of $I_0(x)$

Hence, using (5.22) in (5.20), the optimum decision rule sets $\widehat{m}(\rho(t)) = m_k$ when

$$I_0(\frac{2\xi_i}{N_0}) \exp\left(-\frac{E_i}{N_0}\right) \tag{5.23}$$

is maximum for $k = i$. Recall from (5.17) that the envelope ξ_i is given by

$$\xi_i = \sqrt{z_{ci}^2 + z_{si}^2} \tag{5.24}$$

where z_{ci} and z_{si} are the projections of $\rho_c(t)$ and $\rho_s(t)$ onto the ith signal as given by (5.15a) and (5.15b), respectively. For equal energy signals, the factor $\exp(-E_i/N_0)$ can be ignored and since $I_0(2\xi_i/N_0)$ is a monotone increasing function, we need only to maximize ξ_i or ξ_i^2 where

$$\xi_i^2 = z_{ci}^2 + z_{si}^2 \tag{5.25}$$

The optimum receiver for equal energy signals is depicted in Fig. 5.3 where the received signal is first mixed with sine and cosine waveforms at the incoming frequency and the resulting outputs are correlated with each of the M possible transmitted signals.[3] The receiver then forms the sum of the squares of the sine and cosine correlations and decides on the message with maximum resulting envelope. It is clear that the decision process involves nonlinear operations unlike the coherent receiver discussed in Chapter 4. This is the penalty paid for not knowing or estimating the incoming carrier phase θ. An alternative implementation can be obtained with matched filters with impulse responses

$$h_i(t) = s_i(T - t), \quad i = 0, 1, \cdots, M - 1 \tag{5.26}$$

followed by a sampler at multiples of T seconds as shown in Fig. 5.4. When the signals can be expressed as

$$s_i(t) = \sum_{j=1}^{N} s_{ij}\phi_j(t), \quad i = 0, 1, \cdots, M - 1 \tag{5.27}$$

as given by (4.40), then it is possible to correlate $r_c(t)$ and $r_s(t)$ with the basis functions $\phi_i(t), i = 1, 2, \cdots, N$ and apply the appropriate signal weights to form the various ξ_i^2, $i = 0, 1, \cdots, M - 1$, as shown in Fig. 5.5.

A significantly different approach for optimum detection of equiprobable equal energy signals is possible with envelope detectors. Consider Fig. 5.6, which depicts a filter $h_{bp,k}(t)$ followed by an envelope detector. Since the input to the filter $r(t)$ is centered at f_c, let the filter $h_{bp,k}(t)$ be matched to the signal $s_k(t)$, translated to frequency f_c ("bp" stands for bandpass), that is

$$h_{bp,k}(t) \triangleq \sqrt{2}s_k(T - t) \cos \omega_c t, \quad k = 0, 1, \cdots, M - 1 \tag{5.28}$$

The output of the filter becomes

$$u_k(t) = \int_{-\infty}^{\infty} r(\tau)h_{bp,k}(t - \tau)d\tau$$

$$= \sqrt{2}\int_{-\infty}^{\infty} r(\tau)s_k(T - t + \tau) \cos \omega_c(t - \tau)d\tau \tag{5.29}$$

or

$$u_k(t) = u_{ck}(t) \cos \omega_c t + u_{sk}(t) \sin \omega_c t \tag{5.30}$$

---

3. Note that in Fig. 5.3, lowpass filters have been inserted at the output of the mixers. As long as the bandwidth of these filters is much larger than the data rate, they do not impact the performance of the optimum receiver. It is however good practice to include them in an analog implementation as the mixers are highly nonlinear devices that generate more than the desired product at their output.

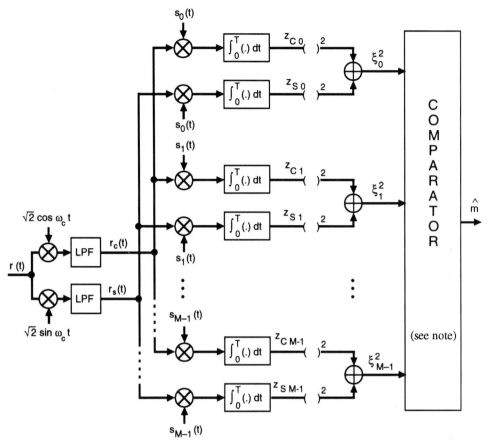

Note: From here on, a "**comparator**" selects the message corresponding to maximum input value or voltage.

Figure 5.3 M-ary noncoherent receiver for equal energy signals

where we define

$$u_{ck}(t) \triangleq \int_{-\infty}^{\infty} r(\tau)s_k(T - t + \tau)\sqrt{2}\cos\omega_c\tau\, d\tau = \int_{-\infty}^{\infty} r_c(\tau)s_k(T - t + \tau)d\tau \quad (5.31a)$$

$$u_{sk}(t) \triangleq \int_{-\infty}^{\infty} r(\tau)s_k(T - t + \tau)\sqrt{2}\sin\omega_c\tau\, d\tau = \int_{-\infty}^{\infty} r_s(\tau)s_k(T - t + \tau)d\tau \quad (5.31b)$$

Since $r_c(\tau)$ and $r_s(\tau)$ are low-frequency signals and $s_k(\tau)$ is also a slowly varying signal, both $u_{ck}(t)$ and $u_{sk}(t)$ remain relatively constant over several cycles of $\cos\omega_c t$. As a result, $u_k(t)$ is a bandpass signal and can thus be expressed as

$$u_k(t) = \text{Re}\left\{a_k(t)e^{-j\omega_c t}\right\} \quad (5.32)$$

where the complex baseband waveform $a_k(t)$ is given by

$$a_k(t) = u_{ck}(t) + ju_{sk}(t) \quad (5.33)$$

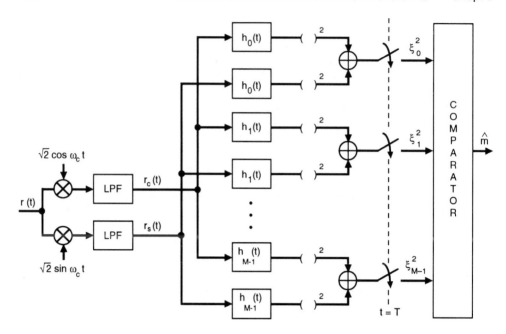

Figure 5.4 Alternate implementation of M-ary noncoherent receivers for equal energy signals with matched filters

The output of the envelope detector becomes

$$e_k(t) = |a_k(t)| \tag{5.34a}$$

or

$$e_k(t) = \sqrt{u_{ck}^2(t) + u_{sk}^2(t)} \tag{5.34b}$$

If we sample the output at $t = T$, we have from (5.34) and (5.31)

$$e_k(T) = \sqrt{u_{ck}^2(T) + u_{sk}^2(T)}$$

$$= \sqrt{\left(\int_0^T r_c(\tau) s_k(\tau) d\tau \right)^2 + \left(\int_0^T r_s(\tau) s_k(\tau) d\tau \right)^2} \tag{5.35}$$

which is identical to ξ_k as given by (5.24) since $u_{ck}(T) = z_{ck}$ and $u_{sk}(T) = z_{sk}$ as given by (5.15a) and (5.15b), respectively. Hence, the optimum M-ary noncoherent detector for equiprobable, equal energy signals can be implemented using matched filters (to the band-pass rather than the baseband signals) and envelope detectors as shown in Fig. 5.7. Receivers that examine only the envelope of the matched filter output are termed *noncoherent* or *incoherent receivers* as they do not exploit the knowledge of the carrier phase; receivers that do exploit knowledge of the carrier phase are called *coherent*.

5.1.1 Optimum *M*-FSK Receivers

A signal set of great interest in noncoherent communications consists of M sinusoids at distinct frequencies and the set is referred to as *Multiple Frequency-Shift-Keying (M-FSK)* signals. In general, the frequency separation can be nonuniform, but equidistant signals are

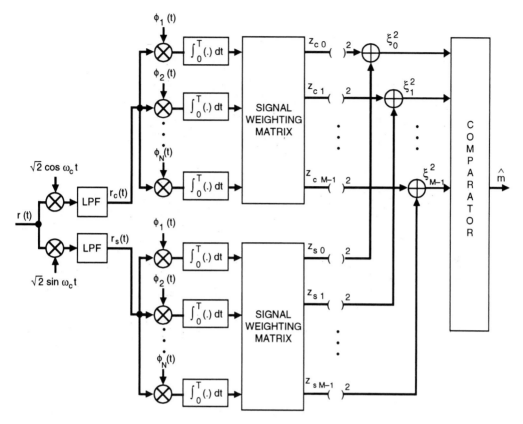

Figure 5.5   *M*-ary noncoherent receiver implementation with the basis functions

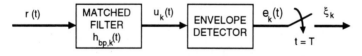

Figure 5.6   Envelope detection

of great practical interest as they occupy the minimum required bandwidth for orthogonal noncoherent communications. The transmitted signals are characterized by

$$s_k(t) = \sqrt{2P}\cos(\omega_c t + \omega_k t + \theta), \quad 0 \le t \le T \tag{5.36}$$

where ω_k denotes the frequency of the kth tone and θ the unknown random phase. The latter is independent from message to message because each signal is typically generated by a unique oscillator whose initial phase is random when the signal is selected. The pairwise signal correlation coefficient, as defined by (4.45), reduces to

$$\gamma_{ij} = \frac{1}{E}\int_0^T s_i(t)s_j(t)dt = \frac{1}{T}\int_0^T \cos(2\pi f_{i,j}t + \theta - \theta')dt \tag{5.37}$$

where $E = PT$, $f_{i,j} = f_i - f_j$ is the frequency separation between the tones, and θ, θ'

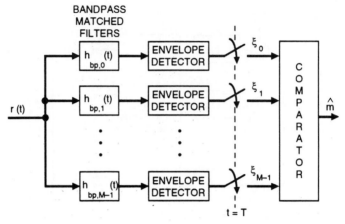

Figure 5.7   Envelope detector noncoherent receiver for equal energy signals

are the random carrier phases when signals $s_i(t)$ and $s_j(t)$ are transmitted. Because of the random phase $\theta - \theta'$, the only way to guarantee zero correlation is for the integral to be carried out over at least a full period of the difference sinusoid. Hence, the minimum frequency separation to guarantee orthogonality is

$$f_{i,j} = \frac{1}{T} \ \text{Hz} \tag{5.38}$$

This separation is equal to twice the minimum separation required for coherent M-FSK transmission, $1/(2T)$, derived earlier in Chapter 4. Given M tones, the minimum bandwidth required for orthogonal noncoherent transmission is M/T Hz achieved by uniformly spacing the tones with $1/T$ Hz separation. The received signal, assuming that message m_k is transmitted, is given by

$$r(t) = \sqrt{2P} \cos(2\pi f_c + 2\pi \left(\frac{k}{T}\right) t + \theta) + n(t), \quad 0 \le t \le T \tag{5.39}$$

Typically, the tones are symmetrically spaced around the carrier f_c, which can easily be modeled by (5.39) by considering negative values of k. The optimum receiver is still as shown in Fig. 5.3 where the various signals are the corresponding sinusoids. The receiver still computes ξ_i^2 for each signal using $\xi_i^2 = z_{ci}^2 + z_{si}^2$ where z_{ci} and z_{si} are given by (5.14) with $a_i(t) = \sqrt{P} \ \forall i$ and $\phi_i(t) = \omega_i t \ \forall i$, that is

$$z_{ci} = \sqrt{2P} \int_0^T r(t) \cos(\omega_c + \omega_i)t\, dt \tag{5.40}$$

and

$$z_{si} = \sqrt{2P} \int_0^T r(t) \sin(\omega_c + \omega_i)t\, dt \tag{5.41}$$

Hence

$$\xi_i^2 = 2P \left\{ \left(\int_0^T r(t) \cos(\omega_c + \omega_i)t\, dt \right)^2 + \left(\int_0^T r(t) \sin(\omega_c + \omega_i)t\, dt \right)^2 \right\} \tag{5.42}$$

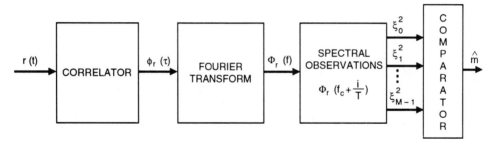

Figure 5.8   Highpass spectrum analyzer receiver

Using the procedure outlined in Appendix 5B, (5.42) is identically equivalent to

$$\xi_i^2 = 4P \int_0^T \int_0^{T-\tau} r(t)r(t+\tau)\cos(\omega_c+\omega_i)\tau d\tau dt \tag{5.43}$$

Defining the finite-time autocorrelation function of $r(t)$ as

$$\phi_r(\tau) \triangleq \int_0^{T-\tau} r(t)r(t+\tau)dt, \quad 0 \le \tau \le T \tag{5.44}$$

then, (5.43) can be expressed as

$$\xi_i^2 = 4P \int_0^T \phi_r(\tau)\cos(\omega_c+\omega_i)\tau d\tau \tag{5.45}$$

Following the argument in Appendix 5B, the quantity ξ_i^2 can be viewed as the Fourier transform of the even function $\phi_{e,r}(\tau)$ [constructed from $\phi_r(\tau)$ as in (5B.10)] evaluated at the discrete frequency $f_c + f_i$; thus

$$\xi_i^2 = 4P\Phi_{e,r}(f_c + f_i) \tag{5.46}$$

where $\Phi_{e,r}(f)$ denotes the Fourier transform [5] of $\phi_{e,r}(\tau)$. This interpretation of the decision process leads to the the highpass spectrum analyzer receiver in Fig. 5.8.[4] In practice, the received signal is first downconverted to an equivalent lowpass waveform $v(t)$ using a reference at the carrier frequency. Thus, any practical implementation must be provided with an estimate of f_c. This heterodyning operation reduces the highpass spectrum analyzer receiver of Fig. 5.8 to its lowpass equivalent illustrated in Fig. 5.9, with the corresponding observations $\{\Phi_v(i/T), \ i=0,1,\cdots,M-1\}$.

5.1.2 Suboptimum M-FSK Receivers

Exact evaluation of the spectral observations is difficult when one elects to implement the spectrum analyzer receivers of Figures 5.8 and 5.9. However, it is possible to approximate the spectral estimate $\Phi_v(f)$ which then results in a receiver performance that is suboptimum. There are various approaches to approximating the spectral observations $\{\Phi_v(k/T), \ k=0,1,\cdots,M-1\}$.

First, one can evaluate the autocorrelation function $\phi_v(\tau)$ directly and then evaluate the Fourier transform to obtain $\Phi_v(f)$. This is referred to as the *autocorrelation function*

---

4. In Figures 5.8 and 5.9, the Fourier transform is operating on the resulting even function, as discussed in Appendix 5B.

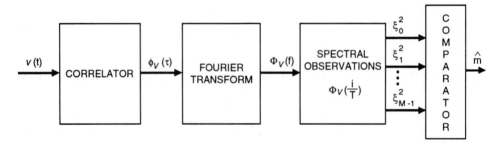

Figure 5.9   Lowpass spectrum analyzer receiver

method. A practical realization of this method that is convenient for digital processing is discussed in [6, 7] and summarized here. Since the Fourier transform of the random process $v(t)$ is essentially zero outside the frequency interval $|f| \le M/T$, then in accordance with the sampling theorem we have, to a good approximation

$$v(t) \simeq v_a(t) \triangleq \sum_{n=1}^{2BT} v\left(\frac{n}{2B}\right) \text{sinc}\left[2B\left(t - \frac{n}{2B}\right)\right] \tag{5.47}$$

where $B \triangleq M/T$ and $v_n = v(n/2B)$ is a sample of $v(t)$ at $t = n/2B$. In terms of the above, the autocorrelation function of $v_a(t)$ is given by

$$\phi_{v_a}(\tau) = \int_0^{T-\tau} v_a(t)v_a(t+\tau)dt = \sum_{n=1}^{2BT}\sum_{m=1}^{2BT} v_n v_m \text{sinc}\left[2B\left(\tau + \frac{n-m}{2B}\right)\right] \tag{5.48}$$

which approximates that of $\phi_v(\tau)$. We observe that for $\tau = k/2B$, k an integer, the above expression simplifies to

$$\phi_{v_a}\left(\frac{k}{2B}\right) = \begin{cases} \frac{1}{2B}\sum_{n=1}^{2BT-k} v_n v_{n+k}, & k = 0, \pm 1, \pm 2, \cdots, \pm(2BT-1) \\ 0, & \text{otherwise} \end{cases} \tag{5.49}$$

Since the spectrum of $\phi_{v_a}(\tau)$ is limited to the frequency range $|f| \le B$, one obtains by means of the sampling theorem

$$\phi_{v_a}(\tau) = \sum_{k=-(2BT-1)}^{2BT-1} \phi_{v_a}\left(\frac{k}{2B}\right) \text{sinc}\left[2B\left(\tau - \frac{k}{2B}\right)\right] \tag{5.50}$$

Taking Fourier transforms, the spectrum of $v(t)$ is approximated by

$$\Phi_v(f) \simeq \Phi_{v_a}(f) = \frac{1}{2B}\phi_{v_a}(0) + \frac{1}{B}\sum_{k=1}^{2BT-1} \phi_{v_a}\left(\frac{k}{2B}\right)\cos\left(\frac{\pi k f}{B}\right), \quad -B \le f \le B \tag{5.51}$$

These expressions for the correlation coefficients and for the approximate spectral observations $\Phi_{v_a}(k/T)$ form the basis for the computational procedure.

Second, one can calculate the spectrum by the discrete Fourier transform approach. The latter takes samples $v(k/2B)$, $k = \{1, 2, \cdots, 2BT\}$ and evaluates

$$X(f) = \sum_{k=1}^{2BT} v\left(\frac{k}{2B}\right)\cos\left(\frac{\pi k f}{B}\right) \quad \text{and} \quad Y(f) = \sum_{k=1}^{2BT} v\left(\frac{k}{2B}\right)\sin\left(\frac{\pi k f}{B}\right) \tag{5.52}$$

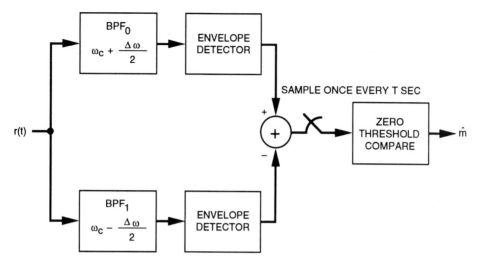

Figure 5.10   Suboptimum binary FSK receiver

and then computes the spectrum

$$\Phi_v(f) \simeq X^2(f) + Y^2(f) \qquad (5.53)$$

which yields the spectral observations $\Phi_v(k/T)$, $k = \{1, 2, \cdots, 2BT\}$. The discrete Fourier transform method permits greatly simplified analysis [8].

Third, the receiver can be implemented with a small computer programmed with the *Cooley-Tukey* Fast Fourier Transform algorithm [9] and processing the output of an analog-to-digital converter. The Fast Fourier Transform (FFT) approach produces results at frequencies that are multiples of $1/T$ provided that the number of samples is a power of two. Although the FFT method does not give better results than the discrete Fourier transform or the autocorrelation function method, it has the advantage that considerably less time is required to perform numerical operations.

Suboptimum detection also occurs when there is a bandpass limiter prior to the detector. In many applications, bandpass limiters are used to provide some form of automatic gain control (AGC) in the receiver and because of their inherent nonlinearity, they degrade the performance of the detector. Suboptimum detection of binary FSK signals in the presence of a bandpass limiter is discussed in [10], where it is shown that a 0.5 dB degradation occurs for $P_s(E) \le 10^{-2}$. Another form of suboptimum detection occurs when the integrate-and-dump filters are replaced by bandpass filters as shown in Fig. 5.10 for binary FSK signals, symmetrically located around the carrier frequency. In this case, the receiver performance is degraded and is a function of the bandpass filter bandwidth, as discussed later.

5.2 PERFORMANCE OF NONCOHERENT RECEIVERS IN RANDOM PHASE CHANNELS

It is of great interest to assess the performance of noncoherent systems and compare it to their coherent counterpart. First, we will derive the performance of orthogonal signals and determine the resulting penalty paid in not tracking the carrier phase at the receiver. We will then generalize the results to any arbitrary set of signals and in particular, consider the effect of frequency and timing synchronization errors. A bound on the performance of orthogonal M-FSK signals is then presented in the presence of frequency error. Finally, the performances of two suboptimum M-FSK receivers are evaluated and the effect of time domain truncation in evaluating the spectral observations is assessed.

5.2.1 Performance of *M*-ary Orthogonal Signals

The M-ary receiver for equal energy signals is depicted in Fig. 5.3 and the receiver picks the message m_k corresponding to maximum ξ_i^2 for $i = 0, 1, \cdots, M - 1$. In order to compute the probability of symbol error, we need to characterize the joint pdf $f_{\xi_0^2, \xi_1^2, \cdots, \xi_{M-1}^2}(\xi_0^2, \xi_1^2, \cdots, \xi_{M-1}^2)$. Assuming that message m_k is transmitted, then $r(t) = \sqrt{2} s_k(t) \cos(\omega_c t + \theta) + n(t)$ with

$$z_{ci} = \int_0^T r_c(t) s_i(t) dt = \int_0^T (s_k(t) \cos \theta + n_c(t)) s_i(t) dt \tag{5.54}$$

and

$$z_{si} = \int_0^T r_s(t) s_i(t) dt = \int_0^T (s_k(t) \sin \theta + n_s(t)) s_i(t) dt \tag{5.55}$$

For a fixed θ, the conditional means become

$$E\{z_{ci}/\theta\} = \cos \theta \int_0^T s_k(t) s_i(t) dt = \begin{cases} E_s \cos \theta, & i = k \\ 0, & i \neq k \end{cases} \tag{5.56}$$

Similarly

$$E\{z_{si}/\theta\} = \sin \theta \int_0^T s_k(t) s_i(t) dt = \begin{cases} E_s \sin \theta, & i = k \\ 0, & i \neq k \end{cases} \tag{5.57}$$

The conditional variances are computed using

$$\sigma_{z_{ci}/\theta}^2 = E\left\{ \left(\int_0^T n_c(t) s_i(t) dt \right)^2 \right\} = \frac{N_0}{2} E_s \tag{5.58a}$$

and

$$\sigma_{z_{si}/\theta}^2 = E\left\{ \left(\int_0^T n_s(t) s_i(t) dt \right)^2 \right\} = \frac{N_0}{2} E_s \tag{5.58b}$$

which are also the unconditional variances. Recall that $n_c(t)$ and $n_s(t)$ are bandlimited processes; however, assuming that $BT \gg 1$, they can be viewed as nonbandlimited using the matched filter argument previously outlined in Chapter 4. Since $n_c(t)$ and $n_s(t)$ are independent, z_{ci} and z_{si} are conditionally uncorrelated and hence independent since they are conditionally Gaussian, that is

$$f_{z_{ci}, z_{si}}(z, y/\theta) = \frac{1}{\pi E_s N_0} \exp \left\{ -\frac{(z - E\{z_{ci}/\theta\})^2 + (y - E\{z_{si}/\theta\})^2}{E_s N_0} \right\}$$

$$= \begin{cases} \frac{1}{\pi E_s N_0} \exp \left\{ -\frac{z^2 + y^2}{E_s N_0} \right\}, & i \neq k \\ \frac{1}{\pi E_s N_0} \exp \left\{ -\frac{z^2 + y^2 + E_s^2 - 2 E_s z \cos \theta - 2 y E_s \sin \theta}{E_s N_0} \right\}, & i = k \end{cases} \tag{5.59}$$

Consider now the pairwise correlation among the various z_{ci} and z_{si}, $i = 0, 1, \cdots$, $M - 1$. Assuming that m_k is transmitted, then

$$E\{z_{ci}z_{cj}\} = E\left\{\left[\int_0^T (s_k(t)\cos\theta + n_c(t))s_i(t)dt\right]\left[\int_0^T (s_k(\tau)\sin\theta + n_c(\tau))s_j(\tau)d\tau\right]\right\}$$

$$= E\left\{\left(\int_0^T n_c(t)s_i(t)dt\right)\left(\int_0^T n_c(\tau)s_j(\tau)d\tau\right)\right\} \tag{5.60}$$

$$= 0, \quad i \neq j, i \neq k, j \neq k$$

Moreover

$$E\{z_{ck}z_{cj}\} = E\left\{\left(E_s\cos\theta + \int_0^T n_c(t)s_k(t)dt\right)\left(\int_0^T n_c(\tau)s_j(\tau)d\tau\right)\right\} \tag{5.61}$$

$$= 0, \quad j \neq k$$

Using similar arguments, it can be easily shown that

$$E\{z_{si}z_{sj}\} = E\{z_{si}\}E\{z_{sj}\} = 0, \quad i \neq j$$
$$E\{z_{ci}z_{sj}\} = E\{z_{ci}\}E\{z_{sj}\} = 0, \quad i = j \tag{5.62}$$

which holds true even when $i = k$ or $j = k$ but not both. Therefore, the variables z_{ci} and z_{si}, $i = 0, 1, \cdots, M - 1$ are uncorrelated and hence independent since they are conditionally Gaussian. Defining $\theta_i = \tan^{-1}(z_{si}/z_{ci})$, then (5.59) becomes

$$f_{z_{ci},z_{si}}(z, y/\theta) = \begin{cases} \frac{1}{\pi E_s N_0}\exp\left\{-\frac{z^2+y^2}{E_s N_0}\right\}, & i \neq k \\ \frac{1}{\pi E_s N_0}\exp\left\{-\frac{z^2+y^2+E_s^2-2E_s\sqrt{z^2+y^2}\cos(\theta-\theta_i)}{E_s N_0}\right\}, & i = k \end{cases} \tag{5.63}$$

Integrating over θ and using the definition of $I_0(x)$ as given by (5.22), (5.63) reduces to

$$f_{z_{ci},z_{si}}(z, y) = \begin{cases} \frac{1}{\pi E_s N_0}\exp\left\{-\frac{z^2+y^2}{E_s N_0}\right\}, & i \neq k \\ \frac{1}{\pi E_s N_0}\exp\left\{-\frac{z^2+y^2+E_s^2}{E_s N_0}\right\}I_0\left(\frac{2\sqrt{z^2+y^2}}{N_0}\right), & i = k \end{cases} \tag{5.64}$$

Making the change of variables $\xi_i = \sqrt{z_{ci}^2 + z_{si}^2}$, then

$$f_{\xi_i}(\xi) = \begin{cases} \frac{2\xi}{E_s N_0}\exp\left\{-\frac{\xi^2}{E_s N_0}\right\}, & i \neq k \\ \frac{2\xi}{E_s N_0}\exp\left\{-\frac{\xi^2+E_s^2}{E_s N_0}\right\}I_0\left(\frac{2\xi}{N_0}\right), & i = k \end{cases} \tag{5.65}$$

The Rayleigh and Rician densities of (5.65) can also be obtained using the results of Appendix 5A, namely, (5A.21) and (5A.27) with $\sigma^2 = N_0 E_s/2$ and $s^2 = E_s^2$. Let us proceed by normalizing ξ_i^2 using

$$A_i = \frac{\xi_i^2}{E_s N_0} \tag{5.66}$$

then $f_{A_i}(a)$ is given by

$$f_{A_i}(a) = \frac{f_{\xi_i}(\sqrt{E_s N_0 a})}{2\sqrt{(a/E_s N_0)}} \tag{5.67}$$

or

$$f_{A_i}(a) = \begin{cases} \exp\left\{-\left(a + \frac{E_s}{N_0}\right)\right\} I_0\left(\sqrt{4\frac{E_s}{N_0}a}\right) u(a), & i = k \\ \exp\{-a\}\, u(a), & i \neq k \end{cases} \tag{5.68}$$

where $u(a)$ is the unit step function. The variables A_i, $i = 0, 1, \cdots, M-1$ are also independent since they are nonlinear functions of independent random variables [11]. Hence

$$f_{A_0, A_1, \cdots, A_{M-1}}(a_0, a_1, \cdots, a_{M-1}) = f_{A_0}(a_0) f_{A_1}(a_1) \cdots f_{A_{M-1}}(a_{M-1}) \tag{5.69}$$

From this joint pdf, we can now evaluate the probability of symbol error $P_s(E)$. Assuming that m_k is transmitted and that the output of the envelope detector matched to $s_k(t)$ is A_k, then the probability of correct symbol detection conditioned on message m_k and A_k is given by

$$P_s(C/m_k, A_k) = \mathrm{Prob}\,\{A_0 \leq A_k, A_1 \leq A_k, \cdots, A_{k-1} \leq A_k, A_{k+1} \leq A_k, \cdots, A_{M-1} \leq A_k/m_k\}$$

$$= \prod_{\substack{i=0 \\ i \neq k}}^{M-1} \int_{-\infty}^{A_k} f_{A_i}(a)da = \left(1 - e^{-A_k}\right)^{M-1} \tag{5.70}$$

Hence

$$P_s(C/m_k) = \int_{-\infty}^{\infty} f_{A_k}(a_k)\left(1 - e^{-a_k}\right)^{M-1} da_k$$

$$= \exp\left\{-\frac{E_s}{N_0}\right\} \int_0^{\infty} e^{-a_k} I_0\left(\sqrt{4\frac{E_s}{N_0}a_k}\right)\left(1 - e^{-a_k}\right)^{M-1} da_k \tag{5.71}$$

Using the expansion

$$\left(1 - e^{-x}\right)^{M-1} = \sum_{m=0}^{M-1} (-1)^m \,_{M-1}C_m e^{-mx} \tag{5.72}$$

where $_{M-1}C_m$ was defined in (4.153), (5.71) reduces to

$$P_s(C/m_k) = \exp\left\{-\frac{E_s}{N_0}\right\} \sum_{m=0}^{M-1} (-1)^m \,_{M-1}C_m$$

$$\times \int_0^{\infty} \exp\{-(m+1)a_k\} I_0\left(\sqrt{4\frac{E_s}{N_0}a_k}\right) da_k \tag{5.73}$$

Using the integral[5]

$$\int_0^{\infty} x \exp\left\{-a^2 x^2\right\} I_0(bx)dx = \frac{1}{2a^2} \exp\left\{\frac{b^2}{4a^2}\right\} \tag{5.74}$$

---

5. This integral is derived from the more general integral [12]

$$\int_0^{\infty} t^{\mu-1} I_\mu(\alpha t) e^{-p^2 t^2} dt = \frac{\Gamma\left(\frac{\mu+\nu}{2}\right)\left(\frac{\alpha}{2p}\right)^\nu}{2p^\mu \Gamma(\nu+1)} \exp\left(\frac{\alpha^2}{4p^2}\right) \,_1F_1\left(\frac{\nu-\mu}{2} + 1, \nu+1; -\frac{\alpha^2}{4p^2}\right)$$

Upon substitution $\mu = 2$, $\nu = 0$ and noting that $_1F_1(0, 1; x) = 1$, the integral of (5.74) is obtained.

or equivalently (by letting $u = x^2$)

$$\int_0^\infty \exp\left\{-a^2 u\right\} I_0\left(b\sqrt{u}\right) du = \frac{1}{a^2} \exp\left\{\frac{b^2}{4a^2}\right\} \tag{5.75}$$

then (5.73) is independent of m_k and the probability of correct symbol detection $P_s(C)$ is given by

$$P_s(C) = \sum_{m=0}^{M-1} (-1)^m {}_{M-1}C_m \frac{1}{m+1} \exp\left\{-\frac{m}{m+1}\left(\frac{E_s}{N_0}\right)\right\} \tag{5.76}$$

The probability of symbol error, $P_s(E)$, is given by $1 - P_s(C)$, or

$$P_s(E) = \sum_{m=1}^{M-1} (-1)^{m+1} \frac{1}{m+1} {}_{M-1}C_m \exp\left\{-\frac{m}{m+1}\left(\frac{E_s}{N_0}\right)\right\} \tag{5.77}$$

The symbol error probability is depicted in Fig. 5.11 versus bit SNR [$E_b/N_0 = (E_s/N_0)/n$ where $n = \log_2 M$] for various values of M. Since the signals are orthogonal, the bit error probability $P_b(E)$ is related to $P_s(E)$ through (4.96) and is plotted in Fig. 5.12 versus E_b/N_0.

Consider now the case of coded versus uncoded transmission. The symbol error probability performance of a noncoherent system that transmits an $n = \log_2 M$ bit symbol by sending one bit at a time (so-called "uncoded transmission") is given by

$$P_s(E) = 1 - \left(1 - P_b(E)|_{M=2}\right)^n \tag{5.78}$$

with

$$P_b(E)|_{M=2} = \frac{1}{2} \exp\left(-\frac{E_b}{2N_0}\right) \tag{5.79}$$

obtained from (5.77) by setting $M = 2$ and $E_s = E_b$. The uncoded symbol error probability, as given by (5.78), is illustrated in Fig. 5.13 for various values of n as a function of bit SNR. Coded and uncoded symbol error probabilities are compared in Figures 5.14 and 5.15 for $n = 5$ ($M = 32$) and 10 ($M = 1024$), respectively. Also shown in these two figures are results given in Chapter 4 for coherent communication employing orthogonal and biorthogonal signals. For orthogonal codes, we observe that the type of detection performed, whether a coherent or noncoherent detector is mechanized, is not very critical for error probabilities less than 10^{-3}. As a matter of fact, the use of 5-bit biorthogonal codes and coherent reception is not appreciably more effective (given a certain energy-to-noise ratio) than the use of orthogonal codes and noncoherent reception. On the other hand, a comparison of coded versus uncoded transmission indicates that coding the transmitted signal into n-bit orthogonal symbols (rather than transmitting an n-bit symbol one bit at a time) is more effective in noncoherent communications. For example, if five bits of information are sent with a symbol error probability of 10^{-3} (typical for telemetry systems), the use of orthogonal codes and coherent detection will reduce the signal-to-noise ratio per bit, that is, E_b/N_0, by approximately 2.5 dB under that required for equivalent performance using bit-by-bit detection. Under the same symbol error probability conditions, the use of orthogonal codes and noncoherent detection will reduce the required E_b/N_0 by approximately 6 dB under that required for equivalent performance using bit-by-bit detection. For ten bits

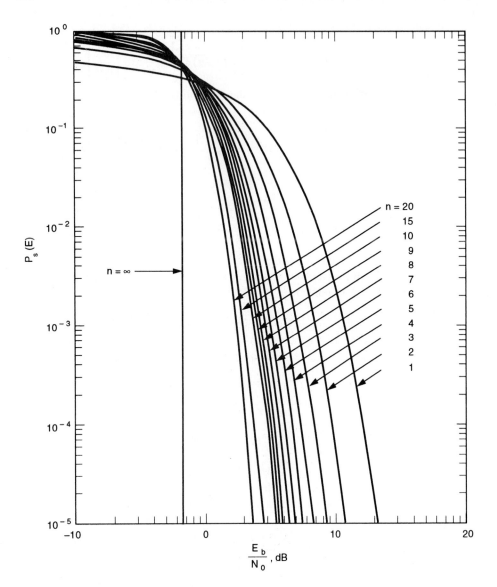

Figure 5.11   Symbol error probability of ideal noncoherent receiver ($M = 2^n$) as a function of bit SNR (Reprinted from [28])

($M = 1024$) and $P_s(E) = 10^{-3}$, orthogonal and biorthogonal coding and coherent reception reduce the required E_b/N_0 by approximately 5 dB under that required without coding. Finally, using noncoherent reception, 10-bit orthogonal codes and $P_s(E) = 10^{-3}$, one realizes an improvement of approximately 8.5 dB over the uncoded noncoherent procedure of bit-by-bit detection.

Finally, by a limiting procedure similar to that employed in Chapter 4, it may be shown (see Problem 5.3) that the asymptotic ($M \rightarrow \infty$) error probability performance of the optimum noncoherent receiver is identical to that for coherent communications as described by (4.182).

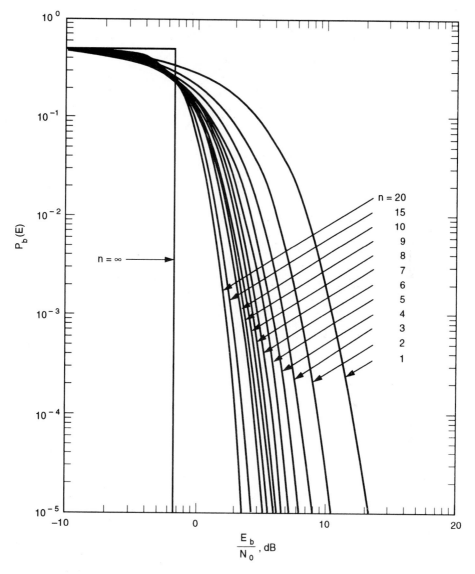

Figure 5.12   Bit error probability of ideal noncoherent receiver ($M = 2^n$) as a function of bit SNR (Reprinted from [28])

5.2.2 Performance of *M*-ary Nonorthogonal Signals

Nonorthogonal signals can occur in either one of two scenarios; the signals can be chosen nonorthogonal at the transmitter or the latter sends orthogonal signals but due to imperfections in the receiver, the signals become correlated and thus nonorthogonal. The second of the two cases occurs as the result of frequency or timing error in the receiver and will be treated in more detail in a later section. The key difference between the two cases in terms of analysis complexity is that when the signals are nonorthogonal at the transmitter, the receiver correlates the observed waveform with the transmitted signals and the noises at the output of the integrate-and-dump filters are correlated and dependent. On the other hand,

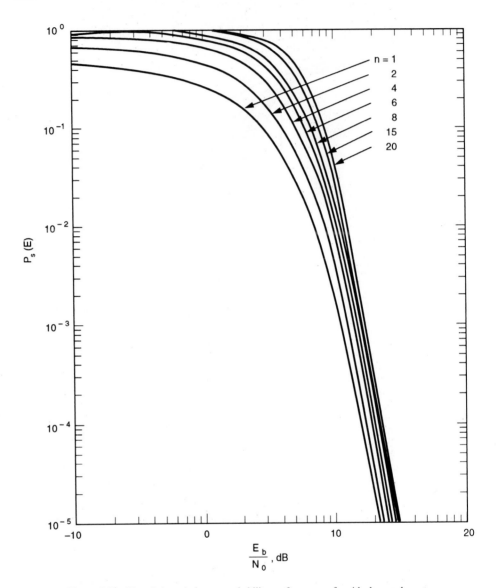

Figure 5.13 Uncoded symbol error probability performance of an ideal noncoherent receiver (Reprinted from [28])

when the signals are nonorthogonal due to frequency or timing errors, the receiver correlates the observed waveform with the original transmitted orthogonal signals and consequently, the noises at the output of the integrate-and-dump filters remain uncorrelated and hence independent in an AWGN channel. Assuming nonorthogonal signals, the received waveform, when message m_i is transmitted, is given by

$$r(t) = \sqrt{2}a_i(t)\cos(\omega_c t + \phi_i(t) + \theta) + n(t), \quad 0 \le t \le T \tag{5.80}$$

where $a_i(t)$ and $\phi_i(t)$ are the amplitude and phase information bearing components of signal

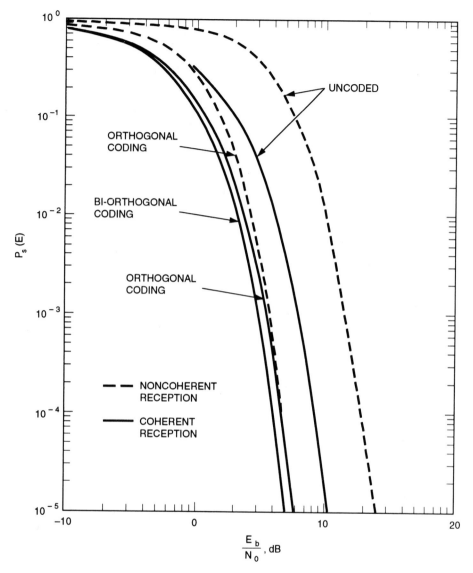

Figure 5.14 A comparison of coded and uncoded, coherent and noncoherent, symbol error probabilities ($M = 32$) (Reprinted from [28])

$s_i(t)$ and $n(t)$ is AWGN[6] with one-sided PSD N_0 Watts/Hz. It is convenient to use complex notation to denote the transmitted signals. Let

$$\tilde{s}_i(t) \triangleq a_i(t)e^{j\phi_i(t)} \tag{5.81}$$

be the complex baseband signal associated with signal $s_i(t)$. Then

---

6. It is easier to derive the results when the noise is white. For bandlimited noise, the results are identical, but the derivation is tedious.

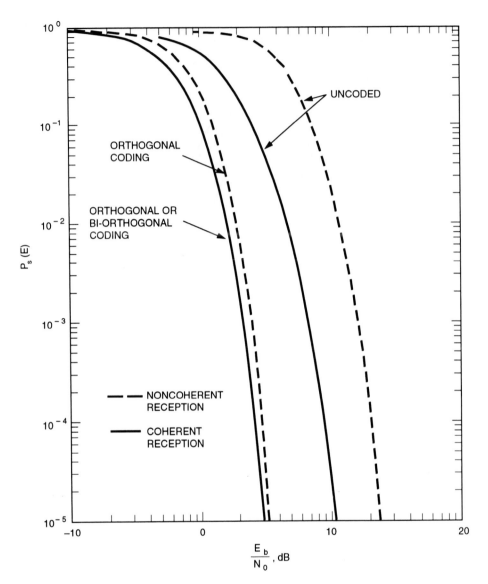

Figure 5.15   A comparison of coded and uncoded, coherent and noncoherent, symbol error probabilities ($M = 1024$) (Reprinted from [28])

$$s_i(t) = \text{Re}\left\{\sqrt{2}\tilde{s}_i(t)e^{j(\omega_c t + \theta)}\right\} \tag{5.82}$$

constitutes the signal component of the received waveform when message m_i is transmitted. The signal $s_i(t)$ can be expressed in an alternative form as

$$s_i(t) = \sqrt{2}\{a_i(t)\cos\phi_i(t)\cos(\omega_c t + \theta) - a_i(t)\sin\phi_i(t)\sin(\omega_c t + \theta)\}$$
$$= \sqrt{2}\{a_{ci}(t)\cos(\omega_c t + \theta) - a_{si}(t)\sin(\omega_c t + \theta)\} \tag{5.83}$$

where

$$a_{ci}(t) \triangleq a_i(t) \cos \phi_i(t)$$
$$a_{si}(t) \triangleq a_i(t) \sin \phi_i(t) \tag{5.84}$$

We will assume that the signals have equal energy E_s, that is

$$E_s \triangleq \int_O^T s_i^2(t)dt = \int_0^T a_i^2(t)dt$$
$$= \int_0^T a_{ci}^2(t)dt + \int_0^T a_{si}^2(t)dt, \quad i = 0, 1, \cdots, M-1 \tag{5.85}$$

Define the complex correlation coefficient $\tilde{\rho}_{ij}$ [13] as the correlation between any two complex baseband signals $\tilde{s}_i(t)$ and $\tilde{s}_j(t)$ as (let $*$ denote complex conjugate)

$$\tilde{\rho}_{ij} \triangleq \frac{1}{E_s} \int_0^T \tilde{s}_i(t)\tilde{s}_j^*(t)dt \tag{5.86}$$

Substituting (5.81) into (5.86), then

$$\tilde{\rho}_{ij} = \alpha_{ij} + j\beta_{ij} \tag{5.87}$$

where the real part of the complex correlation is given by

$$\alpha_{ij} \triangleq \frac{1}{E_s} \int_0^T a_i(t)a_j(t) \cos(\phi_i(t) - \phi_j(t))dt$$
$$= \frac{1}{E_s} \int_0^T \left(a_{ci}(t)a_{cj}(t) + a_{si}(t)a_{sj}(t) \right) dt \tag{5.88}$$

while the imaginary part is

$$\beta_{ij} \triangleq \frac{1}{E_s} \int_0^T a_i(t)a_j(t) \sin(\phi_i(t) - \phi_j(t))dt$$
$$= \frac{1}{E_s} \int_0^T \left(a_{si}(t)a_{cj}(t) - a_{ci}(t)a_{sj}(t) \right) dt \tag{5.89}$$

Note that $\alpha_{ii} = 1$ and $\beta_{ii} = 0$. Furthermore, $\alpha_{ij} = \alpha_{ji}$ and $\beta_{ij} = -\beta_{ji}$. In the special case where the signals are orthogonal, then $\alpha_{ij} = \beta_{ij} = 0$ for $i \neq j$. Recall that the optimum receiver still computes the decision variables ξ_k^2 for $k = 0, 1, \cdots, M-1$ where $\xi_k^2 = z_{ck}^2 + z_{sk}^2$ and

$$z_{ck} \triangleq \int_0^T r(t)\sqrt{2}a_k(t) \cos(\omega_c t + \phi_k(t))dt \tag{5.90a}$$

$$z_{sk} \triangleq \int_0^T r(t)\sqrt{2}a_k(t) \sin(\omega_c t + \phi_k(t))dt \tag{5.90b}$$

It can be easily demonstrated that ξ_k^2 is independent of the carrier phase angle θ of the incoming waveform (see Problem 5.4). Therefore, we will set $\theta = 0$ from here on to simplify the analysis. Substituting (5.80) into (5.90a) and (5.90b), we have

$$z_{ck} = E_s\alpha_{ki} + n_{ck} \tag{5.91a}$$

$$z_{sk} = E_s\beta_{ki} + n_{sk} \tag{5.91b}$$

where

$$n_{ck} \triangleq \int_0^T n(t)\sqrt{2}a_k(t)\cos(\omega_c t + \phi_k(t))dt \qquad (5.92a)$$

$$n_{sk} \triangleq \int_0^T n(t)\sqrt{2}a_k(t)\sin(\omega_c t + \phi_k(t))dt \qquad (5.92b)$$

It is easily shown that n_{ck} and n_{sk} are zero-mean Gaussian random variables with variance $(N_0/2)E_s$ and cross-correlations

$$E\{n_{ck}n_{cj}\} = \frac{N_0}{2}E_s\alpha_{kj} \qquad (5.93a)$$

$$E\{n_{sk}n_{sj}\} = \frac{N_0}{2}E_s\alpha_{kj} \qquad (5.93b)$$

$$E\{n_{ck}n_{sj}\} = \frac{N_0}{2}E_s\beta_{kj} \qquad (5.93c)$$

Alternatively, (5.91a) and (5.91b) can be written in normalized form as

$$z_{ck} = \sqrt{\frac{2E_s}{N_0}}\alpha_{ki} + x_{ck} \qquad (5.94)$$

$$z_{sk} = \sqrt{\frac{2E_s}{N_0}}\beta_{ki} + x_{sk} \qquad (5.94b)$$

where x_{ck} and x_{sk} are normalized unit variance random Gaussian variables with

$$E\{x_{ck}x_{cj}\} = \alpha_{kj} \qquad (5.95a)$$

$$E\{x_{sk}x_{sj}\} = \alpha_{kj} \qquad (5.95b)$$

$$E\{x_{ck}x_{sj}\} = \beta_{kj} \qquad (5.95c)$$

Define now the matrices A and B as

$$A = \begin{pmatrix} 1 & \cdots & \alpha_{ij} \\ \vdots & \ddots & \vdots \\ \alpha_{ji} & \cdots & 1 \end{pmatrix} \quad \text{with} \quad \alpha_{ij} = \alpha_{ji} \qquad (5.96)$$

and

$$B = \begin{pmatrix} 0 & \cdots & \beta_{ij} \\ \vdots & \ddots & \vdots \\ \beta_{ji} & \cdots & 0 \end{pmatrix} \quad \text{with} \quad \beta_{ij} = -\beta_{ji} \qquad (5.97)$$

and the vectors $\mathbf{x}_c$, $\mathbf{x}_s$, $\mathbf{z}_c$, and $\mathbf{z}_s$ as

$$\mathbf{x}_c = \begin{pmatrix} x_{c0} \\ x_{c1} \\ \vdots \\ x_{cM-1} \end{pmatrix}, \quad \mathbf{x}_s = \begin{pmatrix} x_{s0} \\ x_{s1} \\ \vdots \\ x_{sM-1} \end{pmatrix}, \quad \mathbf{z}_c = \begin{pmatrix} z_{c0} \\ z_{c1} \\ \vdots \\ z_{cM-1} \end{pmatrix} \quad \text{and} \quad \mathbf{z}_s = \begin{pmatrix} z_{s0} \\ z_{s1} \\ \vdots \\ z_{sM-1} \end{pmatrix} \qquad (5.98)$$

and $\boldsymbol{\alpha}_i$ and $\boldsymbol{\beta}_i$ as

$$
\boldsymbol{\alpha}_i = \begin{pmatrix} \alpha_{0i} \\ \alpha_{1i} \\ \vdots \\ \alpha_{M-1i} \end{pmatrix} \quad \text{and} \quad \boldsymbol{\beta}_i = \begin{pmatrix} \beta_{0i} \\ \beta_{1i} \\ \vdots \\ \beta_{M-1i} \end{pmatrix} \tag{5.99}
$$

Then the pdf of the augmented vector $(\mathbf{z}_c^T \ \mathbf{z}_s^T)$, assuming message m_i is transmitted, is $2M$-multidimensional Gaussian given by (3.61) as

$$
f_{\mathbf{z}_c, \mathbf{z}_s}(\mathbf{z}, \mathbf{y}/m_i) = G\left(\begin{pmatrix} \mathbf{z} \\ \mathbf{y} \end{pmatrix}; \sqrt{\frac{2E_s}{N_0}} \begin{pmatrix} \boldsymbol{\alpha}_i \\ \boldsymbol{\beta}_i \end{pmatrix}; \mathcal{C} \right) \tag{5.100}
$$

where the $2M \times 2M$ matrix $\mathcal{C}$ is given by

$$
\mathcal{C} \triangleq \begin{pmatrix} \mathcal{A} & \mathcal{B} \\ \mathcal{B}^T & \mathcal{A} \end{pmatrix} \tag{5.101}
$$

and is the covariance matrix of the augmented vector $(\mathbf{z}_c \ \mathbf{z}_s)^T$. The probability of correct decision, assuming that message m_i is transmitted, becomes

$$
P_s(C/m_i) = \int_{-\infty}^{\infty} \int_{-\infty}^{\infty} \int\!\!\int_{R_0} \int\!\!\int_{R_1} \cdots \int\!\!\int_{R_{i-1}} \int\!\!\int_{R_{i+1}} \cdots \int\!\!\int_{R_{M-1}}
$$

$$
G\left(\begin{pmatrix} \mathbf{z} \\ \mathbf{y} \end{pmatrix}; \sqrt{\frac{2E_s}{N_0}} \begin{pmatrix} \boldsymbol{\alpha}_i \\ \boldsymbol{\beta}_i \end{pmatrix}; \mathcal{C} \right) \tag{5.102}
$$

$$
dz_{M-1}dy_{M-1}\cdots dz_{i+1}dy_{i+1}dz_{i-1}dy_{i-1}\cdots dz_1dy_1dz_0dy_0dz_idy_i
$$

where R_i is the region within the circle of radius $\sqrt{z_{ci}^2 + z_{si}^2}$ centered at the origin. The unconditional probability of correct symbol detection $P_s(C)$ becomes

$$
P_s(C) = \sum_{i=0}^{M-1} P_s(C/m_i) P(m_i) \tag{5.103}
$$

where $P(m_i)$ is the a priori probability of sending message m_i. The $2M$-fold integral in (5.102) cannot be simplified any further in general. In the following sections, we will consider scenarios under which $P_s(C)$ can be computed in closed form or in terms of well-known functions.

5.2.3 Performance of *M*-ary Equicorrelated Signals

For M-ary signals, Nuttall [14] derived the probability of error for the case when the signals are equicorrelated and furthermore, the correlation is real and nonnegative, that is

$$
\tilde{\rho}_{ij} = \begin{cases} 1, & i = j \\ \alpha, & i \neq j \text{ and } \alpha \text{ real and nonnegative} \end{cases} \tag{5.104}
$$

Hence, the imaginary component β of the complex correlation is assumed zero. Nuttall [15] provides a heuristic argument (not a proof) that for a given magnitude of the complex correlation, the assumption of zero angle (or zero imaginary component of $\tilde{\rho}$) realizes the minimum error probability. It is shown [14] that the probability of correct symbol detection is given by

$$
P_s(C) = (1-\alpha)\exp\left(-\frac{E_s}{N_0}\right)\int_0^\infty\int_0^\infty xy\exp\left(-\frac{1}{2}\left(x^2+y^2\right)\right)
$$

$$
\times I_0\left(\sqrt{\frac{2E_s}{N_0}(1-\alpha)x}\right)I_0\left(\sqrt{\alpha}xy\right)\left[1-Q\left(\sqrt{\alpha}y,x\right)\right]^{M-1}dxdy \tag{5.105}
$$

If $\alpha = 0$ in (5.105), the integral on y is unity resulting in

$$
P_s(C) = \exp\left(-\frac{E_s}{N_0}\right)\int_0^\infty x\exp\left(-\frac{x^2}{2}\right)I_0\left(\sqrt{\frac{2E_s}{N_0}}x\right)\left[1-\exp\left(-\frac{x^2}{2}\right)\right]^{M-1}dx \tag{5.106}
$$

Using the expansion in (5.72) and the integral of (5.74), (5.106) reduces to (5.76) as expected. If, on the other hand, $\alpha \neq 0$ but $M = 2$, use the integral [15]

$$
\int_0^\infty s\exp\left(-\frac{1}{2}\left(s^2+c^2\right)\right)I_0(cs)Q(as,b)ds = Q\left(\frac{ac}{\sqrt{1+a^2}},\frac{b}{\sqrt{1+a^2}}\right) \tag{5.107}
$$

in (5.105) to obtain

$$
P_s(C) = (1-\alpha)\exp\left(-\frac{E_s}{N_0}\right)\int_0^\infty r\exp\left(-\frac{1}{2}r^2(1-\alpha)\right)
$$

$$
\times I_0\left(r\sqrt{\frac{2E_s}{N_0}(1-\alpha)}\right)\left[1-Q\left(\frac{\alpha r}{\sqrt{1+\alpha}},\frac{r}{\sqrt{1+\alpha}}\right)\right]dr
$$

$$
= (1-\alpha^2)\exp\left(-\frac{E_s}{N_0}\right)\int_0^\infty x\exp\left(-\frac{1}{2}x^2\left(1-\alpha^2\right)\right) \tag{5.108}
$$

$$
\times I_0\left(x\sqrt{\frac{2E_s}{N_0}\left(1-\alpha^2\right)}\right)(1-Q(\alpha x,x))xdx
$$

Helstrom [16] integrated that expression to obtain

$$
P_s(C) = 1 - Q\left(\sqrt{\frac{E_s}{2N_0}\left(1-\sqrt{1-\alpha^2}\right)},\sqrt{\frac{E_s}{2N_0}\left(1+\sqrt{1-\alpha^2}\right)}\right)
$$

$$
+\frac{1}{2}\exp\left(-\frac{E_s}{2N_0}\right)I_0\left(\alpha\frac{E_s}{2N_0}\right) \tag{5.109}
$$

which is exactly $1 - P_s(E)$ where $P_s(E)$ is the probability of error for two nonorthogonal signals and will be given by (5.133).

5.2.4 Performance of Binary Nonorthogonal Signals

For the binary case, the probability of error can be simplified much further as follows. The receiver compares ξ_0 and ξ_1 where $\xi_i = \sqrt{z_{ci}^2+z_{si}^2}$ and z_{ci}, z_{si} for $i = 0, 1$ are still given by (5.94). Mutiplying z_{ci} and z_{si}, $i = 0, 1$, by $\sqrt{2E_s/N_0}$, we have (let $\alpha = \alpha_{10} = \alpha_{01}$ and

$\beta = \beta_{10} = -\beta_{01})$

$$v_{c0} \triangleq \sqrt{\frac{2E_s}{N_0}} z_{c0} = \frac{2E_s}{N_0} + u_{c0}$$

$$v_{s0} \triangleq \sqrt{\frac{2E_s}{N_0}} z_{s0} = u_{s0}$$

(5.110)

and

$$v_{c1} \triangleq \sqrt{\frac{2E_s}{N_0}} z_{c1} = \frac{2E_s}{N_0}\alpha + u_{c1}$$

$$v_{s1} \triangleq \sqrt{\frac{2E_s}{N_0}} z_{s1} = \frac{2E_s}{N_0}\beta + u_{s1}$$

(5.111)

assuming m_0 was transmitted. The Gaussian noise variables, u_{c0}, u_{s0}, u_{c1}, and u_{s1}, are zero-mean, unit variance with $E\{u_{c0}u_{c1}\} = E\{u_{s0}u_{s1}\} = (2E_s/N_0)\alpha$, $E\{u_{c0}u_{s1}\} = -E\{u_{c1}u_{s0}\} = (2E_s/N_0)\beta$. Letting $\xi_{vi} \triangleq \sqrt{v_{ci}^2 + v_{si}^2}$, we know from Appendix 5A that the marginal densities of ξ_{v0} and ξ_{v1} are Rician, but due to the nonzero correlation of the Gaussian noises, ξ_{v0} and ξ_{v1} are not independent. The probability of error, assuming message m_0 is transmitted, is equal to the probability that ξ_{v1} exceeds ξ_{v0} or

$$P_s(E/m_0) = \text{Prob}(\xi_{v1} > \xi_{v0}/m_0) = \int_0^\infty \int_x^\infty f_{\xi_{v0}\xi_{v1}}(x, y)\,dy\,dx \qquad (5.112)$$

which requires the computation of the joint pdf $f_{\xi_{v0}\xi_{v1}}(x, y)$. This approach is outlined by Helstrom [16] who determined the joint pdf and computed the double integral in (5.112). Another approach to the computation of the probability of error, taken by Proakis [17], is based on the observation that $\xi_{v0}^2 - \xi_{v1}^2$ is a special case of a general quadratic form in complex-valued random variables, defined by

$$A = \sum_{k=1}^L \left(A_1|x_k|^2 + A_2|y_k|^2 + A_3x_ky_k^* + A_3^*x_k^*y_k \right) \qquad (5.113)$$

where A_1, A_2, A_3 are constants and $\{x_k, y_k\}$ are pairs of correlated complex-valued Gaussian random variables.

The approach followed in this book (also see Viterbi [18]) is based on whitening the observations v_{c0}, v_{s0}, v_{c1}, v_{s1} so that the resulting variables $v_{w,c0}$, $v_{w,s0}$, $v_{w,c1}$, and $v_{w,s1}$ are uncorrelated. Consequently, the variables $\xi_{w,0} \triangleq \sqrt{v_{w,c0}^2 + v_{w,s0}^2}$ and $\xi_{w,1} \triangleq \sqrt{v_{w,c1}^2 + v_{w,s1}^2}$ are Rician and independent and the probability of error reduces to the probability that one Rician variable exceeds another independent Rician variable. Defining the vector $\mathbf{x}$ as

$$\mathbf{x}^T = (v_{c0} \quad v_{s0} \quad v_{c1} \quad v_{s1})^T \qquad (5.114)$$

then, the mean and covariance of $\mathbf{x}$ are given by

$$\mathbf{m_x} = \begin{pmatrix} (2E_s/N_0) \\ 0 \\ (2E_s/N_0)\alpha \\ (2E_s/N_0)\beta \end{pmatrix} = \left(\frac{2E_s}{N_0}\right) \begin{pmatrix} 1 \\ 0 \\ \alpha \\ \beta \end{pmatrix} \tag{5.115}$$

and

$$\mathcal{K}_\mathbf{x} = \left(\frac{2E_s}{N_0}\right) \begin{pmatrix} 1 & 0 & \alpha & \beta \\ 0 & 1 & -\beta & \alpha \\ \alpha & -\beta & 1 & 0 \\ \beta & \alpha & 0 & 1 \end{pmatrix} \tag{5.116}$$

Due to the symmetry of the problem, it is easily shown that $P_s(E/m_0) = P_s(E/m_1) = P_s(E)$. Hence, the error probability is given by

$$P_s(E) = \mathrm{Prob}(\xi_{v0}^2 < \xi_{v1}^2) = \mathrm{Prob}(v_{c0}^2 + v_{s0}^2 - v_{c1}^2 - v_{s1}^2 < 0) \tag{5.117}$$

or

$$P_s(E) = \mathrm{Prob}\left(\mathbf{x}^T Q \mathbf{x} < 0\right) \tag{5.118}$$

where

$$Q = \begin{pmatrix} 1 & 0 & 0 & 0 \\ 0 & 1 & 0 & 0 \\ 0 & 0 & -1 & 0 \\ 0 & 0 & 0 & -1 \end{pmatrix} \tag{5.119}$$

In order to whiten the vector $\mathbf{x}$, we need to perform a linear nonsingular transformation $\mathcal{T}$ on $\mathbf{x}$ which diagonalizes $\mathcal{K}_\mathbf{x}$, that is

$$\mathbf{x}_w = \mathcal{T}\mathbf{x} \tag{5.120}$$

One such transformation is given by

$$\mathcal{T} = \frac{\sqrt{N_0/E_s}}{2k\sqrt{1+k}} \begin{pmatrix} 1+k & 0 & -\alpha & -\beta \\ 0 & 1+k & \beta & -\alpha \\ -\alpha & \beta & 1+k & 0 \\ -\beta & -\alpha & 0 & 1+k \end{pmatrix} \tag{5.121}$$

where

$$k = \sqrt{1 - \beta^2 - \alpha^2} \tag{5.122}$$

with the result (see Problem 5.5)

$$\mathcal{K}_{\mathbf{x}_w} = \mathcal{T}\mathcal{K}_\mathbf{x}\mathcal{T}^T = \mathcal{I} \tag{5.123}$$

$$\mathbf{x}^T Q \mathbf{x} = \mathbf{x}_w^T \left(\mathcal{T}^{-1}\right)^T Q \left(\mathcal{T}^{-1}\right) \mathbf{x}_w = \left(\frac{2E_s}{N_0}\right) k \left(\mathbf{x}_w^T Q \mathbf{x}_w\right) \tag{5.124}$$

Hence, the probability of error becomes

$$P_s(E) = \mathrm{Prob}(\mathbf{x}^T Q \mathbf{x} < 0) = \mathrm{Prob}(\mathbf{x}_w^T Q \mathbf{x}_w < 0)$$
$$= \mathrm{Prob}(v_{w,c0}^2 + v_{w,s0}^2 < v_{w,c1}^2 + v_{w,s1}^2) \tag{5.125}$$

where $v_{w,c0}$, $v_{w,s0}$, $v_{w,c1}$, and $v_{w,s1}$ are unit variance uncorrelated Gaussian (and hence independent) random variables with means

$$E\{\mathbf{x}_w\} = T E\{\mathbf{x}\} = \frac{\sqrt{N_0/E_s}}{2k\sqrt{1+k}} \begin{pmatrix} 2(E_s/N_0)k(k+1) \\ 0 \\ 2(E_s/N_0)k\alpha \\ 2(E_s/N_0)k\beta \end{pmatrix} \tag{5.126}$$

The optimum receiver compares $\xi_{w,0}^2$ to $\xi_{w,1}^2$ where $\xi_{w,i}^2$ is the sum of the squares of $v_{w,ci}$ and $v_{w,si}$. The pdf of $\xi_{w,i}$ is Rician with parameter s_i, that is

$$f_{\xi_{w,i}}(\xi/m_0) = \xi \exp\left(-\frac{\xi^2 + s_i^2}{2}\right) I_0(s_i\xi) \tag{5.127}$$

where

$$s_0^2 = E^2\{v_{w,c0}\} + E^2\{v_{w,s0}\} = \left(\frac{E_s}{N_0}\right)(1+k)$$
$$s_1^2 = E^2\{v_{w,c1}\} + E^2\{v_{w,s1}\} = \left(\frac{E_s}{N_0}\right)(1-k) \tag{5.128}$$

Hence, the probability of error becomes

$$P_s(E) = \text{Prob}(\xi_{w,1} > \xi_{w,0}/m_0)$$
$$= \int_0^\infty f_{\xi_{w,0}}(x) \int_x^\infty f_{\xi_{w,1}}(y) dy dx \tag{5.129}$$

which is similar to (5.112) except that the variables $\xi_{w,0}$ and $\xi_{w,1}$ are now independent. It can be shown that [19] (see Problem 5.6), given two independent Rician random variables R_0 and R_1, each with parameter s_i and variance σ_i^2 ($i = 0, 1$), then the probability that R_1 exceeds R_0 is given by

$$P = \text{Prob}(R_0 < R_1)$$
$$= Q\left(\sqrt{a}, \sqrt{b}\right) - \frac{\sigma_0^2}{\sigma_0^2 + \sigma_1^2} \exp\left(-\frac{a+b}{2}\right) I_0\left(\sqrt{ab}\right) \tag{5.130}$$

where

$$a = \frac{s_1^2}{\sigma_0^2 + \sigma_1^2} \quad \text{and} \quad b = \frac{s_0^2}{\sigma_0^2 + \sigma_1^2} \tag{5.131}$$

In our case, $\sigma_1^2 = \sigma_0^2 = 1$ and

$$a = \frac{E_s}{2N_0}(1-k) = \frac{E_s}{2N_0}\left(1 - \sqrt{1 - |\tilde{\rho}|^2}\right)$$
$$b = \frac{E_s}{2N_0}(1+k) = \frac{E_s}{2N_0}\left(1 + \sqrt{1 - |\tilde{\rho}|^2}\right) \tag{5.132}$$

with $\tilde{\rho} = \alpha + j\beta$. Hence, the probability of error for two nonorthogonal signals is given by

$$P_s(E) = Q\left(\sqrt{\frac{E_s}{2N_0}\left(1 - \sqrt{1 - |\tilde{\rho}|^2}\right)}, \sqrt{\frac{E_s}{2N_0}\left(1 + \sqrt{1 - |\tilde{\rho}|^2}\right)}\right)$$
$$- \frac{1}{2}\exp\left(-\frac{E_s}{2N_0}\right) I_0\left(\frac{E_s}{2N_0}|\tilde{\rho}|\right)$$
(5.133)

which is identical to $1 - P_s(C)$ where $P_s(C)$ is given by (5.109). The probability of error is illustrated in Fig. 5.16 for several values of $|\tilde{\rho}|$. It is clear that orthogonal signals (that is, $\tilde{\rho} = 0$) minimize $P_s(E)$. In this case, $P_s(E)$ reduces to

$$P_s(E) = Q\left(0, \sqrt{\frac{E_s}{N_0}}\right) - \frac{1}{2}\exp\left(-\frac{E_s}{2N_0}\right)$$
(5.134)

But since $Q(0, x) = \exp(-x^2/2)$, then

$$P_s(E) = \exp\left(-\frac{E_s}{2N_0}\right) - \frac{1}{2}\exp\left(-\frac{E_s}{2N_0}\right) = \frac{1}{2}\exp\left(-\frac{E_s}{2N_0}\right)$$
(5.135)

which is identical to (5.77), for $M = 2$, or equivalently, (5.79) with $E_s = E_b$.

Consider the cross-correlation between two FSK signals at frequencies f_0 and f_1. Then, the cross-correlation between the corresponding complex baseband signals is

$$\tilde{\rho} = \frac{1}{2E_s}\int_0^T \sqrt{2E_s/T}\,e^{j2\pi f_0 t}\sqrt{2E_s/T}\,e^{-j2\pi f_1 t}\,dt$$
$$= \frac{1}{T}\int_0^T e^{j2\pi(f_0 - f_1)t}\,dt = \frac{\sin\pi f_{0,1}T}{\pi f_{0,1}T}e^{j\pi f_{0,1}T}$$
(5.136)

and

$$|\tilde{\rho}| = \left|\frac{\sin\pi f_{0,1}T}{\pi f_{0,1}T}\right|$$
(5.137)

The performance for any frequency separation $f_{0,1}$ can then be computed by first evaluating the corresponding $\tilde{\rho}$ and then picking the corresponding curve in Fig. 5.16. In particular, if $f_{0,1} = 1/T$, then $|\tilde{\rho}| = 0$ and $P_s(E)$ reduces to (5.135).

5.2.5 Effect of Frequency Error on Orthogonal *M*-FSK Detection

When the received frequency is not perfectly known, the observed signal, assuming message m_i is transmitted, is given by

$$r(t) = \sqrt{2P}\cos\left(2\pi(f_c + f_i + \Delta f)t + \theta\right) + n(t), \quad 0 \le t \le T$$
(5.138)

where Δf is the error in carrier frequency. Alternatively, the received signal can be interpreted as a carrier f_c shifted by the appropriate signal frequency $f_i + \Delta f$, rather than f_i. This scenario is illustrated in Fig. 5.17, where the dashed lines denote start and end of integration times, with the depicted frequencies. The inphase integrator output z_{ck}, matched to

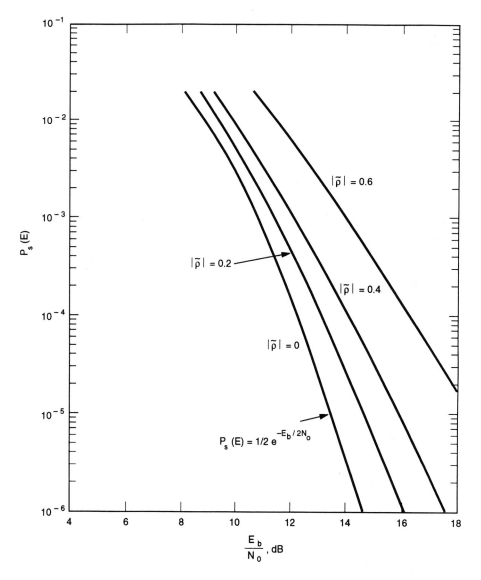

Figure 5.16   Probability of error for noncoherent detection of two nonorthogonal signals (Reprinted from [17])

signal $s_k(t)$, becomes[7]

$$z_{ck} \triangleq \int_0^T r(t)\sqrt{2P}\cos\left(2\pi(f_c + f_k)t\right)dt$$

$$= 2P\int_0^T \cos\left(2\pi(f_c + f_i + \Delta f)t\right)\cos\left(2\pi(f_c + f_k)t\right)dt + n_{ck} \tag{5.139}$$

---

7. We can set θ to zero without loss of generality.

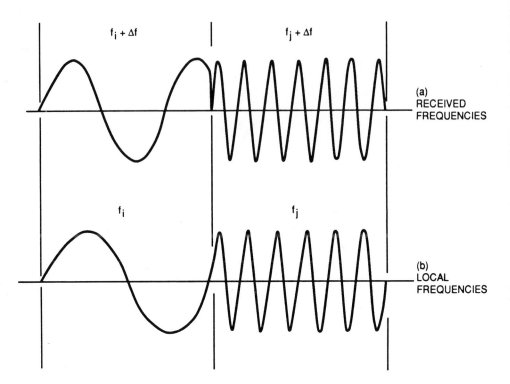

Figure 5.17 Effect of frequency error

where n_{ck} is a zero-mean Gaussian random variable with variance $\sigma^2 = (N_0/2)E_s$. Simplifying, (5.139) reduces to

$$z_{ck} = \frac{P}{2\pi(f_{i,k} + \Delta f)} \sin\left(2\pi(f_{i,k} + \Delta f)T\right) + n_{ck} \qquad (5.140)$$

where again, $f_{i,k}$ denotes the frequency difference between messages i and k, that is

$$f_{i,k} \triangleq f_i - f_k \qquad (5.141)$$

Similarly, the quadrature integrator output z_{sk} is given by

$$
\begin{aligned}
z_{sk} &\triangleq \int_0^T r(t)\sqrt{2P} \sin\left(2\pi(f_c + f_k)t\right) dt \\
&= \frac{P}{2\pi(f_{i,k} + \Delta f)} \left\{\cos\left(2\pi(f_{i,k} + \Delta f)T\right) - 1\right\} + n_{sk}
\end{aligned}
\qquad (5.142)
$$

The envelope statistic $\xi_k = \sqrt{z_{ck}^2 + z_{sk}^2}$ will then be Rician distributed with parameter s_k^2 given by

$$s_k^2 = \left(\frac{P}{2\pi(f_{i,k} + \Delta f)}\right)^2 \left\{\sin^2\left(2\pi(f_{i,k} + \Delta f)T\right) + \left(\cos\left(2\pi(f_{i,k} + \Delta f)T\right) - 1\right)^2\right\}$$

$$= \left(\frac{P}{2\pi(f_{i,k} + \Delta f)}\right)^2 \left\{2 - 2\cos\left(2\pi(f_{i,k} + \Delta f)T\right)\right\} \tag{5.143}$$

$$= \left(\frac{P}{2\pi(f_{i,k} + \Delta f)}\right)^2 \left\{4\sin^2\left(\pi(f_{i,k} + \Delta f)T\right)\right\}$$

Normalizing z_{ck} and z_{sk} by $1/\sigma = \sqrt{2/(N_0 E_s)}$, the parameter s_k^2 is then normalized by $2/(N_0 E_s)$ and since $f_i = i/T$ for orthogonal signals, then

$$s_k^2 = \left(\frac{2E_s}{N_0}\right)\frac{1}{\pi^2(i - k + \rho)^2}\sin^2\left(\pi(i - k + \rho)\right) \tag{5.144}$$

where ρ denotes the normalized frequency error [20] given by

$$\rho \triangleq \Delta f T \tag{5.145}$$

From (5A.27), the density of ξ_k is given by (note that $\sigma^2 = 1$ after the normalization)

$$f_{\xi_k}(\xi) = \xi \exp\left\{-\left(\frac{\xi^2}{2} + \frac{(2E_s/N_0)f_\rho^2(i, k)}{2}\right)\right\} I_0\left(\xi\sqrt{\frac{2E_s}{N_0}f_\rho^2(i, k)}\right) \tag{5.146}$$

where we let

$$f_\rho^2(i, k) \triangleq \frac{1}{\pi^2(i - k + \rho)^2}\sin^2\left(\pi(i - k + \rho)\right) \tag{5.147}$$

First, note that the detector matched to the incoming signal suffers from signal attenuation equal to

$$f_\rho^2(i, i) = \frac{1}{\pi^2\rho^2}\sin^2(\pi\rho) \tag{5.148}$$

which, as expected, reduces to unity if $\rho = 0$. Simultaneously, loss of signal orthogonality occurs as a result of signal spillover into the remaining $M - 1$ detectors; hence, the nonzero means and the resulting Rician densities. Note that for zero frequency error (that is, $\rho = 0$), then

$$f_{\rho=0}^2(i, k) = \frac{1}{\pi^2(i - k)^2}\sin^2\left(\pi(i - k)\right) = \begin{cases} 0, & \text{for } i \neq k \\ 1, & \text{for } i = k \end{cases} \tag{5.149}$$

and signal orthogonality is restored. In this case, the Rician densities reduce to the Rayleigh densities derived earlier. Despite loss of orthogonality, the variables $\xi_0, \xi_1, \cdots, \xi_{M-1}$ remain independent since the Gaussian random variables resulting from the noise integration are still independent as the local signals remain orthogonal. In this case, the probability of correct symbol detection, assuming message m_i is transmitted, is given by

$$P_s(C/m_i) = \text{Prob}\left\{\xi_i = \max_m \xi_m, \quad m = 0, 1, \cdots, M-1\right\}$$

$$= \int_0^\infty f_{\xi_i}(x_i) \underbrace{\int_0^{x_i} \cdots \int_0^{x_i}}_{M-1 \text{ integrals}} f_{\xi_0}(x_0) f_{\xi_1}(x_1) \cdots f_{\xi_{i-1}}(x_{i-1}) f_{\xi_{i+1}}(x_{i+1}) \cdots$$

$$\times f_{x_{M-1}}(\xi_{M-1}) dx_0 dx_1 \cdots dx_{i-1} dx_{i+1} \cdots dx_{M-1} dx_i$$ (5.150)

$$= \int_0^\infty f_{\xi_i}(x_i) \left(\prod_{\substack{m=0 \\ m \neq i}}^{M-1} \int_0^{x_i} f_{\xi_m}(x_m) dx_m \right) dx_i$$

Using the definition of the Marcum Q-function as given by (5A.15), we have

$$\int_0^{x_i} f_{\xi_m}(x_m) dx_m = 1 - Q\left(\frac{2E_s}{N_0} f_\rho^2(i, m), x_i\right)$$ (5.151)

The probability of symbol error, assuming message m_i is transmitted, is then given by

$$P_s(E/m_i) = 1 - \int_0^\infty f_{\xi_i}(x_i) \prod_{\substack{m=0 \\ m \neq i}}^{M-1} \left(1 - Q\left(\frac{2E_s}{N_0} f_\rho^2(i, m), x_i\right)\right) dx_i$$ (5.152a)

and the unconditional probability of symbol error becomes

$$P_s(E) = \frac{1}{M} \sum_{i=0}^{M-1} P_s(E/m_i)$$ (5.152b)

In order to compute the bit error probability, one can assume that the signals remain orthogonal in the presence of frequency error, which facilitates the use of the well-known result, $P_b(E) = (M/2(M-1))P_s(E)$, relating bit and symbol error probabilities of orthogonal M-FSK. Unfortunately however, this assumption is not valid in the presence of frequency error and hence, the bit error probability results found in [20] are approximate. In fact, to properly evaluate the bit error probability in the presence of synchronization errors, one must specify an appropriate mapping, for example a Gray code, of the symbols to bits. In the perfectly synchronized case, the bit error probability performance is completely independent of the symbol-to-bit mapping since all errors are *equally likely* to occur. To compute the average bit error probability, one must first compute the probability of a *particular* symbol error for a given transmitted message. Analogous to (5.150), the probability of choosing m_k when message m_i is transmitted is given by

$$P_{sk}(E/m_i) = \text{Prob}\left\{\xi_k = \max_m \xi_m, \quad m = 0, 1, \cdots, M-1\right\}$$

$$= \int_0^\infty f_{\xi_k}(x_k) \left(\prod_{\substack{m=0 \\ m \neq k}}^{M-1} \int_0^{x_k} f_{\xi_m}(x_m) dx_m \right) dx_k, \quad k \neq i$$ (5.153)

If (k, i) denote the difference between the codewords (bit mapping) assigned to messages (symbols) m_i and m_k, that is, the number of bits in which the two differ, then the average bit error probability is as given by (4.86), that is

$$P_b(E) = \frac{1}{M \log_2 M} \sum_{i=0}^{M-1} \sum_{\substack{k=0 \\ k \neq i}}^{M-1} (k, i) P_{sk}(E/m_i) \tag{5.154}$$

It is clear that if a symbol error occurs, it is more likely to occur in an adjacent frequency than in any other. Thus, a Gray code mapping, in which two codewords from two adjacent symbols differ by only one bit, is appropriate to this type of modulation. Figure 5.18 depicts the average bit error probability versus E_b/N_0 in dB with ρ as a parameter for binary, 4-ary and 8-ary FSK and the conventional Gray code assignment given in Table 4.1.

5.2.6 Effect of Timing Error on Orthogonal *M*-FSK Detection

When the received frequency is known precisely but the symbol epoch is not, the receiver integrates using its own estimate of the symbol epoch which is offset from the true epoch by Δt sec. This phenomenon is depicted in Fig. 5.19 where the integration overlaps two successive symbol intervals. This results in signal attenuation in the detector matched to the incoming frequency and moreover, loss of orthogonality due to signal spillover into the remaining detectors [8]. The received signal can be modeled as

$$r(t) = \begin{cases} \sqrt{2P} \cos(2\pi(f_c + f_i)t + \theta) + n(t), & 0 \leq t \leq T \\ \sqrt{2P} \cos(2\pi(f_c + f_j)t + \theta) + n(t), & T \leq t \leq 2T \end{cases} \tag{5.155}$$

where we assumed that message m_i is transmitted followed by message m_j and that phase continuity is maintained. Since the local epoch estimate is not perfect, the receiver integrators operate from Δt to $\Delta t + T$ to obtain at the kth detector (setting $\theta = 0$ without loss of generality)

$$z_{ck} = \int_{\Delta t}^{\Delta t + T} r(t)\sqrt{2P} \cos(2\pi(f_c + f_k)t)dt$$

$$= 2P \left\{ \int_{\Delta t}^{T} \cos(2\pi(f_c + f_i)t) \cos(2\pi(f_c + f_k)t)dt \right. \tag{5.156}$$

$$\left. + \int_{T}^{\Delta t + T} \cos(2\pi(f_c + f_j)t) \cos(2\pi(f_c + f_k)t)dt \right\} + n_{ck}$$

and

$$z_{sk} = \int_{\Delta t}^{\Delta t + T} r(t)\sqrt{2P} \sin(2\pi(f_c + f_k)t)dt$$

$$= 2P \left\{ \int_{\Delta t}^{T} \cos(2\pi(f_c + f_i)t) \sin(2\pi(f_c + f_k)t)dt \right. \tag{5.157}$$

$$\left. + \int_{T}^{\Delta t + T} \cos(2\pi(f_c + f_j)t) \sin(2\pi(f_c + f_k)t)dt \right\} + n_{sk}$$

Normalizing z_{ck} and z_{sk} by $1/\sigma$, the pdf of ξ_k can be expressed as

$$f_{\xi_k}(\xi) = \xi \exp\left\{ -\left(\frac{\xi^2}{2} + \frac{(2E_s/N_0)f_\lambda^2(i, j, k)}{2} \right) \right\} I_0\left(\xi \sqrt{\frac{2E_s}{N_0}} f_\lambda^2(i, j, k) \right) \tag{5.158}$$

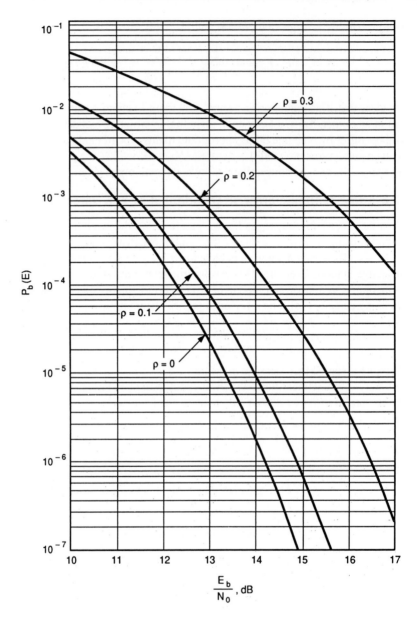

Figure 5.18a   Bit error probability of M-FSK with frequency error, $M = 2$

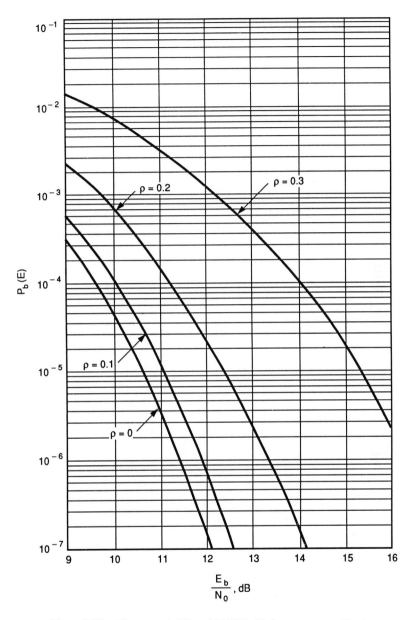

Figure 5.18b   Bit error probability of M-FSK with frequency error, $M = 4$

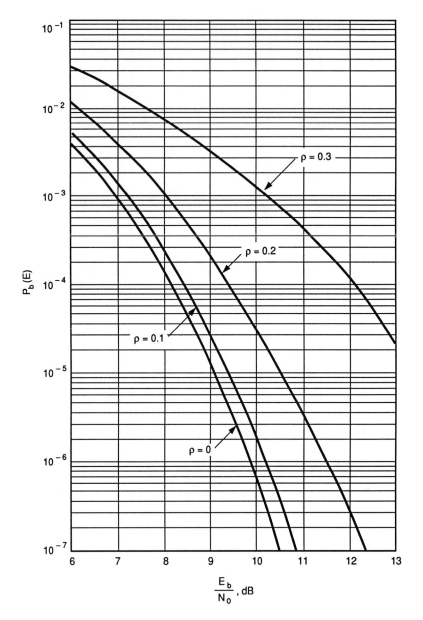

Figure 5.18c   Bit error probability of M-FSK with frequency error, $M = 8$

where

$$f_\lambda^2(i, j, k) = \frac{1}{\pi^2(i-k)^2} \sin^2[\pi(i-k)(1-\lambda)] + \frac{1}{\pi^2(j-k)^2} \sin^2[\pi(j-k)\lambda]$$

$$- \frac{1}{\pi^2(i-k)(j-k)} \sin^2\{\pi[(i-j)-(j-k)\lambda]\}$$

$$- \frac{1}{\pi^2(i-k)(j-k)} \sin^2\{\pi[(i-k)\lambda - (j-k)]\}$$ (5.159a)

$$+ \frac{1}{\pi^2(i-k)(j-k)} \sin^2\{\pi[(k-j)-(j-i)\lambda]\}, \quad i \neq k, j \neq k$$

$$f_\lambda^2(i, i, k) = \frac{1}{\pi^2(i-k)^2} \sin^2 \pi(i-k)$$

$$= \begin{cases} 1, & i = k \\ 0, & i \neq k \end{cases}$$ (5.159b)

$$f_\lambda^2(i, j, i) = \frac{1}{\pi^2(i-j)^2} \left\{ \pi^2(i-j)^2(1-\lambda)^2 + \sin^2(\pi\lambda(i-j)) \right.$$

$$\left. + \pi(i-j)(1-\lambda)\sin(2\pi(i-j)(\lambda+1)) \right\} \quad i \neq j$$ (5.159c)

$$f_\lambda^2(i, k, k) = \frac{1}{\pi^2(i-k)^2} \left\{ \sin^2(\pi(i-k)(1-\lambda)) \right.$$

$$+ \pi^2\lambda^2(i-k)^2$$

$$\left. - \pi\lambda(i-k)\sin(2\pi\lambda(i-k)) \right\}, \quad i \neq k$$ (5.159d)

$$f_\lambda^2(i, i, i) = 1, \quad \forall i$$ (5.159e)

and λ denotes the normalized timing error given by

$$\lambda \triangleq \frac{\Delta t}{T}$$ (5.160)

As a check, we have

$$\lim_{i \to j} f_\lambda^2(i, j, i) = \lim_{x \to 0} \frac{1}{4\pi^2 x^2} \left\{ 2 + 4\pi^2 x^2(1-\lambda)^2 - 2\cos(2\pi x\lambda) \right.$$

$$\left. + 4\pi x(1-\lambda)\sin(2\pi x(\lambda+1)) - 4\pi x(1-\lambda)\sin(2\pi x) \right\}$$ (5.161a)

which is 0/0 ! Using L'Hopital's rule once, we have

$$\lim_{i \to j} f_\lambda^2(i, j, i) = \lim_{x \to 0} \frac{1}{8\pi^2 x} \left\{ 8\pi^2 x(1-\lambda)^2 + 2(2\pi\lambda)\sin(2\pi x\lambda) \right.$$

$$+ 4\pi(1-\lambda)(\sin(2\pi x(\lambda+1)) + x2\pi(\lambda+1)\cos(2\pi x(\lambda+1)))$$ (5.161b)

$$\left. - 4\pi(1-\lambda)(\sin(2\pi x) - x(2\pi)\cos(2\pi x)) \right\}$$

which is still 0/0 ! Applying L'Hopital's rule one more time, we obtain

$$\lim_{i \to j} f_\lambda^2(i, j, i) = \lim_{x \to 0} \frac{1}{8\pi^2} \left\{ 8\pi^2(1 - \lambda)^2 + 4\pi(1 - \lambda) + 4\pi(\lambda + 1) \right.$$

$$\left. - 4\pi(1 - \lambda)4\pi + 8\pi^2\lambda^2 \right\}$$

$$= 1$$

(5.161c)

as expected. Using a similar procedure, it can be shown that

$$\lim_{i \to k} f_\lambda^2(i, k, k) = 1 \tag{5.162}$$

From (5.152), the probability of symbol error, assuming message m_i is sent followed by message m_j, is then given by

$$P_s(E/m_i, m_j) = 1 - \int_0^\infty x_i \exp\left\{ -\left(\frac{x_i^2}{2} + \frac{(2E_s/N_0)f_\lambda^2(i, j, i)}{2} \right) \right\}$$

$$\times I_0\left(x_i \sqrt{\frac{2E_s}{N_0} f_\lambda^2(i, j, i)} \right)$$

$$\times \prod_{\substack{m=0 \\ m \neq i}}^{M-1} \left(1 - Q\left(\frac{2E_s}{N_0} f_\lambda^2(i, j, m), x_i \right) \right) dx_i$$

(5.163)

The unconditional probability of symbol error, $P_s(E)$, is then

$$P_s(E) = \frac{1}{M} \sum_{i=0}^{M-1} P_s(E/m_i)$$

$$= \frac{1}{M} \sum_{i=0}^{M-1} \left[\frac{1}{M} \sum_{j=0}^{M-1} P_s(E/m_i, m_j) \right]$$

(5.164)

As was the case for frequency error, the presence of timing error produces a lack of orthogonality which results in the symbols not being equally likely. Hence to compute the average bit error probability, one must once again compute the probability of a *particular* symbol error for a given transmitted message. Analogous to (5.153), the probability of choosing m_k when message m_i was sent followed by message m_j, is given by

$$P_{sk}(E/m_i, m_j) = \int_0^\infty x_i \exp\left\{ -\left(\frac{x_i^2}{2} + \frac{E_s}{N_0} f_\lambda^2(i, j, k) \right) \right\} I_0\left(x_i \sqrt{\frac{2E_s}{N_0} f_\lambda^2(i, j, k)} \right)$$

$$\times \prod_{\substack{m=0 \\ m \neq k}}^{M-1} \left[1 - Q\left(\frac{2E_s}{N_0} f_\lambda^2(i, j, m), x_i \right) \right] dx_i$$

(5.165)

with $f_\lambda(i, j, m)$ as in (5.159). Finally, the average bit error probability is analogous to (5.154)

$$P_b(E) = \frac{1}{M^2 \log_2 M} \sum_{i=0}^{M-1} \sum_{j=0}^{M-1} \sum_{\substack{k=0 \\ k \neq i}}^{M-1} (k, i) P_{sk}(E/m_i, m_j) \tag{5.166}$$

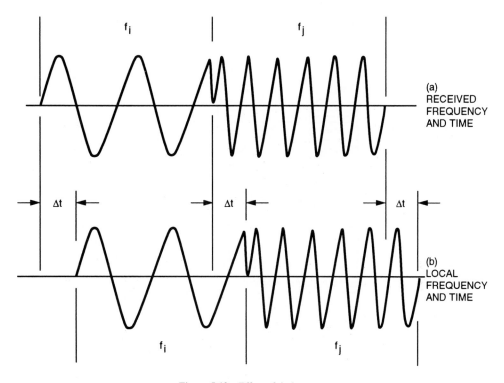

Figure 5.19   Effect of timing error

Here again, the evaluation of (5.166) will depend on the mapping of the symbols to bits. For the conventional Gray code mapping of Table 4.1, Fig. 5.20 plots the average bit error probability versus E_b/N_0 in dB for binary, 4-ary and 8-ary FSK with λ as a parameter. Digital computer simulations were used to confirm some of the results.

5.2.7 Effect of Timing and Frequency Errors on Orthogonal *M*-FSK Detection

When both the incoming frequency and the symbol epoch are not known [21], the received signal is still given by (5.155) with f_c replaced by $f_c + \Delta f$, as depicted in Fig. 5.21. The inphase and quadrature integrator outputs become

$$
\begin{aligned}
z_{ck} = 2P \Bigg\{ &\int_{\Delta t}^{T} \cos(2\pi(f_c + \Delta f + f_i)t)\cos(2\pi(f_c + f_k)t)dt \\
&+ \int_{T}^{\Delta t + T} \cos(2\pi(f_c + \Delta f + f_j)t)\cos(2\pi(f_c + f_k)t)dt \Bigg\} + n_{ck}
\end{aligned}
\tag{5.167}
$$

and

$$
\begin{aligned}
z_{sk} = 2P \Bigg\{ &\int_{\Delta t}^{T} \cos(2\pi(f_c + \Delta f + f_i)t)\sin(2\pi(f_c + f_k)t)dt \\
&+ \int_{T}^{\Delta t + T} \cos(2\pi(f_c + \Delta f + f_j)t)\sin(2\pi(f_c + f_k)t)dt \Bigg\} + n_{sk}
\end{aligned}
\tag{5.168}
$$

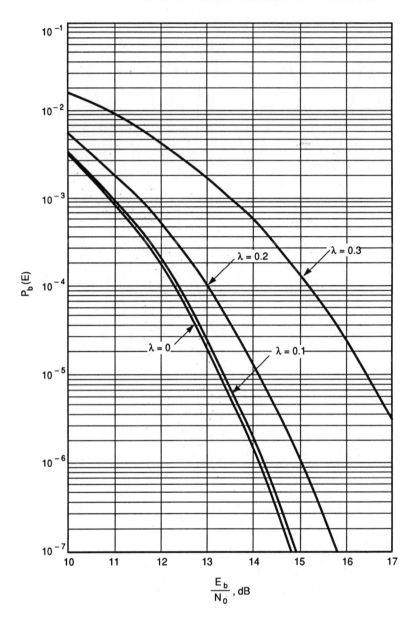

Figure 5.20a   Bit error probability of M-FSK with timing error, $M = 2$

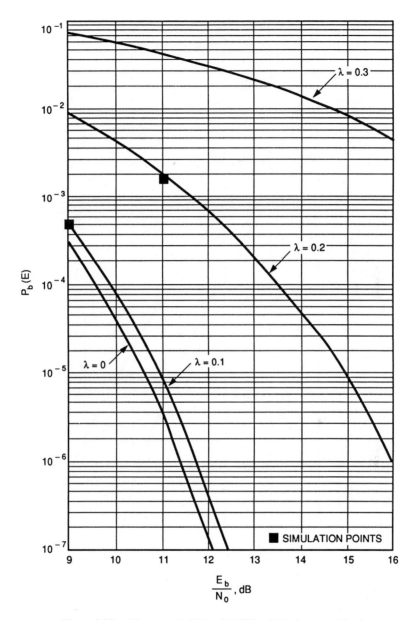

Figure 5.20b   Bit error probability of M-FSK with timing error, $M = 4$

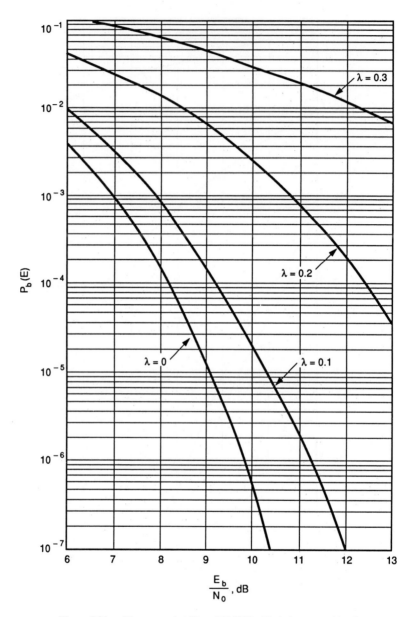

Figure 5.20c   Bit error probability of M-FSK with timing error, $M = 8$

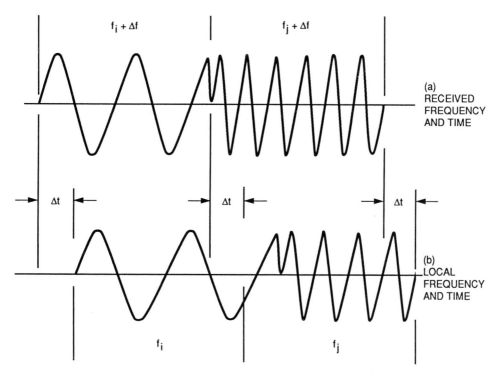

Figure 5.21   Effect of both frequency and timing errors

Normalizing as before and following a similar procedure, we have

$$f_{\xi_k}(\xi) = \xi \exp\left\{-\left(\frac{\xi^2}{2} + \frac{(2E_s/N_0)f_{\rho,\lambda}^2(i,j,k)}{2}\right)\right\} I_0\left(\xi\sqrt{\frac{2E_s}{N_0}f_{\rho,\lambda}^2(i,j,k)}\right) \quad (5.169)$$

where

$$\begin{aligned}
f_{\rho,\lambda}^2(i,j,k) =\ & \frac{1}{\pi^2(i-k+\rho)^2}\sin^2\left[\pi(i-k+\rho)(1-\lambda)\right] \\
& + \frac{1}{\pi^2(j-k+\rho)^2}\sin^2\left[\pi(j-k+\rho)\lambda\right] \\
& + \frac{1}{2\pi^2(i-k+\rho)(j-k+\rho)}\Big\{\cos(2\pi(i-j-(j-k)\lambda-\rho\lambda)) \\
& \quad - \cos(2\pi(i-j)) - \cos(2\pi((i-j)\lambda-j-k-\rho)) \\
& \quad + \cos(2\pi((i-k+\rho)\lambda-(j-k+\rho)))\Big\}, \quad \rho \neq 0, \lambda \neq 0
\end{aligned} \qquad (5.170)$$

If $\rho = 0$, then $f_{\rho,\lambda}^2(i,j,k)|_{\rho=0}$ reduces to $f_\lambda^2(i,j,k)$ as expected. Similarly, if $\lambda = 0$, then $f_{\rho,\lambda}^2(i,j,k)|_{\lambda=0}$ reduces to $f_\rho^2(i,k)$. The probability of symbol error is still given by (5.163) and (5.164) with $f_\lambda^2(i,j,k)$ replaced by $f_{\rho,\lambda}^2(i,j,k)$. The probability of bit error is still given by (5.165) and (5.166) with $f_\lambda^2(i,j,k)$ replaced by $f_{\rho,\lambda}^2(i,j,k)$. For the conventional Gray code mapping, Fig. 5.22 depicts the average bit error probability versus E_b/N_0 in dB for binary, 4-ary and 8-ary FSK with both ρ and λ as parameters. Digital computer simulations were used to confirm some of the results. Note that when timing and frequency errors occur simultaneously, the losses are not additive.

5.2.8 A Bound on the Performance of Orthogonal *M*-FSK Detection in the Presence of Frequency Error

A number of years back, Jim K. Omura developed a Chernoff-type bound on the error probability performance of certain types of M-ary communication systems. This unpublished result has particular application in noncoherent M-FSK communications.

In this section, we apply a slightly generalized version of the bound to predicting the error probability performance of noncoherent M-FSK with frequency error. Upper union bounds for this performance have been previously obtained in [22] in terms of the exact result for the performance of binary ($M = 2$) FSK with frequency error (see Problem 5.7). The latter is expressed in terms of the Marcum's Q function, which in general is cumbersome to compute. Here, we shall derive a simple-to-compute bound on this performance that will enable system comparisons to be readily made. The result is obtained in a form that is similar to the exact error probability performance of noncoherent M-FSK with no frequency error.

Since, as mentioned above, Omura's bound was never published but rather privately communicated to the authors, Appendix 5C presents the derivation of the bound in its generalized form. Assuming that message m_n is transmitted, then the detector matched to f_n produces ξ_n with pdf as given by (5.146), that is

$$f_{\xi_n}(\xi) = \xi \exp\left\{ -\left(\frac{\xi^2}{2} + \frac{(2E_s/N_0)f_\rho^2(n, n)}{2} \right) \right\} I_0\left(\xi\sqrt{\frac{2E_s}{N_0}f_\rho^2(n, n)} \right) \qquad (5.171)$$

while the remaining $M - 1$ detectors produce independent ξ_i's with

$$f_{\xi_i}(\xi) = \xi \exp\left\{ -\left(\frac{\xi^2}{2} + \frac{(2E_s/N_0)f_\rho^2(n, i)}{2} \right) \right\} I_0\left(\xi\sqrt{\frac{2E_s}{N_0}f_\rho^2(n, i)} \right) \qquad (5.172)$$

As required by the results in Appendix 5C, we need to evaluate the characteristic functions of the pdfs in (5.171) and (5.172). In particular, letting $\beta_i = \xi_i^2$, then

$$g_1(x) \triangleq \int_{-\infty}^{\infty} e^{x\xi^2} f_{\xi_n}(\xi)d\xi$$

$$= \frac{1}{1 - 2x} \exp\left\{ \frac{2E_s}{N_0} \frac{f_\rho^2(n, n)x}{1 - 2x} \right\} \qquad (5.173)$$

and

$$g_0\left(x; f_\rho^2(n, i) \right) \triangleq \int_{-\infty}^{\infty} e^{x\xi^2} f_{\xi_i}(\xi)d\xi$$

$$= \frac{1}{1 - 2x} \exp\left\{ \frac{2E_s}{N_0} \frac{f_\rho^2(n, i)x}{1 - 2x} \right\} \qquad (5.174)$$

The bound on $P_s(E/m_n)$ is given by (5C.8) and repeated here for convenience

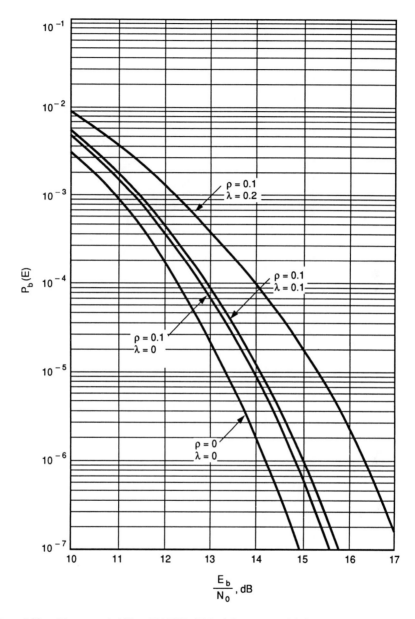

Figure 5.22a   Bit error probability of M-FSK with both frequency and timing errors, $M = 2$ and $\rho = 0.1$

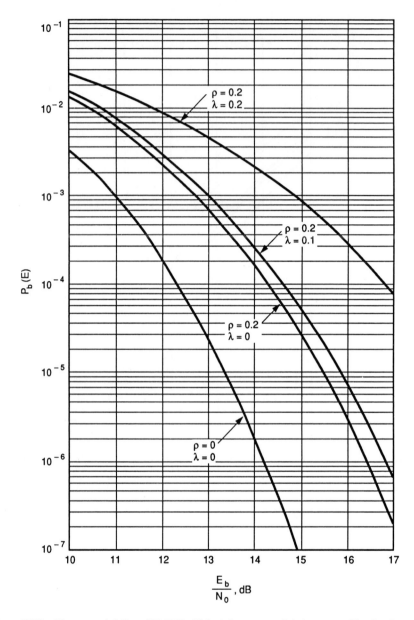

Figure 5.22b   Bit error probability of M-FSK with both frequency and timing errors, $M = 2$ and $\rho = 0.2$

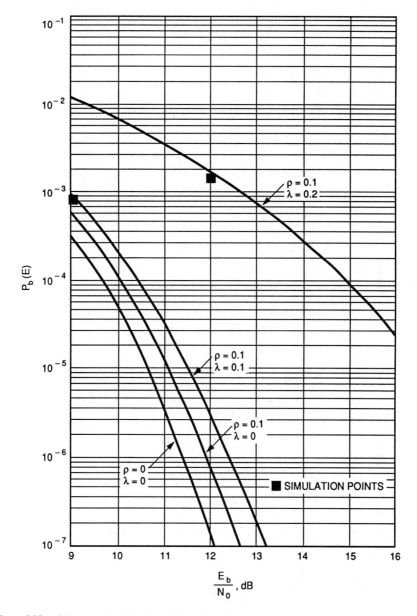

Figure 5.22c   Bit error probability of M-FSK with both frequency and timing errors, $M = 4$ and $\rho = 0.1$

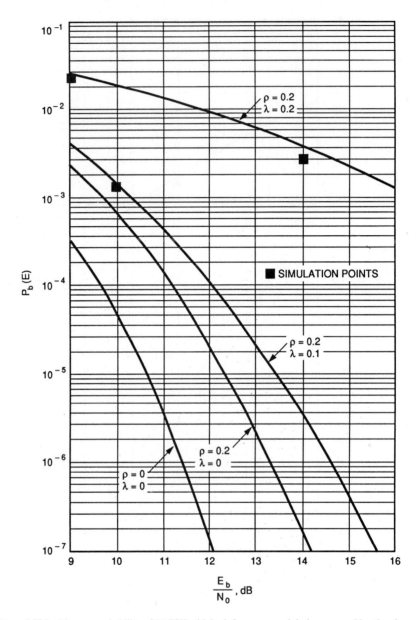

Figure 5.22d Bit error probability of M-FSK with both frequency and timing errors, $M = 4$ and $\rho = 0.2$

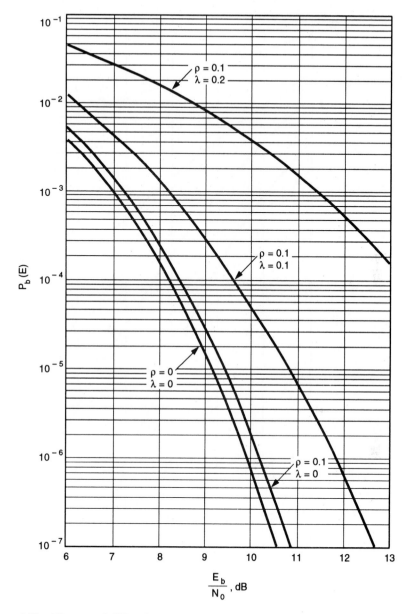

Figure 5.22e   Bit error probability of M-FSK with both frequency and timing errors, $M = 8$ and $\rho = 0.1$

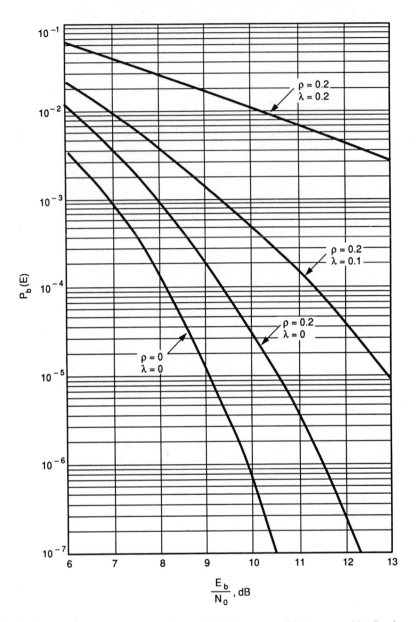

Figure 5.22f   Bit error probability of M-FSK with both frequency and timing errors, $M = 8$ and $\rho = 0.2$

$$P_s(E/m_n) \leq \min_{\lambda_n \geq 0} \sum_{k=1}^{M-1} (-1)^{k+1} g_1(-\lambda_n k)$$

$$\times \underbrace{\sum_{i_1=0}^{M-1} \sum_{i_2=0}^{M-1} \cdots \sum_{i_k=0}^{M-1}}_{\substack{i_1, i_2, \cdots, i_k \neq n \text{ and} \\ i_1 < i_2 < \cdots < i_k}} \prod_{j=1}^{k} g_0(\lambda_n; \zeta_{i_j n}) \tag{5.175}$$

Letting $x_0 = 2x$ and using (5.173) and (5.174) in (5.175) gives after some manipulations

$$P_s(E/m_n) \leq \min_{x_0 \geq 0} \sum_{k=1}^{M-1} (-1)^{k+1} \left(\frac{1}{1-x_0}\right)^k \left(\frac{1}{1+kx_0}\right)$$

$$\times \exp\left\{-\frac{E_s}{N_0} \frac{kx_0 f_\rho^2(n,n)}{1+kx_0}\right\}$$

$$\times \underbrace{\sum_{i_1=0}^{M-1} \sum_{i_2=0}^{M-1} \cdots \sum_{i_k=0}^{M-1}}_{\substack{i_1 < i_2 < \cdots < i_k \text{ and} \\ i_1, i_2, \cdots, i_k \neq n}} \exp\left\{\frac{E_s}{N_0}\left(\frac{x_0}{1-x_0}\right) \sum_{j=1}^{k} f_\rho^2(n, i_j)\right\} \tag{5.176}$$

Finally, the desired upper bound on $P_s(E)$ is given by

$$P_s(E) \leq \frac{1}{M} \sum_{n=0}^{M-1} P_s(E/m_n) \tag{5.177}$$

Figure 5.23 illustrates the upper bound on $P_s(E)$ as given by (5.177) versus bit energy-to-noise ratio E_b/N_0 for $M = 4$ and various values of ρ (the normalized frequency error), assuming minimum spacing for orthogonality. It is to be emphasized that the results in Fig. 5.23 are upper bounds and thus should not be used to predict the true error probability performance. Rather, their value is for making system comparisons and trade-offs since the relative tightness of the bound to the exact result should be about the same in all cases considered. As is true for most Chernoff-type bounds, they are asymptotically loose by about 1 to 1.5 dB. For other bounds, see [29].

5.2.9 Performance of Two Suboptimum Receivers

The spectrum analyzer receiver with autocorrelation function truncation. Here, we consider the error probability performance of the spectrum analyzer receiver when spectral estimates are evaluated by means of time domain truncation of the autocorrelation function. This approach reduces the amount of computation time required but increases the error probability.

The simplest form of time domain truncation of the autocorrelation function is to apply a rectangular window to the function; that is, compute only $\phi_{v_a}(j/2B)$ for $j = 0, 1, 2, \cdots, J$ where $J \leq 2BT - 1$. When the correlation function is truncated in this manner, the computation of the approximate spectral observation $\Phi_{v_a}(k/T)$ is based on the expression

$$\phi_{v_a}(\tau; J) \triangleq \phi_{v_a}(\tau) W(\tau) \tag{5.178}$$

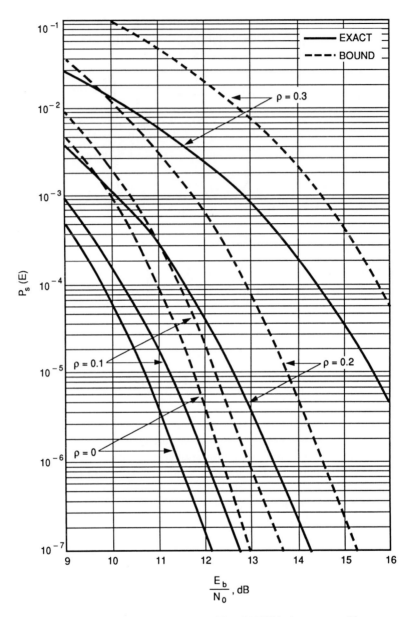

Figure 5.23 A bound on the symbol error probability of 4-FSK in the presence of frequency error

where $W(\tau)$ is the truncating window defined by

$$W(\tau) \triangleq \begin{cases} 1, & |\tau| \le (\frac{J+1}{2BT})T \\ 0, & \text{for all } \tau \end{cases} \tag{5.179}$$

Hence, from (5.51)

$$\Phi_v(f) \simeq \Phi_{v_a}(f; J) \triangleq \frac{\phi_{v_a}(0)}{2B} + \frac{1}{B} \sum_{j=1}^{J} \phi_{v_a}\left(\frac{j}{2B}\right) \cos\left(\frac{\pi j f}{B}\right), \quad |f| \le B \quad (5.180)$$

The quantity $1 - [(J+1)/2BT]$ is a measure of the fractional truncation. With truncation, the frequency separation $f_{i,j}$ (between tones i and j) must be adjusted so that they remain envelope orthogonal at the receiver. Thus

$$f_{i,j} = \frac{1}{T}\left(\frac{2BT}{J+1}\right) \quad (5.181)$$

yields orthogonal signals when truncation is employed. Hence, the minimum total transmission bandwidth B_{tr} required for M tones must now satisfy the inequality

$$B_{tr} = \frac{M}{T}\left(\frac{2BT}{J+1}\right) \ge \frac{M}{T} \quad (5.182)$$

Using a result from the theory of sequences of j-dependent random variables [23], Springett and Charles [7] are able to approximate the pdf of the spectral estimates $\Phi_{v_a}(f_l; J)$, $l = 0, 1, \cdots, M-1$ and hence the error probability performance under the assumption $P/N_0B \ll 1$ and $J + 1 \ll 2BT$. Under these conditions and assuming that the signal corresponding to f_k was in fact transmitted, the pdf of $\Phi_{v_a}(f; J)$ at $f = f_l$ is Gaussian with mean

$$\mu_l \triangleq \{\Phi_{v_a}(f_l; J)\} = \frac{T}{2B}[N_0B + R_{s_k}(0)]$$
$$+ 2\sum_{j=1}^{J}\left[\frac{2BT - j}{(2B)^2}\right] R_{s_k}(j) \cos\left(2\pi f_l\left(\frac{j}{2B}\right)\right) \quad (5.183)$$

where

$$R_{s_k}(j) \triangleq P \cos\left[2\pi f_k\left(\frac{j}{2B}\right)\right] \quad (5.184)$$

The variance of $\Phi_{v_a}(f_j; J)$, that is

$$\sigma_l^2 \triangleq E\{\Phi_{v_a}^2(f_l, J)\} - E^2\{\Phi_{v_a}(f_l; J)\} \quad (5.185)$$

is quite tedious to evaluate and hence we outline only the major steps. The first term in (5.185) can be written as a double sum, namely

$$E\left\{\Phi_{v_a}^2(f_l; J)\right\} = \sum_{j=0}^{J}\sum_{h=0}^{J} \epsilon_j \epsilon_h \left(\frac{1}{2B}\right)^2$$
$$\times E\left\{\phi_{v_a}(j)\phi_{v_a}(h)\right\} \cos 2\pi f_l\left(\frac{j}{2B}\right) \cos 2\pi f_l\left(\frac{h}{2B}\right) \quad (5.186)$$

where $\epsilon_0 = 1$ and $\epsilon_n = 2$ $(n \ne 0)$. The elements $E\{\phi_{v_a}(j)\phi_{v_a}(h)\}$ are now considered for the following four cases: (1) $j = h = 0$; (2) $j = 0, h \ne 0$; (3) $j = h \ne 0$; and (4) $j \ne h \ne 0$,

$j < h$. The results are:

$E\{\phi_{v_a}(j)\phi_{v_a}(h)\}$

$$
= \begin{cases}
\left(\frac{T}{2B}\right)(N_0B)^2\left[2BT\left(1+\frac{E_s}{N_0}\delta\right)^2 + 2\left(1+2\frac{E_s}{N_0}\delta\right)\right], & j=h=0 \\[2mm]
\frac{2BT-h}{(2B)^2}(N_0B)\left[2BT\left(1+\frac{E_s}{N_0}\delta\right)+4\right]R_{s_k}(h), & j=0, h\neq 0 \\[2mm]
\frac{2BT-h}{(2B)^2}(N_0B)^2\left\{1+\left[(2BT-h)\left(\frac{E_s}{N_0}\delta\right)^2+4\frac{E_s}{N_0}\delta\right]\frac{R_{s_k}^2(h)}{R_{s_k}^2(0)}\right. \\[2mm]
\quad \left. -\left(\frac{(E_s/N_0)\delta}{2BT-h}\right)\left[2h\left(\frac{R_{s_k}(2h)}{R_{s_k}(0)}-\frac{(E_s/N_0)\delta}{2}\right)\right]\right\}, & j=h\neq 0 \\[2mm]
\frac{2BT-h}{(2B)^2}(N_0B)^2\left\{\left[(2BT-j)\left(\frac{E_s}{N_0}\delta\right)^2+2\frac{E_s}{N_0}\delta\left(2+\frac{E_s}{N_0}\delta\right)\right]\frac{R_{s_k}(h)R_{s_k}(j)}{R_{s_k}^2(0)}\right. \\[2mm]
\quad \left. -\left(\frac{j(E_s/N_0)\delta}{2BT-h}\right)\left(2+\frac{E_s}{N_0}\delta\right)\frac{R_{s_k}(j+h)}{R_{s_k}(0)}\right\}, & j\neq h\neq 0, j<
\end{cases}
$$

(5.187)

where

$$\delta \triangleq \frac{1}{BT} \tag{5.188}$$

The variance of the spectral estimates can now be obtained by weighing each j, h element of the covariance matrix by $\epsilon_j\epsilon_h(1/2B)^2[\cos 2\pi f_l(j/2B)\cos 2\pi f_l(h/2B)]$, summing the resulting $(J+1)^2$ elements and subtracting the square of the mean. Although closed form expressions can be obtained for both (5.183) and (5.185), it is simpler to evaluate them numerically on a digital computer. Assuming that the M Gaussian distributed spectral estimates $\Phi_{v_a}(f_0; J)$, $\Phi_{v_a}(f_1; J)$, $\cdots$, $\Phi_{v_a}(f_{M-1}; J)$ are uncorrelated, an expression for the probability of correct detection is given by

$$P_s(C) = \frac{1}{\sqrt{\pi}}\int_{-\infty}^{\infty}\exp\left(-x^2\right)\left\{\frac{1}{2}\text{erfc}\left[-\frac{\sigma_k}{\sigma_l}x-\frac{\mu_k-\mu_l}{\sqrt{2}\sigma_l}\right]\right\}^{M-1}dx \tag{5.189}$$

where the μ_k's and σ_k's are defined in (5.183) and (5.185), respectively. Also in (5.189), any value of $l \neq k$ is suitable since μ_l and σ_l are independent of l for $l \neq k$ (see Problem 5.8). Numerical integration of $P_s(E) = 1 - P_s(C)$ was carried out in [7] for various values of the normalized window width $(J+1)/2BT$ in percent. The results are illustrated in Figures 5.24 and 5.25 where $P_s(E)$ is plotted versus E_b/N_0, for $M = 2^n$, $n = 1$, and 5 bits. Although the results were plotted for small percentage truncations (including no truncation), one should be suspect of applying these results to a practical system, since the assumptions regarding the independence and Gaussian nature of the spectral estimates break down when $J \to 2BT$. Comparing the no-truncation curve of Figs. 5.24 and 5.25 with the ideal noncoherent results of Fig. 5.11, one observes that the approximation breaks down in such a way as to give a pessimistic estimate of the correct answer.

The bandpass filter receiver.   Let's consider the performance of the suboptimum receiver (binary case) when the integrate-and-dump filters are replaced by bandpass filters as illustrated in Fig. 5.10. The two possible transmitted signals are given by

$$s_1(t) = \sqrt{2P}\cos\left[(\omega_c - \frac{\Delta\omega}{2})t + \theta\right]; \quad s_0(t) = \sqrt{2P}\cos\left[(\omega_c + \frac{\Delta\omega}{2})t + \theta\right] \tag{5.190}$$

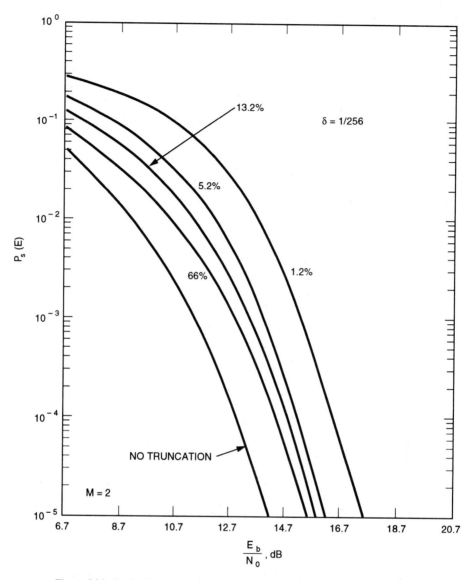

Figure 5.24   Probability of error performance for the spectrum analyzer receiver with autocorrelation function truncation; $n = 1$ (Reprinted from [7])

The bandpass filters in Fig. 5.10 each have a noise bandwidth $B \ll f_c$ and are centered, respectively, at the frequencies corresponding to the two possible transmitted signals. Assuming further that $\Delta\omega$ is large enough (B is small enough) that the BPFs don't spectrally overlap, then the noises at the filter outputs are uncorrelated (see Problem 5.9). Finally, assume that B is also wide enough to assume negligible intersymbol interference (ISI).

We express the noises at the filter outputs as narrowband expansions around the appropriate bandpass frequency. Thus

$$r(t) = s_i(t) + n_i(t), \quad i = 0, 1 \tag{5.191}$$

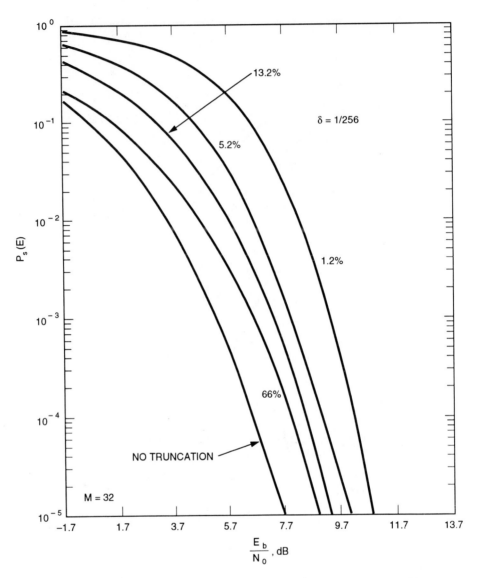

Figure 5.25   Probability of error performance for the spectrum analyzer receiver with autocorrelation function truncation; $n = 5$ (Reprinted from [7])

where

$$n_1(t) = n_{c1}(t) \cos\left[\left(\omega_c - \frac{\Delta\omega}{2}\right)t + \theta\right] - n_{s1}(t) \sin\left[\left(\omega_c - \frac{\Delta\omega}{2}\right)t + \theta\right] \quad (5.192a)$$

$$n_0(t) = n_{c0}(t) \cos\left[\left(\omega_c + \frac{\Delta\omega}{2}\right)t + \theta\right] - n_{s0}(t) \sin\left[\left(\omega_c + \frac{\Delta\omega}{2}\right)t + \theta\right] \quad (5.192b)$$

Rewriting (5.191) in polar form gives

$$r(t) = \xi_i(t) \cos\left[\left(\omega_c \pm \frac{\Delta\omega}{2}\right)t + \theta + \psi_i(t)\right], \quad i = 0, 1 \tag{5.193}$$

where

$$\xi_i(t) = \sqrt{(\sqrt{2P} + n_{ci}(t))^2 + (n_{si}(t))^2} \tag{5.194}$$

$$\psi_i(t) = \tan^{-1} \frac{n_{si}(t)}{\sqrt{2P} + n_{ci}(t)} \tag{5.195}$$

and the "$-$" sign in the "$\pm$" notation is associated with the subscript $i = 1$ and the "$+$" sign with the subscript $i = 0$. The process $\xi_i(t)$ is the envelope of $r(t)$ (assuming message m_i is transmitted) and any sample of it has a Rician pdf, namely

$$f_{\xi_i}(\xi) = \begin{cases} \frac{\xi}{\sigma^2} \exp\left(-\frac{\xi^2 + 2P}{2\sigma^2}\right) I_0(\frac{\sqrt{2P}\xi}{\sigma^2}), & \xi \geq 0 \\ 0, & \text{otherwise} \end{cases} \tag{5.196}$$

where $\sigma^2 = N_0 B$ is the noise power at the output of either BPF. Since the noises $n_1(t)$ and $n_0(t)$ are uncorrelated (and thus independent since they are Gaussian), ξ_1 and ξ_0 are independent random variables, that is, at any sample time, the envelopes of the processes $r_1(t)$ and $r_0(t)$ are independent.

To compute the probability of error, we first compute the conditional probability of error assuming $s_1(t)$ was transmitted. Under this assumption, BPF$_0$ has no signal in it. Thus, $f_{\xi_0}(x)$ should be computed with $P = 0$, yielding the Rayleigh pdf

$$f_{\xi_0}(x) = \begin{cases} \frac{x}{\sigma^2} \exp\left(-\frac{x^2}{2\sigma^2}\right), & x \geq 0 \\ 0, & \text{otherwise} \end{cases} \tag{5.197}$$

The probability of choosing $s_0(t)$ when indeed $s_1(t)$ was transmitted is the probability that $\xi_1 - \xi_0$ is less than zero (the threshold) at the sample time, that is

$$P_s(E/m_1) = \text{Prob}\{\xi_1 - \xi_0 < 0\} = \text{Prob}\{\xi_0 > \xi_1\}$$
$$= \int_0^\infty f_{\xi_1}(y) \int_y^\infty f_{\xi_0}(x)|_{P=0} dx dy \tag{5.198}$$

Similarly if $s_0(t)$ were transmitted, then $f_{\xi_1}(y)$ should be computed with $P = 0$ and the conditional error probability would be identical to (5.198) with the subscripts "1" and "0" reversed. Thus, because of the symmetry of the problem, the two conditional probabilities are equal and the average error probability $P_s(E)$ is equal to either one of them. Substituting (5.196) and (5.197) into (5.198) gives

$$P_s(E) = \int_0^\infty \frac{y}{\sigma^2} \exp\left(-\frac{y^2 + 2P}{2\sigma^2}\right) I_0\left(\frac{\sqrt{2P}\,y}{\sigma^2}\right) \underbrace{\int_y^\infty \frac{x}{\sigma^2} \exp\left(-\frac{x^2}{2\sigma^2}\right) dx}_{\exp(-y^2/2\sigma^2)} dy$$

$$= \exp\left(-\frac{P}{\sigma^2}\right) \underbrace{\int_0^\infty \frac{y}{\sigma^2} \exp\left(-\frac{y^2}{\sigma^2}\right) I_0\left(\frac{\sqrt{2P}\,y}{\sigma^2}\right) dy}_{(1/2)\exp(P/2\sigma^2)} \qquad (5.199)$$

$$= \frac{1}{2} \exp\left(-\frac{P}{2\sigma^2}\right) = \frac{1}{2} \exp\left(-\frac{1}{2}\text{SNR}_{BP}\right)$$

where SNR_{BP} is the ratio of the signal power P to the bandpass (input) noise power $N_0 B$. Comparing (5.199) with (5.79), we observe that the suboptimum implementation gives "essentially" the same performance if the bandpass bandwidth B of the filters in Fig. 5.10 is chosen equal to $1/T$, that is, the bandpass noise bandwidth of an integrate-and-dump filter. By "essentially," we imply that we are ignoring the portion of the data spectrum that falls outside the bandpass bandwidth B.

5.3 NONCOHERENT RECEIVERS IN RANDOM AMPLITUDE AND PHASE CHANNELS

In some communication channels, the amplitude of the received signal is random and unknown to the receiver. In this case, the optimum decision rule maximizes the likelihood of the signal over all possible amplitudes and phases. Sometimes, the amplitude and phase are independent variables. However, in other cases, they are correlated. We will provide a model and a physical intuition of the statistics of the amplitude and phase for both Rayleigh and Rician channels. But first, let's consider the following problem, which will be related to both channels later. Assume, when message m_k is transmitted, that the received signal is given by

$$r(t) = \sum_{i=1}^J g_i s_{i,k}(t) + n(t), \quad 0 \le t \le T \qquad (5.200)$$

where the g_i's are independent Gaussian random variables each with mean μ_i and variance σ_i^2 and $n(t)$ is AWGN with one-sided PSD N_0 Watts/Hz. We further assume that the various signals $\{s_{1,j}(t), s_{2,j}(t), \cdots, s_{J,j}(t)\}$ are orthogonal with equal energy E_j, which is a function of the message m_j, that is

$$\int_0^T s_{i,j}(t) s_{l,j} dt = E_j \delta_{i,l} \qquad (5.201)$$

where $\delta_{i,l}$ is the Kronecker delta function. The MAP decision rule selects message m_j, which maximizes the general likelihood function $\Lambda(r(t))$, given by

$$\Lambda(r(t)) = \underbrace{\int_{-\infty}^\infty \cdots \int_{-\infty}^\infty}_{J \text{ integrals}} f_{G_1}(g_1) \cdots f_{G_J}(g_J) \exp\left\{ \frac{2}{N_0} \int_0^T r(t) \sum_{i=1}^J g_i s_{i,j}(t) dt \right.$$

$$\left. - \frac{1}{N_0} \int_0^T \sum_{i=1}^J \sum_{l=1}^J g_i g_l s_{i,j}(t) s_{l,j}(t) dt \right\} \qquad (5.202)$$

The exponential term is the likelihood function assuming a perfectly known signal in AWGN. The general likelihood function is its average over the unknown parameters, the Gaussian amplitudes in this specific instance. Defining

$$W_{i,j} \triangleq \int_0^T r(t)s_{i,j}(t)dt, \, i = 1, 2, \ldots, J \text{ and } j = 0, 1, \ldots, M - 1 \qquad (5.203)$$

then

$$\Lambda(r(t)) = \prod_{i=1}^J \int_{-\infty}^\infty f_{G_i}(g_i) \exp\left\{\frac{2}{N_0} g_i W_{i,j} - \frac{g_i^2}{N_0} E_j\right\} dg_i \qquad (5.204)$$

Substituting for the Gaussian density, (5.204) can be written as

$$\Lambda(r(t)) = \prod_{i=1}^J \frac{1}{\sqrt{2\pi}\sigma_i} \int_{-\infty}^\infty e^{h(g_i)} dg_i \qquad (5.205)$$

where $h(g_i)$ is the argument of the exponential function and is given by

$$h(g_i) = -\frac{1}{2\sigma_i^2}\left\{g_i^2\left(1 + \frac{2\sigma_i^2}{N_0}E_j\right) - g_i\left(2\mu_i + 4\frac{\sigma_i^2}{N_0}W_{i,j}\right)\right\} - \frac{\mu_i^2}{2\sigma_i^2} \qquad (5.206)$$

Using the integral [12]

$$\int_{-\infty}^\infty \exp\left\{-p^2 x^2 \pm qx\right\} dx = \frac{\sqrt{\pi}}{p} \exp\left\{\frac{q^2}{4p^2}\right\} \qquad (5.207)$$

then (5.205) becomes

$$\Lambda(r(t)) = \prod_{i=1}^J \sqrt{2\pi}\sigma_i \left\{\frac{\sqrt{\pi}}{p_{i,j}} \exp\left(\frac{q_{i,j}^2}{4p_{i,j}}\right)\right\} e^{-\mu_i^2/2\sigma_i^2} \qquad (5.208)$$

where

$$p_{i,j}^2 \triangleq \frac{1}{2\sigma_i^2}\left(1 + \frac{2\sigma_i^2}{N_0}E_j\right) \qquad (5.209)$$

$$q_{i,j} \triangleq \frac{1}{2\sigma_i^2}\left(2\mu_i + 4\frac{\sigma_i^2}{N_0}W_{i,j}\right) \qquad (5.210)$$

Hence, the MAP decision rule selects message m_j, which maximizes

$$\prod_{i=1}^J \left(\frac{1}{p_{i,j}}\right) \exp\left(\frac{q_{i,j}^2}{4p_{i,j}}\right) \qquad (5.211)$$

Assuming that the signals have equal energy, then $E_j = E$ and as a result, $p_{i,j} = p_i$. In this case, we need to only maximize

$$\prod_{i=1}^J \exp\left\{\frac{q_{i,j}^2}{4p_i}\right\} \qquad (5.212)$$

Taking the natural logarithm, we have

$$\sum_{i=1}^{J} \frac{q_{i,j}^2}{4p_i} = \sum_{i=1}^{J} \frac{\left(\frac{1}{2\sigma_i^2}\right)^2 \left(2\mu_i + 4\frac{\sigma_i^2}{N_0} W_{i,j}\right)^2}{4\left(\frac{1}{2\sigma_i^2}\right)\left(1 + 2\frac{\sigma_i^2}{N_0} E\right)} \tag{5.213}$$

or alternatively, the decision rule reduces to maximizing

$$\sum_{i=1}^{J} \sigma_i^2 \frac{\left(\frac{\mu_i}{2\sigma_i^2} + \frac{1}{N_0} W_{i,j}\right)^2}{\left(\frac{N_0}{2E} + \sigma_i^2\right)} \tag{5.214}$$

As a special case, let the means be zero (that is, $\mu_i = 0$ for all i.) Then the MAP rule reduces to maximizing

$$\sum_{i=1}^{J} W_{i,j}^2 \left(\frac{\sigma_i^2}{\frac{N_0}{2E} + \sigma_i^2}\right) \tag{5.215}$$

On the other hand, when the means are nonzero but the variances are equal (that is, $\sigma_i^2 = \sigma^2$), then we need to maximize

$$\sum_{i=1}^{J} \left(\frac{\mu_i}{2\sigma^2} + \frac{1}{N_0} W_{i,j}\right)^2 \tag{5.216}$$

Both rules given by (5.215) and (5.216) will be very useful in deriving the optimum detection structures in Rayleigh and Rician channels, respectively.

5.3.1 Optimum Receivers in Rayleigh Channels

In Rayleigh channels, the received signal, when message m_k is transmitted, is given by

$$r(t) = r\sqrt{2}a_k(t)\cos(\omega_c t + \phi_k(t) + \theta) + n(t), \quad 0 \le t \le T \tag{5.217}$$

where r and θ are independent random variables, r Rayleigh distributed [that is, $f_R(r) = (r/\sigma^2)\exp(-r^2/2\sigma^2), r \ge 0$], and θ uniformly distributed over $(0, 2\pi)$. In real physical channels, the fade $r(t)$ is a random process and its statistics vary with time. In slowly fading channels, the fade is assumed constant over a symbol duration and can thus be denoted by r. At this point, it would be very helpful to examine the scenarios or circumstances under which the signal model given by (5.217) reflects the actual output of a physical channel. Consider the case where a "large" number of scatterers exist in the propagation path and are furthermore randomly located, as depicted in Fig. 5.26. Let the received signal from the lth scatterer be (no noise case)

$$r_l(t) = c_l\sqrt{2}a_k(t - \tau_l)\cos\left(\omega_c(t - \tau_l) + \phi_k(t - \tau_l)\right) \tag{5.218}$$

with attenuation c_l and delay τ_l. The composite received signal is the sum of the signals from all scatterers, that is

$$r(t) = \sum_{\text{all } l} r_l(t) \tag{5.219}$$

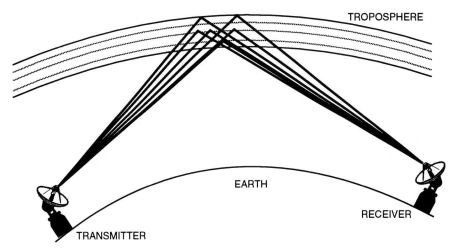

Figure 5.26   Scattering of the transmitted signal by the troposphere

Assuming that the delays $\{\tau_l\}$ are all small relative to the time constant (which is approximately the inverse of the bandwidth) of the complex signals $\{a_k(t)\exp(j\phi_k(t))\}$, then $r(t)$ can be approximated by

$$r(t) \simeq \sqrt{2}a_k(t)\sum_{\text{all } l} c_l \cos\left(\omega_c(t-\tau_l)+\phi_k(t)\right) \qquad (5.220)$$

As an example, consider the case of a baseband BPSK signal. The bandwidth of the signal is roughly $1/T$ sec where T is the symbol duration. Hence, its time constant is approximately the symbol period, which is intuitive since the signal is constant during a symbol period. If the delays from the various scatterers are fractions of the symbol period T (that is, $\tau_l \ll T$), then for all practical purposes, the delays can be ignored without any significant distortion of the signal. The received signal of (5.220) can be alternatively written as

$$r(t) \simeq \sqrt{2}a_k(t)\sum_{\text{all } l} c_l \cos\left(\omega_c t +\phi_k(t)-\phi_l\right) \qquad (5.221)$$

where $\phi_l \triangleq \omega_c \tau_l$. If the delays are comparable to $2\pi/\omega_c$, the phases can not be ignored and are essentially independent (each is a function of its respective path) and uniformly distributed over $(0, 2\pi)$. Expanding the cosine with trigonometric identities, $r(t)$ can be written as

$$r(t) \simeq r_c\sqrt{2}a_k(t)\cos(\omega_c t +\phi_k(t)) + r_s\sqrt{2}a_k(t)\sin(\omega_c t +\phi_k(t)) \qquad (5.222)$$

where

$$r_c = \sum_{\text{all } l} c_l \cos\phi_l \qquad (5.223a)$$

$$r_s = \sum_{\text{all } l} c_l \sin\phi_l \qquad (5.223b)$$

Using the independence of the $\{\phi_l\}$, it is easily shown that

$$E\{r_c\} = E\{r_s\} = 0 \qquad (5.224)$$

$$\sigma_{r_c}^2 = \sigma_{r_s}^2 = \frac{1}{2} \sum_{\text{all } l} E\{c_l^2\} = \sigma^2 \tag{5.225}$$

and

$$E\{r_c r_s\} = 0 \tag{5.226}$$

Assuming that the number of scatterers is "large," it can be argued via the central limit theorem that r_c and r_s are Gaussian, independent with zero-mean and equal variance σ^2. Rewriting $r(t)$ as

$$r(t) = r\sqrt{2}a_k(t) \cos(\omega_c t + \phi_k(t) + \theta) \tag{5.227}$$

where

$$r = \sqrt{r_c^2 + r_s^2} \quad \text{and} \quad \theta = \tan^{-1} \frac{r_s}{r_c} \tag{5.228}$$

then from Appendix 5A, r is Rayleigh distributed and independent of θ, which is uniformly distributed over $(0, 2\pi)$. Thus, the Rayleigh model is the result of a channel that has many independently located scatterers in the propagation path.

In order to derive the optimum decision rule in a Rayleigh channel, consider the signal model of (5.222) in the presence of noise, namely

$$r(t) = r_c\sqrt{2}a_k(t) \cos(\omega_c t + \phi_k(t)) + r_s\sqrt{2}a_k(t) \sin(\omega_c t + \phi_k(t)) + n(t) \tag{5.229}$$

where $n(t)$ is AWGN with one-sided PSD N_0 Watts/Hz. Comparing (5.229) with (5.200), the optimum decision rule for equal energy signals can be written by inspection of (5.216), with $\mu_i = 0$, $J = 2$ and $s_{1,i}(t) = \sqrt{2}a_i(t) \cos(\omega_c t + \phi_i(t))$, $s_{2,i}(t) = \sqrt{2}a_i(t) \sin(\omega_c t + \phi_i(t))$; namely, the receiver sets $\widehat{m}(r(t)) = m_k$ if and only if

$$\xi_i^2 = W_{1,i}^2 + W_{2,i}^2$$

$$= \left(\int_0^T r(t)\sqrt{2}a_i(t) \cos(\omega_c t + \phi_i(t))dt \right)^2 \tag{5.230}$$

$$+ \left(\int_0^T r(t)\sqrt{2}a_i(t) \sin(\omega_c t + \phi_i(t))dt \right)^2$$

is maximum for $i = k$. Note that $W_{1,i} = z_{ci}$ and $W_{2,i} = z_{si}$ where z_{ci}, z_{si} are defined by (5.14). The decision rule of (5.230) is identical to the optimum detector in noncoherent channels with a fixed amplitude. The optimum receiver can thus be alternatively derived by noting that the noncoherent detector with a known amplitude is the optimum noncoherent detector for random amplitude channels, if we condition on a known amplitude. Once we condition on the amplitude, it can be ignored as it does not contribute to the maximization process and as a result, averaging over its distribution does not alter the structure of the optimum detector. Note however that even though the structure of the detector is not altered, its performance will be a function of the amplitude pdf, as will be shown shortly.

5.3.2 Optimum Receivers in Rician Channels

In numerous fading channels, there is a direct line of sight between the transmitter and the receiver. As a result, the observed signal is composed of a fixed (or specular) component in

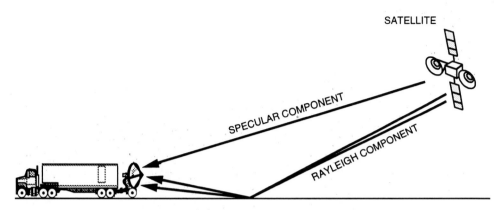

Figure 5.27   Example of a Rician channel

addition to the Rayleigh component, as depicted in Fig. 5.27. Such channels are referred to as Rician and the received signal, assuming message m_k is transmitted, is given by

$$r(t) = \alpha\sqrt{2}a_k(t)\cos(\omega_c t + \phi_k(t) + \delta)$$
$$+ r\sqrt{2}a_k(t)\cos(\omega_c t + \phi_k(t) + \theta) + n(t), \quad 0 \le t \le T \tag{5.231}$$

where α and δ are the amplitude and phase of the specular signal. At this point, we differentiate between two distinct receiver structures, namely, the coherent and noncoherent receivers. In both scenarios, we assume that α is known and fixed. In the coherent case, the phase of the fixed component δ is estimated perfectly while in the noncoherent structure, δ is not tracked. In either case, the phase of the Rayleigh component θ is unknown and assumed uniformly distributed. We will thus derive both coherent and noncoherent receiver structures and compare their performances.

Coherent detection.   When the phase δ is known (that is, coherent detector), the received signal can be expressed as

$$r(t) = r_c\sqrt{2}a_k(t)\cos(\omega_c t + \phi_k(t)) + r_s\sqrt{2}a_k(t)\sin(\omega_c t + \phi_k(t)) + n(t) \tag{5.232}$$

where r_c and r_s are independent Gaussian random variables with nonzero means given by

$$E\{r_c\} = \alpha\cos\delta \quad \text{and} \quad E\{r_s\} = \alpha\sin\delta \tag{5.233}$$

and

$$\text{Var}\{r_c\} = \text{Var}\{r_s\} = \sigma^2 \tag{5.234}$$

Comparing (5.232) with (5.200), the optimum receiver can be written down immediately using (5.216) with $J = 2$, $\mu_1 = \alpha\cos\delta$, $\mu_2 = \alpha\sin\delta$, that is, set $\widehat{m}(r(t)) = m_k$ if and only if

$$\left(\frac{\alpha\cos\delta}{2\sigma^2} + \frac{1}{N_0}W_{1,i}\right)^2 + \left(\frac{1}{N_0}W_{2,i} + \frac{\alpha\sin\delta}{2\sigma^2}\right)^2 \tag{5.235}$$

is maximum for $k = i$, where $W_{1,i} = \int_0^T r(t)\sqrt{2}a_i(t)\cos(\omega_c t + \phi_i(t))dt$ and $W_{2,i} = \int_0^T r(t)\sqrt{2}a_i(t)\sin(\omega_c t + \phi_i(t))dt$. Equivalently, the decision rule can be modified to maximize

$$\frac{\alpha N_0}{\sigma^2}\int_0^T r(t)\sqrt{2}a_i(t)\cos(\omega_c t + \phi_i(t) - \delta)dt + W_{1,i}^2 + W_{2,i}^2 \tag{5.236}$$

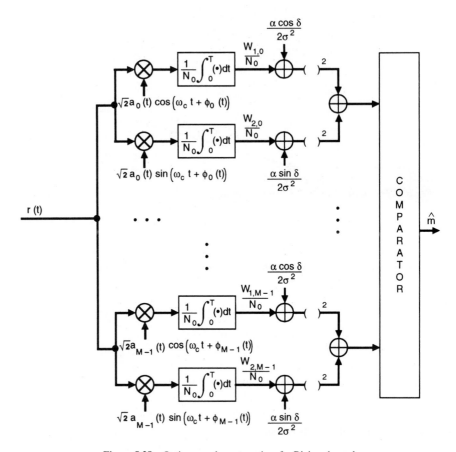

Figure 5.28   Optimum coherent receiver for Rician channels

which clearly shows a coherent and a noncoherent contribution. One implementation of coherent receivers in Rician channels is depicted in Fig. 5.28 and requires knowledge of $\sigma^2/\alpha N_0$. Keep in mind that the term "coherent" refers to the phase of the fixed component only. As expected when $\alpha = 0$ (that is, no specular component), the receiver reduces to the classical noncoherent receiver of Fig. 5.3.

Noncoherent detection.   In the noncoherent receiver, the phase δ is unknown and thus can be assumed uniformly distributed. In this case, the optimum decision rule is derived by expressing the received signal in the form of (5.200), by conditioning on δ. Then, the likelihood function conditioned on δ and message m_j (is transmitted) is given by (5.211) with $J = 2$, that is

$$\Lambda(r(t)/\delta, m_j) = \prod_{i=1}^{2} \left(\frac{1}{p_{i,j}}\right) \exp\left(\frac{q_{i,j}^2}{4p_{i,j}}\right) \tag{5.237}$$

Averaging over a uniform δ, (5.237) reduces to

$$\Lambda(r(t)/m_j) = K_1 e^{K_2\left(W_{1,j}^2 + W_{2,j}^2\right)} I_0\left(K_3\sqrt{W_{1,j}^2 + W_{2,j}^2}\right) \tag{5.238}$$

where K_1, K_2, and K_3 are positive constants and are independent of j for the equal energy, equilikely case. Since the likelihood function is a monotonic increasing function of the envelope (or equivalently, envelope squared), the optimum receiver makes its decision on the basis of samples of the envelope alone, that is, $W_{1,j}^2 + W_{2,j}^2$ (see Problem 5.10). The receiver need not know $\sigma^2/\alpha N_0$, which is a great practical advantage over a coherent receiver.

5.4 PERFORMANCE OF NONCOHERENT RECEIVERS IN RANDOM AMPLITUDE AND PHASE CHANNELS

We shall now evaluate the performance, in terms of probability of error, of the optimum receivers in Rayleigh and Rician channels. Note that since the received signal energy is random itself, we will evaluate the performance as a function of the "average" received energy E_{av}. Moreover, results are presented for orthogonal and, when possible, correlated signals.

5.4.1 Performance in Rayleigh Channels

M-ary orthogonal signals.   Since the optimum receiver in Rayleigh channels is identical to its noncoherent counterpart in AWGN channels (that is, both seek to maximize $\xi_i^2 = z_{ci}^2 + z_{si}^2$), then the performance of the receiver in Rayleigh channels, conditioned on a fixed fade r, is identical to that of noncoherent signaling in AWGN as given by (5.77) with E_s replaced by $r^2 E_s$, that is

$$P_s(E/r) = \sum_{m=1}^{M-1} (-1)^{m+1} {}_{M-1}C_m \left(\frac{1}{m+1} \right) \exp \left\{ -\frac{m}{m+1} \left(r^2 \frac{E_s}{N_0} \right) \right\} \quad (5.239)$$

Averaging over the Rayleigh fade r, we obtain

$$
\begin{aligned}
P_s(E) &= \int_0^\infty P_s(E/r) \frac{r}{\sigma^2} \exp \left(-\frac{r^2}{2\sigma^2} \right) dr \\
&= \sum_{m=1}^{M-1} {}_{M-1}C_m \frac{(-1)^{m+1}}{m+1+m(E_{av}/N_0)}
\end{aligned}
\quad (5.240)
$$

where $E_{av}/N_0 \triangleq 2\sigma^2(E_s/N_0)$. Note that $P_s(E)$ in Rayleigh channels is no longer an exponentially decreasing function of E_{av}/N_0 as in AWGN channels, but is a function of the inverse of E_{av}/N_0. This means that for a fading channel, the transmitter must provide a large amount of power in order to provide a reasonably low probability of error at the receiver. In many applications, the required transmitted power is not permissible for various reasons. An alternative solution is obtained by means of diversity techniques [24, 25], which insert redundancy in the channel. Depending on the nature of the fade, time, frequency, spatial diversity, or combinations thereof may be employed (see Problems 5.11 to 5.17).

For the binary case (that is, $M = 2$), $P_s(E)$ as given by (5.240) reduces to

$$P_s(E) = \frac{1}{2 + (E_{av}/N_0)}, \quad M = 2, \text{ Rayleigh Channel} \quad (5.241)$$

It would be interesting to assess the effect of signal fading, in the presence of perfect carrier coherency. As a result, assume that the receiver performs an exact estimate of the carrier phase and uses that information in the detection process. Then, the conditional probability of error, assuming orthogonal signals, is given by (4.59), that is

$$P_s(E/r) = Q\left(r\sqrt{\frac{E_s}{N_0}}\right) \tag{5.242}$$

and

$$P_s(E) = \int_0^\infty \frac{r}{\sigma^2} \exp\left(-\frac{r^2}{2\sigma^2}\right) \int_{r\sqrt{E_s/N_0}}^\infty \frac{1}{\sqrt{2\pi}} e^{-y^2/2} dy\, dr \tag{5.243}$$

Performing the integration, one obtains

$$P_s(E) = \frac{1}{2}\left[1 - \sqrt{\frac{E_{av}/N_0}{1 + E_{av}/N_0}}\right], \quad M = 2, \text{ Random Amplitude Channel} \tag{5.244}$$

and for comparison, we list the other two previously derived results for orthogonal signals, namely

$$P_s(E) = Q\left(\sqrt{\frac{E_s}{N_0}}\right), \quad M = 2, \text{ Coherent Channel} \tag{5.245}$$

and

$$P_s(E) = \frac{1}{2} \exp\left(-\frac{E_s}{2N_0}\right), \quad M = 2, \text{ Noncoherent Channel} \tag{5.246}$$

The performance of these four channels is depicted in Fig. 5.29. Note that at high E_s/N_0, coherent detection offers about 3 dB advantage over noncoherent detection. Simultaneously, detection in random amplitude channels (Rayleigh channel with perfect phase measurement) offers about a 3 dB gain over detection in random amplitude and phase channel.

Binary correlated signals. Other results have been derived in the literature for correlated signals in Rayleigh channels. Turin [26] has derived the performance of correlated signals for $M = 2$ and is given by

$$P_s(E) = \frac{1}{2}\left(1 - \frac{\mu\sqrt{1 - |\tilde{\rho}|^2}}{\sqrt{1 - \mu^2|\tilde{\rho}|^2}}\right) \tag{5.247}$$

with $\tilde{\rho} = \tilde{\rho}_{10}$ denoting the correlation between the complex baseband signals as given by (5.86) and

$$\mu \triangleq \frac{(E_{av}/N_0)}{(E_{av}/N_0) + 2} \tag{5.248}$$

5.4.2 Performance in Rician Channels

In Rician channels, two different receivers can be employed depending on whether the phase of the specular component is known or not.

Noncoherent detection. Let's consider the noncoherent receiver that selects the signal corresponding to maximum $\xi_i^2 = z_{ci}^2 + z_{si}^2$.

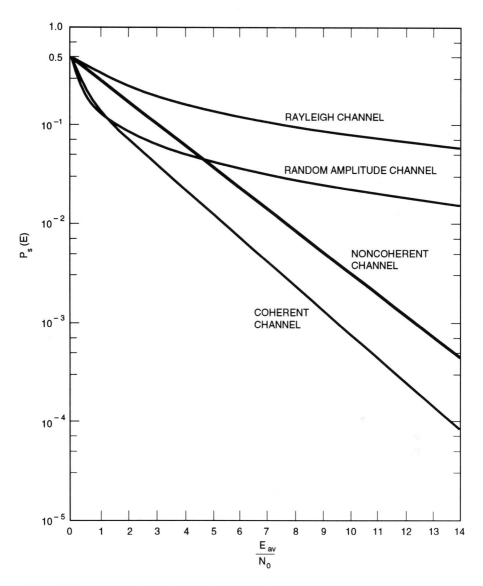

Figure 5.29   Probability of error for the coherent, noncoherent, random amplitude and Rayleigh channels

M-ary Orthogonal Signals.   Here again, if we condition on the fade r, then the performance is identical to (5.77). Averaging over the Rician variable r, one obtains

$$
P_s(E) = \sum_{m=1}^{M-1} (-1)^{m+1} {}_{M-1}C_m \frac{1}{m+1} \int_0^\infty \frac{r}{\sigma^2}
$$

$$
\times \exp\left\{ -\frac{1}{2\sigma^2} \left(r^2 + \alpha^2 \right) \right\} \exp\left\{ -\frac{m}{m+1} \left(r^2 \frac{E_s}{N_0} \right) \right\} I_0\left(\frac{\alpha r}{\sigma^2} \right) dr
$$

(5.249)

Using the integral in (5.74), then $P_s(E)$ becomes

$$P_s(E) = \sum_{m=1}^{M-1} (-1)^{m+1} {}_{M-1}C_m \frac{1}{m+1+m\beta} \exp\left\{-\frac{m\gamma^2\beta}{2(m+1+m\beta)}\right\} \qquad (5.250)$$

where β is the SNR in the scatterlike signal, that is

$$\beta \triangleq 2\sigma^2 \frac{E_s}{N_0} \qquad (5.251)$$

and γ^2 is twice the ratio of the power in the specular component to the power in the Rayleigh component, that is

$$\gamma^2 \triangleq \frac{\alpha^2}{\sigma^2} \qquad (5.252)$$

As a check, if $\gamma = 0$, the Rician channel reduces to a Rayleigh channel and (5.250) reduces to (5.240), as expected. Note that the average symbol SNR in a Rician channel is given by

$$\frac{E_{av}}{N_0} = \left(\alpha^2 + 2\sigma^2\right)\frac{E_s}{N_0} = \left(1 + \frac{\gamma^2}{2}\right)\beta \qquad (5.253)$$

which for $\gamma = 0$ reduces to β, the average symbol SNR in a Rayleigh channel. In the binary case, (5.250) reduces to

$$P_s(E) = \frac{1}{2+\beta} \exp\left\{-\frac{\gamma^2\beta}{2(\beta+2)}\right\}, \qquad M = 2 \text{ Rician Channel (Noncoherent)} \quad (5.254)$$

Here again, when $\gamma = 0$, (5.254) reduces to (5.241) as expected. The performance in Rician channels for $M = 2$ is depicted in Fig. 5.30 as a function of β for various values of γ. For $\gamma = 0$, the performance reduces to that in a Rayleigh channel. On the other hand, when $\gamma = \infty$, it reduces to the performance of binary orthogonal FSK in an AWGN channel.

Binary Correlated Signals. The performance of binary noncoherent reception in the presence of correlated signals has also been derived by Turin [26] and the probability of error is given by [with $\tilde{\rho}$ defined by (5.86)]

$$P_s(E) = Q(uv, wv) - \frac{1}{2}\left[1 + \frac{\mu\sqrt{1-|\tilde{\rho}|^2}}{\sqrt{1-\mu^2|\tilde{\rho}|^2}}\right]$$

$$\times \exp\left\{-\frac{(u^2+w^2)v^2}{2}\right\} I_0\left(uwv^2\right) \qquad (5.255)$$

where

$$u = \sqrt{1 - \frac{\sqrt{(1-|\tilde{\rho}|^2)(1-\mu^2|\tilde{\rho}|^2)}}{1-\mu|\tilde{\rho}|^2}} \qquad (5.256)$$

$$w = \sqrt{1 + \frac{\sqrt{(1-|\tilde{\rho}|^2)(1-\mu^2|\tilde{\rho}|^2)}}{1-\mu|\tilde{\rho}|^2}} \qquad (5.257)$$

$$v = \sqrt{\frac{\mu\gamma^2}{2} \frac{1-\mu|\tilde{\rho}|^2}{1-\mu^2|\tilde{\rho}|^2}} \qquad (5.258)$$

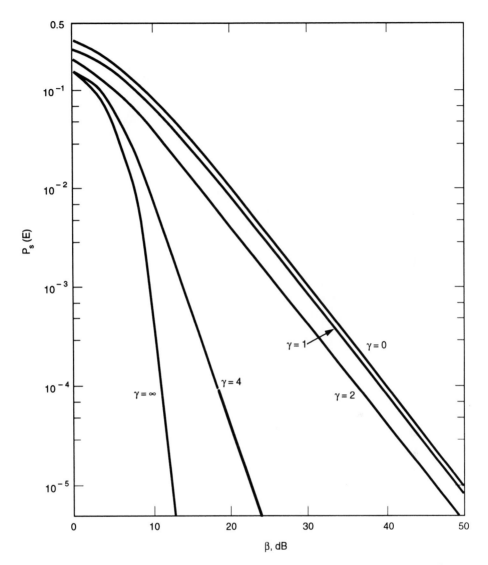

P_s (E)

β, dB

Figure 5.30   Probability of error for binary orthogonal signals, Rician channels (Reprinted from [27])

and $\mu \triangleq \beta/(\beta + 2)$. Note that when $|\tilde{\rho}| = 0$, $u = 0$, $w = \sqrt{2}$ and $v = \gamma\sqrt{\mu/2}$. Since $Q(0, x) = \exp(-x^2/2)$, (5.255) reduces to

$$P_s(E) = \exp\left\{-\frac{\gamma^2\mu}{2}\right\} - \frac{1}{2}(1 + \mu)\exp\left\{-\frac{\gamma^2\mu}{2}\right\}$$

$$= \frac{1}{\beta + 2}\exp\left\{-\frac{\gamma^2\beta}{2(\beta + 2)}\right\}$$

(5.259)

which is identical to (5.254). On the other hand, when $\gamma = 0$, $v = 0$ and (5.255) reduces to

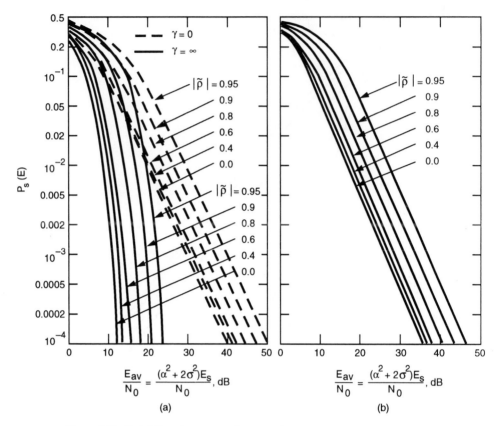

Figure 5.31 Probability of error, noncoherent receiver: (a) $\gamma = 0$ and $\gamma = \infty$, (b) $\gamma = 2$ (Reprinted from [26] © IEEE)

$$P_s(E) = Q(0, 0) - \frac{1}{2}\left[1 + \frac{\mu\sqrt{1 - |\tilde{\rho}|^2}}{\sqrt{1 - \mu^2|\tilde{\rho}|^2}}\right]$$

$$= \frac{1}{2}\left(1 - \frac{\mu\sqrt{1 - |\tilde{\rho}|^2}}{\sqrt{1 - \mu^2|\tilde{\rho}|^2}}\right) \tag{5.260}$$

which is identical to the performance of the Rayleigh channel as given by (5.247). The performance for noncoherent detection is depicted in Fig. 5.31 for $\gamma = 0, \infty$ and $\gamma = 2$ for various values of $|\tilde{\rho}|$, as a function of E_{av}/N_0. Note that orthogonal signals (that is, $|\tilde{\rho}| = 0$) provide the optimum performance for all γ and E_{av}/N_0.

Coherent detection.　In the coherent Rician receiver, the decision rule is modified by the addition of the appropriate biases as given by (5.235). Since the phase δ is known perfectly, it can be set to zero without any loss in generality.

Binary Orthogonal Signals.　For the binary case, the rule reduces to

$$\left(\frac{\alpha}{2\sigma^2} + \frac{W_{1,0}}{N_0}\right)^2 + \left(\frac{W_{2,0}}{N_0}\right)^2 \underset{m_1}{\overset{m_0}{\gtrless}} \left(\frac{\alpha}{2\sigma^2} + \frac{W_{1,1}}{N_0}\right)^2 + \left(\frac{W_{2,1}}{N_0}\right)^2 \tag{5.261}$$

where

$$\begin{pmatrix} W_{1,0} \\ W_{2,0} \end{pmatrix} = \int_0^T r(t)\sqrt{2}a_0(t) \begin{pmatrix} \cos \\ \sin \end{pmatrix} (\omega_c t + \phi_0(t)) dt \qquad (5.262)$$

and

$$\begin{pmatrix} W_{1,1} \\ W_{2,1} \end{pmatrix} = \int_0^T r(t)\sqrt{2}a_1(t) \begin{pmatrix} \cos \\ \sin \end{pmatrix} (\omega_c t + \phi_1(t)) dt \qquad (5.263)$$

Assuming that m_0 is transmitted, then from (5.232), $r(t)$ is given by

$$r(t) = r_c\sqrt{2}a_0(t)\cos(\omega_c t + \phi_0(t)) + r_s\sqrt{2}a_0(t)\sin(\omega_c t + \phi_0(t)) + n(t) \quad (5.264)$$

and $E\{r_c\} = \alpha$, $E\{r_s\} = 0$ with $\text{Var}\{r_c\} = \text{Var}\{r_s\} = \sigma^2$. Assuming that the signals are orthogonal, then

$$W_{1,0} = r_c E_s + n_1 \qquad (5.265a)$$

$$W_{2,0} = r_s E_s + n_2 \qquad (5.265b)$$

$$W_{1,1} = n_3 \qquad (5.265c)$$

$$W_{2,1} = n_4 \qquad (5.265d)$$

where n_i, $i = 1, 2, 3, 4$ are zero-mean Gaussian independent random variables with variances $N_0 E_s/2$. From Appendix 5A, we know that R_i is Rician where

$$R_i^2 \triangleq \left(\frac{\alpha}{2\sigma^2} + \frac{W_{1,i}}{N_0} \right)^2 + \left(\frac{W_{2,i}}{N_0} \right)^2, \quad i = 0, 1 \qquad (5.266)$$

with parameters

$$s_0^2 = \left(\frac{\alpha}{2\sigma^2} + \frac{E_s\alpha}{N_0} \right)^2 = \left(\frac{\alpha}{2\sigma^2} \right)^2 (1 + \beta)^2 \qquad (5.267a)$$

$$s_1^2 = \left(\frac{\alpha}{2\sigma^2} \right)^2 \qquad (5.267b)$$

and

$$\sigma_0^2 = \frac{1}{N_0^2} \left(E_s^2\sigma^2 + \frac{N_0 E_s}{2} \right) = \frac{E_s}{2N_0}(1 + \beta) \qquad (5.268a)$$

$$\sigma_1^2 = \frac{1}{N_0^2} \left(\frac{N_0 E_s}{2} \right) = \frac{E_s}{2N_0} \qquad (5.268b)$$

The conditional probability of error, $P_s(E/m_0)$, is then the probability that $R_1 > R_0$, which is given by (5.130) for two general Rician random variables. In our case, the parameters a and b from (5.131) become

$$a = \frac{s_1^2}{\sigma_0^2 + \sigma_1^2} = \frac{\gamma^2}{\beta(\beta + 2)} \qquad (5.269)$$

$$b = \frac{s_0^2}{\sigma_0^2 + \sigma_1^2} = \frac{\gamma^2(\beta + 1)^2}{\beta(\beta + 2)} \qquad (5.270)$$

and

$$\frac{\sigma_0^2}{\sigma_0^2 + \sigma_1^2} = \frac{\beta + 1}{\beta + 2} \tag{5.271}$$

Substituting all these results in (5.130) and noting that $P_s(E)$ is identical to $P_s(E/m_0)$ because of symmetry, then

$$P_s(E) = Q\left(\frac{\gamma}{\sqrt{\beta(\beta + 2)}}, \frac{\gamma(\beta + 1)}{\sqrt{\beta(\beta + 2)}}\right)$$
$$- \left(\frac{\beta + 1}{\beta + 2}\right) \exp\left\{-\frac{\gamma^2\left(\beta^2 + 2\beta + 2\right)}{2\beta(\beta + 2)}\right\} I_0\left(\gamma^2 \frac{\beta + 1}{\beta(\beta + 2)}\right) \tag{5.272}$$

which was also derived in [27].

Binary Correlated Signals. The performance of binary coherent reception in the presence of correlated signals has been derived by Turin [26] and is given by (5.255) with [$\tilde{\rho}$ defined by (5.86)]

$$u = 1 - \frac{\mu\sqrt{1 - |\tilde{\rho}|^2}}{\sqrt{1 - \mu^2|\tilde{\rho}|^2}} \tag{5.273}$$

$$w = 1 + \frac{\mu\sqrt{1 - |\tilde{\rho}|^2}}{\sqrt{1 - \mu^2|\tilde{\rho}|^2}} \tag{5.274}$$

and

$$v = \sqrt{\frac{\beta\gamma^2\left[\beta(1 - |\tilde{\rho}|^2) + 2(1 - \mathrm{Re}(\tilde{\rho}))\right]}{4\beta^2(1 - |\tilde{\rho}|^2)}} \tag{5.275}$$

Here again when $\tilde{\rho} = 0$, then $u = 1 - \mu = 2/(\beta + 2)$, $w = 1 + \mu = (2\beta + 2)/(\beta + 2)$, and $v = (\gamma/2)\sqrt{(\beta + 2)/\beta}$ and $P_s(E)$ reduces to (5.272) since $uv = \sqrt{a}$, $wv = \sqrt{b}$, and so on. The performance for coherent detection is depicted in Fig. 5.32 for $\gamma = 0, \infty$ and $\gamma = 2$. For comparison purposes, various curves from Fig. 5.31 and Fig. 5.32 are replotted in Fig. 5.33. The results indicate that when fading does exist, there is vary little gain obtained by the use of a coherent receiver. Of course, for a specular signal alone (that is, $\gamma = \infty$), the use of a coherent receiver can result in up to 4 dB improvement over a noncoherent receiver. But the performance of the coherent receiver for $\gamma = \infty$ is already excellent so that the extra advantage obtained from a coherent detection might not be needed.

In conclusion, a system that is to operate through fixed and fading paths will be for all practical purposes optimal if it is designed to use uncorrelated signals and noncoherent reception. The latter has the practical advantages that it does not require estimates of δ or $\sigma^2/\alpha N_0$ at the receiver.

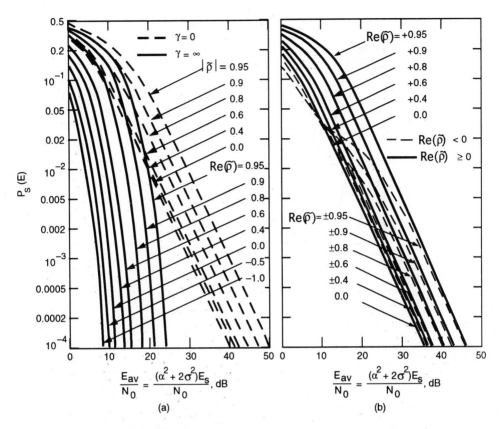

Figure 5.32   Probability of error, coherent receiver: (a) $\gamma = 0$ and $\gamma = \infty$, (b) $\gamma = 2$ (Reprinted from [26] © IEEE)

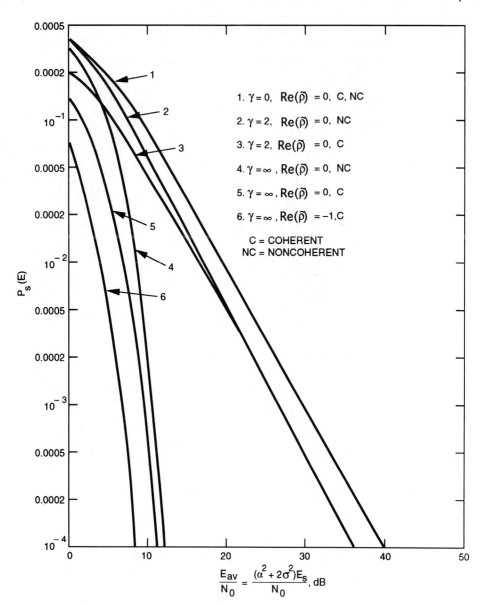

Figure 5.33 Comparison of probability of error curves for several important cases (Reprinted from [26] © IEEE)

Appendix 5A

Useful Probability Density Functions

In various optimum receivers, we encounter nonlinear operations on Gaussian random variables. Specifically, in noncoherent and later on in Chapter 6 partially coherent communications, the optimum receiver requires squaring operations on Gaussian or conditionally Gaussian random variables. In this appendix, we summarize key results in this area and define useful densities, namely, the Chi-square, the noncentral Chi-square, the Rayleigh and finally, the Rician density. A more detailed treatment of this topic can be found in general probability theory books [11]. Here, we merely present the results as taken from Proakis [17].

5A.1 CENTRAL CHI-SQUARE DENSITY

Define the random variable Y as

$$Y = \sum_{i=1}^{n} X_i^2 \tag{5A.1}$$

where the X_i, $i = 1, 2, \cdots, n$ are statistically independent and identically distributed (i.i.d.) Gaussian random variables with zero-mean and variance σ^2. Then, the characteristic function of Y is given by

$$\Psi_Y(j\omega) \triangleq E\left\{e^{j\omega Y}\right\} = \frac{1}{\left(1 - j2\omega\sigma^2\right)^{n/2}} \tag{5A.2}$$

which is the consequence of the statistical independence of the X_i's. The pdf of Y is obtained by the use of the inverse Fourier transform [5], that is

$$f_Y(y) = \frac{1}{\sigma^n 2^{n/2} \Gamma(n/2)} y^{n/2-1} e^{-y/2\sigma^2}, \quad y \geq 0 \tag{5A.3}$$

where $\Gamma(q)$ is the gamma function, defined as

$$\Gamma(q) = \int_0^\infty t^{q-1} e^{-t} dt, \quad q > 0$$

$$\Gamma(q) = (q-1)!, \quad q \text{ an integer and } q > 0 \tag{5A.4}$$

$$\Gamma(\tfrac{1}{2}) = \sqrt{\pi}, \quad \Gamma\left(\frac{3}{2}\right) = \frac{\sqrt{\pi}}{2}$$

This pdf is called the Gamma or Chi-square pdf with n degrees of freedom and is illustrated in Fig. 5A.1 for several values of n. The first and second moments of Y can be computed to give

$$E\{Y\} = n\sigma^2$$

$$E\{Y^2\} = 2n\sigma^4 + n^2\sigma^4 \tag{5A.5}$$

$$\sigma_Y^2 = 2n\sigma^4$$

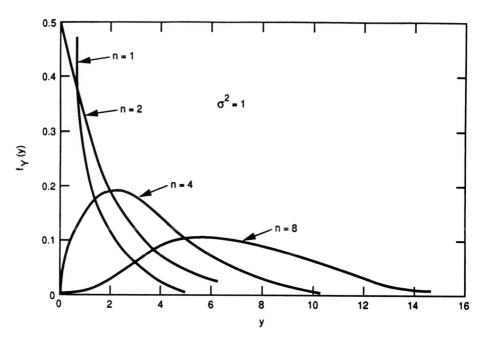

Figure 5A.1   Pdf of a chi-square-distributed random variable for several degrees of freedom

The cumulative distribution function (cdf) of Y is obtained by integrating (5A.3), that is

$$F_Y(y) \triangleq \text{Prob}\{Y \le y\} = \int_0^y \frac{1}{\sigma^n 2^{n/2} \Gamma(n/2)} u^{n/2-1} e^{-u/2\sigma^2} du \qquad (5A.6)$$

For the special case when n is even (let k=n/2), (5A.6) can be computed in closed form to give

$$F_Y(y) = 1 - e^{-y/2\sigma^2} \sum_{j=0}^{k-1} \frac{1}{j!} \left(\frac{y}{2\sigma^2}\right)^j \qquad (5A.7)$$

5A.2 NONCENTRAL CHI-SQUARE DENSITY

When the Gaussian random variables of (5A.1) have nonzero means m_i, $i = 1, 2, \cdots, n$ but still identical variances σ^2, the characteristic function of Y is no longer given by (5A.2) but becomes

$$\psi_Y(j\omega) = \frac{1}{(1 - j2\omega\sigma^2)^{n/2}} \exp\left\{\frac{j\omega \sum_{i=1}^n m_i^2}{1 - j2\omega\sigma^2}\right\} \qquad (5A.8)$$

As expected, (5A.8) reduces to (5A.2) when all the means are zero. The inverse Fourier transform of this characteristic function yields

$$f_Y(y) = \frac{1}{2\sigma^2} \left(\frac{y}{s^2}\right)^{(n-2)/4} \exp\left\{-\frac{s^2 + y}{2\sigma^2}\right\} I_{\frac{n}{2}-1}\left(\sqrt{y}\frac{s}{\sigma^2}\right), \quad y \ge 0 \qquad (5A.9)$$

where s^2 is the sum of means squared, that is

$$s^2 = \sum_{i=1}^{n} m_i^2 \tag{5A.10}$$

and $I_k(x)$ is the kth-order modified Bessel function of the first kind [3] which can be expressed by the infinite series [4]

$$I_k(x) = \sum_{j=0}^{\infty} \frac{(x/2)^{k+2j}}{j!\Gamma(k+j+1)}, \quad x \geq 0 \tag{5A.11}$$

The pdf, as given by (5A.9), is typically referred to as the noncentral Chi-square density with n degrees of freedom. The parameter s^2 is called the noncentrality parameter of the pdf. The cdf is obtained by integration as

$$F_Y(y) = \int_0^y \frac{1}{2\sigma^2} \left(\frac{u}{s^2}\right)^{(n-2)/4} \exp\left\{-\frac{s^2+u}{2\sigma^2}\right\} I_{\frac{n}{2}-1}\left(\sqrt{u}\frac{s}{\sigma^2}\right) du \tag{5A.12}$$

Here again, when n is even (k = n/2), the cdf can be expressed in terms of the generalized Q-function $Q_k(x, y)$ as

$$F_Y(y) = 1 - Q_k\left(\frac{s}{\sigma}, \frac{\sqrt{y}}{\sigma}\right) \tag{5A.13}$$

where

$$Q_k(a, b) \triangleq \int_b^{\infty} x \left(\frac{x}{a}\right)^{k-1} \exp\left\{-\frac{x^2+a^2}{2}\right\} I_{k-1}(ax)dx$$

$$= Q_1(a, b) + \exp\left\{-\frac{a^2+b^2}{2}\right\} \sum_{j=1}^{k-1} \left(\frac{b}{a}\right)^j I_j(ab) \tag{5A.14}$$

The function $Q_1(a, b)$ is typically denoted by $Q(a, b)$ [not to be confused with the Gaussian tail integral $Q(x)$ which is a function of one variable] and is referred to as the Marcum Q-function and is given by

$$Q(a, b) \triangleq \int_b^{\infty} x \exp\left\{-\frac{x^2+a^2}{2}\right\} I_0(ax)dx \tag{5A.15}$$

The first and second moments of Y can be computed in closed form to give

$$E\{Y\} = n\sigma^2 + s^2$$

$$E\{Y^2\} = 2n\sigma^4 + 4\sigma^2 s^2 + \left(n\sigma^2 + s^2\right)^2 \tag{5A.16}$$

$$\sigma_y^2 = 2n\sigma^4 + 4\sigma^2 s^2$$

which reduce to (5A.5) when the means are zero (that is, $s = 0$).

5A.3 RAYLEIGH DENSITY

Suppose we define a new random variable R which is

$$R \triangleq \sqrt{Y} = \sqrt{\sum_{i=1}^{n} X_i^2} \tag{5A.17}$$

where the X_i, $i = 1, 2, \cdots, n$ are as defined earlier and the means are zero. Then, the pdf of R can be easily obtained from the Chi-square pdf with a simple transformation to give

$$f_R(r) = \frac{r^{n-1}}{2^{(n-2)/2}\sigma^n \Gamma(n/2)} \exp\left\{-\frac{r^2}{2\sigma^2}\right\}, \quad r \geq 0 \tag{5A.18}$$

Here again, when n is even (let $k = n/2$), the cdf of R can be obtained in closed form as

$$F_R(r) = 1 - \exp\left\{-\frac{r^2}{2\sigma^2}\right\} \sum_{j=0}^{k-1} \frac{1}{j!} \left(\frac{r^2}{2\sigma^2}\right)^j, \quad r \geq 0 \tag{5A.19}$$

and the mth moment of R is given by

$$E\{R^m\} = \left(2\sigma^2\right)^{m/2} \frac{\Gamma((n+m)/2)}{\Gamma(n/2)}, \quad m \geq 0 \tag{5A.20}$$

for any integer m. As a special case of (5A.18) when $n = 2$, we obtain the familiar Rayleigh pdf

$$f_R(r) = \frac{r}{\sigma^2} \exp\left\{-\frac{r^2}{2\sigma^2}\right\}, \quad r \geq 0 \tag{5A.21}$$

with corresponding cdf

$$F_R(r) = 1 - \exp\left\{-\frac{r^2}{2\sigma^2}\right\}, \quad r \geq 0 \tag{5A.22}$$

5A.4 RICIAN DENSITY

When R is still defined by (5A.17) and the X_i's are nonzero means (each X_i has mean m_i) each with variance σ^2, then the pdf of R becomes

$$f_R(r) = \frac{r^{n/2}}{\sigma^2 s^{(n-2)/2}} \exp\left\{-\frac{r^2 + s^2}{2\sigma^2}\right\} I_{\frac{n}{2}-1}\left(\frac{rs}{\sigma^2}\right), \quad r \geq 0 \tag{5A.23}$$

with corresponding cdf for n even ($k = n/2$)

$$F_R(r) = 1 - Q_k\left(\frac{s}{\sigma}, \frac{r}{\sigma}\right) \tag{5A.24}$$

where s^2 is defined in (5A.10). The mth moment of R is

$$E\{R^m\} = \left(2\sigma^2\right)^{m/2} \exp\left\{-\frac{s^2}{2\sigma^2}\right\} \frac{\Gamma((n+m)/2)}{\Gamma(n/2)} {}_1F_1\left(\frac{n+m}{2}, \frac{n}{2}; \frac{s^2}{2\sigma^2}\right) \tag{5A.25}$$

where ${}_1F_1(\alpha, \beta; x)$ is the confluent hypergeometric function [3] given by

$${}_1F_1(\alpha, \beta; x) \triangleq \sum_{j=0}^{\infty} \frac{\Gamma(\alpha + j)\Gamma(\beta)x^j}{\Gamma(\alpha)\Gamma(\beta + j)j!}, \quad \beta \neq 0, -1, -2, \cdots \tag{5A.26}$$

When $n = 2$, (5A.23) reduces to the familiar Rician pdf

$$f_R(r) = \frac{r}{\sigma^2} \exp\left\{-\frac{s^2 + r^2}{2\sigma^2}\right\} I_0\left(\frac{rs}{\sigma^2}\right), \quad r \geq 0 \tag{5A.27}$$

where $s = \sqrt{m_1^2 + m_2^2}$ and corresponding cdf

$$F_R(r) = 1 - Q\left(\frac{s}{\sigma}, \frac{r}{\sigma}\right) \tag{5A.28}$$

When $s = 0$, the Rician pdf and cdf of (5A.27) and (5A.28) reduce to the Rayleigh pdf and cdf of (5A.21) and (5A.22), respectively.

5A.5 ON THE JOINT DENSITY OF AMPLITUDE AND PHASE

So far, we have been concentrating on the marginal densities of Y or R, basically ignoring any phase information. Let's define

$$R = \sqrt{X_1^2 + X_2^2} \quad \text{and} \quad \tan\Theta = \frac{X_2}{X_1} \tag{5A.29}$$

where X_1 and X_2 are i.i.d. Gaussian random variables with variance σ^2 and $m_1 = m$ but $m_2 = 0$. We already know that R is Rician distributed if $m \neq 0$ but Rayleigh distributed if $m = 0$. It would be interesting to derive the joint pdf $f_{R,\Theta}(r, \theta)$ for both cases. Since R and Θ are two functions of two random variables, it is easily shown [11] that

$$f_{R,\Theta}(r, \theta) = r f_{X_1, X_2}(r\cos\theta, r\sin\theta), \quad r \geq 0, -\pi \leq \theta \leq \pi$$

$$= \frac{r}{2\pi\sigma^2} \exp\left\{-\frac{1}{2\sigma^2}\left((r\cos\theta - m)^2 + (r\sin\theta)^2\right)\right\} \tag{5A.30}$$

or equivalently

$$f_{R,\Theta}(r, \theta) = \frac{r}{2\pi\sigma^2} \exp\left\{-\frac{1}{2\sigma^2}\left(r^2 - 2mr\cos\theta + m^2\right)\right\}, \quad r \geq 0, -\pi \leq \theta \leq \pi \tag{5A.31}$$

To obtain the marginal densities, we integrate the joint density

$$f_R(r) = \int_{-\pi}^{\pi} f_{R,\Theta}(r, \theta)d\theta = \frac{r}{2\pi\sigma^2} \exp\left\{-\frac{r^2 + m^2}{2\sigma^2}\right\} \int_{-\pi}^{\pi} \exp\left\{\frac{2mr\cos\theta}{2\sigma^2}\right\} d\theta$$

$$= \frac{r}{\sigma^2} \exp\left\{-\frac{r^2 + m^2}{2\sigma^2}\right\} I_0\left(\frac{mr}{\sigma^2}\right), \quad r \geq 0 \tag{5A.32}$$

which is the Rician pdf as expected with $s = |m|$. On the other hand, the marginal pdf of Θ becomes

$$f_\Theta(\theta) = \int_0^\infty f_{R,\Theta}(r, \theta)dr$$

$$= \frac{1}{2\pi\sigma^2} \exp\left\{-\frac{m^2}{2\sigma^2}\right\} \int_0^\infty r \exp\left\{-\frac{r^2 - 2mr\cos\theta}{2\sigma^2}\right\} dr \tag{5A.33}$$

Completing the square in the integral and then substituting $x = r - m \cos \theta$, (5A.33) reduces to

$$f_\Theta(\theta) = \frac{1}{2\pi\sigma^2} \exp\left\{-\frac{m^2 \sin^2 \theta}{2\sigma^2}\right\} \int_{-m\cos\theta}^{\infty} (x + m \cos \theta) \exp\left\{-\frac{x^2}{2\sigma^2}\right\} dx \quad (5A.34)$$

or

$$f_\Theta(\theta) = \frac{m \cos \theta}{\sqrt{2\pi}\sigma} \exp\left\{-\frac{m^2 \sin^2 \theta}{2\sigma^2}\right\}\left[1 - Q(\frac{m \cos \theta}{\sigma})\right] + \frac{\exp\left\{-\frac{m^2}{2\sigma^2}\right\}}{2\pi} \quad (5A.35)$$

Note that when $m = 0$, the joint pdf in (5A.31) becomes

$$f_{R,\Theta}(r, \theta) = \frac{r}{2\pi\sigma^2} \exp\left\{-\frac{r^2}{2\sigma^2}\right\} = f_R(r)f_\Theta(\theta) \quad (5A.36)$$

where $f_R(r) = (r/\sigma^2) \exp(-r^2/2\sigma^2), r \geq 0$ and $f_\Theta(\theta) = 1/2\pi, -\pi \leq \theta \leq \pi$. Hence, R and Θ are independent random variables with R Rayleigh distributed and Θ uniformly distributed. On the other hand, when $m \neq 0$, R and Θ are dependent with joint pdf as given by (5A.31). The marginal pdf of R is Rician and $f_\Theta(\theta)$ is given by (5A.35) and is plotted in Fig. 5A.2 as a function of $|m|/\sigma$. As one would expect, the pdf of θ becomes quite peaked as $|m|/\sigma$ increases.

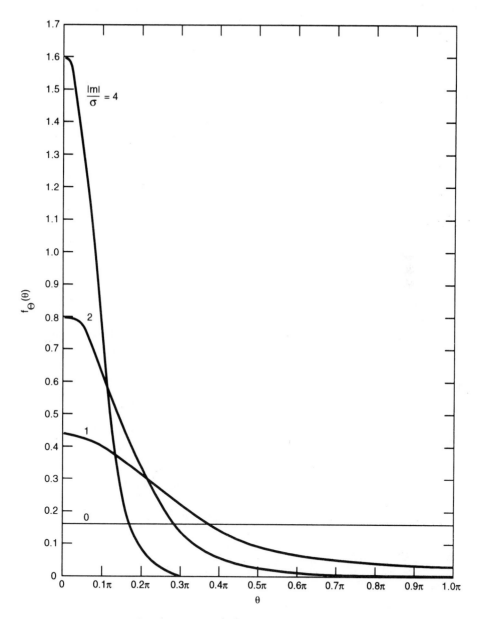

Figure 5A.2   Pdf for phase angle, Rician channel

Appendix 5B

Derivation of Equation (5.43)

From (5.42) (let $\omega_i' \triangleq \omega_c + \omega_i$)

$$\frac{\xi_i^2}{2P} = \int_0^T \int_0^T r(t)r(s)\left\{\cos\omega_i't\cos\omega_i's + \sin\omega_i't\sin\omega_i's\right\}dtds$$

$$= \int_0^T \int_0^T r(t)r(s)\cos\omega_i'(s-t)dtds \tag{5B.1}$$

Let

$$f(t,s) \triangleq r(t)r(s)\cos\omega_i'(s-t) \tag{5B.2}$$

then (5B.1) can be written as

$$\frac{\xi_i^2}{2P} = \int_0^T \int_0^T f(t,s)dtds \tag{5B.3}$$

Divide the integration on T^2 square region in the t, s plane into two integrations on triangular regions as shown in Fig. 5B.1. As such, (5B.3) can be written as the sum of two integrals, namely

$$\frac{\xi_i^2}{2P} = \int_0^T \int_0^t f(t,s)dsdt + \int_0^T \int_0^s f(t,s)dtds$$

$$= \int_0^T \int_0^t f(t,s)dsdt + \int_0^T \int_0^t f(s,t)dsdt \tag{5B.4}$$

where the second integral is obtained by switching t and s. But from (5B.2), $f(t,s) = f(s,t)$, thus

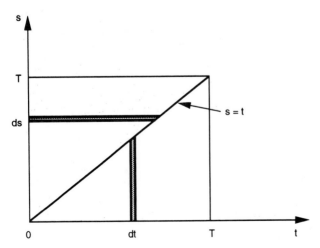

Figure 5B.1 Integration on T^2 region

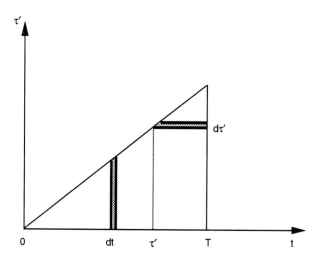

Figure 5B.2   Integration on triangular region

$$\frac{\xi_i^2}{2P} = 2\int_0^T \int_0^t f(t,s)\,ds\,dt$$

$$= 2\int_0^T \int_0^t r(t)r(s)\cos\omega_i'(s-t)\,ds\,dt \tag{5B.5}$$

Letting $s = t + \tau$, then

$$\frac{\xi_i^2}{2P} = 2\int_0^T \int_{-t}^0 r(t)r(t+\tau)\cos\omega_i'\tau\,d\tau\,dt$$

$$= 2\int_0^T \int_0^t r(t)r(t-\tau')\cos\omega_i'\tau'\,d\tau'\,dt \tag{5B.6}$$

This is an integration in the triangular region of the t, τ' plane using a vertical strip of width dt extending from $t = 0$ to $t = T$. If, instead, we choose for our integration in the same triangular region a horizontal strip of height $d\tau$ extending from $t = \tau'$ to $t = T$ as shown in Fig. 5B.2, then we get

$$\frac{\xi_i^2}{2P} = 2\int_0^T \int_{\tau'}^T r(t)r(t-\tau')\cos\omega_i'\tau'\,dt\,d\tau'$$

$$= 2\int_0^T \cos\omega_i'\tau'\,d\tau' \int_{\tau'}^T r(t)r(t-\tau')\,dt \tag{5B.7}$$

Let $v = t - \tau'$ in the inner integral, then

$$\frac{\xi_i^2}{2P} = 2\int_0^T \cos\omega_i'\tau'\,d\tau' \int_0^{T-\tau'} r(v+\tau)r(v)\,dv \tag{5B.8}$$

which is identical to (5.43).

Consider the finite-time autocorrelation function $\phi_r(\tau)$ defined by (5.44) as

$$\phi_r(\tau) = \int_0^{T-\tau} r(t)r(t+\tau)\,dt, \quad 0 \le \tau \le T \tag{5B.9}$$

Note that $\phi_r(\tau)$ is not an even function due to the finite duration. Define the modified function $\phi_{e,r}(\tau)$ as

$$\phi_{e,r}(\tau) = \begin{cases} \phi_r(\tau), & \tau \geq 0 \\ \phi_r(-\tau), & \tau \leq 0 \end{cases} \qquad (5\text{B}.10)$$

Then, $\phi_{e,r}(\tau)$ is an even function and its Fourier transform is given by

$$\Phi_{e,r}(f) = \int_0^\infty \phi_{e,r}(\tau) \cos 2\pi f \tau \, d\tau = \int_0^\infty \phi_r(\tau) \cos 2\pi f \tau \, d\tau \qquad (5\text{B}.11)$$

Hence, the Fourier transform of $\phi_{e,r}(\tau)$ (which is identical to its cosine transform) is identical to the cosine transform of $\phi_r(\tau)$. This technique provides an efficient computational method for evaluating the cosine transform of $\phi_r(\tau)$ using fast Fourier transform (FFT) routines on $\phi_{e,r}(\tau)$.

Appendix 5C

Generalized *M*-ary Symbol Error Probability Bound

Consider an M-ary communication system whose decisions are made based on a relation among M outputs $x_0, x_1, \cdots, x_{M-1}$. Let these outputs be represented by independent random variables with pdfs as follows:

$$
\begin{aligned}
x_n &\longrightarrow f_1(x_n) \\
x_i &\longrightarrow f_0(x_i; \zeta_{in}) \quad \text{for} \quad i = 0, 1, \cdots, n-1, n+1, \cdots, M-1
\end{aligned}
\tag{5C.1}
$$

That is to say, for some particular n, the random variable x_i has a fixed pdf whereas the remaining $M - 1$ r.v.'s x_i, $i \neq n$, all have the identical form pdf (perhaps different than that for x_n)[1] which, however, depend on a parameter ζ_{in} that varies with both i and n. Assuming that message m_n is transmitted, then a correct decision is made at the receiver when $x_n > x_i$ for all $i \neq n$. Then, the conditional probability of a correct decision is

$$
\begin{aligned}
P_s(C/m_n) &= \text{Prob}\{\text{Correct decision}/m_n\} \\
&= \text{Prob}\{x_n > x_i \quad \text{for all } i \neq n/m_n\} \\
&= \int_{-\infty}^{\infty} \text{Prob}\{x_n > x_i \text{ for all } i \neq n/m_n, x_n = \alpha\} \, f_1(\alpha) d\alpha \\
&= \int_{-\infty}^{\infty} \text{Prob}\{x_i < \alpha \text{ for all } i \neq n/m_n\} \, f_1(\alpha) d\alpha \\
&= \int_{-\infty}^{\infty} \prod_{\substack{i=0 \\ i \neq n}}^{M-1} \text{Prob}\{x_i < \alpha/m_n\} f_1(\alpha) d\alpha \\
&= \int_{-\infty}^{\infty} \prod_{\substack{i=0 \\ i \neq n}}^{M-1} [1 - \text{Prob}\{x_i \geq \alpha/m_n\}] f_1(\alpha) d\alpha
\end{aligned}
\tag{5C.2}
$$

If $\text{Prob}\{x_i \geq \alpha/m_n\}$ is hard to evaluate, then use the Chernoff bound

$$
\begin{aligned}
\text{Prob}\{x_i \geq \alpha/m_n\} &\leq E\left\{e^{\lambda(x_i - \alpha)}/m_n\right\} \\
&= e^{-\lambda\alpha} E\left\{e^{\lambda x_i}/m_n\right\} \quad \text{for any } \lambda \geq 0
\end{aligned}
\tag{5C.3}
$$

Define

$$
\begin{aligned}
g_1(\lambda) &\triangleq E\left\{e^{\lambda x_n}/m_n\right\} = \int_{-\infty}^{\infty} e^{\lambda\beta} f_1(\beta) d\beta \\
g_0(\lambda; \zeta_{in}) &\triangleq E\left\{e^{\lambda x_i}/m_n\right\} = \int_{-\infty}^{\infty} e^{\lambda\beta} f_0(\beta; \zeta_{in}) d\beta
\end{aligned}
\tag{5C.4}
$$

---

1. If the two pdfs have identical form, then we shall ignore the "0" and "1" subscripts on them and simply write $f(x)$ or $f(x; \zeta)$, as appropriate. An example of where the two pdfs are, in principle, different in form would correspond to the case of ideal (zero frequency error) noncoherent detection of M-FSK, in which case $f_1(x)$ would be Rician and $f_0(x)$ would be Rayleigh. Also in this ideal situation, $f_0(x)$ would not be dependent on a parameter ζ which varies with the random variable being characterized.

Then

$$\text{Prob}\,\{x_i \geq \alpha/m_n\} \leq e^{-\lambda\alpha} g_0(\lambda; \zeta_{in}) \tag{5C.5}$$

and

$$P_s(C/m_n) \geq \int_{-\infty}^{\infty} \prod_{\substack{i=0 \\ i \neq n}}^{M-1} \left[1 - e^{-\lambda\alpha} g_0(\lambda; \zeta_{in})\right] f_1(\alpha) d\alpha \tag{5C.6}$$

Finally

$$P_s(E/m_n) = 1 - P_s(C/m_n)$$

$$\leq \int_{-\infty}^{\infty} \left\{ 1 - \prod_{\substack{i=0 \\ i \neq n}}^{M-1} \left[1 - e^{-\lambda\alpha} g_0(\lambda; \zeta_{in})\right] \right\} f_1(\alpha) d\alpha$$

$$= \int_{-\infty}^{\infty} \left\{ 1 - \left[1 + \sum_{k=1}^{M-1} (-1)^k e^{-\lambda k \alpha} \right. \right. \tag{5C.7}$$

$$\left. \left. \times \underbrace{\sum_{i_1=0}^{M-1} \sum_{i_2=0}^{M-1} \cdots \sum_{i_k=0}^{M-1}}_{\substack{i_1, i_2, \cdots, i_k \neq n \text{ and} \\ i_1 < i_2 < \cdots < i_k}} \prod_{j=1}^{k} g_0(\lambda; \zeta_{i_j n}) \right] \right\} f_1(\alpha) d\alpha$$

or using (5C.4), simplifying and minimizing over the Chernoff parameter, we get

$$P_s(E/m_n) \leq \min_{\lambda_n \geq 0} \sum_{k=1}^{M-1} (-1)^{k+1} g_1(-\lambda_n k)$$

$$\times \underbrace{\sum_{i_1=0}^{M-1} \sum_{i_2=0}^{M-1} \cdots \sum_{i_k=0}^{M-1}}_{\substack{i_1, i_2, \cdots, i_k \neq n \text{ and} \\ i_1 < i_2 < \cdots < i_k}} \prod_{j=1}^{k} g_0(\lambda_n; \zeta_{i_j n}) \tag{5C.8}$$

Assuming equiprobable signals, then the average probability of symbol error is given by

$$P_s(E) = \frac{1}{M} \sum_{n=0}^{M-1} P_s(E/m_n) \tag{5C.9}$$

Note that if the parameter ζ_{in} is independent of i, that is, all x_i, $i \neq n$, have identical pdfs $f_0(x)$, then

$$\underbrace{\sum_{i_1=0}^{M-1} \sum_{i_2=0}^{M-1} \cdots \sum_{i_k=0}^{M-1}}_{\substack{i_1, i_2, \cdots, i_k \neq n \text{ and} \\ i_1 < i_2 < \cdots < i_k}} \prod_{j=1}^{k} g_0(\lambda; \zeta_{i_j n}) = \binom{M-1}{k} g_0^k(\lambda) \tag{5C.10}$$

where

$$g_0(\lambda) = \int_{-\infty}^{\infty} e^{\beta\lambda} f_0(\beta) d\beta \tag{5C.11}$$

Thus, (5C.9) together with (5C.8) simplify to

$$P_s(E) \leq \min_{\lambda} \sum_{k=1}^{M-1} (-1)^{k+1} \binom{M-1}{k} g_1(-\lambda k) g_0^k(\lambda) \qquad (5C.12)$$

PROBLEMS

5.1 Let $\mathbf{z}$ be a complex Gaussian random vector of dimension N, with

$$\mathbf{z} = \mathbf{x} + j\mathbf{y}$$

where $\mathbf{x}$ and $\mathbf{y}$ are the real and imaginary parts of $\mathbf{z}$. Let $\mathbf{m_z} = E\{\mathbf{z}\}$ denote the mean and $\mathcal{K}_{\mathbf{zz}^*} = E\{\mathbf{zz}^{*T}\}$ denote the covariance matrix. Assume further that $E\{\mathbf{zz}^T\} = [0]$ (the all-zero matrix). Show that the probability density function of the random vectors $\mathbf{x}$ and $\mathbf{y}$ is given by

$$f_{\mathbf{x},\mathbf{y}}(\mathbf{w}_1, \mathbf{w}_2) = \frac{\exp\left\{-(\mathbf{w} - \mathbf{m_z})^{*T} \mathcal{K}_{\mathbf{zz}^*}^{-1}(\mathbf{w} - \mathbf{m_z})\right\}}{\pi^N \det(\mathcal{K}_{\mathbf{zz}^*})}$$

where $\mathbf{w} = \mathbf{w}_1 + j\mathbf{w}_2$.

5.2 Consider the following binary detection problem:

$$H_1: \quad \mathbf{r} = \mathbf{s} + \mathbf{n}$$

$$H_0: \quad \mathbf{r} = \mathbf{n}$$

where $\mathbf{r}$ is the observed vector, $\mathbf{n}$ is a zero-mean, complex Gaussian noise vector with

$$E\left\{\mathbf{nn}^{*T}\right\} = 2N_0 \mathcal{I}$$

$$E\left\{\mathbf{nn}^T\right\} = [0], \quad \text{(all-zeros matrix)}$$

All vectors are N-dimensional and $\mathbf{s}$ is composed of independent, complex random variables r_n, $n = 1, \cdots, N$ of the form

$$r_n = \sqrt{2E} a_n e^{j\phi_n}$$

where a_n is a Rayleigh random variable with unit second moment, that is

$$f_{a_n}(a) = \begin{cases} 2ae^{-a^2}, & a > 0 \\ 0, & \text{otherwise} \end{cases}$$

and ϕ_n is an independent phase, uniformly distributed over 2π, that is

$$f_{\phi_n}(\phi) = \begin{cases} \frac{1}{2\pi}, & 0 \leq \phi \leq 2\pi \\ 0, & \text{otherwise} \end{cases}$$

(a) Show that r_n is a complex Gaussian random variable when H_1 is true. Compute its mean and covariance matrix.

(b) Compute the likelihood ratio

$$\Lambda(\mathbf{r}) = \frac{f_{\mathbf{r}}(\mathbf{r}/H_1)}{f_{\mathbf{r}}(\mathbf{r}/H_0)}$$

and simplify the likelihood test to derive the optimum detector.

(c) Let $z = \sum_{n=1}^{N} |r_n|^2$. What is the pdf of z when H_0 is true? Repeat when H_1 is true.

(d) Derive an expression for the probability of detection.

(e) Derive an expression for the false alarm probability.

5.3 Show that the asymptotic ($M \to \infty$) error probability performance of the noncoherent receiver as given by (5.77) is identical to that for coherent communications as described by (4.182).

5.4 With $r(t)$ given by (5.80), verify that the decision variables ξ_k, $k = 0, 1 \cdots, M - 1$ are independent of the incoming carrier phase θ, as claimed below (5.90b).

5.5 Perform the matrix multiplication $T K_x T^T$ and verify that the whitening transformation T as given by (5.121) does indeed produce $K_{x_w} = \mathcal{I}$.

5.6 Consider two independent Rician random variables R_0 and R_1, each with parameters s_i and variance σ_i^2 ($i = 0, 1$), that is

$$f_{R_i}(r_i) = \frac{r_i}{\sigma_i^2} \exp\left\{-\frac{s_i^2 + r_i^2}{2\sigma_i^2}\right\} I_0\left(\frac{r_i s_i}{\sigma_i^2}\right), \quad r_i \geq 0$$

We wish to evaluate

$$P = \text{Prob}\left\{R_1^2 > R_0^2\right\} = \text{Prob}\{R_1 > R_0\}$$

(a) Show that

$$\text{Prob}\{R_1 > R_0/R_0\} = Q\left(\frac{s_1}{\sigma_1}, \frac{R_0}{\sigma_1}\right)$$

where $Q(\alpha, \beta)$ is the Marcum Q-function. Therefore

$$P = \int_0^\infty Q\left(\frac{s_1}{\sigma_1}, \frac{r_0}{\sigma_1}\right) f_{R_0}(r_0) dr_0$$

(b) Show that the Laplace transform (L.T.) of $I_0(2\sqrt{\beta t})$ is given by $(1/s)\exp(\beta/s)$, that is

$$\int_0^\infty I_0(2\sqrt{\beta t})e^{-st} dt = \frac{1}{s}e^{\beta/s}, \quad \text{Re}(s) > 0$$

or, equivalently

$$I_0(2\sqrt{\beta t}) = \frac{1}{2\pi j} \int_{\sigma - j\infty}^{\sigma + j\infty} e^{ts}\frac{e^{\beta/s}}{s} ds, \quad \sigma > 0$$

(c) By inserting the expression for I_0 in part (b) into the definition of the Q-function, show that

$$Q\left(\sqrt{2\alpha}, \sqrt{2\beta}\right) = e^{-(\alpha + \beta)}\frac{1}{2\pi j} \int_{\sigma - j\infty}^{\sigma + j\infty} \frac{e^{\alpha s + \frac{\beta}{s}}}{s - 1} ds, \quad \sigma > -1$$

(d) By substituting the above expression in (c) for the Q-function into the expression for P in (a), interchanging the order of integration and simplifying, show that

$$P = \exp\left(-\frac{s_1^2}{2\sigma_1^2}\right) \exp\left(-\frac{s_0^2}{2\sigma_0^2}\right) \frac{1}{2\pi j} \int_{\sigma - j\infty}^{\sigma + j\infty} \frac{\exp\left(\frac{s_1^2 s}{2\sigma_1^2} + \frac{s_0^2}{2\sigma_0^2\left[1 + \frac{\sigma_0^2}{\sigma_1^2}\left(1 - \frac{1}{s}\right)\right]}\right)}{(s - 1)\left[1 + \frac{\sigma_0^2}{\sigma_1^2}\left(1 - \frac{1}{s}\right)\right]} ds$$

(e) Define

$$a \triangleq \frac{s_1^2}{\sigma_0^2 + \sigma_1^2} \qquad b \triangleq \frac{s_0^2}{\sigma_0^2 + \sigma_1^2}$$

$$c_0 \triangleq \frac{\sigma_0^2}{\sigma_0^2 + \sigma_1^2} \qquad c_1 \triangleq \frac{\sigma_1^2}{\sigma_0^2 + \sigma_1^2}$$

Use the partial-fraction expansion

$$c_1 \left(\frac{1}{s-1} \right) \left(\frac{s}{s-c_0} \right) = \frac{1}{s-1} - \frac{c_0}{s-c_0}$$

to separate the expression for P in (d) into the sum of two integrals. Make the change of variable $c_1^{-1}(s-1) = (p-1)$ in both integrals. Using the results of (c), simplify the first integral to show that it reduces to $Q(\sqrt{a}, \sqrt{b})$. Using results from (d), show that the second integral is given by

$$-c_1 \exp\left(-\frac{a+b}{a}\right) I_0\left(\sqrt{ab}\right)$$

Therefore

$$P = Q\left(\sqrt{a}, \sqrt{b}\right) - c_1 \exp\left(-\frac{a+b}{2}\right) I_0\left(\sqrt{ab}\right)$$

5.7 Consider the noncoherent reception of an M-FSK signal with a common frequency error Δf in all of the possible input tones. It is desirable to compute an upper union bound on the average symbol error probability in terms of the exact bit error probability of noncoherent binary FSK in the presence of the same frequency error. Let $P_b(E; \rho, m)$ denote the exact bit error probability of noncoherent binary FSK when the tones are spaced by m/T (recall that $1/T$ is the minimum spacing required for orthogonality) where $\rho = \Delta fT$ is the normalized frequency error. Assume that the tones in the M-FSK signaling set are spaced by $1/T$.
(a) For $M = 4$, compute a union bound on the average symbol error probability $P_s(E)$.
(b) Repeat (a) for $M = 8$.
(c) By induction, generalize the results of (a) and (b) to the case of arbitrary M.

5.8 Using the trigonometric identities

$$\sum_{k=0}^{n} \cos^2 k\theta = \frac{n+2}{2} + \frac{\sin n\theta \cos[(n+1)\theta]}{2 \sin \theta}$$

$$\sum_{k=1}^{n-1} k \cos k\theta = \frac{n \sin\{[(2n-1)/2]\theta\}}{2 \sin \theta/2} - \frac{1-\cos n\theta}{4 \sin^2 \theta/2}$$

show that the mean of the spectral estimate $\Phi_{v_a}(f_l; J)$ from (5.183) is given by

$$\mu_l \triangleq E\{\Phi_{v_a}(f_l; J)\} = \begin{cases} \frac{N_0 T}{2} \left\{ 1 + \frac{P}{N_0 B}\left[J - \frac{(J-1)(J+1)}{4BT} \right] \right\}, & f_l = \frac{l}{T}, l = k \\ \frac{N_0 T}{2} \left\{ 1 + \frac{P}{N_0 B}\left[\frac{J+1}{2BT} - 1 \right] \right\}, & f_l = \frac{l}{T}, l \neq k \end{cases}$$

where k/T is the signaling frequency. A similar independence on l for $l \neq k$ can be shown for the variance of $\Phi_{v_a}(f_l; J)$, but the mathematics required is cumbersome.

5.9 Consider a WSS random process $x(t)$ which is input to two different linear filters $G(f)$ and $H(f)$ resulting, respectively, in outputs $y(t)$ and $z(t)$. If $G(f)$ and $H(f)$ do not overlap in frequency, that is, $G(f)H(f) = 0$, show that $y(t)$ and $z(t)$ are uncorrelated.

5.10 Derive the optimum noncoherent receiver for the Rician channel. Follow the approach as outlined by (5.237) and (5.238). Compute the constants K_1, K_2, and K_3 for the unequal energy case. Simplify your results for the equal energy case.

5.11 One form of combating fading is to employ diversity. In this case, multiple channels are used to transmit the signal using different frequencies (frequency diversity), multiple antennas (spatial diversity), different time slots (time diversity), or different polarizations (polarization diversity).

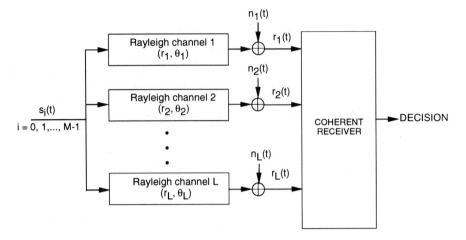

The signal propagates through L channels for L^{th}-order diversity to produce

$$r_i(t) = r_i\sqrt{2}a_k(t)\cos(\omega_c t + \phi_k(t) + \theta_i) + n_i(t), \quad 0 \le t \le T$$

for $i = 1, 2\cdots, L$, when message m_k is transmitted (out of M messages). The parameters (r_i, θ_i) are Rayleigh (with parameter σ^2) and uniform distributed and are independent from each other and from the parameters of other channels, as shown in Fig. P5.11.

(a) Assuming that the channel attenuations $(r_i, i = 1, 2, \cdots, L)$ and phase shifts are perfectly known at the receiver, derive the optimum coherent receiver for $L = 2$ when $n_i(t)$ is white Gaussian noise with one-sided PSD N_0 (independent of i) and the noises are independent.

(b) Generalize your results for an arbitrary L and draw the optimum coherent receiver.

5.12 Using the optimum receiver derived in Problem 5.11

(a) Show that the probability of error for BPSK is given by

$$P_b(E) = \left(\frac{1-\mu}{2}\right)^L \sum_{k=0}^{L-1} \binom{L-1+k}{k}\left(\frac{1+\mu}{2}\right)^k$$

where

$$\mu = \sqrt{\frac{(E_{av}/N_0)}{1 + (E_{av}/N_0)}}$$

and (E_{av}/N_0) is the average SNR per channel and is given by

$$\frac{E_{av}}{N_0} = 2\sigma^2\left(\frac{E}{N_0}\right)$$

(b) When $(E_{av}/N_0) \gg 1$, use the approximations $(1 + \mu)/2 \simeq 1$, $(1 - \mu)/2 \simeq 1/4(E_{av}/N_0)$ and

$$\sum_{k=0}^{L-1}\binom{L-1+k}{k} = \binom{2L-1}{L}$$

to show that

$$P_b(E) \simeq \left(\frac{1}{4(E_{av}/N_0)}\right)^L\binom{2L-1}{L}$$

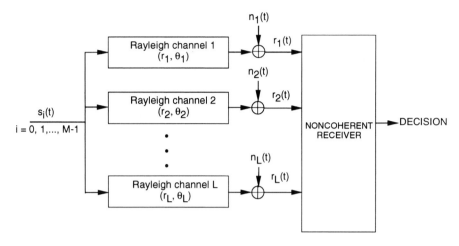

Figure P5.15

5.13 Repeat Problem 5.12 using binary, orthogonal FSK and show that the same results hold, with (E_{av}/N_0) replaced by $(E_{av}/N_0)/2$.

5.14 The transmitted electric field of a communication signal is circular polarized and used to provide dual diversity in a Rayleigh fading channel, using binary orthogonal FSK signal. Assume $\sigma^2 = 1/2$.
  (a) Assuming a coherent receiver, find the error probability if $E_b/N_0 = 20$ dB.
  (b) Compare the result in (a) to the case where no diversity is used.
  (c) Repeat (a) if the channel adds only white noise, that is, does not fade.

5.15 Repeat Problem 5.11, assuming that the phase information about the various channels is not available at the receiver. In this case, the receiver combines the signals noncoherently as shown in Fig. P5.15. Draw the optimum noncoherent receiver.

5.16 Show that the performance of the noncoherent receiver (derived in Problem 5.15) for binary, orthogonal FSK is given by

$$P_s(E) = \left(\frac{1}{2+\beta}\right)^L \sum_{k=0}^{L-1} \binom{L+k-1}{k} \left(\frac{1+\beta}{2+\beta}\right)^k$$

where $\beta = E_{av}/N_0$.
  Hint: See [25].

5.17 A binary noncoherent receiver is required to operate in a Rayleigh fading channel with a bit error probability of 10^{-5}, using binary orthogonal FSK signal.
  (a) What is the required E_b/N_0 assuming $\sigma^2 = 1/2$?
  (b) Design a diversity system such that the 10^{-5} bit error rate can be met if an $E_b/N_0 \le 15$ dB is available.
  (c) What is the required E_b/N_0 in your design and what is the power savings over that required in (a)?

5.18 Consider the M-ary system in Problem 5.11. All the assumptions remain the same except now we assume that the channels are independent Rician [with parameters $(r_i, \theta_i; \alpha_i, \delta_i)$] instead of Rayleigh, as depicted in Fig. P5.18. Assume that the amplitude α_i and phase δ_i of the specular component of channel i is known for $i = 1, 2, \cdots, L$.
  (a) Derive the optimum coherent receiver for $L = 2$.
  (b) Generalize the result to an arbitrary L and draw the optimum coherent receiver.

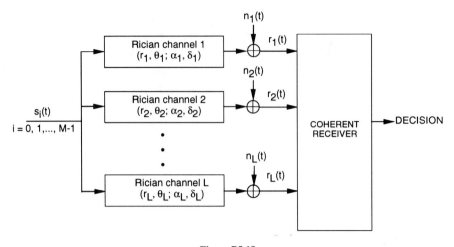

Figure P5.18

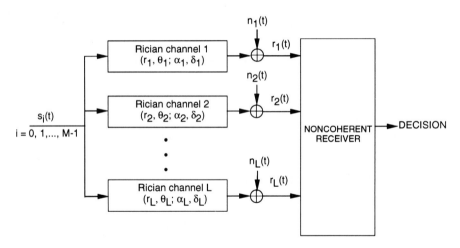

Figure P5.19

5.19 Repeat Problem 5.18 assuming that the phase of the specular component in channel i is not known for all i and that the δ_i's are independent from channel-to-channel. Draw the optimum noncoherent receiver as shown in Fig. P5.19.

5.20 In noncoherent combining,[2] one decision rule in M-ary communications is to decide on message m_j ($j = 0, 1, \cdots, M - 1$) if and only if

$$\sum_{k=1}^{L} \xi_{i,k}^2 = \sum_{k=1}^{L} \left\{ \left(\int_0^T r_k(t)\sqrt{2}a_i(t)\cos(\omega_c t + \phi_i(t))dt \right)^2 \right.$$
$$\left. + \left(\int_0^T r_k(t)\sqrt{2}a_i(t)\sin(\omega_c t + \phi_i(t))dt \right)^2 \right\}$$

---

2. W. C. Lindsey, "Error Probabilities for Rician Multichannel Reception of Binary and N-ary Signal," *IEEE Transactions on Information Theory*, October 1964, pp. 339–50.

is maximum for $i = j$, where $r_k(t)$ is the received signal in channel k. The combining rule is optimum for Rayleigh channels but not for Rician channels. Assuming that the channels are Rician with respective parameters $(r_k, \theta_k; \alpha_k, \delta_k)$ for $k = 1, 2, \cdots, L$

(a) Show that the probability of error for binary $(M = 2)$ orthogonal FSK is given by

$$P_s(E) = \left(\frac{1}{2+\beta}\right)^L \exp\left\{-\frac{\Gamma\beta}{2+\beta}\right\} \sum_{k=0}^{L-1} \binom{L+k-1}{k}$$

$$\times \left(\frac{1+\beta}{2+\beta}\right)^k {}_1F_1\left(-k, L; -\frac{\Gamma\beta}{(1+\beta)(2+\beta)}\right)$$

where $\beta = 2\sigma^2(E_s/N_0)$, $\eta_k = \alpha_k^2(E_s/N_0)$, $\Gamma\beta = \sum_{k=1}^{L}\eta_k$ and ${}_1F_1(a, c; z)$ is the confluent hypergeometric function defined by (5A.26).

(b) Let $\alpha_k = 0$ for $k = 1, 2 \cdots, L$ and show that the result in (a) reduces to that in Problem 5.16, that is

$$P_s(E) = \left(\frac{1}{2+\beta}\right)^L \sum_{k=0}^{L-1} \binom{L+k-1}{k}\left(\frac{1+\beta}{2+\beta}\right)^k$$

where $\beta = E_{av}/N_0$ as all the signal power is in the Rayleigh component [see (5.253)]. Hint: see [25].

5.21 Given the M-FSK signal set

$$s_i(t) = \begin{cases} \sqrt{E}\phi_i(t), & 0 \le t \le T \\ 0, & \text{otherwise} \end{cases}$$

with $\phi_i(t) = \sqrt{2/T}\cos[(\omega_c + \omega_i)t + \theta]$ for $i = 0, 1, \cdots, M-1$ and θ a uniformly distributed random variable. Find the minimum frequency spacing required to obtain orthogonal signals if

(a) Coherent detection is used,

(b) Noncoherent detection is used,

(c) When $T = 10^{-7}$ sec and $M = 1024$, what is the required minimum bandwidth for coherent detection?

(d) Repeat (c) if noncoherent detection is used.

5.22 A BFSK transmitter sends one of two orthogonal signals

$$s_i(t) = \sqrt{2E/T}\cos(\omega_i t + \theta)$$

for $0 \le t \le T$. Here, θ is a uniformly distributed random variable. The receiver knows ω_i, E, T, and N_0 exactly.

(a) Sketch the optimum waveform receiver.

(b) What is the message error probability and the energy per bit required if $E_b/N_0 = 2$ dB and $T = 290°$ Kelvin?

5.23 For the binary signals defined in Problem 5.22, determine and sketch the optimum receiver if

(a) $f(\theta) = \frac{1}{2}\left[\delta(\theta + \frac{\pi}{2}) + \delta(\theta - \frac{\pi}{2})\right]$.

(b) Repeat Problem 5.22(b).

5.24 Assume the same signal model as in Problem 5.22. The channel model is as shown in Fig. P5.24. Assuming the multiplier "r" is a Rayleigh distributed random variable,

(a) Determine the optimum decision rule for noncoherent detection.

(b) Sketch the optimum waveform receiver.

(c) Derive an expression for the average bit error probability.

(d) Sketch the average bit error probability if $\sigma = 1$ (σ is the parameter which characterizes the Rayleigh random variable).

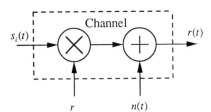

Figure P5.24

REFERENCES

1. Helstrom, C. W., *Statistical Theory of Signal Detection*, 2nd ed., Oxford: Pergamon Press, 1968.

2. Papoulis, A., *Signal Analysis*, New York: McGraw-Hill, 1977.

3. Abramowitz, M. and I. A. Stegun, *Handbook of Mathematical Functions*, New York: Dover Publications, 1970.

4. Watson, G., *Theory of Bessel Functions*, London and New York: Cambridge University Press, 1944.

5. Bracewell, R., *The Fourier Transform and Its Applications*, New York: McGraw-Hill, 1978.

6. Charles, F. and N. Shein, "A Preliminary Study of the Application of Noncoherent Techniques to Low Power Telemetry," Pasadena, CA: Jet Propulsion Laboratory, Technical Memorandum 3341-65-14 (Reorder No. 65-815), November 1965.

7. Charles, F., N. Shein and J. Springett, "The Statistical Properties of the Spectral Estimates Used in the Decision Process by a System Analyzer Receiver," *Proceedings of the National Telemetering Conference*, San Francisco, CA, May 1967.

8. Chadwick, H., "Time Synchronization in an MFSK Receiver," Pasadena, CA: Jet Propulsion Laboratory, SPS 37-48, vol. III, 1967, pp. 252–64. Also presented at the *Canadian Symposium on Communications*, Montreal, Quebec, November 1968.

9. Cooley, J. and J. Tukey, "An Algorithm for the Machine Calculation of Complex Fourier Series," *Math. of Computation*, vol. 19, pp. 297–301, 1965.

10. Simon, M. K. and J. C. Springett, "The Performance of a Noncoherent FSK Receiver Preceded by a Bandpass Limiter," *IEEE Transactions on Communications*, vol. COM-20, no. 6, December 1972, pp. 1128–36.

11. Papoulis, A., *Probability, Random Variables and Stochastic Processes*, New York: McGraw-Hill, 1965.

12. Gradshteyn, I. S. and I. M. Ryshik, *Table of Integrals, Series and Products*, corrected and enlarged edition, Florida: Academic Press, 1980.

13. Weber, C. L., *Elements of Detection and Signal Design*, New York: Springer-Verlag, 1987.

14. Nuttall, A. H., "Error Probabilities for Equicorrelated M-ary Signals under Phase-Coherent and Phase-Incoherent Reception," *IRE Transactions on Information Theory*, July 1962, pp. 305–14.

15. Nuttall, A. N., "Error Probabilities for Nonorthogonal M-ary Signals under Phase-Coherent and Phase-Incoherent Reception," Waltham, MA: Litton Systems Inc., Tech Report TR-61-1-BF, June 1961.

16. Helstrom, C. W., "The Resolution of Signals in White Gaussian Noise," *IRE Proceedings*, vol. 43, September 1955, pp. 1111–18.

17. Proakis, J. G., *Digital Communications*, New York: McGraw-Hill, 1983.

18. Viterbi, A. J., *Principles of Coherent Communications*, New York: McGraw-Hill, 1966.

19. Stein, S., "Unified Analysis of Certain Coherent and Noncoherent Binary Communication Systems," *IEEE Transactions on Information Theory*, vol. IT-10, January 1964, pp. 43–51.

20. Nakamoto, F., R. Middlestead and C. Wolfson, "Impact of Time and Frequency Errors on Processing Satellites with MFSK Modulations," *ICC 81 Conference Record*, Denver, CO, June 1981, pp. 37.3.1–37.3.5.

21. Chadwick, H., "The Error Probability of a Spectrum Analysis Receiver with Incorrect Time and Frequency Synchronization," *International Symposium of Information Theory*, Ellenville, NY, January 1969.

22. Wittke, P. and P. McLane, "Study of the Reception of Frequency Dehopped M-ary FSK," Research Report 83-1, Queen's University, Kingston, Ontario, March 1983.

23. Frasen, D., *Nonparametric Methods in Statistics*, New York: John Wiley and Sons, 1957.

24. Lindsey, W. C., "Error Probabilities for Rician Fading Multichannel Reception of Binary and *N*-ary Signals," *IEEE Transactions on Information Theory*, vol. IT-10, October 1964, pp. 339–50.

25. Lindsey, W. C., "Error Probabilities for Incoherent Diversity Reception," *IEEE Transactions on Information Theory*, vol. IT-11, October 1965, pp. 491–99.

26. Turin, G. L., "Error Probabilities for Binary Symmetric Ideal Reception through Nonselective Slow Fading and Noise," *Proceedings of the IRE*, vol. 46, September 1958, pp. 1603–19.

27. Van Trees, H., *Detection, Estimation and Modulation—Part I: Detection, Estimation and Linear Modulation Theory*, New York: John Wiley and Sons, 1968.

28. Lindsey, W. C. and M. K. Simon, *Telecommunication Systems Engineering*, Englewood Cliffs, N.J.: Prentice Hall, 1973. Reprinted by Dover Press, New York, NY, 1991.

29. Hinedi, S., M. K. Simon and D. Raphaeli, "The Performance of Noncoherent Orthogonal *M*-FSK in the Presence of Timing and Frequency Errors," to be published in *IEEE Transactions on Communications*.

Chapter 6

Partially Coherent Communication with Waveforms

When the incoming carrier phase is estimated in the receiver, the latter can employ coherent detection schemes as discussed in Chapter 4, provided the estimation is performed perfectly. In many practical communication systems, the estimation results in a residual carrier phase error, the amount of which depends on various parameters such as signal-to-noise ratio, frequency stability, the estimation algorithm itself, and so on. Given the statistical description of the phase error, the communication engineer is faced with two choices: design the detector assuming a perfect carrier reference (that is, coherent detection) and then account for any degradation due to the nonideal nature of the estimate, or as a second alternative, implement the optimal detector which exploits the available statistical information on the phase error to obtain an improved detector. Such a scheme, which utilizes the carrier phase error information, is referred to as *partially coherent* detection as it is neither coherent nor noncoherent in the absolute sense. If the estimation is improved to the point where it is perfect, then we expect the partially coherent receiver to approach the coherent receiver of Chapter 4. On the other hand, if the estimation is degraded to the extent of providing no credible information about the phase error, then we expect the partially coherent receiver to reduce to the noncoherent detector, as discussed in Chapter 5.

In partially coherent communications, the structure of the optimal detector depends heavily on the statistics of the phase error. This is to be expected since the optimal detector, by definition, fully exploits the available information. The phase error itself is typically a random process that we assume is slowly varying over a symbol (or multiple symbols) and can thus be modeled as a random variable. Assuming that the communication system transmits an unmodulated carrier for the sole purpose of aiding synchronization in the receiver, a phase-locked loop (PLL) can then be used to track the incoming carrier phase θ. For a first-order PLL, the stationary pdf of the carrier phase error ϕ, defined as $\theta - \widehat{\theta}$ where $\widehat{\theta}$ is the locally generated carrier phase estimate, is given by the Tikhonov density [1, 2], depicted in Fig. 6.1 and can be expressed as

$$f_\phi(\phi) = \begin{cases} \frac{\exp(\rho_\phi \cos \phi)}{2\pi I_0(\rho_\phi)}, & |\phi| \leq \pi \\ 0, & \text{otherwise} \end{cases} \tag{6.1}$$

where ρ_ϕ denotes the carrier loop signal-to-noise ratio (SNR) and is given by

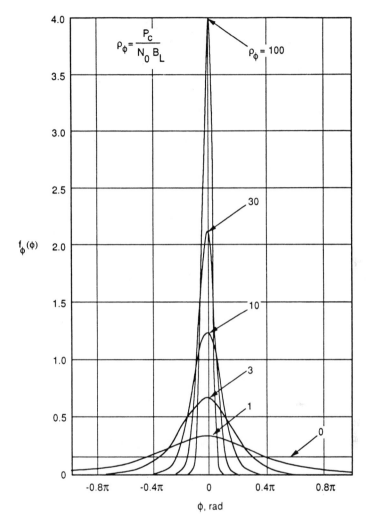

Figure 6.1 Tikhonov pdf

$$\rho_\phi \triangleq \frac{P_c}{N_0 B_L} \tag{6.2}$$

In (6.2), P_c denotes the power of the unmodulated carrier (which is unrelated to the power of the information bearing signal), N_0 is the one-sided PSD of the white Gaussian noise, and B_L is the one-sided bandwidth (in Hz) of the PLL. The synchronization aspect of the receiver will be the topic of Volume II of this book and it suffices to state here the tracking performance as characterized by the pdf of (6.1). Note that (6.1) is the exact pdf for a first-order analog PLL but can also serve as an approximation for a second-order PLL, provided that ρ_ϕ is sufficiently large. Moreover, since the phase reference is arbitrary, we may set $\widehat{\theta} = 0$ without loss in generality so that the received phase θ is identical to ϕ (that is, $\theta = \phi$), as depicted in Fig. 6.2. Partially coherent reception is a more general mode of operation than previously considered since

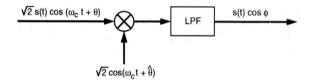

(a) Practical Implementation

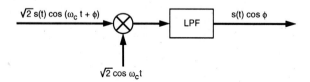

(b) Equivalent Mathmatical Model   **Figure 6.2**   Phase error generation

$$\lim_{\rho_\phi \to 0} \frac{\exp(\rho_\phi \cos \phi)}{2\pi I_0(\rho_\phi)} = \frac{1}{2\pi} \tag{6.3}$$

corresponds to noncoherent reception, while

$$\lim_{\rho_\phi \to \infty} \frac{\exp(\rho_\phi \cos \phi)}{2\pi I_0(\rho_\phi)} = \delta(\phi) \tag{6.4}$$

corresponds to coherent communication as $\phi = 0$ with probability one. Since, in principle, ρ_ϕ can take on any value between 0 and ∞, all intermediate cases are possible.

6.1 OPTIMUM PARTIALLY COHERENT RECEIVERS BASED ON ONE SYMBOL OBSERVATION

In partially coherent channels, the received signal is still given by (5.2) with the understanding that θ is no longer uniformly distributed, but rather distributed according to (6.1) since $\theta = \phi$. In this case, the conditional pdf of $\mathbf{r}$ is still given by (5.19), namely

$$f_{\mathbf{r}/\theta}(\boldsymbol{\rho}/m_i, \theta) \sim \exp\left\{\frac{2\xi_i}{N_0} \cos(\theta + \phi_i)\right\} \exp\left(-\frac{E_i}{N_0}\right) \tag{6.5}$$

Assuming equal energy signals and averaging over the pdf of θ, (6.5) becomes

$$f_{\mathbf{r}}(\boldsymbol{\rho}/m_i) \sim \frac{1}{2\pi I_0(\rho_\phi)} \int_{-\pi}^{\pi} \exp\left\{\frac{2\xi_i}{N_0} \cos(\theta + \phi_i)\right\} \exp(\rho_\phi \cos \theta) d\theta \tag{6.6}$$

Using trigonometric identities, (6.6) reduces to

$$f_{\mathbf{r}}(\boldsymbol{\rho}/m_i) \sim \frac{1}{2\pi I_0(\rho_\phi)} \int_{-\pi}^{\pi} \exp\left\{\chi_i \cos(\theta + \psi_i)\right\} d\theta \tag{6.7}$$

where

$$\chi_i^2 \triangleq \left(\frac{2\xi_i}{N_0} \cos \phi_i + \rho_\phi\right)^2 + \left(\frac{2\xi_i}{N_0} \sin \phi_i\right)^2 \tag{6.8}$$

and

$$\tan \psi_i \triangleq \frac{(2\xi_i/N_0) \sin \phi_i}{(2\xi_i/N_0) \cos \phi_i + \rho_\phi} \tag{6.9}$$

Using the definition of $I_0(x)$ as given by (5.22), (6.7) can then be expressed as

$$f_{\mathbf{r}}(\boldsymbol{\rho}/m_i) \sim \frac{I_0(\chi_i)}{I_0(\rho_\phi)} \tag{6.10}$$

Recalling once again that $I_0(x)$ is a monotonically increasing function of its argument, the MAP receiver sets $\widehat{m}(\boldsymbol{\rho}) = m_i$ if and only if χ_k^2 is maximum when $k = i$. Using z_{ci} and z_{si} as given by (5.17), the rule reduces to maximizing

$$\chi_i^2 = \left(\frac{2}{N_0} z_{ci} + \rho_\phi\right)^2 + \left(\frac{2}{N_0} z_{si}\right)^2 \tag{6.11}$$

For the case where $\phi_i(t) = 0 \ \forall i$ and $a_i(t) = s_i(t)$, then the I-Q projections z_{ci} and z_{si} are given by (5.15) and are repeated here for convenience:

$$z_{ci} = \int_0^T \rho_c(t) s_i(t) dt \tag{6.12a}$$

$$z_{si} = \int_0^T \rho_s(t) s_i(t) dt \tag{6.12b}$$

Alternatively, the rule can be simplified to maximizing

$$\left(\frac{z_{ci}}{N_0}\right)^2 + \left(\frac{z_{si}}{N_0}\right)^2 + \rho_\phi \left(\frac{z_{ci}}{N_0}\right) \tag{6.13}$$

which clearly indicates that the optimum partially coherent detector is simply a linear combination of the optimum coherent and optimum noncoherent detector with the weighing factors ρ_ϕ and 1, respectively.[1] Two alternate implementations of the partially coherent detector are shown in Fig. 6.3. In both cases, knowledge of N_0 as well as ρ_ϕ are required for optimal detection. Clearly, as $\rho_\phi \to 0$, the partially coherent receiver reduces to the familiar noncoherent receiver and as $\rho_\phi \to \infty$, the noncoherent receiver contribution may be ignored, resulting in the coherent receiver.

6.2 PERFORMANCE OF PARTIALLY COHERENT RECEIVERS BASED ON ONE SYMBOL OBSERVATION

It would interesting to compute the performance of partially coherent receivers for various signals and determine their advantages over noncoherent reception. In this section, we will consider binary antipodal, binary orthogonal, binary nonorthogonal, and finally, M-ary orthogonal signals.

---

1. Note [3] that if we define $X_i = (2/N_0)z_{ci}$ and $Y_i = (2/N_0)z_{si}$, then the rule of (6.11) becomes $(X_i + \rho_\phi)^2 + Y_i^2$ or after expansion, $(X_i^2 + Y_i^2) + 2\rho_\phi X_i$, which indicates a weighting of $2\rho_\phi$ and 1 for the coherent and noncoherent contributions.

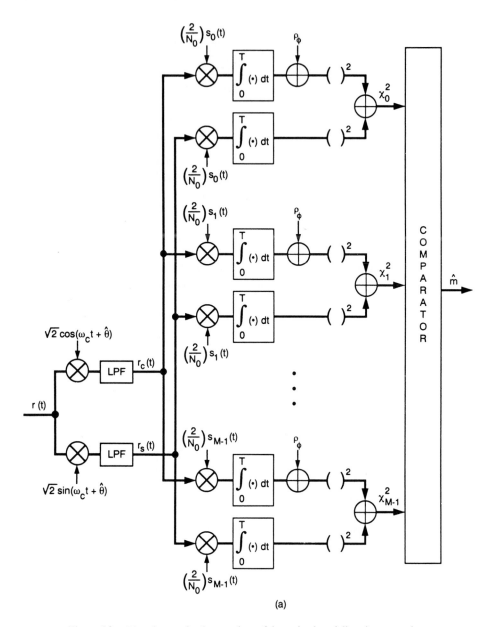

(a)

Figure 6.3a   Two alternate implementations of the optimal partially coherent receiver

6.2.1 Performance of Binary Antipodal Signals

When the signals are antipodal, $s_0(t) = -s_1(t)$ and as a result, the noncoherent contribution (that is, $z_{ci}^2 + z_{si}^2$) to the partially coherent detector is independent of which signal was transmitted since the squaring operation destroys the sign of the signal. Thus, the optimal receiver for antipodal signals is identical to the coherent receiver whose performance in the presence of carrier phase jitter was computed in Chapter 4 and given by (4.363) and (4.364),

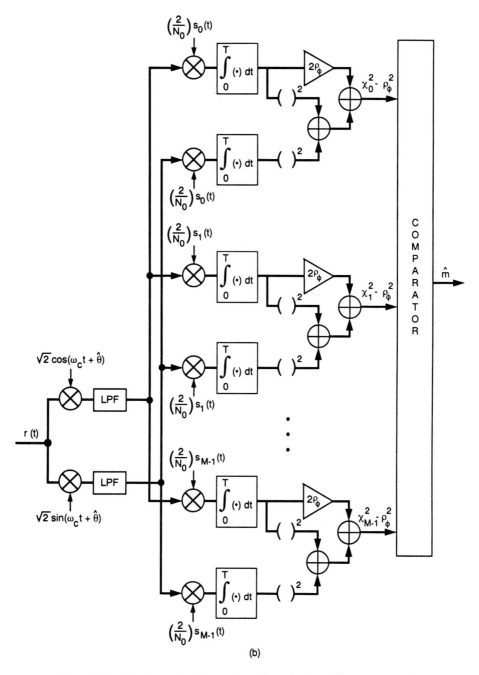

(b)

Figure 6.3b   Two alternate implementations of the optimal partially coherent receiver

that is

$$P_s(E) = \int_{-\pi}^{\pi} \frac{\exp(\rho_\phi \cos \phi)}{2\pi I_0(\rho_\phi)} Q\left(\sqrt{\frac{2E_s}{N_0}} \cos \phi\right) d\phi \tag{6.14}$$

The performance is actually plotted in Fig. 4.46 for various values of loop SNR, ρ_ϕ. When the latter is less than 14 dB, significant degradation starts to occur and gets worse with decreasing ρ_ϕ. For the limiting cases of noncoherent ($\rho_\phi = 0$) and coherent ($\rho_\phi = \infty$) detection, (6.14) reduces to

$$\lim_{\rho_\phi \to 0} P_s(E) = \frac{1}{2} \tag{6.15a}$$

$$\lim_{\rho_\phi \to \infty} P_s(E) = Q\left(\sqrt{\frac{2E_s}{N_0}}\right) \tag{6.15b}$$

the latter being identical to (4.55).

6.2.2 Performance of Binary Orthogonal Signals

For binary orthogonal signals, the receiver compares χ_0 and χ_1 and selects the message corresponding to the largest value. Conditioning on the phase error ϕ, the variables χ_0 and χ_1 become Rician distributed (see Appendix 5A) with parameters s_0^2 and s_1^2, which can be computed using (5.56) and (5.57) to yield (assuming message m_1 is transmitted)

$$\begin{aligned} s_1^2 &= \left(\frac{2E_s}{N_0} \cos \phi + \rho_\phi\right)^2 + \left(\frac{2E_s}{N_0} \sin \phi\right)^2 \\ &= \left(\frac{2E_s}{N_0}\right)^2 + \rho_\phi^2 + 4\left(\frac{E_s}{N_0}\right)\rho_\phi \cos \phi \end{aligned} \tag{6.16}$$

and

$$s_0^2 = \rho_\phi^2 \tag{6.17}$$

Since both z_{ci} and z_{si} have identical variance given by (5.58), then σ_1^2 and σ_0^2 [equivalent to σ^2 in (5A.27)] are equal and given by

$$\sigma_1^2 = \sigma_0^2 = \left(\frac{2}{N_0}\right)^2 \frac{N_0 E_s}{2} = 2\left(\frac{E_s}{N_0}\right) \tag{6.18}$$

Because of the symmetry of the problem, $P_s(E/m_1)$ is identical to $P_s(E/m_0)$. The conditional probability of error, $P_s(E/\phi)$, is then the probability that one Rician variable exceeds another independent Rician variable, which is given by (5.130). Hence

$$\begin{aligned} P_s(E/\phi) &= \text{Prob}\left\{\chi_0 > \chi_1/\phi, m_1\right\} \\ &= Q\left(\sqrt{a}, \sqrt{b}\right) - \frac{1}{2}\exp\left(-\frac{a+b}{-2}\right) I_0\left(\sqrt{ab}\right) \end{aligned} \tag{6.19}$$

where

$$a = \frac{s_0^2}{2\sigma_0^2} = \frac{\rho_\phi^2}{4}\left(\frac{N_0}{E_s}\right) \tag{6.20}$$

and

$$b = \frac{s_1^2}{2\sigma_0^2} = \frac{\left(\frac{2E_s}{N_0}\right)^2 + \rho_\phi^2 + 4\left(\frac{E_s}{N_0}\right)\rho_\phi\cos\phi}{4\left(\frac{E_s}{N_0}\right)} \tag{6.21}$$

The probability of error is then obtained by averaging over the pdf of ϕ, that is

$$P_s(E) = \int_{-\pi}^{\pi} P_s(E/\phi)f_\phi(\phi)d\phi = \int_{-\pi}^{\pi} P_s(E/\phi)\frac{\exp(\rho_\phi\cos\phi)}{2\pi I_0(\rho_\phi)}d\phi \tag{6.22}$$

If we take the limit of (6.22) as ρ_ϕ approaches zero, we obtain the result for noncoherent reception of binary orthogonal signals, namely

$$\lim_{\rho_\phi\to 0} P_s(E) = \frac{1}{2}\exp\left(-\frac{E_s}{2N_0}\right) = \lim_{\rho_\phi\to 0} P_s(E/\phi) \tag{6.23}$$

which is identical to (5.79). To obtain the limit as $\rho_\phi \to \infty$, it is more convenient to return to the optimal detection metric as given by (6.13) and note that we are only interested in z_{c0} and z_{c1}. In this case, it can be shown that (see Problem 6.9)

$$\lim_{\rho_\phi\to\infty} P_s(E) = Q\left(\sqrt{\frac{E_s}{N_0}}\right) \tag{6.24}$$

which is identical to (4.59). Note that for large E_s/N_0, the limiting cases are not very far apart, since asymptotically $Q(\sqrt{E_s/N_0})$ approaches $\exp(-E_s/2N_0)/\sqrt{2\pi E_s/N_0}$ [see Appendix 3B, (3B.12)], so that for transmission with orthogonal signals, the amount of coherence is not a crucial matter for performance, as it is with antipodal signals.

6.2.3 Performance of Binary Nonorthogonal Signals

To compute the performance for arbitrary binary signals, we need to evaluate z_{c0}, z_{s0}, z_{c1}, and z_{s1} in the presence of carrier phase error ϕ. Assuming that m_0 is transmitted, then from (5.86) through (5.90), we have (let $\beta \triangleq \beta_{01}$, $\alpha \triangleq \alpha_{01}$)

$$z_{c0} = E_s\cos\phi + n_{c0} \tag{6.25a}$$

$$z_{s0} = E_s\sin\phi + n_{s0} \tag{6.25b}$$

$$z_{c1} = E_s\alpha\cos\phi - E_s\beta\sin\phi + n_{c1} \tag{6.25c}$$

$$z_{s1} = E_s\beta\cos\phi + E_s\alpha\sin\phi + n_{s1} \tag{6.25d}$$

where $E\{n_{c0}n_{c1}\} = E\{n_{s0}n_{s1}\} = (E_sN_0/2)\alpha$ and $E\{n_{c1}n_{s0}\} = -E\{n_{s1}n_{c0}\} = (E_sN_0/2)\beta$. We will follow the same procedure as in Chapter 5, where the observations are first whitened and then the probability of error is computed using (5.130). Rewriting χ_k^2 in (6.11) as

$$\chi_k^2 = Z_k^2 + Y_k^2 \tag{6.26}$$

where

$$Z_k \triangleq \frac{2}{N_0}z_{ck} + \rho_\phi \tag{6.27a}$$

$$Y_k \triangleq \frac{2}{N_0}z_{sk} \tag{6.27b}$$

then we define the vector $\mathbf{X}$ as

$$\mathbf{X}^T = (Z_0 \; Y_0 \; Z_1 \; Y_1)^T \qquad (6.28)$$

with mean

$$\mathbf{m_X} = \begin{pmatrix} \rho_\phi + (2E_s/N_0)\cos\phi \\ (2E_s/N_0)\sin\phi \\ \rho_\phi + (2E_s/N_0)(\alpha\cos\phi - \beta\sin\phi) \\ (2E_s/N_0)(\beta\cos\phi + \alpha\sin\phi) \end{pmatrix} \qquad (6.29)$$

and covariance matrix

$$\mathcal{K}_\mathbf{X} = \left(\frac{2E_s}{N_0}\right)\begin{pmatrix} 1 & 0 & \lambda & \beta \\ 0 & 1 & -\beta & \lambda \\ \lambda & -\beta & 1 & 0 \\ \beta & \lambda & 0 & 1 \end{pmatrix} \qquad (6.30)$$

As in Chapter 5, we whiten the vector $\mathbf{X}$ using the transformation

$$\mathcal{T} = \frac{\sqrt{N_0/E_s}}{2k\sqrt{1+k}}\begin{pmatrix} 1+k & 0 & -\lambda & -\beta \\ 0 & 1+k & \beta & -\lambda \\ -\lambda & \beta & 1+k & 0 \\ -\beta & -\lambda & 0 & 1+k \end{pmatrix} \qquad (6.31)$$

to get the vector

$$\mathbf{X}_w = \mathcal{T}\mathbf{X} \triangleq \begin{pmatrix} Z_{w,0} \\ Y_{w,0} \\ Z_{w,1} \\ Y_{w,1} \end{pmatrix} \qquad (6.32)$$

with (see Problem 6.10)

$$\mathcal{K}_{\mathbf{X}_w} = \mathcal{T}\mathcal{K}_\mathbf{X}\mathcal{T}^T = \mathcal{I} \qquad (6.33)$$

and

$$\mathbf{X}^T \mathcal{Q}\mathbf{X} = k\left(\frac{2E_s}{N_0}\right)\left(\mathbf{X}_w^T \mathcal{Q}\mathbf{X}_w\right) \qquad (6.34)$$

where k is given by (5.122) and $\mathcal{Q}$ by (5.119). Due to the symmetry of the problem, the conditional probability of error is given by

$$P_s(E/\phi) = \text{Prob}(\mathbf{X}^T \mathcal{Q}\mathbf{X} < 0/m_0) = \text{Prob}(\mathbf{X}_w^T \mathcal{Q}\mathbf{X}_w < 0/m_0) \qquad (6.35)$$

The components of $\mathbf{X}_w$ are conditionally uncorrelated and independent with unit variance and means

$$E\{\mathbf{X}_w\} = \mathcal{T}E\{\mathbf{X}\} = \frac{\sqrt{N_0/E_s}}{2k\sqrt{k+1}}\begin{pmatrix} \rho_\phi(1+k-\alpha) + 2(E_s/N_0)k(k+1)\cos\phi \\ \rho_\phi\beta + (2E_s/N_0)k(1+k)\sin\phi \\ \rho_\phi(1+k-\alpha) + 2(E_s/N_0)k(\alpha\cos\phi - \beta\sin\phi) \\ -\rho_\phi\beta + (2E_s/N_0)k(\alpha\sin\phi + \beta\cos\phi) \end{pmatrix} \qquad (6.36)$$

Letting $\chi_{w,i}^2 \triangleq Z_{w,i}^2 + Y_{w,i}^2$, then $\chi_{w,i}, i = 0, 1$ are Rician variables with parameters

$$s_0^2 = (E\{Z_{w,0}\})^2 + (E\{Y_{w,0}\})^2 \qquad (6.37a)$$

$$s_1^2 = (E\{Z_{w,1}\})^2 + (E\{Y_{w,1}\})^2 \qquad (6.37b)$$

and since the variables have been normalized, $\sigma_0^2 = \sigma_1^2 = 1$. The conditional probability of error is then given by (5.130), that is

$$P_s(E/\phi) = Q\left(\sqrt{a_0}, \sqrt{a_1}\right) - \frac{1}{2}\exp\left(-\frac{a_0 + a_1}{2}\right) I_0\left(\sqrt{a_0 a_1}\right) \tag{6.38}$$

where

$$a_i = \frac{1}{2}\left\{\alpha_i^2 + \beta_i^2 + 2\alpha_i\beta_i\cos(\phi - \zeta_i)\right\}, \quad i = 0, 1 \tag{6.39}$$

and

$$\alpha_0 = \alpha_1 = \frac{\rho_\phi}{k}\sqrt{\frac{1 - \alpha}{(2E_s/N_0)}} \tag{6.40a}$$

$$\beta_0 = \sqrt{(1 + k)(E_s/N_0)} \tag{6.40b}$$

$$\beta_1 = -\sqrt{(1 - k)(E_s/N_0)} \tag{6.40c}$$

$$\zeta_0 = \tan^{-1}\left(\frac{\beta}{1 + k - \alpha}\right) \tag{6.40d}$$

$$\zeta_1 = \tan^{-1}\left(\frac{\beta}{1 - k - \alpha}\right) \tag{6.40e}$$

The unconditional probability of error is then given by

$$P_s(E) = \int_{-\pi}^{\pi} P_s(E/\phi)\frac{\exp(\rho_\phi\cos\phi)}{2\pi I_0(\rho_\phi)}d\phi \tag{6.41}$$

which is evaluated numerically in Figures 6.4 and 6.5 for $\beta = 0$ and α between 0 and -1. The error probability is plotted versus ρ_ϕ where E_s/N_0 is selected so that the error probabilities for coherent reception of antipodal signals are 10^{-3} and 10^{-5}, respectively. These correspond to $E_s/N_0 = 6.8$ dB and 9.58 dB, obtained from Fig. 4.46 with $\rho_\phi = \infty$. The curve $\alpha = -1$ corresponds to antipodal signals while the case $\alpha = 0$ corresponds to orthogonal signals. The effect of nonzero imaginary signal correlation (that is, nonzero β) is illustrated in Fig. 6.6 for $E_s/N_0 = 9.58$ dB. It is clear that the best performance is obtained when $\beta = 0$.

6.2.4 Performance of M-ary Orthogonal Signals

With M-ary orthogonal signals, the receiver selects the message corresponding to the maximum χ_i. Assuming that message m_k is sent, then χ_i, $i = 0, 1, \ldots, M - 1$ are all conditionally Rician distributed with parameters as given by (6.16) and (6.17), that is

$$s_i^2 = \begin{cases} \left(\frac{2E_s}{N_0}\right)^2 + \rho_\phi^2 + 4(\frac{E_s}{N_0})\rho_\phi\cos\phi, & i = k \\ \rho_\phi^2, & i \neq k \end{cases} \tag{6.42}$$

Furthermore, z_{ci} and z_{si} have identical variances for $i = 0, 1, \ldots, M - 1$, given by $\sigma^2 = 2E_s/N_0$. The conditional probability of correct detection is then given by

$$P_s(C/m_k, \phi) = \text{Prob}\{\chi_k = \max_i \chi_i, i = 0, 1, \ldots, M - 1\} \tag{6.43}$$

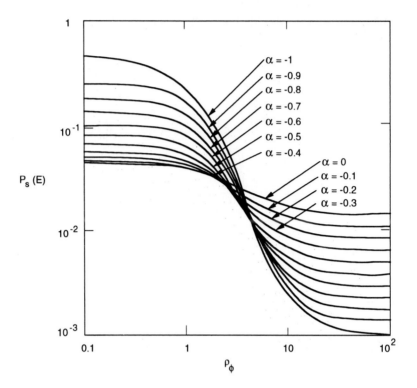

Figure 6.4   Performance of the partially coherent receiver for $M = 2$ and real nonpositive α ($\beta = 0$, $E_s/N_0 = 6.8$ dB) (Reprinted from [3] © IEEE)

Since the signals are all orthogonal, the variables χ_i are all independent. Hence

$$P_s(C/m_k, \phi) = \int_0^\infty f_{\chi_k}(x_k)$$

$$\times \underbrace{\int_0^{x_k} \cdots \int_0^{x_k} f_{\chi_0}(x_0) f_{\chi_1}(x_1) \cdots f_{\chi_{k-1}}(x_{k-1}) f_{\chi_{k+1}}(x_{k+1}) \cdots f_{\chi_{M-1}}(x_{M-1})}_{M-1 \text{ integrals}}$$

$$dx_0 dx_1 \cdots dx_{k-1} dx_{k+1} \cdots dx_{M-1} dx_k \tag{6.44}$$

The variables χ_i, $i = 0, 1, \ldots, M - 1$ and $i \neq k$ are all identically distributed. Therefore, (6.44) becomes

$$P_s(C/m_k, \phi) = \int_0^\infty f_{\chi_k}(x_k) \left[\int_0^{x_k} f_{\chi_0}(x_0) dx_0 \right]^{M-1} dx_k \tag{6.45}$$

Using the definition of $Q(a, b)$ as given by (5A.15), the inner integral reduces to

$$\int_0^{x_k} f_{\chi_0}(x_0) dx_0 = 1 - Q\left(\frac{s_0}{\sigma}, \frac{x_k}{\sigma}\right) \tag{6.46}$$

Substituting (6.46) into (6.45) and simplifying, we have

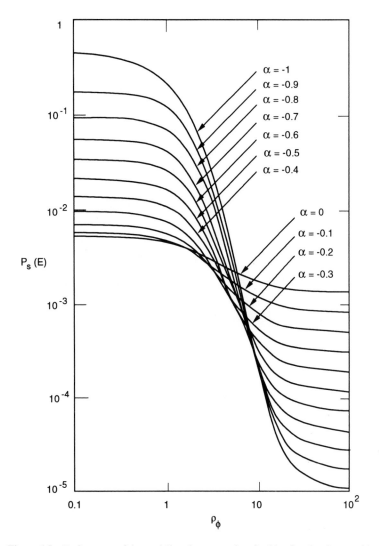

Figure 6.5   Performance of the partially coherent receiver for $M = 2$ and real nonpositive α $(\beta = 0,\ E_s/N_0 = 9.58\ \text{dB})$ (Reprinted from [3] © IEEE)

$$P_s(C/m_k, \phi) = P_s(C/\phi)$$

$$= \int_0^\infty y \exp\left(-\frac{c_1^2 + y^2}{2}\right) I_0(yc_1)\,[1 - Q(c_2, y)]^{M-1}\,dy \qquad (6.47)$$

where

$$c_1^2 = 2(E_s/N_0) + \frac{\rho_\phi^2}{2(E_s/N_0)} + 2\rho_\phi \cos \phi \qquad (6.48)$$

$$c_2^2 = \frac{\rho_\phi^2}{2(E_s/N_0)} \qquad (6.49)$$

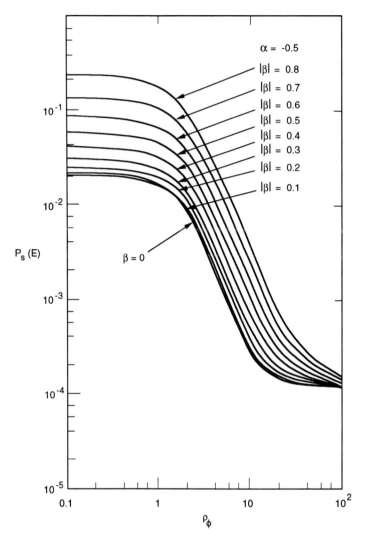

Figure 6.6   Effect of nonzero β ($M = 2$, $E_s/N_0 = 9.58$ dB) (Reprinted from [3] © IEEE)

The unconditional probability of symbol error becomes

$$P_s(E) = 1 - \int_{-\pi}^{\pi} \int_0^\infty y \exp\left(-\frac{c_1^2 + y^2}{2}\right) I_0(yc_1)$$

$$\times [1 - Q(c_2, y)]^{M-1} \frac{\exp(\rho_\phi \cos \phi)}{2\pi I_0(\rho_\phi)} dy d\phi \tag{6.50}$$

The probability of symbol and bit error [obtained using (5.154)] can then be computed numerically for various values of M and loop SNR ρ_ϕ.

6.3 OPTIMUM PARTIALLY COHERENT RECEIVERS BASED ON MULTIPLE SYMBOL OBSERVATION

In this section we consider the structure of the optimum receiver for the partially coherent channel defined by the phase error statistics of (6.1) when the observation interval is not restricted to a single symbol interval [4]. Since the presence of a constant phase error between the incoming carrier phase and the receiver's estimate of it introduces memory into the observable (assumed to be greater than T sec in duration), one should exploit this property in designing an optimum decision algorithm. In particular, one should resort to a form of *sequence* estimation wherein a *joint* decision is made on a group, say N_s, of information symbols, the length of the sequence being proportional to the duration over which the phase error can be assumed constant.

As in Section 6.1, we take a maximum-likelihood (ML) approach to this problem. All our attention will be focused on the case of M-PSK modulation. As such, the analogous results to be presented in Chapter 7 for optimum differential detection of M-PSK can be viewed as a special case of the results derived here corresponding to $\rho_\phi = 0$. We shall say more about this later on.

6.3.1 Maximum-Likelihood Block-by-Block Detection of M-PSK Signals

Since M-PSK is a two-dimensional modulation, it shall be convenient to deal with the complex representation of signals. Thus, the baseband representation of the transmitted signal in the kth transmission interval $kT_s \le t \le (k+1)T_s$ has the complex form[2]

$$\tilde{s}_k = \sqrt{2E_s/T_s}\,e^{j\theta_k} \tag{6.51}$$

where θ_k denotes the transmitted phase in this interval which for M-PSK takes on values from the set $\beta_m = 2\pi m/M$; $m = 0, 1, 2, \ldots, M-1$. Assume that the phase θ introduced by the channel can be assumed constant (independent of time) over a duration of N_s data symbols but is otherwise unknown. The bandpass representation of the received signal in this same time interval is, in accordance with (4B.1), given by

$$r(t) = \mathrm{Re}\,\{\tilde{r}(t)\} = \mathrm{Re}\,\{\tilde{s}_k \exp(j\omega_c t + \theta) + \tilde{n}(t)\} = s_k(t;\theta) + n(t) \tag{6.52}$$

where $\tilde{n}(t)$ is a complex additive white Gaussian process with single-sided PSD $2N_0$. The receiver demodulates $\tilde{r}(t)$ with the complex reference signal $\exp\left\{-j\left(\omega_c t + \hat{\theta}\right)\right\}$ to produce the complex baseband signal $\tilde{s}_k e^{j\phi} + \tilde{n}(t)e^{-j(\omega_c t + \hat{\theta})}$ which is then passed through an I&D, resulting in the complex detected sample[3]

---

2. As in Section 4.1.6, it is necessary to introduce a notation for characterizing discrete time when dealing with observations that span more than one symbol interval. As such, we shall once again use a subscript, for example, k, on a variable to denote the value of that variable in the kth transmission interval. To avoid confusion with the previous notation of θ_k to denote the kth message phase $2\pi k/M$ (k^{th} hypothesis), we introduce a new Greek symbol, that is, β_k, to denote the kth message phase. Whenever it is necessary to represent a variable *both* in discrete time and/or by an associated hypothesis, we shall use a parenthetical superscript for the latter.

3. It is convenient in what follows to introduce a normalization of $1/T_s$ into the definition of the complex I&D output sample. This normalization is denoted by the use of a script typeface.

$$\tilde{\mathcal{V}}_k = \tilde{s}_k e^{j\phi} + \left(\frac{1}{T_s} \int_{kT_s}^{(k+1)T_s} \tilde{n}(t) e^{-j\omega_c t} dt \right) e^{-j\hat{\theta}} \tag{6.53}$$

$$= \tilde{s}_k e^{j\phi} + \tilde{N}_k e^{-j\hat{\theta}}$$

where $\tilde{N}_k$ is a zero-mean complex Gaussian random variable with variance $2N_0/T_s$, or equivalently, a variance $\sigma_n^2 = N_0/T_s$ for the real and imaginary components. For an observation of the received signal over the interval $(k - N_s + 1)T_s \le t \le (k+1)T_s$, the sequence $\tilde{\mathcal{V}} = \left(\tilde{\mathcal{V}}_{k-N_s+1}, \tilde{\mathcal{V}}_{k-N_s+2}, \dots, \tilde{\mathcal{V}}_{k-1}, \tilde{\mathcal{V}}_k \right)$ conditioned on ϕ is a sufficient statistic for making an ML decision on the transmitted sequence $\tilde{s} = \left(\tilde{s}_{k-N_s+1}, \tilde{s}_{k-N_s+2}, \dots, \tilde{s}_{k-1}, \tilde{s}_k \right)$.

For the assumed AWGN model, the a posteriori pdf of $\tilde{\mathcal{V}}$ given $\tilde{s}$ and ϕ is

$$f_{\tilde{\mathcal{V}}|\phi}(\tilde{\mathcal{V}}|\tilde{s}, \phi) = \frac{1}{(2\pi\sigma_n^2)^{N_s}} \exp\left\{ -\frac{\left\| \tilde{\mathcal{V}} - \tilde{s} e^{j\phi} \right\|^2}{2\sigma_n^2} \right\}$$

$$= \frac{1}{(2\pi\sigma_n^2)^{N_s}} \exp\left\{ -\frac{1}{2\sigma_n^2} \left[\sum_{i=0}^{N_s-1} \left[\left| \tilde{\mathcal{V}}_{k-i} \right|^2 + |\tilde{s}_{k-i}|^2 \right] \right. \right. \tag{6.54}$$

$$\left. \left. -2 \left| \sum_{i=0}^{N_s-1} \tilde{\mathcal{V}}_{k-i} \tilde{s}_{k-i}^* \right| \cos(\phi - \alpha) \right] \right\}$$

where

$$\alpha \triangleq \tan^{-1} \frac{\operatorname{Im}\left\{ \sum\limits_{i=0}^{N_s-1} \tilde{\mathcal{V}}_{k-i} \tilde{s}_{k-i}^* \right\}}{\operatorname{Re}\left\{ \sum\limits_{i=0}^{N_s-1} \tilde{\mathcal{V}}_{k-i} \tilde{s}_{k-i}^* \right\}} \tag{6.55}$$

Averaging (6.54) over the pdf in (6.1) gives, upon simplification

$$f_{\tilde{\mathcal{V}}}(\tilde{\mathcal{V}}|\tilde{s}) = \int_{-\pi}^{\pi} f_{\tilde{\mathcal{V}}|\phi}(\tilde{\mathcal{V}}|\tilde{s}, \phi) f_\phi(\phi) d\phi$$

$$= \frac{1}{I_0(\rho_\phi)} \frac{1}{(2\pi\sigma_n^2)^{N_s}} \exp\left\{ -\frac{1}{2\sigma_n^2} \sum_{i=0}^{N_s-1} \left[\left| \tilde{\mathcal{V}}_{k-i} \right|^2 + |\tilde{s}_{k-i}|^2 \right] \right\} \tag{6.56}$$

$$\times I_0 \left(\frac{1}{\sigma_n^2} \sqrt{ \left(\left| \sum_{i=0}^{N_s-1} \tilde{\mathcal{V}}_{k-i} \tilde{s}_{k-i}^* \right| \cos\alpha + \rho_\phi \sigma_n^2 \right)^2 + \left(\left| \sum_{i=0}^{N_s-1} \tilde{\mathcal{V}}_{k-i} \tilde{s}_{k-i}^* \right| \sin\alpha \right)^2 } \right)$$

Since

$$\left| \sum_{i=0}^{N-1} \tilde{\mathcal{V}}_{k-i} \tilde{s}_{k-i}^* \right| \cos\alpha = \operatorname{Re}\left\{ \sum_{i=0}^{N-1} \tilde{\mathcal{V}}_{k-i} \tilde{s}_{k-i}^* \right\},$$

$$\left| \sum_{i=0}^{N-1} \tilde{\mathcal{V}}_{k-i} \tilde{s}_{k-i}^* \right| \sin\alpha = \operatorname{Im}\left\{ \sum_{i=0}^{N-1} \tilde{\mathcal{V}}_{k-i} \tilde{s}_{k-i}^* \right\} \tag{6.57}$$

Equation (6.56) further simplifies to

$$f_{\tilde{\mathcal{V}}}(\tilde{\mathcal{V}}|\tilde{s}) = \frac{1}{I_0(\rho_\phi)} \frac{1}{(2\pi\sigma_n^2)^{N_s}} \exp\left\{ -\frac{1}{2\sigma_n^2} \sum_{i=0}^{N_s-1} \left[\left|\tilde{\mathcal{V}}_{k-i}\right|^2 + |\tilde{s}_{k-i}|^2 \right] \right\} \tag{6.58}$$

$$\times I_0\left(\frac{1}{\sigma_n^2} \sqrt{ \left(\mathrm{Re}\left\{ \sum_{i=0}^{N_s-1} \tilde{\mathcal{V}}_{k-i}\tilde{s}_{k-i}^* \right\} + \rho_\phi\sigma_n^2 \right)^2 + \left(\mathrm{Im}\left\{ \sum_{i=0}^{N_s-1} \tilde{\mathcal{V}}_{k-i}\tilde{s}_{k-i}^* \right\} \right)^2 } \right)$$

Note that for M-PSK, $|\tilde{s}_k|^2$ is constant for all possible transmitted phases β_m. Thus, maximizing $f_{\tilde{\mathcal{V}}}(\tilde{\mathcal{V}}|\tilde{s})$ over $\tilde{s}$ is equivalent to finding

$$\max_{\tilde{s}} \left\{ \left(\mathrm{Re}\left\{ \sum_{i=0}^{N_s-1} \tilde{\mathcal{V}}_{k-i}\tilde{s}_{k-i}^* \right\} + \rho_\phi\sigma_n^2 \right)^2 + \left(\mathrm{Im}\left\{ \sum_{i=0}^{N_s-1} \tilde{\mathcal{V}}_{k-i}\tilde{s}_{k-i}^* \right\} \right)^2 \right\} \tag{6.59a}$$

or equivalently

$$\max_{\tilde{s}} \left\{ \left| \sum_{i=0}^{N_s-1} \tilde{\mathcal{V}}_{k-i}\tilde{s}_{k-i}^* + \rho_\phi\sigma_n^2 \right|^2 \right\}$$

$$= \max_{\tilde{s}} \left\{ \left| \sum_{i=0}^{N_s-1} \tilde{\mathcal{V}}_{k-i}\tilde{s}_{k-i}^* \right|^2 + 2\rho_\phi\sigma_n^2 \, \mathrm{Re}\left\{ \sum_{i=0}^{N_s-1} \tilde{\mathcal{V}}_{k-i}\tilde{s}_{k-i}^* \right\} \right\} \tag{6.59b}$$

Dividing (6.59a) by σ_n^4 and noting that

$$\frac{\tilde{\mathcal{V}}_{k-i}\tilde{s}_{k-i}^*}{\sigma_n^2} = \frac{1}{N_0} \int_{(k-i)T_s}^{(k-i+1)T_s} \tilde{r}(t)\tilde{s}_{k-i}^* \exp\left(-j\omega_c t\right) dt \tag{6.60}$$

where we have arbitrarily set $\hat{\theta} = 0$ and thus $\theta = \phi$, we see immediately that this metric is the generalization of (6.11) for the case of multiple symbol observation ($N_s \geq 1$). As for the single symbol observation case, the first term in (6.59a) represents the component of the decision metric associated with *noncoherent* (differential) detection, whereas the second term represents the component of the decision metric for ideal *coherent* detection. Thus, the partially coherent decision metric represents a hybrid of the two extremes with a weighting of the two terms proportional to ρ_ϕ. Note that for any nonzero value of ρ_ϕ, a decision rule based on this metric is unique. For $\rho_\phi = 0$ (differentially coherent detection), there is a phase ambiguity associated with it since the addition of an arbitrary fixed phase, say θ_a, to all N_s estimated phases $\hat{\theta}_{k-N_s+1}, \ldots, \hat{\theta}_{k-1}, \hat{\theta}_k$ results in the same decision for $\theta_{k-N_s+1}, \ldots, \theta_{k-1}, \theta_k$. In Chapter 7 we shall see that by letting $\theta_a = \theta_{k-N_s+1}$ and differentially encoding the input phases at the transmitter, that is,

$$\theta_k = \theta_{k-1} + \Delta\theta_k \tag{6.61}$$

where now $\Delta\theta_k$ denotes the input data phase corresponding to the kth transmission interval and θ_k the differentially encoded version of it, then we can turn the decision rule into one in which the phase ambiguity is resolved. For the moment, we assume $\rho_\phi \neq 0$ and thus there is no formal requirement for differentially encoding the data phase symbols.

Figure 6.7 is an illustration in complex form of a receiver implemented on the basis of (6.59b). Note that this receiver requires knowledge of the loop SNR ρ_ϕ (as well as the signal power $P = E_s/T_s$ and the noise variance σ_n^2.) The accuracy of this knowledge, which must be obtained by measurement, will have an impact on the ultimate performance of this receiver. Later we shall investigate the sensitivity of the receiver to a mismatch between the true loop SNR and the value supplied to the receiver implementation in Figure 6.7. For the moment, we assume that the receiver has perfect knowledge of ρ_ϕ and thus should outperform a conventional bit-by-bit correlation receiver that does not make use of this knowledge. The next matter under investigation is to determine just how much the optimum partially coherent sequence receiver outperforms the conventional bit-by-bit correlation receiver, which was shown in Section 6.2.1 (at least for binary signals) to be the same as the optimum partially coherent receiver based on a single symbol observation interval.

6.4 PERFORMANCE OF PARTIALLY COHERENT RECEIVERS BASED ON MULTIPLE SYMBOL OBSERVATION

To obtain a simple upper bound on the average bit error probability, $P_b(E)$, of the proposed block-by-block detection scheme, we use a union bound analogous to that used for upper bounding the performance of error correction coded systems (see Chapters 12, 13). In particular, the upper bound on $P_b(E)$ is the sum of the pairwise error probabilities associated with each N_s-symbol error sequence. Each pairwise error probability is then either evaluated directly or itself upper bounded. Mathematically speaking, let $\boldsymbol{\theta} = (\theta_{k-N_s+1}, \theta_{k-N_s+2}, \dots, \theta_{k-1}, \theta_k)$ denote the sequence of N_s data phases and $\hat{\boldsymbol{\theta}} = (\hat{\theta}_{k-N_s+1}, \hat{\theta}_{k-N_s+2}, \dots, \hat{\theta}_{k-1}, \hat{\theta}_k)$ be the corresponding sequence of detected phases. Let $\mathbf{u}$ be the sequence of $b = N_s \log_2 M$ information bits that produces $\boldsymbol{\theta}$ at the transmitter and $\hat{\mathbf{u}}$ the sequence of b bits that result from the detection of $\hat{\boldsymbol{\theta}}$. Then, since M-PSK is a symmetric signaling set, that is, it satisfies a uniform error probability (UEP) criterion,[4] we get the upper bound on conditional (on ϕ) probability of error[5]

$$P_b(E \mid \phi) \le \frac{1}{N_s \log_2 M} \sum_{\boldsymbol{\theta} \ne \hat{\boldsymbol{\theta}}} w(\mathbf{u}, \hat{\mathbf{u}}) \Pr\left\{ \hat{\eta} > \eta \mid \boldsymbol{\theta}, \phi \right\} \tag{6.62}$$

where $w(\mathbf{u}, \hat{\mathbf{u}})$ denotes the Hamming distance between $\mathbf{u}$ and $\hat{\mathbf{u}}$, $\boldsymbol{\theta}$ is any input sequence, (for example, the null sequence $(0, 0, \dots, 0) = \mathbf{0}$), and $\Pr\{\hat{\eta} > \eta \mid \boldsymbol{\theta}, \phi\}$ denotes the conditional pairwise probability that $\hat{\boldsymbol{\theta}}$ is incorrectly chosen when indeed $\boldsymbol{\theta}$ was sent.

Finally, the decision statistic η is defined from (6.51) and (6.59b) by

$$\eta = \left| \sum_{i=0}^{N_s-1} \tilde{V}_{k-i} e^{-j\theta_{k-i}} + \frac{\rho_\phi \sigma_n^2}{\sqrt{2E_s/T_s}} \right|^2 \tag{6.63}$$

and the corresponding error statistic $\hat{\eta}$ is identical to (6.63) with each θ_k replaced by $\hat{\theta}_k$.

---

4. By UEP we mean that the probability of error conditioned on having sent a particular phase is independent of which phase is indeed sent.

5. The result in (6.62) is a special case of the more general result in (4.88) wherein M is replaced by M^{N_s} and because of the UEP condition, the double sum collapses to M times a single sum. Furthermore, a dependence on ϕ is now included in the notation for the bit and pairwise error probabilities.

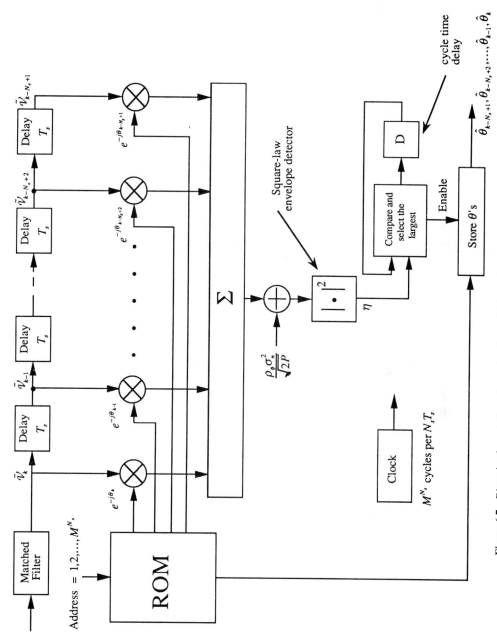

Figure 6.7 Direct implementation of a receiver for multiple symbol detection of partially coherent M-PSK

423

6.4.1 Evaluation of the Pairwise Error Probability

To compute $\Pr\{\hat{\eta} > \eta | \boldsymbol{\theta}, \phi\}$, we use the approach taken in Section 6.2.2 for evaluating the probability of one Rician random variable exceeding another independent Rician random variable. In particular, letting $\eta = |z_1|^2$ and $\hat{\eta} = |z_2|^2$ [see (6.63) and the statement below it for the definitions of z_1 and z_2], then[6]

$$
\Pr\{\hat{\eta} > \eta | \boldsymbol{\theta}, \phi\} = \Pr\{|z_2| > |z_1| \, | \boldsymbol{\theta}, \phi\}
$$

$$
= \frac{1}{2}\left[1 - Q(\sqrt{b}, \sqrt{a}) + Q(\sqrt{a}, \sqrt{b})\right] \triangleq f(a, b) \tag{6.64}
$$

where the arguments of the Marcum-Q function are given by

$$
\left\{\begin{matrix} b \\ a \end{matrix}\right\} = \frac{1}{2N_z}\left\{\frac{S_1 + S_2 - 2|\rho_z|\sqrt{S_1 S_2}\cos(\Theta_1 - \Theta_2 + v)}{1 - |\rho_z|^2} \pm \frac{S_1 - S_2}{\sqrt{1 - |\rho_z|^2}}\right\} \tag{6.65}
$$

with

$$
S_1 \triangleq \frac{1}{2}|\overline{z_1}|^2 = \frac{E_s}{T_s}\left|N_s + \frac{\rho_\phi}{2E_s/N_0}e^{-j\phi}\right|^2 = \frac{E_s}{T_s}\left(N_s^2 + \frac{\rho_\phi}{E_s/N_0}N_s\cos\phi + \left(\frac{\rho_\phi}{2E_s/N_0}\right)^2\right)
$$

$$
S_2 \triangleq \frac{1}{2}|\overline{z_2}|^2 = \frac{E_s}{T_s}\left|\delta + \frac{\rho_\phi}{2E_s/N_0}e^{-j\phi}\right|^2 = \frac{E_s}{T_s}\left(|\delta|^2 + \frac{\rho_\phi}{E_s/N_0}\operatorname{Re}\{\delta e^{j\phi}\} + \left(\frac{\rho_\phi}{2E_s/N_0}\right)^2\right)
$$

$$
N_z \triangleq \frac{1}{2}\overline{|z_1 - \overline{z_1}|^2} = \frac{1}{2}\overline{|z_2 - \overline{z_2}|^2} = N_s\frac{N_0}{T_s} \tag{6.66}
$$

$$
\rho_z \triangleq \frac{1}{2N_z}\overline{(z_1 - \overline{z_1})(z_2 - \overline{z_2})^*} = \frac{\delta}{N_s}; \quad v \triangleq \arg\rho_z = \arg\delta
$$

$$
\Theta_1 \triangleq \arg\overline{z_1} = \arg\left\{N_s e^{j\phi} + \frac{\rho_\phi}{2E_s/N_0}\right\}; \quad \Theta_2 \triangleq \arg\overline{z_2} = \arg\left\{\delta e^{j\phi} + \frac{\rho_\phi}{2E_s/N_0}\right\}
$$

and

$$
\delta \triangleq \sum_{i=0}^{N_s-1} e^{j\left(\theta_{k-i} - \hat{\theta}_{k-i}\right)} \tag{6.67}
$$

Substituting (6.66) into (6.65) results, after considerable simplification, in

$$
\left\{\begin{matrix} b \\ a \end{matrix}\right\} = \frac{E_s}{2N_0}\left\{N_s\left[1 + \frac{1}{N_s}\left(\frac{\rho_\phi}{E_s/N_0}\right)\cos\phi\right]\right.
$$

$$
\left. + \frac{1}{2N_s(N_s^2 - |\delta|^2)}\left(\frac{\rho_\phi}{E_s/N_0}\right)^2(N_s - |\delta|\cos v)\right\} \tag{6.68}
$$

$$
\pm \frac{E_s}{2N_0}\left[\sqrt{N_s^2 - |\delta|^2} + \frac{1}{\sqrt{N_s^2 - |\delta|^2}}\left(\frac{\rho_\phi}{E_s/N_0}\right)(N_s\cos\phi - |\delta|\cos(\phi + v))\right]
$$

We now consider some special cases of practical interest.

---

6. The equivalence between (6.19) and (6.64) is shown in [5].

6.4.2 Special Cases

Case 1. *BPSK with Two-Symbol Observation and Detection ($M = 2$, $N_s = 2$)*
In this case, $E_s/N_0 = E_b/N_0$. There are $M^2 - 1 = 3$ possible error sequences each of length 2. The pertinent results related to the evaluation of (6.67) and (6.68) are given below:

$\theta_k - \hat{\theta}_k$	$\theta_{k-1} - \hat{\theta}_{k-1}$	δ
0	π	0
π	0	0
π	π	-2

For the first two error sequences, (6.68) evaluates to

$$b = \frac{E_b}{2N_0}\left[4 + 2\left(\frac{\rho_\phi}{E_b/N_0}\right)\cos\phi + \frac{1}{4}\left(\frac{\rho_\phi}{E_b/N_0}\right)^2\right]$$

$$a = \frac{E_b}{2N_0}\left[\frac{1}{4}\left(\frac{\rho_\phi}{E_b/N_0}\right)^2\right] \tag{6.69}$$

where $E_b/N_0 = (E_s/N_0)/\log_2 M$ is, as always, the bit energy-to-noise spectral density ratio. For the third error sequence, both a and b approach infinity (the ratio a/b, however, approaches unity) as δ approaches -2. Thus, we must evaluate (6.64) separately for this case. It is straightforward to show that

$$\lim_{\substack{a \to \infty \\ b \to \infty \\ a/b \to 1}} f(a,b) = \frac{1}{2}\,\mathrm{erfc}\left(\sqrt{\frac{b}{2}} - \sqrt{\frac{a}{2}}\right) \tag{6.70}$$

Furthermore, in the general case where $\delta \to -N_s$, (6.70) evaluates to

$$\lim_{\delta \to -N_s} f(a,b) = \frac{1}{2}\,\mathrm{erfc}\left(\sqrt{N_s\frac{E_b}{N_0}}\cos\phi\right) \tag{6.71}$$

which for $N_s = 2$ and $M = 2$ becomes

$$\lim_{\delta \to -2} f(a,b) = \frac{1}{2}\,\mathrm{erfc}\left(\sqrt{\frac{2E_b}{N_0}}\cos\phi\right) \tag{6.72}$$

Finally, noting that the Hamming distance $w(\mathbf{u}, \hat{\mathbf{u}})$ is equal to 1 for the first two error sequences and is equal to 2 for the third sequence, then substituting (6.72) and (6.64) combined with (6.68) and (6.69) into (6.62) gives

$$P_b(E\,|\phi) \le f\left(\frac{E_b}{N_0}\left[\frac{1}{8}\left(\frac{\rho_\phi}{E_b/N_0}\right)^2\right], \frac{E_b}{N_0}\left[2 + \left(\frac{\rho_\phi}{E_b/N_0}\right)\cos\phi + \frac{1}{8}\left(\frac{\rho_\phi}{E_b/N_0}\right)^2\right]\right)$$

$$+ \frac{1}{2}\,\mathrm{erfc}\left(\sqrt{\frac{2E_b}{N_0}}\cos\phi\right) \tag{6.73}$$

Finally, the upper bound on average bit error probability $P_b(E)$ is obtained by averaging the upper bound in (6.73) over the pdf in (6.1). Figures 6.8 and 6.9 are plots of this upper bound

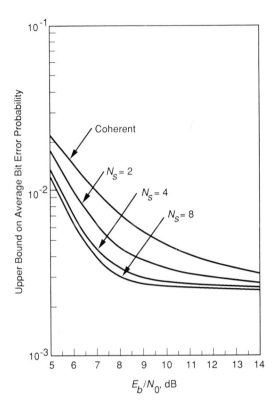

Figure 6.8 Upper bound on average bit error probability versus E_b/N_0 for optimum block-by-block detection receiver with N_s as a parameter; $M = 2$, $\rho_\phi = 7$ dB

on average probability versus E_b/N_0 in dB for values of $\rho_\phi = 7$ dB and 10 dB. For purposes of comparison, we also show the exact results for the conventional ideal coherent metric (or optimum partially coherent metric with $N_s = 1$) operating in a noisy carrier synchronization environment as given by (6.14). We see that even with only one additional observation bit interval, considerable savings in E_b/N_0 can be achieved at a fixed error probability, particularly in the region of the knee of the curve where the system begins approaching its irreducible error probability[7] asymptote.

Case 2.   *BPSK with N_s-Symbol Observation and Detection ($M = 2$, N_s arbitrary)*
      For N_s arbitrary, δ takes on values $-(N_s - 2i)$; $i = 0, 1, 2, \ldots, N_s - 1$. The number of error sequences corresponding to each of these values of δ is binomially distributed, that is, there are $\binom{N_s}{i}$ sequences that yield a value $\delta = -(N_s - 2i)$. Furthermore, the Hamming weight associated with each of the $\binom{N_s}{i}$ sequences that yield a value $\delta =$

---

7. We have seen in Chapter 4 (for example, Fig. 4.46) and also earlier in this chapter that conventional BPSK systems exhibit an irreducible error probability (i.e., a finite error probability in the limit as E_b/N_0 approaches infinity) when a noisy carrier synchronization reference *with fixed power* is used as a demodulation signal. This is observed by examining a curve of $P_b(E)$ vs. E_b/N_0 with loop SNR, ρ_ϕ, held fixed. The value of this irreducible error probability is given by $P_b(E)|_{irr} = \int_{\pi/2}^{\pi} f_\phi(\phi)d\phi$. Unfortunately, increasing the observation interval and using sequence estimation does not reduce this error. Thus, any gain due to performing block detection must take place prior to approaching this asymptotic limit.

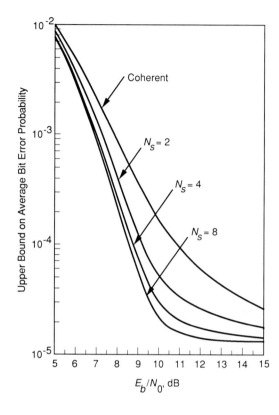

Figure 6.9 Upper bound on average bit error probability versus E_b/N_0 for optimum block-by-block detection receiver with N_s as a parameter; $M = 2$, $\rho_\phi = 10$ dB

$-(N_s - 2i)$ is $w(\mathbf{u}, \hat{\mathbf{u}}) = N_s - i$. Finally then, using the above in (6.68) and (6.71) and substituting the results in (6.62), the conditional bit error probability is upper bounded by

$$P_b(E \mid \phi) \leq \frac{1}{2}\, \mathrm{erfc}\left(\sqrt{N_s \frac{E_b}{N_0}}\cos\phi\right) + \frac{1}{N_s}\sum_{i=1}^{N_s-1}\binom{N_s}{i}(N_s - i)\, f(a_i, b_i)$$

$$= \frac{1}{2}\, \mathrm{erfc}\left(\sqrt{N_s \frac{E_b}{N_0}}\cos\phi\right) + \sum_{i=1}^{N_s-1}\binom{N_s - 1}{i}f(a_i, b_i)$$

(6.74)

where

$$\left\{\begin{matrix} b_i \\ a_i \end{matrix}\right\} = \frac{E_b}{2N_0}\left\{\left[N_s + \left(\frac{\rho_\phi}{E_b/N_0}\right)\cos\phi + \frac{1}{4i}\left(\frac{\rho_\phi}{2E_b/N_0}\right)^2\right]\right.$$

$$\left.\pm\left[2\sqrt{i(N_s - i)} + \sqrt{\frac{N_s - i}{i}}\left(\frac{\rho_\phi}{E_b/N_0}\right)\cos\phi\right]\right\}$$

(6.75)

The upper bound on unconditional average bit error probability is now obtained by averaging (6.74) over the pdf in (6.1). The numerical results are illustrated in Figures 6.8 and 6.9 for values of $N_s = 4$, 6, and 8. We see that as N_s gets large, the curves appear to approach an asymptote. This asymptotic behavior is analytically evaluated as follows.

For N_s large, the first term in (6.74) when integrated over the pdf in (6.1) approaches the irreducible error probability $P_b(E)|_{irr} = \int_{\pi/2}^{\pi} f_\phi(\phi)d\phi$. Also, the dominant term in the summation term of (6.74) corresponds to $i = N_s - 1$, that is, $\delta = N_s - 2$. Thus, for large N_s, the second term of (6.74) approaches $f(a_{N-1}, b_{N-1})$ where

$$\begin{Bmatrix} b_{N_s-1} \\ a_{N_s-1} \end{Bmatrix} \cong \frac{E_b}{2N_0} \left\{ N_s \pm 2\sqrt{(N_s - 1)} \right\} \tag{6.76}$$

Since, from (6.76), $\sqrt{b_{N_s-1}} \gg \sqrt{b_{N_s-1}} - \sqrt{a_{N_s-1}}$, then using the asymptotic form of (6.64) for a and b large (see [5], Appendix A), namely

$$f(a, b) \cong \frac{1}{2} \text{erfc} \left(\frac{\sqrt{b} - \sqrt{a}}{2} \right) \tag{6.77}$$

we obtain for large N_s

$$f(a_{N_s-1}, b_{N_s-1}) \cong \frac{1}{2} \text{erfc} \left\{ \sqrt{\frac{E_b}{N_0}} \right\} \tag{6.78}$$

independent of ϕ. Finally then, for large N_s, the asymptotic behavior of the average bit error probability is approximately upper bounded by

$$P_b \leq \frac{1}{2} \text{erfc} \sqrt{\frac{E_b}{N_0}} + 2 \int_{\pi/2}^{\pi} f_\phi(\phi)d\phi \tag{6.79}$$

namely, *the sum of the performance for ideal coherent detection and the error floor.*

   Equation (6.79) is in very close agreement with the curves for $N_s = 8$ in Figures 6.8 and 6.9.

Case 3.   *QPSK with Two-Symbol Observation and Detection ($M = 4$, $N_s = 2$)*
   In this case, $E_s/N_0 = 2E_b/N_0$. There are now a total of $M^2 - 1 = 15$ possible error sequences, each of length 2. Of these, only 8 produce distinct combinations of $|\delta|$ and ν. These are tabulated below:

Err. Seq.	Case	$\theta_k - \hat{\theta}_k$	$\theta_{k-1} - \hat{\theta}_{k-1}$	$|\delta|$	ν
1, 2, 3, 4	1	$\pi, 0, 3\pi/2, \pi/2$	$0, \pi, \pi/2, 3\pi/2$	0	0
5, 6	2	$0, \pi/2$	$\pi/2, 0$	$\sqrt{2}$	$\pi/4$
7, 8	3	$0, 3\pi/2$	$3\pi/2, 0$	$\sqrt{2}$	$-\pi/4$
9, 10	4	$\pi/2, \pi$	$\pi, \pi/2$	$\sqrt{2}$	$3\pi/4$
11, 12	5	$\pi, 3\pi/2$	$3\pi/2, \pi$	$\sqrt{2}$	$-3\pi/4$
13	6	$\pi/2$	$\pi/2$	2	$\pi/2$
14	7	$3\pi/2$	$3\pi/2$	2	$-\pi/2$
15	8	π	π	2	π

The corresponding values of a and b for each of the first five cases are given as follows:

$$
\left\{ \begin{array}{c} b \\ a \end{array} \right\}_1 = \frac{E_b}{N_0} \left\{ \left[2 + \left(\frac{\rho_\phi}{2E_b/N_0} \right) \cos \phi + \frac{1}{4} \left(\frac{\rho_\phi}{2E_b/N_0} \right)^2 \right] \right.
$$
$$
\left. \pm \left[2 + \left(\frac{\rho_\phi}{2E_b/N_0} \right) \cos \phi \right] \right\}
$$

$$
\left\{ \begin{array}{c} b \\ a \end{array} \right\}_2 = \frac{E_b}{N_0} \left\{ \left[2 + \left(\frac{\rho_\phi}{2E_b/N_0} \right) \cos \phi + \frac{1}{4} \left(\frac{\rho_\phi}{2E_b/N_0} \right)^2 \right] \right.
$$
$$
\left. \pm \left[\sqrt{2} + \frac{1}{\sqrt{2}} \left(\frac{\rho_\phi}{2E_b/N_0} \right) (\cos \phi + \sin \phi) \right] \right\}
$$

$$
\left\{ \begin{array}{c} b \\ a \end{array} \right\}_3 = \frac{E_b}{N_0} \left\{ \left[2 + \left(\frac{\rho_\phi}{2E_b/N_0} \right) \cos \phi + \frac{1}{4} \left(\frac{\rho_\phi}{2E_b/N_0} \right)^2 \right] \right.
$$
$$
\left. \pm \left[\sqrt{2} + \frac{1}{\sqrt{2}} \left(\frac{\rho_\phi}{2E_b/N_0} \right) (\cos \phi - \sin \phi) \right] \right\}
$$

$$
\left\{ \begin{array}{c} b \\ a \end{array} \right\}_4 = \frac{E_b}{N_0} \left\{ \left[2 + \left(\frac{\rho_\phi}{2E_b/N_0} \right) \cos \phi + \frac{3}{4} \left(\frac{\rho_\phi}{2E_b/N_0} \right)^2 \right] \right. \tag{6.80}
$$
$$
\left. \pm \left[\sqrt{2} + \frac{1}{\sqrt{2}} \left(\frac{\rho_\phi}{2E_b/N_0} \right) (3\cos \phi + \sin \phi) \right] \right\}
$$

$$
\left\{ \begin{array}{c} b \\ a \end{array} \right\}_5 = \frac{E_b}{N_0} \left\{ \left[2 + \left(\frac{\rho_\phi}{2E_b/N_0} \right) \cos \phi + \frac{3}{4} \left(\frac{\rho_\phi}{2E_b/N_0} \right)^2 \right] \right.
$$
$$
\left. \pm \left[\sqrt{2} + \frac{1}{\sqrt{2}} \left(\frac{\rho_\phi}{2E_b/N_0} \right) (3\cos \phi - \sin \phi) \right] \right\}
$$

For cases 6 and 7, we have analogous to (6.71)

$$
\lim_{\delta \to \pm jN_s} f(a, b) = \frac{1}{2} \operatorname{erfc} \left(\sqrt{\frac{N_s E_b}{2N_0}} (\cos \phi \pm \sin \phi) \right) \tag{6.81}
$$

which for $N_s = 2$ and $M = 4$ becomes

$$
\lim_{\delta \to \pm j2} f(a, b) = \frac{1}{2} \operatorname{erfc} \left(\sqrt{\frac{2E_b}{N_0}} (\cos \phi \pm \sin \phi) \right) \tag{6.82}
$$

Finally, for case 8, we use (6.71) to obtain

$$
\lim_{\delta \to -2} f(a, b) = \frac{1}{2} \operatorname{erfc} \left(\sqrt{\frac{4E_b}{N_0}} \cos \phi \right) \tag{6.83}
$$

Evaluating the Hamming distances for the 15 error sequences and substituting the above results into (6.62) gives

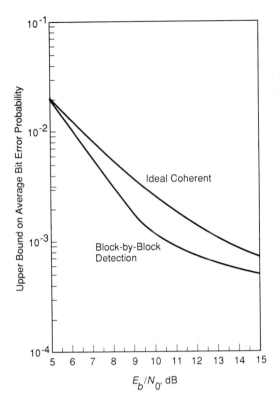

Figure 6.10   Upper bound on average bit error probability versus E_b/N_0 for optimum block-by-block detection receiver and comparison with exact performance of ideal coherent metric; $M = 4$, $N_s = 2$, $\rho_\phi = 13$ dB

$$P_b\left(E\mid\phi\right) \leq \frac{1}{4}\left\{6f\left(a_1, b_1\right) + 2f\left(a_2, b_2\right) + 2f\left(a_3, b_3\right) + 4f\left(a_4, b_4\right) + 4f\left(a_5, b_5\right)\right\}$$

$$+ \frac{1}{4}\left\{\mathrm{erfc}\left(\sqrt{\frac{2E_b}{N_0}}\left(\cos\phi + \sin\phi\right)\right) + \mathrm{erfc}\left(\sqrt{\frac{2E_b}{N_0}}\left(\cos\phi - \sin\phi\right)\right)\right. \tag{6.84}$$

$$\left. + \mathrm{erfc}\left(\sqrt{\frac{4E_b}{N_0}}\cos\phi\right)\right\}$$

Figures 6.10 and 6.11 are the comparable plots to Figures 6.8 and 6.9 for the $M = 4$ (QPSK) case. The analytical exact result corresponding to the ideal coherent metric operating in a noisy carrier synchronization environment is now

$$P_b(E) = \int_{-\pi}^{\pi} \frac{1}{4}\,\mathrm{erfc}\left(\sqrt{\frac{E_b}{N_0}}\left(\cos\phi + \sin\phi\right)\right) f_\phi(\phi)d\phi$$

$$\tag{6.85}$$

$$+ \int_{-\pi}^{\pi} \frac{1}{4}\,\mathrm{erfc}\left(\sqrt{\frac{E_b}{N_0}}\left(\cos\phi - \sin\phi\right)\right) f_\phi(\phi)d\phi$$

6.4.3 Performance Sensitivity to Mismatch

Here we investigate the sensitivity of the average bit error probability (in terms of its upper bound) of Fig. 6.7 to a mismatch between the true loop SNR, ρ_ϕ, and the estimate of it,

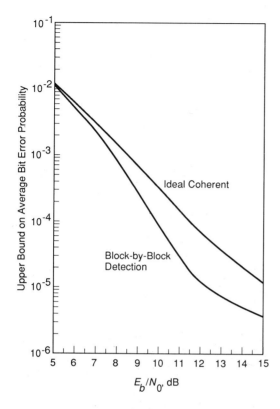

Figure 6.11 Upper bound on average bit error probability versus E_b/N_0 for optimum block-by-block detection receiver and comparison with exact performance of ideal coherent metric; $M = 4$, $N_s = 2$, $\rho_\phi = 16$ dB

$\hat{\rho}_\phi$, supplied to this implementation. In particular, we shall evaluate for the special cases of Section 6.4.2 the upper bound

$$P_b(E) \le \int_{-\pi}^{\pi} P_{bu}\left(E \,|\, \phi, \hat{\rho}_\phi\right) f_\phi(\phi) d\phi \qquad (6.86)$$

where $P_{bu}\left(E \,|\, \phi, \hat{\rho}_\phi\right)$ is given by the upper bound in (6.73) or (6.74) with ρ_ϕ replaced by $\hat{\rho}_\phi = \rho_\phi\left[1 + (\hat{\rho}_\phi - \rho_\phi)/\rho_\phi\right] \triangleq \rho_\phi\left[1 + \varepsilon_\phi\right]$. Figures 6.12 and 6.13 are illustrations of (6.86) for $M = 2$, $\rho_\phi = 10$ dB, and $N_s = 2$ and 8, respectively, with fractional mismatch ε_ϕ as a parameter. We observe that even with mismatches as much as 50% ($\varepsilon_\phi = \pm 0.5$), there is negligible effect on the error probability performance. Thus, we conclude that the block-by-block detection receiver is quite insensitive to mismatch in the parameter ρ_ϕ.

PROBLEMS

6.1 The random phase inserted into the transmitted signal $s_i(t) = a_i(t) \cos(\omega_c t + \theta_i(t) + \phi)$ is governed by the probability density function

$$f(\phi) = \frac{1}{4}|\sin \phi|, \quad |\phi| < \pi$$

for all $i = 0, 1 \ldots, M - 1$. Determine the optimum receiver when white Gaussian noise is present.

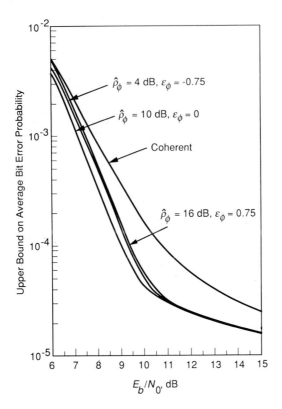

Figure 6.12 Upper bound on average bit error probability versus E_b/N_0 for optimum block-by-block detection receiver in the presence of loop SNR mismatch; $M = 2$, $N_s = 2$, $\rho_\phi = 10$ dB

6.2 Consider the following binary hypothesis testing problem:

$$H_0: \quad r(t) = n(t) \quad 0 \leq t \leq T$$

$$H_1: \quad r(t) = \sqrt{2E}a(t)\cos(\omega_c t + \theta(t) + \phi) + n(t) \quad 0 \leq t \leq T$$

where ϕ is a random variable with the Tikhonov density given by (6.1) and $n(t)$ is AWGN with PSD $N_0/2$.

(a) Derive the decision rule that minimizes the probability of error.

(b) Express the probability of false alarm (choosing H_1 when H_0 is true) in terms of the Marcum Q-function.

(c) Express the probability of detection as an integral of the Marcum Q-function.

6.3 Consider the M-PSK communication system and assume that

$$H_i: \quad r(t) = \sqrt{2P}\cos\left(\omega_c t + i\frac{\pi}{M} + \phi\right) + n(t), \quad 0 \leq t \leq T$$

for $i = 0, 1, \ldots, M - 1$ where ϕ is a random variable with the Tikhonov density given by (6.1) and $n(t)$ is AWGN with PSD $N_0/2$.

(a) Show that the optimum receiver is independent of ρ_ϕ.

(b) What is the minimum number of correlators required in implementing the optimum receiver?

(c) Express the probability of error as a multidimensional integral of the multidimensional Gaussian density.

6.4 (a) Derive the optimum receiver in an AWGN channel (N_0 one-sided PSD) for M-DPSK

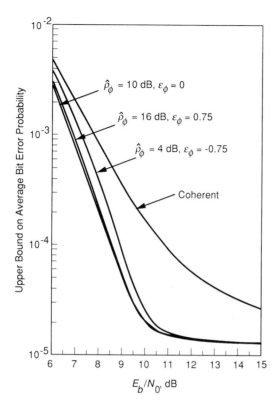

Figure 6.13 Upper bound on average bit error probability versus E_b/N_0 for optimum block-by-block detection receiver in the presence of loop SNR mismatch; $M = 2$, $N_s = 8$, $\rho_\phi = 10$ dB

signaling if the phase uncertainty is given by the Tikhonov density. Assume signal models of the form

$$s_{j-1}(t) = \sqrt{2P}\,\sin(\omega_c t + \theta^{(j-1)} + \phi), \quad 0 \le t \le T$$

$$s_j(t) = \sqrt{2P}\,\sin(\omega_c t + \theta^{(j)} + \phi), \quad T \le t \le 2T$$

where $\theta^{(j)} - \theta^{(j-1)} = 2\pi i/M$ for $i = 0, 1, \ldots, M - 1$.

(b) If $\rho_\phi \to 0$, what is the optimum receiver?

(c) If $\rho_\phi \to \infty$, what is the optimum receiver?

6.5 (a) Derive an expression for the bit error probability when $M = 2$ for the receiver in Problem 6.4(a).

(b) What does the expression in (a) reduce to when $\rho_\phi = 0$? Discuss your answer.

(c) What does the expression in (a) reduce to when $\rho_\phi = \infty$? Discuss your answer.

6.6 A binary source produces equilikely symbols m_0 and m_1. The transmitter emits $s_0(t)$ or $s_1(t)$ defined by

$$s_0(t) = \sqrt{2P}\,\cos \omega_c t, \quad 0 \le t \le T$$

$$s_1(t) = -\sqrt{2P}\,\cos \omega_c t, \quad 0 \le t \le T$$

and the channel randomly phase shifts $s_0(t)$ or $s_1(t)$ through ϕ radians with probability density function

$$f(\phi) = \frac{1}{2}\left[\delta\left(\phi - \frac{\pi}{2}\right) + \delta\left(\phi + \frac{\pi}{2}\right)\right]$$

and then adds white Gaussian noise (N_0 one-sided PSD) to the resulting signal to produce the received signal

$$r(t) = s_k(t, \phi) + n(t), \quad 0 \le t \le T$$

(a) Derive the optimum decision rule.
(b) Illustrate the optimum receiver.
(c) Derive the error probability performance.

6.7 Repeat Problem 6.6 if FSK signals are used

$$s_0(t) = \sqrt{2P} \cos \omega_0 t$$

$$s_1(t) = \sqrt{2P} \cos \omega_1 t$$

where $|f_0 - f_1| = 1/T$ Hz.

6.8 Consider the Rician diversity channel described in Problem 5.19. Assume that the phase of the specular component is not known exactly but is distributed according to

$$f_{\delta_i}(\delta) = \frac{\exp(\rho_\delta \cos \delta)}{2\pi I_0(\rho_\delta)}, \quad |\delta| \le \pi$$

and that the δ_i's are independent from each other and all other random variables in the model.
(a) Derive the optimum receiver when $L = 2$.
(b) Generalize your result to an arbitrary L.
(c) Simplify the optimum receiver if $\rho_\delta = 0$.
(d) Simplify the optimum receiver if $\rho_\delta = \infty$.

6.9 Verify for binary orthogonal signals that the limit of (6.22) as $\rho_\phi \to \infty$ is indeed given by (6.24).

6.10 Verify that the whitening transformation as given by (6.31) produces $\mathcal{K}_{\mathbf{X}_w} = \mathcal{I}$ as claimed in (6.33).

6.11 Consider the received signal of (6.52) corresponding to the transmission of M-PSK over an AWGN channel with unknown phase θ. Assuming binary modulation ($M = 2$), then passing $r(t)$ through an I&D filter produces at time $t = (k + 1)T_b$ the received sample

$$\tilde{r}_k = \sqrt{\frac{2E_b}{T_b}} e^{j(\theta_k + \theta)} + \tilde{n}_k$$

where θ_k denotes the transmitted information phase which takes on values of 0 or π and n_k is a sample of a zero-mean complex Gaussian noise process with variance $\sigma_n^2 = N_0/T_b$. Consider a received sequence $\tilde{\mathbf{r}} = (\tilde{r}_{k-1}, \tilde{r}_{k-2}, \ldots, \tilde{r}_{k-N_s})$ of length N_s and assume that the unknown channel phase θ is constant (in time) over the length of this sequence.

(a) Write an expression, analogous to (6.54), for the conditional pdf of $\tilde{\mathbf{r}}$ given the information phase sequence $\boldsymbol{\theta} = (\theta_{k-1}, \theta_{k-2}, \ldots, \theta_{k-N_s})$ and channel phase θ. Show that this pdf can be expressed in the form

$$f_{\tilde{\mathbf{r}}|\boldsymbol{\theta}}(\tilde{\mathbf{r}}|\boldsymbol{\theta}, \theta) = F \exp \left\{ \alpha \sum_{i=1}^{N_s} \text{Re} \left\{ \tilde{r}_{k-i} e^{-j(\theta_{k-i} + \theta)} \right\} \right\}$$

where F is a constant independent of the data and

$$\alpha = \frac{\sqrt{2E_b/T_b}}{\sigma_n^2} = \frac{\sqrt{2E_b T_b}}{N_0}.$$

(b) Show that the average of the conditional pdf $f_{\tilde{\mathbf{r}}|\boldsymbol{\theta}}(\tilde{\mathbf{r}}|\boldsymbol{\theta}, \theta)$ over the information phase sequence is given by

$$f_{\tilde{\mathbf{r}}|\theta} \left(\tilde{\mathbf{r}} \,|\theta \right) = E_{\boldsymbol{\theta}} \left\{ f_{\tilde{\mathbf{r}}|\theta} \left(\tilde{\mathbf{r}} \,|\boldsymbol{\theta}, \theta \right) \right\} = F \prod_{i=1}^{N_s} \cosh \left(\alpha \operatorname{Re} \left\{ \tilde{r}_{k-i} e^{-j\theta} \right\} \right)$$

For a given observation (received sequence), we would like to determine the maximum-likelihood estimate, $\hat{\theta}_{ML}$, of the channel phase θ and then use this as a demodulation reference phase resulting in a pseudocoherent receiver (a form of partial coherence).

(c) Show that $\hat{\theta}_{ML}$ defined by

$$\hat{\theta}_{ML} = \max_{\theta}{}^{-1} f_{\tilde{\mathbf{r}}|\theta} \left(\tilde{\mathbf{r}} \,|\theta \right) = \max_{\theta}{}^{-1} \ln f_{\tilde{\mathbf{r}}|\theta} \left(\tilde{\mathbf{r}} \,|\theta \right)$$

is the solution to

$$\sum_{i=1}^{N_s} \tanh \left(\alpha \operatorname{Re} \left\{ \tilde{r}_{k-i} e^{-j\theta} \right\} \operatorname{Re} \left\{ j\tilde{r}_{k-i} e^{-j\theta} \right\} \right) = 0$$

(d) For small SNR, we can approximate $\tanh x$ by x. Making this approximation in the result of part (c), show that the maximum-likelihood phase estimate is given by

$$e^{j\hat{\theta}_{ML}} = \left(\frac{\displaystyle\sum_{i=1}^{N_s} \tilde{r}_{k-i}^2}{\left| \displaystyle\sum_{i=1}^{N_s} \tilde{r}_{k-i}^2 \right|} \right)^{1/2}$$

(e) Sketch in block diagram form the complex representation of a pseudo-coherent receiver that implements the above phase estimation and performs detection of θ_k using $\hat{\theta}_{ML}$ (actually $e^{j\hat{\theta}_{ML}}$) as a demodulation reference.

6.12 Redo parts (b)–(e) of Problem 6.11 for the case of arbitrary $M = 2^m$. In particular:

(a) Show that the average of the conditional pdf $f_{\tilde{\mathbf{r}}|\theta} \left(\tilde{\mathbf{r}} \,|\boldsymbol{\theta}, \theta \right)$ over the information phase sequence is now given by

$$f_{\tilde{\mathbf{r}}|\theta} \left(\tilde{\mathbf{r}} \,|\theta \right) = E_{\boldsymbol{\theta}} \left\{ f_{\tilde{\mathbf{r}}|\theta} \left(\tilde{\mathbf{r}} \,|\boldsymbol{\theta}, \theta \right) \right\} = F' \prod_{i=1}^{N_s} \sum_{l=0}^{\frac{M}{2}-1} \cosh \left(\alpha \operatorname{Re} \left\{ \tilde{r}_{k-i} e^{-j(\theta + 2\pi l/M)} \right\} \right)$$

where F' is a constant independent of the data and α is as previously defined.

(b) Show that $\hat{\theta}_{ML}$ is the solution to

$$\sum_{i=1}^{N_s} \frac{\displaystyle\sum_{l=0}^{\frac{M}{2}-1} \sinh \left(\alpha \operatorname{Re} \left\{ \tilde{r}_{k-i} e^{-j(\theta + 2\pi l/M)} \right\} \right) \operatorname{Re} \left\{ j\tilde{r}_{k-i} e^{-j(\theta + 2\pi l/M)} \right\}}{\displaystyle\sum_{l=0}^{\frac{M}{2}-1} \cosh \left(\alpha \operatorname{Re} \left\{ \tilde{r}_{k-i} e^{-j(\theta + 2\pi l/M)} \right\} \right)} = 0$$

(c) For small SNR, we can make a first-order approximation of $\sinh x$ by x and $\cosh x$ by 1. Making these approximations in the result of part (b), show that it reduces to

$$\sum_{l=0}^{\frac{M}{2}-1} \sum_{i=1}^{N_s} \operatorname{Re} \left\{ \tilde{r}_{k-i} e^{-j\left(\hat{\theta}_{ML} + 2\pi l/M \right)} \right\} \operatorname{Re} \left\{ j\tilde{r}_{k-i} e^{-j\left(\hat{\theta}_{ML} + 2\pi l/M \right)} \right\} = 0$$

which does not yield a unique solution for $\hat{\theta}_{ML}$ since the left-hand side of this equation is identically equal to zero for any $\hat{\theta}_{ML}$ and $M > 2$ (show this).

(d) To get an approximate solution for $\hat{\theta}_{ML}$, we must take the next higher order terms in the Taylor series expansions of $\sinh x$ and $\cosh x$, namely $\sinh x = x + x^3/6$ and $\cosh x = 1 + x^2/2$. Using these approximations in the result of part (b) and recognizing that $(1 + x^2/2)^{-1} \cong (1 - x^2/2)$ for small x, show that, for QPSK ($M = 4$), the maximum-likelihood phase estimate is given by

$$e^{j\hat{\theta}_{ML}} = \left(\frac{\sum_{i=1}^{N_s} \tilde{r}_{k-i}^4}{\left| \sum_{i=1}^{N_s} \tilde{r}_{k-i}^4 \right|} \right)^{1/4}$$

which for the case of arbitrary $M = 2^m$ generalizes to

$$e^{j\hat{\theta}_{ML}} = \left(\frac{\sum_{i=1}^{N_s} \tilde{r}_{k-i}^M}{\left| \sum_{i=1}^{N_s} \tilde{r}_{k-i}^M \right|} \right)^{1/M}$$

6.13 Reconsider Problem 6.11 where now the unknown channel phase $\theta(t)$ varies with time in accordance with a frequency offset, Δf, due to Doppler. In particular, $\theta(t)$ at time $t = kT_b$ is related to $\theta(t)$ at time $t = (k - i)T_b$ by $\theta(kT_b) = \theta\left((k - i)T_b\right) + 2\pi i \Delta f T_b$. As before, it is desired to find an ML estimate of $\theta(t)$ now, however, in the presence of frequency offset.

(a) Show that the conditional pdf of the received vector $\tilde{\mathbf{r}} = \left(\tilde{r}_{k-1}, \tilde{r}_{k-2}, \ldots, \tilde{r}_{k-N_s}\right)$ given the information phase sequence $\boldsymbol{\theta} = \left(\theta_{k-1}, \theta_{k-2}, \ldots, \theta_{k-N_s}\right)$, channel phase $\theta(t)$ at time $t = kT_b$, and frequency offset Δf can be expressed in the form

$$f_{\tilde{\mathbf{r}}|\theta,\Delta f}\left(\tilde{\mathbf{r}} | \boldsymbol{\theta}, \theta\left(kT_b\right), \Delta f\right) = F \exp\left\{ \alpha \sum_{i=1}^{N_s} \operatorname{Re}\left\{ \tilde{r}_{k-i} e^{-j\left(\theta_{k-i} + \theta(kT_b) - 2\pi \Delta f i T_b\right)} \right\} \right\}$$

where F is a constant independent of the data.

(b) Show that the average of the conditional pdf $f_{\tilde{\mathbf{r}}|\theta,\Delta f}\left(\tilde{\mathbf{r}} | \boldsymbol{\theta}, \theta\left(kT_b\right), \Delta f\right)$ over the information phase sequence is given by

$$f_{\tilde{\mathbf{r}}|\theta,\Delta f}\left(\tilde{\mathbf{r}} | \theta\left(kT_b\right), \Delta f\right) = E_{\boldsymbol{\theta}}\left\{ f_{\tilde{\mathbf{r}}|\theta,\Delta f}\left(\tilde{\mathbf{r}} | \boldsymbol{\theta}, \theta\left(kT_b\right), \Delta f\right) \right\}$$

$$= F \prod_{i=1}^{N_s} \cosh\left(\alpha \operatorname{Re}\left\{ \tilde{R}_{k-i} e^{-j\theta(kT_b)} \right\} \right)$$

where

$$\tilde{R}_{k-i} = \tilde{r}_{k-i}\, e^{j2\pi i \Delta f T_b}$$

(c) Assume that the frequency offset is completely unknown and as such has a uniform pdf between $-f_{max}$ and f_{max} where f_{max} corresponds to the maximum expected Doppler frequency shift. Compute the conditional pdf $f_{\tilde{\mathbf{r}}|\theta}\left(\tilde{\mathbf{r}} | \theta\left(kT_b\right)\right)$ by averaging the result of part (b) over this uniform distribution.

(d) Show that the ML estimate, $\hat{\theta}_{ML}$, of the channel phase $\theta(t)$ at time $t = kT_b$ as defined by

$$\hat{\theta}\left(kT_b\right)_{ML} = \max_{\theta(kT_b)}{}^{-1} f_{\tilde{\mathbf{r}}|\theta}\left(\tilde{\mathbf{r}} | \theta\left(kT_b\right)\right) = \max_{\theta(kT_b)}{}^{-1} \ln f_{\tilde{\mathbf{r}}|\theta}\left(\tilde{\mathbf{r}} | \theta\left(kT_b\right)\right)$$

is the solution to

$$\int_{-f_{max}}^{f_{max}} \sum_{i=1}^{N_s} \sinh\left(\alpha\, \mathrm{Re}\left\{\tilde{R}_{k-i}e^{-j\theta(kT_b)}\right\}\right)\left(\mathrm{Re}\left\{j\tilde{R}_{k-i}e^{-j\theta(kT_b)}\right\}\right)$$

$$\times \prod_{\substack{m=1 \\ m\neq i}}^{N_s} \cosh\left(\alpha\, \mathrm{Re}\left\{\tilde{R}_{k-m}e^{-j\theta(kT_b)}\right\}\right) df = 0$$

(e) For small SNR, we can approximate $\sinh x$ by x and $\cosh x$ by unity. Making these approximations in the result of part (d), show that the ML phase estimate is given by

$$e^{j\hat{\theta}(kT_b)_{ML}} = \left(\frac{\displaystyle\sum_{i=1}^{N_s} \tilde{r}_{k-i}^2\, \mathrm{sinc}\,(4f_{max}iT_b)}{\left|\displaystyle\sum_{i=1}^{N_s} \tilde{r}_{k-i}^2\, \mathrm{sinc}\,(4f_{max}iT_b)\right|}\right)^{1/2}$$

REFERENCES

1. Lindsey, W. C., *Synchronization Systems in Communication and Control*, Englewood Cliffs, N.J.: Prentice-Hall, 1972.

2. Lindsey, W. C. and M. K. Simon, *Telecommunication Systems Engineering*, Englewood Cliffs, N.J.: Prentice-Hall, 1973. Reprinted by Dover Press, New York, NY, 1991.

3. Viterbi, A., "Optimum Detection and Signal Selection for Partially Coherent Binary Communication," *IEEE Transaction on Information Theory*, vol. IT-11, pp. 239–46, April 1965.

4. Simon, M. K. and D. Divsalar, "Multiple Symbol Partially Coherent Detection of MPSK," submitted to the *IEEE Transactions on Communications*.

5. Schwartz, M., W. Bennett, and S. Stein, *Communication Systems and Techniques*, New York: McGraw-Hill, 1966.

Chapter 7

Differentially Coherent Communication with Waveforms

In Chapter 4, we discussed coherent communication systems wherein the receiver provides a means (e.g., a carrier synchronization loop) for establishing a carrier demodulation reference signal that is synchronized in phase and frequency to the received signal. The implementation and operation of a coherent receiver is, in general, a complex process involving many ancilliary functions associated with the carrier synchronization loop, for example, acquisition, tracking, lock detection, false lock prevention, etc. In many applications, simplicity and robustness of implementation take precedence over achieving the best possible system performance. In these cases, *differential detection*, which obviates the need for implementing the carrier synchronization function, is an attractive alternative to coherent detection. Aside from implementation considerations, it is also possible that the transmission environment may be sufficiently degraded, for example, a multipath fading channel, that acquiring and tracking a coherent demodulation reference signal are difficult if not impossible. Here again, differential detection is a possible solution that still provides better performance than noncoherent detection. While, in principle, differential detection can be used with many different modulation schemes, the most widespread applications pertain to constant envelope modulations. We will focus in this chapter on the most common of these, namely, multiple phase-shift-keying (M-PSK). In Chapter 10, we shall discuss the application of differential detection to a nonconstant envelope modulation, namely, quadrature amplitude modulation (QAM).

Over the years, differential detection of M-PSK traditionally has been accomplished by comparing the received phase in a given symbol interval (of duration T_s seconds) with that in the previous symbol interval and making a multilevel decision on the difference between these two phases. Implicit in this process is the assumption that the phase introduced by the channel is constant over these same two symbol intervals and thus cancels when the above difference is taken. Since, in the absence of noise, the decision is equivalently being made on the difference between two adjacent *transmitted* phases, then in addition a suitable coding must be applied at the transmitter to allow this phase difference to represent the input *information* phase to be communicated. The encoding of the input information phase into the difference between two adjacent transmitted phases was previously referred to as *differential encoding* (see Section 4.1.6 of Chapter 4). There, differential encoding was included in the system solely as a means of resolving the phase ambiguity associated with the demodulation reference signal supplied by the carrier synchronization loop. Stated another way, differential encoding is not an essential element of the coherent detection process provided that another means for resolving the phase ambiguity of the carrier demodulation reference signal is employed. On the other hand, for differential detection, the inclusion of differential encoding at the transmitter is absolutely essential! This will become clear shortly.

Perhaps the earliest known (to the authors) citation of the above ideas, herein referred to as *classical multiple differential phase-shift-keying (M-DPSK)*, is due to Fleck and Trabka and appears in a 1961 document edited by Lawton [1]. The receiver structures presented in [1], which were based on observation of the received signal plus noise over an interval of two-symbol duration, were obtained without regard to their sense of optimality. We begin our discussion by showing that classical *M*-DPSK is indeed optimum (based on maximum-likelihood (ML) considerations) for a two-symbol observation interval. Later in the chapter, we shall show that if the channel phase characteristic can be assumed constant over $N_s T_s$ seconds, $N_s > 2$, then by extending the observation interval to $N_s T_s$ seconds and applying an appropriate maximum-likelihood detection rule, one can further improve the performance of the system. In fact, by making N_s larger and larger, one can approach the performance of a coherent *M*-PSK system with differential encoding (see Section 4.1.6). More about this later on.

7.1 CLASSICAL *M*-DPSK

Consider the desire to communicate in the ith transmission interval $iT_s \leq t \leq (i+1)T_s$ an information phase, $\Delta\theta_i$, which ranges over the set of allowable values $\beta_m = 2\pi m/M; m = 0, 1, \ldots, M-1$. In accordance with the above discussion, the first step is to differentially encode $\Delta\theta_i$ (see Figure 4.29a) into the phase θ_i according to the relation $\theta_i = \Delta\theta_i + \theta_{i-1}$. The phase θ_i is then modulated onto the carrier and transmitted over the channel in the form of an *M*-PSK signal, namely, $s_i(t) = \sqrt{2P} \cos(\omega_c t + \theta_i)$. In addition to additive white Gaussian noise, $n(t)$, of single-sided power spectral density N_0 Watts/Hz, the channel introduces an unknown phase θ that is assumed to be constant for a duration of at least $2T_s$ seconds. Thus, the received signal in the ith transmission interval has the form

$$r(t) = s_i(t; \theta) + n(t) = \sqrt{2P} \cos(\omega_c t + \theta_i + \theta) + n(t) \tag{7.1}$$

For classical symbol-by-symbol differential detection of *M*-PSK, the optimum receiver takes the form of Fig. 7.1. The derivation of the optimality of this receiver follows along the lines of that presented in Chapters 4, 5, and 6 for coherent, noncoherent, and partially coherent detectors. The details of the derivation for the differentially coherent receiver are presented in Appendix 7A. At first glance, Fig. 7.1 resembles Fig. 4.32b. However, there are some important differences. In both figures, an estimate, η_i, is made of the phase[1] associated with the ith transmission interval. In Figure 4.32b, a hard decision $\hat{\theta}_i$ is made based on this phase estimate and then compared (in the differential decoder) with the hard decision $\hat{\theta}_{i-1}$ based on the previous phase estimate, η_{i-1}, to arrive at the final decision on the information phase $\Delta\theta_i$. In Fig. 7.1, the difference of the two phase estimates η_i and η_{i-1} is first taken and then a hard decision is made based on this difference (modulo 2π) to arrive

---

1. It is understood that the $\tan^{-1}$ operation used to define the phase η_i in Fig. 7.1 is meant in the 4-quadrant sense and not in the principal value sense.

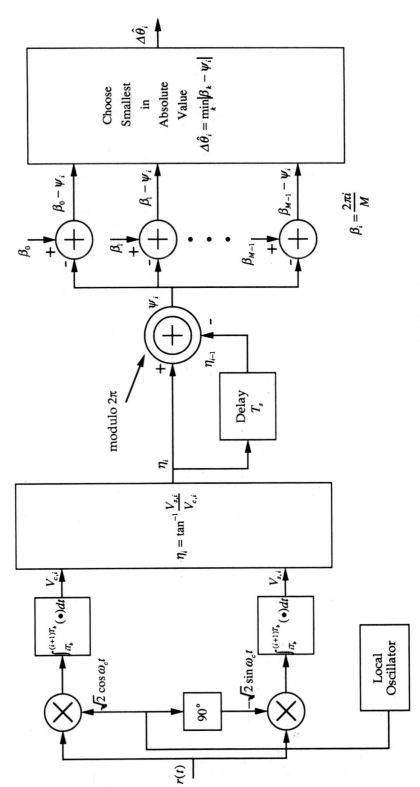

Figure 7.1 Optimum receiver for classical M-DPSK

at the final decision on the information phase $\Delta\theta_i$. As such, the differential detector intrinsically includes the differential decoding operation and thus no separate differential encoder is required.

The receiver of Fig. 7.1 can also be redrawn in a different form as in Fig. 7.2a, which takes on the simpler form of Fig. 7.2b when $M = 2$. This receiver was suggested in [4] as well as several other places and its equivalence with Fig. 7.1 was demonstrated in [5] (see Problem 7.1).

7.1.1 Symbol Error Probability Performance

The decision algorithm used in the receiver in Fig. 7.1 is quite analogous to that in Fig. 4.32b in that the decision regions are still defined as wedges uniformly spaced around a circle (see Fig. 4.22). Here, however, the phase associated with the signal that defines each of these regions is the value of the phase difference $\theta_i + \theta - (\theta_{i-1} + \theta) = \theta_i - \theta_{i-1} = \Delta\theta_i$ and the decision variable that falls into these regions is the phase difference $\psi_i = \eta_i - \eta_{i-1}$ modulo 2π. In particular, if $\theta_i - \theta_{i-1} = \Delta\theta_i = \beta_k$ is the value of the phase communicated in the ith transmission interval, then a correct decision is made when ψ_i falls in the kth wedge defined by the angular interval $(\beta_k - \pi/M, \beta_k + \pi/M)$. In mathematical terms, the conditional probability of a correct symbol decision given that $\theta_i - \theta_{i-1} = \Delta\theta_i = \beta_k$ is

$$P_s(C\,|\,k) = \Pr\left\{\left|(\eta_i - \eta_{i-1})_{\text{mod }2\pi} - \underbrace{(\theta_i - \theta_{i-1})}_{\beta_k}\right| \leq \frac{\pi}{M}\right\} \tag{7.2}$$

$$\triangleq \Pr\left\{|\psi_i - \beta_k| \leq \frac{\pi}{M}\right\} = \int_{-\pi/M}^{\pi/M} f_X(\chi)\,d\chi$$

where $\chi = \psi_i - \beta_k$. Thus, analytical evaluation of symbol error probability reduces to finding the pdf of X.

In [1], the pdf $f_X(\chi)$ was determined by first expressing it in the form of a convolution integral based on the pdf's of η_i and η_{i-1} (since η_i and η_{i-1} are statistically independent) and then evaluating this integral using the method of characteristic functions with the result

$$f_X(\chi) = \begin{cases} \frac{1}{2\pi}\int_0^{\pi/2}(\sin\alpha)\left[1 + \frac{E_s}{N_0}(1 + \cos\chi\sin\alpha)\right] \\ \times\exp\left\{-\frac{E_s}{N_0}(1 - \cos\chi\sin\alpha)\right\}d\alpha; & |\chi| \leq \pi \\ 0; & \text{otherwise} \end{cases} \tag{7.3}$$

Since this pdf is an even function of χ and independent of the particular phase β_k transmitted in the ith interval, then $P_s(C\,|\,k)$ is independent of k and hence the average symbol error probability is given by

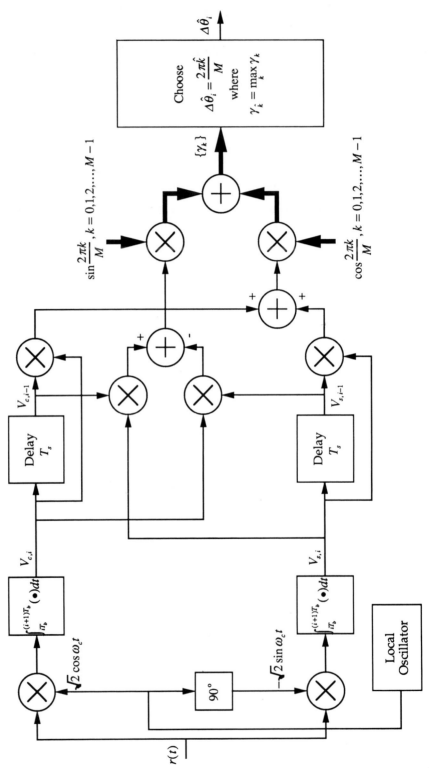

Figure 7.2a An alternate form of Figure 7.1

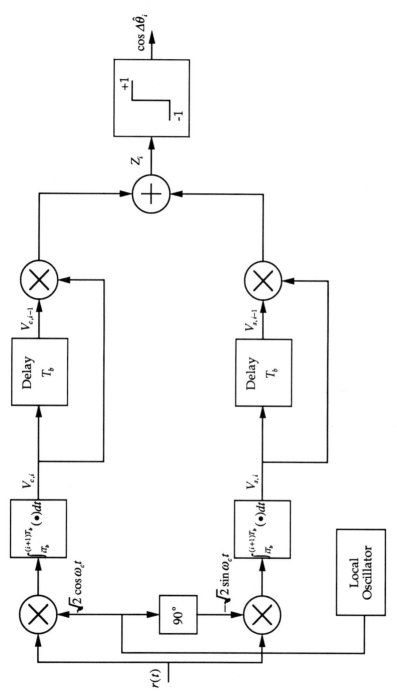

Figure 7.2b   An equivalent receiver to Figure 7.2a for $M = 2$

$$P_s(E) = \frac{1}{M} \sum_{k=1}^{M} P_s(E|k) = 1 - \int_{-\pi/M}^{\pi/M} f_\Psi(\psi) \, d\psi$$

$$= \int_{-\pi}^{\pi} f_\Psi(\psi) \, d\psi - \int_{-\pi/M}^{\pi/M} f_\Psi(\psi) \, d\psi \tag{7.4}$$

$$= 2 \int_{\pi/M}^{\pi} f_\Psi(\psi) \, d\psi$$

where we have arbitrarily assumed $\beta_k = 0$ and hence $\chi = \psi$. Substituting (7.3) into (7.4), we observe that $P_s(E)$ is in the form of a double integral, which must be evaluated numerically.

More recently, the evaluation of (7.2) has been carried out as a special case of the more generic problem of finding the probability distribution of the phase angle between two vectors, each perturbed by additive Gaussian noise. The solution to this problem is discussed in great detail in [4] and summarized here in Appendix 7B. For the M-DPSK receiver of Fig. 7.1, Case 1 applies[2] with $\psi_1 = \pi/M$ and $\psi_2 = \pi$. Thus, from (7B.3) and (7B.4)

$$\int_{\pi/M}^{\pi} f_\Psi(\psi) \, d\psi = F(\pi) - F(\pi/M)$$

$$= 0 - \left(-\frac{\sin\frac{\pi}{M}}{4\pi} \int_{-\pi/2}^{\pi/2} \frac{\exp\left\{ -\frac{A^2}{2\sigma^2}\left[1 - \cos\frac{\pi}{M}\cos t\right] \right\}}{1 - \cos\frac{\pi}{M}\cos t} dt \right) \tag{7.5}$$

$$= \frac{\sin\frac{\pi}{M}}{4\pi} \int_{-\pi/2}^{\pi/2} \frac{\exp\left\{ -\frac{A^2}{2\sigma^2}\left[1 - \cos\frac{\pi}{M}\cos t\right] \right\}}{1 - \cos\frac{\pi}{M}\cos t} dt$$

Since $A^2/2\sigma^2$ represents the symbol SNR, then

$$\frac{A^2}{2\sigma^2} = \frac{E_s}{N_0} = (\log_2 M)\frac{E_b}{N_0} \tag{7.6}$$

Substituting (7.6) into (7.5), the symbol error probability of (7.4) becomes

$$P_s(E) = \frac{\sin\frac{\pi}{M}}{2\pi} \int_{-\pi/2}^{\pi/2} \frac{\exp\left\{ -\frac{E_s}{N_0}\left[1 - \cos\frac{\pi}{M}\cos t\right] \right\}}{1 - \cos\frac{\pi}{M}\cos t} dt \tag{7.7}$$

which is in the form of a single integral of simple functions.

Many approximate expressions have been found for the symbol error probability performance of M-DPSK. One form due to Fleck and Trabka that originally appeared in [1] is accurate for large E_s/N_0 and is given by

$$P_s(E) \cong \text{erfc } \Lambda + \frac{\Lambda \exp\left(-\Lambda^2\right)}{4\sqrt{\pi}\,[1/8 + E_s/N_0]} \tag{7.8}$$

---

2. Since, as mentioned above, $P_s(C|k)$ is independent of k, then for convenience we again take $k = 0$ ($\beta_k = \beta_0 = 0$), which implies that the transmitted signal is the same in both the $(i-1)^{\text{st}}$ and i^{th} intervals.

Table 7.1 Ratio of Symbol Error Probabilities Corresponding to Eq. (7.13)
(Reprinted from [4])

M	Q
3	0.9186
4	0.9768
8	0.9987
16	0.9999

where

$$\Lambda = \sqrt{2E_s/N_0}\,\sin\left(\frac{\pi}{2M}\right) \tag{7.9}$$

Another approximation due to Arthurs and Dym [5] is

$$P_s(E) \cong \operatorname{erfc}\left(\sqrt{\frac{E_s}{N_0}}\,\sin\frac{\pi}{\sqrt{2}M}\right) \tag{7.10}$$

This expression is simpler to evaluate but less accurate than (7.8). Bussgang and Leiter [6] derived an upper bound on $P_s(E)$ which for large E_s/N_0 is itself a good approximation. The result is

$$P_s(E) \cong \operatorname{erfc}\left(\frac{E_s/N_0}{\sqrt{1 + 2E_s/N_0}}\,\sin\frac{\pi}{M}\right) \tag{7.11}$$

When both E_s/N_0 and M are large, (7.11) resembles (7.10) (since for small x, $\sin x \cong x$); however, for smaller values of E_s/N_0, (7.11) is the more accurate of the two.

Although all three of the above approximations are good for large E_s/N_0, none are asymptotic in the sense that the error in using them becomes vanishingly small as E_s/N_0 becomes infinitely large. Pawula et al. [4] obtained a true asymptotic (for large E_s/N_0) formula for symbol error probability. Their result is

$$P_s(E) \cong \sqrt{\frac{1 + \cos \pi/M}{2\cos \pi/M}}\,\operatorname{erfc}\left(\sqrt{\frac{E_s}{N_0}\left[1 - \cos\frac{\pi}{M}\right]}\right); \quad M \geq 3 \tag{7.12}$$

The Fleck and Trabka approximation, that is, (7.8), is closest in form to (7.12). To compare the two, we form the ratio Q of (7.8) to (7.12) by first neglecting the $1/8$ factor in the denominator of (7.8) and then employing the asymptotic expansion of the erfc function in both (7.8) and (7.12). The result is

$$Q = \frac{P_s(E)|_{(7.8)}}{P_s(E)|_{(7.12)}} \cong \frac{\sqrt{\cos\frac{\pi}{M}}\left[1 + \frac{1}{2}\sin^2\frac{\pi}{2M}\right]}{\cos\frac{\pi}{2M}} \tag{7.13}$$

which is independent of E_s/N_0. Table 7.1 tabulates the ratio Q for values of $M = 3, 4, 8,$ and 16. Thus, although Fleck and Trabka's approximation is not, in the strict sense, an asymptotic formula, it nevertheless differs in its leading term from the asymptotic formula by less than 9% in the worst case when $M = 3$ and by less than 3% in the more typical situation when M is a power of 2 ($M > 2$).

The special case of binary ($M = 2$) DPSK, often simply called DPSK, is a case where the error probability performance can be found in closed form (as opposed to an integration). In particular, letting $M = 2$ in (7.7) gives the simple result

$$P_s(E) = P_b(E) = \frac{1}{2} \exp\left(-\frac{E_b}{N_0}\right) \tag{7.14}$$

This result was independently obtained by Lawton [7] and Cahn [8]. Figure 7.3 is an illustration of the exact expression for symbol error probability of M-DPSK as obtained from (7.7) or (7.4) versus E_b/N_0 in dB for $M = 2, 4, 8, 16, 32$, and 64. Superimposed on these curves are the analogous results for the symbol error probability performance of coherent M-PSK as obtained from (4.131).[3] Asymptotically for large M and large E_b/N_0, we see that differential detection of M-PSK requires 3 dB more E_b/N_0 than coherent detection of M-PSK. This result can be obtained quite easily by comparing the effective SNR's for these two detection schemes as determined from the arguments of the erfc functions in (4.139) and (7.13). In particular, for coherent detection the effective symbol SNR is given by $(E_s/N_0)_{\text{eff}} = (E_s/N_0)(\sin^2 \pi/M) \cong (E_s/N_0)(\pi/M)^2$ whereas for differential detection $(E_s/N_0)_{\text{eff}} = (E_s/N_0)(1 - \cos \pi/M) \cong (E_s/N_0)\left[1 - \left(1 - (\pi/M)^2/2\right)\right] = (E_s/2N_0)(\pi/M)^2$.

7.1.2 Bit Error Probability Performance

Assuming a Gray code bit to symbol mapping as was done in Chapter 4, the bit error probability performance of M-DPSK can be related to its symbol error probability performance evaluated in the previous section by the simple approximate relation given in (4.140). It is also possible to obtain an exact expression for the bit error probability performance using the method of Lee [11] combined with the results of Appendix 7B. For example, for 8-DPSK, Lee obtained the result

$$P_b(E) = \frac{2}{3}\left[F\left(\frac{13\pi}{8}\right) - F\left(\frac{\pi}{8}\right)\right] \tag{7.15}$$

where $F(\psi)$ is given by (7B.4) with $U = E_s/N_0$. This result was later generalized by Korn [12] to other values of M. In particular

$$P_b(E) = F\left(\frac{5\pi}{4}\right) - F\left(\frac{\pi}{4}\right); \quad M = 4 \tag{7.16a}$$

$$P_b(E) = \frac{1}{2}\left[F\left(\frac{13\pi}{16}\right) - F\left(\frac{9\pi}{16}\right) - F\left(\frac{3\pi}{16}\right) - F\left(\frac{\pi}{16}\right)\right]; \quad M = 16 \tag{7.16b}$$

---

3. Using the results of Appendix 7B in the degenerate case where one of the two vectors is noise free, Pawula et al. [4] have been able to arrive at a simpler expression for the symbol error probability performance of coherent M-PSK, namely

$$P_s(E) = \frac{1}{\pi}\int_{-\pi/2}^{\pi/2 - \pi/M} \exp\left\{-\frac{E_s}{N_0}\sin^2\frac{\pi}{M}\sec^2\theta\right\} d\theta$$

which is essentially the same result obtained by Weinstein [9].

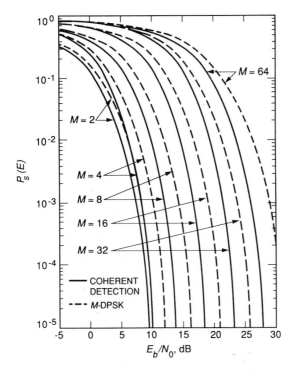

Figure 7.3   Symbol error probability performance of *M*-PSK and *M*-DPSK (Reprinted from [10])

$$P_b(E) = \frac{2}{5}\left[F\left(\frac{29\pi}{32}\right) + F\left(\frac{23\pi}{32}\right) - F\left(\frac{19\pi}{32}\right) - F\left(\frac{17\pi}{32}\right)\right.$$

$$\left. + F\left(\frac{13\pi}{32}\right) - F\left(\frac{9\pi}{32}\right) - F\left(\frac{3\pi}{32}\right) - F\left(\frac{\pi}{32}\right)\right]; \quad M = 32$$

(7.16c)

Note that the last term in each expression in (7.15) and (7.16) is

$$P_b(E)|_1 = -\frac{2}{\log_2 M} F\left(\frac{\pi}{M}\right) = \frac{P_s(E)}{\log_2 M}$$

(7.17)

which represents the Gray code approximation of (4.135). When compared with coherent detection, some typical values of E_b/N_0 degradation at $P_b = 10^{-5}$ are 0.75 dB for $M = 2$, 2.2 dB for $M = 4$, and 2.5 dB for $M = 8$.

7.1.3 Effect of Noise Correlation and Power Imbalance

Suppose now that because of channel bandlimitation, the amplitude of the received sinusoid is time-varying and the power spectral density of the received noise is not flat for all frequencies. In particular, the received signal of (7.1) now takes the form

$$r(t) = A(t) \cos(\omega_c t + \theta_i + \theta) + n'(t)$$

(7.18)

where $n'(t)$ is additive Gaussian noise with power spectral density $[S_n(f - f_0) + S_n(f + f_0)]/2$ and $S_n(f) = \mathcal{F}\{R_n(\tau)\}$ is a symmetric function. After demodulation with inphase and quadrature reference signals as in Fig. 7.1 and passing through the integrate-and-dump (I&D) circuits, we get the lowpass signals

$$V_{c,i} = A_i \cos(\theta_i + \theta) + N'_{c,i} \quad V_{s,i} = A_i \sin(\theta_i + \theta) + N'_{s,i} \tag{7.19}$$

where

$$A_i = \frac{1}{\sqrt{2}} \int_{iT_s}^{(i+1)T_s} A(t)dt$$

$$\tag{7.20}$$

$$N'_{c,i} = \int_{iT_s}^{(i+1)T_s} n'(t) \left(\sqrt{2}\cos\omega_c t\right) dt; \quad N'_{s,i} = \int_{iT_s}^{(i+1)T_s} n'(t) \left(\sqrt{2}\sin\omega_c t\right) dt$$

If, as in (7B.1), we relate the complex form of these I&D outputs to the geometry in Fig. 7B.1, then the association is now

$$A_1 = A_{i-1}; \quad A_2 = A_i; \quad \phi_1 = \theta_{i-1} + \theta; \quad \phi_2 = \theta_i + \theta$$

$$x_1 = N'_{c,i-1}; \quad y_1 = N'_{s,i-1}; \quad x_2 = N'_{c,i}; \quad y_2 = N'_{s,i}$$

$$\sigma_1^2 = \frac{1}{2} \int_{(i-1)T_s}^{iT_s} \int_{(i-1)T_s}^{iT_s} R_n(u-v)dudv = \sigma^2;$$

$$\tag{7.21}$$

$$\sigma_2^2 = \frac{1}{2} \int_{iT_s}^{(i+1)T_s} \int_{iT_s}^{(i+1)T_s} R_n(u-v)dudv = \sigma^2$$

$$r = \frac{\int_{(i-1)T_s}^{iT_s} \int_{iT_s}^{(i+1)T_s} R_n(u-v)dudv}{\sigma^2}; \quad \rho = 0$$

For binary DPSK, Pawula and Roberts [13] have been able to evaluate the bit error probability performance in closed form (in terms of tabulated functions). For the binary case, the error probability can be written in the form

$$P_b(E) = \Pr\{\Delta\phi = 0\} \left(\int_{-3\pi/2}^{-\pi/2} f_\Psi(\psi) d\psi \right)$$

$$+ \Pr\{\Delta\phi = \pi\} \left(\int_{-\pi/2}^{\pi/2} f_\Psi(\psi) d\psi \right) \tag{7.22}$$

$$= P_0 \left(\int_{-3\pi/2}^{-\pi/2} f_\Psi(\psi) d\psi \right) + P_\pi \left(\int_{-\pi/2}^{\pi/2} f_\Psi(\psi) d\psi \right)$$

where P_0 and P_π are the a priori probabilities[4] of the two possible information phases transmitted in the ith interval. Evaluation of the integrals in (7.22) follows the results corresponding to Case 2 in Appendix 7B. For the case where the limits of (7B.3) are separated by π radians, as is true in (7.22), Pawula and Roberts [13] have been able to evaluate the integrals in closed form. In particular

---

4. In their analysis, Pawula and Roberts [13] allow different a priori probabilities for the information phases 0 and π.

$$\Pr\{\psi_0 - \pi \le \psi \le \psi_0\} = \int_{\psi_0-\pi}^{\psi_0} f_\Psi(\psi)\,d\psi$$

$$= \begin{cases} \frac{1}{2}\left[1 - \sqrt{1 - \frac{\beta^2}{\alpha^2}}\, Ie\left(\frac{\beta}{\alpha}, \alpha\right)\right] \\ \quad + \frac{r\sin\psi_0\, e^{-\alpha} I_0(\beta)}{2\sqrt{1 - r^2\cos^2\psi_0}}; \\[2mm] \qquad\qquad\qquad\qquad \Delta\Phi - \pi \le \psi_0 \le \Delta\Phi \\[2mm] \frac{1}{2}\left[1 + \sqrt{1 - \frac{\beta^2}{\alpha^2}}\, Ie\left(\frac{\beta}{\alpha}, \alpha\right)\right] \\ \quad + \frac{r\sin\psi_0\, e^{-\alpha} I_0(\beta)}{2\sqrt{1 - r^2\cos^2\psi_0}}; \\[2mm] \qquad\qquad\qquad\qquad \text{elsewhere} \end{cases} \tag{7.23}$$

where $Ie(k, x)$ is the Rice Ie-function [14] defined by

$$Ie(k, x) \triangleq \int_0^x \exp(-t) I_0(kt)\,dt \tag{7.24}$$

and tabulated in [14, 15], and

$$\alpha = \frac{U - rW\cos\psi_0\cos(\Delta\Phi - \psi_0)}{1 - r^2\cos^2\psi_0}$$

$$\beta = \sqrt{\alpha^2 - \frac{W^2\sin^2(\Delta\Phi - \psi_0)}{1 - r^2\cos^2\psi_0}} \tag{7.25}$$

Since in our application $\Delta\Phi = \Delta\theta_i$, and noting that for $\Delta\Phi = 0$, $\psi_0 = -\pi/2$ while for $\Delta\Phi = \pi$, $\psi_0 = \pi/2$, then combining (7.22) and (7.23) gives the desired result

$$P_b(E) = \frac{1}{2}\left[1 - \frac{W}{U} Ie\left(\frac{V}{U}, U\right)\right] + \frac{r}{2}(P_\pi - P_0) e^{-U} I_0(V) \tag{7.26}$$

where U, V, and W are defined in (7B.2) using (7.21). From (7.26), we observe that the error probability does not depend on the noise correlation r so long as the a priori probabilities P_0 and P_π are equal, *whether or not there is a power imbalance*. When there is no power imbalance, $V = 0$, and (7.26) reduces to

$$P_b(E) = \frac{1}{2}[1 + (P_\pi - P_0) r] e^{-U} \tag{7.27}$$

which was previously obtained by Lee and Miller [16]. When in addition P_0 and P_π are equal, then (7.27) reduces to (7.13) as it should.

7.1.4 Effect of Imperfect Timing

The effect of bit timing error on the bit error probability of an otherwise ideal DPSK receiver was first considered in [17] and later in [18]. We present only the results from [18] since they are based on the approach taken in [4] and are thus in simpler form. In particular, if we denote the bit sync error, as in Section 4.3.3, by $\varepsilon = \lambda T_b$ where λ represents the normalized (with respect to the bit time) error, then the block diagram of Fig. 7.1 applies if we replace the lower and upper limits of the I&D's by $(i - \lambda)T_b$ and $(i + 1 - \lambda)T_b$, respectively. Having done this, the outputs of these circuits become

$$V_{c,i} = \sqrt{PT_b}\left[(1 - \lambda) \cos (\theta_i + \theta) + \lambda \cos (\theta_{i-1} + \theta)\right] + N_{c,i}$$

$$V_{s,i} = \sqrt{PT_b}\left[(1 - \lambda) \sin (\theta_i + \theta) + \lambda \sin (\theta_{i-1} + \theta)\right] + N_{s,i}$$

(7.28)

where

$$N_{c,i} \triangleq \int_{(i-\lambda)T_b}^{(i+1-\lambda)T_b} n(t)\left[\sqrt{2}\cos \omega_c t\right] dt,$$

$$N_{s,i} \triangleq \int_{(i-\lambda)T_b}^{(i+1-\lambda)T_b} n(t)\left[-\sqrt{2}\sin \omega_c t\right] dt$$

(7.29)

are independent zero-mean Gaussian random variables with variance $\sigma^2 = N_0 T_b/2$. The equivalent phase η_i for the ith transmission interval is found from the arctangent of the ratio of $V_{s,i}$ to $V_{c,i}$ and using (7.28) can be expressed in the form

$$\eta_i = \tan^{-1}\left\{\frac{\sqrt{PT_b}\Gamma_i \sin (\xi_i + \theta) + N_{s,i}}{\sqrt{PT_b}\Gamma_i \cos (\xi_i + \theta) + N_{c,i}}\right\}$$

(7.30)

where

$$\xi_i = \tan^{-1}\left\{\frac{(1 - \lambda) \sin \theta_i + \lambda \sin \theta_{i-1}}{(1 - \lambda) \cos \theta_i + \lambda \cos \theta_{i-1}}\right\}$$

$$\Gamma_i = \sqrt{(1 - \lambda)^2 + \lambda^2 + 2\lambda(1 - \lambda) \cos (\theta_i - \theta_{i-1})}$$

$$= \sqrt{(1 - \lambda)^2 + \lambda^2 + 2\lambda(1 - \lambda) \cos \Delta\theta_i}$$

(7.31)

A similar expression can be written for the equivalent phase, η_{i-1}, in the $(i - 1)^{\text{st}}$ transmission interval. Once again the geometry of Fig. 7B.1 applies, where now

$$A_1 = \sqrt{PT_b}\Gamma_{i-1}; \quad A_2 = \sqrt{PT_b}\Gamma_i$$

$$\phi_1 = \xi_{i-1} + \theta; \quad \phi_2 = \xi_i + \theta$$

$$x_1 = N_{c,i-1}; \quad y_1 = N_{s,i-1}; \quad x_2 = N_{c,i}; \quad y_2 = N_{s,i}$$

$$\sigma_1^2 = \sigma_2^2 = \sigma^2$$

$$r = \rho = 0$$

(7.32)

Also, from (7B.2)

$$U = \frac{1}{2}\left(\frac{A_1^2}{2\sigma^2} + \frac{A_2^2}{2\sigma^2}\right) = \frac{E_b}{N_0}\left[\frac{(\Gamma_{i-1})^2 + (\Gamma_i)^2}{2}\right]$$

$$V = \frac{1}{2}\left(\frac{A_2^2}{2\sigma^2} - \frac{A_1^2}{2\sigma^2}\right) = \frac{E_b}{N_0}\left[\frac{(\Gamma_i)^2 - (\Gamma_{i-1})^2}{2}\right] \tag{7.33}$$

$$\Delta\Phi = (\phi_2 - \phi_1) \text{ modulo } 2\pi = (\xi_i - \xi_{i-1}) \text{ modulo } 2\pi$$

Substituting (7.31) into (7.33) and simplifying gives

$$U = \frac{E_b}{N_0}\left[(1-\lambda)^2 + \lambda^2 + \lambda(1-\lambda)(\cos\Delta\theta_i + \cos\Delta\theta_{i-1})\right]$$

$$V = \frac{E_b}{N_0}\left[\lambda(1-\lambda)(\cos\Delta\theta_i - \cos\Delta\theta_{i-1})\right] \tag{7.34}$$

$$\Delta\Phi = \tan^{-1}\left\{\frac{(1-\lambda)^2\sin\Delta\theta_i + \lambda^2\sin\Delta\theta_{i-1} + \lambda(1-\lambda)\sin(\Delta\theta_i + \Delta\theta_{i-1})}{(1-\lambda)^2\cos\Delta\theta_i + \lambda^2\cos\Delta\theta_{i-1} + \lambda(1-\lambda)\left[1 + \cos(\Delta\theta_i + \Delta\theta_{i-1})\right]}\right\}$$

$$\text{modulo } 2\pi$$

For DPSK, the information phase, $\Delta\theta_i$, to be communicated in the ith transmission interval takes on a value of 0 or π. For each of these values we must compute the conditional bit error probability when $\Delta\theta_{i-1}$ takes on these very same values (with equal probability). To do this, we apply the results of Case 2 in Appendix 7B, in particular, the special case of these results given in (7.23)–(7.25). To begin, we observe from (7.34) that for $\Delta\theta_{i-1}$ equal to either 0 or π, $\Delta\Phi = \Delta\theta_i$. Furthermore, for $\Delta\Phi = 0$, we have $\psi_0 = -\pi/2$ while for $\Delta\Phi = \pi$, we have $\psi_0 = \pi/2$. Thus, the above conditional bit error probabilities are evaluated from (7.23) as

$$P_b(E|0,\lambda) = \Pr\left\{-\frac{3\pi}{2} \leq \psi \leq -\frac{\pi}{2}\right\} = \frac{1}{2}\left[1 - \sqrt{1 - \frac{\beta_0^2}{\alpha_0^2}}\, Ie\left(\frac{\beta_0}{\alpha_0}, \alpha_0\right)\right]$$

$$\tag{7.35}$$

$$P_b(E|\pi,\lambda) = \Pr\left\{-\frac{\pi}{2} \leq \psi \leq \frac{\pi}{2}\right\} = \frac{1}{2}\left[1 - \sqrt{1 - \frac{\beta_\pi^2}{\alpha_\pi^2}}\, Ie\left(\frac{\beta_\pi}{\alpha_\pi}, \alpha_\pi\right)\right]$$

where from (7.25)

$$\alpha_0 = \frac{E_b}{N_0} \times \begin{cases} 1; & \Delta\theta_{i-1} = 0 \\ 1 - 2\lambda(1-\lambda); & \Delta\theta_{i-1} = \pi \end{cases}$$

$$\beta_0 = \frac{E_b}{N_0} \times \begin{cases} 0; & \Delta\theta_{i-1} = 0 \\ 2\lambda(1-\lambda); & \Delta\theta_{i-1} = \pi \end{cases}$$

$$\tag{7.36}$$

$$\alpha_\pi = \frac{E_b}{N_0} \times \begin{cases} 1 - 2\lambda(1-\lambda); & \Delta\theta_{i-1} = 0 \\ (1-2\lambda)^2; & \Delta\theta_{i-1} = \pi \end{cases}$$

$$\beta_\pi = \frac{E_b}{N_0} \times \begin{cases} -2\lambda(1-\lambda); & \Delta\theta_{i-1} = 0 \\ 0; & \Delta\theta_{i-1} = \pi \end{cases}$$

The average error probability conditioned on a given timing offset λ is then obtained by averaging (7.35) over equiprobable events $\Delta\theta_{i-1} = 0$ and $\Delta\theta_{i-1} = \pi$, namely

$$P_b\,(E|\lambda) = \frac{1}{4}\,P_b\,(E|0,\lambda)|_{\Delta\theta_{i-1}=0} + \frac{1}{4}\,P_b\,(E|0,\lambda)|_{\Delta\theta_{i-1}=\pi}$$

$$+ \frac{1}{4}\,P_b\,(E|\pi,\lambda)|_{\Delta\theta_{i-1}=0} + \frac{1}{4}\,P_b\,(E|\pi,\lambda)|_{\Delta\theta_{i-1}=\pi}$$

$$= \frac{1}{8}\left[1 - Ie\left(0, \frac{E_b}{N_0}\right)\right] \tag{7.37}$$

$$+ \frac{1}{8}\left[1 - \sqrt{1 - \frac{(2\lambda(1-\lambda))^2}{(1-2\lambda(1-\lambda))^2}}\,Ie\left(\frac{2\lambda(1-\lambda)}{1-2\lambda(1-\lambda)}, (1-2\lambda(1-\lambda))\frac{E_b}{N_0}\right)\right]$$

$$+ \frac{1}{8}\left[1 - \sqrt{1 - \frac{(-2\lambda(1-\lambda))^2}{(1-2\lambda(1-\lambda))^2}}\,Ie\left(\frac{-2\lambda(1-\lambda)}{1-2\lambda(1-\lambda)}, (1-2\lambda(1-\lambda))\frac{E_b}{N_0}\right)\right]$$

$$+ \frac{1}{8}\left[1 - Ie\left(0, (1-2\lambda)^2\frac{E_b}{N_0}\right)\right]$$

Using the identity [13, Eq. (A3-5)]

$$\frac{1}{2}\left[1 - \sqrt{1 - \frac{\beta^2}{\alpha^2}}\,Ie\left(\frac{\beta}{\alpha}, \alpha\right)\right] = \frac{\sqrt{\alpha^2 - \beta^2}}{2\pi}\int_0^\pi \frac{\exp\{-[\alpha - \beta\cos\theta]\}}{\alpha - \beta\cos\theta}\,d\theta \tag{7.38}$$

we can rewrite (7.37) as

$$P_b\,(E|\lambda) = \left[\frac{1}{2}\exp\left(-\frac{E_b}{N_0}\right)\right]\mathcal{D};$$

$$\mathcal{D} = \frac{1}{4}\left[1 + (1-2\lambda)\frac{1}{\pi}\int_0^\pi \frac{\exp\left\{\frac{E_b}{N_0}\left[4\lambda(1-\lambda)\cos^2\frac{\theta}{2}\right]\right\}}{1 - 4\lambda(1-\lambda)\cos^2\frac{\theta}{2}}\,d\theta\right.$$

$$\tag{7.39}$$

$$+ (1-2\lambda)\frac{1}{\pi}\int_0^\pi \frac{\exp\left\{\frac{E_b}{N_0}\left[4\lambda(1-\lambda)\sin^2\frac{\theta}{2}\right]\right\}}{1 - 4\lambda(1-\lambda)\sin^2\frac{\theta}{2}}\,d\theta$$

$$\left. + \exp\left\{\frac{E_b}{N_0}\left[4\lambda(1-\lambda)\right]\right\}\right]$$

which clearly places in evidence the degradation in performance due to bit timing error relative to the ideal (perfectly timed) performance of (7.14). Note that when $\lambda = 0$, $\mathcal{D}$ reduces to unity as expected. Assuming a Tikhonov pdf for the normalized timing error as was done in Chapter 4 [see Eq. (4.324)], the average bit error probability performance of DPSK in the presence of timing error is obtained by averaging (7.39) over this pdf and is illustrated in Fig. 7.4 for various values of rms normalized timing error σ_λ.

7.1.5 Double Error Rate Performance

In designing receivers that employ error detecting and error correction codes, the occurrence of multiple consecutive errors, that is, burst errors, is an important consideration. For

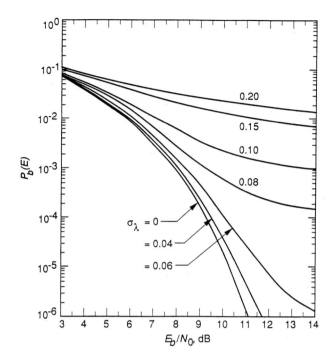

Figure 7.4 Error probability performance of DPSK in the presence of bit timing error (Reprinted from [18]) © IEEE

uncoded coherent receivers that make bit by bit decisions, such as those discussed in Chapter 4, the probability of a burst error is simply the product of the bit error probabilities over the length of the burst, or equivalently, the average bit error probability raised to the power of the burst length in bits. The reason for this is that in a coherent system, the bit decisions are independent of one another. For uncoded DPSK, since in effect the phase of one signal acts as the phase reference for demodulation of the next signal, the decision statistics are not independent from bit to bit and hence the probability of two or more consecutive errors is not the product of the individual bit error probabilities over the length of the error sequence.

The analysis of double error rates in DPSK systems was first considered by Salz and Saltzberg [19]. Although their results were derived for a suboptimum receiver (to be discussed later in this chapter), the application of their approach to the optimum DPSK receiver of Fig. 7.2b follows directly. In particular, let

$$Z_i \triangleq V_{c,i-1} V_{c,i} + V_{s,i-1} V_{s,i}$$

denote the decision variable in the receiver of Fig. 7.2b where from (7.28) and (7.29) with $\lambda = 0$

$$V_{c,i} = \sqrt{P T_b} \cos \theta_i + N_{c,i}, \qquad V_{s,i} = \sqrt{P T_b} \sin \theta_i + N_{s,i}$$

Here again, $N_{c,i}$ and $N_{s,i}$ are independent zero-mean Gaussian random variables with variance $\sigma^2 = N_0 T_b/2$. The decision on the information phase, $\Delta\theta_i$, communicated in the ith transmission interval (or, equivalently, the decision on the binary bit $a_i = \cos \Delta\theta_i$) is then

$$\hat{a}_i = \cos \Delta\hat{\theta}_i = \operatorname{sgn} \ Z_i \tag{7.40}$$

The approach taken in [19] is to first determine the joint pdf of Z_{i-1} and Z_i conditioned on $V_{c,i-1}$ and $V_{s,i-1}$ being fixed. This conditional joint pdf is Gaussian and, since for binary modulation $E\{V_{s,i-1}\} = E\{V_{s,i}\} = 0$, it is given by

$$
f\left(Z_{i-1}, Z_i \mid V_{c,i-1}, V_{s,i-1}\right) = \frac{1}{2\pi\sigma^2\left[\left(V_{c,i-1}\right)^2 + \left(V_{s,i-1}\right)^2\right]}
$$

$$
\times \exp\left\{-\frac{1}{2\sigma^2}\frac{\left(Z_{i-1} - \sqrt{P}T_b\cos\theta_i\,V_{c,i-1}\right)^2 + \left(Z_i - \sqrt{P}T_b\cos\theta_{i-2}V_{c,i-1}\right)}{\left(V_{c,i-1}\right)^2 + \left(V_{s,i-1}\right)^2}\right\}
$$

$$(7.41)$$

Averaging (7.41) over the joint pdf of $V_{c,i-1}$ and $V_{s,i-1}$ which is also Gaussian, namely

$$
f\left(V_{c,i-1}, V_{s,i-1}\right) = \frac{1}{2\pi\sigma^2}\exp\left\{-\frac{1}{2\sigma^2}\left(V_{c,i-1} - \sqrt{P}T_b\cos\theta_{i-1}\right)^2 + \left(V_{s,i-1}\right)^2\right\} \quad (7.42)
$$

we obtain the unconditional joint pdf $f(Z_{i-1}, Z_i)$. This pdf is a function of the three transmitted phases $\boldsymbol{\theta} = (\theta_{i-2}, \theta_{i-1}, \theta_i)$ which for binary modulation takes on eight possible values. Associated with each of these transmitted phase sequences is a double error probability, $P_{bk}(E_{i-1}, E_i)$; $k = 1, 2, \ldots, 8$. For example, if $\boldsymbol{\theta} = (0, 0, 0)$, or equivalently, $\Delta\theta_{i-1} = 0$, $\Delta\theta_i = 0$, then a double error (two consecutive bit errors) occurs when $Z_{i-1} < 0$ and $Z_i < 0$. This double error probability is given by

$$
P_b\left(E_{i-1}, E_i\right)|_{\theta=(0,0,0)} \triangleq P_{b1}\left(E_{i-1}, E_i\right) = \int_{-\infty}^{0}\int_{-\infty}^{0} f\left(Z_{i-1}, Z_i\right) dZ_{i-1}dZ_i \quad (7.43)
$$

which is an integral over the lower left quadrant of the Z_{i-1}, Z_i plane. Salz and Saltzberg [19] show that, with appropriate changes of variables, (7.43) together with the other seven integrals of this type can all be expressed in the form of two integrals each of which integrates over the upper right quadrant of the Z_{i-1}, Z_i plane. Since each phase sequence $\boldsymbol{\theta}$ occurs with probability one-eighth, then averaging $P_{bk}(E_{i-1}, E_i)$; $k = 1, 2, \ldots, 8$ over this equiprobable distribution gives the desired result for average double error probability in the form of a quadruple integral, namely

$$
P_b(E_{i-1}, E_i) = \frac{1}{8}\sum_{k=1}^{8} P_{bk}(E_{i-1}, E_i) = e^{-E_b/N_0}\left(\frac{E_b}{\pi N_0}\right)^2 \int_0^\infty\int_0^\infty\int_{-\infty}^\infty\int_{-\infty}^\infty \frac{1}{x_1^2 + y_1^2}
$$

$$(7.44)$$

$$
\times \exp\left\{-\frac{E_b}{N_0}\left[\frac{(z_1 - x_1)^2 + (z_2 - x_1)}{x_1^2 + y_1^2} + x_1^2 + y_1^2 + 2x_1\right]\right\} dx_1 dx_2 dz_1 dz_2
$$

Finally, it is also shown [19; Appendix I] that this quadruple integral can be simplified to a single integral, which yields the desired result

$$P_b\left(E_{i-1}, E_i\right) = \frac{1}{4} e^{-E_b/N_0} \int_0^\pi \left[1 - \mathrm{erf}\left(\sqrt{\frac{E_b}{N_0}} \cos\theta\right)\right]^2$$

$$+ \sqrt{\frac{\pi E_b}{N_0}} \cos\theta \, e^{(E_b/N_0)\cos^2\theta} \left[1 - \mathrm{erf}\left(\sqrt{\frac{E_b}{N_0}} \cos\theta\right)\right]^2 \quad (7.45)$$

$$\times \left[1 + \mathrm{erf}\left(\sqrt{\frac{E_b}{N_0}} \cos\theta\right)\right] d\theta$$

Oberst and Schilling [20] have also considered the double error probability performance of DPSK and arrived at (7.45) in a somewhat simpler manner. Although they too considered the same suboptimum configuration as Salz and Saltzberg [19], their approach and mathematics is nevertheless applicable to the optimum DPSK receiver of Fig. 7.2b. Oberst and Schilling [20] base their approach on the observation that DPSK can be viewed as the signal in the $(i-1)^{\mathrm{st}}$ transmission interval $(i-1)T_b \le t \le i T_b$ acting as a noisy demodulation reference for the signal in the ith transmission interval $i T_b \le t \le (i+1)T_b$. They referred to these two signals, respectively, as the "reference" and the "data." The clever trick employed in [20] is to switch the names of the "data" and "reference" signals for the intervals $(i-2)T_b \le t \le (i-1)T_b$ and $(i-1)T_b \le t \le i T_b$; that is, one should think of the signal in $(i-2)T_b \le t \le (i-1)T_b$ as the "data" and the signal in $(i-1)T_b \le t \le i T_b$ as the "reference." This allows one to view the two demodulations ordinarily occurring at $t = i T_b$ and $t = (i+1)T_b$ as equivalently having been performed independently (with regard to the additive noise) at $t = (i-1)T_b$ and $t = (i+1)T_b$ by a common reference signal. In mathematical terms, the probability of error for the ith bit conditioned on the phase estimate η_{i-1} associated with the previous bit is given by

$$P_b\left(E_i \mid \eta_{i-1}\right) = \frac{1}{2} \mathrm{erfc}\left(\sqrt{\frac{E_b}{N_0}} \cos\eta_{i-1}\right) \quad (7.46)$$

where

$$f\left(\eta_{i-1}\right) = \frac{1}{2\pi} e^{-E_b/N_0}$$

$$+ \frac{1}{2}\sqrt{\frac{E_b}{\pi N_0}} \cos\eta_{i-1} e^{-(E_b/N_0)\sin^2\eta_{i-1}} \left[1 + \mathrm{erf}\left(\sqrt{\frac{E_b}{N_0}} \cos\eta_{i-1}\right)\right]; \quad (7.47)$$

$$-\pi \le \eta_{i-1} \le \pi$$

Based on the above discussion, the double error probability conditioned on η_{i-1} is then

$$P_b\left(E_{i-1}, E_i \mid \eta_{i-1}\right) = \left[\frac{1}{2} \mathrm{erfc}\left(\sqrt{\frac{E_b}{N_0}} \cos\eta_{i-1}\right)\right]^2 \quad (7.48)$$

Averaging (7.48) over the pdf in (7.47) produces, after some manipulation, the identical result to (7.45).

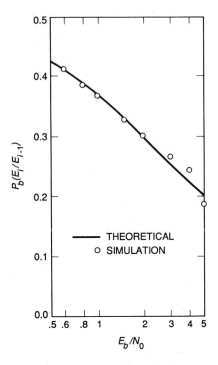

Figure 7.5 The ratio of double to single error probability for DPSK (Reprinted from [20] © IEEE)

The conditional error probability $P_b(E_i | E_{i-1}) = P_b(E_{i-1}, E_i) / P_b(E_{i-1})$ represents the ratio of double to single error probabilities, which, in view of (7.14), is given by

$$P_b(E_i | E_{i-1}) = \frac{P_b(E_{i-1}, E_i)}{P_b(E_{i-1})} = 2e^{E_b/N_0} P_b(E_{i-1}, E_i) \qquad (7.49)$$

Figure 7.5 illustrates this ratio as a function of E_b/N_0 obtained theoretically from (7.49) together with (7.45) as well as by simulation results [20]. The monotonically decreasing nature of $P_b(E_i | E_{i-1})$ with E_b/N_0 indicates that the ratio of double errors to single errors does not increase as SNR increases. As a final note, we also point out that using Oberst and Schilling's viewpoint, it is straightforward to show that nonadjacent decisions are mutually exclusive and thus the conditional probability $P_b(E_i | E_{i-k})$, $k > 1$ is equal to the unconditional bit error probability, namely, (7.14).

7.1.6 Error Probability Performance Comparison of DPSK and Noncoherent FSK

In Chapter 5, we examined the bit error probability performance of noncoherent FSK (NCFSK), in particular, that of the optimum receiver. It was found that whereas optimum coherent detection of FSK (see Chapter 4) resulted in an error probability that varied as a *complementary error* function of $E_b/2N_0$, the comparable behavior for NCFSK was an *exponential* function of $E_b/2N_0$. If we make a similar comparison between optimum coherent PSK (see Chapter 4) and DPSK, we also note that the error probability performance of the former varies as a *complementary error* function of E_b/N_0, whereas that of the latter varies as an *exponential* function of E_b/N_0. Aside from the 3 dB differences in SNR, the performance similarity between optimum coherent detection of PSK and FSK and the performance similarity between optimum differential detection of PSK and noncoherent detection of FSK is more than accidental.

We saw in Chapter 4 that both optimum coherent detection of PSK and optimum coherent detection of FSK resulted in a correlation receiver with the same form of decision variable, the difference between the two being the correlation coefficient, γ, of the signals [see (4.52)] which affects the mean of this variable. In the case of FSK, $\gamma = 0$, whereas for PSK, $\gamma = -1$. Since the effective SNR in both cases is proportional to $(1 - \gamma)$, then the above difference in the value of γ accounts for the 3 dB difference in E_b/N_0 performance. The explanation of the similarity in performance between optimum differential detection of PSK and noncoherent detection of FSK is perhaps not as obvious since indeed the two receiver structures appear to be quite different. In what follows, we shall expose the underlying thread that ties these two different modulation and detection forms together. The basis of this thread is a special case of a unified analysis of certain coherent and noncoherent binary communication systems performed many years ago by Stein [21], which still stands as a classic contribution in the field.

Consider the optimum DPSK receiver of Fig. 7.2b and the optimum NCFSK receiver discussed in Chapter 5, redrawn here as Fig. 7.6 for the purpose of ease of reference and similarity of notation. We are interested in comparing the error probability performances of these receivers for both messages m_0 and m_1, which are defined below for the two different modulation/detection schemes.[5]

$$\text{message } m_0 : \begin{cases} \theta_i - \theta_{i-1} = \Delta\theta_i = 0 & \text{DPSK} \\ f_i = f^{(0)} & \text{NCFSK} \end{cases}$$

$$\text{message } m_1 : \begin{cases} \theta_i - \theta_{i-1} = \Delta\theta_i = \pi & \text{DPSK} \\ f_i = f^{(1)} & \text{NCFSK} \end{cases} \tag{7.50}$$

The rules for deciding between these two messages (see Figs. 7.2b and 7.6) are

$$Z_i = V_{c,i} V_{c,i-1} + V_{s,i} V_{s,i-1} \underset{m_1}{\overset{m_0}{\gtrless}} 0 \qquad \text{DPSK} \tag{7.51a}$$

$$Z_i' = \left[(V_{c1,i})^2 + (V_{s1,i})^2 \right] - \left[(V_{c2,i})^2 + (V_{s2,i})^2 \right] \underset{m_1}{\overset{m_0}{\gtrless}} 0 \qquad \text{NCFSK} \tag{7.51b}$$

where

---

5. We use a parenthetical superscript notation here to characterize the particular message being referred to so as to avoid confusion with the subscript notation used for denoting discrete time.

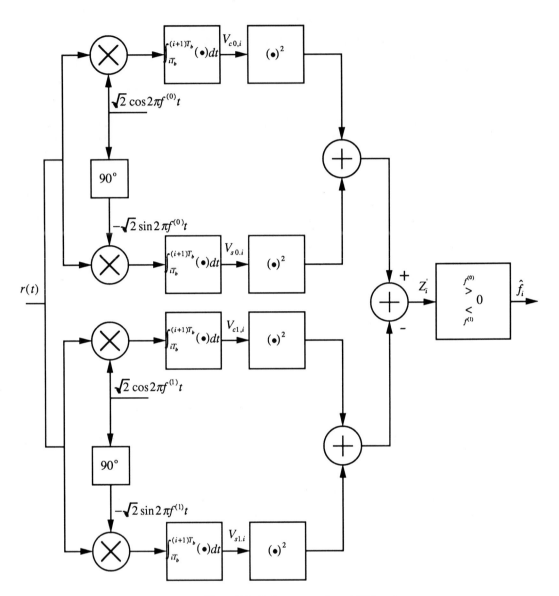

Figure 7.6   Optimum noncoherent FSK receiver

$$V_{c,i} = \sqrt{PT_b} \cos(\theta_i + \theta) + N_{c,i}, \qquad V_{s,i} = \sqrt{PT_b} \sin(\theta_i + \theta) + N_{s,i}$$

$$V_{ck,i} = \sqrt{PT_b} \left[\frac{\sin\left(2\pi i \left(f^{(k)} - f_i\right) T_b + \theta\right) - \sin\left(2\pi (i-1) \left(f^{(k)} - f_i\right) T_b + \theta\right)}{2\pi \left(f^{(k)} - f_i\right)} \right]$$

$$+ N_{ck,i}; \quad k = 0, 1 \tag{7.52}$$

$$V_{sk,i} = \sqrt{PT_b} \left[\frac{-\cos\left(2\pi i \left(f^{(k)} - f_i\right) T_b + \theta\right) + \cos\left(2\pi (i-1) \left(f^{(k)} - f_i\right) T_b + \theta\right)}{2\pi \left(f^{(k)} - f_i\right)} \right]$$

$$+ N_{sk,i}; \quad k = 0, 1$$

For DPSK, $V_{c,i}$ and $V_{s,i}$ are independent Gaussian random variables with variance $\sigma^2 = N_0 T_b / 2$, while for NCFSK, $V_{c1,i}$, $V_{c2,i}$, $V_{s1,i}$, and $V_{s2,i}$ are independent Gaussian random variables also with variance $\sigma^2 = N_0 T_b / 2$. The means of these random variables when messages m_0 and m_1 are transmitted are as follows:

$$E\{V_{c,i}\} = \begin{cases} \sqrt{PT_b} \cos\theta; & \theta_i = 0 \\ -\sqrt{PT_b} \cos\theta; & \theta_i = \pi \end{cases} m_0 \text{ or } m_1$$

$$E\{V_{s,i}\} = \begin{cases} \sqrt{PT_b} \sin\theta; & \theta_i = 0 \\ -\sqrt{PT_b} \sin\theta; & \theta_i = \pi \end{cases} m_0 \text{ or } m_1$$

$$\tag{7.53}$$

$$E\{V_{c1,i}\} = \begin{cases} \sqrt{PT_b} \cos\theta; & m_0 \\ 0; & m_1 \end{cases}, \quad E\{V_{s1,i}\} = \begin{cases} \sqrt{PT_b} \sin\theta; & m_0 \\ 0; & m_1 \end{cases}$$

$$E\{V_{c2,i}\} = \begin{cases} 0; & m_0 \\ \sqrt{PT_b} \cos\theta; & m_1 \end{cases}, \quad E\{V_{s2,i}\} = \begin{cases} 0; & m_0 \\ \sqrt{PT_b} \sin\theta; & m_1 \end{cases}$$

It is at this point that we introduce a clever algebraic trick suggested by Stein [21] that allows (7.51a) and (7.51b) to be more readily identified as being equivalent. In particular, we rewrite (7.51a) as

$$Z_i = \left[\left(\frac{V_{c,i-1} + V_{c,i}}{2} \right)^2 + \left(\frac{V_{s,i-1} + V_{s,i}}{2} \right)^2 \right]$$

$$- \left[\left(\frac{V_{c,i-1} - V_{c,i}}{2} \right)^2 + \left(\frac{V_{s,i-1} - V_{s,i}}{2} \right)^2 \right] \tag{7.54}$$

$$\triangleq \left[(V_{c+,i})^2 + (V_{s+,i})^2 \right] - \left[(V_{c-,i})^2 + (V_{s-,i})^2 \right] \underset{m_1}{\overset{m_0}{\underset{<}{\gtrless}}} 0 \qquad \text{DPSK}$$

The random variables $V_{c+,i}$, $V_{s+,i}$, $V_{c-,i}$, $V_{s-,i}$ are all independent Gaussian with variance $\sigma^2 / 2 = N_0 T_b / 4$ and means as follows:

$$E\left\{V_{c+,i}\right\} = \begin{cases} \sqrt{PT_b}\cos\theta; & \theta_i = 0 \\ -\sqrt{PT_b}\cos\theta; & \theta_i = \pi \\ 0; & \end{cases} \begin{array}{l} \\ \\ m_1, \end{array} \quad \left.\begin{array}{c} \\ \\ \end{array}\right\} m_0$$

$$E\left\{V_{s+,i}\right\} = \begin{cases} \sqrt{PT_b}\sin\theta; & \theta_i = 0 \\ -\sqrt{PT_b}\sin\theta; & \theta_i = \pi \\ 0; & \end{cases} \begin{array}{l} \\ \\ m_1, \end{array} \quad \left.\begin{array}{c} \\ \\ \end{array}\right\} m_0 \qquad (7.55)$$

$$E\left\{V_{c-,i}\right\} = \begin{cases} 0; & \\ \sqrt{PT_b}\cos\theta; & \theta_i = 0 \\ -\sqrt{PT_b}\cos\theta; & \theta_i = \pi \end{cases} \quad m_0 \atop \left.\begin{array}{c} \\ \\ \end{array}\right\} m_1,$$

$$E\left\{V_{s-,i}\right\} = \begin{cases} 0; & \\ \sqrt{PT_b}\sin\theta; & \theta_i = 0 \\ -\sqrt{PT_b}\sin\theta; & \theta_i = \pi \end{cases} \quad m_0 \atop \left.\begin{array}{c} \\ \\ \end{array}\right\} m_1.$$

Comparing (7.53) and (7.55), we observe the similarity between the squares of $V_{c+,i}$ and $V_{c1,i}$, $V_{s+,i}$ and $V_{s1,i}$, $V_{c-,i}$ and $V_{c2,i}$, $V_{s-,i}$ and $V_{s2,i}$. Finally then comparing Z_i of (7.54) with Z_i' of (7.51b), it is clear that they have the same probability statistics except that N_0 should be replaced by $N_0/2$ in the former since, as shown above, each Gaussian component in Z_i has half the variance. Equivalently, DPSK and NCFSK have the same error probability performance except for a factor of two in SNR.

7.2 OPTIMUM WAVEFORM RECEIVERS BASED ON A LARGER THAN TWO SYMBOL OBSERVATION INTERVAL

Thus far, we have discussed symbol-by-symbol differential detection of M-DPSK based on an observation of the received signal plus noise over two symbol intervals. We have observed that although differential detection eliminates the need for carrier acquisition and tracking in the receiver, it suffers from a performance penalty (additional required SNR at a given bit error rate) when compared with ideal (perfect carrier phase reference) coherent detection. The amount of this performance penalty increases with the number of phases, M, and is significant for $M \geq 4$. For example, at a bit error probability $P_b(E) = 10^{-5}$, DPSK requires about 0.75 dB more E_b/N_0 than coherently detected BPSK (see Fig. 7.3). For QPSK ($M = 4$), the difference in E_b/N_0 between differential detection and ideal coherent detection at $P_b(E) = 10^{-5}$ is about 2.2 dB. Finally for 8PSK, the corresponding difference in E_b/N_0 performance between the two is greater than 2.5 dB. The latter two E_b/N_0 penalties may be obtained from Fig. 7.3 and the approximate result of Eq. (4.134).

Thus, it is natural to ask: Is there a way of enhancing classical (two symbol observation) M-DPSK so as to recover a portion of the performance lost relative to that of coherent detection and still maintain the advantages of a simple and robust implementation? Furthermore, if this is possible, what is the trade-off between the amount of performance recovered and the additional complexity added to the classical M-DPSK implementation? The answers to these questions stem from the idea of allowing the obervation interval to be longer than two symbol intervals while at the same time making a *joint* decision on several symbols simultaneously, that is, block-by-block detection, as opposed to symbol-by-symbol detection.

In order for this approach to have meaning, one must extend the previous assumption on the duration of time over which the carrier phase is constant to be commensurate with the extended observation interval. For observations on the order of three or four symbol intervals, this is still a reasonable assumption in many applications.

The theoretical framework in which we shall develop this so-called *multiple-symbol differential detection* technique [22, 23] is once again the ML approach to statistical detection and is analogous to that used in Chapter 6 for deriving the optimum receiver of partially coherent M-PSK based on an observation of N_s symbols. As such, the resulting receiver configurations derived here will be the optimum *block-by-block* detection receivers for differentially coherent M-PSK. In the next section, we derive the appropriate ML algorithm. Clearly, classical M-DPSK must result as a special case of this more general model when $N_s = 2$.

7.2.1 Maximum-Likelihood Differential Detection of *M*-PSK—Block-by-Block Detection

Consider as before the transmission of M-PSK signals over the AWGN channel. The channel introduces the unknown phase, θ, which in the absence of any information is assumed to be uniformly distributed in the interval $(-\pi, \pi)$. The received signal in the kth transmission interval then takes the form of (7.1) (with i replaced by k) which can be expressed in terms of the complex baseband signal $\tilde{s}_k = \sqrt{2P}e^{j\theta_k}$ by [see (6.52)]

$$r(t) = \text{Re}\left\{\tilde{s}_k \exp\left(j\omega_c t + \theta\right) + \tilde{n}(t)\right\} = \text{Re}\left\{\tilde{r}(t)\right\} \tag{7.56}$$

where $\tilde{n}(t)$ is a complex additive white Gaussian process. The receiver demodulates $\tilde{r}(t)$ with the complex reference signal $\exp(-j\omega_c t)$ to produce the complex baseband signal $\tilde{s}_k e^{j\theta} + \tilde{n}(t)e^{-j\omega_c t}$, which is then passed through an I&D, resulting in the complex detected sample[6]

$$\tilde{\mathcal{V}}_k = \tilde{s}_k e^{j\theta} + \frac{1}{T_s}\int_{kT_s}^{(k+1)T_s} \tilde{n}(t)e^{-j\omega_c t}dt = \tilde{s}_k e^{j\theta} + \tilde{N}_k \tag{7.57}$$

Consider now an observation of the received signal over an interval of length $N_s T_s$ (N_s is herein referred to as the *block length*) and assume that the unknown channel phase, θ, is constant over this interval. In particular, for an observation of the received signal over the interval $(k - N_s + 1)T_s \leq t \leq (k + 1)T_s$, the sequence $\tilde{\mathcal{V}} = \left(\tilde{\mathcal{V}}_{k-N_s+1}, \tilde{\mathcal{V}}_{k-N_s+2}, \ldots, \tilde{\mathcal{V}}_{k-1}, \tilde{\mathcal{V}}_k\right)$ conditioned on θ is a sufficient statistic for making a ML decision on the transmitted sequence $\tilde{s} = \left(\tilde{s}_{k-N_s+1}, \tilde{s}_{k-N_s+2}, \ldots, \tilde{s}_{k-1}, \tilde{s}_k\right)$.

For the assumed AWGN model, the a posteriori probability of $\tilde{\mathcal{V}}$ given $\tilde{s}$ and θ is

---

6. We apply the same normalization of $1/T_s$ to the complex I&D output sample as was done in Section 6.3. Again, we denote this normalization by a script typeface.

$$f_{\tilde{\mathcal{V}}|\theta}(\tilde{\mathcal{V}}|\tilde{s}, \theta) = \frac{1}{\left(2\pi\sigma_n^2\right)^{N_s}} \exp\left\{-\frac{\left\|\tilde{\mathcal{V}} - \tilde{s}e^{j\theta}\right\|^2}{2\sigma_n^2}\right\}$$

$$= \frac{1}{\left(2\pi\sigma_n^2\right)^{N_s}} \exp\left\{-\frac{1}{2\sigma_n^2}\left[\sum_{i=0}^{N_s-1}\left[\left|\tilde{\mathcal{V}}_{k-i}\right|^2 + \left|\tilde{s}_{k-i}\right|^2\right]\right.\right. \tag{7.58}$$

$$\left.\left.-2\left|\sum_{i=0}^{N_s-1}\tilde{\mathcal{V}}_{k-i}\tilde{s}_{k-i}^*\right|\cos(\theta - \alpha)\right]\right\}$$

where

$$\alpha \triangleq \tan^{-1}\frac{\operatorname{Im}\left\{\sum_{i=0}^{N_s-1}\tilde{\mathcal{V}}_{k-i}\tilde{s}_{k-i}^*\right\}}{\operatorname{Re}\left\{\sum_{i=0}^{N_s-1}\tilde{\mathcal{V}}_{k-i}\tilde{s}_{k-i}^*\right\}} \tag{7.59}$$

Averaging (7.58) over the uniform pdf of θ and simplifying gives the conditional probability

$$f_{\tilde{\mathcal{V}}}(\tilde{\mathcal{V}}|\tilde{s}) = \int_{-\pi}^{\pi} f_{\tilde{\mathcal{V}}|\theta}(\tilde{\mathcal{V}}|\tilde{s}, \theta) f_\theta(\theta)d\theta \tag{7.60}$$

$$= \frac{1}{\left(2\pi\sigma_n^2\right)^{N_s}} \exp\left\{-\frac{1}{2\sigma_n^2}\sum_{i=0}^{N_s-1}\left[\left|\tilde{\mathcal{V}}_{k-i}\right|^2 + \left|\tilde{s}_{k-i}\right|^2\right]\right\} I_0\left(\frac{1}{\sigma_n^2}\left|\sum_{i=0}^{N_s-1}\tilde{\mathcal{V}}_{k-i}\tilde{s}_{k-i}^*\right|\right)$$

noting as before that for M-PSK, $|\tilde{s}_k|^2$ is constant for all transmitted phases. Then, since $\left|\tilde{\mathcal{V}}_k\right|^2$ is independent of the transmitted signal and since $I_0(x)$ is a monotonic function of x, maximizing $f_{\tilde{\mathcal{V}}}(\tilde{\mathcal{V}}|\tilde{s})$ over $\tilde{s}$ is equivalent to finding $\max_s\left|\sum_{i=0}^{N_s-1}\tilde{\mathcal{V}}_{k-i}\tilde{s}_{k-i}^*\right|^2$, which is identical to (6.59b) with ρ set equal to zero. Defining the transmitted phase sequence $\boldsymbol{\theta} = \left(\theta_{k-N_s+1}, \theta_{k-N_s+2}, \ldots, \theta_{k-1}, \theta_k\right)$, we obtain the decision rule:

$$\text{choose } \hat{\boldsymbol{\theta}} \text{ corresponding to } \max_{\boldsymbol{\theta}}\left|\sum_{i=0}^{N_s-1}\tilde{\mathcal{V}}_{k-i}e^{-j\theta_{k-i}}\right|^2$$

where $\hat{\boldsymbol{\theta}} = \left(\hat{\theta}_{k-N_s+1}, \hat{\theta}_{k-N_s+2}, \ldots, \hat{\theta}_{k-1}, \hat{\theta}_k\right)$ is the decision sequence. Note that the actual complex I&D output in the $(k-i)^{\text{th}}$ interval, $\tilde{V}_k$, is related to $\tilde{\mathcal{V}}_{k-i}$ by (see Fig. 7.2)

$$\tilde{V}_{k-i} = V_{c,k-i} + jV_{s,k-i} = \frac{T_s}{\sqrt{2}}\tilde{\mathcal{V}}_{k-i} = \frac{1}{\sqrt{2}}\int_{(k-i)T_s}^{(k-i+1)T_s}\tilde{r}(t)\exp\left\{-j\omega_c t\right\}dt \tag{7.61}$$

and thus the above decision rule can equivalently be rewritten as

$$\text{choose } \hat{\boldsymbol{\theta}} \text{ corresponding to } \max_{\boldsymbol{\theta}}\left|\sum_{i=0}^{N_s-1}\tilde{V}_{k-i}e^{-j\theta_{k-i}}\right|^2$$

Note that, as for the two symbol observation case in Appendix 7A, the differentially coherent decision rule for an N_s-symbol observation also has a phase ambiguity associated with it since the addition of an arbitrary fixed phase, say θ_a, to all estimated phases $\hat{\theta}_{k-N_s+1}, \hat{\theta}_{k-N_s+2}, \ldots, \hat{\theta}_{k-1}, \hat{\theta}_k$ results in the identical decision rule, that is,

$$\text{choose } \hat{\boldsymbol{\theta}} \text{ corresponding to } \max_{\boldsymbol{\theta}} \left| \sum_{i=0}^{N_s-1} \tilde{V}_{k-i} \exp\{-j(\theta_{k-i} + \theta_a)\} \right|^2$$

Thus, as for classical M-DPSK, we must employ differential encoding at the transmitter to resolve this phase ambiguity. Letting $\theta_a = -\theta_{k-N_s+1}$, then the above decision rule can be rewritten in terms of the input information phases, $\Delta\theta_{k-i} = \theta_{k-i} - \theta_{k-i-1}$; $i = 0, 1, \ldots, N_s - 2$ as:

choose the information sequence $\Delta\boldsymbol{\theta} = (\Delta\theta_{k-N_s+2}, \Delta\theta_{k-N_s+3}, \ldots, \Delta\theta_{k-1}, \Delta\theta_k)$

that maximizes the statistic

$$\begin{aligned} \eta &= \left| \sum_{i=0}^{N_s-1} \tilde{V}_{k-i} \exp\left\{-j\left(\sum_{m=0}^{N_s-i-2} \Delta\theta_{k-i-m}\right)\right\} \right|^2 \\ &= \left| \tilde{V}_{k-N_s+1} + \sum_{i=0}^{N_s-2} \tilde{V}_{k-i} \exp\left\{-j\left(\sum_{m=0}^{N_s-i-2} \Delta\theta_{k-i-m}\right)\right\} \right|^2 \end{aligned} \tag{7.62}$$

A receiver that implements the above decision rule is illustrated in Fig. 7.7. Note that an observation over N_s symbol intervals is used to make a sequence (block) decision on $N_s - 1$ input information phases. As such, the N_s-symbol observation blocks overlap by one symbol since the last symbol of a given block serves as the phase reference for the next block. Thus, for this type of block-by-block detection, the rate at which decisions are made is identical to that of the symbol-by-symbol detection scheme of classical M-DPSK.

Some special cases of (7.62) are of interest. For $N_s = 1$, that is, an observation of the received signal over one symbol interval, (7.62) simplifies to

$$\eta = \left| \tilde{V}_k \right|^2 \tag{7.63}$$

which is completely independent of the input data phases and thus cannot be used for making decisions on differentially encoded M-PSK modulation. In fact, the statistic of (7.63) corresponds to the classical case of noncoherent detection (see Chapter 5) which is not applicable to phase modulation.

Next, let $N_s = 2$, in which case (7.62) becomes

$$\eta = \left| \tilde{V}_{k-1} + \tilde{V}_k e^{-j\Delta\theta_k} \right|^2 = \left| \tilde{V}_{k-1} \right|^2 + \left| \tilde{V}_k \right|^2 + 2\,\text{Re}\left\{ \tilde{V}_k \tilde{V}_{k-1}^* e^{-j\Delta\theta_k} \right\} \tag{7.64}$$

This results in the decision rule

$$\text{choose } \Delta\hat{\theta}_k \text{ corresponding to } \max_{\Delta\theta_k} \text{Re}\left\{ \tilde{V}_k \tilde{V}_{k-1}^* e^{-j\Delta\theta_k} \right\}$$

which has a receiver implementation (see Fig. 7.8) that is the complex form of Fig. 7.2b, as one would expect based on the results in Appendix 7A.

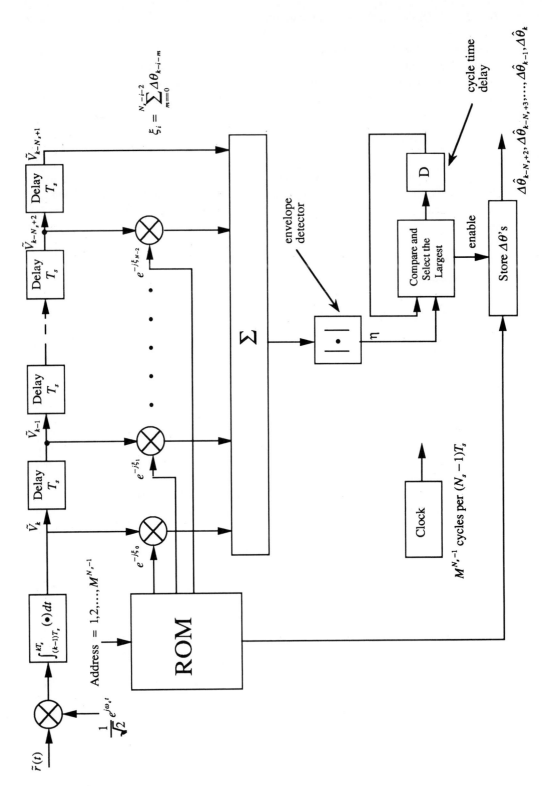

Figure 7.7 Direct implementation of multiple symbol differential detection of *M*-PSK (Reprinted

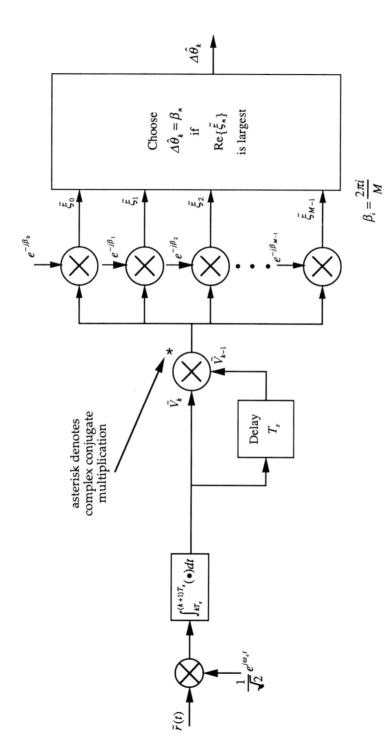

Figure 7.8 Optimum receiver (complex form) for classical M-DPSK (Reprinted from [22] © IEEE)

Next, we consider (7.62) for $N_s = 3$. Here we have

$$\eta = \left| \tilde{V}_{k-2} + \tilde{V}_{k-1} e^{-j\Delta\theta_{k-1}} + \tilde{V}_k e^{-j(\Delta\theta_{k-1}+\Delta\theta_k)} \right|^2$$

$$= \left| \tilde{V}_{k-2} \right|^2 + \left| \tilde{V}_{k-1} \right|^2 + \left| \tilde{V}_k \right|^2 + 2\,\mathrm{Re}\left\{ \tilde{V}_k \tilde{V}_{k-1}^* e^{-j\Delta\theta_k} \right\} \qquad (7.65)$$

$$+ 2\,\mathrm{Re}\left\{ \tilde{V}_{k-1} \tilde{V}_{k-2}^* e^{-j\Delta\theta_{k-1}} \right\} + 2\,\mathrm{Re}\left\{ \tilde{V}_k \tilde{V}_{k-2}^* e^{-j(\Delta\theta_{k-1}+\Delta\theta_k)} \right\}$$

Thus, the decision rule becomes

choose $\Delta\hat{\theta}_k$ and $\Delta\hat{\theta}_{k-1}$ corresponding to

$$\max_{\Delta\theta_{k-1},\Delta\theta_k} \mathrm{Re}\left\{ \tilde{V}_k \tilde{V}_{k-1}^* e^{-j\Delta\theta_k} + \tilde{V}_{k-1} \tilde{V}_{k-2}^* e^{-j\Delta\theta_{k-1}} + \tilde{V}_k \tilde{V}_{k-2}^* e^{-j(\Delta\theta_{k-1}+\Delta\theta_k)} \right\}$$

Note that the first and second terms of the metric used in the decision rule of (7.65) are identical to those used to make successive and independent decisions on $\Delta\theta_k$ and $\Delta\theta_{k-1}$, respectively, in classical M-DPSK. The third term in the optimum metric is a combination of the first two and is required to make an optimum *joint* decision on $\Delta\theta_k$ and $\Delta\theta_{k-1}$.

Clearly, a receiver implemented on the basis of (7.65) will outperform a classical M-DPSK receiver. Before demonstrating the amount of this performance improvement as a function of the number of phases, M, we first present the implementation of the optimum $N_s = 3$ receiver in accordance with the above form of the decision rule. Figure 7.9 is a parallel implementation and Fig. 7.10 is a series implementation that although simpler in appearance requires envelope normalization and additional delay elements. It should also be noted that in the parallel implementation of Fig. 7.9, the apparent M^2 phasors needed to perform the phase rotations of the output $V_k V_{k-2}^*$ can be accomplished with only M phasors since the sum angle $\Delta\theta_k + \Delta\theta_{k-1}$ when evaluated modulo 2π ranges over the same set of M values as does $\Delta\theta_k$ and $\Delta\theta_{k-1}$.

7.2.2 Bit Error Probability Performance

To obtain a simple upper bound on the average bit error probability, $P_b(E)$, of the optimum block-by-block M-DPSK scheme, we use a union bound analogous to that used in Chapter 6 to derive the performance of the optimum partially coherent block detection receiver. In particular, the upper bound on $P_b(E)$ is the sum of the pairwise error probabilities associated with each $(N_s - 1)$-bit error sequence. An exact evaluation of these pairwise error probabilities is carried out in the next section. Mathematically speaking, if $\boldsymbol{\Delta\theta} = \left(\Delta\theta_{k-N_s+2}, \Delta\theta_{k-N_s+3}, \ldots, \Delta\theta_{k-1}, \Delta\theta_k \right)$ is the sequence of $N_s - 1$ information phases and $\boldsymbol{\Delta\hat{\theta}} = \left(\Delta\hat{\theta}_{k-N_s+2}, \Delta\hat{\theta}_{k-N_s+3}, \ldots, \Delta\hat{\theta}_{k-1}, \Delta\hat{\theta}_k \right)$ is the corresponding sequence of detected phases, then if $\mathbf{u}$ denotes the sequence of $b = (N_s - 1)\log_2 M$ information bits that produces $\boldsymbol{\Delta\theta}$ at the transmitter and $\hat{\mathbf{u}}$ the sequence of b bits that result from the detection of $\boldsymbol{\Delta\hat{\theta}}$, then the bit error probability has the upper bound [see (4.88) with M replaced by M^{N_s-1}]

$$P_b(E) \leq \frac{1}{b}\frac{1}{M^{N_s-1}} \sum_{\boldsymbol{\Delta\theta} \neq \boldsymbol{\Delta\hat{\theta}}} \sum w(\mathbf{u},\hat{\mathbf{u}}) \Pr\left\{ \hat{\eta} > \eta \mid \boldsymbol{\Delta\theta} \right\} \qquad (7.66)$$

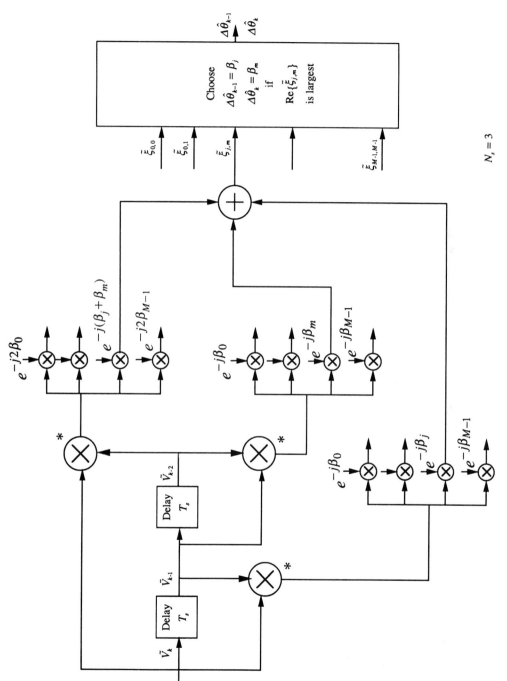

Figure 7.9 Parallel implementation of multiple symbol differential detector; $N_s = 3$ (Reprinted from [22] © IEEE)

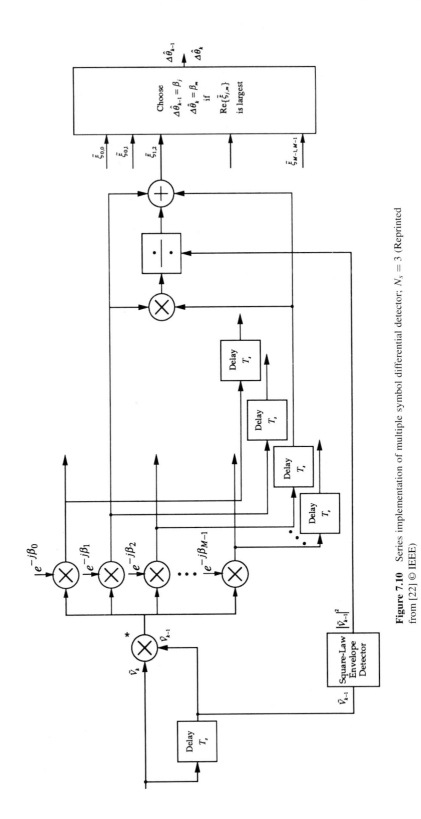

Figure 7.10 Series implementation of multiple symbol differential detector; $N_s = 3$ (Reprinted from [22] © IEEE)

where $w(\mathbf{u}, \hat{\mathbf{u}})$ denotes the Hamming distance between $\mathbf{u}$ and $\hat{\mathbf{u}}$ and $\Pr\{\hat{\eta} > \eta | \boldsymbol{\Delta\theta}\}$ denotes the pairwise probability that $\boldsymbol{\Delta\hat{\theta}}$ is incorrectly chosen when indeed $\boldsymbol{\Delta\theta}$ was sent. The decision statistic η is defined in (7.62) and the corresponding error statistic $\hat{\eta}$ is identical to (7.62) with each $\Delta\theta_k$ replaced by $\Delta\hat{\theta}_k$. For symmetric signaling sets (such as M-PSK), (7.66) satisfies a uniform error probability (UEP) criterion, that is, the probability of error is independent of which input phase sequence $\boldsymbol{\Delta\theta}$ is chosen as the correct sequence. Under these conditions, (7.66) simplifies to [see (6.62) with $\phi = 0$]

$$P_b(E) \le \frac{1}{(N_s - 1)\log_2 M} \sum_{\boldsymbol{\Delta\theta} \ne \boldsymbol{\Delta\hat{\theta}}} w(\mathbf{u}, \hat{\mathbf{u}})\Pr\{\hat{\eta} > \eta | \boldsymbol{\Delta\theta}\} \tag{7.67}$$

where $\boldsymbol{\Delta\theta}$ is any input sequence (e.g., the null sequence $(0, 0, \ldots, 0) = \mathbf{0}$).

Evaluation of the pairwise error probability.   To compute $\Pr\{\hat{\eta} > \eta | \boldsymbol{\Delta\theta}\}$, we take the approach of Section 6.4.1 which, for the particular case under consideration, was evaluated in detail in [22, Appendix A]. In particular, letting $\eta = |z_1|^2$ and $\hat{\eta} = |z_2|^2$, then analogous to (6.64) we have

$$\Pr\{\hat{\eta} > \eta | \boldsymbol{\Delta\theta}\} = \Pr\{|z_2| > |z_1| \,|\, \boldsymbol{\Delta\theta}\} = \frac{1}{2}\left[1 - Q(\sqrt{b}, \sqrt{a}) + Q(\sqrt{a}, \sqrt{b})\right] \tag{7.68}$$

where $Q(\alpha, \beta)$ is as before Marcum's Q-function, whose arguments are given by (6.65) and evaluate to (see Problem 7.8)

$$\begin{Bmatrix} b \\ a \end{Bmatrix} = \frac{E_s}{2N_0}\left[N_s \pm \sqrt{N_s^2 - |\delta|^2}\right] \tag{7.69}$$

with

$$\begin{aligned}
\delta &\triangleq \sum_{i=0}^{N_s-1} \exp\left\{j\sum_{m=0}^{N_s-i-2}\left(\Delta\theta_{k-i-m} - \Delta\hat{\theta}_{k-i-m}\right)\right\} \\
&\triangleq \sum_{i=0}^{N_s-1} \exp\left\{j\sum_{m=0}^{N_s-i-2} \delta\theta_{k-i-m}\right\}
\end{aligned} \tag{7.70}$$

In (7.70), it is understood that the summation in the exponent evaluates to zero if the upper index is negative.

Note that for any given N_s, M, and data sequence $\boldsymbol{\Delta\theta}$, δ can be evaluated for each error sequence $\boldsymbol{\Delta\hat{\theta}}$. We now demonstrate the evaluation of (7.67) and (7.69) for some special cases of interest.

Special cases.

Case 1.   Classical DPSK ($N_s = 2$, $M = 2$)
From (7.70), we immediately get $\delta = 0$ and thus from (7.69)

$$\begin{Bmatrix} b \\ a \end{Bmatrix} = \begin{Bmatrix} \frac{2E_s}{N_0} \\ 0 \end{Bmatrix} \tag{7.71}$$

Substituting (7.71) into (7.68) gives

$$\Pr\{\hat{\eta} > \eta | \boldsymbol{\Delta\theta}\} = \frac{1}{2}\left[1 - Q\left(\sqrt{\frac{2E_s}{N_0}}, 0\right) + Q\left(0, \sqrt{\frac{2E_s}{N_0}}\right)\right] \tag{7.72}$$

From the definition of the Q-function

$$Q(\alpha, 0) = 1; \quad Q(0, \beta) = \exp\left(-\frac{\beta^2}{2}\right) \tag{7.73}$$

Since for the binary case the pairwise error probability is indeed equal to the bit error probability, then we have from (7.72) and (7.73) that

$$P_b(E) = \frac{1}{2} \exp\left(-\frac{E_b}{N_0}\right) \tag{7.74}$$

which is identical with (7.14).

Case 2. DPSK with 3-Symbol Observation ($N_s = 3$, $M = 2$)

Here, there are three possible error sequences of length two. The pertinent results related to the evaluation of (7.70) and (7.71) are given below:

$\Delta\theta_k - \Delta\hat{\theta}_k$	$\Delta\theta_{k-1} - \Delta\hat{\theta}_{k-1}$	δ	$\begin{Bmatrix} b \\ a \end{Bmatrix}$
0	π	-1	$\frac{E_s}{N_0}\left[\frac{3}{2} \pm \sqrt{2}\right]$
π	0	$+1$	$\frac{E_s}{N_0}\left[\frac{3}{2} \pm \sqrt{2}\right]$
π	π	$+1$	$\frac{E_s}{N_0}\left[\frac{3}{2} \pm \sqrt{2}\right]$

Since the Hamming distance $w(\mathbf{u}, \hat{\mathbf{u}})$ is equal to unity for the first two error sequences and is equal to two for the third sequence, then using the above results in (7.69) and (7.70), the upper bound on bit error probability as given by (7.66) is evaluated as

$$P_b(E) \le 1 - Q\left(\sqrt{\frac{E_b}{N_0}\left(\frac{3}{2} + \sqrt{2}\right)}, \sqrt{\frac{E_b}{N_0}\left(\frac{3}{2} - \sqrt{2}\right)}\right)$$
$$+ Q\left(\sqrt{\frac{E_b}{N_0}\left(\frac{3}{2} - \sqrt{2}\right)}, \sqrt{\frac{E_b}{N_0}\left(\frac{3}{2} + \sqrt{2}\right)}\right) \tag{7.75}$$

To see how much performance is gained by extending the observation interval for differential detection from $N_s = 2$ (classical DPSK) to $N_s = 3$, we must compare (7.75) to (7.74). Due to the complex form of (7.75), this comparison is not readily obvious without resorting to numerical evaluation. On the other hand, by examining the asymptotic (large E_s/N_0) behavior of (7.75), we can get an immediate handle on this gain.

When both arguments of the Q-function are large, the following asymptotic approximations are valid [24]:

$$Q(\alpha, \beta) \cong 1 - \frac{1}{\alpha - \beta}\sqrt{\frac{\beta}{2\pi\alpha}} \exp\left\{-\frac{(\alpha - \beta)^2}{2}\right\}; \quad \alpha \gg \beta \gg 1$$
$$Q(\alpha, \beta) \cong \frac{1}{\beta - \alpha}\sqrt{\frac{\beta}{2\pi\alpha}} \exp\left\{-\frac{(\beta - \alpha)^2}{2}\right\}; \quad \beta \gg \alpha \gg 1 \tag{7.76}$$

Using these approximations, (7.68) becomes

$$\Pr\{\hat{\eta} > \eta | \Delta\boldsymbol{\theta}\} = \frac{1}{2}\left[\frac{1}{\sqrt{b} - \sqrt{a}}\left(\sqrt{\frac{\sqrt{a/b}}{2\pi}} + \sqrt{\frac{\sqrt{b/a}}{2\pi}}\right)\exp\left\{-\frac{(\sqrt{b} - \sqrt{a})^2}{2}\right\}\right] \quad (7.77)$$

or, from (7.69)

$$\Pr\{\hat{\eta} > \eta | \Delta\boldsymbol{\theta}\} = \frac{1}{2\sqrt{2\pi\frac{E_s}{N_0}(N_s - |\delta|)}}\left(\left(\frac{N_s - \sqrt{N_s^2 - |\delta|^2}}{N_s + \sqrt{N_s^2 - |\delta|^2}}\right)^{1/4}\right.$$

$$+ \left.\left(\frac{N_s + \sqrt{N_s^2 - |\delta|^2}}{N_s - \sqrt{N_s^2 - |\delta|^2}}\right)^{1/4}\right) \times \exp\left\{-\frac{E_s}{2N_0}(N_s - |\delta|)\right\} \quad (7.78)$$

$$= \frac{1}{2\sqrt{\pi\frac{E_s}{N_0}}}\left(\sqrt{\frac{N_s + |\delta|}{|\delta|(N_s - |\delta|)}}\right)\exp\left\{-\frac{E_s}{2N_0}(N_s - |\delta|)\right\}$$

For $N_s = 3$ and $M = 2$, $|\delta| = 1$. Thus, (7.75) becomes the approximate upper bound

$$P_b(E) \lesssim \frac{2\sqrt{2}}{\sqrt{\pi\frac{E_s}{N_0}}}\left[\frac{1}{2}\exp\left\{-\frac{E_s}{N_0}\right\}\right] \quad (7.79)$$

Comparing (7.79) and (7.74), we observe that the factor in front of the term in brackets in (7.79) represents a bound on the improvement in performance obtained by increasing the memory of the decision by one symbol interval from $N_s = 2$ to $N_s = 3$.

General asymptotic results.   In the general case for arbitrary N_s, the dominant terms in the bit error probability occur for the sequences that result in the minimum value of $N_s - |\delta|$. One can easily show that this minimum value will certainly occur for the error sequence $\Delta\hat{\boldsymbol{\theta}}$ having $N_s - 1$ elements equal to the correct sequence $\Delta\boldsymbol{\theta}$ and one element with the smallest error. Thus

$$\min_{\Delta\boldsymbol{\theta}, \Delta\hat{\boldsymbol{\theta}}}(N_s - |\delta|) = N_s - \left|N_s - 1 + e^{j(\Delta\theta_k - \Delta\hat{\theta}_k)_{\min}}\right|$$

$$= N_s - \sqrt{(N_s - 1)^2 + (N_s - 1)(2 - d_{\min}^2) + 1} \quad (7.80)$$

$$= N_s - |\delta|_{\max}$$

where

$$d_{\min}^2 = 4\sin^2\frac{(\Delta\theta_k - \Delta\hat{\theta}_k)_{\min}}{2} = 4\sin^2\frac{\pi}{M} \quad (7.81)$$

and $(\Delta\theta_k - \Delta\hat{\theta}_k)_{\min}$ denotes the smallest possible value between the kth symbols of the data and error phase sequences, that is, $2\pi/M$. Also note that for $|\delta| = |\delta|_{\max}$, (7.70) reduces to

$$\begin{Bmatrix} b \\ a \end{Bmatrix} = \frac{E_s}{2N_0}\left[N_s \pm 2\sqrt{N_s - 1}\sin\frac{\pi}{M}\right] \quad (7.82)$$

Thus, the average bit error probability is approximately upper bounded by

$$
P_b(E) \lesssim \frac{1}{(N_s - 1)\log_2 M} \left(\sum_{\Delta\boldsymbol{\theta} \neq \Delta\hat{\boldsymbol{\theta}}} w(\mathbf{u}, \hat{\mathbf{u}}) \right)
$$

$$
\times \frac{1}{2\sqrt{\pi \frac{E_s}{N_0}}} \left(\sqrt{\frac{N_s + |\delta|_{\max}}{|\delta|_{\max}(N_s - |\delta|_{\max})}} \right) \exp\left\{ -\frac{E_s}{2N_0}(N_s - |\delta|_{\max}) \right\}
$$

(7.83)

where $w(\mathbf{u}, \hat{\mathbf{u}})$ corresponds only to those error sequences that result in $|\delta|_{\max}$.

For the binary case ($M = 2$), we have from (7.81) that $d_{\min}^2 = 4$ and hence $N_s - |\delta|_{\max} = 2$. Similarly, it is straightforward to show (see [22, Appendix B] and Problem 7.10) that the sum of Hamming distances required in (7.83) is given by

$$
\sum_{\Delta\boldsymbol{\theta} \neq \Delta\hat{\boldsymbol{\theta}}} w(\mathbf{u}, \hat{\mathbf{u}}) = \begin{cases} 2(N_s - 1); & N_s > 2 \\ 1; & N_s = 2 \end{cases}
$$

(7.84)

Thus, (7.83) simplifies to

$$
P_b(E) \lesssim \frac{2}{\sqrt{\pi \frac{E_s}{N_0}}} \left(\sqrt{\frac{N_s - 1}{N_s - 2}} \right) \left[\frac{1}{2} \exp\left\{ -\frac{E_s}{N_0} \right\} \right]
$$

(7.85)

which is the generalization of (7.79) for arbitrary N_s.[7]

Equation (7.85) has an interesting interpretation as N_s gets large. Taking the limit of (7.85) as $N_s \to \infty$, we get

$$
P_b(E) \lesssim \frac{1}{\sqrt{\pi \frac{E_s}{N_0}}} \exp\left\{ -\frac{E_s}{N_0} \right\}
$$

(7.86)

which when expressed in terms of the first term of the asymptotic expansion of the complementary error function, namely

$$
\operatorname{erfc} x \cong \frac{1}{\sqrt{\pi} x} \exp(-x^2)
$$

(7.87)

becomes

$$
P_b \lesssim \operatorname{erfc} \sqrt{\frac{E_s}{N_0}}
$$

(7.88)

From the results in Chapter 4, the exact bit error probability performance for *coherent* detection of BPSK with differential encoding and decoding [see Eq. (4.201)] has an asymptotic upper bound identical to (7.88). Thus, as one might expect, *the performance of multiple symbol differentially detected BPSK approaches that of ideal coherent detection BPSK with differential encoding in the limit as the observation interval (decision memory) approaches infinity.*

---

7. Note that (7.85) is not valid for $N_s = 2$ since in that case $|\delta|_{\max} = 0$ [see (7.80) and (7.81)] and thus the inequalities of (7.76) are not satisfied.

A similar limiting behavior as the above may be observed for other values of M. In particular, it can be shown (see [22, Appendix B] and Problem 7.9) that for $M > 2$ and a Gray code mapping of bits to symbols, the sum of Hamming distances corresponding to $|\delta|_{max}$ is given by

$$\sum_{\Delta\boldsymbol{\theta}\neq\Delta\hat{\boldsymbol{\theta}}} w\left(\mathbf{u},\hat{\mathbf{u}}\right) = \begin{cases} 4(N_s - 1); & N_s > 2 \\ 2; & N_s = 2 \end{cases} \tag{7.89}$$

Using (7.89) in (7.83), we get (for $N_s > 2$)

$$P_b(E) \lesssim \frac{2}{(\log_2 M)\sqrt{\pi \frac{E_s}{N_0}}} \left(\sqrt{\frac{N_s + |\delta|_{max}}{|\delta|_{max}(N_s - |\delta|_{max})}}\right)$$

$$\times \exp\left\{-\frac{E_s}{2N_0}(N_s - |\delta|_{max})\right\} \tag{7.90}$$

where, from (7.80) and (7.81)

$$|\delta|_{max} = \sqrt{(N_s - 1)^2 + 2(N_s - 1)(1 - 2\sin^2 \pi/M) + 1} \tag{7.91}$$

For $N_s = 2$, the asymptotic upper bound on bit error probability becomes

$$P_b(E) \lesssim \frac{1}{(\log_2 M)\sqrt{2\pi \frac{E_s}{N_0}}} \left(\frac{\cos \frac{\pi}{2M}}{\sin \frac{\pi}{2M}\sqrt{\cos \frac{\pi}{M}}}\right)$$

$$\times \exp\left\{-\frac{2E_s}{N_0}\sin^2 \frac{\pi}{2M}\right\} \tag{7.92}$$

As N_s gets large, $|\delta|_{max} \to N_s - 2\sin^2 \pi/M$ and (7.90) reduces to

$$P_b(E) \lesssim \frac{1}{(\log_2 M)\sqrt{\pi \frac{E_s}{N_0}\sin^2 \frac{\pi}{M}}} \exp\left\{-\frac{E_s}{N_0}\sin^2 \frac{\pi}{M}\right\}$$

$$\cong \frac{1}{\log_2 M} \operatorname{erfc}\left(\sqrt{\frac{E_s}{N_0}}\sin \frac{\pi}{M}\right) \tag{7.93}$$

which is identical to the asymptotic bit error probability for *coherent* detection of M-PSK with differential encoding and decoding [see Eq. (4.134), (4.135) together with the asymptotic form of (4.200)].[8]

Figures 7.11, 7.12, and 7.13 are illustrations of the upper bounds of (7.90) and (7.92) for $M = 2$, 4, and 8, respectively. In each figure, the length (in M-PSK symbols) of the observation interval is a parameter varying from $N_s = 2$ (classical M-DPSK) to $N_s = \infty$

---

8. It should be noted that the result in (7.93) can be obtained by observing that, for large N_s, (7.82) satisfies $\sqrt{b} \gg \sqrt{b} - \sqrt{a} > 0$. In this case, (7.69) can be approximated as in (6.67) (also see [25], Appendix A) by

$$\Pr\left\{\hat{\eta} > \eta | \Delta\boldsymbol{\theta}\right\} \cong \frac{1}{2} \operatorname{erfc}\left(\frac{\sqrt{b} - \sqrt{a}}{\sqrt{2}}\right)$$

Using this relation in (7.67) gives the asymptotic bit error probability in (7.93).

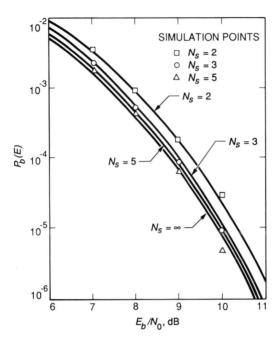

Figure 7.11 Bit error probability versus E_b/N_0 for multiple symbol differential detection of M-PSK; $M = 2$ (Reprinted from [22] © IEEE)

(ideal coherent detection of differentially encoded M-PSK). Also indicated on the figures are computer simulation results corresponding to the exact performance. We observe that, for example, for DPSK, extending the observation interval from $N_s = 2$ to $N_s = 3$ recovers more than half of the E_b/N_0 loss of differential detection versus coherent detection with differential encoding. For $M = 4$, the improvement in E_b/N_0 performance of $N_s = 3$ relative to $N_s = 2$ is more than 1 dB, which is slightly less than half of the total difference between differential detection and coherent detection with differential encoding.

7.2.3 Implementation Algorithms

In the previous section, we showed that for sufficiently large block size, N_s, the bit error probability performance approaches that of coherent detection of differentially encoded M-PSK. Unfortunately, the complexity of a receiver implementation of the type illustrated in Fig. 7.9 for $N_s = 3$ grows exponentially with N_s. To see this, we first observe that, for general N_s, the decision metric [such as that described by (7.65) for $N_s = 3$] must be evaluated for M^{N_s-1} hypotheses (the number of possible $\Delta\boldsymbol{\theta}$ vectors of length $N_s - 1$). For each of these hypotheses there are $\binom{N_s}{2}$ terms in the metric [see, for example, the decision rule below (7.65) which contains $\binom{3}{2} = 3$ terms within the braces]. Thus, we must calculate $\binom{N_s - 1}{2}$ phase angle sums of the form $\Delta\theta_{k-q} + \Delta\theta_{k-q+1} + \ldots + \Delta\theta_{k-p}$ for $0 \leq p < q \leq N_s - 2$, perform $\binom{N_s}{2}$ phase rotations and $\binom{N_s}{2} - 1$ real additions. Since $\binom{N_s}{2}$ is on the order of N_s^2, the total number of operations required to implement the decision rule is on the order of $3N_s^2 \times M^{N_s-1}$.

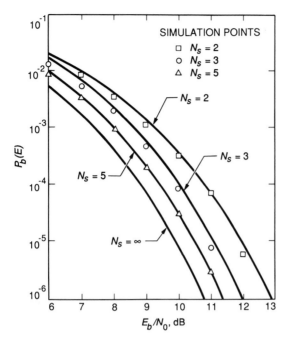

Figure 7.12   Bit error probability versus E_b/N_0 for multiple symbol differential detection of M-PSK; $M = 4$ (Reprinted from [22] © IEEE)

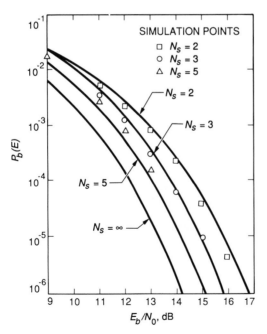

Figure 7.13   Bit error probability versus E_b/N_0 for multiple symbol differential detection of M-PSK; $M = 8$ (Reprinted from [22] © IEEE)

Recently, Mackenthun [25] suggested an algorithm for implementing multiple symbol differential detection of M-PSK, which reduces the number of operations for each N_s-symbol block from order $3N_s^2 \times M^{N_s-1}$ to order $N_s \log_2 N_s$. The impact of this fast computational algorithm is that it allows for a scheme that is a *practical* and viable alternative to coherent detection of differentially encoded QPSK without the need for carrier recovery at the receiver. We present here without proof (see [25] for the details) a description of this algorithm.

Consider the vector of complex I&D samples[9] $\tilde{\mathbf{V}} = \left(\tilde{V}_1, \tilde{V}_2, \ldots, \tilde{V}_{N_s}\right)$ corresponding to a particular observation (block) of N_s symbols. Let $v_i = \tan^{-1}\left(V_{s,i}/V_{c,i}\right)$ denote the argument of each component of this vector [see (7.61)]. For each component in the vector $\tilde{\mathbf{V}}$, define the complex quantity

$$\tilde{Z}_i = \tilde{V}_i e^{-j\tilde{\theta}_i} = \left|\tilde{V}_i\right| e^{j\left(v_i - \tilde{\theta}_i\right)} = \left|\tilde{V}_i\right| e^{j\zeta_i} \tag{7.94}$$

where for each value of $i = 1, 2, \ldots, N_s$ one finds the phase angle $\tilde{\theta}_i \in \{0, 2\pi/M, 4\pi/M, \ldots, 2(M-1)\pi/M\}$ such that $\zeta_i \triangleq \arg\left\{\tilde{Z}_i\right\} = v_i - \tilde{\theta}_i \in [0, 2\pi/M)$. Next, list the values of ζ_i in order of largest to smallest. Define the function $k(i)$ as giving the subscript k of ζ_k for the ith list position, that is,

$$0 \leq \zeta_{k(N_s)}, \zeta_{k(N_s-1)}, \ldots, \zeta_{k(2)}, \zeta_{k(1)} < \frac{2\pi}{M} \tag{7.95}$$

Let

$$\tilde{g}_i = \tilde{Z}_{k(i)} \tag{7.96}$$

The complex quantities $\{\tilde{g}_i; i = 1, 2, \ldots, N_s\}$ are referred to by Mackenthun as the remodulations of the received complex samples $\{\tilde{V}_i; i = 1, 2, \ldots, N_s\}$ in $[0, 2\pi/M)$ ordered by the magnitude of their phase angles. Augment the set of values of $\tilde{g}_i$ by including a $2\pi/M$ phase shift of each as follows:

$$\tilde{g}_i = e^{-j2\pi/M}\tilde{g}_{i-N_s}; \quad i = N_s + 1, N_s + 2, \ldots, 2N_s \tag{7.97}$$

The extended set of $2N_s$ values of $\tilde{g}_i$ represent all possible remodulations of $\{\tilde{V}_i; \ i = 1, 2, \ldots, N_s\}$ in $(-2\pi/M, 2\pi/M)$ indexed by their phase angle, going clockwise around the circle starting at $e^{j2\pi/M}$. Next form the set of N_s sums

$$\eta_q \triangleq \left|\sum_{i=q}^{q+N_s-1} \tilde{g}_i\right|^2; \quad q = 1, 2, \ldots, N_s \tag{7.98}$$

Let q' be the value of q corresponding to the largest value of η_q, i.e.,

$$q' = \max_q^{-1} \eta_q \tag{7.99}$$

and define the phases

$$\hat{\theta}_{k(i)} = \tilde{\theta}_{k(i)}; \quad q' \leq i \leq N_s$$

$$\hat{\theta}_{k(i-N_s)} = \tilde{\theta}_{k(i-N_s)} + \frac{2\pi}{M}; \quad N_s \leq i \leq q' + N_s - 1 \tag{7.100}$$

Arranging the phases $\hat{\theta}_{k(i)}; i = 1, 2, \ldots N_s$ in order of subscript value $k(i)$ produces the joint decision $\hat{\boldsymbol{\theta}} = \left(\hat{\theta}_1, \hat{\theta}_2, \ldots, \hat{\theta}_{N_s}\right)$ on the sequence of transmitted phases. Finally, the decision

---

9. For simplicity of notation, we shall choose the discrete time k equal to $N_s - 1$.

$\boldsymbol{\Delta\hat{\theta}} = \left(\Delta\hat{\theta}_2, \Delta\hat{\theta}_3, \dots, \Delta\hat{\theta}_{N_s}\right)$ on the information data phase vector has components $\Delta\hat{\theta}_i = \hat{\theta}_i - \hat{\theta}_{i-1}$; $i = 1, 2, \dots N_s$ where $\hat{\theta}_i$; $i = 1, 2, \dots N_s$ are the components of $\boldsymbol{\hat{\theta}}$ found in (7.100).

As an example, suppose that $M = 4$, $N_s = 6$ and the ζ_i's are ordered such that $0 \le \zeta_2 < \zeta_6 < \zeta_5 < \zeta_1 < \zeta_3 < \zeta_4 < \pi/2$. Then

$$\tilde{g}_1 = \tilde{Z}_4, \tilde{g}_2 = \tilde{Z}_3, \tilde{g}_3 = \tilde{Z}_1, \tilde{g}_4 = \tilde{Z}_5, \tilde{g}_5 = \tilde{Z}_6, \tilde{g}_6 = \tilde{Z}_2$$

$$\tilde{g}_7 = \tilde{Z}_4 e^{-j\pi/2}, \tilde{g}_8 = \tilde{Z}_3 e^{-j\pi/2}, \tilde{g}_9 = \tilde{Z}_1 e^{-j\pi/2}, \tilde{g}_{10} = \tilde{Z}_5 e^{-j\pi/2}, \qquad (7.101)$$

$$\tilde{g}_{11} = \tilde{Z}_6 e^{-j\pi/2}, \tilde{g}_{12} = \tilde{Z}_2 e^{-j\pi/2}$$

Suppose now for the values of $\tilde{g}_i$ in (7.98) we find that $q' = 4$. Then $\boldsymbol{\hat{\theta}} = (\hat{\theta}_1 + \frac{\pi}{2}, \hat{\theta}_2, \hat{\theta}_3 + \frac{\pi}{2}, \hat{\theta}_4 + \frac{\pi}{2}, \hat{\theta}_5, \hat{\theta}_6)$ and $\boldsymbol{\Delta\hat{\theta}} = (\frac{3\pi}{2}, \frac{\pi}{2}, 0, \frac{3\pi}{2}, 0)$.

7.3 DPSK—NONCOHERENT DETECTION OF OVERLAPPED BINARY ORTHOGONAL MODULATION

Earlier in the chapter, we demonstrated that the decision variables of DPSK and noncoherent FSK are statistically equivalent except for a 3 dB difference in effective E/N_0. Here we show an interpretation of DPSK as noncoherent detection of *overlapped* binary orthogonal signals from which we can immediately glean the 3 dB difference in E/N_0 performance between the two techniques. This interpretation of DPSK also has the advantage that it can be generalized to M-ary $(M > 2)$ modulation.

Consider a pair of complex signals $\tilde{s}_0(t)$ and $\tilde{s}_1(t)$ defined over the interval $0 \le t \le 2T_b$ as follows:

$$\tilde{s}_k(t) = \sqrt{2P} \, p_k(t) e^{j\omega_c t}; \quad k = 0, 1 \qquad (7.102)$$

where

$$p_0(t) = \begin{cases} 1; & 0 \le t \le 2T_b \\ 0; & \text{elsewhere} \end{cases}$$

$$\qquad (7.103)$$

$$p_1(t) = \begin{cases} 1; & 0 \le t \le T_b \\ -1; & T_b \le t \le 2T_b \end{cases}$$

Clearly, $\tilde{s}_0(t)$ and $\tilde{s}_1(t)$ are orthogonal. Suppose now that we were to transmit a DPSK signal corresponding to the binary information sequence $\{a_i\} = 1, -1, 1, 1, -1, 1, -1, -1, -1, 1$ which occurs at a data rate $1/T_b$. The actual transmitted baseband waveform (normalized to unit power)

$$S(t) = \sum_{i=0}^{9} b_i p\left(t - iT_b\right), \quad b_i = a_i b_{i-1} \qquad (7.104)$$

appears as in Fig. 7.14 (assuming a $+1$ as a reference bit, i.e., $b_{-1} = 1$) where $p(t)$ is a unit rectangular pulse of T_b-sec duration. We observe that $S(t)$ of (7.104) can be viewed as a sequence of overlapped $2T_b$-sec pulses of the form given in (7.103). Thus, if $\tilde{s}(t) = \sqrt{2P} \, S(t) e^{j\omega_c t}$ denotes the complex transmitted DPSK signal, then from Fig. 7.14 we see that $\tilde{s}(t)$ can be represented as the overlapped sequence $\tilde{s}_0(t), \tilde{s}_1(t), -\tilde{s}_0(t), -\tilde{s}_0(t), -\tilde{s}_1(t),$

$\tilde{s}_0(t), \tilde{s}_1(t), -\tilde{s}_1(t), \tilde{s}_1(t), -\tilde{s}_0(t)$ where the interval of overlap has duration T_b sec. In short, the DPSK signal has been represented as a sequence of 100% overlapped binary orthogonal signals.

Note that except for the polarity of the $p_k(t)$'s in the overlapped pulse sequence of Fig. 7.14, there is a one-to-one correspondence between the input data bit a_i in the interval $iT_b \le t \le (i+1)T_b$ and the transmitted pulse $p_k(t)$ in the interval $(i-1)T_b \le t \le (i+1)T_b$. In particular, $a_i = +1$ corresponds to transmitting $p_0(t)$ (or $-p_0(t)$) and $a_i = -1$ corresponds to transmitting $p_1(t)$ (or $-p_1(t)$). As such, an optimum decision rule for the information data sequence $\{a_i\}$ can be obtained by noncoherently detecting the sequence of overlapped signals. In particular, based on an observation of the received signal, $r(t)$, over the $2T_b$-sec interval $(i-1)T_b \le t \le (i+1)T_b$, the ±1 decision on a_i is made according to the rule

$$\left| \int_{(i-1)T_b}^{(i+1)T_b} r(t)\tilde{s}_1^*(t)dt \right|^2 \underset{+1}{\overset{-1}{\underset{<}{\gtrless}}} \left| \int_{(i-1)T_b}^{(i+1)T_b} r(t)\tilde{s}_0^*(t)dt \right|^2 \tag{7.105}$$

The decision rule in (7.105) represents an optimum noncoherent decision on a pair of binary orthogonal signals $[\tilde{s}_0(t)$ and $\tilde{s}_1(t)]$ whose duration is $2T_b$ sec and is illustrated in Fig. 7.15. Thus, the error probability associated with this decision is identical to (5.79) with $E_b = PT_b$ replaced by $2E_b = P(2T_b)$, namely

$$P_b(E) = \frac{1}{2}\exp\left(-\frac{E_b}{N_0}\right) \tag{7.106}$$

However, (7.106) is identical to the bit error probability performance of DPSK as given by (7.14). Thus, we have demonstrated the desired equivalence between DPSK and noncoherent detection of overlapped binary orthogonal modulation. Stated another way, Fig. 7.15 is an alternate receiver implementation to Fig. 7.2b.

7.3.1 Generalization to Overlapped *M*-ary Orthogonal Modulation

The notion of signaling with overlapped sequences of orthogonal signals can be generalized to *M*-ary modulation. [10] Here, however, the overlap is less than 100%. In particular, consider a set of *M* complex signals $\{\tilde{s}_k(t); \ k = 0, 1, \ldots, M-1\}$ defined as in (7.102) in terms of a set of *M* pulses $\{p_k(t); \ k = 0, 1, \ldots, M-1\}$ which in addition to being orthogonal over the interval $0 \le t \le \frac{M}{M-1}T_s$ $(T_s = T_b \log_2 M)$ have the following properties. The $p_k(t)$'s are all identical in the interval $0 \le t \le \frac{1}{M-1}T_s$ and differ only in their polarity in the interval $T_s \le t \le \frac{M}{M-1}T_s$. A set of functions that has these properties can be generated from a Hadamard matrix [1] by using the rows of this matrix as generating sequences for rectangular-shaped waveforms. An example of such a set of pulse waveforms for $M = 4$ is illustrated in Fig. 7.16 corresponding to the 4×4 Hadamard matrix

$$H_4 = \begin{bmatrix} 1 & 1 & 1 & 1 \\ 1 & -1 & 1 & -1 \\ 1 & 1 & -1 & -1 \\ 1 & -1 & -1 & 1 \end{bmatrix} \tag{7.107}$$

10. This generalization was suggested by Dan Raphaeli while a Ph.D. student at Caltech and appears in an unpublished document.

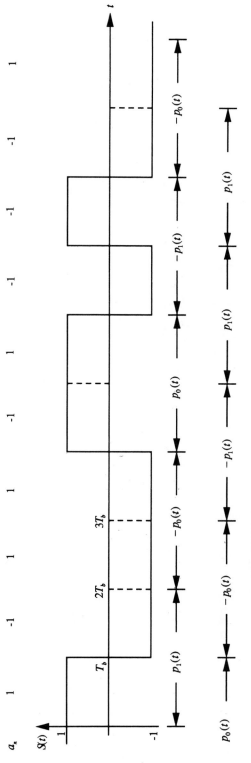

Figure 7.14 Representation of a baseband DPSK waveform in terms of overlapped binary orthogonal pulses

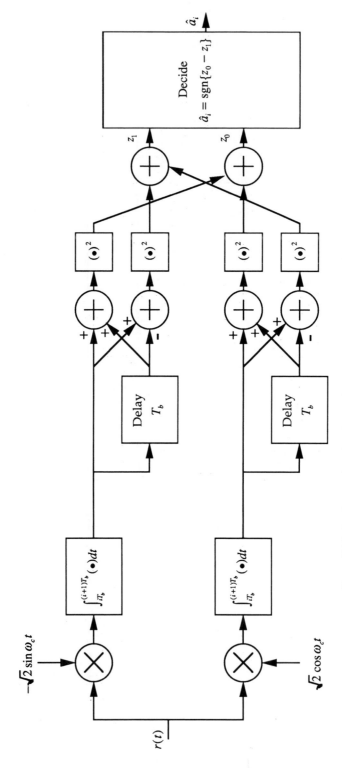

Figure 7.15 Implementation of DPSK receiver as noncoherent detection of binary orthogonal signals

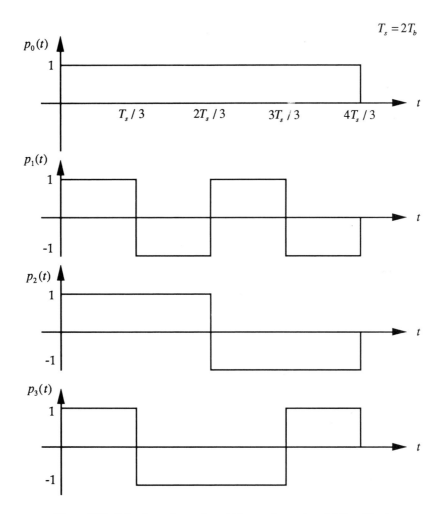

Figure 7.16   Pulse shapes for overlapped binary orthogonal modulation; $M = 4$

Suppose now that we were to transmit the same information data sequence as before (see Fig. 7.14). Consider grouping these bits into symbols consisting of $n = \log_2 M$ bits apiece corresponding to $(a_i, a_{i+1}, \ldots, a_{i+n-1})$ in the interval $iT_b \leq t \leq (i+n)T_b$ and assign $p_k(t)$ (or $-p_k(t)$) as appropriate to the interval $\left(i - \frac{n}{M-1}\right)T_b \leq t \leq (i+n)T_b$ according to the following rule. Let $(k_1, k_2, \ldots, k_n)$ be the binary $(0, 1)$ representation of the integer k where $1 \leq k \leq M$. Then, $p_k(t)$ (or $-p_k(t)$) is assigned to the binary (± 1) sequence $a_i = 1 - 2k_1, a_{i+1} = 1 - 2k_2, \ldots, a_{i+n-1} = 1 - 2k_n$. For the $n = 4$ example, we would have

$$a_i = 1, a_{i+1} = 1 \Rightarrow p_0(t)$$

$$a_i = 1, a_{i+1} = -1 \Rightarrow p_1(t)$$

$$a_i = -1, a_{i+1} = 1 \Rightarrow p_2(t)$$

$$a_i = -1, a_{i+1} = -1 \Rightarrow p_3(t)$$

(7.108)

Using the above assignment, the overlapped pulse sequence for the data sequence of Fig. 7.14 is illustrated in Fig. 7.17. Note that the overlap interval here is $T_s/3 = 2T_b/3$ sec.

An optimum decision rule for choosing groups of n information bits (i.e., data symbols) is again obtained by noncoherently detecting the sequence of overlapped signals. In particular, based on an observation of the received signal, $r(t)$, over the $n\left(\frac{M}{M-1}\right)T_b$-sec interval $\left(i - \frac{n}{M-1}\right)T_b \leq t \leq (i+n)T_b$, the joint ± 1 decisions on a_i and a_{i+1} are made according to the rule. Decide $\hat{a}_i$ and $\hat{a}_{i+1}$ corresponding to $p_{\hat{k}}(t)$ as determined by (7.108) where

$$\hat{k} = \max_k{}^{-1} \left| \int_{\left(i - \frac{n}{M-1}\right)T_b}^{(i+n)T_b} r(t)\tilde{s}_k^*(t)dt \right|^2 \tag{7.109}$$

The above decision rule represents an optimum noncoherent decision on a set of n orthogonal signals, each of duration $n\left(\frac{M}{M-1}\right)T_b = \left(\frac{M}{M-1}\right)T_s$ sec. Thus, the symbol error probability associated with this decision is identical to (5.77) with E_s replaced by $\left(\frac{M}{M-1}\right)E_s$, namely

$$P_s(E) = \sum_{l=1}^{M-1} (-1)^l \frac{1}{l+1} {}_{M-1}C_l \exp\left\{-\left(\frac{l}{l+1}\right)\left(\frac{M}{M-1}\right)\left(\frac{E_s}{N_0}\right)\right\} \tag{7.110}$$

The bit error probability is related to the symbol error probability of (7.110) through (4.96). In view of the above, we see that overlapped noncoherent M-ary orthogonal modulation can achieve an improvement in E/N_0 by a factor of $\left(\frac{M}{M-1}\right)$. Clearly, as M increases, the improvement factor diminishes and is maximum for $M = 2$ where it corresponds to 3 dB as previously shown.

Another way of viewing the $\frac{M}{M-1}$ improvement factor in E/N_0 for overlapped noncoherent orthogonal signaling is to consider the equivalent detection problem for nonoverlapped signals. In particular, suppose that we ignore the overlap region, that is, the region where all transmitted signals are defined to be identical, and just define the complex signals $\tilde{s}_k'(t)$ as in (7.102) but only in the interval $\left(\frac{1}{M-1}\right)T_s \leq t \leq \left(\frac{M}{M-1}\right)T_s$. Then a noncoherent decision on this new signal set $\{\tilde{s}_k'(t)\}$ defined over a T_s-sec time interval should produce the same error probability as the overlapped orthogonal set $\{\tilde{s}_k(t)\}$ defined on a $\left(\frac{M}{M-1}\right)T_s$-sec interval. The difference is that the members of the nonoverlapped set $\{\tilde{s}_k'(t)\}$ are not orthogonal. To see what their correlation properties are we proceed as follows. We can write $\tilde{s}_k'(t)$ in terms of $\tilde{s}_k(t)$ as

$$\tilde{s}_k'(t) = \begin{cases} 0; & 0 \leq t \leq \left(\frac{1}{M-1}\right)T_s \\ \tilde{s}_k(t); & \left(\frac{1}{M-1}\right)T_s \leq t \leq \left(\frac{M}{M-1}\right)T_s \end{cases} \tag{7.111}$$

Using (7.111), it is straightforward to show (see Problem 7.13) that the $\tilde{s}_k'(t)$'s are transorthogonal (see Chapter 4), that is, their off-diagonal normalized correlation is equal to $-\frac{1}{M-1}$. As such, $\{\tilde{s}_k'(t)\}$ represents a simplex set that produces an $\frac{M}{M-1}$ improvement factor in E/N_0 performance relative to an orthogonal signaling set.

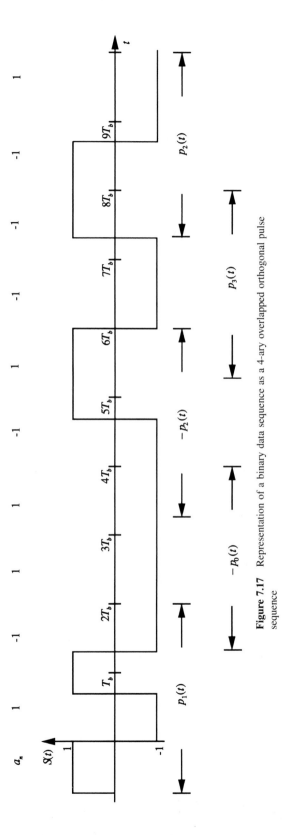

Figure 7.17 Representation of a binary data sequence as a 4-ary overlapped orthogonal pulse sequence

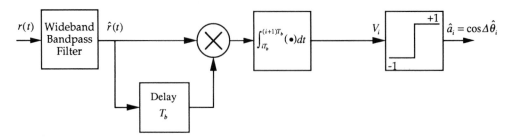

Figure 7.18a   A suboptimum DPSK receiver—wideband IF, I&D detector

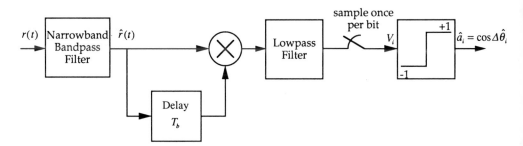

Figure 7.18b   A simpler suboptimum DPSK receiver—narrowband IF, sampling detector

7.4 SUBOPTIMUM WAVEFORM RECEIVERS

To many people, the acronym "DPSK" implies the optimum receiver of Fig. 7.2a while to many others it implies a simpler, but suboptimum, receiver. One such suboptimum receiver is illustrated in Fig. 7.18a. The primary difference between Fig. 7.2a and Fig. 7.18a is the order in which the differential detection and matched filtering (I&D) operations are performed since ordinarily the bandpass filter in Fig. 7.18a is assumed to be sufficiently wideband so that signal distortion and intersymbol interference (ISI) are not an issue. Quite often, however, to simplify the implementation still further, the I&D is replaced by a lowpass filter (to remove second harmonics of the carrier) and sampler (one sample per bit typically taken at the center of the bit). Furthermore, the input bandpass filter is now made narrowband with a bandwidth optimized to trade signal distortion and ISI against noise power reduction. This configuration, which is illustrated in Fig. 7.18b, has been suggested in many places in the literature [26, 27, 28].

The suboptimum receivers in Fig. 7.18 behave similarly in that they both demodulate the received (and filtered) bandpass IF signal in the current transmission interval with a reference corresponding to the same signal in the previous transmission interval. The fact that this demodulation reference is perturbed by additive noise is what distinguishes this form of differential detection from true coherent detection wherein the demodulation reference is assumed to be noise-free. Furthermore, since this noisy demodulation operation is in principle a correlation of the received (and filtered) bandpass signal with itself T_b-sec ago, these implementations are often referred to as *autocorrelation demodulator (ACD)*-type differential detectors.

In what follows, we present approximate analyses of the performances of these two suboptimum receivers and compare them with that of the optimum configuration in

Fig. 7.2b. The details of these analyses follow along the lines given in Park [2]. A more exact evaluation of the performance of ACD-type differential detectors is given in Chapter 8 as a prelude to the discussion of double differential detection.

7.4.1 Wideband IF Filter, Integrate-and-Dump Detector

As stated above, we assume that the IF bandwidth, W ($B = W/2$ is the equivalent lowpass bandwidth) of the input bandpass filter is sufficiently wide that the desired signal will pass through this filter undistorted and with negligible ISI, that is, $1/T_b \ll B \ll \omega_c/2\pi$. Thus, from Fig. 7.18a, and (7.1)

$$\hat{r}(t) = \sqrt{2P} \cos(\omega_c t + \theta_i + \theta) + \hat{n}(t)$$
$$= \sqrt{2P} \cos(\omega_c t + \theta_i + \theta) + \sqrt{2}[n_c(t)\cos(\omega_c t + \theta) - n_s(t)\sin(\omega_c t + \theta)]$$

(7.112)

where the hat denotes filtering by the input bandpass filter. For ease in analysis, we shall assume that the product of the radian carrier frequency and the bit time, $\omega_c T_b$, is an integer multiple of 2π and B is an integer multiple of $1/2T_b$. Under these conditions, the output, V_i, of the I&D is (ignoring second harmonic terms)

$$V_i = \int_{iT_b}^{(i+1)T_b} \hat{r}(t)\hat{r}(t - T_b)dt = PT_b \cos \Delta\theta_i$$

$$+ \sqrt{P} \int_{iT_b}^{(i+1)T_b} \left[n_c(t)\cos\theta_{i-1} + n_c(t - T_b)\cos\theta_i \right] dt$$

(7.113)

$$+ \sqrt{P} \int_{iT_b}^{(i+1)T_b} \left[n_s(t)\sin\theta_{i-1} + n_s(t - T_b)\sin\theta_i \right] dt$$

$$+ \int_{iT_b}^{(i+1)T_b} \left[n_c(t)n_c(t - T_b) + n_c(t)n_c(t - T_b) \right] dt$$

Since for DPSK the transmitted phase only takes on values of 0 and π, then (7.113) simplifies to

$$V_i = \int_{iT_b}^{(i+1)T_b} \hat{r}(t)\hat{r}(t - T_b)dt = PT_b \cos \Delta\theta_i$$

$$+ \sqrt{P} \int_{iT_b}^{(i+1)T_b} [\pm n_c(t) \pm n_c(t - T_b)] dt$$

(7.114)

$$+ \int_{iT_b}^{(i+1)T_b} [n_c(t)n_c(t - T_b) + n_s(t)n_s(t - T_b)] dt$$

where the plus sign is associated with a transmitted phase equal to 0 and the minus sign with a transmitted phase equal to π. Assuming an ideal (unit height rectangular frequency characteristic) filter for the input bandpass filter, then $n_c(t)$ and $n_s(t)$ are statistically independent with autocorrelation function

$$R_n(\tau) = E\{n_c(t)n_c(t + \tau)\} = E\{n_s(t)n_s(t + \tau)\} = N_0 B\left(\frac{\sin 2\pi B\tau}{2\pi B\tau}\right)$$

(7.115)

In view of the assumption that B is an integer multiple of $1/2T_b$, then $R_n(T_b) = 0$ and hence the first two moments of V_i are given by

$$\overline{V_i} = \begin{cases} PT_b; & \Delta\theta_i = 0 \\ -PT_b; & \Delta\theta_i = \pi \end{cases}$$

$$\sigma_{V_i}^2 = 2PBN_0 \int_{iT_b}^{(i+1)T_b} \int_{iT_b}^{(i+1)T_b} \left[\left(\frac{\sin 2\pi B(t-t')}{2\pi B(t-t')} \right) + \left(\frac{\sin 2\pi B(t-t'-T_b)}{2\pi B(t-t'-T_b)} \right) \right] dt\, dt'$$

$$+ 2(BN_0)^2 \int_{iT_b}^{(i+1)T_b} \int_{iT_b}^{(i+1)T_b} \left[\left(\frac{\sin 2\pi B(t-t')}{2\pi B(t-t')} \right)^2 \right.$$

$$\left. + \left(\frac{\sin 2\pi B(t-t'+T_b)}{2\pi B(t-t'+T_b)} \right) \left(\frac{\sin 2\pi B(t-t'-T_b)}{2\pi B(t-t'-T_b)} \right) \right] dt\, dt' \tag{7.116}$$

Since we have assumed that $B \gg 1/T_b$, then the above integrals can be approximately evaluated, resulting in

$$\sigma_{V_i}^2 \cong PT_bN_0 + BT_bN_0^2 \tag{7.117}$$

Both noise terms in (7.114) are zero mean and for sufficiently high SNR, the first of the two, which is Gaussian, dominates. Thus, the bit error probability is evaluated in the same manner as for an antipodal signal set on an AWGN channel, namely

$$P_b(E) = \frac{1}{2} \operatorname{erfc} \sqrt{\frac{(\overline{V_i})^2}{2\sigma_{V_i}^2}} \cong \frac{1}{2} \operatorname{erfc} \sqrt{\frac{E_b}{2N_0 \left(1 + \frac{BT_b}{E_b/N_0}\right)}}$$

$$\cong \frac{1}{2} \sqrt{\frac{2N_0 \left(1 + \frac{N_0 BT_b}{E_b}\right)}{\pi E_b}} \exp\left(-\frac{E_b}{2N_0 \left(1 + \frac{BT_b}{E_b/N_0}\right)} \right) \tag{7.118}$$

Comparing (7.118) with (7.14) we can readily identify the degradation in performance of the suboptimum configuration with respect to that of the optimum one.

We remind the reader that care must be exercised when using (7.118) that all of the above assumptions are met. In particular, if B is made too large, then V_i becomes decidedly non-Gaussian. Similarly, if B is made too small, then the assumption of zero signal distortion becomes inappropriate. In addition to these conditions, it is also important not to let E_b/N_0 become too large relative to BT_b, else the Gaussian assumption again becomes incorrect. In particular, if for BT_b fixed (and large) we were to let $E_b/N_0 \to \infty$, then from (7.118) we reach the erroneous conclusion that the suboptimum differential detector of Fig. 7.18a asymptotically performs 3 dB worse in SNR than the optimum detector of coherent PSK, the latter being asymptotically equivalent in performance to the optimum detector of DPSK (see Fig. 7.2b). The range of BT_b and E_b/N_0 values over which the Gaussian assumption provides approximately valid results for error probabilities of practical interest will be illustrated in Chapter 8, where the performance of this configuration is revisited in the context of a discussion on the behavior of differential detectors in the presence of frequency error.

7.4.2 Narrowband IF Filter, Sampling Detector

Since in this configuration (see Figure 7.18b) there is no I&D to narrowband filter the noise, the input bandpass filter must be used for that purpose. Once again, we will assume an ideal (unit height rectangular frequency characteristic) filter with bandpass bandwidth $W = 2B$, which will now be chosen on the order of half the bit rate. More about that in a moment. It is convenient here to represent the received DPSK signal in the form of a binary rectangular pulse train amplitude modulated on the carrier. Thus, in the ith transmission interval, (7.1) can be rewritten in the form

$$r(t) = \sqrt{2P} b_i\, p\,(t - iT_b) \cos{(\omega_c t + \theta)} + n(t) \tag{7.119}$$

where $b_i = \cos{\theta_i}$ is the binary (± 1) amplitude equivalent of the transmitted phase modulation and $p(t)$ is a unit rectangular pulse in the interval $(0, T_b)$. After passing through the bandpass filter, we get

$$\hat{r}(t) = \sqrt{2P} \sum_{i=-\infty}^{\infty} b_i \hat{p}\,(t - iT_b) \cos{(\omega_c t + \theta)} + \hat{n}(t)$$

$$= \sqrt{2}\left[\sqrt{P} \sum_{i=-\infty}^{\infty} b_i \hat{p}\,(t - iT_b) + n_c(t)\right] \cos{(\omega_c t + \theta)} - \sqrt{2} n_s(t) \sin{(\omega_c t + \theta)} \tag{7.120}$$

where

$$\hat{p}(t) = 2B \int_{t-T_b}^{t} \frac{\sin 2\pi Bt'}{2\pi Bt'} dt' \tag{7.121}$$

Note that because $\hat{p}(t)$ extends before and beyond the T_b-sec duration of the current symbol, ISI exists and we must sum over all past and future data bits. To obtain a detection sample for making a decision on the information bit $a_i = b_{i-1} b_i = \cos \Delta \theta_i$, we multiply $\hat{r}(t)$ by $\hat{r}(t - T_b)$, filter out the second harmonic of the carrier, and sample the remaining lowpass component at $t = (i + 0.5) T_b$, giving

$$V_i = \left[\sqrt{P} \sum_{k=-\infty}^{\infty} b_k \hat{p}\,((i - k + 0.5) T_b) + n_c\,((i + 0.5) T_b)\right]$$

$$\times \left[\sqrt{P} \sum_{k=-\infty}^{\infty} b_k \hat{p}\,((i - k - 0.5) T_b) + n_c\,((i - 0.5) T_b)\right] \tag{7.122}$$

$$+ n_s\,((i + 0.5) T_b)\, n_s\,((i - 0.5) T_b)$$

Computing the bit error probability based on comparing (7.122) (which accounts for the total ISI) with a zero threshold is in general a difficult if not impossible analytical task. Thus, we must make some simplifying assumptions from which we can obtain an approximate result. One such simplification, suggested in [28], assumes that $B = 1/2T_b$ and that the bandpass filter is cleared or reset at the end of each bit time to eliminate ISI. Under this assumption, (7.122) simplifies to

$$V_i = \left[\sqrt{P}b_i\hat{p}(T_b/2) + n_c((i+0.5)T_b)\right]$$

$$\times \left[\sqrt{P}b_{i-1}\hat{p}(T_b/2) + n_c((i-0.5)T_b)\right] \qquad (7.123)$$

$$+ n_s((i+0.5)T_b)\,n_s((i-0.5)T_b)$$

where from (7.121)

$$\hat{p}(T_b/2) = 4B\int_0^{T_b/2}\frac{\sin 2\pi Bt'}{2\pi Bt'}dt' = \frac{2}{\pi}Si(\pi BT_b) = \frac{2}{\pi}Si(\pi/2) \qquad (7.124)$$

with $S(x)$ the sine integral function defined by [30]

$$Si(x) \triangleq \int_0^x \frac{\sin y}{y}dy \qquad (7.125)$$

Recalling from (7.115) that for $B = 1/2T_b$, the filtered noise samples $n_c((i+0.5)T_b)$, $n_c((i-0.5)T_b)$, $n_s((i+0.5)T_b)$, and $n_s((i-0.5)T_b)$ are statistically independent zero-mean Gaussian random variables each with variance $N_0B = N_0/2T_b$, then the decision variable of (7.123) has the same form as that for the optimum DPSK receiver. Hence, we can immediately write the bit error probability in the form of (7.14) where the energy per bit is now degraded by $(\hat{p}(T_b/2))^2$, that is,

$$P_b(E) = \frac{1}{2}\exp\left\{-\frac{E_b}{N_0}\left[\frac{4}{\pi^2}Si^2(\pi/2)\right]\right\} = \frac{1}{2}\exp\left\{-0.76\frac{E_b}{N_0}\right\} \qquad (7.126)$$

Again comparing (7.126) with (7.14), it is a simple matter to see that, under the above assumptions, the degradation in the performance of the suboptimum receiver relative to that of the optimum receiver is a reduction in SNR by the factor $10\log_{10}(4/\pi^2)Si^2(\pi/2) = -1.2$ dB.

A second approach to approximately computing the bit error probability performance of Fig 7.18b is to still make the assumption of neglecting ISI to simplify (7.122) but using a bandwidth $B = 0.57/T_b$, which in [29] is shown to minimize ISI. In this case, the noise variables $n_c((i+0.5)T_b)$ and $n_c((i-0.5)T_b)$ are still zero-mean Gaussian, each with variance $N_0B = 0.57N_0/T_b$, but they are no longer independent and likewise for $n_s((i+0.5)T_b)$, and $n_s((i-0.5)T_b)$. However, the noise variables $n_c((i+0.5)T_b) + n_c((i-0.5)T_b)$ and $n_c((i+0.5)T_b) - n_c((i-0.5)T_b)$ are indeed independent and likewise for $n_s((i+0.5)T_b) + n_s((i-0.5)T_b)$ and $n_s((i+0.5)T_b) - n_s((i-0.5)T_b)$. Thus, remembering the analogy between the noise terms in DPSK and those in noncoherent FSK as discussed in Section 7.1.6, and noting from (7.124) that $\hat{p}(T_b/2) = (2/\pi)Si(0.57\pi)$, we can again immediately write down the bit error probability as

$$P_b(E) = \frac{1}{2}\exp\left\{-\frac{E_b}{N_0}\left[\frac{2}{0.57\pi^2}Si^2(0.57\pi)\right]\right\} = \frac{1}{2}\exp\left\{-0.80\frac{E_b}{N_0}\right\} \qquad (7.127)$$

which represents a degradation of $10\log_{10}(2/0.57\pi^2)Si^2(0.57\pi) = -1.0$ dB in SNR relative to the optimum DPSK detector.

7.4.3 Multiple Decision Feedback Symbol-by-Symbol Detector

A simplification of the optimum multiple symbol block-by-block detector discussed in Section 7.2.1 is suggested in [31, 32] wherein the previous information phases, $\Delta\theta_{k-i}$; $i =$

$0, 1, 2, \ldots, N_s - 2$ in the metric of (7.62), are replaced by past decisions on these quantities, that is,

$$
\eta = \left| \tilde{V}_{k-N_s+1} + \tilde{V}_k \exp \left\{ -j \left(\Delta \theta_k + \sum_{m=1}^{N_s-2} \Delta \hat{\theta}_{k-m} \right) \right\} \right.
$$
$$
\left. + \sum_{i=1}^{N_s-2} \tilde{V}_{k-i} \exp \left\{ -j \left(\sum_{m=0}^{N_s-i-2} \Delta \hat{\theta}_{k-i-m} \right) \right\} \right|^2 \tag{7.128}
$$

Since the only unknown in (7.128) is now the current symbol $\Delta \theta_k$, the detector is once again of the symbol-by-symbol decision type. While suboptimum, this multiple decision feedback type of detector should still improve bit error probability performance relative to that of classical M-DPSK.

As a simple example of the above, consider a three-symbol observation interval ($N_s = 3$). The optimum decision rule for this case is given below (7.65), which for the suboptimum detector now becomes

choose $\Delta \hat{\theta}_k$ corresponding to

$$
\max_{\Delta \theta_k} \mathrm{Re} \left\{ \tilde{V}_k \tilde{V}_{k-1}^* e^{-j \Delta \theta_k} + \tilde{V}_{k-1} \tilde{V}_{k-2}^* e^{-j \Delta \hat{\theta}_{k-1}} + \tilde{V}_k \tilde{V}_{k-2}^* e^{-j \left(\Delta \hat{\theta}_{k-1} + \Delta \theta_k \right)} \right\}
$$

where $\Delta \hat{\theta}_{k-1}$ represents the previous decision that is fed back. Since the middle term in the braces doesn't depend on $\Delta \theta_k$, it can be eliminated, resulting in the yet simpler decision rule

choose $\Delta \hat{\theta}_k$ corresponding to

$$
\max_{\Delta \theta_k} \mathrm{Re} \left\{ \tilde{V}_k \tilde{V}_{k-1}^* e^{-j \Delta \theta_k} + \tilde{V}_k \tilde{V}_{k-2}^* e^{-j \left(\Delta \hat{\theta}_{k-1} + \Delta \theta_k \right)} \right\}
$$

The bit error probability performance of the multiple decision feedback receiver was analytically computed in [32] for DPSK and 4-DPSK under the optimistic assumption of no errors in previous decisions, that is, perfect knowledge of the previous transmitted symbols. The results are given by an expression analogous to (7.68), namely

$$
P_b(E) = \frac{1}{2} \left[1 - Q \left(\sqrt{b}, \sqrt{a} \right) + Q \left(\sqrt{a}, \sqrt{b} \right) \right] \tag{7.129}
$$

where for DPSK

$$
\begin{Bmatrix} b \\ a \end{Bmatrix} = \frac{E_b}{2N_0} \left[N_s \pm 2\sqrt{N_s - 1} \right] \tag{7.130}
$$

and for 4-DPSK

$$
\begin{Bmatrix} b \\ a \end{Bmatrix} = \frac{E_b}{N_0} \left[N_s \pm \sqrt{2 (N_s - 1)} \right] \tag{7.131}
$$

In the limit as $N_s \to \infty$ both performances approach that of coherent M-PSK with differential encoding/decoding. This is the same conclusion as was reached for the optimum ML block-by-block detector.

Figures 7.19 and 7.20 are illustrations of the above theoretical results for $N_s = 2, 3$, and 4. Also shown are two sets of simulation results; one that assumes perfect decision

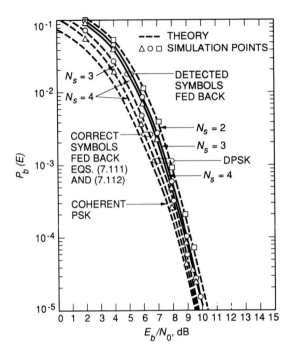

$P_b(E)$

E_b/N_0, dB

Figure 7.19 Bit error probability performance of the multiple decision feedback symbol-by-symbol detector; $M = 2$ (Reprinted from [32] © IEEE)

feedback as does the analysis, and one that takes into account the effects of errors in the past decisions, that is, the true situation. For the case of perfect feedback, the theoretical and simulation results agree exactly as one would hope. For the case of errors in the decisions being fed back, the simulation results, of course, yield a poorer performance than the theoretical results which assume perfect decision feedback. However, over the range of system parameters illustrated in Figs. 7.19 and 7.20, no error propagation was observed in the simulation. Furthermore, the results are quite comparable to those of the optimum ML block-by-block detector. For example, for $N_s = 3$ and $P_b(E) = 10^{-5}$, the difference between the two is only about 0.3 dB.

7.5 SYMMETRIC DPSK

A variation of conventional DPSK that is particularly useful in communication systems possessing a nonlinear channel is *symmetric DPSK (SDPSK)*. This modulation scheme, which is described by Winters [33] and Chethik [34], produces a radian phase shift in the transmitted carrier of $+\pi/2$ or $-\pi/2$ (in contrast to π or 0 of conventional DPSK) depending, respectively, on whether a logical 1 or 0 is to be sent. Mathematically speaking, the transmitted phase θ_i in (7.1) is still the differentially encoded version of the information phase, that is, $\theta_i = \Delta\theta_i + \theta_{i-1}$; however, $\Delta\theta_i$ now ranges over the set $(-\pi/2, \pi/2)$ rather than $(0, \pi)$. As a result, the optimum receiver for SDPSK is similar to that for DPSK except that the complex decision variable is found from a *cross* product rather than a *dot* product of the complex I&D output and a T_b-sec delayed version of itself; that is, the noise-free decision variable for making a ± 1 data decision is proportional to $\sin \Delta\theta_i$ rather than $\cos \Delta\theta_i$. Analogous to Figure 7.2b, an I-Q receiver for demodulating SDPSK is illustrated in Fig. 7.21.

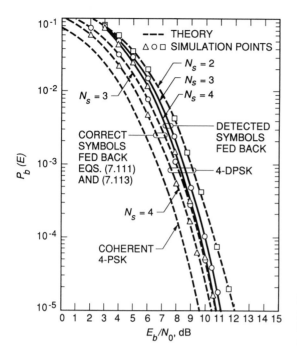

Figure 7.20 Bit error probability performance of the multiple decision feedback symbol-by-symbol detector; $M = 4$ (Reprinted from [32] © IEEE)

By choosing an initial phase θ_0 equal to $\pi/4$, the set of allowable values for θ_i in SDPSK is restricted to $(\pi/4, 3\pi/4, 5\pi/4, 7\pi/4)$ in every symbol interval and since the magnitude of the maximum phase change is limited to $\pi/2$, we see that *either* the I or Q components of the transmitted signal, but not both, must change polarity from symbol to symbol. Thus, the envelope of the transmitted signal never undergoes a complete phase reversal (change of π radians) as is typical of conventional DPSK. The upshot of this modification is that SDPSK suffers from negligible spectrum "regrowth" after being filtered and passed through a nonlinear device. Specifically, laboratory tests [33] show that when SDPSK and DPSK are bandpass filtered and then passed through a hard limiter, spectrum regrowth for DPSK is nearly 100%, whereas for SDPSK it is significantly lower. On a linear channel, unfiltered DPSK and SDPSK have identical bit error probability performance [see (7.14)] since in both cases the two possible symbols are separated by π radians. Furthermore, the two modulations have identical power spectral densities. Finally, Winters [33] shows that SDPSK produces a lower average bit error probability than DPSK in the presence of timing error and frequency uncertainty.

Although the desirable attributes of SDPSK with regard to minimum spectral regrowth after bandpass filtering and hard limiting were demonstrated almost a decade ago [33], it has only recently been suggested by Alexovich, Kossin, and Schonhoff [35] to include the filtering and hard limiting as an inherent part of the modulation scheme itself. Such a *bandpass limited symmetric DPSK (BPL-SDPSK)* produces a bandwidth efficient, constant envelope modulation. By choosing the bandwidth-time product BT_b equal to unity, it was found by simulation that the power spectral density resembled that of minimum-shift-keying (MSK) (see Chapter 2). Furthermore, the bit error probability performance of BPL-SDPSK with $BT_b = 1$ was shown to be comparable to and in some cases better than that of various forms of differentially detected MSK [36], [37], [38], [39]. A detailed discussion of MSK performance will be given in Chapter 10.

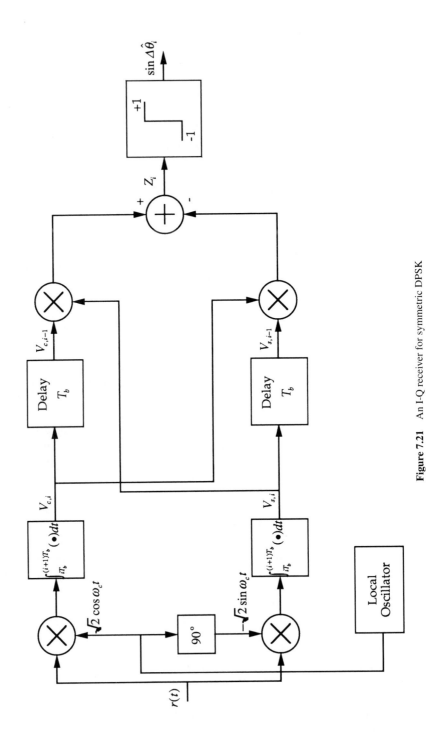

Figure 7.21 An I-Q receiver for symmetric DPSK

Appendix 7A

Derivation of the Optimum Receiver for Classical (2-Symbol Observation) *M*-DPSK

Consider the a priori pdf of the received signal given the transmitted message as given by (4.8). This pdf is the basis for deriving all of the optimum waveform receiver structures in Chapters 4, 5, and 6 and will be once again used here as the starting point for deriving the optimum receiver for differential detection of *M*-PSK (*M*-DPSK) using an observation interval of two symbols. For the purpose of the derivation of the optimum receiver, the observation interval shall be taken as $0 \leq t \leq 2T_s$.

For *M*-DPSK, the transmitted signal waveform in the interval $0 \leq t \leq 2T_s$ corresponding to message m_i can be expressed, analogous to (7.1), as

$$s_i(t; \theta) = \sqrt{2P} \cos(\omega_c t + \theta_i(t) + \theta); \quad 0 \leq t \leq 2T_s \tag{7A.1}$$

where θ is an unknown random phase uniformly distributed in $(-\pi, \pi)$ and

$$\theta_i(t) = \begin{cases} \theta_{i_0}; & 0 \leq t \leq T_s \\ \theta_{i_1}; & T_s \leq t \leq 2T_s \end{cases} \tag{7A.2}$$

is the transmitted data phase waveform with $\theta_{i_0}, \theta_{i_1}$ independently taking on values from the set of *M* possible phases $\beta_m = 2\pi/M; \quad m = 0, 1, \ldots, M - 1$. Since the signal now depends on an additional unknown random variable, that is, θ, the conditional pdf of (4.8) is written as

$$
\begin{aligned}
f_r\left(\boldsymbol{\rho} \,|m_i, \theta\right) &= C \exp\left\{-\frac{\int_0^{2T_s} (\rho(t) - s_i(t; \theta))^2 \, dt}{N_0}\right\} \\
&= C \exp\left\{-\frac{\int_0^{2T_s} \rho^2(t) \, dt + \int_0^{2T_s} s_i^2(t; \theta) \, dt}{N_0}\right\} \\
&\quad \times \exp\left\{\frac{2 \int_0^{2T_s} \rho(t) s_i(t; \theta) \, dt}{N_0}\right\}
\end{aligned} \tag{7A.3}
$$

Since $\int_0^{2T_s} \rho^2(t) \, dt$ is independent of i and $\int_0^{2T_s} s_i^2(t; \theta) \, dt = 2PT_s$, we can rewrite (7A.3) in the equivalent form

$$f_r\left(\boldsymbol{\rho} \,|m_i, \theta\right) = C' \exp\left\{\frac{2 \int_0^{2T_s} \rho(t) s_i(t; \theta) \, dt}{N_0}\right\} \tag{7A.4}$$

where C' is a new normalization constant that is independent of i. Using (7A.1) and (7A.2), the numerator of the exponent in (7A.4) can be expressed in the following convenient form:

$$\int_0^{2T_s} \rho(t)s_i(t;\theta)dt = \sqrt{P}\,\mathrm{Re}\left\{\int_0^{2T_s}\rho(t)\sqrt{2}\exp\{j(\omega_c t + \theta_i(t) + \theta)\}\,dt\right\}$$

$$= \sqrt{P}\,\mathrm{Re}\left\{\left[\left[\underbrace{\int_0^{T_s}\rho(t)\sqrt{2}\exp\{j(\omega_c t + \theta_{i_0})\}\,dt}_{\tilde{U}_0}\right.\right.\right.$$

$$\left.\left.\left.+\underbrace{\int_{T_s}^{2T_s}\rho(t)\sqrt{2}\exp\{j(\omega_c t + \theta_{i_1})\}\,dt}_{\tilde{U}_1}\right]\exp(j\theta)\right]\right\} \tag{7A.5}$$

$$= \sqrt{P}\,\left|\tilde{U}_0 + \tilde{U}_1\right|\cos(\theta + \alpha)$$

where

$$\alpha = \tan^{-1}\frac{\mathrm{Im}\left\{\tilde{U}_0 + \tilde{U}_1\right\}}{\mathrm{Re}\left\{\tilde{U}_0 + \tilde{U}_1\right\}} \tag{7A.6}$$

Substituting (7A.5) into (7A.4) gives

$$f_r(\boldsymbol{\rho}\,|m_i,\theta) = C'\exp\left\{\frac{2\sqrt{P}}{N_0}\left|\tilde{U}_0 + \tilde{U}_1\right|\cos(\theta + \alpha)\right\} \tag{7A.7}$$

Averaging (7A.7) over the uniformly distributed phase, θ, gives

$$f_r(\boldsymbol{\rho}\,|m_i) = \int_{-\pi}^{\pi} f_r(\boldsymbol{\rho}\,|m_i,\theta)\,f_\Theta(\theta)d\theta$$

$$= C'\frac{1}{2\pi}\int_{-\pi}^{\pi}\exp\left\{\frac{2\sqrt{P}}{N_0}\left|\tilde{U}_0 + \tilde{U}_1\right|\cos(\theta + \alpha)\right\}d\theta \tag{7A.8}$$

$$= C'I_0\left(\frac{2\sqrt{P}}{N_0}\left|\tilde{U}_0 + \tilde{U}_1\right|\right)$$

where as before $I_0(x)$ is the zero order modified Bessel function of the first kind with argument x.

For equiprobable messages, the optimum (maximum-likelihood) receiver chooses that message as having been transmitted corresponding to the maximum of $f_r(\boldsymbol{\rho}\,|m_i)$ over i. In particular, letting $\hat{m}(\mathbf{r}) = m_{\hat{i}}$ denote the decision made by the receiver as a result of observing $\mathbf{r}$, then since $I_0(x)$ is a monotonically increasing function of its argument, we have

$$\hat{i} = \max_{i}{}^{-1} f_r(\boldsymbol{\rho}|m_i) = \max_{i}{}^{-1} \left|\tilde{U}_0 + \tilde{U}_1\right|^2$$

$$= \max_{i_0,i_1}{}^{-1} \left| \int_0^{T_s} \rho(t)\sqrt{2}\exp\left\{j\left(\omega_c t + \theta_{i_0}\right)\right\}dt \right. \tag{7A.9}$$

$$\left. + \int_{T_s}^{2T_s} \rho(t)\sqrt{2}\exp\left\{j\left(\omega_c t + \theta_{i_1}\right)\right\}dt \right|^2$$

where $\max_{i}{}^{-1} g(i)$ denotes the value of i that maximizes the function $g(i)$. We note that adding an arbitrary phase, say θ_a, to the decision statistic in (7A.9), that is,

$$\hat{i} = \max_{i_0,i_1}{}^{-1} \left| \int_0^{T_s} \rho(t)\sqrt{2}\exp\left\{j\left(\omega_c t + \theta_{i_0} + \theta_a\right)\right\}dt \right.$$

$$\left. + \int_{T_s}^{2T_s} \rho(t)\sqrt{2}\exp\left\{j\left(\omega_c t + \theta_{i_1} + \theta_a\right)\right\}dt \right|^2$$

$$= \max_{i_0,i_1}{}^{-1} \left| \exp\left(j\theta_a\right)\left[\int_0^{T_s} \rho(t)\sqrt{2}\exp\left\{j\left(\omega_c t + \theta_{i_0}\right)\right\}dt \right.\right.$$

$$\left.\left. + \int_{T_s}^{2T_s} \rho(t)\sqrt{2}\exp\left\{j\left(\omega_c t + \theta_{i_1}\right)\right\}dt \right] \right|^2 \tag{7A.10}$$

$$= \max_{i_0,i_1}{}^{-1} \left| \int_0^{T_s} \rho(t)\sqrt{2}\exp\left\{j\left(\omega_c t + \theta_{i_0}\right)\right\}dt \right.$$

$$\left. + \int_{T_s}^{2T_s} \rho(t)\sqrt{2}\exp\left\{j\left(\omega_c t + \theta_{i_1}\right)\right\}dt \right|^2$$

doesn't change the decisions made on θ_{i_0} and θ_{i_1}. Thus, the decision rule has a phase ambiguity in that the receiver is as likely to choose the pair $\left(\theta_{i_0} + \theta_a, \theta_{i_1} + \theta_a\right)$ as it is to choose the pair $\left(\theta_{i_0}, \theta_{i_1}\right)$. To resolve this ambiguity, we employ differential encoding in which the information (data) phase is the difference of two adjacent transmitted phases. Letting $\Delta\theta_i \triangleq \theta_{i_1} - \theta_{i_0}$ and also setting the arbitrary phase $\theta_a = -\theta_{i_0}$, (7A.10) becomes

$$\hat{i} = \max_{i}{}^{-1} \left| \int_0^{T_s} \rho(t)\sqrt{2}\exp\left(j\omega_c t\right)dt + \int_{T_s}^{2T_s} \rho(t)\sqrt{2}\exp\left\{j\left(\omega_c t + \Delta\theta_i\right)\right\}dt \right|^2$$

$$= \max_{i}{}^{-1} \left| \underbrace{\int_0^{T_s} \rho(t)\sqrt{2}\exp\left(j\omega_c t\right)dt}_{\tilde{V}_0} + \exp\left(j\Delta\theta_i\right)\underbrace{\int_{T_s}^{2T_s} \rho(t)\sqrt{2}\exp\left(j\omega_c t\right)dt}_{\tilde{V}_1} \right|^2 \tag{7A.11}$$

$$= \max_{i}{}^{-1} \left[\left|\tilde{V}_0\right|^2 + \left|\tilde{V}_1\right|^2 + 2\,\mathrm{Re}\left\{\tilde{V}_0\tilde{V}_1^* \exp\left(-j\Delta\theta_i\right)\right\} \right]$$

Since $\left|\tilde{V}_0\right|^2$ and $\left|\tilde{V}_1\right|^2$ are independent of i, we get the equivalent decision rule

$$\hat{i} = \max_{i}{}^{-1} \operatorname{Re}\left\{\tilde{V}_0 \tilde{V}_1^* \exp\left(-j\Delta\theta_i\right)\right\} \tag{7A.12}$$

Finally, letting $\tilde{V}_0 \triangleq \left|\tilde{V}_0\right| e^{-j\eta_0}$ and $\tilde{V}_1 \triangleq \left|\tilde{V}_1\right| e^{-j\eta_1}$, we have

$$\hat{i} = \max_{i}{}^{-1} \operatorname{Re}\left\{\left|\tilde{V}_0\right|\left|\tilde{V}_1^*\right| \exp\left(j\left[\eta_1 - \eta_0 - \Delta\theta_i\right]\right)\right\}$$

$$= \max_{i}{}^{-1} \left|\tilde{V}_0\right|\left|\tilde{V}_1^*\right| \cos\left(\eta_1 - \eta_0 - \Delta\theta_i\right) \tag{7A.13}$$

$$= \min_{i}{}^{-1} \left|\eta_1 - \eta_0 - \Delta\theta_i\right|$$

Figure 7.1 is an implementation of the decision rule in (7A.13) corresponding to the $2T_s$-observation interval $((i-1)T_s \le t \le (i+1)T_s)$.

Appendix 7B

Probability Distribution of the Phase Angle
between Two Vectors Perturbed by Additive Gaussian Noise

Consider the geometry illustrated in Fig. 7B.1. Here A_1, A_2 and ϕ_1, ϕ_2 represent the amplitudes and angles, respectively, of two known signal vectors and x_1, y_1 and x_2, y_2 are the inphase and quadrature components of the complex Gaussian noises n_1, n_2 added to each of these vectors. The noises are each zero mean and, in general, have different variances, that is, $\text{var}(x_1) = \text{var}(y_1) = \sigma_1^2$ and $\text{var}(x_2) = \text{var}(y_2) = \sigma_2^2$. The resulting signal plus noise vectors have amplitudes R_1, R_2 and phases β_1, β_2. The specific application of this geometry to the problem at hand corresponds to the case where the signal plus noise vectors represent the complex form of the integrate-and-dump outputs for the $(i-1)^{\text{st}}$ and i^{th} transmission intervals, namely

$$\tilde{V}_{i-1} \triangleq V_{c,i-1} + jV_{s,i-1} = \left|\tilde{V}_{i-1}\right| e^{j\eta_{i-1}} = R_1 e^{j\beta_1}$$

$$\tilde{V}_i \triangleq V_{c,i} + jV_{s,i} = \left|\tilde{V}_i\right| e^{j\eta_i} = R_2 e^{j\beta_2}$$

(7B.1)

We are interested in the pdf (actually the probability distribution function) of the modulo 2π difference $\psi = (\beta_2 - \beta_1)$ modulo 2π, $-\pi \le \beta_1, \beta_2 \le \pi$.

Defining the parameters[1]

$$\Delta\Phi = (\phi_2 - \phi_1) \text{ modulo } 2\pi$$

(7B.2a)

$$U = \frac{1}{2}\left(\frac{A_1^2}{2\sigma_1^2} + \frac{A_2^2}{2\sigma_2^2}\right), \quad V = \frac{1}{2}\left(\frac{A_2^2}{2\sigma_2^2} - \frac{A_1^2}{2\sigma_1^2}\right)$$

(7B.2b)

---

1. The modulo 2π is to be interpreted so that the resulting $\Delta\Phi$ lies within the 2π interval of interest. Also, $\Delta\Phi$ ordinarily includes the carrier phase shift $\omega_c T$. However, this term is frequently eliminated by receiver processing and thus we do not include it in Fig. 7B.1.

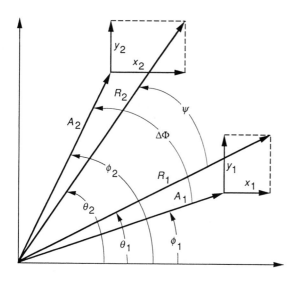

Figure 7B.1 Geometry for angle between vectors perturbed by Gaussian noise (Reprinted from [4] © IEEE)

$$W = \sqrt{U^2 - V^2} \tag{7B.2c}$$

$$r = \frac{E\{x_1 x_2\}}{\sigma_1 \sigma_2} = \frac{E\{y_1 y_2\}}{\sigma_1 \sigma_2}; \quad \rho = \frac{E\{x_1 y_2\}}{\sigma_1 \sigma_2} = \frac{E\{y_1 x_2\}}{\sigma_1 \sigma_2} \tag{7B.2d}$$

then it is shown in [4] that

$$\Pr\{\psi_1 \le \psi \le \psi_2\} = \int_{\psi_1}^{\psi_2} f_\Psi(\psi) d\psi$$

$$= \begin{cases} F(\psi_2) - F(\psi_1) + 1; & \psi_1 < \Delta\Phi < \psi_2 \\ F(\psi_2) - F(\psi_1); & \Delta\Phi > \psi_2 \text{ or } \Delta\Phi < \psi_1 \end{cases} \tag{7B.3}$$

where $F(\psi)$ is an auxiliary function that is periodic with period 2π and discontinuous at the point $\psi = \Delta\Phi$ with a jump discontinuity $F(\Delta\Phi^+) - F(\Delta\Phi^-) = -1$. Specific evaluation of $F(\psi)$ depends on the assumptions made regarding the relative amplitudes and phases of the signal vectors and the correlation between the two noise processes as well as their relative variance. Three different cases are considered in [4]. The results are summarized as follows.

Case 1. Equal Signals ($\Delta\Phi = 0$, $V = 0$) and Uncorrelated Noises ($r = \rho = 0$)

$$F(\psi) = -\frac{\sin\psi}{4\pi} \int_{-\pi/2}^{\pi/2} \frac{\exp\{-U[1 - \cos\psi\cos t]\}}{1 - \cos\psi\cos t} dt \tag{7B.4}$$

Case 2. Unequal Signals and Uncorrelated Noises

$$F(\psi) = \frac{W\sin(\Delta\Phi - \psi)}{4\pi} \int_{-\pi/2}^{\pi/2} \frac{\exp\{-[U - V\sin t - W\cos(\Delta\Phi - \psi)\cos t]\}}{U - V\sin t - W\cos(\Delta\Phi - \psi)\cos t} dt \tag{7B.5}$$

Case 3. Unequal Signals and Correlated Noises

$$F(\psi) = \int_{-\pi/2}^{\pi/2} \frac{e^{-E}}{4\pi} \left[\frac{W\sin(\Delta\Phi - \psi)}{U - V\sin t - W\cos(\Delta\Phi - \psi)\cos t} \right.$$

$$\left. + \frac{r\sin\psi - \rho\cos\psi}{1 - (r\cos\psi + \rho\sin\psi)\cos t} \right] dt \tag{7B.6}$$

where

$$E = \frac{U - V\sin t - W\cos(\Delta\Phi - \psi)\cos t}{1 - (r\cos\psi + \rho\sin\psi)\cos t} \tag{7B.7}$$

Note that for all three cases, the probability distribution function is expressable as a single integral of simple (exponential and trigonometric) functions. Further simplications of the integrands of $F(\psi)$ are considered in [4] at the expense of complicating the integration limits.

PROBLEMS

7.1 The optimum receiver for M-DPSK is illustrated in Fig. 7.1, which has the complex form of Fig. 7.8. With reference to Fig. 7.8, the decision rule can be written as

$$\Delta \hat{\theta}_i = \beta_n$$

where $n = \min_i^{-1} \left\{ \left| \arg \tilde{V}_k \tilde{V}_{k-1}^* e^{-j\beta i} \right| \right\} = \min_i^{-1} \left\{ \left| \arg \tilde{V}_k - \arg \tilde{V}_{k-1} - \beta_i \right| \right\}; \quad \beta_i = \frac{2\pi i}{M}$

An equivalent receiver to the above that makes direct use of the inphase and quadrature outputs $V_{c,k}$ and $V_{s,k}$ is illustrated in Fig. 7.2a. The decision rule for this receiver is given by

$$\Delta \hat{\theta}_i = \beta_n$$

where $n = \max_i^{-1} \left\{ \left(V_{s,k} V_{s,k-1} + V_{c,k} V_{c,k-1} \right) \cos \beta_i + \left(V_{s,k} V_{c,k-1} - V_{c,k} V_{s,k-1} \right) \sin \beta_i \right\}$;

$$\beta_i = \frac{2\pi i}{M}$$

(a) Show that the above two decision rules are completely equivalent. Hint: Note that the value of i that results in $\min_i \{|\Theta_i|\}$ is equal to the value of i that results in $\max_i \{\cos \Theta_i\}$.
(b) Show that the receiver of Fig. 7.2a reduces to that of Fig. 7.2b for binary DPSK ($M = 2$), that is, they implement the same decision rule.

7.2 (Wideband IF Filter, Sampling Detector) A differential detector for binary DPSK is illustrated in Fig. 7P.2. Assume that the IF bandwidth of the input bandpass filter is sufficiently wide that the desired signal will pass through this filter undistorted and with negligible ISI. As such, the output of the bandpass filter can be represented as

$$\hat{r}(t) = \sqrt{2P} \cos \left(\omega_c t + \theta(t) + \theta \right) + n_1(t) \cos \left(\omega_c t + \theta \right) - n_2(t) \sin \left(\omega_c t + \theta \right)$$

where θ is the unknown phase introduced by the channel, $\theta(t) = \theta_i$ is the transmitted phase in the ith bit interval $iT_b \leq t \leq (i+1)T_b$ corresponding to the information phase $\Delta \theta_i = \theta_i - \theta_{i-1}$, and $n_1(t)$, $n_2(t)$ are uncorrelated zero-mean Gaussian noise processes each with variance $\sigma^2 = N_0 B$. Here B is the equivalent lowpass bandwidth of the input bandpass filter. Also assume that the transfer function of this filter is symmetric around its center frequency, ω_c, and that the lowpass equivalent of this transfer function is denoted by $H(\omega)$.
(a) Assuming that the lowpass filter merely eliminates second harmonic terms of the carrier, show that a sample of the output of this filter taken in the ith bit interval can be written in the form

$$z = \frac{1}{2} \operatorname{Re} \left\{ z_1^* z_2 \right\}$$

where

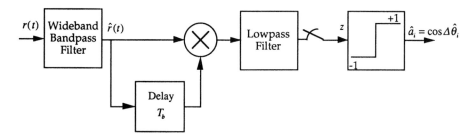

Figure P7.2

$$z_1 = \left(\sqrt{2P}\cos\theta_i + n_1\right) + j\left(\sqrt{2P}\sin\theta_i + n_2\right)$$

$$z_2 = \left(\sqrt{2P}\cos\theta_{i-1} + n_3\right) + j\left(\sqrt{2P}\sin\theta_{i-1} + n_4\right)$$

with $n_3(t) \triangleq n_1(t - T_b)$, $n_4(t) \triangleq n_2(t - T_b)$, and n_1, n_2, n_3, n_4 respectively denoting samples of the noise processes $n_1(t), n_2(t), n_3(t), n_4(t)$. Also, we have further assumed that $\omega_c T_b = 2n\pi$.

(b) Show that the correlation matrix of the noise samples n_1, n_2, n_3, n_4 is given by

$$[\overline{n_i n_j}] = \sigma^2 \begin{bmatrix} 1 & 0 & \rho & 0 \\ 0 & 1 & 0 & \rho \\ \rho & 0 & 1 & 0 \\ 0 & \rho & 0 & 1 \end{bmatrix}$$

where $\sigma^2\rho = \frac{N_0}{4\pi}\int_{-\infty}^{\infty}|H(\omega)|^2 e^{j\omega T_b}d\omega$.

(c) Compute the mean, variance, and cross-correlation of the complex variables z_1 and z_2 defined by

$$\overline{z_k} \triangleq \mu_{c,k} + j\mu_{s,k}; \quad \sigma_k^2 \triangleq \frac{1}{2}\overline{|z_k - \overline{z_k}|^2}; \quad k = 1, 2$$

$$\rho_z \sigma_1 \sigma_2 \triangleq \frac{1}{2}\overline{(z_1 - \overline{z_1})^*(z_2 - \overline{z_2})}; \quad \rho_z = \rho_{cz} + j\rho_{sz}$$

7.3 For threshold decisions made on random variables, z, of the form given in part (a) of Problem 7.2, Stein [21] has shown that the average bit error probability $P_b(E)$ is given by

$$P_b(E) = \frac{1}{2}\left[1 - Q\left(\sqrt{b}, \sqrt{a}\right) + Q\left(\sqrt{a}, \sqrt{b}\right)\right] - \frac{A}{2}\exp\left(-\frac{a+b}{2}\right)I_0\left(\sqrt{ab}\right)$$

where $Q(\alpha, \beta)$ is the Marcum Q-function, $A = \rho_{cz}/\sqrt{1 - \rho_{sz}^2}$, and

$$\begin{Bmatrix} b \\ a \end{Bmatrix} = \frac{1}{2}\left\{ \frac{S_1 + S_2 + (S_1 - S_2)\cos v + 2\sqrt{S_1 S_2}\sin(\Theta_1 - \Theta_2)\sin v}{N_1 + N_2 + \sqrt{(N_1 - N_2)^2 + 4N_1 N_2 \rho_{sz}^2}} \right.$$

$$+ \frac{S_1 + S_2 - (S_1 - S_2)\cos v - 2\sqrt{S_1 S_2}\sin(\Theta_1 - \Theta_2)\sin v}{N_1 + N_2 - \sqrt{(N_1 - N_2)^2 + 4N_1 N_2 \rho_{sz}^2}}$$

$$\left. \mp \frac{2\sqrt{S_1 S_2}\cos(\Theta_1 - \Theta_2)}{\sqrt{N_1 N_2 \left(1 - \rho_{sz}^2\right)}} \right\}$$

with

$$S_k \triangleq \frac{1}{2}|\overline{z_k}|^2; \quad N_k \triangleq \sigma_k^2; \quad \Theta_k \triangleq \arg\{\overline{z_k}\} = \tan^{-1}\frac{\mu_{s,k}}{\mu_{c,k}}; \quad k = 1, 2$$

$$v \triangleq \arg\{\rho_z\} = \tan^{-1}\frac{\rho_{sz}}{\rho_{cz}}$$

Recognizing that $P_b(E) = \frac{1}{2}\Pr\left\{\text{Re}\left\{z_1^* z_2\right\} < 0 \mid \Delta\theta_i = 0\right\} + \frac{1}{2}\Pr\left\{\text{Re}\left\{z_1^* z_2\right\} > 0 \mid \Delta\theta_i = \pi\right\}$, evaluate a, b, and A, and show that $P_b(E)$ simplifies to

$$P_b(E) = \frac{1}{2}\exp\left(-\frac{P}{N_0 B}\right) = \frac{1}{2}\exp\left(-\frac{E_b}{N_0}\frac{1}{BT_b}\right)$$

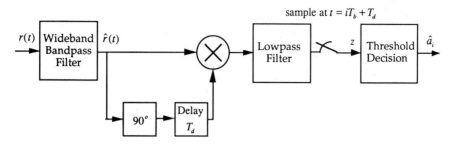

Figure P7.4

7.4 In addition to coherent and noncoherent detection, binary FSK can also be differentially detected. The receiver structures are quite similar to those for DPSK, the primary differences
being: (1) the value of delay, T_d, used to produce the delayed demodulation reference should
be chosen relative to the frequency modulation index, h; and (2) a 90° phase shift should be
applied to the delayed demodulation reference. As such, consider the suboptimum (wideband
IF, sampling) differential detector for binary FSK illustrated in Fig. 7P.4.

Assume that the IF bandwidth of the input bandpass filter is sufficiently wide that the
desired signal will pass through this filter undistorted and with negligible ISI. As such, the
output of the bandpass filter can be represented as

$$\hat{r}(t) = \sqrt{2P} \sin(\omega_c t + \theta(t) + \theta) + n_1(t) \cos(\omega_c t + \theta) + n_2(t) \sin(\omega_c t + \theta)$$

where θ is the unknown phase introduced by the channel, $\theta(t) = \frac{\pi h a_i t}{T_b}$ is the effective phase
modulation transmitted in the ith bit interval $iT_b \le t \le (i+1)T_b$ with a_i denoting the ± 1 i^{th}
data bit, and $n_1(t)$, $n_2(t)$ are uncorrelated zero-mean Gaussian noise processes, each with
variance $\sigma^2 = N_0 B$. Here B is the equivalent lowpass bandwidth of the input bandpass filter.
Also assume that the transfer function of this filter is symmetric around its center frequency, ω_c,
and that the lowpass equivalent of this transfer function is denoted by $H(\omega)$. Finally, we point
out that for FSK, differential encoding is not employed at the transmitter; hence, the transmitted
sequence $\{a_i\}$ is indeed the information data sequence.

Assume that the lowpass filter merely eliminates second harmonic terms of the carrier
and furthermore that $\omega_c T_d = 2k\pi$ (k integer).

(a) Show that, in the absence of noise, the sample of the output lowpass filter taken at $t = iT_b + T_d$ is given by

$$z = a_i P \sin \frac{\pi h T_d}{T_b}$$

(b) Although the noise at the input to the threshold decision device will not be Gaussian,
one would still expect to get the lowest average bit error probability when the separation
between the signal components of z corresponding to $a_i = 1$ and $a_i = -1$ is a maximum.
Using the result of part (a), show that the value of delay, T_d, for which this condition occurs
is given by

$$T_d = \frac{(2n+1)T_b}{2h}; \quad n = 0, 1, 2, \ldots$$

(c) Since the delay must be restricted to lie in the interval $0 \le T_d \le T_b$, show that for $h < 0.5$,
there is no suitable value of delay such that the signal components of z are maximally
separated and thus the performance of differentially detected FSK will be inferior to that of
DPSK.

7.5 Consider the receiver of Problem 7.4.

(a) Show that, in the presence of noise, a sample of the output of the lowpass filter taken at $t = iT_b + T_d$ can be written in the form

$$z = \frac{1}{2} \operatorname{Re} \{z_1^* z_2\}$$

where

$$z_1 = \left(\sqrt{2P} \sin \theta_i + n_1\right) - j \left(\sqrt{2P} \cos \theta_i + n_2\right)$$

$$z_2 = \left(\sqrt{2P} \cos (\theta_i - \xi_i) + n_4\right) + j \left(\sqrt{2P} \sin (\theta_i - \xi_i) + n_3\right); \quad \xi_i \triangleq \frac{\pi h a_i T_d}{T_b}$$

with $n_3(t) \triangleq n_1(t - T_d)$, $n_4(t) \triangleq n_2(t - T_d)$, and $n_1, n_2, n_3, n_4, \theta_i$ respectively denoting samples of the noise processes $n_1(t), n_2(t), n_3(t), n_4(t)$ and the phase process $\theta(t)$ taken at $t = iT_b + T_d$.

(b) Show that the correlation matrix of the noise samples n_1, n_2, n_3, n_4 is given by

$$[\overline{n_i n_j}] = \sigma^2 \begin{bmatrix} 1 & 0 & \rho & 0 \\ 0 & 1 & 0 & \rho \\ \rho & 0 & 1 & 0 \\ 0 & \rho & 0 & 1 \end{bmatrix}$$

where $\sigma^2 \rho = \frac{N_0}{4\pi} \int_0^\infty |H(\omega)|^2 e^{j\omega T_d} d\omega$.

(c) Compute the mean, variance, and cross-correlation of the complex variables z_1 and z_2 defined by

$$\bar{z}_k \triangleq \mu_{c,k} + j\mu_{s,k}; \quad \sigma_k^2 \triangleq \frac{1}{2}\overline{|z_k - \bar{z}_k|^2}; \quad k = 1, 2$$

$$\rho_z \sigma_1 \sigma_2 \triangleq \frac{1}{2}\overline{(z_1 - \bar{z}_1)^* (z_2 - \bar{z}_2)}; \quad \rho_z = \rho_{cz} + j\rho_{sz}$$

7.6 Using the specific results of Problem 7.5 in the general bit error probability results given in the description of Problem 7.3:

(a) Show that for differential detection of FSK, the average bit error probability is given by

$$P_b(E) = \frac{1}{2}\left[1 - Q\left(\sqrt{b}, \sqrt{a}\right) + Q\left(\sqrt{a}, \sqrt{b}\right)\right]$$

where

$$\left\{\begin{matrix} b \\ a \end{matrix}\right\} = \frac{P}{N_0 B}\left[\frac{1 \pm \sqrt{1 - \rho^2}}{1 - \rho^2}\right]$$

(b) Show that for $\rho = 0$, the average bit error probability reduces to

$$P_b(E) = \frac{1}{2} \exp\left(-\frac{P}{N_0 B}\right)$$

which agrees with the comparable result for DPSK given in Problem 7.3.

7.7 A special case of Problem 7.6 corresponding to $h = 0.5$ and a phase modulation that is continuous in time from bit interval to interval is referred to as *minimum-shift-keying* (MSK). (This modulation is discussed in more detail in Chapter 10). In particular, the effective phase modulation transmitted in the ith bit interval $iT_b \le t \le (i + 1)T_b$ is now $\theta(t) = a_i \frac{\pi t}{2T_b} + x_i$ where x_i is a phase constant whose value is determined by the phase continuity requirement at the bit transition instant $t = iT_b$. Consider a modulation $a(t)$ starting at $t = 0$ corresponding to an input data sequence $a_0 = -1, a_1 = 1, a_2 = 1, a_3 = -1, a_4 = -1, a_5 = -1, a_6 = 1, a_7 = -1$, that is, $a(t) = \sum_{i=0}^{7} a_i p\, (t - iT_b)$ where $p(t)$ denotes a unit rectangular pulse of width T_b. Assuming that noise is absent, zero initial phase, that is, $\theta(0) = 0$, and a delay $T_d = T_b$, sketch the following waveforms over the time interval $0 \le t \le 8T_b$:

(a) $a(t)$

(b) $\theta(t)$

(c) $\theta(t - T_b)$

(d) $\Delta\theta(t) = \theta(t) - \theta(t - T_b)$

(e) $\sin \Delta\theta(t)$

Compare $\sin \Delta\theta(t)$ with $a(t)$. How do they differ?

7.8 For maximum-likelihood block-by-block differential detection of M-PSK, the pairwise error probability is given by (7.68) where from (7.62) we have

$$z_1 = \sum_{i=0}^{N_s-1} \tilde{V}_{k-i} \exp\left\{-j\left(\sum_{m=0}^{N_s-i-2} \Delta\theta_{k-i-m}\right)\right\}$$

$$= \tilde{V}_{k-N_s+1} + \sum_{i=0}^{N_s-2} \tilde{V}_{k-i} \exp\left\{-j\left(\sum_{m=0}^{N_s-i-2} \Delta\theta_{k-i-m}\right)\right\}$$

$$z_2 = \sum_{i=0}^{N_s-1} \tilde{V}_{k-i} \exp\left\{-j\left(\sum_{m=0}^{N_s-i-2} \Delta\hat{\theta}_{k-i-m}\right)\right\}$$

$$= \tilde{V}_{k-N_s+1} + \sum_{i=0}^{N_s-2} \tilde{V}_{k-i} \exp\left\{-j\left(\sum_{m=0}^{N_s-i-2} \Delta\hat{\theta}_{k-i-m}\right)\right\}$$

(a) Compute the mean, variance, and cross-correlation of the complex variables z_1 and z_2 defined by

$$\overline{z_k} \triangleq \mu_{c,k} + j\mu_{s,k}; \quad \sigma_k^2 \triangleq \frac{1}{2}\overline{|z_k - \overline{z_k}|^2}; \quad k = 1, 2$$

$$\rho_z \sigma_1 \sigma_2 \triangleq \frac{1}{2}\overline{(z_1 - \overline{z_1})^* (z_2 - \overline{z_2})}; \quad \rho_z = \rho_{cz} + j\rho_{sz}$$

Wherever possible, express your answers in terms of the parameter δ defined in (7.70).

(b) The arguments of the Marcum Q-function in (7.68) are obtained from [see (6.65)]

$$\left\{\begin{matrix} b \\ a \end{matrix}\right\} = \frac{1}{2N_z}\left\{\frac{S_1 + S_2 - 2\,|\rho_z|\,\sqrt{S_1 S_2}\cos\,(\Theta_1 - \Theta_2 + \nu)}{1 - |\rho_z|^2} \pm \frac{S_1 - S_2}{\sqrt{1 - |\rho_z|^2}}\right\}$$

where

$$S_k \triangleq \frac{1}{2}|\overline{z_k}|^2; \quad \Theta_k \triangleq \arg\{\overline{z_k}\} = \tan^{-1}\frac{\mu_{s,k}}{\mu_{c,k}}; \quad k = 1, 2$$

$$\nu \triangleq \arg\{\rho_z\} = \tan^{-1}\frac{\rho_{sz}}{\rho_{cz}}$$

$$N_z \triangleq \sigma_1^2 = \sigma_2^2$$

Using the results found in part (a), show that a and b are given by (7.69).

7.9 In evaluating the upper bound on the bit error probability performance of maximum-likelihood block-by-block differential detection of M-PSK, the sum of Hamming distances required in

(7.83) must be evaluated. To accomplish this, we start by observing that δ of (7.70) can be written in the form

$$\delta = 1 + \sum_{i=1}^{N_s-1} e^{j\alpha_i}; \quad \alpha_i \triangleq \sum_{n=1}^{i} \delta\theta_{k-N_s+1-n}$$

We are interested in determining the various possible solutions for the $\delta\theta$'s such that the maximum value of the magnitude of δ, namely, $|\delta|_{max}$, is achieved. There are four cases to consider.

Case 1: All $N_s - 1$ vectors $e^{j\alpha_i}$; $i = 1, 2, \ldots, N_s - 1$ must be collinear and equal to $e^{j2\pi/M}$.

Case 2: All $N_s - 1$ vectors $e^{j\alpha_i}$; $i = 1, 2, \ldots, N_s - 1$ must be collinear and equal to $e^{-j2\pi/M}$.

Case 3: Any $N_s - 2$ vectors $e^{j\alpha_i}$ must be collinear and equal to 1 and the remaining vector must be equal to $e^{j2\pi/M}$.

Case 4: Any $N_s - 2$ vectors $e^{j\alpha_i}$ must be collinear and equal to 1 and the remaining vector must be equal to $e^{-j2\pi/M}$.

(a) Show that the solutions for the $\delta\theta$'s to Cases 1–4 are as follows:

Case 1: $\delta\theta_{k-N_s+2} = \frac{2\pi}{M}$; $\quad \delta\theta_{k-N_s+i} = 0$; $\quad i = 3, 4, \ldots, N_s$

Case 2: $\delta\theta_{k-N_s+2} = -\frac{2\pi}{M}$; $\quad \delta\theta_{k-N_s+i} = 0$; $\quad i = 3, 4, \ldots, N_s$

Case 3:

$$\delta\theta_{k-N_s+l+1} = \frac{2\pi}{M}; \quad \delta\theta_{k-N_s+l+2} = -\frac{2\pi}{M}; \quad \delta\theta_{k-N_s+i} = 0;$$

$$i = 1, 2, \ldots, l, l+3, l+4, \ldots, N_s, \quad l = 1, 2, \ldots, N_s - 2$$

$$\delta\theta_k = \frac{2\pi}{M}; \quad \delta\theta_{k-N_s+i} = 0; \quad i = 1, 2, \ldots, N_s - 1$$

Case 4:

$$\delta\theta_{k-N_s+l+1} = -\frac{2\pi}{M}; \quad \delta\theta_{k-N_s+l+2} = \frac{2\pi}{M}; \quad \delta\theta_{k-N_s+i} = 0;$$

$$i = 1, 2, \ldots, l, l+3, l+4, \ldots, N_s, \quad l = 1, 2, \ldots, N_s - 2$$

$$\delta\theta_k = -\frac{2\pi}{M}; \quad \delta\theta_{k-N_s+i} = 0; \quad i = 1, 2, \ldots, N_s - 1$$

(b) For the solutions of part (a) show that for a Gray code bit to symbol assignment, the accumulated Hamming distances are as follows:

Case 1: $w(\mathbf{u}, \hat{\mathbf{u}}) = 1$

Case 2: $w(\mathbf{u}, \hat{\mathbf{u}}) = 1$

Case 3: $w(\mathbf{u}, \hat{\mathbf{u}}) = 2(N_s - 2) + 1$

Case 4: $w(\mathbf{u}, \hat{\mathbf{u}}) = 2(N_s - 2) + 1$

(c) From the results of part (b), obtain the result in (7.89) corresponding to $M > 2$.

7.10 For the special case of $M = 2$, show that Cases 1 and 2 of Problem 7.9 are one and the same and likewise for Cases 3 and 4, that is, only half the number of solutions exist. As such, show that (7.84) is now the appropriate solution for the accumulated Hamming distance.

7.11 For maximum-likelihood block-by-block differential detection of M-PSK, the conditional likelihood ratio is given by (7.58). In (7.60), the *average*-likelihood ratio was found by averaging (7.58) over the pdf of the unknown phase, θ. The decision statistic was then found by maximizing this average-likelihood ratio over the signal sequence. Suppose now that we instead first find the *maximum-likelihood* estimate, $\hat{\theta}_{ML}$, of θ, then substitute this value into (7.58) to produce a *maximum*-likelihood ratio, and finally maximize this ratio over the signal sequence.

(a) Find the maximum-likelihood estimate of θ defined by

$$\hat{\theta}_{ML} \triangleq \max_{\theta}{}^{-1} f_{\tilde{\boldsymbol{\nu}}|\theta}\left(\tilde{\boldsymbol{\nu}}\,\middle|\,\tilde{\mathbf{s}}, \theta\right)$$

Note that $\hat{\theta}_{ML}$ will be a function of the signal sequence.

(b) Show that the decision rule obtained by maximizing $f_{\tilde{\boldsymbol{\nu}}|\theta}\left(\tilde{\boldsymbol{\nu}}\,\middle|\,\tilde{\mathbf{s}}, \hat{\theta}_{ML}\right)$ over the signal sequence $\tilde{\mathbf{s}}$ is identical to that obtained by maximizing the average likelihood ratio $f_{\tilde{\boldsymbol{\nu}}}\left(\tilde{\boldsymbol{\nu}}\,\middle|\,\tilde{\mathbf{s}}\right)$ of (7.60) over this same sequence.

7.12 Consider a DPSK system where the signal has a *residual* carrier component. As such, it can be written in the form

$$s(t) = \sqrt{2P} \sin\left(\omega_c t + \theta_m \sum_{n=-\infty}^{\infty} b_n p(t - nT) + \theta\right)$$

where $\theta_m \leq 90°$ is the modulation index, θ is the unknown phase introduced by the channel, $p(t)$ is a unit rectangular pulse, and $\{b_n\}$ is a differentially encoded data sequence, that is, $b_n = a_n b_{n-1}$ with $\{a_n\}$ the true input data sequence. The signal is sent over an AWGN channel and the receiver is as illustrated in Fig. 7.2b.

(a) Find the mean of the sample Z_i under the hypothesis $a_i = 1$ and $a_i = -1$, that is, how does the presence of the residual carrier affect these results relative to the case where it is absent ($\theta_m = 90°$)?

(b) Is it reasonable to expect that the receiver will work even in the presence of the residual carrier? If so, what might you expect to choose for the decision threshold?

(c) How might you expect the value of θ_m to affect the system performance, that is, what is the best choice of θ_m?

7.13 For the equivalent nonoverlapped signal set defined in (7.103), show that the signals $\tilde{s}'_k(t)$ have the normalized correlation matrix given by (4.111). Hint: Write $\tilde{s}'_k(t)$ as $\tilde{s}'_k(t) = \tilde{s}_k(t) - \tilde{s}''_k(t)$ in the interval $0 \leq t \leq \left(\frac{M}{M-1}\right) T_s$ where

$$\tilde{s}''_k(t) = \begin{cases} \tilde{s}_k(t); & 0 \leq t \leq \left(\frac{1}{M-1}\right) T_s \\ 0; & \left(\frac{1}{M-1}\right) T_s \leq t \leq \left(\frac{M}{M-1}\right) T_s \end{cases}$$

and make use of the orthogonality properties of the $\tilde{s}_k(t)$'s over this same time interval.

REFERENCES

1. Fleck, J. T. and E. A. Trabka, "Error Probabilities of Multiple-State Differentially Coherent Phase-Shift-Keyed Systems in the Presence of White, Gaussian Noise," in *Investigation of Digital Data Communication Systems*, J. G. Lawton, ed., Buffalo, NY: Cornell Aeronautical Laboratory, Inc., January 1961. Available as ASTIA Document No. AD 256 584.

2. Park, J. H., Jr., "On Binary DPSK Detection," *IEEE Transactions on Communications*, vol. COM-26, no. 4, April 1978, pp. 484–86.

3. Simon, M. K., "Comments on 'On Binary DPSK Detection'," *IEEE Transactions on Communications*, vol. COM-26, no. 10, October 1978, pp. 1477–78.

4. Pawula, R. F., S. O. Rice and J. H. Roberts, "Distribution of the Phase Angle between Two Vectors Perturbed by Random Noise," *IEEE Transactions on Communications*, vol. COM-30, no. 8, August 1982, pp. 1828–41.

5. Arthurs, E. and H. Dym, "On the Optimum Detection of Digital Signals in the Presence of White Gaussian Noise—A Geometric Interpretation and a Study of Three Basic Data Transmission Systems," *IRE Transactions on Communication Systems*, vol. CS-10, no. 4, December 1962, pp. 336–72. Also TM 3250, Bedford, MA: Mitre Corporation.

6. Bussgang, J. J. and M. Leiter, "Error Rate Approximations for Differential Phase-Shift-Keying," *IEEE Transactions on Communication Systems*, vol. CS-12, no. 1, March 1964, pp. 18–27. Also SR-103, Bedford, MA: Mitre Corporation.

7. Lawton, J. G., "Comparison of Binary Data Transmission Systems," *Proceedings of the National Convention on Military Electronics*, 1958, pp. 54–61.

8. Cahn, C. R., "Performance of Digital Phase-Modulation Communication Systems," *IRE Transactions on Communications Systems*, vol. CS-7, no. 1, May 1959, pp. 3–6.

9. Weinstein, F. S., "A Table of the Cumulative Probability Distribution of the Phase of a Sine Wave in Narrow-Band Normal Noise," *IEEE Transactions on Information Theory*, vol. IT-23, September 1977, pp. 640–43.

10. Lindsey, W. C. and M. K. Simon, *Telecommunication Systems Engineering*, Englewood Cliffs, NJ: Prentice-Hall, 1973. Reprinted by Dover Press, New York, NY, 1991.

11. Lee, P. J., "Computation of the Bit Error Rate of Coherent M-ary PSK with Gray Code Bit Mapping," *IEEE Transactions on Communications*, vol. COM-34, no. 5, May 1986, pp. 488–91.

12. Korn, I. and L. M. Hii, "Relation between Bit and Symbol Error Probabilities for DPSK and FSK with Differential Phase Detection," to appear in the *IEEE Transactions on Communications*.

13. Pawula, R. F. and J. H. Roberts, "The Effects of Noise Correlation and Power Imbalance on Terrestrial and Satellite DPSK Channels," *IEEE Transactions on Communications*, vol. COM-31, no. 6, June 1983, pp. 750–55.

14. Rice, S. O., "Statistical Properties of a Sine Wave Plus Random Noise," *Bell System Technical Journal*, vol. 27, January 1948, pp. 109–57.

15. Roberts, J. H., *Angle Modulation*, England: Peregrinus, 1977.

16. Lee, J. S. and L. E. Miller, "On the Binary Communication Systems in Correlated Gaussian Noise," *IEEE Transactions on Communications*, vol. COM-23, no. 2, February 1975, pp. 255–59.

17. Huff, R. J., "The Imperfectly-Timed Demodulation of Differential Phase Shift Keyed Signals," Report 2738-1, ElectroScience Laboratory, Department of Electrical Engineering, Ohio State University, June 1969.

18. Simon, M. K., "A Simple Evaluation of DPSK Error Probability Performance in the Presence of Bit Timing Error," to appear in the February 1994 issue of the *IEEE Transactions on Communications*.

19. Salz, J. and B. R. Saltzberg, "Double Error Rates in Differentially Coherent Phase Systems," *IEEE Transactions on Communication Systems*, vol. COM-12, June 1964, pp. 202–05.

20. Oberst, J. F. and D. L. Schilling, "Double Error Probability in Differential PSK," *Proceedings of the IEEE*, vol. 56, June 1968, pp. 1099–1100.

21. Stein, S., "Unified Analysis of Certain Coherent and Noncoherent Binary Communications Systems," *IEEE Transactions on Information Theory*, vol. IT-10, no. 1, January 1964, pp. 43–51.

22. Divsalar, D. and M. K. Simon, "Multiple-Symbol Differential Detection of MPSK," *IEEE Transactions on Communications*, vol. COM-38, no. 3, March 1990, pp. 300–08.

23. Wilson, S. G., J. Freebersyser and C. Marshall, "Multi-Symbol Detection of M-DPSK," *GLOBECOM '89 Conference Record*, Dallas, TX, November 27–30, 1989, pp. 47.3.1–47.3.6.

24. Helstrom, C. W., *Statistical Theory of Signal Detection*, England: Pergamon Press, 1960, p. 164.

25. Mackenthun, K., "A Fast Algorithm for Multiple-Symbol Differential Detection of MPSK," submitted for publication in the *IEEE Transactions on Communications*.

26. Schwartz, M., W. R. Bennett and S. Stein, *Communication Systems and Techniques*, New York: McGraw-Hill, 1966.

27. Carlson, A. B., *Communications Systems*, 2nd ed., New York: McGraw-Hill, 1975, pp. 396–404.

28. Downing, J. J., *Modulation, Systems and Noise*, Englewood Cliffs, NJ: Prentice-Hall, 1964, pp. 184–87.

29. Park, J. H., Jr., "Effects of Band Limiting on the Detection of Binary Signals," *IEEE Transactions on Aerospace and Electronic Systems*, vol. AES-5, September 1969, pp. 868–70.

30. Abramowitz, M. and I. A. Stegun, eds., *Handbook of Mathematical Functions with Formulas, Graphs, and Mathematical Tables*, New York: Dover Press, 1965 (originally published by the National Bureau of Standards in 1964).

31. Leib, H. and S. Pasupathy, "The Phase of a Vector Perturbed by Gaussian Noise and Differentially Coherent Receivers," *IEEE Transactions on Information Theory*, vol. IT-34, no. 6, November 1988, pp. 1491–1501.

32. Edbauer, F., "Bit Error Rate of Binary and Quaternary DPSK Signals with Multiple Decision Feedback Detection," *IEEE Transactions on Communications*, vol. COM-40, no. 3, March 1992, pp. 457–60.

33. Winters, J. H., "Differential Detection with Intersymbol Interference and Frequency Uncertainty," *IEEE Transactions on Communications*, vol. COM-32, No. 1, January 1984, pp. 25–33.

34. Chethik, F., "Analysis and Tests of Quadrature Advance/Retard Keying Modulation and Detection," Aeronutronic Ford Corp. (internal technical memorandum), August 1974.

35. Alexovich, J., P. Kossin and T. Schonhoff, "Bandpass-Limited Symmetric DPSK: A Bandwidth-Efficient Modulation for Nonlinear Channels," *MILCOM'92 Conference Record*, San Diego, CA, October 11–14, 1992, pp. 35.4.1–35.4.5.

36. Davarian, F., M. K. Simon and J. Sumida, "DMSK—a Practical 2400 bps Receiver for the Mobile Satellite Service," Pasadena, CA: Jet Propulsion Laboratory, JPL Publication 85-51, June 1985.

37. Simon, M. K. and C. C. Wang, "Two-Bit Differential Detection of MSK," *GLOBECOM'84 Conference Record*, December 1984, Atlanta, GA, pp. 22.3.1–22.3.6.

38. Kaleh, G. K., "Differentially Coherent Detection of MSK," *ICC'88 Conference Record*, Philadelphia, PA, June 12–15, 1988, pp. 23.7.1–23.7.5.

39. Kaleh, G.K., "A Differentially Coherent Receiver for Minimum Shift Keying Signals," *IEEE Journal on Selected Areas in Communications*, vol. SAC-7, no. 1, January 1989, pp. 99–106.

Chapter 8

Double Differentially Coherent Communication with Waveforms

In our discussion of differentially coherent communication in Chapter 7, we were quick to point out that although this form of detection is quite robust in the presence of slowly varying channel phase errors, it degrades quite rapidly when in addition slowly varying channel frequency errors (offsets) are present. A simple example of such a channel is one characterized by communication with a moving vehicle, in which case the frequency offset is that associated with the well-known Doppler effect. As a reminder to the reader, the problem with using classical[1] differential detection in an environment subject to frequency offset is that the phase of the detector output complex signal contains a bias proportional to the product of the frequency offset Δf and the detector delay T. This bias produces an asymmetry (the direction of which depends on the polarity of Δf) in the decision between the two equally likely messages (antipodal signals) which ultimately degrades the average bit error probability performance of the receiver. Since ordinarily the detector delay T is chosen equal to the reciprocal of the data rate, then equivalently we are saying that classical differential detection as discussed in Chapter 7 will yield poor performance if the frequency offset introduced by the channel is an appreciable fraction of the data rate.

One solution to the problem is to include in the receiver a means for estimating, either continuously with a frequency tracking loop or at discrete time intervals with an open loop technique, the frequency offset and using this estimate to cancel the phase bias in the differential detector output. Clearly, the success of this technique depends heavily on the accuracy of the frequency offset estimate, which is ultimately limited by how narrow one can make the tracking loop bandwidth or how long one can smooth the open loop estimate. These system parameters must in turn be commensurate with the rate of variation of the frequency offset itself. An example of a mobile communication receiver that employs such open loop Doppler estimation and phase correction is discussed in [1].

The focus of this chapter is on techniques that do not attempt to estimate the frequency offset but instead modify the classical differential detection scheme in such a way as to make

---

1. Herein, we shall refer to this form of differential detection as *first-order differential detection* (recall that the input information phases are encoded as a first-order difference of two successive transmitted phases) to distinguish it from *second-order* and *higher-order differential detection* schemes, whose definitions will be introduced shortly.

it insensitive to this offset, thereby maintaining the previously discussed advantage of implementation simplicity. The means by which this is achieved lies in encoding the information data phases as a second-order difference phase process[2] and using a two-stage differential detection process. The class of schemes that accomplish this goal are herein referred to as *second-order differential detection* or more simply *double differential detection* [2,3]. When specifically applied to phase-shift-keying, we shall refer to this class of schemes as *double differential phase-shift-keying (DDPSK or D²PSK)*.

From a historical perspective, the Russian literature [4–12] beginning around 1972 was first to introduce the notion of using higher order phase differences for modulation and detection when the channel introduced higher order phase variations. Later in the 1970s, Pent [13] contributed much additional insight to these ideas and indeed produced a result that allowed for analysis of several extensions of the theory given in [2] and a comparison with the work of [3]. Since much of the work in [4–13] is not commonly known in this country, in fact the book (in Russian) by Okunev [2] is, to the authors' knowledge, the only textbook that even mentions higher-order differential detection, we shall make every effort in this chapter to expose the reader to this body of literature, giving due credit where appropriate.

8.1 DEFINITION OF HIGHER ORDER PHASE DIFFERENCE MODULATION (PDM)

Consider an M-ary communication system in which the information in the ith transmission interval $iT_s \le t \le (i+1)T_s$ is conveyed by one of M phase angles. For example, for M-PSK, these phase angles would take on values from the set $\{\beta_m = 2\pi m/M; \ m = 0, 1, 2, \ldots, M - 1\}$. Let θ_i denote the actual signal phase transmitted over the channel in the ith transmission interval. Then a *first-order phase difference modulation (PDM-1)* is one that encodes the input information phase, denoted by $\Delta\theta_i$, in accordance with the operation

$$\Delta\theta_i \triangleq \theta_i - \theta_{i-1} \tag{8.1}$$

namely, the phase difference[3] between the current and previous transmitted signal phases. Turning this relation around, we express the signal phase transmitted in the ith transmission interval as

$$\theta_i = \Delta\theta_i + \theta_{i-1} \tag{8.2}$$

The operation in (8.1) may be viewed as the finite difference representation of the first derivative of $\theta(t)$ at $t = (i+1)T_s$ and we saw in Chapter 7 that it is accomplished by including a differential encoder between the information source and the phase modulator.

Suppose now that we encode the information phases with two back-to-back differential encoders. Since the ouput of the second differential encoder is to be as before, namely, θ_i, then its input (also the output of the first differential encoder) is $\Delta\theta_i$, and thus the input

---

2. If $\Delta^2\theta_i$ denotes the data information phase to be communicated in the ith transmission interval defined by $iT_s \le t \le (i+1)T_s$, then the actual phase θ_i transmitted over the channel in this same interval is related to $\Delta^2\theta_i$ by the second-order difference $\Delta^2\theta_i = (\theta_i - \theta_{i-1}) - (\theta_{i-1} - \theta_{i-2}) = \theta_i - 2\theta_{i-1} + \theta_{i-2}$. This will be expanded upon later in the chapter. For now, it is sufficient to understand this simple definition of the term *second-order difference modulation*.

3. Unless otherwise noted, it is assumed that the result of first and higher order difference and addition operations performed on phase angles are evaluated modulo 2π.

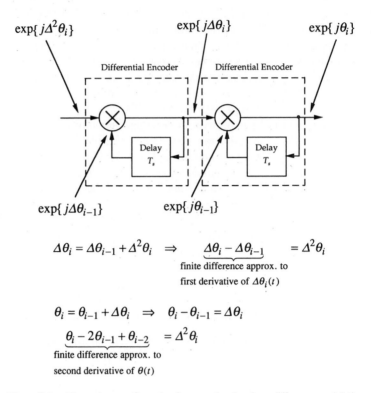

$$\Delta\theta_i = \Delta\theta_{i-1} + \Delta^2\theta_i \quad \Rightarrow \quad \underbrace{\Delta\theta_i - \Delta\theta_{i-1}}_{\substack{\text{finite difference approx. to} \\ \text{first derivative of } \Delta\theta_i(t)}} = \Delta^2\theta_i$$

$$\theta_i = \theta_{i-1} + \Delta\theta_i \quad \Rightarrow \quad \theta_i - \theta_{i-1} = \Delta\theta_i$$

$$\underbrace{\theta_i - 2\theta_{i-1} + \theta_{i-2}}_{\substack{\text{finite difference approx. to} \\ \text{second derivative of } \theta(t)}} = \Delta^2\theta_i$$

Figure 8.1a   Transmitter configuration for second-order phase difference modulation

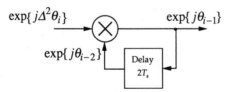

Figure 8.1b   Equivalent transmitter to Figure 8.1a for binary PSK

of the first differential encoder would be

$$\Delta^2\theta_i \triangleq \Delta\theta_i - \Delta\theta_{i-1} = (\theta_i - \theta_{i-1}) - (\theta_{i-1} - \theta_{i-2})$$
$$= \theta_i - 2\theta_{i-1} + \theta_{i-2} \tag{8.3}$$

This defines a *second-order phase difference modulation (PDM-2)* and may be viewed as a finite difference representation of the second derivative of $\theta(t)$ at $t = (i + 1)T_s$. The operation in (8.3) is illustrated in complex form in Figure 8.1a. Note that for binary PSK, wherein the allowable information phases are 0 and π, the term $2\theta_{i-1}$ in (8.3) can only take on values 0 or 2π. Thus, for this case, (8.3) simplifies to

$$\Delta^2\theta_i = \theta_i + \theta_{i-2} = \theta_i - \theta_{i-2} \tag{8.4}$$

where the latter equality in (8.3) again holds because the allowable values for θ_{i-2} are only 0 and π. Figure 8.1b is an illustration in complex form of (8.4).

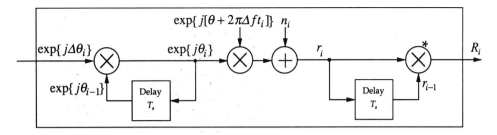

Figure 8.2   Classical differential encoding/detection system based on two-symbol observation

Clearly the above ideas can be extended to higher (than two) order phase differences. In particular, define the nth *order phase difference modulation (PDM-n)* by

$$\Delta^n \theta_i \triangleq \sum_{k=0}^{n} (-1)^{n-k} \binom{n}{k} \theta_{i-n+k} \tag{8.5}$$

where, as in earlier chapters

$$\binom{n}{k} \triangleq \frac{n!}{k!(n-k)!} \tag{8.6}$$

is the binomial coefficient. The modulation of (8.5) is implemented by including n back-to-back differential encoders between the information source and the phase modulator and would be of interest in situations where one is attempting to mitigate the effects of an $(n-1)$-order phase derivative introduced by the channel. When applied to binary PSK, the acronym D^nPSK is used.

What is important in the above is that PDM-n is a modulation with memory in that the current transmitted phase depends on the immediate n previous transmitted phases. This property will be exploited later on in the chapter when consideration is given to optimum detection schemes for such modulations.

8.2 STRUCTURES MOTIVATED BY THE MAXIMUM-LIKELIHOOD (ML) APPROACH

To understand in what sense Fig. 8.1b is an appropriate encoding for channels that introduce frequency offset, consider first a classical differential detection scheme (see Fig. 8.2) that, for a two symbol observation, was shown in Chapter 7 to be optimum (in the ML sense) for a channel with constant (time-invariant) but unknown phase θ. We shall now demonstrate how to augment this scheme to account for the presence of a channel frequency offset Δf.

As before, the complex baseband signal transmitted over the channel in the ith transmission interval is given by (assuming unit amplitude for simplicity)

$$\tilde{s}_i = e^{j\theta_i} \tag{8.7}$$

and the corresponding received complex signal is

$$\tilde{r}_i = \tilde{s}_i e^{j(\theta + 2\pi \Delta f t_i)} + \tilde{n}_i \triangleq \rho_i e^{j(\theta_i + \theta + 2\pi \Delta f t_i + \eta_i)} \tag{8.8}$$

where n_i is a sample of complex Gaussian noise. From Fig. 8.2, we have that the output of the differential detector in the same transmission interval is given by

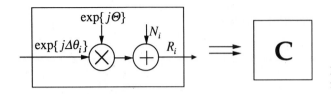

Figure 8.3 Large SNR equivalent channel for system of Figure 8.2

$$\tilde{R}_i = \tilde{r}_i \tilde{r}_{i-1}^* = \left(\tilde{s}_i e^{j(\theta+2\pi\Delta f t_i)} + \tilde{n}_i\right)\left(\tilde{s}_{i-1}^* e^{-j(\theta+2\pi\Delta f t_{i-1})} + \tilde{n}_{i-1}\right) \triangleq \tilde{S}_i e^{j\Theta} + \tilde{N}_i \quad (8.9)$$

where

$$\tilde{S}_i = e^{j(\theta_i-\theta_{i-1})} = e^{j\Delta\theta_i}; \quad \Theta = 2\pi\Delta f(t_i - t_{i-1}) = 2\pi\Delta f T_s$$

$$\tilde{N}_i = \tilde{n}_i \tilde{s}_{i-1}^* e^{-j(\theta+2\pi\Delta f t_{i-1})} + \tilde{n}_{i-1}^* \tilde{s}_i e^{j(\theta+2\pi\Delta f t_i)} + \tilde{n}_i \tilde{n}_{i-1}^* \quad (8.10)$$

For large SNR, we can neglect the term $\tilde{n}_i \tilde{n}_{i-1}^*$ and thus $\tilde{N}_i$ can be approximately modeled as a Gaussian random variable. Assuming that the frequency offset is constant but unknown, then the phase $\Theta = 2\pi\Delta f T_s$ can be modeled as being uniformly distributed in the interval $(-\pi, \pi)$. Hence, we can look at the combination of differential encoder, channel, and differential detector in Fig. 8.2 as an " equivalent" channel with additive white Gaussian noise $\tilde{N}_k$ and random (but time-invariant) phase Θ. This is depicted by the illustration of Fig. 8.3. The problem of finding the optimum receiver for this equivalent channel with input $e^{j\Delta\theta_i}$ is now the same as that previously treated in Chapter 7, namely, apply a differential encoding/detection scheme identical to that used for an AWGN channel with zero frequency offset and random (uniformly distributed) time-invariant phase. For an N_s-symbol observation, the solution is again the maximum-likelihood sequence receiver derived in Section 7.2.1. For $N_s = 2$, we have the classical differential detector based on symbol-by-symbol decisions which requires first-order differential encoding at the transmitter. This simplest case is illustrated in Fig. 8.4. Finally, putting Figures 8.2–8.4 together, we see that for large SNR, the optimum transmitter/receiver configuration equates to back-to-back differential encoders at the transmitter (see Fig. 8.1a) and back-to-back differential detectors at the receiver (Fig. 8.5). The output of the receiver in Fig. 8.5 is from (8.8) and (8.9) given by

$$\tilde{R}_i \tilde{R}_{i-1}^* = \left(\tilde{r}_i \tilde{r}_{i-1}^*\right)\left(\tilde{r}_{i-1} \tilde{r}_{i-2}^*\right)$$

$$= \left(\rho_i \rho_{i-1} e^{j(\Delta\theta_i + 2\pi\Delta f T_s + \Delta\eta_i)}\right)\left(\rho_{i-1}\rho_{i-2} e^{-j(\Delta\theta_{i-1} + 2\pi\Delta f T_s + \eta_{i-1})}\right) \quad (8.11)$$

$$= \rho_i \rho_{i-1}^2 \rho_{i-2} e^{j\left(\Delta^2\theta_i + \Delta^2\eta_i\right)}$$

which is clearly *independent of the frequency offset* Δf as desired. Note that whereas conventional DPSK schemes require that the channel phase be constant over two transmission intervals, D²PSK schemes such as Fig. 8.5 require that the channel phase and frequency be constant over three intervals.

If one reviews the literature, double differential detection schemes fall into two categories depending on whether their two stages are implemented *serially* or with a *parallel* architecture. Furthermore, they can be classified according to whether they have an *autocorrelation demodulator (ACD)* or *inphase-quadrature (I-Q)* type of first stage. Finally,

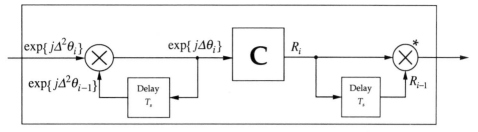

(asterisk on multiplier denotes complex conjugate multiplication)

Figure 8.4   Optimum transmitter/receiver for equivalent channel of Figure 8.3

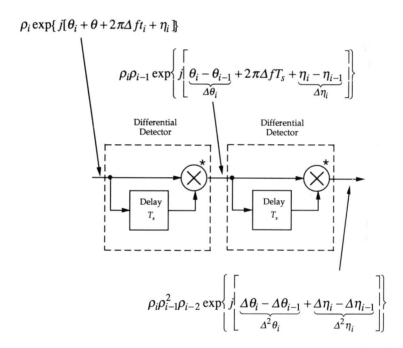

Figure 8.5   Optimum (large SNR) receiver for second-order phase difference modulation

depending on the specific implementation, the final detector output will, in the absence of noise, either be entirely independent of the input frequency (the detection scheme is *absolutely frequency-invariant*) or vary as a function of frequency within some specified small frequency range (the detection scheme is *relatively frequency-invariant*). It is important to emphasize that even though a detector is absolutely frequency-invariant according to the above classification, its error probability performance will in most cases depend on the frequency offset. Typically (as we shall see shortly), this comes about because of the dependence of the signal × noise terms in the detector ouput on the frequency offset.

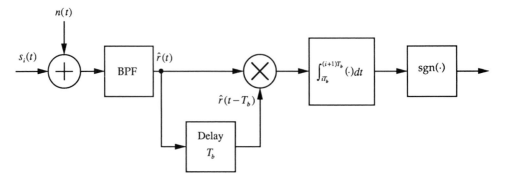

Figure 8.6   Autocorrelation demodulator of DPSK (PDM-1) signals

We begin with an examination of well-known PDM-1 receivers (at least one of which was presented in Chapter 7) with emphasis on their behavior and performance in the presence of carrier frequency offset. The primary purpose of this discourse is to expose the configurations that will later represent the first stage of serial-type receivers for PDM-2 modulation.

8.3 AUTOCORRELATION DEMODULATION (ACD) OF BINARY PDM-1 SIGNALS

Consider the autocorrelation demodulator for binary DPSK illustrated in Fig. 8.6. The input signal $s_i(t)$ in the ith transmission interval $iT_b \leq t \leq (i+1)T_b$ is of the form

$$s_i(t) = A \sin(\omega_c t + \theta + \theta_i) \tag{8.12}$$

where ω_c is the radian carrier frequency, θ_i is the differentially encoded information phase in that interval taking on values 0, π, and θ is the unknown phase introduced by the channel (assumed to be constant over an interval of duration $2T_b$ sec). The sum of $s(t)$ and white Gaussian noise $n(t)$ (with single-sided power spectral density N_0) are passed through a bandpass filter (BPF) of single-sided bandwidth W Hz ($B = W/2$ is the equivalent lowpass bandwidth) producing the signal $x(t)$. Assuming, for convenience, that the center radian frequency, ω_{c0}, of the bandpass filter is an integer multiple of the data rate, that is, $\omega_{c0} = m\Omega$; $\Omega = 2\pi/T_b$, m integer, then the actual carrier radian frequency in (8.12) can be expressed as $\omega_c = m\Omega + \Delta\omega$ where $\Delta\omega = 2\pi\Delta f$ represents the radian frequency uncertainty about the nominal carrier frequency. Note that the bandwidth of the BPF is chosen commensurate with the maximum value of $\Delta\omega$. Assuming that the signal $s(t)$ remains undistorted after passing through the BPF (this is equivalent to assuming a wide enough bandwidth W relative to the data rate $1/T_b$ that intersymbol interference (ISI) can be ignored), then expressing the filtered noise in a truncated Fourier series[4] in the interval $iT_b \leq t \leq (i+1)T_b$, we have

---

4. This form of the noise expansion was used in [10] and in the context of this problem was first introduced in [5].

$$\hat{r}(t) = s_i(t) + \hat{n}_1(t) = A \sin (m\Omega t + \Delta\omega t + \theta + \theta_i)$$

$$+ \sum_{k=k_1}^{k_2} \left(n_{s1,k} \sin (k\Omega t + \Delta\omega t + \theta) + n_{c1,k} \cos (k\Omega t + \Delta\omega t + \theta) \right) \tag{8.13}$$

where the number of harmonics in the sum is given by $k_2 - k_1 + 1 = 2BT_b$ and the carrier frequency integer, m, lies in the range $k_1 \le m \le k_2$. Thus, for the results that follow, $2BT_b$ is assumed to be an integer.

Similarly, in the interval $iT_b \le t \le (i+1)T_b$, the BPF output corresponding to the previous interval is given by

$$\hat{r}(t - T_b) = s_{i-1}(t - T_b) + \hat{n}_2(t) = A \sin (m\Omega t + \Delta\omega (t - T_b) + \theta + \theta_{i-1})$$

$$+ \sum_{k=k_1}^{k_2} \left(n_{s2,k} \sin (k\Omega t + \Delta\omega t + \theta) + n_{c2,k} \cos (k\Omega t + \Delta\omega t + \theta) \right) \tag{8.14}$$

The quantities $n_{s1,k}, n_{s2,k}, n_{c1,k}, n_{c2,k}$ are independent Gaussian random variables with zero mean and variance N_0/T_b.

The ACD algorithm for the DPSK signal is to make the decision $\Delta\hat{\theta}_i$ on the differentially encoded data phase $\Delta\theta_i = \theta_i - \theta_{i-1}$ transmitted in the ith transmission interval according to

$$\hat{d}_i = \text{sgn} \left\{ \int_{iT_b}^{(i+1)T_b} \hat{r}(t)\hat{r}(t - T_b)dt \right\} \tag{8.15}$$

where, as in previous chapters, sgn $\{\cdot\}$ denotes the signum (sign) function of its argument. Using (8.13) and (8.14), the integral in (8.15) can be written as

$$\int_{iT_b}^{(i+1)T_b} \hat{r}(t)\hat{r}(t - T_b)dt = \frac{A^2 T_b}{2} \cos \Delta\theta_i \cos \Delta\omega T_b + \frac{AT_b}{2} n_{s2,m} \cos \theta_i$$

$$+ \frac{AT_b}{2} n_{s1,m} \cos \theta_{i-1} \cos \Delta\omega T_b$$

$$- \frac{AT_b}{2} n_{c1,m} \cos \theta_{i-1} \sin \Delta\omega T_b \tag{8.16}$$

$$+ \frac{T_b}{2} \sum_{k=k_1}^{k_2} \left(n_{s1,k} n_{s2,k} + n_{c1,k} n_{c2,k} \right)$$

Substituting $n_{s1,k} = v_{s1,k} + v_{s2,k}$, $n_{s2,k} = v_{s1,k} - v_{s2,k}$, $n_{c1,k} = v_{c1,k} + v_{c2,k}$, and $n_{c2,k} = v_{c1,k} - v_{c2,k}$ into (8.16) allows it to be expressed as the sum of the squares of Gaussian-distributed independent random variables (the v_s's and the v_c's). Arbitrarily assuming that the data phase $\Delta\theta_i = 0$, then an error occurs when the integral in (8.16) is negative and the error probability is given as

$$P_b(E) = \Pr \left\{ \left(v_{s1,m} + A\cos\theta_{i-1}\cos^2\frac{\Delta\omega T_b}{2} \right)^2 \right.$$

$$+ \left(v_{c1,m} - \frac{A\sin\Delta\omega T_b}{2} \right)^2 + \sum_{\substack{k=k_1\\k\neq m}}^{k_2} \left(v_{s1,k}^2 + v_{c1,k}^2 \right)$$

$$< \left(v_{s2,m} + A\cos\theta_{i-1}\cos^2\frac{\Delta\omega T_b}{2} \right)^2 \tag{8.17}$$

$$\left. + \left(v_{c2,m} + \frac{A\sin\Delta\omega T_b}{2} \right)^2 + \sum_{\substack{k=k_1\\k\neq m}}^{k_2} \left(v_{s2,k}^2 + v_{c2,k}^2 \right) \right\}$$

$$= \Pr\{\lambda_1 < \lambda_2\}$$

where λ_1 and λ_2 have noncentral $\chi 2$ distributions with $4BT_b$ degrees of freedom. The details of the evaluation of (8.18) are carried out in [12] with the following results. Letting $\alpha = \Delta\omega T_b$, that is, the ratio of radian frequency offset to data rate, then

$$P_b(E) = \frac{1}{2^{2BT_b}} \exp\left\{ -\frac{E_b}{N_0}\left(2\sin^2\frac{\alpha}{2} + \cos^2\frac{\alpha}{2} \right) \right\}$$

$$\times \sum_{n=0}^{\infty} \frac{\left(2\frac{E_b}{N_0}\sin^2\frac{\alpha}{2} \right)^n}{n!} \sum_{k=0}^{2BT_b-n-1} \frac{1}{2^k} \sum_{m=0}^{k} \frac{\left(\frac{E_b}{N_0}\cos^2\frac{\alpha}{2} \right)^m}{m!} \binom{k+2BT_b-1}{k-m} \tag{8.18}$$

where $E_b = A^2 T_b/2$ is the energy in the signal and $\binom{n}{k}$ is the combinatorial coefficient defined in (8.6). When the frequency error is absent, that is, $\alpha = 0$, then (8.18) simplifies to a result previously obtained in [4, 11], namely

$$P_b(E) = \frac{1}{2^{2BT_b}} \exp\left\{ -\frac{E_b}{N_0} \right\} \sum_{i=0}^{2BT_b-1} \frac{\left(\frac{E_b}{N_0} \right)^i}{i!} \sum_{j=i}^{2BT_b-1} \frac{1}{2^j} \binom{j+2BT_b-1}{j-i} \tag{8.19}$$

For $2BT_b = 1$, (8.19) becomes

$$P_b(E) = \frac{1}{2}\exp\left(-\frac{E_b}{N_0} \right) \tag{8.20}$$

which is equal[5] to the well-known result for optimum detection of DPSK [see (7.14)]. For $2BT_b = 2, 3, 4$ we have

$$P_b(E)\big|_{2BT_b=2} = \frac{1}{2}\exp\left(-\frac{E_b}{N_0} \right)\left[1 + \frac{1}{4}\frac{E_b}{N_0} \right] \tag{8.21a}$$

---

5. Strictly speaking, for $2BT_b = 1$, the assumption that the signal $s(t)$ remains undistorted after passing through the BPF is not valid. Thus, (8.20) is optimistic for the configuration of Fig. 8.6.

$$P_b(E)\big|_{2BT_b=3} = P_b(E)\big|_{2BT_b=2} + \exp\left(-\frac{E_b}{N_0}\right)\left[\frac{\left(\frac{E_b}{N_0}\right)^2 + 4\left(\frac{E_b}{N_0}\right)}{2!2^5}\right]$$

(8.21b)

$$= \frac{1}{2}\exp\left(-\frac{E_b}{N_0}\right)\left[1 + \frac{3}{8}\left(\frac{E_b}{N_0}\right) + \frac{1}{32}\left(\frac{E_b}{N_0}\right)^2\right]$$

$$P_b(E)\big|_{2BT_b=4} = P_b(E)\big|_{2BT_b=3}$$

$$+ \exp\left(-\frac{E_b}{N_0}\right)\left[\frac{\left(\frac{E_b}{N_0}\right)^3 + 12\left(\frac{E_b}{N_0}\right)^2 + 30\left(\frac{E_b}{N_0}\right)}{3!2^7}\right]$$

(8.21c)

$$= \frac{1}{2}\exp\left(-\frac{E_b}{N_0}\right)\left[1 + \frac{29}{64}\left(\frac{E_b}{N_0}\right) + \frac{1}{16}\left(\frac{E_b}{N_0}\right)^2 + \frac{1}{384}\left(\frac{E_b}{N_0}\right)^3\right]$$

which shows the degradation of performance caused by widening the input BPF bandwidth. Again, these results are somewhat optimistic because of the zero ISI assumption, but improve in their accuracy as $2BT_b$ increases.

Figure 8.7 illustrates $P_b(E)$ as computed from (8.19) versus E_b/N_0 in dB for values of $2BT_b$ equal to 1, 2, 4, 8, 16, and 32. Also illustrated are the approximate values obtained from (7.118) of Chapter 7 corresponding to the analysis of this detector assuming Gaussian statistics at the output of the I&D. We observe that for small values of $2BT_b$, the Gaussian approximation analysis yields very poor results relative to (8.19) over the range of $P_b(E)$ between 10^{-1} and 10^{-6}. As $2BT_b$ increases, the Gaussian approximation analysis becomes better and the range of values of E_b/N_0 values over which (7.118) tends to agree with the exact (assuming zero ISI) result in (8.19) is increased. However, for any fixed and finite value of $2BT_b$, it is clear from Fig. 8.7 that for sufficiently large E_b/N_0, the results in (7.118) and (8.19) deviate significantly from each other and thus it is inappropriate to use (7.118) to predict asymptotic ($E_b/N_0 \to \infty$) performance.

An approximate account of the effect of frequency error on the noise immunity of the ACD is suggested in [4] and is based upon replacing E_b/N_0 by $E_b/N_0 \cos^2\alpha$ in (8.19) (the result for zero frequency error), which yields

$$P_b(E) = \frac{1}{2^{2BT_b}}\exp\left\{-\frac{E_b}{N_0}\cos^2\alpha\right\}\sum_{i=0}^{2BT_b-1}\frac{\left(\frac{E_b}{N_0}\cos^2\alpha\right)^i}{i!}\sum_{j=i}^{2BT_b-1}\frac{1}{2^j}\binom{j+2BT_b-1}{j-i}$$

(8.22)

The advantage of this approximate expression is that, for computational purposes, one does not have to compute an infinite number of terms. The degree of accuracy to which (8.22) approximates (8.18) was investigated in [4] with the following results. For $2BT_b \leq 10$, $E_b/N_0 \geq 4$ (6 dB), and $\alpha = \Delta\omega T_b \leq 45°$, there is less than 1 dB inaccuracy in the assessment of the E_b/N_0 requirement for fixed $P_b(E)$ using the approximate result in (8.22) relative to that in (8.18).

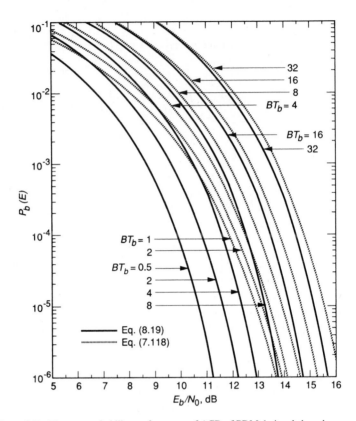

Figure 8.7   Bit error probability performance of ACD of PDM-1 signals based on exact (zero ISI) and Gaussian approximation analyses

8.4 I-Q DEMODULATION OF BINARY PDM-1 SIGNALS

The optimum demodulator for binary DPSK with known carrier frequency is illustrated in Fig. 7.2b and is repeated here in Fig. 8.8 for convenience of reference. As was shown in Chapter 7, the error probability performance of this receiver is given by (8.20). It is tempting to use this configuration in the presence of frequency offset despite the fact that the integrate-and-dump (I&D) circuits are no longer matched filters under these circumstances. We now examine the impact of this mismatch on error probability performance.

Let the signal plus noise in the ith transmission interval be expressed by

$$r(t) = s_i(t) + n(t) = A \sin (m\Omega t + \Delta\omega t + \theta + \theta_i) + n(t) \tag{8.23}$$

The inphase and quadrature demodulation reference signals are given by

$$r_s(t) = 2 \sin m\Omega t; \quad r_c(t) = 2 \cos m\Omega t \tag{8.24}$$

Hence, the I&D outputs (ignoring double harmonic terms) become

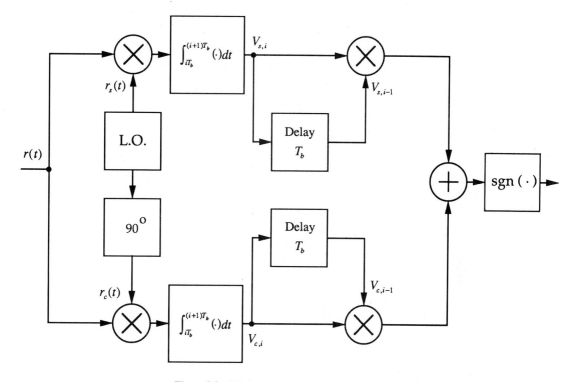

Figure 8.8   I-Q demodulator of DPSK (PDM-1) signals

$$V_{s,i} = \int_{iT_b}^{(i+1)T_b} r(t)r_s(t)\,dt = A\int_{iT_b}^{(i+1)T_b} \cos\left(\Delta\omega t + \theta + \theta_i\right)dt + N_1$$

$$= AT_b \left(\frac{\sin\frac{\Delta\omega T_b}{2}}{\frac{\Delta\omega T_b}{2}}\right)\cos\left(\frac{1}{2}\left[(2i-1)\Delta\omega T_b + 2\theta_i + 2\theta\right]\right) + N_1 \tag{8.25a}$$

and

$$V_{c,i} = \int_{iT_b}^{(i+1)T_b} r(t)r_c(t)\,dt = A\int_{iT_b}^{(i+1)T_b} \sin\left(\Delta\omega t + \theta + \theta_i\right)dt + N_3$$

$$= AT_b \left(\frac{\sin\frac{\Delta\omega T_b}{2}}{\frac{\Delta\omega T_b}{2}}\right)\sin\left(\frac{1}{2}\left[(2i-1)\Delta\omega T_b + 2\theta_i + 2\theta\right]\right) + N_3 \tag{8.25b}$$

Here N_1 and N_2 are independent zero-mean Gaussian random variables with variance $N_0 T_b$. Analogous to (8.15), the decision algorithm here is

$$\hat{d}_i = \text{sgn} \left\{ V_{c,i-1} V_{c,i} + V_{s,i-1} V_{s,i} \right\}$$

$$= \text{sgn} \left\{ A^2 T_b \left(\frac{\sin \frac{\Delta \omega T_b}{2}}{\frac{\Delta \omega T_b}{2}} \right)^2 \cos(\theta_i - \theta_{i-1} + \Delta \omega T_b) + N_{\text{eff}} \right\} \qquad (8.26)$$

$$= \text{sgn} \left\{ A^2 T_b \left(\frac{\sin \frac{\alpha}{2}}{\frac{\alpha}{2}} \right)^2 \cos \alpha \cos \Delta \theta_i + N_{\text{eff}} \right\}$$

where again $\alpha = \Delta \omega T_b$ and we have furthermore made the observation that, for binary phase modulation, $\sin \Delta \phi_i = 0$. Also, N_{eff} is the effective noise consisting of signal $\times$ noise and noise $\times$ noise components as follows:

$$N_{\text{eff}} = A T_b \left(\frac{\sin \frac{\alpha}{2}}{\frac{\alpha}{2}} \right) \left\{ N_2 \cos \left(\frac{1}{2} [(2i-1)\alpha + 2\theta_i + 2\theta] \right) \right.$$

$$+ N_1 \cos \left(\frac{1}{2} [(2i-3)\alpha + 2\theta_{i-1} + 2\theta] \right).$$

$$\left. + N_4 \sin \left(\frac{1}{2} [(2i-1)\alpha + 2\theta_i + 2\theta] \right) + N_3 \sin \left(\frac{1}{2} [(2i-3)\alpha + 2\theta_{i-1} + 2\theta] \right) \right\} \qquad (8.27)$$

$$+ N_1 N_2 + N_3 N_4$$

Here again, N_2, N_4, which represent the noises in $V_{s,i-1}$ and $V_{c,i-1}$, respectively, are independent zero-mean Gaussian random variables with variance $N_0 T_b$ and are independent of N_1, N_3. The effective noise in (8.27) can be written in a more compact form, viz.

$$N_{\text{eff}} = A T_b \left(\frac{\sin \frac{\alpha}{2}}{\frac{\alpha}{2}} \right) N_{S \times N} + N_{N \times N} \qquad (8.28)$$

where $N_{S \times N}$ is a zero-mean Gaussian random variable with variance $2 N_0 T_b$ and $N_{N \times N} = N_1 N_2 + N_3 N_4$ is a non-Gaussian random variable that is uncorrelated with $N_{S \times N}$.

Following the approach taken by Stein [14], the error probability of the decision algorithm given in (8.26) can be shown to be

$$P_b(E) = \frac{1}{2} \left[1 - Q \left(\sqrt{b}, \sqrt{a} \right) + Q \left(\sqrt{a}, \sqrt{b} \right) \right] \qquad (8.29)$$

where

$$a = \frac{2 E_b'}{N_0} \sin^2 \alpha; \quad b = \frac{2 E_b'}{N_0} \cos^2 \alpha \qquad (8.30)$$

with

$$E_b' \triangleq E_b \left(\frac{\sin \frac{\alpha}{2}}{\frac{\alpha}{2}} \right)^2 \qquad (8.31)$$

and $Q(a, b)$ is as before the Marcum Q-function. This result was originally obtained by Henry [15].

Note that for zero frequency error ($\alpha = 0$), we have from (8.30) and (8.31) that $a = 0$ and $b = 2E_b/N_0$. Then, using the special cases of the Marcum Q-function as in (7.76), it is easy to see that (8.29) reduces to (8.20) as it should. On the other hand, there are many values of α, the smallest of these being $\alpha = \pi/2$ ($\Delta f = 1/4T_b$), for which the error probability goes to one-half. Even at $\Delta f = 1/6T_b$, there is already a loss of 3 dB in SNR relative to the case of no frequency error. This rather rapid deterioration of performance with α is a direct consequence of using I&D filters that are not matched to the inphase and quadrature output signals when frequency error is present.

The error probability performance of the I-Q demodulator in Fig. 8.8 can also be obtained in several other equivalent forms. Using a geometric approach [3] to be discussed later in this chapter in connection with the demodulation of higher order PDM, the authors obtained the following integral form which is identical to (8.29)–(8.31), namely

$$P_b(E) = \frac{1}{2} - \frac{1}{2}\sqrt{\frac{E'_b}{\pi N_0}}$$

$$\times \int_0^\pi \left[\mathrm{erf}\left(\sqrt{\frac{E'_b}{N_0}}\cos(\eta - \alpha)\right)\right] \cos\eta \,\exp\left(-\frac{E'_b}{N_0}\sin^2\eta\right) d\eta$$

(8.32)

Also, replacing the error function and exponential functions by their expansions in terms of modified Bessel functions of the first kind, namely

$$\mathrm{erf}(z\cos\phi) = \frac{z\exp(-z^2/2)}{\sqrt{\pi}}\sum_{k=0}^\infty (-1)^k\varepsilon_k I_k\left(\frac{z^2}{2}\right)$$

$$\times \left[\frac{\cos[(2k+1)\phi]}{2k+1} - \frac{\cos[(2k-1)\phi]}{2k-1}\right]$$

(8.33a)

and

$$\exp(z\cos\phi) = \sum_{k=0}^\infty I_k(z)\cos k\phi$$

(8.33b)

the following equivalent series form is obtained [3]:

$$P_b(E) = \frac{1}{2} - \frac{E'_b}{2N_0}\exp\left(-\frac{E'_b}{N_0}\right)$$

$$\times \sum_{k=0}^\infty (-1)^k\left[I_k\left(\frac{E'_b}{2N_0}\right) + I_k\left(\frac{E'_b}{2N_0}\right)\right]^2 \frac{\cos[(2k+1)\alpha]}{2k+1}$$

(8.34)

Miller [16] obtained a similar and equivalent form to (8.32), which is given by

$$P_b(E) = \frac{1}{\pi} \int_0^{\pi/2} \exp\left\{-\frac{E_b'}{N_0}\left(\frac{\cos^2\alpha}{1 - \sin^2\alpha \sin^2\theta}\right)\right\} d\theta$$

$$-\frac{1}{2}\sqrt{E_b'} \int_{-\pi/2}^{\pi/2} \left[\mathrm{erf}\left(\sqrt{\frac{2E_b'}{N_0}}\cos\eta\cos\alpha\right)\right.$$

$$\left.-\mathrm{erf}\left(\sqrt{\frac{2E_b'}{N_0}}\cos(\eta+\alpha)\right)\right]\cos\eta\exp\left(-\frac{E_b'}{N_0}\sin^2\eta\right) d\eta \tag{8.35}$$

It is interesting to study the limit of (8.29) in the limit of large E_b/N_0. In particular, using the asymptotic forms of the Marcum-Q function given in (7.76), we obtain for $\cos\alpha > 0$ (e.g., $0 \le |\alpha| < \pi/2$)

$$P_b(E) = \frac{1}{2\sqrt{2\pi\left(\frac{E_b'}{N_0}\right)}}\sqrt{\frac{1 + |\sin\alpha|}{(|\sin\alpha|)(1 - |\sin\alpha|)}}\exp\left(-E_b'(1 - |\sin\alpha|)\right) \tag{8.36a}$$

and for $\cos\alpha < 0$ (e.g., $\pi/2 < |\alpha| \le \pi$)

$$P_b(E) = 1 - \frac{1}{2\sqrt{2\pi\left(\frac{E_b'}{N_0}\right)}}\sqrt{\frac{1 + |\sin\alpha|}{(|\sin\alpha|)(1 - |\sin\alpha|)}}\exp\left(-\frac{E_b'}{N_0}(1 - |\sin\alpha|)\right) \tag{8.36b}$$

which for $E_b/N_0 \to \infty$ becomes

$$\lim_{E_b/N_0 \to \infty} P_b(E) = \begin{cases} 0; & \cos\alpha > 0 \\ 1; & \cos\alpha < 0 \end{cases} \tag{8.37}$$

8.5 COMPARISON OF ACD AND I-Q DEMODULATION OF PDM-1 SIGNALS

Another way of visualizing the receivers of Figures 8.6 and 8.8 is to observe that the ACD performs differential detection *prior* to "matched filtering" where the I-Q demodulator performs "matched filtering" *prior* to differential detection. The quotations on the words "matched filtering" refer to the fact that for nonzero frequency error, the I&D's are not truly matched filters, as mentioned above.

In the case of zero frequency error, the I-Q demodulator in Fig. 8.8 is the optimum scheme and thus should give the best performance. Nevertheless, we have seen that, for small values of $2BT_b$, the behavior of the ACD of Fig. 8.6 closely approximates the optimum performance under the assumption that ISI is ignored. The assumption of zero ISI is a direct consequence of assuming rectangular pulses for the information modulation (i.e., constant phase over the symbol interval) at the *output* of the BPF in Fig. 8.6, and as such the error probability performance results as given in (8.21) are only approximately valid with the degree of approximation improving as $2BT_b$ increases.

If, instead of rectangular pulses for the phase modulation, one were to use a Nyquist pulse shaping, that is, one that achieves zero ISI (see Chapter 9), then it might be anticipated that the results of (8.21) would *exactly* apply. In [1], this was verified for the case $2BT_b = 2$ where a 100% excess bandwidth raised cosine transfer function was employed for the overall (transmitter plus receiver) Nyquist filter response. Since, for an AWGN channel with no ISI, the average error probability is minimized (with respect to the noise) by splitting the *total* filtering (transmit filter, receive filter, and matched filter) equally between transmitter and receiver (see Chapter 9), then for the above this can be accomplished as follows.

The transmit filter (see Fig. 8.9a) is a *root* raised cosine filter with transfer function

$$P_T(\omega) = \begin{cases} \sqrt{2T_b}\cos\frac{\omega T_b}{4}; & |\omega| \leq \frac{2\pi}{T_b} \\ 0; & \text{otherwise} \end{cases} \tag{8.38}$$

The input receive filter is a brick wall filter whose equivalent lowpass version has a transfer function magnitude $P_R(\omega)$ given by

$$P_R(\omega) = \begin{cases} \sqrt{\frac{T_b}{2}}; & |\omega| \leq \frac{2\pi}{T_b} \\ 0; & \text{otherwise} \end{cases} \tag{8.39}$$

Note that the single-sided bandwidth of this lowpass filter is $B = 1/T_b$.

The combination of $P_T(w)$ and $P_R(w)$ presents a root raised cosine characteristic to the matched filter input, namely

$$P(\omega) = P_T(\omega)P_R(\omega) = \begin{cases} T_b\cos\frac{\omega T_b}{4}; & |\omega| \leq \frac{2\pi}{T_b} \\ 0; & \text{otherwise} \end{cases} \tag{8.40}$$

The impulse response of such an overall characteristic has the property that two of its samples taken at intervals of $T_b/2$ sec are equal and nonzero with the remaining samples all having zero value, that is,

$$p(t) = \frac{4}{\pi}\left[\frac{\cos\left(2\pi\frac{t}{T_b}\right)}{1 - \left(\frac{4t}{T}\right)^2}\right];$$

$$p\left(\frac{T_b}{4}\right) = p\left(-\frac{T_b}{4}\right) = 1 \tag{8.41}$$

$$p\left(\pm(2k+1)\right)\frac{T_b}{4}) = 0; \quad k = 1, 2, \ldots$$

The matched filter (assuming zero frequency offset) for the pulse shape of (8.41) has a transfer function whose magnitude is given by (see Fig. 8.9)

$$|G(\omega)| = \left|2\cos\frac{\omega T_b}{4}\right| \tag{8.42}$$

Thus, we see that the cascade of (8.40) and (8.42) achieves the Nyquist pulse shaping corresponding to a 100% excess bandwidth raised cosine transfer function and also the equal split (except for a scale factor) of the total filtering between transmitter and receiver.

Since zero ISI is achieved by this implementation, the optimum detector (for zero frequency error) would be to perform the matched filtering prior to the differential detection

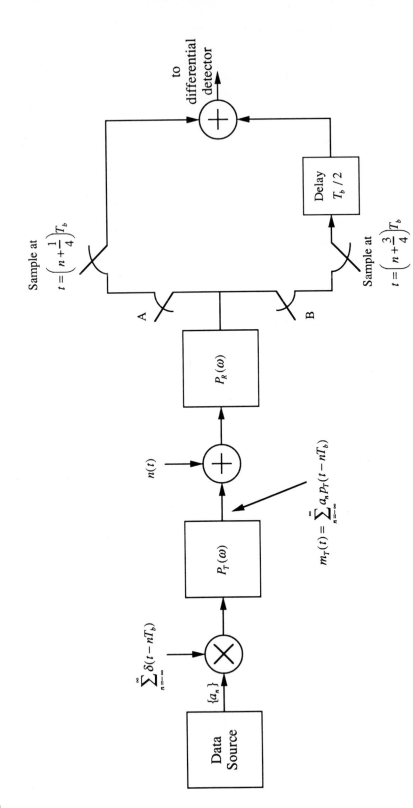

Figure 8.9a Block diagram of the system

(see Fig. 8.9b). When this is done, the error probability performance is indeed given by (8.20). If on the other hand, differential detection is performed prior to matched filtering (see Fig. 8.9c), then it can be shown [17] that the error probability is *exactly* given by $P_b(E)\big|_{2BT_b=2}$ of (8.21) which corresponds to the case $B = 1/T_b$.

8.6 SERIAL (ABSOLUTELY FREQUENCY INVARIANT) STRUCTURES

8.6.1 Autocorrelation Demodulation (ACD) of PDM-2 Signals

When carrier frequency error is present, it is possible to design a differential detection-type receiver whose performance is *absolutely invariant* to this error in the sense that the signal component of the demodulator output is independent of the carrier frequency. There are several possible schemes for accomplishing this, all of which depend on encoding the input information phases as a second-order phase difference modulation (PDM-2) as in (8.3).

An ACD for binary PDM-2 signals in the presence of frequency error is illustrated in Fig. 8.10 (originally shown as Fig. 5 in [8]). This configuration is a hybrid of Figures 8.6 and 8.8 in the following sense. The first stage accomplishes inphase and quadrature ACD of the type illustrated in Fig. 8.6 to convert the frequency error Δf into an equivalent phase $\Delta f T_b$ appearing as $\cos \Delta f T_b$ and $\sin \Delta f T_b$ degradations at the I&D outputs. The second stage then is identical to Fig. 8.8 which, as before, is the optimum processing for signals with unknown but constant phase.

This scheme and others of its type are difficult to analyze because of the successive multiplications of signal plus noise at the first and second stage I and Q multipliers. An approximate analysis of the ACD in Fig. 8.10 is given in [10] by modeling the I&D outputs as Gaussian random variables similar to what was done in Section 7.4.1. In particular, if the decision algorithm is given by

$$
\hat{d}_i = \mathrm{sgn}\left\{ \int_{(i-1)T_b}^{iT_b} \hat{r}(t)\hat{r}(t-T_b)dt \int_{iT_b}^{(i+1)T_b} \hat{r}(t)\hat{r}(t-T_b)dt \right.
$$

$$
\left. + \int_{(i-1)T_b}^{iT_b} \hat{r}(t)\hat{r}_{90}(t-T_b)dt \int_{iT_b}^{(i+1)T_b} \hat{r}(t)\hat{r}_{90}(t-T_b)dt \right\} \tag{8.43}
$$

$$
= \mathrm{sgn}\left\{ V_{c,i-1}V_{c,i} + V_{s,i-1}V_{s,i} \right\}
$$

where $\hat{r}(t), \hat{r}(t - T_b)$ are modeled as in (8.13) and (8.14) and $\hat{r}_{90}(t - T_b)$ is the Hilbert transform of $\hat{r}(t - T_b)$, then the I&D outputs $V_{c,i}$, $V_{c,i-1}$, $V_{s,i}$, and $V_{s,i-1}$ are given by

$$
V_{c,i} = E_b \cos \Delta\theta_i \cos \alpha + N_1; \quad V_{c,i-1} = E_b \cos \Delta\theta_{i-1} \cos \alpha + N_2;
$$

$$
V_{s,i} = E_b \cos \Delta\theta_i \sin \alpha + N_3; \quad V_{s,i-1} = E_b \cos \Delta\theta_{i-1} \sin \alpha + N_4 \tag{8.44}
$$

where from (8.16) and analogous relations

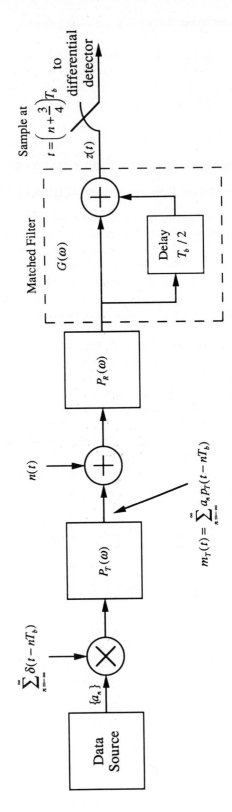

Figure 8.9b Equivalent mathematical model of the system in Figure 8.9a

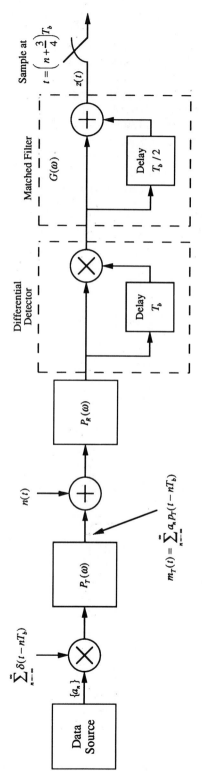

Figure 8.9c Block diagram of the system with differential detector prior to matched filter

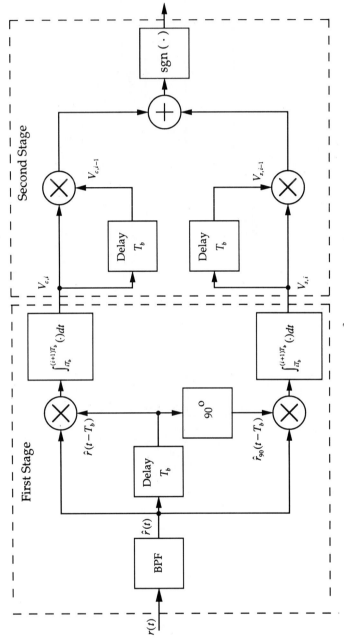

Figure 8.10 Autocorrelation demodulator of D^{2}PSK (PDM-2) signals (Reprinted from [2] © IEEE)

$$N_1 = \frac{AT_b}{2} n_{s2,m} \cos\theta_i + \frac{AT_b}{2} n_{s1,m} \cos\theta_{i-1} \cos\alpha$$

$$- \frac{AT_b}{2} n_{c1,m} \cos\theta_{i-1} \sin\alpha + \frac{T_b}{2} \sum_{k=k_1}^{k_2} \left(n_{s1,k} n_{s2,k} + n_{c1,k} n_{c2,k} \right)$$

$$N_2 = \frac{AT_b}{2} n_{s3,m} \cos\theta_{i-1} + \frac{AT_b}{2} n_{s2,m} \cos\theta_{i-2} \cos\alpha$$

$$- \frac{AT_b}{2} n_{c2,m} \cos\theta_{i-2} \sin\alpha + \frac{T_b}{2} \sum_{k=k_1}^{k_2} \left(n_{s2,k} n_{s3,k} + n_{c2,k} n_{c3,k} \right)$$

$$(8.45)$$

$$N_3 = \frac{AT_b}{2} n_{c2,m} \cos\theta_i + \frac{AT_b}{2} n_{s1,m} \cos\theta_{i-1} \sin\alpha$$

$$+ \frac{AT_b}{2} n_{c1,m} \cos\theta_{i-1} \cos\alpha + \frac{T_b}{2} \sum_{k=k_1}^{k_2} \left(n_{s1,k} n_{c2,k} - n_{c1,k} n_{s2,k} \right)$$

$$N_4 = \frac{AT_b}{2} n_{c3,m} \cos\theta_{i-1} + \frac{AT_b}{2} n_{s2,m} \cos\theta_{i-2} \sin\alpha$$

$$+ \frac{AT_b}{2} n_{c2,m} \cos\theta_{i-2} \cos\alpha + \frac{T_b}{2} \sum_{k=k_1}^{k_2} \left(n_{s2,k} n_{c3,k} - n_{c2,k} n_{s3,k} \right)$$

The noise random variables N_i; $i = 1, 2, 3, 4$ are all zero mean with variance [10]

$$\sigma_N^2 = E_b N_0 \left(1 + \frac{BT_b}{E_b/N_0} \right) \tag{8.46}$$

Furthermore, N_1 and N_3 are uncorrelated and likewise for N_2 and N_4. However, the other pairs are correlated with normalized correlation coefficients[6]

$$\rho_{12} \triangleq \frac{\overline{N_1 N_2}}{\sigma_N^2} = \frac{\overline{N_3 N_4}}{\sigma_N^2} = \frac{\cos\alpha \cos\left(\Delta^2 \theta_i \right)}{2 \left[1 + \frac{BT_b}{E_b/N_0} \right]},$$

$$(8.47)$$

$$\rho_{14} \triangleq \frac{\overline{N_1 N_4}}{\sigma_N^2} = \frac{\overline{N_2 N_3}}{\sigma_N^2} = \frac{\sin\alpha \cos\left(\Delta^2 \theta_i \right)}{2 \left[1 + \frac{BT_b}{E_b/N_0} \right]}$$

For $2BT_b \gg 1$ and E_b/N_0 not too large relative to $2BT_b$, N_1, N_2, N_3, and N_4 can be approximated by Gaussian random variables. Substituting (8.44) into (8.43) gives

$$\hat{d}_i = \text{sgn}\{ E_b^2 \cos\Delta\theta_i \cos\Delta\theta_{i-1} + N_1 E_b \cos\Delta\theta_{i-1} \cos\alpha + N_2 E_b \cos\Delta\theta_i \cos\alpha$$

$$+ N_3 E_b \cos\Delta\theta_{i-1} \sin\alpha + N_4 E_b \cos\Delta\theta_i \sin\alpha + N_1 N_2 + N_3 N_4 \} \tag{8.48}$$

$$= \text{sgn}\{ E_b^2 \cos\Delta^2\theta_i + E_b N_{S \times N} + N_{N \times N} \}$$

---

6. This result was first pointed out by Shouxing Qu while at Technical University of Nova Scotia in a private communication to the authors.

where $N_{S \times N}$ is a zero mean Gaussian random variable with variance $2\sigma_N^2$ and $N_{N \times N} = N_1 N_2 + N_3 N_4$. The error probability may be evaluated using Stein's approach [14]. In particular, it is straightforward to show that

$$P_b(E) = \frac{1}{2}\left[1 - Q\left(\sqrt{b}, \sqrt{a}\right) + Q\left(\sqrt{b}, \sqrt{a}\right)\right] - \frac{A}{2}\exp\left(-\frac{a+b}{2}\right)I_0(ab) \quad (8.49)$$

where

$$\begin{Bmatrix} a \\ b \end{Bmatrix} = \frac{E_b}{2N_0}\left(1 + \frac{BT_b}{E_b/N_0}\right)^{-1}\left\{\left(1 - \frac{1}{2}\sin^2\alpha\left(1 + \frac{BT_b}{E_b/N_0}\right)^{-2}\right)^{-1}\right.$$

$$\left. \mp \left(1 - \frac{1}{2}\sin^2\alpha\left(1 + \frac{BT_b}{E_b/N_0}\right)^{-2}\right)^{-1/2}\right\} \quad (8.50)$$

$$A = \frac{1}{2}\left(1 + \frac{BT_b}{E_b/N_0}\right)^{-1}\left(1 - \frac{1}{2}\sin^2\alpha\left(1 + \frac{BT_b}{E_b/N_0}\right)^{-2}\right)^{-1/2}\cos\alpha$$

For zero frequency offset ($\alpha = 0$), (8.49) simplifies to

$$P_b(E) = \frac{1}{2}\left[1 - \frac{1}{2}\left(1 + \frac{BT_b}{E_b/N_0}\right)^{-1}\right]\exp\left\{-\frac{E_b}{2N_0}\left(1 + \frac{BT_b}{E_b/N_0}\right)^{-1}\right\} \quad (8.51)$$

The result in (8.51) differs from that given in [8], namely

$$P_b(E) = \frac{1}{2}\exp\left\{-\frac{E_b}{2N_0}\left(1 + \frac{BT_b}{E_b/N_0}\right)^{-1}\right\} \quad (8.52)$$

since in (8.52) the correlations of (8.47) were not taken into account. It is felt that this difference between (8.51) and (8.52) will be small over the region of parameter values where the Gaussian assumption is valid.

It is difficult if not impossible to obtain an exact (even assuming zero ISI) expression for $P_b(E)$, analogous to (8.18) and (8.19), for the ACD scheme of Fig. 8.10. Thus, we cannot check the absolute validity of the Gaussian approximation results obtained in (8.49)–(8.51). However, based on the results for PDM-1 in Fig. 8.7, we intuitively expect that for sufficiently large BT_b, over a range of $P_b(E)$ values of practical interest, the above results can be used with a fair degree of accuracy. As a measure of some confidence, the authors in [10] do compare the Gaussian analysis results (not accounting for the correlations of (8.47), however) to those obtained from a computer simulation and show reasonable agreement (see Table 8.1).

The PDM-2 demodulator of Fig. 8.10 can be generalized to apply to M-ary ($M > 2$) D^2PSK (M-D^2PSK) with information phase angles β_m as defined in Section 8.1. The resulting configuration is illustrated in Fig. 8.11. Letting $V_i = V_{c,i} + jV_{s,i}$, then the appropriate decision rule is:

$$\text{Choose } \Delta^2\theta_i = \beta_{\hat{k}} \text{ where } \hat{k} = \max_k{}^{-1}\gamma_k \quad (8.53a)$$

Table 8.1 A Comparison of Simulation and Theoretical Results for
Error Probability Performance of ACD for Binary PDM-2 Signals

FT_b	E_b/N_0 (dB)	$P_b(E)$ Simulation Results	Eq. (8.49)
2	8	4.03×10^{-2}	3.28×10^{-2}
2	10	7.67×10^{-3}	5.31×10^{-3}
2	12	6.37×10^{-4}	2.90×10^{-4}
4	8	5.61×10^{-2}	4.56×10^{-2}
4	10	1.10×10^{-2}	7.75×10^{-3}
4	12	9.08×10^{-4}	4.39×10^{-4}
8	8	8.26×10^{-2}	7.25×10^{-2}
8	10	1.85×10^{-2}	1.41×10^{-2}
8	12	1.64×10^{-3}	8.93×10^{-4}
16	8	1.37×10^{-1}	1.24×10^{-1}
16	10	3.80×10^{-2}	3.11×10^{-2}
16	12	4.02×10^{-3}	2.58×10^{-3}

with

$$\gamma_k = \mathrm{Re}\left\{\xi_k\right\} = \mathrm{Re}\left\{V_i V_{i-1}^* e^{-j\beta_k}\right\}$$

$$= (V_{c,i} V_{c,i-1} + V_{s,i} V_{s,i-1})\cos\beta_k + (V_{c,i} V_{s,i-1} - V_{s,i} V_{c,i-1})\sin\beta_k \qquad (8.53b)$$

Since the exact symbol error probability performance of differentially detected PDM-1 modulation with M signals and no frequency offset is given by (7.7), then by analogy with the relation of (8.53) to the exact bit error probability when $M = 2$, that is, (8.20), the symbol error probability performance of Fig. 8.11 in the presence of frequency offset can be approximated by

$$P_s(E) = \frac{\sin\frac{\pi}{M}}{\pi}\int_0^{\pi/2}\frac{\exp\left\{-\frac{E_s}{2N_0}\left(1 + \frac{BT_s}{E_s/N_0}\right)^{-1}\left[1 - \cos\frac{\pi}{M}\cos\phi\right]\right\}}{1 - \cos\frac{\pi}{M}\cos\phi}\,d\phi \qquad (8.54)$$

In (8.54), E_s denotes symbol energy and T_s denotes symbol time. Furthermore, assuming a Gray code bit-to-symbol mapping, the bit error probability, $P_b(E)$, would approximately be given in terms of the symbol error probability by (4.135).

Finally, in Chapter 7 we also presented an easy-to-evaluate asymptotic (large E_s/N_0) approximation to the symbol error probability performance of M-DPSK, namely, (7.12). For the PDM-2 demodulator of Fig. 8.11, this would become

$$P_s(E) \cong \sqrt{\frac{1 + \cos\frac{\pi}{M}}{2\cos\frac{\pi}{M}}}\left[\frac{1}{2}\,\mathrm{erfc}\,\sqrt{\frac{E_s}{2N_0}\left(1 + \frac{BT_s}{E_s/N_0}\right)^{-1}\left(1 - \cos\frac{\pi}{M}\right)}\right]; \quad M > 3 \; (8.55)$$

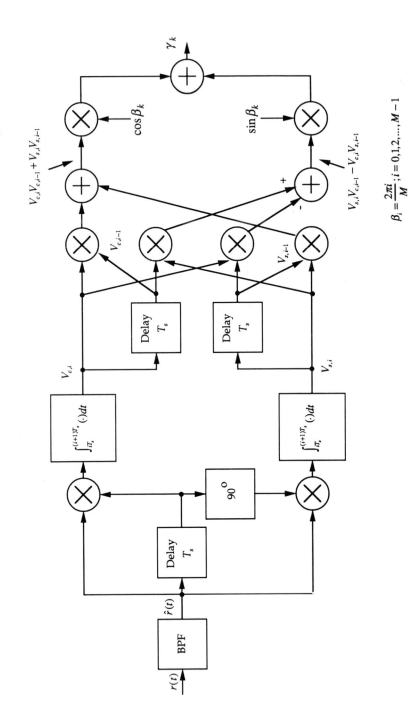

Figure 8.11 Autocorrelation demodulator of M-D²PSK (PDM-2) signals (Reprinted from [2]) © IEEE)

$$\beta_i = \frac{2\pi i}{M}; i = 0,1,2,...,M-1$$

8.6.2 I-Q Demodulation of PDM-2 Signals

An I-Q demodulator for PDM-2 signals is illustrated in Fig. 8.12. An equivalent[7] demodulator [3] which performs the double differencing operation directly on the detected phase angle rather than working with the sine and cosine projections of it is illustrated in Fig. 8.13. The exact error probability performance of this implementation[8] specialized to the case of binary D^2PSK was first obtained in [8] and [13] under the assumption of zero frequency error with the result[9]

$$
P_b(E) = \frac{1}{2} - \frac{\sqrt{\pi}}{4} \left(\frac{E_b}{N_0} \right)^{3/2} \exp \left(-\frac{3}{2} \frac{E_b}{N_0} \right)
$$
$$
\times \sum_{m=0}^{\infty} \frac{(-1)^m}{2m+1} \left\{ I_m \left(\frac{E_b}{2N_0} \right) + I_{m+1} \left(\frac{E_b}{2N_0} \right) \right\}^2 \tag{8.56}
$$
$$
\times \left\{ I_{2m+1/2} \left(\frac{E_b}{2N_0} \right) + I_{2m+1+1/2} \left(\frac{E_b}{2N_0} \right) \right\}
$$

It is also shown in [8] that (8.56) represents a good lower bound on the performance of the ACD of Fig. 8.11 in the presence of frequency offset when $2BT_b$ is small.

For M-DPSK ($M > 2$), an exact expression cannot be found. However, a union bound on symbol error probability $P_s(E)$ in the form of (8.56) can be obtained with the result [13]

$$
P_s(E) \le 1 - \frac{\sqrt{\pi}}{2} \left(\frac{E_s}{N_0} \right)^{3/2} \exp \left(-\frac{3}{2} \frac{E_s}{N_0} \right)
$$
$$
\times \sum_{m=0}^{\infty} \frac{(-1)^m}{2m+1} \left\{ I_m \left(\frac{E_s}{2N_0} \right) + I_{m+1} \left(\frac{E_s}{2N_0} \right) \right\}^2 \tag{8.57}
$$
$$
\times \left\{ I_{2m+1/2} \left(\frac{E_s}{2N_0} \right) + I_{2m+1+1/2} \left(\frac{E_s}{2N_0} \right) \right\} \cos \left\{ (2m+1) \frac{\pi}{2} \left(\frac{M-2}{M} \right) \right\}
$$

The bit error probability upper bound can then be obtained from (4.140) and (8.57).

The exact error probability performance of Fig. 8.13 (specialized to the binary D^2PSK case) in the presence of frequency offset was first carried out in [3] using a geometric approach analogous to that taken by Arthurs and Dym [20] in evaluating the performance of DPSK. The result obtained in [3] is in the form of a double integral and is given by

---

7. Figures 8.12 and 8.13 are equivalent (produce the same decision algorithm) in the same sense that their counterparts for PDM-1 modulation, namely, Figs. 7.1 and 7.2a, are equivalent [19].

8. The actual configuration analyzed in [8] and [13] replaced the I&D's with distortion-free (zero ISI) lowpass filters whose outputs were sampled once per bit. However, due to the zero ISI assumption, the error probability performance results obtained there as a function of the lowpass filter output SNR, $\eta = A^2/\sigma^2$ apply directly to Fig. 8.13 if η is replaced by E_b/N_0.

9. There is a discrepancy between the error probability results obtained in [8] and [13]. Based on a derivation of (8.56) supplied to the authors by Mario Pent, we have concluded that the result in [13] is indeed the correct one and this is the one presented here in (8.56).

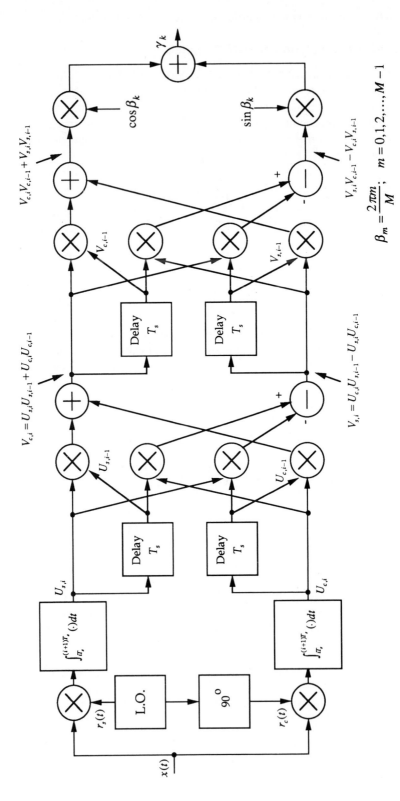

Figure 8.12 I-Q demodulator of M-D^{2}PSK (PDM-2) signals (Reprinted from [2] © IEEE)

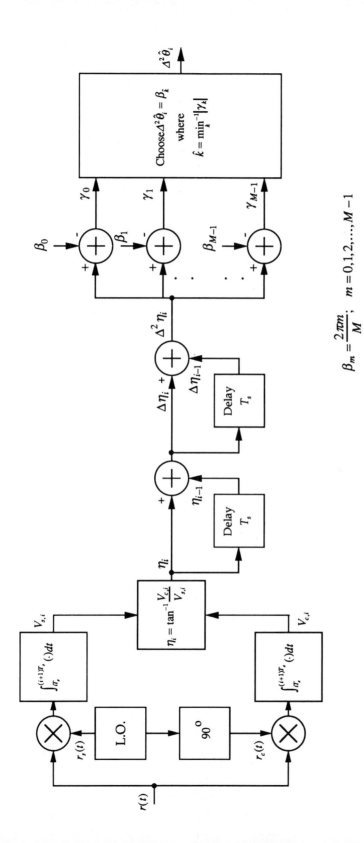

Figure 8.13 Alternate implementation of I-Q demodulator of M-D^{2}PSK (PDM-2) signals (Reprinted from [3] © IEEE)

$$\beta_m = \frac{2\pi m}{M}; \quad m = 0, 1, 2, \ldots, M-1$$

$$P_b(E) = \frac{1}{2} - \frac{E'_b}{2\pi N_0}$$

$$\times \int_0^\pi \int_0^\pi \text{erf}\left(\sqrt{\frac{E'_b}{N_0}} \cos(2\eta_2 - \eta_1)\right) \text{erf}\left(\sqrt{\frac{E'_b}{N_0}} \cos\eta_2\right) \tag{8.58}$$

$$\times \cos\eta_1 \cos\eta_2 \exp\left(-E'_b\left[\sin^2\eta_1 + \sin^2\eta_2\right]\right) d\eta_1 d\eta_2$$

with E'_b defined as before in (8.31). Upon careful examination of the mathematical model used in arriving at (8.56) in the absence of frequency offset and the characterization of the I&D outputs used to obtain (8.58) in the presence of frequency offset, it becomes readily apparent that the two results can be made equivalent if in (8.56) E_b is replaced by E'_b, that is,

$$P_b(E) = \frac{1}{2} - \frac{\sqrt{\pi}}{4}\left(\frac{E'_b}{N_0}\right)^{3/2}$$

$$\times \exp\left(-\frac{3}{2}\frac{E'_b}{N_0}\right) \sum_{m=0}^\infty \frac{(-1)^m}{2m+1}\left\{I_m\left(\frac{E'_b}{2N_0}\right) + I_{m+1}\left(\frac{E'_b}{2N_0}\right)\right\}^2 \tag{8.59}$$

$$\times \left\{I_{2m+1/2}\left(\frac{E'_b}{2N_0}\right) + I_{2m+1+1/2}\left(\frac{E'_b}{2N_0}\right)\right\}$$

Indeed, the equivalence between (8.58) and (8.59) can be formally derived. Thus, we conclude that (8.59) is also an exact result for the error probability performance of Fig. 8.12 or Fig. 8.13 specialized to the binary ($M = 2$) case.

Figure 8.14 illustrates the error probability as given by either (8.56) or (8.58) for the special case of no frequency error $(E'_b = E_b)$. Also plotted in this figure for the purpose of comparison are the corresponding results for coherent BPSK and DPSK as given by (4.55) and (8.20) respectively. We observe that using D^{2}PSK instead of DPSK requires approximately 4 dB more E_b/N_0 and thus a sizable penalty is payed for using a double differential scheme when indeed no frequency offset is present. Figure 8.15 illustrates (8.58) or (8.59) for various normalized frequency offsets $\Delta f T_b$. Also shown for comparison are the corresponding results for DPSK with frequency offset as given by any of the forms (8.29)–(8.31), (8.32), (8.34), or (8.35). Here we see that the D^{2}PSK scheme is quite insensitive to frequency offsets over a large range of values, whereas the DPSK performance degrades rapidly. Thus, it is natural to ask: At what value of $\Delta f T_b$ is D^{2}PSK preferable to DPSK, that is, where do the performance curves of the two intersect? The answer to this question was obtained in [3] and is illustrated in Fig. 8.16 where the values of $\Delta f T_b$ at which $P_b(E)|_{\text{DPSK}} = P_b(E)|_{\text{D}^2\text{PSK}}$ are plotted versus E_b/N_0 in dB. The region above the curve corresponds to the situation where D^{2}PSK would be preferable over DPSK and vice versa for the region below the curve. Thus, for a given E_b/N_0, Fig. 8.16 offers one an idea of how large Δf must be relative to the data rate $1/T_b$ before having to resort to a double differentially coherent detection scheme such as Fig. 8.12 or 8.13.

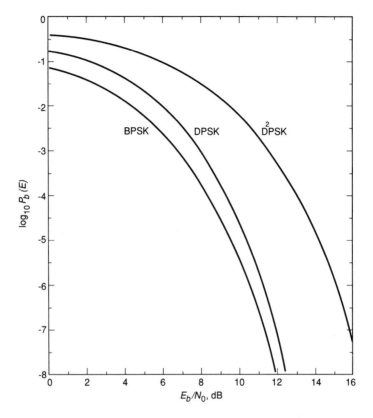

Figure 8.14   A comparison of the error probability performances of BPSK, DPSK, and D^{2}PSK for the case of zero frequency offset (Reprinted from [3] © IEEE)

Before moving on to double differentially coherent detection schemes that make use of longer (than three symbols) observation intervals, we wish to point out that all of the above techniques can readily be extended to n^{th} order differential detection of PDM-n modulation. For example, the scheme of Fig. 8.13 would employ n (rather than two as shown) differential detectors with a corresponding error probability performance given by [3]

$$P_b(E) = \frac{1}{2} \int_{-\pi}^{\pi} \cdots \int_{-\pi}^{\pi} \text{erfc} \left(\sqrt{\frac{E_b'}{N_0}} \cos \left(\eta_n - \Delta^n \eta_{n+1} - \frac{2n+1}{2} \alpha \right) \right)$$

$$\times \prod_{i=1}^{n} f\left(\eta_i \mid 0 \right) d\eta_i$$

(8.60)

where $\Delta^n \eta_{n+1}$ is the nth order differential operator applied to η_{n+1} as defined in (8.5) and

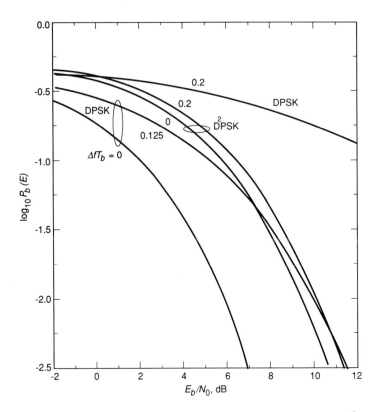

Figure 8.15 A comparison of the error probability performances of DPSK and D²PSK in the presence of frequency offset (Reprinted from [3] © IEEE)

$$
f(\eta_i \mid 0) = \frac{1}{2\pi} \exp\left(-\frac{E_b'}{N_0}\right) + \frac{1}{2}\sqrt{\frac{E_b'}{\pi N_0}} \cos\left(\eta_i - \frac{2i-1}{2}\alpha\right)
$$

$$
\times \exp\left(-\frac{E_b'}{N_0} \sin^2\left(\eta_i - \frac{2i-1}{2}\alpha\right)\right) \tag{8.61}
$$

$$
\times \left[1 + \operatorname{erf}\left(\sqrt{\frac{E_b'}{N_0}} \cos\left(\eta_i - \frac{2i-1}{2}\alpha\right)\right)\right]
$$

is the pdf of η_i given that the message selected for transmission corresponds to zero phase. This result can be evaluated numerically for any n and reduces to (8.32) for $n = 1$ and (8.57) for $n = 2$.

8.6.3 Multiple Symbol Differential Detection of PDM-2 Signals

In Chapter 7, we introduced the notion of *multiple symbol differential detection* [21] which used a multiple (more than two) symbol observation interval for improving the performance of PDM-1 signals relative to that obtained with classical differential detection. Specifically, a metric was derived that made a maximum-likelihood sequence estimate of $N_s - 1$ M-PSK symbols based on an observation over N_s symbol intervals. For $N_s = 2$, it was shown that

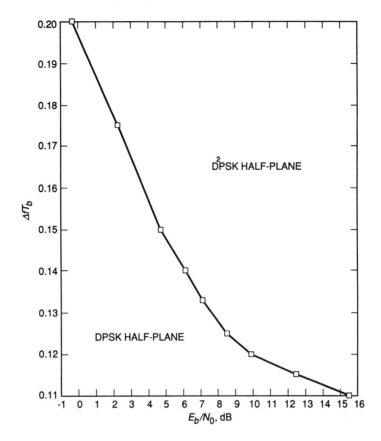

Figure 8.16   Frequency offsets that yield $P_b(E)|_{\text{DPSK}} = P_b(E)|_{\text{D}^2\text{PSK}}$ (Reprinted from [3] © IEEE)

this metric reduced to the classical differential detection algorithm whose performance is given by (7.14) for $M = 2$ or (7.7) for $M > 2$.

It is of interest here to extend these notions to PDM-2 signals. Since the first stage of a PDM-2 demodulator (Figures 8.10–8.13) in effect converts the frequency error Δf to an unknown phase shift $2\pi \Delta f T_s$ for processing by the second stage, then, in principle, one should be able to improve performance by replacing the second stage of the detector with a multiple symbol detection algorithm analogous to that considered for PDM-1 signals in Chapter 7. It is important to emphasize that the resulting total configuration is not truly maximum-likelihood; however, conditioned on the first stage being implemented in the classical way for PDM-1 signals, the remaining portion of the receiver is maximum-likelihood as before.

Mathematically speaking, applying multiple symbol differential detection to the ACD in Fig. 8.11, the decision rule of (8.53) generalizes as follows. We assume that the input phase $\Delta^2\theta_{i-N_s+1}$ now acts as a phase reference for the *sequence* $\boldsymbol{\Delta}^2\boldsymbol{\theta} = (\Delta^2\theta_{i-N_s+2}, \Delta^2\theta_{i-N_s+3}, \dots, \Delta^2\theta_{i-1}, \Delta^2\theta_i)$. Then, second order differentially encoding these phases for transmission over the channel, the appropriate sequence decision rule is:

$$\text{Choose } \boldsymbol{\Delta}^2\boldsymbol{\theta} = \boldsymbol{\Delta}^2\hat{\boldsymbol{\theta}} \qquad (8.62a)$$

where $\Delta^2\hat{\boldsymbol{\theta}}$ is the particular vector, with components ranging over the allowable set $\{\beta_m = 2\pi m/M; m = 0, 1, 2, \ldots, M-1\}$, that maximizes

$$\eta = \left| V_{i-N_s+1} + \sum_{k=0}^{N_s-2} V_{i-k}e^{-j\sum_{m=0}^{N_s-k-2}\Delta^2\hat{\theta}_{i-k-m}} \right|^2 \tag{8.62b}$$

For example, for $N_s = 3$ (one additional symbol observation), (8.62) becomes

Choose $(\Delta^2\theta_{i-1}, \Delta^2\theta_i) = (\Delta^2\hat{\theta}_{i-1}, \Delta^2\hat{\theta}_i)$ that maximizes

$$\eta = \left| V_{i-2} + V_ie^{-j(\Delta^2\hat{\theta}_i+\Delta^2\hat{\theta}_{i-1})} + V_{i-1}e^{-j\Delta^2\hat{\theta}_{i-1}} \right|^2$$

$$= |V_{i-2}|^2 + |V_{i-1}|^2 + |V_i|^2 + 2\operatorname{Re}\left\{ V_iV_{i-2}^*e^{-j(\Delta^2\hat{\theta}_i+\Delta^2\hat{\theta}_{i-1})} \right\} \tag{8.63a}$$

$$+ 2\operatorname{Re}\left\{ V_{i-1}V_{i-2}^*e^{-j\Delta^2\hat{\theta}_{i-1}} \right\} + 2\operatorname{Re}\left\{ V_iV_{i-1}^*e^{-j\Delta^2\hat{\theta}_i} \right\}$$

or, equivalently

Choose $\Delta^2\hat{\theta}_i$ and $\Delta^2\hat{\theta}_{i-1}$ if

$$\tag{8.63b}$$
$$\operatorname{Re}\left\{ V_iV_{i-1}^*e^{-j\Delta^2\hat{\theta}_i} + V_{i-1}V_{i-2}^*e^{-j\Delta^2\hat{\theta}_{i-1}} + V_iV_{i-2}^*e^{-j(\Delta^2\hat{\theta}_i+\Delta^2\hat{\theta}_{i-1})} \right\} \text{ is maximum}$$

An implementation of (8.63b) in complex form is illustrated in Fig. 8.17. It is to be emphasized that the enhancement of Fig. 8.11 with the multiple symbol detection algorithm (as in Fig. 8.17) does not change its status as a demodulator that is absolutely invariant to frequency error.

To determine the performance of multiple symbol detection of PDM-2 signals in the presence of frequency error, we proceed along the lines of the approximate (Gaussian) analysis leading to (8.52) mixed with the approach taken in [21]. In particular, noting that (8.52) is identical to (8.20) if E_b/N_0 is replaced by the "effective E_b/N_0" defined by

$$\left(\frac{E_b}{N_0} \right)_{\text{eff}} = \left(\frac{E_b}{2N_0} \right)\left(1 + \frac{BT_b}{E_b/N_0} \right)^{-1} \tag{8.64}$$

then for large $2BT_b$ and binary PDM-2 signals, we have from (7.85) that

$$P_b(E) \lesssim \frac{2}{\sqrt{\pi\left(\frac{E_b}{N_0}\right)_{\text{eff}}}} \left(\sqrt{\frac{N_s-1}{N_s-2}} \right)\left[\frac{1}{2}\exp\left\{ -\left(\frac{E_b}{N_0}\right)_{\text{eff}} \right\} \right] \tag{8.65}$$

where again $N_s - 1$ denotes the number of simultaneous decisions being made. As $N_s \to \infty$, (8.65) approaches

$$P_b(E) \lesssim \frac{1}{\sqrt{\pi\left(\frac{E_b}{N_0}\right)_{\text{eff}}}} \exp\left\{ -\left(\frac{E_b}{N_0}\right)_{\text{eff}} \right\} \tag{8.66}$$

which, using the asymptotic expansion of the complementary error function [see (7.87)],

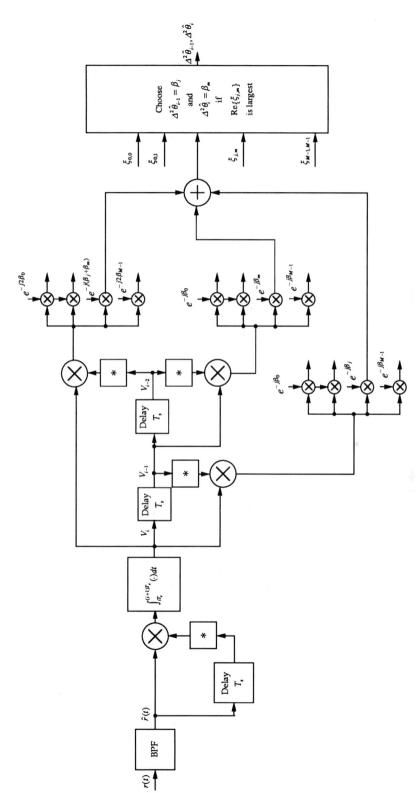

Figure 8.17 Complex form of autocorrelation demodulator of M-D^{2}PSK (PDM-2) signals with multiple symbol detection ($N_s = 3$)

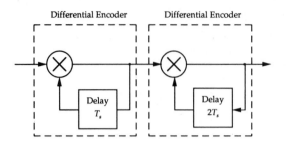

Figure 8.18 Baseband implementation of modified transmitter in complex form

becomes

$$P_b(E) \lesssim \text{erfc} \sqrt{\left(\frac{E_b}{N_0}\right)_{\text{eff}}} \tag{8.67}$$

Thus, in the limit as the observation interval (decision memory) approaches infinity, the large $2BT_b$, large SNR behavior of the performance of an ACD demodulator employing multiple symbol differential detection of PDM-2 signals in the presence of frequency offset approaches that of ideal coherent detection of BPSK (with differential encoding) with an effective E_b/N_0 given by (8.64).

The above can be extended to M-ary PSK, again using the results of Chapter 7 together with (8.64). Analogous to the relation between (8.65) and (4.206) as $N_s \to \infty$, it is straightforward to show that in the limit as the observation interval (decision memory) approaches infinity, the large $2BT_b$, large SNR performance of an ACD demodulator employing multiple symbol differential detection of M-ary D^{2}PSK signals in the presence of frequency offset approaches that of ideal coherent detection of M-PSK (with differential encoding) with an effective E_s/N_0 given in terms of E_s/N_0 by a relation identical to (8.64).

8.6.4 Further Enhancements

The conceptual double differential detection scheme illustrated in Figures 8.1a and 8.5 can be enhanced (i.e., its performance improved) by choosing the delay in the second stage of the transmitter and correspondingly the delay in the first stage of the receiver to be $2T_s$ rather than T_s (see Figures 8.18 and 8.19). The reason for doing this is to eliminate the high noise correlation in the receiver output caused by the sharing of a common noise sample between the two stages of the receiver as in Fig. 8.5. This can be noted in Fig. 8.19 where the receiver output signal consists of four different noise samples each spaced T_s-sec apart. Despite the fact that the transmitter no longer implements a true second-order difference operation, the receiver output is still independent of the frequency offset present in the received input and as such Fig. 8.19 generically represents an implementation of an absolutely frequency invariant demodulator.

To see the amount of improvement in performance gained by the above modification, consider Fig. 8.12 or 8.13 (assuming binary modulation, i.e., $M = 2$, $T_s = T_b$) where the T_b-sec delays in the first stage are now $2T_b$-sec in duration. The exact performance for this modified implementation of Fig. 8.12 can be obtained by the method employed in [13] with the result

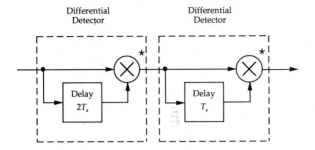

Differential
Detector

Differential
Detector

Delay
$2T_s$

Delay
T_s

(asterisk on multiplier denotes conjugate multiplication)

Figure 8.19   Baseband implementation of
modified receiver in complex form

At output of second differential detector, *the Doppler has been removed*

$$P_b(E) = \frac{1}{2} - \frac{\pi}{8}\left(\frac{E_b'}{N_0}\right)^2 \exp\left(-\frac{2E_b'}{N_0}\right)$$

$$\times \sum_{m=0}^{\infty} \frac{(-1)^m}{2m+1}\left\{I_m\left(\frac{E_b'}{2N_0}\right) + I_{m+1}\left(\frac{E_b'}{2N_0}\right)\right\}^4$$

(8.68)

Alternately, using the geometric approach taken in [3], the bit error probability can be expressed in integral form as

$$P_b(E) = \frac{1}{2} - \frac{1}{2}\left(\frac{E_b'}{\pi N_0}\right)^{3/2} \int_0^\pi \int_0^\pi \int_0^\pi \mathrm{erf}\left(\sqrt{\frac{E_b'}{N_0}}\cos(\eta_3 + \eta_2 - \eta_1)\right)\cos\eta_1 \cos\eta_2 \cos\eta_3$$

$$\times \exp\left\{-\frac{E_b'}{N_0}\left(\sin^2\eta_1 + \sin^2\eta_2 + \sin^2\eta_3\right)\right\}\left[1 + \mathrm{erf}\left(\sqrt{\frac{E_b'}{N_0}}\cos\eta_3\right)\right]d\eta_1 d\eta_2 d\eta_3$$

(8.69)

Figure 8.20 plots $P_b(E)$ of (8.59) and (8.68) versus E_b'/N_0 in dB. The large improvement of the configuration using a combination of T_b-sec and $2T_b$-sec delays over that using T_b-sec delays in both stages is clearly evident. The same kind of improvement in performance can be obtained with an ACD, for example, modifying the autocorrelator (first stage) of Fig. 8.11 to have a $2T_b$-sec delay. Figure 8.21 illustrates simulation results for the error probability performance of Fig. 8.11 and its modification as per the above. Over a wide range of values of $2BT_b$, the advantage of a T_b, $2T_b$ delay combination is strongly demonstrated.

Finally, for $M > 2$, a union bound on the symbol error probability of the modified version of Fig. 8.12 described above can be obtained in a form analogous to (8.57) with the result

$$P_s(E) \le 1 - \frac{\pi}{4}\left(\frac{E_s'}{N_0}\right)^2 \exp\left(-\frac{2E_s'}{N_0}\right)\sum_{m=0}^{\infty}\frac{(-1)^m}{2m+1}\left\{I_m\left(\frac{E_s'}{2N_0}\right) + I_{m+1}\left(\frac{E_s'}{2N_0}\right)\right\}^4$$

$$\times \cos\left\{(2m+1)\frac{\pi}{2}\left(\frac{M-2}{M}\right)\right\}$$

(8.70)

The structures of Figures 8.18 and 8.19 are actually special cases of a form of D²PSK

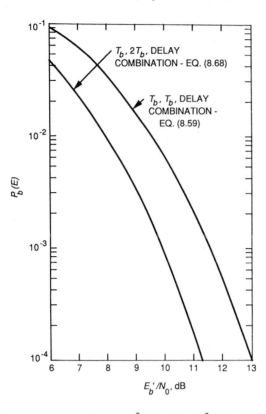

Figure 8.20 Evaluation of Equations (8.59) and (8.68) as a function of E_b/N_0 in dB (Reprinted from [2] © IEEE)

referred to as nonadjacent D^2PSK (NA-D^2PSK) [22]. To see this, we recognize that, analogous to (8.3), the input, $\Delta_1^2\theta_i$, of the first differential encoder in Fig. 8.18 can be written in terms of the transmitted phase, θ_i, as $\Delta_1^2\theta_i = (\theta_i - \theta_{i-2}) - (\theta_{i-1} - \theta_{i-3}) = (\theta_i - \theta_{i-1}) - (\theta_{i-2} - \theta_{i-3}) = \Delta\theta_i - \Delta\theta_{i-2}$, which is recognized as the difference of two first-order phase differences that are separated from each other by a single symbol interval. In the more general case, if one replaces the $2T_s$ delays in Figures 8.18 and 8.19 by LT_s delays ($L \geq 2$), then the input, $\Delta_{L-1}^2\theta_i$, of the first differential encoder and the transmitted phase, θ_i, are related by $\Delta_{L-1}^2\theta_i = (\theta_i - \theta_{i-1}) - (\theta_{i-L} - \theta_{i-L-1}) = \Delta\theta_i - \Delta\theta_{i-L}$. Here the two first order phase differences are separated by $(L-1)$ symbol intervals. With reference to Fig. 8.13 (appropriately modified by replacing the first T_s-sec delay by an LT_s-sec delay), a decision on $\Delta_{L-1}^2\theta_i$ would be made by comparing the phase difference $\Delta_{L-1}^2\eta_i = (\eta_i - \eta_{i-1}) - (\eta_{i-L} - \eta_{i-L-1}) = \Delta\eta_i - \Delta\eta_{i-L}$ with all possible transmitted phases around the circle and choosing the one that is closest. Since for any $L \geq 2$, the random variables $\Delta\eta_i$ and $\Delta\eta_{i-L}$ are statistically dependent, then from the standpoint of error probability performance on the AWGN, there is nothing to be gained[10] by choosing a value of L larger than 2; specifically, the bit error probability performance of NA-D^2PSK is given by (8.68) or (8.69) for all $L \geq 2$. Moreover, Qu [22] shows that asymptotically as E_b/N_0 becomes large, the difference in error probability performance between D^2PSK ($L = 1$) and NA-D^2PSK ($L \geq 2$)

---

10. For fast fading channels, it is shown in [22] that the correlation of the fading process can be utilized to improve error probability performance by choosing L larger than 2. In fact, depending on the fading channel model, e.g., satellite-aircraft, land-mobile, etc., an optimum (in the sense of minimum error probability) value of L can be determined.

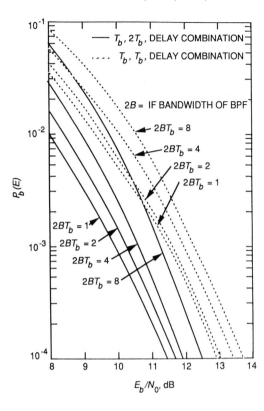

Figure 8.21 Computer simulation results for the error probability performance of Figure 8.10 and its modified version (Reprinted from [2] © IEEE)

is attributed entirely to the removal by the latter of the correlation between the two first-order phase differences making up the decision variable. Specifically, the asymptotic E_b/N_0 degradation of D^{2}PSK relative to NA-D^{2}PSK is given by

$$\Delta E_b/N_0 \ (dB) = 10 \log_{10} \frac{\left(1 - \rho_{\Delta \eta}\right)\big|_{\text{NA}-\text{D}^2\text{PSK}}}{\left(1 - \rho_{\Delta \eta}\right)\big|_{\text{D}^2\text{PSK}}} \tag{8.71}$$

where

$$\rho_{\Delta \eta} \triangleq \frac{E\left\{\Delta \eta_i \Delta \eta_{i-L}\right\}}{\sigma_{\Delta \eta}^2} \tag{8.72}$$

is the normalized correlation between the two first order phase differences making up the decision variable. Since for NA-D^{2}PSK ($L \geq 2$), $\rho_{\Delta \eta} = 0$, whereas for D^{2}PSK ($L = 1$)

$$\rho_{\Delta \eta} = \frac{E\left\{(\eta_i - \eta_{i-1})(\eta_{i-1} - \eta_{i-2})\right\}}{2\sigma_\eta^2} = \frac{E\left\{-\eta_{i-1}^2\right\}}{2\sigma_\eta^2} = -\frac{1}{2} \tag{8.73}$$

then from (8.71) we find that $\Delta E_b/N_0 \ (dB) = 10 \log_{10}(1.5) = 1.77$ dB, which is in full agreement with the numerical results of Fig. 8.20. The asymptotic E_b/N_0 degradation of D^{2}PSK relative to NA-D^{2}PSK as given by (8.71) is also valid for $M > 2$ [22].

8.7 PARALLEL (RELATIVELY FREQUENCY INVARIANT) STRUCTURES— MULTIPHASE ACD

In an effort to improve the noise immunity of autocorrelation demodulators of PDM-2 signals with frequency error, while maintaining to some degree their invariance with respect to the actual signal frequency itself, attention is now directed toward a class of *relatively invariant* ACD's. These ACD's process the received PDM-2 signal plus noise in a parallel bank of L ACD's similar to Fig. 8.6, each of which now introduces an appropriate phase shift $\beta_l = \pi(l - 1)/L$; $l = 1, 2, \ldots, L$ between the direct and delayed signals before multiplication (hence the name *multiphase ACD*). The purpose of this phase shift is to try to match the ACD to the equivalent phase shift $\alpha = \Delta\omega T_b$ introduced by the frequency error in the input.

One configuration of this type is illustrated in Fig. 8.22 and was first suggested in [4, 7]. The outputs $V_i^{(l)}$; $l = 1, 2, \ldots, L$ from the I&D's corresponding to the ith transmission interval are squared[11] (to remove the modulation) and compared to determine the maximum. Letting $\hat{l}$ denote the value of l corresponding to the maximum $\left(V_i^{(l)}\right)^2$, then only the $\hat{l}^{\text{th}}$ switch is closed and as such the $\hat{l}^{\text{th}}$ ACD output, namely, $d_i^{(\hat{l})}$ (which represents a hard decision on the first-order phase difference $\Delta\theta_i = \theta_i - \theta_{i-1}$) is passed on to the differential decoder which performs the second stage of differential detection. Note that the range of values for the L phase shifters in Fig. 8.22 only covers the interval $(0, \pi)$ [rather than $(0, 2\pi)$ as one might first expect]. The reason for this is that a frequency error-induced phase shift of α in the region $(\pi, 2\pi)$ will result in a 180° phase shift at the I&D output of the ACD "matched" to the phase shift $\alpha - \pi$. However, since the "matched" ACD is determined by the *square* of the I&D output, the ACD corresponding to the phase shift $\alpha - \pi$ will ideally be selected. Furthermore, the 180° phase shift at the I&D output of this "matched" ACD, which results in an inversion of the hard decisions at the "sgn" device output, will be resolved by the differential decoder.

Intuitively, one would anticipate that the above scheme would yield an improvement in noise immunity over the absolutely invariant type of ACD since: (1) first stage demodulation is achieved by only a single ACD that is "matched" (within the resolution of the number of ACD's used in the bank) to the input frequency error; and (2) the second phase difference is calculated from the adjacent hard decision outputs of the first stage as is done in coherent detection with differentially encoded input symbols. As such, we would expect that as the number, L, of ACD's in the bank increases and the probability of correctly selecting the one that matches the input approaches unity, the noise immunity of Fig. 8.22 should approach (within approximately a factor of two due to the additional differential encoding operation) that of an ACD of PDM-1 signals with no frequency error (Fig. 8.6). Practically speaking, it has been shown [7] that a bank of $L = 6$ ACD's gives performance quite close to the above limit. This and other conclusions will be demonstrated by the results of the analysis that follows.

Assume that the signal of (8.12) is received (in noise) where θ_i is now the double differentially encoded information phase in the ith transmission interval. The output of the multipler in the lth ACD is (in the absence of noise)

$$e_l(t) = A^2 \cos\left(\Delta\omega T_b + \theta_i - \theta_{i-1} - \beta_l\right) = A^2 \cos\left(\Delta\theta_i\right)\cos\left(\alpha - \beta_l\right) \qquad (8.74)$$

---

11. Equally, one could replace the square-law devices with absolute-value devices (full-wave rectifiers). The choice between the two would depend on the SNR as is typical in situations of this type.

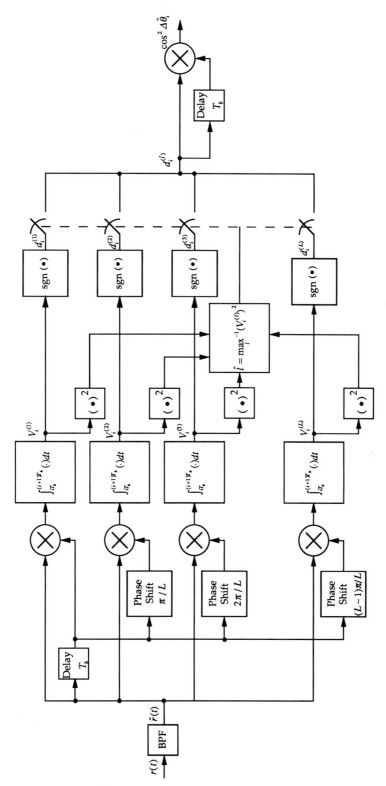

Figure 8.22 A multiphase autocorrelation demodulator of M-D^{2}PSK (PDM-2) signals

where we have again set $\alpha = \Delta\omega T_b$ and recognized the fact that for binary modulation, $\Delta\theta_i = \theta_i - \theta_{i-1}$ only takes on values 0 and π. After passing through the I&D, we get the (noise-free) decision variable

$$V_i^{(l)} = A^2 T_b \cos(\Delta\theta_i) \cos(\alpha - \beta_l) = E_b \cos(\Delta\theta_i) \cos\Phi_l \tag{8.75}$$

where $\Phi_l = \alpha - \beta_l$ denotes the phase error associated with the mismatch of the lth ACD to the input signal. Assuming that based on a comparison of the squared values of $V_i^{(l)}$ the correct ACD (the $\hat{l}^{\text{th}}$ corresponding to the one whose value of Φ_l is minimum) is always chosen,[12] then the hard decision outputs from the correct ACD are described by

$$\begin{aligned} d_i^{(\hat{l})} &= \text{sgn}\left\{ A^2 T_b \cos(\Delta\theta_i) \cos(\alpha - \beta_{\hat{l}}) + N_i^{(\hat{l})} \right\} \\ &= \text{sgn}\left\{ E \cos\Phi_{\hat{l}} \cos(\Delta\theta_i) + N_i^{(\hat{l})} \right\} \end{aligned} \tag{8.76}$$

where $N_i^{(\hat{l})}$ is a noise variable that accounts for $S \times N$ and $N \times N$ degradations. More important, however, is the fact that the correct choice among L ACD's reduces the range of the phase error, $\Phi_{\hat{l}}$, to the interval $(-\pi/2L, \pi/2L)$. As such, the I&D output signal for the correctly selected ACD will dip by no more than a factor of $\cos\pi/2L$ relative to that of the same ACD of PDM-1 signals in the absence of frequency error. For example, for $L = 6$, this dip is less than 4%.

The error probability associated with the hard decisions $d_i^{(\hat{l})}$ is obtained by noting that (8.76) represents the decisions in Fig. 8.6 if the energy E_b is replaced by $E_b \cos\Phi_{\hat{l}}$. As such, this error probability, now denoted by $P_{b1}(E\,|\,\Phi_{\hat{l}})$ to indicate its dependence on $\Phi_{\hat{l}}$, is given by (8.19) with E_b/N_0 replaced by $(E_b/N_0) \cos^2\Phi_{\hat{l}}$. For example, for the special case of $2BT_b = 2$, we would have

$$P_{b1}(E\,|\,\Phi_{\hat{l}}) = \frac{1}{2}\exp\left(-\frac{E_b}{N_0}\cos^2\Phi_{\hat{l}}\right)\left[1 + \frac{1}{4}\frac{E_b}{N_0}\cos^2\Phi_{\hat{l}}\right] \tag{8.77}$$

The relation between the error probability, $P_{b1}(E\,|\,\Phi_{\hat{l}})$, associated with the hard decisions d_i on the first-order phase difference $\Delta\phi_i$ and the error probability, $P_b(E\,|\,\Phi_{\hat{l}})$, associated with the ultimate hard decisions on the second-order phase difference $\Delta^2\theta_i = \Delta\theta_i - \Delta\theta_{i-1}$ is

$$P_b(E\,|\,\Phi_{\hat{l}}) = 2P_{b1}(E\,|\,\Phi_{\hat{l}})\left(1 - P_{b1}(E\,|\,\Phi_{\hat{l}})\right) \tag{8.78}$$

Finally, the average probability of error of the receiver of Fig. 8.22 is given by

$$P_b(E) = \int_{-\pi/2L}^{\pi/2L} P_E(\Phi_{\hat{l}}) f(\Phi_{\hat{l}}) d\Phi_{\hat{l}} \tag{8.79}$$

where $f(\Phi_{\hat{l}})$ is the pdf of $\Phi_{\hat{l}}$.

Assuming that the frequency error is completely unknown and as such the equivalent phase $\alpha = \Delta\omega T_b$ is uniformly distributed in $(-\pi, \pi)$, then the pdf of $\Phi_{\hat{l}}$ is given by

$$f(\Phi_{\hat{l}}) = \begin{cases} \dfrac{L}{\pi}; & -\dfrac{\pi}{2L} \le \Phi_{\hat{l}} \le \dfrac{\pi}{2L} \\ 0; & \text{otherwise} \end{cases} \tag{8.80}$$

12. Practical conditions under which this assumption is valid are discussed in [4, 7].

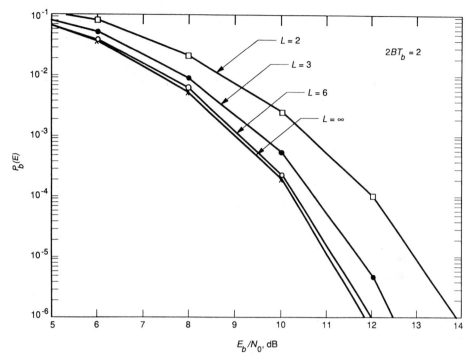

Figure 8.23   Bit error probability performance of the multiphase ACD in Figure 8.22 with the number of ACD's in the bank as a parameter (Reprinted from [2] © IEEE)

Thus, for example, the average error probability performance of the receiver of Fig. 8.22 for $2BT_b = 2$ would be found from[13]

$$P_b(E) = \frac{L}{\pi} \int_{-\pi/2L}^{\pi/2L} \exp\left(-\frac{E_b}{N_0}\cos^2\Phi_{\hat{\imath}}\right) \left[1 + \frac{1}{4}\frac{E_b}{N_0}\cos^2\Phi_{\hat{\imath}}\right]$$

$$\times \left\{1 - \frac{1}{2}\exp\left(-\frac{E_b}{N_0}\cos^2\Phi_{\hat{\imath}}\right)\left[1 + \frac{1}{4}\frac{E_b}{N_0}\cos^2\Phi_{\hat{\imath}}\right]\right\} d\Phi_{\hat{\imath}} \tag{8.81}$$

Evaluation of (8.81) in closed form is difficult if not impossible. Nevertheless, we observe that as L gets large, we can approximate the integral by the value of the integrand at $\Phi_{\hat{\imath}} = 0$ multiplied by the interval of integration π/L. Thus, for large L, we get

$$P_b(E) \cong \exp\left(-\frac{E_b}{N_0}\right)\left[1 + \frac{1}{4}\frac{E_b}{N_0}\right]\left\{1 - \frac{1}{2}\exp\left(-\frac{E_b}{N_0}\right)\left[1 + \frac{1}{4}\frac{E_b}{N_0}\right]\right\} \tag{8.82}$$

which is approximately twice the first result in (8.21). Figure 8.23 is a plot of $P_b(E)$ versus E_b/N_0 (in dB) as determined from (8.81). Also shown is the limiting result of (8.82) which corresponds to L approaching infinity. We observe that a value of $L = 6$ gives performance within a few tenths of a dB of the limiting result of (8.82).

   In addition to the above analytical results, a computer simulation has been written to assess the error probability performance of Fig. 8.22. Numerical results obtained from

---

13. References [4, 7] incorrectly evaluate this error probability since they first average (8.77) over the pdf of $\Phi_{\hat{\imath}}$ and then use a relation analogous to (8.78) to convert this result to the final error probability. Also, Figure 1 in [7] is incorrectly drawn. Figure 4.14 in [4] is the correct figure.

this simulation for the case $2BT_b = 2$ and $L = 6$ lie within 0.1 dB of the corresponding theoretical curve in Fig. 8.23.

Finally, we point out that the configuration of Fig. 8.22 can be clearly generalized to apply to M-DPSK (PDM-2) signals with $M > 2$ in the same manner that Fig. 8.6 generalizes to the first stage of Fig. 8.11. The details are left to the reader.

PROBLEMS

8.1 A binary phase information source whose phases take on values $(0, \pi)$ is to be communicated using a second-order phase difference modulation (PDM-2).
   (a) If the input information sequence $\{\Delta^2\theta_i\}$ is $\pi, \pi, \pi, 0, 0, \pi, 0, \pi$, find the sequence of complex baseband signals $\{e^{j\theta_i}\}$ transmitted in the intervals $(0, T_s), (T_s, 2T_s), \ldots, (9T_s, 10T_s)$. Assume that the first two transmitted phases are given by $\theta_0 = \theta_1 = \pi$ and since the source is binary, $T_s = T_b$.
   (b) If now the information source is 4-ary with phases that take on values $(0, \pi/2, \pi, 3\pi/2)$, repeat part (a) for the information sequence $\pi/2, \pi, \pi, 0, 3\pi/2, \pi/2, 0, 3\pi/2$.

8.2 A binary phase information source whose phases take on values $(0, \pi)$ is to be communicated using an nth-order phase difference modulation (PDM-n). Show that the appropriate encoding scheme relating the information phases $\{\Delta^n\theta_i\}$ to the transmitted phases $\{\theta_i\}$ is given by

$$\Delta^n\theta_i = \theta_i - \theta_{i-n}$$

8.3 A ternary phase information source whose phases take on values $(0, 2\pi/3, 4\pi/3)$ is to be communicated using a third-order phase difference modulation (PDM-3).
   (a) Show that the appropriate encoding scheme relating the information phases $\{\Delta^n\theta_i\}$ to the transmitted phases $\{\theta_i\}$ is given by

$$\Delta^3\theta_i = \theta_i - \theta_{i-3}$$

   (b) For the ternary source, does the relation found in Problem 8.2 also hold, in general, for values of n greater than 3? Explain.

8.4 A binary *amplitude* (±1) information source is to be communicated using a second-order phase difference modulation (PDM-2). Show that the two encoders illustrated in Fig. P8.4 are equivalent, that is, show that the relation between the information sequence $\{a_i\}$ and the transmitted sequence $\{c_i\}$ is the same for both.

8.5 Consider the communication system of Problem 8.4 wherein the data source $\{a_i\}$ is characterized as an independent, identically distributed (i.i.d.) sequence with $\Pr\{a_i = -1\} =$

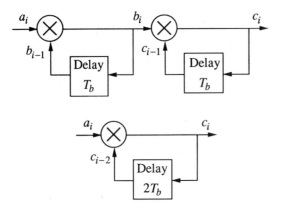

Figure P8.4

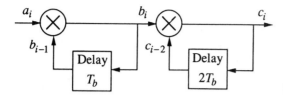

Figure P8.6

p, $\Pr\{a_i = 1\} = 1 - p$. Assuming that the first two bits in the output sequence are $c_0 = 1$, $c_1 = 1$,

(a) Evaluate the statistical mean of the output sequence $\{c_i\}$ for all i.

(b) Evaluate the autocorrelation function of the output sequence $R_c(i, i+m) \triangleq E\{c_i c_{i+m}\}$. Is the sequence $\{c_i\}$ in general, wide sense stationary (WSS)? Explain.

(c) If $p = 1/2$, that is, $\{a_i\}$ is a purely random data sequence, is the sequence $\{c_i\}$ WSS? Is it an uncorrelated sequence, that is, $R_c\{i, i+m\} = 0$ for $m \neq 0$?

8.6 A binary *amplitude* (± 1) information source is to be communicated using a nonadjacent second-order phase difference modulation (NA-PDM-2). The encoder for such a system is illustrated in Fig. 8.18 and is repeated in Fig. P8.6 for convenience. The data source $\{a_i\}$ is characterized as an independent, identically distributed (i.i.d.) sequence with $\Pr\{a_i = -1\} = p$, $\Pr\{a_i = 1\} = 1 - p$. Assuming that the first four bits in the output sequence are $c_0 = 1$, $c_1 = 1$, $c_2 = 1$, $c_3 = 1$,

(a) Evaluate the statistical mean of the output sequence $\{c_i\}$ for all i.

(b) Evaluate the autocorrelation function of the output sequence $R_c(i, i+m) \triangleq E\{c_i c_{i+m}\}$. Is the sequence $\{c_i\}$ in general, wide sense stationary (WSS)? Explain.

(c) If $p = 1/2$, that is, $\{a_i\}$ is a purely random data sequence, is the sequence $\{c_i\}$ WSS? Is it an uncorrelated sequence, that is, $R_c\{i, i+m\} = 0$ for $m \neq 0$?

8.7 A binary phase information source whose phases take on values $(0, \pi)$ is to be communicated using a second-order phase difference modulation (PDM-2). Based on Equations (8.3) and (8.4), an equivalent transmitter implementation is obtained by replacing the pair of back-to-back differential encoders in Fig. 8.1a each having delay T_s by a single differential encoder having delay $2T_s$ (also see Problem 8.4). Is it possible to make the same type of replacement in the receiver of Fig. 8.5, that is, can the two differential detectors each having delay T_s be replaced by a single differential detector having delay $2T_s$? Answer the question separately for the case of $\Delta f = 0$ and $\Delta f \neq 0$.

8.8 Consider a differential PSK system with I-Q detection that transmits an all-zero phase angle sequence of length five, that is, $\theta_0 = \theta_1 = \theta_2 = \theta_3 = \theta_4 = 0$. Assume that there is a channel frequency offset of $\Delta f T_b = 1/8$ and a channel phase offset $\theta = \pi/5$. Also assume for simplicity that there is no additive noise.

(a) Find the sequence of measured angles $\{\eta_i\}$; $i = 0, 1, \ldots, 4$ seen by the phase measurement portion of the I-Q detector as shown in Figures 7.1 and 8.13.

(b) If the system under consideration is binary DPSK (Fig. 7.1), find the receiver's estimate of the sequence of first-order differences $\{\psi_i = \Delta \eta_i\}$ for $i = 1, \ldots, 4$. Also find the receiver's estimate of the sequence of data information angles $\left\{\Delta \hat{\theta}_i\right\}$ for $i = 1, \ldots, 4$. (Assume that $\theta_0 = 0$ is known.) How large can θ and Δf be before the first-order differential decoding results in an error in $\Delta \hat{\theta}_i$?

(c) If the system under consideration is binary D^2PSK (Fig. 8.13), find the receiver's estimate of the sequence of second-order differences $\left\{\Delta^2 \eta_i\right\}$ for $i = 2, 3, 4$. Also find the receiver's estimate of the sequence of data information angles $\left\{\Delta^2 \hat{\theta}_i\right\}$ for $i = 2, 3, 4$. (Assume that $\theta_0 = \theta_1 = 0$ is known.) How large can θ and Δf be before the second-order differential decoding results in an error in $\Delta^2 \hat{\theta}_i$?

8.9 Suppose that signals $s_2(t)$, $s_3(t)$, and $s_4(t)$ as given by Eq. (8.12) are transmitted and received in the absence of noise by the autocorrelation demodulation for D^2PSK signals as illustrated in

Fig. 8.10. Trace the signals through the receiver block diagram to find the summer output (input to the signum operator) $V_{c,4}V_{c,3} + V_{s,4}V_{s,3}$. Assuming that the transmitted phase angles for all three signals have been second-order differentially encoded, explain how the signum operator can be used to make the decision for the binary signal alphabet $(0, \pi)$.

8.10 For a D²PSK system, suppose that, in addition to the frequency offset Δf and phase offset θ, there is a higher-order offset between the received signal carrier and the receiver reference signals given by at_i^2, that is, the complex channel rotation shown in Fig. 8.2 becomes $\exp\left\{ j\left[\theta + 2\pi \Delta f t_i + a t_i^2 \right] \right\}$.

(a) In the absence of noise, find the output of the summer (input to the signum operator) $V_{c,i}V_{c,i-1} + V_{s,i}V_{s,i-1}$ in the ACD of Fig. 8.10. Also find $\Delta^2 \eta_i$ in the I-Q demodulator of Fig. 8.13.

(b) Design a third-order differential encoder and a corresponding demodulator with third-order differential decoding that would, in the absence of noise, eliminate the effect of this higher-order offset.

8.11 It is desired to evaluate the channel capacity (see Chapter 1) of a second-order phase difference modulation (PDM-2) system using a transmitter as in Fig. 8.1a, that is, a double differential encoder, and an I-Q receiver as in Fig. 8.13. Such a system can be modeled as an M-ary symmetric channel (see Chapter 3) for which the channel capacity (in bps) is given by

$$C = \sum_{m=0}^{M-1} \frac{1}{M} \sum_{n=0}^{M-1} P_{sn} (E \,|m) \log_2 P_{sn} (E \,|m) + \log_2 M$$

where $P_{sn} (E \,|m)$ is the conditional probability of choosing the nth symbol (phase) when the mth symbol (phase) was transmitted, that is

$$P_{sn} (E \,|m) \triangleq \Pr\left\{ \Delta^2 \hat{\theta}_i = \beta_n \,\Big|\, \Delta^2 \theta_i = \beta_m \right\}, \quad \beta_m = \frac{2\pi m}{M}$$

Note that $P_{sm} (E \,|m)$ is actually the conditional probability of a correct decision given that the mth symbol was transmitted. Since the β_i's are equiprobable, $P_{sm} (E \,|m)$ is also the average probability of correct symbol detection.

With reference to Figures 8.1a and 8.13, the pdf of the second-order phase difference $\Delta\chi_i \triangleq \Delta^2 \eta_i - \Delta^2 \theta_i$ (see Problem 8.12) is denoted by $f_{\Delta X_i} (\Delta\chi_i)$, which is independent of i and the transmitted information phase $\Delta^2 \theta_i$, and is distributed in the interval $-\pi \le \Delta\chi_i \le \pi$.

(a) Evaluate the channel capacity of the PDM-2 system. Express your answer in terms of the set of probabilities

$$P_l = \int_{(2l-M-1)\pi/M}^{(2l-M+1)\pi/M} f_{\Delta X_i} (\Delta\chi_i)\, d\Delta\chi_i, \quad l = 1, 2, \ldots, M-1$$

$$P_0 = \int_{-\pi}^{-(M-1)\pi/M} f_{\Delta X_i} (\Delta\chi_i)\, d\Delta\chi_i + \int_{(M-1)\pi/M}^{\pi} f_{\Delta X_i} (\Delta\chi_i)\, d\Delta\chi_i$$

(b) Suppose now that we have a PDM-1 system using a single differential encoder as a transmitter and a receiver as in Fig. 8.8 (or Fig. 7.1). Here, the conditional pdf of $\chi_i \triangleq \Delta\eta_i - \Delta\theta_i$, which is denoted by $f_{X_i} (\chi_i)$, is independent of i and the transmitted information phase $\Delta\theta_i$, and is distributed in the interval $-\pi \le \chi_i \le \pi$. Also, denote the probabilities corresponding to the P_l's of part (a) by

$$Q_l = \int_{(2l-M-1)\pi/M}^{(2l-M+1)\pi/M} f_{X_i} (\chi_i)\, d\chi_i, \quad l = 1, 2, \ldots, M-1$$

$$Q_0 = \int_{-\pi}^{-(M-1)\pi/M} f_{X_i} (\chi_i)\, d\chi_i + \int_{(M-1)\pi/M}^{\pi} f_{X_i} (\chi_i)\, d\chi_i$$

In principle, how would your answer found in part (a) change, if at all; that is, express the channel capacity in terms of the Q_l's.

8.12 Consider a second-order phase difference modulation (PDM-2) system composed of the transmitter of Fig. 8.1a and the I-Q receiver of Fig. 8.13. The pdf of the second-order phase difference $\Delta\chi_i \triangleq \Delta^2\eta_i - \Delta^2\theta_i$ is denoted by $f_{\Delta\chi_i}(\Delta\chi_i)$ which is independent of i and the transmitted information phase $\Delta^2\theta_i$, and is distributed in the interval $-\pi \leq \Delta\chi_i \leq \pi$. In terms of the equivalent first-order phase differences $\chi_i \triangleq \Delta\eta_i - \Delta\theta_i$, we have $\Delta\chi_i \triangleq \underbrace{\Delta\eta_i - \Delta\eta_{i-1}}_{\Delta^2\eta_i} - \underbrace{\Delta\theta_i - \Delta\theta_{i-1}}_{\Delta^2\theta_i} = \chi_i - \chi_{i-1}$. The pdf of $\Delta\chi_i$ can, in principle, be obtained if the joint pdf of χ_{i-1} and χ_i, namely $f_{X_{i-1},X_i}(\chi_{i-1}, \chi_i)$, is available. In particular

$$f_{\Delta\chi_i}(\Delta\chi_i) = \int_{-\pi}^{\pi} f_{X_{i-1},X_i}(\chi_{i-1}, \chi_{i-1} + \Delta\chi_i)d\chi_{i-1}$$

Although the joint pdf $f_{X_{i-1},X_i}(\chi_{i-1}, \chi_i)$ can be found and then $f_{\Delta\chi_i}(\Delta\chi_i)$ evaluated as above, the resulting expression becomes quite complicated and is in the form of multiple integrals. In the high SNR case, the matter can be simplified by approximating $f_{X_{i-1},X_i}(\chi_{i-1}, \chi_i)$ by a two-dimensional Gaussian form. In particular

$$f_{X_{i-1},X_i}(\chi_{i-1}, \chi_i) = \frac{1}{2\pi\sigma^2\sqrt{1-\rho^2}} \exp\left\{-\frac{\chi_{i-1}^2 - 2\rho\chi_{i-1}\chi_i + \chi_i^2}{2(1-\rho^2)\sigma^2}\right\}, \quad -\infty \leq \chi_{i-1}, \chi_i \leq \infty,$$

where $\rho = -1/2$ is the correlation between the first-order phase differences in the two adjacent symbol intervals and $\sigma^2 = N_0/ST_s = R_s^{-1}$ denotes the inverse of the symbol SNR.

(a) Show that the second-order phase difference approximate (high SNR) pdf is given by

$$f_{\Delta\chi_i}(\Delta\chi_i) = \frac{1}{2\sigma\sqrt{\pi(1-\rho)}} \sum_{m=-\infty}^{\infty} \exp\left[-\frac{(\Delta\chi_i + 2\pi m)^2}{4\sigma^2(1-\rho)}\right], \quad -\pi \leq \Delta\chi_i \leq \pi$$

Hint: In the case where χ_{i-1} and χ_i are not restricted to values in the interval $(-\pi, \pi)$ as in the approximate pdf above, one can first compute the pdf of $\Delta\chi_{ia} \triangleq \chi_i - \chi_{i-1}$, the "actual" phase difference without the modulo 2π condition by applying the above integral with limits $-\infty$ and ∞ instead of $-\pi$ and π. Then, the pdf of the modulo 2π reduced $\Delta\chi_i$ is given by

$$f_{\Delta\chi_i}(\Delta\chi_i) = \sum_{m=-\infty}^{\infty} f_{\Delta\chi_{ia}}(\Delta\chi_i + 2\pi m)$$

where $f_{\Delta\chi_{ia}}(\Delta\chi_{ia})$ is the pdf of $\Delta\chi_{ia}$.

(b) Using the result found in part (a), evaluate the asymptotic (high SNR) average symbol error probability, $P_s(E)$, of the I-Q D²PSK receiver of Fig. 8.13. Express your answer in terms of "erfc" functions with appropriate arguments.

REFERENCES

1. Divsalar, D. and M. K. Simon, "Doppler-Corrected Differential Detection of MPSK," *IEEE Transactions on Communications*, vol. 37, no. 2, February 1989, pp. 99–109.

2. Divsalar, D. and M. K. Simon, "On the Implementation and Performance of Single and Double Differential Detection," *IEEE Transactions on Communications*, vol. 40, no. 2, February 1992, pp. 278–91.

3. VanAlphen, D. K. and W. C. Lindsey, "Higher-Order Differential Phase Shift Keyed Modulation," to appear in the *IEEE Transactions on Communications*.

4. Okunev, Yu. B., *Theory of Phase-Difference Modulation*, Moscow: Svyaz Press, 1979.

5. Fink, L. M. , *Theory of Digital Data Transmission*, Moscow: Sovetskoye Radio Press, 1970.

6. Korzhik, V. I., L. M. Fink and K. N. Shchelkunov, *Calculation of the Noise Immunity of Digital Data Transmission Systems (A Handbook)*, Moscow: Radio i Svyaz Press, 1981.

7. Okunev, Yu. B., V. A. Pisarev and V. K. Reshemkin, "The Design and Noise-Immunity of Multiphase Autocorrelation Demodulators of Second-Order DPSK Signals," Radiotekhnika, vol. 34, no. 6, 1979 [Telecomm. Radio Engng., Part 2, vol. 34, no. 6, 1979, pp. 60–63].

8. Okunev, Yu. B. and L. M. Fink, "Noise Immunity of Various Receiving Methods for Binary Systems with Second-Order Phase-Difference Modulation," Radiotekhnika, vol. 39, no. 8, 1984 [Telecomm. Radio Engng., vol. 39, no. 8, 1984, pp. 51–56].

9. Okunev, Yu. B., N. M. Sidorov and L. M. Fink, "Noise Immunity of Incoherent Reception with Single Phase-Difference Modulation," Radiotekhnika, no. 11, 1985 [Telecomm. Radio Engng., no. 11, 1985, pp. 103–106].

10. Okunev, Yu. B. and N. M. Sidorov, "Noise Immunity of a Carrier-Frequency-Invariant Demodulator of DPSK-2 Signals," Radiotekhnika, no. 6, 1986 [Telecomm. Radio Engng., no. 6, 1986, pp. 81–83].

11. Goot, R. E., "Noise Immunity of Autocorrelation Reception of Single PSK Signals," Radiotekhnika, no. 10, 1972 [Telecomm. Radio Engng., no. 10, 1972, pp. 120–122].

12. Goot, R. E., Yu. B. Okunev and N. M. Sidorov, "Effect of Carrier Detuning on the Noise Immunity of Autocorrelation Reception of DPSK Signals," Radiotekhnika, no. 7, 1986 [Telecomm. Radio Engng., no. 7, 1986, pp. 96–98].

13. Pent, M., "Double Differential PSK Scheme in the Presence of Doppler Shift," Digital Communications in Avionics, AGARD Proceedings no. 239, 1978, pp. 43-1–43-11.

14. Stein, S., "Unified Analysis of Certain Coherent and Noncoherent Binary Communication Systems," *IEEE Transactions on Information Theory*, January 1964, pp. 43–51.

15. Henry, J. C., III, "DPSK versus FSK with Frequency Uncertainty," *IEEE Transactions on Communications Technology*, vol. COM-18, December 1970, pp. 814–817.

16. Miller, T. W., "Imperfect Differential Detection of a Biphase Modulated Signal—An Experimental and Analytical Study," Ohio State University, Technical Report No. RADC-TR-72-16, February 1972.

17. Divsalar, D. and M. Shahshahani, "The Bit Error Rate (BER) of DPSK with Differential Detector (DD) Followed by Matched Filter (MF)," private correspondence.

18. Pawula, R. F., S. O. Rice and J. H. Roberts, "Distribution of the Phase Angle Between the Vectors Perturbed by Gaussian Noise," *IEEE Transactions on Communications*, vol. COM-30, no. 8, pp. 1828–1841.

19. Simon, M. K., "Comments on 'On Binary DPSK Detection'," *IEEE Transactions on Communications*, vol. COM-26, no. 10, October 1978, pp. 1477–1478.

20. Arthurs, E. and H. Dym, "On the Optimum Detection of Digital Signals in the Presence of White Gaussian Noise—A Geometric Interpretation and a Study of Three Basic Transmission Systems," *IRE Transactions on Communications Systems*, December 1962, pp. 336-372.

21. Divsalar, D. and M. K. Simon, " Multiple Symbol Differential Detection of MPSK," *IEEE Transactions on Communications*, vol. 38, no. 3, March 1990, pp. 300–308. Also see "Multiple Symbol Differential Detection of Uncoded and Trellis Coded MPSK," Pasadena, CA: Jet Propulsion Laboratory, JPL Publication 89-38, November 15, 1989.

22. Qu, S., "New Modulation Schemes: Principles and Performance Analyses," Ph.D. Dissertation, Technical University of Nova Scotia, 1992. Also see S. Qu and S. Fleisher, "Double Differential MPSK on the Fast Rician Fading Channel," submitted for publication in the *IEEE Transactions on Communications*.

Chapter 9

Communication
over Bandlimited Channels

Except in a few instances, all of the communication scenarios we have discussed thus far have corresponded to an AWGN channel with unconstrained bandwidth. As such, the only effect the channel was assumed to have on the transmitted signal (aside from the addition of Gaussian noise) was attenuation of its amplitude and the addition of a phase shift to the carrier, the latter being assumed constant in time over the duration of the data symbol or longer. In reality, the communication channel is never infinite in bandwidth; however, in certain applications, it is considerably wideband when compared to the bandwidth of the modulation being communicated over it. Thus, for all practical purposes, one can in these situations still model the channel's effect on the transmitted waveform as multiplication by a fixed gain (actually an attenuation) and the addition of a constant carrier phase shift. Alternately, the equivalent baseband signal when passed through the channel is multiplied by a fixed gain and delayed by an amount that is constant for all frequencies in the waveform. For all practical purposes, this is the case that we have treated thus far in the book.

In this chapter, we discuss the design of modulation/demodulation schemes for sending digital information over a bandlimited channel whose bandwidth is on the same order as that of the modulation. Our treatment will be brief by comparison with other fine textbooks devoted to this subject [1, 2, 3], and will only focus on the key issues. We shall consider only linear channels, that is, those modeled as a linear filter with bandpass transfer function $F_c(f)$, and linear modulations of the type corresponding to a baseband pulse stream

$$\tilde{m}(t) = A \sum_{i=-\infty}^{\infty} \tilde{c}_i p(t - iT_s) \tag{9.1}$$

modulated onto a carrier where A is the signal amplitude and $\tilde{c}_i = a_i + jb_i$ is, in general, a complex discrete random variable representing the data symbol to be communicated in the ith transmission interval. The key issue here is the design of the pulse shape $p(t)$ to allow communication at the highest possible data rate while achieving the best trade-off between signal power and noise power. As we shall see, this pulse shape design is dictated by the characteristics of $F_c(f)$ over its bandwidth occupancy. In particular, if $F_c(f)$ is an ideal (rectangular transfer characteristic) filter of IF bandwidth W Hz (lowpass bandwidth $B = W/2$), then, at least theoretically, $p(t)$ can be designed to allow distortion-free transmission at a rate equal to W bps with an SNR (E_b/N_0) equivalent to that of the unconstrained bandwidth channel. If the channel is not ideal over its bandwidth occupancy, then signaling at a rate W bps or higher results in a form of distortion known as *intersymbol interference (ISI)* wherein a number of adjacent received pulses interfere with each other in so far as the detection process is concerned, thus producing a degradation in SNR. These last two statements are the basis of the well-known Nyquist criterion [4], which will be reviewed early in this chapter. Even when the channel is not ideal over its passband, it may still satisfy the Nyquist criterion in the sense that it allows distortion (ISI)-free transmission but at a communication rate less than W bps. The pulse shapes that satisfy the Nyquist criterion are referred to as *Nyquist pulses*, and systems that employ such pulse shapes achieve the best level of performance.

Assuming first that the overall pulse response is designed according to the Nyquist criterion, the next issue is to determine how to best apportion it between the transmit and receive filters so as to maximize the detection SNR. We shall see that, under the constraint of a given amount of power available to the channel, the solution to this problem is to equally split the total filtering between the two locations. Once again, this comes about because of the optimization in the trade-off between signal plus distortion power and noise power at the receiver, which is consistent with the notion of matched filtering introduced in Chapter 4.

Quite often, it is not possible or practical to design a Nyquist pulse and thus the system will be corrupted by ISI. When the ISI is a controlled amount, it is possible to exploit this occurrence in the design of the transmitter and receiver. This leads to a class of systems known as *partial response (PR) systems* [5], which have the advantage that they are simple to implement yet produce an efficient trade-off between signal plus distortion and noise powers. The simplest case of a PR system is one where the ISI results only from a single adjacent symbol and is referred to as *duobinary* [6]. We shall discuss this case in detail.

In other instances, one might resort to the more complicated case of designing a receiver that is optimum in the presence of both ISI and additive Gaussian noise. The solution to this problem takes several different forms depending on the optimization criterion specified and the nature of the detection and decision processes. Of particular interest are the structures resulting from application of the maximum-likelihood principle. These structures are based on the notion of *maximum-likelihood sequence estimation (MLSE)* wherein a joint estimate is made on a *sequence* of transmitted information symbols with the goal of minimizing the sequence error probability. The reason for estimating a sequence rather than a single symbol at a time is that ISI has the effect of introducing memory into the modulation in the sense that the detection of a given symbol depends on the knowledge of past (and possibly future) data symbols. This idea of making a joint decision based on an observation of the received signal over more than one symbol interval is not unlike that discussed in several of the previous chapters in connection with optimum block-by-block detectors. We shall discuss an efficient algorithm for performing MLSE, namely, the Viterbi algorithm [7], which was originally proposed in the context of decoding convolutional codes.

The ability to design a receiver as in the above to mitigate the deleterious effects of ISI relies on having complete knowledge of the channel transfer characteristic $F_c(f)$. In certain practical channels, for example, the telephone channel, this characteristic is time-varying, and thus the design of the demodulator must be made adaptive so as to remain optimum as a function of time. The general process by which the receiver attempts to cancel the ISI produced by the channel to restore a Nyquist pulse shape is referred to as *equalization*. In the case of a time-varying channel, the receiver achieves the equalization process by continuously adjusting a number of its parameters based on measurements performed on the channel characteristics. As such, the process is called *adaptive equalization*. A large body of literature has been developed that is devoted to the design and convergence of algorithms for adjusting the equalizer's parameters. Since this is not the major focus of this textbook, we shall be relatively brief in the coverage of this subject.

We begin the chapter with a brief characterization of what is meant by a bandlimited channel and a practical application where it is encountered.

9.1 DEFINITION AND CHARACTERIZATION OF A BANDLIMITED CHANNEL

By "bandlimited channel," we infer a strictly bandlimited channel corresponding to one whose frequency response is nonzero only in a finite portion of the spectrum. If the channel is bandpass with frequency response $F_c(f)$, then we require that $F_c(f) = 0$ for $|f - f_c| > W/2 (f > 0)$ and for $|f + f_c| > W/2 (f < 0)$. If the channel impulse response $f_c(t) = \mathcal{F}^{-1}\{F_c(f)\}$ is written in the form

$$f_c(t) = \mathrm{Re}\left\{ g_c(t)e^{j2\pi f_c t} \right\} \tag{9.2}$$

where $g_c(t)$ is the equivalent lowpass impulse response, then in terms of the lowpass equivalent channel frequency response $G_c(f) = \mathcal{F}\{g_c(t)\} = 2F_c(f + f_c)$, we have that $G_c(f) = 0$ for $|f| > W/2 = B$. The notion of baseband equivalence of an IF function used here is analogous to that discussed for random processes in Appendix 4B. For a baseband channel, the bandlimiting requirement is identical to that imposed on $G_c(f)$. In what follows, we shall find it convenient to work with $G_c(f)$ rather than $F_c(f)$ with the understanding that if we are dealing with a baseband channel they are one and the same. Furthermore, for simplicity, we shall assume that $F_c(f)$ is symmetric around its center frequency, f_c, so that $g_c(t)$ is a real function of t.

In view of the above, a bandlimited AWGN channel is one that, in addition to bandlimiting the signal, also adds white Gaussian noise. Thus, if a bandpass signal $s(t) = \mathrm{Re}\{\tilde{m}(t)e^{j2\pi f_c t}\}$ is transmitted over this channel, the received signal is given by $r(t) = \mathrm{Re}\{\tilde{m}'(t)e^{j2\pi f_c t}\}$ where $\tilde{m}'(t)$ is obtained from the convolution of the lowpass modulation $\tilde{m}(t)$ with the lowpass channel impulse response $g_c(t)$. For the assumed random pulse train form of modulation, $\tilde{m}'(t)$ becomes

$$\tilde{m}'(t) = A \sum_{i=-\infty}^{\infty} \tilde{c}_i p'(t - iT_s);$$

$$p'(t) = \int_{-\infty}^{\infty} p(\tau)g_c(t - \tau)d\tau = \int_{-B}^{B} P(f)G_c(f)e^{j2\pi f t}df \tag{9.3}$$

with $P(f)$ the Fourier transform of $p(t)$.

It is important to understand that both the envelope $|G_c(f)|$ and phase $\theta_c(f) = \arg\{G_c(f)\}$ characteristics of the channel lowpass frequency response are influential in determining the distortion introduced into the transmitted pulse shape. Assuming that $P(f)$ is bandlimited to $(-B, B)$ (since no energy outside that interval will get through the channel anyway), then if $|G_c(f)|$ is constant, that is, $G_{c0} = G_c(0)$, and $\theta_c(f)$ is a linear function of frequency, that is, $-K_\theta f$ over this same interval, then the channel introduces no distortion other than a pure time delay $T_d = K_\theta/2\pi$ and from (9.3) the received pulse shape becomes

$\tilde{p}(t) = G_{c0}p(t - T_d)$. Note that since $P(f)$ is bandlimited, $p(t)$ cannot be timelimited (the uncertainty principle [8] prevents a signal from being limited in both time and frequency) and likewise for $p'(t)$. Thus, the pulses in $\tilde{m}'(t)$ must, by necessity, overlap. This does not, however, mean that ISI is present at the receiver since by proper design of $p(t)$, it may still be possible to separate out the various pulses in the data stream. What is important here is that if $p(t)$ is designed to produce zero ISI at the receiver, then the distortion-free channel above does not change that condition, that is, $p'(t)$ also yields zero ISI.

In the more general case where $|G_c(f)|$ is not constant and $\theta_c(f)$ is not a linear function of frequency over $(-B, B)$, the channel will introduce amplitude and phase distortion into the transmitted pulse shape. It is common to define the *group delay*, $T_d(f)$, by the negative of a scaled derivative of the channel phase characteristic, $\theta_c(f)$, that is,

$$T_d(f) = -\frac{1}{2\pi}\frac{d\theta_c(f)}{df} \tag{9.4}$$

This definition is consistent with the above discussion for the distortion-free channel, whereupon the group delay becomes a constant (fixed) delay, T_d, independent of frequency. For the channel that introduces phase distortion, $T_d(f)$ will be a function of frequency. The effect of amplitude and phase distortion on a transmitted pulse stream modulation, $\tilde{m}(t)$, is to smear the pulses in the received modulation, $\tilde{m}'(t)$, so that they cannot be separated by the receiver. It is indeed this additional smearing of the pulses that accounts for what is referred to as *intersymbol interference*.

Perhaps the most common bandlimited channel that, in addition to many other impairments (e.g., impulse noise, phase jitter, frequency offset, etc.), exhibits both amplitude and phase distortion over its frequency range of definition is the telephone channel, whose usable band is typically in the range of 300 Hz to 3 KHz. A detailed discussion and illustration of the telephone channel characteristics are given in [2]. As an example, an average medium range telephone channel of the switched telecommunications network [9] has an impulse response of about 10 ms; thus, for communication over this channel at a rate on the order of 2.5 Ksps (a symbol duration of 0.4 ms), the ISI would extend over 25 symbols.

9.2 OPTIMUM PULSE SHAPE DESIGN FOR DIGITAL SIGNALING THROUGH BANDLIMITED AWGN CHANNELS

Consider the block diagram of a baseband bandlimited communication system illustrated in Fig. 9.1. This system is also the baseband equivalent of a bandpass bandlimited communication system in the context of our discussion above. The source generates a pulse stream modulation $\tilde{m}(t)$ of the type described by (9.1). This modulation is passed through the transmit filter $G_t(f)$ and the channel $G_c(f)$. The filtered modulation is perturbed by additive complex Gaussian noise $\tilde{n}(t)$ with power spectral density (PSD) $S_n(f)$ and the sum of the two is passed through the receive filter $G_r(f)$.[1] Thus, the detected pulse is shaped by the combination of $G_t(f)$, $G_c(f)$, and $G_r(f)$ whereas the additive noise spectral density is shaped only by $G_r(f)$. Samples of the output of $G_r(f)$ are taken once per symbol interval, T_s, and processed to make decisions on the information sequence $\{\tilde{c}_n\}$. The problem at hand

---

1. Since the noise process $\tilde{n}(t)$ is, in general, complex, its PSD is defined as $S_{\tilde{n}\tilde{n}^*}(f) \triangleq \mathcal{F}\{E\{\tilde{n}(t)\tilde{n}^*(t + \tau)\}\}$. However, to keep the notation simple, we shall define it with only a single subscript, namely, $S_{\tilde{n}}(f)$, as is done for the PSD's of real random processes.

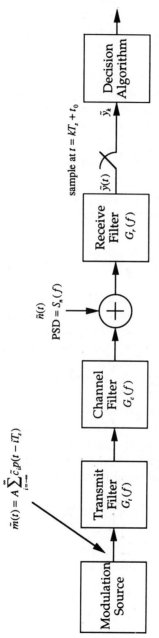

Figure 9.1 A block diagram of a bandlimited communication system

is to design $G_t(f)$ and $G_r(f)$ [for a given $G_c(f)$] to minimize the error probability of the receiver.[2]

The output $\tilde{y}(t)$ of the receive filter can be expressed as

$$\tilde{y}(t) = A \sum_{i=-\infty}^{\infty} \tilde{c}_i g(t - iT_s) + \tilde{\mu}(t) \tag{9.5}$$

where $g(t) = \mathcal{F}^{-1}\{G(f)\} = \mathcal{F}^{-1}\{G_t(f)G_c(f)G_r(f)\}$ is the overall impulse response, that is, the detected pulse shape, and $\tilde{\mu}(t)$ is the receive-filtered complex noise process with PSD $S_{\tilde{\mu}}(f) = S_{\tilde{n}}(f)|G_r(f)|^2$. Suppose that $\tilde{y}(t)$ is now uniformly sampled at times $t = kT_s + t_0$ (k integer) to produce the complex sample values

$$\tilde{y}(kT_s + t_0) = A \sum_{i=-\infty}^{\infty} \tilde{c}_i g((k - i) T_s + t_0) + \tilde{\mu}(kT_s + t_0) \tag{9.6}$$

where t_0 is selected to match the delay of the channel (more about this in a moment). In shorthand notation, (9.6) can be rewritten as

$$\tilde{y}_k = A \sum_{i=-\infty}^{\infty} \tilde{c}_i g_{k-i} + \tilde{\mu}_k = A\tilde{c}_k g_0 + A \sum_{\substack{i=-\infty \\ i \neq k}}^{\infty} \tilde{c}_i g_{k-i} + \tilde{\mu}_k \tag{9.7}$$

where, in general, v_k denotes the sample of the waveform $v(t)$ at time $kT_s + t_0$. The first term in (9.7) represents the desired signal sample corresponding to the kth transmission interval and thus t_0 should be selected to maximize g_0, that is, we should choose our sampling time epoch so that the desired sample corresponds to the peak of the overall pulse shape $g(t)$. The second term represents ISI and ideally should be made equal to zero. The third term is the sample of the filtered complex Gaussian noise process.

Assuming for the moment that we can indeed achieve zero ISI, then the optimum decision rule is identical to that considered in Chapter 4, namely, choose the data symbol $\hat{\tilde{c}}_k$ that is closest in distance to $\tilde{y}_k$. For example, for binary modulation with $\tilde{c}_i$ real ($\tilde{c}_i = a_i$) and taking on equiprobable values ± 1, the decision variable is real, that is, $\tilde{y}_k = y_k$ and the decision rule is implemented as a threshold comparison, that is, $\hat{a}_k = \text{sgn}\{y_k\}$. The corresponding bit error probability is given by

$$P_b(E) = \frac{1}{2} \text{erfc} \left\{ \frac{E\{y_k|_{a_k=1}\}}{\sqrt{2}\sigma_y} \right\} = \frac{1}{2} \text{erfc} \left\{ \frac{Ag_0}{\sqrt{2}\sigma_y} \right\} \tag{9.8}$$

which is the minimum achievable without further specification on $G_t(f)$ or $G_r(f)$. As noted above, the variance of the filtered noise sample depends only on $G_r(f)$ and thus

$$\sigma_y^2 = \int_{-\infty}^{\infty} S_n(f) |G_r(f)|^2 df \tag{9.9}$$

---

2. Although, in a real communication system, the modulation pulse shaping and transmit filtering are separate entities, from a mathematical standpoint, their cascaded effect on the transmitted waveform is represented by the product of their respective transfer functions $P(f)$ and $G_t(f)$. Thus, for analysis purposes, we shall denote the product $P(f)G_t(f)$ by simply $G_t(f)$ with the understanding that in the physical system $G_t(f)$ must be partitioned between the modulation pulse shape and the transmit filter.

Also, as noted above, the desired signal sample g_0 depends on $G_t(f)$, $G_c(f)$, and $G_r(f)$, viz.

$$g_0 = \int_{-\infty}^{\infty} G_t(f)G_c(f)G_r(f)e^{j2\pi f t_0}df \tag{9.10}$$

We will return to a consideration of the selection of $G_t(f)$ and $G_r(f)$ to minimize $P_b(E)$ of (9.8), that is, maximize g_0/σ_y, after first deriving the criteria necessary to achieve zero ISI, that is, the *Nyquist criterion*. For simplicity, we shall let $g_0 = 1$ since an arbitrary gain can be included by scaling $G(f)$ later on. Also, we shall continue to consider the case of real (but not necessarily binary) data, that is, $\tilde{c}_i = a_i$, until further notice.

9.2.1 Nyquist Criterion for Zero ISI

We start out by recalling the well-known sampling theorem, namely, if $X(f) = \mathcal{F}\{x(t)\}$ is bandlimited to $(-B, B)$, then $x(t)$ is uniquely determined by samples taken at $1/2B$, viz.

$$x(t) = \sum_{n=-\infty}^{\infty} x\left(\frac{n}{2B}\right) \frac{\sin 2\pi B\left(t - \frac{n}{2B}\right)}{2\pi B\left(t - \frac{n}{2B}\right)} \tag{9.11}$$

and taking the Fourier transform of $x(t)$

$$X(f) = \begin{cases} \frac{1}{2B} \sum_{n=-\infty}^{\infty} x\left(\frac{n}{2B}\right) \exp\left(-j2\pi f \frac{n}{2B}\right); & |f| \le B \\ 0; & \text{otherwise} \end{cases} \tag{9.12}$$

The lowpass bandwidth B is referred to as the *Nyquist bandwidth* and is equal to one-half the sampling rate.

From (9.7), for zero ISI, we desire a pulse waveform $g(t)$ such that its samples taken at a rate $1/T_s$ (the data symbol rate) satisfy[3]

$$g_n = \int_{-\infty}^{\infty} G(f)e^{j2\pi f n T_s}df = \delta_n = \begin{cases} 1; & n = 0 \\ 0; & n \neq 0 \end{cases} \tag{9.13}$$

We can rewrite the above integral in terms of a partition of adjacent Nyquist bandwidths of width $2B = 1/T_s$, viz.

$$g_n = \sum_{k=-\infty}^{\infty} \int_{(2k-1)/2T_s}^{(2k+1)/2T_s} G(f)e^{j2\pi f n T_s}df = \delta_n \tag{9.14}$$

Using the change of variables $v = f - k/T_s$, (9.14) becomes

---

3. For the purpose of the discussion here, we shall assume without any loss in generality that the sampling epoch t_0 equals zero. Whatever conditions we determine on the overall pulse transfer function $G(f)$ can later be modified by replacing $G(f)$ with $G(f)e^{j2\pi f t_0}$.

$$g_n = \sum_{k=-\infty}^{\infty} \int_{-1/2T_s}^{1/2T_s} G\left(v + \frac{k}{T_s}\right) e^{j2\pi(v+k/T_s)nT_s} dv$$

$$= \sum_{k=-\infty}^{\infty} \int_{-1/2T_s}^{1/2T_s} G\left(v + \frac{k}{T_s}\right) e^{j2\pi vnT_s} dv \tag{9.15}$$

$$= \int_{-1/2T_s}^{1/2T_s} \left(\sum_{k=-\infty}^{\infty} G\left(v + \frac{k}{T_s}\right)\right) e^{j2\pi vnT_s} dv$$

Define the *equivalent Nyquist channel characteristic*

$$G_{eq}(f) = \begin{cases} \sum_{k=-\infty}^{\infty} G\left(f + \frac{k}{T_s}\right); & |f| \le \frac{1}{2T_s} \\ 0; & \text{otherwise} \end{cases} \tag{9.16}$$

that is, the sum of all of the translates of $G(f)$ folded into the interval $(-1/2T_s, 1/2T_s)$. Substituting (9.16) into (9.15) gives

$$g_n = \int_{-1/2T_s}^{1/2T_s} G_{eq}(f) e^{j2\pi fnT_s} df \tag{9.17}$$

But the inverse Fourier transform of $G_{eq}(f)$ is by definition

$$g_{eq}(t) = \int_{-\infty}^{\infty} G_{eq}(f) e^{j2\pi ft} df = \int_{-1/2T_s}^{1/2T_s} G_{eq}(f) e^{j2\pi ft} df \tag{9.18}$$

with samples

$$g_{eq_n} = \int_{-1/2T_s}^{1/2T_s} G_{eq}(f) e^{j2\pi vnT_s} df \tag{9.19}$$

Thus, from (9.18) and (9.19), we see that the Nyquist rate samples of $g(t)$ are also the Nyquist rate samples of $g_{eq}(t)$, that is, $g_{eq_n} = g_n$.

Since, by the definition of (9.16), $G_{eq}(f)$ is a strictly bandlimited function on the interval $(-1/2T_s, 1/2T_s)$, it can be expressed in the form of (9.12), namely

$$G_{eq}(f) = \begin{cases} T_s \sum_{n=-\infty}^{\infty} g_n e^{-j2\pi fnT_s}; & |f| \le \frac{1}{2T_s} \\ 0; & \text{otherwise} \end{cases} \tag{9.20}$$

However, for zero ISI we require that g_n satisfy (9.13). Thus, using (9.13) in (9.20) gives

$$G_{eq}(f) = \begin{cases} T_s; & |f| \le \frac{1}{2T_s} \\ 0; & \text{otherwise} \end{cases} \tag{9.21}$$

that is, an ideal brick wall (rectangular) filter. Finally, combining (9.16) and (9.21), we see that the combined (transmitter, channel, and receiver) overall system frequency response $G(f)$ must satisfy

$$\sum_{k=-\infty}^{\infty} G\left(f + \frac{k}{T_s}\right) = T_s; \quad |f| \le \frac{1}{2T_s} \tag{9.22}$$

that is, the superposition of all of the translates of $G(f)$ in the interval $(-1/2T_s, 1/2T_s)$ must yield a flat response.

It can also be shown (Problem 9.1) that the superposition of all of the translates of $G(f)$ in the interval $((2n - 1)/2T_s \le f \le (2n + 1)/2T_s)$ must also yield a flat response for any integer n. Thus, the zero ISI condition on $G(f)$ as specified by (9.22) generalizes to

$$\sum_{k=-\infty}^{\infty} G\left(f + \frac{k}{T_s}\right) = T_s \quad \text{for all } f \tag{9.23}$$

Note that the criterion for zero ISI does not uniquely specify the pulse spectrum $G(f)$ unless its bandwidth is limited to $(-1/2T_s, 1/2T_s)$, in which case it must be flat since the sum in (9.23) reduces to one term ($k = 0$). In this special case, $G(f)$ itself has a rectangular transfer function in the interval $(-1/2T_s, 1/2T_s)$, or equivalently

$$g(t) = \frac{\sin \pi t/T_s}{\pi t/T_s} \tag{9.24}$$

The interpretation of this result is that for a digital modulation of the form in (9.1), the minimum one-sided lowpass bandwidth needed to transmit $R_s = 1/T_s$ symbols per second (sps) without ISI, is $R_s/2 = 1/2T_s$ Hz. When the data symbols $\{a_n\}$ are independent and the noise samples (spaced T_s-sec apart) are uncorrelated, then each symbol can be recovered without resorting to past history of the waveform, that is, a zero memory system.

Since in the above system R_s sps are transmitted without ISI over a bandwidth $R_s/2$ Hz, then the *throughput* (efficiency of transmission) is R_s sps$/(R_s/2)$ Hz $= 2$ sps/Hz. To achieve this efficiency, one must generate the $\sin x/x$ pulse shape in (9.24). Unfortunately, this pulse shape is impractical because its very slowly decreasing tail will cause excessive ISI if any perturbations from the ideal sampling instants occur.

To reduce this sensitivity to timing (sampling instant) offset and at the same time ease the filter design problem in the transmitter and receiver, we use more practical shapes for $g(t)$ whose Fourier transform $G(f)$ has smoother transition at the band edges yet still satisfies the zero ISI criterion of (9.23). One such class of channels that has achieved a good deal of practical application is the set of *raised-cosine channels* defined by the overall transmission characteristic

$$G(f) = \begin{cases} T_s; & 0 \le |f| \le \frac{1-\alpha}{2T_s} \\ T_s \cos^2\left\{\frac{\pi}{4\alpha}[2|f|T_s - 1 + \alpha]\right\}; & \frac{1-\alpha}{2T_s} \le |f| \le \frac{1+\alpha}{2T_s} \\ 0; & \text{otherwise} \end{cases} \tag{9.25}$$

where $0 \le \alpha \le 1$ is defined as the *excess bandwidth factor*. The special case of $\alpha = 0$ corresponds to the minimum Nyquist bandwidth brick wall (rectangular) filter as discussed above. Values of $\alpha > 0$ trade increased bandwidth (and hence reduced throughput for a fixed data rate) for a reduction in the rate of roll-off of the frequency response. In fact, for the class of systems defined by (9.25), the throughput is reduced to $R_s/[(R_s/2)(1 + \alpha)] = 2/(1 + \alpha)$ sps/Hz. Figure 9.2a is a plot of $G(f)$ for several values of α. The corresponding overall system impulse response is

$$g(t) = \underbrace{\left(\frac{\sin \pi t/T_s}{\pi t/T_s}\right)}_{g_0(t)} \underbrace{\left(\frac{\cos \alpha \pi t/T_s}{1 - (2\alpha t/T_s)^2}\right)}_{g_a(t)} \tag{9.26}$$

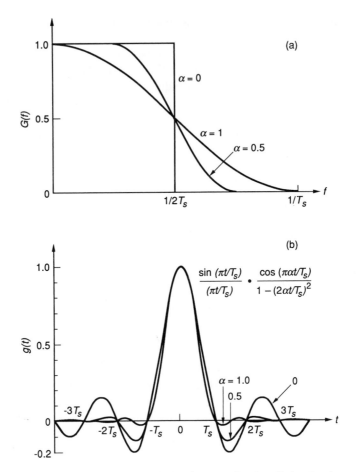

Figure 9.2   The raised cosine pulse: (a) frequency function, (b) time function

which is illustrated in Fig. 9.2b for the same values of α as in Fig. 9.2a. For $\alpha = 0$, we get the time response for the minimum bandwidth brick wall filter as given in (9.24). For values of $\alpha > 0$, the tails of the response decay more rapidly than for the $\sin x/x$ response of (9.24).

Note that (9.26) is written in the form of a product of $g_0(t)$ (the overall system impulse response corresponding to the minimum bandwidth brick wall Nyquist filter) and another response, $g_a(t)$. In fact, (9.26) can be generalized to include any function $g_a(t)$ that is bounded at the zero crossings of $g_0(t)$ since under this condition, the zero crossings of $g(t)$ will be the combined zero crossings of $g_0(t)$ and $g_a(t)$ and thus the zero ISI condition is maintained. Stated in the frequency domain, the convolution of the minimum bandwidth brick wall filter $G_0(f) = \mathcal{F}\{g_0(t)\}$ and an arbitrary filter $G_a(f)$ whose inverse Fourier transform $g_a(t)$ is bounded at the zero crossings of $g_0(t)$ is a Nyquist filter.

9.2.2 Apportionment of Filtering between Transmitter and Receiver

Having just seen how to design the overall system frequency response $G(f)$ to achieve zero ISI, the next issue concerns how to optimally apportion this response between the transmitter and receiver. By optimum, we mean that the apportionment should result in minimum error probability or equivalently maximum detected sample SNR. From (9.8), we see that this is equivalent to maximizing the ratio[4]

$$\frac{g_0^2}{\sigma_y^2} = \frac{\left(\int_{-\infty}^{\infty} G_t(f)G_c(f)G_r(f)e^{j2\pi f t_0}df\right)^2}{\int_{-\infty}^{\infty} S_n(f)\,|G_r(f)|^2df} \tag{9.27}$$

which is identical to the matched filtering problem considered in Chapter 4. Thus, we conclude that the transmit and receive filters are related by (see Problem 9.2)

$$G_r(f) = K\frac{G_t^*(f)G_c^*(f)e^{-j2\pi f t_0}}{S_n(f)} \tag{9.28}$$

where K is a real constant, or, in terms of the overall frequency response, we should choose

$$|G_r(f)| = \left(\frac{KG^*(f)e^{-j2\pi f t_0}}{S_n(f)}\right)^{1/2}$$

$$\tag{9.29}$$

$$|G_t(f)| = \left(\frac{G(f)S_n(f)e^{j2\pi f t_0}}{K\,|G_c(f)|^2}\right)^{1/2}$$

Since $S_n(f)$ is real, then we must have $\arg\{G(f)\} = e^{-j2\pi f t_0}$ in order for (9.29) to have any meaning. Moreover, since $g_0 = 1$, then from (9.27)–(9.29), the constant is evaluated as $K = \left(\int_{-\infty}^{\infty}|G_t(f)|^2|G_c(f)|^2/|S_n(f)df\right)^{-1}$.

If we now assume that $n(t)$ corresponds to white Gaussian noise with PSD $S_n(f) = N_0/2$, then we get

$$|G_r(f)| = \left(\frac{2K\,|G(f)|}{N_0}\right)^{1/2} = \left(\frac{|G(f)|}{\int_{-\infty}^{\infty}|G_t(f)|^2\,|G_c(f)|^2\,df}\right)^{1/2}$$

$$\tag{9.30}$$

$$|G_c(f)|\,|G_t(f)| = \left(\frac{|G(f)|\,N_0}{2K}\right)^{1/2} = \left(|G(f)|\int_{-\infty}^{\infty}|G_t(f)|^2\,|G_c(f)|^2\,df\right)^{1/2}$$

which implies that, except for a scaling constant, the overall pulse response should be split equally between the combined transmit/channel and receive filters. Furthermore, the maximum SNR achieved by this apportionment is from (9.27) given by

$$SNR_{max} = A^2\left(\frac{g_0^2}{\sigma_y^2}\right)_{max} = \frac{A^2}{\frac{N_0}{2}\int_{-\infty}^{\infty}|G_r(f)|^2\,df} = \frac{2A^2\int_{-\infty}^{\infty}|G_t(f)|^2\,|G_c(f)|^2\,df}{N_0}$$

$$\tag{9.31}$$

$$= \frac{2T_s}{N_0}\left(A^2\frac{1}{T_s}\int_{-\infty}^{\infty}|G_t(f)|^2\,|G_c(f)|^2\,df\right) \triangleq \frac{2P_rT_s}{N_0}$$

where P_r is the signal power at the input of the receiver. For binary amplitude modulation,

---

4. Note that for multilevel (M-ary) amplitude modulation, the symbol error probability is given by $P_s(E) = \frac{M-1}{M}\,\mathrm{erfc}\left\{\sqrt{\frac{3}{M^2-1}}\frac{Ag_0}{\sqrt{2}\sigma_y}\right\}$ (see Chapter 4) and thus the optimum apportionment is still obtained by minimizing (9.27).

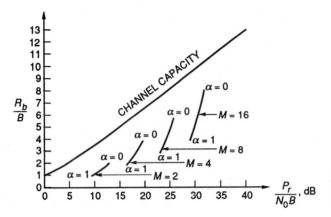

Figure 9.3 Throughput performance of M-ary amplitude modulation on a raised cosine Nyquist channel

the minimum error probability is obtained by using (9.31) in (9.8), resulting in

$$P_b(E)|_{\min} = \frac{1}{2}\,\mathrm{erfc}\,\sqrt{\frac{P_r T_b}{N_0}} \tag{9.32}$$

where we have also substituted T_b for T_s. Comparing (9.32) with the comparable result for the unrestricted bandwidth case (that is, rectangular pulses transmitted over an infinite bandwidth channel) as discussed in Chapter 4, we conclude that the two are completely equivalent if we associate $P_r T_b$ with the energy per bit E_b. For M-level amplitude modulation with a Gray code bit-to-symbol mapping, the result would be approximately given by (9.32) with a factor of $\sqrt{(M^2-1)/3}$ included in the argument of the erfc function. Similarly, the throughput for M-level raised cosine channels would now be $2\log_2 M/(1+\alpha)$ bps/Hz.

It is of interest now to compare the throughput of the various Nyquist systems as a function of the SNR in the Nyquist bandwidth, $P_r/N_0 B$ required to achieve, say, a fixed symbol error rate equal to 10^{-5}. For M-ary amplitude modulation, we have approximately (ignoring the factor of $(M-1)/M$ in front of the erfc function) that $[3/(M^2-1)]P_r T_s/N_0 = 9.12$ (9.6 dB). Thus

$$
\begin{aligned}
\frac{P_r}{N_0 B} &= \left(\frac{P_r T_s}{N_0}\right)\left(\frac{R_s}{B}\right) = \left(\frac{P_r T_s}{N_0}\right)\left(\frac{R_b}{B}\right)\frac{1}{\log_2 M}\\[2mm]
&= \frac{9.12\,(M^2-1)}{3}\,\frac{2}{(1+\alpha)} = \frac{(M^2-1)}{3}\,\frac{18.24}{(1+\alpha)}
\end{aligned}
\tag{9.33}
$$

Figure 9.3 is a plot of throughput $R_b/B = 2\log_2 M/(1+\alpha)$ bps/Hz versus SNR in the Nyquist bandwidth, $P_r/N_0 B$ for $P_s(E) = 10^{-5}$ with α and M as parameters. Also indicated on this plot is the throughput corresponding to channel capacity [see Chapter 1, Eq. 1.30)], which is repeated here (letting $C = R_b$) as

$$\frac{R_b}{B} = \log_2\left(1 + \frac{P_r}{N_0 B}\right) \tag{9.34}$$

As an example, suppose we wanted to achieve a throughput $R_b/B = 2$ bps/Hz. If we were to use a binary ($M = 2$) system, we would require an ideal brick wall Nyquist filter ($\alpha = 0$) with $P_r/N_0 B = 18.24$ (12.6 dB). If instead we were to use a 4-level system ($M = 4$), we could relax the filtering requirement by using a 100% excess bandwidth ($\alpha = 1$) Nyquist

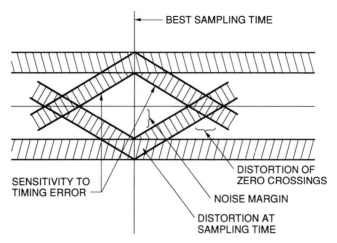

Figure 9.4   The eye pattern and its characteristics

raised cosine filter, but the required P_r/N_0B would now be $18.24(5/2) = 45.6$ (16.6 dB). Thus, the penalty in P_r/N_0B for using higher level modulation with practical filtering is 4 dB.

9.2.3 The Eye Pattern of a Bandlimited Communication System

A useful tool for qualitatively assessing the severity of ISI degradation is the so-called *eye pattern*. To see how this comes about, we proceed as follows. Consider the bandlimited communication system of Fig. 9.1 wherein the transmitted modulation is of the form in (9.1) and a received signal is of the form in (9.3). For a specific data symbol sequence $\{a_n\}$, $\tilde{m}(t)$ takes on a particular waveform. Suppose we consider the superposition of all $\tilde{m}(t)$ waveforms corresponding to *all possible* sequences $\{a_n\}$. The resulting pattern when viewed on an oscilloscope is called an *eye pattern*, so-called because of its resemblance with the shape and characteristics of the human eye (see Fig. 9.4 for a qualitative example corresponding to binary amplitude modulation). Since the above superposition considers all possible sequences (each theoretically of infinite length), the resulting waveform is the same in every T_s-sec interval, that is, *the eye pattern is periodic with period T_s.* Hence, it is sufficient to examine the eye pattern and its properties only in a T_s-sec interval.

The two most important properties of an eye pattern are its *maximum opening* (the largest unobstructed interior vertical distance at any sample time instant) and the *broadness* (horizontally) of the opening in the neighborhood of its maximum. The significance of these two eye pattern measures is as follows.

The maximum eye opening is a measure of the maximum degradation of the *signal* component of the detected sample due to ISI. Equivalently, it is an indication of the ability of the system to protect itself against bit errors when noise is present since, indeed, half the maximum eye opening is the size of the smallest output *noise* sample that will cause an error (assuming that sampling is performed at the point of maximum eye opening). Clearly, the smaller the eye opening, the more ISI is present and the smaller the noise sample need be to cause a bit error.

The broadness of the eye in the neighborhood of the maximum eye opening is an in- dication of how *timing jitter* tolerant the system is. Clearly, the steeper the eye pattern is

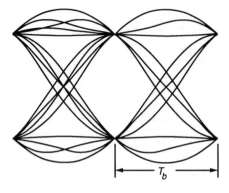

Figure 9.5a An eye pattern for binary amplitude modulation (Reprinted from [3])

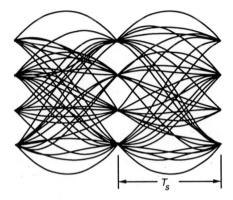

Figure 9.5b An eye pattern for multilevel amplitude modulation (Reprinted from [3])

in this vicinity, the quicker the eye closes with any perturbation around the desired sample point. Thus, a broad eye is desirable so that an offset in the sampling instant will produce minimum eye closure (relative to its maximum value) and minimum error probability degradation. Based on this consideration, it is now possible to clarify a statement made earlier with regard to the reduction in sensitivity to timing offset of the raised cosine channel pulse response relative to the $\sin x / x$ response of the brick wall channel. For the latter, the eye will be much narrower in the vicinity of the peak because of the slowly decreasing tail.

A more realistic eye pattern diagram that results from a digital stream with pulses having a shape that extends over many T_b-sec intervals is illustrated in Fig. 9.5a. We observe that at the point of maximum opening, the eye is "well-focused" with little degradation due to ISI. Figure 9.5b illustrates an eye pattern for multilevel ($M = 4$) pulse amplitude modulation. Here the maximum opening of any of the $M - 1 = 3$ eyes is a measure of the resistance against making an error (due to noise) by choosing a symbol adjacent to the one transmitted.

9.2.4 Systems with Controlled ISI—Partial Response

In many transmission channels, bandwidth is at a premium and thus the question naturally arises: how can we increase the throughput of the system, for example, for a fixed bandwidth, how can we increase the bit rate? If we restrict ourselves to zero ISI systems, that is, channels that satisfy the Nyquist criterion of Section 9.2.1, then the only solution is to increase the number of modulation levels by going to a larger signaling alphabet. While this does indeed produce a higher bit rate for a given symbol rate (or equivalently, a given bandwidth), the penalty paid is an increase in the required transmitter power. On the other hand, if we remove the zero ISI constraint and allow a finite and controlled amount of ISI, then it is possible to achieve a more efficient trade-off between bandwidth and power than that achievable by the above solution.

Lender [6] was the first to show that using binary modulation but introducing *dependencies* or *correlations* between the amplitudes (that is, nonzero ISI) and suitably changing the detection procedure, one could still achieve the maximum efficiency of 2 bps/Hz with, however, realizable and perturbation-tolerant filters. The method was called *duobinary encoded modulation*, the prefix *duo* referring to the fact that the overall system time response $g(t)$ has *two* equal nonzero sample values. As such there exists 100% ISI which stems, however, from only one adjacent symbol. Duobinary is the simplest form of controlled ISI signaling; hence, we start our discussion with a description of this correlative encoding technique.

The duobinary encoding technique.   Consider the block diagram of Fig. 9.1 where the block labeled "modulation source" that generates (9.1) is modeled as a binary data source represented by a uniform train of impulse functions at the data rate $1/T_b$ fed to a pulse shaping filter $P(f)$. Consider now encoding the binary data source prior to pulse shaping with a simple digital filter as shown in Fig. 9.6. The digital filter delays the input and then adds it to itself. Thus, the nth impulse into the pulse shaping filter has a strength proportional to the *algebraic* sum of the nth (current) input bit and the $(n-1)^{\text{st}}$ (preceding) input bit. As such, the effective encoded bit stream $\{a'_n\} = \{a_n + a_{n-1}\}$ at the channel input has dependencies among its members. Note that because of the algebraic addition, the symbol sequence $\{a'_n\}$ is now *ternary* (rather than binary) in accordance with Table 9.1.

Note that when $a_n = -1$, there are two possible levels for a'_n, namely, 0, -2, depending on the value of a_{n-1}. Similarly, when $a_n = 1$, there are again two possible levels for a'_n, namely, 0, 2, also depending on the value of a_{n-1}. We can think of this dependence on the previous symbol as introducing a *controlled* amount of ISI. Thus, whereas Nyquist (zero memory) signaling is based on eliminating ISI, signaling with duobinary or correlative encoding schemes is based on allowing a controlled amount (finite memory) of ISI, that is, a decision on a_n based on a threshold test performed on the sum of a'_n plus noise depends on a_{n-1}.

We can write the signal component of the receive filter output $y(t)$ as

$$y_s(t) = A \sum_{i=-\infty}^{\infty} a'_i g(t - iT_b) = A \sum_{i=-\infty}^{\infty} (a_i + a_{i-1})g(t - iT_b)$$

$$= A \sum_{i=-\infty}^{\infty} [a_i g(t - iT_b) + a_i g(t - (i+1)T_b)] \qquad (9.35)$$

$$= A \sum_{i=-\infty}^{\infty} a_i [g(t - iT_b) + g(t - (i+1)T_b)] = A \sum_{i=-\infty}^{\infty} a_i h(t - iT_b)$$

Thus, as far as the modulation is concerned, the combined transmitter/receiver of Fig. 9.6 can be represented as in Fig. 9.7 where

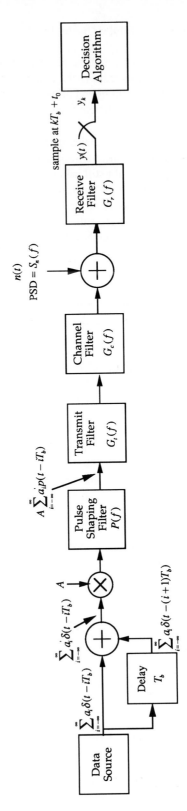

Figure 9.6 Block diagram of bandlimited communication system with duobinary encoded source

Table 9.1   Data Encoding

a_{n-1}	a_n	a'_n
-1	-1	-2
-1	1	0
1	-1	0
1	1	2

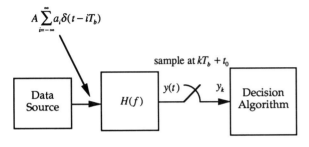

Figure 9.7   Equivalent block diagram to Figure 9.6 for modulation

$$H(f) = G(f) \left[1 + e^{-j2\pi f T_b} \right] = G(f) e^{-j\pi f T_b} \left[e^{j\pi f T_b} + e^{-j\pi f T_b} \right]$$

$$= 2G(f) \cos \pi f T_b \exp \left(-j\pi f T_b \right) \tag{9.36}$$

$$|H(f)| = 2 |G(f)| |\cos \pi f T_b|$$

Suppose that $G(f)$ is an ideal brickwall (minimum bandwidth Nyquist) filter, that is,

$$G(f) = \begin{cases} T_b; & |f| \le \frac{1}{2T_b} \\ 0; & \text{otherwise} \end{cases} \tag{9.37}$$

Then,

$$|H(f)| = \begin{cases} 2T_b \cos \pi f T_b; & |f| \le \frac{1}{2T_b} \\ 0; & \text{otherwise} \end{cases} \tag{9.38}$$

We can implement $H(f)$ directly and thus no separate digital filter is required in front of $P(f)$ as in Fig. 9.6. Rather, as we shall see shortly, we merely split $H(f)$ (equally for additive white noise) into $H_t(f)$ and $H_r(f)$. Note that since no excess (over the minimum Nyquist) bandwidth is used, the efficiency of transmission is 2 bps/Hz. Thus, we are able to achieve the maximum efficiency for binary transmission using, however, practical (perturbation-tolerant) filters.

The impulse response corresponding to (9.38) is

$$h(t) = \frac{4}{\pi} \left(\frac{\sin \frac{\pi t}{T_b}}{1 - 4 \left(\frac{t - \frac{T_b}{2}}{T_b} \right)^2} \right) \tag{9.39}$$

and is illustrated in Fig. 9.8. Note that $h(t)$ has *two* equal nonzero samples, that is, one at $t = 0$ and one at $t = T_b$. Thus, as previously stated, the detected samples will have 100% ISI due to one adjacent pulse.

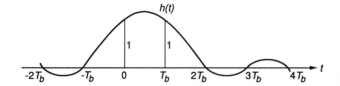

Figure 9.8 Duobinary pulse shape

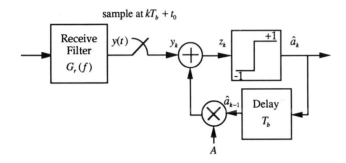

Figure 9.9 Duobinary receiver with perfect decision feedback

If we could always make *perfect* decisions, we could, in principle, subtract out the ISI and use a binary decision scheme with a single threshold. The appropriate receiver for doing this would be as shown in Fig. 9.9. If no error is made on the decision for a_{k-1}, that is, $\hat{a}_{k-1} = a_{k-1}$, then $z_k = y_k - A\hat{a}_{k-1} = Aa_k + \tilde{n}_k$ and the probability of error for a_k would be $\frac{1}{2} \operatorname{erfc}\left(A/\sqrt{2}\sigma_y\right)$ which is the identical result as that for binary Nyquist signaling. On the other hand, if an error was made on the decision for a_{k-1}, that is, $\hat{a}_{k-1} = -a_{k-1}$, then $z_k = y_k - A\hat{a}_{k-1} = A(a_k + 2a_{k-1}) + \tilde{n}_k$. Thus, depending on the relative polarities of a_k and a_{k-1}, the double ISI can help or hurt. The problem with this detection scheme then is that once an error occurs, it tends to *propagate* further errors, that is, if a_{k-1} is in error, then its effect is incorrectly compensated for when deciding a_k; consequently, a_k is likely to be in error also.

To get around the problem of error propagation, Lender [6] suggested a means of avoiding the necessity for ISI compensation (subtracting out the single ISI) at the receiver. His scheme suggested that we *precode* the data $\{a_i\}$ at the transmitter into the sequence $\{d_i\}$ as follows. If we think of the binary data as 0's and 1's, then the precoder is described by $d_n = a_n \oplus d_{n-1}$ where the symbol $\oplus$ denotes modulo-2 addition. Note that a_n now indicates the *change* in the output symbol d_n. If $a_n = 0$, then there is no change in d_n. If $a_n = 1$, then there is a change in d_n. This simple encoding scheme is identical to differential encoding described in Chapter 7 in connection with differential detection of BPSK modulation. The precoder can also be reinterpreted in terms of "1'"s and "−1'"s for the input bits $\{a_n\}$. In particular, the modulo-2 summer is replaced by a multiplier resulting in the relation $d_n = a_n \times d_{n-1}$. Note that d_n is still a binary random sequence with equiprobable 1's and −1's.

Using the ± 1 representation for the a_n's, we obtain the implementation for the duobinary system illustrated in Fig. 9.10. The mean of y_k, the sample of $y(t)$ at $t = kT_b + t_0$, is now equal to $A[d_k + d_{k-1}]$. Table 9.2 gives the values for this mean in relation to the possible values of the input (information) binary symbol a_k.

Thus, we see that when the information bit is a "−1", then the detected signal sample is "0" (independent of a_{k-1}). When the information bit is a "+1", then the detected signal

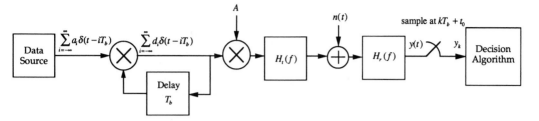

Figure 9.10   Duobinary system implementation

Table 9.2   Possible Values for the Mean of the Detected Sample

a_k	d_{k-1}	d_k	$A[d_k + d_{k-1}] = \bar{y}_k$
1	1	1	$2A$
1	-1	-1	$-2A$
-1	1	-1	0
-1	-1	1	0

sample is $\pm 2A$ with the polarity depending on a_{k-1}. If we examine the *magnitude* of the detected signal sample, then, independent of a_{k-1}, the detected output sample will be $2A$ when the information bit is a "+1". Thus, in the absence of noise, we can decode perfectly according to the following rule:

$$\text{Decide } a_k = 1 \text{ if } |y_k| = 2A \qquad \text{Decide } a_k = -1 \text{ if } |y_k| = 0$$

By choosing a threshold[5] halfway between "0" and "$2A$," then in the presence of noise we have the decision rule $\hat{a}_k = \text{sgn}\{|y_k| - A\}$. This decision rule involves *rectification* of y_k followed by a single threshold binary decision. *Since no knowledge of a sample other than y_k is involved (that is, the decision rule requires no knowledge of previous symbol decisions), error propagation cannot occur.* The threshold comparison and decision logic described above can be implemented as in Fig. 9.11.

What remains is to compute the probability of error for the duobinary scheme with precoding. For a flat AWGN channel, we saw in Section 9.2.2 that the total system response [in this case $H(f)$] should be split equally between the transmitter and receiver. Thus, we choose the two filters $H_t(f)$ and $H_r(f)$ to be identical and equal to the square root of $H(f)$. Under these circumstances, the input power to the receiver is

$$P_r = A^2 \overline{d_k^2} \int_{-\infty}^{\infty} \frac{1}{T_b} |H_t(f)|^2 \, df = A^2 \int_{-(2T_b)^{-1}}^{(2T_b)^{-1}} \frac{1}{T_b} 2T_b \cos \pi f T_b \, df$$

$$= \frac{4}{\pi} \left(\frac{A^2}{T_b} \right)$$

(9.40)

The variance of the detected sample, y_k, is

5. This choice of threshold can be shown (see Prob. 9.5) to minimize the average bit error probability in the limit of large SNR.

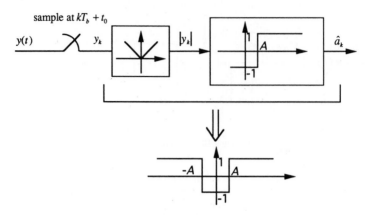

Figure 9.11   Threshold comparison and decision logic

$$\sigma_y^2 = \frac{N_0}{2} \int_{-\infty}^{\infty} |H_r(f)|^2 df = \frac{N_0}{2} \left(\frac{4}{\pi}\right) = \frac{2N_0}{\pi} \tag{9.41}$$

The receiver levels are 0, $2A$, and $-2A$. The level "0" occurs with probability $1/2$ and the levels "$2A$" and "$-2A$" each occur with probability $1/4$. Thus, the probability of error is

$$P_b(E) = \frac{1}{4} \Pr\{-A < y_k < A | a_k = 1 (\bar{y}_k = 2A)\}$$

$$+ \frac{1}{4} \Pr\{-A < y_k < A | a_k = 1 (\bar{y}_k = -2A)\}$$

$$+ \frac{1}{2} \Pr\{y_k > A \text{ or } y_k < -A | a_k = -1 (\bar{y}_k = 0)\}$$

$$= \frac{1}{4} \int_{-A}^{A} \frac{1}{\sqrt{2\pi\sigma_y^2}} \exp\left\{-\frac{(y - 2A)^2}{2\sigma_y^2}\right\} dy$$

$$+ \frac{1}{4} \int_{-A}^{A} \frac{1}{\sqrt{2\pi\sigma_y^2}} \exp\left\{-\frac{(y + 2A)^2}{2\sigma_y^2}\right\} dy \tag{9.42}$$

$$+ \frac{1}{2} \int_{A}^{\infty} \frac{1}{\sqrt{2\pi\sigma_y^2}} \exp\left\{-\frac{y^2}{2\sigma_y^2}\right\} dy + \frac{1}{2} \int_{-\infty}^{-A} \frac{1}{\sqrt{2\pi\sigma_y^2}} \exp\left\{-\frac{y^2}{2\sigma_y^2}\right\} dy$$

$$= \frac{3}{4} \operatorname{erfc}\left(\frac{A}{\sqrt{2}\sigma_y}\right) - \frac{1}{4} \operatorname{erfc}\left(\frac{3A}{\sqrt{2}\sigma_y}\right)$$

In terms of P_r of (9.40), (9.42) becomes

$$P_b(E) = \frac{3}{4} \text{erfc} \left(\sqrt{\frac{\pi^2}{16} \frac{P_r T_b}{N_0}} \right) - \frac{1}{4} \text{erfc} \left(\sqrt{\frac{9\pi^2}{16} \frac{P_r T_b}{N_0}} \right)$$

$$\cong \frac{3}{4} \text{erfc} \left(\sqrt{\frac{\pi^2}{16} \frac{P_r T_b}{N_0}} \right) \quad \text{for large SNR}$$

(9.43)

Thus, when compared with binary amplitude modulation using ideal (brick wall) filters to achieve 2 bps/Hz, the SNR penalty (ignoring the factor of 3/4 versus 1/2) is $-10 \log_{10}(\pi^2/16) = 2.1$ dB.

Generalization of duobinary to raised cosine Nyquist channels.   Suppose now that instead of $G(f)$ being a flat channel as in (9.37), $G(f)$ is the raised cosine channel characterized by (9.25). Then, the magnitude response of the equivalent channel $H(f)$ is given by the product of (9.25) and $2 \cos \pi f T_b$. Since $g(t)$ has a single nonzero sample when sampled at $1/T_b$, then $h(t)$ will once again have two nonzero samples when sampled at the same rate. Hence, the implementation of Fig. 9.10 and the detection scheme of Fig. 9.11 still apply here. In fact, an evaluation of the input power, P_r, and variance of the detected sample, σ_y^2, produce results identical to (9.40) and (9.41) respectively (see Problem 9.6). Thus, the probability of error for the *duobinary raised cosine channel* is also given by (9.42).

What is different then about the raised cosine channel from the flat channel is the trade-off between throughput and signal-to-noise ratio. Ignoring the factor of 3/4 in (9.42), then for $P_b(E) = 10^{-5}$, we require $(\pi^2/16) P_r T_b/N_0 = 9.12$ (9.6 dB). The throughput is, as before, $R_b/B = 2/(1+\alpha)$. Thus

$$\frac{P_r}{N_0 B} = \frac{P_r T_b}{N_0} \left(\frac{R_b}{B} \right) = (9.12) \left(\frac{16}{\pi^2} \right) \left(\frac{2}{1+\alpha} \right) = \frac{16}{\pi^2} \left(\frac{18.24}{1+\alpha} \right)$$

(9.44)

Once again, we see the trade-off between throughput and SNR as a function of the excess fractional bandwidth α. In fact, a plot of R_b/B versus $P_r/N_0 B$ with α as a parameter would be identical to that determined from (9.33) evaluated for $M = 2$, except it would be displaced horizontally (along the $P_r/N_0 B$ axis) by $10 \log_{10}(16/\pi^2) = 2.1$ dB.

Other partial response techniques.   Generalizations of the duobinary encoding scheme to multiple levels is possible. For M-level duobinary, that is, $a_n = \pm 1, \pm 3, \ldots \pm (M-1)$, one would replace the binary differential encoder with an M-ary differential encoder (the modulo-2 summer becomes a modulo-M summer). Asymptotically for large SNR, the resulting symbol error probability performance of this scheme becomes

$$P_s(E) = \left(1 - \frac{1}{M^2} \right) \text{erfc} \left[\sqrt{\frac{\pi^2}{16} \left(\frac{P_r T_s}{N_0} \right) \left(\frac{3}{M^2 - 1} \right)} \right]$$

(9.45)

Thus, ignoring the $(1 - 1/M^2)$ factor, the SNR degradation for any M relative to M-ary amplitude modulation with Nyquist pulses is again $10 \log_{10}(16/\pi^2) = 2.1$ dB.

Aside from extension of duobinary to multiple levels, other generalizations involve correlative encoding techniques that make use of overall pulse responses with more than two nonzero samples. These techniques fall into what is commonly referred to as *partial response* signaling [5, 10, 11]. A generalized partial response digital filter, $T(f)$, has the form of a tapped delay line, that is,

$$T(D) = \sum_{n=0}^{N-1} \gamma_n D^n; \quad T(f) = T(D)|_{D=e^{-j2\pi f T_b}} = \sum_{n=0}^{N-1} \gamma_n e^{-j2\pi f n T_b} \quad (9.46)$$

(see Fig. 9.12) which, when combined with a Nyquist filter, $G(f)$ produces the overall response characteristic $H(f)$. By choosing various combinations of integer values for the coefficients $\{\gamma_k\}$ in $T(f)$, one can obtain a variety of different useful overall responses, $H(f)$, which have a controlled amount of ISI over a span of N bits. Each of these generates a different partial response scheme. As in duobinary, the separation of $H(f)$ into $T(f)$ and $G(f)$ is artificial in that the actual implementation may consist of simply a single transmit filter and a single receive filter corresponding to an equal partition of $H(f)$. The selection of the coefficients $\{\gamma_k\}$ is governed by the following considerations.

If $H(f)$ and its first $K-1$ derivatives are continuous and the Kth derivative is discontinuous, then $h(t)$ decays asymptotically as $1/|t|^{K+1}$, that is, the larger K is, the less energy there exists in the tails of $h(t)$ and thus the smaller the ISI is [12]. Clearly, for minimum bandwidth systems wherein $G(f)$ is the brick wall Nyquist filter, $T(f)$ must have a zero (spectral null) at $f = 1/2T_b$ or equivalently, $T(D)$ must have a zero at $D = -1$ (where $G(f)$ has a discontinuity) in order for $H(f)$ to be a continuous function. Stated another way, for $H(f)$ to be a continuous function, $T(D)$ must contain the factor $(1+D)$. In the more general case, the conditions on $T(D)$ to assure the continuity of $H(f)$ *and* its derivatives are described by the following proposition. The first $K-1$ derivatives of a minimum bandwidth system response $H(f)$ are continuous if and only if $T(D)$ has $(1+D)^K$ as a factor. Note that as the multiplicity of the $(1+D)$ factor increases, the number of output levels in the detected signal sample increases and thus error probability performance degrades. Thus as K increases, a trade-off exists between reduced sharpness of spectral roll-off (easing of practical filter design) and increase in error probability.

In addition to the Kth order spectral null at $f = 1/2T_b$, it is often desirable to have a spectral null at the origin ($f = 0$), for example, for the possibility of inserting a carrier pilot tone. To achieve a spectral null at the origin, $T(D)$ must contain the factor $(1-D)$. Higher order $(1-D)$ factors, such as $(1-D)^L$, cause a more gradual roll-off of frequency components just above dc, which may also be desirable from the standpoint of allowing easier extraction of the pilot tone at $f = 0$ at the receiver. Combining the two spectral null requirements, we see that $T(D)$ should contain the factors $(1+D)$ and $(1-D)$. Most common partial response systems have a $T(D)$ of the form

$$T(D) = (1+D)^K (1-D)^L \quad (9.47)$$

Some common examples are as follows:

Duobinary

$$K = 1, \quad L = 0, \quad T(D) = 1 + D$$

Dicode

$$K = 0, \quad L = 1, \quad T(D) = 1 - D$$

Modified Duobinary (see Prob. 9.7)

$$K = 1, \quad L = 1, \quad T(D) = (1-D)(1+D) = 1 - D^2$$

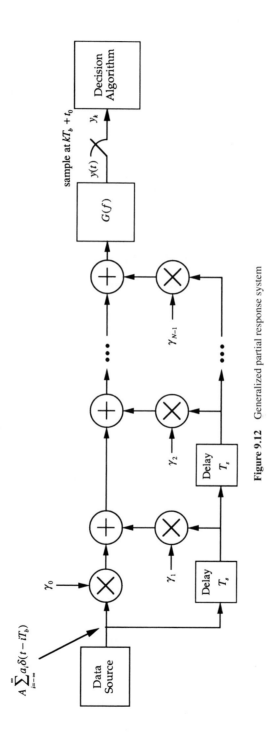

Figure 9.12 Generalized partial response system

Table 9.3 Characteristics of Minimum Bandwidth Partial Response Systems

$T(D)$	$H(f)$ for $\|f\| \le 1/2T_s$	$h(t)$	Number of Output Levels (M-ary Input)
$1+D$	$2T_s \cos \pi f T_s$	$\dfrac{4}{\pi} \dfrac{\cos \pi t/T_s}{1-4(t/T_s)^2}$	$2M-1$
$1-D$	$j2T_s \sin \pi f T_s$	$\dfrac{4t/T_s}{\pi} \dfrac{\cos \pi t/T_s}{4(t/T_s)^2-1}$	$2M-1$
$1-D^2$	$j2T_s \sin 2\pi f T_s$	$\dfrac{2}{\pi} \dfrac{\sin \pi t/T_s}{(t/T_s)^2-1}$	$2M-1$
$(1+D)^2$	$4T_s \cos^2 \pi f T_s$	$\dfrac{2}{\pi t/T_s} \dfrac{\sin \pi t/T_s}{1-(t/T_s)^2}$	$4M-3$
$(1+D)^2(1-D)$	$j4T_s \sin 2\pi f T_s \cos \pi f T_s$	$-\dfrac{64t/T_s}{\pi} \dfrac{\cos \pi t/T_s}{\left(4(t/T_s)^2-9\right)\left(4(t/T_s)^2-1\right)}$	$4M-3$
$(1+D)(1-D)^2$	$-4T_s \sin 2\pi f T_s \sin \pi f T_s$	$\dfrac{16}{\pi} \dfrac{\left(4(t/T_s)^2-3\right)\cos \pi t/T_s}{\left(4(t/T_s)^2-9\right)\left(4(t/T_s)^2-1\right)}$	$4M-3$
$(1+D)^2(1-D)^2$	$-4T_s \sin^2 2\pi f T_s$	$\dfrac{8}{\pi t/T_s} \dfrac{\sin \pi t/T_s}{(t/T_s)^2-4}$	$4M-3$
$2+D-D^2$	$T_s(1+\cos 2\pi f T_s + j3 \sin 2\pi f T)$	$\dfrac{1}{\pi t/T_s} \dfrac{(3t/T_s-1)\sin \pi t/T_s}{(t/T_s)^2-1}$	$4M-3$
$2-D^2-D^4$	$T_s(-1+\cos 4\pi f T_s + j3 \sin 4\pi f T)$	$\dfrac{2}{\pi t/T_s} \dfrac{(2-3t/T_s)\sin \pi t/T_s}{(t/T_s)^2-4}$	$4M-3$

Table 9.3 gives expressions for the minimum bandwidth $H(f)$ and $h(t)$ corresponding to the above examples as well as some others due to Kretzmer [5].[6]

9.3 OPTIMUM DEMODULATION OF DIGITAL SIGNALS IN THE PRESENCE OF ISI AND AWGN

As we have seen in Section 9.2.4, the presence of ISI in a digital pulse stream introduces memory into the decision process in that the detected sample y_k for estimating the data symbol a_k depends on knowledge of the previous data symbols $a_{k-1}, a_{k-2}, \ldots$. If the ISI extends only over a finite number of symbols, then the memory is finite. In the case of small amounts of controlled (intentionally introduced) ISI, we saw that by proper pulse shape design and encoding of the data information symbols, we could still obtain a reasonable error probability performance using partial response techniques. In many instances, the ISI introduced is unintentional and extends over a large number of past symbols, thus potentially causing severe performance degradation. In these cases, it becomes necessary to design an optimum receiver with memory which takes into account the presence of both ISI and the additive noise. The approach to designing such optimum receivers is based on the maximum-likelihood principle and is referred to as *maximum-likelihood sequence estimation (MLSE)* [13]. As the acronym suggests, optimum decisions are made on *sequences* of information symbols rather than *symbol-by-symbol* with the goal of minimizing the sequence error probability.

Consider the block diagram illustrated in Fig. 9.1 which sends a complex digital data sequence $\{c_n\}$ over a bandlimited AWGN channel. As before, the transmitter filtering is accomplished by $G_t(f)$ [impulse response $g_t(t)$] and the channel $G_c(f)$ [impulse response

---

6. In this table, for mathematical convenience, we have shifted the time origin to the center of the nonzero samples, that is, $t = (N-1)T_s/2$ is the new time origin. Thus, for example, the duobinary frequency response $H(f)$ is given by (9.36) multiplied by $e^{-j\pi f T_b}$ and $h(t)$ corresponds to (9.39) with t replaced by $t+T_b/2$.

$g_c(t)$] being bandlimited distorts the pulse shape $g_t(t)$ and introduces ISI.[7] The received pulse shape is given by $h(t) = g_t(t) * g_c(t)$. Thus, the received signal plus noise is expressed as[8]

$$\tilde{r}(t) = \sum_{n=-\infty}^{\infty} \tilde{c}_n h(t - nT_s) + \tilde{n}(t) \tag{9.48}$$

Suppose now we expand $\tilde{r}(t)$ in a Karhunen-Loève series, namely

$$\tilde{r}(t) = \lim_{K \to \infty} \sum_{k=1}^{K} \tilde{r}_k \phi_k(t) \tag{9.49}$$

where $\{\phi_k(t)\}$ is *any* set of orthonormal functions (since the noise is white) and

$$\tilde{r}_k = \int_{-\infty}^{\infty} \tilde{r}(t) \phi_k^*(t) dt = \int_{-\infty}^{\infty} \sum_{n=-\infty}^{\infty} \tilde{c}_n h(t - nT_s) \phi_k^*(t) dt + \int_{-\infty}^{\infty} \tilde{n}(t) \phi_k^*(t) dt$$

$$\tag{9.50}$$

$$= \sum_{n=-\infty}^{\infty} \tilde{c}_n h_{kn} + \tilde{n}_k$$

with

$$h_{kn} = \int_{-\infty}^{\infty} h(t - nT_s) \phi_k^*(t) dt; \quad \tilde{n}_k = \int_{-\infty}^{\infty} \tilde{n}(t) \phi_k^*(t) dt \tag{9.51}$$

The $\tilde{n}_k$'s are in general independent complex zero-mean Gaussian variables with equal variance (because of the white noise assumption all eigenvalues are equal)

$$E\left\{ |\tilde{n}_k|^2 \right\} = 2\sigma^2 \tag{9.52}$$

that is, $\sigma^2 = N_0/2A^2$ represents the variance of either the real or imaginary part of $\tilde{n}_k$.

Consider the K-dimensional observable $\tilde{\mathbf{r}}_K = (\tilde{r}_1, \tilde{r}_2, \ldots, \tilde{r}_K)$ which depends on the data sequence $\tilde{\mathbf{c}} = (\ldots, \tilde{c}_{-2}, \tilde{c}_{-1}, \tilde{c}_0, \tilde{c}_1, \tilde{c}_2, \ldots)$ through (9.48) and (9.49). Since from the above the $\tilde{r}_k$'s are independent complex Gaussian random variables conditioned on $\tilde{\mathbf{c}}$, then the joint pdf of the components of $\tilde{\mathbf{r}}_K$ is given by

$$f(\tilde{\mathbf{r}}_K | \tilde{\mathbf{c}}) = \frac{1}{\left(2\pi\sigma^2\right)^K} \exp\left\{ -\frac{1}{2\sigma^2} \sum_{k=1}^{K} \left| \tilde{r}_k - \sum_{n=-\infty}^{\infty} \tilde{c}_n h_{kn} \right|^2 \right\}$$

$$\tag{9.53}$$

$$= \frac{1}{\left(2\pi\sigma^2\right)^K} \exp\left\{ -\frac{1}{2\sigma^2} \sum_{k=1}^{K} |\tilde{n}_k|^2 \right\}$$

The optimum (in the sense of maximizing the probability of correct detection) decision rule is obtained from maximizing the a posteriori probability of the transmitted *sequence* given the observable, namely, $f(\tilde{\mathbf{c}} | \tilde{\mathbf{r}}_K)$. Using Bayes rule (see Chapter 3), this probability can be expressed in terms of the probability $f(\tilde{\mathbf{r}}_K | \tilde{\mathbf{c}})$ of (9.53). For independent data, maximizing

---

7. As before, we again assume that $G_t(f)$ includes the modulation pulse shaping $P(f)$, and thus $g_t(t)$ is, in reality, the convolution of $g_t(t)$ with $p(t)$.

8. Herein it shall be convenient to renormalize the problem such that the signal amplitude $A = 1$. Thus, the PSD of $\tilde{n}(t)$ now becomes N_0/A^2.

$f(\tilde{c}|\tilde{r}_K)$ is equivalent to maximizing $f(\tilde{r}_K|\tilde{c})$. Alternately, since the optimum (maximum a posteriori) decision rule is: choose the *sequence* $\hat{\tilde{c}}$ that maximizes $f(\tilde{c}|\tilde{r}_K)$ for the given observable $\tilde{r}_K$, then the equivalent optimum (maximum-likelihood) decision rule is:

Choose the *sequence* $\hat{\tilde{c}}$ that maximizes $f(\tilde{r}_K|\tilde{c})$ or equivalently, $\ln f(\tilde{r}_K|\tilde{c})$.

Taking the natural logarithm of (9.53) gives

$$\ln f(\tilde{r}_K|\tilde{c}) = -\ln\left(2\pi\sigma^2\right)^K - \frac{1}{2\sigma^2}\sum_{k=1}^{K}\left|\tilde{r}_k - \sum_{n=-\infty}^{\infty}\tilde{c}_n h_{kn}\right|^2 \tag{9.54}$$

Since the first term of (9.54) is independent of $\tilde{c}$, then equivalently we can maximize the function

$$F_K(\tilde{c}) = -\sum_{k=1}^{K}\left|\tilde{r}_k - \sum_{n=-\infty}^{\infty}\tilde{c}_n h_{kn}\right|^2 = -\sum_{k=1}^{K}|\tilde{n}_k|^2 \tag{9.55}$$

In the limit as $K \to \infty$, we get

$$\int_{-\infty}^{\infty}|\tilde{n}(t)|^2\,dt = \int_{-\infty}^{\infty}\lim_{K\to\infty}\sum_{k=1}^{K}\tilde{n}_k\phi_k(t)\sum_{l=1}^{K}\tilde{n}_l^*\phi_i^*(t)dt = \lim_{K\to\infty}\sum_{k=1}^{K}|\tilde{n}_k|^2 \tag{9.56}$$

Thus, taking the limit of (9.56) as $K \to \infty$ gives

$$F(\tilde{c}) = \lim_{K\to\infty}F_K(\tilde{c}) = -\int_{-\infty}^{\infty}|\tilde{n}(t)|^2\,dt = -\int_{-\infty}^{\infty}\left|\tilde{r}(t) - \sum_{n=-\infty}^{\infty}\tilde{c}_n h(t-nT_s)\right|^2 dt$$

$$= -\int_{-\infty}^{\infty}|\tilde{r}(t)|^2\,dt + 2\operatorname{Re}\left\{\sum_{n=-\infty}^{\infty}\tilde{c}_n^*\int_{-\infty}^{\infty}\tilde{r}(t)h(t-nT_s)dt\right\} \tag{9.57}$$

$$- \sum_{n=-\infty}^{\infty}\sum_{m=-\infty}^{\infty}\tilde{c}_n\tilde{c}_m^*\int_{-\infty}^{\infty}h(t-nT_s)h(t-mT_s)dt$$

The maximum-likelihood estimate of the data sequence $\tilde{c}$ is the particular sequence that maximizes $F(\tilde{c})$ of (9.57). We can simplify this result still further by noting that the first term in (9.57) is independent of $\tilde{c}$; hence, it can be discarded with regard to the maximization operation. For the second term in (9.57), we define

$$\tilde{y}_n \triangleq \tilde{y}(nT_s) = \int_{-\infty}^{\infty}\tilde{r}(t)h(t-nT_s)dt \tag{9.58}$$

where

$$\tilde{y}(t) = \int_{-\infty}^{\infty}\tilde{r}(\tau)h(\tau-t)d\tau \tag{9.59}$$

The interpretation of (9.58) is that it represents uniform samples at $t = nT_s$ of the output, $\tilde{y}(t)$, of a matched filter with impulse response $g_r(t) = h(-t)$ excited by $\tilde{r}(t)$.

In the third term of (9.57), define

$$g\left((m-n)T_s\right) \triangleq g_{m-n} = \int_{-\infty}^{\infty}h(t-nT_s)h(t-mT_s)dt \tag{9.60}$$

where

$$g(t) = \int_{-\infty}^{\infty} h(\tau)h(\tau + t)d\tau = \int_{-\infty}^{\infty} h(\tau' - t)h(\tau')d\tau'$$

$$= \int_{-\infty}^{\infty} h(\tau' - nT_s - t)h(\tau' - nT_s)d\tau' \tag{9.61}$$

The interpretation of (9.60) is that it represents uniform samples at $t = (m - n)T_s$ of the output, $g(t)$, of a matched filter with impulse response $g_r(t) = h(-t)$ excited by $h(t)$, or equivalently, uniform samples at $t = nT_s$ of this same output when the excitation is $h(t - mT_s)$. Thus, the set $\{g_{m-n}\}$ represents the ISI in the output of the receive filter $G_r(f)$ in Fig. 9.1 at the sample time $t = nT_s$. [9] Note from (9.61) that $g(t)$ is an even function of t and is in fact the time autocorrelation of $h(t)$. Thus, the Fourier transform of $g(t)$ is given by $G(f) = |H(f)|^2$ and is a real and even function of frequency. Finally, substituting (9.58) and (9.60) in (9.57) gives the decision rule:

Choose the *sequence* $\hat{\tilde{\mathbf{c}}}$ that maximizes

$$F_0(\tilde{\mathbf{c}}) = 2\,\text{Re}\left\{ \sum_{n=-\infty}^{\infty} \tilde{c}_n^* y_n \right\} - \sum_{n=-\infty}^{\infty} \sum_{m=-\infty}^{\infty} \tilde{c}_n^* \tilde{c}_m g_{m-n} \tag{9.62}$$

We point out that we are not particularly concerned with the noncausality of $h(t)$ since, in practice, we can introduce a sufficiently large delay to make it causal.

Note that for $g(t)$ corresponding to a Nyquist channel (that is, $g(nT_s) = 0$ for $n \neq 0$), (9.62) simplifies to

$$F_0(\tilde{\mathbf{c}}) = 2\,\text{Re}\left\{ \sum_{n=-\infty}^{\infty} \tilde{c}_n^* y_n \right\} - \sum_{n=-\infty}^{\infty} |\tilde{c}_n|^2 g_0 \tag{9.63}$$

Hence, the decision rule becomes:

Choose the *sequence* $\hat{\tilde{\mathbf{c}}}$ that maximizes $\displaystyle\sum_{n=-\infty}^{\infty} \left(\text{Re}\left\{ \tilde{c}_n^* y_n \right\} - \frac{1}{2}|\tilde{c}_n|^2 g_0 \right)$

Furthermore, we can write $\tilde{y}_k = \displaystyle\sum_{m=-\infty}^{\infty} \tilde{c}_m g_{m-k} + \tilde{\mu}_k$ where

$$\tilde{\mu}_k = \int_{-\infty}^{\infty} \tilde{n}(t)g_r(kT_s - t)\,dt = \int_{-\infty}^{\infty} \tilde{n}(t)h(t - kT_s)\,dt \tag{9.64}$$

is the detected Gaussian noise sample at time kT_s with complex correlation function

$$R_{\tilde{\mu}}(l - k) = E\left\{ \tilde{\mu}_k \tilde{\mu}_l^* \right\} = 2\sigma^2 g_{l-k} \tag{9.65}$$

For the Nyquist channel, the noise samples are uncorrelated (also independent because they are Gaussian) and hence the probability of a correct decision for the *sequence* $\tilde{\mathbf{c}}$ is equal to the product of the correct decision probabilities for each symbol detected independently. Thus, the above decision rule results in the same average error probability as symbol-by-symbol detection, that is, *in the absence of ISI on an AWGN channel optimum sequence detection offers no improvement in performance over optimum symbol-by-symbol detection.*

---

9. We assume here that the arbitrary sampling epoch $t_0 = 0$.

9.3.1 Evaluation of the MLSE—The Viterbi Algorithm

The brute force method for determining the optimum sequence $\hat{\tilde{c}}$ is to exhaustively try all possible sequences[10] in (9.62) and then select the one that yields the largest value of $F_0(\tilde{c})$. A more efficient method for processing the sampled complex matched filter outputs $\{\tilde{y}_n\}$ is the so-called *Viterbi algorithm (VA)* [7] which was originally proposed for optimum decoding of convolutional codes. This algorithm performs maximum-likelihood estimation of a digital sequence and thus is often called a maximum-likelihood decoding algorithm [14]. Note again that the optimum decision rule in the presence of ISI and AWGN makes a decision on a *sequence* $\tilde{c}$ rather than symbol-by-symbol detection, which is a suboptimum decision rule. The MLSE technique is capable of producing performance approaching that of an optimum symbol-by-symbol demodulator operating in the presence of AWGN alone.

The VA may be viewed as a solution to the problem of maximum a posteriori (MAP) estimation of the state sequence of a finite-state discrete-time Markov process observed in memoryless noise. Before starting a formal explanation of the VA, we must first cast the MLSE in the presence of ISI and AWGN in this form. The formulation for accomplishing this is due to Forney [13, 15]. For the purpose of our discussion, we shall assume that the ISI exists over a finite time interval corresponding to ν discrete times on either side of g_0, that is, $g_l = 0$ for $|l| > \nu$. Although $g(t)$ is bandlimited and thus cannot also be timelimited, the assumption of a finite length is a reasonable approximation in most practical applications.

We begin by first writing the detected sample $\tilde{y}_k$ of (9.58) as

$$\tilde{y}_k = \sum_{l=-\nu}^{\nu} \tilde{c}_{k-l} g_l + \tilde{\mu}_k \qquad (9.66)$$

The discrete-time process $\{\tilde{y}_k\}$ is called a *shift register process* since it can be modeled by a shift register of length 2ν whose input is the sequence $\tilde{c}$ (see Fig. 9.13a). The problem with the representation of Fig. 9.13a is the correlation of the noise samples [see (9.65)] which makes the performance analysis of the MLSE difficult. To circumvent this problem, we shall introduce a noise whitening discrete filter (see Chapter 4) after the sampling operation in Fig. 9.1 so that the detected samples will take the form of (9.66) with, however, uncorrelated Gaussian noise samples.[11] The noise whitening filter is determined as follows.

Since from (9.60) and (9.61), $g_{m-n} = g_{n-m}$, then g_k can be expressed as the discrete convolution

$$g_k = \sum_{i=0}^{\nu-k} f_i f_{i+k} \qquad (9.67)$$

where $f_i = 0$ for $i < 0$ and $i > \nu$. Note that $g_k = 0$ for $|k| > \nu$ as required above. In terms of the z-transform of these sequences, namely

---

10. This, of course, only has practical meaning if the sequence length and thus the number of possible sequences is finite. We shall be more specific about this shortly.

11. The introduction of a noise whitening filter is made merely for convenience but is not necessarily required to implement the Viterbi algorithm. A description of the operation of the Viterbi algorithm for combined ISI and AWGN channels without using a noise whitening filter is given in [17].

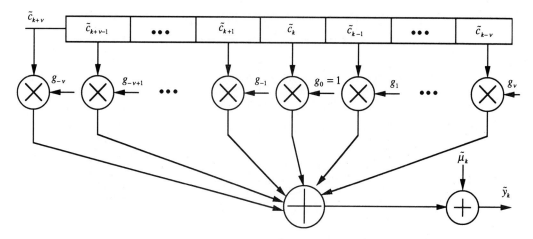

Figure 9.13a   Shift register representation of bandlimited system in the presence of ISI and AWGN

$$G(z) = \sum_{k=-\nu}^{\nu} g_k z^{-k}; \quad F(z) = \sum_{k=0}^{\nu} f_k z^{-k} \tag{9.68}$$

we have $G(z) = F(z)F(z^{-1})$. Since the sequence $\{f_i\}$ is causal, $F(z)$ is a physically realizable digital filter but $F(z^{-1})$ is not. Despite the fact that $F(z^{-1})$ is not physically realizable, we can still choose it to be minimum phase. By doing so, its inverse $1/F(z^{-1})$ is a physically realizable digital filter. By selecting $1/F(z^{-1})$ as the noise whitening digital filter, then passing $\tilde{y}_k$ through this filter produces the sample $\tilde{y}'_k$, which can be expressed in the form

$$\tilde{y}'_k = \sum_{l=0}^{\nu} \tilde{c}_{k-l} f_l + \tilde{\mu}'_k \triangleq \tilde{w}_k + \tilde{\mu}'_k \tag{9.69}$$

where $\tilde{\mu}'_k$ is an element of a zero mean *white* Gaussian noise sequence with variance $E\left\{\left|\tilde{\mu}'_k\right|^2\right\} = 2\sigma^2$. In terms of the sequence $\{\tilde{y}'_k\}$, the decision rule of (9.62) can be rewritten as

Choose the *sequence* $\hat{\tilde{\mathbf{c}}}$ that maximizes

$$F_0(\tilde{\mathbf{c}}) = 2\,\mathrm{Re}\left\{\sum_{n=-\infty}^{\infty} \tilde{\mu}_n^* \tilde{y}'_n\right\} - \sum_{n=-\infty}^{\infty} |\tilde{w}_n|^2 \tag{9.70a}$$

$$= 2\,\mathrm{Re}\left\{\sum_{n=-\infty}^{\infty} \left(\sum_{i=0}^{\nu} \tilde{c}_{n-i}^* f_i\right) \tilde{y}'_n\right\} - \sum_{n=-\infty}^{\infty} \left|\sum_{i=0}^{\nu} \tilde{c}_{n-i} f_i\right|^2$$

or equivalently minimizes

$$F'_0(\tilde{\mathbf{c}}) = \sum_{n=-\infty}^{\infty} |\tilde{y}'_n - \tilde{w}_n|^2 = \sum_{n=-\infty}^{\infty} \left|\tilde{y}'_n - \sum_{i=0}^{\nu} \tilde{c}_{n-i} f_i\right|^2 \tag{9.70b}$$

The $\tilde{y}'_k$'s form a set of sufficient statistics for estimation of the input data sequence $\tilde{\mathbf{c}}$ [15].

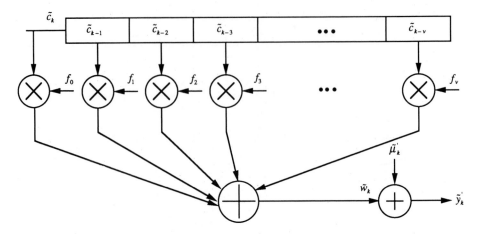

Figure 9.13b   Shift register representation of bandlimited system in the presence of ISI and AWGN (noise whitening filter included)

Figure 9.13b is a model of the shift register process corresponding to (9.69).[12] Note that the length of the register is now equal to ν in accordance with the duration of the f_i sequence. Also, $\{\tilde{y}_k\}$ is now a νth order M-ary Markov process.

The *state*, $\tilde{x}_k$, of the process is defined by

$$\tilde{x}_k = (\tilde{c}_{k-1}, \tilde{c}_{k-2}, \ldots, \tilde{c}_{k-\nu})$$

and the *transition* (change of states), ξ_k, at discrete time k is defined as the pair of states $(\tilde{x}_{k+1}, \tilde{x}_k)$, or in terms of the inputs[13]

$$\tilde{\xi}_k = (\tilde{c}_k, \tilde{c}_{k-1}, \tilde{c}_{k-2}, \ldots, \tilde{c}_{k-\nu})$$

Thus, the maximum number of states is M^ν and the maximum number of transitions is $M^{\nu+1}$. Note that there is a one-to-one correspondence between sequences of states and sequences of transitions. Also if the input sequence "starts" at time zero (that is, $\tilde{c}_0$ is the first transmitted symbol) and "ends" at time $K - \nu - 1$ (that is, $\tilde{c}_{K-\nu-1}$ is the last transmitted symbol), then the contents of the shift register in Fig. 9.13 is zero prior to the start, that is, the state $\tilde{x}_0 = (\tilde{c}_{-1}, \tilde{c}_{-2}, \tilde{c}_{-3}, \ldots, \tilde{c}_{-\nu}) = (0, 0, 0, \ldots, 0)$, and is again zero K symbols later, that is, the state $\tilde{x}_K = (\tilde{c}_{K-1}, \tilde{c}_{K-2}, \tilde{c}_{K-3}, \ldots, \tilde{c}_{K-\nu}) = (0, 0, 0, \ldots, 0)$.

Finally, in the context of the above, we can restate the problem to which the VA is a solution. Given a sequence $\tilde{\mathbf{y}}$ of observations of a discrete-time finite-state Markov process in memoryless noise, find the state sequence $\tilde{\mathbf{x}}$ for which the a posteriori probability $f(\tilde{\mathbf{x}}|\tilde{\mathbf{y}})$ is maximum. Alternately, since $\tilde{\mathbf{x}}$ and the transition sequence $\tilde{\boldsymbol{\xi}}$ have a one-to-one correspondence, we can restate the problem as: find the sequence transition $\tilde{\boldsymbol{\xi}}$ for which $f\left(\tilde{\boldsymbol{\xi}}|\tilde{\mathbf{y}}\right)$ is a maximum. For the shift register model, this is the same as finding the most probable input sequence $\tilde{\mathbf{c}}$ since $\tilde{\mathbf{x}}$ and $\tilde{\mathbf{c}}$ also have a one-to-one correspondence. This MAP

12. Since herein we shall be dealing with the respresentation of (9.69) rather than (9.66), for simplicity of notation, we shall drop the prime on $\tilde{y}'_k$ and $\tilde{\mu}'_k$.

13. Despite the fact that the state and transition are defined by sequences, it is customary not to denote them with a boldface typeface.

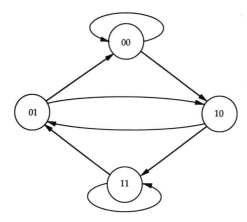

Figure 9.14a   State diagram of a four-state shift register process

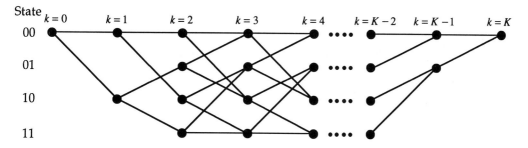

Figure 9.14b   Trellis diagram of a four-state shift register process

decision rule, which minimizes the error probability in detecting the whole sequence, is thus optimum in this sense.

Associated with a discrete-time finite-state Markov process is a *state diagram* in which nodes represent states and branches represent transitions. As discrete time progresses, the process traces a path from state to state in this diagram. An example of such a state diagram is illustrated in Fig. 9.14a for a binary[14] system with ISI memory $\nu = 2$. Since $M = 2$, there are $M^{\nu} = 4$ states and a total of $M^{\nu+1} = 8$ transitions among them. Note that from any given state, there are only two allowable transitions to the next state. For example, if the current (at time k) state is "01" corresponding to $a_{k-1} = 0$ and $a_{k-2} = 1$, then the next state will be "00" if $a_k = 0$ or "10" if $a_k = 1$. Note further that the *structure* of the state diagram depends only on the input sequence **a** and the memory of the ISI, not the values of the ISI (shift register tap weights) themselves.

An alternate and more redundant description of this same process is a *trellis diagram* [14], so-called because of its resemblance to a garden trellis. Figure 9.14b is the trellis diagram corresponding to the state diagram of Fig. 9.14a. The trellis diagram illustrates the progression of the process with time, a property that is lacking in the state diagram representation. Each node in the trellis diagram corresponds to a possible state of the state diagram at that particular instant of discrete time. Thus, the maximum number of states at

---

14. For this example, $\tilde{c}_k$ is real, that is, $\tilde{c}_k = a_k$ and takes on values 0 or 1 with equal probability.

any particular time in the trellis is equal to the number of states in the state diagram. In fact, the only time the trellis will have fewer than this maximum number of states is for the first few time instants at the beginning and at the end of the input sequence. Each branch in the trellis diagram represents a transition from a particular state at a given time to a new state at the next instant of time. The inital state (I) of the trellis corresponds to $\tilde{x}_0$ and the final state (F) corresponds to $\tilde{x}_K$, both of which will be "00" if the initial contents of the shift register was empty. Associated with each sequence of states, $\tilde{\mathbf{x}}$, is a unique path through the trellis and vice versa. Since the most probable sequence is the one for which $f(\tilde{\mathbf{x}}|\tilde{\mathbf{y}})$ [or equivalently $f(\tilde{\mathbf{x}}, \tilde{\mathbf{y}}) = f(\tilde{\mathbf{x}}|\tilde{\mathbf{y}}) f(\tilde{\mathbf{y}})$ is maximum, if we assign to each path $\tilde{\mathbf{x}}$ the quantity $-\ln f(\tilde{\mathbf{x}}, \tilde{\mathbf{y}})$, then since the negative logarithm is a monotonically decreasing function of its argument, the most probable sequence is also the one for which $-\ln f(\tilde{\mathbf{x}}, \tilde{\mathbf{y}})$ is minimum. We shall see in a moment that we can attach to the quantity $-\ln f(\tilde{\mathbf{x}}, \tilde{\mathbf{y}})$ the significance of path "length."

Since the process $\tilde{\mathbf{y}}$ is observed in memoryless noise, then

$$f(\tilde{\mathbf{y}}|\tilde{\mathbf{x}}) = f\left(\tilde{\mathbf{y}}|\tilde{\boldsymbol{\xi}}\right) = \prod_{k=0}^{K-1} f\left(\tilde{y}_k|\tilde{\xi}_k\right) = \prod_{k=0}^{K-1} f(\tilde{y}_k|\tilde{x}_{k+1}, \tilde{x}_k) \tag{9.71}$$

Furthermore, since the states obey a first-order Markov behavior, that is,

$$f(\tilde{x}_{k+1}|\tilde{x}_0, \tilde{x}_1, \ldots, \tilde{x}_k) = f(\tilde{x}_{k+1}|\tilde{x}_k) \tag{9.72}$$

then since the initial state occurs with probability one, that is, $f(\tilde{x}_0) = 1$, the probability of the state sequence becomes

$$f(\tilde{\mathbf{x}}) = \prod_{k=0}^{K-1} f(\tilde{x}_{k+1}|\tilde{x}_k) \tag{9.73}$$

and thus the joint probability $f(\tilde{\mathbf{x}}, \tilde{\mathbf{y}})$ is given by

$$f(\tilde{\mathbf{x}}, \tilde{\mathbf{y}}) = f(\tilde{\mathbf{x}}) f(\tilde{\mathbf{y}}|\tilde{\mathbf{x}}) = \prod_{k=0}^{K-1} f(\tilde{x}_{k+1}|\tilde{x}_k) \prod_{k=0}^{K-1} f(\tilde{y}_k|\tilde{x}_{k+1}, \tilde{x}_k) \tag{9.74}$$

Taking the natural logarithm of (9.74) converts the products into sums; thus

$$-\ln f(\tilde{\mathbf{x}}, \tilde{\mathbf{y}}) = -\sum_{k=0}^{K-1} (\ln f(\tilde{x}_{k+1}|\tilde{x}_k) + \ln f(\tilde{y}_k|\tilde{x}_{k+1}, \tilde{x}_k)) = \sum_{k=0}^{K-1} \lambda\left(\tilde{\xi}_k\right) \tag{9.75}$$

where the weight $\lambda\left(\tilde{\xi}_k\right)$ is assigned to each branch (transition) and represents its length. Hence, the sum in (9.75) represents the length of the entire path.

Finding the shortest length path through the trellis [path with minimum $-\ln f(\tilde{\mathbf{x}}, \tilde{\mathbf{y}})$] in an efficient way is what the VA accomplishes. Consider a segment of the state sequence $\tilde{\mathbf{x}} = (\tilde{x}_0, \tilde{x}_1, \ldots, \tilde{x}_K)$ consisting of the states from discrete time zero to discrete time k, that is, $\tilde{\mathbf{x}}_0^k = (\tilde{x}_0, x_1, \ldots, \tilde{x}_k)$. In the trellis diagram, $\tilde{\mathbf{x}}_0^k$ corresponds to a path from state $\tilde{x}_0$ to state $\tilde{x}_k$. For any of the M^ν (or fewer) nodes at discrete time k, there will be in general several such path segments entering that node, each with a length

$$\lambda\left(\tilde{\mathbf{x}}_0^k\right) = \sum_{i=0}^{k-1} \lambda\left(\tilde{\xi}_i\right) \tag{9.76}$$

that depends on the particular transitions $\tilde{\xi}_i$; $i = 0, 1, \ldots, k - 1$ along that path segment. One of these path segments, $\hat{\tilde{\mathbf{x}}}(\tilde{x}_k)$, entering the node will have the shortest length $\Gamma(\tilde{x}_k) \triangleq$ $\min \lambda(\tilde{\mathbf{x}}_0^k) = \lambda(\hat{\tilde{\mathbf{x}}}(\tilde{x}_k))$. This particular shortest length path segment is referred to as the *survivor* corresponding to that particular node. The reason for this terminology stems from the following observation:

> The shortest state sequence $\hat{\tilde{\mathbf{x}}}$ (the one we are looking for) must at any discrete time k along its path correspond to a survivor path segment.

If this were not true, then at any node along $\hat{\tilde{\mathbf{x}}}$ we could replace the path segment from state x_0 to that node by the survivor and thus achieve a smaller length which leads to a contradiction that $\hat{\tilde{\mathbf{x}}}$ is the shortest length path. The result of all this is that at discrete time $0 < k \leq K$, *it is only necessary to save (keep track of) the survivor paths and their lengths*, one each for the M^ν (or fewer) nodes at that time instant. In proceeding to discrete time $k + 1$, one merely appends the survivors with the branches leading to the next set of states and increments the survivor lengths with values related to the corresponding transitions. At time $k + 1$, one again chooses the M^ν survivors and the process continues recursively. Finally, at time K, the trellis converges to a single node with one survivor that must correspond to the shortest state sequence $\hat{\tilde{\mathbf{x}}}$ being sought after. The length, $\Gamma(\hat{\tilde{\mathbf{x}}}) = \Gamma(\hat{\tilde{\mathbf{x}}}(\tilde{x}_K))$, of this shortest path must from (9.76) be equal to the minimum value of (9.75) as desired.

Figure 9.15 illustrates the above recursive evaluation of the shortest path for the trellis diagram of Fig. 9.14b and a value of $K = 5$. The values assigned to the trellis branches are not intended to be typical of any particular application but are merely selected for simplicity of the distance evaluations. Since $M = 2$ and $\nu = 2$, there are 4 (or fewer) nodes and 8 (or fewer) transitions at each discrete time. At each of these nodes, the solid lines entering the node indicate the survivor path while the dashed lines indicate the nonsurvivor paths. Finally, the shortest path through the trellis has length equal to 3.

Although the VA as described above will theoretically find the shortest length path through the trellis, the presence of very long or infinite state sequences requires a buffer memory beyond practical limits. Hence, in practice, one must invoke a modification of the algorithm wherein survivors are prematurely truncated to a manageable length, say δ. That is, at time k the algorithm makes a final decision on the first $k - \delta$ branches of the shortest path (that is, nodes up to time $k - \delta$) and retains only the length and path information on the most recent δ branches (the truncation depth) of the survivors. The ability to make such a premature decision relies heavily on the extent to which the time-k survivors share a common path for the first $k - \delta$ nodes. For example, in the example of Fig. 9.15 we observe that at $k = 4$, the two surviving paths shared a common path up to time $k = 2$. Since one of these two survivors will ultimately be selected as the shortest path, nothing is lost by making a final decision at time $k = 4$ on the first two branches ($\delta = 2$) and truncating the survivors to two branches. Clearly, for sufficiently large δ, there is a high probability that all survivors at time k will pass through the same set of nodes for the first $k - \delta$ time instants. In the rare cases where survivors disagree in their first $k - \delta$ branches, one must adopt a time-$(k - \delta)$ decision strategy that has negligible effect on the ultimate error probability performance. One such possible strategy is to choose the time-$(k - \delta)$ node corresponding to the shortest time-k survivor hoping, of course, that this survivor will eventually wind up

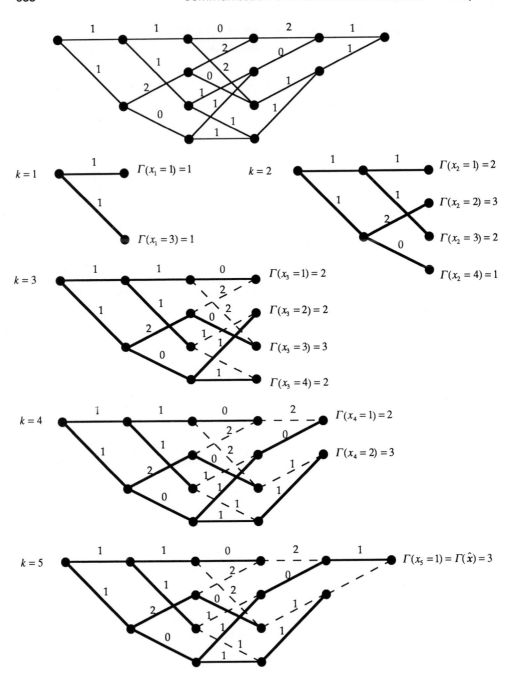

Figure 9.15   Recursive determination of shortest length path for four-state trellis diagram with $K = 5$

to be the shortest length path. Another possibility is to choose the time-$(k - \delta)$ node by a majority vote, that is, the one through which most of the time-k survivors pass.

From a computational point of view, as time k gets large, the values of survivor path length, $\Gamma(\tilde{x}_k)$, for each of the M^ν states also become large and thus from time to time it is necessary to subtract a constant from all of them.

A few words about complexity are in order. Since the algorithm requires M^ν storage locations, one for each state (after the initial transient) capable of storing a survivor path and its length, then from a memory standpoint, the complexity grows exponentially with the length ν of the shift register. The complexity associated with the computation of the incremental lengths, $\lambda\left(\tilde{\xi}_k\right)$, is assessed as follows. Since from (9.74), one component of $\lambda\left(\tilde{\xi}_k\right)$ is equal to $-\ln f(\tilde{x}_{k+1}|\tilde{x}_k)$, then since for a shift register process $f(\tilde{x}_{k+1}|\tilde{x}_k)$ is either equal to $1/M$ or 0 depending upon whether state $\tilde{x}_{k+1}$ is allowable after the occurrence of state $\tilde{x}_k$, all *allowable* transitions have the same value of $-\ln f(\tilde{x}_{k+1}|\tilde{x}_k)$. Thus, this component of $\lambda\left(\tilde{\xi}_k\right)$ may be ignored. The computation of the remaining component of $\lambda\left(\tilde{\xi}_k\right)$, namely, $-\ln f\left(\tilde{y}_k|\tilde{\xi}_k\right)$ depends on the data sequence. However, from Fig. 9.13b and (9.69), we observe that, in general, many different values of $\tilde{\xi}_k$ produce the same value of shift register output $\tilde{w}_k$. Since $\tilde{y}_k = \tilde{w}_k + \tilde{\mu}_k$, then for a given $\tilde{y}_k$, all of these values of $\tilde{\xi}_k$ produce the same value of Gaussian probability $f\left(\tilde{y}_k|\tilde{\xi}_k\right)$ and thus $-\ln f\left(\tilde{y}_k|\tilde{\xi}_k\right)$ need be computed only once. Because of the above, the complexity associated with the computation of the $\lambda\left(\tilde{\xi}_k\right)$'s tends to be insignificant relative to that associated with the memory requirement.

Example 9.1

To illustrate the application of the Viterbi algorithm specifically to the ISI problem, consider a binary PAM ($\tilde{c}_k = a_k = \pm 1$) system with memory $\nu = 2$. The appropriate representation of the system is given in Fig. 9.13b where the shift register now has only two stages. The state, the transition, and the shift register output are real and are given by $x_k = (a_{k-1}, a_{k-2})$, $\xi_k = (a_k, a_{k-1}, a_{k-2})$, and $w_k = f_0 a_k + f_1 a_{k-1} + f_2 a_{k-2}$, respectively. The state transition diagram for this shift register process is identical in form to Fig. 9.14a. We redraw it here in Fig. 9.16a where we have now labeled the branches with the input (a_k)/output (w_k) corresponding to each transition. The conditional pdf $f(y_k|\xi_k)$ is Gaussian and given by

$$f(y_k|\xi_k) = f(y_k|w_k) = \frac{1}{\sqrt{\pi N_0}} \exp\left\{-\frac{(y_k - w_k)^2}{N_0}\right\}$$

$$= \frac{1}{\sqrt{\pi N_0}} \exp\left\{-\frac{[y_k - (f_0 a_k + f_1 a_{k-1} + f_2 a_{k-2})]^2}{N_0}\right\}$$

(9.77)

Taking the natural logarithm of (9.77), we get

$$-\ln f(y_k|\xi_k) = \frac{1}{2}\ln \pi N_0 + \frac{[y_k - (f_0 a_k + f_1 a_{k-1} + f_2 a_{k-2})]^2}{N_0}$$

(9.78)

Since the first term in (9.78) is independent of the data, then dropping the $1/N_0$ scaling constant in the second term, the normalized incremental length (weight) to be assigned to transitions between the kth and $k + 1^{\text{st}}$ state in the trellis diagram is $[y_k - (f_0 a_k + f_1 a_{k-1} + f_2 a_{k-2})]^2$, that is, the squared distance between the output signal plus noise sample and the shift register output. Figure 9.16b is an illustration of the appropriate trellis diagram (identical in form to Fig. 9.14b) where the branches have now been labeled as per the above.

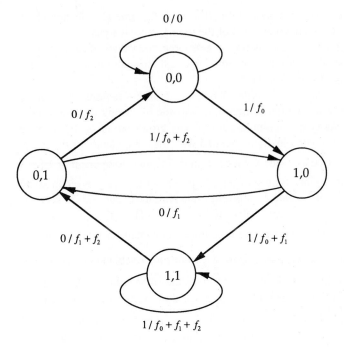

Figure 9.16a   State diagram for binary transmission with ISI memory $\nu = 2$

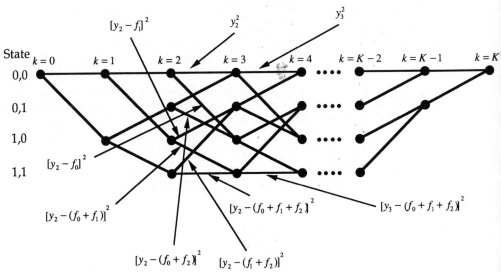

Figure 9.16b   Trellis diagram corresponding to state diagram of Figure 9.16a

9.3.2 Error Probability Performance

To evaluate the error probability performance of the VA, we introduce the notion of an *error event at time k*, which is defined based on the following observation. Consider that the correct path through the trellis (that is, the path corresponding to the transmitted sequence $\tilde{\mathbf{c}}$) is defined by the state sequence $\tilde{\mathbf{x}}$. Suppose, however, that the shortest distance path computed by the VA corresponds to a different state sequence, say $\hat{\tilde{\mathbf{x}}}$. For sufficiently large sequence lengths, the paths $\tilde{\mathbf{x}}$ and $\hat{\tilde{\mathbf{x}}}$ will agree over much of their length with an occasional discrepancy wherein $\hat{\tilde{\mathbf{x}}}$ deviates (diverges) from $\tilde{\mathbf{x}}$ for a number of branches and then returns to (remerges with) $\tilde{\mathbf{x}}$. These deviant paths are referred to as *error events*. Along each of these deviant paths (unmerged segments), the accumulated weight (metric) is larger than that along the correct path. When the time instant at which a deviant path begins corresponds to time k, we refer to these paths as *error events at time k*. Assuming that the error event terminates at time $k + L$, then for such an error event, we have $\tilde{x}_k = \hat{\tilde{x}}_k$, $\tilde{x}_{k+L} = \hat{\tilde{x}}_{k+L}$, and $\tilde{x}_i \neq \hat{\tilde{x}}_i$; $k + 1 \leq i \leq k + L - 1$. The *length* of the error event is defined as being equal to L. Clearly $L \geq \nu + 1$ but has no upper bound, although it can be shown [13] that L is finite with probability 1. Letting $\tilde{\mathbf{e}} = \tilde{\mathbf{x}} - \hat{\tilde{\mathbf{x}}}$ denote the total error event path between $\tilde{\mathbf{x}}$ and $\hat{\tilde{\mathbf{x}}}$, then $\tilde{\mathbf{e}}$ is composed of a number of error events, each of which starts at a different time instant along the path and, in general, have different lengths and number of branches with respect to one another and different probabilities of occurrence. Note that the weight assigned to an error event branch is the difference of the weights assigned to the corresponding branches of $\tilde{\mathbf{x}}$ and $\hat{\tilde{\mathbf{x}}}$.

For a given time instant k, there are many possible paths that diverge from the correct path at the node x_k and remerge for the first time with the correct path L branches later at time $k + L$. When one of these paths corresponds to an error event, the probability of its occurrence relative to the correct path is computed as a *pairwise error probability*, that is, the probability that the segment of $\hat{\tilde{\mathbf{x}}}$ between time k and time $k + L$ is more probable than the corresponding segment along $\tilde{\mathbf{x}}$ between the same two time instants. For a given input sequence, this pairwise error probability can be evaluated as a closed form solution and is a decreasing function of L. Evaluation of the pairwise error event probability is essential to computation of error event probability at time k and also average symbol error probability since, as we shall soon see, upper union bounds on these quantities can be obtained as appropriately weighted sums of all pairwise error probabilities. Thus, we now evaluate the pairwise error probability for the more general case of the correct sequence and *any* incorrect sequence (not necessarily corresponding to an error event).

Evaluation of the pairwise error probability.   Consider two paths $\hat{\tilde{\mathbf{x}}}$ and $\tilde{\mathbf{x}}$ which are identical up to time k and from time $k + L$ on but are not constrained in any other way between times k and $k + L$. We have seen that the metric used for choosing one path in the trellis over another is the sum of the squared Euclidean distances between the detected sample $\tilde{y}_k$ and the shift register output $\tilde{w}_k$ along all of its branches. Let $\eta_k = |\tilde{y}_k - \tilde{w}_k|^2$ and $\hat{\eta}_k = \left|\tilde{y}_k - \hat{\tilde{w}}_k\right|^2$ denote the above squared Euclidean distances at time k where $\tilde{w}_k$ and $\hat{\tilde{w}}_k$ respectively denote the shift register outputs corresponding to the correct and incorrect paths. Then, given that $\tilde{\mathbf{c}}$ is the transmitted sequence, the probability of choosing $\hat{\tilde{\mathbf{x}}}$ rather than $\tilde{\mathbf{x}}$, denoted by $P(\tilde{\mathbf{x}} \to \hat{\tilde{\mathbf{x}}}|\tilde{\mathbf{c}})$, is given by

$$P(\tilde{\mathbf{x}} \to \hat{\tilde{\mathbf{x}}}|\tilde{\mathbf{c}}) = \Pr \left\{ \sum_{n=k}^{k+L-1} \hat{\eta}_n < \sum_{n=k}^{k+L-1} \eta_n \,\middle|\, \tilde{\mathbf{c}} \right\}$$

$$= \Pr \left\{ \sum_{n=k}^{k+L-1} \left[\left|\tilde{y}_n - \hat{\tilde{w}}_n\right|^2 - \left|\tilde{y}_n - \tilde{w}_n\right|^2 \right] < 0 \,\middle|\, \tilde{\mathbf{c}} \right\}$$

$$= \Pr \left\{ \sum_{n=k}^{k+L-1} \left[\mathrm{Re}\left\{ \tilde{y}_n \left(\tilde{w}_n - \hat{\tilde{w}}_n \right)^* \right\} - \frac{1}{2} \left(|\tilde{w}_n|^2 - \left|\hat{\tilde{w}}_n\right|^2 \right) \right] < 0 \,\middle|\, \tilde{\mathbf{c}} \right\} \quad (9.79)$$

$$= \Pr \left\{ \sum_{n=k}^{k+L-1} \left[\mathrm{Re}\left\{ \tilde{y}_n \sum_{i=0}^{\nu} f_i \left(\tilde{c}_{n-i} - \hat{\tilde{c}}_{n-i} \right)^* \right\} \right.\right.$$

$$\left.\left. - \frac{1}{2} \left(\left|\sum_{i=0}^{\nu} f_i \tilde{c}_{n-i}\right|^2 - \left|\sum_{i=0}^{\nu} f_i \hat{\tilde{c}}_{n-i}\right|^2 \right) \right] < 0 \,\middle|\, \tilde{\mathbf{c}} \right\}$$

Note the similarity between (9.79) and the metric used in the decision rule of (9.70).
To evaluate (9.79), let

$$X = \sum_{n=k}^{k+L-1} \mathrm{Re}\left\{ \tilde{y}_n \left(\hat{\tilde{w}}_n - \tilde{w}_n \right)^* \right\} \quad (9.80)$$

The random variable X is Gaussian with mean

$$\bar{X} = \sum_{n=k}^{k+L-1} \mathrm{Re}\left\{ \tilde{w}_n \left(\hat{\tilde{w}}_n - \tilde{w}_n \right)^* \right\} \quad (9.81)$$

and variance

$$\sigma_X^2 = E\left\{ \left(\sum_{n=k}^{k+L-1} \mathrm{Re}\left\{ \tilde{\mu}_n \left(\hat{\tilde{w}}_n - \tilde{w}_n \right)^* \right\} \right)^2 \right\} = \sigma^2 \sum_{n=k}^{k+L-1} \left|\hat{\tilde{w}}_n - \tilde{w}_n\right|^2 \quad (9.82)$$

Hence, the pairwise error probability is evaluated as

$$P(\tilde{\mathbf{x}} \to \hat{\tilde{\mathbf{x}}}|\tilde{\mathbf{c}}) = \Pr\left\{ X > \gamma \,|\, \tilde{\mathbf{c}} \right\} = \frac{1}{2} \mathrm{erfc}\left(\frac{\gamma - \bar{X}}{\sqrt{2\sigma_X^2}} \right) \quad (9.83)$$

where

$$\gamma \triangleq \frac{1}{2} \sum_{n=k}^{k+L-1} \left(\left|\hat{\tilde{w}}_n\right|^2 - |\tilde{w}_n|^2 \right) \quad (9.84)$$

But, from (9.81) and (9.84) we have that

$$\gamma - \bar{X} = \frac{1}{2} \sum_{n=k}^{k+L-1} \left|\hat{\tilde{w}}_n - \tilde{w}_n\right|^2 = \frac{\sigma_X^2}{2\sigma^2} \quad (9.85)$$

Thus, (9.83) becomes

$$P(\tilde{\mathbf{x}} \to \hat{\tilde{\mathbf{x}}}|\tilde{\mathbf{c}}) = \frac{1}{2} \operatorname{erfc} \left(\sqrt{\frac{\sum\limits_{n=k}^{k+L-1} \left| \hat{\tilde{w}}_n - \tilde{w}_n \right|^2}{8\sigma^2}} \right)$$

$$= \frac{1}{2} \operatorname{erfc} \left(\sqrt{\frac{\sum\limits_{n=k}^{k+L-1} \left| \sum\limits_{i=0}^{\nu} f_i \left(\hat{\tilde{c}}_{n-i} - \tilde{c}_{n-i} \right) \right|^2}{8\sigma^2}} \right)$$

(9.86)

Since the correct and incorrect paths share the same set of states for discrete times less than or equal to k and greater than or equal to $k + L$, that is, $\tilde{x}_i = \hat{\tilde{x}}_i$; $i \le k$, $i \ge k + L$, then from the definition of the system state in terms of the input data sequence, we have that $\hat{\tilde{c}}_i - \tilde{c}_i = 0$; $i \le k - 1$, $i \ge k + L - \nu$. Then, in terms of the definition of the actual ISI coefficients given in (9.67), we can rewrite (9.86) as

$$P(\tilde{\mathbf{x}} \to \hat{\tilde{\mathbf{x}}}|\tilde{\mathbf{c}}) = \tag{9.87a}$$

$$\frac{1}{2} \operatorname{erfc} \left(\sqrt{\frac{1}{8\sigma^2} \sum_{n=k}^{k+L-1} \left[g_0 \left| \hat{\tilde{c}}_n - \tilde{c}_n \right|^2 + 2 \operatorname{Re} \left\{ \sum_{i=1}^{\nu} g_i \left(\hat{\tilde{c}}_n - \tilde{c}_n \right)^* \left(\hat{\tilde{c}}_{n-i} - \tilde{c}_{n-i} \right)^* \right\} \right]} \right)$$

which in view of the above properties of $\hat{\tilde{c}}_i - \tilde{c}_i$ can be rewritten as

$$P(\tilde{\mathbf{x}} \to \hat{\tilde{\mathbf{x}}}|\tilde{\mathbf{c}}) = \tag{9.87b}$$

$$\frac{1}{2} \operatorname{erfc} \left(\sqrt{\frac{1}{8\sigma^2} \left[\sum_{n=k}^{k+L-\nu-1} g_0 \left| \hat{\tilde{c}}_n - \tilde{c}_n \right|^2 + 2 \operatorname{Re} \left\{ \sum_{i=1}^{\nu} g_i \sum_{n=k+i}^{k+L-\nu-1} \left(\hat{\tilde{c}}_n - \tilde{c}_n \right)^* \left(\hat{\tilde{c}}_{n-i} - \tilde{c}_{n-i} \right)^* \right\} \right]} \right)$$

Despite the fact that (9.86) is in closed form, it is convenient from the standpoint of average symbol error probability evaluation to find an upper bound on $P(\tilde{\mathbf{x}} \to \hat{\tilde{\mathbf{x}}}|\tilde{\mathbf{c}})$ in the form of a *product* of terms. To achieve this desired form, we overbound the complementary function by $\operatorname{erfc} y < \exp\{-y^2\}$ which when used on the right-hand side of (9.86) gives

$$P(\tilde{\mathbf{x}} \to \hat{\tilde{\mathbf{x}}}|\tilde{\mathbf{c}}) < \frac{1}{2} \prod_{n=k}^{k+L-1} \exp \left\{ -\frac{\left| \hat{\tilde{w}}_n - \tilde{w}_n \right|^2}{8\sigma^2} \right\}$$

$$= \frac{1}{2} \prod_{n=k}^{k+L-1} \exp \left\{ -\frac{\left| \sum\limits_{i=0}^{\nu} f_i \left(\hat{\tilde{c}}_{n-i} - \tilde{c}_{n-i} \right) \right|^2}{8\sigma^2} \right\}$$

(9.88)

This result (except for the factor of $1/2$) can also be obtained by applying a Chernoff bound directly to (9.79).

Evaluation of an upper bound on the average symbol error probability for PAM. We consider here the case of PAM wherein c_i and $\hat{c}_i$ are real, that is, a_i and $\hat{a}_i$, respectively, and take on values from the set $\pm 1, \pm 3, \ldots \pm (M-1)$ where M is an even integer. Also, the real sequences $\hat{\mathbf{x}}$ and $\mathbf{x}$ are now such that they correspond to a single error event in the time interval k to $k + L$. We define the input error sequence $\boldsymbol{\varepsilon}_c$ associated with the error event by $\boldsymbol{\varepsilon}_c = \left(\varepsilon_{c,k-\nu}, \varepsilon_{c,k-\nu+1}, \ldots, \varepsilon_{c,k}, \varepsilon_{c,k+1}, \ldots, \varepsilon_{c,k+L-1}\right)$ where $\varepsilon_{c,i} \triangleq \left(\hat{a}_i - a_i\right)/2$ and $\varepsilon_{c,i} = 0$; $i \leq k - 1$, $i \geq k + L - \nu$. $\boldsymbol{\varepsilon}_c$ has dimension[15] $L + \nu$ and has the property that it contains no sequence of ν consecutive zeros (else we would have $x_i = \hat{x}_i$ for some intermediate value of i between k and $k + L$ resulting in two distinct error events). Also, the elements of $\boldsymbol{\varepsilon}_c$ take on values from the set $0, \pm 1, \pm 2, \ldots, \pm(M-1)$. Associated with the *input* error sequence $\boldsymbol{\varepsilon}_c$ is a *signal* error sequence $\boldsymbol{\varepsilon}_w$ defined by the corresponding sequences of shift register outputs, that is, $\boldsymbol{\varepsilon}_w = \left(\varepsilon_{w,k}, \varepsilon_{w,k+1}, \ldots, \varepsilon_{w,k+L-2}, \varepsilon_{w,k+L-1}\right)$ where $\varepsilon_{w,i} \triangleq \left(\hat{w}_i - w_i\right)/2$. $\boldsymbol{\varepsilon}_w$ has dimension L. The squared Euclidean weight of an error event, $d^2(\mathcal{E})$, is defined as the squared norm of the associated signal error sequence, that is,

$$d^2(\mathcal{E}) = \sum_{n=k}^{k+L-1} \varepsilon_{w,n}^2 = \sum_{n=k}^{k+L-1} \left(\frac{\hat{w}_n - w_n}{2}\right)^2 = \sum_{n=k}^{k+L-1} \left(\sum_{i=0}^{\nu} f_i \varepsilon_{c,n-i}\right)^2$$

$$= \sum_{n=k}^{k+L-1} \left[g_0 \varepsilon_{c,n}^2 + 2 \sum_{i=1}^{\nu} g_i \varepsilon_{c,n} \varepsilon_{c,n-i}\right] \tag{9.89}$$

The number of errors in $\boldsymbol{\varepsilon}_c$, that is, the number of nonzero elements in $\boldsymbol{\varepsilon}_c$, is defined as the Hamming weight $w_H(\mathcal{E})$ of the error event.

Following Forney [13], we consider a particular error event $\mathcal{E}$ as being composed of three subevents, $\mathcal{E}_1$, $\mathcal{E}_2$, and $\mathcal{E}_3$, all of which must occur in order for $\mathcal{E}$ to occur. Subevent $\mathcal{E}_1$ corresponds to the requirement that both the incorrect and correct path must have the same state at time k, that is, $x_k = \hat{x}_k$. Subevent $\mathcal{E}_2$ requires that for $0 \leq i \leq L - \nu - 1$, the input sequence must be such that, for the particular error sequence assumed, $\hat{c}_i = 2\varepsilon_{c,i} + c_i$ is an allowable error sequence, that is, it takes on values from the set $\pm 1, \pm 3, \ldots \pm (M-1)$. Finally, subevent $\mathcal{E}_3$ corresponds to having noise terms $\tilde{n}_i$; $k \leq i \leq k + L - 1$ such that the likelihood of the segment of $\hat{\mathbf{x}}$ over this interval is greater than for the corresponding segment of $\mathbf{x}$. The probability of $\mathcal{E}_3$ occurring, $\Pr\{\mathcal{E}_3\}$, is precisely the pairwise error probability discussed in the previous section applied now to the specific case where the incorrect path corresponds to an error event.

Based on the above partitioning of $\mathcal{E}$ into $\mathcal{E}_1$, $\mathcal{E}_2$, and $\mathcal{E}_3$, we can write the probability of the particular event $\mathcal{E}$ occurring as

$$\Pr\{\mathcal{E}\} = \Pr\{\mathcal{E}_1 | \mathcal{E}_3\} \Pr\{\mathcal{E}_3\} \Pr\{\mathcal{E}_2 | \mathcal{E}_1, \mathcal{E}_3\} \tag{9.90a}$$

However, since $\mathcal{E}_2$ is independent of $\mathcal{E}_1$, and $\mathcal{E}_3$ being only dependent on the message ensemble, then (9.90a) can be simplified to

$$\Pr\{\mathcal{E}\} = \Pr\{\mathcal{E}_1 | \mathcal{E}_3\} \Pr\{\mathcal{E}_3\} \Pr\{\mathcal{E}_2\} \tag{9.90b}$$

As mentioned above, the subevent probability $\Pr\{\mathcal{E}_3\}$ may be computed from the exact result in (9.86) or the upper bound in (9.88). In particular, using (9.89) in these

---

15. Actually, $\boldsymbol{\varepsilon}_c$ has a true dimension equal to $L - \nu$ in terms of the number of elements that may or may not be zero. However, it is convenient to define the extended sequence $\boldsymbol{\varepsilon}_c$ as we have done.

equations gives

$$\Pr\{\mathcal{E}_3\} = P(\mathbf{x} \to \hat{\mathbf{x}}|\boldsymbol{\varepsilon}_c) = \frac{1}{2}\,\mathrm{erfc}\left(\sqrt{\frac{d^2(\mathcal{E})}{2\sigma^2}}\right) \tag{9.91}$$

and

$$\Pr\{\mathcal{E}_3\} = P(\mathbf{x} \to \hat{\mathbf{x}}|\boldsymbol{\varepsilon}_c) < \frac{1}{2}\prod_{n=k}^{k+L-1}\exp\left\{-\frac{\varepsilon_{w,n}^2}{2\sigma^2}\right\} = \frac{1}{2}\prod_{n=k}^{k+L-1}\exp\left\{-\frac{\left(\sum\limits_{i=0}^{\nu}f_i\varepsilon_{c,n-i}\right)^2}{2\sigma^2}\right\}$$

$$= \frac{1}{2}\prod_{n=k}^{k+L-1}\exp\left\{-\frac{g_0\varepsilon_{c,n}^2 + 2\sum\limits_{i=1}^{\nu}g_i\varepsilon_{c,n}\varepsilon_{c,n-i}}{2\sigma^2}\right\} \tag{9.92}$$

where we have now changed the conditioning event to the error event sequence. It is convenient to normalize (9.91) and (9.92) in terms of the received signal-to-noise ratio. From (9.60), we have

$$g_0 = \int_{-\infty}^{\infty} h^2(t - nT_s)\,dt = \int_{-\infty}^{\infty} |H(f)|^2\,df = \int_{-\infty}^{\infty} |G_t(f)|^2\,|G_c(f)|^2\,df \tag{9.93}$$

Analogous to (9.31), the average received power for M-ary PAM is

$$P_r = \left(\frac{M^2-1}{3}\right)\frac{A^2}{T_s}\int_{-\infty}^{\infty}|G_t(f)|^2\,|G_c(f)|^2\,df = \left(\frac{M^2-1}{3}\right)\frac{A^2 g_0}{T_s} \tag{9.94}$$

Since $\sigma^2 = N_0/2A^2$, then solving for A^2 from (9.94) and substituting into (9.91) and (9.92) gives

$$\Pr\{\mathcal{E}_3\} = P(\mathbf{x} \to \hat{\mathbf{x}}|\boldsymbol{\varepsilon}_c) = \frac{1}{2}\,\mathrm{erfc}\left(\sqrt{\left(\frac{3}{M^2-1}\right)\frac{P_r T_s}{N_0}d^2(\mathcal{E})}\right) \tag{9.95}$$

and

$$\Pr\{\mathcal{E}_3\} = P(\mathbf{x} \to \hat{\mathbf{x}}|\boldsymbol{\varepsilon}_c) < \frac{1}{2}\prod_{n=k}^{k+L-1}\exp\left\{-\left(\frac{3}{M^2-1}\right)\frac{P_r T_s}{N_0}\left(\sum_{i=0}^{\nu}f_i\varepsilon_{c,n-i}\right)^2\right\}$$

$$= \frac{1}{2}\prod_{n=k}^{k+L-1}\exp\left\{-\left(\frac{3}{M^2-1}\right)\frac{P_r T_s}{N_0}\left[\varepsilon_{c,n}^2 + 2\sum_{i=1}^{\nu}g_i\varepsilon_{c,n}\varepsilon_{c,n-i}\right]\right\} \tag{9.96}$$

where, as earlier in the chapter, we have assumed $g_0 = 1$ without loss in generality.

To compute $\Pr\{\mathcal{E}_2\}$, we observe that the subevent $\mathcal{E}_2$ is independent of $\mathcal{E}_1$ and $\mathcal{E}_3$. Specifically, for any particular input error sequence of length $L - \nu$, in each of the $L - \nu - w_H(\mathcal{E})$ positions where $\varepsilon_{c,i} = 0$, there are M possible values for the correct sequence element and likewise for the incorrect sequence element. In each of the $w_H(\mathcal{E})$ positions where $\varepsilon_{c,i} \neq 0$, there are $M - |\varepsilon_{c,i}|$ possible combinations for the values of the correct and incorrect sequence elements. Thus, if I_{w_H} denotes the set of values of i where $\varepsilon_{c,i} \neq 0$, then the number of input data sequences that produce a particular error sequence with Hamming weight $w_H(\mathcal{E})$ is $M^{L-v-w_H(\mathcal{E})} \prod_{i \in I_{w_H}} (M - |\varepsilon_{c,i}|)$. Since the total number of possible input data sequences of length $L - v$ is M^{L-v}, then the probability of subevent $\mathcal{E}_2$ occurring is

$$\Pr\{\mathcal{E}_2\} = M^{-w_H(\mathcal{E})} \prod_{i \in I_{w_H}} (M - |\varepsilon_{c,i}|)$$

$$= \prod_{i=k}^{k+L-v-1} \left(\frac{M - |\varepsilon_{c,i}|}{M} \right) = \prod_{i=k}^{k+L-1} \left(\frac{M - |\varepsilon_{c,i}|}{M} \right) \tag{9.97}$$

Finally, computing the conditional probability $\Pr\{\mathcal{E}_1|\mathcal{E}_3\}$ is difficult because of the possible dependence of $\mathcal{E}_1$ on the noise terms involved in $\mathcal{E}_3$. However, the unconditional probability that $\mathcal{E}_1$ does not occur, namely, $1 - \Pr\{\mathcal{E}_1\}$ is on the order of the symbol error probability. Thus, in the normal operating region, the conditional probability $\Pr\{\mathcal{E}_1|\mathcal{E}_3\}$ is closely approximated as well as overbounded by 1. Substituting this upper bound on $\Pr\{\mathcal{E}_1|\mathcal{E}_3\}$ together with (9.97) and (9.95) [or (9.96)] into (9.90b) gives the desired upper bound on the probability of the particular event $\mathcal{E}$, namely

$$\Pr\{\mathcal{E}\} < \frac{1}{2} \, \text{erfc} \left(\sqrt{\left(\frac{3}{M^2 - 1} \right) \frac{P_r T_s}{N_0} d^2(\mathcal{E})} \right) \prod_{i=k}^{k+L-1} \left(\frac{M - |\varepsilon_{c,i}|}{M} \right)$$

$$< \frac{1}{2} \prod_{n=k}^{k+L-1} \left(\frac{M - |\varepsilon_{c,n}|}{M} \right) \exp\left\{ -\left(\frac{3}{M^2 - 1} \right) \frac{P_r T_s}{N_0} \left(\sum_{n=0}^{v} f_i \varepsilon_{c,n-i} \right)^2 \right\} \tag{9.98}$$

$$= \frac{1}{2} \prod_{n=k}^{k+L-1} \left(\frac{M - |\varepsilon_{c,n}|}{M} \right) \exp\left\{ -\left(\frac{3}{M^2 - 1} \right) \frac{P_r T_s}{N_0} \left[\varepsilon_{c,n}^2 + 2 \sum_{i=1}^{v} g_i \varepsilon_{c,n} \varepsilon_{c,n-i} \right] \right\}$$

For binary PAM where $\varepsilon_{c,i}$ can only take on values $0, \pm 1$, (9.97) simplifies to

$$\Pr\{\mathcal{E}_2\} = \frac{1}{2^{w_H(\mathcal{E})}} = \prod_{n=k}^{k+L-1} \frac{1}{2^{w_H(\varepsilon_{c,n})}} \tag{9.99}$$

where $w_H(\varepsilon_{c,n})$ is the Hamming weight of $\varepsilon_{c,n}$, that is, $w_H(\varepsilon_{c,n}) = 1$ if $\varepsilon_{c,n} \neq 0$ and $w_H(\varepsilon_{c,n}) = 0$ if $\varepsilon_{c,n} = 0$. Thus, (9.98) becomes

$$\Pr\{\mathcal{E}\} < \frac{1}{2} \, \text{erfc} \left(\sqrt{\frac{P_r T_b}{N_0} d^2(\mathcal{E})} \right) \frac{1}{2^{w_H(\mathcal{E})}}$$

$$< \frac{1}{2} \prod_{n=k}^{k+L-1} \frac{1}{2^{w_H(\varepsilon_{c,n})}} \exp\left\{ -\frac{P_r T_b}{N_0} \left(\sum_{i=0}^{v} f_i \varepsilon_{c,n-i} \right)^2 \right\} \tag{9.100}$$

$$= \frac{1}{2} \prod_{n=k}^{k+L-1} \frac{1}{2^{w_H(\varepsilon_{c,n})}} \exp\left\{ -\frac{P_r T_b}{N_0} \left[\varepsilon_{c,n}^2 + 2 \sum_{i=1}^{v} g_i \varepsilon_{c,n} \varepsilon_{c,n-i} \right] \right\}$$

Since the probability of a union of events is less than or equal to the sum of their individual probabilities, if E denotes the set of all possible error events $\mathcal{E}$ starting at time k, then the probability of *any* error event starting at time k is union bounded by

$$P_e(E) < \sum_{\mathcal{E} \in E} \frac{1}{2} \operatorname{erfc} \left(\sqrt{\left(\frac{3}{M^2 - 1} \right) \frac{P_r T_s}{N_0} d^2 (\mathcal{E})} \right) \prod_{i=k}^{k+L-1} \left(\frac{M - |\varepsilon_{c,i}|}{M} \right)$$

$$< \sum_{\mathcal{E} \in E} \frac{1}{2} \prod_{n=k}^{k+L-1} \left(\frac{M - |\varepsilon_{c,n}|}{M} \right) \exp \left\{ -\left(\frac{3}{M^2 - 1} \right) \frac{P_r T_s}{N_0} \left(\sum_{i=0}^{v} f_i \varepsilon_{c,n-i} \right)^2 \right\}$$

$$= \sum_{\mathcal{E} \in E} \frac{1}{2} \prod_{n=k}^{k+L-1} \left(\frac{M - |\varepsilon_{c,n}|}{M} \right) \qquad (9.101)$$

$$\times \exp \left\{ -\left(\frac{3}{M^2 - 1} \right) \frac{P_r T_s}{N_0} \left[\varepsilon_{c,n}^2 + 2 \sum_{i=1}^{v} g_i \varepsilon_{c,n} \varepsilon_{c,n-i} \right] \right\}$$

Similarly, the average probability of a symbol error is obtained by weighting each error event probability by the number of decision errors $w_H (\mathcal{E})$ resulting in the upper bound

$$P_s (E) < \sum_{\mathcal{E} \in E} w_H (\mathcal{E}) \frac{1}{2} \operatorname{erfc} \left(\sqrt{\left(\frac{3}{M^2 - 1} \right) \frac{P_r T_s}{N_0} d^2 (\mathcal{E})} \right) \prod_{i=k}^{k+L-1} \left(\frac{M - |\varepsilon_{c,i}|}{M} \right)$$

$$< \sum_{\mathcal{E} \in E} w_H (\mathcal{E}) \frac{1}{2} \prod_{n=k}^{k+L-1} \left(\frac{M - |\varepsilon_{c,n}|}{M} \right)$$

$$\times \exp \left\{ -\left(\frac{3}{M^2 - 1} \right) \frac{P_r T_s}{N_0} \left(\sum_{i=0}^{v} f_i \varepsilon_{c,n-i} \right)^2 \right\} \qquad (9.102)$$

$$= \sum_{\mathcal{E} \in E} w_H (\mathcal{E}) \frac{1}{2} \prod_{n=k}^{k+L-1} \left(\frac{M - |\varepsilon_{c,n}|}{M} \right)$$

$$\times \exp \left\{ -\left(\frac{3}{M^2 - 1} \right) \frac{P_r T_s}{N_0} \left[\varepsilon_{c,n}^2 + 2 \sum_{i=1}^{v} g_i \varepsilon_{c,n} \varepsilon_{c,n-i} \right] \right\}$$

Evaluation of (9.101) or (9.102) is facilitated by the use of a so-called *error-state diagram*. The error-state diagram is similar to the conventional state diagram, the major difference being the labeling of the transitions. In particular, we define an error state, $\boldsymbol{\varepsilon}_i$, at time i as the difference between the incorrect and correct states $\hat{\mathbf{x}}_i$ and $\mathbf{x}_i$ at the same time instant, namely, $\boldsymbol{\varepsilon}_i \triangleq \hat{\mathbf{x}}_i - \mathbf{x}_i = (\varepsilon_{c,i-1}, \varepsilon_{c,i-2}, \ldots, \varepsilon_{c,i-v})$. Since $\varepsilon_{c,i}$ takes on values $0, \pm 1, 2, \ldots \pm (M - 1)$, then the maximum number of possible error states is $(2(M - 1) + 1)^v = (2M - 1)^v$. The initial and final states are the all zeros state since by definition the correct path and incorrect path start and end at identical states. There are $2(M - 1)$ branches leaving the initial all zeros state (at time k). To each of these branches, we assign the weight $a_{0,j} I (M - |\varepsilon_{c,k}|) / M$ where $\varepsilon_{c,k} = j$; $j = \pm 1, \pm 2, \ldots, (M - 1)$ is the input error that causes the transition to the next state, I is an indicator variable whose presence indicates that the next state is not the all zeros state, and $a_{0,j}$ is the exponential term in the product of (9.102) corresponding to $\boldsymbol{\varepsilon}_k = (\varepsilon_{c,k-1}, \varepsilon_{c,k-2}, \ldots, \varepsilon_{c,k-v}) =$

$(0, 0, \ldots, 0)$, that is,

$$a_{0,j} = \exp\left\{-\left(\frac{3}{M^2 - 1}\right)\frac{P_r T_s}{N_0}\varepsilon_{c,k}^2\right\} \tag{9.103}$$

There are $2(M - 1)$ branches entering the final all zeros state (at time $k + L$). To each of these branches, we assign the weight 1 since the input error $\varepsilon_{c,k+L-1}$ that causes the transition to this final state is zero (thus $\left(M - |\varepsilon_{c,k+L-1}|\right)/M = 1$), I is absent (since the next state is the all zeros state), and the exponential term in the product of (9.102) corresponding to $\varepsilon_{c,k+L-1} = 0$ is unity. To the remaining branches between error states, we assign the weight $a_{n-k,j}I\left(M - |\varepsilon_{c,n}|\right)/M$ where $\varepsilon_{c,n} = j$ is the input error that causes the transition and $a_{i,n}$ is again the exponential term in the product of (9.102) corresponding to $\varepsilon_{c,n} = j$ and the current state $\boldsymbol{\varepsilon}_n = \left(\varepsilon_{c,n-1}, \varepsilon_{c,n-2}, \ldots, \varepsilon_{c,n-\nu}\right)$, that is,

$$a_{n-k,j} = \exp\left\{-\left(\frac{3}{M^2 - 1}\right)\frac{P_r T_s}{N_0}\left[\varepsilon_{c,n}^2 + 2\sum_{i=1}^{\nu} g_i\varepsilon_{c,n}\varepsilon_{c,n-i}\right]\right\} \tag{9.104}$$

A pictorial representation of the error-state diagram rapidly becomes complicated as ν grows. Even for $M = 4$ and $\nu = 1$, there are already 7 states in the diagram. Thus, for the purpose of example, we shall restrict ourselves to the binary case with $\nu = 1$, in which case the number of states is 3. In terms of the above, there are two branches leaving the initial all zeros state each of which is assigned the weight $a_0 I/2$ where now

$$a_0 = \exp\left\{-\frac{P_r T_b}{N_0}\right\} \tag{9.105}$$

There are two branches entering the all zeros final state each of which is assigned the weight 1. The branches between other pairs of states are assigned weights $a_\pm I/2$ where

$$a_\pm = \exp\left\{-\frac{P_r T_b}{N_0}[1 \pm 2g_1]\right\} \tag{9.106}$$

where the plus sign corresponds to a transition from a given state to the same state, that is, 1 to 1 or -1 to -1, and the minus sign corresponds to a transition from a given state to the opposite state, that is, 1 to -1 or -1 to 1. The complete error-state diagram is illustrated in Fig. 9.17.

To evaluate the error event error probability in (9.101) or symbol error probability in (9.102), we must compute the generating function of the error-state diagram, which can also be regarded as the transfer function of a signal flow graph with unity input. This function can be most directly computed by simultaneous solution of the state equations obtained from the error-state diagram itself. In particular, from Fig. 9.17 letting ξ_b, ξ_c, and ξ_d represent dummy variables for the partial paths to the intermediate nodes, and $T(a_0, a_-, a_+, I) = \xi_d$ the desired generating function, then with $\xi_a = 1$ (since the graph has unity input) the set of state equations is given by

$$T(a_0, a_-, a_+, I) = \xi_b + \xi_c$$

$$\xi_b = a_0\frac{I}{2} + a_+\frac{I}{2}\xi_b + a_-\frac{I}{2}\xi_c \tag{9.107}$$

$$\xi_c = a_0\frac{I}{2} + a_+\frac{I}{2}\xi_c + a_-\frac{I}{2}\xi_b$$

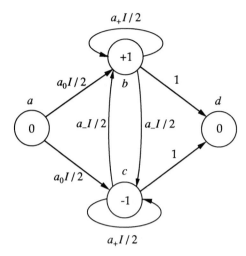

Figure 9.17 Error-state diagram of a binary system with memory one

The solution of (9.107) is

$$\xi_b = \xi_c = \frac{a_0 I/2}{1 - (a_+ + a_-)\,I/2}$$

$$T(a_0, a_-, a_+, I) = \frac{a_0 I}{1 - (a_+ + a_-)\,I/2}$$

(9.108)

Finally, the desired upper bounds on $P_e(E)$ and $P_b(E)$ are obtained from $T(a_0, a_-, a_+, I)$ as [16]

$$P_e(E) < T(a_0, a_-, a_+, I)\big|_{I=1} = \frac{a_0/2}{1 - (a_+ + a_-)/2}$$

$$= \frac{\frac{1}{2}\exp\left\{-\frac{P_r T_b}{N_0}\right\}}{1 - \exp\left\{-\frac{P_r T_b}{N_0}\right\}\cosh\left(\frac{2P_r T_b}{N_0}g_1\right)}$$

(9.109)

$$P_b(E) < \frac{1}{2}\frac{\partial T(a_0, a_-, a_+, I)}{\partial I}\bigg|_{I=1} = \frac{a_0/2}{\left[1 - (a_+ + a_-)/2\right]^2}$$

$$= \frac{\frac{1}{2}\exp\left\{-\frac{P_r T_b}{N_0}\right\}}{\left[1 - \exp\left\{-\frac{P_r T_b}{N_0}\right\}\cosh\left(\frac{2P_r T_b}{N_0}g_1\right)\right]^2}$$

An upper bound on the exact bit error probability performance of binary PAM *with no ISI*, namely, $P_b(E) = (1/2)\,\mathrm{erfc}\left(\sqrt{P_r T_b/N_0}\right)$, is given by the numerator of the upper bound on $P_b(E)$ in (9.109). Thus, we see that as long as the "eye" is open, that is, $g_1 < g_0 = 1$, the upper bound on the performance of the Viterbi algorithm for binary PAM *with ISI of memory 1* asymptotically approaches the no ISI bit-by-bit detection result as $P_r T_b/N_0$ gets large. Clearly, the larger the value of g_1, the slower will be the approach to the asymptotic result. Figure 9.18 is an illustration of this behavior where the upper bound on $P_b(E)$ as given in (9.109) is plotted versus $P_r T_b/N_0$ in dB with g_1 as a parameter.

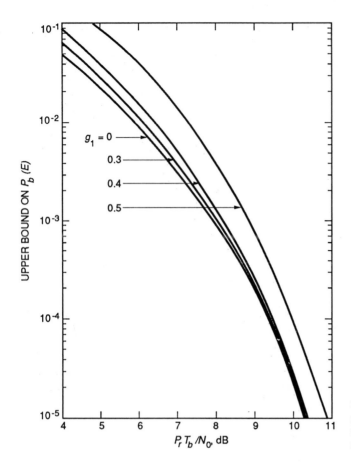

Figure 9.18 Upper bound on the bit error probability performance of the Viterbi algorithm for binary PAM with ISI of memory one

9.4 EQUALIZATION TECHNIQUES

In the broadest sense, the term "equalization" has been used to describe any technique employed by the receiver to deal with ISI. As such, one would include the MLSE scheme described in the previous section in the class of equalization techniques. More often than not, however, equalization refers to the introduction of a linear digital filter between the sampler and the decision algorithm of Fig. 9.1 that compensates for the non-Nyquist behavior of the overall system response. Such a *linear equalizer* tries to undo the ISI distortion introduced by the channel. It is in this narrower context that we shall now discuss the process of equalization.

In the simplest of terms, a linear equalizer tries to introduce gain at those frequencies where the overall system frequency response has a loss so that the resulting cascade of the two is either identically or approximately constant. By introducing gain at the receiver, the total noise at the equalizer output is greater than if the equalizer were not present, resulting in what is commonly called *noise enhancement*. Clearly then, the challenge is to design the equalizer so as to produce as flat a total frequency response as possible for the overall system while at the same time keeping the noise enhancement to a minimum. The manner in which this is accomplished, that is, the criterion used for optimization, defines the type of equalizer. Needless to say, the most desired criterion would be to minimize

D = unit delay

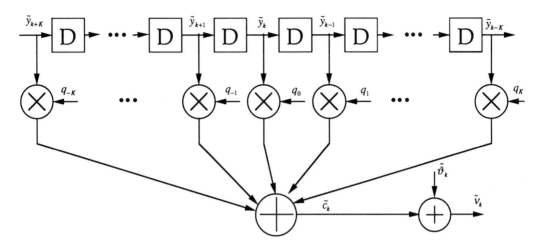

Figure 9.19   Tapped delay line representation of zero-forcing linear equalizer

the probability of error. Since however, the ISI is not Gaussian distributed, designing an equalizer to satisfy this criterion is a difficult task and is typically not attempted. Instead, one considers optimization criteria that lead to analytically tractable yet practically meaningful results. In discussing the various equalizer configurations, we shall not place any constraint on their implementation complexity.

The simplest (but not necessarily the best in performance) class of equalizers are the so-called *linear equalizers*, which have a z-transform of the form

$$Q(z) = \sum_{k=-K}^{K} q_k z^{-k} \tag{9.110}$$

which is implemented as either an infinite impulse response (IIR) filter ($K \to \infty$) or a finite impulse response (FIR) filter (K finite). In the latter case, the filter is referred to as a *transversal filter* or *tapped delay line*[16] and the coefficients $\{q_n\}$ are called the *tap weights* (see Fig. 9.19). Determination of the coefficients (tap weights) is made based on the optimization criterion selected. The input to the filter is the sequence of detected samples $\{\tilde{y}_k\}$ in (9.66) and the output of the filter is the sequence $\{\tilde{v}_k\}$ representing estimates of the complex input data symbols. Figure 9.20 is a discrete-time representation of the entire system including the equalizer.

The two most common types are the *zero-forcing linear equalizer (LE-ZF)* and the *mean-squared error linear equalizer (LE-MSE)*. The former completely removes ISI by identically creating a flat overall system frequency response without regard to the resulting amount of noise enhancement produced. The latter attempts to mitigate the noise enhancement by allowing a residual amount of ISI. We begin with a presentation of the LE-ZF and its performance.

---

16. We note that, from the standpoint of a mathematical characterization, the tapped delay line representation of a digital filter as in Fig. 9.19 is no different than the shift register representation such as that in Fig. 9.13.

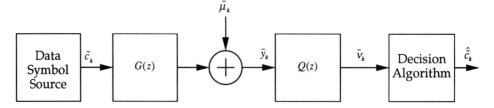

Figure 9.20   A discrete-time block diagram of a linearly equalized communication system

9.4.1 Zero-Forcing Linear Equalizer

As mentioned above, the optimization criterion used by a zero-forcing equalizer is to completely remove the ISI without regard to the noise enhancement that results. Thus, if $G(z)$ of (9.68) represents the overall system response,[17] then to obtain an ISI-free signal sequence $Q(z)$ must be chosen to produce a flat channel response for all frequencies, that is,

$$Q(z) = \frac{1}{G(z)} \tag{9.111}$$

The output of such an equalizer then has the form

$$\tilde{v}_k = \tilde{c}_k + \tilde{\vartheta}_k \tag{9.112}$$

where $\left\{\tilde{\vartheta}_k\right\}$ is a zero-mean complex Gaussian noise sequence with discrete power spectral density

$$S_{\tilde{\vartheta}}(z) = \sum_{m=-\infty}^{\infty} R_{\tilde{\vartheta}}(m) z^{-m} = \frac{2\sigma^2}{G(z)} = 2\sigma^2 Q(z); \quad R_{\tilde{\vartheta}}(m) = E\left\{\tilde{\vartheta}_k \tilde{\vartheta}_{k+m}^*\right\} \tag{9.113}$$

Note that although we have completely removed the ISI, the noise sequence is not white. Thus, symbol-by-symbol detection is not the optimum decision algorithm; nevertheless, it is still the most common algorithm used in receivers employing an LE-ZF because of its simplicity.

The mean-squared error at the equalizer output is simply the noise power in $\tilde{\vartheta}_k$ or equivalently the contour integral of (9.113) around the unit circle defined by $|z| = 1$. If $G(z)$ is zero over some interval, then the equalizer $Q(z)$ becomes infinite over that same interval and hence the mean-squared error at the equalizer output also becomes infinite. Thus, a condition on the existence of the LE-ZF is that $1/G(z)$ be integrable. From (9.110), (9.111), and (9.113), the above mean-squared error is given by

$$\overline{\varepsilon_{LE-ZF}^2} \triangleq E\left\{|\tilde{v}_k - \tilde{c}_k|^2\right\} = E\left\{\left|\tilde{\vartheta}_k\right|^2\right\}$$

$$= R_{\tilde{\vartheta}}(0) = 2\sigma^2 q_0 = 2\sigma^2 \frac{T_s}{2\pi} \int_{-\pi/T_s}^{\pi/T_s} \frac{1}{G\left(e^{j\omega T_s}\right)} d\omega \tag{9.114}$$

where q_0 is the zero[th] equalizer tap.

17. We shall assume here the most general case where the memory ν can be infinite.

Example 9.2

By way of example, we shall consider a received pulse shape, $h(t)$, that is causal and of the following form:

$$h(t) = \sqrt{2}\alpha e^{-\alpha t} \sin \beta t \; u(t) \tag{9.115}$$

From (9.60) the ISI coefficients can be obtained as

$$g_k = \frac{\beta}{2} \left[\frac{\alpha \sin |k| \beta T_s + \beta \cos |k| \beta T_s}{\alpha^2 + \beta^2} \right] e^{-|k|\alpha T_s} \tag{9.116}$$

Letting $\beta T_s = 2\pi$, then (9.116) simplifies to

$$g_k = \frac{\beta^2}{2(\alpha^2 + \beta^2)} e^{-\alpha |k| T_s} \tag{9.117}$$

with transfer function

$$G(z) = \frac{\beta^2}{2(\alpha^2 + \beta^2)} \left[\frac{1 - \eta^2}{(1 - \eta z^{-1})(1 - \eta z)} \right]; \quad \eta = e^{-\alpha T_s} \tag{9.118}$$

The corresponding equalizer is described by

$$Q(z) = \frac{2(\alpha^2 + \beta^2)}{\beta^2 (1 - \eta^2)} \left(1 - \eta z^{-1}\right)(1 - \eta z) = \frac{2(\alpha^2 + \beta^2)}{\beta^2 (1 - \eta^2)} \left(-\eta z^{-1} + 1 + \eta^2 - \eta z\right) \tag{9.119}$$

that is, a finite impulse response (FIR) filter. The corresponding mean-squared error is from (9.114) given by

$$\overline{\varepsilon^2_{LE-ZF}} = \frac{4\sigma^2 (\alpha^2 + \beta^2) \left(1 + e^{-2\alpha T_s}\right)}{\beta^2 \left(1 - e^{-2\alpha T_s}\right)} \tag{9.120}$$

which approaches infinity in the limit as α approaches zero.

9.4.2 Mean-Squared Error Linear Equalizer

The mean-squared error (MSE) linear equalizer allows for a residual amount of ISI at its output as a trade against minimizing the mean-squared error between the equalizer output and the transmitted symbol. Thus, $Q(z)$ of (9.110) is chosen so as to minimize $E\left\{|\tilde{v}_k - \tilde{c}_k|^2\right\}$. Let $\tilde{C}(z) = \sum\limits_{k=-\infty}^{\infty} \tilde{c}_k z^{-k}$ denote the z-transform of the transmitted data symbol sequence and $\tilde{M}(z) = \sum\limits_{k=-\infty}^{\infty} \tilde{\mu}_k z^{-k}$ the z-transform of the noise sequence at the equalizer input. Since the data symbols are independent of one another, that is, the sequence $\{\tilde{c}_k\}$ is a zero mean "white" sequence with variance $E\left\{|\tilde{c}_k|^2\right\} \triangleq \sigma_c^2$, then since

$$R_{\tilde{c}}(m) = E\left\{\tilde{c}_k \tilde{c}^*_{k+m}\right\} = \begin{cases} \sigma_c^2; & m = 0 \\ 0; & m \neq 0 \end{cases} \tag{9.121}$$

the power spectral density of $\{\tilde{c}_k\}$ is flat and is given by

$$S_{\tilde{c}}(e^{j\omega T_s}) = \sum\limits_{m=-\infty}^{\infty} R_{\tilde{c}}(m) e^{-j\omega m T_s} = \sigma_c^2 \tag{9.122}$$

Furthermore, from the correlation properties of the noise sequence $\{\tilde{\mu}_k\}$ given in (9.65), the power spectral density of this sequence becomes

$$S_{\tilde{\mu}}(e^{j\omega T_s}) = \sum_{m=-\infty}^{\infty} R_{\tilde{\mu}}(m)e^{j\omega m T_s} = 2\sigma^2 \sum_{m=-\infty}^{\infty} g_m e^{j\omega m T_s} = 2\sigma^2 G(e^{j\omega T_s}) \quad (9.123)$$

Finally, writing the z-transform of the equalizer output sequence $V(z) = \sum_{m=-\infty}^{\infty} v_m z^{-m}$ as $V(z) = \tilde{C}(z)G(z)Q(z)$, then the power spectral density of the error sequence $\{\tilde{\varepsilon}_k\} = \{\tilde{v}_k - \tilde{c}_k\}$ is given by

$$S_{\tilde{\varepsilon}}\left(e^{j\omega T_s}\right) = S_{\tilde{c}}\left(e^{j\omega T_s}\right)\left|G(e^{j\omega T_s})Q(e^{j\omega T_s}) - 1\right|^2 + S_{\tilde{\mu}}\left(e^{j\omega T_s}\right)\left|Q(e^{j\omega T_s})\right|^2$$

$$= \sigma_c^2 \left|G(e^{j\omega T_s})Q(e^{j\omega T_s}) - 1\right|^2 + 2\sigma^2 G(e^{j\omega T_s})\left|Q(e^{j\omega T_s})\right|^2 \quad (9.124)$$

and hence the mean-squared error is

$$\overline{\varepsilon_{\text{LE-MSE}}^2} = \frac{T_s}{2\pi} \int_{-\pi/T_s}^{\pi/T_s} \left[\sigma_c^2 \left|G(e^{j\omega T_s})Q(e^{j\omega T_s}) - 1\right|^2 \right.$$

$$\left. + 2\sigma^2 G(e^{j\omega T_s})\left|Q(e^{j\omega T_s})\right|^2\right] d\omega \quad (9.125)$$

Completing the square in (9.125) with respect to $G(e^{j\omega T_s})$ and, for simplicity of notation, omitting the argument $e^{j\omega T_s}$ in the various transfer functions, we get

$$\overline{\varepsilon_{\text{LE-MSE}}^2} = \frac{T_s}{2\pi} \int_{-\pi/T_s}^{\pi/T_s} \left[\left(\sigma_c^2 |G|^2 + 2\sigma^2 G\right)\left|Q - \frac{\sigma_c^2 G}{\sigma_c^2 |G|^2 + 2\sigma^2 G}\right|^2 \right.$$

$$\left. - \frac{\left(\sigma_c^2 G\right)^2}{\sigma_c^2 |G|^2 + 2\sigma^2 G} + \sigma_c^2\right] d\omega \quad (9.126)$$

Since $G(z)$ is real and even (since $g_k = g_{-k}$), then $G(e^{j\omega T_s}) = G^*(e^{j\omega T_s})$, that is, $|G|^2 = G^2$ and (9.126) simplifies to

$$\overline{\varepsilon_{\text{LE-MSE}}^2} = \frac{T_s}{2\pi} \int_{-\pi/T_s}^{\pi/T_s} \left[G\left(\sigma_c^2 G + 2\sigma^2\right)\left|Q - \frac{\sigma_c^2}{\sigma_c^2 G + 2\sigma^2}\right|^2 \right.$$

$$\left. + \frac{2\sigma^2 \sigma_c^2}{\sigma_c^2 G + 2\sigma^2}\right] d\omega \quad (9.127)$$

Since both terms in the integrand of (9.127) are positive, the mean-squared error is minimized by choosing Q so as to force the first term equal to zero at every value of ω. This is accomplished by selecting

$$Q(z) = \frac{\sigma_c^2}{\sigma_c^2 G(z) + 2\sigma^2} = \frac{1}{G(z) + 2\sigma^2/\sigma_c^2} \quad (9.128)$$

which results in the minimum mean-squared error

$$\left(\overline{\varepsilon^2_{\text{LE-MSE}}}\right)_{\min} = \frac{T_s}{2\pi} \int_{-\pi/T_s}^{\pi/T_s} \frac{2\sigma^2\sigma_c^2}{\sigma_c^2 G\left(e^{j\omega T_s}\right) + 2\sigma^2} d\omega$$

$$= 2\sigma^2 \frac{T_s}{2\pi} \int_{-\pi/T_s}^{\pi/T_s} \frac{1}{G\left(e^{j\omega T_s}\right) + 2\sigma^2/\sigma_c^2} d\omega \qquad (9.129)$$

Comparing (9.129) with (9.114), we clearly see that

$$\left(\overline{\varepsilon^2_{\text{LE-MSE}}}\right)_{\min} \leq \overline{\varepsilon^2_{\text{LE-ZF}}} \qquad (9.130)$$

as would be expected. At high signal-to-noise ratio as measured by σ_c^2/σ^2, the LE-ZF and the LE-MSE give approximately the same performance. Also, for the LE-MSE the noise enhancement is finite, and hence this equalizer always exists.

For Example 9.2, the LE-MSE would be

$$Q(z) = \frac{\left(\alpha^2 + \beta^2\right)\left(\eta z^{-1} + 1 + \eta^2 + \eta z\right)}{\beta^2\left(1 - \eta^2\right)/2 + 2\left(\sigma^2/\sigma_c^2\right)\left(\alpha^2 + \beta^2\right)\left(\eta z^{-1} + 1 + \eta^2 + \eta z\right)} \qquad (9.131)$$

9.4.3 Decision-Feedback Equalization

The performance of the linear equalizers considered in the previous two sections can be improved upon by introducing nonlinearity into the design. One such structure is the *decision-feedback equalizer (DFE)* [17, 18] which uses decisions on the previous data symbols in an attempt to cancel out a portion of the ISI affecting the decision on the current symbol. The primary advantage of the DFE is that it achieves its purpose with substantially less noise enhancement than its linear counterpart. The main disadvantage of the DFE is that it suffers from error propagation caused by feeding back incorrect decisions. Fortunately, however, the error propagation effect is usually far less significant than the benefit obtained from the reduction in noise enchancement.

The structure of a DFE is composed of two parts: a *forward filter*, $Q(z)$, which typically includes one of the two linear equalizers previously discussed, and a *feedback filter*, $P(z)$, which attempts to further reduce the ISI (see Fig. 9.21). Each of these filters has a distinct purpose in the way it handles the ISI. The forward filter acts on the *precursor ISI* (the samples of the overall system response occurring prior to the present time) in a manner similar to that of the linear equalizer discussed previously. It makes no attempt to equalize the *postcursor ISI* (the samples of the overall system response occurring after the present time). Since the feedback filter operates on past decisions, it must be causal. Thus, it has a discrete-time representation with only positive delay terms

$$P(z) = \sum_{k=1}^{K} p_k z^{-k} \qquad (9.132)$$

and as such aids in cancelling the postcursor ISI. Since the forward filter only attempts to equalize the precursor ISI, it can accomplish this task with less noise enhancement that if it tried to equalize the entire system response. On the other hand, assuming that the decision algorithm produces virtually error-free decisions, then the feedback filter produces no noise at its output and thus the overall effect is a reduction in total noise enhancement.

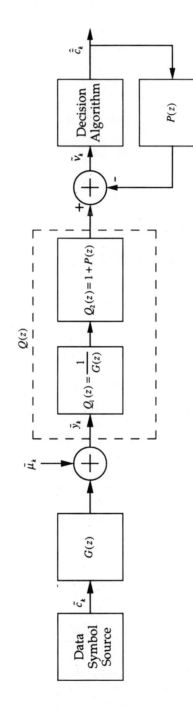

Figure 9.21 A discrete-time block diagram of a zero-forcing decision feedback equalized communication system

DFE's are classified in much the same way as LE's, that is, in accordance with the optimization criterion applied to the forward filter. In this regard, we shall discuss the so-called *zero-forcing decision feedback equalizer (DFE-ZF)* and the *mean-squared error decision feedback equalizer (DFE-MSE)*.

Zero-forcing decision feedback equalizer (DFE-ZF).   The idea behind the DFE-ZF is for the forward filter to accomplish what the LE-ZF did but only for the precursor ISI. Conceptually, the way it does this is to equalize the entire overall response (precursor and postcursor ISI) and then reinsert the postcursor ISI. Thus, the forward filter may be viewed as a cascade of two filters, that is, $Q(z) = Q_1(z)Q_2(z)$, the first being identical to the LE-ZF, namely, $Q_1(z) = 1/G(z)$ [see (9.111)] and the second defined in terms of the feedback filter by $Q_2(z) = 1 + P(z)$. The idea here is that at the output of $Q_1(z)$ [conceptually just an intermediate point within $Q(z)$] there is no ISI (the signal component of these samples is precisely c_k) while at the output of $Q(z)$, there is once again ISI but just due to the postcursor. If now the decision algorithm were to produce totally noise-free decisions, that is, $\hat{c}_k = c_k$, then the output of the feedback filter $P(z)$ would exactly cancel the postcursor ISI introduced by $Q_2(z)$ leaving once again a totally ISI-free set of samples on which to make a decision.

To see how the reduction in noise enhancement comes about, we can write the input to the decision algorithm in Fig. 9.21 as

$$\tilde{v}_k = \hat{\tilde{c}}_k + \tilde{\zeta}_k \tag{9.133}$$

where $\hat{\tilde{c}}_k$ has the z-transform

$$\hat{\tilde{C}}(z) = \tilde{C}(z)G(z)Q(z) - \hat{\tilde{C}}(z)P(z) = \tilde{C}(z)(1 + P(z)) - \hat{\tilde{C}}(z)P(z) \tag{9.134}$$

with $\hat{\tilde{C}}(z)$ denoting the z-transform of the symbol decision sequence $\{\hat{\tilde{c}}_k\}$. If we assume error-free decisions, then (9.134) simplifies to $\hat{\tilde{C}}(z) = \tilde{C}(z)$ and likewise (9.133) becomes

$$\tilde{v}_k = \tilde{c}_k + \tilde{\zeta}_k \tag{9.135}$$

that is, the input to the decision algorithm is once again ISI-free as predicted above. The noise sequence $\{\tilde{\zeta}_k\}$, however, now has the discrete power spectral density [see (9.123) and Fig. 9.21]

$$S_{\tilde{\zeta}}(z) = \frac{2\sigma^2}{G(z)}(1 + P(z))\left(1 + P(z^{-1})\right) \tag{9.136}$$

where we recall that $G(z) = G(z^{-1})$ since g_k is real and even. Comparing (9.135) and (9.136) with (9.112) and (9.113), respectively, we see that the DFE-ZF has an identical signal component to the LE-ZF, that is, it is independent of the feedback filter $P(z)$; however, the noise component of the DFE-ZF is clearly affected by this filter. As such, we should be able to choose the causal filter $P(z)$ to reduce the noise at the input to the decision algorithm relative to that of the LE-ZF wherein $P(z) = 0$.

The mean-squared error of the DFE-ZF at the decision algorithm input, namely, $\overline{\varepsilon^2_{\text{DFE}-\text{ZF}}} = E\left\{|\tilde{v}_k - \tilde{c}_k|^2\right\}$, is from (9.135) simply given by $\overline{\varepsilon^2_{\text{DFE}-\text{ZF}}} = E\left\{\left|\tilde{\zeta}_k\right|^2\right\}$. To determine the optimum $P(z)$, that is, the one that minimizes this mean-squared error, we first reinterpret (9.136) as the power spectrum of the output $\tilde{\zeta}_k$ of a filter $Q_2(z) = 1 + P(z)$ whose input is $\tilde{\vartheta}_k$ with power spectrum as in (9.113). Based on the orthogonality principle [19, Chapter 13], it can be shown that the particular filter $P(z)$ that minimizes $\overline{\varepsilon^2_{\text{DFE-ZF}}}$ has the property that the output of $Q_2(z)$ is uncorrelated with the past of its input, that is,

$$E\left\{\tilde{\zeta}_k \tilde{\vartheta}^*_{k-m}\right\} = 0; \quad m = 1, 2, \ldots \tag{9.137}$$

Furthermore, since from the above interpretation

$$\tilde{\zeta}_{k-m} = \tilde{\vartheta}_{k-m} + \sum_{l=1}^{K} d_l \tilde{\vartheta}^*_{k-m-l} \tag{9.138}$$

then using (9.137), we see that

$$R_{\tilde{\zeta}}(m) \triangleq E\left\{\tilde{\zeta}_k \tilde{\zeta}^*_{k-m}\right\} = 0; \quad m = 1, 2, \ldots \tag{9.139}$$

and since $R_{\tilde{\zeta}}(m) = R^*_{\tilde{\zeta}}(-m)$, it follows that (9.139) is true for all $m \neq 0$, that is,

$$R_{\tilde{\zeta}}(m) \triangleq E\left\{\left|\tilde{\zeta}_k\right|^2\right\} \delta(m) \tag{9.140}$$

where again $\delta(m)$ is the Dirac delta function. Thus, we conclude that *the equalizer output noise sequence that minimizes* $\overline{\varepsilon^2_{\text{DFE-ZF}}} = E\left\{\left|\tilde{\zeta}_k\right|^2\right\}$ *is a "white" sequence, that is, its discrete power spectrum is a constant given by* $S_{\tilde{\zeta}}(z) = \sum_{m=-\infty}^{\infty} R_{\tilde{\zeta}}(m)z^{-m} \triangleq E\left\{\left|\tilde{\zeta}_k\right|^2\right\}$.

In our discussion of noise whitening filters in Section 9.3.1, we considered partitioning the overall system response $G(z)$ into a minimum phase, physically realizable part $F(z)$ and a nonminimum phase, nonphysically realizable part $F(z^{-1})$, the former having all its zeros and poles inside the unit circle and the latter having all its zeros and poles outside the unit circle. To find the optimum forward filter, we again consider this same partitioning of $G(z)$ (actually a normalized version of it, as we shall describe shortly) and then associate the filter $(1 + P(z))$ with the normalized minimum phase, physically realizable part and the filter $(1 + P(z^{-1}))$ with the normalized nonminimum phase, nonphysically realizable part. In particular, let

$$G(z) = F(z)F(z^{-1}) = \sqrt{K}\,\frac{\prod_l (1 - \lambda_l z^{-1})}{\prod_m (1 - \chi_m z^{-1})}\,\sqrt{K}\,\frac{\prod_l (1 - \lambda_l z)}{\prod_m (1 - \chi_m z)} \tag{9.141}$$

where the λ_l's and the χ_m's respectively represent the zeros and poles of $G(z)$. Since from the definition of $P(z)$ in (9.132), $(1 + P(z))|_{z=\infty} = 1$, then the correct association referred to above would be

$$1 + P(z) = \frac{\prod_l (1 - \lambda_l z^{-1})}{\prod_m (1 - \chi_m z^{-1})} \tag{9.142}$$

and thus from (9.136), the sequence $\left\{\tilde{\zeta}_k\right\}$ has the power spectrum

$$S_{\tilde{\zeta}}(z) = \frac{2\sigma^2}{K} \tag{9.143}$$

which is white in accordance with the above discussion. Finally then, the minimum mean-squared error is

$$\left(\overline{\varepsilon_{\text{DFE}-\text{ZF}}^2}\right)_{\min} = \frac{2\sigma^2}{K} \tag{9.144}$$

Note that (9.142) also determines the feedback filter by simply subtracting 1 from the right-hand side of this equation.

Example 9.3

Considering Example 9.2, we immediately see from (9.118) that

$$K = \frac{\beta^2\left(1-\eta^2\right)}{2\left(\alpha^2+\beta^2\right)}; \quad 1+P(z) = \frac{1}{1-\eta z^{-1}}; \quad \eta = e^{-\alpha T_s} \tag{9.145}$$

corresponding to the case of no zeros and one pole. Thus, the forward and feedback filters of the DFE-ZF are given by

$$Q(z) = \frac{2\left(\alpha^2+\beta^2\right)}{\beta^2\left(1-\eta^2\right)}(1-\eta z); \quad P(z) = \frac{\eta z^{-1}}{1-\eta z^{-1}} \tag{9.146}$$

and the minimum mean-squared error is

$$\left(\overline{\varepsilon_{\text{DFE}-\text{ZF}}^2}\right)_{\min} = 4\sigma^2 \frac{\left(\alpha^2+\beta^2\right)}{\beta^2\left(1-e^{-2\alpha T_s}\right)} \tag{9.147}$$

When compared with (9.120) for the LE-ZF we see that we have achieved a reduction in mean-squared error by the factor $1+e^{-2\alpha T_s}$. The fact that $\left(\overline{\varepsilon_{\text{DFE}-\text{ZF}}^2}\right)_{\min} \leq \overline{\varepsilon_{\text{LE}-\text{ZF}}^2}$ can be shown [3] to be true in a more general context than just this example.

Mean-squared error decision feedback equalizer (DFE-MSE). The discrete-time block diagram of a communication system employing a mean-squared error decision-feedback equalizer (DFE-MSE) [20] can be represented as in Fig. 9.21 except that now $Q_1(z)$ is not constrained to be equal to $1/G(z)$ as in the DFE-ZF case but rather is to be determined along with the feedback filter $P(z)$ to minimize the mean-squared error at the decision algorithm input, namely, $E\left\{|\tilde{v}_k - \tilde{c}_k|^2\right\}$. By inspection of Fig. 9.21, the z-transform of the signal component of $\tilde{v}_k$, namely, $\hat{\tilde{c}}_k$, can be written as

$$\hat{\tilde{C}}(z) = \tilde{C}(z)G(z)Q_1(z)\left(1+P(z)\right) - \hat{\tilde{C}}(z)P(z) \tag{9.148}$$

which, assuming again that there are no errors in the decision process, that is, $\hat{\tilde{c}}_k = \tilde{c}_k$, becomes

$$\hat{\tilde{C}}(z) = \left\{\tilde{C}(z)\left[G(z)Q_1(z) - \frac{P(z)}{(1+P(z))}\right]\right\}(1+P(z)) \tag{9.149}$$

The z-transform of the signal component, $\tilde{\varepsilon}_{sk} \triangleq \hat{\tilde{c}}_k - \tilde{c}_k$, of the total error signal $\tilde{\varepsilon}_k$ is then

$$\tilde{E}_s(z) = \hat{\tilde{C}}(z) - \tilde{C}(z) = \tilde{C}(z)\left[G(z)Q_1(z) - 1\right](1+P(z)) \tag{9.150}$$

Finally, since the signal and noise components of $\tilde{\varepsilon}_k$ are uncorrelated, then analogous to (9.124), the power spectral density of the total error signal is

$$S_{\tilde{\varepsilon}}\left(e^{j\omega T_s}\right) = \left\{\sigma_c^2 \left|G(e^{j\omega T_s})Q_1(e^{j\omega T_s}) - 1\right|^2 + 2\sigma^2 G(e^{j\omega T_s})\left|Q_1(e^{j\omega T_s})\right|^2\right\}$$

$$\times \left|1 + P\left(e^{j\omega T_s}\right)\right|^2 \tag{9.151}$$

and hence the mean-squared error to be minimized is

$$\overline{\varepsilon_{\text{DFE-MSE}}^2} = \frac{T_s}{2\pi} \int_{-\pi/T_s}^{\pi/T_s} \left\{\sigma_c^2 \left|G(e^{j\omega T_s})Q_1(e^{j\omega T_s}) - 1\right|^2 \right.$$

$$\left. + 2\sigma^2 G(e^{j\omega T_s})\left|Q_1(e^{j\omega T_s})\right|^2\right\} \left|1 + P\left(e^{j\omega T_s}\right)\right|^2 d\omega \tag{9.152}$$

Because of the way we have partitioned the integral in (9.152) into a part (between braces) dependent only on $Q_1\left(e^{j\omega T_s}\right)$ and a part dependent only on $P\left(e^{j\omega T_s}\right)$, we can perform the minimization of $\overline{\varepsilon_{\text{DFE-MSE}}^2}$ in two steps, that is, we first choose $Q_1\left(e^{j\omega T_s}\right)$ to minimize at each frequency the power spectrum between braces and then choose $P\left(e^{j\omega T_s}\right)$ to minimize the total power spectrum corresponding to the error signal itself. Comparing the power spectrum between braces in (9.152) with the integrand in (9.125), we see that we are dealing with identical minimization problems. Thus, without further ado, we can immediately write the desired solution for the first section of the forward filter as [see (9.128)]

$$Q_1(z) = \frac{1}{G(z) + 2\sigma^2/\sigma_c^2} \tag{9.153}$$

which results in the term between braces in (9.152) being equal to the integrand in (9.129), or equivalently

$$\overline{\varepsilon_{\text{DFE-MSE}}^2} = \frac{T_s}{2\pi} \int_{-\pi/T_s}^{\pi/T_s} \left[\frac{2\sigma^2}{G(e^{j\omega T_s}) + 2\sigma^2/\sigma_c^2}\right] \left|1 + P\left(e^{j\omega T_s}\right)\right|^2 d\omega \tag{9.154}$$

The second step involves a spectral factorization similar to that performed for the DFE-ZF, namely, we factor $G'(z) \triangleq G(z) + 2\sigma^2/\sigma_c^2$ into

$$G'(z) = F'(z)F'(z^{-1}) = \sqrt{K'} \frac{\prod_l (1 - \lambda_l' z^{-1})}{\prod_m (1 - \chi_m' z^{-1})} \sqrt{K'} \frac{\prod_l (1 - \lambda_l' z)}{\prod_m (1 - \chi_m' z)} \tag{9.155}$$

and then make the association

$$1 + P(z) = \frac{\prod_l (1 - \lambda_l' z^{-1})}{\prod_m (1 - \chi_m' z^{-1})} \tag{9.156}$$

from which we can determine the feedback filter $P(z)$. Finally, the minimum mean-squared error for the DFE-MSE is given as

$$\left(\overline{\varepsilon_{\text{DFE-MSE}}^2}\right)_{\min} = \frac{2\sigma^2}{K'} \tag{9.157}$$

Example 9.4

Considering again Example 9.2, we have that

$$G'(z) = \frac{\beta^2 \left(1 - \eta^2\right)}{2 \left(\alpha^2 + \beta^2\right)} \frac{1}{\left(1 - \eta z^{-1}\right) \left(1 - \eta z\right)} + \frac{2\sigma^2}{\sigma_c^2}$$

$$= K' \frac{\left(1 - \mu z^{-1}\right) \left(1 - \mu z^{-1}\right)}{\left(1 - \eta z^{-1}\right) \left(1 - \eta z\right)}$$

(9.158)

where

$$K' = \frac{\frac{\beta^2(1-\eta^2)}{2(\alpha^2+\beta^2)} + \frac{2\sigma^2}{\sigma_c^2}\left(1 + \eta^2\right)}{1 + \mu^2} = \frac{K + \frac{2\sigma^2}{\sigma_c^2}\left(1 + \eta^2\right)}{1 + \mu^2}$$

$$\mu = \frac{\beta^2 \left(1 - \eta^2\right)}{8\eta \left(\alpha^2 + \beta^2\right)} \left(\frac{\sigma_c^2}{\sigma^2}\right) + \frac{1}{2} \left(\frac{1 + \eta^2}{\eta}\right)$$

(9.159)

$$- \sqrt{\left[\frac{\beta^2 \left(1 - \eta^2\right)}{8\eta \left(\alpha^2 + \beta^2\right)} \left(\frac{\sigma_c^2}{\sigma^2}\right) + \frac{1}{2} \left(\frac{1 + \eta^2}{\eta}\right)\right]^2 - 1}$$

It can be shown [3] that in general $\left(\overline{\varepsilon_{\text{DFE-MSE}}^2}\right)_{\min} \leq \left(\overline{\varepsilon_{\text{DFE-ZF}}^2}\right)_{\min}$ (that is, $K' \geq K$) and furthermore $\left(\overline{\varepsilon_{\text{DFE-MSE}}^2}\right)_{\min} \leq \left(\overline{\varepsilon_{\text{LE-MSE}}^2}\right)_{\min}$.

9.5 FURTHER DISCUSSION

In our discussion of equalization, we have made several idealistic assumptions that, although theoretically satisfying, may not be met in practice. First, we implicitly assumed that the channel impulse response is completely known to the receiver and furthermore is fixed, that is, it does not vary with time. This is tantamount to assuming that the ISI is completely known and time-invariant and as such the filters that made up the various equalizer configurations have fixed coefficients determined in accordance with the particular optimization criterion that defined the equalizer type. Second, we placed no constraint on the complexity (number of coefficients) in the equalizer; that is, we assumed, in general, an implementation in the form of an IIR rather than an FIR (transversal filter).

The first of the above two idealizations is handled by making the equalizer dynamic. In particular, the channel impulse response is continuously (in discrete time increments) estimated and the equalizer coefficients are then adjusted to minimize ISI and noise in accordance with these estimates. This combined process is referred to as *adaptive equalization* and was first introduced by Lucky [21] in connection with the telephone channel application. A simple block diagram of an adaptive linear equalized receiver is illustrated in Fig. 9.22. The error signal used to drive the filter coefficient adjustment algorithm is obtained from the difference of the data symbol decisions and the equalizer output. For example, in the absence of noise and ISI, the two would be identical and the error signal would be zero, resulting in no change in the equalizer coefficients. When there is noise but no ISI, then the error signal is zero whenever a correct decision is made and nonzero whenever an error (due to noise) occurs. However, since the equivalent additive noise at the decision algorithm input is zero mean, the effect of decision errors on the adaptation of the equalizer

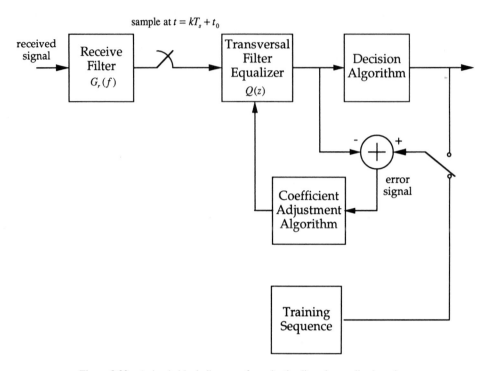

Figure 9.22 A simple block diagram of an adaptive linearly equalized receiver

will be averaged (smoothed) out by the large time constants typically built into the adjustment algorithm resulting in no *net* change in the equalizer coefficients. Finally, when both noise and ISI are present, the adaptive equalizer will function on the average to bring the input and output of the symbol decision algorithm into agreement.

The error signal produced from the difference of the data symbol decisions and the equalizer output will be successful in adapting the equalizer provided that the data symbol estimates are reasonably accurate, which in turn requires that the ISI not be excessively large. While this condition is typical of the steady-state (tracking) behavior of an adaptive equalizer, it is not typical during initial acquisition when the equalizer is learning the state of the channel. Thus, at the outset of transmission (and perhaps periodically for reacquisition purposes), it is common to send a training data sequence (assumed to be known at the receiver) over the channel during which time the adaptation error signal is formed as the difference of the known training symbol sequence and the equalizer output (see Fig. 9.22). Upon conclusion of the training interval, actual data transmission begins, whereupon the training sequence is replaced by the decision algorithm output for the formation of the error signal.

Since the complexity of the equalizer and thus the speed with which it adapts is directly related to the number of coefficients in its z-transform representation [for example, the parameter K in (9.110)], it is only meaningful to talk about adaptive equalizers in the context of an FIR. Hence, the second idealization mentioned above, namely, the lack of a constraint on complexity, is automatically handled as a consequence of the need for adaptation. However, one must re-solve the problem of finding the filter coefficients under the various optimization criteria discussed in Section 9.4 subject now to the constraint of a finite number of degrees of freedom. The solutions to these constrained complexity equalizer problems require methods quite similar to those used in designing other adaptive systems such as adaptive antenna arrays [22]. For example, designing an LE subject to an MSE criterion requires a so-called *mean-squared error gradient (MSEG) algorithm* [3]. This and other methods of its type lead directly to methods for designing the adaptation algorithm. For example, by substituting a time average for the statistical (ensemble) average in the MSEG leads to an adaptation algorithm called the *stochastic gradient (SG) algorithm*. Since, as mentioned in the introduction of this chapter, these subjects are not the main focus of this textbook, we shall not continue their discussion beyond this point. A detailed presentation of the subject of constrained-complexity equalizers and their corresponding adaptation algorithms may be found in such excellent references as [2] , [3], and [23].

PROBLEMS

9.1 Following an approach analogous to (9.14)–(9.21), show that the Nyquist condition on $G(f)$ can alternately be expressed as

$$\sum_{k=-\infty}^{\infty} G\left(f + \frac{k}{T_s}\right) = T_s; \quad \frac{(2n-1)}{2T_s} \le f \le \frac{(2n+1)}{2T_s}$$

for any integer n [including $n = 0$ which corresponds to (9.22)].

9.2 For a Nyquist (zero ISI) overall system frequency response, show that the choice of transmit and receive filters as per (9.28) maximizes the detected sample SNR of (9.27), or equivalently minimizes the average bit error probability of (9.8).

9.3 Let $G(f)$ characterize the overall channel response of a bandlimited system sending data at a rate $1/T_s$. Then, if $g(t) = \mathcal{F}^{-1}\{G(f)\}$ is to have nonzero sample values *only* at $t = 0, T_s$, and both of these values are required to be equal to unity, show that $G(f)$ must satisfy

$$\left|\sum_{k=-\infty}^{\infty} G\left(f + \frac{k}{T_s}\right)\right| = 2T_s \cos \pi f T_s; \quad |f| \le \frac{1}{2T_s}$$

9.4 Consider a frequency response $G(f)$ characterized by the convolution of an arbitrary (not necessarily bandlimited but bounded impulse response) filter $G_a(f)$ and a strictly bandlimited filter

$$G_b(f) = \begin{cases} 2T_s \cos \pi f T_s e^{-j\omega T_s/2}; & |f| \le \frac{1}{2T_s} \\ 0; & \text{otherwise} \end{cases}$$

that is, $G(f) = G_a(f) * G_b(f)$. Show that using $G(f)$ for the overall channel response of a communication system sending data at a rate $1/T_s$ results in zero ISI from all but adjacent symbols.

9.5 For the duobinary system illustrated in Fig. 9.10, show that choosing a detection threshold halfway between "0" and "2A" minimizes the average bit error probability in the limit of large SNR.

9.6 Show that, for the raised cosine class of Nyquist channels as described by the overall system frequency response of (9.25), the received power P_r and amplitude A are related by (9.40), and the variance of the detected sample, y_k, is given by (9.41).

9.7 A modified duobinary system is illustrated in Fig. P9.7a. As in the conventional duobinary case, this system has an equivalent, as illustrated in Fig. P9.7b.
(a) Letting $P(\omega) = P_1(\omega) P_2(\omega)$, and $H(\omega) = H_1(\omega) H_2(\omega)$, find $H(\omega)$ in terms of $P(\omega)$ such that the above equivalence is established, that is, in the absence of noise, $y(t)$ is the same in both.

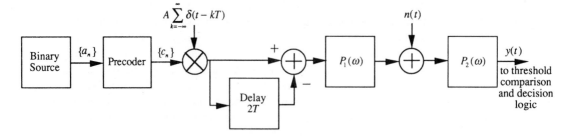

Figure P9.7a

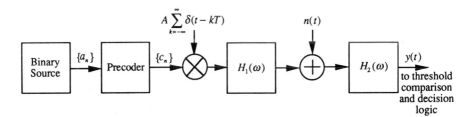

Figure P9.7b

(b) If $P(\omega)$ is the ideal brick wall Nyquist filter, that is, $P(\omega) = T$; $|\omega| \leq \pi/T$ and $P(\omega) = 0$; otherwise, find the overall channel response $h(t) = \mathcal{F}^{-1}\{H(\omega)\}$. [Hint: make use of the fact that $p(t) = \mathcal{F}^{-1}\{P(\omega)\}$ is a $\sin x/x$ function and then use the relation between $H(\omega)$ and $P(\omega)$ found in part (a)]. Sketch $H(\omega)$ and $h(t)$.

(c) What are the values of $h(t)$ at the sampling points $t = nT$, that is, how many have nonzero value and what are these values?

(d) Based on the result of part (c), in the absence of precoding, which data bit(s) contribute ISI to the detection of a_0?

(e) As in duobinary, error propagation will also occur for modified duobinary unless precoding is used. Find the appropriate precoder and sketch its configuration.

(f) For the precoder of part (e), what is the appropriate decision rule for determining $\{a_n\}$ from the samples of $y(t)$?

(g) Assuming $H(\omega)$ is split equally between $H_1(\omega)$ and $H_2(\omega)$, find the error probability for the modified duobinary system using the results found for parts (b)–(f). Assume $n(t)$ is white Gaussian noise with single-sided spectral density N_0 Watts/Hertz.

9.8 A system is designed for a *quadrature partial response* (QPR) modulation, in particular, *quadrature modified duobinary*, that is, modified duobinary (Problem 9.7) on quadrature carriers transmitted over a common channel. The transmitter for such a modulation scheme is illustrated in Fig. P9.8a. Binary (± 1) data at a rate $1/T_b$ is split into its even and odd components, a_n and b_n, respectively, each occurring at a rate $1/T = 1/2T_b$. These two data bit streams are precoded and formed into baseband streams with pulse shaping equal to that produced by a modified duobinary transmit filter, that is

$$H_1(\omega) = \begin{cases} \sqrt{2T}\sin\omega T\, e^{-j\left(\frac{\omega T}{2} - \frac{\pi}{4}\right)}; & |\omega| \leq \frac{\pi}{T} \\ 0; & \text{otherwise} \end{cases}$$

The two baseband data bit streams are amplitude modulated onto quadrature carriers and the sum of the two sent over the channel as the transmitted signal

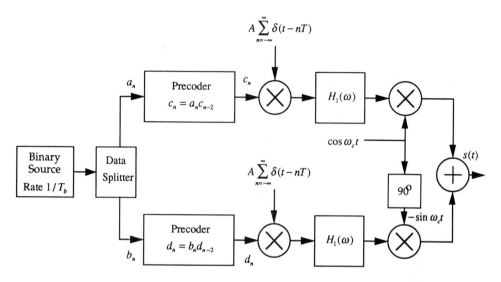

Figure P9.8a

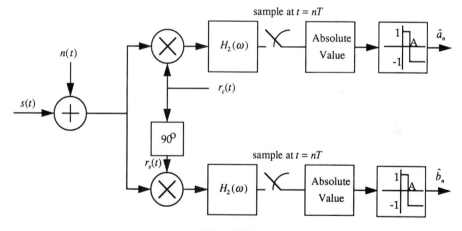

Figure P9.8b

$$s(t) = A \left[\sum_{n=-\infty}^{\infty} c_n h_1(t - nT) \right] \cos \omega_c t - A \left[\sum_{n=-\infty}^{\infty} d_n h_1(t - nT) \right] \sin \omega_c t$$

Assume an additive white Gaussian noise channel with single-sided power spectral density N_0 Watts/Hz. The received signal plus noise is first demodulated with quadrature carriers

$$r_c(t) = 2 \cos \omega_c t; \quad r_s(t) = -2 \sin \omega_c t$$

and then passed through the modified duobinary receive filters $H_2(\omega) = H_1(\omega)$. The outputs of these filters are sampled at times $t = nT$ (n integer) and passed through full-wave recti-fied threshold comparison devices typical of modified duobinary baseband modulation. The detected bits are denoted by $\hat{a}_n$ and $\hat{b}_n$, respectively, as shown in Fig. P9.8b.

(a) Assuming $\omega_c T \gg 1$ (that is, ignore all second harmonic carrier frequency terms), find the bit error probabilities for the in-phase and quadrature data sequences, that is, $P(E)|_a =$

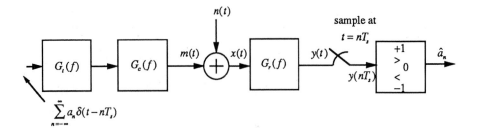

Figure P9.9a

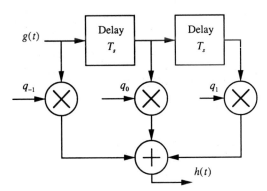

Figure P9.9b

$\Pr\{\hat{a}_n \neq a_n\}$ and $P(E)|_b = \Pr\{\hat{b}_n \neq b_n\}$. Express your answer in terms of the bit energy-to-noise ratio E_b/N_0 of the system where $E_b = P_{in}T_b$ is the bit energy with P_{in} the input power to the channel, that is, the average power of $s(t)$.

(b) If a pair of bits a_n and b_n is thought of as having sent the complex symbol $x_n = a_n + jb_n$, evaluate the *symbol* error probability of the system, that is, $P(E)|_x = \Pr\{\hat{x}_n \neq x_n\}$. Express your answer in terms of the bit energy-to-noise ratio E_b/N_0 of the system.

9.9 A pulse $g_t(t) = \mathcal{F}^{-1}\{G_t(f)\}$ is designed for transmission over a given channel $G_c(f)$ perturbed by additive white Gaussian noise so that the overall system response $g_0(t) = \mathcal{F}^{-1}\{G_0(f)\} = \mathcal{F}^{-1}\{G_t(f)G_c(f)G_r(f)\}$ has no ISI. In particular, let $G_0(f)$ satisfy the Nyquist condition of Eq. (9.22). Furthermore, assume that $G_r(\omega) = (G_t(\omega)G_c(\omega))^*$ so that the error probability is minimized. Suppose now that the channel has a somewhat different transfer function, say $\hat{G}_c(f)$, than that expected in the design such that the overall system response $g(t) = \mathcal{F}^{-1}\{G(f)\} = \mathcal{F}^{-1}\{G_t(f)\hat{G}_c(f)G_r(f)\}$ now contains ISI at the sample time instants. It is desired to pass $y(t)$ through a zero-forcing linear equalizer (LE-ZF) with transfer function $Q(f)$ to restore the zero ISI condition at a *finite* number of sample times in the equalizer filter's output pulse response $h(t) = \mathcal{F}^{-1}\{G(f)Q(f)\}$. The LE-ZF has the form of a transversal filter with three taps as illustrated in Fig. P9.9b. Samples of the pulse shapes are illustrated in Fig. P9.9c.

(a) Find a matrix relation that when solved yields the tap coefficients q_{-1}, q_0, q_1 in terms of the samples $h_0 = h(0)$, $h_1 = h(T_s)$, $h_2 = h(2T_s)$ and an appropriate set of samples of $g(t)$ taken at integer multiples of T_s, namely, $g_i = g(iT_s)$. How many and which samples of $g(t)$ and $h(t)$ are needed?

(b) If the equalized pulse is to have the same nominal delay as the equalizer, that is, $h(t) = g_0(t - T_s)$ at the three sample points $t = 0, T_s, 2T_s$, solve the matrix relation found in

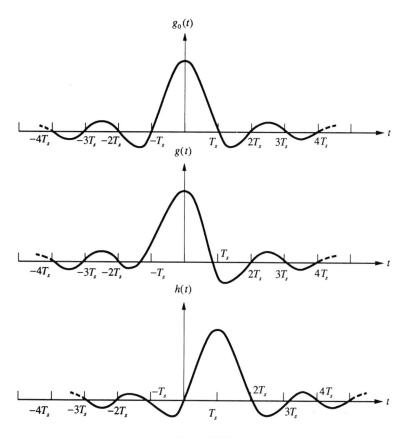

Figure P9.9c

part (a) for q_{-1}, q_0, q_1 when $g_{-1} \neq 0$, $g_0 = 1$, $g_1 \neq 0$ and $g_i = 0$; $i \neq -1, 0, 1$. Express your answers entirely in terms of g_{-1} and g_1.

(c) Numerically evaluate the tap coefficients found in part (b) when $g_{-1} = .2$, $g_0 = 1$, $g_1 = -.1$.

9.10 The design of an LE-MSE equalizer can be carried out in the discrete time domain as opposed to the frequency domain (z-transform) approach described in Section 9.4.2. In particular, consider a linear equalizer described by (9.110) whose input is the sequence $\{y_k\}$ and output is the sequence $\{v_k\}$ where $v_k = \sum_{n=-K}^{K} q_n y_{k-n}$. For simplicity of notation, we shall assume that all sequences are real. Define the MSE for the kth equalizer output sample by $\mathcal{E} = E\left\{(v_k - a_k)^2\right\}$ where a_k is the desired k^{th} data symbol. Note that if a_k is binary (± 1), then $\sigma_a^2 = 1$. Since $\mathcal{E}$ is a concave function of the tap weights $\{q_k\}$, a set of sufficient conditions for minimizing $\mathcal{E}$ is

$$\frac{\partial \mathcal{E}}{\partial q_m} = 0, \quad m = 0, \pm 1, \pm 2, \ldots, \pm K$$

(a) Show that the above set of conditions leads to the following set of relations:

$$R_{yv}(m) = R_{ya}(m), \quad m = 0, \pm 1, \pm 2, \ldots, \pm K$$

where $R_{yv}(m) \triangleq E\{y_{k-m}v_k\}$ is the cross-correlation of the input and output equalizer sequences and $R_{ya}(m) \triangleq E\{y_{k-m}a_k\}$ is the cross-correlation of the equalizer input sequence
with the data symbol sequence.

(b) Using the convolution relation between the input and output equalizer sequences in terms
of the equalizer tap weights, show that the set of conditions of part (a) can be expressed in
the following matrix form:

$$\mathbf{R}_y\mathbf{q} = \mathbf{R}_{ya}$$

where

$$\mathbf{q} = \begin{bmatrix} q_{-K} \\ q_{-K+1} \\ \vdots \\ q_{K-1} \\ q_K \end{bmatrix}, \quad \mathbf{R}_{ya} = \begin{bmatrix} R_{ya}(-K) \\ R_{ya}(-K+1) \\ \vdots \\ R_{ya}(K-1) \\ R_{ya}(K) \end{bmatrix},$$

$$\mathbf{R}_y = \begin{bmatrix} R_y(0) & R_y(1) & \cdots & R_y(2K) \\ R_y(-1) & R_y(0) & & R_y(2K-1) \\ \vdots & \vdots & \ddots & \vdots \\ R_y(-2K) & R_y(-2K+1) & & R_y(0) \end{bmatrix}$$

with $R_y(m) \triangleq E\{y_{k-m}y_k\}$ equal to the autocorrelation function of the equalizer input sequence.

(c) Assume now that the equalizer input sequence is as specified in (9.66), namely $y_k = \sum_{l=-v}^{v} g_l a_{k-l} + \mu_k$, where $2v$ is the number of ISI components), $\{a_k\}$ is an uncorrelated data
sequence with variance σ_a^2, and $\{\mu_k\}$ is a zero-mean Gaussian correlated noise sequence
with correlation function $R_\mu(m) = E\{\mu_{k-m}\mu_k\} = \sigma^2 g_m$. Also, assume that $g_0 = 1$. Show
that the elements of the matrix $\mathbf{R}_y$ and the column vector $\mathbf{R}_{ya}$ are given by

$$R_{ya}(m) = \sigma_a^2 g_{-m}$$

$$R_y(m) = \sigma_a^2 \sum_{l=-v}^{v} g_l g_{l+m} + \sigma^2 g_m$$

from which the tap weight vector $\mathbf{q}$ can be uniquely found using the result in part (b),
that is

$$\mathbf{q} = \mathbf{R}_y^{-1}\mathbf{R}_{ya}$$

9.11 Consider the design of an LE-MSE in accordance with the development of Problem 9.10.
Letting $K = v = 1$, $\sigma_a^2 = 1$, $\sigma^2 = .2$ and an ISI channel described by $g_{-1} = .2$, $g_0 = 1$, $g_1 = -.1$, find the optimum tap weights q_{-1}, q_0, q_1.

9.12 Consider a zero-forcing linear equalizer (LE-ZF) for use in a system with an overall system
response of the form

$$G(z) = g_{-1}z^{-1} + g_0 + g_1z$$

that is, ISI is contributed only by the pulses adjacent to the one of interest. Letting $g_{-1} = g_1$,
we define

$$\rho \triangleq \frac{g_1}{g_0}$$

as the normalized ISI contributed by each of the two adjacent pulses.

(a) Show that the necessary Fourier transform condition $G\left(e^{j\omega T_s}\right) \geq 0$ for all ω implies that $|\rho| \leq 1/2$.

(b) Show that $G(z)$ can be factored in the form

$$G(z) = \rho g_0 \frac{(z - \alpha)\left(z - \alpha^{-1}\right)}{z}$$

where $\rho \triangleq -\left(\alpha + \alpha^{-1}\right)^{-1}$.

(c) Assuming without loss in generality that $|\alpha| \leq 1$, find a partial fraction expansion for the equalizer filter $Q(z)$.

(d) Expanding each of the partial fraction terms in the answer of part (c) in a power series in z or z^{-1} as appropriate, show that the zeroth equalizer tap q_0 can be expressed as

$$q_0 = -\frac{\alpha}{\rho g_0 \left(1 - \alpha^2\right)}$$

(e) Evaluate the mean-squared error of the LE-ZF using your answer found in part (d). Express your answer only in terms of g_0, σ^2, and ρ.

(f) What is the value of the mean-squared error as $\rho \to \pm 1/2$ or equivalently $\alpha \to \pm 1$? What happens to the gain of $Q(z)$ at these limiting values and what does this result in in terms of noise enhancement?

9.13 Consider now a mean-squared error linear equalizer (LE-MSE) for use in a system with the same overall system response as in Problem 9.12 and the same ISI condition, namely $g_{-1} = g_1$.

(a) Show that the equalizer $Q(z)$ is now given by the reciprocal of the equivalent overall system response

$$G'(z) = \rho g_0 \frac{(z - \alpha')\left(z - \alpha'^{-1}\right)}{z}$$

where $\rho' \triangleq \rho \left(1 + \frac{2\sigma^2}{g_0 \sigma_c^2}\right)^{-1} \triangleq -\left(\alpha' + \alpha'^{-1}\right)^{-1}$.

(b) Expressing $Q(z)$ as a partial fraction expansion and expanding each of the partial fraction terms in power series in z or z^{-1} as appropriate, show that the zeroth equalizer tap q_0 can be expressed as

$$q_0 = -\frac{\alpha'}{\rho g_0 \left(1 - \alpha'^2\right)} = \frac{\rho'}{\rho g_0 \sqrt{1 - 4\rho'^2}}$$

(c) Evaluate the mean-squared error of the LE-MSE using your answer found in part (b). Express your answer only in terms of g_0, σ^2, σ_c^2, and ρ'.

(d) What is the value of the mean-squared error now as $\rho \to \pm 1/2$? Is it bounded? Explain.

REFERENCES

1. Lucky, R. W., J. Salz, and E. J. Weldon, Jr., *Principles of Data Communication*, New York: McGraw-Hill, 1968.

2. Proakis, J., *Digital Communications*, 2nd ed., New York: McGraw-Hill, 1989.

3. Messerschmitt, D. G., and E. A. Lee, *Digital Communication*, 1st ed., Boston: Kluwer Academic Publishers, 1988.

4. Nyquist, H., "Certain Topics in Telegraph Transmission Theory," *AIEE Transactions*, vol. 47, 1928, pp. 617–44.

5. Kretzmer, E. R., "Generalization of a Technique for Binary Data Communication," *IEEE Transactions on Communication Technology*, vol. COM-14, no. 1, February 1966, pp. 67–68.

6. Lender, A., "The Duobinary Technique for High-Speed Data Transmission," *IEEE Transactions on Communication Electronics*, vol. 82, May 1963, pp. 214–18.

7. Viterbi, A. J., "Error Bounds for Convolutional Codes and an Asymptotically Optimum Decoding Algorithm," *IEEE Transactions on Information Theory*, vol. IT-13, April 1967, pp. 260–69.

8. Papoulis, A., *The Fourier Integral and its Applications*, New York: McGraw-Hill, 1962, Chapter 4.

9. Duffy, F. P. and T. W. Tratcher, "Analog Transmission Performance on the Switch Telephone Communications Network," *Bell System Technical Journal*, vol. 50, April 1971, pp. 1311–47.

10. Pasupathy, S., "Correlative Encoding: A Bandwidth Efficient Signaling Scheme," *IEEE Communications Magazine*, July 1976, pp. 4–11.

11. Kabal, P. and S. Pasupathy, "Partial-Response Signaling," *IEEE Transactions on Communications*, vol. COM-23, September 1975, pp. 921–34.

12. Bennett, W. R., *Introduction to Signal Transmission*, New York: McGraw-Hill, 1970, p. 16.

13. Forney, G. D., Jr., "Maximum-Likelihood Sequence Estimation of Digital Sequences in the Presence of Intersymbol Interference," *IEEE Transactions on Information Theory*, vol. IT-18, May 1972, pp. 363–78.

14. Forney, G. D., Jr., "Coding System Design for Advanced Solar Missions," Final Report on Contract NAS 2-3637 for NASA Ames Research Center, Watertown, MA: Codex Corp., December 1967.

15. Forney, G. D., Jr., "The Viterbi Algorithm," *Proceedings of the IEEE*, vol. 61, no. 3, March 1973, pp. 268–78.

16. Viterbi, A. J., and J. K. Omura, *Principles of Digital Communication and Coding*, New York: McGraw-Hill, 1979.

17. Austin, M. E., *Decision Feedback Equalization for Digital Communication over Dispersive Channels*, Lexington, MA: MIT Lincoln Laboratory, August 1967.

18. Belfiore, C. A. and J. H. Park, "Decision Feedback Equalization," *Proceedings of the IEEE*, vol. 67, no. 8, August 1979, pp. 1143–56.

19. Papoulis, A., *Probability, Random Variables, and Stochastic Processes*, 2nd ed., New York: McGraw-Hill, 1984.

20. Salz, J., "Optimum Mean-Square Decision Feedback Equalization," *Bell System Technical Journal*, vol. 52, October 1973, pp. 1341–73.

21. Lucky, R. W., "Automatic Equalization for Digital Communications," *Bell System Technical Journal*, no. 44, April 1965, pp. 547–88.

22. Widrow, B. and S. D. Stearns, *Adaptive Signal Processing*, Englewood Cliffs, NJ: Prentice-Hall, 1985.

23. Qureshi, S. U. H., "Adaptive Equalization," *Proceedings of the IEEE*, vol. 73, no. 9, September 1985, pp. 1349–87.

Chapter 10

Demodulation and Detection of Other Digital Modulations

Thus far in the book, we have focused our attention on the characterization of traditional digital modulation schemes and their performance over the AWGN channel. By traditional modulations we mean those that are treated in most textbooks on the subject of digital communications, although not perhaps in the depth to which they are covered here. While in many of today's applications these schemes still provide adequate performance, there exist other applications for which modulations having greater flexibility and sophistication are absolutely essential. An example of an application requiring increased flexibility might be a satellite channel over which it is desired to simultaneously send two independent and generically different types of information. In this instance, we would need a modulation scheme capable of handling two asynchronous data streams with, in general, different rates, different powers, and different formats. Both the Tracking and Data Relay Satellite System (TDRSS) and Space Shuttle make widespread use of such modulation schemes for communicating various combinations of operational, scientific, and telemetry data. An example of an application requiring increased sophistication of design is today's commercial modem market where baud rates as high as 19.2 Kbps are being transmitted over a nominal 5 KHz bandwidth telephone channel. Here the focus is on modulation schemes that achieve a high degree of bandwidth conservation with as little expenditure of additional power as possible.

For the most part, all other digital modulations that meet the above requirements fall into two primary categories: inphase-quadrature (I-Q) modulations and continuous phase modulations (CPM), the latter most often being represented by a constant envelope signal with digital frequency modulation. In certain special cases, CPM also has an I-Q representation and thus the two categories are not always divorced from one another. Nevertheless, it is convenient to separate them this way since the generic implementation for conventional CPM is in the form of a single carrier which is frequency modulated by a waveform that maintains phase continuity from symbol interval to symbol interval. More about this later on.

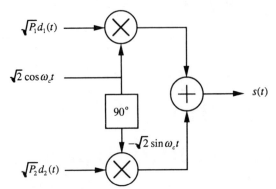

Figure 10.1a An unbalanced QPSK modulator

10.1 I-Q MODULATIONS

10.1.1 Unbalanced QPSK (UQPSK)

Following the form of the generic I-Q modulation introduced in Chapter 4, an unbalanced QPSK modulation (see Fig. 10.1a) can be written as

$$s(t) = \sqrt{2}\left[\sqrt{P_1}d_1(t)\cos\omega_c t - \sqrt{P_2}d_2(t)\sin\omega_c t\right] \tag{10.1}$$

where ω_c is the radian carrier frequency, P_1, P_2 are, respectively, the powers of the inphase and quadrature components of $s(t)$, and $d_1(t)$, $d_2(t)$ are unit power inphase and quadrature binary (± 1) modulations. In general, $d_1(t)$ and $d_2(t)$ are allowed to have different data rates, that is, $\mathcal{R}_1 = 1/T_1$ and $\mathcal{R}_2 = 1/T_2$, different data formats (e.g., NRZ, Manchester), and different data types (e.g., coded or uncoded). Furthermore, the two data streams may be asynchronous or synchronous the latter requiring that their data rates be integer related. In the special case where $P_1 = P_2$, $T_1 = T_2$, the data streams are synchronous, and both are uncoded NRZ waveforms, (10.1) characterizes balanced QPSK modulation as discussed in Chapter 4.

The channel introduces the bandpass Gaussian noise process

$$n(t) = \sqrt{2}\left[n_c(t)\cos\omega_c t - n_s(t)\sin\omega_c t\right] \tag{10.2}$$

where $n_c(t)$ and $n_s(t)$ are lowpass bandlimited "white" Gaussian noise processes with flat spectral density $S_n(f) = N_0/2$ in the frequency interval $-B \le f \le B$ where $B \ll f_c = \omega_c/2\pi$. Also, assume that the channel introduces a constant (time independent) but unknown phase shift θ into the transmitted signal resulting in

$$s(t,\theta) = \sqrt{2}\left[\sqrt{P_1}d_1(t)\cos(\omega_c t + \theta) - \sqrt{P_2}d_2(t)\sin(\omega_c t + \theta)\right] \tag{10.3}$$

Assuming initially the case of ideal coherent detection, then at the receiver, the sum of (10.2) and (10.3), namely, $r(t) = s(t,\theta) + n(t)$, is first demodulated[1] by the quadrature pair of unmodulated carriers

$$r_c(t) = \sqrt{2}\cos(\omega_c t + \theta); \quad r_s(t) = -\sqrt{2}\sin(\omega_c t + \theta) \tag{10.4}$$

---

1. We assume as before that the inphase and quadrature demodulators (multipliers) include lowpass filters that eliminate second harmonics of the carrier.

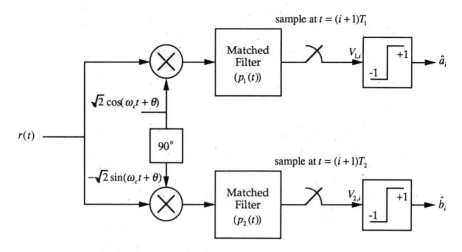

Figure 10.1b   An optimum coherent receiver for unbalanced QPSK

to produce the I and Q lowpass signals

$$\varepsilon_1(t) = \sqrt{P_1}d_1(t) + n_c(t)\cos\theta + n_s(t)\sin\theta$$
$$\varepsilon_2(t) = \sqrt{P_2}d_2(t) + n_c(t)\cos\theta - n_s(t)\sin\theta \tag{10.5}$$

Representing $d_1(t)$ and $d_2(t)$ as binary pulse streams, that is,

$$d_1(t) = \sum_{i=-\infty}^{\infty} a_i p_1(t - iT_1 - v_1); \quad d_2(t) = \sum_{i=-\infty}^{\infty} b_i p_2(t - iT_2 - v_2) \tag{10.6}$$

where $p_1(t)$ and $p_2(t)$ are unit power pulses of duration T_1 and T_2, respectively, $\{a_i\}$ and $\{b_i\}$ are ± 1 independent sequences at rates $1/T_1$ and $1/T_2$, respectively, and v_1, v_2 are random independent timing epochs to allow for asynchronization, then after passing through the I and Q matched filters and assuming $1/T_1, 1/T_2 \ll B$, we obtain the output samples (at times $t = (i + 1)T_1$ and $t = (i + 1)T_2$)

$$V_{1,i} = \sqrt{P_1}a_i + N_c; \quad V_{2,i} = \sqrt{P_2}b_i + N_s \tag{10.7}$$

where N_c and N_s are independent zero-mean Gaussian noise samples with variances $N_0/2T_1$ and $N_0/2T_2$, respectively. Suppose now that the sequences $\{a_i\}$ and $\{b_i\}$ correspond to uncoded random data. Then, making independent hard decisions on the I and Q channels (see Fig. 10.1b), results in the bit error probabilities

$$P_{b1}(E) = \frac{1}{2}\operatorname{erfc}\sqrt{\frac{E_{b1}}{N_0}}; \quad P_{b2}(E) = \frac{1}{2}\operatorname{erfc}\sqrt{\frac{E_{b2}}{N_0}} \tag{10.8}$$

where $E_{b1} = P_1T_1$ and $E_{b2} = P_2T_2$ are the respective energies per bit on the I and Q channels. Thus, in the ideal phase coherent case, the bit error probability performance per channel for UQPSK is identical in form to that of balanced QPSK.

We now consider the case where a carrier phase error exists between the received signal and the I and Q demodulation reference signals. In particular, assume now that the quadrature pair of demodulation reference signals of (10.4) are described by

$$r_c(t) = \sqrt{2}\cos\left(\omega_c t + \hat{\theta}\right); \quad r_s(t) = -\sqrt{2}\sin\left(\omega_c t + \hat{\theta}\right) \tag{10.9}$$

where $\hat{\theta}$ is an estimate of the received signal phase θ such as that produced by a carrier synchronization subsystem. Demodulating $r(t)$ with the I and Q references of (10.9) produces

$$\varepsilon_1(t) = \sqrt{P_1}d_1(t)\cos\phi - \sqrt{P_2}d_2(t)\sin\phi + n_c(t)\cos\hat{\theta} + n_s(t)\sin\hat{\theta}$$

$$\varepsilon_2(t) = \sqrt{P_2}d_2(t)\cos\phi + \sqrt{P_1}d_1(t)\sin\phi + n_s(t)\cos\hat{\theta} - n_c(t)\sin\hat{\theta} \tag{10.10}$$

where $\phi \triangleq \theta - \hat{\theta}$ is the carrier phase error assumed to be constant[2] over the observation interval pertinent to making decisions on $d_1(t)$ and $d_2(t)$.

Consider representing $d_1(t)$ and $d_2(t)$ as in (10.6). After passing through the normalized I and Q matched filters, we obtain the output samples [at time $t = (i+1)T$]

$$V_{1,i} = \sqrt{P_1}a_i\cos\phi - \sqrt{P_2}d_{2,1}\sin\phi + N_{c,1}\cos\hat{\theta} + N_{s,1}\sin\hat{\theta}$$

$$= \sqrt{P_1}a_i\cos\phi - \sqrt{P_2}d_{2,1}\sin\phi + N_1$$

$$V_{2,i} = \sqrt{P_2}b_i\cos\phi + \sqrt{P_1}d_{1,2}\sin\phi + N_{s,2}\cos\hat{\theta} - N_{c,2}\sin\hat{\theta} \tag{10.11}$$

$$= \sqrt{P_2}b_i\cos\phi + \sqrt{P_1}d_{1,2}\sin\phi + N_2$$

where

$$d_{2,1} = \frac{1}{T_1}\int_{iT_1+v_1}^{(i+1)T_1+v_1} d_2(t)p_1(t - iT_1 - v_1)dt$$

$$d_{1,2} = \frac{1}{T_2}\int_{iT_2+v_2}^{(i+1)T_2+v_2} d_1(t)p_2(t - iT_2 - v_2)dt$$

$$N_{c,k} = \frac{1}{T_k}\int_{iT_k+v_k}^{(i+1)T_k+v_k} n_c(t)p_k(t - iT_i - v_k)dt; \quad k = 1, 2 \tag{10.12}$$

$$N_{s,k} = \frac{1}{T_k}\int_{iT_k+v_k}^{(i+1)T_k+v_k} n_s(t)p_k(t - iT_i - v_k)dt; \quad k = 1, 2$$

Here N_1 and N_2 are zero-mean independent noise samples with variances $\sigma_1^2 = N_0/2T_1$ and $\sigma_2^2 = N_0/2T_2$, respectively. We note from (10.11) that, as for balanced QPSK, the presence of carrier phase error introduces *crosstalk* into the detection process, that is, the data in one channel affects the detection of the data in the other channel. The nature of the crosstalk, however, now depends on both the powers in the I and Q channels as well as the I and Q data formats through the definitions of $d_{1,2}$ and $d_{2,1}$ in (10.12). To assess the effect of this crosstalk on uncoded bit error probability performance, we proceed as follows.

---

2. In Volume II of the book, we shall introduce other phase error variation models wherein the phase error can vary in time over the duration of a data decision. For now, we consider only the so-called *slowly varying phase error* case.

The data estimates $\hat{a}_i$ and $\hat{b}_i$ are found from

$$\hat{a}_i = \operatorname{sgn}\{V_{1,i}\}; \quad \hat{b}_i = \operatorname{sgn}\{V_{2,i}\} \tag{10.13}$$

Hence, the I and Q bit error probabilities conditioned on a fixed phase error ϕ are given by

$$P_{b1}(E\,|\,\phi) = \Pr\{\hat{a}_i \neq a_i\,|\,\phi\} = \frac{1}{4}\operatorname{erfc}\overline{\left[\sqrt{\frac{P_1 T_1}{N_0}}\cos\phi + \sqrt{\frac{P_2 T_1}{N_0}}d_{2,1}\sin\phi\right]}$$

$$+ \frac{1}{4}\operatorname{erfc}\overline{\left[\sqrt{\frac{P_1 T_1}{N_0}}\cos\phi - \sqrt{\frac{P_2 T_1}{N_0}}d_{2,1}\sin\phi\right]} \tag{10.14a}$$

and

$$P_{b2}(E\,|\,\phi) = \Pr\left\{\hat{b}_i \neq b_i\,\Big|\,\phi\right\} = \frac{1}{4}\operatorname{erfc}\overline{\left[\sqrt{\frac{P_2 T_2}{N_0}}\cos\phi + \sqrt{\frac{P_1 T_2}{N_0}}d_{1,2}\sin\phi\right]}$$

$$+ \frac{1}{4}\operatorname{erfc}\overline{\left[\sqrt{\frac{P_2 T_2}{N_0}}\cos\phi - \sqrt{\frac{P_1 T_2}{N_0}}d_{1,2}\sin\phi\right]} \tag{10.14b}$$

where the overbar denotes averaging over the data sequences and uniformly distributed random epochs ν_1 and ν_2 in $d_{1,2}$ and $d_{2,1}$.

In principle, (10.14a) and (10.14b) can be evaluated exactly for given pulse shapes $p_1(t)$ and $p_2(t)$, that is, data formats on the I and Q channels. This, however becomes a tedious calculation when, for example, the ratio of data rates $\mathcal{R}_2/\mathcal{R}_1$ is large[3] since the statistical average over the data sequence required in $P_{b1}(E\,|\,\phi)$ of (10.14a) will then require evaluation of $d_{2,1}$ for each possible sequence $d_2(t)$ in the interval $nT_1 + \nu_1 \leq t \leq (n+1)T_1 + \nu_1$. Furthermore, in order to ultimately calculate the I-channel average probability of error [by averaging (10.14a) over the probability density function $f(\phi)$ of ϕ], these statistical averages over $d_2(t)$ must be computed for each value of ϕ in the domain of $f(\phi)$.

To get around this computational bottleneck, a scheme was proposed in [1] that expands $P_{bi}(E\,|\,\phi)$; $i = 1, 2$ into a power series in ϕ whereupon maintaining only the first few terms, the average error probability P_{bi} for the I and Q channels is easily evaluated as a function of E_{b1}/N_0, E_{b2}/N_0, the mean-squared crosstalks $\overline{d_{2,1}^2}$ and $\overline{d_{1,2}^2}$, and the mean-squared phase error, σ_ϕ^2. For small values of σ_ϕ^2, as is typically the case of interest, this scheme gives credible results. We now proceed to describe this simplified evaluation technique.

From (10.14a) and (10.14b), we note that $P_{bi}(E\,|\,\phi)$; $i = 1, 2$ is of the form

$$P_{bi}(E\,|\,\phi) = \frac{1}{4}\overline{\operatorname{erfc}[A_i\cos\phi + \xi_i\sin\phi]} + \frac{1}{4}\overline{\operatorname{erfc}[A_i\cos\phi - \xi_i\sin\phi]} \tag{10.15}$$

where A_i is a constant and ξ_i is a random variable in the sense that it is both data sequence

---

3. With no loss in generality, we shall assume that the data rate $\mathcal{R}_2 = 1/T_2$ is arbitrarily chosen to be greater than or equal to the data rate $\mathcal{R}_1 = 1/T_1$.

and random epoch dependent. Differentiating $P_{bi}(E|\phi)$ twice with respect to ϕ and evaluating the results at $\phi = 0$ gives

$$\left.\frac{dP_{bi}(E|\phi)}{d\phi}\right|_{\phi=0} = \frac{1}{\sqrt{\pi}}\overline{\xi_i}\exp\left(-A_i^2\right) = 0$$

$$\left.\frac{d^2P_{bi}(E|\phi)}{d\phi^2}\right|_{\phi=0} = \frac{A_i\exp\left(-A_i^2\right)}{\sqrt{\pi}}\left[1 + 2\overline{\xi_i^2}\right]$$

(10.16)

Furthermore, from (10.15)

$$P_{bi}(E|\phi)|_{\phi=0} = \frac{1}{2}\operatorname{erfc} A_i$$

(10.17)

which corresponds to the bit error probability performance in the absence of carrier phase error. Thus, expanding (10.15) in a Taylor series about ϕ and keeping only the first three terms gives, upon substitution of (10.16) and (10.17)

$$P_{bi}(E|\phi) \cong P_{bi}(E|\phi)|_{\phi=0} + \left.\frac{dP_{bi}(E|\phi)}{d\phi}\right|_{\phi=0}\phi + \frac{1}{2}\left.\frac{d^2P_{bi}(E|\phi)}{d\phi^2}\right|_{\phi=0}\phi^2$$

$$= \frac{1}{2}\operatorname{erfc} A_i + \frac{A_i\exp\left(-A_i^2\right)}{\sqrt{\pi}}\left[1 + 2\overline{\xi_i^2}\right]\phi^2$$

(10.18)

Comparing (10.15) with (10.14) and identifying A_i and ξ_i in terms of the receiver parameters, (10.18) becomes

$$P_{b1}(E|\phi) \cong \frac{1}{2}\operatorname{erfc}\sqrt{\frac{E_{b1}}{N_0}} + \frac{1}{2}\sqrt{\frac{E_{b1}}{\pi N_0}}\exp\left(-\frac{E_{b1}}{N_0}\right)\left[1 + 2\frac{E_{b2}}{N_0}\gamma_T\overline{d_{2,1}^2}\right]\phi^2$$

$$P_{b2}(E|\phi) \cong \frac{1}{2}\operatorname{erfc}\sqrt{\frac{E_{b2}}{N_0}} + \frac{1}{2}\sqrt{\frac{E_{b2}}{\pi N_0}}\exp\left(-\frac{E_{b2}}{N_0}\right)\left[1 + 2\frac{E_{b1}}{N_0}\frac{\overline{d_{2,1}^2}}{\gamma_T}\right]\phi^2$$

(10.19)

where

$$\gamma_T \triangleq \frac{T_1}{T_2} = \frac{\mathcal{R}_2}{\mathcal{R}_1} \geq 1$$

(10.20)

denotes the ratio of the I and Q data rates.

Finally, averaging (10.19) over the phase error probability density function $f(\phi)$ gives the desired result for I and Q channel average bit error probability performance, namely

$$P_{b1}(E) \cong \frac{1}{2}\operatorname{erfc}\sqrt{\frac{E_{b1}}{N_0}} + \frac{1}{2}\sqrt{\frac{E_{b1}}{\pi N_0}}\exp\left(-\frac{E_{b1}}{N_0}\right)\left[1 + 2\frac{E_{b2}}{N_0}\gamma_T\overline{d_{2,1}^2}\right]\sigma_\phi^2$$

$$P_{b2}(E) \cong \frac{1}{2}\operatorname{erfc}\sqrt{\frac{E_{b2}}{N_0}} + \frac{1}{2}\sqrt{\frac{E_{b2}}{\pi N_0}}\exp\left(-\frac{E_{b2}}{N_0}\right)\left[1 + 2\frac{E_{b1}}{N_0}\frac{\overline{d_{2,1}^2}}{\gamma_T}\right]\sigma_\phi^2$$

(10.21)

Note that (10.21) consists of the sum of a term representing perfect carrier synchronization performance [see (10.8)] and a term proportional to the mean-squared crosstalk and mean-squared phase error. What remains is to evaluate the mean-squared crosstalks $\overline{d_{2,1}^2}$ and $\overline{d_{1,2}^2}$

Table 10.1 Evaluation of Mean-Squared Crosstalk for NRZ and Manchester Data Formats

		$d_2(t)$	
		Manchester	NRZ
$d_1(t)$	Manchester	$\frac{1}{6}\left[\gamma_T^2 - \frac{5}{\gamma_T} + 12 - 6\gamma_T\right];$ $\mathcal{R}_1 \le \mathcal{R}_2 \le 2\mathcal{R}_1$	$\gamma_T - \frac{1}{6}\gamma_T^2 - 1 + \frac{1}{3\gamma_T};$ $\mathcal{R}_1 \le \mathcal{R}_2 \le 2\mathcal{R}_1$
		$\frac{1}{2\gamma_T};\quad \mathcal{R}_2 \ge 2\mathcal{R}_1$	$1 - \frac{1}{\gamma_T};\quad \mathcal{R}_2 \ge 2\mathcal{R}_1$
	NRZ	$\frac{1}{6\gamma_T}$	$1 - \frac{1}{3\gamma_T}$

for various combinations of data formats in the I and Q channels. From the definitions of $d_{2,1}$ and $d_{1,2}$ in (10.12) and the approach taken in Chapter 2, Section 2.1, it is straightforward to show that

$$\overline{d_{1,2}^2} = \frac{1}{T_2} \int_{-\infty}^{\infty} S_{m1}(f) S_{m2}(f) df$$

$$\overline{d_{2,1}^2} = \frac{1}{T_1} \int_{-\infty}^{\infty} S_{m1}(f) S_{m2}(f) df = \frac{\overline{d_{1,2}^2}}{\gamma_T} \tag{10.22}$$

where $S_{m1}(f)$ and $S_{m2}(f)$ are the power spectral densities of the modulations $d_1(t)$ and $d_2(t)$, respectively. In [1], closed form expressions for $\overline{d_{1,2}^2}$ are evaluated for all combinations of Manchester and NRZ coding on the I and Q channels. The results are summarized in Table 10.1.

10.1.2 Quadrature Amplitude Modulation (QAM)

An I-Q modulation formed by quadrature carrier multiplexing two M-AM modulations is called *quadrature amplitude modulation (QAM)*.[4] The signal constellation corresponding to such a signal takes the form of a rectangular array (grid) of points in the form of a square lattice (see Fig. 10.2 for $M = 16$). If each M-AM modulation corresponds to a signal set of size 2^m (m integer), then the QAM signal set contains $M = (2^m)^2 = 4^m$ signals. In accordance with the above description, a QAM signal can be mathematically represented by (see Fig. 10.3a)

$$s(t) = \sqrt{2}A\left[d_c(t) \cos \omega_c t - d_s(t) \sin \omega_c t\right] \tag{10.23}$$

where A is the signal amplitude (which will shortly be related to the average signal power), and $d_c(t)$, $d_s(t)$ are, respectively, the inphase and quadrature data pulse streams, that is,

$$d_c(t) = \sum_{i=-\infty}^{\infty} a_i p(t - iT); \quad d_c(t) = \sum_{i=-\infty}^{\infty} b_i p(t - iT) \tag{10.24}$$

In (10.24), $p(t)$ is a unit power pulse of duration T_s, T_s is the symbol time which is related to the bit time by $T_s = 2mT_b$, and the data amplitudes $\{a_i\}$ and $\{b_i\}$ range over the set of values $\pm 1, \pm 3, \ldots, \pm\sqrt{M} - 1$. Note that for $M = 4$, QAM is identical to QPSK.

---

4. In the special case where the I and Q amplitude modulations correspond to amplitude-shift-keying (ASK), the quadrature multiplexed form is often referred to as *quadrature amplitude-shift-keying (QASK)*.

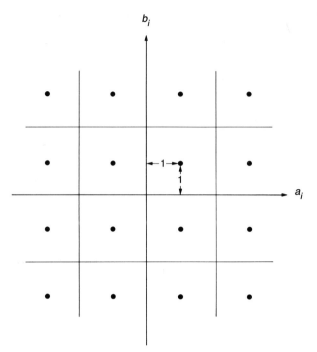

Figure 10.2   The signal constellation for QAM with $M = 16$

The average power P of $s(t)$ is easily computed as

$$P = A^2 \left(\overline{a_i^2} + \overline{b_i^2} \right) = A^2 \frac{1}{M} \left[\sum_{k=\pm1,\pm3,\ldots}^{\pm(\sqrt{M}-1)} \sum_{l=\pm1,\pm3,\ldots}^{\pm(\sqrt{M}-1)} (k^2 + l^2) \right] = \frac{2}{3} (M-1) A^2 \quad (10.25)$$

Assume now that (10.26) is transmitted over an AWGN as characterized by (10.2) and in addition the channel introduces a constant phase into $s(t)$ producing

$$s(t, \theta) = A \left[d_c(t) \cos (\omega_c t + \theta) - d_s(t) \sin (\omega_c t + \theta) \right] \quad (10.26)$$

Thus, the received signal is the sum of (10.2) and (10.26), namely, $r(t) = s(t, \theta) + n(t)$.

Coherent detection.   In the case of ideal coherent detection, the received signal is first demodulated (see Fig. 10.3) by the pair of quadrature carriers in (10.4) producing the lowpass signals

$$\varepsilon_c(t) = A d_c(t) + n_c(t) \cos \theta + n_s(t) \sin \theta$$
$$\varepsilon_s(t) = A d_s(t) + n_s(t) \cos \theta - n_c(t) \sin \theta \quad (10.27)$$

After passing through the I and Q matched filters and assuming $1/T_s \ll B$, we obtain the output samples [at time $t = (i+1)T_s$]

$$V_{c,i} = A a_i + N_{c,i}$$
$$V_{s,i} = A b_i + N_{s,i} \quad (10.28)$$

where $N_{c,i}$ and $N_{s,i}$ are independent zero-mean Gaussian noise samples with variance $N_0/2T_s$. A multilevel threshold decision (see Fig. 10.3b) is made on $V_{c,i}$ and $V_{s,i}$ to produce the I and Q data estimates $\hat{a}_i$ and $\hat{b}_i$. A correct QAM symbol decision then corresponds to

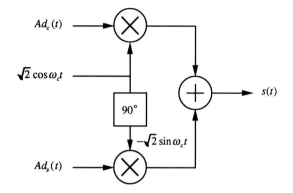

Figure 10.3a   A QAM modulator

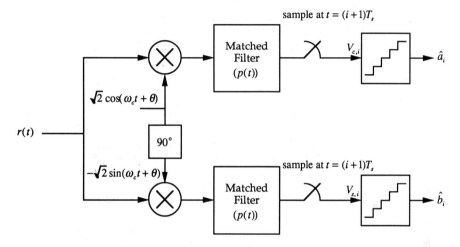

Figure 10.3b   An optimum coherent receiver for QAM

$\hat{a}_i = a_i$ and $\hat{b}_i = b_i$. The average probability of a correct symbol decision is computed as follows.

For a QAM signal constellation with M points, there are 4 corner points, $4\left(\sqrt{M} - 2\right)$ exterior points, and $\left(\sqrt{M} - 2\right)^2$ interior points. Because of the symmetry of the constellation, it is sufficient to examine only the points in the first quadrant. For each of the interior points, the probability of a correct decision is

$$P_{sA}(C) = \Pr\left\{A(a_i - 1) \le V_{c,i} \le A(a_i + 1); A(b_i - 1) \le V_{s,i} \le A(b_i + 1)\right\}$$

$$= \Pr\left\{-A \le N_{c,i} \le A; -A \le N_{s,i} \le A\right\}$$

$$= \left[\frac{1}{2}\,\mathrm{erfc}\left(-\sqrt{\frac{A^2 T_s}{N_0}}\right) - \frac{1}{2}\,\mathrm{erfc}\left(\sqrt{\frac{A^2 T_s}{N_0}}\right)\right]^2 \tag{10.29}$$

$$= \left[1 - \mathrm{erfc}\left(\sqrt{\frac{A^2 T_s}{N_0}}\right)\right]^2 = (1 - 2P_{s1}(E))^2$$

For each of the exterior points, the probability of a correct decision is

$$P_{sB}(C) = \Pr\left\{A(a_i - 1) \leq V_{c,i} \leq \infty; A(b_i - 1) \leq V_{s,i} \leq A(b_i + 1)\right\}$$

$$= \Pr\left\{-A \leq N_{c,i} \leq \infty; -A \leq N_{s,i} \leq A\right\}$$

$$= \left[\frac{1}{2}\operatorname{erfc}\left(-\sqrt{\frac{A^2 T_s}{N_0}}\right)\right]\left[\frac{1}{2}\operatorname{erfc}\left(-\sqrt{\frac{A^2 T_s}{N_0}}\right) - \frac{1}{2}\operatorname{erfc}\left(\sqrt{\frac{A^2 T_s}{N_0}}\right)\right] \quad (10.30a)$$

$$= (1 - P_{s1}(E))\,(1 - 2P_{s1}(E))$$

or

$$P_{sB}(C) = \Pr\left\{A(a_i - 1) \leq V_{c,i} \leq A(a_i + 1); A(b_i - 1) \leq V_{s,i} \leq \infty\right\}$$

$$= \Pr\left\{-A \leq N_{c,i} \leq A; -A \leq N_{s,i} \leq \infty\right\}$$

$$= \left[\frac{1}{2}\operatorname{erfc}\left(-\sqrt{\frac{A^2 T_s}{N_0}}\right) - \frac{1}{2}\operatorname{erfc}\left(\sqrt{\frac{A^2 T_s}{N_0}}\right)\right]\left[\frac{1}{2}\operatorname{erfc}\left(-\sqrt{\frac{A^2 T_s}{N_0}}\right)\right] \quad (10.30b)$$

$$= (1 - 2P_{s1}(E))\,(1 - P_{s1}(E))$$

Finally, for the single corner point, the probability of a correct decision is

$$P_{sC}(C) = \Pr\left\{A(a_i - 1) \leq V_{c,i} \leq \infty; A(b_i - 1) \leq V_{s,i} \leq \infty\right\}$$

$$= \Pr\left\{-A \leq N_{c,i} \leq \infty; -A \leq N_{s,i} \leq \infty\right\} \quad (10.31)$$

$$= \left[\frac{1}{2}\operatorname{erfc}\left(-\sqrt{\frac{A^2 T_s}{N_0}}\right)\right]^2 = (1 - P_{s1}(E))^2$$

The average probability of symbol error is then

$$P_s(E) = 1 - P_s(C)$$

$$= 1 - \frac{1}{M}\left\{4(1 - P_{s1}(E))^2 + 4\left(\sqrt{M} - 2\right)(1 - P_{s1}(E))(1 - 2P_{s1}(E))\right.$$

$$\left. + \left(\sqrt{M} - 2\right)^2 (1 - 2P_{s1}(E))^2\right\}$$

$$= 4\left(1 - \frac{1}{\sqrt{M}}\right)P_{s1}(E)\left[1 - \left(1 - \frac{1}{\sqrt{M}}\right)P_{s1}(E)\right] \quad (10.32)$$

$$= 2\left(1 - \frac{1}{\sqrt{M}}\right)\operatorname{erfc}\left(\sqrt{\left(\frac{3}{2(M-1)}\right)\frac{E_s}{N_0}}\right)$$

$$\times \left[1 - \frac{1}{2}\left(1 - \frac{1}{\sqrt{M}}\right)\operatorname{erfc}\left(\sqrt{\left(\frac{3}{2(M-1)}\right)\frac{E_s}{N_0}}\right)\right]$$

where we have substituted for A^2 in terms of P from (10.25) and as in previous chapters $E_s/N_0 \triangleq PT_s/N_0$ denotes the average symbol energy-to-noise ratio.

It is interesting to observe that (10.32) can also be expressed in terms of the average error probability performance of $\sqrt{M}$-AM, that is, the individual modulations on the I and Q channels. Since a correct QAM decision is made only when a correct decision is made on both the I and Q channels, then

$$P_s(C)|_{M-QAM} = \left[P_s(C)|_{\sqrt{M}-AM} \right]^2 \tag{10.33}$$

or, equivalently

$$P_s(E)|_{M-QAM} = 1 - \left[1 - P_s(E)|_{\sqrt{M}-AM} \right]^2$$

$$= 2P_s(E)|_{\sqrt{M}-AM} \left(1 - \frac{1}{2} P_s(E)|_{\sqrt{M}-AM} \right) \tag{10.34}$$

which, using the result for the symbol error probability of M-AM [see (4.167)][5] and replacing M by $\sqrt{M}$ and E_s by $E_s/2$, demonstrates the equivalence.

To obtain the average bit error probability of QAM from the average symbol error probability of (10.32), we observe that QAM can be assigned a perfect 2-dimensional Gray code [2], that is, for each symbol error in the set, all adjacent symbol errors cause only a single bit error. Thus, since the number of bits/symbol is $\log_2 M$, then for large SNR we have the simple approximate result that

$$P_b(E) \cong \frac{P_s(E)}{\log_2 M} = \frac{P_s(E)}{2m} \tag{10.35}$$

where in addition the symbol energy-to-noise ratio E_s/N_0 is equal to $2m$ times the bit energy-to-noise ratio E_b/N_0. The values of E_b/N_0 corresponding to $P_b = 10^{-5}$ are tabulated in Table 10.2.

An exact evaluation of bit error probability can be obtained by considering the number of bit errors that are associated with the QAM symbols having each of the symbol error probability types, $P_{sA}(E)$, $P_{sB}(E)$, and $P_{sC}(E)$ and then averaging over all possible cases. For arbitrary M, it is tedious to express the result in closed form. However, for any particular value of M, an answer can be obtained. For example, for $M = 16$ and $M = 64$ and a Gray code mapping (see Chapter 4), it can be shown that

$$P_b(E) = \frac{3}{8} \, \mathrm{erfc} \left(\sqrt{\frac{2}{5} \frac{E_b}{N_0}} \right) + \frac{1}{4} \, \mathrm{erfc} \left(3\sqrt{\frac{2}{5} \frac{E_b}{N_0}} \right) - \frac{1}{8} \, \mathrm{erfc} \left(5\sqrt{\frac{2}{5} \frac{E_b}{N_0}} \right) \tag{10.36a}$$

---

5. The reader is cautioned that (4.167) applies to asymmetric M-AM where all of the signal points lie on the positive real line. For symmetric M-AM (the signal points are symmetrically located about the origin) as referred to here in drawing our analogy with QAM, the equivalent symbol error probability expression would be

$$P_s(E)|_{M-AM} = \left(\frac{M-1}{M} \right) \mathrm{erfc} \left(\sqrt{\left(\frac{3}{(M^2-1)} \right) \frac{E_s}{N_0}} \right)$$

where we have furthermore replaced E_{avg} by E_s for consistency with the usage in this chapter.

Table 10.2 Bandwidth and Power Efficiencies of Various Modulation Schemes ($P_b = 10^{-5}$, Values of E_b/N_0 are in dB)

	QFPM		Constant Envelope QFPM (N-FSK/QPSK)		CPQFPM		CPFPM (MFMSK)		N-FSK/ BPSK		QAM		M-PSK	
M	δ	$\frac{E_b}{N_0}$	δ	$\frac{E_b}{N_0}$	δ	$\frac{E_b}{N_0}$	δ	$\frac{E_b}{N_0}$	δ	$\frac{E_b}{N_0}$	δ	$\frac{E_b}{N_0}$	δ	$\frac{E_b}{N_0}$
4	1.0	9.60	1.0	9.60	0.67	9.60	0.67	9.60	0.76	9.60	1.0	9.6	1.0	9.6
8	—	—	1.0	8.31	—	—	0.60	8.31	0.83	8.31	—	—	1.5	12.9
16	1.33	9.60	0.8	7.36	0.80	9.60	0.44	7.36	0.71	7.36	2.0	13.4	2.0	17.4
32	—	—	0.55	6.71	—	—	0.29	6.71	0.52	6.71	—	—	2.5	22.3
64	1.2	8.31	0.35	6.16	0.67	8.31	0.18	6.16	0.34	6.16	3.0	17.8	3.0	27.5
256	0.89	7.36	0.12	5.36	0.47	7.36	—	—	—	—	4.0	22.5	4.0	38.1
1024	0.59	6.71	0.04	4.80	0.30	6.71	—	—	—	—	5.0	27.5	5.0	49.1
4096	0.36	6.16	0.01	4.40	0.18	6.16	—	—	—	—	6.0	32.6	6.0	60.2

and

$$P_b(E) = \frac{7}{24}\,\mathrm{erfc}\left(\sqrt{\frac{E_b}{7N_0}}\right) + \frac{1}{4}\,\mathrm{erfc}\left(3\sqrt{\frac{E_b}{7N_0}}\right)$$

$$- \frac{1}{24}\,\mathrm{erfc}\left(5\sqrt{\frac{E_b}{7N_0}}\right) + \frac{1}{24}\,\mathrm{erfc}\left(9\sqrt{\frac{E_b}{7N_0}}\right) - \frac{1}{24}\,\mathrm{erfc}\left(13\sqrt{\frac{E_b}{7N_0}}\right)$$

$$(10.36b)$$

respectively, both of which simplify to (10.35) for large SNR.

An alternate and simpler procedure is to observe[6] that because the I and Q channels operate independently, the average bit error probability for M-ary QAM is equal to the average bit error probability for $\sqrt{M}$-AM. Thus, substituting $\sqrt{M}$ for M and $E_s/2$ for E_s (the same substitutions as were used to demonstrate the relation between the symbol error probabilities of QAM and $\sqrt{M}$-AM) in the expression for average bit error probability of M-AM, we again obtain (10.36). Figure 10.4 is an illustration of (10.36) together with the results for $M = 4$, that is, QPSK. The type of relation in (10.34) and (10.36) between the average error probability performance of an I-Q modulation and that of its I and Q channels individually is typical of quadrature carrier multiplexed modulations.

We now consider the case where a carrier phase error exists between the received signal and the I and Q demodulation reference signals. Demodulating $r(t)$ with the I and Q reference signals of (10.9) and passing the results through the I and Q matched filters, we obtain the output samples [at time $t = (i + 1)T_s$]

$$V_{c,i} = A\left[a_i \cos\phi - b_i \sin\phi\right] + N'_{c,i}$$

$$V_{s,i} = A\left[b_i \cos\phi + a_i \sin\phi\right] + N'_{s,i}$$

$$(10.37)$$

where $N'_{c,i}$ and $N'_{s,i}$ are independent zero-mean Gaussian noise samples with variances $N_0/2T_s$.

---

6. This observation was made in [5] for the special case $M = 16$.

To compute the conditional probability of symbol error we again consider the probability of a correct decision (conditioned on ϕ) for each of the three different types of points in the first quadrant of the constellation. Defining

$$P_s\left(C \mid j, k, \phi\right) \triangleq \Pr\left\{\hat{a}_i = a_i, \hat{b}_i = b_i \mid a_i = j, b_i = k\right\} \tag{10.38}$$

then for each of the interior points, the conditional probability of a correct symbol decision is

$$P_s(C|j, k, \phi) = \Pr\left\{A(j-1) \le V_{c,i} \le A(j+1); A(k-1) \le V_{s,i} \le A(k+1)\right\}$$

$$= \Pr\left\{A\left[-1 + j(1 - \cos\phi) + k\sin\phi \le N'_{c,i} \le 1 + j(1 - \cos\phi) + k\sin\phi\right];\right.$$

$$\left. A\left[-1 + k(1 - \cos\phi) - j\sin\phi \le N'_{s,i} \le 1 + k(1 - \cos\phi) - j\sin\phi\right]\right\}$$

$$= \left[\frac{1}{2}\operatorname{erfc}\left(\sqrt{\frac{A^2 T_s}{N_0}}\left(-1 + j(1 - \cos\phi) + k\sin\phi\right)\right)\right.$$

$$\left. -\frac{1}{2}\operatorname{erfc}\left(\sqrt{\frac{A^2 T_s}{N_0}}\left(1 + j(1 - \cos\phi) + k\sin\phi\right)\right)\right] \tag{10.39}$$

$$\times \left[\frac{1}{2}\operatorname{erfc}\left(\sqrt{\frac{A^2 T_s}{N_0}}\left(-1 + k(1 - \cos\phi) - j\sin\phi\right)\right)\right.$$

$$\left. -\frac{1}{2}\operatorname{erfc}\left(\sqrt{\frac{A^2 T_s}{N_0}}\left(1 + k(1 - \cos\phi) - j\sin\phi\right)\right)\right];$$

$$j, k = 1, 3, \ldots, \left(\sqrt{M} - 3\right)$$

For each of the exterior points, the conditional probability of a correct decision is

$$P_s(C|j, k, \phi) = \Pr\left\{A(\sqrt{M} - 2) \le V_{c,i} \le \infty; \quad A(k-1) \le V_{s,i} \le A(k+1)\right\}$$

$$= \left[\frac{1}{2}\operatorname{erfc}\left(\sqrt{\frac{A^2 T_s}{N_0}}\left(-1 + \left(\sqrt{M} - 1\right)(1 - \cos\phi) + k\sin\phi\right)\right)\right]$$

$$\times \left[\frac{1}{2}\operatorname{erfc}\left(\sqrt{\frac{A^2 T_s}{N_0}}\left(-1 + k(1 - \cos\phi) - \left(\sqrt{M} - 1\right)\sin\phi\right)\right)\right. \tag{10.40a}$$

$$\left. -\frac{1}{2}\operatorname{erfc}\left(\sqrt{\frac{A^2 T_s}{N_0}}\left(1 + k(1 - \cos\phi) - \left(\sqrt{M} - 1\right)\sin\phi\right)\right)\right];$$

$$k = 1, 3, \ldots, \left(\sqrt{M} - 3\right)$$

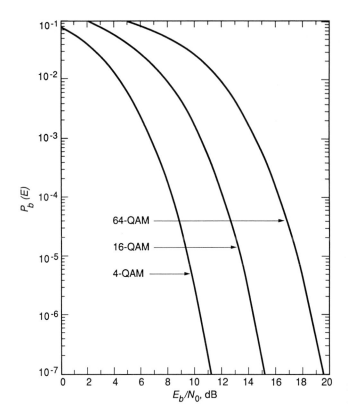

Figure 10.4   Exact bit error rate of 4, 16, and 64 QAM

or

$$P_s\left(C\,|\,j, k, \phi\right) = \Pr\left\{A(j-1) \le V_{c,i} \le A(j+1); \; A(\sqrt{M}-2) \le V_{s,i} \le \infty\right\}$$

$$= \left[\frac{1}{2}\,\text{erfc}\left(\sqrt{\frac{A^2 T_s}{N_0}}\left(-1 + j(1-\cos\phi) + \left(\sqrt{M}-1\right)\sin\phi\right)\right)\right.$$

$$\left. -\frac{1}{2}\,\text{erfc}\left(\sqrt{\frac{A^2 T_s}{N_0}}\left(1 + j(1-\cos\phi) + \left(\sqrt{M}-1\right)\sin\phi\right)\right)\right] \quad (10.40\text{b})$$

$$\times \left[\frac{1}{2}\,\text{erfc}\left(\sqrt{\frac{A^2 T_s}{N_0}}\left(-1 + \left(\sqrt{M}-1\right)(1-\cos\phi) - j\sin\phi\right)\right)\right];$$

$$j = 1, 3, \ldots, \left(\sqrt{M}-3\right)$$

Finally, for the single corner point, the conditional probability of a correct decision is

$$P_s\left(C\mid j, k, \phi\right) = \Pr\left\{A(\sqrt{M}-2) \le V_{c,i} \le \infty; \quad A(\sqrt{M}-2) \le V_{s,i} \le \infty\right\}$$

$$= \left[\frac{1}{2}\,\mathrm{erfc}\left(\sqrt{\frac{A^2 T_s}{N_0}\left(-1+\left(\sqrt{M}-1\right)\left(1-\cos\phi+\sin\phi\right)\right)}\right)\right]$$

$$\times \left[\frac{1}{2}\,\mathrm{erfc}\left(\sqrt{\frac{A^2 T_s}{N_0}\left(-1+\left(\sqrt{M}-1\right)\left(1-\cos\phi-\sin\phi\right)\right)}\right)\right] \tag{10.41}$$

From the symmetry of the signal set and the decision regions, the following properties hold:

$$P_s\left(C\mid j, k, \phi\right) = P_s\left(C\mid -j, -k, \phi\right)$$

$$P_s\left(C\mid j, k, \phi\right) = P_s\left(C\mid j, -k, -\phi\right) \tag{10.42}$$

$$P_s\left(C\mid j, k, \phi\right) = P_s\left(C\mid k, j, -\phi\right)$$

The average probability of correct detection (conditioned on ϕ) is given by

$$P_s(C\mid\phi) = \frac{1}{M}\sum_{j=\pm1,\pm3,\dots}^{\pm\left(\sqrt{M}-1\right)}\sum_{k=\pm1,\pm3,\dots}^{\pm\left(\sqrt{M}-1\right)} P_s\left(C\mid j, k, \phi\right) \tag{10.43}$$

Using the properties of (10.42), we can simplify (10.43) to

$$P_s(C\mid\phi) = \frac{2}{M}\sum_{\substack{j=1,3,\dots \\ j\le k}}^{\sqrt{M}-1}\sum_{k=1,3,\dots}^{\sqrt{M}-1} \varepsilon_{k-j}\left[P_s\left(C\mid j, k, \phi\right) + P_s\left(C\mid j, k, -\phi\right)\right] \tag{10.44}$$

where ε_n is the Neumann factor defined by

$$\varepsilon_n = \begin{cases} 1; & n = 0 \\ 2; & n > 0 \end{cases} \tag{10.45}$$

Substituting (10.40a), (10.41), and (10.42) into (10.44) and combining like terms gives the final form of the desired result, namely [3]

$$P_s(E|\phi) = 1 - P_s(C|\phi) = \frac{4}{M} \sum_{j=\pm1,\pm3,\dots}^{\pm(\sqrt{M}-1)} \sum_{l=0,\pm2,\pm4,\dots}^{\pm(\sqrt{M}-2)} \frac{1}{2}$$

$$\times \operatorname{erfc}\left(\sqrt{\left(\frac{3}{2(M-1)}\right)\frac{E_s}{N_0}}\,(l + (1-l)\cos\phi - j\sin\phi)\right)$$

$$-\frac{4}{M} \sum_{k=0,\pm2,\pm4,\dots}^{\pm(\sqrt{M}-2)} \sum_{l=0,\pm2,\pm4,\dots}^{\pm(\sqrt{M}-2)} \frac{1}{2} \qquad (10.46)$$

$$\times \operatorname{erfc}\left(\sqrt{\left(\frac{3}{2(M-1)}\right)\frac{E_s}{N_0}}\,(k + (1-k)\cos\phi + (l-1)\sin\phi)\right)$$

$$\times \frac{1}{2}\operatorname{erfc}\left(\sqrt{\left(\frac{3}{2(M-1)}\right)\frac{E_s}{N_0}}\,(l + (1-l)\cos\phi - (k-1)\sin\phi)\right)$$

The conditional error probability of (10.46) is plotted in Fig. 10.5 for $M = 16$ with $\Delta = \sqrt{\frac{3}{M-1}\frac{E_s}{N_0}}$ as a parameter. The limiting cases of (10.46), namely, $\Delta \to 0$ and $\Delta \to \infty$, are also illustrated in this figure and are mathematically described by

$$\lim_{\Delta \to 0} P_s(E|\phi) = 1 - \frac{1}{M} \qquad (10.47)$$

$$\lim_{\Delta \to \infty} P_s(E|\phi) = \frac{4}{M} \sum_{j=\pm1,\pm3,\dots}^{\pm(\sqrt{M}-1)} \sum_{l=0,\pm2,\pm4,\dots}^{\pm(\sqrt{M}-2)} \left[\frac{1}{2} - \frac{1}{2}\operatorname{sgn}(l + (1-l)\cos\phi - j\sin\phi)\right]$$

$$-\frac{4}{M} \sum_{k=0,\pm2,\pm4,\dots}^{\pm(\sqrt{M}-2)} \sum_{l=0,\pm2,\pm4,\dots}^{\pm(\sqrt{M}-2)} \left[\frac{1}{2} - \frac{1}{2}\operatorname{sgn}(k + (1-k)\cos\phi + (l-1)\sin\phi)\right]$$

$$\times \left[\frac{1}{2} - \frac{1}{2}\operatorname{sgn}(l + (1-l)\cos\phi - (k-1)\sin\phi)\right]$$

Note that the limiting curve for $\Delta \to \infty$ exhibits discontinuities at values of phase error at which the four signal points in the first quadrant move across a decision boundary (see Fig. 10.6), namely, $\phi_1 = 16°53'$, $\phi_2 = 18°26'$, $\phi_3 = 20°48'$, and $\phi_5 = 45°$. The absence of a discontinuity at $\phi_4 = 32°20'$ is explained as follows. For a phase error of value ϕ_3, the signal point at $(3A, A)$ moves across a decision boundary and into the decision region of the signal at $(3A, 3A)$, thus resulting in an error in its detection (see Fig. 10.6). When the phase error increases to ϕ_4, the signal point at $(3A, A)$ moves across another decision boundary and into the decision region of the signal at $(A, 3A)$. However, this still results in an error insofar as detecting the transmission of the signal point at $(3A, A)$; thus, there is no discontinuity in the conditional error probability at ϕ_4.

In order to calculate the *average* error probability performance of QAM in the presence of carrier phase error, we must characterize the pdf $f(\phi)$ of this phase error. Since this characterization in turn depends on the particular carrier synchronization loop used, we shall delay our discussion of average error probability performance until Volume II of this book,

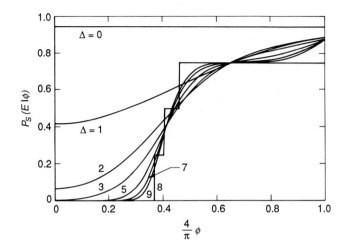

Figure 10.5   Conditional error probability performance of 16-ary QAM (Reprinted from [3] © IEEE)

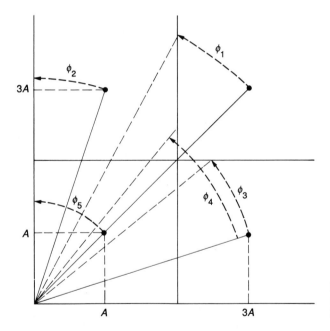

Figure 10.6   Values of phase error at which quadrant signal points cross decision boundaries (Reprinted from [3] © IEEE)

where we shall present carrier synchronization loops for QAM and their associated $f(\phi)$'s. At that time we shall be in a position to average (10.44) over these pdf's.

Differential detection.   Ordinarily, one thinks of differential detection only in the context of constant amplitude modulations such as M-PSK. However, by applying maximum-likelihood decision theory in the presence of unknown channel phase to combined amplitude-phase modulations, it is possible to develop a differential detection decision metric for the special case of QAM. As was done for M-PSK in Chapter 7, the results will be obtained for the general case of observation over $N_s(N_s \geq 2)$ symbol intervals. This leads to an MLSE algorithm that performs multiple symbol differential detection of QAM, herein referred to as *DQAM*.

Without reference to maximum-likelihood theory, one might think that, at least in principle, differential detection of combined amplitude-phase modulations could be accomplished by treating the phase and amplitude of each signal point as separate entities and as such applying differential detection to their phase and noncoherent (envelope) detection to their amplitudes. Such a scheme was proposed for DQAM in [4] in the context of its application in frequency-hopped spread spectrum systems. Unfortunately, because this scheme is quite suboptimum from a maximum-likelihood standpoint, its performance is very poor. Without further ado, we now present the maximum-likelihood approach to DQAM, followed by an evaluation of its performance over the AWGN.

If we assume that the receiver has no knowledge of the channel phase, θ, then without loss in generality we can assume that the quadrature demodulation references have zero phase. This is equivalent to setting $\hat{\theta} = 0$ in the reference signals of (10.9). Then, using complex notation, we obtain analogous to (10.37)

$$V_k \triangleq V_{c,k} + jV_{s,k} = A(a_k + jb_k)\cos\theta + A(-b_k + ja_k)\sin\theta + N'_{c,k} + jN'_{s,k}$$

$$\triangleq Ac_k e^{\theta_k}e^{j\theta} + N'_k \tag{10.48}$$

where $c_k e^{j\theta_k} \triangleq a_k + jb_k$ is the amplitude-phase complex representation of the transmitted QAM signal in the kth transmission interval and $N'_k \triangleq N'_{c,k} + jN'_{s,k}$ is a sample of a zero-mean complex Gaussian noise process with variance $\sigma_n^2 = N_0/T_s$. Consider now a sequence of V_k's of length N_s. This sequence, represented as a row vector, $\mathbf{V}$, is expressed as

$$\mathbf{V} = \mathbf{s}e^{j\theta} + \mathbf{N}' \tag{10.49}$$

where $V_{k-i}, s_{k-i} = Ac_{k-i}e^{j\theta_{k-i}}$, and N'_{k-i} are, respectively, the ith elements[7] $i = 0, 1, 2, \ldots, N_s - 1$ of the N_s-length row vectors $\mathbf{V}$ and $\mathbf{N}'$. For the assumed constant (but unknown) phase channel, the conditional probability of $\mathbf{V}$ given $\mathbf{s}$ and $\boldsymbol{\theta}$ is

$$f_{\mathbf{V}|\mathbf{s},\theta}(\mathbf{V}|\mathbf{s},\theta) = \frac{1}{(2\pi\sigma_n^2)^{N_s}}\exp\left\{-\frac{1}{2\sigma_n^2}|\mathbf{V} - \bar{\mathbf{V}}|^2\right\} \tag{10.50}$$

Using the definition of $\mathbf{V}$, the argument of the exponential in (10.50) can be expressed as

$$|\mathbf{V} - \bar{\mathbf{V}}|^2 = |\mathbf{V}|^2 + |\mathbf{s}|^2 - 2\left|\mathbf{V}\mathbf{s}^{*T}\right|\cos(\theta - \alpha) \tag{10.51}$$

where

$$\alpha \triangleq \tan^{-1}\frac{\operatorname{Im}\left\{\mathbf{V}\mathbf{s}^{*T}\right\}}{\operatorname{Re}\left\{\mathbf{V}\mathbf{s}^{*T}\right\}} \tag{10.52}$$

Substituting (10.51) into (10.50) and averaging over the uniform pdf of θ, we get

$$f_{\mathbf{V}|\mathbf{s}}(\mathbf{V}|\mathbf{s}) = \int_{-\pi}^{\pi} f_{\mathbf{V}|\mathbf{s},\theta}(\mathbf{V}|\mathbf{s},\theta)f_{\theta}(\theta)d\theta$$

$$= \frac{1}{(2\pi\sigma_n^2)^{N_s}}\exp\left\{-\frac{1}{2\sigma_n^2}\left(|\mathbf{V}|^2 + |\mathbf{s}|^2\right)\right\}I_0\left(\sigma_n^{-2}\left|\mathbf{V}\mathbf{s}^{*T}\right|\right) \tag{10.53}$$

---

7. For notational convenience here, we number the elements of an N_s-length sequence vector by $0, 1, \ldots, N_s - 1$ rather than by $1, 2, \ldots, N_s$. A similar notation applies in two dimensions to matrices.

where as before $I_0(x)$ is the zero order modified Bessel function of the first kind. Since the natural logarithm is a monotonically increasing function of its argument, then maximizing $f_{\mathbf{V}|\mathbf{s}}(\mathbf{V}|\mathbf{s})$ over $\mathbf{s}$ is equivalent to maximizing $\ln f_{\mathbf{V}|\mathbf{s}}(\mathbf{V}|\mathbf{s})$ over $\mathbf{s}$. Hence, eliminating all irrelevant terms (that is, those that are independent of the transmitted data), the following decision rule results: Choose the sequence $\mathbf{s}$ that maximizes the metric

$$\eta = -\frac{1}{2\sigma_n^2}|\mathbf{s}|^2 + \ln I_0\left(\sigma_n^{-2}\left|\mathbf{V}\mathbf{s}^{*T}\right|\right)$$

$$= -\frac{1}{2\sigma_n^2}\sum_{i=0}^{N_s-1}c_{k-i}^2 + \ln I_0\left(\sigma_n^{-2}\left|\sum_{i=0}^{N_s-1}V_{k-i}c_{k-i}e^{-j\theta_{k-i}}\right|\right) \tag{10.54}$$

Unfortunately, as was true for the analogous result for M-PSK derived in Chapter 7, the metric of (10.54) is ambiguous in phase in the sense that the argument of the Bessel function is unchanged by a fixed phase rotation. Thus, even in the absence of noise, it is possible to make an erroneous sequence decision. The simplest case of this to envision occurs where the phase rotation is a multiple, say J, of $90°$ in which case a particular sequence of QAM symbols is mistaken for another sequence whose symbols correspond to a rotation of the true symbols by J quadrants. To resolve this ambiguity, we again propose the use of differential encoding at the transmitter. Thus, if $\Delta\theta_k$ now denotes the phase of the kth transmitted complex QAM symbol, that is, $s_k = c_k e^{j\Delta\theta_k}$, then the actual phase transmitted over the channel in this same interval is given by the usual differential encoding relationship

$$\theta_k = \theta_{k-1} + \Delta\theta_k \tag{10.55}$$

Note that unlike M-PSK, the phases $\{\theta_k\}$ do not necessarily take on values corresponding to QAM symbols. This, however, is of no consequence since indeed (10.54) holds for any AM-PM constellation. Finally, then, substituting (10.55) into (10.54) gives the desired unambiguous metric

$$\eta = -\frac{1}{2\sigma_n^2}\sum_{i=0}^{N_s-1}c_{k-i}^2 + \ln I_0\left(\sigma_n^{-2}\left|\sum_{i=0}^{N_s-1}V_{k-i}c_{k-i}e^{-j\sum_{m=0}^{N_s-i-2}\Delta\theta_{k-i-m}}\right|\right) \tag{10.56}$$

Using the approximation (valid for large arguments, that is, high SNR) $\ln I_0(x) \cong x$, then after normalization by σ_n^2, (10.56) becomes

$$\eta = -\frac{1}{2}\sum_{i=0}^{N_s-1}c_{k-i}^2 + \left|\sum_{i=0}^{N_s-1}V_{k-i}c_{k-i}e^{-j\sum_{m=0}^{N_s-i-2}\Delta\theta_{k-i-m}}\right| \tag{10.57}$$

Analogous to Fig. 7.7, a direct implementation of a receiver that employs a decision rule based on the metric in (10.57) is illustrated in Fig. 10.7. Note that the receiver uses one extra complex amplitude, namely, $c_{k-N_s} + 1$, to make its decision on $c_{k-i}, \Delta\theta_{k-i}$; $i = 1, 2, \ldots, N_s - 2$.

Analysis of the average bit error probability performance of (10.57) is difficult. Thus, we present in Fig. 10.8 computer simulation results typical of this performance for the case of 16-ary DQAM. Also illustrated for comparison is the performance of ideal coherent detection of QAM as given by (10.36). We see that despite the use of a maximum-likelihood

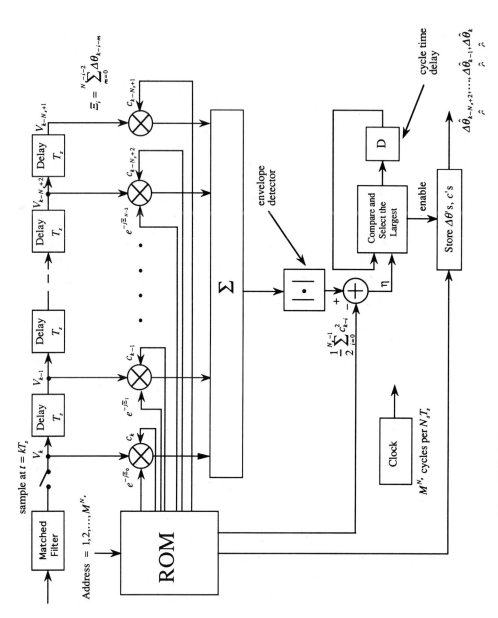

Figure 10.7 A direct implementation of a receiver based on the metric of Equation (10.57)

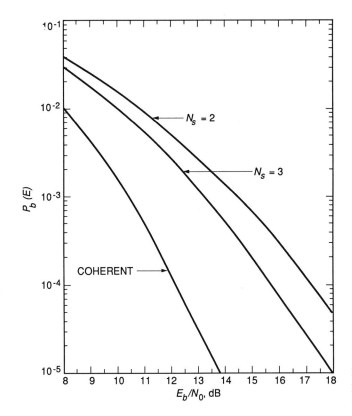

Figure 10.8   Average bit error probability performance of multiple symbol detection of 16-ary DQAM

receiver, the cost (in E_b/N_0) of using DQAM relative to coherent QAM is quite significant, although one can achieve a good reduction in this cost by increasing the observation interval from two to three symbols times. Further increase in the observation interval at the expense of increased complexity of the receiver processing brings one closer and closer to the ideal coherent performance.

10.1.3 Quadrature Biorthogonal Modulation

In Chapter 4, we defined a biorthogonal modulation as one that generates an N-dimensional signal set of size $M = 2N$ consisting of $M/2 = N$ pairs of antipodal signals, each pair being orthogonal to all other pairs. We also discussed a particular realization of such a scheme as a combined frequency/phase modulation (FPM) wherein N frequencies uniformly separated by $2f_d = k/2T_s$ (k an integer, T_s the signaling interval) are each BPSK modulated, that is, N-FSK/BPSK. Typically, $N = 2^n$ (n integer) and the frequencies are symmetrically placed around a carrier $f_c(=\omega_c/2\pi)$, that is, $f_c \pm f_d, f_c \pm 3f_d, \ldots, f_c \pm (N-1)f_d$. Also, the minimum frequency spacing required to guarantee orthogonality, namely, $2f_d = 1/2T_s(k = 1)$, corresponds to a modulation index $h = 2f_dT_s = 0.5$.

Consider now a modulation scheme that results from quadrature carrier multiplexing two N-dimensional biorthogonal signal sets as characterized above. This scheme is referred to as *quadrature biorthogonal modulation (QBOM)* [7] and yields a $2N$-dimensional signal set of size $M = (2N)^2$. Analogous to FPM, a particular embodiment of QBOM results from modulating the orthogonal carriers $\cos \omega_c t$ and $\sin \omega_c t$ with N-FSK/BPSK modulations. Here the minimum frequency spacing required for orthogonality corresponds to selecting $h = 1(f_d T_s = 0.5)$ for each of the two N-FSK/BPSK modulations and the resulting modulation is referred to as *quadrature frequency/phase modulation (QFPM)* [7]. A mathematical representation of a QFPM signal and its bit error probability performance on an AWGN are presented in the following section.

Quadrature frequency/phase modulation (QFPM).

Consider as a data source a binary random sequence $\{d_i\}$ of rate $1/T_b$ [equivalently, a rectangular pulse data stream $d(t)$]. Next subdivide this sequence into two subsequences $\{d_{c,i}\}$ and $\{d_{s,i}\}$ corresponding, respectively, to the even-numbered and odd-numbered bits in $\{d_i\}$ and generating rectangular pulse data streams $d_c(t)$ and $d_s(t)$. Since for each QFPM symbol of duration T_s there are $\log_2 M = \log_2(2N)^2 = 2(1 + \log_2 N)$ bits of duration T_b, then blocks of $\lambda = 1 + \log_2 N$ bits of $\{d_{c,i}\}$ and $\{d_{s,i}\}$ are respectively assigned to the inphase and quadrature biorthogonal components of this symbol. The first bits in these blocks are assigned to binary (± 1) sequences $\{a_{c,i}\}$ and $\{a_{s,i}\}$, respectively, which are used to define the BPSK portions of the biorthogonal modulations. The remaining $\lambda - 1 = \log_2 N$ bits in each block are assigned to N-ary $[\pm 1, \pm 3, \ldots, \pm(N-1)]$ sequences $\{b_{c,i}\}$ and $\{b_{s,i}\}$, respectively, which are used to define the N-FSK portions of the biorthogonal modulations.[8] Defining random rectangular pulse trains $a_c(t)$, $a_s(t)$, $b_c(t)$, and $b_s(t)$ from the sequences $a_{c,i}$, $a_{s,i}$, $b_{c,i}$, and $b_{s,i}$, for example, $a_c(t) = \sum_{i=-\infty}^{\infty} a_{c,i} p(t - iT_s)$, then a QFPM signal is characterized by

$$s(t) = \sqrt{P} \left\{ a_c(t) \cos \left[(\omega_c + b_c(t)\omega_d) t \right] + a_s(t) \sin \left[(\omega_c + b_s(t)\omega_d) t \right] \right\}$$
$$= s_c(t) + s_s(t) \tag{10.58}$$

where P is the average signal power and $\omega_d = 2\pi f_d$ is the radian frequency deviation about the carrier radian frequency ω_c. Note that (10.58) represents a constant energy ($E_s = PT_s$) but nonconstant envelope signal. To see this, we rewrite (10.58) as

$$s(t) = \sqrt{P} a_c(t) \left\{ \cos \left[(\omega_c + b_c(t)\omega_d) t \right] + \frac{a_s(t)}{a_c(t)} \cos \left[(\omega_c + b_s(t)\omega_d) t - \frac{\pi}{2} \right] \right\}$$

$$= 2\sqrt{P} a_c(t) \cos \left[\frac{b_c(t) - b_s(t)}{2} \omega_d t \pm \frac{\pi}{4} \right] \tag{10.59}$$

$$\times \cos \left[\left(\omega_c + \frac{b_c(t) + b_s(t)}{2} \omega_d \right) t \mp \frac{\pi}{4} \right]$$

where the upper signs correspond to $a_s(t)/a_c(t) = 1$ and the lower signs to $a_s(t)/a_c(t) = -1$. The cosine factor in (10.59) having a radian frequency $[(b_c(t) - b_s(t))/2] \omega_d$ represents the envelope of $s(t)$, whereas the second cosine factor of radian frequency $\omega_c + [(b_c(t) + b_s(t))/2] \omega_d$ represents the carrier. Clearly, the envelope is constant only when $b_c(t) = b_s(t)$, that is, the inphase and quadrature component frequencies in (10.58) are identical. In this special case, the $(2N)^2$-ary QFPM signal reduces to a $4N$-ary N-FSK/QPSK signal.

---

8. The coding rule used to map the data sequence $\{d_{c,i}\}$ into the binary amplitude and N-ary frequency sequences $\{a_{c,i}\}$ and $\{b_{c,i}\}$, and likewise the data sequence $\{d_{s,i}\}$ into the binary amplitude and N-ary frequency sequences $\{a_{s,i}\}$ and $\{b_{s,i}\}$, is identical to that given in Chapter 4 for realizing a biorthogonal signal set with an N-FSK/BPSK modulation (see Example 4.6).

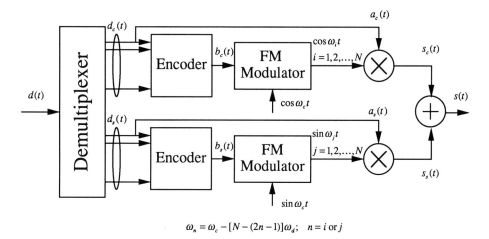

$$\omega_n = \omega_c - [N - (2n-1)]\omega_d; \quad n = i \text{ or } j$$

Figure 10.9a   A block diagram of a modulator for $(2N)^2$-ary QFPM

Block diagrams of modulators for $(2N)^2$-ary QFPM and $4N$-ary N-FSK/QPSK are illustrated in Figures 10.9a and 10.9b, respectively. Since for the former, the quadrature components of $s(t)$, namely, $s_c(t)$ and $s_s(t)$, are orthogonal over the signaling interval T_s, they can be demodulated and detected independently. As such, an optimum coherent receiver for a QFPM signal (see Fig. 10.10) is a parallel combination of two optimum coherent receivers for biorthogonal signals of the type illustrated in Fig. 4.17. Exact bit error probability results for N-FSK/BPSK biorthogonal modulation are given by Eq. (4.107) together with (4.102), (4.105), and (4.106). Since the two constituent N-FSK/BPSK signals in the M-ary QFPM signal of (10.58) are demodulated and detected independently, then the bit error probability performance of the latter is also given by Eq. (4.107), keeping in mind that now $N = \sqrt{M}/2$ and thus $E_s = (1 + \log_2 N) E_b = \left(\frac{1}{2} \log_2 M\right) E_b$. This equivalence between the average bit error probability of the resulting modulation and that of its individual inphase and quadrature components is typical of quadrature multiplexed modulations, for example, QPSK and QAM.

An upper bound on the bit error probability performance of QFPM that is simple to evaluate can be obtained from the same upper bound on the performance of N-FSK/BPSK and the equivalences noted above. In particular, since the probability of error of the first kind, $P_{s1}(E)$, of (4.105) can be upper bounded by the error probability for antipodal signaling, that is, $\frac{1}{2} \operatorname{erfc} \sqrt{E_s/N_0}$, then from (4.107) together with (4.103) and (4.106) we have

$$P_b(E) = \frac{1}{2} P_{s1}(E) + \frac{1}{2} P_s(E)$$

$$\leq \frac{1}{4} \operatorname{erfc} \sqrt{\frac{E_s}{N_0}} + \frac{1}{4} \left[(M-2) \operatorname{erfc} \sqrt{\frac{E_s}{2N_0}} + \operatorname{erfc} \sqrt{\frac{E_s}{N_0}} \right] \tag{10.60}$$

$$= \frac{1}{4} (M-2) \operatorname{erfc} \sqrt{\frac{E_s}{2N_0}} + \frac{1}{2} \operatorname{erfc} \sqrt{\frac{E_s}{N_0}}$$

where $M = 2N$. For $(2N)^2$-ary QFPM with the same number of frequencies as the N-FSK/BPSK scheme, the bit error probability has the same upper bound as in (10.60), which,

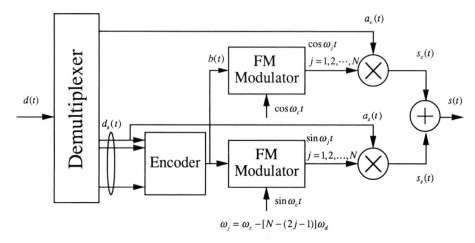

Figure 10.9b   A block diagram of a modulator for 4N-ary N-FSK/QPSK

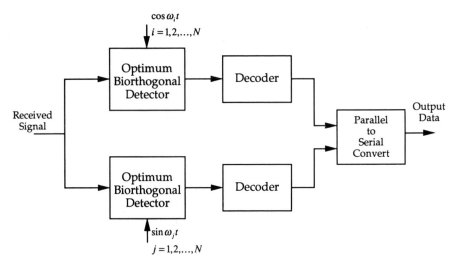

Figure 10.10   Block diagram of an optimum receiver for $(2N)^2$-ary QFPM

when written explicitly in terms of N, becomes

$$P_b(E) \leq \frac{N-1}{2} \, \mathrm{erfc} \sqrt{(1 + \log_2 N) \frac{E_b}{2N_0}} + \frac{1}{2} \, \mathrm{erfc} \sqrt{(1 + \log_2 N) \frac{E_b}{N_0}} \qquad (10.61)$$

For the constant envelope 4N-ary N-FSK/QPSK modulation, the receiver in Fig. 10.10 is not optimum. The reason for this is that the two biorthogonal components $s_c(t)$ and $s_s(t)$ are no longer independent of each other (they share the same frequency) and thus should not be detected as if they were. More specifically, in Fig. 10.10, the common frequency of $s_c(t)$ and $s_s(t)$ is redundantly detected twice (once in each biorthogonal detector) each time using only half the total power available. Rather, an optimum receiver (see Fig. 10.11) for this special case of QFPM should coherently combine both quadrature signal components prior to detection of the common frequency thus utilizing the entire signal

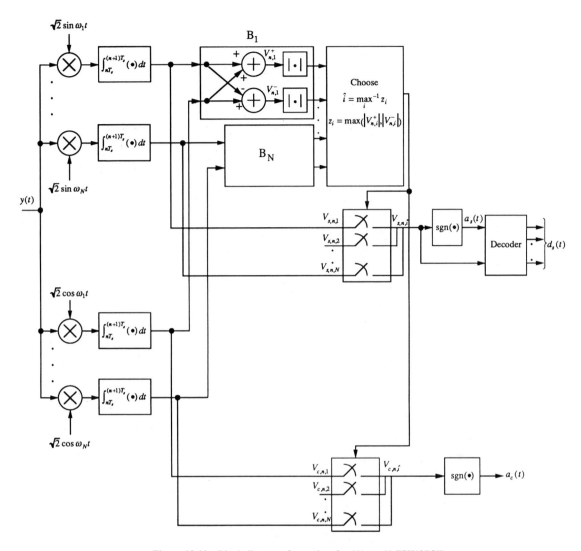

Figure 10.11   Block diagram of a receiver for $4N$-ary N-FSK/QPSK

power for this purpose. The idea behind such a configuration is generically the same as that used in constructing the optimum biorthogonal (N-FSK/BPSK) receiver, namely, remove the phase modulation for detecting the frequency, and then use the frequency decision to determine which of the N parallel channels the phase modulation should be extracted from. Removal of the QPSK modulation in Fig. 10.11 is accomplished at a given frequency by coherently adding and subtracting the I&D output samples (projections of the signal onto the quadrature coordinate axes) and choosing the larger in absolute value of these two quantities. The maximum of the N largest absolute values so produced then provides the desired frequency decision, and a binary hard decision on the amplitude in each of the quadrature components corresponding to this frequency channel provides the QPSK phase decision.

To compute an upper bound on the bit error probability performance of Fig. 10.11, we observe that for the N-FSK/QPSK signal of (10.58) with $b_s(t) = b_c(t) = b(t)$, the sum and difference I&D outputs of the ith frequency channel (at frequency ω_i), namely, $V_{n,i}^+$ and $V_{n,i}^-$, corresponding to an input signal at the same frequency, that is, $\omega_0 + b(t)\omega_d = \omega_i$, are given by

$$V_{n,i}^+ = \sqrt{\frac{P}{2}}\, (a_{s,n} + a_{c,n}) + \overbrace{\left(n_{c,n}^+ + n_{s,n}^-\right)}^{n^+}$$

$$V_{n,i}^- = \sqrt{\frac{P}{2}}\, (a_{c,n} - a_{s,n}) - \overbrace{\left(n_{c,n}^- - n_{s,n}^+\right)}^{n^-}$$

$$(10.62)$$

where for the noise representation of (10.2)

$$n_{c,n}^+ = \sqrt{2} \int_{nT}^{(n+1)T} n_c(t) \sin\left[(\omega_i - \omega_c)t + \pi/4\right] dt;$$

$$n_{s,n}^+ = \sqrt{2} \int_{nT}^{(n+1)T} n_s(t) \sin\left[(\omega_i - \omega_c)t + \pi/4\right] dt$$

$$(10.63)$$

$$n_{c,n}^- = \sqrt{2} \int_{nT}^{(n+1)T} n_c(t) \sin\left[(\omega_i - \omega_c)t - \pi/4\right] dt;$$

$$n_{s,n}^- = \sqrt{2} \int_{nT}^{(n+1)T} n_s(t) \sin\left[(\omega_i - \omega_c)t - \pi/4\right] dt$$

In (10.62), the noise components n^+ and n^- are zero mean and uncorrelated (also independent since they are Gaussian) and have equal variance $\sigma_n^2 = N_0/T$. The sum and difference I&D outputs, $V_{n,j}^+$ and $V_{n,j}^-$, of the jth frequency channel ($j \neq i$), have zero signal components (due to the orthogonality of the signaling frequencies over the symbol interval) and noise components analogous to those in (10.62). Note that depending on whether $a_{s,n} = a_{c,n}$ or $a_{s,n} \neq a_{c,n}$, the signal component of $V_{n,j}^+$ ($V_{n,j}^-$) will have a magnitude equal to $\sqrt{2P}$. Thus, the effective SNR of the signal used for frequency detection is $2PT_s/N_0$, which is a 3 dB improvement over the equivalent signal in Fig. 10.10. Once again, this improvement comes about because of the coherent combining of the I and Q signals prior to frequency detection which is possible when the two quadrature components of $s(t)$ share a common frequency.

An upper union bound on the symbol error probability performance of the receiver in Fig. 10.11 can be obtained by observing that a symbol error results when any of the following occur: a wrong frequency is detected, the sign of $a_c(t)$ is erroneously detected, the sign of $a_s(t)$ is erroneously detected. As such

$$P_s(E) \le \frac{1}{2}(M-4)\,\text{erfc}\sqrt{(2 + \log_2 N)\frac{E_b}{2N_0}} + \frac{1}{2}\,\text{erfc}\sqrt{(2 + \log_2 N)\frac{E_b}{2N_0}}$$

$$+ \frac{1}{2}\,\text{erfc}\sqrt{(2 + \log_2 N)\frac{E_b}{2N_0}} = (2N-1)\,\text{erfc}\sqrt{(2 + \log_2 N)\frac{E_b}{2N_0}}$$

$$(10.64)$$

The first term of the first line in (10.64) corresponds to the detection of an incorrect frequency, while the second and third terms correspond to erroneous detection of the signs of $a_s(t)$ and $a_c(t)$.

An upper bound on the bit error probability performance is obtained by employing reasoning analogous to that used in relating bit error probability to symbol error probability for biorthogonal (in particular, N-FSK/BPSK) signals. In particular, with reference to Table 4.4, if a frequency error occurs, then on the average, half of the bits are in error. If an error occurs in detecting the sign of $a_s(t)$, then all but one [that corresponding to $a_c(t)$] of the total number of bits are in error. Finally, if an error occurs in detecting the sign of $a_c(t)$, then only one bit is in error. Since the total number of bits per symbol is $2 + \log_2 N$, then from (10.64) and the above arguments, the average bit error probability has the upper bound

$$
\begin{aligned}
P_b(E) \leq & \left(\frac{1}{2}\right) \frac{1}{2}\, (M-4)\, \mathrm{erfc}\, \sqrt{(2+\log_2 N)\frac{E_b}{2N_0}} \\[2mm]
& + \left(\frac{1+\log_2 N}{2+\log_2 N}\right) \frac{1}{2}\, \mathrm{erfc}\, \sqrt{(2+\log_2 N)\frac{E_b}{2N_0}} \\[2mm]
& + \left(\frac{1}{2+\log_2 N}\right) \frac{1}{2}\, \mathrm{erfc}\, \sqrt{(2+\log_2 N)\frac{E_b}{2N_0}} \\[2mm]
= & \frac{1}{2}\, (2N-1)\, \mathrm{erfc}\, \sqrt{(2+\log_2 N)\frac{E_b}{2N_0}}
\end{aligned}
\tag{10.65}
$$

which is simply one-half of the symbol error probability upper bound in (10.64). Finally, we shall see later in this chapter that the structure of the receiver in Fig. 10.11 forms the basis of the optimum receiver for continuous phase FPM, which is constant envelope by construction.

Using the methods presented in Chapter 2, it is straightforward to show that the equivalent lowpass two-sided power spectral density (PSD) of QFPM is given by [7]

$$
\begin{aligned}
S_s(f) &= \frac{PT_s}{N} \sum_{i=0}^{N-1} \mathrm{sinc}^2 \left(fT_s + i - \frac{N-1}{2} \right) \\[2mm]
&= \frac{2\left(1+\log_2 N\right) PT_b}{N} \sum_{i=0}^{N-1} \mathrm{sinc}^2 \left(2\left(1+\log_2 N\right) fT_b + i - \frac{N-1}{2} \right)
\end{aligned}
\tag{10.66}
$$

which is illustrated in Fig. 10.12 for various values of N. Note that since QFPM corresponds to a modulation index $h=1$, its PSD is not[9] simply a frequency scaled (by a factor of two) version of the PSD for N-FSK/BPSK ($h=0.5$). This is readily seen by comparing (10.66) with the comparable result in Example 4.6 and Fig. 10.12 with the comparable figure in that example, that is, Fig. 4.45.

The null-to-null bandwidth, W_{nn}, ($B_{nn} = W_{nn}/2$ is the single-sided lowpass null bandwidth) of QFPM is easily obtained from (10.66) by setting the argument of the sinc

9. The exception to this statement occurs for the trivial case of $N = 1$ whereupon QFPM becomes QPSK and N-FSK/BPSK becomes simply BPSK.

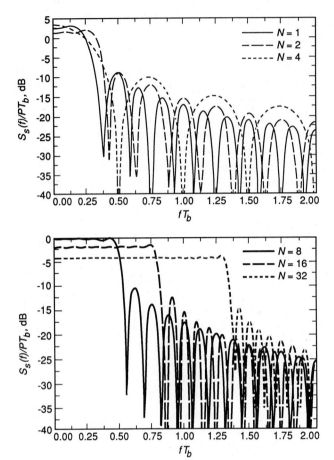

Figure 10.12 Power spectral density of QFPM with N as a parameter (Reprinted from [7] © IEEE)

function equal to unity when $i = 0$ and letting $f = W_{nn}/2$ with the result that

$$W_{nn} = (N+1)\frac{1}{T_s} = \frac{N+1}{2(1+\log_2 N)}\frac{1}{T_b} \qquad (10.67)$$

Table 10.2 illustrates the bandwidth efficiency $\delta = 1/W_{nn}T_b = 2(1+\log_2 N)/(N+1)$ of QFPM for various values of $M = (2N)^2$. Also tabulated are the exact values of E_b/N_0 (in dB) for a bit error probability $P_b = 10^{-5}$ as determined from the exact N-FSK/BPSK results of Chapter 4, that is, Eq. (4.107). We observe that in comparison with QAM and MPSK, the bandwidth efficiency of QFPM is significantly lower for $M > 16$ as one might expect; however, the power efficiency is significantly better. In particular, QFPM yields a power savings of 3.8 dB for $M = 16$, 9.5 dB for $M = 64$, and 15.1 dB for $M = 256$ when compared with QAM. When compared with M-PSK, the power savings are even larger; however, the reader is reminded that M-PSK is a constant envelope modulation whereas, in general, QFPM is not. For the special case of constant envelope QFPM (N-FSK/QPSK) as discussed above, Table 10.2 also gives the exact values of E_b/N_0 (in dB) required for a bit error probability $P_b = 10^{-5}$.

If instead of using null bandwidth, we were to use a fractional out-of-band power definition of bandwidth (see Chapter 1) for computing bandwidth efficiency, then QFPM shapes

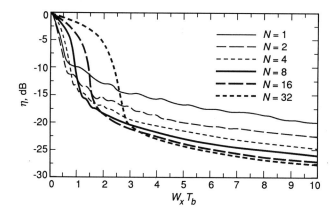

Figure 10.13 Fractional out-of-band power versus normalized bandwidth $W_x T_b$ for QFPM with N as a parameter (Reprinted from [7] © IEEE)

Table 10.3 A Comparison of the Bandwidth Efficiencies of Various Modulation Schemes Using a 90% Power Containment Bandwidth Definition ($\delta = 1/W_{90}T_b$)

M	QFPM	CPQFPM	CPFPM (MFMSK)	QAM M-PSK	CPFSK ($h = 0.5$)	N-FSK/2PSK ($h = 0.5$)
2	—	—	—	0.58	1.25	0.58
4	1.16	1.25	1.25	1.16	1.18	1.00
8	—	—	0.93	1.74	—	1.11
16	1.74	1.24	0.59	2.31	0.56	0.95
32	—	—	0.36	2.90	—	0.65
64	1.52	0.89	0.21	3.47	0.21	0.41
128	—	—	0.12	4.07	—	
256	1.07	0.58	—	4.62	—	
1024	0.68	0.35	—	5.78	—	
4096	0.42	0.21	—	6.93	—	

up considerably better when compared with QAM and M-PSK. Figure 10.13 illustrates the fractional out-of-band power [see (2.125)] $\eta = (1 - x/100)$ in dB of QFPM versus normalized bandwidth $W_x T_b$ where x is the percentage of the total power that lies in the bandwidth W_x. Tables 10.3 and 10.4 [7] present the comparable data to the column labeled δ in Table 10.2 for 90% and 99% power containments, that is, $\delta = 1/W_{90}T_b$ and $\delta = 1/W_{99}T_b$ where

$$\frac{\int_{-W_x/2}^{W_x/2} S_s(f)\,df}{\int_{-\infty}^{\infty} S_s(f)\,df} = \frac{x}{100} \qquad (10.68)$$

In Section 10.2 we will discuss other forms of QBOM that exploit the desirable aspects of carrier phase continuity.

Quadrature-quadrature phase-shift-keying (Q^2PSK). In a 1987 patent [8] and later in the open literature [9], Saha described an I-Q modulation scheme that, using a pair of quadrature (orthogonal) carriers and a pair of quadrature (orthogonal) pulse shapes, aimed at forming a signal set that would make more efficient spectral utilization of the

Table 10.4 A Comparison of the Bandwidth Efficiencies of Various Modulation Schemes Using a 99% Power Containment Bandwidth Definition ($\delta = 1/W_{99}T_b$)

M	QFPM	CPQFPM	CPFPM (MFMSK)	QAM M-PSK	CPFSK ($h = 0.5$)	N-FSK/2PSK ($h = 0.5$)
2	—	—	—	0.05	0.83	0.05
4	0.1	0.83	0.83	0.1	0.79	0.1
8	—	—	0.73	—	—	0.15
16	0.2	0.97	0.51	0.2	0.47	0.20
32	—	—	0.32	—	—	0.23
64	0.3	0.77	0.19	0.3	—	0.25
128	—	—	0.11	—	—	—
256	0.37	0.51	—	0.4	—	—
1024	0.43	0.32	—	0.5	—	—
4096	0.35	0.19	—	0.6	—	—

available signal dimensions than conventional two-dimensional signal sets such as QPSK.[10] The modulation scheme dubbed *quadrature-quadrature phase-shift-keying* (Q²PSK) places two QPSK signals in quadrature to form a four-dimensional signal set that, in a given bandwidth, ideally allows for an increase of the bit rate by a factor of two (i.e., twice the throughput) relative to QPSK. Mathematically speaking, a Q²PSK signal takes the form

$$s(t) = \text{Re}\left\{ \sqrt{P}\left[\sum_{i=-\infty}^{\infty} p_1(t - iT_s)e^{j\theta_{1,i}} + \sum_{i=-\infty}^{\infty} p_2(t - iT_s)e^{j\theta_{2,i}} \right] e^{j\omega_c t} \right\} \quad (10.69)$$

where $T_s = 4T_b$ is the signaling interval, $p_1(t)$ and $p_2(t)$ are unit power real pulse shapes of duration T_s and orthogonal to one another over the interval $0 \leq t \leq T_s$, and $\theta_{n,i}$; $n = 1, 2$ denotes the nth data phase (constant over the nth signaling interval $nT_s \leq t \leq (n + 1)T_s$) from each of the two QPSK signals with values ranging over the set $\beta_m = (2m + 1)\pi/4$; $m = 0, 1, 2, 3$. A particular case of interest is where $p_1(t) = \sqrt{2}\cos(\pi t/T_s)$ and $p_2(t) = \sqrt{2}\sin(\pi t/T_s)$, whereupon (10.69) becomes (in the zeroth signaling interval)

$$s(t) = \sqrt{2P}\left[\cos\left(\frac{\pi t}{T}\right)\cos(\omega_c t + \theta_{1,0}) + \sin\left(\frac{\pi t}{T}\right)\cos(\omega_c t + \theta_{2,0}) \right]$$

$$= \sqrt{P}\left[a_{1,0}\cos\left(\frac{\pi t}{T}\right)\cos\omega_c t - b_{1,0}\cos\left(\frac{\pi t}{T}\right)\sin\omega_c t \right. \quad (10.70)$$

$$\left. + a_{2,0}\sin\left(\frac{\pi t}{T}\right)\cos\omega_c t - b_{2,0}\sin\left(\frac{\pi t}{T}\right)\sin\omega_c t \right]$$

with $a_{n,0}, b_{n,0}$; $n = 1, 2$ representing the equivalent ±1 inphase and quadrature pairs of data bits corresponding to the data phases $\theta_{n,0}$; $n = 1, 2$, that is, $a_{n,0} = \sqrt{2}\cos\theta_{n,0}$, $b_{n,0} = \sqrt{2}\sin\theta_{n,0}$; $n = 1, 2$. The structure of (10.70) clearly identifies this particular form of Q²PSK as the sum of two sinusoidally pulse-shaped QPSK signals that are in quadrature (with respect to their pulse shaping) to each other.

---

10. By "QPSK" here and in the discussion that follows, we mean the more generic case that allows for arbitrary (rather than rectangular) pulse shaping.

It is interesting to note that Q^2PSK described by (10.70) is a special case of QFPM discussed in the previous section corresponding to $N = 2$. To see this, we note that for $N = 2$, $b_c(t)$ and $b_s(t)$ of (10.58) are binary (± 1) waveforms. Then, for the minimum frequency to guarantee orthogonality, namely, $\omega_d = \pi/T_s$, we have

$$s(t) = \sqrt{P}\left[a_c(t)\cos\left(\frac{\pi t}{T}\right)\cos\omega_c t - a_c(t)b_c(t)\sin\left(\frac{\pi t}{T}\right)\sin\omega_c t\right]$$

$$+ \sqrt{P}\left[a_s(t)\cos\left(\frac{\pi t}{T}\right)\sin\omega_c t + a_s(t)b_s(t)\sin\left(\frac{\pi t}{T}\right)\cos\omega_c t\right] \qquad (10.71)$$

Since $a_c(t)$, $a_c(t)b_c(t)$, $a_s(t)$, and $a_s(t)b_s(t)$ are all binary waveforms, that is, rectangular pulse data streams of rate $1/T_s = 1/4T_b$, then letting $a_c(t) = a_1(t)$, $a_c(t)b_c(t) = b_2(t)$, $a_s(t) = -b_1(t)$, and $a_s(t)b_s(t) = a_2(t)$, we obtain the desired equivalence with (10.70). Thus, Q^2PSK of (10.70) can be generated, demodulated, and detected as a QFPM signal with $N = 2$ in accordance with Figures 10.9 and 10.10.

10.2 CONTINUOUS PHASE MODULATION (CPM)

Continuous phase modulation (CPM) refers to a generic class of modulations wherein the carrier phase is maintained continuous from symbol interval to symbol interval. Traditionally, this acronym has been assigned to schemes that are constant envelope and convey their information in the form of a digital frequency modulation represented as a doubly infinite random pulse train. Recently, new nonconstant envelope CPM schemes have been developed (see Section 10.2.3) which convey their information in the form of combined frequency and phase modulation and have certain advantages over their predecessors. Thus, to allow for this distinction, in this book we shall assign the acronym *continuous phase frequency modulation (CPFM)* to what has in the past been referred to as simply CPM.

By allowing the phase to remain continuous from symbol interval to symbol interval, one introduces memory into the transmitted signal which can be exploited at the receiver by making decisions on an observation of the signal longer than one symbol interval. Perhaps the first recognition of this idea was in the 1971 conference paper by Pelchat, Davis, and Luntz [10] which considered coherent detection of continuous phase binary FSK (CPFSK) signals and showed that the high SNR error probability performance improved with increasing observation interval. Since that time, there has been a myriad of publications on the subject of coherent and noncoherent detection of CPM with consideration of both spectral and power efficiencies. An excellent documentation of much of this work appears in a textbook by Anderson et al. [11] that is entirely dedicated to this subject. Since the focus of our book is on a considerably broader coverage of modulation schemes, the treatment of CPM given here shall by comparison be quite brief. For a more extensive expose on the subject, the reader is referred to [11] and the host of references cited therein.

10.2.1 Continuous Phase Frequency Modulation (CPFM)

In the most general case, a CPFM signal is described by

$$s(t) = \sqrt{2P}\cos\left[\omega_c t + 2\pi\sum_{i=-\infty}^{n}\alpha_i h_i q(t - nT_s)\right]; \quad nT_s \le t \le (n+1)T_s \quad (10.72)$$

where α_n denotes the nth transmitted data symbol which takes on values from the M-ary alphabet $\{\Delta_k = -M + (2k+1); \quad k = 0, 1, 2, \ldots, M-1\}$, h_n is the modulation index, which may vary from symbol interval to symbol interval, and $q(t)$ is the phase pulse, which has length LT_s (L integer) and is expressed as the integral of the normalized frequency pulse $g(t)$ by

$$q(t) = \int_0^t g(\tau)d\tau; \quad q(LT_s) = \int_0^{LT_s} g(t)dt = \frac{1}{2} \tag{10.73}$$

CPFM schemes characterized by (10.72) fall into several categories depending on the variation of h and the shape and duration of $g(t)$. Typical shapes for $g(t)$ are rectangular and raised cosine which lead, respectively, to the modulation types LREC and LRC. *Multi-h* modulations refer to the class of CPFM's where h_n varies from interval to interval (typically the variation is cyclic through a set of indices). *Full response CPFM* is one where the frequency pulse duration is one symbol, that is, $L = 1$. Otherwise the modulation is referred to as *partial response CPFM*. *Continuous phase FSK (CPFSK)* is the acronym associated with a full response CPM having a rectangular frequency pulse, that is, 1REC. For our purposes in this chapter, we shall consider only the case where h_i is independent of i, that is, the modulation index is constant over the duration of the transmitted signal, and also $L = 1$. For this case, (10.72) reduces to

$$s(t) = \sqrt{2P} \cos \left[\omega_c t + 2\pi h \alpha_n q(t - nT_s) + \pi h \sum_{i=-\infty}^{n-1} \alpha_i \right] \tag{10.74}$$

$$= \sqrt{2P} \cos \left[\omega_c t + \theta(t, \boldsymbol{\alpha}) \right]; \quad nT_s \leq t \leq (n+1)T_s$$

The first term in $\theta(t, \boldsymbol{\alpha})$ represents the contribution to the total phase in the nth signaling interval due solely to the nth data symbol α_n. The second term represents the contribution due to all previous data symbols α_{n-i}; $i = 1, 2, \ldots, \infty$ and as such is responsible for the memory associated with this form of modulation.

Analogous to (10.74), the excess phase in the $(n-1)^{\text{st}}$ transmission interval $(n-1)T_s \leq t \leq nT_s$ is given by

$$\theta(t, \boldsymbol{\alpha}) = 2\pi h \alpha_{n-1} q(t - (n-1)T_s) + \pi h \sum_{i=-\infty}^{n-2} \alpha_i \tag{10.75}$$

Comparing (10.75) with (10.74), we observe that $\theta(t, \boldsymbol{\alpha})$ is continuous at the symbol transition instant $t = nT_s$ [since $q(0) = 0$] which is required for CPFM.

Transmitting $s(t)$ over a constant phase AWGN channel results in the received signal $r(t) = s(t, \theta) + n(t) = A \cos(\omega_c t + \theta(t, \boldsymbol{\alpha}) + \theta) + n(t)$ where as in previous sections $n(t)$ denotes the additive white Gaussian noise process.

Since, as previously mentioned, the continuity of phase from symbol interval to symbol interval introduces memory into the modulation, the optimum receiver (coherent or noncoherent) should base its decisions on an observation interval, say $N_s T_s$, commensurate with the memory duration. In this regard, two types of optimum receivers have been considered in the literature. One type calculates an estimate of the data sequence in the observation interval and performs *block-by-block* (N_s-*symbol*) detection. The second type decides on only one symbol (e.g., the first one) per N_s-symbol observation by averaging over the remaining

$N_s - 1$ symbols in the sequence, thus performing *symbol-by-symbol* detection. In what follows, we shall present the structure and performance of both receiver types for the special case of CPFSK modulation.

10.2.2 Continuous Phase FSK (CPFSK)

As previously mentioned, continuous phase FSK[11] (CPFSK) or, as it is sometimes, called 1REC CPM, is a full response continuous phase modulation with a rectangular frequency pulse defined by

$$g(t) = \begin{cases} \frac{1}{2T_b}; & 0 \le t \le T_b \\ 0; & \text{otherwise} \end{cases} \tag{10.76}$$

Thus from (10.73), the phase response pulse, $q(t)$, is a linear function of t over its T_b-sec duration and likewise the excess phase in the nth transmission interval $nT_b \le t \le (n+1)T_b$ is also linear, that is,

$$\theta(t, \boldsymbol{\alpha}) = \frac{\pi h \alpha_n (t - nT_b)}{T_b} + \pi h \sum_{i=0}^{n-1} \alpha_i \tag{10.77}$$

where we now assume that the observation interval starts at $t = 0$.

A plot of $\theta(t, \boldsymbol{\alpha})$ versus t resembles a trellis diagram (see Fig. 10.14) where the number of possible states [values of $\theta(t, \boldsymbol{\alpha})$] at any integer multiple of the symbol time T_b depends on the value of h. For example, if $h = 0.5$ [later we shall refer to this special case as *minimum-shift-keying (MSK)*], then there are four possible states (reduced modulo 2π) for $\theta(nT_b, \boldsymbol{\alpha})$, namely, $0, \pm\pi/2$, and π. In general, for arbitrary h, there are an infinite number of states.

Optimum bit-by-bit detection.   An important follow-on to the work of Pelchat et al. [10] appeared in the early paper by Luntz and Osborne [12], who derived structures for optimum bit-by-bit coherent and noncoherent detection of CPFSK. Furthermore, the performance of these structures was determined (in the form of upper and lower bounds) for low SNR as well as high SNR. What follows here is a brief presentation of the results in [12].

Coherent Detection.   For ideal coherent detection, the channel phase θ is assumed to be absolutely known and thus the received signal becomes simply $r(t) = s(t) + n(t)$. In what follows, it will be important to distinguish between signals $s(t)$ observed in the interval $0 \le t \le N_s T_b$ whose first bit (α_0) is a $+1$ and those whose first bit is a -1. To this end, we replace $s(t)$ with the notation $s\left(t, \alpha_0, \boldsymbol{\alpha}'\right)$ where $\boldsymbol{\alpha}'$ denotes the data sequence $(\alpha_1, \alpha_2, \ldots, \alpha_{N_s-1})$, that is, the sequence $\boldsymbol{\alpha}$ excluding the first data bit. Thus, a signal transmitted in the interval $0 \le t \le N_s T_b$ corresponding to a data sequence whose first bit is a $+1$ would be denoted by $s(t, 1, \boldsymbol{\alpha}')$ and likewise a signal transmitted in the same interval corresponding to a data sequence whose first bit is a -1 would be denoted by $s(t, -1, \boldsymbol{\alpha}')$. Representing $r(t)$ in its Karhunen-Loève expansion with coefficient vector $\mathbf{r}$, then in terms of the

---

11. Unless otherwise noted, we shall consider only the case of binary FSK and thus the terminology "symbol" is replaced by "bit." Also, the generalization of the results in this section to continuous phase M-FSK is given by Schonhoff in [13].

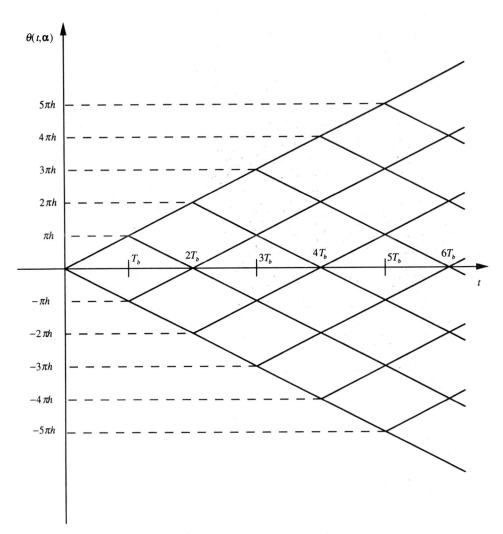

Figure 10.14   Excess phase trellis diagram for CPFSK

above, the conditional probability of $\mathbf{r}$ given $s(t, \alpha_0, \boldsymbol{\alpha}')$, denoted simply by $f(\mathbf{r} | \alpha_0, \boldsymbol{\alpha}')$, is

$$f(\mathbf{r} | \alpha_0, \boldsymbol{\alpha}') = \frac{1}{\sqrt{\pi N_0}} \exp\left\{-\frac{1}{N_0} \int_0^{N_s T_b} \left[r(t) - s(t, \alpha_0, \boldsymbol{\alpha}')\right]^2 dt\right\}$$

$$= C \exp\left(\frac{2}{N_0} \int_0^{N_s T_b} r(t) s(t, \alpha_0, \boldsymbol{\alpha}') dt\right)$$

(10.78)

where C is a constant that is independent of $s(t, \alpha_0, \boldsymbol{\alpha}')$. Averaging (10.78) over the pdf of $\boldsymbol{\alpha}'$, namely

$$f(\boldsymbol{\alpha}') = \prod_{i=1}^{N_s-1} f(\alpha_i) = \prod_{i=1}^{N_s-1} \left[\frac{1}{2}\delta(\alpha_i - 1) + \frac{1}{2}\delta(\alpha_i + 1)\right]$$

(10.79)

gives the conditional probability

$$f(\mathbf{r}\,|\alpha_0) = \int\limits_{\alpha_1}\int\limits_{\alpha_2}\cdots\int\limits_{\alpha_{N-1}} C\exp\left(\frac{2}{N_0}\int_0^{N_sT_b} r(t)s(t,\alpha_0,\boldsymbol{\alpha}')\,dt\right) f(\boldsymbol{\alpha}')\,d\boldsymbol{\alpha}'$$

$$= \frac{1}{2^{N_s-1}}\sum_{i=1}^{2^{N_s-1}} C\exp\left(\frac{2}{N_0}\int_0^{N_sT_b} r(t)s(t,\alpha_0,\boldsymbol{\alpha}'_i)\,dt\right) \tag{10.80}$$

where $\boldsymbol{\alpha}'_i$; $i = 1, 2, \ldots, 2^{N_s-1}$ represents the set of possible sequences for $\boldsymbol{\alpha}'$. A decision between $\alpha_0 = 1$ and $\alpha_0 = -1$ is then made by comparing the likelihood ratio

$$\Lambda \triangleq \frac{f(\mathbf{r}\,|1)}{f(\mathbf{r}\,|-1)} = \frac{\displaystyle\sum_{i=1}^{2^{N_s-1}}\exp\left(\frac{2}{N_0}\int_0^{N_sT_b}r(t)s(t,1,\boldsymbol{\alpha}'_i)\,dt\right)}{\displaystyle\sum_{i=1}^{2^{N_s-1}}\exp\left(\frac{2}{N_0}\int_0^{N_sT_b}r(t)s(t,-1,\boldsymbol{\alpha}'_i)\,dt\right)} = \frac{\displaystyle\sum_{i=1}^{2^{N_s-1}}x_{1,i}}{\displaystyle\sum_{i=1}^{2^{N_s-1}}x_{-1,i}} \tag{10.81}$$

with unity, that is, if $\Lambda > 1$ then decide $\alpha_0 = 1$, otherwise decide $\alpha_0 = -1$. Figure 10.15 is an illustration of a receiver that implements this decision rule. Unfortunately, the exact error probability performance of this receiver cannot be evaluated analytically. However, it is possible to approximate and bound this performance at low and high SNR.

At low SNR, we may use the approximation $e^x \cong 1 + x$, in which case the decision rule dictated by (10.81) and Fig. 10.15 becomes

$$\hat{\alpha}_0 = \mathrm{sgn}\left\{\sum_{i=1}^{2^{N_s-1}}\left(1 + \frac{2}{N_0}\int_0^{N_sT_b}r(t)s(t,1,\boldsymbol{\alpha}'_i)\,dt\right)\right.$$

$$\left. -\sum_{i=1}^{2^{N_s-1}}\left(1 + \frac{2}{N_0}\int_0^{N_sT_b}r(t)s(t,-1,\boldsymbol{\alpha}'_i)\,dt\right)\right\}$$

$$= \mathrm{sgn}\left\{\sum_{i=1}^{2^{N_s-1}}\left(\int_0^{N_sT_b}r(t)s(t,1,\boldsymbol{\alpha}'_i)\,dt - \int_0^{N_sT_b}r(t)s(t,-1,\boldsymbol{\alpha}'_i)\,dt\right)\right\} \tag{10.82}$$

$$= \mathrm{sgn}\left\{\int_0^{N_sT_b}r(t)\left[\sum_{i=1}^{2^{N_s-1}}s(t,1,\boldsymbol{\alpha}'_i) - \sum_{i=1}^{2^{N_s-1}}s(t,-1,\boldsymbol{\alpha}'_i)\right]dt\right\}$$

$$\triangleq \mathrm{sgn}\left\{\int_0^{N_sT_b}r(t)\left[\bar{s}(t,1) - \bar{s}(t,-1)\right]dt\right\}$$

Here, $\bar{s}(t,1)$ and $\bar{s}(t,-1)$ denote the *average* transmitted waveform given that the first bit is a "1" or a "−1", respectively. A block diagram implementation of (10.82) is illustrated in Fig. 10.16, which takes the form of a conventional matched filter receiver for binary communication over the AWGN. This implementation is referred to as an *average matched filter (AMF)* receiver in [11]. Other variations of the AMF are also discussed in Chapter 8 of [11]. Using the results of Chapter 4 and recognizing the symmetry of the present system,

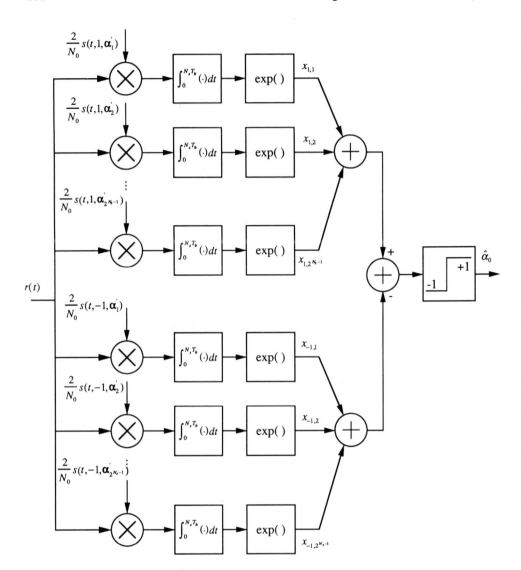

Figure 10.15   Optimum receiver for symbol-by-symbol coherent detection of CPFSK

we immediately have that the average bit error probability performance for low SNR is

$$P_b(E) = \frac{1}{2^{N_s-1}} \sum_{i=1}^{2^{N_s-1}} P_{bi}(E) \tag{10.83}$$

where

$$P_{bi}(E) = \frac{1}{2} \operatorname{erfc}\left\{\frac{\mu_i}{\sqrt{2}\sigma}\right\}; \quad \mu_i = \int_0^{N_s T_b} s(t, 1, \boldsymbol{\alpha}_i')[\bar{s}(t, 1) - \bar{s}(t, -1)]\, dt$$

$$\sigma^2 = \frac{N_0}{2} \int_0^{N_s T_b} [\bar{s}(t, 1) - \bar{s}(t, -1)]^2\, dt$$

(10.84)

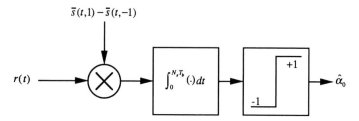

Figure 10.16   An equivalent receiver to Figure 10.15 for low SNR

Using (10.74) together with (10.77), it can be shown [13] that

$$\sigma^2 = \frac{N_0 T_b}{2}\left[1 - \text{sinc }2h + \frac{(1-\cos 2\pi h)\,(1+\text{sinc }2h)\,(\cos^{2N_s-2}\pi h - 1)}{2\left(\cos^2\pi h - 1\right)}\right] \quad (10.85a)$$

and

$$\mu_i = \frac{AT_b}{2}\left\{1 - \text{sinc }2h + \sin\pi h \sum_{k=2}^{N_s}\left(\cos^{k-2}\pi h\right)\right.$$

$$\left. \times\left[\sin\Theta_k + \frac{\alpha_{i,k}}{2\pi h}\left(\cos\Theta_k - \cos\left(2\pi h\alpha_{i,k}+\Theta_k\right)\right)\right]\right\}; \quad \Theta_k \triangleq \sum_{j=1}^{k-1}\pi h\alpha_{i,j} \quad (10.85b)$$

where $\alpha_{i,k}$; $k = 2, 3, \ldots, N_s$ is the kth component of the data vector $\boldsymbol{\alpha}'_i$.

At high SNR, we can upper union bound the error probability performance by first finding the pairwise error probability between each output of the upper group and each output of the lower group of $x_{i,j}$'s (see Fig. 10.15) and then averaging over all of them. Again recognizing the symmetry of the problem, we can write

$$P_b(E) \le \frac{1}{2^{N_s-1}}\sum_{i=1}^{2^{N_s-1}}\sum_{j=1}^{2^{N_s-1}} P_{i,j} \quad (10.86)$$

where

$$P_{i,j} = \Pr\left\{x_{1,i} \le x_{-1,j}|s\left(t, 1, \boldsymbol{\alpha}'_i\right)\right\}$$

$$= \Pr\left\{\exp\left(\frac{2}{N_0}\int_0^{N_s T_b} r(t)s\left(t, 1, \boldsymbol{\alpha}'_i\right)dt\right)\right.$$

$$\le \exp\left(\frac{2}{N_0}\int_0^{N_s T_b} r(t)s\left(t, -1, \boldsymbol{\alpha}'_i\right)dt\right)\left|s\left(t, 1, \boldsymbol{\alpha}'_i\right)\right\} \quad (10.87)$$

$$= \Pr\left\{\int_0^{N_s T_b} r(t)s\left(t, 1, \boldsymbol{\alpha}'_i\right)dt \le \int_0^{N_s T_b} r(t)s\left(t, -1, \boldsymbol{\alpha}'_i\right)dt\left|s\left(t, 1, \boldsymbol{\alpha}'_i\right)\right\}\right.$$

$$= \frac{1}{2}\text{erfc}\sqrt{\frac{N_s E_b}{N_0}\left(1 - \rho_{i,j}\right)}$$

with $E_b/N_0 = PT_b/N_0$ and

$$\rho_{i,j} \triangleq \frac{1}{N_s E_b} \int_0^{N_s T_b} s(t, -1, \alpha_i') \, s(t, 1, \alpha_i') \, dt$$

$$= \frac{1}{N_s} \left\{ \text{sinc } h \cos \pi h + \sum_{k=2}^{N_s} \text{sinc} \left(\frac{h(\alpha_{i,k} - \alpha_{j,k})}{2} \right) \right. \tag{10.88}$$

$$\left. \times \cos \left[\frac{\pi h(\alpha_{i,k} - \alpha_{j,k})}{2} + \sum_{l=1}^{k-1} \pi h(\alpha_{i,l} - \alpha_{j,l}) \right] \right\}$$

A lower bound on average bit error probability for all SNR can be obtained by supposing that for each transmitted sequence the receiver needs only decide between that sequence and its nearest neighbor, that is, the sequence most correlated with it. Then

$$P_b(E) \geq \frac{1}{2^{N_s-1}} \sum_{i=1}^{2^{N_s-1}} P_{i,j_{\max}} \tag{10.89}$$

where

$$P_{i,j \max} = \frac{1}{2} \text{erfc} \sqrt{\frac{N_s E_b}{N_0} \left(1 - \rho_{i,j \max}\right)}; \quad \rho_{i,j \max} = \max_j \rho_{i,j} \tag{10.90}$$

The lower bound of (10.89) and the upper bound of (10.86) asymptotically approach each other at high SNR and thus represent the true bit error probability performance in that region. Thus, recalling that (10.83)–(10.85) represents an approximation of the true performance at low SNR, then a composite curve formed from (10.83)–(10.85) and (10.86)–(10.88) can be looked upon as an approximation of the performance for all SNR. Figure 10.17 is an illustration of the behavior of these various bit error probability bounds and the composite approximation for an observation interval of five bits ($N_s = 5$) and a modulation index $h = 0.715$. This value of modulation index was chosen since in [10] it was shown that it achieves the maximum value of the minimum distance between all pairs of signals $s(t, 1, \alpha_i')$ and $s(t, -1, \alpha_j')$, $i, j = 1, 2, \ldots, 16$ which dominates performance at high SNR. By distance between signals $s(t, 1, \alpha_i')$ and $s(t, -1, \alpha_j')$, we mean

$$d_{i,j} = \left(\int_0^{5T_b} \left[s(t, 1, \alpha_i') - s(t, -1, \alpha_j') \right]^2 dt \right)^{1/2} \tag{10.91}$$

It can also be shown [12] that little or no further improvment in performance can be achieved by extending the observation interval beyond five symbols. This behavior appears to be typical of CPFSK systems independent of modulation index.

Noncoherent Detection. Here we assume that the channel phase θ is unknown and uniformly distributed in the interval $(-\pi, \pi)$. Also, in contrast with the coherent detection case, we shall be making decisions here on the middle bit of the observed sequence as explained in [12]. Thus, to simplify notation, the length of the observed sequence N_s shall be equated to $2n_s + 1$ and the decision will be made on the bit α_{n_s}. If α' now denotes the transmitted sequence excluding the bit α_{n_s}, and $s(t)$ is notationally replaced by

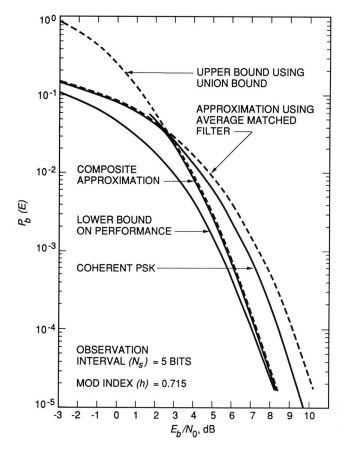

Figure 10.17   Approximations to and bounds on the average bit error probability performance of CPFSK (Reprinted from [12] © IEEE)

$s(t, \alpha_{n_s}, \boldsymbol{\alpha}')$ and $s(t, \theta)$ by $s(t, \alpha_{n_s}, \boldsymbol{\alpha}', \theta)$, then analogous to (10.81), the likelihood ratio is now

$$
\Lambda = \frac{\sum\limits_{i=1}^{2^{2n_s}} \frac{1}{2\pi} \int_{-\pi}^{\pi} \exp\left(\frac{2}{N_0} \int_0^{(2n_s+1)T_b} r(t) s\left(t, 1, \boldsymbol{\alpha}_i', \theta\right) dt\right) d\theta}{\sum\limits_{i=1}^{2^{2n_s}} \frac{1}{2\pi} \int_{-\pi}^{\pi} \exp\left(\frac{2}{N_0} \int_0^{(2n_s+1)T_b} r(t) s\left(t, -1, \boldsymbol{\alpha}_i', \theta\right) dt\right) d\theta}
$$

$$
= \frac{\sum\limits_{i=1}^{2^{2n_s}} I_0\left(z_{1,i}\right)}{\sum\limits_{i=1}^{2^{2n_s}} I_0\left(z_{-1,i}\right)}
$$

(10.92)

where

$$z_{1,i}^2 = \left(\frac{2}{N_0} \int_0^{(2n_s+1)T_b} r(t)s(t, 1, \boldsymbol{\alpha}_i', 0)\, dt \right)^2$$

$$+ \left(\frac{2}{N_0} \int_0^{(2n_s+1)T_b} r(t)s(t, 1, \boldsymbol{\alpha}_i', \pi/2)\, dt \right)^2$$

$$z_{-1,i}^2 = \left(\frac{2}{N_0} \int_0^{(2n_s+1)T_b} r(t)s(t, -1, \boldsymbol{\alpha}_i', 0)\, dt \right)^2 \qquad (10.93)$$

$$+ \left(\frac{2}{N_0} \int_0^{(2n_s+1)T_b} r(t)s(t, -1, \boldsymbol{\alpha}_i', \pi/2)\, dt \right)^2$$

A block diagram of a receiver that implements the decision on α_{n_s} based on (10.92) is illustrated in Fig. 10.18. As was true for the coherent receiver of Fig. 10.15, the exact error probability performance of Fig. 10.18 cannot be evaluated analytically. Thus, once again we resort to suitable approximations and bounds for high and low SNR.

At low SNR, the Bessel functions in (10.92) can be approximated by $I_0(x) \cong 1 + x^2/4$. Making this approximation in (10.92) results in the simplified decision rule

$$\hat{\alpha}_{n_s} = \mathrm{sgn} \left\{ \sum_{i=1}^{2^{2n_s}} \left(z_{1,i}^2 - z_{-1,i}^2 \right) \right\} \qquad (10.94)$$

If we now switch to a complex signal representation, that is,

$$s(t, \alpha_{n_s}, \boldsymbol{\alpha}', 0) + js(t, \alpha_{n_s}, \boldsymbol{\alpha}', \pi/2) \triangleq S(t, \alpha_{n_s}, \boldsymbol{\alpha}') = A \exp\left(j\left[\omega_c t + \theta(t, \boldsymbol{\alpha}) \right] \right) \quad (10.95)$$

then (10.94) may be rewritten as

$$\hat{\alpha}_{n_s} = \mathrm{sgn} \left\{ \sum_{i=1}^{2^{2n_s}} \left(|Z_{1,i}|^2 - |Z_{-1,i}|^2 \right) \right\};$$

$$Z_{\alpha_{n_s}, i} = \frac{2}{N_0} \int_0^{(2n_s+1)T_b} r(t) S(t, \alpha_{n_s}, \boldsymbol{\alpha}')\, dt \qquad (10.96)$$

In order to calculate the error probability performance of a receiver that implements the decision rule of (10.96), it is convenient to rewrite it in a different form. Suppose that, analogous to the coherent detection case, we define the *average* complex transmitted waveform

$$\bar{S}(t, \alpha_{n_s}) = \frac{1}{2^{2n_s}} \sum_{i=1}^{2^{2n_s}} S(t, \alpha_{n_s}, \boldsymbol{\alpha}_i') \qquad (10.97)$$

Then

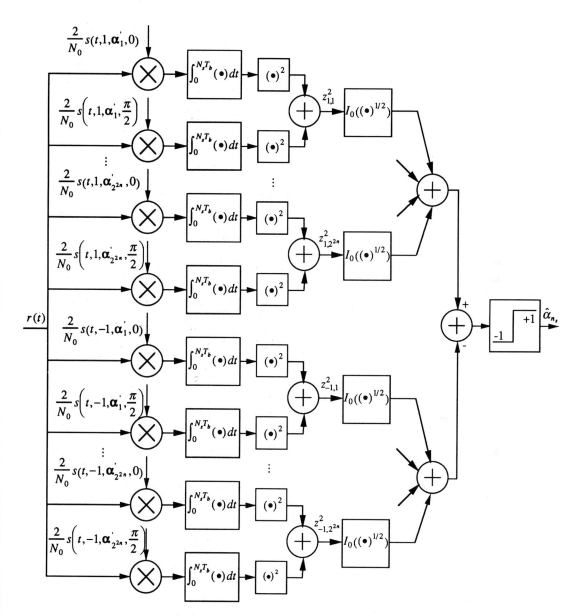

Figure 10.18   Optimum receiver for symbol-by-symbol noncoherent detection of CPFSK

$$\left(2^{2n_s}\right)^2 \left\{ \left| \frac{2}{N_0} \int_0^{(2n_s+1)T_b} r(t)\bar{S}(t,\,1)\,dt \right|^2 - \left| \frac{2}{N_0} \int_0^{(2n_s+1)T_b} r(t)\bar{S}(t,\,-1)\,dt \right|^2 \right\}$$

$$= \sum_{i=1}^{2^{2n_s}} \left[\left| \frac{2}{N_0} \int_0^{(2n_s+1)T_b} r(t)S\big(t,\,1,\,\boldsymbol{\alpha}_i'\big)\,dt \right|^2 \right.$$

$$\left. - \left| \frac{2}{N_0} \int_0^{(2n_s+1)T_b} r(t)S\big(t,\,-1,\,\boldsymbol{\alpha}_i'\big)\,dt \right|^2 \right]$$

$$+ \sum_{\substack{i=1 \\ i\neq j}}^{2^{2n_s}} \sum_{j=1}^{2^{2n_s}} \left[\left(\frac{2}{N_0} \int_0^{(2n_s+1)T_b} r(t)S\big(t,\,1,\,\boldsymbol{\alpha}_i'\big)\,dt \right) \right. \tag{10.98}$$

$$\left. \times \left(\frac{2}{N_0} \int_0^{(2n_s+1)T_b} r(t)S^*\big(t,\,1,\,\boldsymbol{\alpha}_j'\big)\,dt \right) \right]$$

$$- \sum_{\substack{i=1 \\ i\neq j}}^{2^{2n_s}} \sum_{j=1}^{2^{2n_s}} \left[\left(\frac{2}{N_0} \int_0^{(2n_s+1)T_b} r(t)S\big(t,\,-1,\,\boldsymbol{\alpha}_i'\big)\,dt \right) \right.$$

$$\left. \times \left(\frac{2}{N_0} \int_0^{(2n_s+1)T_b} r(t)S^*\big(t,\,-1,\,\boldsymbol{\alpha}_j'\big)\,dt \right) \right]$$

For any given sequence $\boldsymbol{\alpha}_i$ in the double summations of (10.98), the complement (negative) of that sequence, say $\boldsymbol{\alpha}_j = -\boldsymbol{\alpha}_i$, also exists. Hence, it can be shown that these double summation terms cancel each other and what is left (that is, the single summation) is the argument of the sgn function in (10.97). Finally, we conclude that the decision rule of (10.97) can alternately be written as (eliminating the $2/N_0$ scaling)

$$\hat{\alpha}_{n_s} = \mathrm{sgn} \left\{ \left| \int_0^{(2n_s+1)T_b} r(t)\bar{S}(t,\,1)\,dt \right|^2 - \left| \int_0^{(2n_s+1)T_b} r(t)\bar{S}(t,\,-1)\,dt \right|^2 \right\}$$

$$\triangleq \mathrm{sgn} \left\{ |Z_2|^2 - |Z_1|^2 \right\} \tag{10.99}$$

which can be implemented by the simple binary noncoherent AMF receiver of Fig. 10.19.

To obtain the average bit error probability performance of Fig. 10.19, we observe that (10.99) is in a form that allows application of the results in [14] for the solution of the general binary noncoherent problem. In particular, the bit error probability given the transmitted sequence $\boldsymbol{\alpha}_i$ is

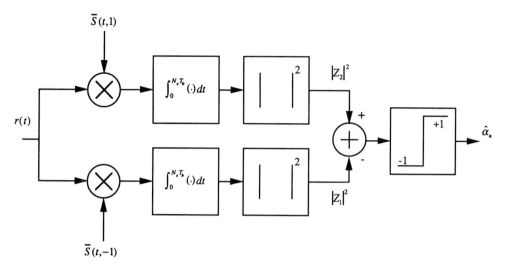

Figure 10.19   An equivalent receiver to Figure 10.18 for low SNR

$$P_{bi}(E) = \Pr\left\{|Z_1|^2 > |Z_2|^2 \,\big|\, \alpha_{n_s} = 1, \boldsymbol{\alpha}_i'\right\}$$

$$= \frac{1}{2}\left[1 - Q\left(\sqrt{b_i}, \sqrt{a_i}\right) + Q\left(\sqrt{a_i}, \sqrt{b_i}\right)\right] \tag{10.100}$$

where $Q(x, y)$ is again the Marcum Q-function and

$$\left\{\begin{array}{c} a_i \\ b_i \end{array}\right\} = \frac{1}{2\sigma^2}\left[\frac{S_1 + S_2 - 2\,|\rho|\,\sqrt{S_1 S_2}\cos\left(\Theta_2 - \Theta_1 + \Phi\right)}{1 - |\rho|^2} \pm \frac{S_1 - S_2}{\sqrt{1 - |\rho|^2}}\right] \tag{10.101}$$

with

$$S_k = \frac{1}{2}\,|E\,\{Z_k\}|^2; \quad k = 1, 2$$

$$\sigma^2 = \frac{1}{2}E\left\{|Z_1 - E\,\{Z_1\}|^2\right\} = \frac{1}{2}E\left\{|Z_2 - E\,\{Z_2\}|^2\right\}$$

$$\rho = \frac{1}{2\sigma^2}E\left\{(Z_1 - E\,\{Z_1\})\,(Z_2 - E\,\{Z_2\})^*\right\} \tag{10.102}$$

$$\Theta_k = \arg\{E\,\{Z_k\}\}; \quad k = 1, 2$$

$$\Phi = \arg\rho$$

In (10.102), the values of the various parameters are computed conditioned on the transmitted middle bit, α_{n_s}, being a $+1$ and the remaining bits belonging to the vector $\boldsymbol{\alpha}_i'$. The evaluation of these parameters is carried out in [12] under the assumption that all signals have zero phase at the beginning of the middle bit interval. The results are as follows:

$$\sigma^2 = \frac{N_0 A^2 T_b}{4} \left\{ 1 + (1 + \text{sinc } 2h) \left[\frac{1 - \cos^{2n} \pi h}{1 - \cos^2 \pi h} \right] \right\}$$

$$\rho\sigma^2 = \frac{N_0 A^2 T_b}{4} \left\{ \frac{1}{2} (1 + \text{sinc } 2h) \left(1 + e^{-j2\pi h} \right) \left[\frac{1 - \cos^{2n} \pi h}{1 - \cos^2 \pi h} \right] + e^{-j\pi h} \text{ sinc } h \right\} \tag{10.103}$$

$$S_1 = \frac{A^4 T_b^2}{8} \left| A_1 + e^{-j\pi h} \text{ sinc } h + e^{-j2\pi h} A_2 \right|^2; \quad S_2 = \frac{A^4 T_b^2}{8} |A_1 + 1 + A_2|^2$$

$$\Theta_1 = \arg \left\{ A_1 + e^{-j\pi h} \text{ sinc } h + e^{-j2\pi h} A_2 \right\}; \quad \Theta_2 = \arg \{ A_1 + 1 + A_2 \}$$

where

$$A_1 = \frac{1}{2} \sum_{l=1}^{n_s} \exp \left(-j\pi h \sum_{k=1}^{l-1} \alpha_{i,n_s-k} \right) \left(\cos^{l-1} \pi h \right) \left(1 + e^{-j\pi h \alpha_{i,n_s-l}} \text{ sinc } h \right)$$

$$\tag{10.104}$$

$$A_2 = \frac{1}{2} \sum_{l=1}^{n_s} \exp \left(-j\pi h \sum_{k=1}^{l-1} \alpha_{i,n_s+k} \right) \left(\cos^{l-1} \pi h \right) \left(1 + e^{-j\pi h \alpha_{i,n_s+l}} \text{ sinc } h \right)$$

Finally, the average error probability is found from (10.83) with $N_s = 2n_s + 1$.

As a simple example of the above, consider the case of $N_s = 1(n_s = 0)$. For this case

$$A_1 = A_2 = 0; \quad S_1 = \frac{A^4 T_b^2}{8} \text{ sinc}^2 h; \quad S_2 = \frac{A^4 T_b^2}{8}$$

$$\sigma^2 = \frac{N_0 A^2 T_b}{4}; \quad \rho\sigma^2 = \frac{N_0 A^2 T_b}{4} e^{-j\pi h} \text{ sinc } h \tag{10.105}$$

$$\Theta_1 = -\pi h; \quad \Theta_2 = 0; \quad \Phi = -\pi h$$

Substituting (10.105) into (10.101) gives

$$\left\{ \begin{matrix} a \\ b \end{matrix} \right\} = \frac{A^2 T_b}{4N_0} \left[1 \mp \sqrt{1 - \text{sinc}^2 h} \right] = \frac{E_b}{2N_0} \left[1 \mp \sqrt{1 - \text{sinc}^2 h} \right] \tag{10.106}$$

for all sequences $\boldsymbol{\alpha}_i$. Combining (10.106) with (10.100) gives the well-known result [14, p. 207] for the error probability performance of noncoherent FSK with modulation index h. If in addition the signals are orthogonal, that is, $h = 1$, then this result simplifies to

$$P_b(E) = \frac{1}{2} \exp \left(-\frac{E_b}{2N_0} \right) \tag{10.107}$$

which is identical to (5.79).

As a second example, let $N_s = 3$ ($n_s = 1$) and $h = 1$. Then

$$A_1 = A_2 = \frac{1}{2}; \quad S_1 = \frac{A^4 T_b^2}{8}; \quad S_1 = \frac{A^4 T_b^2}{2}$$

$$\sigma^2 = \frac{N_0 A^2 T_b}{2}; \quad \rho\sigma^2 = \frac{N_0 A^2 T_b}{4} \tag{10.108}$$

$$\Theta_1 = \Theta_2 = \Phi = 0$$

and

$$\begin{Bmatrix} a \\ b \end{Bmatrix} = \frac{E_b}{N_0} \left[1 \mp \frac{\sqrt{3}}{2} \right] \tag{10.109}$$

for all sequences $\boldsymbol{\alpha}_i$. Combining (10.109) with (10.100) results in an improved error probability performance over the case $N_s = 1$.

For high SNR, we may use the approximation

$$\sum_{i=1}^{2^{2n_s}} I_0(x_i) \cong I_0(x_{\max}) \tag{10.110}$$

where $x_{\max}$ is the largest of the set $\{x_i\}$. With this approximation, the decision rule corresponding to (10.92) becomes

$$\hat{\alpha}_{n_s} = \mathrm{sgn}\left\{ I_0\left(z_{1,\max}\right) - I_0(z_{-1,\max}) \right\} \tag{10.111}$$

or, since $I_0(x)$ is a monotonic function of x

$$\hat{\alpha}_{n_s} = \mathrm{sgn}\left\{ z_{1,\max} - z_{-1,\max} \right\} \tag{10.112}$$

As in the coherent case, we resort to a union bound for characterizing the error probability performance of the receiver. In particular, (10.86) applies with $N_s = 2n_s + 1$ and $P_{i,j}$ now defined by

$$P_{i,j} = \mathrm{Pr}\left\{ z_{1,j} < z_{-1,i} \right\}$$

$$= \mathrm{Pr}\left\{ \left| \frac{2}{N_0} \int_0^{(2n_s+1)T_b} r(t) S(t, 1, \boldsymbol{\alpha}_i') \, dt \right| \right. \tag{10.113}$$

$$\left. < \left| \frac{2}{N_0} \int_0^{(2n_s+1)T_b} r(t) S(t, -1, \boldsymbol{\alpha}_j') \, dt \right| \right\}$$

The pairwise error probability of (10.113) can be expressed in the form of (10.100) where

$$\begin{Bmatrix} a_{i,j} \\ b_{i,j} \end{Bmatrix} = \frac{(2n_s + 1)E_b}{2N_0} \left[1 \mp \sqrt{1 - |\rho_{i,j}|^2} \right] \tag{10.114}$$

with

$$\rho_{l,m} = \frac{1}{2n_s + 1} \sum_{k=1}^{2n_s+1} \exp\left(j\pi h \sum_{v=1}^{k-1} (\alpha_{v,l} - \alpha_{v,m}) \right) \tag{10.115}$$

$$\times \exp\left(j\frac{\pi h}{2} (\alpha_{k,l} - \alpha_{k,m}) \right) \mathrm{sinc}\left(\frac{h}{2} (\alpha_{k,l} - \alpha_{k,m}) \right)$$

where $\alpha_{v,l}$ is the vth bit in the sequence $\boldsymbol{\alpha}_l$ whose middle bit is a -1 and $\alpha_{v,m}$ is the vth bit in the sequence $\boldsymbol{\alpha}_m$ whose middle bit is a $+1$. Note that for $N_s = 1$, (10.114) together with (10.115) reduces to (10.106) and hence the high SNR upper bound of (10.86) agrees with the low SNR approximation, both of which yield the exact performance.

Figure 10.20 is an illustration of the behavior of the composite performance approximation made up of a combination of the low SNR performance of (10.83) together with

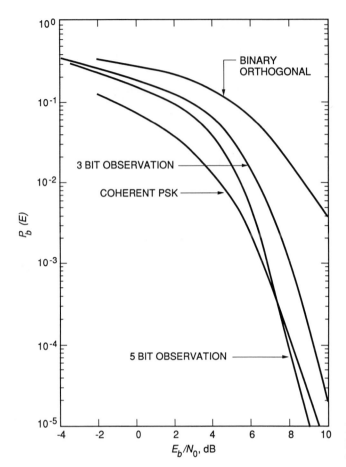

Figure 10.20 Error probability performance of noncoherent CPFSK receiver (Reprinted from [12] © IEEE)

(10.100) and the high SNR approximate bound of (10.86) together with (10.113). The numerical results are plotted for 3-bit ($n_s = 1$) and 5-bit ($n_s = 2$) observations and $h = .715$. Also, plotted for the purpose of comparison is the performance of noncoherent orthogonal ($h = 1$) FSK [Eq. (10.107)] and coherent PSK. We observe that for E_b/N_0 greater than about 7.3 dB, noncoherent CPFSK with 5-bit observation outperforms coherent PSK. For bit error probabilities below 10^{-3}, the performance of noncoherent CPFSK with 3-bit observation is within 1 dB of that corresponding to coherent PSK.

Optimum block-by-block detection.   CPFSK receivers that perform optimum block-by-block detection derive their structure based on a maximum-likelihood estimate of the data sequence in the observation interval. In principle, for an N_s-bit observation this entails an implementation that calculates the correlation of the received signal with each of the possible 2^{N_s} transmitted sequences and selects the largest among these correlations. For the case of coherent detection, one further requires knowledge of the signal phase at the beginning of each block. Approximate knowledge of this can be obtained from the data decisions on the previous blocks; however, the resulting configuration would then be suboptimum. In the noncoherent case, knowledge of the signal phase at the beginning of each block is not required and thus the resulting configuration is indeed optimum from a sequence detection standpoint. Hence, we shall consider only the noncoherent case and furthermore show that considerable simplification of the generic implementation with 2^{N_s} correlators can be achieved. The results to be presented here for CPFSK are taken from [15] where the more general case of full response CPM is considered.

Noncoherent Detection.   Since we are now interested in detecting a sequence rather than just one bit in the sequence, we start by rewriting the conditional probability of (10.78) as

$$f(\mathbf{r}\,|\boldsymbol{\alpha}, \theta) = \exp\left\{-\frac{1}{N_0}\int_0^{N_s T_b}[r(t) - s(t, \boldsymbol{\alpha}, \theta)]^2\,dt\right\}$$

$$= C\exp\left(\frac{2}{N_0}\int_0^{N_s T_b} r(t)s(t, \boldsymbol{\alpha}, \theta)\,dt\right)$$

(10.116)

where as before we have incorporated into the notation the arbitrary (uniformly distributed) phase θ introduced by the channel. For the CPFSK signal of (10.74) together with (10.77), the correlation in the argument of the exponential in (10.116) can be expressed as

$$\int_0^{N_s T_b} r(t)s(t, \boldsymbol{\alpha}, \theta)\,dt = \left(\int_0^{N_s T_b} r(t)s(t, \boldsymbol{\alpha}, 0)\,dt\right)\cos\theta$$

$$-\left(\int_0^{N_s T_b} r(t)s(t, \boldsymbol{\alpha}, \pi/2)\,dt\right)\sin\theta$$

(10.117)

Substituting (10.74) into (10.117), we get

$$\int_0^{N_s T_b} r(t)s(t, \boldsymbol{\alpha}, \theta)\,dt = \left[A\int_0^{N_s T_b} r(t)\cos(\omega_c t + \theta(t, \boldsymbol{\alpha}))\,dt\right]\cos\theta$$

$$-\left[A\int_0^{N_s T_b} r(t)\sin(\omega_c t + \theta(t, \boldsymbol{\alpha}))\,dt\right]\sin\theta$$

(10.118)

$$= A\,|\beta|\cos(\arg\beta - \theta)$$

where the complex baseband correlation β is defined by

$$\beta = \int_0^{N_s T_b} r(t)\exp[-j(\omega_c t + \theta(t, \boldsymbol{\alpha}))]\,dt$$

(10.119)

Evaluating (10.119) separately in each bit interval we get

$$\beta = \int_0^{T_b} r(t)e^{-j\omega_c t}e^{-j\pi h\alpha_0 t/T_b}\,dt + \left(e^{-j\pi h\alpha_0}\right)\int_{T_b}^{2T_b} r(t)e^{-j\omega_c t}e^{-j\pi h\alpha_1(t-T_b)/T_b}\,dt$$

$$+\cdots + \left(\prod_{k=0}^{N_s-2} e^{-j\pi h\alpha_k}\right)\int_{(N_s-1)T_b}^{N_s T_b} r(t)e^{-j\omega_c t}e^{-j\pi h\alpha_{N_s-1}(t-(N_s-1)T_b)/T_b}\,dt$$

(10.120)

Substituting (10.119) into (10.116), averaging over θ, and taking the natural logarithm gives

$$\ln f(\mathbf{r}\,|\boldsymbol{\alpha}) = \ln\left[\frac{1}{2\pi}\int_{-\pi}^{\pi} C\exp\left(\frac{2A}{N_0}|\beta|\cos\left(\arg\beta - \theta\right)\right)d\theta\right]$$

$$= \ln C + \ln I_0\left(\frac{2A}{N_0}|\beta|\right) \tag{10.121}$$

Note that any signal phase (due to prior data bits) that is present at the start of the observation interval can be included in (added to) θ and because of the averaging performed in (10.121) will have no bearing on the decision rule. This affirms a previous statement that for noncoherent detection, the structure of the optimum receiver does not require knowledge of the signal phase at the beginning of each block.

Since $\ln I_0(x)$ is a monotonic function of x, then maximizing $\ln f(\mathbf{r}\,|\boldsymbol{\alpha})$ leads to the appropriate decision rule as follows. For each finite sequence (block) $\boldsymbol{\alpha}_i = (\alpha_{i,0}, \alpha_{i,1}, \ldots, \alpha_{i,N_s-1})$, the optimum receiver computes

$$\beta_i = \sum_{k=0}^{N_s-1} \Gamma_{\alpha_{i,k},N_s-k-1} C_{i,k} \tag{10.122}$$

where

$$\Gamma_{1,m} = \int_{(N_s-1-m)T_b}^{(N_s-m)T_b} r(t)e^{-j\omega_c t}e^{-j\pi h(t-mT_b)/T_b}dt$$

$$\Gamma_{-1,m} = \int_{(N_s-1-m)T_b}^{(N_s-m)T_b} r(t)e^{-j\omega_c t}e^{j\pi h(t-mT_b)/T_b}dt; \quad m = 0, 1, \ldots, N_s - 1 \tag{10.123}$$

and the complex constants $C_{i,k}$ are defined recursively by

$$C_{i,0} = 1, \quad C_{i,k} = C_{i,k-1}e^{-j\pi h\alpha_{i,k-1}}; \quad k = 1, 2, \ldots N_s - 1 \tag{10.124}$$

A decision as to which sequence was indeed transmitted is then made in accordance with $\max_i |\beta_i|$. An implemention (in complex form) of this decision rule is illustrated in Fig. 10.21. We note that only two correlators are required, each of which is followed by a tapped delay line and suitable processing to make the sequence decision.

To evaluate the performance of the receiver in Fig. 10.21, we make use of the technique used in the discussion of multiple symbol detection of MPSK in Chapters 6 and 7 to obtain an upper bound on the bit error probability, $P_b(E)$. In particular, we use a union bound analogous to that used for upper bounding the performance of error correction coded systems. This bound is expressed as the sum of the pairwise error probabilities associated with each N_s-bit error sequence. Exact evaluation of the pairwise error probabilities is again made possible by the results in [14] analogous to (10.100)–(10.102).

Mathematically speaking, let $\boldsymbol{\alpha} = (\alpha_0, \alpha_1, \ldots, \alpha_{N_s-1})$ denote the sequence of N_s information bits and $\hat{\boldsymbol{\alpha}} = (\hat{\alpha}_0, \hat{\alpha}_1, \ldots, \hat{\alpha}_{N_s-1})$ be the corresponding sequence of detected bits. Then[12]

$$P_b(E) \le \frac{1}{2^{N_s}N_s} \sum_{\boldsymbol{\alpha}\neq\hat{\boldsymbol{\alpha}}}\sum w(\boldsymbol{\alpha}, \hat{\boldsymbol{\alpha}})\,\mathrm{Pr}\left\{|\hat{\beta}| > |\beta|\,\Big|\,\boldsymbol{\alpha}\right\} \tag{10.125}$$

---

12. The result in (10.125) is a special case of (4.88) corresponding to $M = 2^{N_s}$.

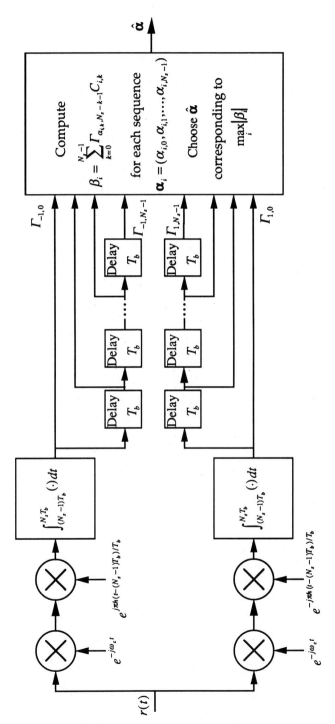

Figure 10.21 . An implementation of a maximum-likelihood sequence estimator for noncoherent detection of CPFSK based on an N_s-symbol observation (Reprinted from [15]© IEEE)

where $w(\boldsymbol{\alpha}, \hat{\boldsymbol{\alpha}})$ denotes the Hamming distance between $\boldsymbol{\alpha}$ and $\hat{\boldsymbol{\alpha}}$ and $\Pr\left\{|\hat{\beta}| > |\beta|\,\big|\,\boldsymbol{\alpha}\right\}$ denotes the pairwise probability that $\hat{\boldsymbol{\alpha}}$ is incorrectly chosen when indeed $\boldsymbol{\alpha}$ was sent. The decision statistic $|\beta|$ is defined in (10.120) and the corresponding error statistic $|\hat{\beta}|$ is identical to (10.120) with each α_k replaced by $\hat{\alpha}_k$. Unlike M-PSK, CPFSK is not a symmetric signaling set and thus it does satisfy a uniform error probability (UEP) criterion where the probability of error is independent of which input phase sequence $\boldsymbol{\alpha}$ is chosen as the correct sequence. Hence, without any additional characterization of $\Pr\left\{|\hat{\beta}| > |\beta|\,\big|\,\boldsymbol{\alpha}\right\}$, further simplification of (10.125) is not possible. Later we shall show that $\Pr\left\{|\hat{\beta}| > |\beta|\,\big|\,\boldsymbol{\alpha}\right\}$ depends only on the error sequence $\boldsymbol{\varepsilon} = \boldsymbol{\alpha} - \hat{\boldsymbol{\alpha}}$ and thus further simplification is possible.

To compute $\Pr\left\{|\hat{\beta}| > |\beta|\,\big|\,\boldsymbol{\alpha}\right\}$, we use the approach taken in [14] for evaluating the probability that one Rician random variable exceeds another. In particular, analogous to (10.100)

$$\Pr\left\{|\hat{\beta}| > |\beta|\,\big|\,\boldsymbol{\alpha}\right\} = \frac{1}{2}\left[1 - Q(\sqrt{b}, \sqrt{a}) + Q(\sqrt{a}, \sqrt{b})\right] \tag{10.126}$$

where

$$\begin{Bmatrix} b \\ a \end{Bmatrix} = \frac{E_b}{2N_0}\left[N_s \pm \sqrt{N_s^2 - |\delta|^2}\right] \tag{10.127}$$

with

$$\delta = c(\boldsymbol{\alpha}, \hat{\boldsymbol{\alpha}}) + js(\boldsymbol{\alpha}, \hat{\boldsymbol{\alpha}}) \tag{10.128}$$

In (10.128), $c(\boldsymbol{\alpha}, \hat{\boldsymbol{\alpha}})$ and $s(\boldsymbol{\alpha}, \hat{\boldsymbol{\alpha}})$ are defined analogous to [13] by

$$c(\boldsymbol{\alpha}, \hat{\boldsymbol{\alpha}}) = \frac{1}{T_b}\int_0^{N_sT_b} \cos\left[\theta(t, \boldsymbol{\alpha}) - \theta(t, \hat{\boldsymbol{\alpha}})\right]dt,$$

$$s(\boldsymbol{\alpha}, \hat{\boldsymbol{\alpha}}) = \frac{1}{T_b}\int_0^{N_sT_b} \sin\left[\theta(t, \boldsymbol{\alpha}) - \theta(t, \hat{\boldsymbol{\alpha}})\right]dt \tag{10.129}$$

Reinterpreting $c(\boldsymbol{\alpha}, \hat{\boldsymbol{\alpha}})$ and $s(\boldsymbol{\alpha}, \hat{\boldsymbol{\alpha}})$ as sums of N_s integrals each corresponding to a single data bit interval T_b, then using (10.77) in (10.129) and substituting the result into (10.128), we obtain after some simplification

$$\delta = \sum_{n=0}^{N_s-1}\left(\frac{1}{T_b}\int_0^{T_b} e^{j\pi h(\alpha_n - \hat{\alpha}_n)t/T_b}dt\right)e^{j\pi h\sum_{i=0}^{n-1}(\alpha_i - \hat{\alpha}_i)}$$

$$= \sum_{n=0}^{N_s-1} e^{j\frac{\pi h}{2}(\alpha_n - \hat{\alpha}_n)}\operatorname{sinc}\left(\frac{h}{2}(\alpha_n - \hat{\alpha}_n)\right)e^{j\pi h\sum_{i=0}^{n-1}(\alpha_i - \hat{\alpha}_i)} \tag{10.130}$$

Before proceeding to the evaluation of (10.126)–(10.130) for specific cases, we first present an expression for the pairwise error probability under the conditions $a, b \gg 1$ which from (10.127) corresponds to the case of large E_b/N_0. Using the appropriate asymptotic expressions for the Marcum Q-function and the complementary error function, (10.126) can be written as

$$\Pr\left\{|\hat{\beta}| > |\beta| \,\Big|\, \boldsymbol{\alpha}\right\} \cong \left(\sqrt{\frac{N_s + |\delta|}{2|\delta|}}\right) \frac{1}{2}\,\mathrm{erfc}\left\{\sqrt{\frac{E_b}{2N_0}(N_s - |\delta|)}\right\} \tag{10.131}$$

Case 1. $N_s = 2$, h arbitrary

Since in evaluating (10.127) or (10.131) we are interested in the magnitude of δ (rather than δ itself), there are only four cases of interest. A tabulation of these evaluations is given below:

$\alpha_1 - \hat{\alpha}_1$	$\alpha_0 - \hat{\alpha}_0$	$\left\{\begin{array}{c} b \\ a \end{array}\right\}$
2	0	$\dfrac{E_b}{N_0}\left(1 \pm \sqrt{1 - \left\|\frac{1+e^{j\pi h}\ \mathrm{sinc}\,h}{2}\right\|^2}\right)$
0	2	$\dfrac{E_b}{N_0}\left(1 \pm \sqrt{1 - \left\|\frac{1+e^{-j\pi h}\ \mathrm{sinc}\,h}{2}\right\|^2}\right)$
2	2	$\dfrac{E_b}{N_0}\left(1 \pm \sqrt{1 - |\cos \pi h\ \mathrm{sinc}\,h|^2}\right)$
2	-2	$\dfrac{E_b}{N_0}\left(1 \pm \sqrt{1 - |\mathrm{sinc}\,h|^2}\right)$

The first two cases in the above tabulation correspond to error sequences having a Hamming distance $w(\boldsymbol{\alpha}, \hat{\boldsymbol{\alpha}}) = 1$ while the third and fourth cases have $w(\boldsymbol{\alpha}, \hat{\boldsymbol{\alpha}}) = 2$. We note further that the values of a and b for cases 1 and 2 are identical. Finally, computing the number of times that each entry in the above table occurs corresponding to the $4 \times 3 = 12$ possible cases for which $\boldsymbol{\alpha} \neq \hat{\boldsymbol{\alpha}}$, then from (10.125)–(10.127) and the above tabulation we arrive at the desired upper bound on bit error probability given by

$$P_b(E) \leq P_1 + \frac{1}{2}(P_3 + P_4) \tag{10.132}$$

where

$$P_i = \frac{1}{2}\left[1 - Q(\sqrt{b_i}, \sqrt{a_i}) + Q(\sqrt{a_i}, \sqrt{b_i})\right]; \quad i = 1, 3, 4 \tag{10.133}$$

with

$$\left\{\begin{array}{c} b_1 \\ a_1 \end{array}\right\} = \frac{E_b}{N_0}\left(1 \pm \sqrt{1 - \left|\frac{1 + e^{j\pi h}\ \mathrm{sinc}\,h}{2}\right|^2}\right) \tag{10.134}$$

$$\left\{\begin{array}{c} b_3 \\ a_3 \end{array}\right\} = \frac{E_b}{N_0}\left(1 \pm \sqrt{1 - |\cos \pi h\ \mathrm{sinc}\,h|^2}\right); \quad \left\{\begin{array}{c} b_4 \\ a_4 \end{array}\right\} = \frac{E_b}{N_0}\left(1 \pm \sqrt{1 - |\mathrm{sinc}\,h|^2}\right)$$

Figure 10.22 is a plot of the upper bound on $P_b(E)$ [as given by (10.132)] versus E_b/N_0 in dB for various values of h between 0 and 1. Figure 10.23 is the associated plot which shows the optimization of $P_b(E)$ as a function of h with E_b/N_0 in dB now as the parameter. We note that the optimum value of h [corresponding to minimum $P_b(E)$] is quite insensitive to E_b/N_0 over the large range of values considered.

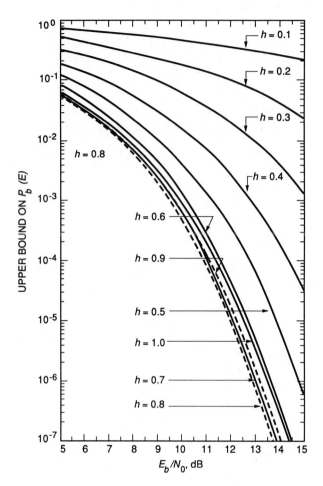

Figure 10.22 Upper bound on bit error probability versus E_b/N_0 with modulation index as a parameter; binary modulation with rectangular frequency pulses; two-symbol observation (Reprinted from [15] © IEEE)

Case 2. $N_s = 3$, h arbitrary

Again recognizing that we are only interested in the magnitude of δ, we can identify certain symmetries and arrive at a table analogous to that for the case $N_s = 2$ which will contain only eight distinct entries. Finally, computing the number of times that each of the eight entries in this table occurs corresponding to the $8 \times 7 = 56$ possible cases for which $\alpha \neq \hat{\alpha}$, then from (10.125)–(10.127) we arrive, after much simplification, at the desired upper bound on bit error probability given by

$$P_b(E) \leq P_1 + \frac{1}{3}(2P_2 + P_3 + 2P_4 + P_5) + \frac{1}{4}(P_6 + 2P_7 + P_8) \qquad (10.135)$$

where P_i; $i = 1, 2, \ldots, 8$ is given by (10.133) with

$$\left\{ \begin{matrix} b_i \\ a_i \end{matrix} \right\} = \frac{3E_b}{2N_0}\left(1 \pm \sqrt{1 - \frac{|\delta|^2}{9}}\right) \qquad (10.136)$$

and

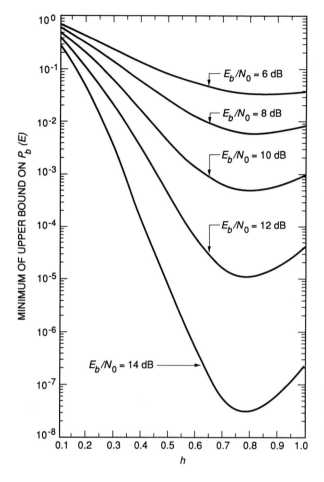

Figure 10.23 Minimum of upper bound on bit error probability versus modulation index with E_b/N_0 as a parameter; binary modulation with rectangular frequency pulses; two-symbol observation (Reprinted from [15] © IEEE)

i	$|\delta|$
1	$\left|1 + 2e^{j\pi h} \operatorname{sinc} h\right|$
2	$\left|1 + \left(e^{j\pi h} + e^{j3\pi h}\right) \operatorname{sinc} h\right|$
3	$\left|1 + \left(e^{j\pi h} + e^{-j\pi h}\right) \operatorname{sinc} h\right|$
4	$\left|2 + e^{j\pi h} \operatorname{sinc} h\right|$
5	$\left|1 + e^{j2\pi h} + e^{j\pi h} \operatorname{sinc} h\right|$
6	$\left|\left(1 + e^{j2\pi h} + e^{j4\pi h}\right) \operatorname{sinc} h\right|$
7	$\left|\left(1 + 2e^{j2\pi h}\right) \operatorname{sinc} h\right|$
8	$\left|3 \operatorname{sinc} h\right|$

Figure 10.24 illustrates the analogous results to Fig. 10.22 where the upper bound on $P_b(E)$ is now computed from (10.135) together with (10.133), (10.136), and the above table. Comparing these two figures, one can assess the improvement in performance in going from a two-symbol to a three-symbol observation interval.

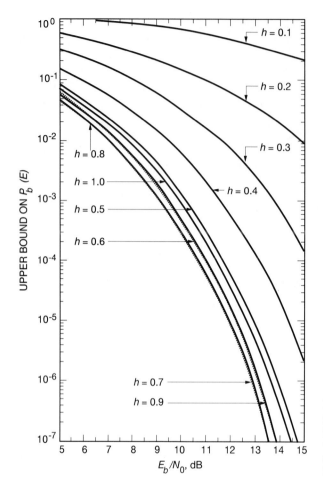

UPPER BOUND ON $P_b(E)$

$h = 0.1$

$h = 0.2$

$h = 0.8$

$h = 0.3$

$h = 1.0$

$h = 0.5$ $h = 0.4$

$h = 0.6$

$h = 0.7$

$h = 0.9$

E_b/N_0, dB

Figure 10.24   Upper bound on bit error probability versus E_b/N_0 with modulation index as a parameter; binary modulation with rectangular frequency pulses; three-symbol observation (Reprinted from [15] © IEEE)

To extend these results to values of $N_s > 3$, we make use of the fact that from (10.130), $|\delta|$ is only a function of the error sequence $\boldsymbol{\varepsilon} = \boldsymbol{\alpha} - \hat{\boldsymbol{\alpha}}$ rather than $\boldsymbol{\alpha}$ and $\hat{\boldsymbol{\alpha}}$ individually. Thus, from (10.126) and (10.127), $\Pr\left\{|\hat{\beta}| > |\beta| \big| \boldsymbol{\alpha}\right\}$ is likewise dependent only on the error sequence $\boldsymbol{\varepsilon} = \boldsymbol{\alpha} - \hat{\boldsymbol{\alpha}}$. Hence, analogous to [16, Eq. (4.9.21)], we can write (10.125) in the simpler form

$$P_b(E) \leq \frac{1}{N_s} \sum_{\boldsymbol{\varepsilon}} \frac{w(\boldsymbol{\varepsilon})}{2^{w(\boldsymbol{\varepsilon})}} \Pr\left\{|\hat{\beta}| > |\beta| \big| \boldsymbol{\varepsilon}\right\} \tag{10.137}$$

where we have now indicated the dependencies of the Hamming weights and conditional pairwise error probabilities on the error sequence. Since the number of terms in the double sum of (10.125) is $2^{N_s} \times (2^{N_s} - 1)$ and the number of possible error sequences in the single sum of (10.137) is $3^{N_s} - 1$, we see that we have indeed achieved a simplification. Also, since from (10.130), $\delta(-\boldsymbol{\varepsilon}) = \delta^*(\boldsymbol{\varepsilon})$, then in (10.137) we need only sum over half the number of possible error sequences and double the result.

Case 3. N_s arbitrary, $h = 1$ (orthogonal signals)
In the general case of arbitrary N_s, the upper bound on bit error probability is dominated by the terms that correspond to the sequences resulting in the maximum value of $|\delta|$

(minimum value of equivalent squared distance $N_s - |\delta|$) provided that the other values of $|\delta|$ generated by the remaining sequences are sufficiently smaller than $|\delta|_{max}$. For arbitrary h, one cannot show that this always occurs. In fact, for certain values of h, the values of $|\delta|$ may indeed be bunched around $|\delta|_{max}$. However, for $h = 1$, the values of δ are all real and integer and there is sufficient separation between them that one can evaluate the upper bound on error probability based on the above approximation. Thus, we proceed with an analysis of the behavior for this special case.

For $h = 1$ (orthogonal CPFSK), it is easy to show that the maximum value of $|\delta|$ occurs for each error sequence $\hat{\boldsymbol{\alpha}}$ having $N_s - 1$ elements equal to the correct sequence $\boldsymbol{\alpha}$ and one element in error. From (10.130), $|\delta|_{max}$ is easily shown to have value $N_s - 1$. Since the element in error can occur in any one of the N_s positions of the sequence, then the total number of pairs of sequences yielding $|\delta|_{max}$ is $N_s \times 2^{N_s}$. For each of these sequence pairs, the Hamming distance is $w(\boldsymbol{\alpha}, \hat{\boldsymbol{\alpha}}) = 1$. Thus, from (10.125)–(10.127), an approximate upper bound on the performance of orthogonal CPFSK with N_s-symbol observation ($N_s \geq 2$) is given by

$$P_b(E) \lesssim \Pr\left\{|\hat{\beta}| > |\beta|\,\Big|\,\boldsymbol{\alpha}\right\} = \left[1 - Q(\sqrt{b_{max}}, \sqrt{a_{max}}) + Q(\sqrt{a_{max}}, \sqrt{b_{max}})\right] \quad (10.138)$$

where

$$\left\{\begin{matrix} b_{max} \\ a_{max} \end{matrix}\right\} = \frac{E_b}{2N_0}\left[N_s \pm \sqrt{N_s^2 - |\delta|_{max}^2}\right] = \frac{E_b}{2N_0}\left[N_s \pm \sqrt{2N_s - 1}\right] \quad (10.139)$$

For large SNR, we can use (10.131) instead of (10.126), which gives the asymptotic upper bound

$$P_b(E) \lesssim \left(\sqrt{\frac{N_s + |\delta|_{max}}{2|\delta|_{max}}}\right)\frac{1}{2}\,\text{erfc}\left\{\sqrt{\frac{E_b}{2N_0}(N_s - |\delta|_{max})}\right\}$$

$$= \left(\sqrt{\frac{2N_s - 1}{2N_s - 2}}\right)\frac{1}{2}\,\text{erfc}\left\{\sqrt{\frac{E_b}{2N_0}}\right\}; \quad N_s \geq 2 \quad (10.140)$$

As $N_s \to \infty$, (10.140) reduces to

$$P_b(E) \lesssim \frac{1}{2}\,\text{erfc}\left\{\sqrt{\frac{E_b}{2N_0}}\right\} \quad (10.141)$$

which is identical to the performance of orthogonal FSK with *coherent* detection (see Chapter 4).

10.2.3 Minimum-Shift-Keying (MSK)

The special case of CPFSK corresponding to a modulation index $h = .5$, or equivalently a frequency spacing $\Delta f = 1/2T_b$, is referred to as *minimum-shift-keying (MSK)*. The reason for this acronym stems from the fact that this value of Δf is the *minimum* frequency spacing that allows for coherent detection of *orthogonal* signals. As mentioned in the introduction to this chapter, certain forms of CPM also have an I-Q representation. MSK is one such case that is partially responsible for the popularity of this form of modulation. To demonstrate this alternate representation, we proceed as follows.

Substituting $h = .5$ in (10.77), the excess phase for MSK in the nth transmission interval $nT_b \leq t \leq (n+1)T_b$ becomes

$$\theta(t, \boldsymbol{\alpha}) = \frac{\pi \alpha_n (t - nT_b)}{2T_b} + \pi h \sum_{i=0}^{n-1} \alpha_i = \alpha_n \left(\frac{\pi t}{2T_b} \right) + x_n \qquad (10.142)$$

where x_n is a phase constant required to keep the phase continuous at the transition points $t = nT_b$ and $t = (n+1)T_b$. In the previous interval, $(n-1)T_b \leq t \leq nT_b$ the excess phase is given by

$$\theta(t, \boldsymbol{\alpha}) = \frac{\pi \alpha_{n-1}(t - (n-1)T_b)}{2T_b} + \pi h \sum_{i=0}^{n-2} \alpha_i = \alpha_{n-1} \left(\frac{\pi t}{2T_b} \right) + x_{n-1} \quad (10.143)$$

For phase continuity at $t = nT_b$, we require that

$$\alpha_n \left(\frac{\pi}{2T_b} \right)(nT_b) + x_n = \alpha_{n-1} \left(\frac{\pi}{2T_b} \right)(nT_b) + x_{n-1} \qquad (10.144)$$

or

$$x_n = x_{n-1} + \frac{\pi n}{2}(\alpha_{n-1} - \alpha_n) \qquad (10.145)$$

Equation (10.145) is a recursive relation for x_n in terms of the data in the nth and the $n-1^{st}$ intervals.

We observe that $(\alpha_{n-1} - \alpha_n)/2$ takes on values $0, \pm 1$. Thus, for $\alpha_n = \alpha_{n-1}$, we have $x_n = x_{n-1}$ while for $\alpha_n = -\alpha_{n-1}$, we have $x_n = x_{n-1} \pm n\pi$. If we arbitrarily set x_0 equal to zero, then we see that x_n takes on values of 0 or π (when reduced modulo 2π). Note from (10.142) that x_n represents the y-intercept (reduced modulo 2π) of the line representing $\theta(t, \boldsymbol{\alpha})$ in the interval $nT_b \leq t \leq (n+1)T_b$. This is easily verified by letting $h = .5$ in the trellis diagram of Fig. 10.14.

Let us now apply trigonometric identities to (10.74) with $\theta(t, \boldsymbol{\alpha})$ defined as in (10.142). In particular, we rewrite (10.74) as

$$s(t) = \sqrt{2P} \left[\cos \omega_c t \cos \theta(t, \boldsymbol{\alpha}) - \sin \omega_c t \sin \theta(t, \boldsymbol{\alpha}) \right] \qquad (10.146)$$

Now

$$\cos \theta(t, \boldsymbol{\alpha}) = \cos \frac{\pi t}{2T_b} \cos x_n - \alpha_n \sin \frac{\pi t}{2T_b} \sin x_n \qquad (10.147)$$

Recalling that x_n only takes on values 0 or π (modulo 2π), then $\sin x_n = 0$ and (10.147) simplifies to

$$\cos \theta(t, \boldsymbol{\alpha}) = a_n \cos \frac{\pi t}{2T_b}; \quad a_n = \cos x_n = \pm 1 \qquad (10.148)$$

Similarly

$$\sin \theta(t, \boldsymbol{\alpha}) = \sin \left(\alpha_n \frac{\pi t}{2T_b} + x_n \right) = \alpha_n \sin \frac{\pi t}{2T_b} \cos x_n + \cos \frac{\pi t}{2T_b} \sin x_n$$

$$\qquad (10.149)$$

$$= b_n \sin \frac{\pi t}{2T_b}; \quad b_n = \alpha_n \cos x_n = \alpha_n a_n = \pm 1$$

Finally then

$$s(t) = \sqrt{2P}\,[a_n C(t)\cos\omega_c t - b_n S(t)\sin\omega_c t]; \quad nT_b \le t \le (n+1)T_b$$

$$a_n = \cos x_n; \; b_n = \alpha_n a_n \; \Rightarrow \; \text{effective I\&Q binary data sequences} \tag{10.150}$$

$$C(t) = \cos\frac{\pi t}{2T_b}; \quad S(t) = \sin\frac{\pi t}{2T_b} \; \Rightarrow \; \text{effective I\&Q waveshapes}$$

To relate this representation back to the notion of a *frequency modulation*, we observe that

$$C(t)\cos\omega_c t = \frac{1}{2}\cos\left[\left(\omega_c + \frac{\pi}{2T_b}\right)t\right] + \frac{1}{2}\cos\left[\left(\omega_c - \frac{\pi}{2T_b}\right)t\right]$$

$$\tag{10.151}$$

$$S(t)\sin\omega_c t = -\frac{1}{2}\cos\left[\left(\omega_c + \frac{\pi}{2T_b}\right)t\right] + \frac{1}{2}\cos\left[\left(\omega_c - \frac{\pi}{2T_b}\right)t\right]$$

Thus

$$s(t) = \sqrt{2P}\left\{\left(\frac{a_n + b_n}{2}\right)\cos\left[\left(\omega_c + \frac{\pi}{2T_b}\right)t\right] + \left(\frac{a_n - b_n}{2}\right)\cos\left[\left(\omega_c - \frac{\pi}{2T_b}\right)t\right]\right\}$$

$$a_n = b_n \;(\alpha_n = 1) \Rightarrow s(t) = Aa_n\cos\left[\left(\omega_c + \frac{\pi}{2T_b}\right)t\right] \tag{10.152}$$

$$a_n \ne b_n \;(\alpha_n = -1) \Rightarrow s(t) = Aa_n\cos\left[\left(\omega_c - \frac{\pi}{2T_b}\right)t\right]$$

Note that in (10.150), $C(t)$ is $S(t)$ shifted by T_b seconds. Thus, it appears that $s(t)$ might be in the *form* of offset QPSK with half sinusoidal pulse shaping. To justify that this is indeed the case, we must examine more carefully the effective data sequences $\{a_n\}$ and $\{b_n\}$. Since the actual data sequence α_n can change every T_b seconds, it might appear that a_n and b_n can also change every T_b seconds. To the contrary, it can be shown that as a result of the phase continuity constraint [Eq. (10.145)], $a_n = \cos x_n$ can change only at the zero crossings of $C(t)$ and $b_n = \alpha_n \cos x_n$ can change only at the zero crossings of $S(t)$. Since the zero crossings of $C(t)$ and $S(t)$ are each spaced $T_s = 2T_b$ apart, then a_n and b_n are constant over $2T_b$-sec intervals (see Fig. 10.25 for an illustrative example). Noting that $C(t)$ and $S(t)$ alternate in sign every $2T_b$ sec, we can strengthen the analogy with sinusoidally pulse-shaped offset QPSK by defining the basic pulse shape

$$p(t) = \begin{cases} \sin\frac{\pi t}{2T_b}; & 0 \le t \le 2T_b \\ 0; & \text{otherwise} \end{cases} \tag{10.153}$$

and then rewriting $s(t)$ of (10.150) in the form

$$s(t) = \sqrt{2P}\,[d_c(t)\cos\omega_c t - d_s(t)\sin\omega_c t] \tag{10.154}$$

where

$$d_c(t) = \sum_n c_n p(t - (2n-1)T_b), \quad d_s(t) = \sum_n d_n p(t - 2nT_b) \tag{10.155}$$

with

k	α_k	$x_k \pmod{2\pi}$	a_k	b_k	time interval
0	1	0	1	1	$0 \le t \le T_b$
1	-1	π	-1	1	$T_b \le t \le 2T_b$
2	-1	π	-1	1	$2T_b \le t \le 3T_b$
3	1	0	1	1	$3T_b \le t \le 4T_b$
4	1	0	1	1	$4T_b \le t \le 5T_b$
5	1	0	1	1	$5T_b \le t \le 6T_b$
6	-1	0	1	-1	$6T_b \le t \le 7T_b$
7	1	π	-1	-1	$7T_b \le t \le 8T_b$
8	-1	π	-1	1	$8T_b \le t \le 9T_b$

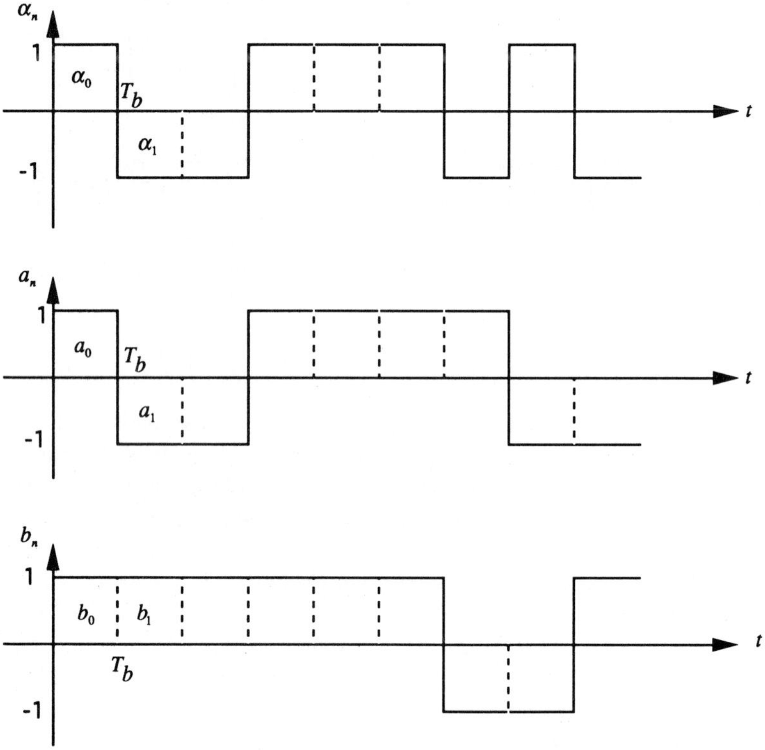

Figure 10.25　An example of the equivalent I&Q data sequences represented as rectangular pulse streams

$$c_n = (-1)^n a_{2n-1}, \quad d_n = (-1)^n b_{2n} \tag{10.156}$$

To complete the analogy, we point out that the sequences $\{a_{2n-1}\}$ and $\{b_{2n}\}$ are the odd/even split of a sequence $\{v_n\}$ which is the *differentially encoded* version of $\{\alpha_n\}$. This is easily verified (see Fig. 10.26 for an illustrative example) by recalling that for differential encoding, $v_n = \alpha_n v_{n-1}$ and then noting from (10.150) that $b_n = \alpha_n a_n$. The alternate representation of MSK as in (10.152) can also be expressed in terms of the differentially encoded bits, v_n. In particular

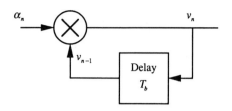

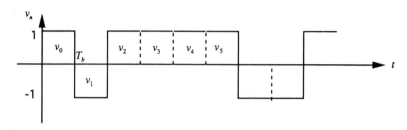

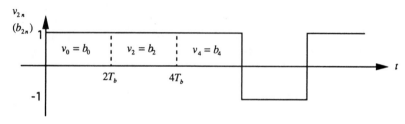

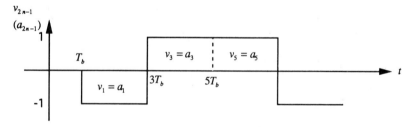

Figure 10.26   Equivalence between differentially encoded input bits and effective I&Q bits

For n odd

$$s(t) = \sqrt{2P} \left\{ \left(\frac{v_n + v_{n-1}}{2} \right) \cos \left[\left(\omega_c + \frac{\pi}{2T_b} \right) t \right] \right.$$
$$\left. + \left(\frac{v_n - v_{n-1}}{2} \right) \cos \left[\left(\omega_c - \frac{\pi}{2T_b} \right) t \right] \right\} \qquad (10.157a)$$

For n even

$$s(t) = \sqrt{2P} \left\{ \left(\frac{v_{n-1} + v_n}{2} \right) \cos \left[\left(\omega_c + \frac{\pi}{2T_b} \right) t \right] \right.$$
$$\left. + \left(\frac{v_{n-1} - v_n}{2} \right) \cos \left[\left(\omega_c - \frac{\pi}{2T_b} \right) t \right] \right\} \qquad (10.157b)$$

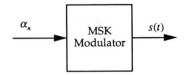

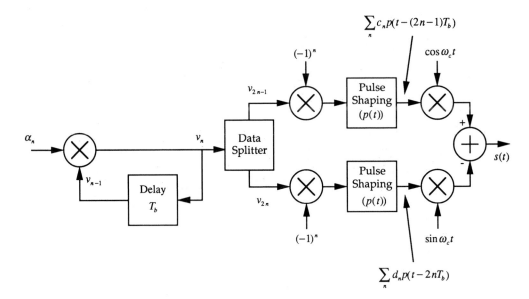

Figure 10.27   I-Q implementation of MSK

Combining these two results we get

$$
\begin{aligned}
s(t) = \sqrt{2P} \left\{ \left(\frac{v_n + v_{n-1}}{2} \right) \cos\left[\left(\omega_c + \frac{\pi}{2T_b} \right) t \right] \right. \\
\left. + (-1)^{n+1} \left(\frac{v_n - v_{n-1}}{2} \right) \cos\left[\left(\omega_c - \frac{\pi}{2T_b} \right) t \right] \right\}; \quad nT_b \le t \le (n+1)T_b
\end{aligned}
\tag{10.158}
$$

Finally, the I-Q implementation of MSK is as shown in Fig. 10.27. We observe that this figure resembles a transmitter for offset QPSK except that here the pulse shaping is half-sinusoidal [see (10.153)] rather than rectangular and in addition a differential encoder is applied to the input data sequence prior to pulse shaping and modulation onto the quadrature carriers.

Coherent detection.   Based on the above I-Q representation of MSK, an optimum receiver would consist of an I-Q coherent receiver for offset QPSK with I&D filters replaced by matched (to the half-sinusoidal pulse shape) filters and followed by a differential decoder (see Fig. 10.28). Because of the differential encoding/decoding operation, the bit error probability of the combination of Figures 10.27 and 10.28 has a penalty of approximately a factor of two [see (4.202) for the exact result] relative to the same performance of

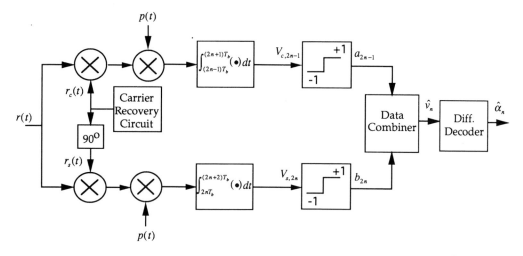

Figure 10.28   An I-Q receiver implementation of MSK

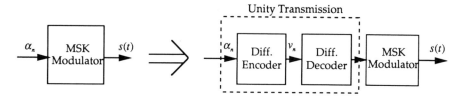

Figure 10.29   Two equivalent MSK transmitters

offset QPSK (or QPSK or BPSK). Thus, the often quoted statement that "optimal coherent detection of MSK and optimal coherent detection of BPSK yield the same bit error probability performance" is, strictly speaking, incorrect. Again we remind the reader that the reason behind this inaccuracy is that MSK is *defined* as a frequency modulation as per (10.74) together with (10.142) and its equivalent I-Q representation requires a differential encoder at the transmitter and hence a differential decoder at the receiver.

It is possible, however, to modify MSK so that the above statement becomes valid. Since a differential encoder followed by a differential decoder is equivalent to a unity transmission device, that is, input = output, then preceding the MSK modulator with this combination (see Fig. 10.29) still produces $s(t)$ as an output. Comparing Figures 10.27 and 10.29, we then have the equivalence illustrated in Fig. 10.30, namely, that preceding an MSK modulator with a differential *decoder* has an I-Q representation identically equivalent to that of offset QPSK with, of course, half-sinusoidal pulse shaping. The modulation implemented in Fig. 10.30 is referred to as *precoded MSK* and as such its receiver requires no differential decoder following the data detection. *It is this modulation that has the exact same bit error probability performance as BPSK, QPSK, or offset QPSK.* While all of this may seem like a moot point, it is nonetheless important to point out the discrepancy. In this text, we shall be careful to make the distinction between MSK, and precoded MSK both of which have the identical power spectral density given in Chapter 2.

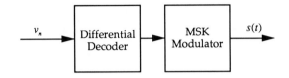

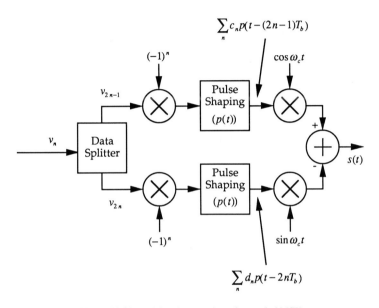

Figure 10.30   I-Q implementation of precoded MSK

Noncoherent detection.   Since MSK is nothing more than a special case of CPFSK corresponding to $h = .5$, then the noncoherent techniques discussed in Section 10.2.2 also apply here with, however, some simplification in implementation. For example, consider the simplification of the optimum block-by-block noncoherent receiver of Fig. 10.21 when the transmitted signal is MSK and for example, $N_s = 2$. Under these conditions, the complex weights [see (10.124)] needed to compute the decision statistic from the matched filter (and delayed matched filter) outputs are:

$$C_{i,0} = 1, \quad C_{i,1} = (-j)^{\alpha_{i,0}}; \quad i = 1, 2, 3, 4 \tag{10.159}$$

which leads to the simplified complex receiver of Fig. 10.31.

A real implementation of Fig. 10.31 in the form of an I-Q structure can be obtained as follows. Letting $\omega_m = \pi/2T_b$ denote the radian frequency of the equivalent I and Q half-sinusoidal baseband pulse shaping, then the four values of β_i can be expressed as:

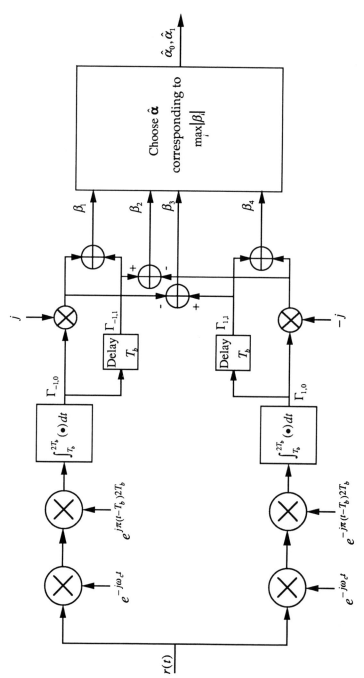

Figure 10.31 An implementation of a block detector for noncoherent MSK based on a two-symbol observation (Reprinted from [15] © IEEE)

$$\beta_1 = \int_0^{T_b} r(t) e^{-j\omega_c t} e^{j\omega_m t} dt + j \int_{T_b}^{2T_b} r(t) e^{-j\omega_c t} e^{j\omega_m (t - T_b)} dt$$

$$\beta_2 = \int_0^{T_b} r(t) e^{-j\omega_c t} e^{j\omega_m t} dt + j \int_{T_b}^{2T_b} r(t) e^{-j\omega_c t} e^{-j\omega_m (t - T_b)} dt$$

$$\text{(10.160)}$$

$$\beta_3 = \int_0^{T_b} r(t) e^{-j\omega_c t} e^{-j\omega_m t} dt - j \int_{T_b}^{2T_b} r(t) e^{-j\omega_c t} e^{j\omega_m (t - T_b)} dt$$

$$\beta_4 = \int_0^{T_b} r(t) e^{-j\omega_c t} e^{-j\omega_m t} dt - j \int_{T_b}^{2T_b} r(t) e^{-j\omega_c t} e^{-j\omega_m (t - T_b)} dt$$

Taking the squared[13] magnitudes of the four complex quantities in (10.152) gives terms of the form:

$$|\beta_i|^2 = (a_1 R_s + a_2 I_c + a_3 R_{cT} + a_4 I_{sT})^2 + (a_5 R_c + a_6 I_s + a_7 R_{sT} + a_8 I_{cT})^2 \quad \text{(10.161)}$$

where

$$R_s = \int_{T_b}^{2T_b} r(t) \cos \omega_c t \sin \omega_m (t - T_b) \, dt, \quad R_c = \int_{T_b}^{2T_b} r(t) \cos \omega_c t \cos \omega_m (t - T_b) \, dt$$

$$I_s = \int_{T_b}^{2T_b} r(t) \sin \omega_c t \sin \omega_m (t - T_b) \, dt, \quad I_c = \int_{T_b}^{2T_b} r(t) \sin \omega_c t \cos \omega_m (t - T_b) \, dt$$

$$\text{(10.162)}$$

$$R_{sT} = \int_0^{T_b} r(t) \cos \omega_c t \sin \omega_m t \, dt, \quad R_{cT} = \int_0^{T_b} r(t) \cos \omega_c t \cos \omega_m t \, dt$$

$$I_{sT} = \int_0^{T_b} r(t) \sin \omega_c t \sin \omega_m t \, dt, \quad I_{cT} = \int_0^{T_b} r(t) \sin \omega_c t \cos \omega_m t \, dt$$

and the a_i's take on values of ± 1 depending on the value of i. For example

$$|\beta_1|^2 = (R_{cT} + I_{sT} - R_s + I_c)^2 + (R_{sT} - I_{cT} + R_c + I_s)^2 \quad \text{(10.163)}$$

Figure 10.32 illustrates a portion of the I-Q receiver implementation, namely that which generates $|\beta_1|^2$. The remainder of the receiver that generates the other three $|\beta_i|^2$'s makes use of the same front end structure (the part enclosed in a box in Fig. 10.32) which is, in fact, identical to that found at the front end of a *coherent* receiver of MSK.

The bit error probability performance of the receiver in Fig. 10.32 can be obtained from (10.132) simply by substituting the value $h = .5$ in (10.130). In particular, we get

$$\begin{Bmatrix} b_1 \\ a_1 \end{Bmatrix} = \frac{E_b}{N_0} \left(1 \pm \sqrt{\frac{3 - 4/\pi^2}{4}} \right); \quad \begin{Bmatrix} b_3 \\ a_3 \end{Bmatrix} = \frac{E_b}{N_0} \begin{Bmatrix} 2 \\ 0 \end{Bmatrix};$$

$$\text{(10.164)}$$

$$\begin{Bmatrix} b_4 \\ a_4 \end{Bmatrix} = \frac{E_b}{N_0} \left(1 \pm \sqrt{1 - 4/\pi^2} \right);$$

13. Note that a decision based on the maximum $|\beta_i|$ is equivalent to one based on the maximum $|\beta_i|^2$.

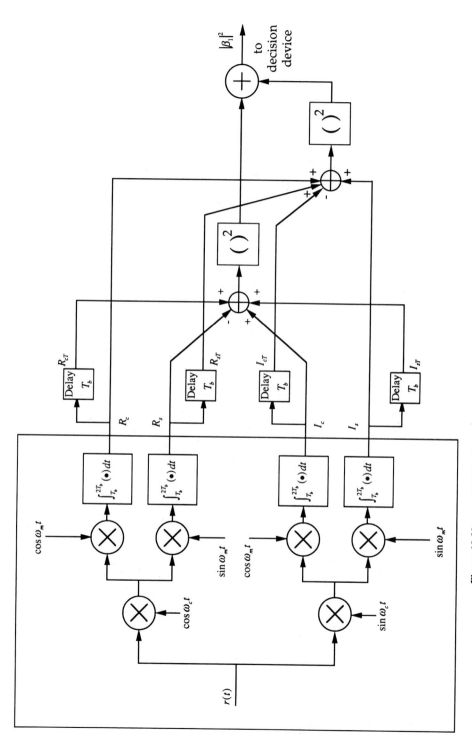

Figure 10.32 A real I-Q implementation of a block detector for noncoherent MSK based on a two-symbol observation (Reprinted from [15] © IEEE)

Since

$$Q(\alpha, 0) = 1; \quad Q(0, \beta) = \exp\left(-\frac{\beta^2}{2}\right) \tag{10.165}$$

we get the simplified result

$$P_b(E) \leq \frac{1}{2}\left[1 - Q(\sqrt{b_1}, \sqrt{a_1}) + Q(\sqrt{a_1}, \sqrt{b_1})\right]$$

$$+ \frac{1}{4}\left[1 - Q(\sqrt{b_4}, \sqrt{a_4}) + Q(\sqrt{a_4}, \sqrt{b_4})\right] + \frac{1}{4}\exp\left(-\frac{E_b}{N_0}\right) \tag{10.166}$$

where a_1, a_4 and b_1, b_4 are as defined in (10.164).

If the large SNR approximation of (10.131) is applied to the first two terms of (10.166), then we get

$$P_b(E) \leq \sqrt{\frac{2 + \sqrt{1 + 4/\pi^2}}{2\sqrt{1 + 4/\pi^2}}}\, \frac{1}{2}\, \mathrm{erfc}\left(\sqrt{\frac{E_b}{2N_0}\left(2 - \sqrt{1 + 4/\pi^2}\right)}\right)$$

$$+ \sqrt{\frac{2 + 4/\pi}{8/\pi}}\, \frac{1}{4}\, \mathrm{erfc}\left(\sqrt{\frac{E_b}{2N_0}(2 - 4/\pi)}\right) + \frac{1}{4}\exp\left(-\frac{E_b}{N_0}\right) \tag{10.167}$$

What we see from (10.167) is that the bit error probability performance (at least in terms of its asymptotic upper bound) of noncoherent MSK with two symbol observation and block-by-block detection is a scaled mixture of the performance of DPSK and coherent PSK's with effective E/N_0's determined from the arguments of the erfc functions. Figure 10.33 illustrates the upper bound of (10.167) as well as the accompanying results for $N_s > 2$ as determined from (10.137). From this figure, one can see the dramatic improvement in performance obtainable with moderate values of N_s.

10.2.4 Sinusoidal Frequency-Shift-Keying (SFSK)

As mentioned at the beginning of Section 10.2.1, another common pulse shape for full response CPM is the raised cosine (RC) pulse, which is a special case of LRC corresponding to $L = 1$. In particular, the frequency pulse $g(t)$ has the form

$$g(t) = \begin{cases} \frac{1}{2T}\left(1 - \cos\frac{2\pi t}{T}\right); & 0 \leq t \leq T \\ 0; & \text{otherwise} \end{cases} \tag{10.168}$$

For $h = .5$, this modulation scheme has independently been referred to as *sinusoidal frequency-shift-keying* by Amoroso [17]. The primary advantage of this form of CPM over, say, MSK is its increased rate of spectral roll-off as discussed in Chapter 2. It also possesses an I-Q representation in the form of offset QPSK with I and Q pulse shapes as in (2.122). As such, coherent and noncoherent detection techniques are quite similar to those for MSK and will not be discussed here in detail. As an example, for noncoherent block-by-block detection, the upper bounds on bit error probability given by (10.132) [together with (10.133)]

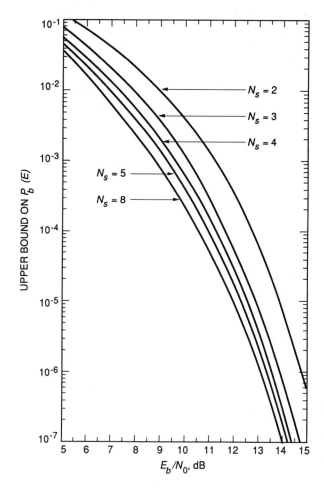

Figure 10.33 Upper bound on bit error probability versus E_b/N_0 with the number of observation symbols, N_s, as a parameter; $h = 0.5$, binary modulation with rectangular frequency pulses (Reprinted from [15] © IEEE)

for $N_s = 2$, (10.135) [together with (10.136)] for $N_s = 3$, and (10.137) for arbitrary N_s still apply with δ now given by [15]

$$\delta = \sum_{n=0}^{N_s-1} \xi_n e^{j\frac{\pi}{2} \sum_{i=0}^{n-1} (\alpha_i - \hat{\alpha}_i)} \tag{10.169}$$

where

$$\xi_i = \frac{1}{2} \sum_{k=0}^{\infty} \varepsilon_k J_{2k}\left(\frac{h}{2}(\alpha_i - \hat{\alpha}_i)\right) \left[A\left(\frac{h}{2}(\alpha_i - \hat{\alpha}_i) + 2k\right) + A\left(\frac{h}{2}(\alpha_i - \hat{\alpha}_i) - 2k\right) \right]$$

$$- \sum_{k=1}^{\infty} J_{2k-1}\left(\frac{h}{2}(\alpha_i - \hat{\alpha}_i)\right) \tag{10.170}$$

$$\times \left[A\left(\frac{h}{2}(\alpha_i - \hat{\alpha}_i) + 2k - 1\right) - A\left(\frac{h}{2}(\alpha_i - \hat{\alpha}_i) - 2k + 1\right) \right]$$

and

$$A(x) = e^{j\pi x} \operatorname{sinc} x \tag{10.171}$$

Also in (10.170), ε_k is again the Neumann factor, that is, $\varepsilon_k = 1$ if $k = 0$ and $\varepsilon_k = 2$ if $k \neq 0$, and $J_k(x)$ is again the kth order Bessel function of the first kind with argument x. As such the values of a and b, that is, the arguments of the Marcum-Q function, given previously for CPFSK still apply to SFSK if $\operatorname{sinc} h$ is replaced by $\operatorname{sinc} h \times f(h)$ where

$$f(h) \triangleq \sum_{k=0}^{\infty} \varepsilon_k J_{2k}(h) \left[\frac{h^2}{h^2 - (2k)^2} \right] - 2 \sum_{k=1}^{\infty} J_{2k-1}(h) \left[\frac{h(2k-1)}{h^2 - (2k-1)^2} \right] \tag{10.172}$$

and then evaluated at $h = .5$.

10.2.5 Continuous Phase FPM and QFPM

Continuous phase QFPM (CPQFPM) [7] is a modification of the QFPM scheme discussed in Section 10.1.3 that allows for sinusoidal pulse shaping and an offset of $T_s/2$ in the relative alignment of the baseband modulating waveforms for the two quadrature components. Analogous to the I-Q representation of MSK as an offset QPSK modulation with sinusoidal pulse shaping, the above modification of QFPM produces a continuous phase modulation. *Continuous phase FPM (CPFPM)* or alternately *Multifrequency Minimum-Shift-Keying (MFMSK)* [18] is a special case of CPQFPM wherein the frequencies of the two quadrature components are constrained to be identical, thereby producing a constant envelope signal set. In both schemes, the minimum frequency spacing required for orthogonality corresponds to selecting $h = 2$ ($f_d T_s = 1$).

A $(2N)^2$-ary CPQFPM signal is represented as

$$
\begin{aligned}
s(t) = \sqrt{2P} \Big\{ & a_c(t) \cos\left(\frac{\pi t}{T_s}\right) \cos\left[(\omega_c + b_c(t)\omega_d)\,t\right] \\
& + a_s(t) \sin\left(\frac{\pi t}{T_s}\right) \sin\left[(\omega_c + b_s(t)\omega_d)\,t\right] \Big\}
\end{aligned}
\tag{10.173}
$$

$$= s_c(t) + s_s(t)$$

where the even- and odd-numbered data waveforms $d_c(t)$ and $d_s(t)$ that generate the amplitude-frequency pairs $a_c(t)$, $b_c(t)$ and $a_s(t)$, $b_s(t)$, respectively, are offset by $T_s/2$ in their relative alignments.[14] This staggered alignment between $a_c(t)$ and $a_s(t)$, and similarly between $b_c(t)$ and $b_s(t)$ together with the sinusoidal pulse shaping by $\cos(\pi t/T_s)$ and $\sin(\pi t/T_s)$ results in the phase of $s(t)$ at data transition instants $t = (2i - 1)T_s/2$ and $t = iT_s$ being continuous. To see this, we rewrite (10.173) as follows:

---

14. The data streams $a_c(t)$ and $b_c(t)$ have data transition instants at $t = (2i - 1)T_s/2$ and the data streams $a_s(t)$ and $b_s(t)$ have data transition instants at $t = iT_s$.

$$s(t) = \sqrt{\frac{P}{2}} a_c(t) \left\{ \cos\left[\left(\omega_c + b_c(t)\omega_d + \frac{\pi}{T_s}\right)t\right] + \cos\left[\left(\omega_c + b_c(t)\omega_d - \frac{\pi}{T_s}\right)t\right]\right.$$

$$+ \frac{a_s(t)}{a_c(t)} \cos\left[\left(\omega_c + b_s(t)\omega_d - \frac{\pi}{T_s}\right)t\right]$$

$$\left. - \frac{a_s(t)}{a_c(t)} \cos\left[\left(\omega_c + b_s(t)\omega_d + \frac{\pi}{T_s}\right)t\right]\right\}$$

$$= \sqrt{2P} a_c(t) \left\{ \cos\left(\frac{b_s(t) - b_c(t)}{2}\omega_d t\right)\right.$$

$$\times \cos\left[\left(\omega_c + \frac{b_s(t) + b_c(t)}{2}\omega_d \mp \frac{\pi}{T_s}\right)t\right]$$

$$\left. + \sin\left(\frac{b_s(t) - b_c(t)}{2}\omega_d t\right) \sin\left[\left(\omega_c + \frac{b_s(t) + b_c(t)}{2}\omega_d \pm \frac{\pi}{T_s}\right)t\right]\right\}$$

(10.174)

where the upper signs correspond to $a_s(t)/a_c(t) = 1$ and the lower signs correspond to $a_s(t)/a_c(t) = -1$. Evaluating (10.174) at $t = (2i - 1)T_s/2$, we get

$$s\left((2i-1)\frac{T_s}{2}\right) = \pm(-1)^{i-1}\sqrt{2P} a_c\left((2i-1)\frac{T_s}{2}\right)$$

$$\times \sin\left\{\left[\omega_c + b_s\left((2i-1)\frac{T_s}{2}\right)\omega_d\right]\left((2i-1)\frac{T_s}{2}\right)\right\}$$

(10.175)

$$= (-1)^{i-1}\sqrt{2P} a_s\left((2i-1)\frac{T_s}{2}\right)$$

$$\times \sin\left\{\left[\omega_c + b_s\left((2i-1)\frac{T_s}{2}\right)\omega_d\right]\left((2i-1)\frac{T_s}{2}\right)\right\}$$

Since $a_s(t)$ and $b_s(t)$ cannot make a transition at $t = (2i-1)T_s/2$, then $s(t)$ will be continuous at these time instants. Similarly, at $t = iT_s$, (10.174) evaluates to

$$s(iT_s) = (-1)^i \sqrt{2P} a_c(iT_s) \cos\{[\omega_c + b_c(iT_s)\omega_d] iT_s\}$$ (10.176)

Since $a_c(t)$ and $b_c(t)$ cannot make a transition at $t = iT_s$, then $s(t)$ will be continuous at these time instants.

The special case of (10.173) corresponding to $N = 1$, that is, $b_c(t) = b_s(t) = 0$, yields

$$s(t) = \sqrt{2P} \left\{ a_c(t) \cos\left(\frac{\pi t}{T_s}\right) \cos\omega_c t + a_s(t) \sin\left(\frac{\pi t}{T_s}\right) \sin\omega_c t\right\}$$ (10.177)

which is in the form of an offset QPSK signal with sinusoidal pulse shaping, or equivalently, precoded MSK.

A block diagram of a modulator for $(2N)^2$-ary CPQFPM is obtained from the block diagram of a $(2N)^2$-ary QFPM modulator (Fig. 10.9a) by introducing a $T_s/2$ offset between the alignments of $d_c(t)$ and $d_s(t)$ and adding sinusoidal pulse shaping to the $s_c(t)$ and $s_s(t)$ signals prior to their summation. The result is illustrated in Fig. 10.34a. Similarly, the block

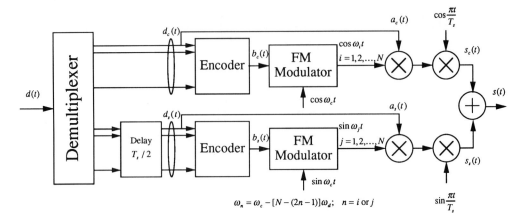

Figure 10.34a   A block diagram of a modulator for $(2N)^2$-ary CPQFPM

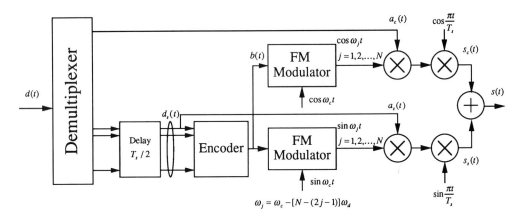

Figure 10.34b   A block diagram of a modulator for $4N$-ary CPFPM

diagram of a modulator for $4N$-ary CPFPM is the same modification of the modulator for $4N$-ary N-FSK/QPSK (Fig. 10.9b). The result is illustrated in Fig. 10.34b. The optimum receiver for CPQFPM is a modification of the optimum receiver for QFPM (Fig. 10.10) wherein the I and Q integrate-and-dump filters of the optimum biorthogonal detector blocks are replaced by matched filters, that is, amplitude weighting by $\cos(\pi t/T_s)$ and $\sin(\pi t/T_s)$ followed by I&D's with integration limits as well as the sampling instants staggered by $T_s/2$ with respect to one another (see Fig. 10.35). It should be clear from Fig.10.35 that the ideal coherent error probability performance of CPQFPM on the AWGN is identical to that of QFPM (see Table 10.2).

Using the methods presented in Chapter 2, it is straightforward to show that the equivalent lowpass two-sided PSD of CPQFPM is given by [7]

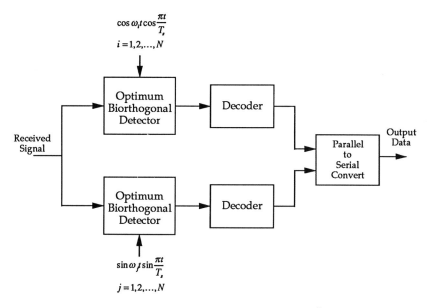

Figure 10.35 Block diagram of an optimum receiver for $(2N)^2$-ary CPQFPM

$$S_s(f) = \frac{PT_s}{2N} \sum_{i=0}^{N-1} \left[\mathrm{sinc}\left(fT_s + 2i - N + \frac{1}{2} \right) + \mathrm{sinc}\left(fT_s + 2i - N + \frac{3}{2} \right) \right]^2$$

$$= \frac{(1 + \log_2 N)}{N} PT_b \sum_{i=0}^{N-1} \left[\mathrm{sinc}\left(2\left(1 + \log_2 N\right) fT_b + 2i - N + \frac{1}{2} \right) \right. \tag{10.178}$$

$$\left. + \mathrm{sinc}\left(2\left(1 + \log_2 N\right) fT_b + 2i - N + \frac{3}{2} \right) \right]^2$$

which is illustrated in Fig. 10.36 for various values of N. The curve labeled $N = 1$ is identical to the PSD of MSK. Comparing Fig. 10.36 with Fig. 10.12 we observe that, analogous to the PSD's of QPSK and MSK, the main lobe of the CPQFPM spectrum is wider than that of QFPM; however, the PSD of CPQFPM falls off at a faster rate than that of QFPM. This can also be seen by comparing the fractional out-of-band behavior of CPQFPM (Fig. 10.37) with that of QFPM (Fig. 10.13). Finally, the power and bandwidth efficiencies of CPQFPM are included in Tables 10.2–10.4 along with the comparable data for the other I-Q modulations previously discussed.

As previously mentioned, CPFPM is a special case of CPQFPM wherein the I and Q signal component frequencies are identical. With reference to (10.173), it is mathematically modeled as a $4N$-ary constant envelope signal of the form[15]

---

15. We remind the reader that, as for N-FSK/QPSK, the number of bits per symbol, i.e., T_s/T_b, is now $\log_2 4N = 2 + \log_2 N$ since two bits are required for the pair of binary amplitudes $a_c(t)$ and $a_s(t)$ and $\log_2 N$ bits to determine the frequency common to both channels.

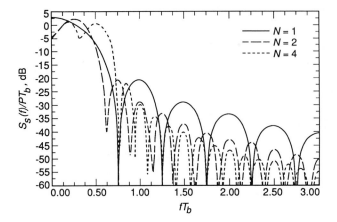

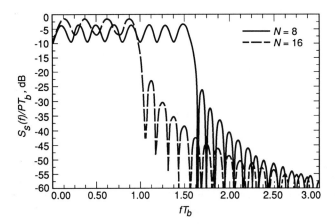

Figure 10.36 Power spectral density of CPQFPM with N as a parameter (Reprinted from [18] © IEEE)

$$s(t) = \sqrt{2P} \left\{ a_c(t) \cos\left(\frac{\pi t}{T_s}\right) \cos\left[(\omega_c + b(t)\omega_d) t\right] \right.$$

$$\left. + a_s(t) \sin\left(\frac{\pi t}{T_s}\right) \sin\left[(\omega_c + b(t)\omega_d) t\right] \right\} \tag{10.179}$$

$$= s_c(t) + s_s(t)$$

Here $a_c(t)$ and $a_s(t)$ are offset in their relative alignment by $T_s/2$, that is, $a_s(t)$ has transitions at $t = iT_s$ and $a_c(t)$ has transitions at $t = (2i-1)T_s/2$. However, $b(t)$ being common to both I and Q signal components now has the same transitions as $b_s(t)$ of CPQFPM, that is, at $t = iT_s$. Also, $s_c(t)$ and $s_s(t)$ are orthogonal over the interval $(iT_s/2, (i+1)T_s/2)$ of duration $T_s/2$ sec, which allows for demodulating and detecting them separately.

An equivalent representation to (10.179) is

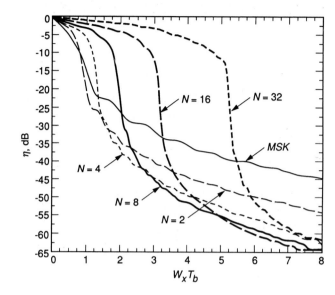

Figure 10.37 Fractional out-of-band power versus normalized bandwidth $W_x T_b$ for CPQFPM with N as a parameter (Reprinted from [18] © IEEE)

$$s(t) = \sqrt{2P} a_c(t) \cos\left[(\omega_c + b(t)\omega_d) t - a_c(t)a_s(t)\frac{\pi t}{T_s}\right]$$

$$= \sqrt{2P} a_c(t) \cos\left[\omega_c t + (2b(t) - a_c(t)a_s(t))\frac{\pi t}{T_s}\right]$$

(10.180)

which clearly elucidates the constant envelope nature of the CPFPM signal. Moreover, the phase continuity of $s(t)$ is easily identified since a transition in $b(t)$ and/or $a_s(t)$ at $t = iT_s$ or a transition in $a_c(t)$ at $t = (2i - 1)T_s/2$ produces a phase change equal to 2π (modulo 2π).

An optimum receiver for coherently demodulating and detecting CPFPM is given in [18] and is illustrated here in Fig. 10.38. Its structure resembles the optimum receiver for constant envelope QFPM (see Fig. 10.11) . Because of the coincident alignment of $b(t)$ and $a_s(t)$, that is, both transition at integer multiples of T_s, the bit stream, $d_s(t)$, corresponding to the signal frequency is estimated only in the receiver block, D_s, associated with $s_s(t)$. This estimation, however, is aided by the receiver block, D_c, associated with $s_c(t)$ so as to improve performance. This will be explained shortly. Thus, at the output of block D_s, $a_s(t)$ and $d_s(t)$ [or equivalently, $b(t)$] are recovered.

In the block D_c, only the binary amplitude data stream $a_c(t)$ (or a delayed version of it) needs to be recovered since as mentioned above no frequency decisions are made in this block. The particular correlator in D_c, that is, the $\hat{i}$th, from which $a_c(t)$ is estimated should be chosen in accordance with the frequency $\omega_{\hat{i}}$ decided on in the D_s block. However, since the time at which this frequency estimate is made, namely, $t = (n + 1)T_s$, occurs in the middle of the pulse $a_c(t) \cos \pi t/T_s$ associated with $s_c(t)$, then it is possible that a frequency transition may occur at this very same time instant. Stated another way, the first half of the pulse $a_c(t) \cos \pi t/T_s$ may correspond to one frequency while the second half of this pulse may correspond to another frequency. Because of this, the integration interval of the I&D's associated with the correlators in block D_c must be limited to a duration of $T_s/2$ sec so that the appropriate correlator for the

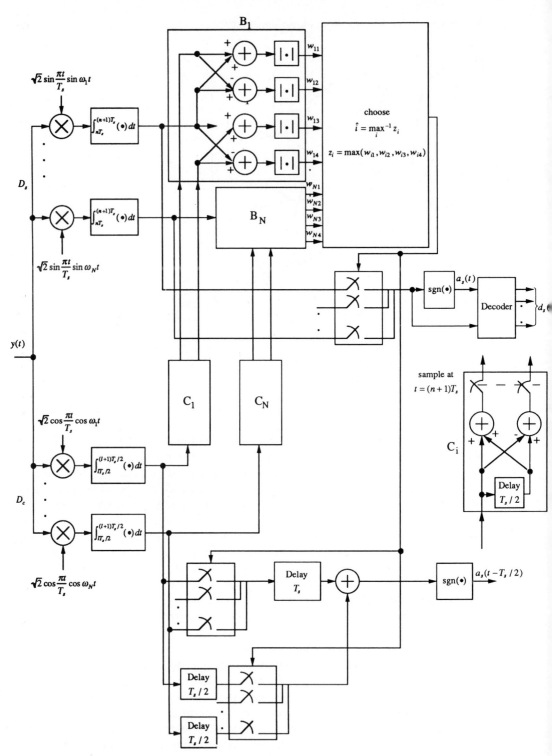

Figure 10.38 Block diagram of the optimum receiver for $4N$-ary CPFPM (Reprinted from [18] © IEEE)

first and second halves of the symbol $a_c(t)$ can be *separately* chosen in accordance with the frequency decisions at $t = (n + 1)T_s$ and $t = (n + 2)T_s$, respectively. For example, in order to detect $a_c(t)$ [the sign of $s_c(t)$] in the interval $T_s/2 \leq t \leq 3T_s/2$, the particular correlator corresponding to I&D integration in the interval $T_s/2 \leq t \leq T_s$ should be selected by the frequency decision at $t = T_s$ whereas the particular correlator corresponding to I&D integration in the interval $T_s \leq t \leq 3T_s/2$ should be selected by the frequency decision at $t = 2T_s$. To effectively utilitize both of these components in making the decision on $a_c(t)$, the first-half symbol correlator output should be delayed by T_s sec and then added to the second-half symbol correlator output (see the lower portion of block D_c in Figure 10.38). By taking the sign of this sum, we obtain an estimate of $a_c(t - T_s/2)$.

As mentioned above, an improvement in detection of the signal frequency can be obtained by utilizing the correlator outputs corresponding to the quadrature signal component $s_c(t)$. Since midway in the detection interval for $s_s(t)$ a transition in $a_c(t)$ may occur, we must separately add and subtract the first- and second-half symbol correlator outputs corresponding to $s_c(t)$ (this is accomplished in the D_c block of Fig. 10.38 by the blocks C_i; $i = 1, 2, \ldots, N$) to obtain pairs of signals that can be used to aid in the formation of the block D_s output magnitudes, $w_{i,j}$; $i = 1, 2, \ldots, N$, $j = 1, 2, 3, 4$. As in Fig. 10.11, the pair of signals from each of the blocks C_i; $i = 1, 2, \ldots, N$ is coherently combined (added and subtracted) with the corresponding I&D output of the D_s block to produce a 3 dB improvement in SNR relative to the case where the frequency estimate is obtained without aiding from the D_c block.

In determining upper bounds on the symbol and bit error probability performances of the optimum $4N$-ary CPFPM receiver of Fig. 10.38, we follow the same arguments as used to arrive at upper bounds on the performance of the optimum M-ary receiver for $4N$-ary N-FSK/QPSK (see Fig. 10.11). In particular, it is straightforward to show that the symbol and bit error probabilities of Fig. 10.38 are upper bounded oy (10.64) and (10.65), respectively.

Figure 10.39 is an illustration of the upper bound of (10.65) for values of $N = 1, 2, 4, 8$, and 16. We observe from these numerical results that the power efficiency of CPFPM is a monotonically increasing function of N and that relative to MSK ($N = 1$) significant power savings (on the order of 1.5 dB at $P_b(E) = 10^{-5}$) can be achieved with only $N = 2$. A more detailed numerical comparison is made in Table 10.2.

10.3 FURTHER DISCUSSION

In this chapter we have attempted to cover the demodulation and detection of a variety of digital modulations that have achieved a certain degree of popularity by finding their way into modern communication system design. Needless to say, there exist a host of other digital modulations that for one reason or another have received attention in the literature. Our purpose here is to mention some of these and briefly discuss their merits.

In our discussion of I-Q modulations we have focused entirely on full response schemes, that is, those whose pulse duration is equal to a single symbol interval. Analogous to partial response CPFM schemes, such as, LREC and LRC, it is possible to achieve additional bandwidth compression by considering I-Q modulations whose pulse duration extends over more than one symbol interval. An example of this is so-called *Quadrature Overlapped Raised Cosine (QORC)* modulation [19] wherein the I and Q pulse shape is a raised cosine pulse that extends over two symbol (four bit) intervals. Mathematically speaking, QORC is described by

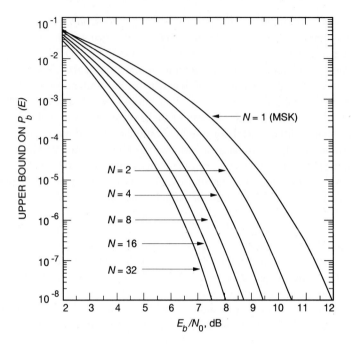

Figure 10.39 Upper bound on the bit error probability performance of CPQFPM with N as a parameter (Reprinted from [18] © IEEE)

$$s(t) = A \left[\left(\sum_{n=-\infty}^{\infty} a_n p\,(t - 2nT_b) \right) \cos \omega_c t - \left(\sum_{n=-\infty}^{\infty} b_n p\,(t - 2nT_b) \right) \sin \omega_c t \right] \quad (10.181)$$

where

$$p(t) = \begin{cases} \frac{1}{2} \left(1 - \cos \frac{\pi t}{2T_b} \right); & 0 \le t \le 4T_b \\ 0; & \text{otherwise} \end{cases} \quad (10.182)$$

Note that for QORC, the amplitude A is related to the average power of the modulation P by $A = \sqrt{4P/3}$, whereas for QPSK the relationship is $A = \sqrt{P}$ and for precoded MSK, $A = \sqrt{2P}$. While the power performance of QORC is quite comparable to that of QPSK and precoded MSK, the real advantage of QORC lies in its spectral occupancy. Indeed it can be shown [19] that, except for a normalization constant, the equivalent lowpass PSD of QORC is the *product* of the equivalent lowpass PSD for QPSK and MSK (or precoded MSK), that is,

$$S_s(f) = 2PT_b \left[\frac{\sin 2\pi f T_b}{2\pi f T_b} \right]^2 \left[\frac{\cos 2\pi f T_b}{1 - (4fT_b)^2} \right]^2 \quad (10.183)$$

Thus, asymptotically the PSD of QORC rolls off at a rate f^{-6} as compared with that of MSK, which rolls off as f^{-4}, and that of QPSK, which rolls off as f^{-2}. Offset (staggered) versions of QORC have also been considered for application on nonlinear amplification channels [20].

Another I-Q technique that has merit on nonlinear amplification channels is a variation of conventional QPSK referred to as $\pi/4$-*QPSK* and was initially described by Baker [21].

The idea here is to reduce the maximum 100% envelope fluctuations due to possible 180° phase reversals in QPSK without resorting to an offset form of modulation. While indeed offset QPSK reduces the envelope fluctuation by limiting carrier phase changes to 90°, its performance suffers relative to conventional QPSK when differential detection is employed. By limiting the maximum phase change to 135°, $\pi/4$-QPSK achieves a compromise between QPSK and offset QPSK in the spectral spread caused by nonlinear amplification. The following brief description of $\pi/4$-QPSK explains how this reduction in maximum phase change is accomplished without offsetting the I and Q channels.

The basic difference between $\pi/4$-QPSK and conventional QPSK is the set of values assigned to the phases representing the four possible two-bit information sequences. In conventional QPSK, the information phases used are $0, \pm\pi/2, \pi$ whereas in $\pi/4$-QPSK we use $\pm\pi/4, \pm3\pi/4$. Since with differential encoding (which is required for differential detection) the information phases represent the phase *changes* between adjacent transmitted symbols, then we see that whereas with QPSK a maximum phase change of π radians is possible, with $\pi/4$-QPSK the magnitude of the maximum phase change is limited to $3\pi/4$ radians. In terms of the actual phases transmitted over the channel, with QPSK the allowable set remains the same in every transmission interval, namely, $0, \pm\pi/2, \pi$ whereas with $\pi/4$-QPSK it alternates between $0, \pm\pi/2, \pi$ and $\pm\pi/4, \pm3\pi/4$.

Finally, we bring the reader's attention to *multi-amplitude minimum-shift-keying (MAMSK)*, a generalization of MSK introduced in the late 1970's [22, 23]. As the name of the modulation indicates, MAMSK is a multilevel version of MSK in that the effective inphase and quadrature data sequences are multilevel rather than binary. Specifically, MAMSK assumes the form of (10.154) where the data sequences c_n and d_n take on values $\pm1, \pm3, \ldots, \pm\sqrt{M} - 1$. An alternate representation of MAMSK is in the form of the sum of $n = \log_2\left(\log_2 M\right)$ MSK signals whose powers are distributed such that the ith component of the sum, $i = 1, 2, \ldots, n$ has an amplitude that is 2^{i-1} times that of the first component [23]. For example, for $M = 16$, the signal

$$s(t) = A \cos\left(\omega_c t + \alpha_n\left(\frac{\pi t}{2T_b}\right) + x_n\right) + 2A \cos\left(\omega_c t + \beta_n\left(\frac{\pi t}{2T_b}\right) + y_n\right);$$

$$nT_b \leq t \leq (n+1)T_b \tag{10.184}$$

is identically equivalent to

$$s(t) = A \cos \omega_c t \cos \frac{\pi t}{2T_b} \left[\cos x_n + 2 \cos y_n\right]$$

$$- A \sin \omega_c t \sin \frac{\pi t}{2T_b} \left[\alpha_n \cos x_n + 2\beta_n \cos y_n\right] \tag{10.185}$$

$$= A c_n \cos \omega_c t \cos \frac{\pi t}{2T_b} - A d_n \sin \omega_c t \sin \frac{\pi t}{2T_b}$$

where the effective inphase and quadrature data sequences c_n and d_n take on values $\pm1, \pm3$.

The probability of error performance of precoded MAMSK is identical to that of QASK with the same value of M in the same sense that the probability of error performance of MSK is identical to that of PSK. The advantage of the former in both cases is, of course, the faster spectral roll-off achieved by using I and Q half-sinusoidal pulse shaping.

Table 10.5   A Table of Various Modulation Types and Their Properties

Modulation Type	Description and Relation to Other Modulation Types	Continuous or Discontinuous Phase	Constant Envelope
Unbalanced Quadrature Phase-Shift-Keying (UQPSK)	QPSK with, in general, different power, data rate, and data format on I&Q channels	N/A	Yes
Quadrature Amplitude Modulation (QAM)	Independent M-AM modulations of the same power, data rate, and data format on I&Q channels	N/A	No
Quadrature Amplitude-Shift-Keying (QASK)	Independent M-ASK modulations (rectangular pulse shape implied) of the same power, data rate, and data format on I&Q channels	N/A	No
Quadrature Biorthogonal Modulation (QBOM)	Independent N-dimensional biorthogonal modulations of the same power, data rate, and data format on I and Q channels	Discontinuous	No
Quadrature Frequency/Phase Modulation (QFPM)	Particular embodiment of QBOM using, for the biorthogonal modulations, N-FSK/BPSK with minimum frequency spacing required for orthogonality	Discontinuous	No
Quadrature-Quadrature Phase-Shift-Keying (Q^2PSK)	Independent QPSK modulations on I&Q channels (special case of QPFM when $N = 2$, pulse shaping is sinusoidal, and I&Q channels are offset by half a symbol)	Discontinuous	No
Continuous Phase Modulation (CPM) or Continuous Phase Frequency Modulation (CPFM)	Digital frequency modulation with continuous phase from symbol interval to symbol interval	Continuous	Yes
Full Response CPFM	CPFM wherein frequency pulse is one symbol in duration	Continuous	Yes
Partial Response CPFM	CPFM wherein frequency pulse is greater than one symbol in duration	Continuous	Yes
Continuous Phase Frequency-Shift-Keying (CPFSK)	Special case of full response CPFM using a rectangular frequency pulse	Continuous	Yes
Multi-h CPFM	CPFM wherein the modulation index varies from symbol interval to symbol interval	Continuous	Yes
Minimum-Shift-Keying (MSK)	Special case of CPFSK wherein the modulation index equals 0.5. Also equivalent to precoded offset QPSK with sinusoidal pulse shaping	Continuous	Yes

Table 10.5   A Table of Various Modulation Types and Their Properties, *continued*

Modulation Type	Description and Relation to Other Modulation Types	Continuous or Discontinuous Phase	Constant Envelope
Continuous Phase QFPM (CPQFPM)	Modification of QFPM wherein pulse shaping is sinusoidal and I&Q channels are offset by half a symbol	Continuous	No
Continuous Phase FPM (CPFPM) or Multifrequency Minimum-Shift-Keying (MFMSK)	Special case of CPFPM wherein frequencies of the two quadrature components are constrained to be identical	Continuous	Yes
Quadrature Overlapped Raised Cosine (QORC)	QPSK with raised cosine pulse shapes of two-bit duration	N/A	No
$\pi/4$-Quadrature Phase-Shift-Keying ($\pi/4$-QPSK)	QPSK where the signal phases take on values $\pm\pi/4, \pm3\pi/4$ (rather than $0, \pm\pi/2, \pi$)	N/A	Yes
Multi-amplitude Minimum-Shift-Keying (MAMSK)	Multi-amplitude version of I-Q form of MSK	Continuous	No

We conclude this chapter by presenting, for quick reference purposes, a table listing the various modulations considered in this chapter together with their acronyms, relation to other modulation types, and key properties such as discontinuous versus continuous phase and constant versus nonconstant envelope. These results are presented in Table 10.5.

PROBLEMS

10.1 Consider a digital communication system transmitting an offset QPSK modulation. As such, the received signal is of the form

$$r(t) = \sqrt{P}\,[d_c(t)\cos\omega_c t - d_s(t)\sin\omega_c t] + n(t)$$

where P is the received power, ω_c is the carrier radian frequency, $n(t)$ is an additive white Gaussian noise source with single-sided power spectral density, N_0 W/Hz, and $d_c(t)$, $d_s(t)$ are, respectively, the in-phase (I) and quadrature (Q) data streams defined by

$$d_c(t) = \sum_{i=-\infty}^{\infty} d_{ci}\, p\,(t - iT_s)$$

$$d_s(t) = \sum_{i=-\infty}^{\infty} d_{si}\, p\left(t - \left(i + \frac{1}{2}\right)T_s\right)$$

with $p(t)$ a unit amplitude rectangular pulse of duration T_s sec. The two data streams $\{d_{ci}\}$ and $\{d_{si}\}$ are independent, identically distributed (iid) ± 1 sequences.

The carrier reference signals generated at the receiver (see Fig. P10.1) are not ideal in that their phase differs from that of the received signal by ϕ radians. That is

$$r_c(t) = 2\cos(\omega_c t - \phi), \quad r_s(t) = -2\sin(\omega_c t - \phi)$$

(a) Find the conditional error probability, $P_I(E; \phi)$, associated with detecting the sequence $\{d_{ci}\}$ in the presence of the phase error ϕ. For simplicity, consider only the detection of the data bit d_{c0} and ignore any second harmonics of the carrier frequency.

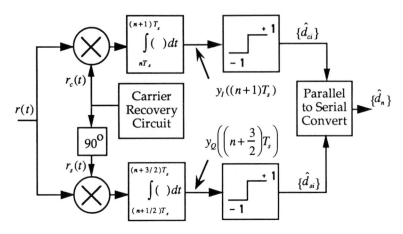

Figure P10.1

(b) If the I and Q data sequences $\{d_{ci}\}$ and $\{d_{si}\}$ were originally obtained as the odd and even bits of an iid ± 1 data sequence $\{d_i\}$ occurring at a bit rate of $1/T_b = 1/2T_s$, what is the conditional bit error probability, $P_b(E;\phi)\big|_{\text{OQPSK}}$, of the sequence $\{d_i\}$? Express your answer in terms of the conditional bit error probability of QPSK (non-offset), $P_b(E;\phi)\big|_{\text{QPSK}}$, and the conditional bit error probability of PSK, $P_b(E;\phi)\big|_{\text{PSK}}$, in the presence of the same phase error, ϕ. That is, write $P_b(E;\phi)\big|_{\text{OQPSK}}$ as a function of $P_b(E;\phi)\big|_{\text{QPSK}}$ and $P_b(E;\phi)\big|_{\text{PSK}}$.

(c) For any $\phi \neq 0$, how does $P_b(E;\phi)\big|_{\text{OQPSK}}$ compare with $P_b(E;\phi)\big|_{\text{PSK}}$? Is it better or worse? How does $P_b(E;\phi)\big|_{\text{OQPSK}}$ compare with $P_b(E;\phi)\big|_{\text{QPSK}}$? Is it better or worse? How do these conditional bit error probability performances compare at $\phi = 0$?

10.2 Consider an unbalanced QPSK receiver with imperfect carrier phase synchronization. Because of the presence of a carrier phase error, ϕ, between the received modulation $s(t,\theta)$ of (10.3) and the I and Q demodulation reference signals, $r_c(t)$ and $r_s(t)$ of (10.9), the I and Q integrate-and-dump outputs will each contain a term due to crosstalk [see (10.11)]. Normalized versions of these crosstalk terms denoted as $d_{1,2}$ and $d_{2,1}$ are defined in (10.12). Using the approach taken in Chapter 2, Section 2.1 to derive the PSD of a random pulse train, show that the mean-squared values of $d_{1,2}$ and $d_{2,1}$ can be expressed in terms of the PSDs of the I and Q data streams as in (10.22).

10.3 Using (10.22), derive the mean-squared crosstalk $\overline{d_{1,2}^2}$ for the case where $d_1(t)$ is an NRZ waveform of rate $1/\mathcal{R}_1$ and $d_2(t)$ is an NRZ waveform of rate $1/\mathcal{R}_2$.

10.4 Using (10.22), derive the mean-squared crosstalk $\overline{d_{1,2}^2}$ for the case where $d_1(t)$ is a Manchester waveform of rate $1/\mathcal{R}_1$ and $d_2(t)$ is an NRZ waveform of rate $1/\mathcal{R}_2$.

10.5 (Gray Code for QAM) In Chapter 4 we defined the notion of a Gray code as applied to M-PSK modulation. In particular, a bit assignment was made to each symbol (signal point on the circular constellation) in such a way that adjacent symbol errors resulted in only a single bit error. Since each point on the circle has only two adjacent neighbors, when viewed in polar (amplitude-phase) coordinates, the Gray code for M-PSK is one-dimensional, that is, it relates only to the phase coordinate. For QAM modulation, a two-dimensional Gray code can be assigned in such a way that for each of the two rectangular coordinates, an adjacent symbol error results in only a single bit error. Construct such a 2-dimensional Gray code for

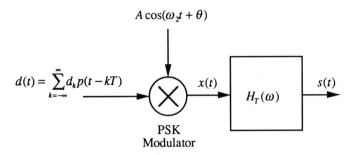

$$d(t) = \sum_{k=-\infty}^{\infty} d_k p(t - kT)$$

$A \cos(\omega_2 t + \theta)$

$x(t)$

$H_T(\omega)$

$s(t)$

PSK
Modulator

Figure P10.6

16-QAM, that is, show the appropriate bit to symbol assignments for each of the 16 points in the constellation.

10.6 (Serial MSK) Consider the transmitter implementation illustrated in Fig. P10.6. Binary (± 1) data $\{d_k\}$ is PSK modulated onto a carrier $A \cos(\omega_2 t + \theta)$ to produce the signal $x(t) = A \sum_{k=-\infty}^{\infty} d_k p(t - kT) \cos(\omega_2 t + \theta)$ where $p(t)$ is a unit rectangular pulse shape and θ is an arbitrary phase. The PSK output is then passed through a linear filter $H_T(\omega)$ with finite impulse response

$$h_T(t) = \begin{cases} \frac{\pi}{T} \sin \omega_1 t; & 0 \le t \le T \\ 0; & \text{otherwise} \end{cases}$$

Suppose that we choose the carrier frequency $\omega_2 = \omega_c + \Delta\omega/2$ and the filter impulse response frequency $\omega_1 = \omega_c - \Delta\omega/2$ where $\Delta\omega T = \pi$.

(a) Show that the output signal $s(t)$ in the kth transmission interval $kT \le t \le (k+1)T$ depends only on the input bits d_k and d_{k-1}, that is, write $s(t)$ as a convolution of $x(t)$ with $h_T(t)$ and express the result as two integrals with appropriate limits involving only d_k and d_{k-1}. Ignore double harmonic terms.

(b) Evaluate your result of part (a) and show that it can be written in the form of an MSK-type signal, that is

$$s(t) = A \left\{ u_k \cos \omega_2 t + (-1)^k v_k \cos \omega_1 t \right\}; \quad kT \le t \le (k+1)T$$

where $\{u_k\}$ and $\{v_k\}$ are ($-1, 0, +1$)-valued sequences that depend on d_k and d_{k-1} such that $u_k = 0$ when $v_k = \pm 1$ and vice versa. Indicate your choice of θ.

10.7 In addition to coherent detection, MSK can be *differentially* detected as illustrated in Fig. P10.7. Assume that $n(t)$ is white Gaussian noise with PSD $N_0/2$, and $s(t)$ is the usual MSK input, that is

$$s(t) = A \cos(\omega_c t + \theta(t)); \quad \theta(t) = d_k \frac{\pi t}{2T} + x_k; \quad kT \le t \le (k+1)T$$

The MSK signal plus noise is multiplied by itself delayed one bit and phase shifted 90°. The resulting product is passed through a lowpass filter $H(f)$ which, for simplicity, we shall assume does nothing more than remove second harmonics of the carrier frequency terms, that is, assume that $H(f)$ is wide enough not to introduce any distortion into the signal $s(t)$. Also assume that the carrier frequency and data rate are related such that $\omega_c T = 2k\pi$ (k an integer).

(a) Write expressions for $\Delta\theta(t) = \theta(t) - \theta(t - T)$ and the statistical mean of $y(t)$ in terms of the appropriate data bits $\{d_k\}$ that are valid over the interval $kT \le t \le (k+1)T$.

(b) Show that a single sample of $y(t)$ taken at $t = (k+1)T$ yields a *suitable* decision variable

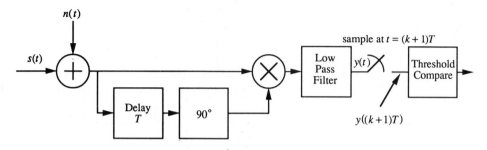

Figure P10.7

for detecting d_k, that is, express the statistical mean of $y((k+1)T)$ in terms of d_k and give its values for $d_k = -1$ and $d_k = 1$.

(c) Consider an input data sequence $d_0 = -1$, $d_1 = 1$, $d_2 = 1$, $d_3 = -1$, $d_4 = -1$, $d_5 = -1$, $d_6 = 1$, and $d_7 = -1$. Letting $p(t)$ denote a unit rectangular pulse of width T, and *assuming that noise is absent*, sketch, over the interval $0 \le t \le 8T$, the following waveforms:

1. $d(t) = \sum_{k=-\infty}^{\infty} d_k p(t - kT)$ (the input pulse stream)

2. $\theta(t)$

3. $\theta(t - T)$

4. $\Delta\theta(t) = \theta(t) - \theta(t - T)$

5. $\sin \Delta\theta(t)$

6. $\hat{d}(t - T) = \sum_{k=-\infty}^{\infty} y((k+1)T) \, p(t - (k+1)T)$ [the sample and held values of $y(t)$].

You may assume $A = 1$ for the purpose of illustration. Assume an initial phase value $x_0 = 0$. Compare $d(t)$ and $\hat{d}(t - T)$ and justify your previous results.

10.8 Consider an offset QPSK signal with half sinusoidal pulse shaping, that is, precoded MSK. The conventional receiver for this modulation (see Fig. 10.28 without the differential encoder) uses I and Q matched filters that are implemented as follows. The I and Q demodulated signals are multiplied by the half sinusoidal pulse shape $p(t) = \sin \pi t/2T_b$ and then passed through I&D filters spanning the appropriate T_s-sec time intervals. An approximation to the above matched filter design that avoids the multiplication by $p(t)$ is a *windowed* I&D filter, namely, one whose integration time is shortened by an amount ε at the beginning and at the end of the integration interval. In particular, with reference to Fig. 10.28, the multiplications by $p(t)$ in the I and Q channels are removed and the remaining I&D filters have integration intervals respectively given by $((2n - 1)T_b + \varepsilon, (2n + 1)T_b - \varepsilon)$ and $(2nT_b + \varepsilon, (2n + 2)T_b - \varepsilon)$.

(a) Evaluate the error probability of the resulting (suboptimum) receiver in terms of the *relative* gate parameter ε/T_b.

(b) Find the value of ε/T_b that minimizes the error probability of part (a).

10.9 Redo part (a) of Problem 10.8 for the case where the I and Q carrier reference signals are not ideal and are given as in Problem 10.1. Express your answer in terms of the carrier phase error, ϕ, and ε/T_b.

10.10 An MFMSK signal can be viewed as an MSK signal with carrier frequency which may vary from symbol to symbol and can take on one of N equiprobable values. Consequently, the PSD of the MFMSK signal can be derived as the average of the PSD's of N MSK signals with carrier frequencies $\omega_i = \omega_c + [2i - (N + 1)]\pi/T_s$, $i = 1, 2, \ldots, N$. Starting with the equivalent lowpass PSD for an MSK signal as given by (2.148), derive the equivalent lowpass PSD for an MFMSK signal in a form analogous to (10.178), namely

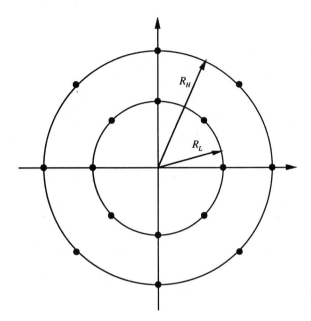

Figure P10.12

$$S_s(f) = \frac{PT_s}{2N} \sum_{i=0}^{N-1} \left[\text{sinc} \left(fT_s + i - \frac{N}{2} \right) + \text{sinc} \left(fT_s + i + 1 - \frac{N}{2} \right) \right]^2$$

$$= \frac{(2 + \log_2 N)}{2N} PT_b \sum_{i=0}^{N-1} \left[\text{sinc} \left((2 + \log_2 N) fT_b + i - \frac{N}{2} \right) \right.$$

$$\left. + \text{sinc} \left((2 + \log_2 N) fT_b + i + 1 - \frac{N}{2} \right) \right]^2$$

10.11 For MAMSK modulation with $M = 16$, show the equivalence between the I-Q representation of (10.185) and the multiple MSK representation of (10.184).

10.12 Consider a 16-signal amplitude-phase-shift-keyed (APSK) modulation whose signal point constellation is composed of two circles (one having radius R_L and one having radius R_H) each containing eight uniformly spaced points (see Fig. P10.12). The phase angles associated with the points on each circle are identical, that is, for each of the eight possible phase angles, $\beta_i = (i - 1)\pi/4$, $i = 1, 2, \ldots, 8$, the signal can be at one of two amplitude levels. As such, it is possible, in principle, to *independently* detect the amplitude and phase of the modulation.

It is desired to detect this modulation differentially both in amplitude and phase, hence the acronym 16DAPSK (sometimes called 16 star QAM). Assume an ISI-free environment, for example, Nyquist filtering equally shared between transmitter and receiver. Then the complex samples of the receive filter output (taken once every T_s sec) are given by $\tilde{r}_k = \tilde{s}_k e^{j\theta} + \tilde{n}_k$ where $\tilde{s}_k = \sqrt{2P} a_k e^{j\theta_k} \triangleq \sqrt{2P} \tilde{S}_k$ is the transmitted amplitude-phase signal in the kth signaling interval, θ is the unknown channel phase, and $\tilde{n}_k$ is the AWGN sample with variance σ_n^2. Writing $\tilde{s}_k$ in differential form, namely $\tilde{s}_k = \tilde{s}_{k-1} \left(\frac{a_k}{a_{k-1}} \right) \exp\{j (\theta_k - \theta_{k-1})\} = \tilde{s}_{k-1} \Delta a_k \exp(j \Delta\phi_k)$, then the differential phase $\Delta\phi_k$ ranges over the set of values $\{\beta_i = (i - 1)\pi/4, \quad i = 1, 2, \ldots, 8\}$ and the differential amplitude, Δa_k, ranges over the set of values $\{\gamma^{-1}, 1, \gamma\}$ where $\gamma = R_H/R_L > 1$ denotes the ratio of the circle radii.

(a) Show that if $P = \frac{1}{16} \sum\limits_{k=1}^{16} \frac{1}{2} |\tilde{s}_k|^2$ denotes the average power of the signal set, then the radii of the circle must satisfy the relations

$$R_L = \sqrt{\frac{2}{1+\gamma^2}}, \quad R_H = \sqrt{\frac{2\gamma^2}{1+\gamma^2}}$$

(b) Show that the conditional pdf of $\tilde{r}_k$ given $\Delta a_k, \Delta\phi_k, \tilde{S}_{k-1}, \theta$ is expressed by

$$f\left(\tilde{r}_k \,\middle|\, \Delta a_k, \Delta\phi_k, \tilde{S}_{k-1}, \theta\right) = \frac{1}{2\pi\sigma_n^2} \exp\left\{ -\frac{\left| \tilde{r}_k - \Delta a_k e^{j\Delta\phi_k}\left(\sqrt{2P}\tilde{S}_{k-1}e^{j\theta}\right)\right|^2}{2\sigma_n^2} \right\}$$

(c) It would be desirable to detect the information pair $\Delta a_k, \Delta\phi_k$ using maximum-likelihood detection, namely, select that particular pair of values, $\Delta\hat{a}_k, \Delta\hat{\phi}_k$, that maximize the pdf in part (a). However, $\tilde{S}_{k-1}$ and θ are both unknown. Thus, to obtain an implementable but suboptimum solution, the term $\sqrt{2P}\tilde{S}_{k-1}e^{j\theta}$ in the argument of the exponential is first replaced by an estimate, $\hat{\tilde{r}}_{k-1}$ (see Problem 10.13 for the characterization of the estimate) and then the maximization of the pdf is performed. With this replacement, show that the joint decision can be decomposed as follows:

$$\Delta\hat{\phi}_k = \max\limits_{\Delta\phi_k}{}^{-1} \operatorname{Re}\left\{ \tilde{r}_k \hat{\tilde{r}}_{k-1}^* e^{-j\Delta\phi_k} \right\}$$

$$\Delta\hat{a}_k = \min\limits_{\Delta a_k}{}^{-1} \left| \Delta a_k - \operatorname{Re}\left\{ \frac{\tilde{r}_k \hat{\tilde{r}}_{k-1}^*}{\left|\hat{\tilde{r}}_{k-1}\right|^2} e^{-j\Delta\hat{\phi}_k} \right\} \right|$$

Note that the decision rule for $\Delta\hat{\phi}_k$ follows the form of that corresponding to pure M-DPSK, whereas the decision rule for $\Delta\hat{a}_k$ depends on the decision $\Delta\hat{\phi}_k$.

10.13 In Problem 10.12 the unknown received signal in the $k - 1^{\text{st}}$ interval, $\tilde{s}_k e^{j\theta} = \sqrt{2P}\tilde{S}_k e^{j\theta}$, is replaced by an estimate, $\hat{\tilde{r}}_{k-1}$, before maximum-likelihood detection of Δa_k and $\Delta\phi_k$ is performed. One possibility for $\hat{\tilde{r}}_{k-1}$ is the maximum-likelihood estimate of $\tilde{s}_k e^{j\theta}$ using the set of L past samples $\tilde{r}_{k-1}, \tilde{r}_{k-2}, \ldots, \tilde{r}_{k-L}$ wherein the past decisions $\Delta\hat{\phi}_{k-l}, \Delta\hat{a}_{k-l}, \; l = 1, 2, \ldots, L$ arrived at by the decision rules in Problem 10.12 are substituted for the true past values $\Delta\phi_{k-l}, \Delta a_{k-l}, \; l = 1, 2, \ldots, L$.

(a) Show that $\tilde{S}_{k-l}$ and $\tilde{S}_{k-1}$ are related by

$$\tilde{S}_{k-l} = \tilde{S}_{k-1} \prod\limits_{i=1}^{l-1} \Delta a_{k-i}^{-1} e^{-j\Delta\phi_{k-i}}$$

(b) Using the relation in part (a), and defining the past received complex sample vector $\tilde{\mathbf{r}} \triangleq (\tilde{r}_{k-1}, \tilde{r}_{k-2}, \ldots, \tilde{r}_{k-L})$, show that the joint conditional pdf of $\tilde{\mathbf{r}}$ given $\Delta\mathbf{a}, \Delta\boldsymbol{\phi}_k, \tilde{S}_{k-1}, \theta$ is expressed by

$$f\left(\tilde{\mathbf{r}} \,\middle|\, \Delta\mathbf{a}, \Delta\boldsymbol{\phi}_k, \tilde{S}_{k-1}, \theta\right) = \frac{1}{\left(2\pi\sigma_n^2\right)^L}$$

$$\times \exp\left\{ -\frac{1}{2\sigma_n^2} \sum\limits_{l=1}^{L} \left| \tilde{r}_{k-l} - \sqrt{2P}\tilde{S}_{k-1} e^{j\theta} \prod\limits_{i=1}^{l-1} \Delta a_{k-i}^{-1} e^{-j\Delta\phi_{k-i}} \right|^2 \right\}$$

(c) Substituting $\Delta\hat{\phi}_{k-l}, \Delta\hat{a}_{k-l}, \; l = 1, 2, \ldots, L$ for $\Delta\phi_{k-l}, \Delta a_{k-l}, \; l = 1, 2, \ldots, L$ in the pdf of part (b), show that the maximum-likelihood estimate of $\tilde{s}_{k-1}e^{j\theta} = \sqrt{2PS}_{k-1}e^{j\theta}$, that is, the value that maximizes the joint pdf is given by

$$\hat{\tilde{r}}_{k-1} = \text{ML estimate of } \tilde{s}_{k-1}e^{j\theta} = \frac{\sum\limits_{l=1}^{L} \tilde{r}_{k-l} \prod\limits_{i=1}^{l-1} \Delta\hat{a}_{k-i}^{-1} e^{j\Delta\hat{\phi}_{k-i}}}{\sum\limits_{l=1}^{L} \prod\limits_{i=1}^{l-1} \Delta\hat{a}_{k-i}^{-2}}$$

Upon substitution of this result in the decision rules of part (c) of Problem 10.12, one obtains a suboptimum detection scheme for 16DAPSK based on decision feedback.

10.14 An approximation (asymptotically tight for high SNR) to the average symbol error probability performance of an M-ary two-dimensional (amplitude-phase) modulation scheme transmitted over an AWGN channel is given as

$$P_s(E) \cong \frac{1}{M} \sum_{i=0}^{M-1} N_{i_{\max}} P_{ji_{\max}}$$

where $P_{ji_{\max}}$ denotes the maximum of the pairwise error probabilities between signals $s_i(t)$ and $s_j(t)$, $j = 0, 1, 2, \ldots, M-1$, $i \neq j$ which is expressed in terms of the phasor representation of these signals by

$$P_{ji_{\max}} = \frac{1}{2} \text{erfc} \left(\sqrt{\frac{E_s}{N_0}} \left[\frac{\min\limits_{i,j} |s_i - s_j|}{2\sqrt{\frac{1}{M} \sum\limits_{i=0}^{M-1} |s_i|^2}} \right] \right)$$

E_s is the average symbol energy, and $N_{i_{\max}}$ is the number of signals that result in $P_{ji_{\max}}$ for each value of i, that is, the number of signals at the minimum distance from signal point i.

Evaluate the approximate average symbol error probability for the 8-ary signal constellations illustrated in Fig. P10.14.

10.15 A quadrature *overlapped* raised cosine (QORC) signaling scheme is illustrated and defined as shown in Fig. P10.15 where $p_1(t)$ is a unit rectangular pulse of duration T_b and $p_2(t)$ is a unit rectangular pulse of duration $2T_b$. The output can be expressed as:

$$s(t) = A \left[\left(\sum_{n=-\infty}^{\infty} a_n p(t - 2nT_b) \right) \cos \omega_c t - \left(\sum_{n=-\infty}^{\infty} b_n p(t - 2nT_b) \right) \sin \omega_c t \right]$$

with $p(t)$ a raised cosine pulse, that is

$$p(t) = \begin{cases} \frac{1}{2}\left(1 - \cos \frac{\pi t}{2T_b}\right) = \sin^2 \frac{\pi t}{4T_b}; & 0 \leq t \leq 4T_b \\ 0; & \text{otherwise} \end{cases}$$

Note that the transmission rate on the in-phase and quadrature channels is $1/2T_b$ but the pulse width is $4T_b$, that is, the adjacent pulses overlap in time; hence the name of the modulation QORC.

(a) If we consider the pulse shape $p(t)$ as being generated by passing a rectangular unit pulse

$$p_2(t) = \begin{cases} 1; & 0 \leq t \leq 2T_b \\ 0; & \text{otherwise} \end{cases}$$

through a filter $H(f)$, find the impulse response $h(t) = \mathcal{F}\{H(f)\}$ of the filter.

(b) Show that the average power P of $s(t)$ is related to the amplitude A by $P = 3A^2/4$.

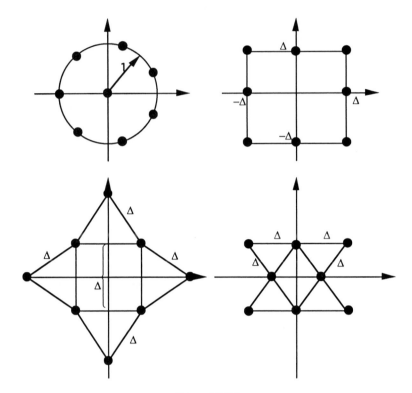

Figure P10.14

(c) From your result of part (a), which choice, when filled in the blank space, correctly completes the following sentence: A raised cosine pulse of duration $4T_b$ is obtained by convolving a square pulse of duration $2T_b$ with the pulse shape of _____ modulation.

      (a) PSK   (b) QPSK   (c) MSK   (d) OQPSK   (e) SFSK

(d) Using the results of parts (a) and (c), derive the two-sided equivalent low-pass PSD $S_s(f)$ of $s(t)$ as given by Eq. (10.183).

(e) If the modulation is now changed to an offset QORC scheme, which is now characterized by

$$s(t) = A\left[\left(\sum_{n=-\infty}^{\infty} a_n p\,(t - 2nT_b)\right)\cos\omega_c t - \left(\sum_{n=-\infty}^{\infty} b_n p\,(t - 2(n+1)T_b)\right)\sin\omega_c t\right]$$

how does the result of part (d) change if at all?

10.16 A generalized modulation scheme that includes MSK and offset QPSK as special cases, and has the practical advantage of not requiring any RF filtering, is described as follows. Consider a modulator in the form of a sequence transducer followed by an RF carrier selector as illustrated in Fig. P10.16a. The sequence transducer (see Fig. P10.16b) converts the binary (±1) random data sequence $\{d_i\}$ into a pair of ternary $(0, \pm1)$ sequences $\{a_{Ii}\}$ and $\{a_{Qi}\}$ in accordance with the relations

$$a_{Ii} = \frac{1}{2}\,(d_i + d_{i-1}), \quad a_{Qi} = \frac{1}{2}(-1)^{i+1}\,(d_i - d_{i-1})$$

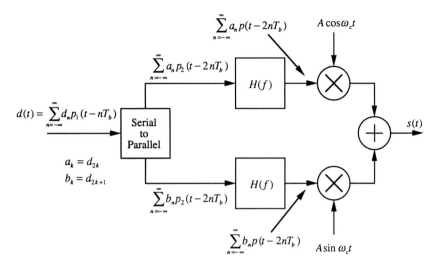

Figure P10.15

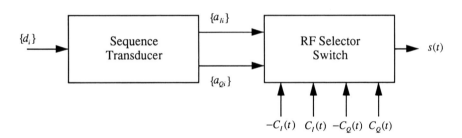

Figure P10.16a

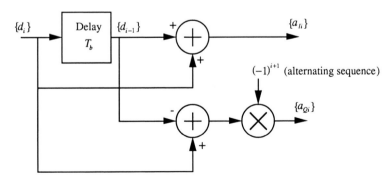

Figure P10.16b

(a) Show that for each i, either $a_{Ii} = 0$ or $a_{Qi} = 0$, but not both; that is, in each bit interval, one of the two transducer outputs is zero while the other takes on value $+1$ or -1.

(b) Show that the transducer illustrated in Fig. P10.16c, whose outputs satisfy the relations

$$a_{Ii} = \frac{1}{2}(d_i + d_{i-1}), \quad a_{Qi} = \frac{1}{2}(d'_i + d'_{i-1}), \quad d'_i = (-1)^{i+1} d_i$$

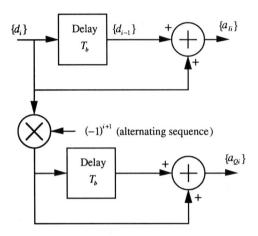

Figure P10.16c

and thus have identical statistics, is equivalent to the transducer shown in Fig. 10.16b.

The inputs $C_I(t)$ and $C_Q(t)$ to the RF selector are carrier waveforms that are assigned to the modulator output $s(t)$ in accordance with the following rule:

$$s(t) = \begin{cases} C_I(t) & \text{if } a_{Ii} = 1 \\ -C_I(t) & \text{if } a_{Ii} = -1 \\ C_Q(t) & \text{if } a_{Qi} = 1 \\ -C_Q(t) & \text{if } a_{Qi} = -1 \end{cases}, \quad iT_b \le t \le (i+1)T_b$$

If the carrier waveforms are selected as

$$C_I(t) = \sqrt{2P} \sin\left[\left(\omega_c + \frac{\pi}{2T_b}\right)t + \theta\right], \quad C_Q(t) = \sqrt{2P} \sin\left[\left(\omega_c - \frac{\pi}{2T_b}\right)t + \theta\right]$$

(c) Show that the phase of $s(t)$ is continuous at the potential carrier switching transitions, that is, at times $t = iT_b$ for all i and for all choices of the data sequence $\{d_i\}$. Consider even and odd values of i separately.

(d) With the choice of carrier waveforms as in part (c), show that the modulator realizes differentially encoded MSK, that is, show that the output signal, $s(t)$, is equivalent to what one would get from an MSK modulator preceded by a differential encoder.

10.17 For the modulator of Problem 10.16, assume that the carrier waveforms are selected as

$$C_I(t) = \sqrt{2P} \cos(\omega_c t + \theta), \quad C_Q(t) = \sqrt{2P} \sin(\omega_c t + \theta)$$

(a) Show that the phase of $s(t)$ can never have a transition of π radians. Equivalently, show that $s(t)$ can never make a transition from $C_I(t)$ to $-C_I(t)$, or $C_Q(t)$ to $-C_Q(t)$.

(b) Show that the modulator realizes differentially encoded offset QPSK, that is, show that the output signal, $s(t)$, is equivalent to what one would get from an offset QPSK modulator preceded by a differential encoder.

REFERENCES

1. Simon, M. K., "Error Probability Performance of Unbalanced QPSK Receivers," *IEEE Transactions on Communications*, vol. COM-26, no. 9, September 1978, pp. 1390–97.

2. Weber, W. J., III, "Differential Encoding for Multiple Amplitude and Phase Shift Keying Systems," *IEEE Transactions on Communications*, vol. COM-26, no. 3, March 1978, pp. 385–91.

3. Simon, M. K. and J. G. Smith, "Carrier Synchronization and Detection of QASK Signal Sets," *IEEE Transactions on Communications*, vol. COM-22, no. 2, February 1974, pp. 98–106.

4. Simon, M. K., G. K. Huth and A. Polydoros, "Differentially Coherent Detection of QASK for Frequency-Hopping Systems—Part I: Performance in the Presence of a Gaussian Noise Environment," *IEEE Transactions on Communications*, vol. COM-30, no. 1, January 1982, pp. 158–164.

5. Simon, M. K. and A. Polydoros, "Coherent Detection of Frequency-Hopped Quadrature Modulations in the Presence of Jamming—Part 1: QPSK and QASK Modulations," *IEEE Transactions on Communications*, vol. COM-29. no. 11, November 1981, pp. 1644–60.

6. Divsalar, D. and M. K. Simon, "Maximum-Likelihood Differential Detection of Uncoded and Trellis-Coded Amplitude-Phase Modulation over AWGN and Fading Channels—Metrics and Performance," *IEEE Transactions on Communications*, vol. 42, no. 1, January 1994, pp. 76–89.

7. Fleisher, S. and S. Qu, "Quadrature Frequency/Phase Modulation," submitted for publication in the *IEEE Transactions on Communications*. Also see S. Fleisher, S. Qu and S. Periyalwar, "Quadrature Biorthogonal Modulation," Canadian Conference on Electrical and Computer Engineering, September 25–27, 1992, Quebec, Canada, pp. 61.2.1–61.2.4.

8. Saha, D., "Quadrature-Quadrature Phase Shift Keying," U.S. Patent 4,680,777, July 1987.

9. Saha, D. and T. G. Birdsall, "Quadrature-Quadrature Phase Shift Keying," *IEEE Transactions on Communications*, vol. 37, no. 5, May 1989, pp. 437–48.

10. Pelchat, M. G., R. C. Davis and M. B. Luntz, "Coherent Demodulation of Continuous Phase Binary FSK Signals," *Proceedings of the International Telemetry Conference*, Washington, D.C., 1971.

11. Anderson, J. B., T. Aulin and C.-E. Sundberg, *Digital Phase Modulation*, New York and London: Plenum Press, 1986.

12. Osborne, W. P. and M. B. Luntz, "Coherent and Noncoherent Detection of CPFSK," *IEEE Transactions on Communications*, vol. COM-22, no. 8, August 1974, pp. 1023–36.

13. Schonhoff, T. A, "Symbol Error Probabilities for M-ary CPFSK: Coherent and Noncoherent Detection," *IEEE Transactions on Communications*, vol. COM-24, no. 6, June 1976, pp. 644–52.

14. Proakis, J., *Digital Communications*, New York: McGraw-Hill, 1983.

15. Simon, M. K. and D. Divsalar, "Maximum-Likelihood Sequence Detection of Noncoherent Continuous Phase Modulation," *IEEE Transactions on Communications*, vol. 41, no. 1, January 1993, pp. 90–98.

16. Viterbi, A. J. and J. K. Omura, *Principles of Digital Communications and Coding*, New York: McGraw-Hill, 1979.

17. Amoroso, F., "Pulse and Spectrum Manipulation in the Minimum (Frequency) Shift Keying (MSK) Format," *IEEE Transactions on Communications*, vol. COM-24, no. 3, March 1976, pp. 381–84.

18. Fleisher, S. and S. Qu, "Multifrequency Minimum Shift Keying," *IEEE Journal on Selected Areas in Communications*, vol. SAC-10, no. 8, October 1992, pp. 243–53. Also see S. Qu and S. Fleisher, "Continuous Phase Frequency/Phase Modulation," AIAA 14th International Communications Satellite Systems Conference, Washington, D.C., March 22–24, 1992, pp. 500–06.

19. Austin, M. C. and M. U. Chang, "Quadrature Overlapped Raised-Cosine Modulation," *ICC'80 Conference Record*, Seattle, WA, June 1980, pp. 26.7.1–26.7.5.

20. Simon, M. K., J. K. Omura and D. Divsalar, "Performance of Staggered Quadrature Modulations over Nonlinear Satellite Channels with Uplink Noise and Intersymbol Interference," *GLOBECOM'82 Conference Record*, Miami, FL, December 1982, pp. A7.1.1–A7.1.8.

21. Baker, P. A., "Phase-Modulation Data Sets for Serial Transmission at 2000 and 2400 bits per Second, Part I," *AIEE Transactions on Communication and Electronics*, no. 61, July 1962, pp. 166-71.

22. Weber, W. J., III, P. H. Stanton and J. Sumida, "A Bandwidth Compressive Modulation System Using Multi-Amplitude Minimum Shift Keying (MAMSK)," *NTC'77 Conference Record*, Los Angeles, CA, December 1977, pp. 05:5-1–05:5-7.

23. Simon, M. K., "An MSK Approach to Offset QASK," *IEEE Transactions on Communications*, vol. COM-24, no. 8, August 1976, pp. 921–23.

Chapter 11

Coded Digital Communications

In previous chapters, we studied the reliability and communications efficiency of various M-ary modulation-demodulation and signal detection techniques. For the most part, the techniques that have been expounded upon are rather exhaustive of those that are of practical interest. In particular, in Chapters 3–8, we studied communications reliability and efficiency without a channel bandwidth constraint. On the other hand, Chapters 9 and 10 consider communication via channels where constraints are placed on bandwidth occupancy and information throughput is maximized. In other words, *spectrum utilization efficiency* was considered to be a design constraint. The question naturally arises as to what else can be done to improve communications reliability and efficiency. What is the ultimate limit, the so-called "thermodynamic limit?" This question can be answered by introducing the information coding applica at the transmitter and decoding of information applica at the receiver. As we shall see, a coded[1] digital communication system is a good example of a complex system.

This chapter is organized so as to introduce the systems engineer to the notion of *coded communications, top-down system engineering*, and *design*. We begin by introducing channel coding and decoding using the level 2 and 3 system architectures (see Figs. 1.3, 1.6, and 1.7). This has the advantage that all signal processing functions in the signal path are combined (levels 4 through 8) into the *baseband channel function* (see Figs. 1.3, 1.6, and 1.7). A clear picture of the coded information flow can be presented to the reader from this system perspective. Next, we introduce the need for a *communications carrier* which requires that the *modulation* and *demodulation* functions be introduced into the coded system architecture; hence, the level 5 system architecture, seen in Figs. 1.6 and 1.7, must be considered. From an analysis perspective, this function differs from that in an uncoded system in that it is mathematically convenient to combine the *signal generation-modulation functions* and the *signal detection-demodulation functions* into the *coded-modulation* and *coded-demodulation function*. At level 5, the IF channel model includes channel features realized in hardware as well as the propagation media. This IF channel model is time and

---

1. Did you know? The first code to be used in digital communications was the Morse Code. The Morse Code System lasted as the primary communication code until the end of the nineteenth century. The insightful Samuel B. Morse used what is called a "variable-length code"; however, the "variable-length bit pattern" of the Morse Code was a problem to interpret so a new code was developed. This was done by the brilliant Frenchman Jean Maurice Baudot. Baudot's code was based on five bits of information with each alphabet character represented by combinations of binary digits "1" and "0"; there are 32 such combinations (codewords). It is from Baudot's name that the speed (information rate) of information transmission called the baud rate evolved. It is frequently used in computer communications. Donald A. Huffman was apparently the first to qualitatively investigate variable-length codes (see *Proc. IRE*, 40, 1952).

amplitude continuous, but the demodulator output is continuous in amplitude at discrete points in time. Since decoders are realized in digital hardware, some form of *quantization* of the demodulator output samples is required. This process is characterized in terms of the *quantization levels* and a *quantization interval* for use in the chapters that follow. The quantization converts the *continuous IF channel model* into an equivalent *discrete IF channel model*.

Next, the question of *channel memory* is addressed. To mitigate the deleterious effects that channel memory has on *communications efficiency*, the concept of an *interleaver-deinterleaver* system is introduced; hence, the notion of interleaved coding for burst noise channels is introduced. From this, we are then able to understand the notions of a *continuous memoryless* and a *discrete memoryless channel* (CMC and DMC). Using the approach of top-down engineering, the systems engineer thereby gains the insight needed to design an error control system for use in a coded digital communication system.

We now proceed by evaluating the reliability and efficiency of transmitting information in packets of K_c bits; bandwidth is not constrained. This allows us to introduce the notion of sequential message transmission (bit-by-bit) with the passage of time; hence, the requirement for a stable clock to mark the boundaries of the information bits at both the transmitting and receiving end.

11.1 INFORMATION TRANSFER USING BIT-BY-BIT BINARY SIGNALING

Previous chapters have dealt with the problem of transmitting a single message from the M_c-ary alphabet[2] $(m_0, m_1, \cdots, m_{M_c-1})$. This has been dubbed the "one-shot" approach to information transmission in communication theory. In practice, information is generally transmitted serially in time so systems that communicate a sequence of messages—one right after the other—are of practical interest.

Consider now the problem of transmitting a message from the M_c-ary alphabet by transmitting bit-by-bit sequentially in time. We need $K_c = \log_2 M_c$ one-bit transmissions for each message. Let T_b denote the transmission time per information bit. Then, the transmission time per message is $T = K_c T_b$ seconds. The information *data rate* can be defined as

$$\mathcal{R}_b \triangleq \frac{1}{T_b} = \frac{\log_2 M_c}{T} = \frac{K_c}{T} \text{ bits/sec} \tag{11.1}$$

Now the K_c-bit binary sequence which represents message m_m can be represented by the "carrier waveform"

$$s_m(t) = \sum_{k=0}^{K_c-1} p_i(t - kT_b); \quad m = 0, \cdots, M_c - 1 \tag{11.2}$$

where i is a random variable taking on values "0" and "1" with equal probability and $p_i(t), i = 0, 1$, are the corresponding pulse waveforms. When viewed as vectors in a space of K_c dimensions, the M_c points represent the vertices of a hypercube. Assuming the use of the most efficient modulation-demodulation and detection system (BPSK) operating in the

---

2. For coded digital communications, we use M_c as the number of messages to be coded and reserve M for the number of distinct waveforms the modulator generates. This is consistent with the notation used in the first ten chapters.

face of no bandwidth constraint, we have then that $p_0(t) = -p_1(t) = \sqrt{2E_b/T_b}\sin\omega_c t$ for $0 \le t \le T_b$ and zero elsewhere. With equal energy signals given by

$$E_b = PT_b = \int_0^{T_b} s_m^2(t)dt = \int_0^{T_b} p_i^2(t)dt \text{ joules/bit} \tag{11.3}$$

we know from Chapter 4 that the bit error probability is given by

$$P_b(E) = Q\left(\sqrt{\frac{2E_b}{N_0}}\right) = Q\left(\sqrt{\frac{2P}{N_0\mathcal{R}_b}}\right) \triangleq p_b \tag{11.4}$$

where we have assumed the AWGN channel and P is the average received power. Since the errors in the K_c-bit message occur independently, then the probability of correct symbol or message detection is

$$P_s(C) = (1 - p_b)^{K_c} = (1 - p_b)^{\mathcal{R}_b T} \tag{11.5}$$

Therefore, the *message or symbol error probability* (MEP) is

$$P_s(E) = 1 - (1 - p_b)^{K_c} \tag{11.6}$$

Since $p_b < 1/2$ for all values of E_b/N_0, then from (11.4) and (11.6), we have

$$\lim_{K_c \to \infty} P_s(E) = \lim_{T \to \infty} P_s(E) \to 1 \tag{11.7}$$

which says that the MEP tends to one as the amount of information transmitted (K_c bits) becomes large. The reason for this behavior is that the Euclidean distance $\sqrt{2E_b}$ between nearest neighbors remains fixed as K_c increases, whereas the number of nearest neighbors and the number of dimensions occupied by the signal set increase linearly with K_c. The probability that at least one of the K_c relevant noise components will carry the received vector closer to a neighbor than the transmitted signal vector becomes large as K_c increases, that is, there are K_c chances for this to happen. On the other hand for a fixed T and N_0, the MEP can be made small by either decreasing the data rate $\mathcal{R}_b$ or by increasing the *received power* of P Watts. This fate of an uncoded communication system performance can be alleviated by employing encoding and decoding techniques at the transmitter and receiver respectively; we now investigate methodologies of practical interest.

11.2 CODED DIGITAL COMMUNICATION SYSTEM ARCHITECTURES

Our journey into the application of error-correcting coding and decoding of information as a technique to further improve communications reliability, $P_b(E)$, and efficiency, E_b/N_0, begins by establishing the coding-decoding viewpoint from the system engineering perspective. In 1948, the brilliant and insightful Claude E. Shannon demonstrated, via his celebrated work *Mathematical Theory of Communication* [1], that by proper encoding and decoding of information, errors induced by a noisy channel could be reduced to any desired level, subject to a code rate constraint. Since the appearance of that work, a great deal of effort has been expended on the problem of designing encoding and decoding techniques for error control in a communications channel. Documentation of these efforts has led to the development of the subjects of Information and Coding Theory.

Our treatment of this subject begins by introducing coded communications from the systems perspective; this serves to take advantage of the system architecture presented in

Chapter 1. In particular, level 3 and level 5 coded-system architectures will now be used to further our investigation from a systems point of view (see Fig. 1.6).

11.2.1 Coding-Decoding System Architecture (Level 3)

Figure 11.1 illustrates the level 3 system architecture discussed in Chapter 1; this architecture will lead us to our mathematical model of coded information flow between the *information source* and *information sink* (user). As discussed earlier, we will not treat the problem of source encoding and decoding[3] from an information redundancy perspective, that is, the techniques associated with *data compression* should not be confused with channel coding which is used for error control.

For most practical purposes, the channel encoder (see Figs. 11.1a and 11.1b) is characterized by the three parameters M_c, N_c, and r_c. Here, M_c serves to characterize the number of *codewords* in the *codeword dictionary*. These M_c words are frequently called the *code*. The parameter N_c serves to characterize the *dimensionality* (codeword length) of the code while r_c specifies the *code rate* defined as $r_c = \log_2 M_c/N_c$ bits/dimension. Assuming that the source encoder output is characterized by a Q-ary alphabet and that the messages that are generated from this alphabet contain K_c letters, then we require that $M_c = Q^{K_c}$. This assures that the source encoder and channel encoder[4] are matched with respect to the information to be transmitted. Clearly, each message contains $\log_2 M_c = K_c \log_2 Q$ bits. In the binary case, a case of great practical interest, $Q = 2$ and $M_c = 2^{K_c}$ so there are $K_c = \log_2 M_c$ bits per message. In general, we will consider the class of (M_c, N_c) Q-ary codes and, in particular, the class of (N_c, K_c) *binary codes*.[5] In the binary case, the *code rate* is $r_c = K_c/N_c$ bits/dimension. As we shall later see, r_c^{-1} serves as the parameter which characterizes the *bandwidth expansion* due to coding; both r_c and r_c^{-1} play key roles in what follows.

In the case where we are working with Q-ary codes, we will find it convenient to refer to the codeword elements as *channel or coded symbols*. When we talk about binary codes, we will refer to the N_c elements which make up the codewords as *coded message bits, message bits,* or *codeword symbols* (there seems to be no standard in the literature). Thus, the $\log_2 M_c$ *information bits*, each of duration T_b sec, inputted into the encoder are converted into N_c coded message bits, message bits, or codeword symbols. The exact prescription for this mapping defines the encoding algorithm (the type of code) and constitutes the subject of *Coding Theory*. Later, in Section 11.6, the classification for the error correction and detection coding techniques of greatest practical interest is presented.

When time is introduced, we require that each *codeword duration* T sec satisfy the relationship $T = T_b \log_2 M_c = T_b K_c \log_2 Q$ sec. For a binary code, this means that $T = K_c T_b = N_c T_c$ where T_c is the duration of the coded message bit or code symbol. From this relationship, we see that the message bit duration is related to the bit duration via the code rate, viz., $T_c = K_c T_b/N_c = r_c T_b$ sec. Thus, the *channel symbol rate* $\mathcal{R} = 1/T_c = \mathcal{R}_b/r_c$ is r_c^{-1} times the *data rate* $\mathcal{R}_b$. This is why the inverse code rate r_c^{-1} is referred to as the *bandwidth expansion*. For example, a *binary block encoder* accepts data in blocks of K_c

3. The term *codec* used in Fig. 11.1 was defined in Chapter 1 and refers to the channel coder-decoder pair. Similarly, the term *modem* refers to the channel modulator-demodulator pair.

4. In general, $N_c > K_c$ and the encoder adds redundancy in the data stream for error correction purposes. When $N_c < K_c$, the data is compressed to remove redundancy in the information and thus, the term *data compression*. For $N_c = K_c$, the encoder acts like a scrambler for secure communications.

5. We use two different notations for binary and Q-ary codes to emphasize the different alphabets.

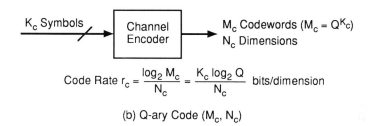

(a) Binary Code (N_c, K_c)

(b) Q-ary Code (M_c, N_c)

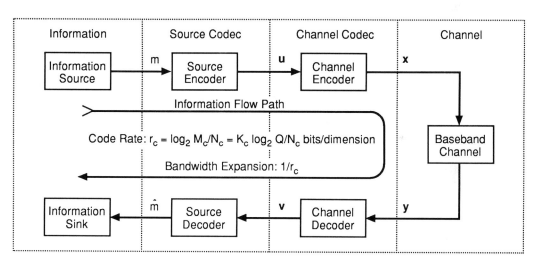

(c) System Functional Architecture (Level 3)

Figure 11.1 Simplified coded-digital communication system functional model: (a) Binary code (N_c, K_c), (b) Q-ary code (M_c, N_c), (c) System functional architecture (level 3)

bits and outputs coded message bits in blocks of N_c coded bits (see Chapter 12). Since *redundancy* $r = N_c - K_c$ bits is added, N_c will be larger than K_c. On the other hand, we shall see in Chapter 13 that *binary convolutional coders* can also be described as providing N_c output coded message bits for each group of K_c input bits. The major difference is that any channel symbol output consisting of N_c coded message bits depends, not only on the last set of K_c input bits, but also on several preceding sets of input bits. Furthermore, with binary convolutional encoders, K_c and N_c are usually much smaller than their block code counterparts (see Chapters 12 and 13). In either case, we will see that low-rate codes have a greater *error-detection-and-correction* potential than high-rate codes; as a consequence, they also require a greater bandwidth expansion.

Refer now to the coded baseband system illustrated in Fig. 11.1. We assume that the *information source* is discrete and outputs a message symbol m from the M_c-ary alphabet $m = 0, 1, \cdots, M_c - 1$. The *source encoder* transforms the information source output into the random sequence $\mathbf{u} = (u_1, u_2, \cdots, u_{K_c})$; in particular, if message symbol m is transmitted then $\mathbf{u} = \mathbf{u}_m = (u_{m1}, u_{m2}, \cdots, u_{mK_c})$. When the source output symbol alphabet is composed of the binary digits u_i="0" or "1", the symbols are generally called *information bits*; this is the case of greatest interest. We will think of these binary digits as corresponding to "+1" volt or "−1" volt.

The *channel encoder*, according to known rules, transforms the source encoder output sequence $\mathbf{u}$ into N_c channel input symbols (coded message bits) denoted by the vector $\mathbf{x} = (x_1, x_2, \cdots, x_{N_c})$; $\mathbf{x}$ is called a *codeword* for $\mathbf{u}$. This transformation, $\mathbf{u} \to \mathbf{x}$, is called *channel encoding*.[6] In particular, if message m is transmitted then $m \to \mathbf{u}_m \to \mathbf{x}_m = (x_{m1}, x_{m2}, \cdots, x_{mN_c})$. When the elements of $\mathbf{x}_m$, say x_{mn}, are drawn from a Q-ary alphabet, then the number of possible codewords in the vector space is Q^{N_c} words; only $M_c = Q^{K_c}$ of these are needed and are included in the codeword dictionary. We also note that each channel input symbol $x_n \in \mathbf{x}$ can depend on the source encoder output symbols $u_i \in \mathbf{u}$ in any desired way. The channel symbols (coded message bits) are denoted by x_{mn} during the known n^{th} symbol time $(n - 1)T_c \leq t \leq nT_c$; $m = 0, \cdots, M_c - 1$; $n = 1, 2, \cdots, N_c$. We assume that the source and channel clocks at the transmitter and receiver are synchronized.

Example 11.1

> Consider a 4-ary message source ($M_c = 4$) whose output is to be encoded using a binary ($Q = 2$) rate 2/5 code ($K_c = \log_2 M_c = 2$, $N_c = 5$). The four possible messages m_0, m_1, m_2, m_3 are respectively represented by the four possible binary sequences 00, 01, 10, 11 which serve as the input to the channel encoder in each T-sec interval. Of the $2^{N_c} = 2^5 = 32$ possible binary sequences of length 5, we choose the following four sequences to act as channel encoder outputs corresponding to the four possible input sequences.

Message	Input Sequence	Output Sequence (Coded Bits)
m_0	00	00000
m_1	01	01010
m_2	10	10100
m_3	11	11111

> This one-to-one mapping between input and output channel encoder sequences defines the code. Note that the above mapping produces a *nonlinear* code, that is, the modulo-2 sum of two output sequences (codewords) does not necessarily generate another possible codeword.

Example 11.2

> Consider a 16-ary message source ($M_c = 16$) whose output is to be encoded using a 4-ary ($Q = 4$) rate 2/3 code ($K_c = \log_Q M_c = 2$, $N_c = 3$). The 16 possible messages $m_0, m_1, m_2, \ldots, m_{15}$ are respectively represented by the 16 possible 4-ary sequences 00, 01, 02, 03, 10, 11, 12, 13, 20, 21, 22, 23, 30, 31, 32, 33 which serve as the input to the channel encoder in each T-sec interval. Of the $Q^{N_c} = 4^3 = 64$ possible 4-ary sequences of length 3, we choose the

---

6. The codes used in the examples in this chapter are for illustration purposes only and are not necessarily "good" codes.

following 16 sequences to act as channel encoder outputs corresponding to the 16 possible input sequences.

Message	Input Sequence	Output Sequence (Coded Bits)
m_0	00	000
m_1	01	010
m_2	02	202
m_3	03	030
m_4	10	102
m_5	11	020
m_6	12	302
m_7	13	110
m_8	20	002
m_9	21	300
m_{10}	22	133
m_{11}	23	210
m_{12}	30	321
m_{13}	31	111
m_{14}	32	322
m_{15}	33	213

Once again the above mapping represents a nonlinear code since the modulo-4 sum of any two output sequences (codewords) does not necessarily result in another codeword.

Based on the coded system architecture discussed thus far, we observe that, as N_c gets large, the decision regarding the message is delayed for $T = N_c T_c = K_c T_b \log_2 Q$ sec. This delay is called *coding delay* and represents a constraint which must be considered by the system engineer. Moreover, since $M_c = Q^{K_c}$ grows exponentially with K_c we would expect the complexity of the encoder and decoder to grow rapidly; this represents another issue to be faced by the system engineer. In practice, there are other issues which affect the system engineer's choice of a particular coding-decoding technique, for example, it is driven by implementation complexity, performance requirements, and cost. We will not be able to address these issues from a theoretical point of view.

Baseband channel characterization.   The physics of the baseband channel is that, for analysis and synthesis purposes, it can be modeled as a CMC. For decoding purposes (using digital hardware), this channel can be converted into a DMC. The baseband channel maps the input vector $\mathbf{x}$ into the output vector (observables) $\mathbf{y} = (y_1, y_2, \cdots, y_{N_c}) \in R^{N_c}$ where R^{N_c} is the N_c-dimensional Euclidean space. In practice, the vector components y_n are continuous random variables.

In general, the continuous baseband channel is characterized by the known transition probability density function $f(\mathbf{y}|\mathbf{x})$; the *continuous memoryless baseband channel* is characterized by

$$f(\mathbf{y}|\mathbf{x}) = \prod_{n=1}^{N_c} f(y_n|x_n). \qquad (11.8)$$

Furthermore for the AWGN channel, if codeword $\mathbf{x}_m$ is transmitted, then $x_n = x_{mn}$ and

$$f(y_n|x_{mn}) = \frac{1}{\sqrt{\pi N_0}} \exp\left[-\frac{(y_n - x_{mn})^2}{N_0}\right] \tag{11.9}$$

for all $n = 1, 2, \cdots, N_c$ and $m = 0, 1, \cdots, M_c - 1$.

Now, the channel output sequence $\mathbf{y} = (y_1, y_2, \cdots, y_{N_c})$ serves as the input to the *channel decoder* whose purpose is to produce the output sequence $\mathbf{v} = (v_1, v_2, \cdots, v_{K_c})$ such that the effects of the channel disturbances are minimized, that is, the number of errors between the output sequence $\mathbf{v}$ and input sequence $\mathbf{u}$ is minimized. The transform $\mathbf{y} \to \mathbf{v}$ is called *channel decoding*. From a system performance perspective, we are interested in *block errors*, that is, errors that occur at the receiver in the transmitted block $(x_1, \cdots, x_{N_c})$. Thus, one can conveniently drop the sequences $\mathbf{u}$ from consideration and regard the channel encoding simply as a mapping from the m integers, $m = 0, \cdots, M_c - 1$, onto the codewords $\mathbf{x}_0, \cdots, \mathbf{x}_{M_c-1}$. If message m enters the source encoder, $\mathbf{x}_m$ is transmitted, and on the basis of the channel output sequence $\mathbf{y}$, the channel decoder outputs $\mathbf{v} = (v_1, \cdots, v_{K_c})$ and the source decoder produces $\hat{m}$. An error occurs if and only if $m \neq \hat{m}$. We will demonstrate the conversion of the CMC into the DMC after we introduce the reader to the Level 5 coded system architecture (see Fig. 1.6) which incorporates a communications carrier. This level of modeling requires careful consideration of the interfaces between the coder and the modulator and between the demodulator and the decoder.

11.2.2 Coded-Modulation and Demodulation-Decoding System Architecture (Level 5)

Level 5 of our information transportation system architecture (see Fig. 1.6), serves as the coded system model when a communications carrier of intermediate frequency (IF) f_c Hz is introduced to transport the information (see Fig. 11.2). Introducing the carrier into the system picture requires the use of a **Modulator** and a **Demodulator** and this requires separate attention. Unlike the uncoded system, for analysis and synthesis purposes only, the *modulator* function in a coded system can be viewed as incorporating the two functions of signal generation and modulation. Similarly the **Demodulator** in a coded system can also be viewed as incorporating the two functions of demodulation and detection (see Fig. 11.3). We adopt this point of view here.[7] In practice, the demodulator projects the received IF signal to baseband; this projection is accomplished by a *carrier synchronization circuit*.

In one simple form, the modulator can transform, via a known technique, the codeword $\mathbf{x}$ from the channel encoder output into the transmitted signal vector $\mathbf{S}_m$ [waveform $S_m(t)$] for presentation to the IF channel. In yet another simple form, the modulator can transform each coded bit or symbol from the codeword into the transmitted waveform. In between, the modulator can transform a group of coded bits or symbols into the transmitted waveform. In general, this process is tedious and is best accomplished in two steps using a signal mapper which we now discuss.

---

7. Demodulation involves four major tasks: (1) synchronization of the local communication carrier with the received carrier component used at the modulator, (2) use of this reference to convert the received modulation to baseband, (3) extraction of channel symbol timing so that the "matched filtered" version of this baseband signal can be sampled and the recovered clock synchronization maintained with the transmit clock, and (4) detection of the demodulated channel symbols and decoding them in accordance with the appropriate decoding algorithm.

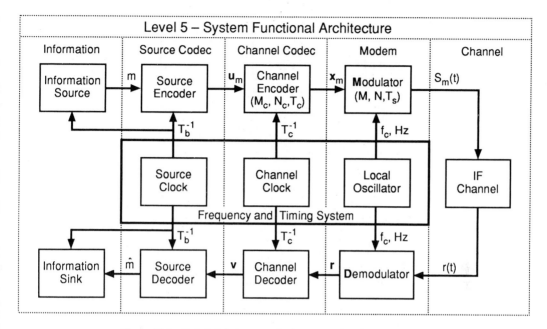

Figure 11.2   Coded-digital communication system functional architecture (level 5)

Modulator characterization.   For our purposes, a digital modulator is characterized by the three parameters M, N, and T_s. Here M represents the number of distinct waveforms in the set $\{s_m(t); m = 0, 1, \cdots, M - 1\}$ which the modulator can generate and T_s represents the time duration of each waveform (signal). We emphasize that the set $\{s_m(t); m = 0, 1, \cdots, M - 1\}$ of modulator waveforms is not in general equal to the set $\{S_m(t); m = 0, 1, \cdots, M_c - 1\}$ of coded modulation waveforms. As demonstrated in Chapter 4, these waveforms can be decomposed into N pieces using the orthonormal expansion

$$s_m(t) = \sum_{n=1}^{N} s_{mn}\varphi_n(t), \quad 0 \le t \le T_s = NT_N \tag{11.10}$$

and zero elsewhere; we will refer to $s_m(t)$ as the transmitted signal when codeword $\mathbf{x}_m$ is to be transmitted if and only if $N = N_c$ and $M = M_c$. When this is not the case, we will consider mapping methods to provide codeword transmission. From the above representation, it is clear that the waveform set $\{s_m(t), m = 0, \cdots, M - 1\}$ can be represented by the N-dimensional vector $\mathbf{s}_m$ of its coefficients, that is, $\mathbf{s}_m = (s_{m1}, s_{m2}, \cdots, s_{mN})$, $m = 0, \cdots, M - 1$. In other words, there exists a one-to-one correspondence between the waveform $s_m(t)$ and the vectors $\mathbf{s}_m$. As we shall see, $s_m(t) = S_m(t)$ if and only if $M = M_c$ and $N = N_c$.

It is convenient to identify $T_N \triangleq T_s/N$ as the time used per dimension by the modulator and $D = 1/T_N$ as the number of dimensions used per second by the modulator. In Chapter 12, we will need to introduce the notion of bandwidth as it relates to signal dimensionality. The interconnect will be made (Chapter 12) via the *Shannon-Nyquist sampling rate* defined as $B = 1/(2T_N)$ Hz; there, we identify dimensionality $D = W = 2B$ Hz as the required *channel bandwidth*.

There are two modulator rate parameters of practical interest. They are defined using

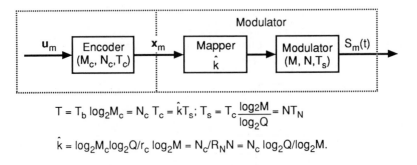

$$T = T_b \log_2 M_c = N_c \, T_c = \hat{k} T_s; \; T_s = T_c \frac{\log_2 M}{\log_2 Q} = NT_N$$

$$\hat{k} = \log_2 M_c \log_2 Q / r_c \log_2 M = N_c / R_N N = N_c \log_2 Q / \log_2 M.$$

(a) Modulator Model for Coded System Design

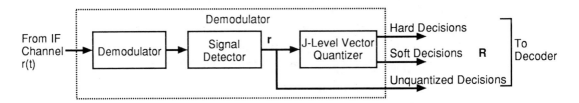

(b) Demodulator Model for Coded System Design

Figure 11.3 Combining transmitter and receiver functions in a coded-digital communication system: (a) Modulator model for coded systen design, (b) Demodulator model for coded system design

the parameters which characterize a digital modulator. These are the *channel rates* defined by $\mathcal{R} \triangleq \log_2 M / T_s$ bits/sec or by $R_N \triangleq \log_2 M / N$ bits/dimension. Thus, the modulator output rate is $\mathcal{R} = R_N / T_N = D R_N$ bits/sec. These two rates will play a key role in what follows. We will be interested in coded digital systems which are implemented using the (M_c, N_c, T_c) code and the (M, N, T_s) modulator. Thus, the modulation/demodulation techniques studied in the earlier chapters are applicable here. In the coded digital communication system architecture of Fig. 1.6, the signal generator and signal detector serve as bridges which respectively connect the channel encoder to the modulator and the demodulator to the decoder. We will describe these bridges next and present combined coding-modulation waveform designs which can support the transmission of binary and multilevel codes using the modulation-demodulation techniques (BPSK, QPSK, M-PSK, M-FSK, QASK, QAM, and so on) presented in Chapters 3 through 10. In this chapter, we deal with rectangular baseband pulses; generalization to arbitrary pulse shapes is straightforward (see Chapter 9).

Finally, the energy in the mth waveform output of the (M, N, T_s) modulator is

$$E_{s_m} = \int_0^{T_s} s_m^2(t) dt \qquad (11.11)$$

for $m = 0, \cdots, M - 1$. For equal energy signals, $E_{s_m} = E_s$ and the energy generated per dimension is $E_N = E_s / N$ joules. Since the time per dimension is $T_N = T_s / N$, then $E_s = P T_s = P N T_N = N E_N$ joules. If the (M, N, T_s) modulator is used $\hat{k}$ times per codeword transmission and we denote this composite waveform by $S_m(t)$, then the energy required per codeword is $E = PT = \hat{k}E_s = \hat{k}PT_s = N_c E_{N_c} = (K_c \log_2 Q)PT_b = (K_c \log_2 Q)E_b$ Joules. These various energy relationships will be useful in what follows.

Example 11.3

Consider mapping the channel encoder output into a ternary ($M = 3$) signal set of dimensionality $N = 2$. Such a signal set can be generated by an M-PSK modulation with normalized signal points located on the perimeter of a unit radius circle at $\theta = 0, 2\pi/3$, and $4\pi/3$ radians. The orthonormal basis functions in (11.10) are defined by $\phi_1(t) = \sqrt{2/T_s} \cos \omega_c t, \quad \phi_2(t) = \sqrt{2/T_s} \sin \omega_c t$ and the coefficients, s_{mn}, for each signal are tabulated below:

Signal	θ	Coefficients	
$s_0(t)$	0	$s_{01} = 1,$	$s_{02} = 0$
$s_1(t)$	$2\pi/3$	$s_{11} = -1/2,$	$s_{12} = \sqrt{3}/2$
$s_2(t)$	$4\pi/3$	$s_{21} = -1/2,$	$s_{22} = -\sqrt{3}/2$

Signal generator mapper characterization.   Clearly, the output of a (M_c, N_c, T_c) channel encoder cannot directly serve as the input to a (M, N, T_s) modulator. The problem is that, in general, the codewords in the code dictionary cannot be mapped one-to-one with the modulator output waveforms $s_m(t)$ because $M_c \neq M$; a matching problem occurs if $M_c > M$. In this sense, we will use the signal generator in a coded digital communication system as a *mapper*, viz., one that maps codewords (vectors) onto the signal vectors or modulator output waveforms; coding theorists frequently call this a mapper. For example, the mapper could be a demultiplexer which partitions the coded message bits into two coded message bit streams. One stream could be used to modulate the I-channel and the other to modulate the Q-channel of the modulator.

We begin our discussion regarding this mapping process by considering the class of (M_c, N_c, T_c) Q-ary codes. In this class, $\log_2 M_c = K_c \log_2 Q$ information bits are mapped into one of the M_c codewords containing N_c coded symbols (or coded bits). Since T_b is the time per information bit, then $K_c \log_2 Q$ bits require $T = (K_c \log_2 Q)T_b$ sec. These bits must be mapped into the N_c coded symbols in the same amount of time, that is, $T = N_c T_c$ sec. At this time, we define also the coded bit duration $T_{cb} = T_c/\log_2 Q$. Note that T_{cb} is not necessarily a physical parameter and differs from the information bit T_b. For purposes of discussion, let us assume that the modulator is capable of producing M points (vectors) in N-dimensional Euclidean space; each point contains $\log_2 M$ bits of information. If $M > M_c$, the modulator produces enough waveforms to map one-to-one with the codewords. The problem then is to select the best subset of waveforms. If $M < M_c$, a problem exists and something must be done to match the modulator output to the channel encoder output. In particular, let us assume that we group the $N_c \log_2 Q$ coded message bits into $\hat{k}$ groups of $\log_2 M$ coded bits such that $N_c \log_2 Q = \hat{k} \log_2 M$. In the special case of binary codes, we group N_c coded bits into $\hat{k}$ groups of $\log_2 M$ coded bits such that $N_c = \hat{k} \log_2 M$. Physically speaking, $\hat{k} = N_c \log_2 Q/\log_2 M = (\log_2 M_c)(\log_2 Q)/r_c \log_2 M = (N_c/N)(\log_2 Q/R_N)$ characterizes the mapper. The parameter $\hat{k}$ is the ratio of the number of coded message bits per codeword to the number of bits per message that the modulator can represent in T_s sec. Then $T = N_c T_c = \hat{k}T_c(\log_2 M/\log_2 Q)$ which means that, in T sec, we must use the modulator $\hat{k}$ times. This says that within T sec, the modulator can produce one of $M^{\hat{k}} \geq M_c$ waveforms (vectors); we only need M_c which the encoder output selects. Since we required that $T = \hat{k}T_s = \hat{k}T_c \log_2 M/\log_2 Q$, then $T_s = T_c \log_2 M/\log_2 Q$. Thus, the *channel rate* $\mathcal{R} = (\log_2 M^{\hat{k}})/T = \log_2 M/T_s = \log_2 Q/T_c = \log_2 Q(\mathcal{R}_b/r_c)$ bits/sec is matched to

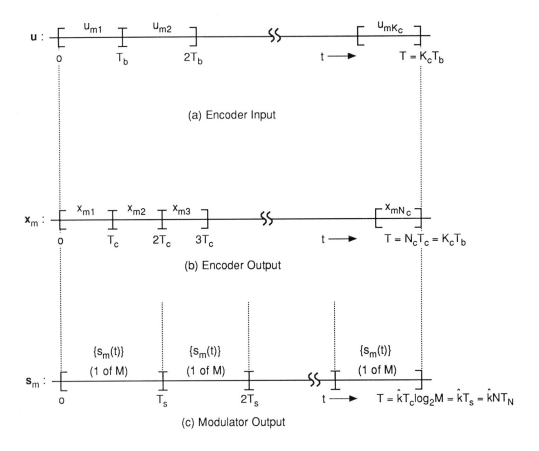

(a) Encoder Input

(b) Encoder Output

(c) Modulator Output

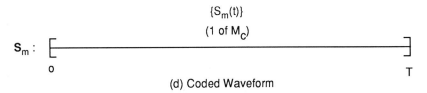

(d) Coded Waveform

Figure 11.4   Encoder-Modulator mapping and timing diagram when $Q = 2$

the information rate as required. In this sense, the (M_c, N_c, T_s) code is mapped onto the (M, N, T) modulator.

More generally, Table 11.1 defines all time intervals, rates, energies, bandwidths, and so on. Table 11.2 summarizes the coder-modulator time-interval and rate relationships, while Fig. 11.4 demonstrates the time scale mapping and summarizes the encoder-modulator mapping for some examples which follow. Some examples will clarify the role of this mapper (see Table 11.2)

Coded BPSK: Consider first the (N_c, K_c) binary code and the $(2, 1)$ BPSK modulator. Here $M = 2$ and $N = 1$ so that $\hat{k} = N_c$, with $T = N_c T_N$ and $T_N = T_c$. In principle the nth coded message bit is used to modulate the basis function $\varphi_n(t) = \sqrt{2/T_N} \cos \omega_c t$, $(n-1)T_N \leq t \leq nT_N$, which is used serially in time for all $n = 1, 2, \cdots, N_c$. In other words, the set of

Table 11.1 Fundamental Parameters of a Coded Digital Communication System

Time intervals	T_c—Time used per codeword dimension, sec T_s—Modulator waveform duration; $T_s = NT_N$ sec T_b—Time per message bit, sec T—Codeword duration; $T = (\log_2 M_c)T_b = \hat{k}T_s = N_cT_c$, $T_c = r_cT_b$, sec T_N—Time used per modulator dimension by modulator, sec

Codec/modem	$\hat{k}$—Number of times modulator is used per codeword transmission
	 $M_c = Q^{K_c}$—Number of codewords in code dictionary
	 K_c—Number of information symbols per codeword (message
	     length)
	 N_c—Number of coded message symbols per codeword
	     (Codeword dimensionality = block length)
	 M—Number of waveforms modulator generates per duration T_s sec
	 N—Dimensionality of modulator; $N = WT_s = WT/\hat{k}$

Rates	$\mathcal{R}_b = (\log_2 M_c)/T = r_c/T_c = 1/T_b$, bits/sec $\mathcal{R} = \log_2 M/T_s = R_N/T_N$; modulator output rate, bits/sec $r_c = \log_2 M_c/N_c$; code rate, bits/dimension $R_N = (\log_2 M)/N$; modulator output rate, bits/dimension R_0—Channel cutoff rate, bits/dimension $\mathcal{R}_0 = R_0 W$—Channel cutoff rate, bits/sec
Energy (joules)	$E = PT$; received energy per codeword $E_s = PT_s$; modulator waveform received energy per use of modulator $E_{N_c} = PT_c$; received energy per codeword dimension $E_b = PT_b$; received energy per information bit $E_N = PT_N$; received energy per modulator dimension used $N_0 = kT^\circ$; noise energy or single-sided PSD $P = E/T$; received power, Watts
Bandwidths (Hz)	$W = N_c/T = 2B$; two-sided IF Nyquist bandwidth $B = 1/2T_c$; single-sided Nyquist bandwidth $D = 1/T_N = N/T_s$ number of dimensions used per sec by modulator $D = W = 2B$ for binary coded BPSK and binary coded QPSK r_c^{-1} and R_N^{-1}—Coder and modulator bandwidth expansion, respectively
Normalized throughputs	$\dfrac{\mathcal{R}}{W} = \dfrac{\log_2 M}{WT_s} = \dfrac{\log_2 M}{NWT_N} = \dfrac{R_N}{WT_N}$ bits/sec/Hz = bits/dimension $\dfrac{\mathcal{R}_b}{W} = r_c$, bits/sec/Hz = bits/dimension
Frequency	f_c - IF channel center frequency, Hz

basis functions are time-orthogonal within the transmission interval if ω_c is chosen to be some multiple of π/T_N. If $\mathbf{x}_m$ is to be transmitted, then the BPSK *coded waveform* model becomes, for $m = 0, 1, \cdots, M_c - 1$

$$S_m(t) = \sqrt{2/T_N} \sum_{n=1}^{\hat{k}} s_{mn} \cos \omega_c t; \quad 0 \le t \le T = N_c T_N \qquad (11.12)$$

Table 11.2 The (M_c, N_C, T_c) Code Mapped into the (M, N, T_s) Modulator at a Bit Rate of $\mathcal{R}_b = 1/T_b$ bits/sec

Parameter	Coder	Modulator
Rate (bits/dim)	$r_c = \log_2 M_c / N_c$ $\quad = K_c \log_2 Q / N_c$ when $M_c = Q^{K_c}$	$R_N = \log_2 M/N$
Time intervals	$T = (\log_2 M_c) T_b$ $\quad = (K_c \log_2 Q) T_b$ $\quad = N_c T_c$ so $T_c = r_c T_b$	$T_s = N T_N = (\log_2 M) T_c / \log_2 Q$ $T_c = N T_N \log_2 Q / \log_2 M$ $\quad = T_N \log_2 Q / R_N$ so $T_N = r_c R_N T_b / \log_2 Q$
Rates (bit/sec)	Code rate: $\mathcal{R} = \log_2 M_c / T$ $\quad = \log_2 M_c / N_c T_c$ $\quad = r_c / T_c = 1/T_b$ $\quad = \mathcal{R}_b$	Channel rate: $\mathcal{R} = \log_2 M / T_s$ $\quad = \log_2 M / N T_N = \log_2 Q / T_c$ or $\mathcal{R} = R_N / T_N \log_2 Q / T_c$ $\quad = (\mathcal{R}_b / r_c) \log_2 Q$
Coder—modulator time interval matching	$T = \hat{k} T_s = \hat{k} T_c (\log_2 M / \log_2 Q) = N_c T_c = (K_c \log_2 Q) T_b$ so $T_s = T_c \log_2 M / \log_2 Q = r_c T_b \log_2 M / \log_2 Q$. Therefore $\hat{k} = \frac{T}{T_s} = \frac{1}{r_c} \left(\frac{\log_2 M_c}{\log_2 M} \right) \log_2 Q = \frac{\log_2 Q}{R_N} \left(\frac{N_c}{N} \right) = \frac{N_c \log_2 Q}{\log_2 M}$	

and $s_{mn} = \sqrt{E_N} x_{mn}$ for $(n-1)T_N \le t \le nT_N$ and zero elsewhere. The parameter E_N represents the energy per dimension. This coded waveform is generated using the (2, 1) BPSK modulator $\hat{k} = N_c$ times. If one now identifies the coded message bits x_{mn} with the voltage levels $+1$ or -1, we have a waveform carrying N_c coded message bits (see Table 11.2 and Fig. 11.4). Figure 11.5(a) demonstrates the coded waveform mapping. Needless to say, binary phase-shift-keying combined with coherent detection can always be used to transport coded information. As we shall see shortly, QPSK can also be used.

Coded QPSK: An alternate approach to transmitting the (N_c, K_c) code is to use the (4, 2) QPSK modulator. Here $M = 4$, $N = 2$ and one uses the modulator $\hat{k} = N_c / \log_2 M = N_c / 2$ times to generate the coded waveform $S_m(t)$ corresponding to one codeword. In this case we introduce an I-channel and a Q-channel which supports the transmission of the coded message bits. Thus, $T = \hat{k} T_s = 2\hat{k} T_N = N_c T_c$ and $T_s = 2T_N = 2T_c$. In order to map the codewords onto the (4, 2) QPSK modulator, we demultiplex the coded message bits out of the encoder for codeword $\mathbf{x}_m$ into the I and Q channel components $\mathbf{x}_{mI} = (x_{m2}, x_{m4}, \cdots, x_{mN_c})$ and $\mathbf{x}_{mQ} = (x_{m1}, x_{m3}, \cdots, x_{mN_c-1})$. The orthonormal basis functions are $\phi_I(t) = \phi_{2n}(t) = \sqrt{2/T_N} \cos \omega_c t$ and $\phi_Q(t) = \phi_{2n-1}(t) = \sqrt{2/T_N} \sin \omega_c t$ with ω_c a multiple of $2\pi/T_N$ and $(n-1)T_N \le t \le nT_N; n = 1, 2, \cdots, \hat{k}$. The resulting QPSK coded waveform model becomes (for N_c even)

$$S_m(t) = \sqrt{2/T_N} \left(\underbrace{\sum_{n \text{ even}}^{\hat{k}} s_{mn} \cos \omega_c t}_{I \text{ channel}} + \underbrace{\sum_{n \text{ odd}}^{\hat{k}} s_{mn} \sin \omega_c t}_{Q \text{ channel}} \right) \quad 0 \le t \le T \quad (11.13)$$

where $s_{mn} \triangleq \sqrt{E_N} x_{mn}$ for QPSK and x_{mn} takes on value $+1$ or -1 volts. Figure 11.5 illus-

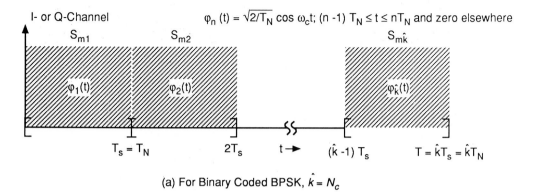

(a) For Binary Coded BPSK, $\hat{k} = N_c$

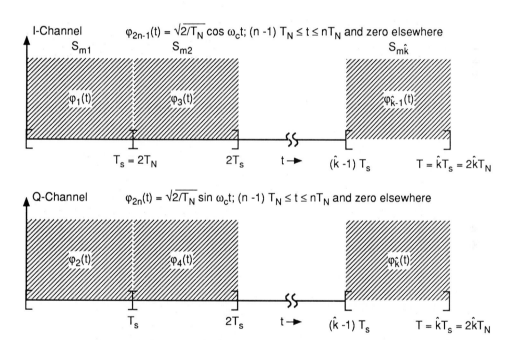

(b) For Binary Coded QPSK $\hat{k} = N_c/2$, 8-PSK $\hat{k} = N_c/3$, and M-PSK $\hat{k} = N_c/\log_2 M$

Figure 11.5   Codeword transmission using a set of time-orthogonal functions

trates the use of time-orthonormal and time-orthonormal quadrature functions to transport coded information.

Coded 8-ary PSK: Finally, consider the transmission of the (N_c, K_c) code using the $(8, 2)$ 8-ary PSK modulator. Then $\log_2 M = 3$, $\hat{k} = N_c/3$ and $T = \hat{k}T_s = 2\hat{k}T_N = 3\hat{k}T_c = K_cT_b$. The coded message bits are sequentially mapped three at a time onto the eight phase states of the modulator. Table 11.3 summarizes various coded modulation mappers for the class of binary and multilevel codes. Table 11.4 summarizes the corresponding coded modulation waveforms $S_m(t)$ corresponding to the transmission of codeword $\mathbf{x}_m$, $m = 0, 1, \cdots, M_c - 1$. The coded modulation waveforms needed for transmission of nonbinary codes will be given in the next chapter, that is, aside from the use of time-orthonormal functions one can use frequency-orthonormal functions. Table 11.5 summarizes numerous communication carriers of great practical interest; these carriers represent basic building blocks needed in the generation of $\{S_m(t)\}$.

Example 11.4

Consider a 4-ary message source ($M_c = 4$) whose output is to be encoded using a binary ($Q = 2$) rate 2/3 code ($K_c = \log_2 M_c = 2$, $N_c = 3$) and transmitted over the channel using an M-PSK modulation. Define the mapping of the four possible messages m_0, m_1, m_2, m_3, which are respectively represented by the four possible binary sequences 00, 01, 10, 11, into codewords as below:

Message	Input Sequence	Output Sequence
m_0	00	000
m_1	01	001
m_2	10	010
m_3	11	111

(a) Coded BPSK Modulation ($M = 2, N = 1$)

Here the modulator is used $\hat{k} = N_c/\log_2 M = 3$ times to generate the coded waveform $S_m(t)$ corresponding to each codeword. That is, each of the $N_c = 3$ coded bits per codeword is assigned to one BPSK waveform. These waveforms are transmitted sequentially over the channel every $T_s = T_N = T/3$ sec. Also note that $T_b = 3T_N/2 = 3T_s/4$. Letting $s_0(t) = \sqrt{2/T_b} \cos \omega_c t$ be assigned to a "0" coded bit and $s_1(t) = -\sqrt{2/T_b} \cos \omega_c t$ be assigned to a "1" code bit, then the mapping from a typical message sequence to the transmitted BPSK modulation would appear as below:

Message Sequence: $m_1, m_0, m_3.m_1, m_2, m_0, \ldots$

Coded Bit Sequence: 001000111001010000 . . .

Modulator Output : $\underbrace{s_0(t), s_0(t), s_1(t),}_{S_1(t)}$ $\underbrace{s_0(t), s_0(t), s_0(t),}_{S_0(t)}$ $\underbrace{s_1(t), s_1(t), s_1(t),}_{S_3(t)}$

Coded Modulation Waveform

$\underbrace{s_0(t), s_0(t), s_1(t),}_{S_1(t)}$ $\underbrace{s_0(t), s_1(t), s_0(t),}_{S_2(t)}$ $\underbrace{s_0(t), s_0(t), s_0(t),}_{S_0(t)} \ldots$

(b) Coded QPSK Modulation ($M = 4, N = 2$)

Here the modulator is used $\hat{k} = N_c/\log_2 M = 3/2 = 1.5$ times to generate the coded waveform $S_m(t)$ corresponding to each codeword. That is, two of the $N_c = 3$ coded bits per codeword are assigned to one QPSK waveform. These waveforms are transmitted sequentially over the channel every $T_s = 2T_N = 2T/3$ sec. Also note that $T_b = 3T_N/2 = 3T_s/4$. Letting $s_0(t) = \sqrt{2/T_s} \cos \omega_c t$ be assigned to the coded bit sequence "00," $s_1(t) = -\sqrt{2/T_s} \sin \omega_c t$ be assigned to the coded bit sequence "01," $s_2(t) = -\sqrt{2/T_s} \cos \omega_c t$ be assigned to the coded bit sequence "10," and $s_3(t) = \sqrt{2/T_s} \sin \omega_c t$ be assigned to the coded bit sequence "11," then the mapping from a typical message sequence to the transmitted QPSK modulation would appear as below:

Message Sequence: $m_1, m_0, m_3.m_1, m_2, m_0, \ldots$

Coded Bit Sequence: 00 10 00 11 10 01 01 00 00 . . .

Modulator Output : $s_0(t), s_2(t), s_0(t),$ $s_3(t), s_2(t), s_1(t),$ $s_1(t), s_0(t), s_0(t), \ldots$

Coded Modulation Waveform $S_1(t)$ $S_0(t)$ $S_3(t)$ $S_1(t)$ $S_2(t)$ $S_0(t)$

(c) Coded 8-PSK Modulation ($M = 8, N = 2$)

Table 11.3 Coded Modulation Waveform Mappers for Binary and Multilevel Codes

Modulator	Codec-Modem Parameters	Communication Carrier	Coded Message Bit to Modulator
BPSK ($M=2$, $N=1$)	$\hat{k} = N_c \log_2 Q$ $T_s = T_N = \frac{T_c}{\log_2 Q}$	$\varphi_n(t) = \sqrt{2/T_N}\cos\omega_c t$ $= \varphi(t)$ for all $n=1,2,\ldots,\hat{k}$ and $(n-1)T_N \le t \le nT_N$; ω_c is a multiple of π/T_N	$s_{mn} = \sqrt{E_N}x_{mn}$ with $x_{mn} = +1$ or -1 for all m and $n=1,2,\ldots,\hat{k}$
QPSK ($M=4$, $N=2$)	$\hat{k} = \frac{N_c \log_2 Q}{2}$ $T_s = 2T_N = \frac{2T_c}{\log_2 Q}$	I-channel: $\varphi_{2n-1}(t)$ $= \sqrt{2/T_N}\cos\omega_c t$ Q-channel: $\varphi_{2n}(t)$ $= \sqrt{2/T_N}\sin\omega_c t$, $(n-1)T_N \le t \le nT_N$, $n=1,2,\ldots,\hat{k}$ and ω_c is a multiple of $2\pi/T_N$	$s_{mn} = \sqrt{E_N}x_{mn}$ with $x_{mn} = +1$ or -1 for all m and $n=1,2,\ldots,\hat{k}$
8-PSK ($M=8$, $N=2$)	$\hat{k} = \frac{N_c \log_2 Q}{3}$ $T_s = 2T_N = \frac{3T_c}{\log_2 Q}$	I-channel: Same as QPSK Q-channel: Same as QPSK and ω_c is a multiple of $2\pi/T_N$	I-channel: (n odd) $S_{mn} = \sqrt{E_N}\cos(\pi m/4)$ Q-channel: (n even) $S_{mn} = \sqrt{E_N}\sin(\pi m/4)$ for all m and $n=1,2,\ldots,\hat{k}$
M-PSK (M, $N=2$)	$\hat{k} = \frac{N_c \log_2 Q}{\log_2 M}$ $T_s = 2T_N$ $= \frac{T_c \log_2 M}{\log_2 Q}$	Same as QPSK	I-channel: (n odd) $S_{mn} = \sqrt{E_N}\cos(\pi m/M)$ Q-channel: (n even) $S_{mn} = \sqrt{E_N}\sin(\pi m/M)$ for all m and $n=1,2,\ldots,\hat{k}$
ASK (M, $N=1$) $M \ge 2$	$\hat{k} = \frac{N_c \log_2 Q}{\log_2 M}$ $T_s = T_N = \frac{T_c \log_2 M}{\log_2 Q}$	$\varphi_n(t) = \sqrt{2/T_N}\cos\omega_c t$, $(n-1)T_N \le t \le nT_N$, and ω_c is a multiple of π/T_N	$s_{mn} = \sqrt{E_{N_n}}x_{mn}$ with $x_{mn} = +1$ or -1 for all m and $n=1,2,\ldots,\hat{k}$
QASK (M, $N=2$) $M \ge 4$	$\hat{k} = \frac{N_c \log_2 Q}{\log_2 M}$ $T_s = 2T_N$ $= \frac{T_c \log_2 M}{\log_2 Q}$	Same as QPSK	I-channel: (n odd) $S_{mn} = \sqrt{E_{N_n}}x_{mn}$ Q-channel: (n even) $S_{mn} = \sqrt{E_{N_n}}x_{mn}$ $x_{mn} = +1$ or -1 for all m and $n=1,2,\ldots,\hat{k}$
Quadrature Amplitude Modulation (QAM)	Same as QASK	Same as QASK	Same as QASK except nonrectangular pulse shapes are used in I- and Q-channels (see Chapter 10)

Table 11.4 Coded Modulation Waveforms for Transmission of Binary and Multilevel Codes

Modulator	Coded Modulation Waveform $0 \leq t \leq T = \hat{k} T_s$	Communication Carrier Coded Message Bit Map
(2, 1) BPSK or $(M, 1)$ M-ASK	$\mathbf{x}_m:\ S_m(t) = \sum_{n=1}^{\hat{k}} s_{mn}(t) \varphi_n(t)$ for $m = 0, 1, \ldots, M_c - 1;\ \hat{k} = N_c$	See Table 11.3, columns 2 and 3
(4, 2) QPSK, $(M, 2)$ QASK, M-PSK M-FSK	$\mathbf{x}_m:\ S_m(t) = \underbrace{\sum_{n\ \text{odd}}^{\hat{k}} s_{mn} \varphi_{2n-1}(t)}_{\text{I-channel}} + \underbrace{\sum_{n\ \text{even}}^{\hat{k}} s_{mn} \varphi_{2n}(t)}_{\text{Q-channel}}$ for $m = 0, 1, \ldots, M_c - 1;\ \hat{k} = N_c/2$	See Table 11.3, columns 2 and 3

The expansions assume unit energy rectangular pulse shapes; if other baseband pulse shapes are used these models can be readily modified (see Chapters 9 and 10) to accommodate arbitrary I and Q pulse shapes

Table 11.5 Communication Carriers of Greatest Practical Interest

Communication Carrier	Channel Waveform
Time-orthogonal (ω_c a multiple of π/T_N)	$\varphi_n(t) = \begin{cases} \sqrt{2/T_N} \sin \omega_c t, & (n-1)T_N \leq t \leq nT_N \\ 0, & \text{else} \end{cases}$
Time-orthogonal (I-Q modulation) (ω_c a multiple of $2\pi/T_N$)	$\varphi_{2n}(t) = \begin{cases} \sqrt{2/T_N} \sin \omega_c t, & (n-1)T_N \leq t \leq nT_N \\ 0, & \text{else} \end{cases}$ $\varphi_{2n+1}(t) = \begin{cases} \sqrt{2/T_N} \cos \omega_c t, & (n-1)T_N \leq t \leq nT_N \\ 0, & \text{else} \end{cases}$
Frequency-orthogonal (ω_c a multiple of π/T)	$\varphi_n(t) = \begin{cases} \sqrt{2/T} \sin[(\omega_c + \pi n/T)t], & 0 \leq t \leq T \\ 0, & \text{else} \end{cases}$
Frequency-orthogonal (I-Q modulation) (ω_c a multiple of $2\pi/T$)	$\varphi_{2n}(t) = \sqrt{2/T} \sin[(\omega_c + 2\pi n/T)t], \quad 0 \leq t \leq T$ $\varphi_{2n+1}(t) = \sqrt{2/T} \cos[(\omega_c + 2\pi n/T)t], \quad 0 \leq t \leq T$

Here the modulator is used $\hat{k} = N_c/\log_2 M = 3/3 = 1$ times to generate the coded waveform $S_m(t)$ corresponding to each codeword. That is, all three of the $N_c = 3$ coded bits per codeword are assigned to one 8-PSK waveform. These waveforms are transmitted sequentially over the channel every $T_s = 2T_N = T$ sec. Also note that $T_b = T_N = T_s/2$. Letting $S_0(t) = \sqrt{2/T_s} \cos \omega_c t$ be assigned to the coded bit sequence "000," $S_1(t) = \sqrt{2/T_s} \cos(\omega_c t + \pi/4) = \sqrt{1/T_s} \cos \omega_c t - \sqrt{1/T_s} \sin \omega_c t$ be assigned to the coded bit sequence "001," $S_2(t) = \sqrt{2/T_s} \cos(\omega_c t + \pi/2) = -\sqrt{2/T_s} \sin \omega_c t$ be assigned to the coded bit sequence "010," etc., then the mapping from a typical message sequence to the transmitted 8-PSK modulation would appear as below:

Message Sequence: $m_1, m_0, m_3.m_1, m_2, m_0, \ldots$

Coded Bit Sequence: 001 000 111 001 010 000...

Modulator Output : $s_1(t), s_0(t), s_3(t),\ s_1(t), s_2(t), s_0(t), \ldots$

Coded Modulation Waveform $S_1(t)\quad S_0(t)\quad S_3(t)\quad S_1(t)\quad S_2(t)\quad S_0(t)$

(d) Coded 16-PSK Modulation ($M = 16$, $N = 2$)

Here the modulator is used $\hat{k} = N_c / \log_2 M = 3/4 = .75$ times to generate the coded waveform $S_m(t)$ corresponding to each codeword. That is, all three of the coded bits of one codeword and one bit of the next codeword (or two coded bits of one codeword and two coded bits of the next codeword) are assigned to one 16-PSK waveform. These waveforms are transmitted sequentially over the channel every $T_s = 2T_N = 4T/3$ sec. Also note that $T_b = 3T_N/4 = 3T_s/8$. Letting $s_0(t) = \sqrt{2/T_s} \cos \omega_c t$ be assigned to the coded bit sequence "0000," $s_1(t) = \sqrt{2/T_s} \cos(\omega_c t + \pi/8) = (\sqrt{2/T_s} \cos \pi/8) \cos \omega_c t - (\sqrt{2/T_s} \sin \pi/8) \sin \omega_c t$ be assigned to the coded bit sequence "0001," $s_2(t) = \sqrt{2/T_s} \cos(\omega_c t + \pi/4) = \sqrt{1/T_s} \cos \omega_c t - \sqrt{1/T_s} \sin \omega_c t$ be assigned to the coded bit sequence "0010," etc., then the mapping from a typical message sequence to the transmitted 16-PSK modulation would appear as below:

Message Sequence:	$m_1, m_0, m_3.m_1, m_2, m_0, \ldots$
Coded Bit Sequence:	0010 0011 1001 0100 00 $\ldots$
Modulator Output :	$s_1(t)$, $s_3(t)$, $s_9(t)$, $s_4(t)$, $\ldots$
Coded Modulation Waveform	$S_1(t)$ $S_0(t)$ $S_3(t)$ $S_1(t)$ $S_2(t)$

Example 11.5

Consider a 2^k-ary message source ($M_c = 2^k$) whose output is to be encoded using a binary ($Q = 2$) rate $k/k + 1$ code ($K_c = \log_2 M_c = k$, $N_c = k + 1$) and transmitted over the channel using an M-PSK modulation where $M = 2^{k+1}$. This is a generalization of case (c) in Example 11.4. Here the modulator is used $\hat{k} = N_c / \log_2 M = (k + 1)/(k + 1) = 1$ times to generate the coded waveform $S_m(t)$ corresponding to each codeword. That is, all $k + 1$ of the $N_c = k + 1$ coded bits per codeword are assigned to one M-PSK waveform. These waveforms are transmitted sequentially over the channel every $T_s = T$ sec.

Example 11.6

Consider a 16-ary message source ($M_c = 16$) whose output is to be encoded using a quaternary ($Q = 4$) rate 2/3 code ($K_c = \log_Q M_c = 2$, $N_c = 3$) and transmitted over the channel using an M-PSK modulation. Define the mapping of the 16 possible messages $m_0, m_1, m_2, \ldots, m_{15}$, which are respectively represented by the 16 possible quaternary sequences 00, 01, 02, 03, 10, 11, $\ldots$, 33, into codewords as in Example 11.2.

(a) Coded BPSK Modulation ($M = 2$, $N = 1$)

Here the modulator is used $\hat{k} = N_c \log_2 Q / \log_2 M = (3)(2)/1 = 6$ times to generate the coded waveform $S_m(t)$ corresponding to each codeword. That is, each of the $N_c = 3$ coded 4-ary symbols per codeword is converted to two coded bits [the mapping from coded symbols to coded bits is assumed here to be binary coded decimal (BCD)] each of which is assigned to one BPSK waveform. These waveforms are transmitted sequentially over the channel every $T_s = T/6$ sec. Also note that $T_b = 3T_N/2 = 3T_s/2$. Letting $s_0(t) = \sqrt{2/T_b} \cos \omega_c t$ be assigned to a "0" coded bit and $s_1(t) = -\sqrt{2/T_b} \cos \omega_c t$ be assigned to a "1" code bit, then the mapping from a typical message sequence to the transmitted BPSK modulation would appear as below:

Message Sequence:	$m_1, m_0, m_3, m_6, \ldots$
Coded Symbol Sequence:	010 000 030 302 $\ldots$
Coded Bit Sequence:	000100000000001100110010
Modulator Output :	$s_0(t), s_0(t), s_0(t),\ s_1(t), s_0(t), s_0(t),$
Coded Modulation Waveform	$S_1(t)$
	$s_0(t), s_0(t), s_0(t),\ s_0(t), s_0(t), s_0(t),$
	$S_0(t)$

$$\underbrace{s_0(t), s_0(t), s_1(t), \; s_1(t), s_0(t), s_0(t),}_{S_3(t)}$$

$$\underbrace{s_1(t), s_1(t), s_0(t), \; s_0(t), s_1(t), s_0(t),}_{S_6(t)} \ldots$$

(b) Coded QPSK Modulation ($M = 4$, $N = 2$)

Here the modulator is used $\hat{k} = N_c \log_2 Q / \log_2 M = (3)(2)/2 = 3$ times to generate the coded waveform $S_m(t)$ corresponding to each codeword. That is, each of the $N_c = 3$ coded 4-ary symbols per codeword is directly assigned to one QPSK waveform without the need for first converting it to coded bits. These waveforms are transmitted sequentially over the channel every $T_s = T/3$ sec. Also note that $T_b = 3T_N/2 = 3T_s/4$. Letting $S_0(t) = \sqrt{2/T_s} \cos \omega_c t$ be assigned to the coded symbol "0," $S_1(t) = -\sqrt{2/T_s} \sin \omega_c t$ be assigned to the coded symbol "1," $S_2(t) = -\sqrt{2/T_s} \cos \omega_c t$ be assigned to the coded symbol "2," and $S_3(t) = \sqrt{2/T_s} \sin \omega_c t$ be assigned to the coded symbol "3," then the mapping from a typical message sequence to the transmitted QPSK modulation would appear as below:

Message Sequence: $m_1, m_0, m_3.m_6, \ldots$

Coded Symbol Sequence: 010 000 030 302 ...

Modulator Output : $\underbrace{s_0(t), s_1(t), s_0(t),}_{S_1(t)} \; \underbrace{s_0(t), s_0(t), s_0(t),}_{S_0(t)}$

Coded Modulation Waveform

$$\underbrace{s_0(t), s_3(t), s_0(t),}_{S_3(t)} \; \underbrace{s_3(t), s_0(t), s_2(t),}_{S_6(t)} \ldots$$

(c) Coded 8-PSK Modulation ($M = 8$, $N = 2$)

Here the modulator is used $\hat{k} = N_c \log_2 Q / \log_2 M = (3)(2)/3 = 2$ times to generate the coded waveform $S_m(t)$ corresponding to each codeword. That is, each of the $N_c = 3$ coded 4-ary symbols per codeword is converted to two coded bits [the mapping from coded symbols to coded bits is assumed here to be binary coded decimal (BCD)] and each three coded bits is assigned to one 8-PSK waveform. Thus two 8-PSK waveforms are generated from each codeword. These waveforms are transmitted sequentially over the channel every $T_s = T/2$ sec. Also note that $T_b = T_N = T_s/2$. Using the same coded bit to waveform mapping as in Example 11.4, part (c), then the mapping from a typical message sequence to the transmitted 8-PSK modulation would appear as below:

Message Sequence: $m_1, m_0, m_3.m_6, \ldots$

Coded Symbol Sequence: 010 000 030 302 ...

Coded Bit Sequence: 000 100 000 000 001 100 110 010 ...

Modulator Output : $\underbrace{s_0(t), s_4(t),}_{S_1(t)} \; \underbrace{s_0(t), s_0(t),}_{S_0(t)} \; \underbrace{s_1(t), s_4(t),}_{S_3(t)} \; \underbrace{s_6(t), s_2(t),}_{S_6(t)} \ldots$

Coded Modulation Waveform

(d) Coded 16-PSK Modulation ($M = 16$, $N = 2$)

Here the modulator is used $\hat{k} = N_c \log_2 Q / \log_2 M = (3)(2)/4 = 1.5$ times to generate the coded waveform $S_m(t)$ corresponding to each codeword. That is, each of the $N_c = 3$ coded 4-ary symbols per codeword is converted to two coded bits [the mapping from coded symbols to coded bits is assumed here to be binary coded decimal (BCD)] and each four coded bits is assigned to one 16-PSK waveform. Thus three 16-PSK waveforms are generated from each pair of codewords. These waveforms are transmitted sequentially over the channel every $T_s = 2T/3$ sec. Also note that $T_b = 3T_N/4 = 3T_s/8$. Using the same coded bit to waveform mapping as in Example 11.4, part (d), then the mapping from a typical message sequence to the transmitted 16-PSK modulation would appear as below:

Message Sequence: $m_1, m_0, m_3.m_6, \ldots$

Coded Symbol Sequence: 010 000 030 302...

Coded Bit Sequence: 0001 0000 0000 0011 0011 0010...

Modulator Output : $s_1(t), s_0(t), s_0(t),\ \ s_3(t),\ s_3(t), s_2(t), \ldots$

Coded Modulation Waveform $S_1(t)$ $S_0(t)$ $S_3(t)$ $S_6(t)$

(e) Coded 16-QASK Modulation ($M = 16, N = 2$)

Here the modulator is used $\hat{k} = N_c \log_2 Q / \log_2 M = (3)(2)/4 = 1.5$ times to generate the coded waveform $S_m(t)$ corresponding to each codeword. That is, each of the $N_c = 3$ coded 4-ary symbols per codeword is converted to two coded bits [the mapping from coded symbols to coded bits is assumed here to be binary coded decimal (BCD)] and each four coded bits is assigned to one QASK waveform. Thus three QASK waveforms are generated from each pair of codewords. These waveforms are transmitted sequentially over the channel every $T_s = 2T/3$ sec. Also note that $T_b = 3T_N/4 = 3T_s/8$. Using the $\pm 1, \pm 3$ coordinates of a 16-point QASK grid in two-dimensional signal space, we let $s_0(t) = \sqrt{2/T_s} \cos \omega_c t - \sqrt{2/T_s} \sin \omega_c t$ be assigned to the coded bit sequence "0000," $s_1(t) = \sqrt{2/T_s} \cos \omega_c t - 3\sqrt{2/T_s} \sin \omega_c t$ be assigned to the coded bit sequence "0001," $s_2(t) = 3\sqrt{2/T_s} \cos \omega_c t - \sqrt{2/T_s} \sin \omega_c t$ be assigned to the coded bit sequence "0010," $s_3(t) = 3\sqrt{2/T_s} \cos \omega_c t - 3\sqrt{2/T_s} \sin \omega_c t$ be assigned to the coded bit sequence "0011," etc. Then, the mapping from a typical message sequence to the transmitted 16-PSK modulation would appear as below:

Message Sequence: $m_1, m_0, m_3.m_6, \ldots$

Coded Symbol Sequence: 010 000 030 302...

Coded Bit Sequence: 0001 0000 0000 0011 0011 0010...

Modulator Output : $s_1(t), s_0(t), s_0(t),\ \ s_3(t),\ s_3(t), s_2(t), \ldots$

CodedModulationWaveform $S_1(t)$ $S_0(t)$ $S_3(t)$ $S_6(t)$

This mapping is identical to that in part (d) of this example. The reason for this is that the code to waveform mapping is only dependent on the values of M and N associated with the modulation and not on the actual form of the signal constellation, that is, the values of s_{mn} in its orthonormal representation.

Example 11.7

In this example, we consider the use of a ternary M-PSK ($M = 3, N = 2$) modulator in combination with binary ($Q = 2$) and ternary ($Q = 3$) rate 2/3 codes.

(a) Binary Code ($Q = 2$)

Since each symbol of the ternary PSK modulator would have to be assigned to $\log_2 M = 1.585$ coded bits, the use of a binary code does not lend itself naturally to this sort of mapping since the conversion of coded symbols to coded bits (in accordance with BCD) has an integer relationship.

(b) Ternary Code ($Q = 3$)

Here it is not necessary to convert the coded symbols to coded bits since the ternary code alphabet represents a natural match to the ternary modulator alphabet. In particular, each coded symbol of the ternary codeword, regardless of the value of N_c, would be assigned to one of the three signals in the modulator of Example 11.3.

IF channel characterization.   This *channel* transforms the coded waveform $S_m(t)$ into the received noise-corrupted waveform $r(t) = S_m(t) + n(t)$; $0 \le t \le T = \hat{k}T_s$. As we have seen in the previous section, $S_m(t)$ can be rewritten as a linear combination of a set of N_c orthonormal functions; furthermore, $S_m(t)$ is representable by the vector $\mathbf{S}_m =$

$(S_{m1}, S_{m2}, \cdots, S_{mN_c})$ of its coefficients in this expansion for $0 \le t \le T$. The *demodulation-detection* process in Fig. 11.3 reduces $r(t)$ to its projection $\mathbf{r} = (r_1, r_2, \cdots, r_{N_c})$; this process is covered in Chapter 3 and this chapter. The vector $\mathbf{r}$ characterizes the output of the demodulator (demodulation processor followed by matched filter detection). The *channel decoder* then transforms $\mathbf{r}$ into the *codeword* output $\mathbf{v} = (v_1, v_2, \cdots, v_{K_c})$ from the codeword dictionary. The *source decoder*, based on the rules of source encoding, transforms $\mathbf{v}$ into an estimate $\hat{m}$ of the information source output. *Source decoding (data compression)* techniques are not covered herein.

The IF channel model consists of modeling numerous subsystems; the subsystems which make up the model are demonstrated in levels 6 through 8 of the system architecture shown in Fig. 1.6 of Chapter 1. For the continuous IF CMC it is characterized by the known channel transition probability density $f(\mathbf{r}|\mathbf{S})$ and, if the channel is memoryless, then

$$f(\mathbf{r}|\mathbf{S}_m) = \prod_{n=1}^{N_c} f(r_n|s_{mn}), \qquad (11.14)$$

when message m is transmitted, that is, message $m \to \mathbf{x}_m \to \mathbf{S}_m$. For the AWGN channel

$$f(r_n|s_{mn}) = \frac{1}{\sqrt{\pi N_0}} \exp\left[-\frac{(r_n - s_{mn})^2}{N_0}\right] \qquad (11.15)$$

for all m and n. Frequently, $s_{mn} = \sqrt{E_{N_c}}x_{mn}$ for all n and m so that the IF channel input-output is directly related to the baseband channel input-output via the energy E_{N_c} used per dimension. So much for the generic baseband and IF channel models of a coded digital communication system. They are both referred to as CMCs.

11.2.3 Quantizing the Demodulator/Detector Outputs

Practical demodulator outputs are defined on the continuum, that is, $\mathbf{r} \in R^{N_c}$ where R^{N_c} denotes the N_c-dimensional Euclidean space. However, before the demodulator output can be processed by a digital signal processor (decoder), some form of *amplitude quantization* on the observable $\mathbf{r}$ must be performed. The quantization process is accomplished in practice using some form of vector quantization circuit (see Fig. 11.6). In essence, the *quantizing circuit* converts the CMC into the DMC which is then an approximation to the CMC. We will discuss the details of this probability transformation shortly. For the moment however, we note that this circuit transforms each observable $r_n \in \mathbf{r}$ from a random variable (r.v.) of the continuous type into a discrete r.v., say R_n. Strictly speaking, we will need N_c quantizing circuits; however, this can be avoided by using BPSK modulation, requiring one quantizer, and using the modulator and demodulator $\hat{k} = N_c$ times to create $\mathbf{r}$ (see Fig. 11.7). In this case, the components of $\mathbf{r}$ are output serially in time every T_N sec. If M-PSK or QAM modulation is used, we need two quantizing circuits; one for the I-channel and one for the Q-channel. In addition, one will need to compute the $\tan^{-1}(r_{2n}/r_{2n-1})$ sequentially in time for M-PSK, $M > 4$. The QASK or QPSK modulator and demodulator will only need to be used $\hat{k} = N_c/2$ times to transmit the coded message bits serially in time (see Fig. 11.7). The quantizer converts the continuous random variable r_n into the discrete random variable R_n. In particular, a J-point *scalar quantizer* Q_1 is a mapping $Q_1 : r_n \to R_n$ where r_n is the real tine in the nth direction of an N_c-dimensional vector space and $R_n = Q_1(r_n) \in \{R_{n0}, R_{n1}, \cdots, R_{n,J-1}\}$ in the quantizer output set of size J for $n = 1, 2, \cdots, N_c$. The output values are referred to as output levels, Associated with each J-level quantizer is a partition of the real line r_n into J cells of width Δ_j for $j = 0, 1, \cdots, J-1$.

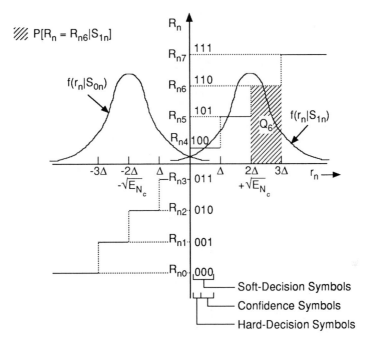

Figure 11.6   Quantization of the observable r_n using a uniform eight-level (3-bit) quantizing circuit; $n = 1, 2, \ldots, N_c$

The *vector quantizer* of dimension N_c and size J illustrated in Fig. 11.3 is a mapping from a vector $\mathbf{r} = (r_1, r_2, \cdots, r_{N_c})$ in N_c-dimensional Euclidean space R^{N_c} into a finite set $\mathbf{R} = (R_1, R_2, \cdots, R_{N_c})$ where each component takes on one of J output levels; $R_n \in \{R_{n0}, R_{n1}, \cdots, R_{n,J-1}\}$ for all $n = 1, 2, \cdots, N_c$. Thus, $\mathbf{R} = Q_{N_c}(\mathbf{r}) : R^{N_c} \to \mathbf{R}$.

The design, that is, selection of the number of *quantization levels*, of the quantizing circuit is an important part of the demodulator design. Assume that R_n takes on one of J values (a $J = 2^q$ level quantizer), say $(R_{n0}, R_{n1}, \cdots, R_{nJ-1})$. For example, an 8-level ($q = 3$ bit) quantizing circuit is illustrated in Fig. 11.6; this figure demonstrates the *quantization regions*, and the input-output transformation law.

In order to discuss the process further, two more concepts involved in error control system design must be presented, viz., the concept of *hard decisions* and *soft decisions*. From the point of view of codec design, creation of the DMC from the CMC outputs is most important. When the demodulator outputs r_n and this value is quantized into two levels ($J = 2$), for all n, the demodulator is said to make *hard decisions* and the decoding process is termed *hard-decision decoding*. When the demodulation outputs are quantized into more than two levels ($J > 2$), the demodulator is said to make *soft decisions* and the decoding process is called *soft-decision decoding*. A soft-decision demodulator is more complex to implement when compared with a hard-decision demodulator. Soft-decision demodulators require automatic gain control in the signal path and $q = \log_2 J$ bits have to be manipulated for every channel symbol interval of duration T_N sec. Infinite bit quantization is frequently referred to by system engineers as *no quantization*, that is, perfect representation of the amplitude voltages associated with $\mathbf{r}$ is retained, no quantizer is used, and no information is lost at this point in the decoding process. We defer until later the presentation of the details regarding codec performance as a function of the quantizer design parameter $J = 2^q$ and the

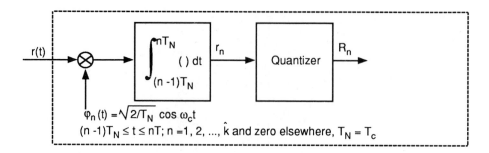

(a) BPSK and ASK Demodulator

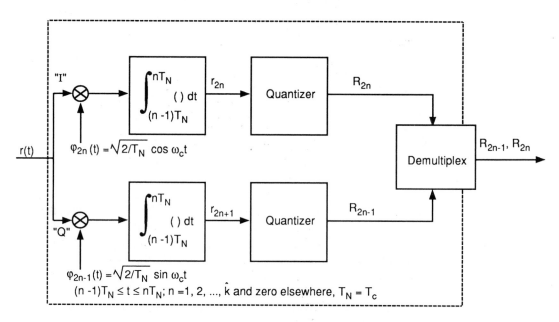

(b) I-Q Demodulator for QPSK, *M*-PSK, QASK, or QAM

Figure 11.7   Optimum demodulation of the coded message bits: (a) BPSK and ASK demodulator, (b) I-Q demodulator for QPSK, *M*-PSK, QASK, or QAM

chosen coding-decoding technique. Suffice it to say, we shall show in the next chapter that 3-bit soft-decision demodulators improve the communication efficiency by approximately 2.2 dB for coherent detection and 4.7 dB for noncoherent detection when compared with hard-decision demodulators. On the other hand, 3-bit quantizers loose little in efficiency when compared with the unquantized demodulator output (see Fig. 11.6). Figure 11.8 demonstrates the DMC model for the quantizer shown in Fig. 11.6. The confidence thresholds illustrated in Fig. 11.6 will be useful in creating *erasures* in Reed-Solomon decoding (see Chapter 12).

As we shall see later, there is only a small difference in the communication performance efficiency achievable between eight-level quantization and the unquantized case. The eight-level quantized outputs involve one *decision threshold* and three pairs of *confidence*

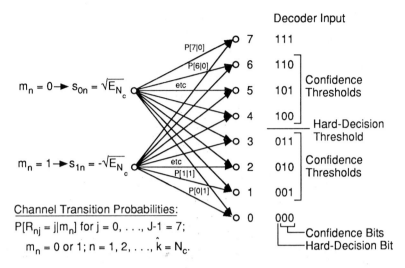

Figure 11.8   DMC for BPSK demodulation using 8-ary (3-bit) soft-decision demodulation (for QPSK one has an I-channel and a Q-channel; however, $n = 1, 2, \ldots, N_c/2$)

thresholds for the ($M = 2, N = 1$) BPSK **M**odulator-**D**emodulator of a (N_c, K_c) code; this 3-bit quantization process is illustrated in Fig. 11.8.

QAM or QASK. To accommodate *multilevel codes*, that is, the case where the coded message bits x_{mn} can take on more than two values, *quadrature amplitude modulation* (QAM) can be used and the coefficients s_{mn} can be adjusted to give the appropriate mapping. The telephone channel is a good example of this. Here the data rate requirement is 9,600 bits/sec and the modulator can support the transmission of 2,400 symbols/sec (bauds). Without channel coding, a 16-point QAM modulator (Chapter 10) can be used to accommodate the data rate. With coding and a code of rate $r_c=4/5$, a 32-point signal structure will support the 9,600 baud data rate. In practice, the CCITTV.32 convolutional encoder is used in combination with a QAM modulator. We consider the transmission of multilevel codes in Chapter 12.

11.2.4 Creating the DMC for Channel Decoding

Now the *channel transition probabilities* $\{P[R_n = R_{nj}|m]$ for $j = 0, 2, \cdots, J - 1; m = 0$ or $1; n = 1, 2, \cdots, N_c\}$ which define the IF DMC are found from the IF CMC *transition probability densities* $f(r_n|s_{mn})$ defined in (11.15). In particular, using (11.15), we have

$$P[R_n = R_{nj}|s_{mn}] = \int_{r_n \in Q_j} f(r_n|s_{mn}) dr_n \triangleq P(j|m) \tag{11.16}$$

for $j = 0, 1, \cdots, J - 1; m = 0$ or $1; n = 1, 2, \cdots, N_c$. Here Q_j is the jth quantization interval of width Δ (see Fig. 11.4). Selection of the design values for Δ and J will be considered in the next chapter.

The simplest DMC which can be created using the demodulator output is the *binary symmetric channel* (BSC) for which the coded message bits are transmitted using BPSK, with $J = 2$ (hard-decision demodulation). We only need to consider quantizing the observable $r_n \triangleq r$ into $R_n \triangleq R$ for all n where $R = 0$ or 1. For the BSC with inputs

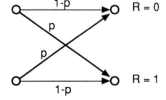

(a) BSC Model

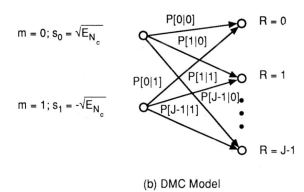

(b) DMC Model

Figure 11.9 BSC and DMC model for hard and J-ary soft-decision demodulation of BPSK coded message bits: (a) BSC model, (b) DMC model

$m_0 = 0 \rightarrow s_0 = \sqrt{E_{N_c}}$ or $m_1 = 1 \rightarrow s_1 = -\sqrt{E_{N_c}}$, for all coded message bits, the channel transition probabilities are given respectively by

$$P[r \leq 0 | s_0] = \int_{-\infty}^{0} f(r | s_0) dr = p \qquad (11.17)$$

$$P[r \geq 0 | s_1] = \int_{0}^{\infty} f(r | s_1) dr = p \qquad (11.18)$$

where $p = Q\left(\sqrt{2 E_{N_c}/N_0}\right)$. Figure 11.9 demonstrates the BSC channel model and the J-ary DMC for $M = 2$; this channel is used $\hat{k} = N_c$ times per message transmission. These two quantized channel models of the IF channel of Fig. 11.2 are of greatest interest in what follows.

When the physical channel is not memoryless, it can usually be made so by adding an interleaver-deinterleaver system into the information flow path. Interleaving-deinterleaving techniques are discussed next.

11.3 INTERLEAVING-DEINTERLEAVING TO MITIGATE CHANNEL MEMORY

We have completed our discussion regarding the problem of quantizing the demodulator output so as to create the DMC needed to accomplish the channel decoding function using digital hardware. In doing so, we tacitly assumed that the channel is *memoryless*, that is, the received channel symbols are independently contaminated by noise. In practice, many channels exist such that this is not the case; *noise bursts* are common. Examples include multipath, interference, jammed, and fading channels. Telephone channels typically have impulses on the line which last for approximately 10 ms.

We now address techniques whereby channels with memory can be engineered such that they appear to the channel decoder as memoryless. *Interleaving* or *scrambling* is the common technique used to make the channel appear to the decoder as memoryless when burst noise is present. In practice, the system designer begins by specifying some maximum *noise burst length* which the system must tolerate—the so-called *channel memory time*. Once this is determined, the design of the interleaving system can proceed.

To mitigate channel memory, thereby taking full advantage of coding, an *interleaver-deinterleaver* (I-D) system is inserted into the signal path. We will discuss the placement or logical position of the I-D system in the functional system architecture in the next section. For the moment, we present various interleaver types and discuss the trade-offs between their performance measures of *interleaving depth (span)*, *interleaving (channel) delay*, and *interleaving memory size*. The I-D system is critical to being able to realize the performance gain provided by coding in channels with memory. The function of the I-D system is to effectively reduce the degradation in bit error probability by randomizing the distribution of the burst errors (or correlated bits) caused by channels with memory. The *interleaving process* serves to scramble the order of the code symbols x_{mn} in time while the *deinterleaving process* unscrambles the recovered symbol stream into the original sequence for application to the decoder. Error bursts which would overwhelm the decoder are spread out by the deinterleaver over a number of channel symbols longer then the duration of the error burst. Thus for purposes of error control, the burst or memory characteristics of the channel must be specified.

There are at least five I-D system techniques of practical interest. They include: (1) the *periodic block* I-D system, (2) the *helical* I-D system, (3) the *periodic convolutional* I-D system, (4) the *pseudorandom* I-D system, and (5) the *hybrid* I-D system (see [2], [4], [6], [7], and [8]).

11.3.1 Periodic Block Interleavers-Deinterleavers

Figure 11.10 depicts a block of length $L = 4$ symbols which is block interleaved to depth $D = 3$. In essence, an (L, D) periodic block interleaver is characterized by the *block interleaver length L* and *interleaver depth D*. Symbols moving across the channel are obtained by scanning rows from left to right; symbols leaving the channel encoder, or entering the decoder, are obtained by scanning succesive columns from top to bottom. The codewords occupy the positions labeled ABCD, EFGH, IJKL, ... The transmitted code symbols are AEI, BFJ, CGK, DHL, and so on. Suppose a burst of noise "wipes out" the transmitted code symbols AEI, and converts these into XXX (see Fig. 11.10) . The decoder input consists of (XBCD, XFGH, XJKL). A single error correcting code which corrects one error will alleviate this error. Also a single error detecting code will detect but not correct this error. Thus, an interleaving depth of D units means that a noise burst must have length $(D + 1)T_N$ sec in order to strike any codeword more than once.

Figure 11.11 demonstrates a block of length L symbols which is block-interleaved to a depth of D units of time. We follow the convention that the block symbols x_{ij} to be transmitted over the channel are read into the interleaver columns and then transmitted over the channel row by row from left to right. On the receiver side, the action is reversed. The deinterleaver inputs are the rows of the interleaver output. The rows are read into the

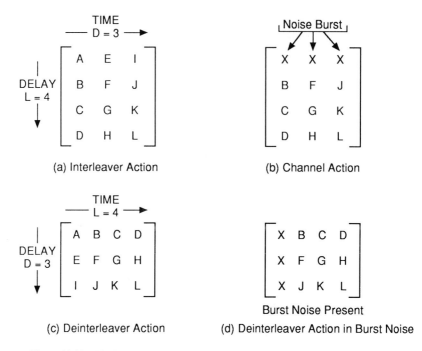

Figure 11.10   Block I-D system of length $L = 4$ and depth $D = 3$: (a) Interleaver action, (b) Channel action, (c) Deinterleaver action, (d) Deinterleaver action in burst noise

columns of the deinterleaver and are read out of the deinterleaver (decoder inputs) by rows as shown in Fig. 11.10(b). Thus, the $D \times L$ deinterleaving matrix is the transpose of the $L \times D$ interleaving matrix.

In practice, the encoder output symbols are written into L vertically contiguous memory cells in a column and the outgoing symbols are read from D horizontally contiguous memory cells in a row. The *interleaver span* is a measure of the time across which previously contiguous data is spread. It serves to characterize the ability of the interleaver to randomize burst errors or spread out adjacent symbols over time intervals greater than the noise burst durations. The span of the block interleaver is given by

$$\tau_s = LDT_N \text{ sec} \tag{11.19}$$

where T_N is the time duration of each channel code symbol. The *interleaving* (channel) *delay* introduced into the recovered code symbols is

$$\tau_d = 2LDT_N \text{ sec} \tag{11.20}$$

whereas the required memory at the transmit and receive end is $M_r = 2LD$ symbols for hard-decision demodulator/decoders.

11.3.2 The Helical Interleaver

Figure 11.12 demonstrates a helical interleaver for a block of length $L = 4$ with a depth $D = 3$. The codewords are seen from this figure to be ABfa, CDbc, and EFde. As a function of time the output of the encoder is ABfaCDbcEFdeABfa, the output of the interleaver (channel input) is abFAcdBCefDEabFA, and the input to the decoder (deinterleaver output)

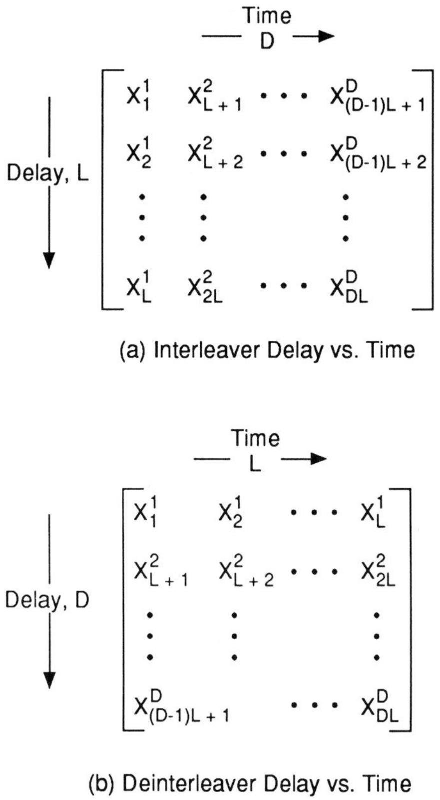

(a) Interleaver Delay vs. Time

(b) Deinterleaver Delay vs. Time

Figure 11.11 Periodic block interleaver-deinterleaver: (a) Interleaver delay versus time, (b) Deinterleaver delay versus time

is the same as the encoder output; no errors are assumed. The total interleaving delay is seen to be $4 \times 3 = 12$ characters. If a burst of noise in the channel were to convert the transmitted codeword abFA into the word XXXX, then the deinterleaver action would scramble the X symbols such that the input to the decoder would be the codeword sequence ABfXCDXcEXdeABfX. Again a single error correcting code would be able to correct these errors or a single error detecting code would be able to detect the errors.

Figure 11.13 depicts a code of length 8 which is helical-interleaved to a depth of 7. This means that a noise burst must have a length of at least 8 in order to hit any code block more than once. Each codeword lies in a column and the symbols are fed into the interleaver by columns and then read out by rows. A block interleaver requires the receiver to know the synchronization modulo LD; however, the helical interleaver requires the receiver to know synchronization modulo L. In general, a code of length L can be helical-interleaved to a depth of $L - 1$. The helical-interleaver requires a memory size about one-half the size of the corresponding block interleaver of the same depth.

11.3.3 The Periodic Convolutional Interleaver-Deinterleaver

Variable delay is the essence of interleaving. Hence, it is natural to view an interleaver as a collection of delay lines of various lengths with multiplexers shuffling data through successively different routes. An (L, D) periodic convolutional I-D system which uses this implementation concept is shown in Fig. 11.14 for the case $L = 30$, $D = 116$. When the

—Channel Time—➤

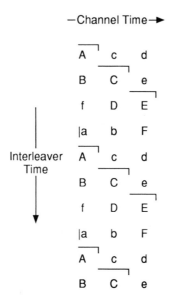

Interleaver
Time

Figure 11.12   A (4,3) helical interleaver

delay lines are tens of thousands of symbols (bits) long, then they are most effectively imple-mented with random access memory (RAM). We note from Fig. 11.14 that the convolutional interleaver has a staircase geometry, as opposed to the rectangular geometry of the block type. This staircase geometry may be thought of as a series of shift registers of graduated lengths, as indicated in Fig. 11.14. Actual implementations are more likely to use a random access memory (RAM) and stepped address pointers as opposed to shift registers.

The shift register model facilitates discussion of the operation. In contrast to the block case where the data was stationary and a two-dimensional commutator was used, a one-dimensional commutator is used in the convolutional interleaver and the data in each row of the interleaver and deinterleaver is periodically shifted to the right. The commutator at the interleaver input is stepped row by row from top to bottom inserting one bit in each row. After the commutators reach the bottom position, each data bit in every row is shifted one bit toward the output, and the commutators are reset to the lowest row. The output or last bit in each row is overwritten after it is read. In contrast with the block interleaver, there is no conflict between input and output, so long as both input and output precede shifting on a consistent basis for all rows.

The convolutional interleaver has several advantages over the block interleaver. The memory and delay required are lower and it lends itself to rate buffering of the data. When inserted in the encoded symbol sequence, the system guarantees separation (of any two symbols within LT_N sec of each other in the interleaver symbol sequence) to be at least $[LD/(D-1)]T_N$ sec between each other in the deinterleaved sequence. Thus, the *inter-leaver span* is given by

$$\tau_s = [LD/(D-1)]T_N \text{ sec} \qquad (11.21)$$

and the *interleaving delay* is $\tau_d = 2\tau_s$ sec. The memory required for the convolutional interleaver is the sum of the row lengths. For a typical interleaver, it becomes

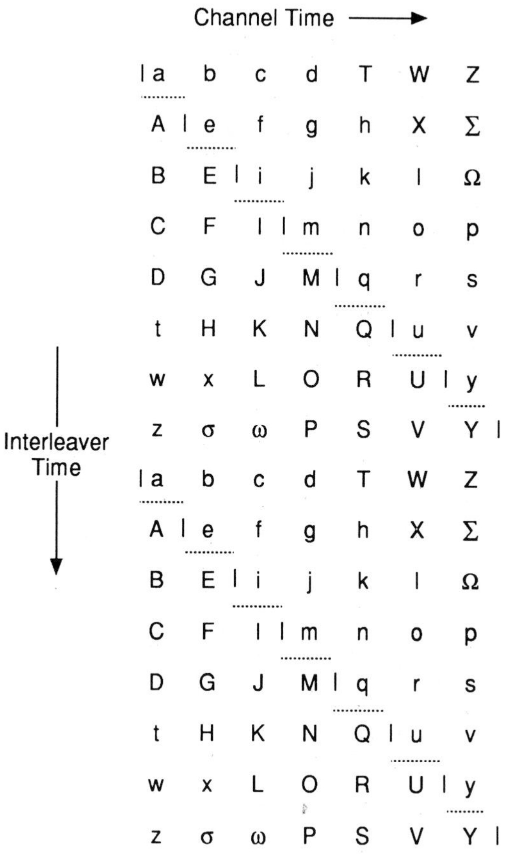

Channel Time ⟶

I a	b	c	d	T	W	Z
A	I e	f	g	h	X	Σ
B	E	I i	j	k	I	Ω
C	F	I	I m	n	o	p
D	G	J	M I	q	r	s
t	H	K	N	Q I	u	v
w	x	L	O	R	U I	y
z	σ	ω	P	S	V	Y I
I a	b	c	d	T	W	Z
A	I e	f	g	h	X	Σ
B	E	I i	j	k	I	Ω
C	F	I	I m	n	o	p
D	G	J	M I	q	r	s
t	H	K	N	Q I	u	v
w	x	L	O	R	U I	y
z	σ	ω	P	S	V	Y I

Interleaver
Time

Figure 11.13 A (8,7) helical interleaver

$$M_r = \frac{(D+1)}{L} \sum_{k=1}^{L} (k-1) \qquad (11.22)$$

characters or symbols.

11.3.4 Other Interleavers-Deinterleavers

McEliece [2] discusses *pseudo-random interleavers*; it is also possible to consider *concatenated* I-D systems. Berlekamp [7] discusses concatenating block and helical interleavers to obtain *deep-staggered* and *shallow-staggered* I-D systems; the description and operation of such systems are outside the scope of our presentation. The *hybrid interleaver* [4] was perceived for use in coded systems employing information packets.

11.4 SYNCHRONIZATION OF I-D SYSTEMS

Coded digital communication systems that employ interleaving require a level of synchronization to operate the I-D commutators. In particular, decoders for block codes need to know when codewords begin and end; convolutional decoders need to know when the nodes in the code tree begin and end. It is also necessary to ensure that the receiver's deinterleaver is synchronized with the transmit side interleaver. For the most part, the methodologies of

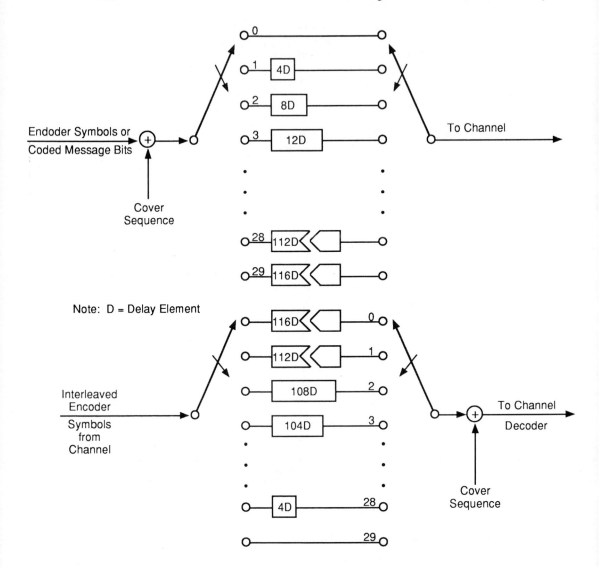

Note: D = Delay Element; The cover sequence can be used to provide data privacy.

Figure 11.14   Periodic convolutional interleaver-deinterleaver

synchronization and resynchronization of I-D systems which accept hard or soft decisions from the demodulator largely remain the trade secrets of organizations who design and build error control systems.

11.5 CODED SYSTEM ARCHITECTURE FOR CHANNELS WITH MEMORY

As we have just discussed, the common strategy for eliminating the deleterious effects of noise bursts on coded system performance is to introduce into the information flow path an interleaver-deinterleaver system. Figure 11.15 shows the logical position of interleavers in

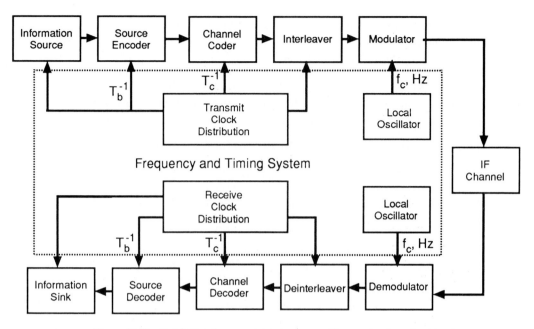

Figure 11.15   Coded-digital communication system architecture for channels with memory

the system architecture. The interleaver lies between the channel encoder and the modulator; the deinterleaver is inserted between the demodulator output and the channel decoder. Both units frequently require substantial memory.

Suffice it to say, block decoders tend to be less complex when driven by hard decisions from the demodulator; convolutional decoders can perform using either hard- or soft-decision demodulator outputs. In order to gain the advantages of both error control methods in the design of an overall error control system, a convolutional decoder is driven by the soft decisions from quantized demodulator outputs. This decoder then outputs hard decisions to a block decoder for further error reduction. Combining the attributes of block and convolutional codes into cascaded codes is known as *concatenated coding*.

Concatenated coding, introduced by Forney [5], represents an alternate technique to mitigate noise bursts when channel bandwidth is not constrained. Figure 11.16 demonstrates the architecture of a system that employs an *inner code*, an *outer code*, and interleavers and deinterleavers designed to randomize the location of the burst errors in the symbol stream seen by the channel decoders.

Typically, the inner code, that is, the code that interfaces with the channel, is designed to correct most of the channel errors while the higher rate outer code reduces the error rate to the desired level. The purpose of concatenated coding is to obtain a small error probability with an end-to-end coded system implementation complexity which is less than that which would be required by a single codec. Such coding techniques could also be applied to improve the error probability performance of an existing coding system. One of the most successful concatenated coded systems is one that employs a *convolutional inner code* and a *Reed-Solomon* outer code with interleaving-deinterleaving [8]. The inner codec can take maximum advantage of soft decisions from the demodulator output while the outer Reed-Solomon codec can easily deal with the hard-decision output from the inner convolutional codec.

In many cases, the analog signals being processed in the front end of the receiver or the confidence symbols shown in Fig. 11.6 can be used to generate a "fade detector" or "erasure indicator" signal (shown dotted in Fig. 11.16). When an error-correcting code is used as the outer code, this erasure signal may enable the decoder to correct more severe noise bursts than it could if there were no such control signal. We will discuss the error and erasure correction capacity for RS codes in the next chapter. This will be followed by presenting the performance of certain convolutional and concatenated coding systems.

11.5.1 Coding and Modulation Matching for Concatenated Coding Systems

Consider the binary concatenated coding-modulation system depicted in Fig. 11.17 which cascades the binary codes (N_{ci}, K_{ci}) of code rates r_{ci} with $M_{ci} = 2^{K_{ci}}$ for all $i = 1, 2, \cdots,$ L. At each encoder output, we must have (let $T_{c1} = T_b$)

$$T = K_{c1}T_{c1} = K_{c2}T_{c2} = \cdots = K_{cL}T_{cL} \text{ sec} \tag{11.23}$$

where T_{ci} are the duration of the coded bits at the output of each coder. Now the signal generator must map N_{cL} coded bits of the Lth coder output to one of the M available modulation waveforms (signals). Then

$$T = (\log_2 M)T_{cL} \tag{11.24}$$

and since the channel rate $\mathcal{R} = (\log_2 M)/T_s$, we can use (11.23) and (11.24) to write

$$T = (\log_2 M)T_b \prod_{i=1}^{L} r_{ci} \tag{11.25}$$

so that the channel rate becomes $\mathcal{R} = \mathcal{R}_b/r_e$ bits/sec where

$$r_e \triangleq \prod_{i=1}^{L} r_{ci} \tag{11.26}$$

is the equivalent *concatenated code rate*. The *bandwidth expansion* of the coded system over that of an uncoded system is r_e^{-1}, that is

$$\frac{\mathcal{R}}{\mathcal{R}_b} = \frac{1}{r_e} = r_e^{-1} \tag{11.27}$$

A BPSK modulator can be used to transmit the codewords one coded bit at a time, while demultiplexing the coded message bits; a QPSK modulator can transmit them two at a time, one in the I-channel and one in the Q-channel (see Fig. 11.5). In the next chapter, we investigate block-coded digital communication techniques; this is followed by looking at convolutional coded digital communication techniques.

11.6 CLASSIFICATION OF ERROR CORRECTION AND DETECTION CODES

For our purposes, the class of error correction/detection codes to be considered in the next two chapters can be broken into two types—*tree codes* and *block codes*. Both code types can be further classified as linear or nonlinear depending on the hardware mechanization required to generate them. Herein we are concerned with the class of *linear codes*. Figure 11.18 depicts the Venn diagram for error correction/detection codes of interest. The major characteristic of linear codes is that they have a block or a tree architecture with an optional systematic versus nonsystematic structure. As seen from Fig. 11.18, cyclic codes

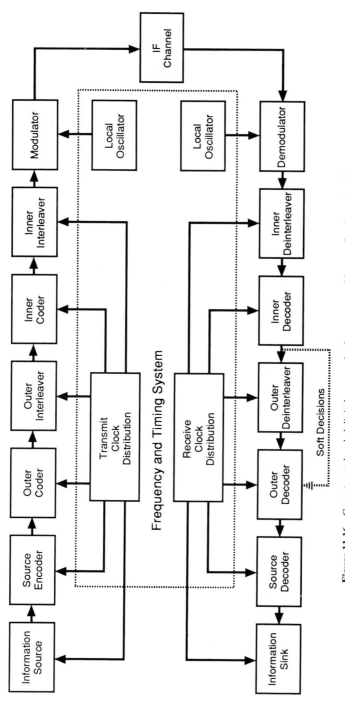

Figure 11.16 Concatenated-coded digital communication system architecture for channels with memory

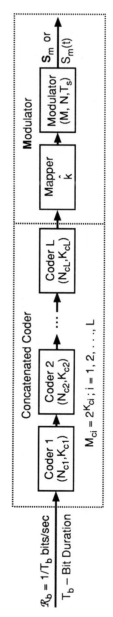

Figure 11.17   Concatenated coding and modulation system model

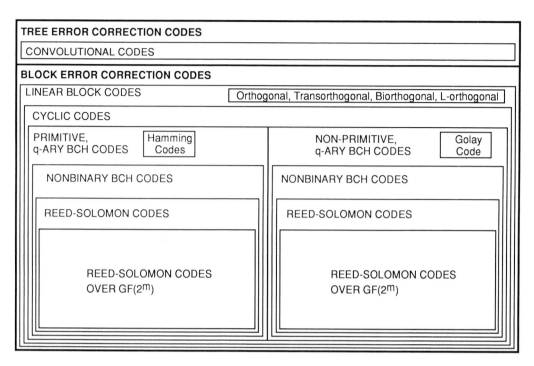

Figure 11.18   Venn diagram for error-correction codes

are a subset of linear block codes [9, 10]. The text book by MacWilliams and Sloane [10] probably contains the most comprehensive discussion of block codes to date while Michelson and Levesque [11] gives a clear discussion of the details of codec design using block or convolutional codes.

Bose, Chaudhuri, and *Hocquenghem* (BCH) codes are a subset of cyclic codes. BCH codes can be classified as P-ary or binary ($P = 2$). P-ary BCH codes have codeword symbols and codeword generator polynomial coefficients which come from Galois field (GF) algebra, viz., GF(P) where P is prime. BCH codeword generator roots are from GF(q)=GF(P^m) where q is the order of the field and m is an integer greater than one. Binary BCH codes are most often used in practice; they are from GF(2). The codeword generators of binary BCH codes are nonbinary; they come from GF(q) = GF(P^m) = GF(2^m). These codes have inherent error detection capability without sacrificing any of the error-correction capability. Berlekamp [12], Lin [13], Lin and Costello [14], Peterson [15], and Peterson and Weldon [16] treat the Galois field algebra of linear block codes and present numerous decoding algorithms. This important material lies outside the scope of this book.

Reed-Solomon (R-S) codes are special cases of q-ary BCH codes (see Fig. 11.18). The code symbols of a binary R-S code are also nonbinary; they are generated from GF(q) = GF(P^m). R-S codes are very powerful burst and random error correcting codes. An excellent tutorial on R-S error correcting coding can be found in Geisel [17]. Discussions of current applications of codes and codec design can be found in Odenwalder [18], Berlekamp [19], and Berlekamp, Peile, and Pope [20]. Odenwalder [18] presents a clean presentation from a system engineering and design point of view. Finally, Chapter 12 covers block coded communication system design and performance, while Chapter 13 presents the use of convolutional codes and their performance.

PROBLEMS

11.1 **(a)** For the one-bit quantizer characteristic illustrated in Fig. P11.1, characterize the channel transition probabilities when the input channel symbols are given by $s_0 = +\sqrt{E_{N_c}}$ and $s_1 = -\sqrt{E_{N_c}}$ with equal probability. Assume additive white Gaussian noise and BPSK signaling.

(b) Sketch the DMC (binary input–binary output) model and label all input to output connections.

11.2 Repeat Problem 11.1 for the quantizer characteristic illustrated in Fig. P11.2. The DMC model for this case represents the binary erasure channel (BEC), discussed in Chapter 3. Here, $M = 2$ and $J = 3$.

11.3 We are given the demodulator output random variable r to quantize. The pdf for r is given by

$$f_r(\rho) = \begin{cases} A - \rho & 0 \le \rho \le 1 \\ A + \rho & -1 \le \rho \le 0 \\ 0 & \text{otherwise} \end{cases}$$

Assuming a 2-level (1-bit) quantizer for this input signal with output levels $R_1 = 1/2$ and $R_2 = -1/2$

(a) Find the constant A.

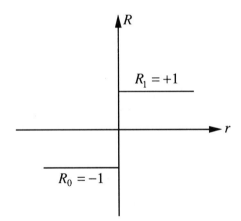

Figure P11.1 Quantizer characteristic required to create the BSC

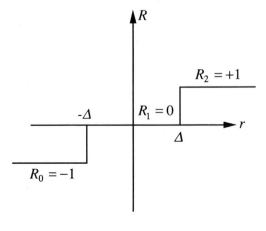

Figure P11.2 Quantizer characteristic required to create the BEC

(b) Find the pdf $f_e(\rho)$ of the quantizer error $e(r) = Q(r) - r = \text{sgn}(r)$, where $\text{sgn}(r) = 1$ if $r \geq 0$ and $\text{sgn}(r) = -1$ if $r < 0$.

(c) Find the distortion $D = E\{e^2(r)\}$.

(d) Now redesign the quantizer so that $R_1 = 1/3$ and $R_2 = -1/3$. Find the new distortion $D = E\{e^2(r)\}$. Compare this to the result obtained in (c).

11.4 The span of an interleaver is a measure of the time duration over which previously continuous bits are spread, and is a metric of the ability of the interleaver to randomize burst errors. For the (10, 20) block interleaver, determine the span for symbol rates of 2.4, 4.8, and 9.6 kb/sec.

11.5 (a) For the block interleaver, show that the memory required at the transmitter or receiver is given by

$$M_r = 2LD\Delta q$$

where $\Delta q = 1$ for block interleavers and $\Delta q = 1$ for deinterleavers with **Demodulators** using hard decisions and $\Delta q = 3$ for deinterleavers with **Demodulators** using 3-bit soft decisions.

(b) For the (10, 20) block interleaver, evaluate the required transmitter and receiver memory for both hard- and soft-decision **Demodulation**.

(c) Repeat (b) for the (39, 116) interleaver.

11.6 (a) Given a coded symbol rate of $\mathcal{R} = 1/T_N = 19.2$ kb/sec, find the span for a block interleaver of length $L = 8$ and depth $D = 8$.

(b) Determine the channel delay τ_d.

(c) Determine the required memory size M_r for hard-decision demodulators.

(d) Using Fig. 11.11 and the numbers 1, 2, 3, ... to fill the interleaver input rows, demonstrate the interleaver output matrix and the deinterleaver output matrix.

11.7 Repeat Problem 11.6 assuming a convolutional interleaver.

11.8 Show that the memory M_r required for the convolutional interleaver is given by

$$M_r = \frac{\Delta q(D+1)}{L} \sum_{k=1}^{L} (k-1)$$

where Δq is defined in Problem 11.5.

11.9 Repeat Problem 11.5 (b) and (c) for a convolutional interleaver-deinterleaver system. Compare the results found in Problem 11.5 with those obtained for the convolutional I-D system.

11.10 For the (4, 3) helical interleaver illustrated in Fig. 11.12

(a) Specify the symbol sequence versus time across the channel.

(b) Specify the symbol sequence versus time out of the decoder assuming no channel errors.

(c) Repeat (a) and (b) for the (8, 7) helical interleaver specified in Fig. 11.13.

11.11 (a) Assuming coded BPSK modulation, develop the channel transition probability mappings for the hard- and soft-decision demodulator thresholds illustrated in Fig. P11.11.

(b) For each of the scalar quantization methods, illustrate the equivalent DMC model for use in the decoding operation.

11.12 (a) Given a noise burst of 20 msec duration, design a block I-D system to accommodate a bit rate of 9, 600 bits/sec such that all errors into the decoder due to this burst noise are independent.

(b) What is the interleaver span, delay, and memory required?

(c) Repeat (a) and (b) for a convolutional interleaver.

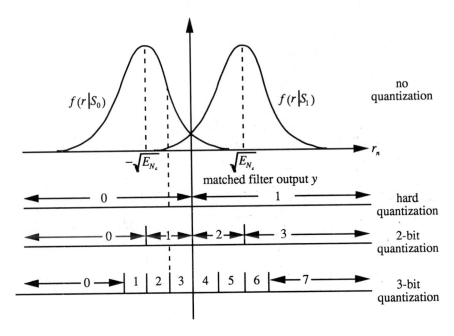

Figure P11.11   Soft decision demodulator thresholds

REFERENCES

1. Shannon, C. E. and W. Weaver, *The Mathematical Theory of Communication*, Urbana, IL: University of Illinois Press, 1949.

2. McEliece, R. J., "Communication in the Presence of Jamming—An Information-Theoretic Approach," Chapter 3 in *Secure Digital Communications*, ed. G. Longo, Wien, NY: Springer-Verlag, 1983.

3. Berlekamp, E. R., *Algebraic Coding Theory*, New York: McGraw-Hill, 1968.

4. Olson, D., "Hybrid Interleaver," *International Telemetering Conference*, vol. 19, San Diego, CA, pp. 703–04, 1983.

5. Forney, G. D., "Burst-Correcting Codes for the Classic Bursty Channel," *IEEE Transactions on Communication Technology*, vol. 19, pp. 772–81, 1971.

6. Ramsey, J. L., "Realization of Optimum Interleavers," *IEEE Transactions on Information Theory*, vol. IT-16, pp. 338–45, 1970.

7. Berlekamp, E. R. et. al., *Interleaved Coding for Bursty Channels*, Cyclotomics Technical Report, April 12, 1983.

8. Berlekamp, E. R. and J. L. Ramsey, "Readable Erasures Improve the Performance of Reed-Solomon Codes," *IEEE Transactions on Information Theory*, vol. IT-24, pp. 632–33, 1978.

9. Clark, Jr., G. C. and J. B. Cain, *Error-Correction Coding for Digital Communications*, New York, NY: Plenum Press, 1981.

10. MacWilliams, F. J. and N. J. A. Sloane, *The Theory of Error-Correcting Codes*, Amsterdam: North Holland, 1977.

11. Michelson, A. M. and A. H. Levesque, *Error-Control Techniques For Digital Communication*, New York: John Wiley and Sons, 1985.

12. Berlekamp, E. R., *Algebraic Coding Theory*, New York: McGraw-Hill, 1968.

13. Lin, S., *An Introduction to Error-Correcting Codes*, Englewood Cliffs, N.J.: Prentice Hall, 1970.

14. Lin, S. and D. J. Costello, *Error Control Coding: Fundamentals and Applications*, Englewood Cliffs, N.J.: Prentice Hall, 1983.

15. Peterson, W. W., *Error-Correcting Codes*, New York: MIT Press, 1961.

16. Peterson, W. W. and E. J. Weldon, *Error-Correcting Codes*, New York: MIT Press, 1972.

17. Geisel, W. A., "NASA Technical Memorandum 102162," Houston, TX: Lyndon B. Johnson Space Center, June 1990.

18. Bartee, T. C., *Data Communications, Networks, and Systems*, Chapter 10 by J. P. Odenwalder, IN: Howard W. Sams & Co., 1985.

19. Berlekamp, E. R., "The Technology of Error-Correcting Codes," *Proceedings of the IEEE*, vol. 68, no. 5, pp. 564–93, May 1980.

20. Berlekamp, E. R., R. E. Peile and S. P. Pope, "The Application of Error Control to Communications," *IEEE Communications Magazine*, vol. 25, no. 4, pp. 44–57, April 1987.

Chapter 12

⊓⌐

Block-Coded Digital Communications

⊓⌐

In previous chapters, we have optimized the design of a digital communication system from the viewpoint of the receiver; that is, we have determined optimum receiver structures and evaluated their performance for an arbitrary set of modulation waveforms which represent the messages in the channel. The following two chapters are primarily concerned with the problem of optimizing the transmitter design, that is, determining the set of coded modulation waveforms that minimizes the bit error probability. Optimizing the transmitter design for a given channel characterization represents the *signal design* or *coding and modulation* selection problem. We shall equivalently pose this problem as one of finding the best codec and modem design which minimizes the bit error probability when the receiver used is optimum. This leads us to presenting the Shannon capacity theory.

In Chapter 11, the functional architecture of a coded digital communication system was presented. In this chapter, we shall exercise this model by developing the performance of block-coded digital communication systems. The "R_0-criterion" for modem and codec design will be introduced as the methodology for separating the design of the codec from that of the modem; hence, the material presented in Chapters 3 through 10 can be used in the modulation-demodulation design process for uncoded systems, while the theory given in Chapters 11 through 13 can be used in the design of the coder-decoder (codec) and Modulator-Demodulator (modem).

The analysis revolves around the fundamental parameter called the *cutoff rate*, R_0 ("R zero"). As we shall see, R_0 (in bits/dimension) is a Shannon capacity-like quantity defined for any CMC or DMC, whose value is less than the *Shannon capacity*, C_{N_c}, by approximately 3 dB in the energy-efficient region where coding is of interest. From a system engineer's perspective, R_0 sets the practical limit on the rate at which information can be reliably transmitted through the channel. Apparently, Wozencraft and Kennedy [1] were the first to propose the use of the R_0 criterion for modulation-demodulation system design; however, as pointed out later by Massey [2], their proposal fell on deaf ears. They observed that a modem for use in a coded system should be designed to achieve the highest possible value of R_0 rather than the lowest value of post-demodulation bit error probability for the comparable uncoded system. In this context, R_0 is optimized not only with respect to the probability distribution of data symbols entering the channel and the modulation-demodulation choice, but also with respect to the quantization boundaries which define the decision regions at the quantized demodulator output; that is, for a given number, J, of quantized demodulator out-

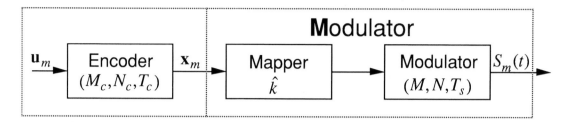

$$T = T_b \log_2 M_c = N_c T_c = \hat{k} T_s; \quad T_s = T_c \log_2 M / \log_2 Q = N T_N$$

$$\frac{T}{T_s} = \frac{1}{r_c}\left(\frac{\log_2 M_c}{\log_2 M}\right)\log_2 Q = \frac{\log_2 Q}{R_N}\left(\frac{N_c}{N}\right) = \frac{N_c \log_2 Q}{\log_2 M}$$

Figure 12.1a Block-coded digital communication system model: Coded modulator model for coded system design

puts (quantization intervals—see Chapter 11), the decision boundaries will be chosen so as to maximize the value of R_0 subject to a hardware complexity constraint. This criterion is also applicable when convolutional and/or concatenated codes are used for error control. In fact, in the development and analysis of powerful decoding algorithms for long constraint-length convolutional codes, R_0 has been given a more descriptive name—the *computational cutoff rate*, sometimes denoted R_{COMP}. We proceed by developing an analytic understanding of this important parameter with respect to the coding and modulation process.

12.1 BLOCK-CODED DIGITAL COMMUNICATION SYSTEM ARCHITECTURE

The functional architecture of a block-coded communication system is illustrated in Fig. 12.1. System engineers accept the perspective that coding and modulation are both components of the signal design problem and that demodulation and decoding are likewise components of the signal detection problem. The (M_c, N_c, T_c) encoder, mapper, and (M, N, T_s) modulator of Fig. 12.1a work with a binary alphabet as follows: every T sec $\log_2 M_c$ *message bits* are encoded into N_c *coded message bits*, and every $T_s = T/\hat{k}$ sec, the mapper maps $\log_2 M$ of these N_c coded message bits onto the M-ary modulator waveform $s_m(t)$; thus, in $T = \hat{k} T_s$ sec, the *coded modulation* waveform $S_m(t), m = 0, \ldots, M_c - 1, 0 \le t \le T$, contains $\log_2 M_c$ bits of information. This process is functionally illustrated in Fig. 12.1a and described in detail in Chapter 11. Numerous coded modulation examples using BPSK, QPSK, QASK, QAM, and so on, are given in Chapter 11. See Tables 11.2, 11.3, 11.4, and 11.5 for the details of $S_m(t)$.

Block-coded communications proceed by taking blocks $\mathbf{u}_m = (u_{m1}, \ldots, u_{mK_c})$ of K_c source output symbols and presenting these to the channel encoder as one of M_c possible messages. Here $M_c = Q^{K_c}$ and Q represents the alphabet size of the encoder input; that is, $u_i \in \mathbf{u}$ can take on one of Q values while $\mathbf{u}$ can take on one of Q^{K_c} values. The baseband channel codeword vectors $\{\mathbf{x}_m = (x_{m1}, x_{m2}, \ldots, x_{mN_c}), m = 0, \ldots, M_c - 1\}$ are then mapped, one-to-one, by the modulator onto the IF coded modulation vectors $\{\mathbf{S}_m = (s_{m1}, s_{m2}, \ldots, s_{mN_c})\}$ or, equivalently, the IF channel input waveform $S_m(t)$. For purposes

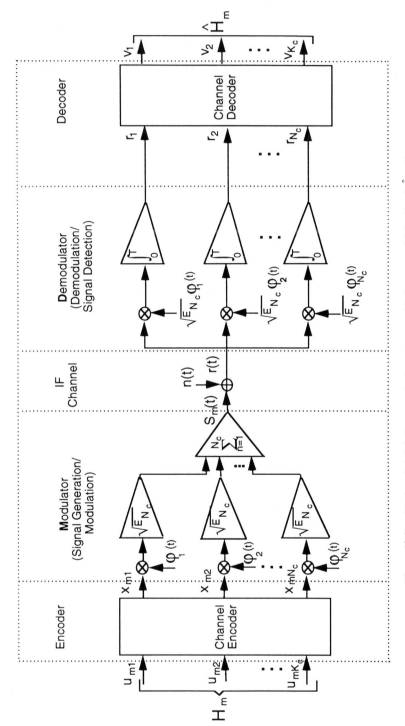

Figure 12.1b Block-coded digital communication system model: System model for $N = 1$, $\hat{k} = N_c$

of being precise, in this section we assume a **M**odulator which produces the coded modulation waveform

$$S_m(t) = \sum_{n=1}^{N_c} s_{mn}\varphi_n(t), \quad m = 0, 1, \ldots, M_c - 1, \quad 0 \leq t \leq T \tag{12.1}$$

so that $\hat{k} = N_c$ and $T_c = T_s$. As discussed in Chapters 3, 4, and 11, the building blocks $\{\varphi_n(t)\}$ form a known orthonormal set over the time interval T sec. Each of these coded signals has energy

$$E_m = P_m T = \int_0^T S_m^2(t)dt = |\mathbf{S}_m|^2 \tag{12.2}$$

for all $m = 0, \ldots, M_c - 1$. Notice that $s_{mn} = \sqrt{E_{N_c}}x_{mn}$ where E_{N_c} is the peak energy available per codeword dimension. For constant envelope signals (for example, M-PSK), E_{N_c} is also the average energy used per codeword dimension.

A particular mechanization of the encoder-mapper-modulator is shown in Fig. 12.1b for the special case $\hat{k} = N_c$, $N = 1$. Two cases of particular interest are binary coded BPSK and Q-ary coded Q-ary ASK. We note that while the actual coded message bits assumed in coding theory for BPSK are $x_{mn} = 0, 1$, we often write $x_{mn} = +1, -1$ corresponding to voltage levels realized in hardware. Thus, for binary coded BPSK, we have $x_{mn} = \pm 1$, while for Q-ary coded Q-ary ASK we have $x_{mn} = q_i = (-1) + (2i - 2)/(Q - 1)$ with probability p_i, $i = 1, \ldots, Q$, where Q is even, and $s_{mn} = \sqrt{E_{N_c}}x_{mn}$. For both of these coded modulation techniques,

$$\varphi_n(t) = \begin{cases} \sqrt{2/T_s}\cos\omega_0 t, & 0 \leq t \leq T_s \\ 0, & \text{else} \end{cases}$$

for all n.

12.1.1 Coding from the Baseband and IF Channel Model Perspectives

When the IF channel model is assumed to add white Gaussian noise (N_0 watts/Hertz single-sided) to the transmitted signal, it was shown in Chapters 3 and 4 that the optimum transmitter-receiver configuration for equally probable transmitted messages reduces to that illustrated in Fig. 12.1. Notice that the signal generator and modulator functions and the demodulation and signal detection functions in the uncoded system have been combined into the **M**odulator and **D**emodulator functions of the coded system; see Chapter 11. We use bold, capital letters in "**M**odulator" when referring to the combined mapper/modulator, and in "**D**emodulator" when referring to the combined demodulator/detector. In addition, the interleaver and deinterleaver functions (if needed) in a coded system are also combined into the channel encoder and decoder functions respectively.

Referring now to Fig. 12.1, we note that if hypothesis H_m is true, that is, if message m_m is to be transmitted, then the code vector $\mathbf{x}_m$ implies coded modulation vector $\mathbf{S}_m$, which implies the IF channel waveform $S_m(t)$. We then observe $r(t) = S_m(t) + n(t)$, from which the received signal plus noise projection $\mathbf{r} = (r_1, \ldots, r_{N_c})$ is generated by the demodulator; $r_n = s_{mn} + n_n$ for all n. The decoded vector $\mathbf{v} = (v_1, \ldots, v_{K_c})$ is generated from $\mathbf{r}$ and transformed into the recovered message $\hat{m}$ corresponding to hypothesis $\hat{H}_m$. Table 12.1 summarizes several sets of time and frequency orthonormal functions (communication carriers) of great practical interest. The I-Q time-orthonormal functions are of greatest interest in the

Table 12.1 Communication Carriers of Greatest Practical Interest

Modulation	Carriers	Communication Basis Functions ($n = 1, 2, \ldots, \hat{k}$)
BPSK or ASK	Time orthonormal (ω_c a multiple of π/T_N)	$\varphi_n(t) = \begin{cases} \sqrt{2/T_N} \cos \omega_c t, & (n-1)T_N \le t \le nT_N \\ 0, & \text{else} \end{cases}$
M-PSK, M-DPSK, QASK, or QAM	I-Q modulation: time orthonormal (ω_c a multiple of $2\pi/T_N$)	$\varphi_{2n}(t) = \begin{cases} \sqrt{2/T_N} \sin \omega_c t, & (n-1)T_N \le t \le nT_N \\ 0, & \text{else} \end{cases}$ $\varphi_{2n-1}(t) = \begin{cases} \sqrt{2/T_N} \cos \omega_c t, & (n-1)T_N \le t \le nT_N \\ 0, & \text{else} \end{cases}$
Coherent M-FSK	Frequency orthonormal (ω_c a multiple of π/T)	$\varphi_n(t) = \begin{cases} \sqrt{2/T} \cos \left[(\omega_c + \pi n/T)t\right], & 0 \le t \le T \\ 0, & \text{else} \end{cases}$
Noncoherent M-FSK	I-Q modulation: frequency orthonormal (ω_c a multiple of $2\pi/T$)	$\varphi_{2n}(t) = \sqrt{2/T} \sin \left[(\omega_c + 2\pi n/T)t\right], \, 0 \le t \le T$ $\varphi_{2n-1}(t) = \sqrt{2/T} \cos \left[(\omega_c + 2\pi n/T)t\right], \, 0 \le t \le T$

implementation of coherent, partially coherent and differentially coherent systems, while the I-Q frequency-orthonormal functions are used as building blocks in noncoherent communication systems.

IF channel perspective. When the IF channel is considered, we wish to select the coded modulation vector set $\{\mathbf{S}_m, m = 0, \ldots M_c - 1\}$ so as to minimize the probability of error. This points immediately to the problem of optimally mapping the vector space generated by the encoder onto the IF waveform space (generated at the modulator output) such that the distance properties of the code are not degraded. In what follows, we shall address this coded modulation problem from a system engineer's perspective. We shall consider the coded communication design problem assuming that the mapping of the encoder output onto the modulator output is isometric when properly normalized. A metric space (X, d) is *isometric* [19] to a metric space (Y, e) if there exists a one-to-one, onto function $f : X \to Y$ which preserves distances; that is, for all $p, q \in X$, $d(p, q) = e(f(p), f(q))$. Here d is a "distance function" defined on the Cartesian Product $X \times X$, so that $d(p, q)$ is the distance (as defined by d) between two points $p, q \in X$. Similarly, $e(f(p), f(q))$ is the distance (as defined by e) between two points $f(p), f(q) \in Y$.

Baseband channel perspective. For the baseband channel model (see Chapter 11), we can equivalently pose the problem of communicating with block codes from a geometric point of view. A block code consists of a subset of M_c vectors (points or codewords) chosen from an N_c-dimensional vector space. It may be thought of as a way of transmitting an integer $m = 0$ to $M_c - 1$ to the receiver by sending the corresponding codeword $\mathbf{x}_m$. A *decoding system* for such a code is a partitioning of the N_c-dimensional observation space into M_c subsets or disjoint decision regions Λ_m. If the received signal plus noise projection $\mathbf{r}$ belongs to Λ_m, the transmitted message is taken to be message m. See Chapters 3 and 4 for all the details. An *optimal decoding system* for a code is one which minimizes the probability of error for the code. Since the Gaussian density is monotone decreasing about its mean, an optimal decoding system for a given code is one which decodes any received signal as the integer m corresponding to the geometrically nearest codeword, assuming equally probable, equal energy signals. If there are several codewords at the same minimum distance, any of these may be used without affecting the probability of error. As shown in Chapter 3, a decoding system of this sort is called *minimum distance* decoding or *maximum-likelihood* decoding; it amounts to partitioning the N_c-dimensional space into N_c-dimensional polyhedra or decision regions "centered" around the different signal points. Each polyhedron is bounded by a finite number (not more than $M_c - 1$) of $(N_c - 1)$-dimensional hyperplanes. In this context, there are $M_c = Q^{K_c}$ different codewords to be assigned to points in our N_c-dimensional signal space, in which there are Q^{N_c} points available; it follows that there are $Q^{N_c M_c}$ different ways (codes) to assign the M_c codeword vectors. This is the usual approach to studying coding system design; that is, the baseband channel model is assumed, and the modem design problem does not explicitly enter into the system model. The question of isometric mapping of the encoder output onto the modulator output does not arise in the baseband model; the modem remains hidden in the baseband channel and is assumed to be ideal. In this chapter, we work with the IF CMC and DMC models which take the modem outside the channel.

12.1.2 Fundamental Parameters of a Block-Coded Communication System

Before proceeding, it seems appropriate to summarize the communication parameters and fundamental relationships of our coded system model.[1] It is highly recommended that the reader gain a good physical understanding of each; they not only play a key role in the theory to be presented, but they also enter into the specification of the subsystems illustrated in the functional models shown in Figures 11.2 and 12.1. Understanding of these models and their associated parameters will be required if one wishes to comprehend the system engineering implications associated with Shannon's random coding bounds, his channel capacity theorem, and the physical meaning of efficient and reliable communications.

Tables 12.2 and 12.3 provide a summary of the basic parameters. **Study** the contents of these two tables; they are of great importance to system engineers. For convenience, the parameters are divided into several categories; some are physical, yet others take on numerical values. The first category in Table 12.2 consists of time intervals. These intervals are generated in practice by means of frequency generators called clocks when dealing with baseband waveforms and called local oscillators when generating the IF and RF waveforms at their respective nominal center frequencies. Without loss of generality, we shall assume the existence of a single clock which can be used to generate the various time intervals defined in Table 12.2. We introduce bandwidth via the *sampling theorem* (or the so-called Nyquist interval of T_c sec), that is, $B \triangleq 1/2T_c$ Hertz—the single-sided *Nyquist* bandwidth.

12.2 PERFORMANCE OF BLOCK-CODED COMMUNICATION SYSTEMS

The exact performance expression of an uncoded M-ary, coherent communication system with equal energy signals was given by (4.80). Here we wish to establish the interplay between the channel, the codec, and the modem illustrated in Figures 11.2 and 12.1. This is best accomplished using the union bound result developed in Section 3.2.7, which allows

---

1. A coded digital communication system is complex and consists of many parameters involving different time intervals, rates, energy, energy-to-noise ratios, and bandwidth.

Table 12.2   Fundamental Parameters of a Coded Digital Communication System

Time intervals	T_c—Time used per codeword dimension, sec T_s—Modulator waveform duration; $T_s = NT_N$ sec T_b—Time per message bit, sec T—Codeword duration; $T = (\log_2 M_c)T_b = \hat{k}T_s = N_cT_c$, $T_c = r_cT_b$, sec T_N—Time used per modulator dimension by modulator, sec
Codec/modem	$\hat{k}$—Number of times modulator is used per codeword transmission $M_c = Q^{K_c}$—Number of codewords in code dictionary K_c—Number of information symbols per codeword (message length) N_c—Number of coded message symbols per codeword (Codeword dimensionality = block length) M—Number of waveforms modulator generates per duration T_s sec N—Dimensionality of modulator; $N = WT_s = WT/\hat{k}$
Rates	$\mathcal{R}_b = (\log_2 M_c)/T = r_c/T_c = 1/T_b$, bits/sec $\mathcal{R} = \log_2 M/T_s = R_N/T_N$; modulator output rate, bits/sec $r_c = \log_2 M_c/N_c$; code rate, bits/dimension $R_N = (\log_2 M)/N$; modulator output rate, bits/dimension R_0—Channel cutoff rate, bits/dimension $\mathcal{R}_0 = R_0 W$—Channel cutoff rate, bits/sec
Energy (joules)	$E = PT$; received energy per codeword $E_s = PT_s$; modulator waveform received energy per use of modulator $E_{N_c} = PT_c$; received energy per codeword dimension $E_b = PT_b$; received energy per information bit $E_N = PT_N$; received energy per modulator dimension used $N_0 = kT^\circ$; noise energy or single-sided PSD $P = E/T$; received power, Watts
Bandwidths (Hz)	$W = N_c/T = 2B$; two-sided IF Nyquist bandwidth $B = 1/2T_c$; single-sided Nyquist bandwidth $D = 1/T_N = N/T_s$ number of dimensions used per sec by modulator $D = W = 2B$ for binary coded BPSK and binary coded QPSK r_c^{-1} and R_N^{-1}—Coder and modulator bandwidth expansion, respectively
Normalized throughputs	$\frac{\mathcal{R}}{W} = \frac{\log_2 M}{WT_s} = \frac{\log_2 M}{NWT_N} = \frac{R_N}{WT_N}$ bits/sec/Hz = bits/dimension $\frac{\mathcal{R}_b}{W} = r_c$, bits/sec/Hz = bits/dimension
Frequency	f_c—IF channel center frequency, Hz

us to introduce the reader to the important concept called the *channel cutoff rate*, R_0. As we shall see, this rate is a simple function of the modem and the communication parameter E_{N_c}/N_0. In Section 3.2.7 and in (4.81), we showed that the conditional message error probability performance of an M_c-ary communication system with equally likely messages over an AWGN channel is union bounded from above by a sum of the *pairwise error probabilities* (PEP's) discussed in Appendix 3D; that is,

Table 12.3 Processing Gains and Energy-to Noise Ratios

Processing Gains	$PG_N = WT_N = 2BT_N$ $PG_b = WT_b = 2BT_b$ $PG = WT = 2BT = N_c$
Energy-to-noise ratios	$\frac{E_b}{N_0}$—Received energy per bit-to-noise density ratio $\frac{E_N}{N_0}$—Received energy per modulator dimension-to-noise density ratio $\frac{E}{N_0}$—Received energy per message-to-noise density ratio $\frac{E_{N_c}}{N_0}$—Received energy per coded message bit-to-noise density ratio
Received signal power-to-noise density ratio for constant envelope modulation, Hertz	$\frac{P}{N_0} = \frac{E_{N_c}}{N_0}\frac{1}{T_c} = \frac{E_N}{N_0}\frac{1}{T_N} = \frac{E_s}{N_0}\frac{1}{T_s}$ $= \frac{E_b}{N_0}\frac{1}{T_b} = \frac{E}{N_0}\frac{1}{T}$

$$P(E \mid S_i) \leq \sum_{\substack{j=0;i\neq j}}^{M_c-1} \underbrace{Q\left(\frac{d_{ij}}{\sqrt{2N_0}}\right)}_{(PEP)} \tag{12.3}$$

where the Euclidean distance between the coded modulation waveforms $S_i(t)$ and $S_j(t)$ is given by

$$d_{ij} = |\mathbf{S}_i - \mathbf{S}_j| = \sqrt{\int_0^T [S_i(t) - S_j(t)]^2 dt}$$

$$= \sqrt{E_i + E_j - 2\sqrt{E_i E_j}\cos\theta_{ij}} = \sqrt{E_i + E_j - 2\sqrt{E_i E_j}\gamma_{ij}} \tag{12.4a}$$

From (12.3) we see that good coded modulation techniques employ coded modulation vectors such that the Euclidean distance between any pair is large enough to ensure that the channel noise will not cause the message that was sent to be confused with another message.

The *Hamming distance* $D(\mathbf{x}_i, \mathbf{x}_j) \triangleq D_{ij}$ between the two binary codewords $\mathbf{x}_i$ and $\mathbf{x}_j$ is the number of positions in which the two codewords differ. For example, the Hamming distance between $(0, 0, 1, 1, 1)$ and $(1, 0, 1, 0, 1)$ is 2. Considerable insight into the coded communications problem is obtained by writing (12.3) in terms of the *Hamming distance* D_{ij} when the coded modulation waveforms have equal energy E. In Section 12.8.1, we will show that for binary coded BPSK modulation, the squared Euclidean distance is related to the Hamming distance through

$$d_{ij}^2 = 2E(1 - \gamma_{ij}) = 4E_{N_c}D_{ij} = 4ED_{ij}/N_c \tag{12.4b}$$

Apparently, for most other coded modulation techniques, the relationship between the Hamming distance between codewords and the corresponding Euclidean distance between coded modulation vectors cannot be readily identified.

Using (12.4b), (12.3) can be rewritten as

$$P(E \mid S_i) \leq \sum_{j=0,i\neq j}^{M_c-1} Q\left(\sqrt{\frac{2E}{N_0}\left(\frac{D_{ij}}{N_c}\right)}\right) \tag{12.5}$$

If we denote the *minimum Hamming distance* by D_{min} and the *minimum Euclidean distance* by d_{min}, then (12.5) can be used to upper bound the message error probability for a block-coded system with equally probable messages which uses binary coded BPSK modulation. The resulting bound is

$$P(E) < (M_c - 1)Q\left(\frac{d_{min}}{\sqrt{2N_0}}\right) = (M_c - 1)Q\left(\sqrt{\frac{2E}{N_0}\left(\frac{D_{min}}{N_c}\right)}\right)$$

$$= (M_c - 1)Q\left(\sqrt{2r_c D_{min}\frac{E_b}{N_0}}\right)$$

(12.6)

where we have used the facts that $E = PT = P(N_c T_c) = P N_c(r_c T_b) = E_b(N_c r_c)$ and the definitions

$$d_{min} \triangleq \min_{\substack{i,j \\ i \neq j}} d_{ij} \text{ and } D_{min} \triangleq \min_{\substack{i,j \\ i \neq j}} D_{ij}$$

(12.7)

The result in (12.6) will be very useful in what follows because the minimum Hamming distance D_{min} plays a key role in error-correction code selection. From (12.6), we also see that the use of binary coding has effectively enhanced the value of E_b/N_0 by the factor $r_c D_{min}$ over that of an uncoded BPSK system.

12.2.1 Transmission of Orthogonal Modulation Vectors

Since (4.80) cannot in general be simplified, we first illustrate key features of (12.3) by assuming transmission of N_c-dimensional, equal energy, orthogonal coded modulation vectors, namely

$$\mathbf{S}_i = \sqrt{E}\boldsymbol{\varphi}_i$$

(12.8)

with $\boldsymbol{\varphi}_i^T \cdot \boldsymbol{\varphi}_j = \delta_{ij}$ for all $i, j = 0, 1, \ldots, M_c - 1$ (these codes will be considered further in Section 12.3.1). In this case, $d_{ij} = \sqrt{2E}$. Thus (12.3), when averaged over equal prior probabilities, reduces to

$$P(E) \leq (M_c - 1)Q\left(\sqrt{E/N_0}\right)$$

(12.9)

Using the fact that $Q(x) < \exp(-x^2/2)$, then

$$P(E) < M_c \exp(-E/2N_0)$$

(12.10)

If we assume a binary code with $Q = 2$, then $M_c = 2^{K_c}$. Since the bit rate $\mathcal{R}_b = \log_2 M_c/T = K_c/T$ or $K_c = \mathcal{R}_b T$ (see Table 12.2), we can write (12.10) as

$$P(E) < 2^{K_c} \exp(-E/2N_0) = 2^{\mathcal{R}_b T} \exp(-PT/2N_0)$$

(12.11)

or

$$P(E) < \exp\left[-T\left(\frac{P}{2N_0} - \mathcal{R}_b \ln 2\right)\right]$$

(12.12)

so

$$P(E) < \exp\left[-K_c(E_b/2N_0 - \ln 2)\right]$$

(12.13)

From (12.12) we see that the decoded message error probability can be made to *approach zero exponentially* with increasing T as long as the bit rate $\mathcal{R}_b$ satisfies the bound

$$\mathcal{R}_b < \frac{P}{N_0}\left(\frac{1}{2\ln 2}\right) \simeq .72\,P/N_0 \text{ bits/sec} \tag{12.14}$$

Alternatively, from (12.13) we see that $P(E)$ approaches zero exponentially as K_c increases, as long as the *communication efficiency* parameter

$$\frac{E_b}{N_0} > 2\ln 2 \simeq 1.4 \text{ dB} \tag{12.15}$$

Since $N_0 = kT^\circ$, it follows that the minimum received energy per bit per degree Kelvin is

$$E_b/^\circ K = 2k\ln 2 \simeq 1.91 \times 10^{-23} \text{ joules}/^\circ K \tag{12.16}$$

where $k = 1.38 \times 10^{-23}$ joules$/^\circ K$ is *Boltzmann's* constant. At room temperature ($T^\circ = 290^\circ K$), this requires an energy per bit of $E_b > 5.55 \times 10^{-21}$ joules! The results established in (12.12) and (12.13) differ from the bit-by-bit signaling results given in (11.7) for large K_c when $E_b/N_0 > 2\ln 2$. We next consider the reason for such contrasting behavior.

In the bit-by-bit signaling (signal points on the vertices of a hypercube) of K_c bits using uncoded BPSK, increasing T (or K_c) forced $P(E)$ toward unity regardless of how large the value for E_b/N_0. In the case of signaling with orthogonal codes, by increasing K_c we can force $P(E)$ to approach zero (perfectly reliable communication) provided that E_b/N_0 exceeds 1.4 dB! The reason for this interesting asymptotic behavior is that the minimum Euclidian distance between the signals (vectors) used to accomplish bit-by-bit signaling is $d = 2\sqrt{E_b}$, no matter how large we make K_c or T. In the case of block-orthogonal coded modulation, $d = \sqrt{2K_cE_b}$, so that as K_c increases, so does the distance to the nearest coded modulation vector. Finally, we note that we have exploited two different signal designs; however, in both cases, we use the optimum matched filter receiver for signal detection (demodulation and decoding). The important role which signal design has on communication reliability and efficiency is clearly manifested by comparing these two signaling techniques. In the next section, we consider the generation of three types of binary codeword dictionaries.

12.3 SPECIAL TYPES OF BINARY BLOCK CODES

In this section, we describe methods of generating binary orthogonal, biorthogonal, and transorthogonal (simplex) codes. The codewords are assumed to be represented by sequences of "0s" and "1s", the case of greatest interest. Block codes have the interesting feature that the minimum Hamming distance between any two codewords is related to the *error-correcting capability*, say t (not to be confused with time) of the code. In general, it is known [3] that a code with minimum Hamming distance

$$D_{\min} \triangleq \min_{\substack{i,j \\ i \neq j}} D_{ij} \tag{12.17}$$

has the t error-correcting capability

$$t = \lceil (D_{\min} - 1)/2 \rceil \tag{12.18}$$

where the brackets denote the largest integer no greater than $(D_{\min} - 1)/2$. It is also true (see Section 12.8.1) that the cross-correlation coefficient between any two binary codewords, say $\mathbf{x}_i$ and $\mathbf{x}_j$, of length N_c, is given by

$$\gamma_{ij} = \frac{A_{ij} - D_{ij}}{N_c} = \frac{A_{ij} - D_{ij}}{A_{ij} + D_{ij}} \tag{12.19}$$

where A_{ij} represents the number of term-by-term agreements and D_{ij} represents the number of term-by-term disagreements (or the Hamming distance) between the two codewords. We shall use these facts in what follows by considering the transmission of specific block codes using BPSK. For this case, we note that $R_N = 1$ and that the coding to modulation mapping is isometric.

12.3.1 Orthogonal Binary Codes

Let $\mathbf{H}_{K_c}$ be a $2^{K_c} \times 2^{K_c}$ matrix defined inductively by

$$\mathbf{H}_{K_c} \triangleq \begin{bmatrix} \mathbf{H}_{K_c-1} & \vdots & \mathbf{H}_{K_c-1} \\ \cdots & \vdots & \cdots \\ \mathbf{H}_{K_c-1} & \vdots & \overline{\mathbf{H}}_{K_c-1} \end{bmatrix} = \begin{bmatrix} \mathbf{x}_0 \\ \mathbf{x}_1 \\ \vdots \\ \mathbf{x}_{N_c-1} \end{bmatrix} \tag{12.20}$$

where we have assumed $N_c = M_c$ and

$$\mathbf{H}_1 \triangleq \begin{bmatrix} 0 & 0 \\ 0 & 1 \end{bmatrix}. \tag{12.21}$$

Here the overbar denotes the complement of the "**H**" matrix, wherein the ones are replaced by zeros and vice versa. Note that the rows of $\mathbf{H}_{K_c}$ represent the M_c orthogonal codewords $\{\mathbf{x_m} = (x_{m1}, x_{m2}, \ldots, x_{mN_c})\}$ (see Figures 11.3 and 12.1). Such a matrix is called a Hadamard matrix [4] and the codeword dictionary generated by (12.20) is known as a first-order $(2^{K_c}, K_c)$ *Reed-Muller* code. For an orthogonal code, $A_{ij} = D_{ij}$ so that $\gamma_{ij} = 0$, for all $i \neq j$. Thus, from (12.4b), $\gamma_{ij} = 0 = 1 - 2D_{ij}/N_c$ since $A_{ij} = N_c - D_{ij}$, so

$$D_{ij} = D_{\min} = N_c/2 = 2^{K_c-1}, \quad i \neq j \tag{12.22}$$

and

$$t = \lceil \left(2^{K_c-1} - 1 \right) /2 \rceil \tag{12.23}$$

errors are correctable.

   Now that the codeword dictionary has been formed, all that remains to completely specify the code is to define the mapping between encoder input vectors $\mathbf{u}_m = (u_{m1}, \ldots, u_{mK_c})$ and codeword vectors $\mathbf{x}_m$. Define a matrix $\mathbf{G}_{K_c}$ to be a $2^{K_c} \times K_c$ matrix whose m^{th} row is the binary representation of the integer $(m - 1)$. The columns of $\mathbf{G}_{K_c}$, that is, $\mathbf{g}_1, \mathbf{g}_2, \ldots, \mathbf{g}_{K_c}$, represent the *code generators* of the codeword dictionary $\mathbf{H}_{K_c}$. Also define the $2^{K_c} \times K_c$ *data symbol matrix* $\mathbf{I}_{K_c}$, whose i^{th} row $i = 1, 2, \ldots, N_c = M_c$ is the mth data vector $\mathbf{u}_m$. Since the elements of $\mathbf{u}_m$ correspond to the binary representation of the integer $(m - 1)$, we notice that $\mathbf{I}_{K_c}$ is identical to $\mathbf{G}_{K_c}$. In terms of the above matrix definitions, the codeword dictionary is easily generated from

$$\mathbf{H}_{K_c} = \mathbf{G}_{K_c} \mathbf{I}_{K_c}^t \tag{12.24}$$

where the superscript "t" denotes a matrix transpose. It is understood that the multiplication and addition operations involved in the matrix multiplication are reduced modulo-2.

Example 12.1

(8,3) Reed-Muller code

$$\mathbf{G}_3 = \begin{bmatrix} 0 & 0 & 0 \\ 0 & 0 & 1 \\ 0 & 1 & 0 \\ 0 & 1 & 1 \\ 1 & 0 & 0 \\ 1 & 0 & 1 \\ 1 & 1 & 0 \\ 1 & 1 & 1 \end{bmatrix} ; \quad \mathbf{I}_3^t = \begin{bmatrix} 0 & 0 & 0 & 0 & 1 & 1 & 1 & 1 \\ 0 & 0 & 1 & 1 & 0 & 0 & 1 & 1 \\ 0 & 1 & 0 & 1 & 0 & 1 & 0 & 1 \end{bmatrix} = \begin{bmatrix} \mathbf{u}_0^t & \vdots & \cdots & \vdots & \mathbf{u}_7^t \end{bmatrix}$$

so that

$$\mathbf{H}_3 = \begin{bmatrix} 0 & 0 & 0 & 0 & \vdots & 0 & 0 & 0 & 0 \\ 0 & 1 & 0 & 1 & \vdots & 0 & 1 & 0 & 1 \\ 0 & 0 & 1 & 1 & \vdots & 0 & 0 & 1 & 1 \\ 0 & 1 & 1 & 0 & \vdots & 0 & 1 & 1 & 0 \\ \cdots & \cdots & \cdots & \cdots & \vdots & \cdots & \cdots & \cdots & \cdots \\ 0 & 0 & 0 & 0 & \vdots & 1 & 1 & 1 & 1 \\ 0 & 1 & 0 & 1 & \vdots & 0 & 1 & 0 & 1 \\ 0 & 0 & 1 & 1 & \vdots & 1 & 1 & 0 & 0 \\ 0 & 1 & 1 & 0 & \vdots & 1 & 0 & 0 & 1 \end{bmatrix} = \begin{bmatrix} \mathbf{x}_0 \\ \mathbf{x}_1 \\ \vdots \\ \mathbf{x}_7 \end{bmatrix}$$

The message and bit error probabilities have been derived and are given by (4.93) and (4.96), respectively. We note that for coded communications, there exists a one-to-one correspondence between messages and codewords, so that the phrases "message error probability (MEP)" and "codeword error probability" are synonymous.

12.3.2 Biorthogonal Binary Codes

The codeword dictionary of a $(2^{K_c-1}, K_c)$ biorthogonal code is easily defined in terms of the dictionary of an orthogonal code through the partitioned matrix

$$\mathbf{B}_{K_c} = \begin{bmatrix} \mathbf{H}_{K_c-1} \\ \overline{\mathbf{H}}_{K_c-1} \end{bmatrix}$$

The message and bit error probabilities have been derived and are given by (4.102) and (4.107), respectively. Here $M_c = 2N_c$ and for any particular codeword $\mathbf{x}_i$, $D_{ij} = N_c/2$ for $M_c - 2$ of the codewords $\mathbf{x}_j (i \neq j)$, and $D_{ij} = N_c$ for the complementary codeword $\mathbf{x}_j = \overline{\mathbf{x}_i}$.

12.3.3 Transorthogonal Binary Codes

The codeword dictionary of a $(2^{K_c} - 1, K_c)$ transorthogonal code can be generated from an orthogonal codeword dictionary by deleting the first column in $\mathbf{H}_{K_c}$. In practice, this can be accomplished by deleting the first digit of each codeword. The message and bit error probabilities are given in (4.116) and (4.96) respectively. The minimum Hamming distance

Table 12.4 Comparison of the MEP for Various Binary Coding Techniques

Coding Technique	$D_{\min}$	Message Error Probability (Union Bound Assuming Coded BPSK)
$\perp$-codes	$N_c/2$	$(M_c - 1)Q\left(\sqrt{K_c E_b/N_0}\right)$
Bi-$\perp$ codes	$N_c/2$	$(M_c - 2)Q\left(\sqrt{K_c E_b/N_0}\right) + Q\left(\sqrt{K_c E_b/N_0}\right)$
Trans-$\perp$ codes	$M_c/2$	$(M_c - 1)Q\left(\sqrt{(K_c E_b/N_0)[M_c/(M_c - 1)]}\right)$
Uncoded BPSK (signal points on hypercube)	1	$(M_c - 1)Q\left(\sqrt{2E_b/N_0}\right)$

is $D_{\min} = M_c/2$ with $N_c = M_c - 1$. Also note from (12.19) that $\gamma_{ij} = -1/(M_c - 1)$ since $A_{ij} = D_{ij} - 1$ and $N_c = M_c - 1 = A_{ij} + D_{ij}$.

A comparison of the message error probability bounds, as found from the union bound result given in (12.5), is given in Table 12.4 for various binary coding techniques. The case of bit-by-bit signaling (uncoded BPSK) is also given for comparison purposes. For this case, we note that $r_c = 1$ and $D_{\min} = 1$ since the signal coordinates disagree in at least one place.

12.3.4 The Bandwidth Issue for Binary Coded BPSK Modulation

In Chapters 1 and 2, definitions and techniques for assessing the bandwidth requirements of a communication system were presented. Although no constraint was placed on the bandwidth of the binary coded waveforms discussed in the previous section, it is necessary in practice to be able to allocate and control bandwidth. Assuming that the transmitter uses the $(2, 1, T_s)$ binary coded BPSK modulator, then $D \triangleq N/T_s = 1/T_N = 1/T_c = 1/T_s$ dimensions per sec; however, we know from the sampling theorem that a signal of bandwidth B Hertz can be uniquely represented by $N_c = 2BT$ samples taken at the Nyquist sampling rate of $2B = 1/T_c = 1/T_N = D$ samples per second [15]. Thus

$$B = 1/2T_c = D/2 \text{ Hz} \tag{12.25}$$

serves as the relationship which we will adopt between the parameters D, T_c, and B. We will later see that (12.25) also holds for Q-ary coded Q-ary ASK.

Since $N_c = 2^{K_c}$ for orthogonal binary codes, we observe that $B = 2^{K_c-1}/T$ Hertz, which demonstrates that B expands exponentially with $K_c = R_b T$! This is undesirable because the bandwidth required also expands exponentially with T. However, the orthogonal binary codes constitute a small subset of the class of codes, binary or otherwise. This leads one to ask if there exist codes whose bandwidth requirements do not expand exponentially with T and yet have the property that the message error probability (MEP) asymptotically goes to zero as T gets large. This question is answered in the next few sections.

12.3.5 Channel Throughput and Time-Bandwidth Requirements

In selecting modulation and coding techniques, one important consideration is *channel throughput*. This is usually defined as the ratio of the bit rate $\mathcal{R}_b$ to the two-sided Nyquist bandwidth $W = 2B$ Hz. From Tables 11.1 and 12.2, we have the *modulator throughput*

$$\frac{\mathcal{R}}{W} = \frac{\log_2 M}{W T_s} = \frac{\log_2 M}{W T_N N} = \frac{R_N}{W T_N} \text{ bits/sec/Hz} \tag{12.26}$$

Table 12.5 Channel Throughput Comparison for Binary Coded
Modulation Techniques

Coding	Modulation	R_N	M_c	N_c	$\mathcal{R}_b/W$
None	BPSK or QPSK	1	—	—	1
None	M-PSK ($M = M_c$)	$\frac{K_c}{2}$	—	—	$\frac{K_c}{2}$
$\perp$-codes	BPSK or QPSK	1	2^{K_c}	2^{K_c}	$\frac{K_c}{2^{K_c}}$
Bi-$\perp$ codes	BPSK or QPSK	1	2^{K_c}	2^{K_c-1}	$\frac{K_c}{2^{K_c-1}}$
Trans-$\perp$ codes	BPSK or QPSK	1	2^{K_c}	$2^{K_c} - 1$	$\frac{K_c}{2^{K_c}-1}$

and for coded systems we have the normalized *channel throughput*

$$\frac{\mathcal{R}_b}{W} = \log_2 M_c/WT = \log_2 M_c/N_c = r_c \text{ bits/sec/Hz} \tag{12.27}$$

Equations (12.26) and (12.27) demonstrate that the units "bits/sec/Hz" and "bits/dimension" are equivalent since WT_N is unitless. Also note that for binary coded BPSK, $T_c = T_N$ and $WT_N = 1$. For the case of an uncoded system, we recall that $\mathcal{R}_b = \log_2 M/T$, so $\mathcal{R}_b/W = \log_2 M/WT = (\log_2 M)/N = R_N$ bits/sec/Hz.

Table 12.5 provides a comparison of channel throughputs achievable for various uncoded and binary coded modulation techniques. Notice that as K_c (the number of bits per message) increases, the normalized channel throughput $\mathcal{R}_b/W$ increases linearly with K_c for M-PSK without coding, while for the three binary coded modulation techniques, the normalized throughput approaches zero exponentially as K_c increases. This effect is due to the fact that the bandwidth requirement $W = N_c/T$ grows exponentially with K_c (since N_c is proportional to 2^{K_c}). This severely limits the applicability of block-orthogonal codes in practice.

12.4 THE ENSEMBLE OF BLOCK-CODED MODULATION SYSTEMS

As we observed in the previous section and in Chapter 3, it is difficult to evaluate the decoded MEP for an arbitrary set of code vectors, binary or otherwise. Yet a more difficult signal design (coding) problem is to find the matrix of correlation coefficients $\Gamma = [\gamma_{ij}]$ which minimizes the MEP. Even if this matrix were known, it would still be difficult to specify the design rules for the channel encoder and decoder. Furthermore, recall from Section 12.3.4 that, while binary orthogonal codes have MEP that can be made to approach zero, the bandwidth $W = 2^{K_c}/T$ grows exponentially with message length K_c. Such a bandwidth requirement quickly exceeds any bandwidth allocated by the Federal Communication Commission (FCC) for any physical channel. Perhaps Shannon [5] recognized that nature has hidden good codes (channel codec algorithms) with MEP as demonstrated in (12.10) to (12.16), and devised an astute argument to show that, as T becomes large, many codes exist such that bandwidth does not increase exponentially with T even though the MEP decreases exponentially with T. Shannon's derivation of the *exponential error bound* was based on an analysis called the *random coding argument*. The bound is obtained by averaging the message error probability over an ensemble of all possible codes, as we shall subsequently do.

We begin by considering the ensemble of block coded systems illustrated in Fig. 12.2. As we shall presently see, the Q-ary block code ensemble consists of the $Q^{N_c M_c}$ distinct block codes $\mathbf{x} \triangleq (\{\mathbf{x}_i\}_1, \{\mathbf{x}_i\}_2, \ldots, \{\mathbf{x}_i\}_{Q^{N_c M_c}})$ which we now define. Each code in the ensemble consists of M_c codewords, not necessarily distinct, such that each message m ($m = 0, \ldots, M_c - 1$) can be mapped to a codeword. The $Q^{N_c M_c}$ coded systems in the ensemble operate in parallel as follows. If message m is sent, then $\mathbf{u} = \mathbf{u}_m = (u_{m1}, \ldots, u_{m K_c})$ for all $m = 0, \ldots, M_c - 1$. We further assume a Q-ary code with $u_{mk} \in \{q_1, \ldots, q_Q\}$ for all $m = 0, \ldots, M_c - 1$ and $k = 1, 2, \ldots, K_c$. Thus, there exist $M_c = Q^{K_c}$ possible source encoder outputs for M_c distinct input messages.

Now the ensemble of block encoders maps $\mathbf{u}_m$ into $\{\mathbf{x}\}_\ell = \{\mathbf{x}_0, \mathbf{x}_1, \ldots, \mathbf{x}_{M_c-1}\}_\ell$ consisting of codewords $\mathbf{x}_m$, $m = 0, \ldots, M_c - 1$ for $\ell = 1, 2, \ldots, Q^{N_c M_c}$. These codes are now described along with the operation of the system ensemble in Fig. 12.2. If message m is sent, then $\mathbf{u} = \mathbf{u}_m$ for all ℓ and $\{\mathbf{x}_m\}_l = \{(x_{m1}, \ldots, x_{mN_c})\}_l$. Thus, there are Q^{N_c} different code vectors available and $M_c = Q^{K_c} = Q^{R_b T} = Q^{r_c N_c}$ messages to be assigned to code vectors; it follows that there are $(Q^{N_c})^{M_c} = Q^{N_c M_c}$ distinct ways (codes) to assign the messages to codewords; thus, the communication system consists of $Q^{N_c M_c}$ block-coded systems operating in parallel; see Figures 12.2 and 12.3.

Assume that the ensemble of modulators isometrically maps the ensemble of block codes $\mathbf{x}$ into the ensemble of coded modulation vectors $\mathbf{S} \triangleq (\{\mathbf{S}_m\}_1, \{\mathbf{S}_m\}_2, \ldots, \{\mathbf{S}_m\}_{Q^{N_c M_c}})$ or waveforms $\mathbf{S}(t)$ (see Figures 12.2 and 12.3). The ensemble of channels maps the transmitted coded modulation ensemble $\mathbf{S}(t)$ into the received waveform ensemble $\mathbf{r}(t)$. The ensemble of demodulators maps the received waveform ensemble into the demodulator output vector ensemble of projections $\mathbf{r} = (\{\mathbf{r}\}_1, \ldots, \{\mathbf{r}\}_{Q^{N_c M_c}})$. The continuous random vector ensemble $\mathbf{r}$ emerging from the IF CMC ensemble can be mapped into the quantized random vector ensemble $\mathbf{R} = (\{\mathbf{R}\}_1, \ldots, \{\mathbf{R}\}_{Q^{N_c M_c}})$ using a J-level vector quantizer (see Fig. 11.3). Here the ensemble of IF CMC's is characterized by the ensemble of conditional channel transition probability density functions $\{f(\mathbf{r}|\mathbf{S}_m)\} \triangleq \{f_{\mathbf{r}}(\rho_1, \ldots, \rho_{N_c}|\mathbf{S}_m)\}$ and the quantized IF CMC (that is, the IF DMC) is characterized by the ensemble of transition probabilities $\{P(R_n = R_{nj}|s_{mn} = q_i) = p_{ij} \triangleq P(j|i)\}$ for all $i = 1, \ldots, Q$, for all $j = 0, \ldots, J - 1$, for all $n = 1, \ldots, N_c$, and for all $m = 0, \ldots, M_c - 1$ (see Section 3.1.1). Later in this section we will show how the IF CMC is reduced to the IF DMC. Finally, the ensemble of block decoders maps ensemble $\mathbf{r}$ (or $\mathbf{R}$) into the ensemble vector $\hat{\mathbf{u}}$, the estimate of $\mathbf{u}_m$; and the source decoder maps $\hat{\mathbf{u}}$ into the recovered message $\hat{m}$. Thus, our ensemble of block-coded systems can be conceptually and functionally represented as in Fig. 12.3. Notice, the ensemble of AWGN channels in Fig. 12.2 has been replaced by the ensemble of IF channels. In this chapter, we will only consider "matched" coding and modulation techniques. By this we mean that the Hamming distance between codewords is preserved in the Euclidean sense at the modulator output, or that the mapping from codewords to waveforms is isometric. Binary coded BPSK and QPSK are examples of isometric modulation techniques.

12.4.1 Performance of the Ensemble of Binary Coded Systems Using BPSK Modulation

Consider now the binary coding case ($Q = 2$) and assume that $m = m_m$. This implies that the code vector is $\mathbf{x}_m$, and the coded modulation vector is $\mathbf{S}_m = (s_{m1}, \ldots, s_{mN_c})$. With BPSK modulation we assume that s_{mn} takes on $+\sqrt{E_{N_c}}$ or $-\sqrt{E_{N_c}}$ with equal probability for all $m = 0, \ldots, M_c - 1$, and $n = 1, \ldots, N_c$. For these equal energy signals, the received signal energy is

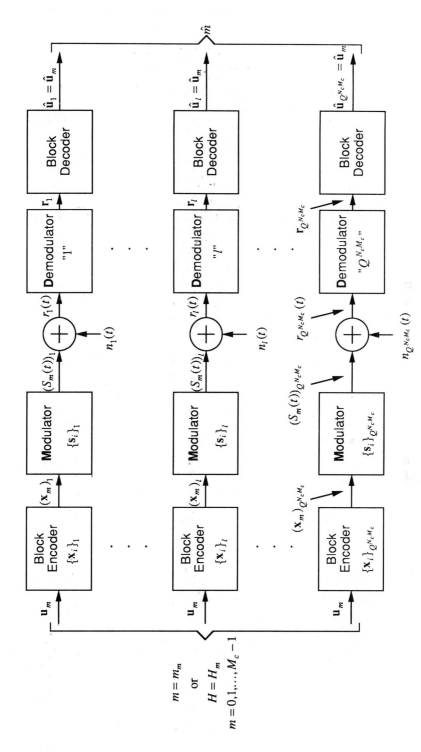

Figure 12.2 Ensemble of block-coded digital communication systems

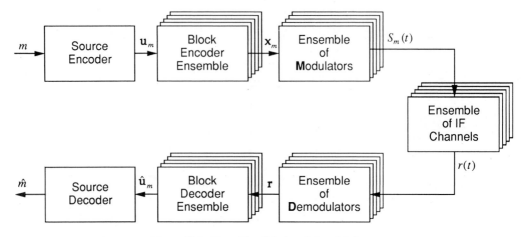

Figure 12.3   Ensemble of block-coded modulation systems

$$E = PT = \sum_{n=1}^{N_c} s_{mn}^2 = N_c E_{N_c} \text{ joules} \tag{12.28}$$

and since $T = N_c T_c$, we have that $E_{N_c} = PT_c$ joules. Since the calculation of the decoded message error probability (MEP) is not feasible for an arbitrary block code, we consider instead the problem of upper bounding the ensemble average MEP for the class of binary block codes. Examples of binary codes in this class include the Hamming and Golay codes, while examples of Q-ary codes are the Reed-Solomon and BCH codes. Applications of these coding techniques to the error-control problem will be given later on in this chapter.

Now the conditional MEP, $P(E|\{\mathbf{S}_m\}_\ell)$, for the ℓth system in the ensemble can be expressed using (4.80) as a function of E/N_0 and the ℓth code cross-correlation matrix $\Gamma_\ell = [\gamma_{jk}]_\ell$; however, as we have seen, it is difficult to find the global minimum of $P(E)$ as a function of Γ_ℓ. In principle, the ensemble average MEP

$$\overline{P(E)} = E[P(E|\{\mathbf{S}_m\}_\ell)] \tag{12.29}$$

can be upper bounded using the union bound result (4.81) for the conditional MEP (assuming message m is sent and all messages are equally likely) given by

$$P(E|\{\mathbf{S}_m\}_\ell, \mathbf{S}_m) \le \sum_{k \ne m, k=0}^{M_c-1} Q\left(\frac{\{d_{mk}\}_\ell}{\sqrt{2N_0}}\right) \tag{12.30}$$

where $\{d_{mk}\}_\ell$ denotes the Euclidean distance between the coded modulation vectors $\mathbf{S}_m$ and $\mathbf{S}_k$ in the ℓth codeword dictionary, $\ell = 1, \ldots, 2^{N_c M_c}$. In the context of random coding, we note that d_{mk} is a random variable since the components s_{mn} and $s_{kn}, n = 1, 2, \ldots, N_c$, of the vectors $\mathbf{S}_m$ and $\mathbf{S}_k$ are binary random variables taking on values $\sqrt{E_{N_c}}$ or $-\sqrt{E_{N_c}}$ with equal probability over the ensemble. To further characterize the random variable d_{mk}, let D_{mk} be a random variable denoting the Hamming distance between $\mathbf{S}_m$ and $\mathbf{S}_k$, taking on value h. If $\mathbf{S}_m$ and $\mathbf{S}_k$ differ from each other in h coordinates, then

$$\left\{d_{mk}^2\right\}_\ell = \left\{|\mathbf{S}_m - \mathbf{S}_k|^2\right\}_\ell = \sum_{n=1}^{N_c}(s_{mn} - s_{kn})^2 = h\left(2\sqrt{E_{N_c}}\right)^2 = 4hE_{N_c} \quad (12.31)$$

for all $\ell = 1, 2, \ldots, 2^{N_c M_c}$ where $h = 0, 1, 2, \ldots, N_c$.

In general, we point out that the Euclidean distance d_{mk} is the variable of interest in the design of coded modulation systems, while the Hamming distance has been used as a measure of the goodness of a code in coding theory. For binary coded BPSK, the Hamming distance is related to the Euclidean distance via $d_{mk}^2 = 4E_{N_c}D_{mk}$ from (12.4b). Now substitution of this relationship into (12.30) leads to

$$P(E|\{\mathbf{S}_m\}_\ell, \mathbf{S}_m) \le \sum_{k=0,k\neq m}^{M_c-1} Q\left(\sqrt{\frac{2\{D_{mk}\}_\ell E_{N_c}}{N_0}}\right) \quad (12.32)$$

Since D_{mk} counts the number of places in which $\mathbf{S}_m$ and $\mathbf{S}_k$ differ, it takes on value h with the same probability as that of getting h heads in N_c tosses of a fair coin; that is

$$P(\{D_{mk}\}_\ell = h) = \binom{N_c}{h}\left(\frac{1}{2}\right)^{N_c} \quad (12.33)$$

for $h = 0, 1, \ldots, N_c$ and $\ell = 1, 2, \ldots, 2^{N_c M_c}$. Using (12.33) to perform the ensemble average over the Hamming distances in order to eliminate the conditioning on $\{\mathbf{S}_m\}_\ell$ in (12.32) produces

$$\overline{P(E|\mathbf{S}_m)} \le \sum_{h=0}^{N_c}\sum_{k=0,k\neq m}^{M_c-1} Q\left(\sqrt{\frac{2hE_{N_c}}{N_0}}\right)\binom{N_c}{h}\left(\frac{1}{2}\right)^{N_c}$$

$$\le \left(\frac{1}{2}\right)^{N_c}\sum_{k=0,k\neq m}^{M_c-1}\sum_{h=0}^{N_c} e^{-\frac{hE_{N_c}}{N_0}}\binom{N_c}{h} \quad (12.34)$$

where the last line follows by changing the order of summation and upper bounding $Q(x)$ by $e^{-x^2/2}$. Note that the inner sum is independent of k and reduces to $(1 + e^{-E_{N_c}/N_0})^{N_c}$. Upper bounding $M_c - 1$ by M_c and summing over h and k yields the closed form expression

$$\overline{P(E|\mathbf{S}_m)} < M_c[1 + \exp(-E_{N_c}/N_0)]^{N_c}/2^{N_c} \quad (12.35)$$

Then, since the right-hand side is independent of $\mathbf{S}_m$, we have the exponential MEP bound for the ensemble,

$$\overline{P(E)} < M_c 2^{-N_c R_0} = M_c 2^{-T\mathcal{R}_0} \quad (12.36)$$

where we have introduced the binary coded BPSK *channel cutoff rate*

$$R_0 \triangleq \log_2\{2/[1 + \exp(-E_{N_c}/N_0)]\} \text{ bits/dimension} \quad (12.37)$$

and $\mathcal{R}_0 \triangleq R_0/T_c = WR_0$, the normalized channel cutoff rate in bits/sec. Since $M_c = 2^{K_c} = 2^{\mathcal{R}_b T} = 2^{r_c N_c}$, (12.36) can be rewritten as

$$\overline{P(E)} < 2^{-N_c[R_0-r_c]} = 2^{-K_c[R_0/r_c-1]} = 2^{-T(\mathcal{R}_0-\mathcal{R}_b)} \quad (12.38)$$

For the system engineer this is a very striking result: it says that as long as the code rate $r_c < R_0$, the ensemble average error probability can be made arbitrarily small by taking the *block length* N_c (and hence *message length* K_c) sufficiently large. Equivalently, as long as $\mathcal{R}_b < \mathcal{R}_0$, the MEP can be made arbitrarily small by taking the codeword duration T sufficiently large. Since the ensemble of binary codes satisfies (12.38), it follows that at least one code in the ensemble must have an error probability $P(E)$ no greater than the ensemble average, $\overline{P(E)}$. Notice from (12.37) that R_0 approaches zero for small E_{N_c}/N_0 and unity for large E_{N_c}/N_0.

Now let us determine the form of R_0 in the energy-limited region, $E_{N_c}/N_0 \ll 1$, where, with unlimited bandwidth, coding can be used to reduce the required signal energy to the lowest possible level. To do this, we differentiate (12.37) with respect to $x = E_{N_c}/N_0$ and take the limit as x approaches 0 to obtain

$$\lim_{x \to 0} [d R_0/dx] = 1/2 \ln 2$$

Thus, $R_0 \simeq (E_{N_c}/N_0)/2 \ln 2 = (r_c E_b/N_0)/2 \ln 2$. When the coded system operates at a code rate $r_c = R_0$ and $E_{N_c}/N_0 \ll 1$, then $R_0 \simeq R_0 (E_b/N_0)/2 \ln 2$. Again, this says that the minimum value of E_b/N_0 to achieve the cutoff rate is $E_b/N_0 = 2 \ln 2 = 1.4$ dB.

12.4.2 MEP Performance Comparison of Orthogonal Codes with the Ensemble of Binary Codes Using BPSK

For binary orthogonal codes, (12.13) can be rewritten as

$$P(E) < 2^{-K_c[(E_b/N_0)/2 \ln 2 - 1]} \tag{12.39}$$

by noting that $e^x = 2^{x/\ln 2}$. We observe that the bound on the MEP tends to zero with increasing message length K_c if $E_b/N_0 > 2 \ln 2 = 1.4$ dB.

Since $T_c = r_c T_b$ (see Table 11.2), then for binary coded BPSK

$$\frac{R_0}{r_c} = \frac{T_b}{T_c} R_0 = \frac{E_b}{E_{N_c}} R_0 = \frac{E_b/N_0}{E_{N_c}/N_0} R_0 \tag{12.40}$$

and we can use this in (12.38) to write

$$\overline{P(E)} < 2^{-K_c[(E_b/N_0)/\beta - 1]} \tag{12.41}$$

where

$$\beta \triangleq \frac{E_{N_c}/N_0}{R_0} \tag{12.42}$$

Thus, the bound on $\overline{P(E)}$ goes to zero with increasing K_c if $E_b/N_0 > \beta$ for the ensemble of binary coded BPSK versus $E_b/N_0 > 2 \ln 2$ for orthogonal modulation vectors (signals). Notice that for small E_{N_c}/N_0, where coding is most useful, we have shown that $R_0 \simeq (E_{N_c}/N_0)/2 \ln 2$, so that β in (12.42) approaches $2 \ln 2$. This says that for very noisy channels, binary coded BPSK is essentially equivalent to binary coded orthogonal signals, which (from Chapter 4) are equivalent to binary coded transorthogonal signals for large M, which are known to be optimum for the AWGN channel.

The binary coded BPSK result can also be obtained another way. Since $E_{N_c}/N_0 = r_c(E_b/N_0)$, then we can use this in (12.37) to write

$$R_0 = \mathcal{R}_0/W = 1 - \log_2[1 + \exp(-r_c E_b/N_0)] \text{ bits/sec/Hz} \tag{12.43}$$

This says that operating the system at $r_c = R_0$ requires the communications efficiency of

$$\frac{E_b}{N_0} = -\ln[2^{1-r_c} - 1]/r_c \tag{12.44}$$

Using the restriction $r_c \leq R_0$ corresponds to requiring that

$$\frac{E_b}{N_0} \geq -\ln{(2^{1-r_c} - 1)}/r_c \tag{12.45}$$

which says that, as the code rate r_c approaches zero (N_c approaches infinity or the band-width expansion $1/r_c$ approaches infinity), that

$$\frac{E_b}{N_0} \geq 2\ln{2} = 1.4 \text{ dB} \tag{12.46}$$

for $r_c \leq R_0$. Can one do better? If yes, then the question is how much better? We proceed to answer this question.

12.4.3 Performance of the Ensemble of Q-ary Codes Using Q-ary ASK and QASK

In Section 12.4.1, we used the $(2, 1, T_s)$ BPSK modulator to generate the binary coded BPSK waveform

$$S_m(t) = \sum_{n=1}^{N_c} s_{mn}\varphi_n(t), \quad 0 \leq t \leq T \tag{12.47}$$

in which s_{mn} is a binary-valued random variable taking on values $+\sqrt{E_{N_c}}$ or $-\sqrt{E_{N_c}}$ with equal probability. We then demonstrated that over the AWGN channel the ensemble error probability delivered by an optimum receiver approaches zero exponentially with block length N_c for rates R_0 greater than r_c. There are two issues of concern: (1) the bandwidth $W = N_c/T$ required to achieve this rate grows with the block length $N_c = WT$, and (2) the cutoff rate R_0 saturates at one bit/sec/Hz for large E_{N_c}/N_0 due to the two-level nature of the BPSK coded waveform $S_m(t)$. One would intuitively expect that large values of E_{N_c}/N_0 would permit reliable communication at rates above one bit per dimension so that binary codes (binary-waveform sequences) will not be exponentially optimum when E_{N_c}/N_0 is large. As we shall see in this section, both of these concerns can be significantly alleviated using Q-ary Coded Q-ary Amplitude-Shift-Keying (Q-ary Coded ASK).

Consider now the $(M, 1, T_s)$ Q-ary coded ASK modulator used $\hat{k} = N_c$ times per codeword (that is, $T_c = T_s$) to transmit the class of Q-ary codes where $M = Q$. The coded message symbols x_{mn} are determined from the Q-ary alphabet $Q = \{q_1, q_2, \ldots, q_Q\}$; that is, x_{mn} is a random variable taking on value q_i from Q with probability $P(q_i) \triangleq p_i$. We further assume $q_i^2 \leq 1$ for all i. The modulator waveform coefficients s_{mn} in (12.47) are then given by $s_{mn} = \sqrt{E_{N_c}}x_{mn}$. Figure 12.4a demonstrates the possible coded message symbols for the case $Q = M = 8$, $N = 1$, and $T_c = T_s$, assuming the points are equally spaced on $(-1, 1)$.

The first concern mentioned above is addressed by comparing the channel through-puts $\mathcal{R}_b/W$ for Q-ary and binary coded $(M, 1, T_s)$ modulators. For Q-ary coded ASK, the throughput is given by $\mathcal{R}_Q/W_Q = \log_2{M_c}/W_QT = K_c\log_2{Q}/N_{cQ}$; for binary coded BPSK, the throughput is given by $\mathcal{R}_2/W_2 = K_c/N_{c2}$ bits/sec/Hz. Thus, the throughput ratio is given by $(\mathcal{R}_Q/W_Q)/(\mathcal{R}_2/W_2) = N_{c2}\log_2{Q}/N_{cQ}$. For a fixed data rate requirement, $\mathcal{R}_Q = \mathcal{R}_2$, and equal block lengths, $N_{c2} = N_{cQ}$, the throughput ratio reduces to bandwidth ratio $W_2/W_Q = \log_2{Q}$ or $W_2 = W_Q\log_2{Q}$. On the other hand, for a fixed bandwidth re-quirement, $W_2 = W_Q$, and equal block lengths, the throughput ratio reduces to the data

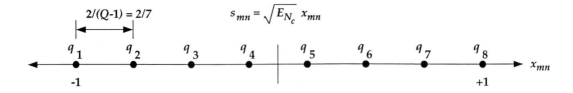

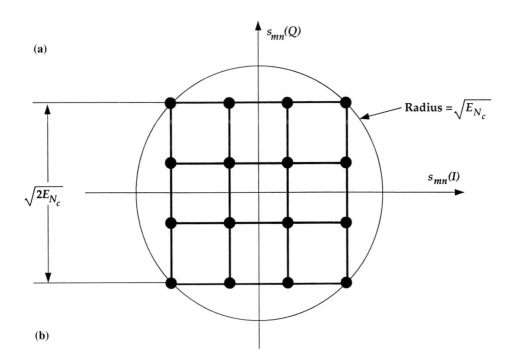

Figure 12.4 Q-ary ASK and Q^2-ary QASK signal vector models: (a) Possible set of values permitted the s_{mn}; $M = Q = 8$, $N = 1$, (b) Possible set of values permitted the $(s_{mn}(I), s_{mn}(Q))$; $M = Q^2 = 16$, $N = 2$

rate ratio $\mathcal{R}_Q/\mathcal{R}_2 = \log_2 Q$ or $\mathcal{R}_Q = \mathcal{R}_2 \log_2 Q$. These relationships define the throughput-bandwidth trade-off of interest to the systems engineer.

The second concern mentioned above is addressed by obtaining a bound on the ensemble MEP and examining the resulting R_0. Our approach is identical with that used for the case of binary codes; that is, we determine the R_0 (as a function of Q and $\overline{E_{N_c}/N_0}$) which exponentially bounds the ensemble of codes average error probability $\overline{P(E)}$. There are Q^{N_c} signal points available (potential codewords), and M_c messages to be assigned to the available points. Thus, the total number of ways to make the assignment, and hence the total number of possible codeword dictionaries in the ensemble, is $Q^{N_c M_c}$. This says that the ensemble of communication systems illustrated in Fig. 12.3 is now $Q^{N_c M_c}$ systems; each system uses different coded modulation vectors $\{\mathbf{S}_m\}_\ell$, $\ell = 1, \ldots, Q^{N_c M_c}$, together with the optimum receiver.

Notice from Fig. 12.4a that the end values $q_8 = 1$ and $q_1 = -1$ have neighbors only on one side. In the presence of noise these values have a higher noise immunity (or are more distinguishable) and it is intuitively clear that the error probability will be smaller for systems which use the coded modulation vectors $\{\mathbf{S}_m\}_\ell$ in which the values q_i near the ends are used more frequently than the interior values.

If we let N_i count the number of times that q_i is used in the entire ensemble of encoder-modulator pairs and if we assume that the message symbols occur independently, then to the coded modulation vectors $\{\mathbf{S}_m\}_\ell$ we assign the a priori probability

$$P\{\mathbf{S}_m\}_\ell = p_1^{N_1} p_2^{N_2} \cdots p_Q^{N_Q}, \quad \ell = 1, \ldots, Q^{N_c M_c} \tag{12.48}$$

where $p_1 + p_2 + \ldots + p_Q = 1$. Since each code yields M_c coded modulation vectors $\mathbf{S}_m$, made up of N_c symbols s_{mn}, then $N_1 + N_2 + \ldots + N_Q = N_c M_c$. In order to bound the ensemble MEP we use the union bound as before. Using the fact that $Q(x) < e^{-x^2/2}$, we note that the pairwise error probabilities in (12.3), conditioned on using the ℓth code, become

$$\{P_2(\mathbf{S}_m, \mathbf{S}_k)\}_\ell = Q\left(\frac{\{d_{mk}\}_\ell}{\sqrt{2N_0}}\right)$$

$$= Q\left(\frac{\{|\mathbf{S}_m - \mathbf{S}_k|\}_\ell}{\sqrt{2N_0}}\right) < \prod_{n=1}^{N_c} \exp\left[-(s_{mn} - s_{kn})^2/4N_0\right] \tag{12.49}$$

Since the coefficients s_{mn} of $\mathbf{S}_m$ are random variables, we will need to average over all possible values in order to get the ensemble error probability, $\overline{P(E)}$. If $s_{mn} = q_i\sqrt{E_{N_c}}$ and $s_{kn} = q_j\sqrt{E_{N_c}}$ then we have that the coordinate distances are $|s_{mn} - s_{kn}| = |q_i - q_j|\sqrt{E_{N_c}} \triangleq q_{ij}\sqrt{E_{N_c}}$. Moreover, the squared coordinate distance $(s_{mn} - s_{kn})^2 = q_{ij}^2 E_{N_c}$ occurs with probability $p_i p_j$ since the levels q_i and q_j are chosen independently and at random from the Q-ary alphabet. Thus

$$E\left\{\exp\left[-(s_{mn} - s_{kn})^2/4N_0\right]\right\} = \sum_{i=1}^{Q}\sum_{j=1}^{Q} p_i p_j \exp\left(-q_{ij}^2 E_{N_c}/4N_0\right) \tag{12.50}$$

and the conditional PEP in (12.49) can be averaged over all codes in the ensemble to give

$$\overline{P_2(\mathbf{S}_m, \mathbf{S}_k)} < \left[\sum_{i=1}^{Q}\sum_{j=1}^{Q} p_i p_j \exp\left(-q_{ij}^2 E_{N_c}/4N_0\right)\right]^{N_c} \tag{12.51}$$

Averaging the union bound over the ensemble of codes, we have

$$\overline{P(E|\mathbf{S}_m)} \leq \sum_{k=0,k\neq m}^{M_c-1} \overline{P_2(\mathbf{S}_m, \mathbf{S}_k)} = (M_c - 1)\overline{P_2(\mathbf{S}_m, \mathbf{S}_k)} \tag{12.52}$$

where the last equality holds since the right-hand-side of (12.52) is independent of k. Since it is also independent of m, the MEP for the ensemble of codes is $\overline{P(E)} = \overline{P(E|\mathbf{S}_m)}$. Bounding $M_c - 1$ by $M_c = Q^{K_c}$, and assuming equally likely messages, we obtain

$$\overline{P(E)} < 2^{-N_c(R_0 - r_c)} \tag{12.53}$$

where

$$R_0 \triangleq -\log_2\left[\sum_{i=1}^{Q}\sum_{j=1}^{Q} p_i p_j \exp\left(-q_{ij}^2 E_{N_c}/4N_0\right)\right] \tag{12.54}$$

and where $r_c = K_c \log_2 Q / N_c$. Since $R_0 = T_c \mathcal{R}_0 = \mathcal{R}_0 / W$ and $r_c = \mathcal{R}_b / W$, we can rewrite (12.53) as

$$\overline{P(E)} < 2^{-N_c[(\mathcal{R}_0 - \mathcal{R}_b)/W]} = 2^{-T(\mathcal{R}_0 - \mathcal{R}_b)}$$

Equation (12.53) demonstrates that the error probability can be forced to zero by using long block lengths as long as $R_0 > r_c$. Furthermore, there exists at least one code in the ensemble of codes with $P(E) \leq \overline{P(E)}$.

If we further assume that the levels q_i and q_j occur with equal probability, then $p_i = 1/Q$ for all $i = 1, \ldots, Q$, so the channel cutoff rate in (12.54) reduces to

$$R_0 = \mathcal{R}_0 / W = - \log_2 \left[\frac{1}{Q^2} \sum_{i=1}^{Q} \sum_{j=1}^{Q} \exp\left(-q_{ij}^2 E_{N_c} / 4N_0\right) \right] \text{ bits/sec/Hz} \quad (12.55)$$

For $Q = 2$, $q_{ij}^2 = 4$ for $i \neq j$ and $q_{ii} = 0$; substitution of these conditions in (12.55) leads to (12.37) for coded BPSK as it should. Upper and lower bounds for the cutoff rate R_0 can be derived from (12.55) by noting that $4E_{N_c}/(Q-1)^2 \leq q_{ij}^2 E_{N_c} \leq 4E_{N_c}$ for all $i \neq j$. Using these conditions, we find from (12.55) that

$$- \log_2 \left\{ 1 + (Q-1) \exp \left[\frac{-E_{N_c}/N_0}{(Q-1)^2} \right] \right\} \leq (R_0 - \log_2 Q)$$

$$\leq - \log_2 \left[1 + (Q-1) \exp\left(-E_{N_c}/N_0\right) \right]$$

with equality holding for $Q = 2$. This says that for large E_{N_c}/N_0, we have $R_0 \leq \log_2 Q$ bits/sec/Hz.

Given any choice of the amplitude levels $\{q_i\}$, (12.55) can be evaluated numerically. For equally probable levels and equally spaced amplitudes, as shown in Fig. 12.4a, the curves represented by (12.55) for R_0 are illustrated in Fig. 12.5 as a function of E_{N_c}/N_0 for various values of Q [7]. Notice that as E_{N_c}/N_0 approaches infinity, R_0 approaches $\log_2 Q$, which represents the source entropy for equally probable message symbols; equivalently, $\mathcal{R}_0$ approaches $W \log_2 Q$ bits/sec. This alleviates the concern that existed for binary codes, where R_0 saturated at 1 bit/sec/Hz. Furthermore, as E_{N_c}/N_0 approaches infinity, there is no need for adding redundancy in the codewords to achieve error control; hence, M_c approaches Q^{N_c} and the code rate $r_c = \mathcal{R}_b / W$ also approaches the source entropy, $\log_2 Q$.

At lower values of E_{N_c}/N_0 (the region where coding is of greatest practical interest), we see the effects that the interior signal points ($q_2, \ldots, q_7$ in Fig. 12.4a) have on R_0 when $Q > 2$. The curves for $Q > 2$ bend over later in the range of values for E_{N_c}/N_0 and also saturate at larger values of E_{N_c}/N_0. Note that R_0 increases as the number Q of amplitude levels increases and that the noise immunity of the Q-ary code vectors that lie to the interior of the outermost points is reduced. As Q increases, so does the number of interior points. The R_0 curves represented by (12.54) for equally spaced amplitude levels using an optimum probability assignment are shown in Fig. 12.6 for various values of Q. We note that the improvement in R_0 due to the optimality of the probability assignment is negligible.

In Section 12.7.6, we will employ (12.54) to demonstrate how the R_0 criterion can be used to simultaneously optimize the design choice for the coding and modulation. For low E_{N_c}/N_0, where coding is most useful, we will show that the optimum code is the $(M_c = 2^{K_c}, N_c = 2^{K_c} - 1, T_c)$ binary transorthogonal code, and the optimum choice for the modem is compatible with the $(M = Q = 2, 1, T_s)$ coded ASK or BPSK modulator.

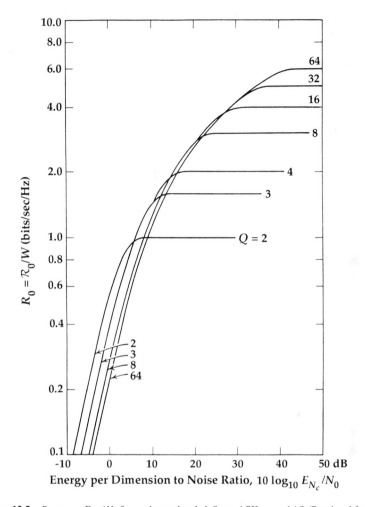

Figure 12.5 R_0 versus E_{N_c}/N_0 for equispaced coded Q-ary ASK; $p_i = 1/Q$ (Reprinted from [7])

To complete the discussion of Q-ary coded ASK, we revisit the bandwidth considerations in Section 12.3.4 pertaining to use of the $(2, 1, T_s)$ binary coded BPSK modulator. We note that the argument that $D = 1/T_N = 1/T_s = 1/T_c$ is true as long as $\hat{k} = N_c$ and $N = 1$. This condition holds for the $(M, 1, T_s)$ Q-ary coded ASK modulator of this section. Thus (12.25) holds for Q-ary coded ASK.

Q^2-ary coded QASK.   The cutoff rate for Q^2-ary coded QASK can be derived easily starting with the I-Q generalization of (12.49). Figure 12.4b demonstrates the extension of the Q-ary coded ASK signal design to the Q^2-ary coded QASK signal design using the $(16, 2, T_s)$ modulator. For the Q^2-ary coded $(M = Q^2, 2, T_s)$ QASK modulator, the modulation vectors $\mathbf{S}_m$ are broken into in-phase, $s_{mn}(I)$, and quadrature, $s_{mn}(Q)$, components with $s_{mn}(I) = \sqrt{E_{N_c}/2}\,x_{mn}(I)$ and $s_{mn}(Q) = \sqrt{E_{N_c}/2}\,x_{mn}(Q)$, where $x_{mn}(I)$ and $x_{mn}(Q)$ take on values $q_i = (-1) + (2i - 2)/(Q - 1)$, $i = 1, \ldots, Q$ for Q even, independently and at random with probability p_i. In other words, we transmit one of $M = Q^2$

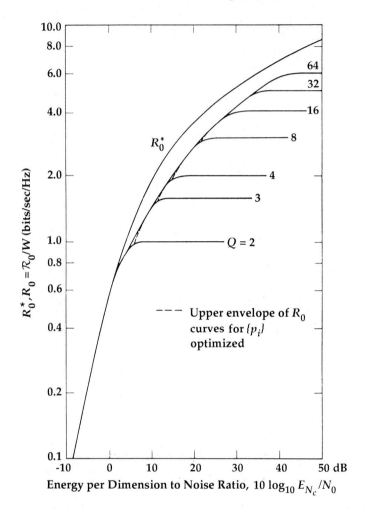

Figure 12.6 R_0 versus E_{N_c}/N_0 for equispaced coded Q-ary ASK; $\{p_i\}$ optimized (Reprinted from [7])

QASK signal vectors $(s_{mn}(I), s_{mn}(Q))$ for each of the $\hat{k} = N_c$ codeword coordinates x_{mn}, with energy constraint $|(s_{mn}(I), s_{mn}(Q))|^2 < E_{N_c}$.

Thus, the I-Q generalization of (12.49) becomes

$$\{P_2(\mathbf{S}_m, \mathbf{S}_k)\}_\ell < \prod_{n=1}^{N_c} \exp\left\{\left([s_{mn}(I) - s_{kn}(I)]^2 + [s_{mn}(Q) - s_{kn}(Q)]^2\right)/(-4N_0)\right\}$$

Furthermore, since the I and Q channel random variables are independent,

$$E\left(\exp\left\{\left([s_{mn}(I) - s_{kn}(I)]^2 + [s_{mn}(Q) - s_{kn}(Q)]^2\right)/(-4N_0)\right\}\right)$$

$$= \left[\sum_{i=1}^{Q}\sum_{j=1}^{Q} p_i p_j \exp\left(-q_{ij}^2 E_{N_c}/4N_0\right)\right]^2$$

where the expected value is computed over the ensemble. We use the above expression to find the ensemble average PEP, $\overline{P_2(\mathbf{S}_m, \mathbf{S}_k)}$, which when substituted into the union bound yields

$$\overline{P(E|\mathbf{S}_m)} \le (M_c - 1) \left[\sum_{i=1}^{Q} \sum_{j=1}^{Q} p_i p_j \exp\left(-q_{ij}^2 E_{N_c}/4N_0\right) \right]^{2N_c}$$

Bounding $(M_c - 1)$ by $M_c = Q^{2K_c}$ and assuming equally likely messages yields

$$\overline{P(E)} < 2^{-N_c(R_0 - r_c)}$$

where $R_0 = \mathcal{R}_0/W = -2\log_2 \left[\sum_{i=1}^{Q} \sum_{j=1}^{Q} p_i p_j \exp\left(-q_{ij}^2 E_{N_c}/4N_0\right) \right]$ bits/sec/Hz denotes the

cutoff rate for Q^2-ary coded QASK and where $r_c = 2K_c \log_2 Q/N_c$. We note that both r_c and R_0 are twice the values obtained for Q-ary coded ASK, as we would expect.

12.4.4 Comparison of the Ensemble Performance of *Q*-ary Codes with Energy Constrained Codes

Using a derivation we shall not repeat here, Shannon [6,7] considers the AWGN channel and the N_c-dimensional coded modulation vectors $\{\mathbf{S}_i\}$ constrained only in energy, that is, $|\mathbf{S}_i|^2 \le N_c E_{N_c} = E$ joules. He shows that

$$\overline{P(E)} < 2^{-N_c(R_0^* - r_c)} = 2^{-K_c \log_2 Q(R_0^*/r_c - 1)} \tag{12.56}$$

where the channel cutoff rate is defined by

$$R_0^* \triangleq \frac{1}{2\ln 2} \left[1 + \frac{E_{N_c}}{N_0} - \sqrt{1 + (E_{N_c}/N_0)^2} \right]$$

$$+ \frac{1}{2}\log_2 \left[\frac{1}{2}\left(1 + \sqrt{1 + (E_{N_c}/N_0)^2} \right) \right] \text{ bits/dimension} \tag{12.57}$$

Here R_0^* denotes the channel cutoff rate for a continuum of possible transmitted signal points on or within the hypersphere of radius $\sqrt{E} = \sqrt{N_c E_{N_c}}$. This derivation assumes that the vector space generated by the modulator is isometric to the vector space generated by the encoder. For small values of E_{N_c}/N_0, where coding is most useful, we note that $R_0^* \simeq (E_{N_c}/N_0)/2\ln 2 = (r_c E_b/N_0)/2\ln 2$, which is the same approximation obtained for the channel cutoff rate, R_0, for binary coded BPSK in Section 12.4.1. Operation of the system at code rate $r_c = R_0^*$ in this energy-efficient region requires $E_b/N_0 > 2\ln 2$.

Figure 12.6 plots R_0^* versus E_{N_c}/N_0 together with the R_0 of (12.54) for the ensemble of Q-ary codes with equally spaced amplitude levels and an optimum probability assignment. We note that the richness of the continuum of possible signal points from the transmitter results in a larger channel cutoff rate; that is, $R_0^* \ge R_0$ for all $Q \ge 2$. This in turn leads to a tighter bound on the ensemble error probability, as seen by comparing (12.56) and (12.53).

Some insight into the physics of the parameter E_{N_c}/N_0 will be useful before we further discuss Fig. 12.6. This can be obtained by using the bandwidth defined via the Nyquist sampling rate of $2B = 1/T_c$. Thus, $2E_{N_c}/N_0 = P/N_0 B \triangleq \text{CNR}$, which can be interpreted as the *receiver input signal-to-noise ratio* in a single-sided Nyquist bandwidth of

B Hz. For low input CNR's, we see from Fig. 12.6 that R_0 nearly coincides with R_0^* for all Q, but for CNR's greater than zero dB the binary codes are less desirable. In fact for CNR ≥ 10 dB binary codes are exceedingly undesirable; however, for high input CNR's, coding is generally not required to achieve the desired performance. At low input CNR's (power limited channels) where coding is usually applied, binary codes are essentially optimum. The increase in achievable R_0 as Q increases for large CNR's is due to the fact that the signal vectors (codewords) that lie inside the signal hypersphere of radius $\sqrt{N_c E_{N_c}}$ can be detected without error, while at low input CNR's only the signal vectors (codewords) that lie on or near the surface of this sphere can be detected without error. This behavior tends to justify the fact that *binary codes are optimum for use in channels that are power limited*, that is, those with low values of CNR. For large values of CNR, it should be clear why $R_0 \simeq \log_2 Q$ bits/dimension can be made to approach R_0^* for large Q. In this region of operation, those codewords in the Q-ary codes which lie inside the hypersphere have enough signal-to-noise ratio to be detected with essentially the same reliability as the set of codewords constrained only in energy. In fact, as Q approaches infinity, the discrete constrained-energy signal points approach a continuum, and R_0 approaches R_0^*. For large CNR's, the channel throughput is usually limited by bandwidth so that the application of coding (which tends to expand bandwidth) is of lesser interest.

12.4.5 *Q*-ary Coded *M*-PSK

As discussed in Chapter 4, M-PSK modulation is frequently used to reduce the bandwidth requirement. Consider now the use of the $(M, 2, T_s)$ modulator $\hat{k} = N_c$ times with $T_s = T_c = 2T_N$ and $M \geq 3$. Assume further that the coded message symbols x_{mn} are M-ary; that is, $Q = M$. Then the modulator maps each codeword $\mathbf{x}_m$ onto the coded M-PSK waveform

$$S_m(t) = \sum_{n=1}^{N_c} s_{mn}[t - (n-1)T_s], \quad 0 \leq t \leq T$$

where

$$s_{mn}(t) = \begin{cases} \sqrt{2E_s/T_s} \cos(\omega_0 t + 2\pi i/M), & (n-1)T_s \leq t \leq nT_s \\ 0, & \text{else} \end{cases}$$

if the random variable $x_{mn} = i\,(i = 0, \ldots, M-1)$, and where $E_s = E_{N_c}$. For example, if $M = Q = 4$, $s_{mn}(t)$ models one of the four QPSK signals for each one of the coded message symbols $x_{mn} \in \{0, \ldots, 3\}$ in the codeword $(x_{m1}, \ldots, x_{mN_c})$.

Following the approach used for Q-ary coded ASK, we again have $M_c = Q^{K_c}$ messages to be assigned to Q^{N_c} available signal points, for a total of $Q^{N_c M_c}$ possible codes in the ensemble. Assuming equally probable codewords and transmission of the coded M-PSK waveform over the AWGN channel, it is possible to show that the ensemble message error probability is bounded by

$$\overline{P[E]} < 2^{-N_c(R_0 - r_c)} = 2^{-N_c[(\mathcal{R}_0 - \mathcal{R}_b)/W]} = 2^{-T(\mathcal{R}_0 - \mathcal{R}_b)} \tag{12.58}$$

where

$$R_0 = \mathcal{R}_0/W \triangleq -\frac{1}{2}\log_2\left\{\frac{1}{M}\sum_{k=1}^{M}\left[\exp\left(-\frac{E_{N_c}}{N_0}\sin^2\frac{k\pi}{M}\right)\right]\right\} \text{ bits/sec/Hz} \tag{12.59}$$

for $M \geq 3$. For small values of E_b/N_0, it is easy to show from (12.59) that

$$R_0 \simeq \frac{1}{2 \ln 2} \left(\frac{E_{N_c}}{N_0} \right) = \frac{r_c}{2 \ln 2} \left(\frac{E_b}{N_0} \right) \tag{12.60}$$

If we insist on operating the system at $R_0 \geq r_c$, then this requires a communications efficiency of

$$\frac{E_b}{N_0} \geq (2 \ln 2) \tag{12.61}$$

equivalent to that given for block orthogonal codes in (12.15).

It was shown in Chapter 4 that the M-PSK symbol error probability is well approximated by

$$P_s(E) \simeq 2Q \left(\sqrt{\frac{2E_s}{N_0}} \sin \frac{\pi}{M} \right) \tag{12.62}$$

for $E_s/N_0 \gg 1$. For large M and E_s/N_0, we have, using the small-angle formula and the fact that $E_s = E_{N_c}$ for this model, that

$$P_s(E) \simeq 2Q \left(\sqrt{\frac{2E_{N_c}}{N_0}} \frac{\pi}{M} \right) \tag{12.63}$$

so that the coded symbol energy-to-noise ratio required to maintain a given error probability must increase as the square of the number of modulation states, M.

12.4.6 Further Discussion Regarding R_0 and R_0^*

Except for the special cases of M-PSK, orthogonal, biorthogonal, and transorthogonal coded modulations, we have seen that the problem of expressing the message error probability performance of a coherent communication system in closed form and evaluating the expression is a formidable problem. The problem of optimum code selection centers around being able to determine that signal cross-correlation matrix, namely Γ of equation (4.78), which minimizes $P(E)$. Shannon no doubt recognized this and in order to gain insight into the coding problem proposed the astute method of bounding the MEP by using the *random coding techniques* presented in the previous sections.

This method uses the pairwise error probability and the union bound to evaluate the MEP of an ensemble of communication systems which use amplitude modulation and the class of all possible Q-ary codes to communicate over the AWGN channel. The significant result, an existence proof, is that this ensemble of communication systems has the ensemble message error probability union bounded for isometric coded modulation mapping by

$$\overline{P(E)} < 2^{-N_c(R_0 - r_c)} = 2^{-N_c[(\mathcal{R}_0 - \mathcal{R}_b)/W]} = 2^{-PG(\mathcal{R}_0 - \mathcal{R}_b/W)} = 2^{-T(\mathcal{R}_0 - \mathcal{R}_b)} \tag{12.64}$$

where we have used the facts that $PG = WT = N_c$ (see Table 12.3). This says that the ensemble MEP approaches zero as N_c approaches infinity provided that $R_0 = \mathcal{R}_0/W > r_c = \mathcal{R}_b/W$ or as T approaches infinity provided that $\mathcal{R}_0 > \mathcal{R}_b$. From this, we can conclude that there exists at least one codeword dictionary among the ensemble of Q-ary codes with MEP less than or equal to the average value. Indeed, such code(s) are of great practical interest; however, it appears that nature has hidden them and the search for such codes has led to the development of Coding Theory. Later in this chapter, we investigate specific block-coding techniques for error control from a system design perspective. In the next chapter, we will investigate the use of convolutional codes as a coding alternative.

12.4.7 The Bhattacharyya Bound and Distance for Isometric Coded Modulation

When message m is sent over an AWGN channel, the *pairwise error probability* has been shown (in Appendix 3D) to be given by

$$P_2(\mathbf{S}_m, \mathbf{S}_k) = Q\left(\frac{d_{mk}}{\sqrt{2N_0}}\right) \tag{12.65}$$

for all $m \neq k = 0, 1, \ldots, M_c - 1$. More generally, it can be shown [8, 20] that, for the arbitrary IF CMC

$$P_2(\mathbf{S}_m, \mathbf{S}_k) \leq \underbrace{\int_{-\infty}^{\infty} \cdots \int_{-\infty}^{\infty}}_{N_c-\text{fold}} \sqrt{f(\mathbf{r}|\mathbf{S}_m) f(\mathbf{r}|\mathbf{S}_k)} d\mathbf{r} \tag{12.66}$$

when the coder output is isometrically mapped onto the modulator output. For the arbitrary IF DMC which outputs one of J symbols, $R_{nj}(j = 0, \ldots, J - 1)$, we have

$$P_2(\mathbf{S}_m, \mathbf{S}_k) \leq \sum_{R_1=R_{10}}^{R_{1(J-1)}} \cdots \sum_{R_{N_c}=R_{N_c0}}^{R_{N_c(J-1)}} \sqrt{P(\mathbf{R}|\mathbf{S}_m) P(\mathbf{R}|\mathbf{S}_k)} \tag{12.67}$$

under the same assumptions, where a vector quantizer produces the quantized random vector $\mathbf{R} = (R_1, \ldots, R_{N_c})$ whose coordinates are the quantized random variables R_n taking on values R_{nj}. Here $P(\mathbf{R}|\mathbf{S}_m)$ and $f(\mathbf{r}|\mathbf{S}_k)$ denote, respectively, IF channel transition probabilities and probability densities. The bounds in (12.66) and (12.67) are called *Bhattacharyya bounds* and the negative logarithms are the *Bhattacharyya distances*, say B_{mk}. For the BSC with crossover probability p, we have from (12.67)

$$B_{10} = -\ln\left(\sqrt{4p(1-p)}\right) \tag{12.68}$$

while for the AWGN channel we have, from (12.66)

$$P_2(\mathbf{S}_m, \mathbf{S}_k) \leq \left(\frac{1}{\pi N_0}\right)^{N_c/2} \underbrace{\int_{-\infty}^{\infty} \cdots \int_{-\infty}^{\infty}}_{N_c-\text{fold}} \exp\left(\frac{|\mathbf{r} - \mathbf{S}_m|^2 + |\mathbf{r} - \mathbf{S}_k|^2}{-2N_0}\right) d\mathbf{r} \tag{12.69}$$

$$= \exp\left(-\frac{d_{mk}^2}{4N_0}\right)$$

Thus, application of the total error probability formula, the union bound, and the PEP result from the AWGN channel yields

$$P(E) \leq (1/M_c) \sum_{k=0}^{M_c-1} \sum_{\substack{m=0 \\ m \neq k}}^{M_c-1} \exp\left(-d_{mk}^2/4N_0\right) \tag{12.70}$$

for equally likely messages.

Now let $d_{\min}$ denote the *minimum Euclidean distance* as defined in (12.7). Using this in (12.70) clearly demonstrates the important role $d_{\min}$ plays in the design of error-correcting codes, that is, $P(E) < M_c \exp(-d_{\min}^2/4N_0)$. Recall further from (12.4b) that the Hamming distance is proportional to the square of the Euclidean distance for binary coded BPSK.

12.5 SHANNON'S CHANNEL CODING THEOREM

We now present and discuss the most basic theorem of information theory, the achievability of the ultimate in channel capacity. The basic argument was first stated by Shannon in his original 1948 paper [5]. The result is very striking as well as counter-intuitive, for if the channel introduces errors randomly, how does one correct all of them? Shannon used a number of key ideas to prove that information can be sent reliably (error-free) over a channel at all rates up to what he called the *channel capacity*. These ideas include: (1) allowing for an arbitrarily small, say ϵ, nonzero MEP; (2) using the channel many times in succession; (3) evaluating the average MEP over a set of all possible codeword dictionaries (the ensemble); and (4) showing the existence of at least one good code.

Shannon's Channel Coding Theorem.   The rate r_c of an (M_c, N_c, T_c) code is

$$r_c = \frac{\log_2 M_c}{N_c} \text{ bits/dimension} \tag{12.71}$$

and all code rates below the *channel capacity* C_{N_c} are achievable. Specifically, for Gaussian channels and M_c sufficiently large, an optimum receiver, and code rate

$$r_c < C_{N_c} = \frac{1}{2} \log_2 \left(1 + \frac{2E_{N_c}}{N_0}\right) = \frac{1}{2} \log_2[1 + 2r_c(E_b/N_0)] \text{ bits/dimension} \tag{12.72}$$

there *exists* an (M_c, N_c, T_c) code with MEP approaching zero. The restriction $r_c < C_{N_c}$ implies that $E_b/N_0 > (2^{2r_c} - 1)/2r_c$ for all $r_c < C_{N_c}$. If $r_c > C_{N_c}$ and M_c is sufficiently large then $P(E)$ approaches one for every possible (M_c, N_c, T_c) code. It can be shown [7] also that $\frac{1}{2}C_{N_c} \le R_0^* \le C_{N_c}$. The code rate inverse $r_c^{-1} \triangleq 1/r_c$ is frequently identified as the *bandwidth expansion* factor by system engineers.

The proof of this famous theorem is given in numerous books on information theory [7, 8, 9] and in Shannon's original paper [5]; however, it is not our purpose to present this important proof here since it is too lengthy to reproduce. Shannon's approach to proving this famous theorem included the modulator and demodulator as a part of the channel; that is, he used the baseband channel model to prove the theorem. This is equivalent to assuming an isometric mapping from coder output to modulator output when the IF channel model is used. In this sense a codec and modem exist such that the theorem is true.

12.5.1 Shannon Limits and Information Transfer Physics

In Shannon's famous paper [5], it was shown that all information sources, for example, produced by people speaking, television cameras, telegraphy key, and so on, have a rate (r_c) associated with them which can be measured in bits/sec. Communication channels have a *capacity* (C_{N_c}) that can be measured in the same units. Shannon's theorem states that a codec-modem exists such that information can be transmitted essentially error-free over the channel (baseband or IF) if and only if the rate does not exceed the channel capacity. To gain further insight, let us investigate the communication efficiency requirement implied by (12.72) for the AWGN channel. Since $E_{N_c}/N_0 = r_c E_b/N_0$, we can write

$$r_c < C_{N_c} = \frac{1}{2} \log_2(1 + 2r_c E_b/N_0) \tag{12.73}$$

Thus

$$\frac{E_b}{N_0} = (2^{2C_{N_c}} - 1)/2r_c \text{ for } r_c < C_{N_c} \tag{12.74}$$

Operation at $r_c = C_{N_c}$ implies that

$$\frac{E_b}{N_0} > (2^{2C_{N_c}} - 1)/2C_{N_c} \tag{12.75}$$

and in the limit as the code rate r_c approaches zero (N_c approaches infinity), we have

$$\frac{E_b}{N_0} > \ln 2 = -1.6 \text{ dB} \qquad \text{for } r_c < C_{N_c} \tag{12.76}$$

This compares with

$$\frac{E_b}{N_0} \geq 2 \ln 2 = 1.4 \text{ dB} \qquad \text{for } r_c \leq R_0 \tag{12.77}$$

as given in Section 12.4.2 for binary orthogonal codes; note the 3 dB difference. From a system perspective, this says that the best codec-modem technique will require an efficiency of $E_b/N_0 > -1.6$ dB, a large coding delay, and complex coder-decoder (codec) implementation; but for small error rates of practical interest and a reasonable coding delay and codec complexity, $E_b/N_0 \geq 1.4$ dB is required. In conclusion, since $N_0 = kT^\circ$, we see from (12.76) that

$$E_b/^\circ K > k \ln 2 \qquad \text{for } r_c < C_{N_c} \tag{12.78}$$

and, we see from (12.77) that

$$E_b/^\circ K \geq 2k \ln 2 = 1.91 \times 10^{-23} \text{ joules}/^\circ K \text{ for } r_c \leq R_0 \tag{12.79}$$

so that a 3 dB difference exists between R_0 for binary orthogonal codes and C_{N_c} for all codes in the region of small code rates.

We can gain further insight into information transport by transforming the units of C_{N_c} from bits/dimension to the more useful physical units of bits/sec. Since

$$E_{N_c} = E/N_c = PT_c = PT/N_c \text{ joules} \tag{12.80}$$

then, with $N_c = 2BT$ per our bandwidth discussion of Section 12.3.4, we write

$$\frac{2E_{N_c}}{N_0} = \frac{2PT}{N_0(2BT)} = \frac{P}{N_0 B} \triangleq \text{CNR} \tag{12.81}$$

Thus, we have a physical interpretation of the parameter $2E_{N_c}/N_0$ as merely the input *signal power-to-noise power ratio*! Substituting the above relationship into (12.72), we see that the temporal channel capacity is given by

$$C \triangleq C_{N_c}/T_c = B \log_2\left(1 + \frac{P}{N_0 B}\right) \text{ bits/sec} \tag{12.82}$$

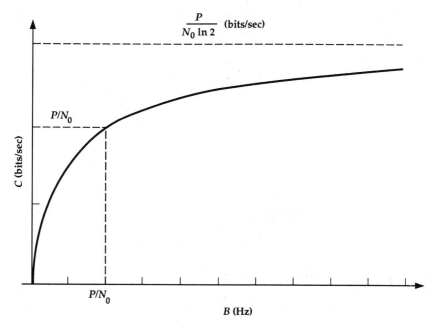

Figure 12.7   AWGN channel capacity versus bandwidth $B = 1/2T_c$

Equivalently, $C_{N_c} = C/W$ bits/dimension = bits/sec/Hz. A plot of the capacity C versus B is illustrated in Fig. 12.7.

It is now interesting to see what happens as the single-sided bandwidth $B = 1/2T_c$ Hertz approaches infinity, that is, $T_c \to 0$. Now

$$C_\infty \triangleq \lim_{B \to \infty} C = \lim_{B \to \infty} \log_2 \left(1 + \frac{P}{N_0} \frac{1}{B} \right)^B$$

$$= \log_2 e^{P/N_0} = \frac{P}{N_0} \frac{1}{\ln 2} \text{ bits/sec} \tag{12.83}$$

The parameter C_∞ is the so-called *infinite bandwidth Shannon capacity limit* for the AWGN channel. Furthermore, the maximum amount of information that can be transported error-free in T sec is

$$I_\infty = C_\infty T = \frac{E}{N_0} \frac{1}{\ln 2} \text{ bits} \tag{12.84}$$

The parameter I_∞ is the so-called *infinite bandwidth Shannon information transport* limit! In practice, most radio communication systems are designed to operate with 10 dB-Hz $\leq P/N_0 \leq 120$ dB-Hz. Thus, with $P/N_0 = 120$ dB-Hz, $I_\infty = 10^{12}/\ln 2$ bits of information can be delivered error-free in one second; this is more information than stored in the Library of Congress.

One question that remains as a consequence of the existence proof is to specify the ultimate codec, modem and their implementation. Unfortunately, nature seems to have hidden the particular coded modulations; they remain to be the slippery eel which motivates the further development of coding theory from the perspective of coded modulation.

12.5.2 The Channel Capacity, Matched Filter, and Spread Spectrum Communications Interconnect

For the bandlimited AWGN IF channel, there exist a codec and a modem which allow one to communicate error-free provided that the rate $r_c = \log_2 M_c / N_c = \log_2 M_c / WT = \mathcal{R}_b / W$ is less than the capacity

$$C_{N_c} = \frac{1}{2} \log_2 \left(1 + \frac{2E_{N_c}}{N_0} \right) \tag{12.85}$$

in bits/dimension where the available energy-per-dimension to noise density ratio for each coded message bit is

$$\frac{2E_{N_c}}{N_0} = \frac{P}{N_0} 2T_c = \frac{2P}{N_0 W} = \frac{P}{N_0 B} \equiv \text{CNR} \tag{12.86}$$

and W is the double-sided *Nyquist bandwidth*. Thus, (12.85) can also be written as

$$C_{N_c} = \frac{1}{2} \log_2 (1 + \text{CNR}) \tag{12.87}$$

As before, converting C_{N_c} to bits/sec requires that one normalize C_{N_c} by T_c; that is, define $C \triangleq C_{N_c} / T_c$, so that

$$C \triangleq \frac{C_{N_c}}{T_c} = B \log_2 (1 + \text{CNR}) \text{ bits/sec} \tag{12.88}$$

since $WT_c = 2BT_c = 1$.

Let us now relate the capacity theorem to the matched filter theory developed in Chapter 4. We note that the input signal power to noise power ratio can be written as

$$\text{CNR} = \frac{2P}{N_0 W} = 2 \frac{PT_b}{N_0} \frac{1}{WT_b} = 2 \frac{E_b}{N_0} \frac{1}{WT_b} \tag{12.89}$$

Now, with the T_b-sec processing gain defined by $PG_b \triangleq WT_b$, and noting that the output signal power-to-noise power ratio of a matched filter is $\rho_{out} = 2E_b / N_0$, then

$$\rho_{out} = \frac{2E_b}{N_0} = (PG_b)(\text{CNR}) = (PG_b)\rho_{in} \tag{12.90}$$

where $\rho_{in} = \text{CNR}$ is the input signal-power-to-noise power ratio to the matched filter. Thus

$$C = B \log_2 (1 + \rho_{out}/PG_b) \tag{12.91}$$

serves to relate the "matched-filter-theory" to the channel capacity formula. Moreover

$$\lim_{B \to \infty} C \triangleq C_\infty = \frac{P}{N_0 \ln 2} \text{ bits/sec} \tag{12.92}$$

so that in T sec, the maximum amount of information that can be transported error-free is

$$I_\infty = C_\infty T = \frac{E}{N_0 \ln 2} \text{ bits!} \tag{12.93}$$

In summary, there exists a codec-modem design such that error-free communications can take place (even in the presence of noise) provided that the channel throughput

$$r_c = \frac{\mathcal{R}_b}{W} < \frac{C}{W} = C_{N_c} = \frac{1}{2} \log_2 \left(1 + \frac{2E_b}{N_0} \frac{\mathcal{R}_b}{W} \right) \tag{12.94}$$

This requires that the communication efficiency E_b/N_0 satisfy

$$\frac{E_b}{N_0} > \frac{(2^{2\mathcal{R}_b/W} - 1)}{2\mathcal{R}_b/W} \qquad (12.95)$$

while operation at $\mathcal{R}_b = C$ requires that

$$\frac{E_b}{N_0} > \frac{(2^{2C/W} - 1)}{2C/W} \qquad (12.96)$$

Thus, three basic parameters which define the ability to communicate error-free over the bandlimited AWGN channel are E_b/N_0, $\mathcal{R}_b/W$, and C/W. The region of $C/B < 1$ is called the *power-limited region* of operation and the region $C/B > 1$ is called the *bandwidth-limited region*. The boundary between the two regions occurs at CNR = 1. The greatest communication efficiency (minimum E_b/N_0) is achieved when the bandwidth can be made large in comparison with the information rate $\mathcal{R}_b$. In the limiting case of very large bandwidth, and operating at $C = \mathcal{R}_b = 1/T_b$, then from (12.94) and (12.96), C/B approaches zero and E_b/N_0 approaches the *Shannon limit*, ln 2 = −1.6 dB. Thus, by invoking the coding theorem, one can say that for $E_b/N_0 > -1.6$ dB, the probability of message error in information delivered to the user can theoretically be made arbitrarily close to zero by use of a suitably chosen error-control code.

The fact that $\rho_{out} = PG_b \cdot \text{CNR} = PG_b\rho_{in}$ in (12.90) is very interesting. For example, if the channel input signal power to noise power ratio is CNR = −20 dB (signal power one hundred times smaller than the noise power) with $PG_b = 1000$, which is easily attainable, then $2E_b/N_0 = 10$ dB! Using this together with (12.88), we can relate the notion of information transfer efficiency, C/B, to the communication efficiency, E_b/N_0. In fact, for small CNR, we see that $C/B \simeq CNR/\ln 2 = (P/N_0B)/\ln 2 = 2(E_b/N_0)/(PG_b \ln 2)$.

Shannon's formula for the capacity of the bandlimited AWGN channel (baseband or IF) was one of the earliest results found in information theory [5]. This remarkable theorem demonstrated to system engineers that enormous improvements in the performance of digital communications systems could be achieved. The initial era of excitement (circa 1950) was followed by an era of discouragement since practical schemes which realized the improvement were difficult to find. This feeling was summarized in an often quoted folk theorem [16] which has been paraphrased as: "All codes are good, except those that we know of." For example, voice grade telephone channels can be modeled as bandlimited AWGN channels. The capacity of typical channels exceeds 20,000 bits/sec and the commercial incentive for achieving the highest possible data rates has a high dollar value. Yet 20 years after Shannon, the upper practical limit was about 2,400 bits/sec. Thirty years after Shannon, the upper practical limit was about 9,600 bits/sec and this advancement had little to do with Shannon. Ironically enough, this improvement over telephone channels was primarily due to the implementation of digital adaptive equalizers.

12.5.3 Spread Spectrum Communications

Spread spectrum communications is motivated by the need for reliable communications in the heat of battle. The basic concept can be summarized as follows: Suppose we transmit an N_c-dimensional signal in the presence of a jammer with finite power J Watts. If this signal is disguised as an $N_{c'}$-dimensional signal, $N_{c'} > N_c$, then the jammer must distribute his power over $N_{c'}$ dimensions and his power will be effectively reduced by the factor $N_{c'}/N_c$. If the jammer's power is J Watts, the average single-sided jamming noise power density is $N_J = J/B$ Watts/Hz where $W = 2B$ Hz is the spread spectrum bandwidth. The jamming power is effectively added to the noise N_0 from the channel. Now since in the jamming channel, $N_J \gg N_0$, we usually approximate the spread spectrum communication efficiency $E_b/(N_0 + N_J)$ by E_b/N_J. Since $E_b = PT_b = P/\mathcal{R}_b$, we have the relationship $J/P = (B/\mathcal{R}_b)/(E_b/N_J) = PG_b/(2E_b/N_J)$. The quantity J/P represents the received jamming power to signal power ratio. In antijam systems, it is desirable to have J/P as large as possible so as to minimize the vulnerability to jamming. From the relationship for J/P we see that this can be accomplished in two ways: one can increase the processing gain, $PG_b = WT_b$, which in essence increases the dimensionality of the transmitted signal, or decrease the required E_b/N_J for a particular bit error probability. In practice, the jamming margin can be 30 to 50 dB, or more.

We now return to the concept of channel capacity and relate it to spread spectrum communications in the presence of jamming. From (12.82) we have

$$\frac{C}{B} = \log_2 \left(1 + \frac{P}{J} \right) \text{ bits} \tag{12.97}$$

which says that $C/B \simeq (P/J)/\ln 2$ if the jamming margin is large ($J/P \gg 1$), a case of great practical interest.

12.5.4 The Union Bound for Binary Orthogonal Codes Written in Terms of the Shannon Capacity Limit

Using the fact that $2^x = e^{x \ln 2}$, (12.12) can be rewritten in terms of the Shannon capacity limit, namely

$$P(E) < 2^{-T\left(\frac{C_\infty}{2} - \mathcal{R}_b\right)} \tag{12.98}$$

which says that, as long as $0 < \mathcal{R}_b < C_\infty/2$, then $P(E)$ can be made to approach zero by making T large. This remarkably simple expression serves to upperbound (4.80); unfortunately this bound is not very tight.

12.6 GALLAGER'S UPPER BOUND ON THE ENSEMBLE ERROR PROBABILITY

One major purpose of information theory has been that of understanding Shannon's coding theorem by developing upper and lower bounds on the decoded error probability. Tremendous insight into the problem of transmitter design (coding trade-off) has been obtained using various error bounding methodologies on the ensemble of block codes defined in Section 12.4. It is not our purpose to develop this detailed theory here; rather, we will present Gallager's bound [8] and discuss its meaning from a system perspective. The reader who is interested in the details associated with the development of this bound should consult Gallager [8]. A special case of his bound is the *Bhattacharyya bound*.

We wish to gain insight into the minimum probability of decoding error as a function of the code rate r_c, the block length N_c, and the channel, by analyzing the performance achievable with the ensemble of block codes discussed in Section 12.4. This peculiar approach is dictated by the fact that there appears to be no systematic way to find block codes that minimize the probability of decoding error, $P(E)$, perhaps with the exception of those treated earlier in the chapter.

Gallager has developed a tight bound on the ensemble error probability $\overline{P(E)}$ for block codes. In particular, consider the case of the Q-ary coded $(M, 1, T_s)$ modulator used $\hat{k} = N_c$ times per codeword to transmit the class of Q-ary codes where $M = Q$, as discussed in Section 12.4.3. For this case, he has shown that

$$\overline{P(E)} < 2^{-N_c[E_r(r_c)]} \tag{12.99}$$

where $E_r(r_c)$ is the *random coding exponent* defined via

$$E_r(r_c) \triangleq \max_{0 \leq \rho \leq 1} \max_{\mathbf{Q}} [E_0(\rho, \mathbf{Q}) - \rho r_c] \text{ bits/dimension} \tag{12.100}$$

and where $\mathbf{Q}$ denotes an arbitrary probability assignment for the coordinates of the coded modulation vector, which are chosen independently with probability $P[s_{mn} = q_k\sqrt{E_{N_c}}] \triangleq p_k$. If $Q = M = 2$ (binary coded BPSK), for example, we have

$$Q(k) = \begin{cases} P\left(s_{mn} = \sqrt{E_{N_c}}\right) = p_0, & k = 0 \\ P\left(s_{mn} = -\sqrt{E_{N_c}}\right) = p_1, & k = 1, \end{cases}$$

where the probability is computed over the ensemble.

For the IF CMC model defined in Section 11.2.2

$$E_0(\rho, \mathbf{Q}) = -\log_2 \left(\int_{-\infty}^{\infty} \left[\sum_{k=1}^{Q} f\left(r_n | s_{mn} = q_k\sqrt{E_{N_c}}\right)^{\frac{1}{1+\rho}} p_k \right]^{1+\rho} dr_n \right) \tag{12.101}$$

for all observables r_n, $n = 1, \ldots, N_c$, and all messages $m = 0, \ldots, M_c - 1$. Here $f(r_n | s_{mn} = q_k\sqrt{E_{N_c}})$ denotes the known channel transition probability density function (which characterizes the channel) for the nth received observable, given that message m is sent.

For the IF DMC (quantized IF CMC) model defined in Section 3.1.1, Section 11.2, and used in Section 11.2.4

$$E_0(\rho, \mathbf{Q}) = -\log_2 \left(\sum_{j=0}^{J-1} \left[\sum_{k=1}^{Q} P\left[R_n = R_{nj} | s_{mn} = q_k\sqrt{E_{N_c}} \right]^{\frac{1}{1+\rho}} p_k \right]^{1+\rho} \right) \tag{12.102}$$

for all quantized observables R_n, $n = 1, \ldots, N_c$, and all messages $m = 0, \ldots, M_c - 1$. Here $P(R_n = R_{nj} | s_{mn} = q_k\sqrt{E_{N_c}}) = p_{kj} \triangleq P(j|k)$ denotes the known channel transition probability for the nth received quantized observable (that is, the quantized version of r_n), given that message m is sent.

For modem and codec design purposes, it is of interest to connect $E_r(r_c)$ with the channel cutoff rate R_0 for the IF CMC model and with the quantized R_0 for the IF DMC model. This is accomplished using Gallager's function $E_0(\rho, \mathbf{Q})$ and setting $\rho = 1$, that is $R_0 = E_0(1, \mathbf{Q})$. In regions of low E_{N_c}/N_0, then $E_r(r_c)$ is maximized at $\rho = 1$ [20], so $E_r(r_c) \simeq (R_0 - r_c)$ for equally likely, independent message symbols. For the IF CMC we have, from (12.101)

$$R_0 = \mathcal{R}_0/W = E_0(1, \mathbf{Q})$$

$$= -\log_2 \left(\int_{-\infty}^{\infty} \left[\sum_{k=1}^{Q} p_k \sqrt{f\left(r_n | s_{mn} = q_k\sqrt{E_{N_c}}\right)} \right]^2 dr_n \right) \text{ bits/sec/Hz}$$

while for the IF DMC model, we have, from (12.102)

$$R_0 = \mathcal{R}_0/W = E_0(1, \mathbf{Q})$$

$$= -\log_2 \left(\sum_{j=0}^{J-1} \left[\sum_{k=1}^{Q} p_k \sqrt{P\left(R_n = R_{nj}|s_{mn} = q_k\sqrt{E_{N_c}}\right)} \right]^2 \right) \quad \text{bits/sec/Hz}$$

where the channel transition probabilities for the quantized observables R_n are related to the channel transition probability density functions $f(r_n|s_{mn} = q_k\sqrt{E_{N_c}})$ through

$$P\left(R_n = R_{nj}|s_{mn} = q_k\sqrt{E_{N_c}}\right) = p_{kj} = \int_{\Delta_j} \frac{1}{\sqrt{\pi N_0}} \exp\left(-\frac{(r_n - q_k\sqrt{E_{N_c}})^2}{N_0}\right) dr_n$$

for Q-ary coded ASK and Δ_j is the observation region which gets mapped into the observable R_j for all $j = 0, 1, \ldots, J-1$ (see Fig. 11.6). This defines the connection between the IF CMC and the quantized IF CMC (IF DMC).

Properties of the random coding exponent $E_r(r_c)$ have been studied with much success; from our system perspective, one important feature is that $E_r(r_c) > 0$ for all $0 \le r_c < C_{N_c}$. Thus, (12.99) says that for appropriately chosen codes and modems, the probability of decoding error can be made to approach zero with increasing block length N_c for any rate less than capacity. In expression (12.100), $E_r(r_c)$ is a positive function of the channel transition probability via $E_0(\rho, \mathbf{Q})$. This function yields a fundamental characterization of the channel for coding purposes. It brings the relationship between decoding error probability, data rate, block length, and channel conditions into focus. In fact, as we shall later see, the question of what modulation-demodulation techniques one should use when coding is applied to system design can be addressed on the basis of $E_r(r_c)$. Moreover, since the ensemble of block codes satisfies (12.99), *it follows that at least one block code in the ensemble must have $P(E)$ no greater than the ensemble average $\overline{P(E)}$*. From a system design perspective, this bound is not useful in accessing design point requirements with respect to communications efficiency (E_b/N_0), supportable data rate ($\mathcal{R}_b$), or power requirements which achieve a specified error probability. The insight provided is philosophical in the sense that a code(s) exist(s) such that if $r_c < C_{N_c}$ then the error probability can be made arbitrarily small by employing large block lengths N_c. Unfortunately, the complexity of the encoder and decoder which accommodate long (or short) block lengths is not addressed by this interesting bound. In fact, the complexity issue remains to be the important unsolved problem in information theory. The *random coding exponent* versus rate curve is of interest to the system engineer, especially for comparing the R_0 values corresponding to specific codec-modem pairs. We will consider an example of how such a comparison can be useful in the design of the demodulator in Section 12.7.5, for binary coded DPSK. The general characteristics of R_0 are discussed extensively in Gallager. However, for our purposes, we are particularly interested in its behavior for very noisy channels, the case of greatest practical interest.

12.6.1 Block versus Convolutional Coding for the Very Noisy Channel

Without elaborating the details, let us consider the class of very noisy channels (low values of E_b/N_0, where coding is of interest) and present the random coding exponent versus rate curve $E_r(r_c)$ which results and discuss the implications from a coding viewpoint. For block

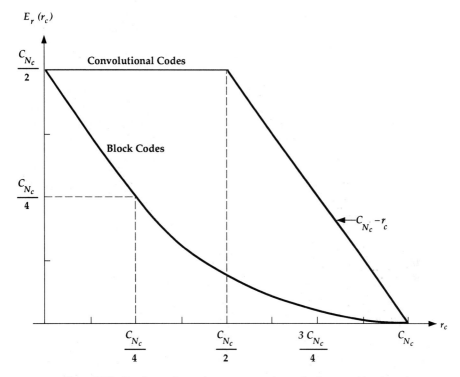

Figure 12.8   Random coding exponent versus code rate for the very noisy channel

codes, it can be shown that [8]

$$E_r(r_c) \simeq \begin{cases} \left(\sqrt{C_{N_c}} - \sqrt{r_c}\right)^2, & \frac{C_{N_c}}{4} \le r_c \le C_{N_c} \\ \frac{C_{N_c}}{2} - r_c, & 0 \le r_c < \frac{C_{N_c}}{4} \end{cases} \qquad (12.103)$$

Figure 12.8 demonstrates plots of the random coding exponent given in (12.103) versus code rate. For comparison purposes we also demonstrate the random coding exponent of convolutional codes (to be discussed in Chapter 13) found by Viterbi [10]. From a systems perspective, we conclude from Fig. 12.8 that for a given code rate, there exist convolutional codes which yield a lower error probability than that of block codes. (This statement assumes a bandwidth unconstrained, memoryless channel.) Notice too that the maximum value of $E_r(r_c)$ predicted by this bound is $C_{N_c}/2$. The result in (12.103) was derived using the baseband system architecture which absorbs the modulator and demodulator as parts of the channel model.

12.7 DEMODULATOR DESIGN USING THE R_0 CRITERION

We now make use of Gallager's bound (12.99) and develop a philosophy for the design of the modulation-demodulation system in a coded BPSK system with equally likely messages. Practical demodulator outputs, **r**, are continuous random vectors, that is, the projections of $r(t)$ onto the signal set basis functions by the combined demodulator-detector. Moreover, digital signal processing algorithms require that the observable **r** be vector-quantized into say $J = 2^L$ levels.

To develop a quantizer design philosophy for processing the demodulator output, we again consider the special case of $E_0(\rho, \mathbf{Q})$ when $\rho = 1$. Then, with $R_0 = E_0(1, \mathbf{Q})$, the ensemble error probability bound in (12.99) reduces to

$$\overline{P(E)} < 2^{-N_c(R_0-r_c)} = 2^{-T(R_0-\mathcal{R}_b)} \tag{12.104}$$

when we use (12.100). The number R_0 not only establishes a limit on the rate at which one can communicate with arbitrarily small error probability, but also gives an *exponent* of error probability. No other communication parameter gives such insight into the communication process—not even the channel capacity, C_{N_c}, which only establishes a limit on the rates at which reliable communication is possible. The "unquantized R_0" given by (12.101) with $\rho = 1$ is also the *computational cutoff rate* of sequential decoding (see [7], Chapter 6), that is, the rate above which the average number of decoding steps per decoded symbol becomes infinity.

The idea that the quantizer design used in the demodulation system can be based on R_0 for a coded digital communication system was first suggested by Wozencraft and Kennedy [1]. A modem design based on the fact that $\mathbf{r}$ is the input to the decoder is impractical because operation of digital hardware requires that $\mathbf{r}$ be quantized. Therefore, the unquantized $R_0 = E_0(1, \mathbf{Q})$ found from (12.101) yields an upper bound for the cut-off rate that is unachievable by any practical demodulator. On the other hand, the quantized $R_0 = E_0(1, \mathbf{Q})$ found from (12.102) can yield an upper bound for the case of a finite demodulator output alphabet which is almost as tight. In other words, the choice of the modulation-demodulation quantizer design to be used in a coded system is not based on the error probability (symbol, bit or message) derived in Chapters 3 through 10; this error probability serves only as input to the computation of R_0. The choice of the isometric modulation-demodulation technique and quantizer design should proceed on the basis of maximizing R_0 [as seen from (12.104)], as opposed to minimizing the M-ary symbol error probability of an uncoded system. In the next section we will demonstrate this system engineering design philosophy for coded BPSK for hard-decision and soft-decision demodulators assuming that the IF CMC has been converted to the IF DMC using the methodology presented in Sections 11.2.3, 11.2.4, and 12.6.

Finally, from a system engineer's perspective, the R_0 parameter can be considered to be characteristic of both the codec (since it is a function of r_c) and the modem. For hard decisions, we will see from (12.107) that R_0 is a function of the coded message bit error probability, p, given by $Q\left(\sqrt{2E_{N_c}/N_0}\right)$ for binary coded BPSK. In this sense, the bit error probabilities derived in Chapters 3 through 10 for uncoded systems serve as a tool in the design of coded systems.

12.7.1 Design of Hard-Decision Binary Coded BPSK Demodulators

Consider first the IF DMC with binary coded BPSK and with equally probable messages $m_0 = 0$ and $m_1 = 1$. The rate R_0, found from (12.102) with $\rho = 1$ for *soft-decision demodulation*, reduces to

$$R_0 = -\log_2 \left\{ \frac{1}{2} \right. \tag{12.105}$$

$$\left. \times \left[1 + \sum_{j=0}^{2^L-1} \sqrt{P\left(R_n = R_{nj}|s_{mn} = \sqrt{E_{N_c}}\right) P\left(R_n = R_{nj}|s_{mn} = -\sqrt{E_{N_c}}\right)} \right] \right\}$$

for $J = 2^L$ output quantization levels or L bits of quantization. This compares with the capacity of the same IF DMC [8]

$$C_{N_c} = \sum_{j=0}^{2^L-1} \left\{ P\left(R_n = R_{nj} | s_{mn} = -\sqrt{E_{N_c}} \right) \right. \tag{12.106}$$

$$\left. \times \log_2 \left[\frac{2P(R_n = R_{nj} | s_{mn} = -\sqrt{E_{N_c}})}{P(R_n = R_{nj} | s_{mn} = -\sqrt{E_{N_c}}) + P(R_n = R_{nj} | s_{mn} = \sqrt{E_{N_c}})} \right] \right\}$$

For *hard-decision demodulators* ($L = 1$), (12.105) reduces to

$$R_0 = -\log_2 \left[\frac{1}{2} \left(1 + \sqrt{4p(1-p)} \right) \right] \tag{12.107}$$

for the BSC with crossover probability p of the uncoded message bits. The quantized R_0 given in (12.107) for hard-decision decoding compares with the channel capacity of [8]

$$C_{N_c} = 1 + p \log_2 p + (1-p) \log_2(1-p) \tag{12.108}$$

for the same BSC. Figure 12.9 demonstrates the channel symbol error probability p required to operate at rates of $r_c = R_0$ and $r_c = C_{N_c}$ using (12.107) and (12.108) versus code bandwidth expansion $r_c^{-1} = 1/r_c$. For example, from Fig. 12.9, we observe that for hard-decision demodulators (one-bit quantizers) and a code rate r_c, it appears that coding is only useful when the channel symbol error probability p is less than 0.11.

For BPSK modems operating in the AWGN channel,

$$p = Q(\sqrt{2E_{N_c}/N_0}) = Q(\sqrt{2r_c E_b/N_0}).$$

Notice here that p not only depends on the modulation but also on the code via the code rate. For this case, the limiting value (small E_{N_c}/N_0) of r_c can be increased from that of (12.107) to the limiting value of the unquantized error exponent given by R_0 in (12.37), with a $10\log_{10}(\pi/2) \simeq 2$ dB improvement in the E_{N_c}/N_0 efficiency of energy utilization. This is the 2 dB loss due to the two-level quantization process required for hard-decision decoding.

12.7.2 Design of Soft-Decision Coded BPSK Demodulators

As we have mentioned, the R_0 criterion is natural for use in the design of the modem as it can be optimized as a function of the number J of quantization levels, allowing the system engineer to trade off system performance with the implementation complexity of the decoder. For example, in making this trade, the engineer would examine the quantized R_0 attainable with $J = 2, 4, 8, 16$, and so on, until he has obtained an R_0 sufficiently close to the unquantized ($J = \infty$) R_0 which justifies the decoder cost and complexity versus the improved bit error probability performance. As we have already pointed out, the absolute upper limit on the code rate r_c is the channel capacity C_{N_c}, and the upper limit for all practical purposes (considering cost and implementation complexity) is the computational cutoff rate R_0; the unquantized R_0 appears to be approximately 3 dB smaller than C_{N_c} for the cases of greatest interest.

The procedure used in the early design of modems for selecting the quantization cell-size parameter Δ demonstrated in Fig. 11.6 (see Sections 11.2.3, 11.2.4, and 12.6) was to choose it to minimize the E_b/N_0 required to operate at a code rate of R_0 bits/dimension.

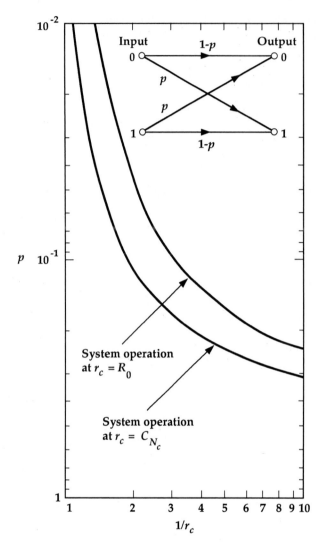

Figure 12.9 Coding limits for the BSC using hard-decision demodulation of binary coded BPSK or QPSK modulation (Reprinted from [17])

This philosophy is intuitively justified since one is attempting to maximize the *coding gain* (the reduction in E_b/N_0 required when coding is used to that required in an uncoded BPSK system for the same BEP) when operating at rate r_c near R_0. Usually computer simulations of the coding system are required to make this trade, and the results have shown good agreement with those determined using the theoretical considerations due to Massey [2] which we now describe.

Consider the case of binary codes with equally likely message symbols using BPSK modulators, that is, $s_{mn} = \sqrt{E_{N_c}}$ or $-\sqrt{E_{N_c}}$, and J output symbols; see Fig. 11.6 for the case $L = 3$ and $J = 8$. Recall that each received observable r_n must be assigned one of J quantization levels $R_{nj} \in \{R_{n0}, \ldots, R_{n(J-1)}\}$. To do this, the quantizer needs $(J - 1)$ thresholds $T_j, j = 0, \ldots, J - 2$. Here T_j denotes the threshold between two quantizer

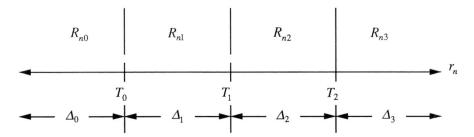

Figure 12.10   Quantization regions and levels for the case $J = 4$

output levels R_{nj} and $R_{n(j+1)}$. The sketch in Fig. 12.10 illustrates the decision regions or quantization levels for the case $J = 4$.

Our goal is to select these threshold levels in an optimum way; specifically, we seek a selection that will maximize the quantized cutoff rate R_0, given by (12.105) and repeated below for convenience:

$$R_0 = -\log_2 \left\{ \frac{1}{2} \right.$$

(12.109)

$$\left. \times \left[1 + \sum_{j=0}^{2^L-1} \sqrt{ P\left(R_n = R_{nj} | s_{mn} = \sqrt{E_{N_c}} \right) P\left(R_n = R_{nj} | s_{mn} = -\sqrt{E_{N_c}} \right) } \right] \right\}$$

Hence the demodulator design problem is to choose each threshold T_j such that

$$\frac{\partial R_0}{\partial T_j} = 0$$

(12.110)

Applying the above equation, Massey found that the optimum threshold levels were given by

$$T_j = \sqrt{ \Lambda(R_{nj}) \Lambda(R_{n,j+1}) }$$

(12.111)

where the likelihood ratio is given by

$$\Lambda(R_{nj}) \triangleq \frac{ P\left(R_n = R_{nj} | s_{mn} = \sqrt{E_{N_c}} \right) }{ P\left(R_n = R_{nj} | s_{mn} = -\sqrt{E_{N_c}} \right) }$$

(12.112)

for a given output quantization level R_{nj}, $j = 0, \ldots, J - 1$. This says that each optimum threshold is the geometric mean of the likelihood ratios for the quantizer outputs whose decision regions the threshold value separates.

Odenwalder [17] considers the use of equally spaced thresholds at $0, \pm T, \ldots,$ $\pm(2^{L-1} - 1)T$, with the spacing $\Delta_j = T$ chosen to maximize R_0. For $L = 3$ (3-bit quantization) and $E_{N_c}/N_0 \simeq 1.5$ dB, we obtain threshold spacing T of roughly .5 or .6 times the standard deviation of the unquantized demodulator outputs [17]. Decreasing (or increasing) E_{N_c}/N_0 results in only a minor decrease (or increase) in T. Thus, a fixed value for T can be used over the entire dynamic range of E_{N_c}/N_0, and an automatic gain control (AGC) circuit can be used to estimate the noise variance $N_0/2$. System performance is not sensitive to small variations in the AGC output, and the value of T is usually selected for the worst case value for E_{N_c}/N_0 or E_b/N_0.

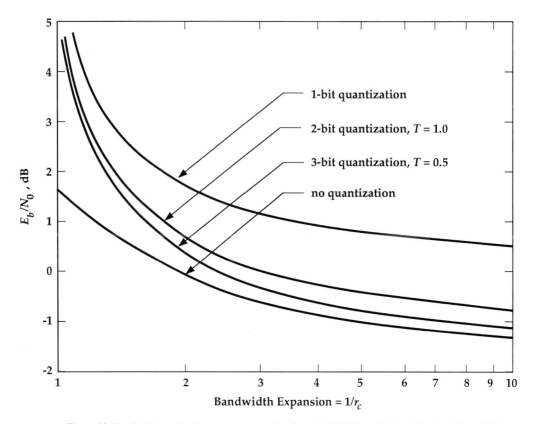

Figure 12.11 E_b/N_0 required to operate at $r_c = C_{N_c}$ for coded BPSK modulation (Reprinted from [17])

12.7.3 Choice of the Number of Quantization Levels

The results presented so far in this section are now combined to demonstrate the effects of the number of quantization levels on demodulator performance. We assume that coherent demodulation of coded BPSK or QPSK is used in the AWGN channel. First, we consider the problem of demodulator design when we wish to operate at code rates $r_c = C_{N_c}$. Equation (12.72) for infinite quantization and (12.108) for 1-bit quantization establish this performance by relating the bandwidth expansion $r_c^{-1} = 1/r_c$ and the required communication efficiency to operate at $r_c = C_{N_c}$. Plots of this relationship are demonstrated in Fig. 12.11. In addition, comparable results found from (12.106) for soft-decision demodulators are demonstrated for 2- and 3-bit quantizers with equally-spaced threshold levels.

Figure 12.12 gives the corresponding curves of required E_b/N_0 versus $1/r_c$ for system operation at $r_c = R_0$ using (12.37), (12.105), (12.107) and equally spaced threshold levels. From Figures 12.11 and 12.12 we see that 3-bit soft quantization is almost as good as infinite quantization ($L = \infty$), and that 1-bit quantization (hard-decision) requires about 2 dB more E_b/N_0 than infinite quantization for comparable performance. It can also be concluded by comparison of Figures 12.11 and 12.12 that for small values of required E_b/N_0 corresponding to large bandwidth expansions, approximately 3 dB more E_b/N_0 is required to operate the system at $r_c = R_0$ than is required to operate at $r_c = C_{N_c}$.

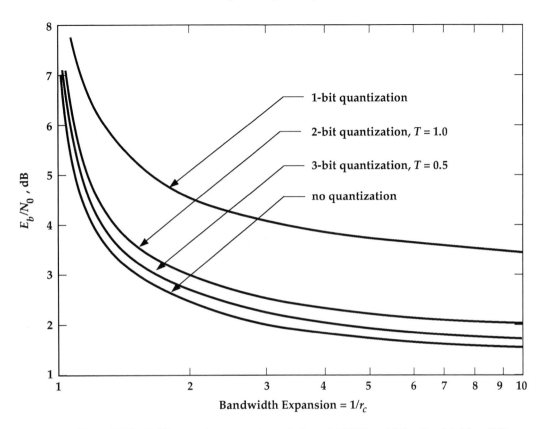

Figure 12.12 E_b/N_0 required to operate at $r_c = R_0$ for coded BPSK modulation (Reprinted from [17])

12.7.4 Coded DPSK and DDPSK Modems with Hard Decisions

For binary coded DPSK modulation, $p = (1/2) \exp(-r_c E_b/N_0)$ is the coded message bit error probability as given by (7.14). For binary coded double differential phase shift keying (DDPSK), the expression for p is given by (8.58). In both cases, we have assumed optimum demodulators and no frequency offsets. Furthermore, the R_0 for both coded DPSK and coded DDPSK with hard decisions is given by (12.107). Figure 12.13 illustrates the E_b/N_0 required to operate at code rate $r_c = R_0$ versus bandwidth expansion $1/r_c$ for coded BPSK, DPSK, and DDPSK. These curves reveal the minimum value of E_b/N_0 for which coding is useful for each of the coded modulation techniques. From the DDPSK curve, for example, we see that if the E_b/N_0 is less than 10.45 dB, no practical code exists at any rate that will allow us to achieve arbitrarily small bit error probabilities. Figure 12.13 also illustrates the E_b/N_0 needed to achieve a bit error probability of 10^{-5} without coding for each of the three modulation schemes; these curves enable us to compute achievable coding gains. From the DDPSK curve, for example, we see that if $r_c = 1/2$, then the needed E_b/N_0 is 10.45 dB. Thus, the maximum coding gain achievable (for BER $= 10^{-5}$ and hard-decision decoding) is 3.55 dB. To achieve a greater gain we must increase the complexity of the receiver to allow for soft decisions.

We further note from Fig. 12.13 that for both coded DPSK and DDPSK it is not the

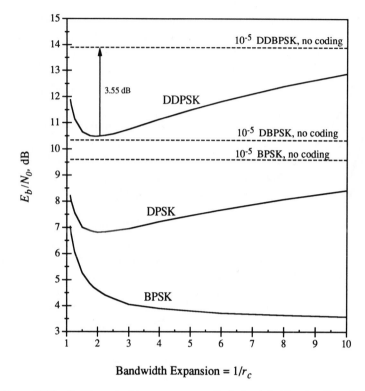

Figure 12.13 E_b/N_0 required to operate at $r_c = R_0$ for hard decision decoding of various coded modulations (Courtesy of Debbie van Alphen)

case that decreasing the code rate (say by increasing the redundancy of the code) automatically increases the achievable coding gain, as was the case for BPSK. This is because the energy in each coded message bit is given by $E_{N_c} = r_c E_b$; thus, decreasing the code rate r_c decreases the channel symbol energy, effectively making the demodulation phase reference noisier. This effect is not evident in the case of BPSK because a perfect phase reference is assumed [17]. Both DPSK and DDPSK appear to achieve maximum coding gain with code rates near $r_c = 1/2$.

12.7.5 Binary Coded DPSK with Infinite Quantization

We now consider the case of binary coded DPSK with infinite quantization ($L = \infty$); that is, we assume that the demodulator output random variable is handed over to the decoder unquantized. For this case, Odenwalder [17] has shown that

$$R_0 = -\log_2\left[\frac{1}{2}\left(1 + \int_{-\infty}^{\infty}\sqrt{f(\psi|\theta^i = 0)f(\psi|\theta^i = \pi)}d\psi\right)\right] \quad (12.113)$$

where $f(\psi|\theta^i = \theta_0)$ denotes the conditional probability density function for the demodulator output random variable ψ, given that the information phase angle $\theta^i = \Delta^n\theta_i$ is equal to θ_0. (See Chapters 7 and 8.) Clearly, the form of the conditional probability density function depends on the detector implementation, via ψ.

In order to illustrate our thesis that the R_0 criterion should be used to design the modem in a coded system versus using the post-demodulation bit error probability suggested for the comparable uncoded system, we shall consider two demodulation techniques, namely the optimum I-Q demodulator and the suboptimum *autocorrelation detector* (ACD) demodulator (delay received signal one coded message bit and multiply; see Fig. 7.18). In the case of the autocorrelation detector, the demodulator output random variable ψ is equal to the sampled output of the lowpass filter, and the resulting R_0 is given in [17]. For the optimum I-Q detector the random variable ψ_i is equal to the difference of the two adjacent transmitted phase angle measurements, $\eta_i - \eta_{i-1}$. The conditional probability density function $f(\psi|\theta^i = 0)$ is given by (7.3), while $f(\psi|\theta^i = \pi)$ is simply a shift of the same density function by $\pi \pmod{2\pi}$.

In Fig. 12.14, we plot the E_b/N_0 necessary to operate at rate $r_c = R_0$ versus bandwidth expansion for binary coded DPSK with infinite quantization using both the ACD and I-Q demodulators. For comparison, we also show the curve for binary coded DPSK with hard decisions using the I-Q demodulator. It is very intriguing to note that, while in the absence of coding the I-Q demodulator is known to be optimum, with coding the autocorrelation detector (ACD) actually gives better R_0 results. Thus, for binary coded DPSK, it is preferable to use an autocorrelation detector rather than an I-Q detector. This confirms the thesis that, from a communication efficiency perspective, the R_0 criterion is better to use in the design of a modem for coded systems; the post-demodulation bit error probability is better to use in the design of an uncoded system.

12.7.6 Optimization of Codes and Modem Design Using the R_0 Criterion

In (12.105) through (12.108), we have presented the cutoff rate R_0 and capacity C_{N_c} for soft- and hard-decision binary coded BPSK demodulators. We know from Shannon's theorem that channel coding will be of no value unless $r_c < C_{N_c}$; we also assert that, for most practical purposes, operation at $r_c < R_0$ will be required in order to realize the decoder mechanization with acceptable hardware complexity.

As in Section 12.4.3, we now consider the use of the $(M, 1, T_s)$ modulator with $\hat{k} = N_c$ to transmit $M = Q$-ary (M_c, N_c, T_c) codes. This implies that the Q-ary coded ASK modulator is used $\hat{k} = N_c$ times for the transmission of a codeword. If we use (12.54) and optimize over the a priori probabilities of the coded message symbols, we have

$$R_0 \le - \min_P \log_2 \left[\sum_{j=1}^{Q} \sum_{k=1}^{Q} p_j p_k \exp\left(-q_{jk}^2 E_{N_c}/4N_0\right) \right] \qquad (12.114)$$

where the minimization is taken over all possible probability distributions $P = \{p_j\}$, $j = 1, \ldots, Q$, on the coded message symbols. (Notice we have assumed an isometric mapping between the coder and modulator.) Since the minimization occurs when $p_j = 1/Q$ for all j, then one can show [2] that (12.114) reduces to

$$R_0 \le \log_2 Q - \log_2 \left[1 + (Q-1) \exp\left(-\frac{Q\overline{E_{N_c}}}{2(Q-1)N_0}\right) \right] \qquad (12.115)$$

where the equality holds only for equally likely Q-ary transorthogonal (simplex) signals and where $\overline{E_{N_c}}$ denotes the average energy per codeword dimension:

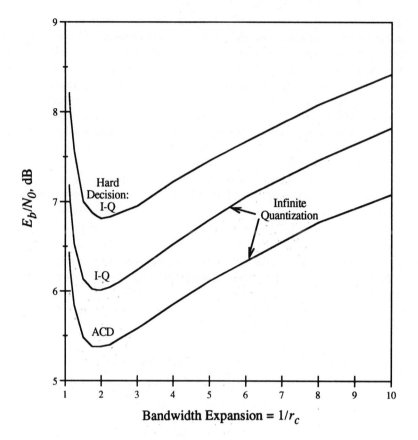

Figure 12.14 E_b/N_0 required to operate coded DPSK at $r_c = R_0$ (Courtesy of Debbie van Alphen)

$$\overline{E_{N_c}} = \frac{1}{Q} \sum_{i=1}^{Q} q_i^2$$

This inequality can be used to gain considerable insight into the design of coded modulation systems for low and high values of $\overline{E_{N_c}}/N_0$.

Now for transorthogonal (simplex) signals, the equality in (12.115) holds and we differentiate with respect to $x = \overline{E_{N_c}}/N_0$ to obtain

$$\frac{dR_0}{dx} = (Q/2 \ln 2) \left[\exp\left(\frac{Q(\overline{E_{N_c}}/N_0)}{2(Q-1)} \right) + Q - 1 \right]^{-1} \tag{12.116}$$

which decreases monotonically with increasing x. Thus, for small values of x, where coding is most useful, (12.116) yields

$$\frac{dR_0}{dx} \simeq 1/2 \ln 2 \text{ or } R_0 \simeq (\overline{E_{N_c}}/N_0)/2 \ln 2 \tag{12.117}$$

which is independent of Q! This demonstrates that one should use binary coded BPSK ($Q = M = 2$) in the design of coded modulation systems since this technique is easiest to implement. In this sense, the optimum value for Q is $Q_{opt} = 2$. From a systems engineering perspective, this value for Q simultaneously implies two things: (1) it says that $M = 2 = Q_{opt}$ which implies that the $(2, 1, T_s)$ binary coded ASK or BPSK modulator is optimum, and (2) that the binary transorthogonal (simplex) code ($M_c = 2^{K_c}$, $N_c = 2^{K_c} - 1$) is simultaneously optimum! Further, with $Q = 2$ in (12.115), we observe for transorthogonal codes and the $(2, 1, T_s)$ coded BPSK modulator that

$$R_0 = \log_2[2/(1 + \exp(-E_{N_c}/N_0)]$$

which agrees with (12.37), as it should.

Finally, from (12.117) and (12.83) we see that as T_c approaches zero (B approaches infinity)

$$\lim_{T_c \to 0} \mathcal{R}_0 \triangleq \lim_{T_c \to 0} \frac{R_0}{T_c} = C_\infty/2 \text{ bits/sec}$$

This says that $\mathcal{R}_0$ is approximately 3 dB less than the infinite bandwidth channel capacity, C_∞. This 3 dB backoff from channel capacity is frequently taken to be the achievable region of operation by practicing system engineers.

For high values of $\overline{E_{N_c}}/N_0$, then we readily observe from (12.115) that $R_0 \leq \log_2 Q$ and $dR_0/dx = 0$. This serves to demonstrate why one might choose a large Q in system designs which require a large value of $\overline{E_{N_c}}/N_0$, where (12.115) holds with near equality.

In conclusion, in this section we have demonstrated how the R_0-criterion can be used to accomplish signal design, that is, the simultaneous choice of both the coding and modulation techniques.

12.8 LINEAR BLOCK CODES

In the past several sections of this chapter, we have demonstrated the implications of a very striking theorem regarding the transmission of information over a noisy channel, namely, Shannon's coding theorem. This theorem says that every communication channel has a definite capacity C_{N_c} (or C) and that for any rate $r_c < C_{N_c}$ (or $\mathcal{R}_b < C$), there exist codes (and modems) of rate r_c which, with maximum likelihood decoding, render an arbitrarily small probability of codeword decoding error $P(E)$. Shannon's theorem only demonstrates the existence of such codes; unfortunately, it does not indicate how these codes can be constructed. The system engineer who wishes to use error control coding is faced with three problems: (1) finding good codes, (2) finding practical encoding methods (encoders), and (3) finding practical decoding methods (decoders). By practical we mean small coding delay and minimum hardware complexity. Transmission errors created in the channel are frequently referred to as *random errors*. Codes devised by coding theorists to combat independent errors are called *random error-correcting codes*. Sometimes channel errors cluster together and occur in bursts. Codes devised to correct burst errors are called *burst error-correcting codes*. In this section, we will discuss and present, from a systems perspective, coding techniques for error correction—in particular, the Hamming, Golay, Reed-Solomon, and BCH techniques. See the Venn diagram of Fig. 11.18. We will not present the algebraic theory leading to the structure of such codes; this can be found in textbooks of algebric coding theory [3, 11, 13] and information theory [8, 9].

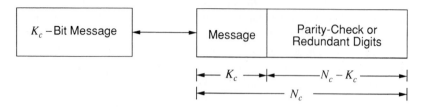

Figure 12.15 The codeword for a systematic block code

We will primarily be concerned with the class of (N_c, K_c) codes.[2] An (N_c, K_c) block code is a code in which each codeword contains N_c symbols; K_c of these are *information symbols* and $r = N_c - K_c$ are *redundant parity check symbols*. In this context, the dimensionality of the codeword space refers to the number of encoder output symbols for each block of K_c encoder input symbols. It is possible to encode the K_c-symbol message into a codeword such that the first K_c symbols of the codeword are exactly the same as the message block and the last $N_c - K_c$ symbols are redundant parity check symbols which are functions of the information symbols (see Fig. 12.15). A code of this form is called a *systematic* code. The redundant symbols are used to combat (control) errors which occur during transmission. Systematic codes are of greatest practical interest and we shall concentrate on the use of such codes for error control in what follows. Linear block codes can be described by a $K_c \times N_c$ generator matrix **G**. If the K_c information symbols into the encoder are represented by $\mathbf{u} = (u_1, u_2, \ldots, u_{K_c})$, the encoder output word is represented by the N_c-dimensional row vector **x**. The encoder input-output relationship is characterized by

$$\mathbf{x} = \mathbf{uG} \tag{12.118}$$

so that the N_c-symbol encoder output blocks (that is, the codewords) are linear combinations of the rows of the generator matrix. For the binary case, the output blocks are bit-by-bit modulo-2 sums of the appropriate rows of **G**. Usually block codes are decoded using algebraic decoding techniques; the important and elegant details can be found by the system engineer in many coding and information theory books [3, 11, 12, 13]. The Venn diagram for error-correcting codes is illustrated in Fig. 11.18. In what follows we shall assume that the mapping of code vectors **x** to coded modulation vectors **S** is isometric.

12.8.1 The Union Bound Expressed in Terms of the Hamming Distance for Binary Codes

From the coding perspective it is very instructive to relate the Euclidean distance, d_{ij}, to the Hamming distance, D_{ij}, and use this result in the union bound result given in (12.3) to gain further insight into block coding. To do so we must make sure that the modulation technique chosen to transmit the coded message bits does not destroy the Hamming distance properties of the code; otherwise, the effectiveness of the coding scheme to minimize the required E_b/N_0 is lost. As was demonstrated in Chapter 11, coded BPSK and QPSK modulators can be used since they map the coded message bits into the modulator output waveform space without modifying the effective distance properties of the code. In this section we will quantify this fact for the class of binary codes.

---

2. In the coding literature, it has been fashionable to refer to such codes as (n, k) codes. We prefer to use (N_c, K_c) here so as to minimize notational confusion; that is, k is Boltzmann's constant, and n and k are summing indices.

Recall that the *Hamming distance* $D(\mathbf{x}_i, \mathbf{x}_j) \triangleq D_{ij}$ between the two binary codewords $\mathbf{x}_i$ and $\mathbf{x}_j$ is the number of positions in which the two codewords differ. On the other hand, the *Hamming weight* of a codeword (N_c-tuple) $\mathbf{x}$ is defined as the number of nonzero components in $\mathbf{x}$. Using coded BPSK, the Euclidean distance d_{ij} between two binary coded signal vectors $(\mathbf{S}_i)$ and $(\mathbf{S}_j)$ can be written in terms of the *Hamming distance* D_{ij}. Now

$$d_{ij}^2 = |\mathbf{S}_i - \mathbf{S}_j|^2 = E_i + E_j - 2\sqrt{E_i E_j}\gamma_{ij} \tag{12.119}$$

where

$$\gamma_{ij} = \frac{\mathbf{S}_i^T \cdot \mathbf{S}_j}{\sqrt{E_i E_j}} = \cos\theta_{ij} \tag{12.120}$$

measures the normalized cross-correlation between the two coded modulation vectors. For binary codewords

$$\gamma_{ij} = \frac{A_{ij} - D_{ij}}{A_{ij} + D_{ij}} \tag{12.121}$$

where $A_{ij}(D_{ij})$ represents the number of term-by-term agreements (disagreements) between the two binary vectors.

For binary coded BPSK we have from (12.119), (12.120), and (12.121)

$$d_{ij}^2 = 2E(1 - \gamma_{ij}) = 2N_c E_{N_c}(1 - \gamma_{ij}) \tag{12.122}$$

and using the facts that $N_c = A_{ij} + D_{ij}$ and $A_{ij} = N_c - D_{ij}$ in (12.121) we find that $1 - \gamma_{ij} = 2D_{ij}/N_c$. Substitution of this into (12.122) yields

$$d_{ij}^2 = 2N_c E_{N_c}(1 - \gamma_{ij}) = 4E_{N_c}D_{ij} \tag{12.123}$$

Equation (12.123) implies that $D_{ij} = N_c/2$ for binary orthogonal codes since $\gamma_{ij} = 0$. This result does not readily generalize to other coded modulation techniques.

Using (12.123) in the union bound result given in (4.81), we have

$$P(E|\mathbf{S}_j) < \sum_{\substack{i \neq j, i=0}}^{M_c-1} Q\left(\sqrt{\frac{2E_{N_c}D_{ij}}{N_0}}\right) \tag{12.124}$$

This demonstrates the key role the Hamming distance plays in specifying good error control codes. With

$$D_{\min} = \min_{\substack{i,j \\ i \neq j}}\{D_{ij}\} \tag{12.125}$$

we can bound the decoded error probability for block length N_c by

$$P(E) < M_c Q\left(\sqrt{\frac{2E_{N_c}}{N_0}D_{\min}}\right) = M_c Q\left(\sqrt{2r_c\frac{E_b}{N_0}D_{\min}}\right) \tag{12.126}$$

since $E_{N_c}/N_0 = r_c E_b/N_0$. This result clearly demonstrates the importance of selecting codes with large values for the minimum Hamming distance. As we shall subsequently see, the minimum Hamming distance $D_{\min}$ plays a key role in defining the error-correcting and detecting capability of a good block code. Notice from (12.126) that the minimum Hamming distance parameter serves as a multiplier on the energy-to-noise ratio, thereby increasing the effective communication efficiency E_b/N_0. From this discussion, we see that one must be careful in selecting the modulation-demodulation technique when coding is used to improve system performance. The main reason for this is that the modem can modify the distance properties of the code in such a way that the coding gain is destroyed. Fortunately, coded BPSK and QPSK modems do not change the distance properties of the code. This combined problem of designing the coding and modulation so as to maintain the codec distance properties and minimize the bandwidth expansion due to coding falls under the topic of *Coded Modulation*.

Finally, since $M_c = 2^{r_c N_c}$ and $Q(x) < \exp(-x^2/2)$ then (12.126) can be written as

$$P(E) < 2^{-N_c[R_0 - r_c]} = 2^{-WT[R_0 - R_b/W]} \tag{12.127}$$

where

$$N_c R_0 \triangleq \frac{D_{\min}}{\ln 2} \left(\frac{E_{N_c}}{N_0} \right) = r_c \frac{D_{\min}}{\ln 2} \frac{E_b}{N_0} \tag{12.128}$$

and, as N_c approaches infinity, $P(E)$ approaches zero if $R_0 > r_c$. From (12.128) we can write the ratio

$$r_c^{-1} R_0 \triangleq \frac{R_0}{r_c} = \frac{1}{\ln 2} \frac{E_b}{N_0} \frac{D_{\min}}{N_c} > 1 \tag{12.129}$$

which says that the product of the bandwidth expansion factor and the cutoff rate is proportional to both the relative minimum Hamming distance, $D_{\min}/N_c$, for any particular binary code and the communication efficiency, E_b/N_0. Thus, the communication efficiency must be

$$\frac{E_b}{N_0} > \ln 2/(D_{\min}/N_c) \tag{12.130}$$

in order to achieve arbitrarily small $P(E)$ with large N_c. For orthogonal codes, $D_{\min} = N_c/2$, so that (12.130) simplifies to (12.15) as it should.

12.8.2 The Plotkin Bound for Binary Codes

Now Plotkin has shown [8] that the minimum Hamming distance is bounded by

$$D_{\min} \leq \frac{N_c M_c}{2(M_c - 1)} \tag{12.131}$$

for binary codes with M_c codewords and block length N_c. This is called the Plotkin bound. Thus

$$\frac{D_{\min}}{N_c} \leq \frac{M_c}{2(M_c - 1)} \tag{12.132}$$

Direct substitution of (12.132) into (12.130) demonstrates the requirement that $E_b/N_0 > (2 \ln 2)(M_c - 1)/M_c$ in order for $P(E)$ to approach zero for large N_c. Moreover, for large M_c, we see from (12.132) that $D_{\min} \leq N_c/2$ with the equality being achieved by orthogonal codes.

12.8.3 The Role of the Minimum Hamming Distance in Specifying Good Error Control Codes

Much insight into the problem of selecting a good error control code can be obtained by the systems engineer by noting that $D_{\min}$ is directly related to the error correcting and error detecting capability of any block code. In particular, it can be shown [3] that a block code with minimum Hamming distance D_{min} has t *error-correcting capability*

$$t = \lceil (D_{min} - 1)/2 \rceil \qquad (12.133)$$

This means that any occurrence of t or fewer symbol errors per codeword will be corrected. On the other hand, it can be shown [3] that if a block code with minimum Hamming distance D_{min} is used for *error detection* only, the decoder can detect all binary error patterns of $(D_{min} - 1)$ or fewer errors per codeword.

  As previously discussed, codes can be characterized as *systematic* or *nonsystematic*. For a systematic code, the encoding process takes the K_c-tuple of bits and maps them into the N_c-tuple by appending the additional redundant bits for error correction and/or detection; see Fig. 12.15. If the error-correcting capability of the code is not exceeded, then the decoder is guaranteed to correctly decode the N_c-tuple into the transmitted K_c-tuple. If the error-correction capability is exceeded, then the decoder will usually do one of two things. It will either decode the received N_c-tuple into incorrect information bits or it will detect that the error-correction capability was exceeded. If the decoder error cannot be detected then it is an *undetectable decoder error*. If the number of errors detected is in excess of the codes error-correction capability, then when a feedback channel is available, the decoder might be designed to send an *automatic repeat request* (ARQ) [12]. Otherwise, the decoder will output the errors to the user or source decoder. With an ARQ approach, the decoder detects errors and makes use of a feedback channel to the transmitter to have the incorrectly received codewords (frame) repeated. Presumably these requests can be repeated until the codewords are received correctly. This procedure has the effect of reducing the information rate. In an ARQ system which employs error-correcting codes, a bit error occurs when the decoder incorrectly accepts a frame of data with errors. Lin and Costello [12] give a good summary of numerous approaches to error control design using ARQ approaches. We now investigate the error-correcting and detecting performance characteristics of other block codes of great practical interest.

12.8.4 Error Probability Performance for Error-Correcting Block Codes

As we have observed in Section 12.8.1 for the AWGN channel, the performance of block codes can be estimated in terms of the minimum Hamming distance if **x** is mapped onto **S** isometrically. Furthermore, for a fixed code rate r_c and block length N_c, the goal is to choose a code with a large minimum Hamming distance. Usually, block codes are decoded using algebraic decoding techniques which require a demodulator-detector to make a *hard decision* on each received channel symbol, that is, "1" or "0." Let p denote the probability of a *channel symbol (coded message bit) error* on the resulting binary symmetric channel. The performance of block codes using hard decisions (one-bit quantization of the demodulator outputs occurring every $T_N = T_c$ sec) on the observables $r_1, r_2, \ldots, r_{N_c}$ can be characterized in terms of their t error-correcting ability. In particular, suppose that the decoder can correct any combination of t or fewer channel symbol errors, and no combination of more than t errors. Such a decoder is called a *hard-decision bounded distance decoder*. Thus, the codeword error probability is

$$P(E) = \text{Prob}[\mathcal{E} \geq t + 1] \qquad (12.134)$$

where $\mathcal{E}$ is a random variable representing the number of coded message bit errors in the codeword. Since there are $\binom{N_c}{k}$ different ways of having k errors in N_c detected coded message bits, the hard-decision codeword error probability is

$$P(E) = \sum_{k=t+1}^{N_c} \binom{N_c}{k} p^k(1-p)^{N_c-k} \qquad (12.135)$$

On the other hand, when hard-decision decoding techniques are not used (that is, with finer quantization of the observables $\mathbf{r}$; see Fig. 11.6), the decoder implementation complexity is usually much greater than that of hard-decision (one-bit) quantization. Forney [13] has investigated such block decoder quantization techniques; they are sometimes called *erasure-and-error decoders*. For infinite quantization and binary coded BPSK, (12.126) upper bounds the codeword error probability for the AWGN channel in terms of D_{min} and E_b/N_0.

The bit error probability for systematic block codes can be found from (12.135) to be [17,18]

$$P_b(E) = \frac{1}{N_c} \sum_{k=t+1}^{N_c} e_k \binom{N_c}{k} p^k(1-p)^{N_c-k} \qquad (12.136)$$

where e_k is the average number of channel symbol errors that remain in the corrected N_c-tuple when the channel caused k symbol errors. In practice, it has been fashionable to use $e_k \simeq k$ for $k > t$, which appears to be a good approximation for the majority of codes of greatest interest, thereby making (12.136) easy to evaluate numerically.

If the block code is designed for the sole purpose of *error detection*, the decoder fails to detect an error in the observed block if and only if the channel transforms the encoder output into another valid codeword. The *undetected error probability* $P(E_u)$ is given by

$$P(E_u) = \sum_{k=t+1}^{N_c} w_k p^k(1-p)^{N_c-k} \qquad (12.137)$$

where w_k is the number of codewords of Hamming weight k.

12.8.5 Hamming Codes and Their Error Probability Performance

Hamming codes are perhaps the simplest class of block codes. The (N_c, K_c) Hamming codes are characterized by the parameters defined in Table 12.6. Using this table, (12.135), (12.136), and (12.137), one can numerically evaluate the codeword error probability, the bit error probability, and the undetected error probability. Figures 12.16, 12.17, and 12.18 demonstrate these probabilities versus the hard-decision coded message bit error probability, p, for $m = 3$, 4, and 5 respectively. Figure 12.19 demonstrates the bit error probability versus E_b/N_0 for various Hamming codes on an AWGN channel, assuming a BPSK modulator so that $p = Q(\sqrt{2r_c E_b/N_0})$. This figure also demonstrates the *coding gain* as a function of parameter m. Coding gain, G, is the difference between the E_b/N_0 (in decibels) required for uncoded communication and the E_b/N_0 (in decibels) required for coded communication to achieve the same bit error probability; that is, $G = (E_b/N_0)_{uncoded} - (E_b/N_0)_{coded}$.

Table 12.6 Hamming Code Parameters

Property	Parameter
Block length	$N_c = 2^m - 1; m = 2, 3, \ldots$
Number of parity check symbols	$m = N_c - K_c \equiv r$
Number of information symbols	$K_c = 2^m - (m + 1)$
Error correcting capability	$t = 1$
Minimum distance	$D_{\min} = 3$

12.8.6 Golay Codes and Their Error Probability Performance

Golay codes are more powerful than the Hamming codes discussed in the previous section; however, the decoder complexity and the bandwidth expansion factor are greater. The ubiquitous Golay (23,12) code [3] is the only known multiple-error correcting binary perfect code that can correct any combination of three or fewer errors per 23-bit codeword. From a system perspective, the *extended Golay code* is more useful. This code is formed by adding one parity bit to the (23,12) Golay code to form the (24,12) *extended Golay code*. This parity bit increases the minimum Hamming distance of the code from seven to eight and yields a code rate 1/2 which is easier to decode in hardware than the (23,12) Golay code. The codeword, bit, and undetected error probabilities for hard-decision decoding can be determined from (12.135), (12.136), and (12.137) to be [17,18]

$$P(E) = \sum_{k=4}^{24} \binom{24}{k} p^k (1 - p)^{24-k} \tag{12.138}$$

$$P_b(E) = \frac{1}{24} \sum_{k=4}^{24} e_k \binom{24}{k} p^k (1 - p)^{24-k} \tag{12.139}$$

and

$$P(E_u) = (1 - p)^{24} \left[A\left(\frac{p}{1 - p}\right) - 1 \right] \tag{12.140}$$

where function $A(z)$ is defined in detail in [17]. The codeword, bit, and undetected error probabilities versus p are illustrated in Fig. 12.20 for the extended (24,12) Golay code.

12.8.7 Cyclic Codes

Cyclic codes are a subset of linear block codes (see the Venn diagram in Fig. 11.18) with the additional property that every cyclic shift of a codeword is also a codeword. If $\mathbf{x} = (x_1, \ldots, x_{N_c})$ is a codeword, then the shifted version $\mathbf{x} = (x_{N_c}, x_1, \ldots, x_{N_c-1})$ is also a codeword. Cyclic codes are usually easier to encode and decode than all other linear block codes. The encoding operation is easily performed using shift registers; furthermore, implementation of the decoder becomes more practical due to their simpler algebraic structures [12]. Cyclic block codes require that synchronization words be attached to the beginning of the transmitted codewords.

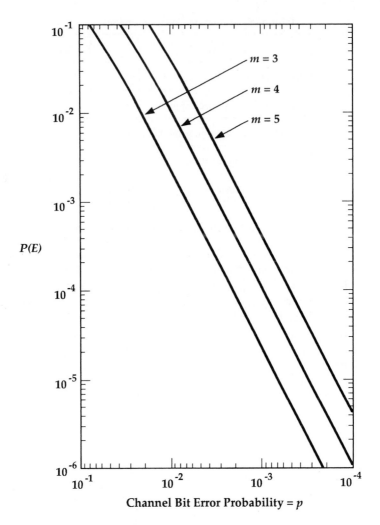

Figure 12.16   Codeword error probability versus channel bit error probability for block
length $N_c = 2^m - 1$ Hamming codes with $m = 3, 4,$ and 5 (Reprinted from [17])

12.8.8 BCH Codes and Their Error Probability Performance

Bose-Chaudhuri-Hocquenghem (BCH) codes are a special case of powerful cyclic codes
which have well-defined decoding algorithms [3, 12, 18]. In this section, we will consider
P-ary (P prime), binary, and Q-ary BCH codes. For each type, we will identify characteristics and present typical achievable performance results.

   ***P*-ary BCH codes.** Here the codeword symbols belong to the Galois Field
$GF(P).$[3] Every codeword is associated with a code polynomial with coefficients in $GF(P)$.
The roots of the generator polynomial lie in an extension field $GF(P^m)$ of $GF(P)$ for some
integer $m \geq 1$. BCH codes can either be *primitive* or *nonprimitive*; see the Venn diagram in

---

 3. An introduction to the algebra of Galois Fields (also known as finite fields) may be found in Lin and Costello
[12].

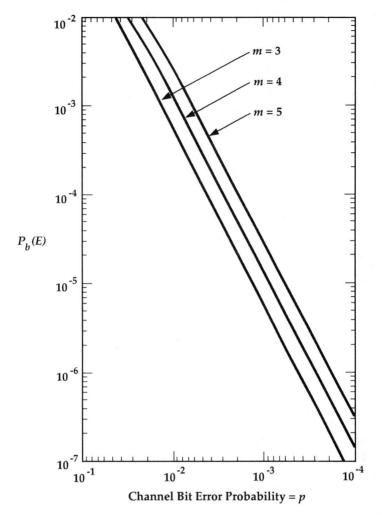

Figure 12.17   Bit error probability versus channel bit error probability for block length $N_c = 2^m - 1$ Hamming codes with $m = 3, 4,$ and 5 (Reprinted from [17])

Fig. 11.18. Primitive BCH codes are defined as codes whose block length is $N_c = P^m - 1$. Given any error-correction capability t, there exist choices of the message length (K_c symbols) and block length (N_c symbols) for which a t-error correcting P-ary (N_c, K_c) code exists [8, 12, 18]; see Table 12.7 with $Q = P$.

Binary BCH codes.   Binary BCH codes are a special case of P-ary BCH codes with $P = 2$. The codeword symbols (coded message bits) of binary BCH codes are from the Galois Field $GF(2)$. The roots of the generator polynomial of the code lie in the extension field $GF(2^m)$ for some integer $m \geq 1$. In particular, a t-error correcting ($t < 2^{m-1}$) primitive, binary BCH code having generator polynomial with roots in $GF(2^m)$ can be found which achieves the parameters listed in Table 12.8 for any $m \geq 3$. There is no simple formula for the system engineer to use for calculating $N_c - K_c$; however, $N_c - K_c$ is less than or equal to mt. The parameters for all binary BCH codes of length $N_c = 2^m - 1$ with $m \leq 10$

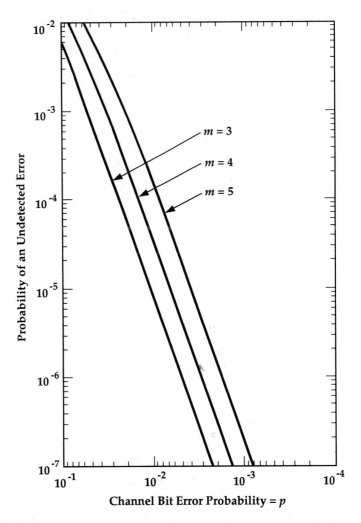

Figure 12.18   Probability of an undetected error versus channel bit error probability for block length $N_c = 2^m - 1$ Hamming codes with $m = 3, 4$, and 5 (Reprinted from [17])

are summarized in Table 6.1 of [3] and Table 4.6 in [18]. The codeword error probability performance versus channel symbol (coded message bit) error probability, p, is illustrated in Figures 12.21, 12.22 and 12.23 for several binary BCH codes. Figure 12.24 provides a bit error probability comparison versus E_b/N_0 for various BCH codes.

These codes often have some inherent error detection capability beyond their error-correction capability. If some error-correcting capability is sacrificed to gain some additional error detection capability, say t_d, then the resultant t error correcting, t_d additional error detecting (t_d is an even number), primitive, binary BCH code is characterized by the parameters summarized in Table 12.9.

Q-ary BCH codes. Q-ary BCH codes are a generalization of the P-ary codes. The codeword symbols of a Q-ary BCH code come from $GF(Q) = GF(P^c)$, for some integer $c \geq 1$, while the roots of the generator polynomial lie again in an extension field

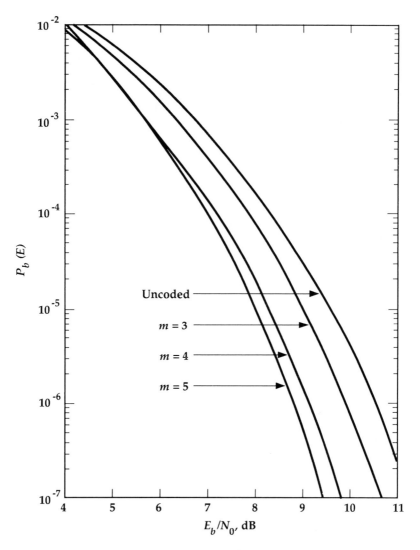

Figure 12.19 Bit error probability versus E_b/N_0 for block length $N_c = 2^m - 1$
Hamming codes with $m = 3, 4$, and 5 on an AWGN channel (Reprinted from [17])

$GF(Q^m) = GF(P^{mc})$, for some $m \geq 1$. If $c = 1$, Q-ary codes become P-ary. There exist Q-ary BCH codes having the code parameters listed in Table 12.7. A discussion of the theory of finite fields underlying Q-ary BCH codes can be found in [3, 8, 11, and 18].

12.8.9 Reed-Solomon Codes and Their Error Probability Performance

Reed-Solomon (RS) codes have found widespread applications beginning with the 1977 Voyager's Deep Space Communications System [21]. At the time of Voyager's launch, efficient encoders existed, but accurate decoding methods were not available! Jet Propulsion Laboratory scientists and engineers gambled that by the time Voyager II would reach Uranus in 1986, decoding algorithms and equipment would both be available for decoding of the

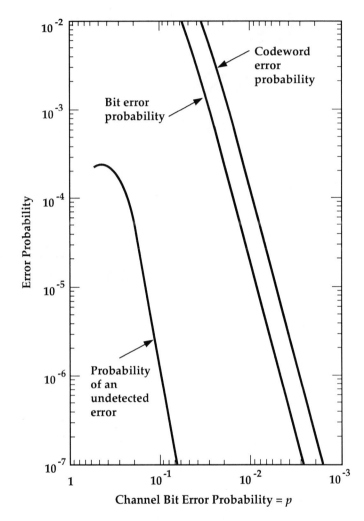

Figure 12.20   Codeword, bit, and undetected error probabilities versus channel symbol error probability with extended Golay coding (Reprinted from [17])

data transmitted back to Earth. Fortunately, they were correct! Voyager's communications system was able to transmit at a data rate of 21,600 bits per sec from 2 billion miles away and successfully extract the data from a signal whose received power was 100 billion times weaker than that generated by an ordinary wrist watch battery!

Other areas of application of RS codes include: (1) communications over telephone systems with such applications as advanced facsimile machines or faxes (which send and receive imaging data) and high-speed modems (which usually send and receive computer data); (2) satellite communications with such applications as the Hubble Space Telescope, Mobile Satellite Terminals, and the 300 megabits per sec (Mbps) return link of the Space Station Freedom via the Tracking Data Relay Satellite System (SF/TDRSS); (3) Deep Space Communications with such applications as Voyager, the Galileo Orbiter, the Mars Observer,

Table 12.7 Parameters for Primitive Q-ary BCH Codes

Block length	$N_c = Q^m - 1$ code symbols
Number of parity checks	$N_c - K_c$ code symbols
Minimum distance	$D_{\min} \geq 2t + 1$ code symbols

Table 12.8 Parameters for Binary BCH Error-Correcting Codes

Block length	$N_c = 2^m - 1$ code symbols
Number of parity checks	$N_c - K_c \leq mt$ code symbols
Minimum distance	$D_{\min} \geq 2t + 1$ code symbols

Table 12.9 Parameters for Binary BCH Error Detecting Codes

Block length	$N_c = 2^m - 1$ code symbols
Number of parity checks	$N_c - K_c \leq m[t + (t_d/2)]$ code symbols
Minimum distance	$D_{\min} \geq 2t + t_d + 1$ code symbols

and the Cassini Titan Orbiter/Saturn Probe; (4) Space Shuttle Communications; (5) entertainment, with applications such as compact disks (CDs) and CD-ROM, etc.

In general, Reed-Solomon (RS) codes may be regarded as a special subclass of Q-ary BCH codes that results when parameter $m = 1$. Thus, the coefficients of the code polynomials and the roots of the generator polynomial all lie in $GF(P^c)$. A primitive RS code has block length $N_c = P^c - 1$. For any RS code, the error-correction capability t and the message length K_c are simply related by $K_c = N_c - 2t$. For example, if $Q = P^c = 2^c = 2^3 = 8$, then this can either be a (7,5) or a (7,3) or a (7,1) RS code depending on whether it is desired to correct $t = 1, 2$, or 3 symbols respectively. Table 12.10 summarizes the achievable paramaters for a t error-correcting, primitive RS code.

Binary RS codes are most frequently used in practice. A "binary RS code" is simply an RS code where $P = 2$. In the nonbinary case, P is an odd prime. Note that the code symbols of a binary RS code are not in fact binary, but belong to the field $GF(2^c)$. The parameters for a binary, t error-correcting, primitive RS code are summarized in Table 12.10 with $P = 2$.

RS codes meet the Singleton bound $(D_{\min} \leq N_c - K_c + 1)$ because for an RS code, $D_{\min} = 2t + 1 = N_c - K_c + 1$. Therefore, RS codes are optimal with respect to this bound, and the term *maximum distance separable* (MDS) is used to describe optimality in this

Table 12.10 Q-ary RS Code Parameters

Block length	$N_c = Q - 1 = P^c - 1$ code symbols
Number of parity checks	$N_c - K_c = 2t$ code symbols
Minimum distance	$D_{\min} = 2t + 1$ code symbols

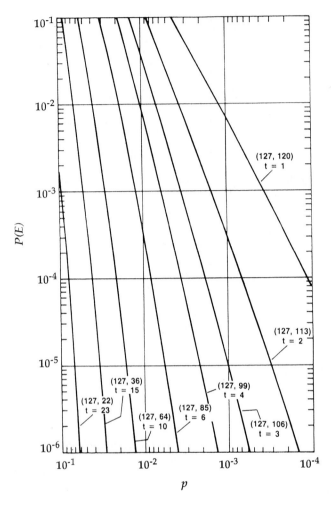

Figure 12.21 Codeword error probability versus bit error probability for some binary BCH codes of block length 127 (Reprinted from [18])

sense. It is also worth pointing out that the block length N_c of an RS code over $GF(Q) = GF(P^c)$ can be extended to either Q or $Q + 1$ while still maintaining the MDS condition. It can also be shown [17, 18] that the block (codeword) and 2^m-ary symbol error probabilities for RS codes are given respectively by

$$P(E) = \sum_{k=1+t}^{2^m-1} \binom{2^m - 1}{k} p^k (1 - p)^{2^m - 1 - k} \tag{12.141}$$

and

$$P_s(E) \simeq \frac{1}{2^m - 1} \sum_{k=1+t}^{2^m-1} k \binom{2^m - 1}{k} p^k (1 - p)^{2^m - 1 - k} \tag{12.142}$$

where we have used the approximation $e_k \simeq k$ for $k > t$, and where p denotes the channel symbol error probability. The bit error probability can sometimes be upper bounded by $P_s(E)$. For certain cases, namely for orthogonal codes, the relationship is

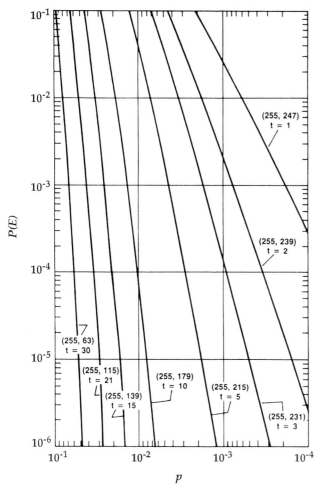

Figure 12.22 Codeword error probability versus bit error probability for some binary BCH codes of block length 255 (Reprinted from [18])

$$P_b(E) = \left(\frac{2^{K_c-1}}{2^{K_c}-1} \right) P_s(E) \tag{12.143}$$

where K_c is the number of bits per codeword; see Chapter 4. Figures 12.25, 12.26, 12.27, and 12.28 demonstrate the codeword error probability (MEP) versus p profiles for various RS codes of practical interest.

Interleaving of the channel encoder output is required when the channel errors occur in bursts; see Section 11.3 in Chapter 11. Inserting an interleaving-deinterleaving system into the signal path introduces another new system error control system parameter, namely the *interleaving length or depth*. The value of this parameter is specified in terms of the prespecified *channel memory length* (sec), say τ_c. Physical considerations which lead to specifying τ_c are the expected burst-error duration and the multipath duration which are frequently characterized in terms of the *temporal correlation time* of the channel.

RS codes also have applications in error control system designs as *outer codes* in concatenated coding systems (see Section 11.5 and Chapter 13). In such designs, an *inner code* is also used to provide a portion of the error control, thus the name *concatenated code*

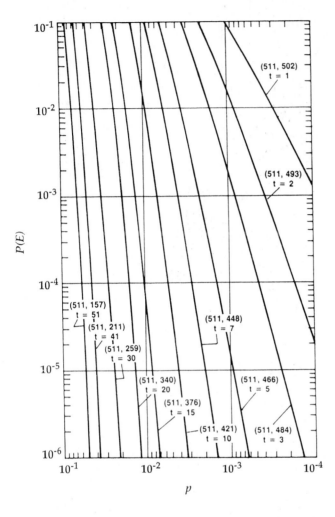

Figure 12.23   Codeword error probability versus bit error probability for some binary BCH codes of block length 511 (Reprinted from [18])

[13]. The inner code is usually a binary code which maps each symbol in the RS alphabet to a unique binary codeword. Decoding of a concatenated code typically proceeds as follows: First, the demodulator-detector outputs are soft quantized (soft-decision) and presented as inputs to the inner channel decoder. The output of this decoder are hard decisions which serve to drive the outer RS decoder. The inner code protects the RS code against random channel errors, leaving the correction of burst errors to the outer RS code.

The performance of concatenated coding systems which use an RS outer code are given in the next chapter. Concatenated codes were first introduced by Forney [13], and their performance for special coding techniques will be presented in Chapter 13.

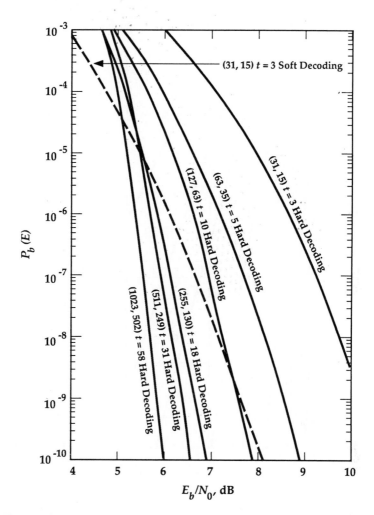

Figure 12.24   Bit error probability versus E_b/N_0 for coherently demodulated BPSK over a Gaussian channel using BCH codes (Reprinted from [22])

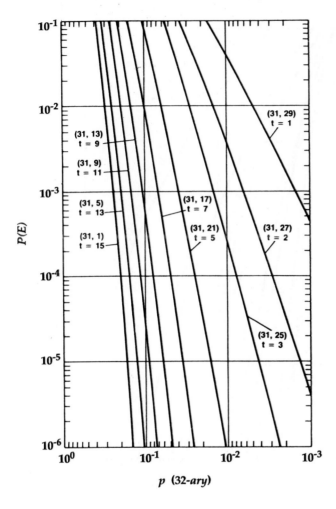

Figure 12.25 Codeword error probability versus channel symbol error probability for various RS codes defined on a 32-ary alphabet (Reprinted from [18])

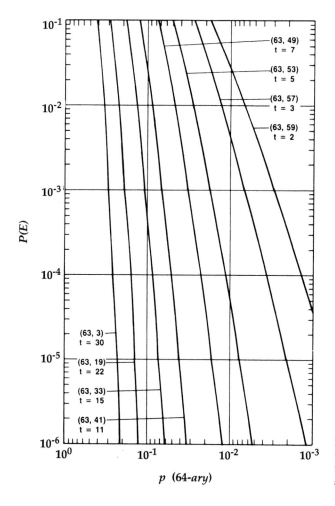

Figure 12.26 Codeword error probability versus channel symbol error probability for various RS codes defined on a 64-ary alphabet (Reprinted from [18])

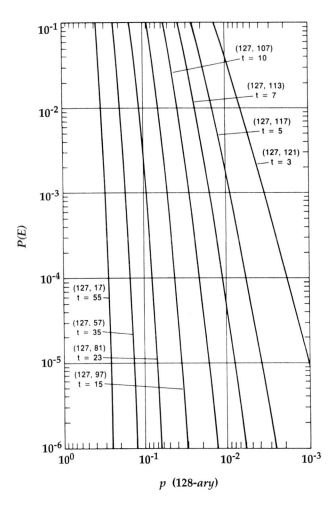

Figure 12.27 Codeword error probability versus channel symbol error probability for various RS codes defined on a 128-ary alphabet (Reprinted from [18])

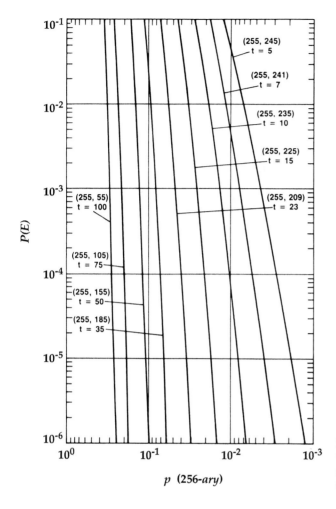

Figure 12.28 Codeword error probability versus channel symbol error probability for various RS codes defined on a 256-ary alphabet (Reprinted from [18])

PROBLEMS

12.1 Consider the discrete memoryless channel (DMC) modeled by the diagram in Fig. P12.1. Write an expression for the cutoff rate R_0 and the channel capacity C_{N_c} assuming equally probable input messages.

12.2 For the discrete memoryless channels illustrated in Fig. P12.2, write expressions for the channel cutoff rate R_0 and capacity C_{N_c}. Assume equally probable input symbols.

12.3 **(a)** Develop an explicit expression for the achievable R_0 performance for coded QPSK, $M = Q = 4$.

(b) Compare the result in (a) with that for coded BPSK and comment on your results.

12.4 Consider the M-ary signal set

$$s_i(t) = \sum_{n=1}^{N} s_{in}\phi_n(t)$$

for $0 \le t \le T$ and $i = 0, 1, \ldots, M - 1$. Assume an AWGN channel with spectral density $N_0/2$. Suppose we randomly select codewords by choosing each s_{in} independently from the ensemble of random variables. Using the union bound, one can show that the ensemble message error

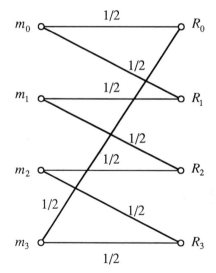

Figure P12.1   4-ary input with four output levels

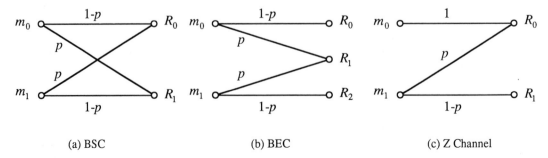

(a) BSC   (b) BEC   (c) Z Channel

Figure P12.2   DMC models

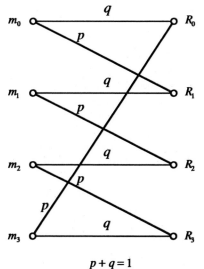

$$p + q = 1$$ **Figure P12.5**

probability is bounded by

$$\overline{P_E(M)} < M2^{-NR_0}$$

(a) Determine R_0 when s_{in} is a zero-mean Gaussian random variable with variance E.

(b) Repeat (a) if s_{in} is a binary random variable taking on values $\sqrt{E}$ or $-\sqrt{E}$ with equal probability.

12.5 (a) Calculate the quantized cutoff rate R_0 (in bits/per channel use or per clock cycle) for the DMC model shown in Fig. P12.5.

(b) Derive a way to use this DMC at rate $r_c = 1$ with $P(E) = 0$.

12.6 For block orthogonal signals, the union bound states that

$$P(E) < \exp\left[-K_c\left(\frac{E_b}{2N_0} - \ln 2\right)\right]$$

Calculate how large the number of messages M_c must be to guarantee that $P(E) < 10^{-6}$ when $E_b/N_0 = 0, 3,$ and 6 dB.

12.7 (a) For M equilikely transorthogonal signals disturbed by additive white Gaussian noise, derive a bound similar to (12.12) and (12.13) which is valid for orthogonal signals.

(b) Repeat (a) for M equiprobable biorthogonal signals.

(c) Using the bounds, discuss the relative advantages of the three signaling systems from a systems engineering viewpoint.

   Hint: See Chapter 3 for the union bound result and Table 12.4.

12.8 Consider communicating over the AWGN channel using binary codes and antipodal signals.

(a) Using the fact that $2BT_c = 1$, write an expression for R_0 as a function of P/N_0B.

(b) Show that as B increases, R_0 in bits/dimension decreases for a fixed P/N_0.

(c) Now consider the channel throughput $\mathcal{R}_0 = R_0/T_c = 2BR_0$ in bits/sec and show that $\mathcal{R}_0$ increases monotonically with increasing bandwidth B.

(d) Explain these results.

12.9 Show that the channel capacity C in bits/sec increases monotonically toward its maximum value C_∞ as B increases. Show that the capacity C_{N_c} in bits/dimension decreases monotonically as B increases. Explain this phenomenon.

12.10 Show for block-orthogonal codes that (12.12) can be written as

$$P(E) < \exp\left[-T \ln 2 \left(\frac{C_\infty}{2} - \mathcal{R}_b\right)\right]$$

so that as T approaches infinity, $P(E)$ approaches zero if $\mathcal{R}_b < C_\infty/2$. This bound is not tight but we know that orthogonal signals actually achieve C_∞ in the limit (see Chapter 4).

12.11 Derive an expression for the channel cutoff rate for the ensemble of binary codes when orthogonal, equal energy, equiprobable signal waveforms (BFSK) are used to transport information.

    (a) Derive the channel cutoff rate for the ensemble of binary codes when equal energy, equiprobable waveforms having cross-correlation coefficient γ are used.

    (b) Compare and discuss the results found in (a) and (b) with the channel cutoff rate in (12.37) for antipodal signals (BPSK).

12.12 Consider a voice-grade telephone circuit with a bandwidth of $B = 3$ kHz. Assume that the circuit can be modeled as an AWGN channel.

    (a) What is the capacity of such a circuit if the CNR is 30 dB?

    (b) What is the minimum CNR required for a data rate of 4,800 bits/sec on such a voice-grade circuit?

    (c) Repeat part (b) for a data rate of 19,200 bits/sec.

12.13 Consider that a 100-kbits/sec data stream is to be transmitted on a voice-grade telephone circuit (having a bandwidth of 3 kHz). Is it possible to achieve error-free transmission with a CNR of 10 dB? Justify your answer. If it is not possible, suggest system modifications that might be made.

12.14 Telephone modems operating at 19.2 kbits/sec are available in the market place.

    (a) Assuming AWGN and an available $E_b/N_0 = 10$ dB, calculate the theoretically available capacity in the 2,400-Hz single-sided bandwidth.

    (b) What is the required E_b/N_0 that will enable a 2,400-Hz bandwidth to have a capacity of 19.2 kbits/sec?

12.15 Consider the set of J orthonormal waveforms $\{\phi_k(t)\}$, $k = 1, 2, \ldots, J$ over the interval $[0, T_c]$. These waveforms are used to construct the signal set $\{s_i(t)\}$ of the form [7]

$$s(t) = \sqrt{E_{N_c}} \sum_{i=1}^{N} \phi_{k_i}(t - iT_c)$$

in which the $\{k_i\}$ are integers between 1 and J. Thus, each signal in the set $\{s_i(t)\}$ is specified by a vector of the form $(k_1, k_2, \ldots, k_J)$.

    (a) Assuming $N = 5$, $J = 4$, and

$$\phi_k(t) = \sqrt{2/T_c} \sin(2\pi kt/T_c), \quad 0 \le t \le T_c$$

    and zero elsewhere, sketch the signal specified by the vector $(2, 1, 4, 2, 3)$.

    (b) How many waveforms are there in the set $\{s_i(t)\}$ for arbitrary N and J? Are all these waveforms orthogonal?

    (c) Using the union bound, show that the ensemble MEP is bounded by

$$\overline{P(E)} < \exp\left\{-K_c\left[\frac{E_b}{2N_0}\frac{1}{x}\ln\frac{J}{1 + (J-1)e^{-x}} - \ln 2\right]\right\}$$

    where $x = PT_c/2N_0$.

12.16 An upper bound for the performance of linear block codes is given by (12.127). Notice from (12.127) that the performance depends on both code parameters r_c and D_{min} as well as the energy per bit-to-noise ratio. Assume that the bit error probability and message error probability are approximately equal.

(a) With $M_c = 2^{K_c}$, show that the coding gain G, defined in Section 12.8.5, is given by

$$G \triangleq 10 \log_{10} [(E_b/N_0)_{\text{uncoded}}/(E_b/N_0)_{\text{coded}}]$$
$$= 10 \log_{10} \{r_c D_{\min} - K_c/[\ln 2(E_b/N_0)_{\text{coded}}]\}$$

if one uses uncoded BPSK (antipodal) transmission. Use the asymptotic expansion for $Q(x)$, (4.55), and (12.126).

(b) What is the limiting value for G as E_b/N_0 approached infinity?

(c) Assuming the use of a Hamming code defined in Table 12.6, write an expression for the code rate r_c and plot the asymptotic gain achieved with $m = 2, 3, 4, 5, \ldots, 10$.

12.17 Using the coding gain formula provided in Problem 12.16

(a) Develop an explicit expression for the coding gain achievable for the class of binary BCH codes defined in Table 12.8.

(b) From Problem 12.16, we see that the asymptotic coding gain for large E_b/N_0 is given by $G = 10 \log_{10} r_c D_{min}$. Repeat (a) and write an expression for G in terms of the error-correcting capability t, N_c, and K_c.

12.18 Repeat Problem 12.17 for the binary Reed-Solomon codes defined in Table 12.10.

12.19 Consider the coding gain formula given in Problem 12.16.

(a) For the binary primitive (15,9) RS code, write an expression for coding gain.

(b) Plot this gain as a function of E_b/N_0.

(c) What is the maximum achievable gain using this code?

(d) Repeat (a), (b), and (c) for the (154,130) RS code.

12.20 (a) Repeat Problem 12.19 (b) and (c) for the (255,235) RS code with $t = 10$.

(b) Repeat Problem 12.19 (b) and (c) for the binary (255,179) BCH code with $t = 10$.

12.21 Figure 12.5 shows a family of channel cutoff rate (R_0) curves for the set of equispaced, equally probable Q-ary ASK signal points as shown in Fig. 12.4a.

(a) Find the expression for the distance q_{ij} in terms of Q, i, j, and $\sqrt{E_{N_c}}$.

(b) Use the expression for the distance q_{ij} found in (a) to find R_0 for the special case of $Q = 4$.

(c) Graph your result as a function of E_{N_c}/N_0.

(d) Verify your answer by checking for agreements with Figure 12.5.

12.22 Consider a set of QPSK signal points assigned to the coded message symbols $\{\mathbf{x_m}\} = \{00, 01, 10, 11\}$.

(a) Find an isometric mapping from coded message symbols to coded modulation vectors $\{\mathbf{S_m}\}$. For each pair of points in your signal set, compare the Hamming distance to the Euclidean distance.

(b) Find a nonisometric mapping from coded message symbols to coded modulation vectors $\{\mathbf{S_m}\}$. For each pair of points in your signal set, compare the Hamming distance to the Euclidean distance.

12.23 Heuristically, justify use of the signal space dimensionality to bandwidth relationship $N_c = WT = 2BT$ where T is the signal duration and B (in Hz) is the single-sided bandwidth measure.

Hint: See Chapter 5 of [7], Nuttall and Amoroso, *IT*, July 1965, and Landau and Pollak, *Bell System Tech. Journal*, 41, July 1962, pp. 1295–1336.

12.24 A customer wants to send information through a channel with single-sided bandwidth 10 kHz. Channel noise is white and Gaussian with single-sided power spectral density $N_0 = 10^{-19}$ Watts/Hz. He/she can achieve a received signal power of 1 pW.

(a) What is the maximum theoretical rate at which error-free transmission is possible?

(b) If the customer decides to send at rate 100 kbps using uncoded quaternary PSK, what will the actual symbol probability be?

12.25 The received signal power in a coherent AWGN channel is -102 dBm and the power spectral density of the noise is -228 dBm.
   (a) What is the maximum rate (bits/sec) at which one can send information error-free?
   (b) What is the corresponding value of E_b/N_0?
   (c) What is the maximum amount of error-free information that can be sent in two seconds?

12.26 For QPSK signaling and $M = 4$
   (a) Evaluate R_0 using the appropriate Q-ary coded M-PSK model from Section 12.4.5.
   (b) Evaluate R_0 using the appropriate QASK model from Section 12.4.3.
   (c) Compare your answers from parts (a) and (b). If they agree, explain why; if they disagree, explain why.

REFERENCES

1. Wozencraft, J. M. and R. S. Kennedy, "Modulation and Demodulation for Probabilistic Coding," *IEEE Trans. Information Theory*, IT-12, July 1966, pp. 291–97.

2. Massey, J. L., "Coding and Modulation in Digital Communications," Zurich Seminar, vol. E2, 1974, pp. 1–4.

3. Lin, S., *An Introduction to Error-Correcting Codes*, Englewood Cliffs, NJ: Prentice Hall, 1970.

4. Lindsey, W. C. and M. K. Simon, *Telecomunication Systems Engineering*, Englewood Cliffs, NJ: Prentice Hall, 1973. Republished by Dover Publications, 0-486-66838-x, 1991.

5. Shannon, C. E., "A Mathematical Theory of Communication," *Bell System Tech. J.*, 27, 1948, pp. 379–423 (Part I), pp. 623–56 (Part II). Reprinted in book form with postscript by W. Weaver, Urbana, IN: Univ. of Illinois Press, 1949; also see *Claude E. Shannon Collected Papers*, edited by N. A. Sloane and A. D. Wyner, IEEE Press, 1992.

6. Shannon, C. E., "Probability of Error for Optimal Codes in a Gaussian Channel," *Bell System Tech. Journal*, vol. 38, May 1959, pp. 611–56.

7. Wozencraft, J. M. and I. M. Jacobs, *Principles of Communication Engineering*, New York: John Wiley and Sons, 1965.

8. Gallager, R. G., *Information Theory and Reliable Communication*, New York: John Wiley and Sons, 1968.

9. Cover, T. and J. A. Thomas, *Elements of Information Theory*, New York: John Wiley and Sons, 1991.

10. Viterbi, A. J., "Error Bounds for Convolutional Codes and an Asymptotically Optimum Decoding Algorithm," *IEEE Trans. Information Theory*, vol. IT-13, April 1967, pp. 260–69.

11. Berlekamp, E. R., *Algebraic Coding Theory*, New York: McGraw-Hill, 1968.

12. Lin, S. and D. J. Costello Jr., *Error Control Coding: Fundamentals and Applications*, Englewood Cliffs, NJ: Prentice Hall, 1983.

13. Forney, G. D. Jr., *Concatenated Codes*, Cambridge, MA: MIT Press, 1966.

14. "Troposcatter Interleaver Study Report," CNR, Inc., Report Number RADC-TR-75-19 for Rome Air Dev. Center, Air Force Systems Command, Griffiss AFB, NY, February 1975.

15. Nyquist, H., "Certain Topics on Telegraph Transmission Theory," *Trans. AIEEE*, vol. 47, April 1928, pp. 617–44.

16. Wozencraft, J. M. and B. Reiffen, *Sequential Decoding*, Cambridge, MA: MIT Press, 1961.

17. Odenwalder, Joseph P., *Error Control Coding Handbook*, Final Report under contract No. F44620-76-C-0056, Linkabit Corporation, July 1976. Also see Chapter 10 on *Error Control in Data Communications, Netwoks and Systems* by Thomas C. Bartee, ed., Indianapolis: Howard W. Sams, 1985.

18. Michelson, A. M. and A. H. Levesque, *Error-Correcting Techniques for Digital Communications*, New York: John Wiley and Sons, 1985.

19. Lipschutz, Seymour, *Schaum's Outline of Theory and Problems of General Topology*, New York: McGraw-Hill, 1965.

20. Viterbi, Andrew J., and Omura, Jim K., *Principles of Digital Communication and Coding*, New York: McGraw-Hill, 1979.

21. Wicker, Stephen and Vijay, Bhargava, editors, *Reed-Solomon Codes and Their Applications*, New York, NY: IEEE Press, 1993.

22. Weng, L. J., "Soft and Hard Decoding Performance Comparisons for BCH Codes," *ICC'79 Conference Record*, Boston, MA, pp. 25.5.1–25.5.5, 1979.

Chapter 13

Convolutional-Coded Digital Communications

In the previous chapter, we saw that a linear (N_c, K_c) *block code* can be used for error control purposes by partitioning the data stream into blocks containing K_c bits each. Then, using an encoding rule to associate the $r = N_c - K_c$ parity check bits with each K_c data bits, an N_c-symbol codeword was formed for transmission. However, with *convolutional codes* (originally called *recurrent codes*), the encoded data do not have a simple block structure, and, as we shall see, a convolutional encoder operates on the input bit stream such that each information bit can affect a finite number of consecutive symbols (coded message bits) in the encoder output.

Apparently convolutional codes were first introduced by Elias in 1955 [1] as an alternative to block codes. In this chapter, we shall introduce convolutional codes, methods for representing such codes, and present the maximum-likelihood (ML) decoding algorithm due to Viterbi [2]. Detailed performance results achievable with maximum-likelihood decoding of convolutional codes are given for a set of particular convolutional codes and key design issues are treated. The chapter concludes by presenting the applications and performance of *concatenated codes* introduced by Forney [3] for error control. Punctured convolutional codes are also discussed.

Other decoding techniques such as threshold decoding, sequential decoding, etc., lie outside the scope of this book; however, the details of these techniques can be found in Gallager [4], Michelson and Levesque [5], Berlekamp [6], Anderson and Mohan [7], and Lin and Costello [8].

13.1 CONVOLUTIONAL ENCODERS

As we observed in Chapter 12, a *linear block code* is characterized by two integers N_c and K_c and a generator polynomial. The integer K_c is the number of data bits that form an input to a block encoder while the integer N_c represents the *block length* of the encoder output. The codeword N_c-tuple is uniquely determined by the input message K_c-tuple and the ratio $r_c = K_c/N_c$ is called the *code rate* which measures the amount of added redundancy; r_c^{-1} also serves to measure the *bandwidth expansion* due to coding.

A *convolutional code* is characterized by three integers N_c, K_c, and K where the ratio $r_c = K_c/N_c$ has the same physical significance that it has for block codes; however, N_c does not define the block length as it does in block codes. Moreover, the parameter K is used to denote the *constraint length* of the convolutional code. The parameter K_c plays the same role for block or convolutional codes. An important difference between binary block codes and binary convolutional codes is that the encoder has memory. As we shall see, the N_c-tuple output of a convolutional coder is not only a function of the input K_c-tuple but is also a function of the previous $K_c - 1$ input K_c-tuples. Usually K_c and N_c assume small integer values and K is varied to control redundancy in the transmitted data stream.

The functional architecture of a (K_c, N_c, K) convolutional encoder is mechanized with a kK-stage linear shift register and N_c modulo-2 adders (see Fig. 13.1). From this figure, we observe that a convolutional code is defined by the number, kK, of stages in the shift register, the number of modulo-2 adders (number of outputs), and the connections between the shift register and the modulo-2 adders. The *encoder state* is defined to be the contents of the rightmost $kK - 1$ stages in Fig. 13.1. Knowledge of the next K_c input bits is therefore necessary and sufficient to determine the next output sequence. In other words, an input sequence of k data bits ("ones" and "zeros") of time duration $T = K_cT_b$ sec are shifted into the first k stages of the kK register and all bits in the register are shifted k stages to the right, thereby establishing a new *state* of the kK-stage register. The outputs of the N_c modulo-2 adders are sequentially sampled to yield the N_c *output channel code symbols* (coded message bits) which are identified as the *codeword branch* vectors $\mathbf{x}_j = (x_{j1}, x_{j2}, \ldots, x_{jN_c})$ of N_c channel symbols. In what follows, we shall consider the most commonly used code for which $k = 1$. Generalization to higher-order alphabets can be found in [8]; however, treatment of such alphabets here lies outside the scope of this book. The N_c code symbols occurring at the encoder output during the time interval $t_{j+1} - t_j = N_cT_c$ comprise the jth *codeword branch* $\mathbf{x}_j = (x_{j1}, \ldots, x_{jN_c})$. Here $x_{ji}(i = 1, \ldots, N_c)$ is the ith binary channel symbol in the jth codeword branch and is assumed to take on values "0" or "1." When $K_c = 1$, the convolutional coder of rate $r_c = 1/N_c$ is implemented with a $(K - 1)$-stage register and the constraint length K can be referred to in units of bits.

13.2 REPRESENTATION OF CONVOLUTIONAL CODES

Several methods can be used for representing convolutional codes. These include: (1) the *connection diagram*, (2) *generator polynomials*, (3) the *tree diagram*, and (4) the *trellis diagram*. In what follows, these techniques will be described using a rate $r_c = 1/2$ convolutional code with $K_c = 1$, $N_c = 2$, and with constraint length $K = 3$. Fortunately, convolutional codes are most easily described by example and any generalizations should become intuitively obvious. Figure 13.2 shows a shift-register circuit that generates this code. Input bits are clocked into the circuit from left to right at a rate of $\mathcal{R}_b = 1/T_b$ Hz. After each input is received, the coder output is generated by sampling and multiplexing outputs of the two mod-2 adders. Thus, for a rate $1/2$ code, we observe that $T_b = 2T_c$ sec.

13.2.1 Polynomial Representation

The encoder of Fig. 13.2 performs two polynomial multiplications. The output of the upper mod-2 adder is the product of the degree-2 polynomial $g_1(x) = 1 + x + x^2$ and the input sequence $u(x) = u_0 + u_1x + u_2x^2 + \ldots + u_jx^j + \ldots$, where $u_j \in \{0, 1\}$ represents the input information bit in the jth bit time interval $(j - 1)T_c < t \leq jT_c$. The output of the lower modulo-2 adder is the product of $u(x)$ and $g_2(x) = 1 + x^2$. The output sequence is represented by the two (in this case) semi-infinite polynomials $X_1(x) = x_{10} + x_{11}x +$

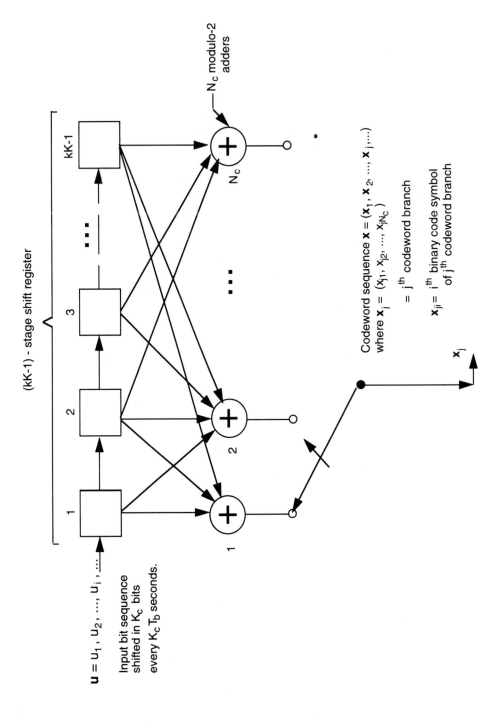

Figure 13.1 Convolutional encoder of constraint length kK and rate $r_c = K_c/N_c$

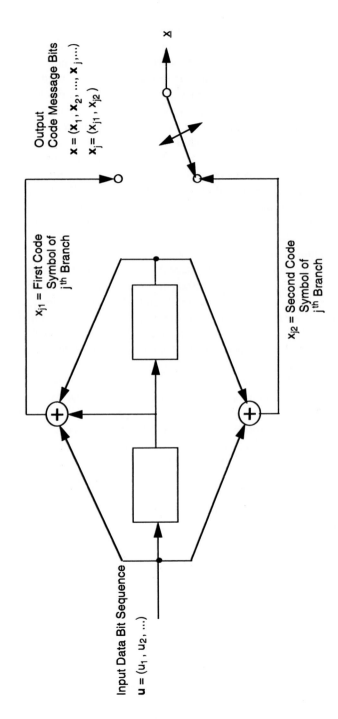

Figure 13.2 Rate 1/2 convolutional encoder with $K = 3$

$x_{12}x^2 + \ldots + x_{1j}x^j + \ldots = g_1(x)u(x)$ and $X_2(x) = x_{20} + x_{21}x + x_{22}x^2 + \ldots + x_{2j}x^j + \ldots = g_2(x)w(x)$ which are interleaved to produce the encoder output sequence. The convolutional code is completely defined by the set of *generator polynomials* $g_1(x)$ and $g_2(x)$. These polynomials define the length of the shift register, the connections to the modulo-2 adders, and the number of modulo-2 adders. For example, the encoder of Fig. 13.2 maps each information bit into $N_c = 2$ coded message bits using a rule which depends on the previous two information bits. Specifically, an input 1 produces the outputs 11, 01, 00, or 10 when the previous inputs are 00, 10, 01, or 11 respectively. The *state* of the encoder in Fig. 13.2 corresponds to the contents in the shift register, viz., the previous two input bits. The *constraint length* K of the code is one plus the number of shift registers required to implement the encoder. The number of stages in the shift register required to implement the encoder is $K - 1$ and the number of *encoder states* is 2^{K-1}. The encoder of Fig. 13.2 has constraint length $K = 3$ since the current output pair is a function of the current input plus the two previous inputs. Note that different definitions of constraint length can be found in the literature on convolutional coding. In all cases, however, constraint length is a measure of the memory required to implement the encoder.

In general, convolutional codes produce N_c output channel symbols for each K_c input symbols and have rate $r_c = K_c/N_c$. Although this would appear to make convolutional codes identical to block codes, they are significantly different. For instance, the number of past inputs which affect the mapping of the current K_c inputs into N_c outputs is an important parameter of convolutional codes. This parameter affects convolutional code performance and complexity much as the block length affects block code performance and complexity. As already pointed out, consideration of the more complex (K_c, N_c, K) codes lies outside the scope of this book and the reader is referred to Lin and Costello [8] for a through treatment. With this limitation, however, convolutional codes can be defined by a set of N_c generator polynomials $\{g_1(x), g_2(x), \ldots, g_{N_c}(x)\}$, where there is one polynomial for each of the N_c output channel symbols. The constraint length K of the code is equal to one plus the degree of the highest-degree generator polynomial. The length of the shift register for the encoder is equal to the degree of the highest-degree generator polynomial. Table 13.1 summarizes the *code generator vectors* $\mathbf{g}_i, i = 1, \ldots, N_c$, for a set of optimum rate 1/2 and 1/3 codes. The *free distance*, d_f, indicated in this table will be further discussed in Section 13.4. For example, if the code generator vector $\mathbf{g}_1 = 111$ then $g_1(x) = 1 + x + x^2$ and if $\mathbf{g}_2 = 101$ then $g_2(x) = 1 + x^2$ as required in the implementation of the encoder illustrated in Fig. 13.2.

Finally, we point out that convolutional codes can be characterized as *systematic* or *nonsystematic* depending on whether or not the information sequence appears directly within the code sequence. For a convolutional code to be systematic, at least one of the code generator polynomials must equal 1 (for example, $g_1(x) = 1$) so that one of the N_c outputs is simply the current information bit. Unlike block codes, the performance of the best systematic convolutional codes is *not* equivalent to the performance of the best nonsystematic convolutional codes. For the same code performance, a systematic code will have a longer constraint length and be more complex to generate than the performance-equivalent nonsystematic code. For this reason, most of the convolutional codes used in practice are nonsystematic.

Table 13.1 Optimum Short Constraint Length
Convolutional Codes (Rate 1/2 and Rate 1/3).
Source: J. P. Odenwalder, *Error Control Coding
Handbook*, San Diego, Calif.: Linkabit Corp.,
July 15, 1976

Rate	Constraint Length	Free Distance	Code Vector
1/2	3	5	111 101
1/2	4	6	1111 1011
1/2	5	7	10111 11001
1/2	6	8	101111 110101
1/2	7	10	1001111 1101101
1/2	8	10	10011111 11100101
1/2	9	12	110101111 100011101
1/3	3	8	111 111 101
1/3	4	10	1111 1011 1101
1/3	5	12	11111 11011 10101
1/3	6	13	101111 110101 111001
1/3	7	15	1001111 1010111 1101101
1/3	8	16	11101111 10011011 10101001

13.2.2 The State Diagram

In order to introduce the reader to the state diagram, it is necessary to understand the concept of *state*. The *state* of a rate $1/N_c$ convolutional encoder is defined as the contents of the shift register (in this case two). Since $2^{K-1} = 4$ for the encoder of Fig. 13.2, there are four *encoder states* $\mathbf{A} = 00$, $\mathbf{B} = 10$, $\mathbf{C} = 01$, and $\mathbf{D} = 11$. One way to represent simple encoders is with a *state diagram*. Such a representation for the encoder in Fig. 13.2 is shown in Fig. 13.3. The states, shown in the circles of the diagram, represent the possible contents of the shift register, and the paths between the states represent the output branch words (codeword branches) resulting from such state transitions. The diagram shown in Fig. 13.3 illustrates all the state transitions that are possible for the encoder in Fig. 13.2. There are *only two transitions* emanating from each state, corresponding to the two possible input bits. Next to each path between states is written the coder output-branch codeword associated with the state transition. In drawing the path, we adopt the convention that a solid line denotes a path associated with an input bit, zero, and a dashed line denotes a path associated with an input bit, one. Notice that it is *not possible* in a single transition to move from a given state to *any arbitrary state*. As a consequence of shifting-in one bit at a time, there are only two possible state transitions that the register can make at each input bit time. For example, if the present encoder state is 00, the *only possibilities* for the state at the next shift are 00 or 10, whereas if the state is 11 then the next state can either be 11 or 01. Thus, the state diagram depicts the output branch codeword as a function of the encoder input data bit stream and demonstrates that convolutional coders can be thought of as finite state machines that change states as a function of the input bit sequence.

13.2.3 The Tree Diagram

The disadvantage of the state diagram as a method for representing a convolutional encoder is that it suppresses the time scale. The *tree diagram*, although a little more complex, adds the time dimension to the state diagram. Figure 13.4 illustrates the tree diagram for the encoder shown in Fig. 13.2. At each successive input bit time, the encoding procedure can be described by traversing the diagram from left to right. Here, in Fig. 13.2, each *tree branch* depicts the encoder output codeword branch $\mathbf{x}_j$. The approach used to generate the tree diagram is described as follows: If the input bit is a zero, its associated codeword branch output is found by moving to the next rightmost branch in the upward direction. If the input bit is a one, its codeword branch output is found by moving to the next rightmost branch in the downward direction. Assuming that the initial contents of the encoder is all zeros, the diagram shows that if the first input bit is a zero, the output branch word is 00 and, if the first input bit is a one, the output branch word is 11. Similarly, if the first input bit is a one and the second input bit is a zero, the second output branch word is 10. Or, if the first input bit is a one and the second input bit is a one, the second output branch word is a 01. Following this procedure, it is easy to see that the input data bit sequence $\mathbf{u} = 11011$ traces the heavy line illustrated in Fig. 13.4. This path corresponds to the output codeword sequence $\mathbf{x} = (\mathbf{x}_1, \mathbf{x}_2, \mathbf{x}_3, \mathbf{x}_4, \mathbf{x}_5)$ where $\mathbf{x}_1 = 11$, $\mathbf{x}_2 = 01$, $\mathbf{x}_3 = 01$, $\mathbf{x}_4 = 00$, and $\mathbf{x}_5 = 01$. From this tree diagram, we notice that the tree repeats itself (period K in general) after three input bits which corresponds to the constraint length; however, for large K one readily sees the limitation of using the tree diagram to represent the encoder output. Noting this periodic feature of the tree diagram suggests an alternate representation, viz., the use of a *trellis diagram* first suggested by Forney [10].

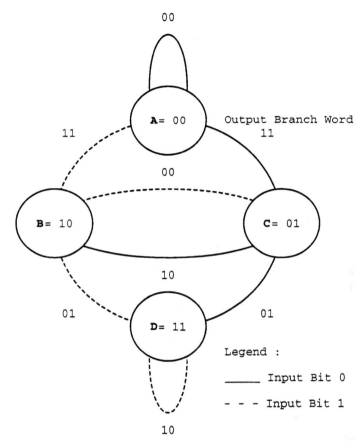

Figure 13.3   State transition diagram for rate $1/2$, $K = 3$ encoder

13.2.4 The Trellis Diagram

To produce the trellis diagram for the encoder shown in Fig. 13.2, we take advantage of the fact that the tree structure repeats itself after K branchings, that is, it is periodic with period K. We therefore begin by labeling the tree nodes in Fig. 13.4 to correspond to the four possible code states **A, B, C**, and **D** defined by the state diagram illustrated in Fig. 13.3. To incorporate time in the diagram, let t_j denote the time epochs of the codeword branches. In this case $t_{j+1} = t_j + 2T_c$ sec, for all $j \geq 1$. We arbitrarily set $t_1 = 0$. More generally, for a code rate $r_c = 1/N_c$, then $t_{j+1} = t_j + N_cT_c$ sec. We now develop the trellis diagram shown in Fig. 13.5 for the encoder shown in Fig. 13.2.

Assume that the encoding operation begins in state $\mathbf{A} = 00$ shown on the far left of the trellis at $t = t_1 = 0$. If the first information bit is a zero, the encoder moves along the solid line out of state $\mathbf{A} = 00$ arriving again at state $\mathbf{A} = 00$ at $t = t_2$. The encoder output is the symbol pair 00 which is labeled on the *trellis branch* between the two states. On the other hand, if at $t = t_1$ the first encoder input bit were a one, the encoder would move along the dashed branch out of state $\mathbf{A} = 00$ arriving at state $\mathbf{B} = 10$ at time $t = t_2 = 2T_c$. In this case, the encoder output is 11 which is indicated on the trellis branch connecting states $\mathbf{A}$ and $\mathbf{B}$. The second possible branching takes place at $t = t_2$ where the encoder is in either

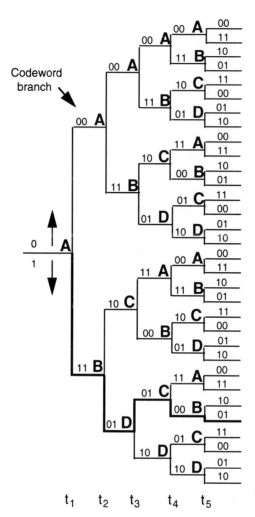

Figure 13.4 Tree diagram representation for encoder shown in Figure 13.2

state **A** or **B**. If state **A** exists and a zero is inputted to the encoder, then it moves along the solid branch outputting 00 and arriving in state **A** at $t = t_2$; if a one is input, the encoder moves to state **B** along the dashed line outputting 11 at $t = t_3$. Similar comments apply if at $t = t_2$ the encoder is in state **B**. This process of moving from left to right every bit time through the trellis and outputting the trellis branch labels continues as long as the encoder input is clocked with data bits. Using our representation, input information bits which equal zero cause the encoder to move along the solid trellis branches to the next state while input bits which equal one cause the encoder to move along the dashed branches. The codeword branch output sequence $\mathbf{x} = (\mathbf{x}_1, \mathbf{x}_2, \ldots, \mathbf{x}_j, \ldots)$ consists of the sequence of branch labels $\mathbf{x}_j = (x_{j1}, x_{j2})$ encountered as the encoder follows the path from left to right. A typical input output sequence and associated modulator input waveform for this convolutional code is shown in Fig. 13.6. For L input bits into the encoder, there are $2^{N_c L}$ possible output sequences that can be generated.

Since convolutional encoding is best understood using illustrative examples, we further demonstrate the concepts of the state and trellis diagrams using a rate $r_c = 1/3$ convolu-

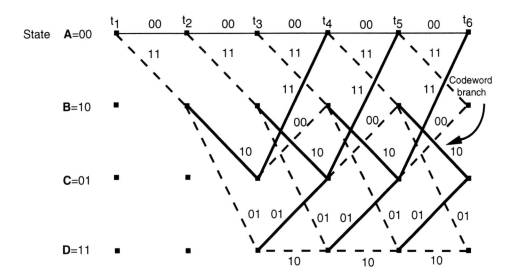

Figure 13.5   Trellis diagram for the rate $1/2$, $K = 3$ convolutional encoder of Figure 13.2

tional of constraint length $K = 4$. Figure 13.7 illustrates the encoder mechanization for the code generator polynomials defined by

$$g_1(x) = 1 + x + x^2 + x^3$$
$$g_2(x) = 1 + x + x^3 \qquad (13.1)$$
$$g_3(x) = 1 + x^2 + x^3.$$

For this case, three channel output symbols defined by the output branch word $\mathbf{x}_j = (x_{j1}, x_{j2}, x_{j3})$, are produced at multiples of $T_b = 3T_c$ sec for each input bit. The output from the 3 mod-2 adders in Fig. 13.7 are sampled to produce the components of $\mathbf{x}_j$. The specific registers summed to produce an output are defined by the generater polynomials in (13.1). None of the generators equal 1 so that the code is nonsystematic. Since all powers of x between 0 and 3 occur in the generator $g_1(x)$, the first output symbol is the modulo-2 sum of the input bit and all three previous inputs stored in the shift register. The number of stages of the shift register is equal to 3, which is the degree of the largest-degree generator polynomial (the degrees of all generators are the same). The constraint length of the code is $K = 4$ which is 1 plus the degree of any of the generator polynomials. Thus, an input bit influences the output due to the next three input bits. The number of states in the trellis representation of the code is $2^{K-1} = 8$. The state transition diagram for this convolutional encoder is illustrated in Fig. 13.8 while the corresponding trellis diagram is shown in Fig. 13.9. The trellis is found using the procedure discussed earlier for the rate $r_c = 1/2$ code

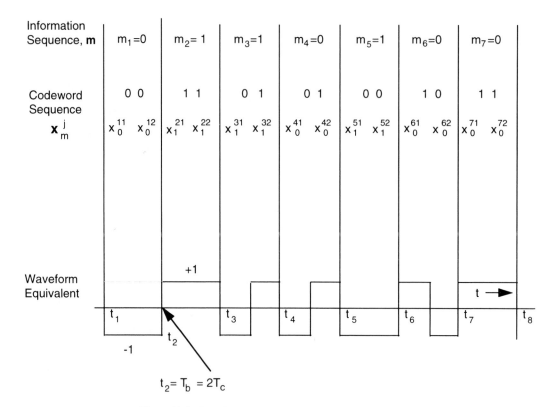

Figure 13.6 Encoder input and output sequences and waveform for $L = 7$

where the trellis state is the contents of the shift-register in Fig. 13.7. The three components of a trellis branch label represent the mod-2 sums defined by the generator polynomials or by the operations indicated in Fig. 13.7. Since $x^0 = 1$ is a component of all three generator polynomials, the labels on the two branches leaving the trellis state are complements of each other. The careful reader should take time to verify the state and trellis diagrams shown in Figs. 13.8 and 13.9.

13.3 DECODING CONVOLUTIONAL CODES

Consider now the convolutional coded digital communication system model illustrated in Fig. 13.10. This model is a special case of the coded system model discussed in Chapter 11, Fig. 11.2. In order to discuss convolutional decoding techniques, we must introduce the reader to the notation required to keep track of the codeword branches, codeword (channel) symbols, and codeword paths that are generated by a random (equal probable ones and zeros) binary data source. In addition, we must be specific about the output of the demodulator which drives the convolutional decoder. We tacitly assume that the decoder and demodulator have acquired the various levels of synchronization needed in accomplishing the decoding task, viz., the demodulator is coherent with respect to the carrier modulation process and has achieved channel symbol, bit, and *codeword branch (node)* synchronization. This knowledge, together with that of the convolutional encoder choice, is equivalent to providing the decoder with the knowledge contained in the trellis diagram. The transition from Fig. 11.2

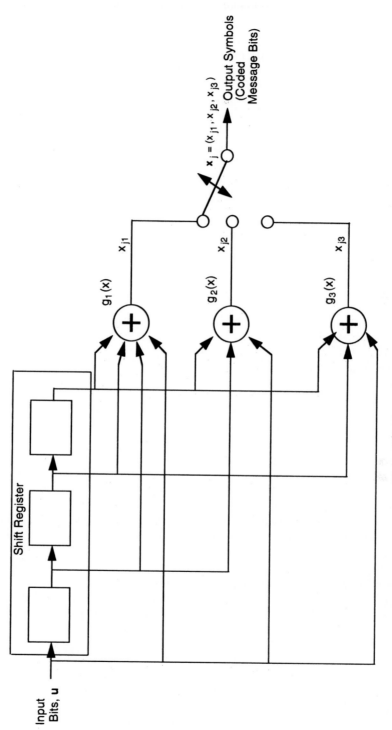

Figure 13.7 Rate $1/3$, $K = 4$ convolutional encoder

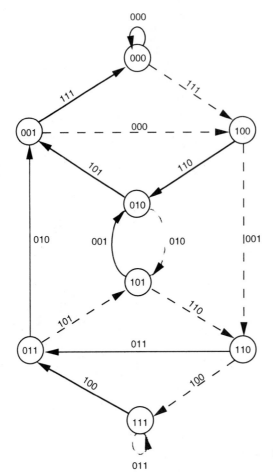

Figure 13.8   State diagram for $r_c = 1/3$, $K = 4$ convolutional encoder

to 13.10 is accomplished by adding the sub or superscript "j" to binary source bits, channel codeword branches, modulator and demodulator outputs to denote that the jth branch in the trellis diagram is being processed in the decoder. This branch occurs during the jth time interval $t_{j+1} - t_j = N_c T_c$, $j \geq 1$, where knowledge of t_j corresponds to codeword *branch or node* synchronization we assume $t_1 = 0$. Assuming further that the IF channel in Fig. 13.10 is memoryless, that is, a CMC, then the demodulator functions to create the output vector $\mathbf{r}_j = (r_{j1}, \ldots, r_{jN_c})$ whose components are the projections of received waveform $r_j(t)$ onto the basis functions of the coded modulator output waveform $S_m^j(t)$; see Chapters 3 and 4 and Fig. 13.10. Furthermore, by quantizing the demodulator outputs the modulator, IF channel, and demodulator can be easily converted into the DMC discussed in Section 11.2 of Chapter 11. This model will be useful when we wish to operate the decoder using *hard-* or *soft-decision* outputs from the demodulator. This important design issue will be discussed further when we present the so-called *Viterbi algorithm*.

13.3.1 Maximum-Likelihood Decoding of Coded BPSK Modulation

For purposes of being precise, consider now the class of binary convolutional codes of rate $r_c = 1/N_c$; specifically consider encoding a binary message sequence of L message bits. In

State

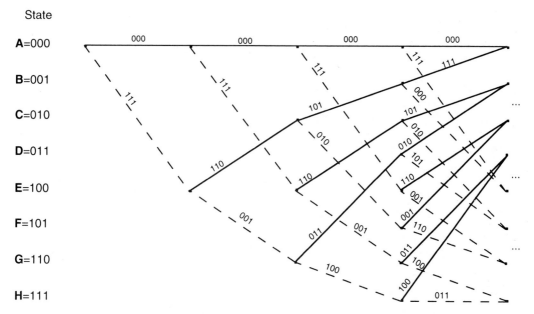

Figure 13.9   Trellis diagram for $r_c = 1/3$, $K = 4$ convolutional encoder

this context, one of $M_c = 2^L$ messages sequences exist and can be represented by the binary L-tuple $\mathbf{m} = (m_1, \ldots, m_L)$. Here $m \triangleq m_i$ is a binary random variable taking on values one and zero. Let u_m^j (with $m = 0$ or 1) characterize the source encoder output of Fig. 13.10 during the jth known data bit interval $(j-1)T_b \leq t \leq jT_b = jN_cT_c$, $j = 1, 2, \ldots, L$. In other words, if $m = 0$ then $u_0^j = 0$ and if $m = 1$ then $u_1^j = 1$. The convolutional encoder in Fig. 13.10 maps the input bit u_m^j into the jth *codeword branch* $\mathbf{x}_m^j = (x_m^{j1}, \ldots, x_m^{jN_c})$ of the trellis diagram where x_m^{ji} is a binary random variable taking on values $x_0^{ji} = 0$ or $x_1^{ji} = 1$ during the ith channel symbol interval of the jth bit interval. To connect the codeword branches to the physics of the modulator, then the vectors $\mathbf{x}_m^j$ are mapped one-to-one onto the coded BPSK modulation vectors $\mathbf{S}_m^j = (s_m^{j1}, \ldots, s_m^{jN_c})$ which are equivalently represented by the coded modulation waveform

$$S_m^j(t) = \sum_{i=1}^{N_c} s_m^{ji} p_j(t - iT_c) \tag{13.2}$$

for $(j-1)T_b \leq t \leq jT_b$ sec. Here the ith coded message bit (channel symbol) coefficient s_m^{ji} takes on values $s_1^{ji} = \sqrt{E_{N_c}}$ if $m = 1$ during the jth codeword time interval $t_{j+1} - t_j = N_cT_c$ and $s_0^{ji} = -\sqrt{E_{N_c}}$ if $m = 0$ during the same time interval; E_{N_c} represents the "energy" of the BPSK channel symbols and $T_N = T_c$. Furthermore, the pulse shape $p_j(t - iT_c)$ of unit energy serves to characterize the modulator output waveform during the known i^{th} channel symbol time of the jth codeword branch and is zero elsewhere. Thus, the IF channel coded modulation waveform for a sequence of L input bits (the message $\mathbf{m}$) can be represented as

$$S_{\mathbf{m}}(t) \triangleq \sum_{j=1}^{L} S_m^j(t), \quad 0 \leq t \leq N_cLT_{N_c} = LT_b \tag{13.3}$$

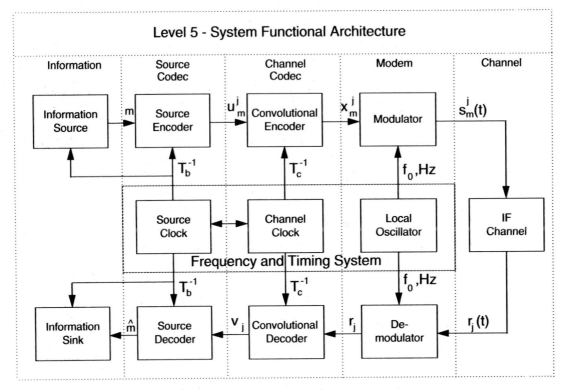

Figure 13.10   Convolutional coded-digital communication system model

where for each j, $m = 0$ or 1 with equal probability and $\mathbf{m}$ is one of 2^L binary message vectors. Now $S_{\mathbf{m}}(t)$ encodes L bits into $N_c L$ binary channel symbols s_m^{ji} or one of $2^{N_c L}$ binary sequences, viz., the L trellis codeword branches containing N_c channel symbols $\mathbf{x}_m^j = (x_m^{j1}, \ldots, x_m^{jN_c})$. Substitution of (13.2) into (13.3) leads to the channel waveform

$$S_{\mathbf{m}}(t) = \sum_{j=1}^{L} \underbrace{\sum_{i=1}^{N_c} s_m^{ji} p_j(t - iT_c)}_{j^{\text{th}} \text{ codeword branch}} \tag{13.4}$$

which is observed to be the sum of L waveforms which represent the codeword branches in the trellis diagram. The vector equivalent to the coded modulation waveform $S_{\mathbf{m}}(t)$ is defined by the coded modulation vectors $\mathbf{S_m} = (\mathbf{S}_m^1, \mathbf{S}_m^2, \ldots, \mathbf{S}_m^L)$ where $\mathbf{S}_m^j = (s_m^{j1}, \ldots, s_m^{jN_c})$. If $r_j(t)$ denotes the noise corrupted received waveform occurring during the time interval $t_{j+1} - t_j = N_c T_c$, $j = 1, \ldots, L$, then the demodulator creates the output projection $\mathbf{r}_j = (r_{j1}, \ldots, r_{jN_c})$ of this waveform using the equivalent basis functions for the modulator output waveform $s_m^j(t)$.

We are now in a position to establish the maximum-likelihood *decision function* for the jth trellis codeword branch. Assuming infinite quantization, if bit m (equals 0 or 1) is to be transmitted, then the codeword branch $\mathbf{x}_m^j$ implies $\mathbf{S}_m^j \triangleq (s_m^{j1}, \ldots, s_m^{jN_c})$ so that the decision function can be written as[1]

$$p\left(\mathbf{r}_j | \mathbf{S}_m^j\right) = \underbrace{\prod_{i=1}^{N_c} p\left(\mathbf{r}_{ji} | s_m^{ji}\right)}_{j^{\text{th}} \text{ trellis branch metric}} \tag{13.5}$$

where we have assumed the noise samples in $\mathbf{r}_j$ are independent and $p(\mathbf{r}_j | \mathbf{S}_m^j) \triangleq f_{\mathbf{r}_j}(\rho_j | \mathbf{S}_m^j)$ characterizes the conditional joint probability density function of the demodulator output, or equivalently, the IF channel transition probabilities. If we assume the L bits transmitted consist of the message $\mathbf{m} = (m_1, m_2, \ldots, m_L)$, then the corresponding maximum-likelihood decision function (conditional upon $\mathbf{m}$ being transmitted) becomes

$$\Lambda(\mathbf{m}) = \prod_{j=1}^{L} p\left(\mathbf{r}_j | \mathbf{S}_m^j\right) = \prod_{j=1}^{L} \underbrace{\left[\prod_{i=1}^{N_c} p\left(r_{ji} | s_m^{ji}\right)\right]}_{j^{\text{th}} \text{ trellis branch metric}} \tag{13.6}$$

Since the logarithm is a monotone increasing function of an increasing argument, the decoder could equivalently use the decision metric $\ln \Lambda(\mathbf{m})$. In this case, the decoder metric (decision function) becomes

$$\ln \Lambda(\mathbf{m}) = \sum_{j=1}^{L} \ln \left[\prod_{i=1}^{N_c} p\left(r_{ji} | s_m^{ji}\right)\right] \tag{13.7}$$

and the maximum-likelihood decoder outputs the vector $\hat{\mathbf{m}} = \mathbf{m}$ which maximizes $\ln \Lambda(\mathbf{m})$. When the IF CMC is modeled as the AWGN channel, then

$$p\left(r_{ji} | s_m^{ji}\right) = \frac{1}{\sqrt{\pi N_0}} \exp\left[\frac{\left(r_{ji} - s_m^{ji}\right)^2}{-N_0}\right] \tag{13.8}$$

where $s_m^{ji} = \sqrt{E_{N_c}}$ if $m = 1$ and $s_m^{ji} = -\sqrt{E_{N_c}}$ if $m = 0$. From (13.8), it is clear that the probability of making an error in the channel BPSK symbol s_m^{ji} is

$$p = Q\left(\sqrt{\frac{2E_{N_c}}{N_0}}\right) = Q\left(\sqrt{\frac{2r_c E_b}{N_0}}\right) \tag{13.9}$$

for all i, j, and m. Substitution of (13.8) into (13.7) leads to a path (decision) metric (function) which can be written equivalently in terms of the squared Euclidean distances $d_{jm}^2 = |\mathbf{r}_j - \mathbf{S}_m^j|^2$, that is

$$d_L^2(\mathbf{m}) \triangleq \sum_{j=1}^{L} d_{jm}^2 + C = \sum_{j=1}^{L} |\mathbf{r}_j - \mathbf{S}_m^j|^2 + C \tag{13.10}$$

---

1. Note again that $m = 0$ or 1 for each $j = 1, 2, \ldots, L$; hence, no reason to subscript m in the probability densities.

This says that one selects the message sequence $\hat{\mathbf{m}}$ which minimizes the sum of the squared Euclidean distances, that is, set $\hat{\mathbf{m}} = \mathbf{m}$ if

$$\min_{\mathbf{m}}[d_L(\mathbf{m})] = d_L(\hat{\mathbf{m}}) \qquad (13.11)$$

where $\hat{\mathbf{m}}$ is now one of the 2^L binary vectors representing the binary message sequence. This says that the maximum likelihood decoder evaluates the Euclidean distance between the received vector $\mathbf{r} = (\mathbf{r}_1, \mathbf{r}_2, \ldots, \mathbf{r}_L)$ and all possible coded modulation vectors $\mathbf{S_m} \triangleq (\mathbf{S}_m^1, \mathbf{S}_m^2, \ldots, \mathbf{S}_m^L)$ and selects that binary vector $\hat{\mathbf{m}} = \mathbf{m} = (m_1, m_2, \ldots, m_L)$ which minimized $d(\mathbf{m})$. We are using the fact that maximum likelihood decoding of convolutional codes is equivalent to *minimum distance decoding*; see Chapter 3. Equivalently, for the AWGN channel the minimum distance decoding can be shown to reduce to that of maximizing a set of $2^{N_c L}$ cross-correlation operations; see Chapters 3 and 4. Further simplification of (13.10) leads to the equivalent decision rule of selecting $\hat{\mathbf{m}} = \mathbf{m}$ if and only if $\hat{\mathbf{m}}$ maximizes the cross-correlations

$$\max_{\mathbf{m}} \left\{ \sum_{j=1}^{L} \sum_{i=1}^{N_c} r_{ji} s_m^{ji} \right\} + C' \qquad (13.12)$$

where for each value of i and j, $m = 0$ or 1 and C' is a constant bias. This says that to perform maximum-likelihood decoding we see that the test in (13.7) requires that one evaluate the $2^{N_c L}$ cross-correlations defined by (13.12). For large L this algorithmic approach to decoding becomes too complex to implement, that is, exponential growth in decoder complexity occurs. To avoid this exponential buildup in decoder complexity and still achieve the performance offered by maximum-likelihood decoding, Viterbi [2] recognized two key facts: first, that the tree diagram representation of a convolutional code was periodic with period K; and second, that paths in the tree diagram actually merge and hence this fact could be used to simplify the decoding process. Forney [10] clarified this point by introducing the trellis diagram from which it is clear that certain codeword paths remerge as a function of the state of the encoder. In what follows, the *Viterbi algorithm* for use in maximum-likelihood decoding of convolutional codes is presented.

13.3.2 Maximum-Likelihood Decoding Using the Viterbi Algorithm

As the L-bit message grows in length we observed that the numbers of correlations (decoder complexity) required for a rate $r_c = 1/N_c$ code grows exponentially. The Viterbi algorithm (VA) is an elegant and efficient method for performing ML decoding of convolutional codes. The algorithm was first described mathematically by Viterbi [2] in 1967. Since that time the algorithm has been described many times by many authors; most notably Forney [10] provides a highly readable and insightful description of the algorithm and its performance.

Recall from the previous section that the function of a ML decoder is to find the codeword vector sequence $\mathbf{S}_m = (\mathbf{S}_m^1, \mathbf{S}_m^2, \ldots, \mathbf{S}_m^L)$ which was most likely to have been transmitted given the received channel output vector sequence $\mathbf{r} = (\mathbf{r}_1, \ldots, \mathbf{r}_L)$ (see Fig. 11.2). Corresponding to each $\mathbf{S_m}$ there is a binary message vector $\mathbf{m} = (m_1, \ldots, m_L)$, which corresponds to a path through the trellis that gives the largest value for the log-likelihood function $\ln \Lambda(\mathbf{m})$ established by (13.7). Maximizing this is equivalent to selecting the binary vector $\mathbf{m}$, say $\hat{\mathbf{m}}$, such that the Euclidean distance

$$d_L(\mathbf{m}) = |\mathbf{r} - \mathbf{S_m}| = \sqrt{\sum_{j=1}^{L} |\mathbf{r}_j - \mathbf{S}_m^j|^2} = \sqrt{\sum_{j=1}^{L} d_{jm}^2} \qquad (13.13)$$

is minimum at $\hat{\mathbf{m}} = \mathbf{m}$. Here $d_{jm} \triangleq |\mathbf{r}_j - \mathbf{S}_m^j|$ is the distance between the demodulator output $\mathbf{r}_j$ during the jth bit interval and the hypothesized coded modulation vector $\mathbf{S}_m^j = (s_{j1}^m, \ldots, s_{jN_c}^m)$ transmitted during the jth bit interval. In the context that follows, the variable L is used to truncate the "length" of the cross-correlation operations which in practice will be a few constraint lengths.

13.3.3 Hard-Decision and Soft-Decision Viterbi Decoding

In order to illustrate this decoding procedure, it is simplest to consider the case of *hard-decision decoding*, that is, the case where $\mathbf{R}_j = (R_{j1}, \ldots, R_{jN_c})$ is one of 2^{N_c} binary vectors with components taking on values $R_{ji} = 1$ or 0. The vector $\mathbf{R}_j$ is generated from $\mathbf{r}_j$ by making hard decisions on each component of $\mathbf{r}_j$, that is, $R_{ji} = \text{sgn}(r_{ji})$; see Figs. 11.3, 11.6, and 11.7. As opposed to $\mathbf{S}_j^m$, these hard-decision vectors are to be compared with all possible encoder output codeword branch vectors $\mathbf{x}_m^j = (x_m^{j1}, \ldots, x_m^{jN_c})$, generated in the trellis. From (12.122) we know that the Euclidean distance squared is proportional to the Hamming distance; this important fact will now be used to rewrite (13.13) and illustrate the Viterbi algorithm. Thus, for hard-decision decoding, the ML decoding algorithm in (13.13) uses the decoding components $d_{jm}^2 = |\mathbf{r}_j - \mathbf{S}_m^j|^2 = 4E_{N_c}D_j(m)$ where $D_j(m) \triangleq |\mathbf{R}_j - \mathbf{x}_m^j|$ for all $j = 1, 2, \ldots, L$ with $m = 0$ or 1. Here the distance measurement is now made in the Hamming sense, that is, $D_j(m) \triangleq |\mathbf{R}_j - \mathbf{x}_m^j|$ is the number of places in which the binary vector $\mathbf{R}_j$ differs from $\mathbf{x}_m^j$. *In other words, for hard-decision decoding of coded BPSK, the Euclidean distance measure is "equivalent" to the Hamming distance.* For example, if the hard-decision BPSK demodulator output during the jth codeword branch is $\mathbf{R}_j = 11$ and the jth codeword branch is $\mathbf{x}_m^j = 00$ then the distance $D_j(m) = |\mathbf{R}_j - \mathbf{S}_m^j| = 2$. In this context, the ML decision rule for hard-decision decoding is to select that message vector $\mathbf{m} = (m_1, \ldots, m_L)$, say $\hat{\mathbf{m}}$, selected such that the accumulated Hamming distance

$$D_L(\mathbf{m}) = \sum_{j=1}^{L} D_j(m) = \sum_{j=1}^{L} |\mathbf{R}_j - \mathbf{x}_m^j| \qquad (13.14)$$

is minimum. The Hamming distances $D_j(m) = |\mathbf{R}_j - \mathbf{x}_m^j|$ are sometimes called *branch metrics* and $D_L(\mathbf{m})$ is called the accumulated path metric (consisting of the sum of the branch metrics) due to transmission of data vector $\mathbf{m}$. Specifically, each codeword branch vector $\mathbf{x}_m^j$ in the trellis diagram is at a Hamming distance

$$D_j(m) = |\mathbf{R}_j - \mathbf{x}_m^j| \qquad (13.15)$$

away from the demodulated codeword vector and as such serves as the *branch metric* for $\mathbf{m}$. The sum of the branch metrics create the so-called *path metric value* $D_j(\mathbf{m})$ at time $t = t_{j+1}$ for $j \geq 1, 2, \ldots, L$ and for each path in the trellis diagram. Using the sum of the branch metric values to create the path metric value as a function of the state of the encoder and the demodulator outputs $\mathbf{R}_j$, then ML decoding can be greatly simplified using the Viterbi algorithm. We now demonstrate the Viterbi algorithm using the encoder of Fig. 13.2 and its corresponding trellis diagram shown in Fig. 13.5.

Hard-decision Viterbi decoding.   Since hard-decision decoding requires that the IF CMC be converted into the BSC (binary symmetric channel) with transition probability given by (13.9), it is convenient to convert the encoder trellis in Fig. 13.5 into the decoder trellis shown in Fig. 13.11. Here we have assumed that $\mathbf{m} = 1\ 1\ 0\ 1\ 1$. The corresponding values for the transmitted codeword branches $\mathbf{x}_m^j$ and assumed hard-decision demodulator output vectors $\mathbf{R}_j$ are illustrated in Fig. 13.11; here the hard-decision demodulator output vectors are called the "received sequence." As we shall see, the VA essentially performs maximum-likelihood decoding by reducing the cross-correlation computational load by taking advantage of the special structure in the code trellis. The advantage of Viterbi decoding is that the complexity of a Viterbi decoder is not a function of the number of symbols in the codeword sequence. The algorithm involves calculating the distance (branch metric) between the demodulator output time t_j and all the trellis paths entering each state of the encoder at time t_j. The Viterbi algorithm removes from consideration (eliminates cross-correlation paths) those trellis paths entering each state that cannot possibly be candidates for the maximum-likelihood choice. This is accomplished as follows: Consider the two paths which enter the same state. The one having the largest path metric is chosen; this path is called *the surviving path*. This selection of surviving paths is performed for all the states for each $t_j = t_{j-1} + N_c T_c$. The decoder continues in this way to advance deeper into the trellis, making decisions by eliminating the least likely paths. The early rejection of the unlikely paths reduces the decoding complexity, that is, the number of required cross-correlations which have to be maintained every bit time.

The decoding procedure discussed above can be best understood by simultaneously considering the encoder trellis diagram in Fig. 13.5 and the decoder trellis diagram illustrated in Fig. 13.11. Along with the hypothesized input message sequence $\mathbf{m}$, the encoder codeword output sequence $\mathbf{x} = \{\mathbf{x}_m^j = (x_m^{j1}, x_m^{j2}),\ j = 1, 2, \ldots, 5\}$ and the hard-decision outputs $\mathbf{R} = (\mathbf{R}_1, \mathbf{R}_2, \ldots, \mathbf{R}_5)$ of the demodulator; see Figs. 11.3, 11.6, and 11.7 for a system perspective. The encoder codeword branch outputs for each state (see Fig. 13.5) are known a priori by the encoder and decoder. The labels on the *decoder trellis* of Fig. 13.11 at time $t_{j+1} = t_j + 2T_c$ are the accumulated Hamming distances $D_j(\mathbf{m})$ between the transmitted sequence and the received sequence, viz. with $L = j$ in (13.14) then

$$D_j(\mathbf{m}) = \sqrt{\sum_{i=1}^{j} |\mathbf{R}_i - \mathbf{x}_m^i|^2}. \tag{13.16}$$

At $t = t_1$ we assume that the encoder is initialized to the state $\mathbf{A} = 00$. In fact, (13.16) serves to define the cumulative Hamming distance (path metric) as the sum of the branch Hamming distances relative to the demodulator outputs up to time $t = t_j$. For example, at $t = t_2$ the demodulator output is $\mathbf{R}_2 = 11$ and the Hamming distance between $\mathbf{R}_2$ and the encoder output of $\mathbf{x}_0^2 = 00(m = 0)$ is 2, while for $\mathbf{x}_1^2 = 11(m = 1)$ and the distance is 0. These accumulated distances are labeled on the decoder branches at $t = t_2$ in Fig. 13.11, that is, going from encoder state "$\mathbf{A}$" at $t = t_1$ to state "$\mathbf{A}$" at $t = t_2$ we label the decoder trellis branch with 2, while going from state "$\mathbf{A}$" at t_1 to state "$\mathbf{B}$" at t_2 we label the decoder trellis with an accumulated distance of 0. One continues in this manner by labeling the accumulated distances on the decoder trellis branches as the demodulator outputs occur

Input data sequence	m	1	1	0	1	1	...
Transmitted codeword	x	11	01	01	00	01	...
Received sequence	R	11	01	01	10	01	...

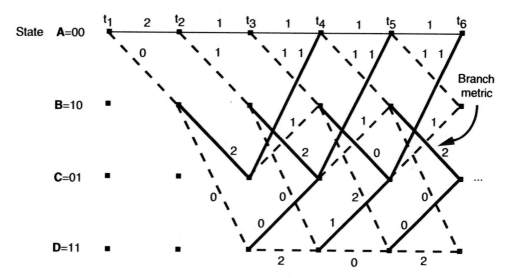

Figure 13.11   Trellis diagram and channel input-output sequences

through $t = t_6$. The VA uses these accumulated Hamming distances to find the accumulated path metric $D_j(\mathbf{m})$ which gives rise to the minimum distance, that is, the shortest path through the trellis relative to the demodulator outputs $\mathbf{R} = (\mathbf{R}_1, \ldots, \mathbf{R}_5)$.

Viterbi made the following creative and innovative observation which in essence eliminates the need to create all cross-correlations required in (13.12). If any two paths in the trellis merge to a single state at any time $t = t_j$, one of them can always be eliminated in the search for the shortest (minimum distance) path. For example, in Fig. 13.12 two paths merge at $t = t_5$. The upper path has the accumulated distance (path metric) $D_5 = 4$ while the lower accumulated path metric $D_5 = 1$. Thus, the upper path need not be given further consideration with respect to finding the shortest path through the trellis because the lower path, which enters the same decoder state, has a lower path metric. This lower path is called the *survivor path* for this state. Needless to say, this was Viterbi's astute observation and is primarily responsible for eliminating the need to compute, store, and accumulate all $2^{N_c L}$ correlations required in (13.12). Technically speaking, this is justified on the basis of the Markov nature of the encoder state, that is, the present encoder state summarizes the encoder history in the sense that past encoder states do not affect future trellis branch outputs. More generally, we can state that a (N_c, K_c, K) convolutional encoder has 2^{K-1} states in the trellis at time $t = t_{j+1} = t_j + N_c T_c$, $j \geq 1$. Each of these states can be entered by one of two paths. The VA consists of computing the accumulated branch metrics for the two paths entering each state and *eliminating one of them*. This computation is done for each of

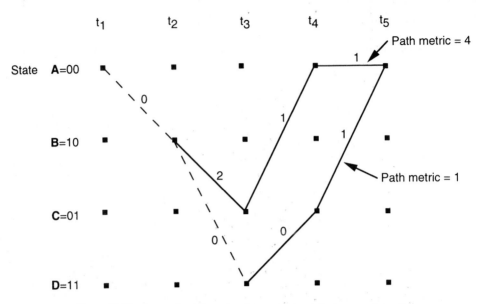

Figure 13.12   Accumulated Hamming distances at $t = t_5$ for two merging paths

the 2^{K-1} states at time t_j and the survivors are retained; then the decoder moves to time t_{j+1} and repeats the process of updating the accumulated path metrics and creating the survivors.

     To further demonstrate the process, we consider decoding the message for the demodulator outputs and trellis diagram given in Fig. 13.11. Following Sklar [27], we assume that the decoder is synchronized with the encoder and knows that the initial state at $t = t_1$, is $\mathbf{A} = 00$. In practice, knowledge of the initial state is not necessary; however, for purposes of discussion there is no loss in generality. From state $\mathbf{A} = 00$ the only possible transitions are to state $\mathbf{A} = 00$ or $\mathbf{B} = 10$; see Fig. 13.13(a). At time $t = t_2$ state "A to A" has branch metric $\lambda_\mathbf{A} = D_2(\mathbf{A}) = 2$ while state "A to B" has branch metric $\lambda_\mathbf{B} = D_2(\mathbf{B}) = 0$. At $t = t_2$ there are two possible branches leaving states $\mathbf{A} = 00$ and $\mathbf{B} = 10$; see Fig. 13.13(b); the cumulative path metrics of the codeword branches are labeled $\lambda_\mathbf{A} = D_3(\mathbf{A})$, $\lambda_\mathbf{B} = D_3(\mathbf{B})$, $\lambda_\mathbf{C} = D_3(\mathbf{C})$, $\lambda_\mathbf{D} = D_3(\mathbf{D})$ terminating at $t = t_3$. At time $t = t_3$ in Fig. 13.13(b) there will be two possible trellis branches diverging from each state. As a result there are two paths entering each state at $t = t_4$; see Fig. 13.13(c). For each state, the path giving rise to the largest cumulative can be eliminated with regard to further computation and the resulting decoding trellis is illustrated in Fig. 13.13(d). Here we observe that $\lambda_\mathbf{A} = D_4(\mathbf{A}) = D_4(\mathbf{B}) = \lambda_\mathbf{B}$ (accumulated path metrics) equal at $t = t_4$. When this occurs in the decoding process, one path can be chosen at random for elimination without affecting the average error probability. The surviving path for each state is illustrated in Fig. 13.13(d).

     Thus far in the decoding process, the cumulated branch distance metrics have been the major concern and no mention has been made regarding the decoded bit stream. With respect to this we note from Fig. 13.13(d) that there is only a single surviving path at time t_2 and this path (shown dashed) has been produced by the input bit $m = 1$; the decoder outputs a 1 as its first decoded bit. Here we can see how the decoding of the surviving branch is facilitated by having drawn the lattice branches with solid lines for input zeros and dashed lines for input ones. Note that the first bit was not decoded until the path metric computation had proceeded three bit times in practice into the trellis. This so-called *coding delay* can be as much as five times the constraint lenght in bits, that is, $5KT_b$ sec.

Now that we have demonstrated the decoder bit decision process and the concept of coding delay, let us continue with the decoding process. At each succeeding step ($t \geq t_4$) in the decoding process there will always be two possible paths entering each state of the encoder and decoder; one of the two will be eliminated by comparing cumulated path metrics leaving the *survivor paths*. At $t = t_4$, Fig. 13.13(e) demonstrates the two departing paths which arrive at the four states at time t_5. At $t = t_5$ there are two paths entering each state and one of each pair can be eliminated; Fig. 13.13(f) shows the survivors. At this point we note that we cannot make a decision on the second message bit because there still are two paths leaving the state 10 node at time t_2. At time t_6 in Fig. 13.13(g) we see the pattern of remerging paths, and in Fig. 13.13(h) we see the survivors at time t_6. Also, in Fig. 13.13(h) the decoder outputs one as the second decoded bit, corresponding to the single surviving path between t_2 and t_3. The decoder continues in this way to advance deeper into the trellis and to make decisions on the message bits by eliminating all paths but one.

Soft-decision Viterbi decoding.   For soft-decision decoding the VA described here, using the Hamming distance decoding metric, works in exactly the same manner using soft-decision decoding metrics [18]. A minor difference is that for soft-decision decoding the path with the *largest* log-likelihood metric, $\ln \Lambda(\mathbf{m})$ given in (13.12), is being found rather than the path with the *smallest* Hamming distance. Thus, discarded path segments are the segments with the smaller (rather than the larger) of the two metrics. In both cases, the path most likely to have been followed by the encoder is found.

13.3.4 Path Memory and Synchronization Requirements

The storage requirements of the Viterbi decoder grow exponentially with constraint length K. For a code with rate $1/N_c$, the decoder retains a set of 2^{K-1} paths after each decoding step. With high probability, these paths will not be mutually disjoint very far back from the present decoding depth. All of the 2^{K-1} paths tend to have a common stem which eventually branches to the various states. Thus, if the decoder stores enough of the history of the 2^{K-1} paths, the oldest bits on all paths will be the same. A simple decoder implementation, then, contains a *fixed amount of path history* and outputs the oldest bit on an arbitrary path each time it steps one level deeper into the trellis. The amount of path storage required, S, is [13] $S = h2^{K-1}$ where h is the length of the information bit path history per state. A refinement, which minimizes the value of h, uses the oldest bit on the most likely path as the decoder output, instead of the oldest bit on an arbitrary path. It has been demonstrated, [13], that a value of h of 4 or 5 times the code constraint length is sufficient for near-optimum decoder performance. The storage requirement, S, is the basic limitation on the implementation of Viterbi decoders. The current state of the art in microelectronics technology limits decoders to a constraint length of about $K = 15$. For $K > 10$ such decoders have been called *big Viterbi decoders*.

Codeword branch synchronization is the process of determining the beginning of a branch word in the received sequence. Such synchronization can take place without new information being added to the transmitted symbol stream because the received data appear to have an excessive error rate when the branch synchronization is incorrect. Therefore, a simple way of accomplishing synchronization is to monitor some concomitant indication of this large error rate, that is, the rate at which the path metrics are increasing or the rate at

Path Metric

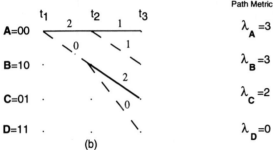

$\lambda_A = 2$

$\lambda_B = 0$

(a)

Path Metric

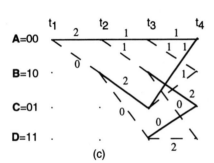

$\lambda_A = 3$

$\lambda_B = 3$

$\lambda_C = 2$

$\lambda_D = 0$

(b)

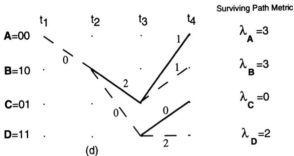

(c)

Surviving Path Metric

$\lambda_A = 3$

$\lambda_B = 3$

$\lambda_C = 0$

$\lambda_D = 2$

(d)

Figure 13.13 Selection of survivor paths: (a) Survivors at $t = t_2$, (b) Survivors at $t = t_3$, (c) Metric comparisons at t_4, (d) Survivors at $t = t_4$ (Reprinted from [27])

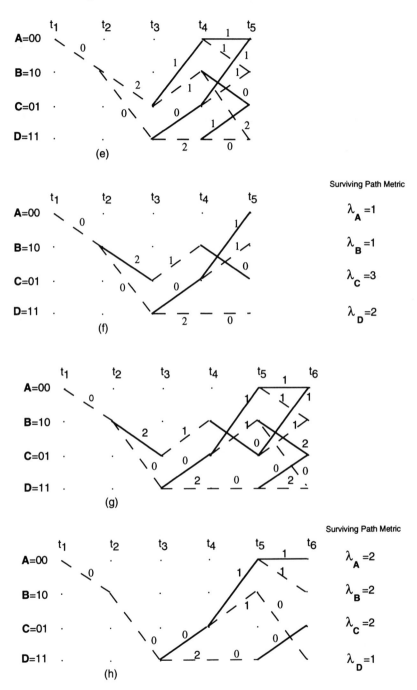

Figure 13.13 *continued*   Selection of survivor paths: (e) Metric comparisons at t_5, (f) Survivors at $t = t_5$, (g) Metric comparisons at t_6, (h) Survivors at $t = t_6$ (Reprinted from [27])

which the surviving paths in the trellis merge. The monitored parameters are compared to a threshold, and synchronization is acquired accordingly. The important details of this process lie outside the scope of this book and are usually best understood in practice where decoders are built to work in practical digital communication systems.

13.4 PROBABILITY OF DECODING ERROR FOR CONVOLUTIONAL CODES

Analogous with block codes, the probability of error in decoding an L-branch Q-ary convolutional tree code transmitted over the equivalent IF DMC (see Chapter 12) is upper bounded by [2]

$$P(E) < \frac{L(Q - 1)}{1 - Q^{\epsilon/r_c}} \exp\left[-N_c E(r_c)\right] \tag{13.17}$$

where

$$E(r_c) = \begin{cases} R_0; & 0 \le r_c = R_0 - \epsilon < R_0 \\ E(\rho, \mathbf{Q}); & R_0 - \epsilon \le R_r = \frac{E(\rho, \mathbf{Q}) - \epsilon}{\rho} < C_{N_c} \end{cases} \tag{13.18}$$

for $0 \le \rho \le 1$ and $R_0 = E_0(1, \mathbf{Q})$ is the quantized channel cutoff rate defined in (12.102).

Analogous with block codes, the probability of error in decoding convolutional codes can be made arbitrarily small by allowing N_c to approach infinity in the region where $E(r_c) > 0$.

Figure 12.8 illustrates the random coding exponent $E(r_c)$ achievable in very noisy channel for convolutional and block codes. The relative improvement increases with code rate; in particular for $r_c = R_0 = C_{N_c}/2$, the exponent for convolutional codes is almost six times that for block codes. Since exact expressions for probability of error are difficult to come by, we consider alternate approaches in what follows.

13.4.1 Distance Properties of Convolutional Codes

Like block codes, the distance properties of convolutional codes play a key role in understanding the error-correcting capability of the code. This point can best be made using the encoder shown in Fig. 13.2 to demonstrate the distance between all possible pairs of codeword sequences that can be generated by a random bit stream. In block codes, it was the minimum distance $D_{\min}$ that determined the t error-correcting capability of the code. For convolutional codes, $D_{\min}$ is replaced by the so-called *minimum free distance* D_f and related to t via

$$t = \left\lceil \frac{D_f - 1}{2} \right\rceil \tag{13.19}$$

where $\lceil x \rceil$ denotes the largest integer which does not exceed the bracketed ratio.

To demonstrate the meaning of D_f, we make use of the fact that a convolutional code is a linear code; hence, there is no loss in generality in simply finding the minimum distance between each codeword sequence (a cascade of code branch word) and the all-zeros sequence. First, we redraw the trellis diagram in Fig. 13.5 and relabel the branch outputs by the Hamming distance from the all-zeros codeword. The resulting trellis is shown in Fig. 13.14. Assuming that the all-zeros input sequence was transmitted, the trellis paths of interest are those that start and end in state $\mathbf{A} = 00$ anywhere in between. According to the VA discussed in the previous section, at least one bit error will occur whenever the

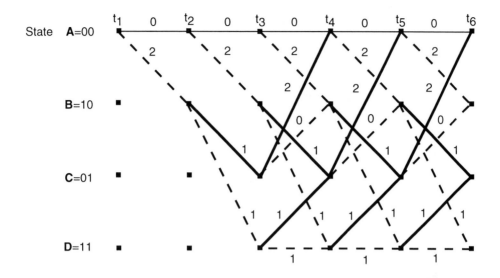

Legend

———————   Input bit 0

— — —   Input bit 1

Figure 13.14   Trellis diagram labeled with distances from the all-zeros path

distance of any other path that merges with state $\mathbf{A} = 00$ at time $t = t_j$ is less than that of the all-zeros path up to time t_j, thereby causing the all-zeros path to be eliminated in the decoding process. The minimum distance for making such an error can be found by exhaustively evaluating the distance between all paths from the 00 state to the 00 state in trellis relative to the all-zeros path. For those paths that remerge to the 00 state that are shorter in time length than the longest path, the sequence will be made of equal length by appending the necessary number of zeros so as to make all paths have equal length. Using Fig. 13.14, consider then all paths that diverge from the all-zeros path and then merge for the first time at some arbitrary trellis node ending at t_k. There is one path that departs at $t = t_1$ and remerges to the all-zeros path at $t = t_4$; this path has distance 5 from the all-zero path and corresponds to the input data sequence of 100. Similarly, there are two paths at distance 6 from the all-zeros path; one departs at t_1 and merges at time t_5, while the other departs at t_1 and merges again at $t = t_6$, etc. The input message bits for these two paths are, respectively, 110 and 10100. Each of the message vectors differs in two bits from the all-zeros path. Thus, the minimum distance is seen to be 5 since the distances for all arbitrarily long paths in greater than or equal 5. Setting $D_f = 5$ in (13.19) verifies that Viterbi decoder corresponding to the encoder in Fig. 13.2 can correct any two channel symbol errors.

For an arbitrary convolutional coder of rather short constraint length, the distance properties can be examined qualitatively using the so-called *code generating function or transfer function*; however, for long constraint lengths the complexity of the transfer

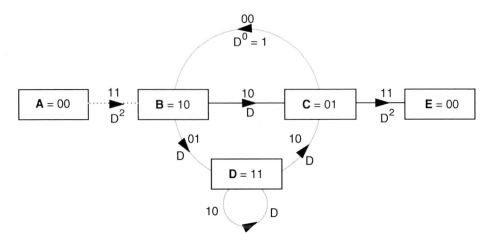

Figure 13.15   State diagram labeled with distances from the all-zeros path

function increases exponentially and other methods must be explored [8], [12]. Table 13.1 summarizes the free distance for several optimum short constraint length rate 1/2 and 1/3 convolutional codes. The details associated with using the transfer function to study the distance properties of convolutional codes lie outside the scope of this chapter; however, in the next section we will present one example to demonstrate the approach.

13.4.2 The Transfer Function Approach

Usually the best way to find D_f for larger constraint lengths is by computer simulation methodology. The analytical tool is to use the *transfer function* which can in principle be found from the state diagram. In addition, a rich by-product of establishing the transfer function is that it can also be used as a tool to calculate the bit error probability performance for convolutional codes. First, we illustrate the methodology for finding the transfer function for the encoder shown in Fig. 13.2 and its corresponding state transition diagram shown in Fig. 13.3. This state transition diagram is redrawn in Fig. 13.15 so as to clearly demonstrate the all-zeros state into two states which correspond to the state from which the trellis path of interest leaves and the state to which the path of interest returns. In this figure, the states **A**, **B**, **C**, and **D** are identified with **A** = **E**. Additionally, the loop from state **A** = 00 to itself has been discarded since paths that follow this loop fall outside the span of interest.

Now redraw the trellis diagram of Fig. 13.11 by labeling each codeword branch with its Hamming distance from the all-zero codeword sequence; see Fig. 13.14. By redrawing the trellis diagram in this way, we can clearly depict and determine the Hamming distance from the all-zeros path and the number of information one bits associated with all trellis paths through the modified bit equivalent state diagram. It is recognized that there are an infinite number of paths through this state diagram; however, we will enumerate the most important paths for application to the bit error probability analysis.

To accomplish this, we label the modified state diagram in Fig. 13.16 with the letters D, L, and N so as to account for additional information contained in the trellis. The letter D raised to power n is used to characterize the number of ones in the codeword symbols associated with the particular trellis branch. In particular, the power of D represents the number of ones. The letter L is associated with every branch and is used to characterize the

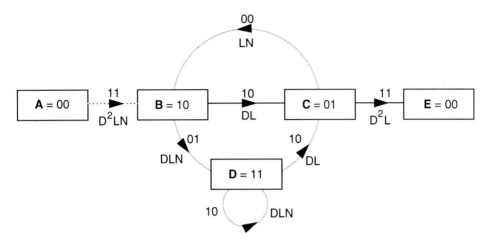

Figure 13.16   State diagram labeled according to distance, length, and number of input ones

number of branches traversed by a path. Letter N characterizes the number of ones in a path. The branch label includes N only if the information bit associated with that path is a one. To demonstrate how this notation accounts for a codeword sequence ones, information bit ones, and path lengths consider the path passing through the state sequence 00, 10, 01, 10, 11, 01, 00. The product of the labels on this path is $D^7L^6N^3 = D^2LN \cdot DL \cdot LN \cdot DLN \cdot DL \cdot D^2L$ and the codeword sequence for this path is 11, 10, 00, 01, 11 which contains seven ones corresponding to the power of D in the overall path product. The corresponding transmitted bit sequence for this path is 1 0 1 1 0 which contains three ones corresponding to the power of N in the path product. On the other hand, the total number of branches traversed is seen in the power of L and the path product is six. Consequently, we conclude that the path label products for the parameters D, L, and N for all paths through the state diagram characterize the desired features from which we will now take advantage.

Let $S_s(00)$, $S(10)$, $S(01)$, $S(11)$, and $S_e(00)$ denote a sum of the path label products for all paths starting with 00 and proceding with 10, 01, 11 and ending with 00. Then the generating function, which also characterizes the transfer function of a signal flow graph, can be found by simultaneous solution of the state equations written from Fig. 13.16, viz.

$$
\begin{aligned}
S_s(00) &= 1 \\
S(10) &= S_s(00) \cdot D^2LN + S(01)LN \\
S(11) &= S(10) \cdot DLN + S(11)DLN \\
S(01) &= S(11)DL + S(10)DL \\
S_e(00) &= S(01) \cdot D^2L
\end{aligned}
\tag{13.20}
$$

so that the transfer function becomes

$$
\begin{aligned}
T(D, N, L) \triangleq \frac{S_e(00)}{S_s(00)} &= \frac{D^5L^3N}{1 - DNL(1 + L)} \\
&= D^5L^3N + D^6L^4(1 + L)N^2 + D^7L^5(1 + L)^2N^3 + \dots \\
&\quad + D^{k+5}L^{k+3}(1 + L)^kN^{k+1} + \dots
\end{aligned}
\tag{13.21}
$$

Each term in the above sum corresponds to the branch label sequence for a path which diverges and then merges with the all-zero path. The first term corresponds to the shortest path which goes through the state sequence 00, 10, 01, 00. This path has length three corresponding to L^3 with Hamming weight five corresponding to D^5 and a message sequence Hamming weight of one corresponding to N. The second and third terms in (13.21) enumerate the two paths with Hamming weight of six and a message Hamming weight of two. This analysis methodology which has just been described by example is applicable in principal to all convolutional codes; however, for large K it is difficult to write down.

13.5 ERROR PROBABILITY PERFORMANCE OF CONVOLUTIONAL CODED BPSK MODULATION

The decoded error probability and bit error probability associated with the Viterbi algorithm does not lend itself to an exact analysis; usually for codes of practical interest the error probability performance must be established using the method of computer simulation. However, like maximum-likelihood decoding of block codes it is possible to establish bounds on error probability performance. In what follows, the *decoded error probability* and the *bit error probability* are given in terms of the transfer function $T(D, L, N)$. Further details of the methodology used to establish error probability bounds can be found in [2, 12, 18].

13.5.1 Hard- and Soft-Decision Decoded Error Probability

For convolutional codes with Viterbi decoding, the appropriate performance measure is the probability that the correct path through the trellis is discarded at depth j (time $t = t_j$). That is, the performance measure is the probability of making an error at each opportunity for decoding error. An opportunity to discard the correct path exists each time the decoder moves one step deeper into the trellis. This measure is comparable to the measure used for block codes where an opportunity for error exists each time a new block is decoded and the appropriate performance measure is the block decoding error probability. Note that the probability of selecting an incorrect path over the full extent of the trellis is an inappropriate performance measure. That is, it would not be correct to evaluate the probability that the total semi-infinite decoder output sequence is not continuously merged with the correct sequence. Over all time, the probability that the decoded path diverges from and remerges with the correct path at least once approaches unity. Since convolutional codes are linear, overall average decoding error probability is then equal to the decoding error probability calculated assuming that the all-zero sequence is transmitted. For the remainder of this discussion assume, therefore, that the correct path through the trellis is the all-zero path. The decoding error probability is then the probability that a nonzero path entering state 00 at depth j of the trellis is selected as the survivor. In order to evaluate decoding error probability, all potential decoding error events at depth j are enumerated, and their error probabilities are calculated and summed. The summation is the result of a union bounding argument similar to that used for the error probability bound for block codes. The reader is referred to [2, 12, 18] for details of the error proability derivation. There it is shown that the decoding error probability $P(E)$ is overbound by

$$P(E) < \sum_{k=D_f}^{\infty} a_k P_k \qquad (13.22)$$

where a_k is the number of nonzero paths with Hamming weight k that merge with the all-zero path at depth j and diverge once from the all-zero path at some previous depth, and P_k is the probability of incorrectly selecting a path with Hamming weight k. The values a_k can be found by differentiation from the generating function $T(D, L, N)$ for the code with $L = 1$ and $N = 1$. Specifically

$$T(D, 1, 1) = \sum_{k=D_f}^{\infty} a_k D^k \tag{13.23}$$

where

$$a_k = \frac{1}{k!} \frac{d^k T(0, 1, 1)}{d D^k} \Big|_{D=0}$$

The probability of incorrectly selecting a path which is Hamming weight k from the all-zero path is a function of the channel and the received signal-to-noise ratio. For the binary symmetric channel, P_k is equal to the probability that symbol errors in the k positions where the correct and incorrect paths differ cause the received sequence to be closer in Hamming distance to the incorrect path than to the correct path. In particular, it can be shown that [12]

$$P_k = \sum_{e=(k/2)+1}^{k} \left[\binom{k}{e} p^e (1-p)^{k-e} + \frac{1}{2} \binom{k}{k/2} p^{k/2} (1-p)^{k/2} \right] \tag{13.24}$$

for k even and

$$P_k = \sum_{e=(k+1)/2}^{k} \binom{k}{e} p^e (1-p)^{k-e} \tag{13.25}$$

for k odd. Furthermore, for infinite quantization and convolutional encoded BPSK modulation it can be shown that [2, 12]

$$P_k = Q\left(\sqrt{\frac{2k r_c E_b}{N_0}} \right). \tag{13.26}$$

For soft-decision decoding, the calculation of P_k is much more complex and the reader is referred to [2, 12] for the details.

13.5.2 Bit Error Probability

Bit error probability is calculated using essentially the same techniques as used for the decoding error probability. To calculate bit error probability, each decoding error event is weighted by the number of bit errors associated with that event. Since the all-zero path is the correct path, the number of bit errors associated with a nonzero path is equal to the number of information ones associated with that path. The generating function is used to enumerate error events along with the number of attendant bit errors. Recall that the number of information ones on a nonzero path through the trellis is equal to the power of N in the generating function term corresponding to that path. In order to change the exponent of N into a multiplier to weight the path by the number of bit errors, the generating function is differentiated with respect to N and N is then set to unity.

Consider the generating function given in (13.21). This function enumerates all nonzero paths through the trellis of Fig. 13.5. Differentiating $T(D, L, N)$ with respect to

N and setting $N = 1$ and $L = 1$ results in

$$\frac{dT(D, 1, 1)}{dN} = D^5 + (2 \cdot 2 \cdot D^6) + (3 \cdot 4 \cdot D^7) + \ldots + (j + 1) \cdot 2^j \cdot D^{j+5} + \ldots$$

$$= \sum_{k=D_f}^{\infty} c_k D^k \tag{13.27}$$

since the path length does not affect the result. This summation accounts for all possible nonzero paths through the trellis which merge with the all-zero path at time t_j. The first term indicates that there is a single bit error associated with all nonzero paths which are Hamming distance five from the all-zero path. The second term indicates that there are four bit errors associated with all nonzero paths which are Hamming distance six from the all-zero path. These four bit errors come from two different paths, each of which is associated with two bit errors. In general, there are a total of c_k bit errors associated with all paths which are Hamming distance k from the all-zero path. Using the generating function to weight error events by the number of bit errors, an upper bound on the bit error probability is

$$P_b(E) < \sum_{k=D_f}^{\infty} c_k P_k \tag{13.28}$$

where P_k is calculated using (13.24) for the hard-decision channel and (13.26) for the continuous output additive white Gaussian noise channel. The coefficients c_k can be found from (13.27) by differentiation. Unfortunately, the performance analysis of a decoder which uses the VA does not lend itself to closed form analysis and computer simulation techniques have to be used. We shall present several such results in the next section for a set of convolutional codes of greatest practical interest to date.

13.6 ERROR PROBABILITY PERFORMANCE OF CONVOLUTIONAL CODES USING THE VITERBI ALGORITHM

Using the methodology discussed in the previous section, Ziemer and Peterson [18] evaluate the bounds on bit error probability performance. Figures 13.17 and 13.18 demonstrate the hard- and soft-decision bit error probability performance versus E_b/N_0 for various rate 1/2 convolutional codes. Figures 13.19 and 13.20 demonstrate the bit error probability performance for rate 1/3 convolutional codes. While both upper bounds in (13.22) and (13.28) for error probability are reasonably good for large signal-to-noise ratios, computer simulation methods must be used to get results over the signal-to-noise ratio region of interest. Indeed, extensive computer simulation results have been published for use in practice and can be found in the frequently used work of Heller and Jacobs [13] and Odenwalder [14]. Typical results are demonstrated in Figs. 13.21 and 13.22 for rate 1/2 codes and soft-decision decoding. Both upper bounds (solid curve) and simulation results (dotted lines) are compared.

13.7 CONCATENATED REED-SOLOMON AND CONVOLUTIONAL CODES

Powerful error-correction capability can be obtained by concatenating a Viterbi decoded convolutional code with a Reed-Solomon block code. Figures 11.16 illustrates a block diagram of a coded communications system which uses both a convolutional and a block code.

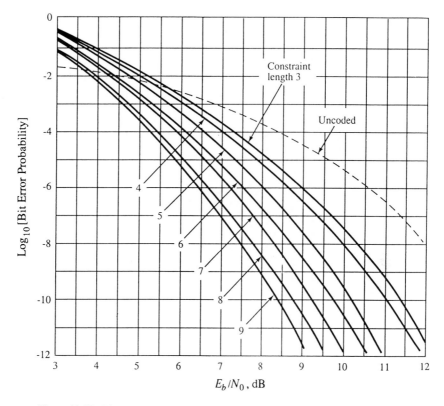

Figure 13.17   Bit error probability performance for various rate 1/2 convolutional codes using hard-decision decoding (Reprinted from [18])

Information bits from the data source are grouped into blocks of K_c to create the 2^{K_c}-ary symbol alphabet used by the Reed-Solomon code. The coding system which appears first in the chain, in this case the RS code, is called the *outer code* of a concatenated code system. The RS output symbols are converted to their K_c-bit binary representations and input to a symbol interleaver. The purpose of this interleaver is to enable the spreading of bursts of channel symbol errors over RS codewords as uniformly as possible. As discussed in Chapter 11, the interleaver changes the order in which the RS symbols are transmitted over the channel. The symbols are put back in their original order in the deinterleaving buffer in the receiver. This process of changing the transmission order, transmitting, and reestablishing the original order causes contiguous bursts of channel errors to be spread uniformly over a number of RS codewords rather than occurring all in a single codeword. RS codes, as well as all of the other codes which have been studied, are designed to perform best when channel errors occur independently from one symbol to the next and do not occur in bursts. For our purposes, we assume that the interleaver/deinterleaver combination may be considered processors which take bursts of channel errors and spread these errors over some number of codewords. The outer interleaver output of Fig. 11.16 serves as the input to a convolutional coder whose output is sent over the channel. The convolutional code and Viterbi decoder operate exactly as discussed previously; the convolutional code is called the *inner code* of the system. The output of the Viterbi decoder is an estimate of the binary sequence which was

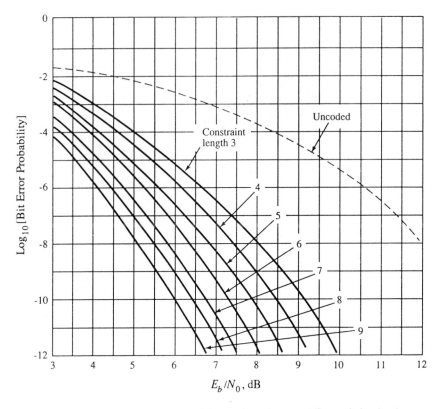

Figure 13.18 Bit error probability performance for various rate 1/2 convolutional codes using soft-decision decoding (Reprinted from [18])

input to the encoder. The Viterbi decoder output is input to the outer deinterleaver. Detailed measurements and simulations of Viterbi decoder performance have shown that output errors tend to occur in bursts. It is for this reason that the interleaving scheme is required. Without the interleaver, the channel would not appear to be memoryless. The outer deinterleaver output binary symbols are grouped into blocks of K_c to construct the 2^{K_c}-ary symbols for input to the RS decoder. The RS decoder performs the processing required to estimate the data source sequence. The RS decoder is able to correct some of the decoding errors of the Viterbi inner decoder, thus improving communication reliability. Figure 13.23 illustrates the bit error probability performance of the concatenated coding system of Fig. 11.16. The convolutional code used for Fig. 13.28 has $r_c = 1/2$ and constraint length $K = 7$. Results are shown for a number of different RS codes with $K = 6, 7, 8$, and 9. Observe the coding gain at the bit error probability of $P_b(E) = 10^{-5}$ of aproximately 7.0 dB relative to an uncoded BPSK system.

Concatenated codes consisting of an inner convolutional code and an outer Reed-Solomon (RS) code are used in several current and planned deep-space missions. The Voyager spacecraft, for example, employs a concatenated coding system based on a (7, 1/2) convolutional code as the inner code and an 8-bit (255, 223) RS code as the outer code. The Galileo mission will use an experimental (15, 1/4) convolutional code concatenated with the same 8-bit RS code. Future missions may use the recently discovered (15, 1/6) convolutional code together with a 10-bit (1023, 959) RS code [19].

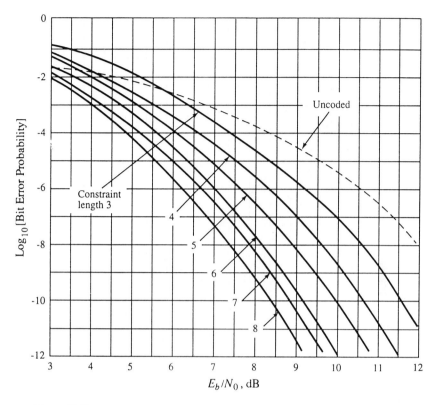

Figure 13.19 Bit error probability performance for various rate $1/3$ convolutional codes using hard-decision decoding (Reprinted from [18])

Figures 13.24 to 13.32 show the performance of concatenated codes with ideal and nonideal interleaver, and point out the difference in performance between 8-bit and 10-bit RS codes [19]. The E_b/N_0 (bit SNR) shown in these figures is the bit signal-to-noise ratio of the concatenated system, which includes a penalty of 0.58 dB due to the rate of the 8-bit RS code, or 0.28 dB for the 10-bit RS code.

Figures 13.24 and 13.25 show the bit error rate of the (7, 1/2) code concatenated with the 8-bit and 10-bit RS codes, respectively. Interleaving depths $I = 2$ and $I = 4$ are shown along with ideal interleaving. Higher values of I had nonmeasurable performance degradation relative to ideal interleaving. Larger constraint length codes suffer more from shallow interleaving since the average length of bursts grows with the constraint length. It must be observed that results for nonideal interleaving need very large amounts of data (decoded bits) and are not very accurate even with runs of 10 million or more bits. Figure 13.26 shows a comparison of the bit error rates of the 8-bit and 10-bit codes concatenated with the (7, 1/2) code and ideal interleaving. The advantage of the 10-bit RS code becomes apparent only at very low bit error rates. A different 10-bit RS code specifically optimized for concatenation with the (7, 1/2) code could offer a larger improvement over the 8-bit RS code than the (1023, 959) RS code, which was optimized for the (15, 1/6) code.

Figures 13.27 and 13.28 show the bit error rate of the (15, 1/4) code concatenated with the 8-bit and 10-bit RS codes, respectively. Figure 13.29 shows a comparison of the bit error rates of the 8-bit and 10-bit codes concatenated with the (15, 1/4) code and ideal inter-

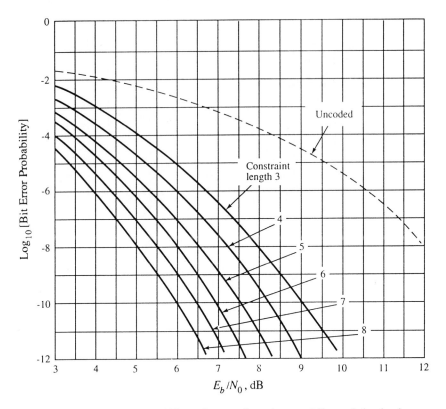

Figure 13.20   Bit error probability performance for various rate 1/2 convolutional codes using soft-decision decoding (Reprinted from [18])

leaving. Now the advantage of the 10-bit RS code over the 8-bit RS code is approximately 0.2 dB at a bit error probability of 10^{-6}. This advantage grows to approximately 0.3 dB at 10^{-12}.

Finally, Figs. 13.30 and 13.31 show the bit error rate of the (15, 1/6) code concatenated with the 8-bit and 10-bit RS codes, respectively. Figure 13.32 shows a comparison of the bit error probability of the 8-bit and 10-bit codes concatenated with the (15, 1/6) code and ideal interleaving. The advantage of the 10-bit code is slightly less than 0.2 dB at a bit error probability of 10^{-6}, and approximately 0.3 dB at a bit error probability of 10^{-12}.

13.8 OTHER CONVOLUTIONAL DECODING TECHNIQUES

The scope of this book places a constraint on the ability to cover the details of all decoding algorithms that have been proposed and reported upon in the literature. In particular, prior to the 1971 paper by Viterbi convolutional codes there were two other decoding algorithms that were of greatest interest and used in decoding convolutional codes, viz., the *sequential decoding algorithm* suggested by Wozencraft [15] and subsequently modified by Fano [16], and the *threshold decoding* algorithm suggested by Massey [17]. The two algorithms are discussed briefly in what follows; however, the reader who is interested in the details should see [5], [12], [15], and [16].

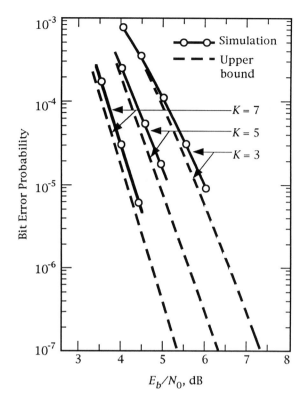

Figure 13.21   Bit error probability versus E_b/N_0 for rate 1/2 convolutional codes— simulation with 8-level quantization and 32-bit path memory (solid lines); upper bounds for unquantized AWGN channel (dashed lines) (Reprinted from [13] © IEEE)

13.8.1 Sequential Decoding

A sequential decoder works [15], [16] by generating hypotheses about the transmitted code-word sequence; it computes a metric between these hypotheses and the received signal. It goes forward as long as the metric indicates that its choices are likely; otherwise, it goes beckward, changing hypotheses until, through a systematic trial-and-error search, it finds a likely hypothesis. Sequential decoders can be implemented to work with hard or soft decisions, but soft decisions are usually avoided because they greatly increase the amount of the required storage and the complexity of the computations.

While the error probability delivered by the VA decreases exponentially with K, the number of states of the decoder complexity grows exponentially with constraint length. Yet the computational complexity of the VA is independent of characteristics in the sense that, compared to hard-decision decoding, soft-decision decoding requires a small increase in the number of computations. The error probability associated with sequential decoding algorithm is asymptotically equivalent to that of maximum-likelihood decoding and does not require searching all possible codeword paths; in fact, the number of paths searched is essentially independent of the constraint length, thereby making it possible to use large constraint lengths ($K = 4$). This important factor provides for the ability to operate at very small values of error probability. The major disadvantage of sequential decoding is that the number of state metrics is a random variable, while the average number of poor hypotheses and backward searches is a function of channel symbol energy-to-noise ratio, $E_{N_c}/N_0 =$

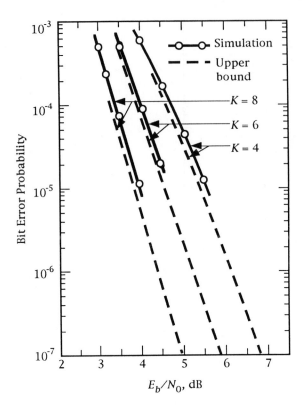

Figure 13.22 Bit error probability versus E_b/N_0 for rate 1/2 convolutional codes—simulation with 8-level quantization and 32-bit path memory (solid lines); upper bounds for unquantized AWGN channel (dashed lines) (Reprinted from [13] © IEEE)

$r_c E_b/N_0$. When E_{N_c}/N_0 is small, more hypotheses must be tried than with high values of E_{N_c}/N_0. Because of this variability in computational load, buffers must be provided to store the arriving sequences. For small E_{N_c}/N_0 the received sequences must be buffered while the decoder is laboring to find a likely hypothesis. If the average symbol arrival rate exceeds the average symbol decode rate, the buffer will overflow, no matter how large it is, causing a loss of data. The sequential decoder typically puts out error-free data until the buffer overflows, at which time the decoder has to go through a recovery procedure. The buffer overflow threshold is a very sensitive function of E_b/N_0. Therefore, an important part of a sequential decoder specification is the *probability of buffer overflow*, see [5], [15], and [16].

Figure 13.33 compares the bit error probability achievable versus E_b/N_0 when Viterbi and sequential decoding are used to decode convolutional codes. Their comparative performance is made with BPSK operating in the AWGN channel. The curves compare Viterbi decoding (rates $\frac{1}{2}$ and $\frac{1}{3}$ soft decision, $K = 41$). One can see from Fig. 13.33 that coding gains of approximately 8 dB at $P_B = 10^{-6}$ can be achieved with sequential decoders. Relative to the approximate 11 dB of coding gain in E_b/N_0 promised by the "Shannon limit" relative to uncoded BPSK, one might conclude that the major portion of that has already been achieved. Sequential decoding has been used in various early (circa 1970) important deep space missions.

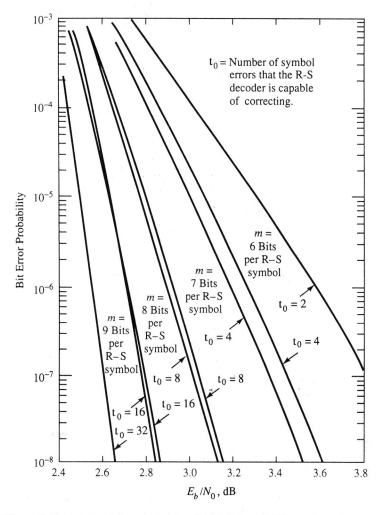

Figure 13.23 Summary of concatenated coding bit error probability performance with a $K = 7, r = 1/2$ convolutional inner code and various RS outer codes (Reprinted from [18])

13.8.2 Threshold Decoding

Certain convolutional codes can be decoded using circuitry similar in complexity to that used to decode Hamming codes. In order to be decoded so simply, the codes must have additional structure as defined in detail by Massey [17] and by Lin and Costello [8]. The simplicity of threshold decoders makes their implementation at very high speeds possible. Although the performance of these special convolutional codes is not outstanding for the AWGN channel, the structure is particularly convenient for applications where interleaving is required; see Chapter 11. For channels where the interference is of the burst type, large values of coding gain can be achieved.

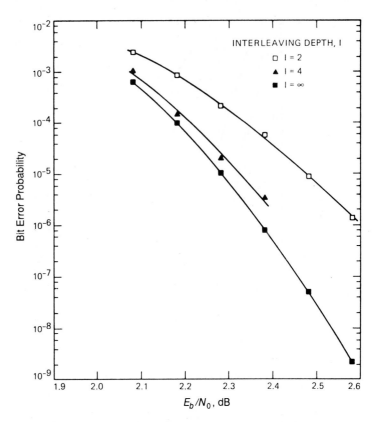

Figure 13.24   Concatenated code performance (7, 1/2) convolutional code and 8-bit (255, 223) RS code (Reprinted from [19])

13.9 PUNCTURED CONVOLUTIONAL CODES

The class of high-rate, $r_c \approx 1$, punctured convolutional codes [20] are of interest in bandlimited channels, for example, transport of high definition television. These special high-rate codes are more easily decoded using either Viterbi decoding, [21], [22], [23], or sequential decoding, [24], [25], because they have the underlying structure of low-rate codes. They also lend themselves to an easy implementation of variable-rate coding-decoding, [22], [25], as well as rate-compatible coding-decoding [26]. Begin and Haccoun [24] present some properties of punctured codes in addition to providing an upper bound on the free distance which indicates that high-rate punctured codes are good codes. The reader is referred to the reference material since the details fall outside the scope of this book.

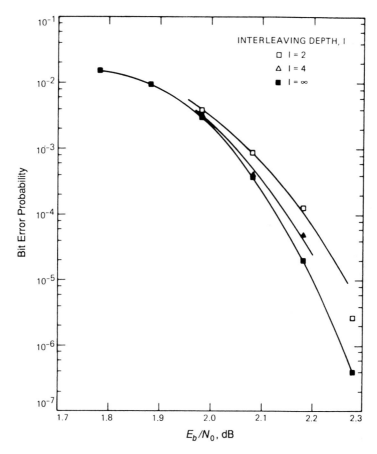

Figure 13.25   Concatenated code performance (7, 1/2) convolutional code and 10-bit
(1023, 959) RS code (Reprinted from [19])

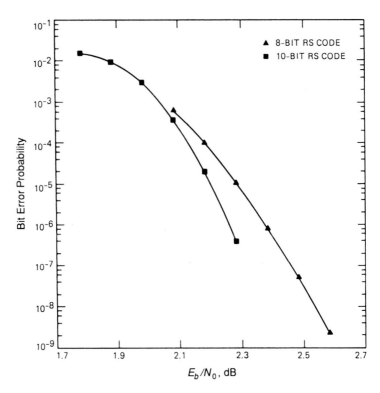

Figure 13.26   Concatenated code performance comparison: 8-bit (255, 223) and 10-bit (1023, 959) RS code concatenated with (7, 1/2) convolutional code (Reprinted from [19])

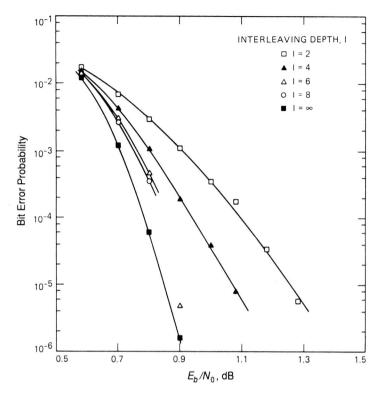

Figure 13.27   Concatenated code performance (15, 1/4) convolutional code and 8-bit (255, 223) RS code (Reprinted from [19])

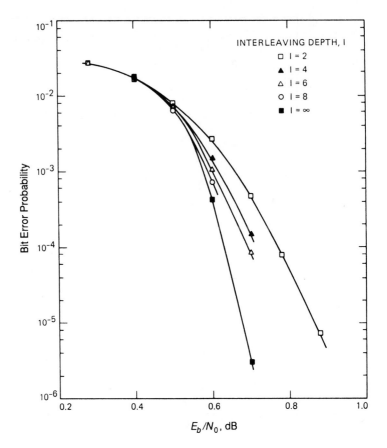

Figure 13.28   Concatenated code performance (15, 1/4) convolutional code and 10-bit
(1023, 959) RS code (Reprinted from [19])

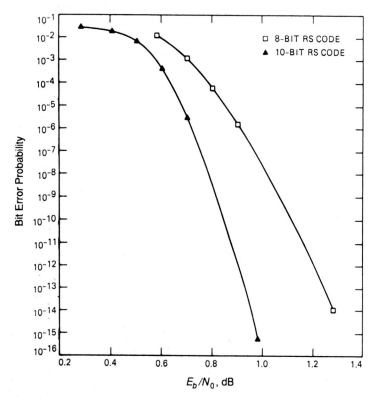

Figure 13.29   Concatenated code performance comparison: 8-bit (255, 223) and 10-bit (1023, 959) RS codes concatenated with (15, 1/4) convolutional code (Reprinted from [19])

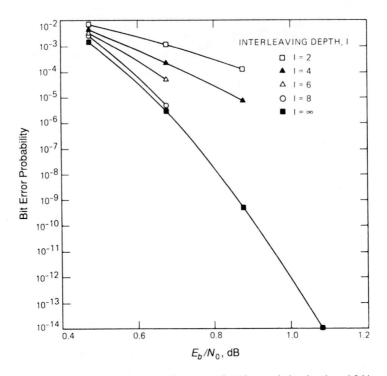

Figure 13.30   Concatenated code performance (15, 1/6) convolutional code and 8-bit (255, 223) RS code (Reprinted from [19])

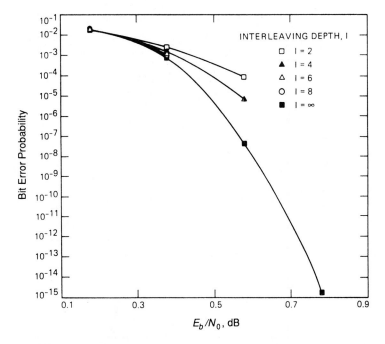

Figure 13.31   Concatenated code performance (15, 1/6) convolutional code and 10-bit (1023, 959) RS code (Reprinted from [19])

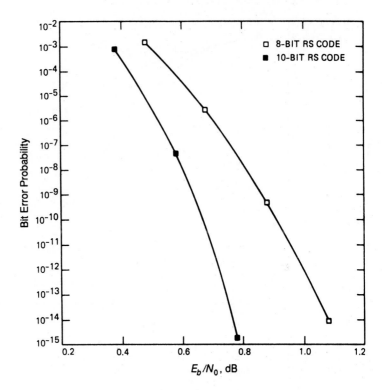

Figure 13.32 Concatenated code performance comparison: 8-bit (255, 223) and 10-bit (1023, 959) RS codes concatenated with (15, 1/6) convolutional code (Reprinted from [19])

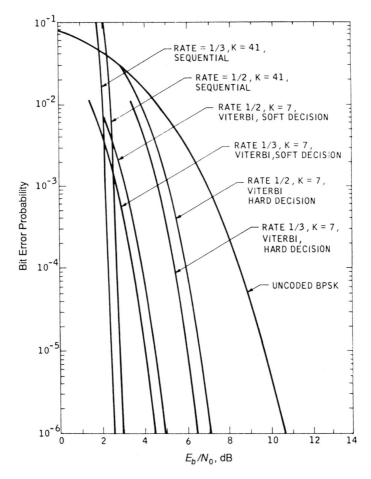

Figure 13.33   Bit error probability performance for various Viterbi and sequential decoding schemes using coherent BPSK over an AWGN channel (Reprinted from [28] © IEEE)

PROBLEMS

13.1 Assuming one input bit per clock cycle:
  (a) What is the constraint length K and code rate for the convolutional encoder illustrated in Fig. P13.1?
  (b) Draw the tree, trellis, and state diagram generated by this encoder.

13.2 Assuming one input bit per clock cycle:
  (a) What is the constraint length K and code rate for the convolutional encoder illustrated in Fig. P13.2?
  (b) Draw the tree, trellis, and state diagram generated by this encoder.

13.3 Assuming one input bit per clock cycle:
  (a) What is the constraint length K and code rate for the convolutional encoder illustrated in Fig. P13.3?
  (b) Draw the state diagram generated by this encoder.

13.4 Given the encoder illustrated in Fig. P13.2 and suppose the decoder employs hard decisions to create the binary symmetric channel. Suppose that the received sequence for the first eight branches is [12]

00 01 10 00 00 00 10 10

Trace the decisions on a trellis diagram labeling the surviving path's Hamming distance metric at each node. When a tie occurs in the decision metric, always choose the upper path.

13.5 Given the $K = 3, r = 1/2$ code in Fig. P13.2, suppose the code is used on a BSC and the transmitted sequence for the first eight branches is [12]

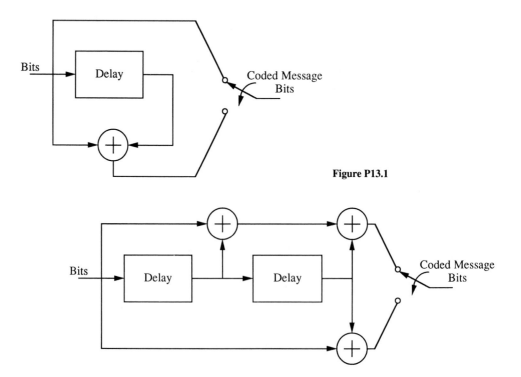

Figure P13.1

Figure P13.2

$$11\ 01\ 01\ 00\ 10\ 11\ 00$$

Suppose this sequence is received *error free*, but the first bit is lost and the received sequence is incorrectly synchronized giving the assumed received sequence

$$10\ 10\ 10\ 01\ 01\ 10$$

This is the same sequence except the first bit is missing and there is now incorrect bit synchronization at the decoder. Trace the decisions on a trellis diagram, labeling the survivor's Hamming distance metric at each node.

Hint: Note that when there is incorrect synchronization, path metrics tend to remain relatively close together. This fact is used to detect incorrect bit synchronization.

13.6 Draw the state diagram, tree diagram, and trellis diagram for the $K = 3$, rate $1/3$ convolutional code generated by

$$g_1(x) = x + x^2$$

$$g_2(x) = 1 + x$$

$$g_3(x) = 1 + x + x^2$$

13.7 Suppose that you were trying to find the quickest way to get from London to Vienna by boat or train. The diagram in Fig. P13.7 was constructed from various schedules. The labels on each path are travel times. Using the Viterbi algorithm, find the fastest route from London to Vienna. In a general sense, explain how the algorithm works, what calculations must be made, and what information must be retained in the memory used by the algorithm.

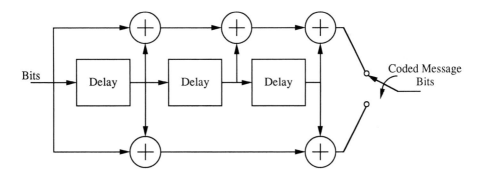

Figure P13.3

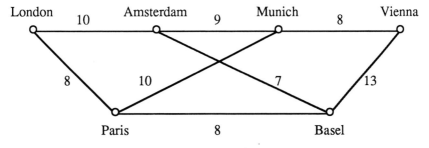

Figure P13.7

13.8 Assuming the coded system operates at capacity C_{N_c}, use Fig. 12.11 and determine the maximum coding gain (relative to uncoded BPSK operated at a bit error rate of 10^{-5}) achievable for a rate 1/2 code. Obtain this gain for hard-decision decoding ($J = 2$), 3-bit quantization ($J = 4$), and infinite quantization. Recall that the communication efficiency required for BPSK is $E_b/N_0 = 9.6$ dB.

13.9 Assuming that the coded system operates at R_0, repeat Problem 13.8 using Fig. 12.12. Compare your results with those found in Problem 13.8.

13.10 What is the free distance, D_f, achievable with the rate 1/3 convolutional coder illustrated in Fig. 13.7? The free distance D_f is the minimum Hamming distance between two distinct trellis paths through the code trellis.

13.11 Assuming rate 1/2 convolutional coded BPSK and operation at a bit error probability of 10^{-5}, what is the coding gain relative to uncoded BPSK if:
 (a) a rate 1/2, $K = 7$ convolutional code and hard-decision decoding is used.
 (b) a rate 1/2, $K = 7$ convolutional code and 3-bit soft-decision decoding is used.
 (c) Repeat (a) and (b) assuming a rate 1/3, $K = 7$ convolutional code is used.
 (d) Compare the coding gains achievable for the two code rates and comment on why the results are different.

13.12 A concatenated Reed-Solomon/convolutional code is used to improve system performance in the AWGN channel at a bit error probability of 10^{-5}.
 (a) Assuming a $K = 7$, rate 1/2 convolutional code and coded BPSK, use Fig. 13.23 and find the coding gain achievable (relative to uncoded BPSK) when the 2^n-ary RS symbols are constructed from n-bit blocks of input data bits, $n = m = 6, 8$.
 (b) Compare your results with those found in Problem 13.11.

REFERENCES

1. Elias, P., "Error-Free Coding," *IRE Trans. Information Theory*, IT-4, 1954, pp. 29–37.

2. Viterbi, A. J., "Error Bounds for Convolutional Codes and an Asymptotically Optimum Decoding Algorithm," *IEEE Trans. Information Theory*, vol. IT-13, April 1967, pp. 260–69.

3. Forney, G. D., Jr., *Concatenated Codes*, Cambridge, MA: MIT Press, 1966.

4. Gallager, R. G., "Information Theory and Reliable Communication," New York: John Wiley and Sons, 1968.

5. Michelson, A. M. and A. H. Levesque, *Error-Control Techniques for Digital Communications*, New York: John-Wiley and Sons, 1985.

6. Berlekamp, E. R., *Algebriac Coding Theory*, New York: McGraw-Hill, 1968.

7. Anderson, J. B. and S. Mohan, *Source and Channel Coding,* Boston, MA: Klumer Academic Publishers, 1991.

8. Lin, S. and D. J. Costello, Jr., *Error Control Coding: Fundamentals and Applications*, Englewood Cliffs, NJ: Prentice Hall, 1983.

9. Forney, G. D., Jr., "Convolutional Codes I: Algebraic Structure," *IEEE Trans. Information Theory*, vol. IT-16, 1970, pp. 720–38.

10. Forney, G. D., Jr., "The Viterbi Algorithm," *Proceedings of the IEEE*, vol. 61, March 1973, pp. 268–78.

11. Forney, G. D., Jr., "Convolutional Codes, II: Maximum Likelihood Decoding," Information and Control, vol. 25, July 1974, pp. 222–26.

12. Viterbi, A. J. and J. K. Omura, *Principles of Digital Communication and Coding*, New York: McGraw-Hill Book Company, 1979.

13. Heller, J. A. and I. M. Jacobs, "Viterbi Decoding for Satellite and Space Communication," *IEEE Trans. Communication Technology*, vol. COM-19, no. 5, pp. 835–48, October 1971.

14. Odenwalder, J. P., "Error Control" in *Data Communications, Networks, and Systems*, Thomas Bartee, ed., Indianapolis, IN: Howard W. Sams, 1985. Also see *Error Control Coding Handbook*, San Diego, CA: Linkabit Corp., July 1976.

15. Wozencraft, J. M. "Sequential Decoding for Reliable Communication Record," vol. 5, part 2, 1957, pp. 11–15.

16. Fano, R. M., *Transmission of Information*, New York: MIT Press and Wiley, 1961.

17. Massey, J. L., *Threshold Decoding*, Cambridge, MA: MIT Press, 1963.

18. Ziemer, R. E. and R. L. Peterson, *Introduction to Digital Communication*, New York: Macmillan Publishing Company, 1992.

19. Pollara, F. and K. M. Cheung, "Performance of Concatenated Codes Using 8-Bit and 10-Bit Reed-Solomon Codes," Telecommunications and Data Acquisition Progress Report 42-97, Jan.– Mar. 1989, pp. 194–201, Pasadena, CA: Jet Propulsion Laboratory.

20. Begin, G. and D. Haccoun, "High-Rate Punctured Convolutional Codes: Structure Properties and Construction Technique," *IEEE Trans. on Communications*, vol. 37, no. 12, December 1989, pp. 1381–85.

21. Cain, J. B., G. C. Clark and J. Geist, "Punctured Convolutional Codes of Rate $(n-1)/n$ and Simplified Maximum Likelihood Decoding," *IEEE Trans. Inform. Theory*, vol. IT-25, January 1979, pp. 97–100.

22. Yasuda, Y., Y. Hirata, K. Nakamura and O. Otani, "Development of a Variable-Rate Viterbi Decoder and Its Performance Characteristics," 6th Int. Conf. Digital Satellite Commun., Phoenix, AZ, September 1983.

23. Yasuda, Y., K. Kashiki and Y. Hirata, "High-Rate Punctured Convolutional Codes for Soft Decision Viterbi Decoding," *IEEE Trans. Commun.*, vol. COM-32, March 1984, pp. 315–19.

24. Begin, G. and D. Haccoun, "Decoding of Punctured Convolutional Codes by the Stack Algorithm," Abstracts of Papers, 1986 IEEE Int. Symp. Inform. Theory, Ann Arbor, MI, October 1986, p. 159.

25. Haccoun, D. and G. Begin, "High-Rate Punctured Convolutional Codes for Viterbi and Sequential Decoding," *IEEE Trans. Commun.*, vol. COM-37, November 1989, pp. 1113–25.

26. Hagenover, J., "Rate Compatible Punctured Convolutional Codes and Their Applications," *IEEE Trans. Commun.*, vol. COM-36, April 1988, pp. 389–400.

27. Sklar, B., *Digital Communications*, Englewood Cliffs, NJ: Prentice Hall, 1988.

28. Omura, J. K. and B. K. Levitt, "Coded Error Probability Evaluation for Antijam Communication Systems," *IEEE Trans. Commun.*, vol. COM-30, no. 5, May 1982, pp. 896–903.

Index

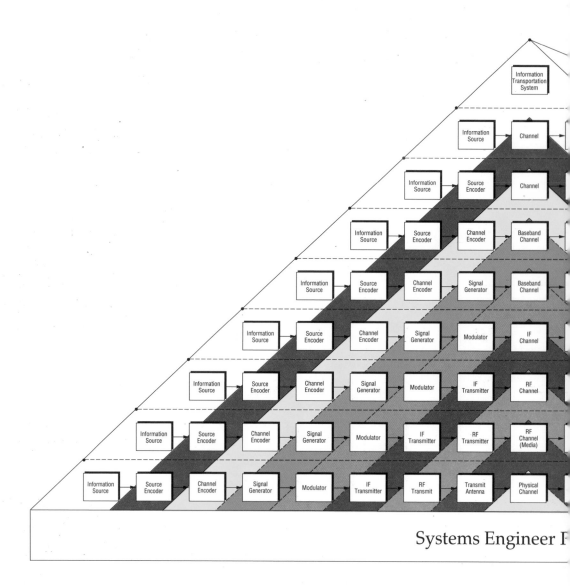

Figure 1.6 Information Transportation System (ITS) pyramid; coded digital communication system functional architectu